Amino Acid Abbreviations*

Amino Acid	Three-Letter Abbreviation	One-Letter Abbreviation	Amino Acid	Three-Letter Abbreviation	One-Letter Abbreviation
Alanine	Ala	A	Leucine	Leu	L
Arginine	Arg	R	Lysine	Lys	K
Asparagine	Asn	N	Methionine	Met	M
Aspartic acid	Asp	D	Phenylalanine	Phe	F
Cysteine	Cys	C	Proline	Pro	P
Glutamic acid	Glu	E	Serine	Ser	S
Glutamine	Gln	Q	Threonine	Thr	T
Glycine	Gly	G	Tryptophan	Trp	W
Histidine	His	H	Tyrosine	Tyr	Y
Isoleucine	Ile	I	Valine	Val	V

*This is Table 3.3 on page 66.

Greek Alphabet

Name of Letter	Greek Alphabet		Name of Letter	Greek Alphabet	
Alpha	Α	α	Nu	Ν	ν
Beta	Β	β	Xi	Ξ	ξ
Gamma	Γ	γ	Omicron	Ο	ο
Delta	Δ	δ ∂	Pi	Π	π
Epsilon	Ε	ϵ	Rho	Ρ	ρ
Zeta	Ζ	ζ	Sigma	Σ	σ ς
Eta	Η	η	Tau	Τ	τ
Theta	Θ	θ ϑ	Upsilon	Υ	υ
Iota	Ι	ι	Phi	Φ	ϕ φ
Kappa	Κ	κ	Chi	Χ	χ
Lambda	Λ	λ	Psi	Ψ	ψ
Mu	Μ	μ	Omega	Ω	ω

BIOCHEMISTRY
A Foundation

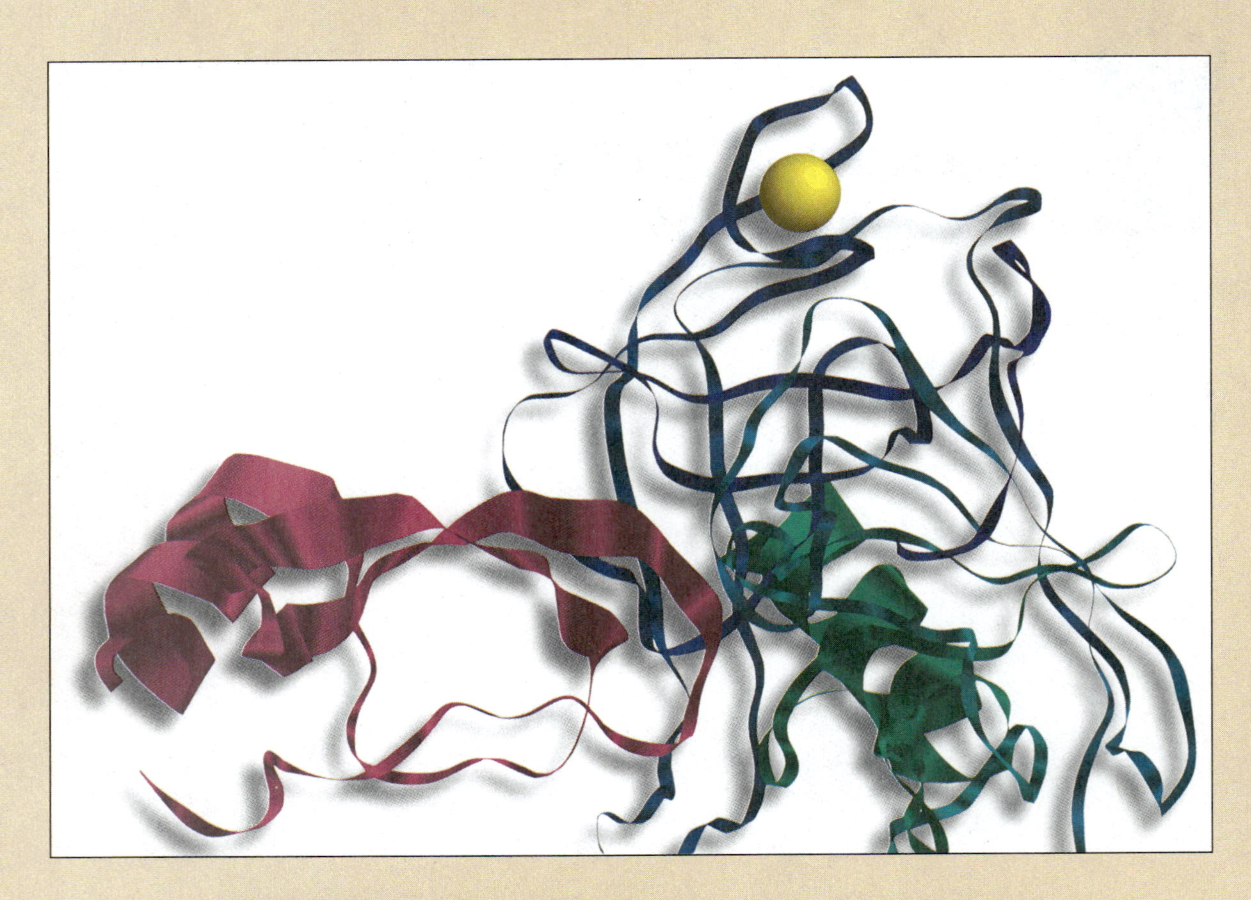

COVER IMAGE Computer-generated Richardson (ribbon) model of the digestive enzyme trypsin (blues and greens) complexed with pancreatic trypsin inhibitor (magenta). A calcium ion (yellow sphere) within the enzyme confers structural stability and might influence its activity. Trypsin is produced in its inactive form, trypsinogen, by the pancreas. Upon entry into the duodenum, trypsinogen is converted into active trypsin, which aids in the digestion of proteins. Trypsin is later inactivated by trypsin inhibitor, also produced by the pancreas. The inactivated complex is shown here. Structural coordinates of this model were determined by R. Huber and J. Deisenhofer.

BIOCHEMISTRY
A Foundation

PECK RITTER
Eastern Washington University

Brooks/Cole Publishing Company
ITP™ An International Thomson Publishing Company

Pacific Grove • Albany • Bonn • Boston • Cincinnati • Detroit • London • Madrid • Melbourne • Mexico City • New York • Paris • San Francisco • Singapore • Tokyo • Toronto • Washington

SPONSORING EDITOR: Harvey Pantzis
PROJECT DEVELOPMENT EDITOR: Sue Ewing
MARKETING TEAM: Connie Jirovsky, Romy Fineroff
MARKETING REPRESENTATIVE: Hester Winn
EDITORIAL ASSOCIATE: Beth Wilbur
PRODUCTION EDITOR: Nancy L. Shammas
PRODUCTION SERVICE: Graphic World Publishing Services
MANUSCRIPT EDITOR: Lindsay Ardwin
PERMISSIONS EDITOR: May Clark
INTERIOR DESIGN: Jeanne Wolfgeher
INTERIOR ILLUSTRATION: Academy Art Works, Inc., PageCrafters, Graphic World Illustration Services
COVER DESIGN: E. Kelly Shoemaker
COVER ILLUSTRATION: Ken Eward/BioGrafx
ART COORDINATOR & PHOTO EDITOR: Marcia Craig
PHOTO RESEARCHER: Clare Maxwell
INDEXER: Kevin Mulrooney
TYPESETTING: Graphic World
COVER PRINTING: Color Dot Graphics, Inc.
PRINTING AND BINDING: R. R. Donnelley & Sons Company

For more information, contact:

BROOKS/COLE PUBLISHING COMPANY
511 Forest Lodge Road
Pacific Grove, CA 93950
USA

International Thomson Publishing Europe
Berkshire House 168-173
High Holborn
London WC1V 7AA
England

Thomas Nelson Australia
102 Dodds Street
South Melbourne, 3205
Victoria, Australia

Nelson Canada
1120 Birchmount Road
Scarborough, Ontario
Canada M1K 5G4

International Thomson Editores
Campos Eliseos 385, Piso 7
Col. Polanco
11560 México D. F. México

International Thomson Publishing GmbH
Königswinterer Strasse 418
53227 Bonn
Germany

International Thomson Publishing Asia
221 Henderson Road
#05-10 Henderson Building
Singapore 0315

International Thomson Publishing Japan
Hirakawacho Kyowa Building, 3F
2-2-1 Hirakawacho
Chiyoda-ku, Tokyo 102
Japan

Printed in the United States of America

10 9 8 7 6 5 4 3 2 1

Library of Congress Cataloging-in-Publication Data

Ritter, Peck.
Biochemistry : a foundation / Peck Ritter.
p. cm.
Includes bibliographical references and index.
ISBN 0-534-33865-8
1. Biochemistry. I. Title.
QP514.2.R55 1996 95-25490
574.19′2--dc20 CIP

In memory of my mother,
Olive B. Ritter,
a humanitarian, librarian, and scholar.

To good health for
Sarah, Chris, Katie, Tasha, and
all children everywhere.

Brief Contents

Contents

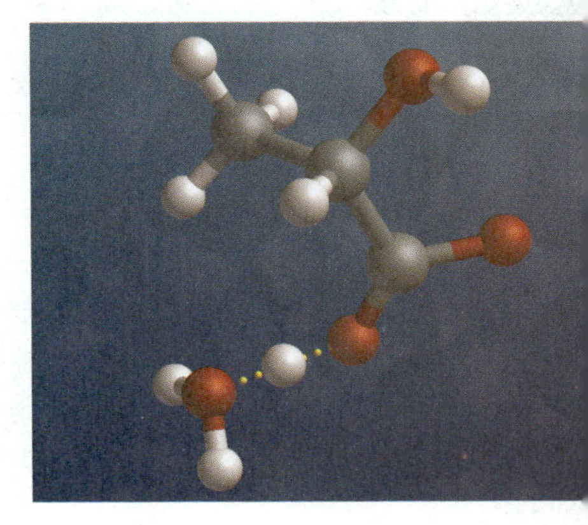

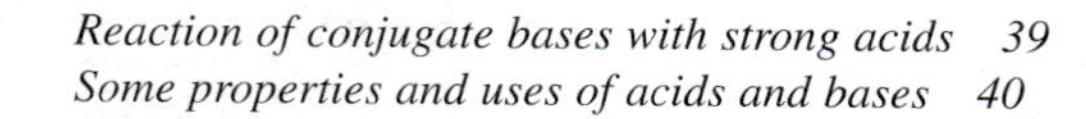

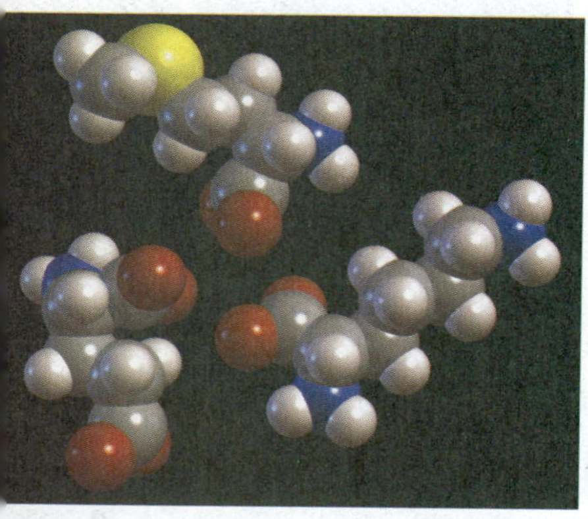

CHAPTER 4

Proteins 112

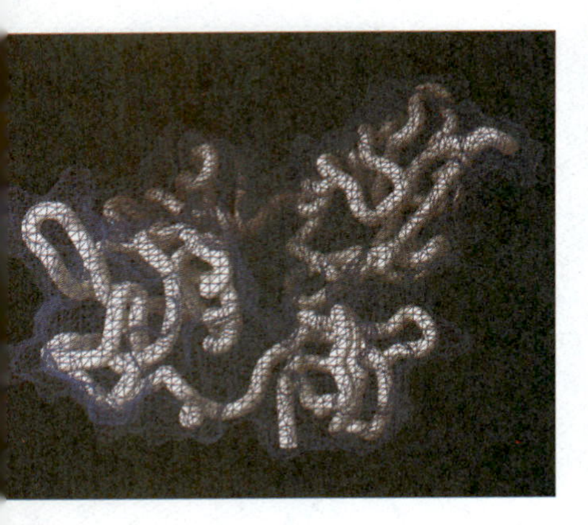

CHAPTER 5

Enzymes 169

CHAPTER 6

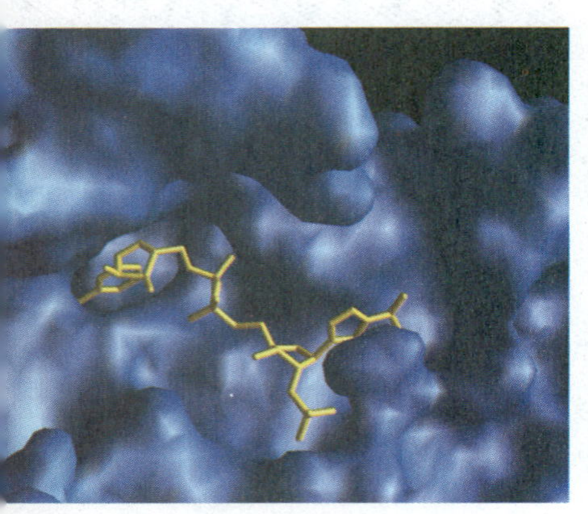

CHAPTER 9

Lipids and Biological Membranes 324

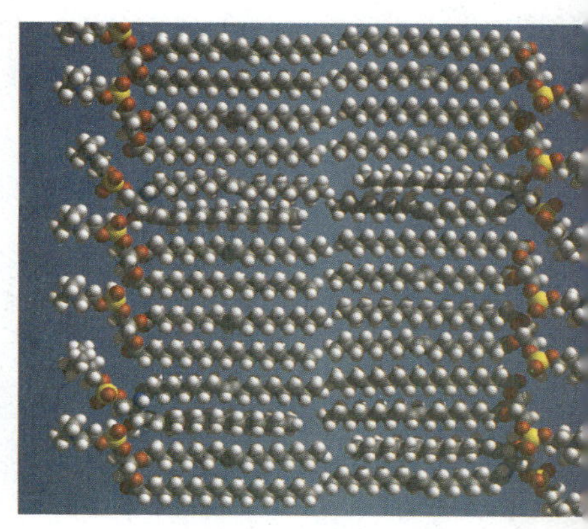

CHAPTER 10

Vitamins, Hormones, and an Introduction to Metabolism 375

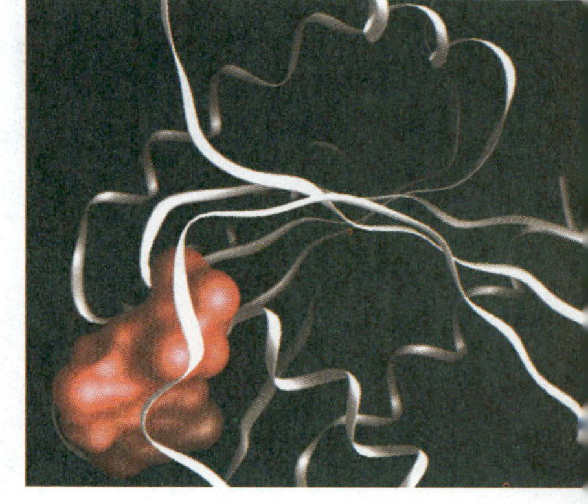

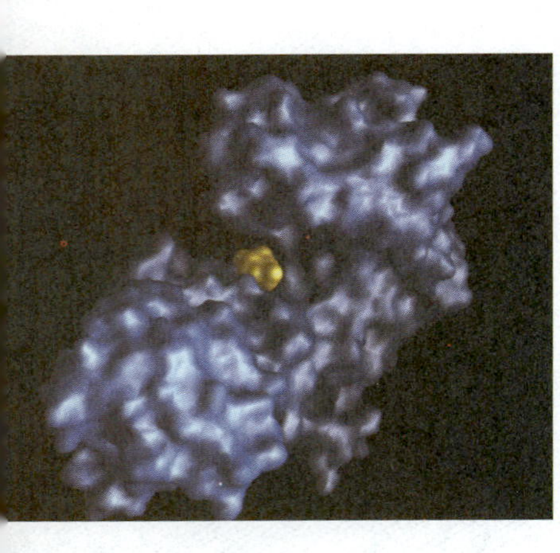

CHAPTER 12

Additional Topics in Carbohydrate Catabolism 479

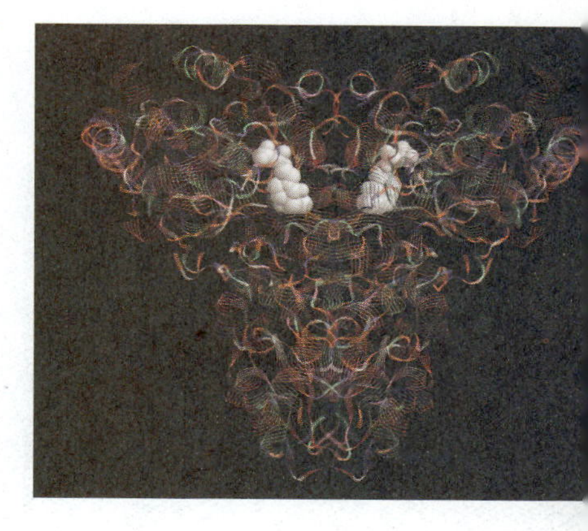

CHAPTER 13

Catabolism of Fats and Proteins 502

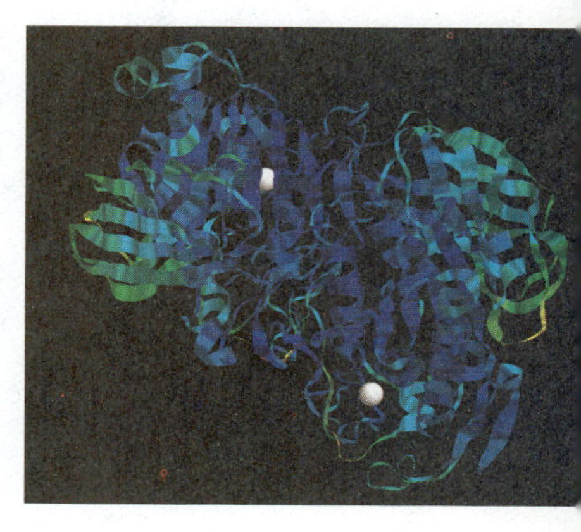

FAT CATABOLISM

PROTEIN AND AMINO ACID CATABOLISM

CHAPTER 14

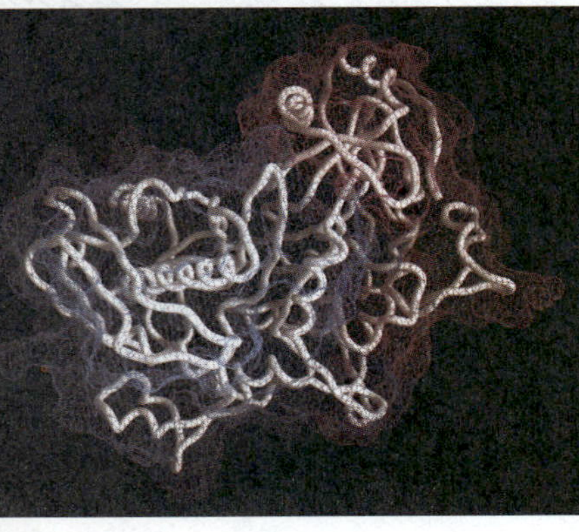

CHAPTER 15

Photosynthesis and Nitrogen Fixation 591

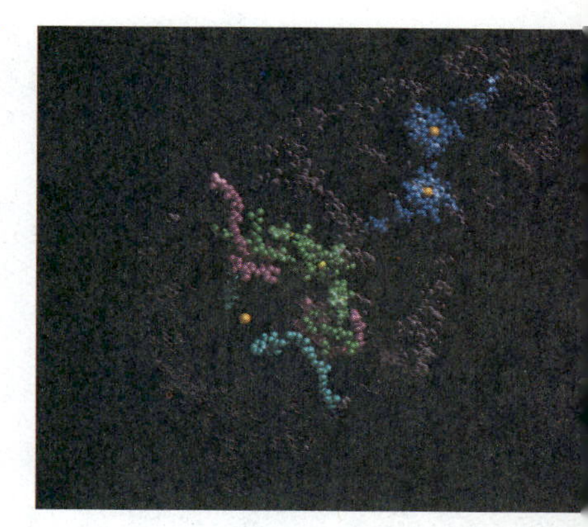

CHAPTER 16

Nucleic Acids 642

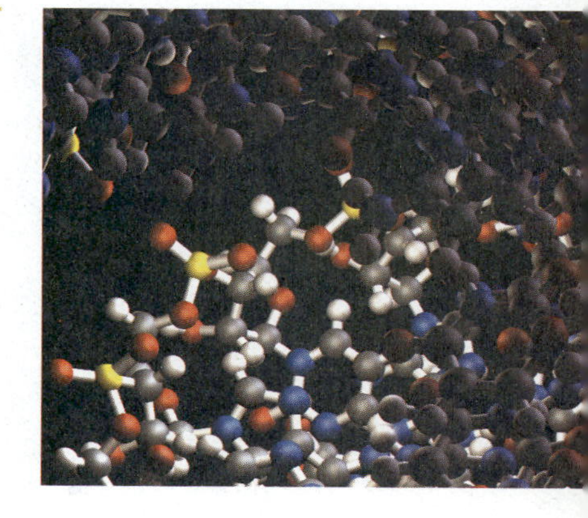

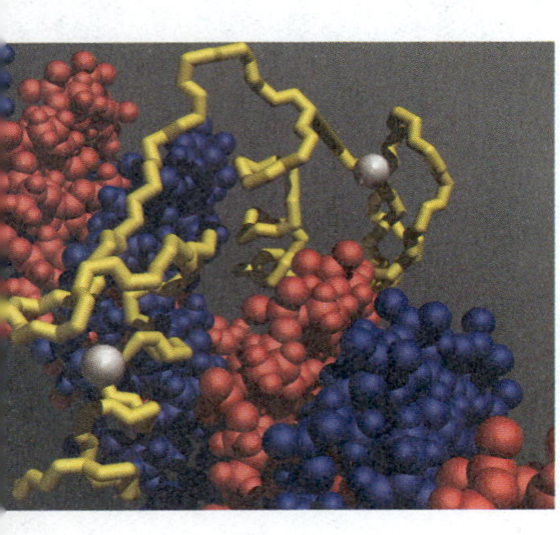

CHAPTER 17

Nucleic Acid Biosynthesis 689

Preface

Interest has never been higher in biochemistry—the study of compounds, chemical transformations, and molecular interactions involved in the development, maintenance, and reproduction of living organisms. Today, biochemical knowledge is expanding at an exponential rate and setting the pace for advances in medicine, dentistry, genetics, immunology, microbiology, agriculture, nutrition, psychology, exercise science, and related disciplines. No mystery or adventure novel is more intriguing than the current biochemical literature. Over the next several decades, biochemistry will have a greater impact on our lives and our society than any other area of the natural sciences.

MRI is used to detect certain biochemically-linked abnormalities.

After teaching survey courses in biochemistry from over a dozen different texts during a 20-year period, it became clear to me that there was a need for a user-friendly introductory text with a unique combination of pedagogical features including:

- a large number of problems with detailed solutions,
- an approach that actively helps students review and use the essential skills and factual information presented in prerequisite courses,
- a balanced coverage of topics that accurately reflects the expanding and changing nature of biochemistry,
- clear, concise explanations that emphasize themes and general principles,
- an approach that assists students in formulating an integrated overview of biochemistry, and
- a writing style and emphasis that fully capture the wonder and excitement of biochemistry.

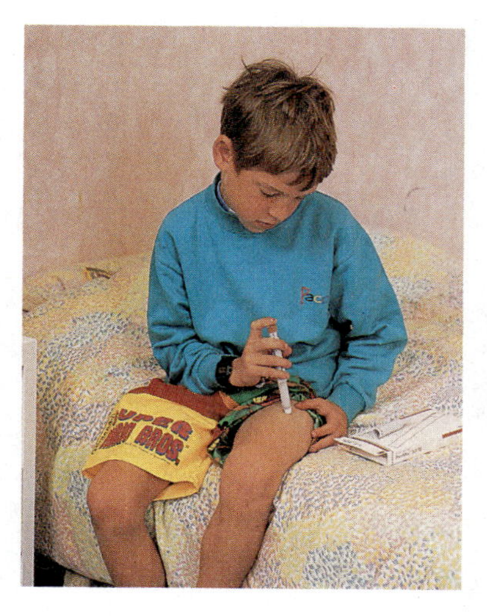

Many therapies are based on our knowledge of biochemistry, such as insulin injections to manage diabetes.

In writing *Biochemistry: A Foundation,* I have used these features to develop a pedagogical system specifically designed for one-quarter, two-quarter, and one-semester introductory survey courses in biochemistry. (The text can also be used for a two-semester course if the instructor chooses to cover most sections of the text.) Topic emphasis is particularly well suited for a class containing students majoring in both chemistry and biology because it caters to the needs and interests of each of these groups. For example, the routine use of basic chemical principles, organic chemistry, thermodynamics, and molecular-level discussions of structure–function relationships will appeal to students majoring in chemistry, while those majoring in biology will appreciate the many sections that involve cell biology, evolution, genetics, physiology, and nutrition. Biochemistry is built on both chemistry and biology, a key concept for all students. Although the text addresses the biochemistry of both simple (prokaryotic) and complex (eukaryotic) organisms, including both plants and animals, emphasis is

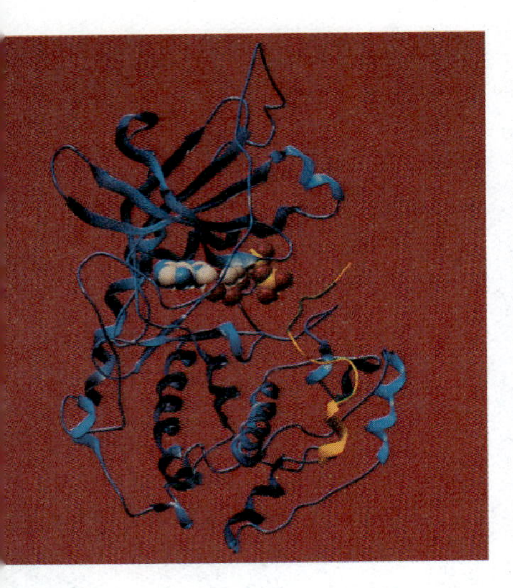
cAMP-dependent protein kinase

placed on human biochemistry and the applications of biochemistry in medicine and other health-related disciplines.

Up-to-Date and Flexible with a Balanced Survey of Biochemistry

Rapid advances in biochemistry make it essential that each new generation of one-semester survey texts adds new topics and alters the relative emphasis placed on some traditional topics. This text places a greater relative emphasis on major biochemical themes, applications of biochemical knowledge, genetic diseases, regulatory strategies, signal transduction, membrane function, diet and health, and the tools of biochemistry. At the same time, it provides complete coverage of traditional core topics and offers individual instructors breadth, depth, and topic-sequence flexibility. An instructor, for example, can cover nucleic acids, gene expression, and related topics prior to the discussion of more traditional metabolism. I favor this approach because genes and the regulation of gene expression are at the heart of all metabolism. After the first eight chapters, the remaining material can be presented in a variety of sequences. Many sections can be omitted with little, if any, impact on other chapters or sections. Enzymes, for example, can be covered at any of several levels of depth and sophistication with the appropriate selection of sections and subsections, and Chapter 15, *Photosynthesis and Nitrogen Fixation,* can be omitted completely.

A Wealth of Integrated Problems

The text includes over 1,200 problems and exercises carefully integrated with chapter material. I consider in-chapter problems an integral part of the text, and their placement enhances their pedagogical value (see Try Problem prompts in each chapter). The end-of-chapter exercises provide the reinforcement needed for students to acquire a working knowledge of introductory biochemistry. Some of the problems and exercises are designed to emphasize key points and to help students direct and concentrate their studies (for example, Problems 5.54, 9.40, 14.28, and Exercises 3.6, 4.23, and 16.17). Some allow students to hone their computational skills (for example, Problems 4.31, 8.12, 9.10 and Exercises 3.13 and 13.8) and others help students recognize and appreciate how their growing biochemical knowledge relates to their bodies, daily lives, and careers (for example, Problems 3.54, 4.37, 6.42, 9.30, and Exercises 4.40 and 14.30). Still other problems give users the opportunity to analyze and apply what they have learned (for example, Problems 3.38, 10.17, 19.16, and Exercises 5.45, 14.9, and 19.23). I have prepared detailed solutions for all problems. The solutions usually discuss the significance of problems and expand on text material; I consider them a key component of my problem-oriented pedagogical approach.

A Review of Material from Prerequisite Courses

The text actively helps students review and use their accumulating chemistry knowledge through in-text discussions (for example, Sections 2.3, 3.3, 5.16, 7.1 [tautomerism] and 8.7), problems, and detailed solutions to problems (for example, Problems 3.2, 3.10, 3.27, and 3.28 and the solutions to these problems). Bonding,

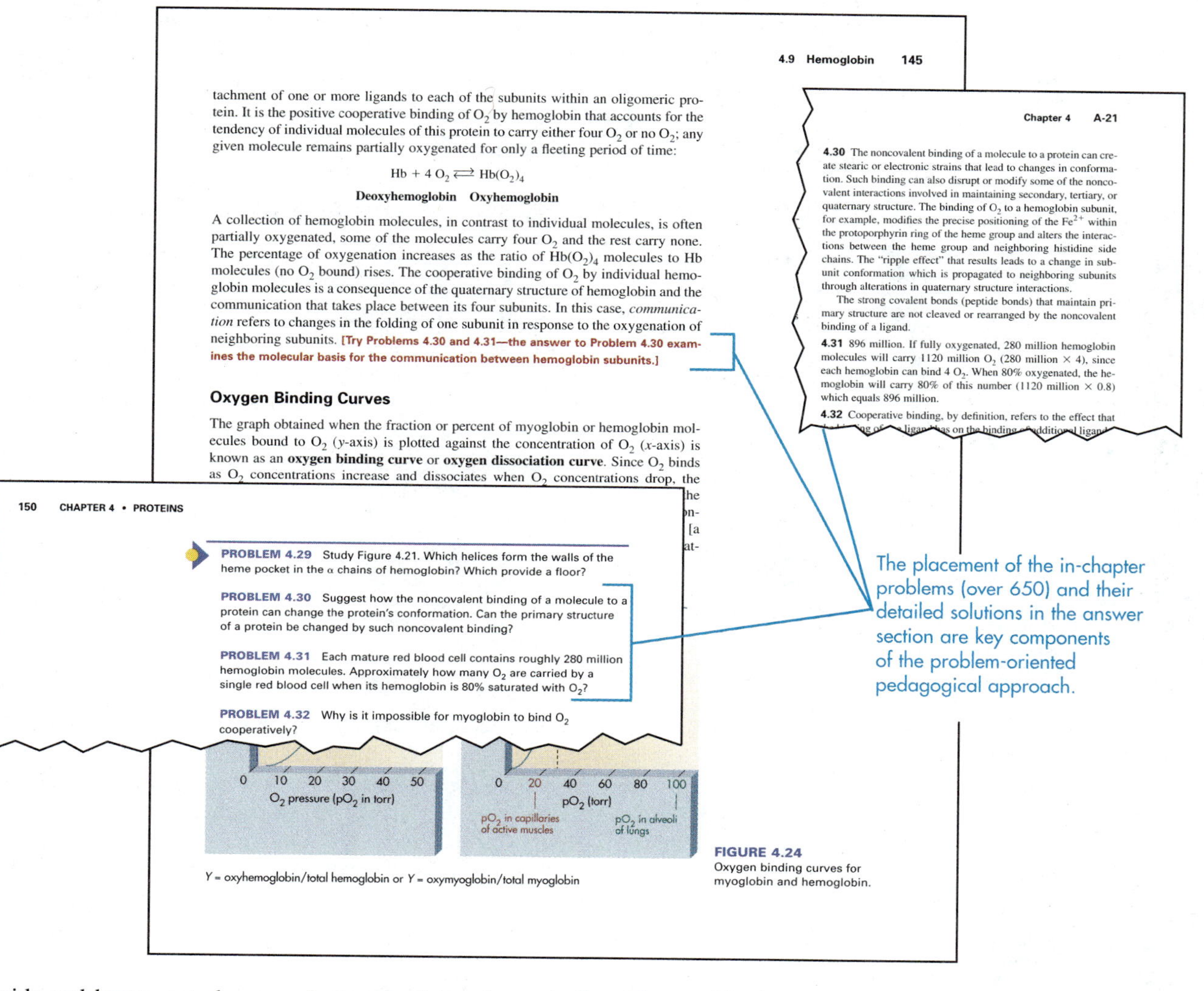

4.9 Hemoglobin 145

tachment of one or more ligands to each of the subunits within an oligomeric protein. It is the positive cooperative binding of O_2 by hemoglobin that accounts for the tendency of individual molecules of this protein to carry either four O_2 or no O_2; any given molecule remains partially oxygenated for only a fleeting period of time:

$$Hb + 4\,O_2 \rightleftharpoons Hb(O_2)_4$$

Deoxyhemoglobin Oxyhemoglobin

A collection of hemoglobin molecules, in contrast to individual molecules, is often partially oxygenated, some of the molecules carry four O_2 and the rest carry none. The percentage of oxygenation increases as the ratio of $Hb(O_2)_4$ molecules to Hb molecules (no O_2 bound) rises. The cooperative binding of O_2 by individual hemoglobin molecules is a consequence of the quaternary structure of hemoglobin and the communication that takes place between its four subunits. In this case, *communication* refers to changes in the folding of one subunit in response to the oxygenation of neighboring subunits. **[Try Problems 4.30 and 4.31—the answer to Problem 4.30 examines the molecular basis for the communication between hemoglobin subunits.]**

Oxygen Binding Curves

The graph obtained when the fraction or percent of myoglobin or hemoglobin molecules bound to O_2 (*y*-axis) is plotted against the concentration of O_2 (*x*-axis) is known as an **oxygen binding curve** or **oxygen dissociation curve**. Since O_2 binds as O_2 concentrations increase and dissociates when O_2 concentrations drop, the

Chapter 4 A-21

4.30 The noncovalent binding of a molecule to a protein can create stearic or electronic strains that lead to changes in conformation. Such binding can also disrupt or modify some of the noncovalent interactions involved in maintaining secondary, tertiary, or quaternary structure. The binding of O_2 to a hemoglobin subunit, for example, modifies the precise positioning of the Fe^{2+} within the protoporphyrin ring of the heme group and alters the interactions between the heme group and neighboring histidine side chains. The "ripple effect" that results leads to a change in subunit conformation which is propagated to neighboring subunits through alterations in quaternary structure interactions.

The strong covalent bonds (peptide bonds) that maintain primary structure are not cleaved or rearranged by the noncovalent binding of a ligand.

4.31 896 million. If fully oxygenated, 280 million hemoglobin molecules will carry 1120 million O_2 (280 million × 4), since each hemoglobin can bind 4 O_2. When 80% oxygenated, the hemoglobin will carry 80% of this number (1120 million × 0.8) which equals 896 million.

4.32 Cooperative binding, by definition, refers to the effect that

150 CHAPTER 4 • PROTEINS

PROBLEM 4.29 Study Figure 4.21. Which helices form the walls of the heme pocket in the α chains of hemoglobin? Which provide a floor?

PROBLEM 4.30 Suggest how the noncovalent binding of a molecule to a protein can change the protein's conformation. Can the primary structure of a protein be changed by such noncovalent binding?

PROBLEM 4.31 Each mature red blood cell contains roughly 280 million hemoglobin molecules. Approximately how many O_2 are carried by a single red blood cell when its hemoglobin is 80% saturated with O_2?

PROBLEM 4.32 Why is it impossible for myoglobin to bind O_2 cooperatively?

FIGURE 4.24 Oxygen binding curves for myoglobin and hemoglobin.

acids and bases, net charges, electronegativity, the periodic table, equilibrium, kinetics, thermodynamics, major classes of organic compounds, tautomerism, and resonance are among the many topics reviewed.

An Integrated Picture of the Chemistry of Life

I have written each chapter to provide students with an integrated picture of the chemistry of life. The major themes in biochemistry are identified in Chapter 1 (Exhibit 1.4) and then are carefully followed as they weave through each of the remaining chapters. The 12 major themes include experimentation and hypothesis construction, storage and transfer of information, transformation and use of energy, structure–function relationships, catalysis and the regulation of reaction rates, specificity, communication, and the association of like with like. The introduction to each chapter lists the dominant themes within the chapter and identifies the chapter's re-

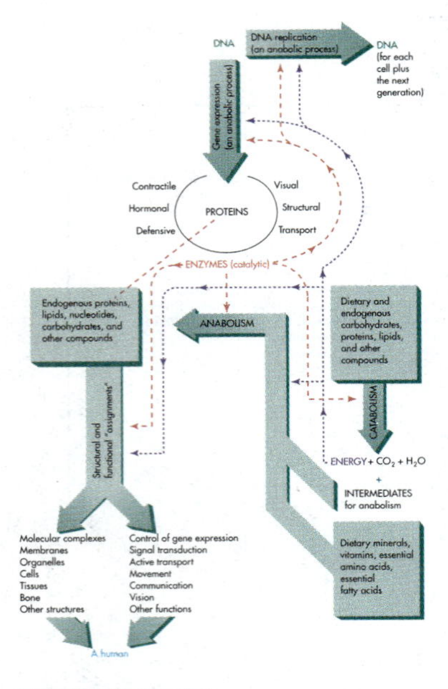

FIGURE 19.20
Molecular basis for human life.

lationship to the preceding chapter and the chapter that follows. In-text discussions, figures, tables, problems, and cross references (38 in Chapter 11 and 50 in Chapter 14, for example) are used to emphasize themes, interrelationships, integrating principles, and the big picture. The text concludes with a global and integrated summary of the chemistry of life (see Figure 19.20).

A Philosophy that Reflects the Needs of Students

I have found that students are both intrigued and motivated by the bigger picture of biochemistry, and so I often examine the present uses, implications, and future promises of our rapidly expanding biochemical knowledge. I ask and then answer frequently raised questions (for example, pages 210, 311, 443, and 539) and examine the fallacies behind common misconceptions (for example, pages 44, 122, and 294). I emphasize major classes of compounds, central metabolic processes, and central themes; key concepts and principles are revisited time and time again. The central role of DNA in life processes, for example, is introduced in Chapter 1, reviewed in Chapters 4 and 10, and examined in detail in Chapters 16 through 19. In addition, virtually every chapter contains problems or discussions that deal with gene expression. Protein kinases and protein phosphatases provide an additional example; they are introduced in Chapter 5, examined in depth in Chapter 10, and repeatedly encountered in later discussions.

A Complete Ancillary Package

- *Solutions Manual:* Here I have included complete solutions to all end-of-chapter exercises.
- *Instructor's Manual with Test Items:* Along with W. Anthony Oertling of Eastern Washington University, I have written a manual that includes lecture outlines, teaching guidelines, a guide to biochemistry on the Internet (see also Internet Resources on pages 821–822 in this text), an expanded reading list, and over 450 short-answer and multiple-choice test items.
- *Electronic Test Items:* The test items are available for DOS, Windows, and Macintosh platforms.
- *Transparencies:* 100 four-color transparencies are available.

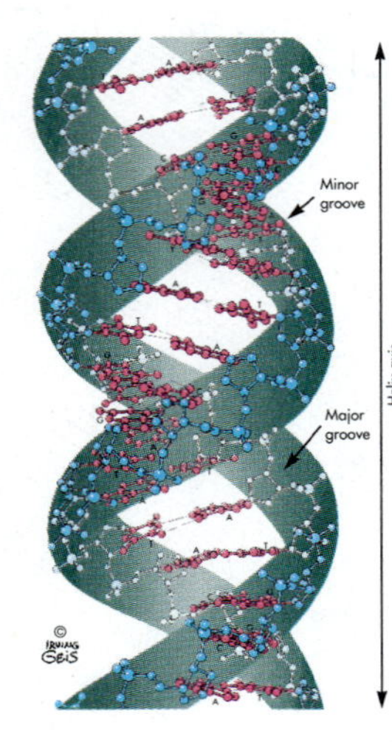

A-DNA

Acknowledgments

The publication of a text is too big a job for one person; it is definitely a team project. The co-captains are Maureen Allaire (former Chemistry Editor, Brooks/Cole), Lindsay Ardwin (copy editor), Marcia Craig (Managing Editor, Graphic World Publishing Services), May Clark (Permissions Editor, Brooks/Cole), John Kenneth Eward (artist for chapter openers and the cover), Sue Ewing (developmental editor), Romy Fineroff (Marketing Communications, Brooks/Cole), Irving Geis (artist for select illustrations), Connie Jirovsky (Senior Marketing Manager, Brooks/Cole), Malcolm Klein (Tripos, Inc.), Clare Maxwell (photo researcher), Lisa Moller (former Chemistry Editor, Brooks/Cole), Harvey Pantzis (Executive Editor, Brooks/Cole), Elizabeth Rammel (Assistant Editor, Brooks/Cole), Nancy Shammas (Senior Production Editor, Brooks/Cole), Kelly Shoemaker (Senior Designer,

Brooks/Cole), and Beth Wilbur (Editorial Associate, Brooks/Cole). A number of artists, proofreaders, and other dedicated individuals made significant contributions as they worked behind the scenes. I am particularly indebted to Lindsay, Marcia, and Nancy, who put the greatest amount of time, energy, and themselves into the project. Marcia Craig spent close to nine months coordinating and managing the final production process. Her eye for details, quest for perfection, and organizational and design skills were invaluable. Special thanks to Beth Wilbur for the guidance, sound advice, moral support, and continuity that she provided throughout the project.

I am indebted to the colleagues and students at Eastern Washington University who provided encouragement, support, and advice: Kenneth W. Raymond, Ernie McGoran, and Donald R. Lightfoot for reviewing specific chapters during the early stages of manuscript development, and two former students, Susan J. Huntley (now a medical technologist) and Charles F. Wandler (now a community college instructor), for critiquing an initial version of the entire text. My deepest thanks and appreciation go to the other numerous reviewers: Hugh Akers, Lamar University; Ronald Bentley, University of Pittsburgh; Kerry L. Cheesman, Capital University; Clyde Denis, University of New Hampshire; Don Dennis, University of Delaware; Francis DeToma, Mount Holyoke College; Walther Ellis, University of Utah; John Hess, Virginia Polytech and State University; Marjorie Jones, Illinois State University; Albert Light, Purdue University; Paul Ludden, University of Wisconsin, Madison; Jerome Maas, Oakton Community College; Scott Mohr, Boston University; Charlotte Otto, University of Michigan, Dearborn; Richard Paselk, Humboldt State University; David Sadava, The Claremont Colleges—Joint Science Center; David E. Saleeby, Florence-Darlington Technical College; Mary Scott, Wheeling Jesuit College; Ming Tien, Pennsylvania State University; Jerry Wilson, California State University, Sacramento; and Beulah Woodfin, University of New Mexico. Utmost thanks go to W. Anthony Oertling, who prepared the outstanding test bank that accompanies the text and checked the accuracy of the answers to all text problems and exercises. I also extend appreciation to Prakash H. Bhuta for writing a section on the Internet for the Instructor's Manual as well as the Internet resources list in this book and to my students, who, collectively, have made teaching fun and rewarding, have allowed me to test and retest pedagogical strategies, and have given me a reason to tackle this textbook project.

A final word of thanks to my wife, Darlene, for her support and understanding as I struggled with what she called *Biochemistry: A Never-Ending Story.*

As I prepare for a second edition of *Biochemistry: A Foundation,* I would like the assistance of everyone who has reviewed or used my text. I can be reached at:

Department of Chemistry and Biochemistry
EWU, MS#74
Cheney, WA 99004
E-mail: pritter@ewu.edu
FAX: 509-359-6973
Phone: 509-359-7901

Your comments and recommendations would be greatly appreciated.

Peck Ritter

Index

Page references in **boldface** refer to text passages in which a compound or concept is defined or introduced; page references followed by an "f" indicate figures and page references followed by a "t" indicate tables.

1 Introduction

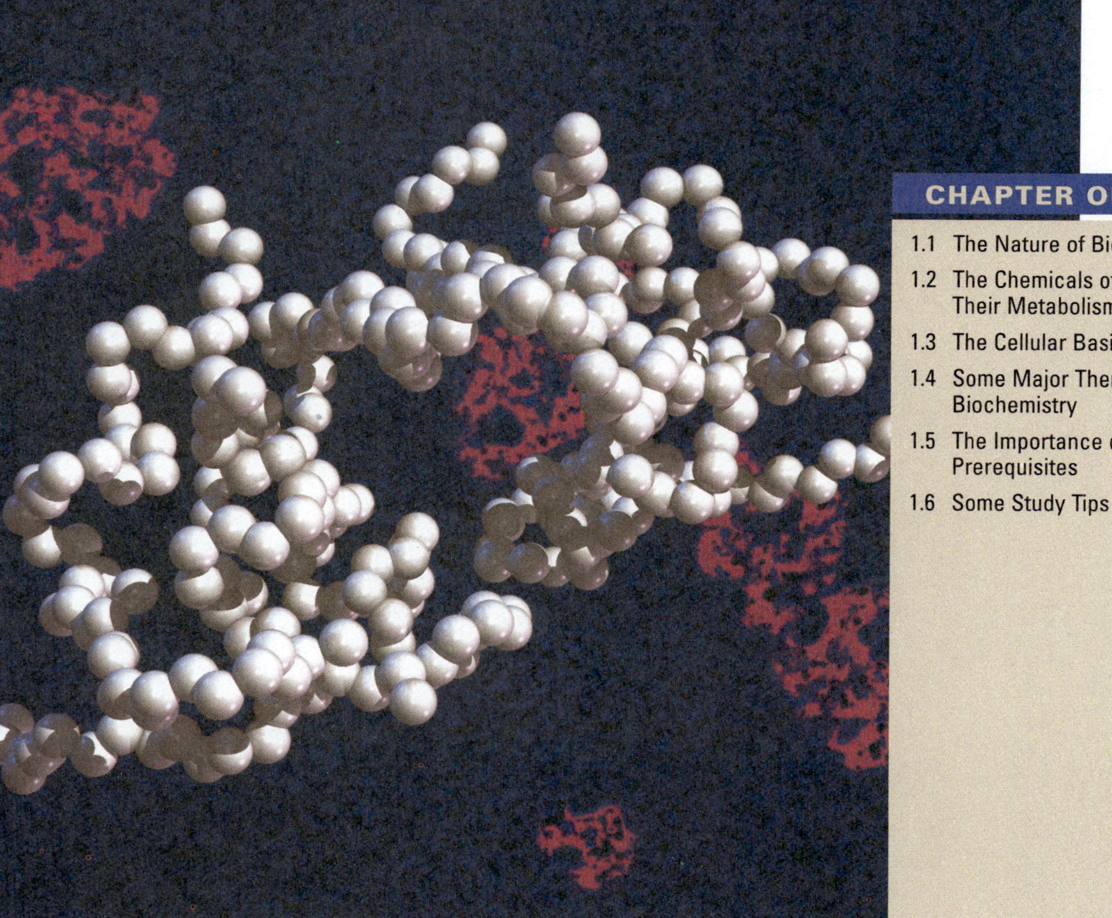

CHAPTER OUTLINE

The human body develops from a single fertilized egg cell, called a **zygote,** that contains thousands of different compounds. Among these compounds one finds **deoxyribonucleic acid (DNA),** the genetic material contributed by the egg cell and the sperm cell that fused to form the zygote. Within this DNA lies the inherited blueprints for a man or a woman.

The human zygote divides into two daughter cells. Each daughter cell divides to produce a total of four cells that, in turn, divide again, and so on. The DNA in each cell is replicated before division, and each daughter cell normally receives a complete copy of the **human genome** (all of the genetic information originally contained in the zygote). As the rounds of cell division continue, individual cells start to specialize and to assume different functions; some cells become liver cells, some muscle cells, some skin cells, and so on. This process, known as **differentiation,** is brought about by the orderly reading of the blueprints carried in the cell's DNA, for it is the precisely timed expression of this genetic information that leads to the production of those compounds that orchestrate development.

The entire process of growth and development is initiated by molecules present in the zygote. These molecules, many of which have not yet been identified, act by directly or indirectly turning on or turning off the expression of specific genes. **Genes** are segments of DNA that carry information for the production of specific molecules. The molecules produced from the expression of particular genes keep growth and development going once it has been initiated. If all goes well, the ultimate product is a miracle—a normal, healthy individual.

The production and maintenance of other organisms involve similar events and processes. DNA is the genetic material in every instance, and growth and cell division are guided by those compounds whose structures are encoded in the DNA. Many of the same compounds, chemical reactions, and molecular interactions play a central role in all forms of life.

1.1 THE NATURE OF BIOCHEMISTRY

The study of the compounds, chemical reactions, and molecular interactions that are involved in the production, maintenance, and reproduction of living organisms is defined as **biochemistry.** Central to this study is the analysis of energy and energy transformations, for the order and the support systems that sustain life require a constant input of energy from the environment. Biochemistry is a vast, largely unexplored frontier. A significant but unknown fraction of the molecular events that provide the basis for life has not yet been identified, and few molecular processes have been fully characterized. The existing gaps in knowledge help explain the lack of answers, the partial answers, the changing answers, and the controversies that are all part of biochemistry today. To avoid frustration when dealing with this dynamic discipline, it is essential to understand fully the scientific method and the basic characteristics of the natural sciences.

The **natural sciences,** including biochemistry, chemistry, physics, and biology, operate on the assumption that all of the changes and interactions that occur in the universe are governed by certain fixed rules. This may or may not be the case, but present evidence supports this view. Natural scientists work to discover these rules. Once discovered, the rules contribute to our understanding of natural processes and can be used by humans to manipulate and modify natural systems, both living and nonliving. Since some rules make no common sense from a typically human perspective and there is no evidence that the rules can be changed, it may be counterproductive to dwell on the question "Why does nature operate the way that it does?".

Natural scientists use the **scientific method** to discover the rules of nature and learn how to employ these rules. The process begins with the observation of natural systems and the changes that go on within them. The systems are then dissected and their components are analyzed and studied. Scientists can also experiment with nature by observing how systems and their components respond to human-induced alterations. After pondering their observations, scientists construct **hypotheses.** Hypothesis construction is strictly a creative, mental exercise, and the final product is a possible explanation for what has been observed or a possible rule of nature. A good hypothesis can be used to make predictions and is, therefore, testable. If predictions are found to be valid, a hypothesis gains in stature. If a hypothesis passes enough tests, if no observations are made that are inconsistent with the hypothesis, and if its ramifications are broad enough, it may be elevated to the status of a **theory** or **law** (Figure 1.1). That does not necessarily mean that it represents an absolute truth. Further studies and observations may challenge its validity.

Biochemistry, like all of the natural sciences, grows through the efforts of thousands of dedicated individuals around the globe. It grows through a careful and crit-

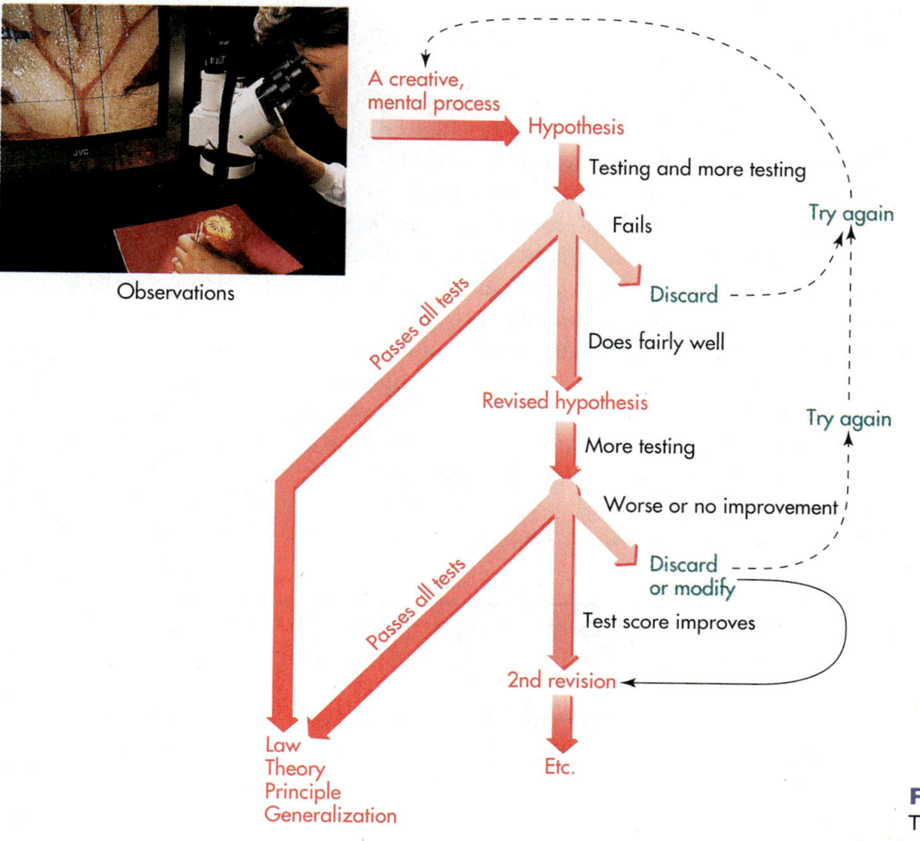

FIGURE 1.1
The scientific method.

ical examination of the compounds, reactions, and interactions within living organisms. It grows through the construction of hypotheses and the testing of these hypotheses. It grows through the realization that some hypotheses and some currently accepted beliefs need to be modified or discarded as more and more is learned about the molecular basis of life. Occasionally, serendipity (accidental discovery) also plays a significant role in the advancement of scientific knowledge. You should remember these facts, and keep an open mind, as this text describes how hormones bring about changes in target cells, how oncogenes convert normal cells into tumor cells, how enzymes catalyze reactions, how cholesterol contributes to heart disease, how aspirin lowers body temperature, how AZT inhibits the growth of the AIDS virus, and numerous other biochemical events and processes.

If the past is any predictor of the future, the uncharted paths to the molecular secrets of life will be full of pitfalls and surprises. Evolution has led to millions of diverse life forms, and much of their chemistry has proved unpredictable. Multiple life processes have already been found to proceed in unanticipated and wondrous ways.

Although no one knows what path biochemistry will follow into the future, it is certain that the health sciences will be drastically impacted along the way. As biochemical phenomena are examined in greater and greater molecular detail, biochemistry will continue to be a central component of all the health sciences, including microbiology, genetics, immunology, physiology, nutrition, medicine, dentistry, exercise science, and psychology. Progress in biochemistry will largely set the pace for future advances in these and related disciplines. Growing biochemical knowledge will, for example, lead to improvements in the prevention and treatment of heart disease, cancer, genetic diseases, periodontal disease, tooth decay, autoimmune diseases, nutritional deficiencies, mental illness, infectious diseases, and other health problems. That knowledge may also allow health scientists to start grappling directly with the ultimate health problem facing advanced societies, biological aging.

Along with the ability to prevent and treat health-related problems comes the ability to manipulate and modify life forms. The latter ability raises many moral and ethical questions, and it holds perils as well as promises. Foreign genes have been introduced into dozens of different organisms, including humans. Hundreds of patents have been issued on genetically modified life forms, including mammals. Could the mythological chimera (a monster with a lion's head, a goat's body, and a serpent's tail) become a reality? Is Huxley's "Brave New World" upon us?

This text will provide a glimpse at biochemistry by providing partial answers to a number of basic questions, including those listed in Exhibit 1.1. At the same time, the text will continually illustrate and emphasize the central role of biochemistry in the health sciences and will capture the excitement that is biochemistry.

[Try Problems 1.1 and 1.2]

PROBLEM 1.1 Add six questions to those already listed in Exhibit 1.1. Design questions that relate directly to your chosen career.

PROBLEM 1.2 Are there any limits on what can be accomplished with science and technology (the application of scientific knowledge)? If so, what ultimately determines or sets these limits?

EXHIBIT 1.1
Some Basic Questions Related to Biochemistry

1. What are the major classes of compounds found within living organisms?
2. How do organisms build these compounds?
3. What are the biological uses and functions of these compounds?
4. What are the chemical and physical properties of these compounds and how do these properties contribute to their biological activities?
5. How is the genetic blueprint in DNA read, and how is this reading regulated?
6. How is DNA replicated and passed from generation to generation?
7. What are genetic diseases?
8. What are the promises and perils of genetic engineering?
9. How do organisms coordinate and control the thousands of reactions occurring within them?
10. How do organisms acquire energy, and what are the major uses of this energy?
11. How does the human body digest food, and how is food utilized by the body?
12. What are vitamins and hormones, and what roles do these compounds play within the human body?
13. What are some of the tools of biochemistry?

1.2 THE CHEMICALS OF LIFE AND THEIR METABOLISM

The study of biochemistry normally begins with an examination of those compounds that are major structural or functional components within living organisms, and continues with an exploration of how these compounds are synthesized and utilized. Along the way, the molecular basis for a multitude of life-sustaining processes are analyzed. This section provides a background for this pedagogical strategy by introducing the major classes of compounds found within living organisms.

All living organisms are predominantly constructed from carbon, oxygen, and hydrogen; water and organic compounds account for a majority of their mass. Nitrogen, sodium, magnesium, phosphorus, sulfur, chlorine, potassium, and calcium complete the list of **bulk elements**, those chemical elements required in relatively large amounts. Most organisms need significantly smaller amounts of over a dozen additional elements, known as **trace elements** (Section 10.1). The trace elements include iron, zinc, and iodine. Although many of the bulk and trace elements are components of organic compounds, a variety of biologically important compounds are strictly, or at least predominantly, inorganic in nature. Bone and teeth, for example, are predominantly constructed from calcium, phosphorus, oxygen, and hydrogen. Nitric oxide (NO) is a biologically important chemical messenger.

DNA has been identified as one of the major classes of carbon-containing compounds that play key roles in life processes; it is the genetic material within cells. *Proteins, ribonucleic acid* (*RNA*), *carbohydrates,* and *lipids* are the other major classes of carbon-containing compounds (Table 1.1). Cells contain, in addition, a

TABLE 1.1

The Major Classes of Biochemicals

Major Class	Some Biological Functions*
DNA	Storage of genetic information
RNA	Expression of genetic information, catalysts
Proteins	Catalysts, structural elements, chemical messengers, transport agents, protective barriers
Carbohydrates	Fuels, structural elements, lubricants, cellular communication agents
Lipids	Fuels, structural elements, membrane components, transport agents, chemical messengers

*The list of functions is not comprehensive.

large number of substances that do not fit into any of these major categories. Many of these substances are intermediates in the synthesis or breakdown of members of the major classes.

The vast majority of the information carried in DNA is information for the production of **proteins**, polymers produced from small building blocks known as amino acids. Most of the proteins within a cell function as biological catalysts, **enzymes**. In humans, enzymes control the rates of the 50,000 to 100,000 reactions that appear to be involved in the production and maintenance of a mature person. Other proteins constitute part of the material used to construct skin, hair, connective tissue, bones, membranes, muscles, and other body components. Still other proteins help defend the body against infectious agents, serve as transport vehicles, help provide vision, function as chemical messengers, and so on. The proteins encoded in DNA play more roles, and more diverse roles, than any other class of compounds in the body.

RNA is primarily involved in the expression of the genetic information carried in DNA. Both **carbohydrates** and **lipids** serve as fuel molecules and play numerous additional roles as well. Like proteins, carbohydrates and lipids are membrane components and are participants in recognition processes, cellular communication, molecular transport, and signal transduction. Some aid in digestion, provide lubrication for bone joints, provide structural support for plant cell walls, and occupy other biochemical niches. In their functional state, carbohydrates are frequently bound covalently to proteins or lipids, and lipids are sometimes joined covalently to proteins. Lipids include fats, steroids, eicosanoids, and several other categories of nonpolar compounds.

All of the enzyme catalyzed reactions within a living organism, including those that involve the major classes of organic compounds, constitute its **metabolism**. Metabolism is divided into **anabolism** (those reactions involved in assembling substances) and **catabolism** (those reactions that participate in degrading substances). Any compound that is a reactant, product, or intermediate during metabolism is known as a **metabolite**. In humans and many other organisms, the catabolism of dietary organic compounds, including carbohydrates and lipids, provides the energy

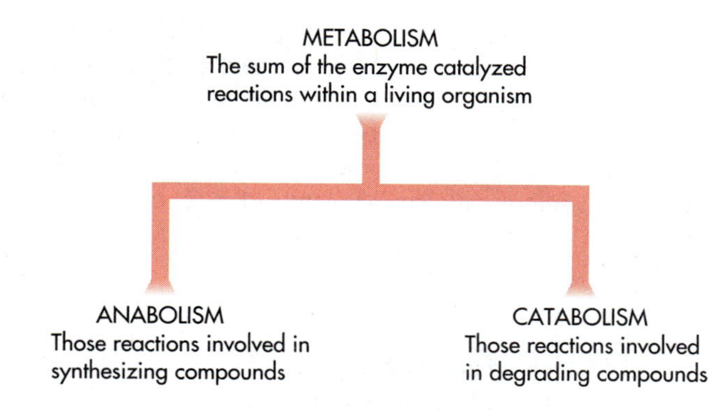

and starting materials for most anabolism. In plants and other photosynthetic organisms, sunlight provides the energy for anabolism and atmospheric CO_2 provides the carbon. A sequence of enzyme catalyzed reactions that leads from initial reactants to specific final products is called a **metabolic pathway**. Although metabolism will not be examined in any detail in the early chapters of this text, the vocabulary introduced in this paragraph will be encountered continuously in most chapters.

1.3 THE CELLULAR BASIS OF LIFE

Biochemistry entails the study of all of the numerous, unique life forms that share our planet. Even the simplest of these employ thousands of distinct reactions, and the more complex ones are genetically programmed to perform tens of thousands of reactions. It is easier to deal with this diversity and complexity if we step back and search for the common features that are shared by all living organisms. Exhibit 1.2 provides a partial listing of such shared features. It is the cellular basis of life that will be examined in this section. The many common characteristics are consistent with the widely held belief that all existing life evolved from a single primitive cell.

All **cells** are distinct morphological units that are surrounded by a **cell membrane**

EXHIBIT 1.2

Some Characteristics Shared By All Living Organisms

1. DNA is the genetic material.
2. Evolution is driven by changes (mutations) in DNA and responses of organisms to their environment.
3. DNA is replicated and passed to all progeny during reproduction.
4. The same major classes of compounds are synthesized and degraded and, for the most part, the same or similar reactions and processes are involved.
5. A constant supply of energy is required.
6. Catalysts are used to control rates of reactions.
7. Organisms consist of a single cell or an aggregate of cells.

(plasma membrane) approximately 7 nm (70 Å) thick. A nanometer (nm) is 10^{-9} meters. Some cells are further encapsulated with a **cell wall,** a **cell coat,** or some other protective barrier. Cells are classified as either **prokaryotic** or **eukaryotic** based on their structural and functional properties. Since prokaryotes and eukaryotes share many biochemical characteristics, the study of even the simplest life forms can provide valuable insights into the molecular secrets of more complex forms of life, including humans.

Prokaryotes

Prokaryotes are single-celled organisms. Most range in size from 1 to 10 μm. A micrometer (μm) is 10^{-6} meters. Most biologists place all of these life forms in the **Monera** (bacteria) kingdom, which includes both **archaebacteria** and **eubacteria.** However, evidence is accumulating that archaebacteria are more closely related to eukaryotes than to prokaryotes, and many experts now recommend that this group of organisms be called **Archaea** and be classified separately from bacteria. Further information on **taxonomy** (the study of the classification of living organisms) can be found in the article by Woese et al. listed in the Selected Readings.

Most prokaryotes lack any detectable internal compartments, and many are surrounded by a rigid cell wall. The plasma membrane may fold into a cell to form a multilayered structure known as a **mesosome.** The contents of the cell, the **cytoplasm,** contains multiple types of particles including a "packaged" form of DNA (known as a **nucleoid**) and **ribosomes** ("factories" for the assembly of proteins). **Pili** are the short protrusions on the outside of the plasma membrane that help some prokaryotes attach to surfaces. The whipping action of the long, thin filaments (**flagella**) found on many prokaryotes provide propulsion. Figure 1.2 (a) illustrates some of the structural features of a typical prokaryote.

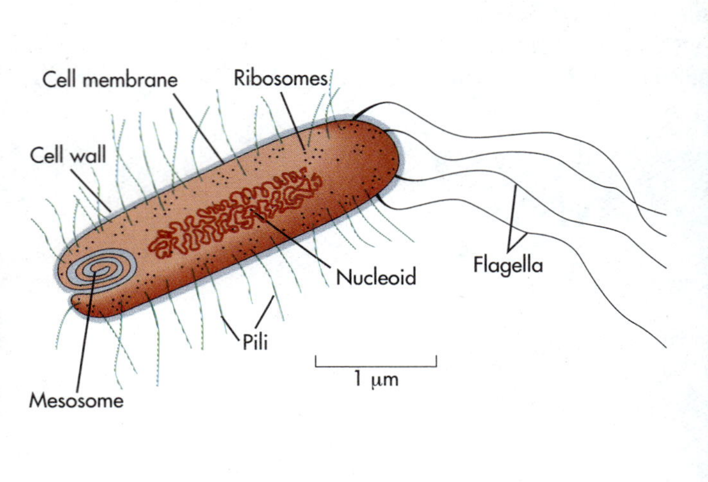

(a)

(b)

FIGURE 1.2
Some structural features of a typical prokaryotic cell. (a) Schematic diagram. (b) Electron micrograph of a dividing *E. coli* cell.

From a biochemical standpoint, the prokaryote *Escherichia coli* [abbreviated *E. coli;* Figure 1.2 (b)] is the most thoroughly studied organism on Earth. It is a rod-shaped bacterium about 2 μm long and 1 μm in diameter. It is a normal inhabitant of the human intestine and is capable of dividing once every 20 minutes. *E. coli* possesses a single, large DNA molecule that carries the blueprints for the construction of approximately 4000 different proteins. Some *E. coli* also encompass one or more small DNAs known as **plasmids.** A typical member of this class of bacteria probably contains multiple copies of several thousand different molecules; the exact number will depend upon numerous factors, including its nutritional state and its environment.

Those concerned with human biochemistry and human welfare have multiple reasons to study prokaryotes. Some of these microorganisms are normal, beneficial inhabitants of our bodies. Others lead to tooth decay, illness, or death. Clever genetic engineers have even "programmed" prokaryotes to produce scarce human proteins, to destroy environmental toxins, and to protect strawberry plants from frost.

Animal Eukaryotes

Most eukaryotic cells are 10 to 100 μm long and contain hundreds to thousands of times more genetic information than prokaryotes. The over 250 varieties of human cells, for example, vary greatly in size depending on their role and function, but all are considerably larger than a typical prokaryote and all contain about 700 times as much DNA as *E. coli.* In many eukaryotes, only a minor fraction of the DNA carries the blueprints for the production of proteins. Although the role of most of the remainder of the DNA is still uncertain, there is some evidence that at least part of it may be **"selfish DNA"** (the ultimate parasite?), which serves no useful function. However, future research is likely to prove that the body derives some benefit from a significant chunk of this genetic material. Stay tuned for the most recent developments.

Eukaryotic cells contain multiple membrane-enclosed bodies and membranous structures known as **organelles** (Table 1.2, Figure 1.3) that contribute to the division of labor within cells. Only a fraction of the eukaryotic organelles are examined in

TABLE 1.2

The Major Functions of Eukaryotic Organelles

Organelle	Major Functions in Eukaryotic Cells
Nucleus	Store DNA; involved in the expression of genetic information and ribosome assembly
Mitochondrion	Oxidation of fuel; energy production
Endoplasmic reticulum (ER)	Synthesis and transport of certain proteins and lipids
Golgi apparatus	Modification and transport of proteins and lipids from the ER; lysosome production
Lysosome	Digestion of materials brought into cells; breakdown and recycling of cellular components

FIGURE 1.3
A typical animal cell. (a) An electron micrograph of a human lymphocyte (white blood cell): N, nucleus; NE, nuclear envelope; ER, endoplasmic reticulum; Go, Golgi apparatus (Golgi complex); PM, plasma membrane; arrows, nuclear pore complexes; M, mitochondria. (b) Schematic diagram of a thin-sectioned animal cell (Based on Wolfe, 1993).

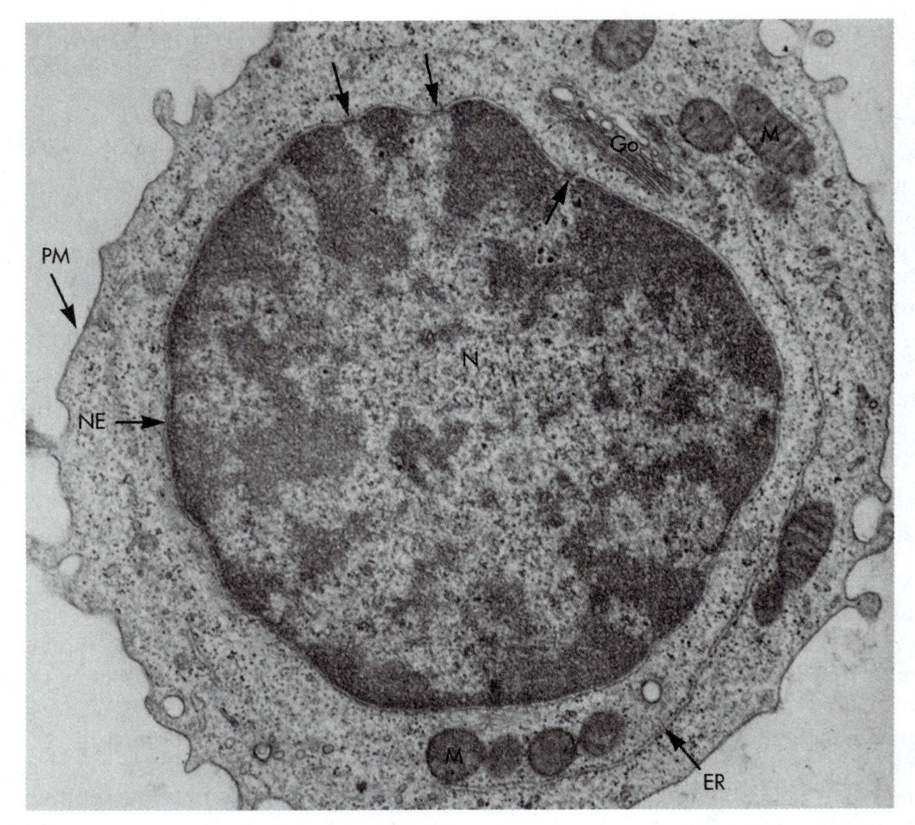

(a)

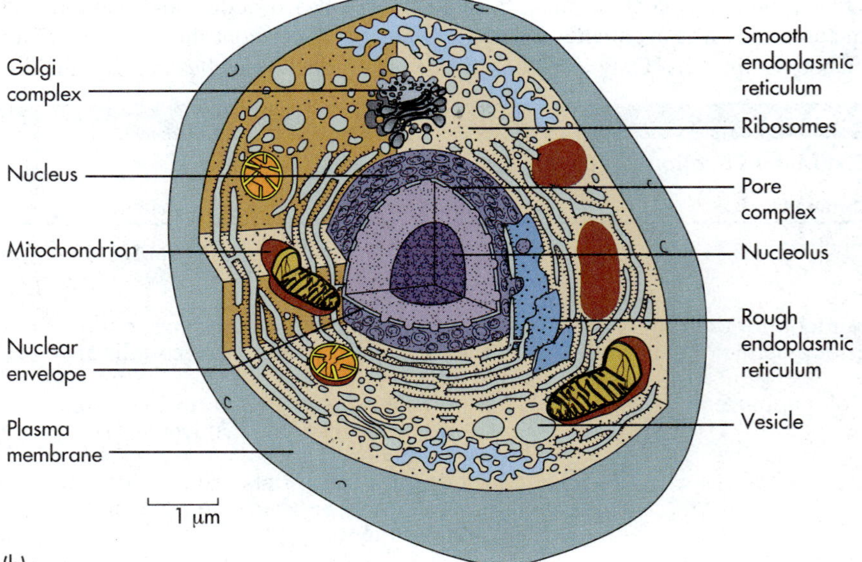

(b)

this text. Although all cellular membranes have certain structural features in common, the membranes associated with each organelle have some unique structural and functional properties that distinguish them from all other membranes. In addition to serving as highly specific chemical and physical barriers, membranes, including the plasma membrane, possess numerous biologically active agents and participate directly in a wide variety of processes. They are not inert, static systems; instead, they are dynamic, ever-changing cellular components.

The most conspicuous organelle is the **nucleus,** which contains the vast majority of a cell's DNA. Within the nucleus there is normally a dark-staining body, known as the **nucleolus** (Figure 1.4), which is involved in the assembly of ribosomes. In eu-

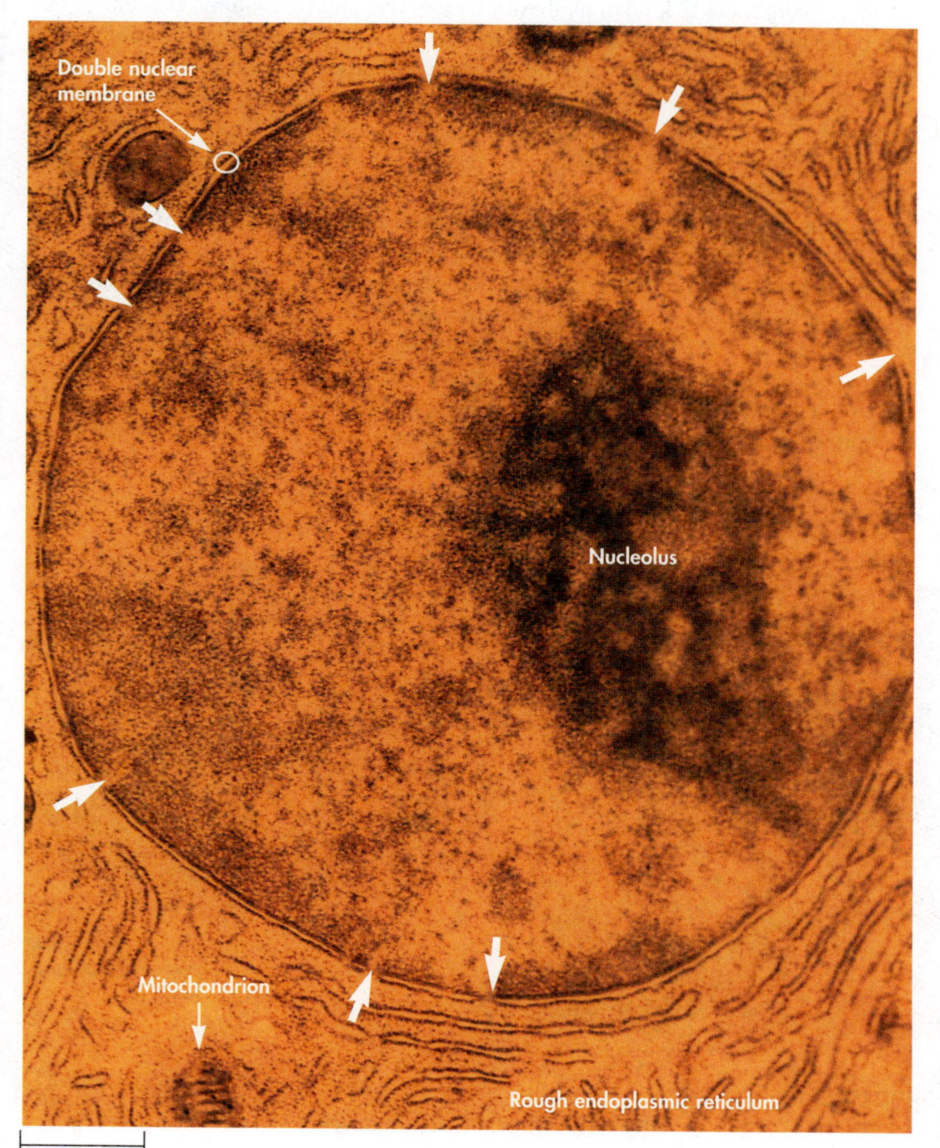

FIGURE 1.4
An electron micrograph of a nucleus with a nucleolus. The small thick arrows identify pores in the nuclear envelope.

karyotes, the term **cytoplasm** refers to the contents of a cell outside the nucleus, and the term **cytosol** identifies that part of the cytoplasm which exists outside of organelles. The double membrane that surrounds the nucleus, along with membrane-associated molecules, is called the **nuclear envelope.** This envelope is breached by numerous **nuclear pore complexes** that rivet the two membrane layers together and serve as channels for the movement of molecules between the nucleus and the cytoplasm.

Mitochondria (Figure 1.5), the powerhouses of most eukaryotic cells, are bacterium-sized organelles that, like the nucleus, are enclosed in a double membrane. The inner membrane folds in upon itself to form **cristae**. The volume enclosed by the inner membrane is known as the **matrix space,** and the volume that exists between the inner membrane and the outer membrane is described as the **intermembrane space**. This distinction will be important when mitochondria-linked metabolism and chemiosmotic theory are discussed in subsequent chapters. Mitochondria contain their own DNA, which encodes some of the proteins found within them. A single eukaryotic cell may contain well over 1000 mitochondria. According to a generally accepted hypothesis, mitochondria evolved from prokaryotic cells which, in the distant past, entered eukaryotic cells and then established a **symbiotic relationship** (one where both live-together partners benefit).

The **endoplasmic reticulum (ER),** which is usually classified as another organelle, is an intricate and winding membrane system that runs throughout a eukaryotic cell (Figure 1.6). The **rough endoplasmic reticulum (RER)** acquired its name from its rough appearance in electron micrographs. The rough appearance is due to the presence of ribosomes that are used by the cell to assemble select groups of proteins. The **smooth endoplasmic reticulum (SER),** which lacks ribosomes,

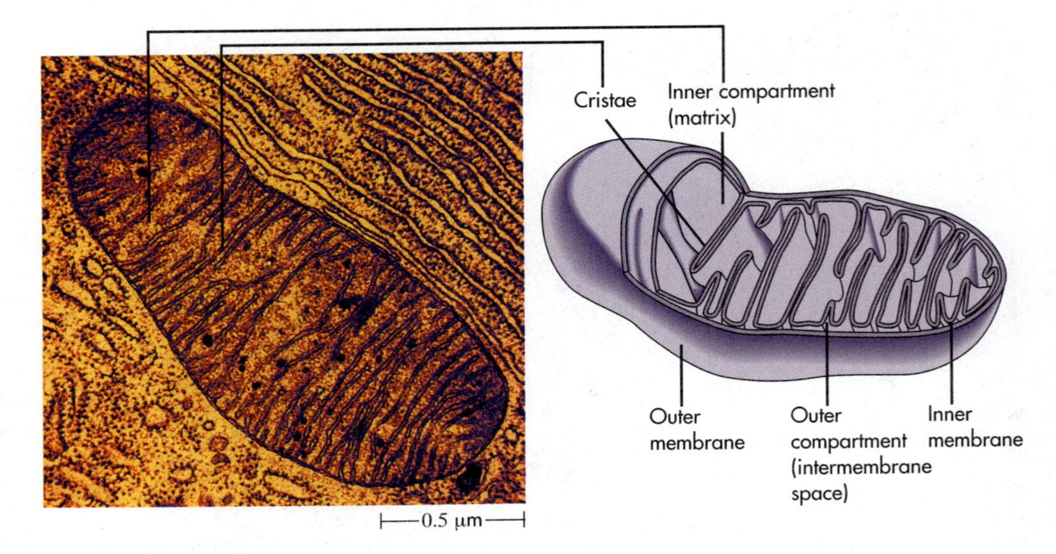

FIGURE 1.5
An electron micrograph (left) and a schematic drawing (right) of a mitochondrion.

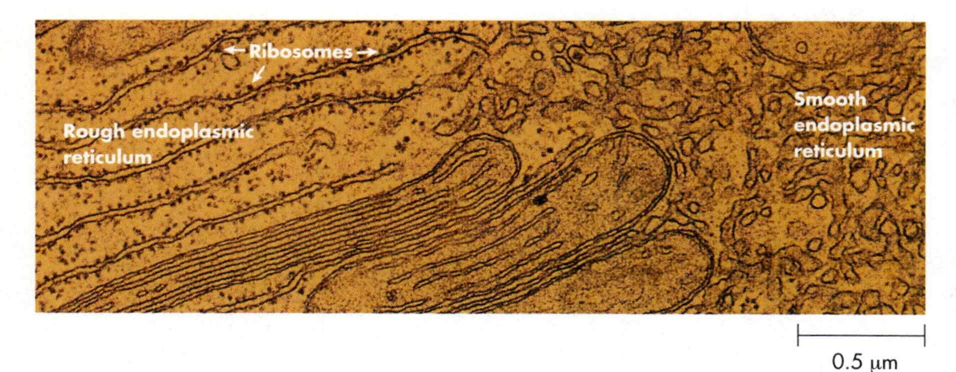

FIGURE 1.6
An electron micrograph of the endoplasmic reticulum.

produces some of the cell's lipids. The ER also participates in the processing, transport, and packaging of the compounds it produces.

The **Golgi apparatus** (**Golgi complex;** Figure 1.7) is a fourth organelle found within eukaryotic cells, including human cells. It contains stacks of flattened, membrane-enclosed bodies called **cisternae,** and its functions include the modification and transport of some of the compounds produced within the ER. The Golgi apparatus is also involved in the production of another organelle, called a **lysosome,** which is basically a membrane-enclosed bag of hydrolytic, digestive enzymes. Lysosomes are responsible for the intracellular digestion of a wide variety of materials, including compounds taken in by the cell and obsolete or damaged parts of the cell itself. Since cells often reuse the compounds produced during the digestion of cellular components, lysosomes can be viewed as miniature recycling centers.

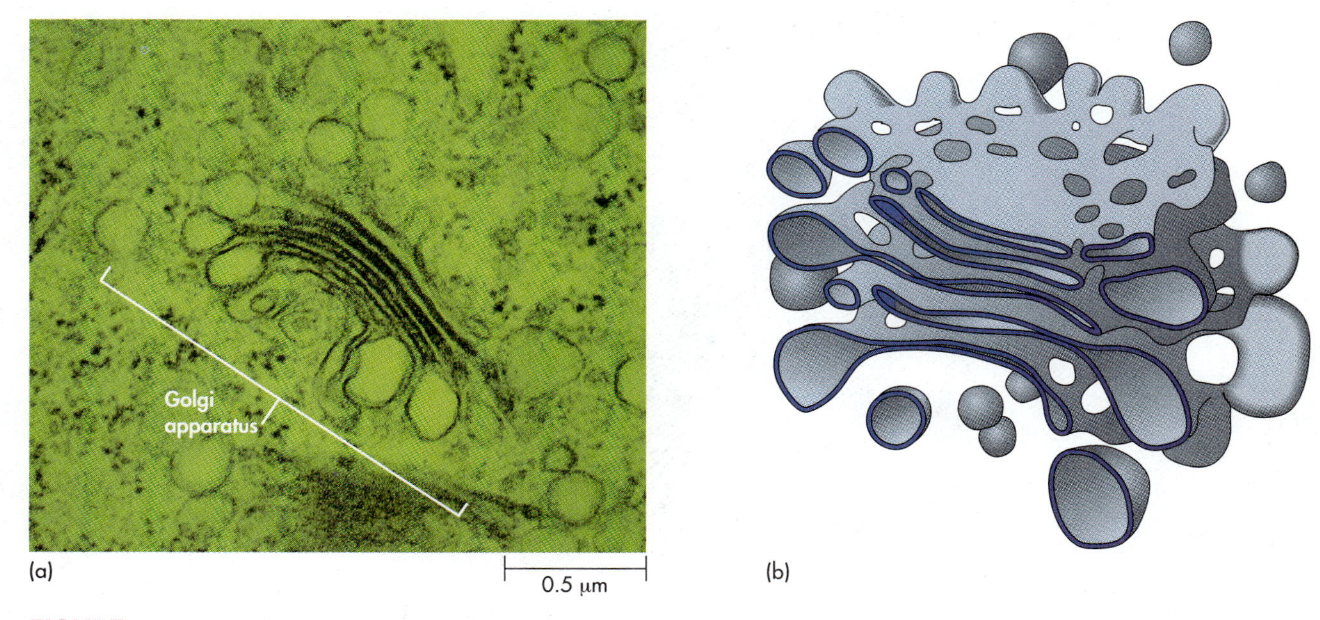

FIGURE 1.7
Golgi apparatus. (a) Electromicrograph of a Golgi apparatus in a plant cell. (b) Schematic diagram. (Based on Wolfe, 1993)

Plant Eukaryotes

A typical eukaryotic plant cell differs from a typical animal cell in several ways: it has a cell wall; it harbors unique double-membraned organelles called **plastids;** and it contains one or more conspicuous membrane-enclosed vacuoles (Figure 1.8). The cell wall provides support and protection. The functions of plastids include photo-

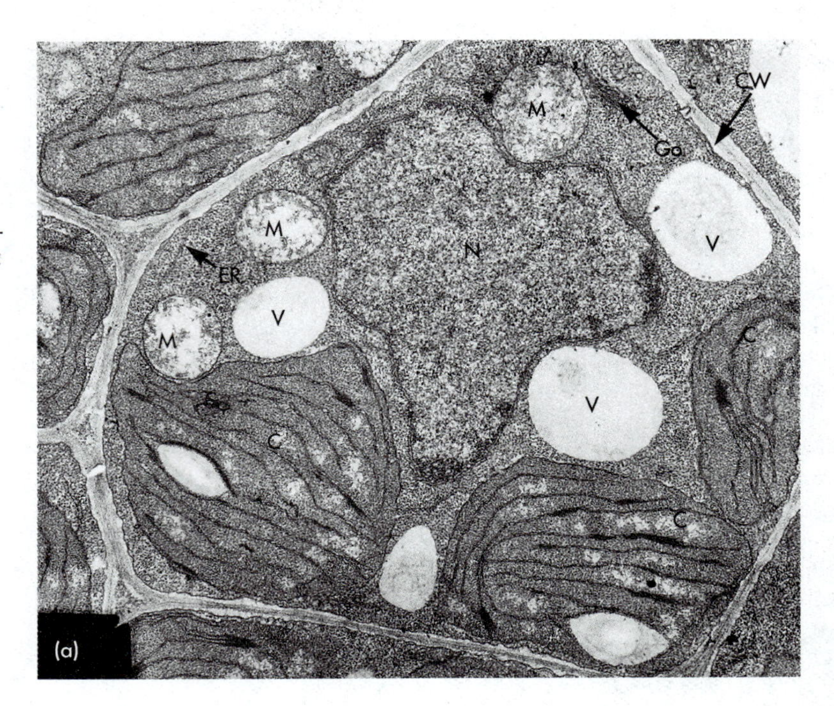

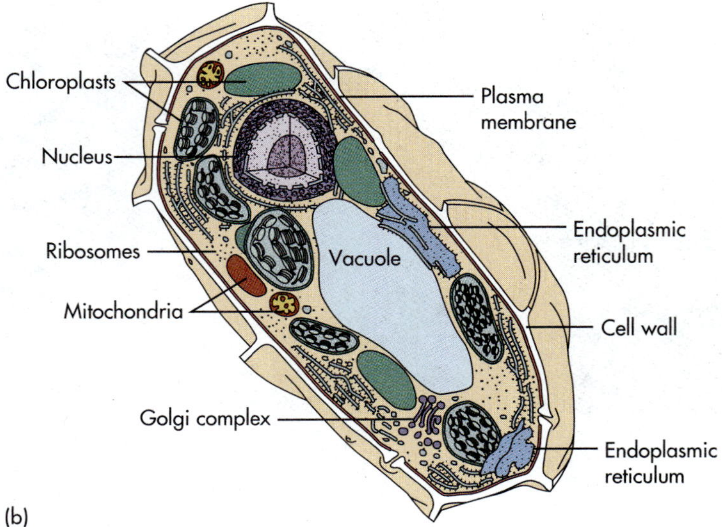

FIGURE 1.8
A typical plant cell. (a) Bean seedling (*Phaseolus vulgaris*) cell, CW, cell wall; C, chloroplasts; V, vacuoles; N, nucleus; M, mitochondria; ER, endoplasmic reticulum; Go, Golgi apparatus (Golgi complex). (b) Schematic thin-sectioned diagram of a typical plant cell (Based on Wolfe, 1993).

synthesis and the assembly and storage of starch. **Photosynthesis** is the process through which atmospheric CO_2 is captured to produce carbohydrates. The plastids that carry out photosynthesis are known as **chloroplasts.** Plastids, like mitochondria and nuclei, contain their own DNA. A major function of the fluid-filled vacuoles is the creation and maintenance of an osmotic pressure, which forces the host cell against the surrounding cell wall. In a mature plant cell, vacuoles may occupy over 90% of the total volume of the cell.

EXHIBIT 1.3

Some Properties of Eukaryotic Cells That Distinguish Them from Prokaryotic Cells

Eukaryotic cells:
- Contain hundreds to thousands of times more DNA than prokaryotic cells,
- Contain organelles,
- Contain a cytoskeleton,
- Are typically 5 to 10 times larger than prokaryotic cells.

Differences Between Prokaryotes and Eukaryotes

Exhibit 1.3 summarizes some of the major differences between prokaryotic and eukaryotic cells. One difference not yet mentioned is the existence of a **cytoskeleton** (Figure 1.9) in eukaryotes. The cytoskeleton is an internal network of protein elements that are constructed from a variety of specific proteins, including **actin, tubulin,** and **keratin.** In addition to providing some structural stability, this network participates in a variety of processes, including the movement of cells and the more or

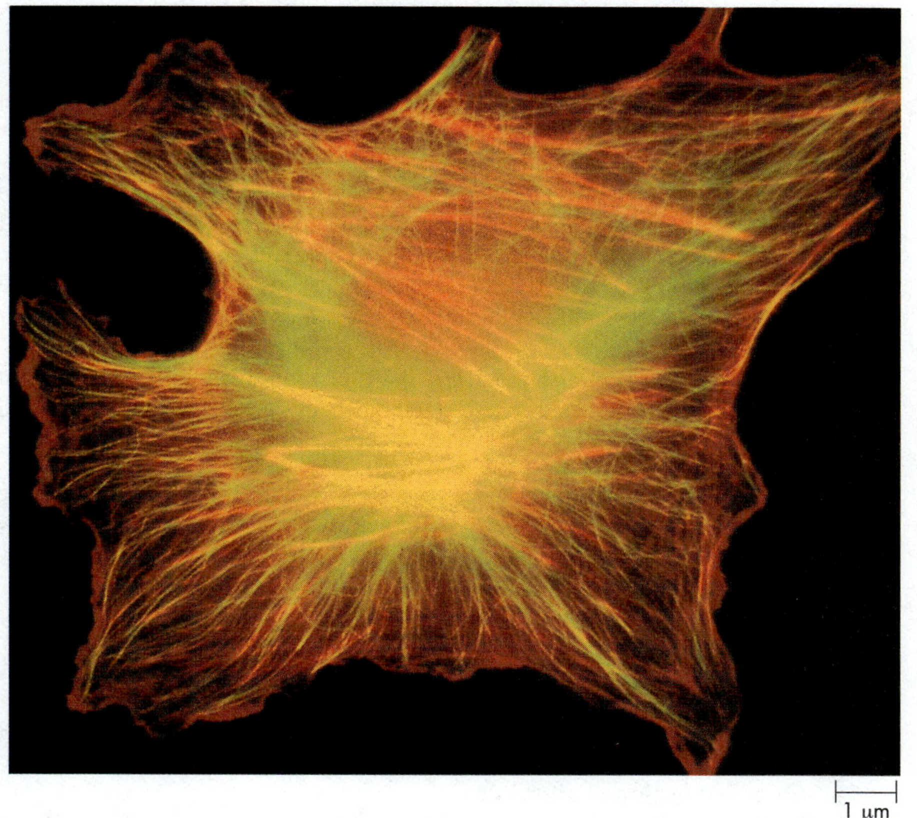

FIGURE 1.9
A micrograph of part of the cytoskeleton within a hamster kidney cell. Actin is red and tubulin is green.

less continuous internal repositioning of organelles and other intracellular components. In some instances, changes in the cytoskeleton may even play a central role in the transformation of normal cells into cancer cells. Because it lacks membranes, the cytoskeleton is not usually considered an organelle.

Viruses

It has been indicated that all living organisms consist of single cells or aggregates of cells. What about viruses? They are not cells and do not contain cells. Aren't they alive? Most biochemists would answer no, because viruses are unable to reproduce outside of living cells. Viruses are examined in more detail in later chapters.

1.4 SOME MAJOR THEMES OF BIOCHEMISTRY

There are a number of basic themes that underlie both prokaryotic and eukaryotic biochemistry. Some of these themes, including experimentation and hypothesis construction, the storage and transfer of information, replication and reproduction, structure–function relationships, structural hierarchy and compartmentation, catalysis and the regulation of reaction rates, and specificity are listed in Exhibit 1.4. Experimentation and hypothesis construction have been described and illustrated, and common sense allows us to comprehend the general nature of some of the other listed themes. *Specificity*, for example, refers to the ability of living organisms to distinguish similar but unique compounds (even stereoisomers) from one another. The specificity theme is closely linked to structure–function relationships, since slight changes in the structure of a compound can have enormous consequences from the standpoint of its biological activity. And it makes sense that the rates of most of the reactions within an organism be carefully regulated, a feat accomplished predominantly with enzymes.

The nature and the central biological significance of some of the other listed themes may not be so obvious, because some biochemistry background is required

EXHIBIT 1.4

Some Major Themes of Biochemistry

1. Experimentation and hypothesis construction
2. Storage and transfer of information
3. Regulation of gene expression
4. Replication and reproduction
5. Evolution, diversity, and change
6. Transformation, storage, and use of energy
7. Structural hierarchy and compartmentation
8. Structure–function relationships
9. Catalysis and the regulation of reaction rates
10. Specificity
11. Communication
12. Association of like with like; polar with polar, nonpolar with nonpolar

to appreciate and understand these themes. All of the themes will be fully characterized and developed as they are encountered time and time again throughout this text. In many instances, they are identified where they are encountered. In other cases, they surface and are utilized without being specifically identified. Watch for them. As they weave together, these themes create the fabric of biochemistry and provide the basis for understanding life processes.

1.5 THE IMPORTANCE OF PREREQUISITES

At one point in history it was widely accepted that some "vital force" set living organisms apart from inanimate matter. No such force has yet been identified. The molecular transformations that provide the basis for life appear to adhere to the same set of chemical laws and principles that guide the transformation of all matter. More than anything else, the ability to simultaneously coordinate and control the rates of a vast number of separate chemical reactions is what distinguishes the living from the nonliving. Although none of the individual reactions is truly unique, collectively they create and maintain life.

An understanding of the molecular basis for life must be built on some prior knowledge of basic chemistry and the chemical and physical properties of those classes of compounds that support life. It is particularly important to have a working knowledge of the prerequisite topics listed in Exhibit 1.5. This basic knowledge is usually acquired by taking at least one quarter or semester of general chemistry plus at least one quarter or semester of organic chemistry. Proficiency in high school algebra is required for general chemistry, and it is also a prerequisite for biochemistry.

Biochemistry is the first course that many students encounter in which they must recall and directly use a vast amount of material from several specific prerequisites. This tends to make biochemistry one of the more difficult courses in a science curriculum. A timely review of basic chemical principles as they are encountered can help, and it may make the difference between success and failure. This text uses both the discussion of review topics and review problems to assist you in the review process.

EXHIBIT 1.5

Some Key Topics That Are Prerequisites to Biochemistry

1. Atomic theory
2. Bonding
3. Shapes of molecules
4. Chemical equations
5. Stoichiometry
6. Equilibrium
7. Chemical kinetics
8. Energy transformations (thermodynamics)
9. Solubility
10. Oxidation and reduction
11. Acids and bases
12. The physical states of matter and those interactions that determine melting point, boiling point, solubility, and density
13. The characteristic reactions of the major classes of organic compounds
14. Isomerism

1.6 SOME STUDY TIPS

The study of biochemistry requires time, energy, self-discipline, and concentration, and is facilitated by a natural curiosity and a desire to know, to understand, and to explain. The central role of biochemistry in the biological and medical sciences should provide ample motivation. Much initial knowledge about biochemistry is acquired through memorization and the analysis of examples. We then gain confidence and proficiency through the repeated application of what we have learned. This means working problems; the more, the better.

Most of the problems within each chapter are designed to help you direct and concentrate your studies, to provide worked examples, and to allow you to test your understanding. The in-chapter problems are an integral part of each chapter, and they should ideally be answered as they are encountered. Although solutions to these problems are provided at the end of the book, it is important that you make an honest effort to solve a problem before turning to the published answer. Solutions to the end-of-chapter exercises are available in a separate Solutions Manual.

Some of the problems are specifically designed to help you review prerequisites. *Review problems that cannot be solved solely on the basis of information presented in this text are identified with an asterisk (*).* If you are unable to answer a review question, turn immediately to the published answer. Detailed discussions and explanations are provided when appropriate.

The importance of prerequisites, problem solving, motivation, time, and self-discipline has been emphasized, but success in biochemistry will also depend upon the development of study skills and some effective study habits. Exhibit 1.6 provides some recommendations on how this can be accomplished. What works well for one may not work for another. If one approach fails, try another.

EXHIBIT 1.6

Some Study Tips

1. Find a quiet place to study and eliminate distractions.
2. Work problems.
3. Study some each day, not just the days before exams.
4. Experiment with different study schedules and then try establishing a routine. Listen to your mind and your body.
5. Eat well and get plenty of rest.
6. Study while alert and rested.
7. Work problems.
8. Look over each chapter to get a feel for where it is going and what it covers before you actually start reading it.
9. Read a chapter both before and after your instructor has covered the material in lecture.
10. Summarize and restate important concepts and principles.
11. Look for relationships between different concepts and topics.
12. Try flash cards. Just the preparation of flash cards can be a learning experience.
13. Prepare a vocabulary list for each chapter.
14. Work problems.
15. Locate related texts in the library and use outside reading to expand on and clarify difficult topics.
16. Use exercise (walking, swimming, tennis, and so on) to clear the mind, relax the body, and break up long periods of intensive study.
17. Relax! Have fun! Go for it! Just do it!
18. Try organizing and recopying lecture notes.
19. Don't lose sight of where you are going and why.
20. Work problems.
21. Try preparing problems on your own, and then working with another student who is willing to do the same. Select a partner who is motivated, self-disciplined, and able to concentrate on studies.
22. List all of the terms that appear in bold print. Define each term and identify its chemical nature, biological function, or significance.
23. Use molecular models when appropriate.

SUMMARY

Biochemistry is the study of those compounds, chemical reactions, and molecular interactions that occur within living organisms. It is one of the most exciting and rapidly advancing natural sciences. Its phenomenal growth is due to the careful and critical application of the scientific method.

The major classes of biologically important compounds are DNA, proteins, RNA, carbohydrates, and lipids. DNA, the genetic material in all cells, carries within its genes information for the production of proteins, including enzymes. Enzymes, by catalyzing reactions, determine what chemical transformations proceed at a biologically significant rate within living organisms. Most RNAs are involved directly or indirectly in protein production. Proteins, carbohydrates, and lipids are components of membranes and other structural elements and are participants in recognition processes, cellular communication, molecular transport, and numerous other life-sustaining processes.

Living organisms are divided into prokaryotes and eukaryotes. Prokaryotes encompass the simpler life forms. Size, DNA content, a cytoskeleton, and organelles are some of the features of eukaryotic cells that distinguish them from prokaryotic cells. Organelles contribute to structural and functional compartmentation, one of the central themes of biochemistry. Other biochemical themes shared by all living organisms include structure–function relationships, catalysis and the regulation of reaction rates, transformation and use of energy, information transfer, and specificity.

Success in a biochemistry course requires a working knowledge of general and organic chemistry plus some good study skills. An active curiosity and an enthusiasm for life and learning are also beneficial.

EXERCISES

1.1 Prepare a list of all the terms presented in **bold** type within this chapter. Define or describe each item in this list and then, when appropriate, summarize its biological significance or importance.

1.2 What is the basic difference between the natural sciences and the social sciences?

1.3 Which of the following terms and statements accurately describe "scientific laws"?

- **a.** Untestable.
- **b.** Unchanging.
- **c.** Evolve from hypotheses.
- **d.** Absolute truths.
- **e.** Based on mystery and magic.
- **f.** Based on careful, critical observations.
- **g.** Statements about how nature appears to operate.

1.4 Summarize both the similarities and the differences between prokaryotes and eukaryotes.

1.5 Which of the following terms and statements accurately describe "membranes"?

- **a.** Found within all living organisms.
- **b.** Contain DNA.
- **c.** Are rigid, inert barriers.
- **d.** Separate the nucleus from the cytoplasm.
- **e.** Are used to build mesosomes.
- **f.** Contribute to the division of labor within eukaryotic cells.

SELECTED READINGS

Alberts, B., D. Bray, J. Lewis, M. Raff, K. Roberts, and J.D. Watson, *Molecular Biology of the Cell,* 3rd ed. New York: Garland Publishing, 1994.

An up-to-date and encyclopedic textbook on the structure and function of cells.

Luna, E.J., and A.L. Hitt, Cytoskeleton-plasma membrane interactions, *Science*, 258, 955–964 (1992).

Analyzes the association of the cytoskeleton with the plasma membrane.

Orgel. L.E., The origin of life on the earth, *Sci. Am.* 271(4), 77–83 (1994).

Presents evidence supporting the idea that the emergence of catalytic RNA was a crucial early step in the events leading to the first living organism.

Rothman, J.E., The compartmental organization of the Golgi apparatus, *Sci. Am.* 253(3), 74–89 (1985).

Examines the structure and function of the Golgi apparatus.

Rothman, J.E., and L. Orci, Molecular dissection of the secretory pathway, *Nature* 355, 409–415 (1992).

Focuses on the roles of the ER and Golgi apparatus during the secretion of substances by cells.

Sedava, D.E., *Cell Biology: Organelle Structure and Function.* Boston: Jones and Bartlett Publishers, 1993.

Places an emphasis on organelle structure and function. There is a separate chapter on each of the major eukaryotic organelles.

Sleigh, M.A., *Protozoa and Other Protists.* London: Edward Arnold, 1989.

Analyzes the structure and function of some simple organisms.

Weber, K., and M. Osborn, The molecules of the cell matrix, *Sci. Am.* 244(6), 100–120 (1985).

Examines the structure and function of the cytoskeleton.

Woese, C.R., O. Kandler, and M.L. Wheelis, Towards a natural system of organisms: proposal for the domains Archaea, Bacteria, and Eucarya. *Proc. Natl. Acad. Sci. USA, 87*, 4576–4579 (1990).

Recommends a new scheme for the classification of organisms.

Yam, P., Talking trash: linguistic patterns show up in junk DNA, *Sci. Am.* 272(3), 24 (1995).

Examines evidence that the regions of DNA that do not carry blueprints for the production of proteins may contain a language. The function, if any, of most of this DNA is unknown.

2 Water, Acids, Bases, and Buffers

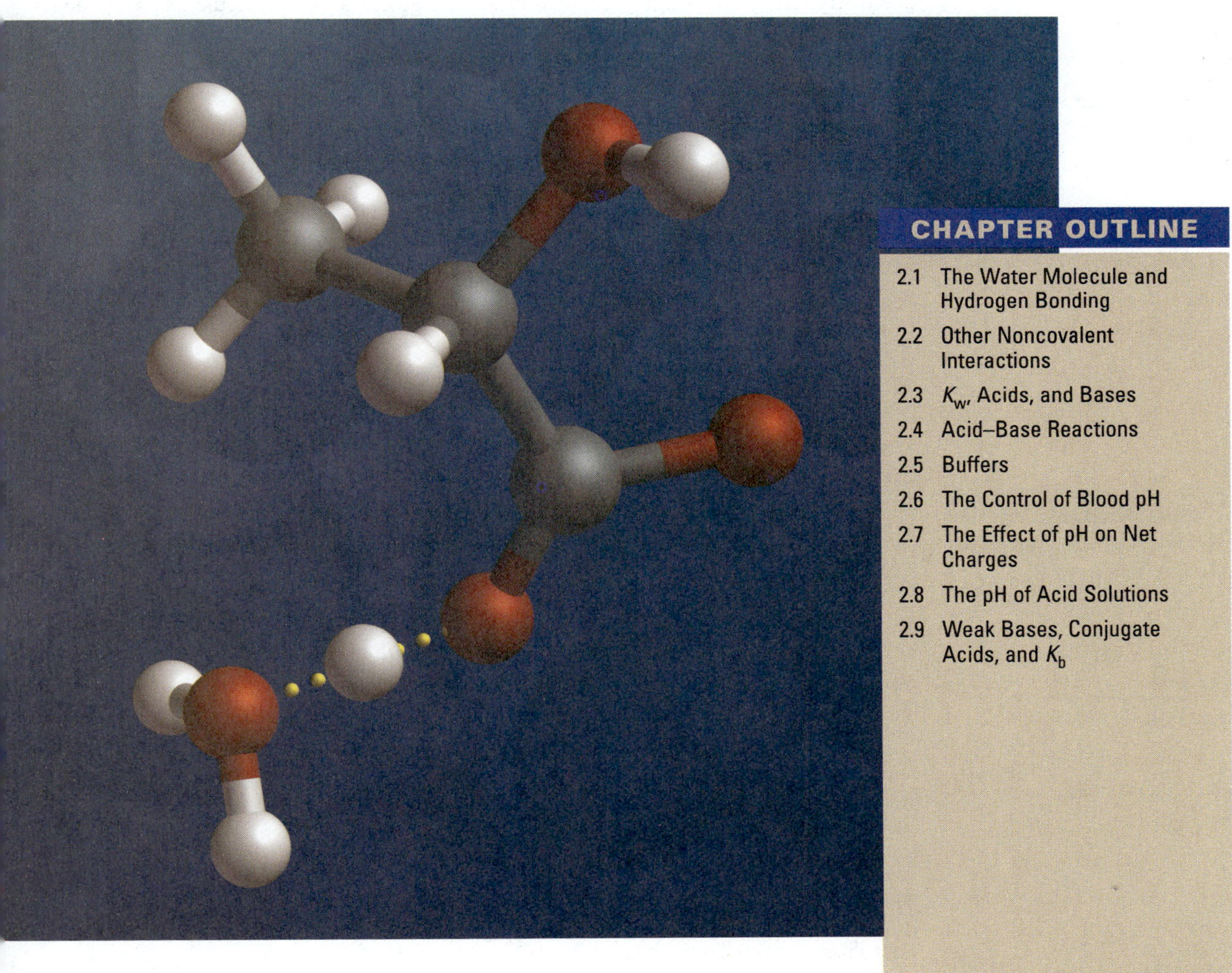

CHAPTER OUTLINE

According to a long popular hypothesis, life originated in a primordial soup in which a complex mixture of organic and inorganic substances were continually colliding, reacting, and interacting in accordance with the laws of nature. Although many of the characteristics of the first life form are uncertain, it undoubtedly consisted predominantly of water. Individual molecules were primarily held together by covalent bonds, as they are today, and the "organism" as a whole was given shape and form then as now by neighbor–neighbor contacts involving electrostatic interactions, van der Waals forces, and hydrophobic forces. The first life form gradually evolved into the prokaryotes and eukaryotes described in Chapter 1.

This chapter reviews chemical bonding and examines a portion of the unique properties of water that contributed to the initial generation of life and presently contribute to the continuing evolution of new life forms. The water story conveniently leads into a discussion of acids, bases, buffers, and salts, topics fundamental to biochemistry. The information presented in this chapter is used during the examination of numerous biologically important compounds and processes in Chapter 3 and subsequent chapters.

The association of like with like is the only major biochemical theme featured in this chapter, since most of the chapter is devoted to a review of some essential chemical principles. It should be pointed out, however, that all of the noncovalent interactions discussed in this chapter play a central role in catalysis, specificity, structure–function relationships, and certain other themes. Water, acids, bases, and salts are key participants in most of the same themes. These claims are documented in the chapters that lie ahead.

2.1 THE WATER MOLECULE AND HYDROGEN BONDING

Water is without a doubt one of the most important chemicals within all living organisms. In humans, it accounts for approximately 70% of a typical cell, over 80% of blood, and usually between 60 to 70% of the body as a whole (all percentages by weight) (Table 2.1). Most of the nucleic acids (RNAs and DNAs), proteins, carbohydrates, and other compounds in the body are designed to function in an **aqueous** (water) environment, and water plays a key role in determining the folding of these compounds and the interactions between them. Most life-sustaining reactions occur in the presence of water, and water molecules participate directly or indirectly in many of these chemical transformations. Water, for example, is a reactant in each of the vast number of hydrolysis reactions that occur in all living organisms. Water carries nutrients, hormones, metabolic wastes, and numerous other substances throughout the body, and it plays a major role in the regulation of body temperature. The average person could live several weeks without food but could live very few days without water.

TABLE 2.1

Percent By Weight of Water in Muscular Tissue and Organs of the Human Body

Tissue or Organ	Percent By Weight of Water[a]
Skeletal muscle	79[b]
Heart	83[b]
Liver	71
Kidney	81
Spleen	79
Lung	79
Brain	77

[a] In adults.
[b] Fat-free tissue.

The Polarity of Water

The polarity of water molecules accounts for many of their biologically important properties. This polarity is explained by water's bent molecular geometry and the

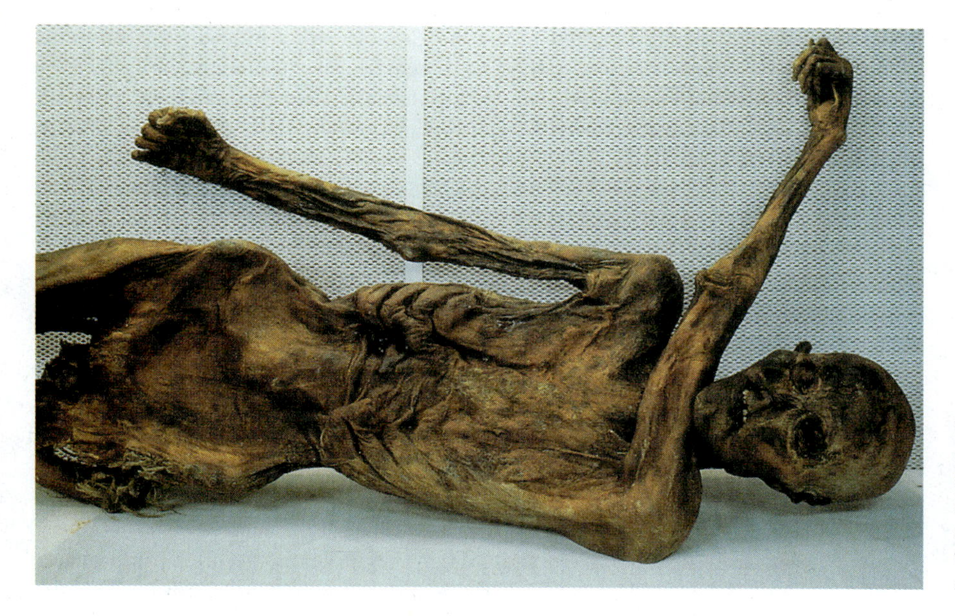

Water accounts for a large fraction of the volume occupied by your body, a fact clearly illustrated by comparing your body with the dehydrated 5,000 year old body of the "iceman" discovered in the Austrian Alps in 1991.

electronegativity (a measure of an atom's electron attracting ability) difference between H and O (Table 2.2). The more electronegative O partially pulls the shared electrons away from the two H atoms and acquires a partial negative charge in the process. Each H atom is left with a partial positive charge. Because a water molecule has a bond angle of approximately 105°, the polarities of the two covalent H—O bonds do not completely cancel one another. As a consequence, the oxygen side of a water molecule is partially negative relative to the hydrogen side of the molecule.

TABLE 2.2

Some Electronegativity Values

Element	Electronegativity
Hydrogen	2.1
Carbon	2.5
Nitrogen	3.0
Oxygen	3.5
Fluorine	4.0
Phosphorus	2.1
Sulfur	2.5

The polarity of a bond or molecule is indicated by putting a δ− (an abbreviation for partial negative) at its negative end or side and a δ+ at its positive end or side. An arrow running along a bond and pointing toward the more electronegative atom (the δ− atom) is also commonly used to depict bond polarity:

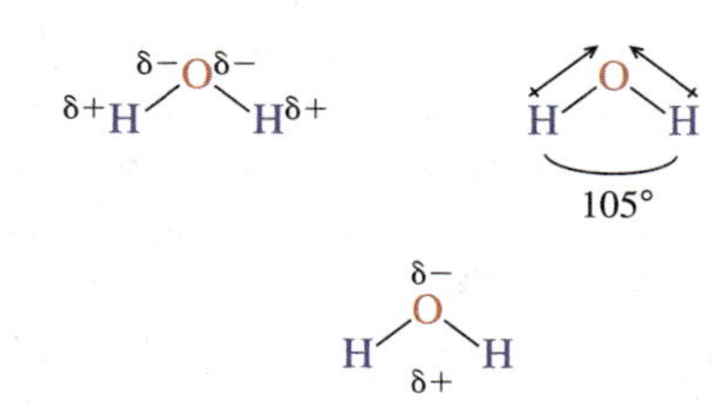

[Try Problems 2.1 through 2.3 and Problem 3.9—the answer to Problem 3.9 provides an in-depth review of electronegativity and polarity.]

The polarity of water makes it a good solvent for both polar and ionic compounds. The basic rule for solubility is "**like associates with like**" or "**like dissolves like**"; polar molecules tend to dissolve in polar solvents, nonpolar molecules in nonpolar solvents. Polar and nonpolar molecules do not readily mix to form solutions. Although water cannot dissolve nonpolar molecules, it can disperse **amphipathic compounds,** compounds with an ionic or highly polar "head" and a large nonpolar "tail." The interactions of water with amphipathic compounds are examined in more detail in Section 2.2.

Hydrogen Bonds

nonbonding pair of electrons

δ− δ+ δ− δ+
Y—H--- :X—Z

dashed line (---) = H bond
solid line (—) = covalent bond
Y = O or N
X = O or N
Z = most often C or H

Although not shown, O normally forms two covalent bonds, N three covalent bonds, and C four covalent bonds.

One of the key properties of water is its ability to form hydrogen bonds, an ability linked to its polarity. **Hydrogen bonds (H bonds)** are attractions between H's with a significant partial positive charge and other atoms with a significant partial negative charge. Within living organisms, the only H's capable of H bonding are those covalently attached to O or N, the two most electronegative elements with the exception of F (Table 2.2). All such H's can H bond. Similarly, in living organisms, an H's partner in H bonding will normally be either a N or an O with a significant partial negative charge. Thus, most biologically important H bonding can be summarized as shown in the diagram in the margin. Although a majority of the N and O atoms in organic compounds have a significant negative charge, the N in a quaternary ammonium salt [$(CH_3)_4N^+Cl^-$, for example] will have a formal positive charge that will repel rather than attract an H capable of H bonding. **[Try Problem 2.4]**

Both of the H atoms in a water molecule can H bond with O atoms in other water molecules and with most of the O or N atoms in those organic compounds from which the body is constructed. Similarly, hydrogen atoms that are covalently bonded to O or N atoms in organic compounds can H bond with O atoms in water molecules and with most O and N atoms in other organic compounds. Figure 2.1 illustrates some of the H bonding possibilities in a sample of pure water and in an aqueous solution of ethanol. Keep in mind that the figure is two dimensional, whereas a liquid sample is three dimensional. A molecule of water in the plane of the page could H bond with water or ethanol molecules behind or in front of this plane. Ethanol itself is not planar since there is a tetrahedral arrangement of bonds about each C in this molecule. **[Try Problem 2.5]**

Water has a concentration of 55.5 mol/L (moles per liter, or *M*), and each molecule in a sample of water at 25°C is, on the average, H bonded to approximately 3.4 other water molecules. The H bonds involved have a **half-life** (the time required for one half of the bonds to be broken) of approximately 10^{-10} seconds. Most H bonds existing in the human body at 37°C have similar half-lives. Once an H bond to an atom is disrupted, that atom is likely to reform immediately a new H bond with the same atom to which it was initially bonded or with a different atom. The thermal energy associated with matter at 25°C to 37°C is responsible for the continuous and repeated rupturing of the relatively weak H bonds. Although it may be difficult to conceptualize and visualize the extremely rapid dynamics (constant changes) that surround H bonding within biological systems, it is well worth the time and effort to do so.

Most H bonds have a **bond energy** (energy required to break a bond; usually expressed as kilojoule (kJ) required to break one mole of bonds) between 2 to 40 kJ/mol (0.5 to 10 kcal/mol). In contrast, most single covalent bonds have bond en-

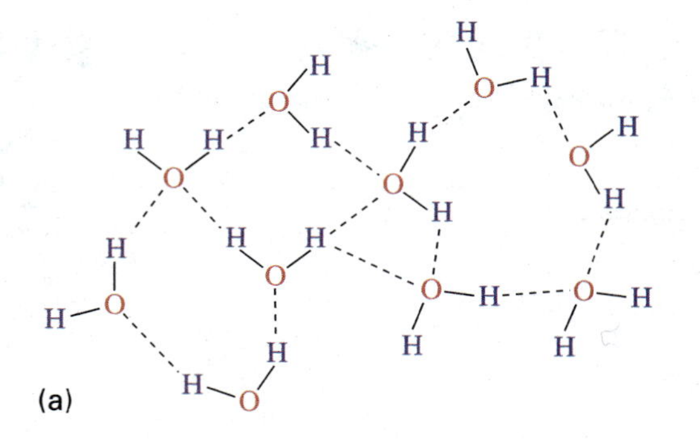

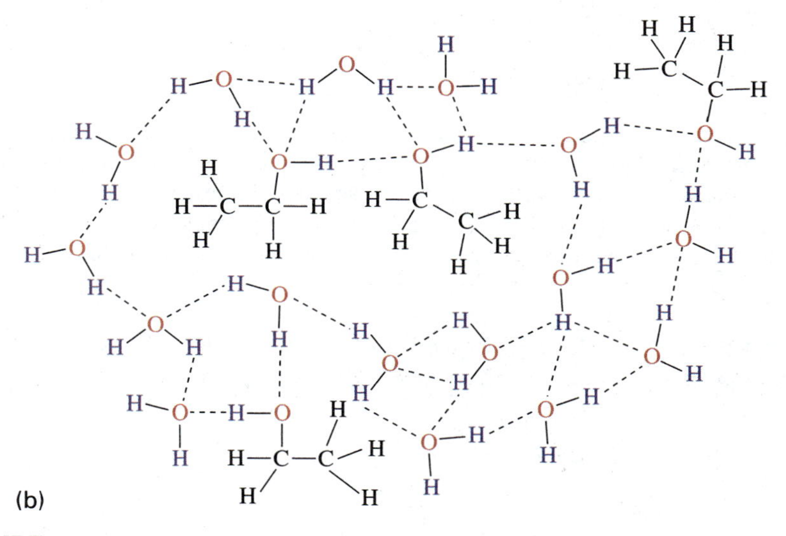

FIGURE 2.1
Hydrogen bonding possibilities in (a) water and (b) aqueous ethanol. Solid lines are covalent bonds and dashed lines are H bonds.

ergies between 250 and 500 kJ/mol (60 to 120 kcal/mol). Although a single H bond is relatively weak when compared to a covalent bond, it is relatively strong when compared to most other noncovalent interactions (Section 2.2).

Multiple H bonds can play a major role in holding separate molecules, or separate parts of the same molecule, together. H bonds, for example, account for the fact that water has a higher melting point and higher boiling point than other covalent compounds of similar molecular weight (Table 2.3). Water's melting point and boiling point contribute to its life-supporting abilities. The H bonding that exists within and between molecules also accounts for many of the biologically important properties of proteins, nucleic acids, and other classes of compounds. Hydrogen bonding is said to be **cooperative** when multiple H bonds lead to a total bond energy from H bonding that is greater than the sum of the bond energies of the individual H bonds. The concept of cooperative bonding will be revisited in Chapters 4 and 5.

TABLE 2.3

The Melting Points and Boiling Points of Some Small Covalent Compounds

Compound	Molecular Weight (amu)	Melting Point (°C)	Boiling Point (°C)
H_2O	18	0	100
CH_4	16	−184	−162
CO_2	44	—	−79 (sublimes)
H_2S	34	−83	−62
CS_2	76	−109	46
CH_2O	30	−92	−21
NH_3	17	−78	−33

***PROBLEM 2.1** *If* water had a 180° bond angle, would it be a polar or a nonpolar molecule? Explain. Note: An asterisk (*) indicates a review problem that you may have difficulty answering based solely on the information presented in this text. The published answers (see end of this book) provide a review, if needed.

***PROBLEM 2.2** Use an accepted convention to illustrate clearly the polarity, if any, of each of the following covalent bonds.

a. C—H
b. C—C
c. N—H
d. C—O
e. C—F

***PROBLEM 2.3** Which one of the five bonds listed in Problem 2.2 is the most polar bond? Explain.

PROBLEM 2.4 In each of the following compounds circle each hydrogen atom that is capable of H bonding. Put a square around each atom to which a hydrogen in a water molecule could H bond.

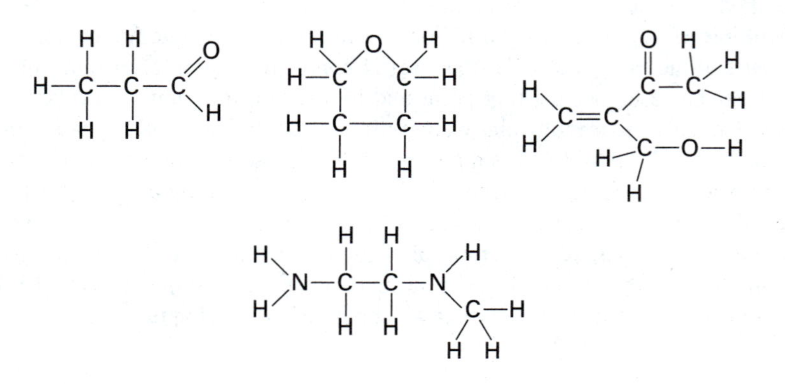

PROBLEM 2.5 Using dashed lines to represent H bonds, illustrate clearly, the H bonding possibilities in an aqueous solution of (a) acetic acid (CH_3COOH) and (b) methylamine (CH_3NH_2).

2.2 OTHER NONCOVALENT INTERACTIONS

Noncovalent interactions play a central role in numerous biochemical processes, including the folding of proteins and nucleic acids into biologically active conformations and the attachment of **ligands** (small substances that bind specifically to larger molecules or particles) to their receptors. They also hold neighboring compounds together in all samples of matter, including the human body. Noncovalent interactions can be divided into three major classes: *electrostatic interactions, van der Waals forces,* and *hydrophobic forces.*

Neighboring divers are held together by bonds (clasped hands) that are readily cleaved and reformed. Similarly, easily disrupted noncovalent interactions hold neighboring molecules together within samples of matter.

Electrostatic Interactions

The term *electrostatic* refers to stationary charges of the type associated with ions and polar molecules. A polar molecule is commonly described as a **dipole,** since it possesses two poles, one partially negative and a second partially positive. From the standpoint of chemical bonding, electrostatic interactions are attractions between oppositely charged ends (sides) of compounds or oppositely charged functional groups within compounds. They include attractions between oppositely charged ions (called **ionic bonds, ion pairs,** or **salt bridges**), attractions between oppositely charged ends of polar molecules (termed **dipole–dipole interactions** or **polar forces**) and attractions between polar molecules and ions **(ion–dipole interactions)** (Table 2.4).

TABLE 2.4

A Summary of Electrostatic Attractions

Class of Electrostatic Attraction	Example
Ion–ion (ion pair, salt bridge, ionic bond)	$CH_3—CH_2—CH_2—\overset{+}{N}H_3$ —— $^{-}O—C(=O)—CH_2—CH_2—CH_2—CH_3$
Ion–dipole	$CH_3—CH_2—CH_2—\overset{+}{N}H_3$ —— $^{\delta-}O{=}C^{\delta+}(CH_3)_2$
Dipole–dipole (polar force)	$^{\delta-}O{=}C^{\delta+}(CH_3)_2$ —— $^{\delta-}O{=}C^{\delta+}(CH_3)_2$
H bond (special class of dipole–dipole attraction)	$^{\delta+}H—\overset{\delta-}{O}—H^{\delta+}$ —— $^{\delta-}O{=}C^{\delta+}(CH_3)_2$

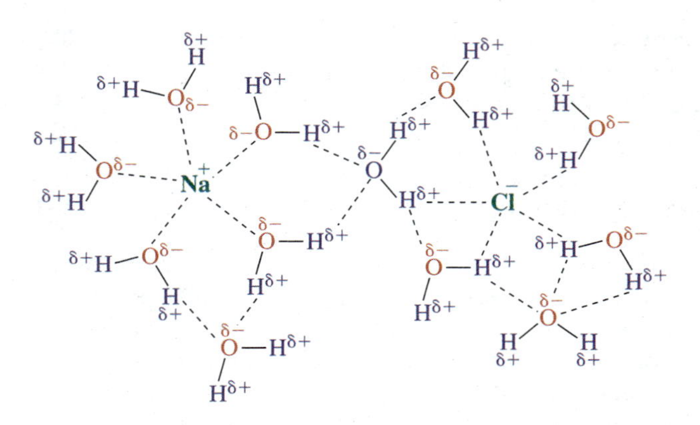

FIGURE 2.2
The hydration of ions.

In general, the greater the magnitude of the attracted opposite charges, the stronger the electrostatic bond. For this reason, ion pairs exhibit the strongest electrostatic attractions. However, such interactions are weakened in an aqueous environment because polar water molecules engulf and interact with individual ions. The ions are said to become **hydrated** (bound to water) or **solvated** (bound to solvent). It is the hydration of ions that causes ionic compounds to dissociate into their component ions when they dissolve in water. The solvent shell around each ion shields it from neighboring ions and reduces the strength of ion–ion interactions (Figure 2.2).

Although H bonds (Section 2.1) are, admittedly, dipole–dipole interactions, they are commonly placed in a unique category and given special emphasis. Hydrogen bonds play a key role in water chemistry and numerous biological processes. They also represent the strongest class of dipole-dipole interactions.

van der Waals Forces

van der Waals forces are those relatively weak interactions (usually less than 2.0 kJ/mol [0.5 kcal/mol]) that tend to bind nonpolar groups or molecules to their neighbors. These forces are explained by the existence of temporary polarities or dipoles. The temporary dipoles are a consequence of the continual movement of the negatively charged electrons within a molecule. At any given instant in time, the electrons may be distributed in favor of one side of the molecule. The favored side temporarily acquires a greater than average electron density and a partial negative charge. The instantaneous distribution of electrons may favor another side of the molecule at the next instant in time. Since the fluctuating polarities quickly average out, the molecule is nonpolar over any extended period of time. However, neighboring temporary dipoles are drawn to one another in the same manner that permanent dipoles are attracted. van der Waals forces are the weakest noncovalent interactions, since the separation of charge within temporary dipoles is normally very small and any given dipole exists for only a fleeting period of time.

$$\overset{\delta-}{\mathrm{H}}\!-\!\overset{\delta+}{\mathrm{H}} \longleftrightarrow \mathrm{H}\!-\!\mathrm{H} \longleftrightarrow \overset{\delta+}{\mathrm{H}}\!-\!\overset{\delta-}{\mathrm{H}}$$

Temporary polarities in H_2.

Hydrophobic Forces

Hydrophobic forces are the weak, nondirectional interactions responsible for the clustering or packing together of nonpolar groups in an aqueous environment. In contrast to other noncovalent interactions, hydrophobic forces are not a result of attractions between those groups or molecules that associate with one another. For this reason, hydrophobic forces are not bonds in a classical sense. A brief review of some **thermodynamics** (the study of energy and natural change) is in order before the nature of hydrophobic forces is examined. A more in-depth review is provided in Chapter 8.

The second law of thermodynamics is of central importance at the moment. This law states that the entropy of the universe as a whole increases during all physical and chemical changes. **Entropy** is a quantitative measure of the randomness or disorder in a system. An increase in entropy during a chemical or physical change tends to make the change spontaneous (favorable). Viewed from a slightly different perspective, an increase in entropy contributes in a positive fashion to the driving force for a reaction.

Hydrophobic interactions are driven by an associated increase in entropy. A mixture of water and nonpolar molecules is more stable and possesses greater entropy when the nonpolar molecules are clustered together rather than uniformly mixed with the water. A random distribution of the nonpolar molecules would disrupt and constrain the H bonding options between neighboring water molecules. Water surrounding nonpolar molecules is in a more ordered state than water far removed from nonpolar neighbors. The increase in entropy associated with hydrophobic interactions accounts for the "bond" energy associated with hydrophobic forces. The energy required to disrupt the nonpolar aggregates is typically close to 3 kJ/mol (0.7 kcal/mol) per interaction, somewhat greater than that of van der Waals forces but considerably less than that for most H bonds. The concept of hydrophobic bonding is tightly coupled to the major biochemical theme "like associates with like." **[Try Problem 2.6]**

Table 2.5 on the next page summarizes the nature and bond energies of noncovalent interactions. When comparing bond energies, keep in mind that most biologically important single covalent bonds have bond energies between 250 and 500 kJ/mol. **[Try Problems 2.7 and 2.8]**

Amphipathic Compounds

Section 2.1 noted that water can disperse **amphipathic compounds.** This dispersal phenomenon is closely related to hydrophobic forces. When amphipathic compounds mix with water they tend to aggregate (clump together) to form **micelles,** tiny spheres in which the nonpolar tails of many individual molecules are buried inside. The polar heads of the amphipathic compounds coat the surface of the micelles and allow these particles to mix readily with the polar solvent.

If the individual amphipathic compounds with their nonpolar tails were uniformly mixed with water (instead of associated in micelles), a higher energy, lower entropy state would result. Thus, an increase in entropy provides the driving force for micelle formation. Hydrophobic forces have a similar origin.

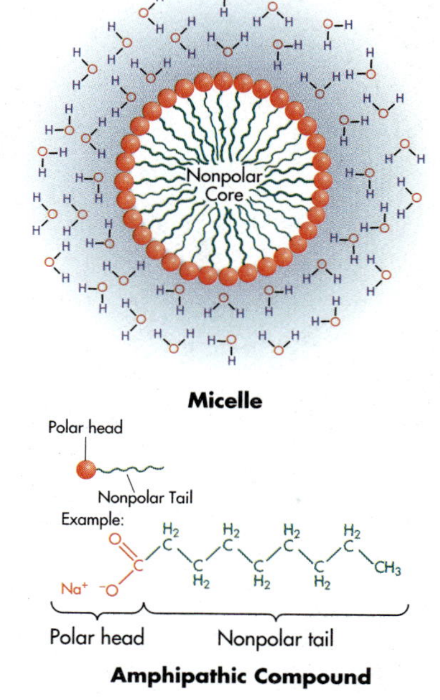

Micelle

Amphipathic Compound

TABLE 2.5

The Nature and Bond Energies of Noncovalent Interactions

Type of Interaction	Nature of Interaction	Typical "Bond" Energy (kJ/mole)
Ion-ion (ion pair) 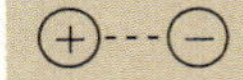 ⊕---⊖	Attraction between opposite charges	400–4000 in the absence of solvent; much smaller in an aqueous solution;---usually less than 50 kJ/mol
Ion-dipole 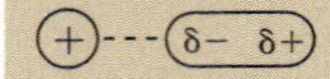 ⊕---(δ− δ+)	Attraction between opposite charges	40–600
Hydrogen bond (special dipole–dipole) 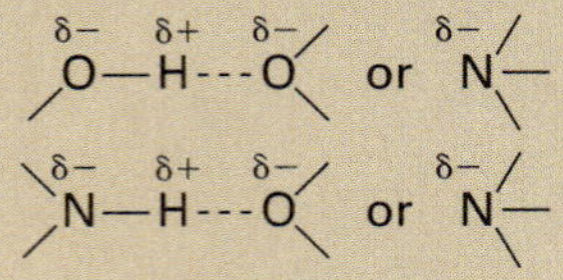 $\overset{\delta-}{O}$—$\overset{\delta+}{H}$---$\overset{\delta-}{O}$ or $\overset{\delta-}{N}$ $\overset{\delta-}{N}$—$\overset{\delta+}{H}$---$\overset{\delta-}{O}$ or $\overset{\delta-}{N}$	Attraction between opposite charges	2–40; normally within the lower half of this range in proteins and nucleic acids
Other dipole–dipole 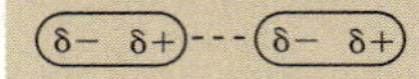 (δ− δ+)---(δ− δ+)	Attraction between opposite charges	1–25
van der Waals (temporary dipole–dipole) 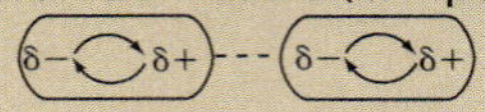 (δ− δ+)---(δ− δ+)	Attraction between opposite charges	0.1–4
Hydrophobic (X = nonpolar molecule) 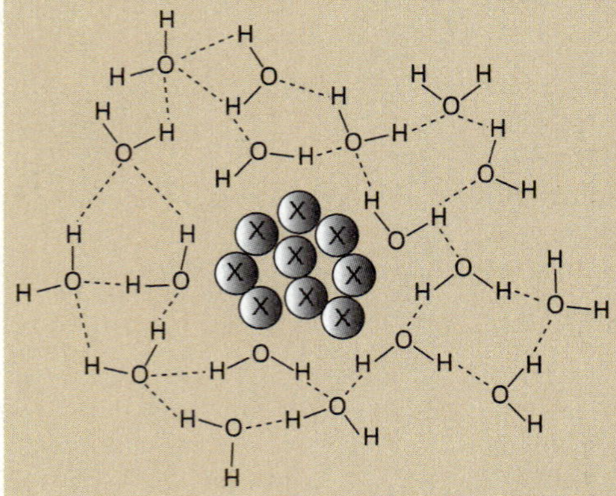	Increase in entropy (randomness)	Normally close to 3

PROBLEM 2.6 Diffusion (the movement of a substance from a region of higher concentration to a region of lower concentration) is a physical process driven by an associated increase in entropy. Explain why entropy increases during diffusion.

***PROBLEM 2.7** List those forces that would tend to hold separate molecules of cyclohexane together in (a) A pure sample of cyclohexane and (b) An aqueous mixture of cyclohexane.

***PROBLEM 2.8** What specific noncovalent interactions hold neighboring molecules of dimethyl ether (CH_3OCH_3) together in a pure sample of dimethyl ether? What noncovalent interactions bind dimethyl ether and water molecules together in an aqueous solution of dimethyl ether?

2.3 K_W, ACIDS, AND BASES

This section moves from water and noncovalent interactions into acids and bases, additional review topics of central importance. The human body is virtually filled with acids and bases, and all life depends upon a careful balance between these biologically active substances.

The acid–base story begins with water. Two water molecules can react to form H_3O^+ (a **hydronium ion**) and OH^- (a **hydroxide ion**). This reaction is reversible and normally exists at equilibrium in aqueous solutions.

$$H_2O + H_2O \rightleftharpoons H_3O^+ + OH^-$$

It is common practice to talk about the dissociation of single water molecules into H^+ and OH^-, and biochemists frequently speak of H^+ concentration (abbreviated $[H^+]$; brackets, in general, are used to abbreviate molar concentration) in aqueous systems.

$$H_2O \rightleftharpoons H^+ + OH^-$$

The true $[H^+]$ in aqueous systems is negligible, since virtually all the H^+ formed reacts with water molecules to form H_3O^+. When $[H^+]$ is encountered during the discussion of acids and bases, it normally refers to $[H_3O^+]$.

K_W

The equilibrium constant expression for the dissociation of water is $K_{eq} = [H^+][OH^-]/[H_2O]$. Multiplying both sides of this equation by $[H_2O]$, we obtain $K_{eq}[H_2O] = [H^+][OH^-]$. In dilute aqueous solutions, the $[H_2O]$ is approximately 55.5 *M* and can, for many purposes, be considered constant. When $[H_2O]$ is assumed to be constant, $K_{eq}[H_2O]$ becomes the product of two constants and equals a third constant known as $\boldsymbol{K_w}$:

$$K_{eq}[H_2O] = K_w = [H^+][OH^-]$$

Experimentally, $K_w = 1 \times 10^{-14}$ at 25°C, and it is approximately the same at **body temperature** (37°C). Knowing the K_w for aqueous solutions, we can calculate the $[H^+]$ given the $[OH^-]$ and vice versa:

$$\text{Since } K_w = [H^+][OH^-] = 1 \times 10^{-14}$$

$$[H^+] = 1 \times 10^{-14}/[OH^-] \quad \text{and} \quad [OH^-] = 1 \times 10^{-14}/[H^+]$$

[Try Problems 2.9 and 2.10]

Acids and Bases

The $[H^+] = [OH^-]$ in a sample of pure water, since each time a water molecule dissociates one H^+ and one OH^- are produced. The $[H^+]$ and the $[OH^-]$ will differ if a solute is present that is an acid or a base. From a biological standpoint, an **acid** is usually defined as a **proton** (an H^+) **donor** and a **base** is defined as a **proton acceptor.** Acids dissociate when dissolved in water to yield H^+ (more accurately, they react with water to give H_3O^+), whereas bases dissolve in water to yield, either directly or indirectly, OH^-.

ACIDS: dissociate in water
to yield H^+ (H_3O^+)

$$AH \longrightarrow A^- + H^+$$

or, more accurately,

$$AH + H_2O \longrightarrow A^- + H_3O^+$$

Examples: $HCl \longrightarrow H^+ + Cl^-$
$HCl + H_2O \longrightarrow Cl^- + H_3O^+$

BASES: dissolve in water to yield OH^-

$$BOH \longrightarrow B^+ + OH^-$$

$$B + H_2O \longrightarrow BH^+ + OH^-$$

Examples: $NaOH \longrightarrow Na^+ + OH^-$
$NH_3 + H_2O \longrightarrow NH_4^+ + OH^-$

An **acidic solution** is one in which the $[H^+]$ is greater than the $[OH^-]$; a **basic solution** is one in which the $[OH^-]$ is greater than the $[H^+]$; and a **neutral solution** is one in which the $[OH^-] = [H^+]$.

pH

The $[H^+]$ and the $[OH^-]$ in aqueous solutions are commonly expressed in terms of **pH** ($-\log [H^+]$; the negative log of the $[H^+]$) and **pOH** ($-\log [OH^-]$; the negative log of the $[OH^-]$), respectively. Since the $[H^+]$ times the $[OH^-]$ equals 1×10^{-14}, pH + pOH = 14. Although the use of logarithms may seem unnecessary to a beginning student, the definitions of pH and pOH do, believe it or not, simplify some calculations. In addition, the use of logarithms provides a convenient means to describe and illustrate certain relationships. The associations among $[H^+]$, $[OH^-]$, pH, and pOH can be summarized as follows:

pH line

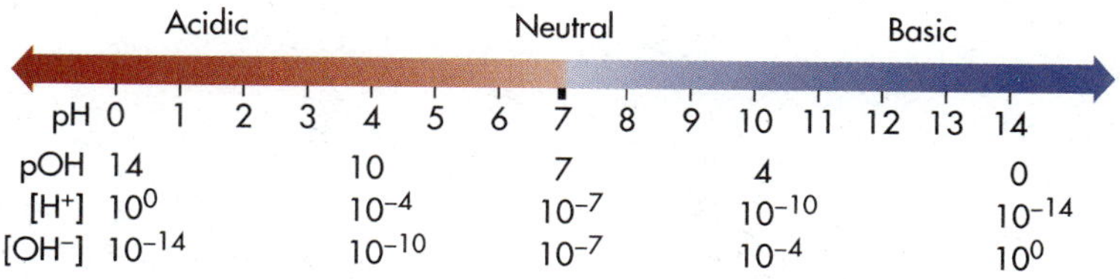

Acidic — Neutral — Basic

pH	0	1	2	3	4	5	6	7	8	9	10	11	12	13	14
pOH	14				10			7			4				0
$[H^+]$	10^0				10^{-4}			10^{-7}			10^{-10}				10^{-14}
$[OH^-]$	10^{-14}				10^{-10}			10^{-7}			10^{-4}				10^0

The pH line given immediately above indicates that it is possible to have pH values below zero and greater than 14. A negative pH (pH < 0) means that the $[H^+]$ is greater than 1 *M,* whereas a pH above 14 (pH > 14) denotes that the $[H^+]$ is less than 10^{-14} *M*. One does not often encounter such extremes in pH, and they are definitely incompatible with life. The approximate pH values of some common liquids are summarized in Table 2.6. **[Try Problem 2.11]**

It is important to distinguish "acids" and "acidic solutions." An acidic solution contains more acid than base, has a pH below 7, and has a $[H^+]$ that is both greater than its $[OH^-]$ and greater than 1×10^{-7} *M*. Since an acid is undissociated unless it is in solution, it does not, in a pure state, have an $[H^+]$ or a pH. An acid, when added to a basic solution, will lower the pH, but in small amounts, it may not make the solution acidic. Acid–base reactions are discussed in Section 2.4.

Weak Versus Strong Acids

Most acids can be classified clearly as either strong or weak. A **strong acid** is normally defined as one that dissociates completely when dissolved in water. Theoretically, no acid dissociates completely, but from a practical standpoint the equilibrium constant for some dissociations is so large that the dissociation can be considered total. **Hydrochloric acid** (HCl), **nitric acid** (HNO_3), **sulfuric acid** (H_2SO_4), and **perchloric acid** ($HClO_4$) are examples of strong acids. **Weak acids** do not dissociate completely when dissolved in water; the equilibrium constant for the dissociation of a weak acid (abbreviated K_a) usually has a numerical value less than one. The **pK_a** for a weak acid is defined as the negative log of its K_a (p$K_a = -\log K_a$). In the expressions pH, pOH, and pK_a, the "p" means negative log.

One can imagine a group of acids in which it would be difficult to decide whether or not, from a practical standpoint, the dissociation was complete. At the present time, few such acids have been discovered. All of the acids encountered in this text can clearly be classified as either strong or weak. Among the weak acids, however, there is a wide range of K_a values, so some weak acids are much weaker than others. The larger the K_a of a weak acid the stronger the acid.

Phosphoric acid and **carbonic acid** are biologically important examples of inorganic weak acids. **Carboxylic acids, phenols, thiols (mercaptans, thioalco-**

TABLE 2.6

The Approximate pH's of Some Common Liquids

Liquid	pH
Gastric juices	1.6–1.8
Lemon juice	2.3
Vinegar	2.4–3.4
Soft drinks	2.0–4.0
Urine	4.4–8.0
Saliva	6.2–7.4
Blood	7.35–7.45
Bile	7.8–8.6
Household ammonia	11

TABLE 2.7

Some Classes of Organic Compounds That Are Weak Acids

Class of Organic Compound	Specific Example of Class	Functional Group	Dissociation Reaction[a] Weak Acid		Dissociation Reaction[a] Conjugate Base
Carboxylic acids	$H_3C{-}C(=O){-}O{-}H$	$-C(=O){-}O{-}H$	$-C(=O){-}O{-}H$	$\rightleftharpoons$	$-C(=O){-}O^- + H^+$
Phenols	$CH_3{-}C_6H_4{-}O{-}H$	Aromatic $-O-H$	$-O-H$	$\rightleftharpoons$	$-O^- + H^+$
Thiols	$H_3C{-}CH_2{-}S{-}H$	$-S-H$	$-S-H$	$\rightleftharpoons$	$-S^- + H^+$
Protonated amines[b]	$H_3C{-}\overset{+}{N}H_3$	$-\overset{+}{N}{-}H$	$-\overset{+}{N}{-}H$	$\rightleftharpoons$	$-N + H^+$

[a]In each case, the weak acid is converted to its conjugate base when it loses a proton (H^+). In the reverse reaction, the conjugate base accepts a proton to form the weak acid.

[b]The protonated form of an amine is commonly classified as the conjugate acid of the nonprotonated form (a weak base).

hols), and the **protonated forms of amines** provide examples of organic weak acids (Table 2.7). Table 2.8 lists the names, formulas, K_a's and pK_a's for some specific weak acids. This information will be needed to solve several of the problems in this and later chapters. **[Try Problems 2.12 through 2.15]**

When a weak acid dissociates in water, the product formed in addition to H^+ is known as its **conjugate base**:

$$\text{weak acid} \rightleftharpoons H^+ + \text{conjugate base}$$

The conjugate base is a true base since it accepts an H^+ during the reverse reaction. The conjugate base of a weak acid is usually either uncharged or negatively charged, as illustrated in Table 2.8. Some conjugate bases, including several listed in Table 2.8, contain a dissociable H^+ and can function as acids as well as bases. The structure of a conjugate base always differs from the structure of its weak acid by a single H^+. **[Try Problems 2.16 and 2.17]**

TABLE 2.8

Names, Formulas, K_a's, and pK_a's for Some Common Weak Acids

Acid Name	Formula	Dissociation Reaction	K_a Expression	K_a at 25°C	pK_a
Phosphoric acid	H_3PO_4	$H_3PO_4 \rightleftharpoons H^+ + H_2PO_4^-$	$K_a = \frac{[H^+][H_2PO_4^-]}{[H_3PO_4]}$	7.08×10^{-3}	2.15
Dihydrogen phosphate	$H_2PO_4^-$	$H_2PO_4^- \rightleftharpoons H^+ + HPO_4^{2-}$	$K_a = \frac{[H^+][HPO_4^{2-}]}{[H_2PO_4^-]}$	6.31×10^{-8}	7.20
Hydrogen phosphate	HPO_4^{2-}	$HPO_4^{2-} \rightleftharpoons H^+ + PO_4^{3-}$	$K_a = \frac{[H^+][PO_4^{3-}]}{[HPO_4^{2-}]}$	4.17×10^{-13}	12.4
Carbonic acid	H_2CO_3	$H_2CO_3 \rightleftharpoons H^+ + HCO_3^-$	$K_a = \frac{[H^+][HCO_3^-]}{[H_2CO_3]}$	4.46×10^{-7}	6.35
Hydrogen carbonate or bicarbonate	HCO_3^-	$HCO_3^- \rightleftharpoons H^+ + CO_3^{2-}$	$K_a = \frac{[H^+][CO_3^{2-}]}{[HCO_3^-]}$	4.68×10^{-11}	10.3
Acetic acid	CH_3COOH	$CH_3COOH \rightleftharpoons H^+ + CH_3COO^-$	$K_a = \frac{[H^+][CH_3COO^-]}{[CH_3COOH]}$	1.74×10^{-5}	4.76
Lactic acid	$CH_3CHOHCOOH$	$CH_3CHOHCOOH \rightleftharpoons H^+ + CH_3CHOHCOO^-$	$K_a = \frac{[H^+][CH_3CHOHCOO^-]}{[CH_3CHOHCOOH]}$	1.38×10^{-4}	3.86
Ammonium	NH_4^+	$NH_4^+ \rightleftharpoons H^+ + NH_3$	$K_a = \frac{[H^+][NH_3]}{[NH_4^+]}$	5.62×10^{-10}	9.25
Phenol	C_6H_5–OH	$C_6H_5\text{–OH} \rightleftharpoons H^+ + C_6H_5\text{–O}^-$	$K_a = \frac{[H^+][C_6H_5\text{–O}^-]}{[C_6H_5\text{–OH}]}$	1.00×10^{-10}	10.0
Protonated form of methylamine	$CH_3NH_3^+$	$CH_3NH_3^+ \rightleftharpoons H^+ + CH_3NH_2$	$K_a = \frac{[H^+][CH_3NH_2]}{[CH_3NH_3^+]}$	3.16×10^{-11}	10.5

PROBLEM 2.9 What is the $[H^+]$ in a solution if the $[OH^-]$ is (a) 2.0×10^{-11} *M* (b) 6.3×10^{-5} *M*, and (c) 8.4×10^{-8} *M*?

PROBLEM 2.10 What is the $[OH^-]$ in a solution if the $[H^+]$ is (a) 1.4×10^{-10} *M*, (b) 7.9×10^{-7} *M*, and (c) 6.6×10^{-5} *M*?

PROBLEM 2.11 Complete the following table:

Solution	$[H^+]$	$[OH^-]$	pH	pOH	Acidic, Basic or Neutral?
1	2.5 *M*				
2		1×10^{-7} *M*			
3			12.3		
4				6.8	

PROBLEM 2.12 Which one of the weak acids in Table 2.8 is the strongest acid? Which is the weakest acid?

PROBLEM 2.13 Write an equation for the dissociation of the acidic hydrogen in aspirin. Provide structural formulas for all reactants and products. Write the K_a expression for aspirin.

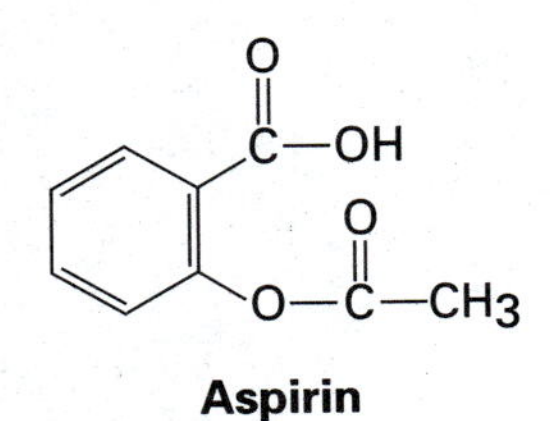

Aspirin

PROBLEM 2.14 If the equilibrium constant for the dissociation of an acid were 3.7×10^{17}, would the acid be classified as weak or strong? Explain.

PROBLEM 2.15 Write the structural formula for a single compound that is simultaneously a thiol, a phenol, and an amine. How many acidic hydrogens does the fully protonated form of this compound contain?

PROBLEM 2.16 Clearly identify the conjugate base of each weak acid given in Table 2.8. Put a circle around those conjugate bases that are also weak acids.

PROBLEM 2.17 Determine which of the following pairs of compounds represent an acid:conjugate base pair. What information would be needed to determine the strength of the acids involved?

a. HCN : CN^-

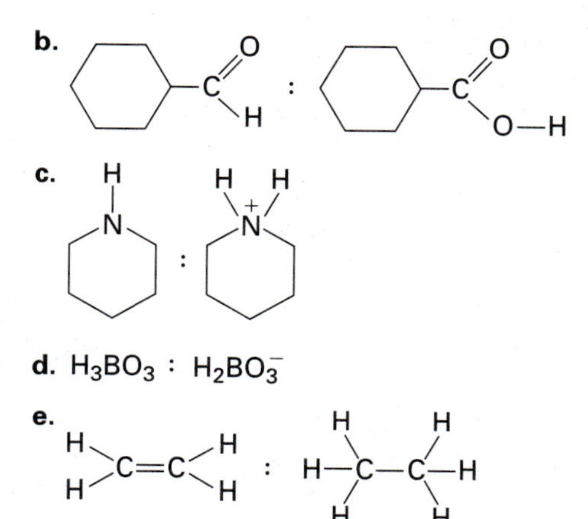

2.4 ACID–BASE REACTIONS

Acid(proton donor)–base(proton acceptor) reactions occur continuously within all living organisms and their environment. Many of these reactions are essential to survival. This section focuses on three categories of reactions: the reaction of strong acids with strong bases; the reaction of strong bases with weak acids; and the reaction of strong acids with the conjugate bases of weak acids (Table 2.9). All three classes of reactions have large K_{eq}'s, so they tend to go to completion. Unless clearly specified otherwise, you should assume that all the acid–base reactions discussed in this text occur in an aqueous environment. **Sodium hydroxide (NaOH)** and **potassium hydroxide (KOH)** are classical examples of strong bases, compounds that dissociate completely in water to yield OH^-. Some common strong acids, weak acids, and conjugate bases were examined in the previous section.

Reaction of Strong Acids with Strong Bases

When a strong acid transfers a proton to a strong base the product is a salt plus water.

$$\text{Strong acid(aq)} + \text{strong base(aq)} \rightleftharpoons \text{salt(aq)} + \text{water}$$

A **salt** is any compound that contains ions other than H^+ and OH^-. The (aq)'s in the equation indicate that the reaction is occurring in an aqueous solution. The reaction of HNO_3 with KOH provides a specific example.

$$HNO_3(aq) + KOH(aq) \rightleftharpoons KNO_3(aq) + H_2O$$

TABLE 2.9

Three Categories of Acid–Base Reactions

Category of Reaction	Overall Reaction	Net Reaction
Strong acid + strong base	Strong acid + strong base $\rightleftharpoons$ H_2O + salt	$H^+ + OH^- \rightleftharpoons H_2O$
	Examples: $HCl + NaOH \rightleftharpoons H_2O + NaCl$ $HNO_3 + KOH \rightleftharpoons H_2O + KNO_3$	
Strong acid + conjugate base (CB) of weak acid (WA) or salt containing CB	Strong acid + CB or salt containing CB $\rightleftharpoons$ WA or salt containing WA + (salt)	$H^+ + CB \rightleftharpoons WA$
	Examples: $HCl + CH_3COONa \rightleftharpoons CH_3COOH + NaCl$ $HCl + NH_3 \rightleftharpoons NH_4Cl$	Examples: $H^+ + CH_3COO^- \rightleftharpoons CH_3COOH$ $H^+ + NH_3 \rightleftharpoons NH_4^+$
Strong base + weak acid or salt containing WA	Strong base + WA or salt containing WA $\rightleftharpoons$ CB or salt containing CB + H_2O + (salt)	$OH^- + WA \rightleftharpoons CB + H_2O$
	Examples: $NaOH + CH_3COOH \rightleftharpoons CH_3COONa + H_2O$ $NaOH + NH_4Cl \rightleftharpoons NH_3 + H_2O + NaCl$	Examples: $OH^- + CH_3COOH \rightleftharpoons CH_3COO^- + H_2O$ $OH^- + NH_4^+ \rightleftharpoons NH_3 + H_2O$

Salts, strong acids, and strong bases all dissociate into ions when dissolved in water. The reaction of HNO_3 with KOH can be rewritten to illustrate this fact.

$$H^+(aq) + NO_3^-(aq) + K^+(aq) + OH^-(aq) \rightleftharpoons K^+(aq) + NO_3^-(aq) + H_2O$$

The water molecule is written in a nondissociated form since it has a very small dissociation constant and is considered a **covalent compound** (contains only covalent bonds).

Any ions or molecules that appear in identical forms and amounts on both sides of a balanced equation can be deleted from the equation. In the equation immediately above, the K^+ and the NO_3^- fall into this category. Such ions are sometimes called **spectator ions** since they simply watch the reactions that go on around them, but they do not react themselves. Leaving out the spectator ions, the reaction between HNO_3 and KOH can be summarized as follows:

$$H^+ + OH^- \rightleftharpoons H_2O$$

This is the **net reaction** that occurs between all strong acids and strong bases. An ion that is a spectator during an acid–base reaction may not be a spectator during other biologically important reactions or processes. This key point must be kept in mind. **[Try Problems 2.18 and 2.19]**

Reaction of Weak Acids with Strong Bases

When one adds a strong base to any weak acid, water is formed along with the conjugate base of the weak acid. Spectator ions will also be present. The conjugate base is commonly an ion that, along with its counter ion(s), represents a salt. The reaction between NaOH and acetic acid (CH_3COOH) illustrates these facts.

$$CH_3COOH(aq) + NaOH(aq) \rightleftharpoons H_2O + CH_3COONa(aq)$$

$$\text{Weak acid} + \text{strong base} \rightleftharpoons \text{water} + \text{salt with conjugate base}$$

or

$$CH_3COOH(aq) + Na^+(aq) + OH^-(aq) \rightleftharpoons H_2O + CH_3COO^-(aq) + Na^+(aq)$$

Ionic form of equation

or

$$CH_3COOH(aq) + OH^-(aq) \rightleftharpoons H_2O + CH_3COO^-(aq)$$

Net reaction; spectator ions omitted

Since CH_3COOH has a small dissociation constant, it is always written in its nondissociated form. **[Try Problems 2.20 and 2.21]**

Reaction of Conjugate Bases with Strong Acids

The addition of a strong acid to a conjugate base of a weak acid leads to the conversion of the conjugate base to the weak acid and, once again, we encounter spectator ions. The reaction of sodium acetate (CH_3COONa), a source of the conjugate base for acetic acid, with HCl provides an example.

$$\underset{\text{Salt with conjugate base}}{CH_3COONa(aq)} + \underset{\text{strong acid}}{HCl(aq)} \rightleftharpoons \underset{\text{weak acid}}{CH_3COOH(aq)} + \underset{\text{salt}}{NaCl(aq)}$$

or

$$CH_3COO^-(aq) + Na^+(aq) + H^+(aq) + Cl^-(aq) \rightleftharpoons CH_3COOH(aq) + Na^+(aq) + Cl^-(aq)$$

Ionic form

or

$$CH_3COO^-(aq) + H^+(aq) = CH_3COOH(aq)$$

Net reaction; spectator ions omitted

Weak acids can react with conjugate bases of other weak acids, but these reactions do not go to completion. **[Try Problem 2.22]**

Some Properties and Uses of Acids and Bases

Both concentrated strong acids and concentrated strong bases are extremely hazardous and can lead to serious **chemical burns.** Even more dilute solutions can readily eat holes in a pair of jeans, as many beginning chemistry students can testify. The chemical burns and the holes in jeans are due to the participation of acids and bases, as either reactants or catalysts, in numerous chemical reactions, including the hydrolysis of many biologically important organic compounds. Some concentrated acids are also powerful dehydrating agents. Since the average American regularly uses products that contain high concentrations of strong acids or strong bases [automobile batteries (contain H_2SO_4) and drain cleaners (some contain NaOH pellets),

A wide variety of household products contain acids and bases. Some drain cleaners contain particularly high concentrations of strong bases.

for example], it is not uncommon to have patients report to emergency rooms with acid or base burns. **[Try Problems 2.23 and 2.24]**

Acids and bases are occasionally used for therapeutic purposes. The treatment of peptic ulcers with **antacids** is an example. The active ingredient in Alka-Seltzer is sodium bicarbonate ($NaHCO_3$), a salt containing the conjugate base of carbonic acid (H_2CO_3). The antacid in sodium free TUMS is calcium carbonate ($CaCO_3$), a salt containing the conjugate base of HCO_3^-. **[Try Problem 2.25]**

PROBLEM 2.18 Write two balanced equations for the reaction of $HClO_4$ with NaOH. In the first equation, include spectator ions. In the second, omit them.

PROBLEM 2.19 Write a balanced equation for the reaction of 2 mol of KOH with 1 mol of H_2SO_4. Omit spectator ions. Note: HSO_4^- is a weak acid. For this problem, however, assume that both hydrogens in H_2SO_4 dissociate completely.

PROBLEM 2.20 Write a balanced equation for the reaction of KOH with (a) phenol and (b) lactic acid (see Table 2.8 for structural formulas).

PROBLEM 2.21 Write a balanced equation for the net reaction (omit spectator ions) that will occur if (a) 1 mol of NaOH is added to 1 mol of H_3PO_4, (b) 2 mol of NaOH is added to 1 mol of H_3PO_4, and (c) 3 mol of NaOH are added to 1 mol of H_3PO_4.

PROBLEM 2.22 Write a balanced equation for the reaction of HCl with $NaHCO_3$. Write the equation both with and without spectator ions.

***PROBLEM 2.23** Hydrolysis, the splitting of a bond through reaction with water, is of major importance in life processes. Most digestive reactions, for example, are hydrolysis reactions. List three different classes of organic compounds that can be hydrolyzed, and then write the expanded structural formula for the functional group of each of these three classes.

***PROBLEM 2.24** Define "catalyst" and "dehydrating agent."

PROBLEM 2.25 Write a balanced equation for the reactions that will occur if excess HNO_3 is added to (a) Alka-Seltzer; and (b) sodium–free TUMS. First write the equations including spectator ions, and then rewrite them without spectator ions.

2.5 BUFFERS

Living organisms must regulate the concentrations of H^+ and OH^- because of the acid–base reactions and other reactions in which these ions participate; H^+ and OH^-

functionally alter many biologically important compounds. Relatively small shifts in the pH of virtually any body fluid can have drastic consequences. A drop in the pH of blood from 7.4 (normal) to 7.0, for example, can lead to a **coma,** a state of deep unconsciousness. An increase in blood pH to 7.8 can lead to **tetany,** a rigid, uncontrolled contraction of skeletal muscles. Severe or uncorrected **acidosis** (a decrease in blood pH) or **alkalosis** (an increase in blood pH) can be lethal.

The production of lactic acid (Table 2.8) by bacteria growing around a tooth can lead to a localized drop in pH, the major cause of tooth decay. **Fluoride** reduces the risk of cavities by allowing the body to build more acid-resistant tooth enamel. Muscles fatigue during strenuous exercise as a consequence of lactic acid production and an associated increase in $[H^+]$. Localized pH gradients play a central role in the production of ATP (the energy currency of cells) within mitochondria (Section 11.9), and chloroplasts (Section 15.4), and localized shifts in pH help regulate photosynthesis and may even be responsible for the spontaneous regression of some tumors. An increase in pH as one moves from lysosomes to the cytosol (Section 1.3) ensures that leaky lysosomes will not digest those cells that contain them. While a lysosome's digestive enzymes are most active at the relatively low pH within these organelles, they harbor little catalytic activity at the pH of the cytosol.

Buffers play a central role in the biological regulation of pH. A **buffer** is a **weak acid:conjugate base pair** (a weak acid along with its conjugate base) that, when in solution, *resists* major changes in pH. Most body fluids are highly buffered.

How Buffers Operate

How are buffers able to resist changes in pH? To understand the answer to this question, we must recognize that essentially all changes in pH are a consequence of the addition of acids or bases. Any weak acid:conjugate base pair (all buffers consist of such a pair) contains one component, the weak acid, which can react with **(neutralize)** added OH^- (from bases) and a second component, the conjugate base, which can react with (neutralize) added H^+ (from acids) (Figure 2.3). Thus, buffers *resist* pH changes by consuming *part* of any H^+ or OH^- that is added to a solution. **[Try Problem 2.26]**

To effectively resist both increases and decreases in pH, a buffer must have an appreciable amount of both a weak acid and its conjugate base. If a solution has a large amount of weak acid, but very little conjugate base, it will be unable to consume

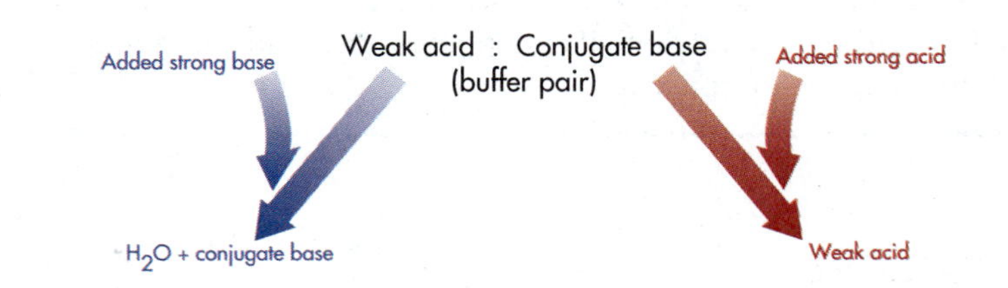

FIGURE 2.3
Operation of buffers. The weak acid in a buffer pair consumes added base while its conjugate base consumes added acid.

much added acid. Consequently, there tends to be a significant drop in pH when acid is added. Similarly, a solution that contains a large amount of conjugate base but very little weak acid will be unable to consume much added base. The addition of relatively little base can lead to a jump in pH.

Under what conditions will a weak acid and its conjugate base both exist at an appreciable concentration within a solution? One can best answer this question by employing the **Henderson–Hasselbalch equation,** an algebraic rearrangement of the equilibrium constant expression for the dissociation of a weak acid.

$$K_{eq} = \frac{[\text{conjugate base}][\text{H}^+]}{[\text{weak acid}]}$$

The Keq for the dissociation of a weak acid

Algebraic rearrangement ↓

$$\text{pH} = \text{p}K_a + \log \frac{[\text{conjugate base}]}{[\text{weak acid}]}$$

The Henderson–Hasselbalch Equation

If the concentration of conjugate base equals the concentration of weak acid, the Henderson–Hasselbalch equation becomes:

$$\text{pH} = \text{p}K_a + \log \frac{[\text{weak acid}]}{[\text{weak acid}]} = \text{p}K_a + \log 1 = \text{p}K_a$$

Restated, whenever the pH of a solution equals the pK_a of a weak acid in that solution, the concentration of the weak acid equals the concentration of its conjugate base. When the concentration of a conjugate base is ten times greater than the concentration of its weak acid ([conjugate base] = 10[weak acid]):

$$\text{pH} = \text{p}K_a + \log \frac{10[\text{weak acid}]}{[\text{weak acid}]} = \text{p}K_a + \log 10 = \text{p}K_a + 1$$

Viewed from a different angle, whenever the pH of a solution is one greater than the pK_a of a weak acid in that solution, the concentration of the conjugate base of that acid will be ten times the concentration of the weak acid. Similar calculations show that the concentration of a weak acid is ten times the concentration of its conjugate base whenever the pH of a solution is one less than the pK_a of the weak acid. These and related generalizations are summarized in Table 2.10.

Table 2.10 clearly reveals that any solution of a weak acid contains some of its conjugate base, and any solution of a conjugate base contains some of its weak acid. Although the exact ratio of conjugate base to weak acid depends on pH, there is always more weak acid than conjugate base at pH's below the pK_a and more conjugate base than weak acid at pH's above the pK_a. Normally (exceptions do exist), the pH of a solution must be within one unit of the pK_a of a weak acid (pH = pK_a ± 1) in order for that weak acid, along with its conjugate base, to serve as a buffer. At pH values outside this range, a solution will have over ten times more conjugate base than weak acid (leads to a limited ability to react with added base) or over ten times more weak acid than conjugate base (leads to an insufficient ability to consume

TABLE 2.10

Effect of pH on Conjugate Base to Weak Acid Ratio ([CB]/[WA])

pH	[CB]/[WA]		
$pK_a + 4$	$\frac{10,000}{1}$	↑	Over 90% of buffer pair exists as conjugate base
$pK_a + 3$	$\frac{1,000}{1}$		
$pK_a + 2$	$\frac{100}{1}$		
$pK_a + 1$	$\frac{10}{1}$	Buffer Range ↕	Appreciable amount of *both* weak acid and conjugate base
pK_a	$\frac{1}{1}$		
$pK_a - 1$	$\frac{1}{10}$		
$pK_a - 2$	$\frac{1}{100}$		Over 90% of buffer pair exists as weak acid
$pK_a - 3$	$\frac{1}{1,000}$		
$pK_a - 4$	$\frac{1}{10,000}$	↓	

added acid). A buffer must be able to resist pH changes effectively in both directions. **[Try Problems 2.27 and 2.28]**

The **buffer capacity** of a buffer is a quantitative measure of its ability to consume added acid and added base. Buffer capacity increases as buffer concentration increases and as the pH of the buffer moves closer to the pK_a of the weak acid in the buffer pair. The concentration of a buffer is defined as the concentration of its weak acid plus the concentration of its conjugate base. It is usually possible to add more acid or more base than a buffer can consume, even when its buffer capacity is very high. A solution then loses its ability to resist pH changes and it is no longer buffered. The concepts of buffer capacity and buffering range can be clearly illustrated with **titration curves** where pH (y-axis) is plotted against amount of strong acid or strong base added (x-axis) (Figure 2.4). **[Try Problems 2.29 and 2.30]**

Common Misconceptions About Buffers

One common misconception is that a buffered solution is always a neutral solution. This is false. One can buffer a solution at any desired pH. A pH 10.3 buffer will tend to hold the pH of a solution near pH 10.3, whereas a pH 2.7 buffer will tend to hold

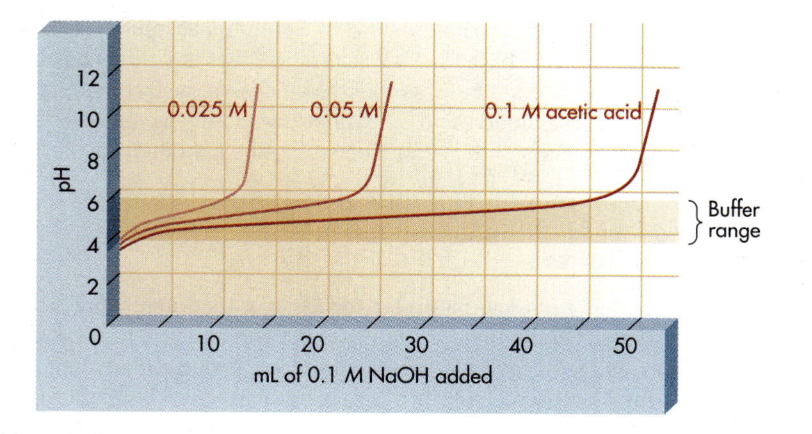

FIGURE 2.4
Titration curves for acetic acid. Three 50 mL acetic acid solutions, each a unique concentration, are titrated. During the titrations, the acetic acid ($pK_a = 4.76$) effectively resists pH changes between approximately pH 3.76 and pH 5.76, the buffer range for the acetic acid:acetate buffer pair. As the concentration of acetic acid increases, the buffer capacity increases within the buffer range and it takes additional base to move through this buffer range.

the pH of a solution near pH 2.7. Normally, we must select a weak acid with a $pK_a = 10.3 \pm 1$ to prepare a pH 10.3 buffer and we must select a weak acid with a $pK_a = 2.7 \pm 1$ to prepare a pH 2.7 buffer. Since, in general, any given weak acid:conjugate base pair can buffer only within a 2 unit pH range, different weak acid:conjugate base pairs are usually required to prepare buffers that differ by more than 2 units in pH.

A second common misconception is that a buffer will hold the pH of a solution absolutely constant. The addition of even small amounts of a strong acid or a strong base to any solution, buffered or not, will lead to a change in pH. The important concept is that the shift in pH will be much less when an effective buffer is present. This fact is documented in Table 2.11 and with the titration curves in Figure 2.4. To un-

TABLE 2.11

A Comparison of pH Shifts in Buffered and Nonbuffered Solutions

Sample	Approximate pH
1 L H_2O	7.0
1 L H_2O + 0.01 mol NaOH	12.0
1 L H_2O + 0.01 mol HCl	2.0
1 L Buffer[a]	7.2
1 L Buffer[a] + 0.01 mol NaOH	7.3
1 L Buffer[a] + 0.01 mol HCl	7.1

[a]Buffer = 0.1 *M* HPO_4^{2-} + 0.1 *M* $H_2PO_4^-$

derstand why a buffer is unable to hold the pH of a solution absolutely constant, you must recall that the conjugate base to weak acid ratio shifts any time an acid or base is added to a buffer pair. Added base (converts some weak acid to conjugate base) increases this ratio, whereas added acid (converts some conjugate base to weak acid) decreases it. The ratio cannot shift without changing the pH (study the Henderson–Hasselbalch equation).

Preparation of Buffers

There are several ways to prepare a buffer for use in experiments. We can use the Henderson–Hasselbalch equation to calculate the conjugate base to weak acid ratio necessary to generate the desired pH, and then mix weak acid and conjugate base in the calculated ratio. The weak acid must, of course, have a pK_a within one unit of the desired pH. A second approach involves titrating an appropriate weak acid or conjugate base with a strong base or a strong acid to the desired pH. The titration will automatically bring the conjugate base to weak acid ratio to that calculated using the first approach. The various approaches to buffer preparation do not lead to identical buffer solutions, because each approach has a unique impact on the concentration of spectator ions. Although such ions are usually inert from the standpoint of acid–base reactions, they may have a significant impact on the overall activity or behavior of a biological system. **[Try Problem 2.31]**

PROBLEM 2.26 For each of the folowing buffer pairs: (a) write a balanced equation for the reaction that will occur if NaOH is added to a buffer containing this pair; and (b) write a balanced equation for the reaction that will occur if HCl is added to a buffer containing this pair.

i. $HPO_4^{2-}:PO_4^{3-}$
ii. $CH_3NH_3^+:CH_3NH_2$
iii. $HCN:CN^-$
iv. $H_2CO_3:HCO_3^-$

PROBLEM 2.27 For each weak acid in Table 2.8, clearly indicate within what pH range this weak acid, along with its conjugate base, can serve as a buffer.

PROBLEM 2.28 The $H_2PO_4^-:HPO_4^{2-}$ pair is a blood buffer. What is the whole number ratio of $[HPO_4^{2-}]$ to $[H_2PO_4^-]$ in blood at pH 7.4? at pH 6.8? at pH 7.8? at pH 9.2?

PROBLEM 2.29 What specific variable primarily determines the amount of added acid that a buffer is able to consume (i.e. accept a proton from)? Explain.

PROBLEM 2.30 What is the concentration of a buffer that contains 0.12 *M* acetic acid (CH_3COOH) and 0.37 *M* sodium acetate (CH_3COONa)? If the pKa for acetic acid is 4.76, what is the pH of this buffer?

PROBLEM 2.31 What conjugate base to weak acid ratio is required to produce a pH 4.0 acetate buffer? An acetate buffer is one in which the acetate ion (CH_3COO^-) is the conjugate base and acetic acid (CH_3COOH) is the weak acid.

2.6 THE CONTROL OF BLOOD pH

The major **blood buffers,** in order of importance, are summarized in Exhibit 2.1. Collectively, they play a central role as the body continuously attempts to hold the pH of blood very close to 7.4. Since $H_2PO_4^-$ has a pK_a of 7.2 and proteins contain weak acid functional groups with pK_a's near 7.4, these buffers are typical buffers that function through the mechanism described in Section 2.5. You may wonder, however, how the H_2CO_3:HCO_3^- pair can buffer at pH 7.4 when the effective pK_a of H_2CO_3 is around 6.35 (Table 2.8). To understand this exception to our rule of thumb (pH must equal pK_a ± 1), we must examine the carbon dioxide story.

EXHIBIT 2.1
Major Blood Buffers

1. H_2CO_3:HCO_3^-
2. Protein—H^+:Protein
3. $H_2PO_4^-$:HPO_4^{2-}

Carbon dioxide, CO_2, is produced by virtually all cells in the body as they oxidize nutrients to acquire the energy that is essential for their survival. This CO_2 enters the blood which carries it to the lungs to be exhaled (Figure 2.5). Since the CO_2 in the blood is in equilibrium with gaseous CO_2 in the lungs, depth (how deeply one inhales) and rate of respiration influence the efficiency with which CO_2 is eliminated from the blood as it passes through the lungs. An increase in depth or rate of respiration can reduce the amount of CO_2 in the lungs and stimulate the flow of CO_2 from the blood into the lungs. A decrease in depth or rate of respiration has the opposite effect. CO_2 affects blood pH because it exists in equilibrium with carbonic acid ($CO_2 + H_2O \rightleftharpoons H_2CO_3$). The higher the concentration of CO_2, the higher the concentration of H_2CO_3. Since H_2CO_3 partially dissociates to produce H^+, the higher the concentration of CO_2, the greater the [H^+] and the lower the pH. Conversely, as CO_2 concentration drops, H_2CO_3 concentration decreases and the pH is elevated.

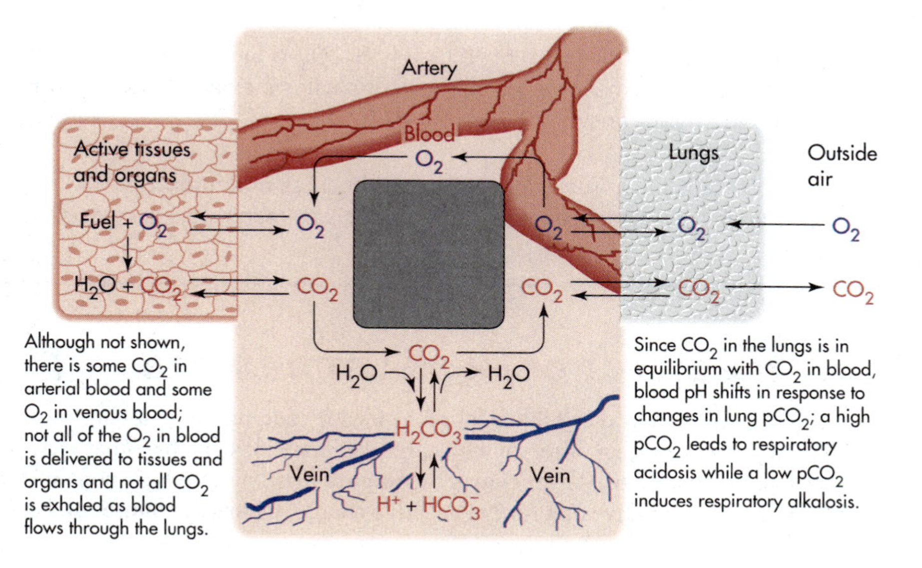

FIGURE 2.5
Effect of CO_2 on blood pH.

Under normal circumstances, the body eliminates CO_2 as rapidly as it is generated, so the concentration of CO_2 (usually expressed as partial pressure of CO_2, abbreviated as pCO_2) in the lungs is relatively constant. However, if a person has impaired lung function resulting from neuromuscular disorders, central nervous system diseases, or other causes, CO_2 accumulates (pCO_2 increases) and can lead to potentially lethal **respiratory acidosis.** Similarly, hyperventilation (rapid and deep breathing) brought on by anxiety, central nervous system injury, aspirin poisoning, fever, or other causes can lead to **respiratory alkalosis.**

Although fluctuations in CO_2 concentration can lead to significant shifts in blood pH, carbon dioxide's relatively constant concentration under normal conditions allows it to play a prominent role in the buffering of blood; it is the equilibrium between CO_2 and H_2CO_3 that allows the H_2CO_3:HCO_3^- pair to buffer outside of its normal buffering range. At the pH of blood, it is the H_2CO_3 that is the deficient member of this pair. In the absence of CO_2, the small amount of H_2CO_3 existing at pH 7.4 is quickly consumed if any significant amount of base enters the blood. With CO_2 present, however, the reaction between CO_2 and H_2O ($CO_2 + H_2O \rightleftharpoons H_2CO_3$) shifts to the right (Le Châtelier's principle) as the H_2CO_3 is consumed. At a near constant CO_2 concentration, this shift replaces most of the H_2CO_3 lost. **[Try Problems 2.32 through 2.34]**

The regulation of blood pH is so important that this responsibility has not been left solely in the hands of respiration and buffers. The kidneys also play a central role. In making urine, these vital organs filter blood and then selectively reabsorb water and many solutes. Excess nonvolatile acids or excess bicarbonate (HCO_3^-; a conjugate base) are, however, not reabsorbed; these substances are excreted from the body. The pH of urine ranges from about 4.4 to 8.0 depending on diet and other variables. A detailed discussion of the design and functioning of the kidneys can be found in most physiology texts.

***PROBLEM 2.32** Summarize Le Châtelier's principle.

PROBLEM 2.33 If the ratio of HCO_3^- to H_2CO_3 in blood is 50 to 1, is the pH of the blood normal or does acidosis or alkalosis exist? Describe how you arrived at your answer.

PROBLEM 2.34 Do respiratory diseases such as pneumonia and emphysema tend to cause respiratory acidosis or respiratory alkalosis? Explain.

2.7 THE EFFECT OF pH ON NET CHARGES

H^+ and OH^- primarily alter biological molecules by adding or removing protons and, in the process, adding or removing charges. Compounds can be either charged or uncharged (Figure 2.6). An **uncharged compound** is one in which none of the atoms has a formal positive or a formal negative charge. **Charged compounds** con-

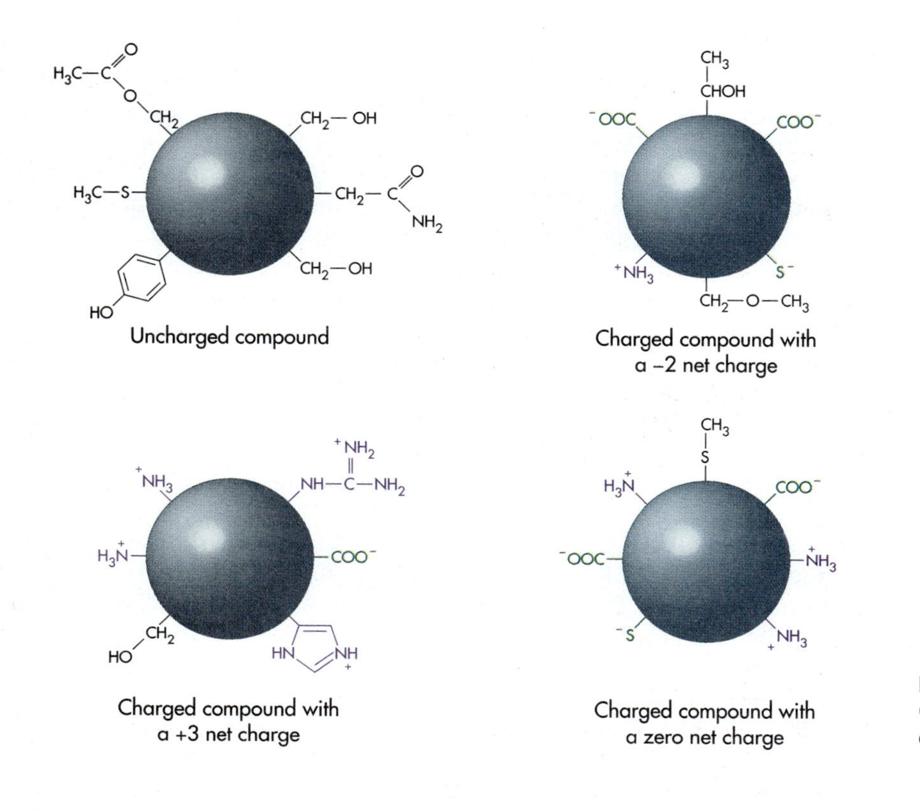

FIGURE 2.6 Charged and uncharged compounds.

tain one or more atoms with formal charges. Such compounds are really ions. The **net charge** on a compound is defined as the sum of its positive charges minus the absolute number of negative charges that it contains. Thus, a compound with 5 positive charges and 4 negative charges has a net charge of +1 (5 minus 4). A compound that contains 6 positive charges and 12 negative charges will have a net charge of −6 (6 minus 12). When the number of positive charges equals the number of negative charges, the net charge on a compound will be zero. It is important to recognize that a compound with a zero net charge is not uncharged; the "zero net" implies that there are both positive and negative charges present.

Net Charge Calculations

The charge on any compound with one or more weak acid functional groups is a function of pH. Acetic acid (CH_3COOH) will be used to illustrate this fact. A fraction of the acetic acid molecules in any acetic acid solution exists in the dissociated or conjugate base form (acetate). The numerical value of this fraction is determined by the pH of the solution (Table 2.10). When the pH equals the pK_a of acetic acid (4.76), $[CH_3COO^-] = [CH_3COOH]$. At any time, half the compounds will have a 0 charge and the other half a −1 charge. But since the dissociation of acetic acid is rapid and reversible, a single molecule that is nondissociated and uncharged at one instant in time may be dissociated and have a −1 charge at the next instant in time.

At pH 4.76, each molecule will have a 0 charge half the time and a −1 charge half the time. Over a finite period of time, this averages out to a −1/2 charge. At pH values far above 4.76, acetic acid exists almost exclusively in its conjugate base form (Table 2.10), and its average charge is very close to −1. For the charge on acetic acid to be precisely −1, it must exist in its dissociated form all the time. Theoretically, this situation never occurs, but for most purposes the small fraction of nondissociated acid present at high pH values can be neglected. At pH values far below 4.76, the average (over time) charge on acetic acid approaches 0 since each molecule exists in its nondissociated or **protonated form** the vast majority of the time (Table 2.10). Thus, the average charge on acetic acid, which ranges from near 0 to approximately −1, depends on pH. *All future references to the charge on a weak acid functional group refers to its average charge over time.*

In this text, the calculation of charges is carried out to the nearest half of a charge unit. There is nothing magical about the one half; it has been selected to simplify calculations. More precise calculations of charges are definitely possible, but such calculations are a bit more complex mathematically.

When we are working with charge calculations to the nearest half charge unit, we assign a weak acid functional group the charge of its protonated form whenever the pH of a solution is 0.5 or more units below the pK_a of the weak acid. A functional group is assigned the charge of its dissociated form when the pH is 0.5 or more units above its pK_a. A functional group is assigned a charge that is the average of the charges on its weak acid and conjugate base forms when the pH is within 0.5 units of the pK_a. These charge assignment rules (Table 2.12) are based on the Henderson–Hasselbalch equation. They allow us to quickly determine specific charges and net charges to the nearest one half of a charge unit with minimal mathematical calculations. The application of these rules is illustrated later in this section.

The charge on the weak acid in any weak acid:conjugate base pair is one unit more positive than the charge on the conjugate base, because the weak acid is produced by adding one H^+ (a proton) to its conjugate base. If the conjugate base has a 0 charge, the weak acid will have a +1 charge. If the conjugate base has a −3 charge, the weak acid will have a −2 charge (study Table 2.8). It follows that the charge on any weak acid functional group will move in a positive direction (it may still be negative, but it will be less negative) as the pH drops and will move in a negative direction as the pH increases. **[Try Problems 2.35 through 2.37]**

TABLE 2.12

Charge Assignment Rules

pH	Charge Assigned to Weak Acid Functional Group
>0.5 below pK_a	Charge associated with weak acid form
$pK_a \pm 0.5$	1/2 (Charge on weak acid form + charge on conjugate base form)
>0.5 above pK_a	Charge associated with conjugate base form

Net Charges on Multifunctional Compounds

To calculate the net charge on a compound with multiple weak acid functional groups, we simply calculate the charge contributed by each individual functional group and then subtract the total number of negative charges from the total number of positive charges. The calculation of the net charge on cysteine at pH 8.2 will be used to illustrate this approach.

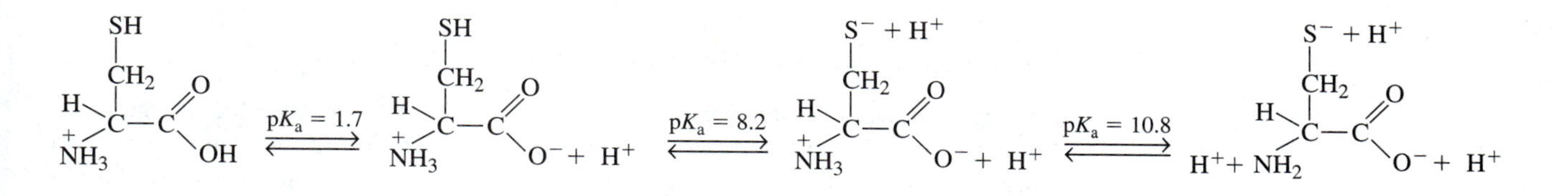

The carboxyl functional group (—COOH) of cysteine has a pK_a of 1.7. Since pH 8.2 is more than 0.5 units above this pK_a and the conjugate base form of the functional group (—COO^-) has a −1 charge, we can use the charge assignment rules summarized in Table 2.12 to assign this functional group immediately a −1 charge. When considering the sulfhydryl functional group (—SH) of cysteine, the pH is within 0.5 of the pK_a and the weak acid has a 0 charge and its conjugate base a −1 charge. The assigned charge is $^1/_2\,(0 - 1) = -^1/_2$. Since the pK_a for the protonated amino group (—NH_3^+) of cysteine is more than 0.5 units above the pH, this functional group is assigned a charge of +1. The approximate net charge on cysteine at pH 8.2 is the sum of the charges assigned to each of its three weak acid functional groups.

$$\text{Net charge} = -1\,[\text{—COO}^-] + (-^1/_2)\,[\text{—SH/—S}^-] + (+1)\,[\text{—NH}_3^+] = -^1/_2$$

"Approximate" should be emphasized since all calculated charges have been rounded off to the nearest half charge unit.

Given the K_a or pK_a for each weak acid functional group in a compound, we should be able to calculate the net charge on that compound at any pH. We should also be able to write the structure of a compound as it will primarily exist at any given pH. To accomplish the latter task, simply write the structure of the compound with each weak acid functional group in the form that will predominate at the specified pH. If the pH is below the pK_a of the functional group, the weak acid form will predominate (study Table 2.10). If the pH is above the pK_a of the functional group, the conjugate base form will predominate. Cysteine will primarily exist in the following form at pH 1.0:

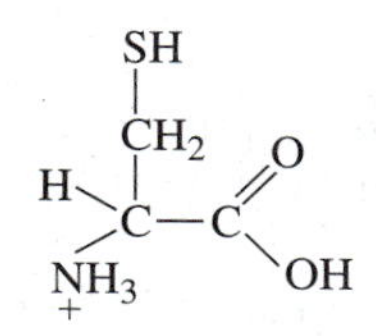

It has a different prevailing structure at pH 9.0:

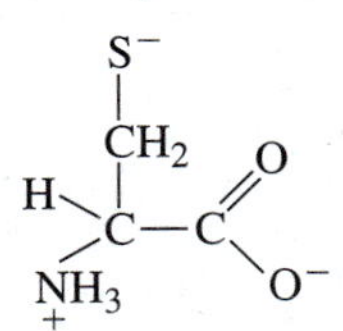

[Try Problems 2.38 through 2.40]

The Importance of Net Charges

Charge calculations are emphasized throughout this text because the charges on a compound or particle normally have a marked impact on its biological activity. Hydrogen bonding and charge–charge interactions (like charges repel, opposite charges attract) play a key role in determining the shapes, the chemical characteristics, and the physical properties of compounds. As pH increases, for example, acidic hydrogens (often involved in H bonding) are removed from weak acid functional groups. This loss tends to destroy specific H bonds and to create or destroy charged centers. Marked changes in neighbor–neighbor interactions result. As a consequence, many compounds, including most proteins, are active within only a relatively narrow pH range, a pH range in which they have a unique distribution of charges within their molecular architecture.

Counter Ions

Isolated charges are usually very unstable; in a solution or compound, the number of positive charges tends to equal the number of negative charges. An ion of opposite charge that helps "balance" the charge on another ion is said to be a **counter ion.** Together, a charged compound and its counter ions represent an ionic compound (usually a salt). Na^+, K^+, Mg^{2+}, Ca^{2+}, Cl^-, NO_3^-, $H_2PO_4^-$, HPO_4^{2-} and SO_4^{2-} are common counter ions for biologically important organic ions.

PROBLEM 2.35 Give an example of a weak acid:conjugate base pair in which the weak acid has a +1 charge and the conjugate base has a zero charge.

PROBLEM 2.36 Give an example of a weak acid that has a −1 charge. What is the charge on the conjugate base of this weak acid?

PROBLEM 2.37 Which of the weak acids in Table 2.8 will exist primarily in their dissociated form at pH 7.4?

PROBLEM 2.38 To the nearest half of a charge unit, calculate the net charge on cysteine at (a) pH 7.0, (b) pH 1.0, and (c) pH 12.

PROBLEM 2.39 To the nearest half of a charge unit, calculate the net charge on lysine at (a) pH 1.0, (b) pH 7.0, and (c) pH 12.

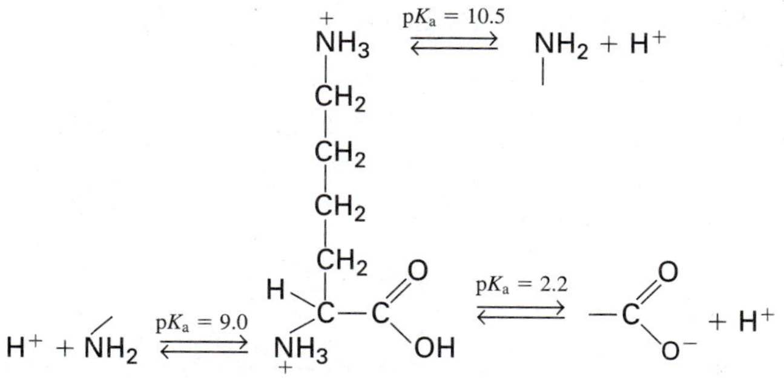

$pK_a = 10.5$: $\overset{+}{N}H_3 \rightleftharpoons NH_2 + H^+$

CH_2
CH_2
CH_2
CH_2

$pK_a = 9.0$: $H^+ + NH_2 \rightleftharpoons$ H–C–C(=O)OH with $\overset{+}{N}H_3$ $\overset{pK_a = 2.2}{\rightleftharpoons}$ –C(=O)O^- + H^+

PROBLEM 2.40 Write the formula for lysine as it would primarily exist at (a) pH 1.0, (b) pH 7.0, (c) pH 9.0, (d) pH 12, and (e) pH 14.

2.8 THE pH OF ACID SOLUTIONS

Since biological systems are so sensitive to changes in pH, biochemists need to be able to calculate the pH of various solutions, including solutions of commonly encountered weak and strong acids. The calculation of the pH of a solution of a strong acid is extremely simple if we assume that the acid does indeed dissociate completely. A 0.01 *M* HCl solution, for example, is assumed to contain 0.01 *M* H^+, 0.01 *M* Cl^-, and no nondissociated HCl. To calculate the pH of this solution, just punch 0.01, the $[H^+]$, into a scientific pocket calculator and then push the log button. A −2 appears on the screen. The −2 is changed to a +2 because pH is defined as the log of the $[H^+]$ times −1.

The situation is a bit more complicated for weak acid solutions. Since weak acids do not dissociate completely, the $[H^+]$ is not equal to the concentration of the weak acid. The $[H^+]$ must be calculated from the K_a expression and published values for K_a's. To calculate the pH of a 0.100 *M* acetic acid solution, for example, first write the K_a expression for acetic acid:

$$K_a = \frac{[CH_3COO^-][H^+]}{[CH_3COOH]} = 1.74 \times 10^{-5}$$

Next, let $X = [H^+]$. Since one acetate ion is formed for every H^+ that is formed when acetic acid dissociates, $[CH_3COO^-] = [H^+] = X$. This equality statement neglects the very small amount of H^+ that is present in aqueous solutions as a consequence of the dissociation of water molecules themselves. The final $[CH_3COOH]$ must equal 0.100 − X, since for each CH_3COO^- and H^+ formed, one molecule of CH_3COOH must have dissociated. That is, the initial $[CH_3COOH]$ is reduced by an amount equal to the final $[CH_3COO^-]$.

$$\underset{0.100 - X}{CH_3COOH} \rightleftharpoons \underset{X}{H^+} + \underset{X}{CH_3COO^-}$$

Next, plug the values for $[H^+]$, $[CH_3COO^-]$, and $[CH_3COOH]$ into the K_a expression.

$$K_a = \frac{X \times X}{0.100 - X} = \frac{X^2}{0.100 - X} = 1.74 \times 10^{-5}$$

To solve this equation precisely for X, we must use the **quadratic equation**:

$$X = \frac{-b \pm \sqrt{b^2 - 4ac}}{2a}$$

The a, b, and c in the quadratic equation are defined by the following algebraic expression:

$$aX^2 + bX + c = 0$$

Fortunately, a simplifying assumption will often provide an acceptable estimate of X without the use of the quadratic equation. In the present example, the simplifying assumption is that $0.100 - X = 0.100$. That is, we assume that X is so small compared to 0.100 that it can be neglected. With this assumption, the K_a expression becomes:

$$K_a = \frac{X^2}{0.100} = 1.74 \times 10^{-5}$$

To solve for X, first multiply both sides of this equation by 0.100 to isolate X^2, and then find the square root of the right-hand side of the equation to arrive at a value for X.

$$X^2 = 0.100 \times 1.74 \times 10^{-5}$$
$$X^2 = 1.74 \times 10^{-6}$$
$$X = 1.32 \times 10^{-3}$$

Since $X = [H^+]$, the pH $= -\log(1.32 \times 10^{-3}) = 2.88$. At this point it is wise to reexamine the simplifying assumption. Is 1.32×10^{-3} (0.00132) really negligible compared to 0.100? It is if we can live with a 1 to 2% error. The error would be greater for acids that are stronger than acetic acid and less for acids that are weaker than acetic acid. **[Try Problem 2.41]**

PROBLEM 2.41 Use the information given in Table 2.8 to calculate the pH of (a) a 0.10 *M* solution of phenol and (b) a 1.0 *M* solution of NH_4Cl.

2.9 WEAK BASES, CONJUGATE ACIDS, AND K_b

Most biochemists approach the topic of weak acid reactions from the viewpoint of weak acids and their conjugate bases. An alternative approach is possible since the conjugate base of a weak acid is a weak base. A **weak base** is a proton acceptor that does not react completely with water to generate OH^- (Figure 2.7). The weak acid that is formed when a weak base accepts a H^+ is known as its **conjugate acid.** Thus, when CH_3COONa is dissolved in water, it first dissociates completely to give Na^+

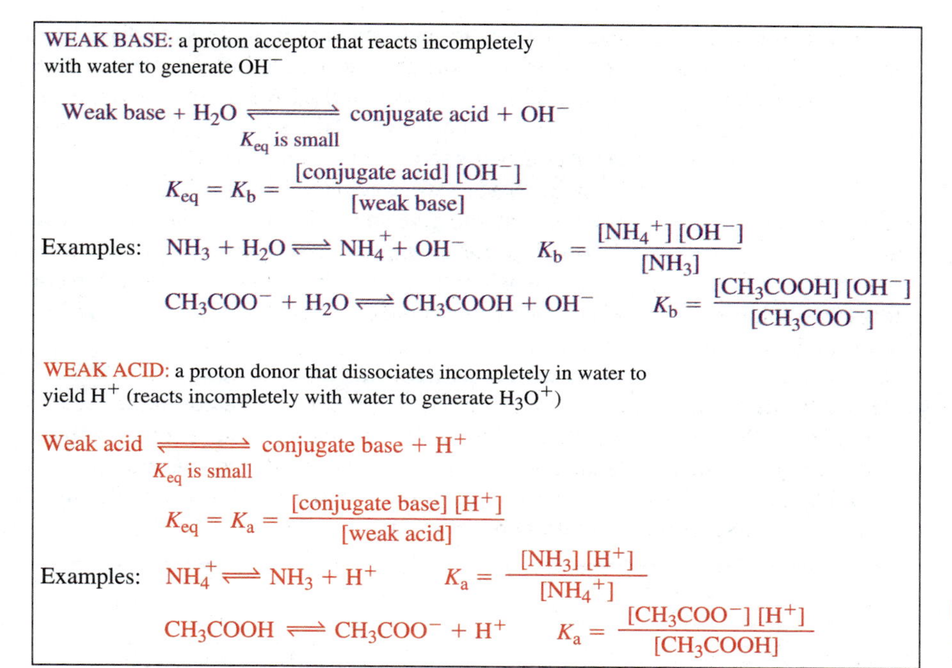

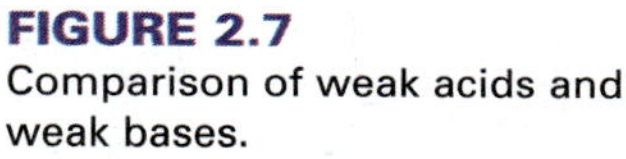
FIGURE 2.7 Comparison of weak acids and weak bases.

and CH_3COO^- (a weak base whose conjugate acid is acetic acid). The CH_3COO^- then reacts with water molecules to generate OH^-:

$$CH_3COO^- + H_2O \rightleftharpoons CH_3COOH + OH^-$$

This reaction does not go to completion. The equilibrium constant for this reaction is:

$$K_{eq} = \frac{[CH_3COOH][OH^-]}{[CH_3COO^-][H_2O]}$$

When the $[H_2O]$ (assumed to be constant in dilute aqueous solutions) is combined with the value of K_{eq}, the K_b expression is obtained:

$$K_{eq}[H_2O] = K_b = \frac{[CH_3COOH][OH^-]}{[CH_3COO^-]}$$

The pK_b is defined as the negative log of the K_b (**pK_b = −log K_b**). The product of the K_b of a weak base and the K_a of its conjugate acid is equal to the K_w for water.

$$\frac{[CH_3COOH][OH^-]}{[CH_3COO^-]} \times \frac{[CH_3COO^-][H^+]}{[CH_3COOH]} = [H^+][OH^-]$$

$$K_b \times K_a = K_w$$

$$pK_b + pK_a = 14$$

K_b values and pK_b values can readily be calculated from K_a values or pK_a values, and vice versa.

Weak bases and weak acids allow us to understand why some salts dissolve to yield acidic or basic solutions, a fact of more than academic interest because such salts can change the pH of blood or a laboratory solution. A salt, in general, is defined as an ionic compound that contains no H^+ or OH^-. Salts may, however, contain weak acids or weak bases. **Acidic salts** ($NH_4^+NO_3^-$ and $CH_3NH_3^+Cl^-$, for examples) are ones that contain an ion that is a weak acid (NH_4^+ and $CH_3NH_3^+$, in the present examples); they dissolve in water to give an acidic solution. Similarly, **basic salts** ($CH_3COO^-Na^+$ and $CH_3S^-K^+$, for example) possess an ion that is a weak base (CH_3COO^- and CH_3S^-, in the examples given); they dissolve in water to create a basic solution. None of the ions within **neutral salts** (Na^+Cl^- and $K^+NO_3^-$, for example) are either weak acids or weak bases. [Try Problem 2.42]

PROBLEM 2.42 (a) Write the structural formulas for the conjugate acids of lactate ($CH_3CHOHCOO^-$) and NH_3. (b) Write balanced equations for the reaction of water with lactate and NH_3. (c) Write the K_b expressions, for both lactate and NH_3. (d) Use the information in Table 2.8 to calculate the K_b and the pK_b of both lactate and NH_3.

SUMMARY

Water is, by far, the most abundant compound in most living organisms, and it usually accounts for 60 to 70% (by weight) of the human body. It is a highly polar molecule that participates directly or indirectly in a vast number of life-sustaining reactions and processes, including the transport of nutrients, the control of body temperature, and the folding of proteins, nucleic acids, and certain other compounds into biologically active conformations.

The noncovalent interactions that take place between neighboring molecules and separate parts of individual molecules account for many of the biologically important properties of water and other substances. These interactions fall into three major categories: electrostatic interactions (attractions between permanent stationary charges), van der Waals forces (attractions between temporary dipoles), and hydrophobic forces (the entropy-driven aggregation of nonpolar molecules in an aqueous environment). Hydrogen bonds are technically a special class of electrostatic interactions, but they are of such great biological importance that they are commonly discussed separately and placed in a category by themselves. Although individual noncovalent interactions tend to be relatively weak, multiple interactions can, collectively, possess a substantial bond energy. Cooperative bonding often contributes to this bond energy.

The water story leads into a discussion of acids, bases, buffers and related topics:

Acid: a proton (H^+) donor
Base: a proton acceptor
Acidic solution: $[H^+] > [OH^-]$; $pH < 7.0$
Basic solution: $[OH^-] > [H^+]$; $pH > 7.0$
Neutral solution: $[H^+] = [OH^-]$; $pH = 7.0$
$\mathbf{K_w} = [H^+][OH^-] = 1 \times 10^{-14}$
pH $= -\log [H^+]$
pOH $= -\log [OH^-]$
Strong acid: dissociates completely to yield H^+
Weak acid: dissociates incompletely to yield H^+
$\mathbf{K_a}$: the equilibrium constant for the dissociation of a weak acid
pK_a $= -\log K_a$
Conjugate base: what remains after a weak acid loses its H^+

Net reaction of a strong acid with a strong base:

$$H^+ + OH^- \rightleftharpoons H_2O$$

Net reaction of a weak acid with a strong base:

$$\text{Weak acid} + OH^- \rightleftharpoons H_2O + \text{conjugate base}$$

Net reaction of a conjugate base with a strong acid:

$$\text{Conjugate base} + H^+ \rightleftharpoons \text{weak acid}$$

Buffer: a weak acid (can consume added OH^-):conjugate base (can consume added H^+) pair that, in solution, resists changes in pH when an acid or base is added.

Buffer range: that pH range in which a buffer pair can effectively consume both added acid (H^+) and added base (OH^-); normally, $pK_a \pm 1$.

Buffer capacity: a quantitative measure of a buffer's ability to consume added H^+ and added OH^-. It is normally determined by buffer concentration ([weak acid] + [conjugate base]) and the proximity of the pH to the pK_a of the weak acid.

Since slight shifts in $[H^+]$ can be lethal to humans, evolution has led to the selection of multiple mechanisms for regulating the pH of human blood. Buffers, depth and rate of breathing, and the kidneys are all involved. The primary blood buffer is the H_2CO_3:HCO_3^- pair in equilibrium with a relatively constant concentration of CO_2:

$$CO_2\,(aq) + H_2O \rightleftharpoons H_2CO_3\,(aq) \rightleftharpoons HCO_3^-\,(aq) + H^+\,(aq)$$

The reaction of H_2O with CO_2 (produced from the oxidation of fuels within cells) tends to replace (Le Châtelier's principle) any H_2CO_3 or HCO_3^- consumed when a base enters the blood. HCO_3^- reacts with added acid.

The charges on a compound are a function of pH if that compound has one or more functional groups that can exist in a weak acid form. Since the charges on a compound normally have a major impact on its biological activity, many of the compounds within the body are active only within a relatively narrow pH range. To calculate the net charge on a compound to the nearest half of a charge unit, one employs three rules derived from the Henderson–Hasselbalch equation ($pH = pK_a + \log$ ([conjugate base]/[weak acid])):

Rule 1. A functional group is assigned the charge of its weak acid form if the pH is more than 0.5 units below its pK_a.

Rule 2. A functional group is assigned the charge of its conjugate base if the pH is more than 0.5 units above its pK_a.

Rule 3. A functional group is assigned a charge that is the average of the charge of its weak acid and conjugate base forms whenever the pH is within 0.5 of the pK_a.

In polyfunctional compounds, the charge on each individual functional group is calculated separately, and then net charge is determined by subtracting the absolute number of negative charges from the total number of positive charges.

The $[H^+]$ within a solution of a strong acid is always a whole number multiple (determined by the number of completely dissociable H^+'s in the acid) of the concentration of the strong acid itself. The $[H^+]$ within a weak acid solution can only be calculated precisely with the employment of the quadratic equation. However, calculations can be simplified if the $[H^+]$ can be neglected when compared with the concentration of the nondissociated weak acid. In such cases:

$$[H^+] = \sqrt{[\text{weak acid}] \times K_a}$$

Since the dissociation of a weak acid is readily reversible, a weak acid can be considered the conjugate acid of the weak base that forms when it dissociates. If we view a weak acid from this perspective, we encounter several new definitions and concepts that follow logically from those associated with the more classical approach to viewing a weak acid:

Weak base: reacts incompletely with water to generate OH^-

Conjugate acid: formed when a weak base accepts a proton

K_b: the $[H_2O]$ multiplied by the equilibrium constant for the reaction of a weak base with water to generate OH^-.

$$\mathbf{pK_b = -\log K_b}$$

$$\mathbf{K_aK_b = K_w}$$

$$\mathbf{pK_a + pK_b = 14}$$

EXERCISES

2.1 Given the structural formulas shown at the top of the next page:

a. In each compound, circle each hydrogen atom that is capable of H bonding.

b. In each compound, put a square around each atom to which a hydrogen in a water molecule could H bond.

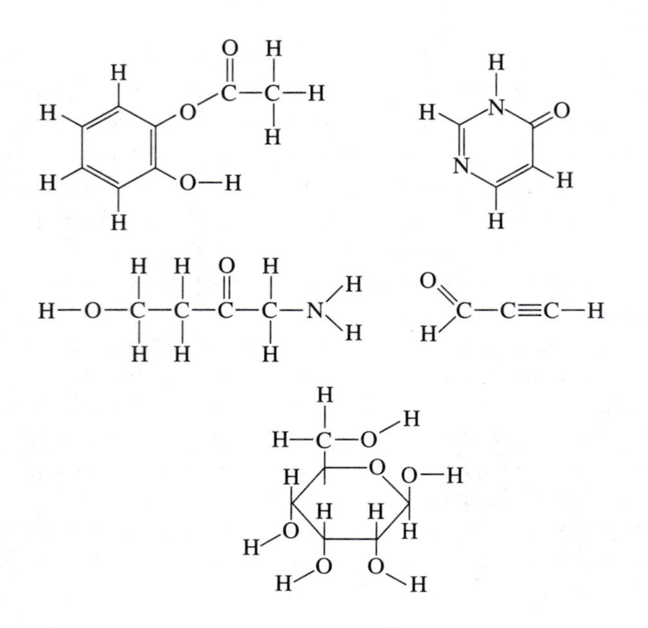

c. Draw formulas for water molecules around each compound, then use dashed lines to illustrate the H bonding possibilities that exist between water and each compound.

2.2 List the classes of noncovalent interactions (forces). Which of these classes are a consequence of electrostatic attractions?

***2.3** List two classes of compounds, in addition to salts of long-chain carboxylic acids, that tend to form micelles when mixed with water.

2.4 What predominant noncovalent force holds neighboring formula units of $^{+}NH_3CH_2COO^{-}$ together in a pure sample of this compound.

***2.5** List six major classes of organic compounds whose members are capable of H bonding with water.

2.6 Complete the following table.

Solution	$[H^+]$	$[OH^-]$	pH	pOH	Acidic, Basic, or Neutral?
0.02 *M* HNO_3					
0.18 *M* KOH					
0.18 *M* H_2CO_3 and 0.11 *M* HCO_3^-					
0.37 *M* $H_2PO_4^-$ and 0.02 *M* H_3PO_4					
0.90 *M* NH_4^+ and 0.08 *M* NH_3					
0.25 *M* NaCl					
0.25 *M* NaCl and 0.25 *M* HCl					
0.15 *M* lactate and 0.75 *M* lactic acid					
1.0 *M* ethanol					
1.2 M NaOH					

(Continued)

2.7 Complete each of the following reactions by writing the formulas for the products that will be formed. Balance each equation.

$H_2SO_4 + KOH \rightleftharpoons$

$HNO_3 + NaHCO_3 \rightleftharpoons$

$CH_3NH_2 + HCl \rightleftharpoons$

$K_2HPO_4 + NaOH \rightleftharpoons$

$K_2HPO_4 + HCl \rightleftharpoons$

$CH_3CH_2SH + NaOH \rightleftharpoons$

$CH_3OPO_3H^- + NaOH \rightleftharpoons$

$CH_3OPO_3H^- + HCl \rightleftharpoons$

$CH_3CH_2CH_2COOH + KOH \rightleftharpoons$

$CH_3CH_2COOK + HNO_3 \rightleftharpoons$

2.8 Write the formula for the conjugate base of each of the following weak acids.

a. HCOOH $pK_a = 3.75$

b. $(CH_3)_2AsO_2H$ $pK_a = 6.27$

c. $C_6H_5CO_2H$ $pK_a = 4.20$

d. $(CH_3)_2NH_2^+$ $pK_a = 10.8$

e. HF $pK_a = 3.18$

f. H_2S $pK_a = 6.89$

g. HCN $pK_a = 9.31$

2.9 Write the equilibrium constant expression for the dissociation of each of the weak acids given in Exercise 2.8.

2.10 List the acids given in Exercise 2.8 in order of increasing strength.

2.11 For each weak acid listed in Exercise 2.8, clearly indicate the pH range within which that weak acid, along with its conjugate base, can serve as a buffer.

2.12 Calculate the K_a for each weak acid given in Exercise 2.8.

2.13 Write structural formulas for three specific weak acids whose formulas have not been given in this text.

2.14 For each of the following weak acid:conjugate base pairs, write, (a) a balanced equation for the reaction that will occur if HNO_3 is added to the pair, and (b) a separate balanced equation for the reaction that will occur if KOH is added to the same pair. Omit all spectator ions in your equations.

i. $HF : F^-$

ii. $H_2SO_3 : HSO_3^-$

iii. $H_2S : HS^-$

iv. $CHCl_2COOH : CHCl_2COO^-$

v.

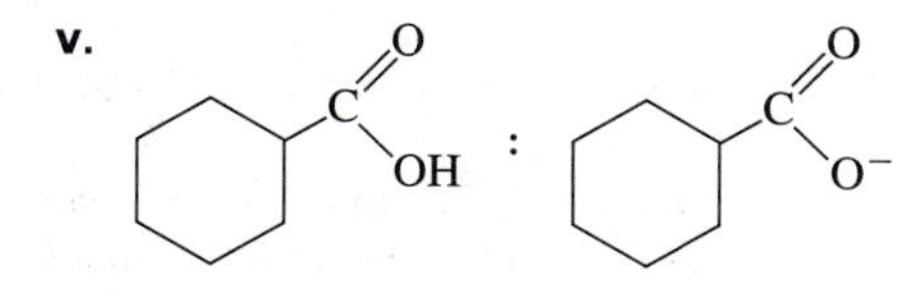

vi. $CH_3CHOHCH_2SH : CH_3CHOHCH_2S^-$

vii. $CH_3CH{=}CHCH_2\overset{+}{N}H_3 : CH_3CH{=}CHCH_2NH_2$

viii.

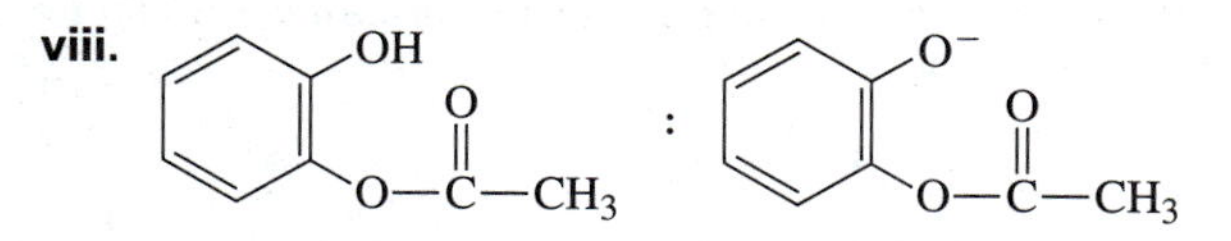

2.15 Calculate the net charge to the nearest half charge unit on each of the following compounds at (a) pH 1, (b) pH 6, (c) pH 10, and (d) pH 14.

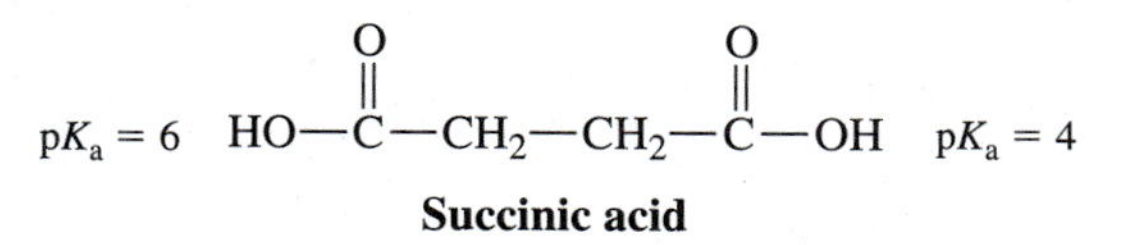

Succinic acid

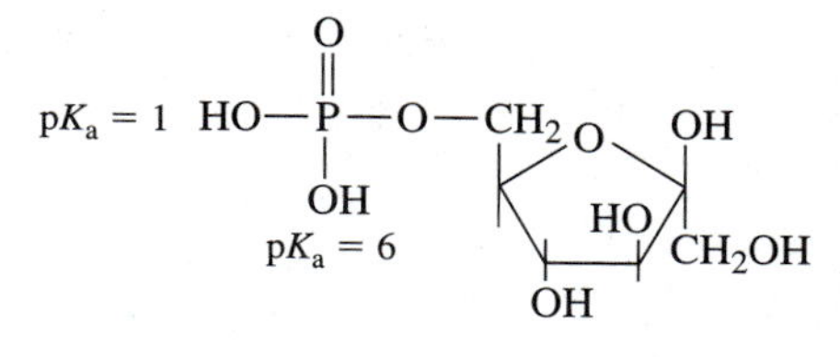

Fructose 6-phosphate

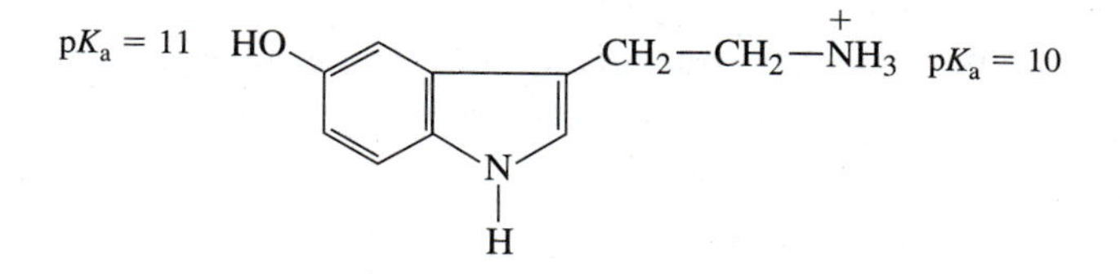

5-Hydroxytryptamine

$$pK_a = 11\quad H_3\overset{+}{N}{-}CH_2{-}CH_2{-}CH_2{-}CH(\overset{+}{N}H_3)\,(pK_a = 9){-}COOH\quad pK_a = 2$$

Ornithine

2.16 Write the structural formula for each of the compounds given in Exercise 2.15 as it primarily exists at pH 10. Rewrite the structure for each compound as it primarily exists at pH 1.

2.17 What is the ratio of HPO_4^{-2} to $H_2PO_4^-$ in saliva at pH 6.8?

2.18 What is the pH of a 1 L solution containing 1 mol of HCl and 2 mol of CH_3COONa?

2.19 If you held your breath, would you tend to develop acidosis or alkalosis? Explain.

2.20 What is the pH of blood if $[HCO_3^-] = 30[H_2CO_3]$? Show your calculations.

2.21 Why is the pH of venous blood slightly lower than the pH of arterial blood?

2.22 What is the pH of a solution prepared by mixing 0.15 mol of sodium acetate with 0.37 mol of acetic acid and then diluting the mixture to 1 L with pure water?

2.23 What is the pH of a solution prepared by mixing 1 mol of HCl with 1.2 mol of NaOH and then diluting the resultant solution to 1 L with pure water?

2.24 Make a list of the terms that appear in bold print in this chapter. Define or describe each term.

2.25 Use the information given in Exercise 2.8 to calculate the approximate pH of:

a. 0.72 M HCN

b. 1.2 M $(CH_3)_2NH_2^+Cl^-$

c. 0.36 M H_2S

d. 0.48 M HCOOH

e. 0.25 M $(CH_3)_2AsO_2H$

2.26 Write the structural formula for the conjugate acid of each of the following weak bases.

a. $HCOO^-$

b. $(CH_3)_2NH$

c. F^-

d. HS^-

e. CN^-

2.27 Use the information given in Exercise 2.8 to calculate the K_b and the pK_b for each weak base given in Exercise 2.26.

2.28 Write the K_b expression for each weak base given in Exercise 2.26.

2.29 List the weak bases given in Exercise 2.26 in order of increasing strength.

SELECTED READINGS

Fersht, A. R., The hydrogen bond in molecular recognition, *Trends Biochem. Sci.* 12, 301–304 (1987).

Emphasizes the role of H bonding within proteins and nucleic acids.

Good, N. E., G. D. Winget, W. Winter, T. N. Conolly, S. Izawa, and R. M. M. Singh, Hydrogen ion buffers for biological research, *Biochemistry* 5, 467–477 (1966).

Examines a number of synthetic buffers used in the biochemistry laboratory.

Honig, B., and A. Nicholls, Classical electrostatics in biology and chemistry, *Science* 268, 1144–1149 (1995).

Describes a major revival in the use of classical electrostatics as an approach to the study of charged and polar molecules in aqueous solutions.

Lehninger, A. L., D. L. Nelson, and M. M. Cox, *Principles of Biochemistry.* New York: Worth Publishers, 1993.

Chapter 4 provides another perspective of the topics covered in this chapter.

Practical Handbook of Biochemistry and Molecular Biology. Fasman, G. D., ed., Cleveland: CRC Press, 1989.

Contains sections on the preparation of buffers and pH measurement.

Stenesh, J., *Core Topics in Biochemistry.* Kalamazoo, MI: Cogno Press, 1993.

A thorough and somewhat advanced analysis of acids, bases and buffers. Two chapters and over 70 pages are devoted to these topics.

Wiggins, P. M., Role of water in some biological processes, *Microbiol. Rev.* 54, 432–449 (1990).

Reviews the structure of water and its importance within living organisms.

3 Amino Acids, Peptides, and Four Laboratory Techniques

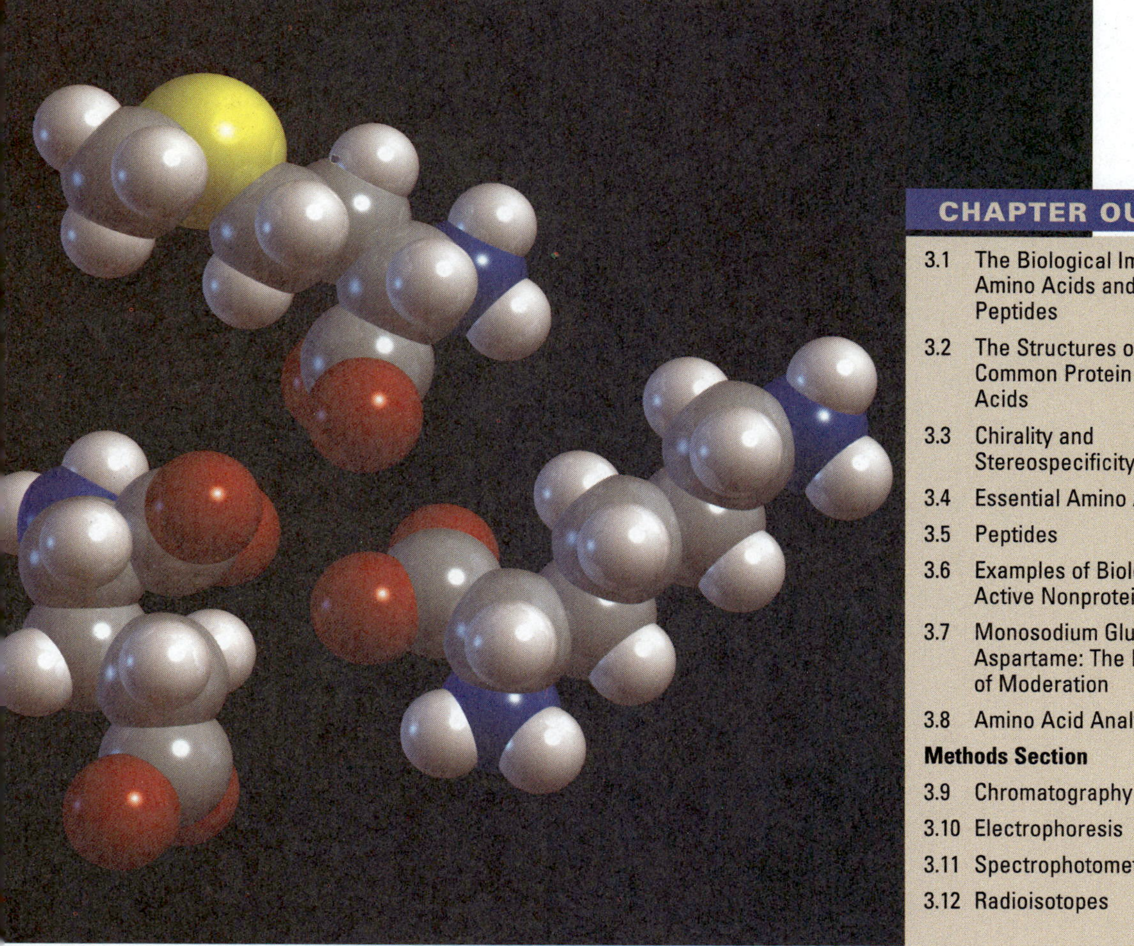

CHAPTER OUTLINE

TABLE 3.1

A Classification System for Peptides

Class	Number of Amino Acids
Dipeptide	2
Tripeptide	3
Tetrapeptide	4
Pentapeptide	5
Hexapeptide	6
Heptapeptide	7
Octapeptide	8
Nonapeptide	9
Decapeptide	10
Oligopeptide	2 to 11
Polypeptide	>11
Protein[a]	>50[b]

[a]Must attain a specific conformation to be biologically active.
[b]Usually

This chapter moves from a review of water, acids, bases, and buffers into an examination of amino acids and those substances, called **peptides,** that are created when amino acids are joined covalently in continuous chains. Since amino acids and peptides contain both acidic and basic functional groups and they usually function in an aqueous environment, the concepts discussed in Chapter 2 are put to good use in this chapter.

Peptides are classified as dipeptides, tripeptides, tetrapeptides, and so on, on the basis of the number of amino acid units they contain (Table 3.1). The prefix *oligo,* which means "several" or "a few," is used in this text to specify 2 to 11. The prefix *poly,* as used in Table 3.1, means over 11. Polypeptides that contain a chain of 50 (approximately) or more amino acids and that must attain a specific conformation to be biologically active are called **proteins.** Thus, all proteins are peptides, but only the larger, more complex peptides are classified as proteins. Other peptides are classified as **nonprotein peptides.**

Proteins are analyzed in detail in Chapters 4 and 5, and they appear as central figures in virtually all of the chapters that follow. This chapter concentrates on the structure and properties of the **protein amino acids** (those used to assemble both nonprotein peptides and proteins) and nonprotein peptides. Several major biochemical themes surface along the way. They include specificity, structure–function relationships, and communication. It is important to remember that the scientific method is the basis for the acquisition of all biochemical knowledge; indirectly, experimentation and hypothesis construction are an integral part of every chapter.

Because the study of amino acids and peptides involves numerous laboratory techniques, four of the most widely employed tools of biochemistry are included in this chapter: chromatography, electrophoresis, spectrophotometric analysis, and radioisotopes. The same four tools play a central role in the study of most other classes of compounds as well.

To comprehend and interpret the characteristics of amino acids and peptides, a substantial amount of organic chemistry must be recalled and utilized. Some essential organic chemistry is reviewed here directly, and some is examined in answers to specific problems. In this and other chapters, the in-chapter problems are an integral part of the chapter. If you cannot solve the review problems relatively quickly and confidently, study the published answers and associated discussions (see end of book). For a more in-depth review, an organic chemistry textbook can be consulted.

3.1 THE BIOLOGICAL IMPORTANCE OF AMINO ACIDS AND NONPROTEIN PEPTIDES

Amino acids are the building blocks for both proteins and nonprotein peptides. They are also used as fuels by cells and as precursors (starting materials) for the construction of certain other compounds. Some possess important physiological activities as well. Glutamate (the salt form of glutamic acid), for example, is the major excitatory **neurotransmitter** (chemical messenger involved in nerve cell communication) in the mammalian central nervous system (Section 10.9). Aspartate (the salt form of aspartic acid) and glycine function in a similar capacity. Tryptophan is the precursor to another neurotransmitter, **serotonin,** as well as the plant growth factor, **auxin.**

Tryptophan's association with neurotransmitters and the central nervous system has led to its use as a food supplement by some individuals seeking relief from depression, insomnia, and certain other nervous system disorders. In 1989, a particular batch of tryptophan was implicated in at least 38 deaths due to **eosinophilia-myalgia syndrome (EMS).** The incident attracted considerable press, in part, because the tryptophan involved was produced with the assistance of a genetically engineered organism (see Chapter 19). It is still uncertain whether the tryptophan itself, or contaminant(s), or both were responsible. Excess tryptophan may be harmful to certain sensitive individuals. The Canadian government already treats amino acid food supplements as drugs, and there is growing pressure to have the Food and Drug Administration in the United States do the same.

A variety of nonprotein peptides are found in virtually every fluid and tissue within the human body (Table 3.2). Most of these were long dismissed as inactive breakdown products of proteins. Although some of these peptides undoubtedly fall into this category, many have potent, unique, and biologically important activities; a large and growing number are known to function as traditional **hormones** (chemical messengers), **growth factors** (compounds that control cell division and tissue growth), or neurotransmitters. Nonprotein peptides have physiological roles in most organs including kidneys, heart, gonads, pancreas, liver, lungs, and brain. Each of

TABLE 3.2

Some Physiologically Active Nonprotein Peptides

Peptide	Number of Amino Acid Units	Role, Function, or Activity
Angiotensin II	8	Regulates blood pressure
Bradykinin	9	A vasodilator and diuretic
Gastrin	17	Stimulates acid secretion by the stomach
Glucagon	29	Regulates glucose metabolism
Endothelin 1	21	Regulates blood pressure; may help control differentiation during embryonic development
Methionine enkephalin	5	An opiate peptide and neurotransmitter; a natural pain reliever
Oxytocin	9	Stimulates lactation and uterine contraction
Substance P	10	A neurotransmitter
Thyrotropin-releasing factor	3	Hypothalamus hormone that stimulates thyrotropin release by the pituitary
Vasopressin (Antidiuretic hormone [ADH])	9	Regulates blood pressure and reabsorption of water by the kidneys; a neurotransmitter

the "messenger" peptides plays a part in the central theme of communication and many are at least as important as better characterized nonpeptide neurotransmitters such as acetylcholine, histamine, dopamine, and noradrenaline. A significant number of nonprotein peptides and their derivatives are already being employed as therapeutic agents.

3.2 THE STRUCTURES OF THE COMMON PROTEIN AMINO ACIDS

The term **amino acid** refers to any compound that contains both an **amino group** (—NH_2) and a **carboxyl group** (—COOH). Dozens of such compounds have been discovered in one or more living organisms. In this text, however, the term *amino acid,* unless specified otherwise, refers to one of the 20 very special amino acids that are most often used to assemble proteins and other peptides. With one rare exception, these 20 amino acids are the only compounds known to be encoded in the **genetic code** (Section 16.5). Since amino acids are sometimes covalently modified after they have been incorporated into a peptide chain (Section 18.11), a "mature" peptide may contain a variety of amino acid units in addition to the 20 common ones.

The generalized structure for a protein amino acid at pH 7.0 can be depicted as follows:

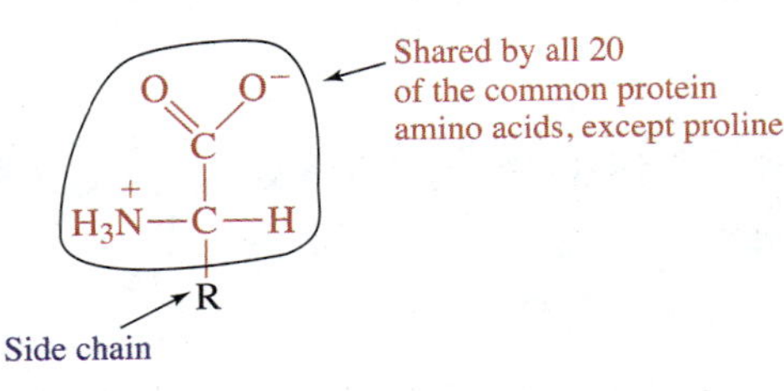

The circled grouping of atoms is shared by all of the common amino acids, except one, known as proline. A unique **side chain,** represented with an R in the generalized structure, distinguishes any one amino acid from all the others. Figure 3.1 provides the name, structural formula, three-letter abbreviation and one-letter abbrevi-

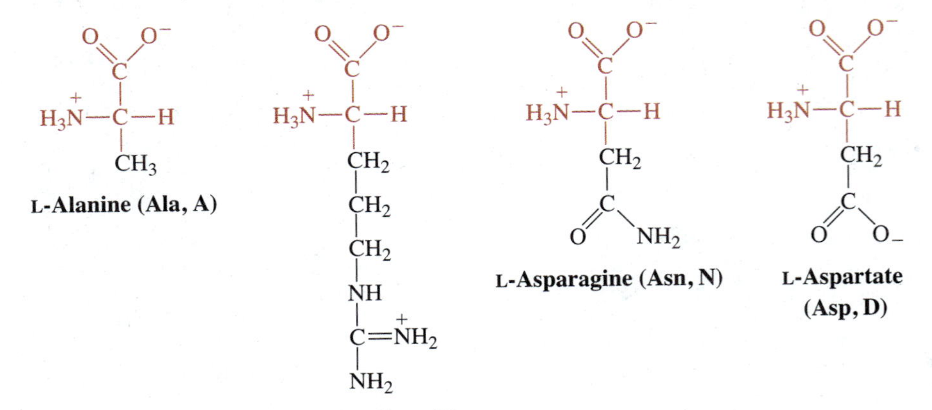

FIGURE 3.1
Structures of the 20 common protein amino acids as they predominantly exist at pH 7.0.

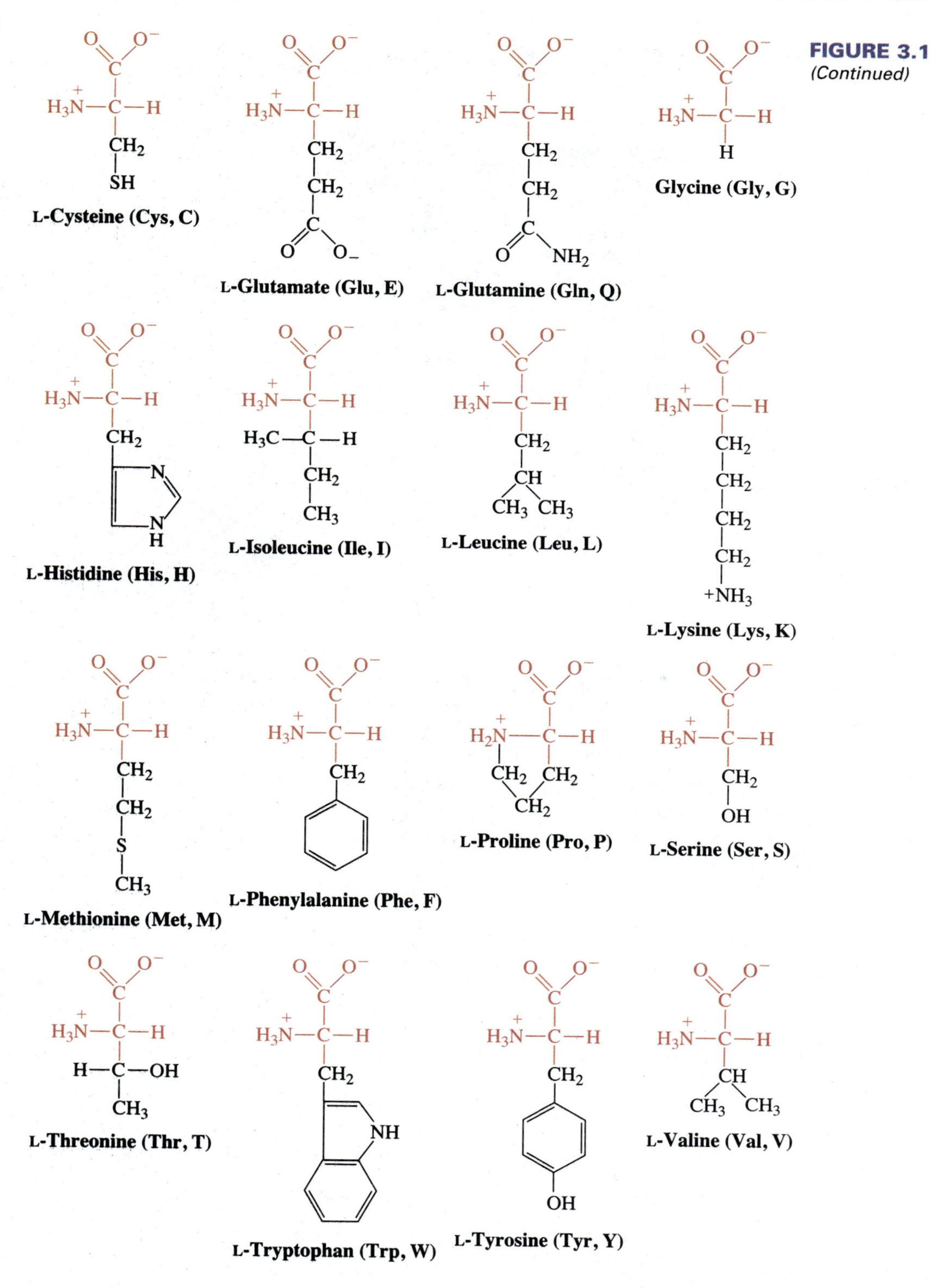

FIGURE 3.1
(Continued)

TABLE 3.3

Amino Acid Abbreviations

Amino Acid	Three-Letter Abbreviation	One-Letter Abbreviation
Alanine	Ala	A
Arginine	Arg	R
Asparagine	Asn	N
Aspartic acid	Asp	D
Cysteine	Cys	C
Glutamic acid	Glu	E
Glutamine	Gln	Q
Glycine	Gly	G
Histidine	His	H
Isoleucine	Ile	I
Leucine	Leu	L
Lysine	Lys	K
Methionine	Met	M
Phenylalanine	Phe	F
Proline	Pro	P
Serine	Ser	S
Threonine	Thr	T
Tryptophan	Trp	W
Tyrosine	Tyr	Y
Valine	Val	V

ation for each amino acid. The abbreviations for the amino acids are also summarized in Table 3.3. **[Try Problems 3.1. and 3.2. Problem 3.2 provides a review of some basic chemistry.]**

All of the common protein amino acids, except proline, are **α-amino acids**; they contain an amino group which is attached to the carbon (called an **α-carbon**) immediately adjacent to the carboxyl group. The stepwise movement of the amino group to carbons further from the carboxyl group generates β, γ, δ and ϵ-amino acids, respectively.

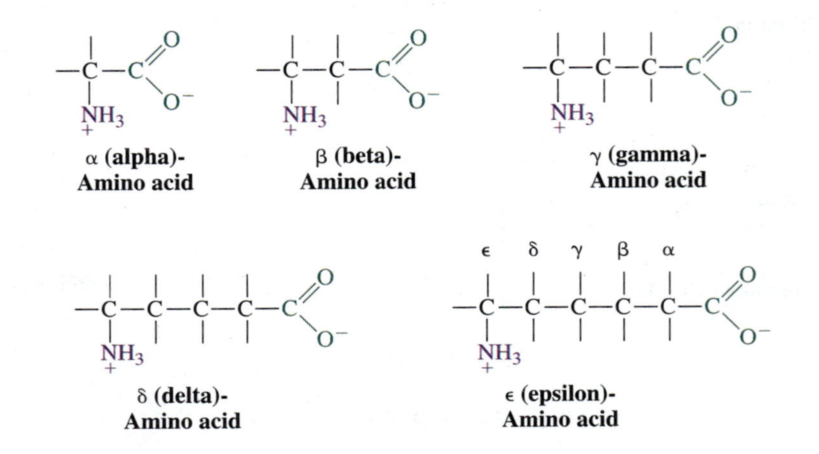

Thus, lysine is both an α and an ϵ-amino acid:

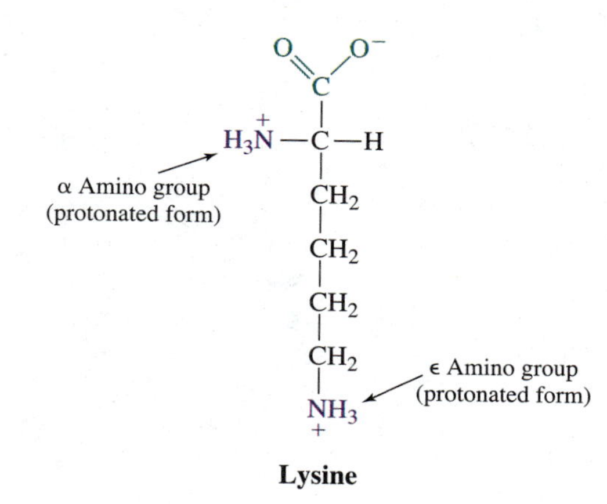

Lysine

The same Greek letters are routinely employed to designate the location of any substituent relative to a reference functional group. Although commonly called an α-amino acid, proline is actually an imino acid; its side chain and α-nitrogen have been covalently joined to form a heterocyclic ring. Chapter 4 describes how the unique structure of proline allows it to play a central role in directing the folding of many proteins.

Table 3.4 summarizes the functional groups found in amino acids and identifies those functional groups that are weak acids. The pK_a's given in this table are average values for the functional groups as they exist in the 20 common amino acids. The precise pK_a for each weak acid functional group in each amino acid is given in Table 3.5. Since the pK_a for a functional group is partially determined by its proximity to neighboring groups, the pK_a's for the protonated α-amino groups range from 8.8 in asparagine to 10.3 in cysteine. The pK_a of the side chain carboxyl group of aspartic acid differs from that of its α-carboxyl group and from the pK_a's of both carboxyl groups in glutamic acid. Similarly, the pK_a's for amino acid functional groups normally shift when amino acids are incorporated into peptides, and the pK_a's for peptide functional groups fluctuate when the environments of these groups are altered as a consequence of chain folding. The latter fact is of more than academic interest. In Section 4.9, conformation-linked changes in pK_a's are used to help explain one of the physiologically important properties of hemoglobin (an oxygen transport protein). **[Try Problems 3.3 and 3.4]**

The 20 common protein amino acids have been divided into a variety of categories on the basis of the similarities and differences in their side chains. Table 3.6 summarizes a frequently employed classification scheme that is based on the polarity and charge of side chains at pH 7.0. Cysteine, glycine, and tryptophan are placed in a borderline category, because some authors classify them as polar and others label them nonpolar. Since the thiol group possesses a relatively large sulfur atom and normally has a pK_a between 8 and 9, the side chain in a cysteine is protonated (uncharged) over 90% of the time at pH's below 7.0 and is compatible with a nonpolar environment. At higher pH's, the cysteine side chain acquires a polar character: Its

TABLE 3.4

Summary of the Functional Groups in the Common Protein Amino Acids

Functional Group	Name	Weak Acid Form	Conjugate Base Form	Average pK_a Value in Free Amino Acids
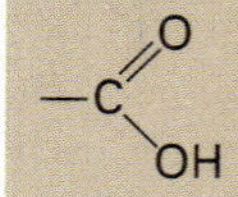 $-C(=O)OH$	Carboxyl	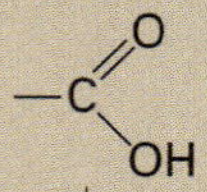 $-C(=O)OH$	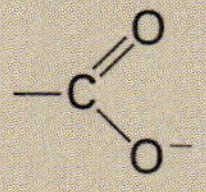 $-C(=O)O^-$	2.4
$-NH_2$	Amino	$-\overset{+}{N}H_3$	$-NH_2$	9.5
$-SH$	Sulfhydryl	$-SH$	$-S^-$	8.2
$-OH$	Hydroxyl	—	—	—
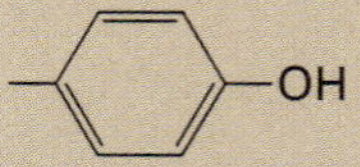 $-C_6H_4-OH$	Phenolic hydroxyl	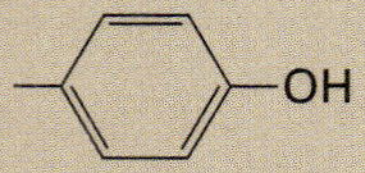 $-C_6H_4-OH$	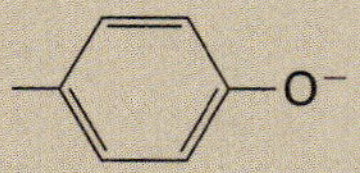 $-C_6H_4-O^-$	10.1
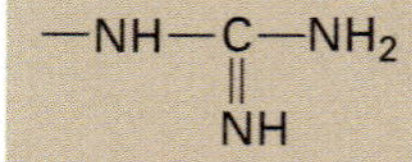 $-NH-C(=NH)-NH_2$	Guanidino	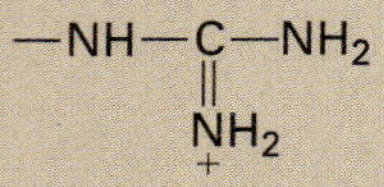 $-NH-C(=\overset{+}{N}H_2)-NH_2$	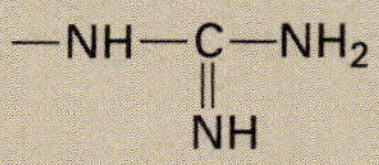 $-NH-C(=NH)-NH_2$	12.5
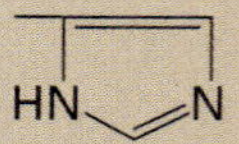 HN–N imidazole ring	Imidazole	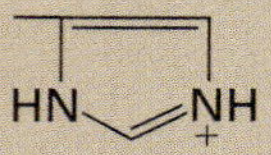 HN–$\overset{+}{N}$H imidazolium ring	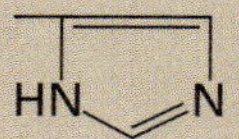 HN–N imidazole ring	6.0
$-S-$	Sulfide	—	—	—
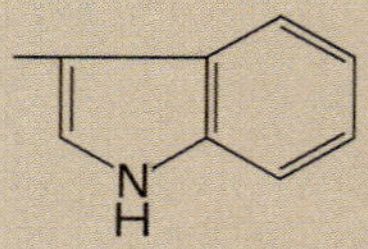 Indole ring (N–H)	Indole	—	—	—
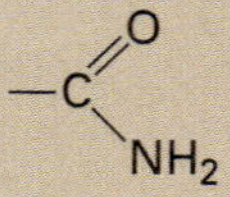 $-C(=O)NH_2$	Amido	—	—	—

time averaged charge approaches −1 at pH's well above the pK_a for the thiol group (Section 2.7). Glycine is actually in a class by itself, because the single hydrogen in its side chain does not have a significant degree of either polar or nonpolar character. Although the tryptophan side chain definitely contains a small polar functional group (—NH—), overall, it has more nonpolar than polar character. **[Try Problems 3.5 through 3.7]**

The side chain of an amino acid has a major impact on its chemistry. Serine and threonine side chains, for example, can undergo reactions characteristic of alcohols, including ester and ether formation. Acidic side chains can react with bases and can provide protons for acid catalysis. Cysteine side chains are readily oxidized to disulfides, whereas the acylation of the amino group in the lysine side chain generates an amide. Charged side chains can form salt bridges and, at pH 7.0, arginine, asparagine, glutamine, histidine, lysine, serine, threonine, tryptophan, and tyrosine side chains contain hydrogens that are capable of H bonding. In water, nonpolar side

TABLE 3.5

The pK_a Values for the Weak Acid Functional Groups in Amino Acids

	pK_a (at 25°C in aqueous solution)		
Amino Acid	**α-COOH**	**α-NH_3^+**	**Side Chain Functional Group**
Ala	2.3	9.7	
Arg	2.2	9.0	12.5
Asn	2.0	8.8	
Asp	1.9	9.6	3.7
Cys	2.0	10.3	8.2
Gln	2.2	9.1	
Glu	2.2	9.7	4.3
Gly	2.3	9.6	
His	1.8	9.2	6.0
Ile	2.4	9.6	
Leu	2.4	9.6	
Lys	2.2	9.0	10.5
Met	2.3	9.2	
Phe	1.8	9.1	
Pro[a]	2.0	10.6[a]	
Ser	2.2	9.2	
Thr	2.1	9.1	
Trp	2.8	9.4	
Tyr	2.2	9.1	10.1
Val	2.3	9.6	

[a]Contains an α-imino group rather than an α-amino group

TABLE 3.6

The Classification of Amino Acids on the Basis of the Polarity and Charge of Their Side Chains at pH 7.0

Nonpolar	Borderline (Polar/Nonpolar)	Polar (Uncharged)	Polar (+ charged)	Polar (− charged)
Ala	Cys	Asn	Arg	Asp
Ile	Gly	Gln	Lys	Glu
Leu	Trp	His		
Met		Ser		
Phe		Thr		
Pro		Tyr		
Val				

chains tend to aggregate through hydrophobic forces. All of these facts will be revisited as the complete amino acid and protein story gradually unfolds. **[Try Problems 3.8 and 3.9]**

PROBLEM 3.1 Based on Figure 3.1, how many atoms does the smallest amino acid contain? The largest amino acid?

***PROBLEM 3.2** List all the elements that would be required to build the 20 common protein amino acids. Which of these belong to the A-family elements (see the periodic table on the back inside cover)? Which are metals? Which are nonmetals? What is the most common valence of each listed element? Arrange these elements in order of increasing electronegativity. Arrange these elements in order of increasing size.

PROBLEM 3.3 Which functional group in Table 3.4 represents the strongest acid? The weakest acid?

***PROBLEM 3.4** Why does a carboxyl group that is close to an α-NH_3^+ group have a lower pK_a than a carboxyl group that is further removed (see glutamic acid and aspartic acid in Table 3.5)?

PROBLEM 3.5 Which, if any, amino acids have side chains that will exist predominantly in a polar, positively charged form at pH 1.0? at pH 12?

PROBLEM 3.6 Which, if any, amino acids have side chains that will exist predominantly in a polar, negatively charged form at pH 1.0? at pH 12?

PROBLEM 3.7 Calculate the net charge (to the nearest half of a charge unit—see Section 2.7) on tyrosine and histidine at both pH 1.0 and pH 12.

PROBLEM 3.8 Circle each hydrogen atom in Figure 3.1 that is capable of hydrogen bonding (reviewed in Section 2.1).

***PROBLEM 3.9** Classify each of the following compounds as polar or nonpolar:

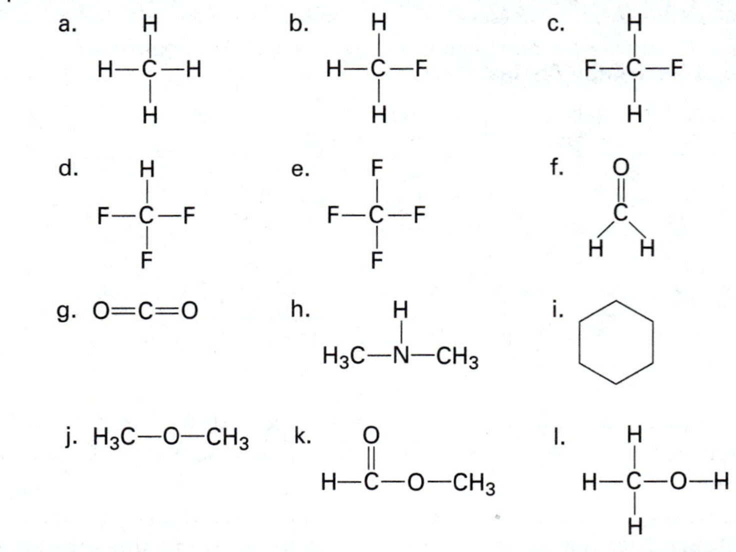

3.3 CHIRALITY AND STEREOSPECIFICITY

All of the common protein amino acids, except glycine, possess one or more chiral centers and are chiral compounds. A **chiral center** is any atom (usually carbon) that has four different groups attached to it in a nonplanar geometry and, consequently, has no plane of symmetry. A **chiral compound** is one that lacks a plane of symmetry. Each chiral compound has one enantiomer. **Enantiomers** are **stereoisomers** that are nonsuperimposable mirror images.

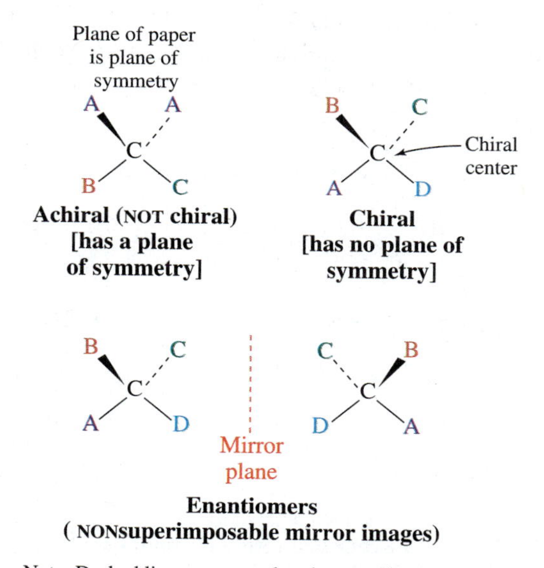

Enantiomers have the same melting point, boiling point, and solubility. They also exhibit the same chemical characteristics, except when interacting with other chiral compounds. Enantiomers rotate the plane of **plane-polarized light** equally but in opposite directions. Their ability to rotate the plane of plane-polarized light makes them **optically active.** **[Try Problems 3.10 through 3.13. They provide a review of stereoisomerism and related concepts]**

When using structural formulas to illustrate and distinguish the enantiomers of an amino acid, it is necessary to follow certain standard conventions. The carbon chain is placed in a vertical column with the carboxyl group up. The α-carbon is assumed to be in the plane of the paper with the two carbons to which it is directly bonded behind that plane. When the carbon chain is locked in space in this fashion, an amino group written to the left identifies the **L-enantiomer** and an amino group written to the right represents the **D-enantiomer** as shown at the top of the next page:

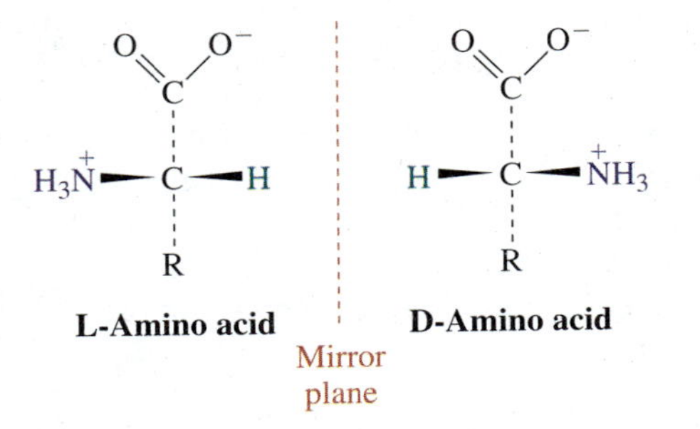

Note: When depicting D and L configurations, the orientation of atoms about achiral centers is unimportant. Thus, it is acceptable to draw the carboxyl group in the same orientation in both enantiomers

The L and D specify the **absolute configuration** about the α-carbon. The absolute configuration is the configuration relative to D- and L-glyceraldehyde, a pair of enantiomers that have been internationally accepted as reference compounds for assigning configurations.

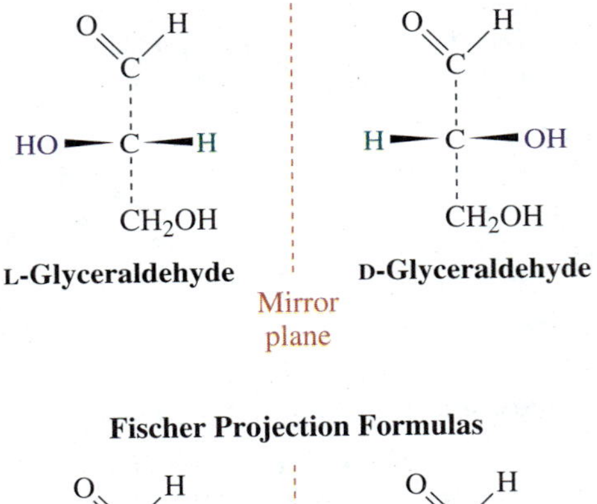

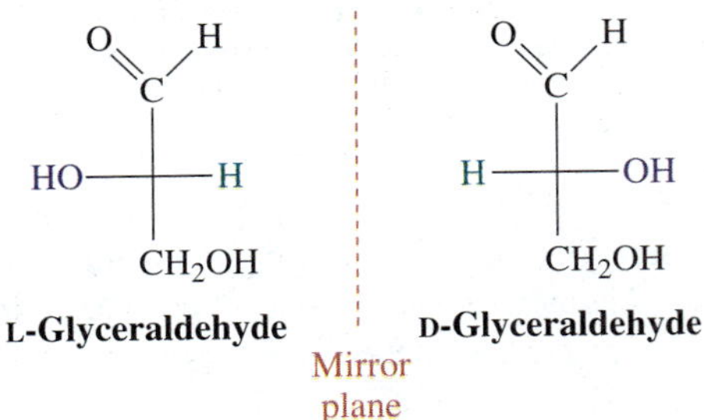

In **Fischer projection formulas,** vertical lines represent bonds that extend below the plane of the paper, and horizontal lines identify bonds that extend above this plane. A carbon atom exists at each point where two lines intersect. **[Try Problem 3.14]**

All living organisms, including humans, appear to use only L-amino acids to assemble polypeptides, and many organisms are not genetically programmed to synthesize or utilize their D enantiomers. However, some organisms produce enzymes that catalyze the interconversion of D- and L-configurations about the α-carbon of specific amino acid units within certain already assembled polypeptides. Remarkable *specificity*! Some bacteria use particular D-amino acids to construct cell walls and to synthesize a number of unique **antibiotics** (organic compounds that are produced by certain microorganisms and plants and are toxic to other organisms), including **valinomycin, actinomycin** (Section 17.1) and **gramicidin S.** Compounds that selectively block D-amino acid production are being used to treat certain bacterial infections in humans, including urinary tract infections and tuberculosis. *Any compound that selectively blocks the production of a substance that is essential for the survival of an infectious agent, a parasite, or a cancer cell, but is not required by the human body, is a potential chemotherapeutic agent.*

Since living organisms can distinguish stereoisomers, they are said to be **stereospecific.** The ability to discriminate between chiral isomers lies in the chiral nature of the protein catalysts that control the rates of most life-sustaining reactions. This fact is examined in detail in Section 5.4. Stereospecificity is of primary importance to anyone attempting to synthesize drugs or other biologically active compounds. One enantiomer of the drug thalidomide, for example, is a sedative, whereas its partner leads to birth defects, a fact tragically discovered some years ago (Figure 3.2).

The configuration about a chiral center can be designated ***R*** or ***S*** (rather than D or L) based on a classification system routinely employed in organic chemistry. When applied to the common protein amino acids, some α-carbons have an *S* configuration and others an *R* configuration. In contrast, all of these amino acids are assigned an L configuration when the D–L system is employed. Only the D–L system is used in this text.

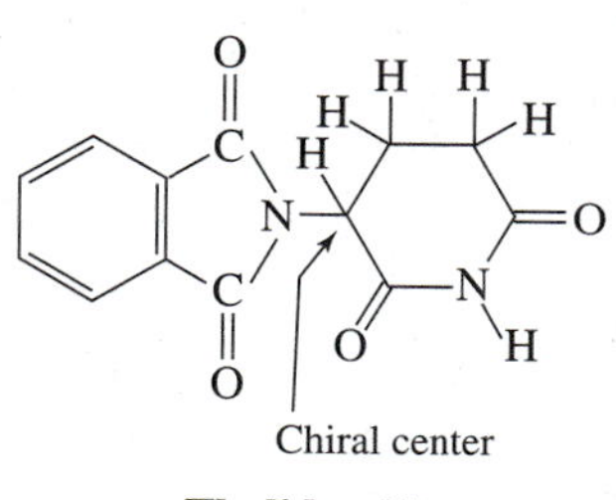

Thalidomide

FIGURE 3.2
A woman who was affected by the drug thalidomide *in utero.* Phocomelia (seal limb) is a common result of the action of one enantiomer of thalidomide on a fetus.

***PROBLEM 3.10** Define or describe each of the following terms:

a. Plane of symmetry
b. Plane-polarized light
c. Chiral center
d. Chiral compound
e. Achiral compound
f. Isomers
g. Stereoisomers
h. Geometric isomers
i. Structural isomers
j. Enantiomers
k. Diastereomers

***PROBLEM 3.11** Circle each chiral center in each of the following compounds:

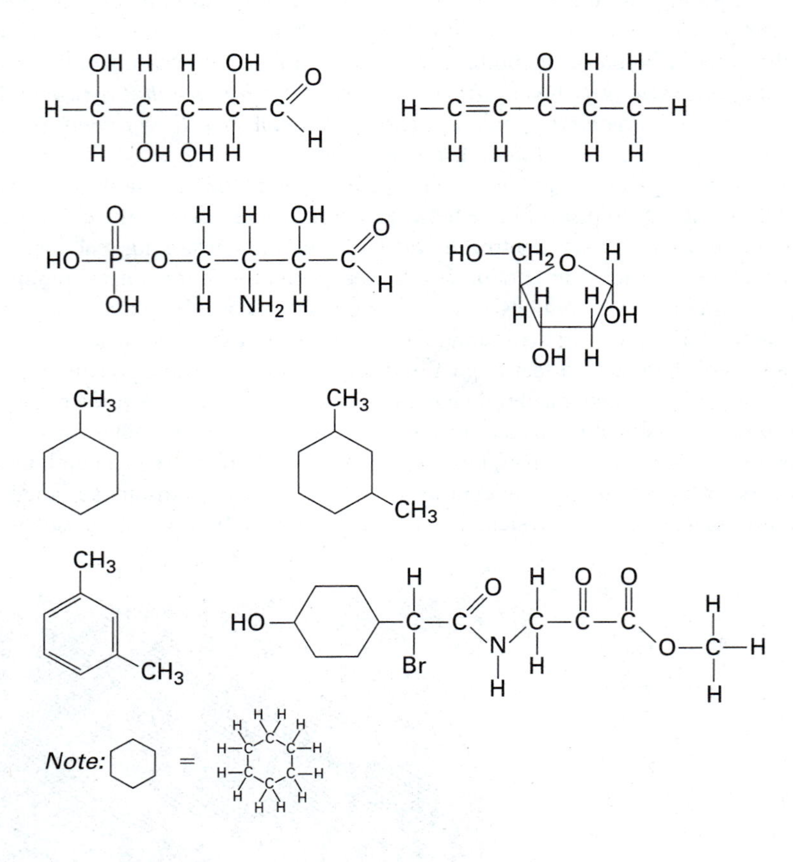

PROBLEM 3.12 Which amino acids have more than one chiral center?

***PROBLEM 3.13** Write the structural formula for a compound that has multiple chiral centers but is achiral. What term is used to describe such a compound?

PROBLEM 3.14 Write a three-dimensional structural formula for the enantiomer of each of the following compounds:

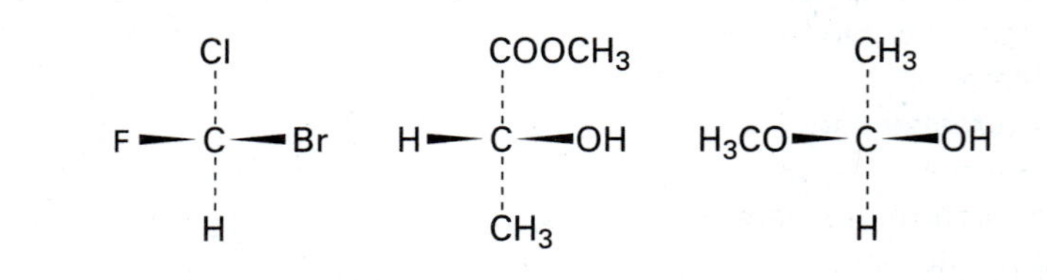

3.4 ESSENTIAL AMINO ACIDS

The human body absolutely requires all 20 of the common protein amino acids, but it is genetically programmed to synthesize only a fraction of these. Those 9 amino acids that are not produced in adequate amounts within the body are described as **essential amino acids** (Table 3.7); they must be obtained in the diet. Although arginine can be assembled in adequate amounts by most adults, it is now thought to be a tenth essential amino acid for children. Diets deficient in essential amino acids are common in many parts of the world, and they frequently lead to serious physical and mental disabilities. **Kwashiorkor** is a potentially lethal deficiency disease linked to the inadequate consumption of protein, the major source of dietary amino acids.

The division of amino acids into essential and nonessential categories is based solely on nutritional requirements in humans. The nonessential amino acids are, in fact, just as vital as the essential ones from the standpoint of protein production and survival. [Try Problem 3.15]

TABLE 3.7

Essential and Nonessential Amino Acids

Essential	Nonessential
Histidine	Alanine
Isoleucine	Arginine[a]
Leucine	Asparagine
Lysine	Aspartic acid
Methionine	Cysteine
Phenylalanine	Glutamic acid
Threonine	Glutamine
Tryptophan	Glycine
Valine	Proline
	Serine
	Tyrosine

[a]Nonessential for most adults but essential for normal growth and development in children.

PROBLEM 3.15 Amino acid mixtures are sometimes sold as dietary supplements. Some mixtures contain only L-amino acids and others contain roughly equal amounts of both the D- and the L-enantiomers. If a health food store offered 50g of L-amino acids for $15 and 75g of D-,L-amino acids for $10, which would represent the better buy? Assume that both mixtures contain the same ratios of each of the 20 common protein amino acids.

3.5 PEPTIDES

Peptides, as defined in the introduction to this chapter, are linear oligomers and polymers of amino acids. Our picture of peptides has been focused by the discussion of the structure of amino acids in Sections 3.2 and 3.3. To further sharpen that picture, it is necessary to examine the covalent link that joins amino acids.

Peptide Bonds

The α-amino group of one amino acid can react with the α-carboxyl group of a second to form an **amide**:

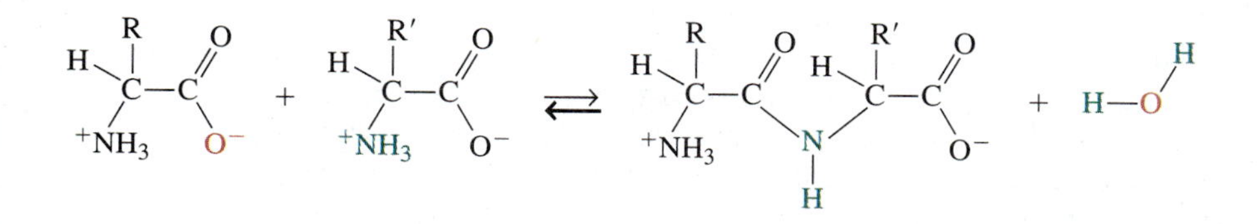

The product is a peptide, and the amide link is called a **peptide bond.** Thus, peptide bonds are a subclass of amide bonds, and all peptides are amides. In the chemical equation given above, the heavy arrow in the reverse direction indicates that the

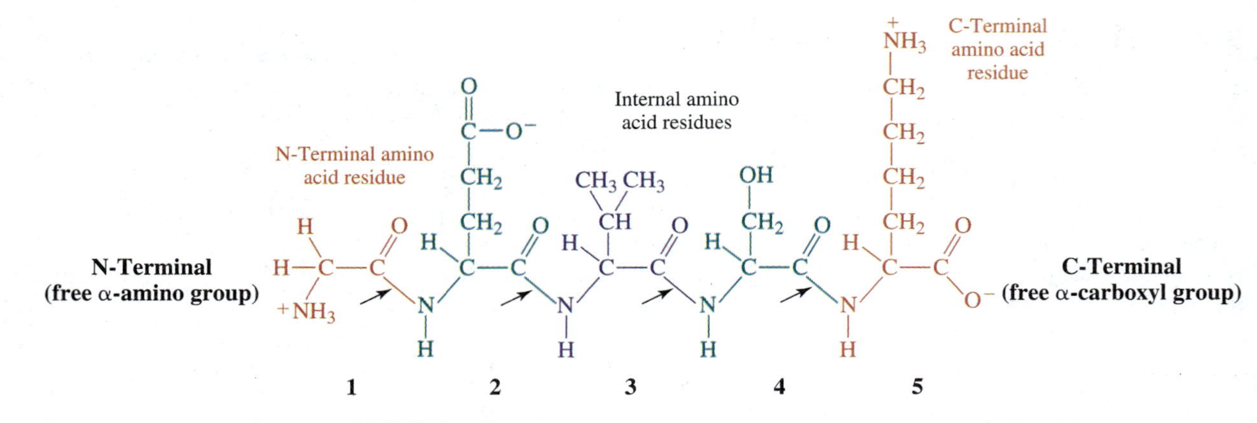

FIGURE 3.3
Structural formula for a pentapeptide. The arrows identify peptide bonds. A pentapeptide contains 4 peptide bonds, a hexapeptide 5, heptapeptide 6, and so on.

equilibrium constant for this reaction is relatively small. For this reason, both organic chemists and living organisms use **activated acid derivatives,** rather than free acids, to synthesize amides. The activated derivatives increase the free energy content of reactants relative to products and provide a thermodynamic driving force for the reaction. Chapter 8 expands on this observation and Chapter 16 examines those activated amino acid derivatives used by living organisms to help drive the coupling of amino acids. **[Try Problems 3.16 and 3.17. They review some organic chemistry and thermodynamics.]**

The term **amino acid residue** refers to an amino acid as it resides within a peptide. Each residue contains all of the atoms contributed to the peptide by a single amino acid. The total number of amino acid residues within a peptide is equal to the number of amino acids that were joined to form that peptide. Since both the α-amino groups and the α-carboxyl groups of internal amino acid residues have reacted to form amide bonds, only the terminal amino acid residues possess a free α-amino or a free α-carboxyl group. Although branching is chemically possible, peptides are usually nonbranched. The structural formula for a specific pentapeptide is given in Figure 3.3. **[Try Problems 3.18 through 3.21]**

Acid–Base Properties of Peptides

Most peptides contain at least two functional groups that can exist in both a weak acid and a conjugate base form: the α-amino group at the **N-terminus** and the α-carboxyl group at the **C-terminus** (Figure 3.3). This statement is qualified with a "most" since the terminal amino and carboxyl groups in linear peptides are occasionally modified (blocked) after the peptides have been assembled and cyclic peptides (rare, but do exist) lack terminal residues. Side chains commonly contribute additional weak acid functional groups. At very high pH values, each of these functional groups is nonprotonated (dissociated) and either negatively charged or uncharged (see Table 3.4); the compound as a whole contains only negative charges (Figure 3.4). At very low pH values, each weak acid functional group is protonated

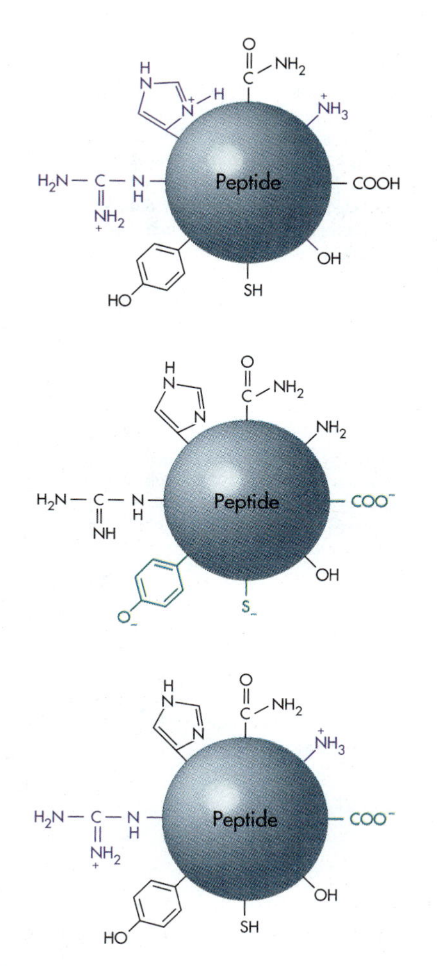

At pH 1, each peptide functional group is uncharged or positively charged; the net charge is positive.

At pH 13, each peptide functional group is uncharged or negatively charged; the net charge is negative.

At most intermediate pH values some functional groups are positively charged, some are negatively charged, and some are uncharged; the net charge is determined by the precise pH and the number of copies of each functional group (Section 2.7). The isoelectric point (pI) is the pH at which the net charge is zero.

FIGURE 3.4
Effect of pH on the net charge of a peptide.

and is either positively charged or uncharged; the compound as a whole contains only positive charges. At intermediate pH values, peptides usually contain both positive and negative charges, since some weak acid functional groups are protonated and others are nonprotonated (Section 2.7). There is a unique pH at which the number of negative charges on a peptide exactly equals the number of positive charges. This pH is known as its **isoelectric point** or **pI.** Although the net charge on a peptide is zero at its isoelectric point, it may still be coated with many charged functional groups. **[Try Problems 3.22 through 3.25]**

Restricted Rotation About Peptide Bonds

The carbon-to-nitrogen link in a peptide bond exhibits restricted rotation because of its partial double bond character, a character that can be explained by **delocalized pi (π) bonding, tautomerism,** or **resonance:**

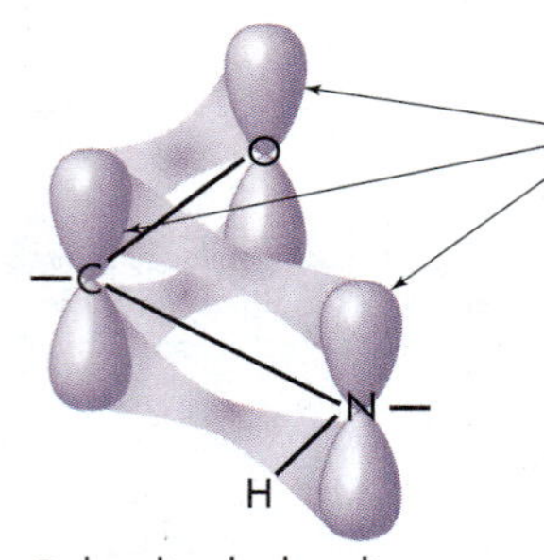

Delocalized π bonding

Three adjacent nonhybridized *p* orbitals simultaneously overlap to form a molecular orbital that encompasses three atoms — it is not localized between two atoms. Pi (π) bonding normally leads to restricted rotation, since rotation destroys the *p* orbital – *p* orbital overlaps that constitute a π bond.

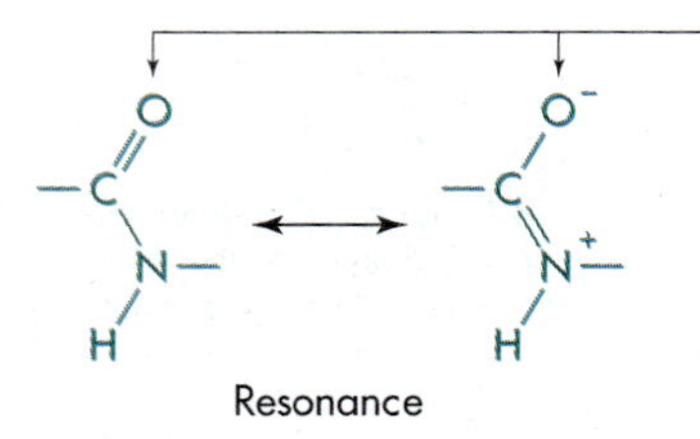

Resonance

Resonance forms differ only in the location of electrons. A molecule is, in reality, a hybrid of its resonance forms.

O
C
N
H
OH
C
N

Tautomerism

Tautomers are structural isomers that exist in equilibrium and differ in the location of a single hydrogen and a single double bond.

These three phenomena are reviewed further in the answer to Problem 3.27 and in Chapter 8. The restricted rotation leads to **geometric isomerism,** so a peptide bond can have either a **cis** or a **trans configuration** (Figure 3.5 a). In the body, most peptide bonds have a trans configuration, another example of stereospecificity. **[Try Problems 3.26 and 3.27. Problem 3.26 reviews geometric isomerism.]**

The six starred (*) atoms shown in the *cis*- and *trans*-peptide bonds (Figure 3.5 a) are all in the same plane (flat surface), known as an **amide plane.** A peptide can be visualized as a sequence of amide planes in which neighboring planes intersect or merge at α-carbons (Figure 3.5 b). With the exception of proline (α-carbon is part of a heterocyclic ring), there is free rotation about all of the bonds to these α-carbons. The degree of rotation (sometimes called a **torsion angle**) about the C_α—N bond is known as ϕ **(phi)** and the degree of rotation about the C_α—C bond is labeled ψ **(psi).** Both ϕ and ψ are assigned values of zero when the amide planes that share an α-carbon are both in a common plane and oriented as depicted in Figure 3.6. Rotation is said to be positive when it is clockwise as viewed from the α-carbon. Rotation in the opposite direction is labeled negative. In theory, the values for both ϕ and ψ can range from $-180°$ to $+180°$.

Only certain combinations of ϕ and ψ are allowed within a peptide. Other combinations are prohibited because of a phenomenon known as **steric interference** or

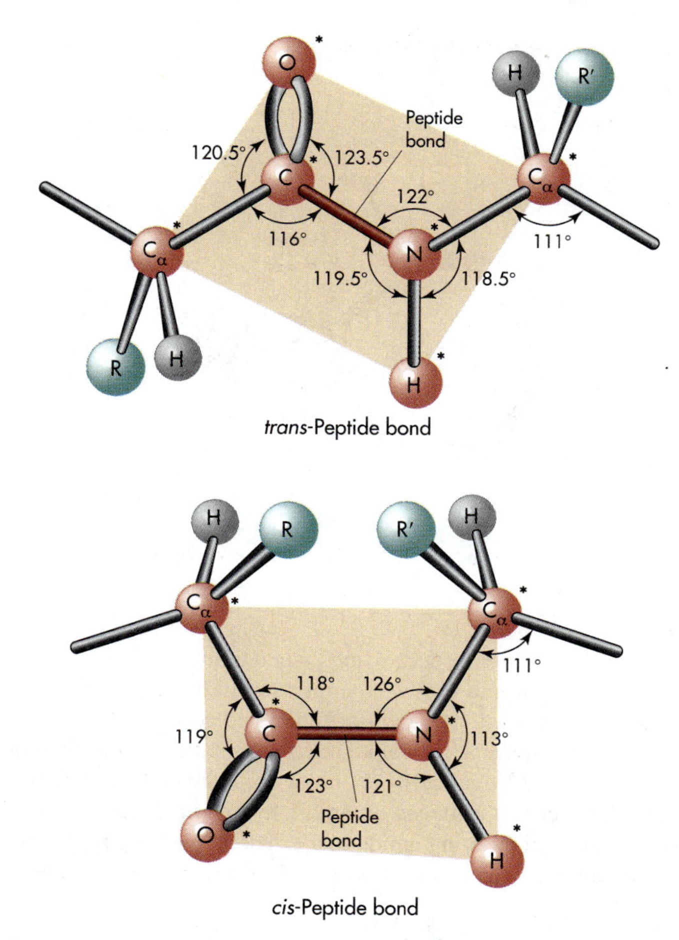

FIGURE 3.5 a
Diagram of *cis*- and *trans*-peptide bonds.

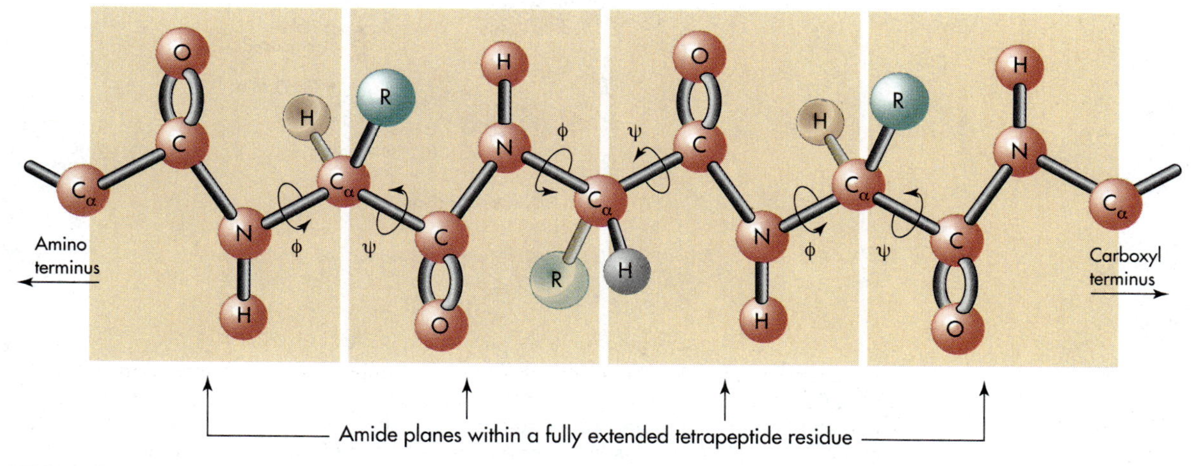

FIGURE 3.5 b
Amide planes within a peptide. (Based on Lehninger et al., 1993.)

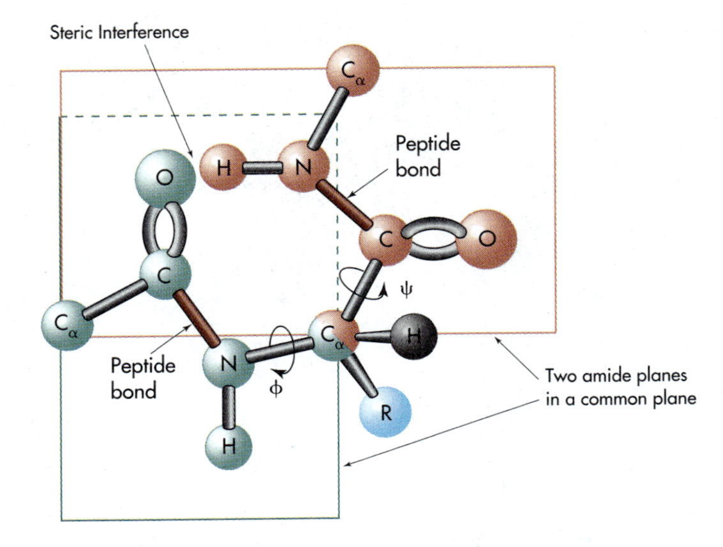

FIGURE 3.6
Reference structure for the assignment of values to φ and ψ. In this conformation, both φ and ψ are assigned a rotation of zero. (Based on Lehninger et al., 1993.)

steric hindrance. Steric interference results when atoms at different sites along the peptide chain are brought into close proximity and like-charged electron clouds repel one another. Viewed from a slightly different perspective, two atoms cannot occupy the same point in space at the same time. The rotational combination shown in Figure 3.6 (both φ and ψ equal zero) is one of the excluded combinations. It leads to severe steric interference between an amide hydrogen in one amide plane and a carbonyl oxygen in a neighboring amide plane. The allowed combinations of φ and ψ, which are commonly summarized with a Ramachandran plot, determine what

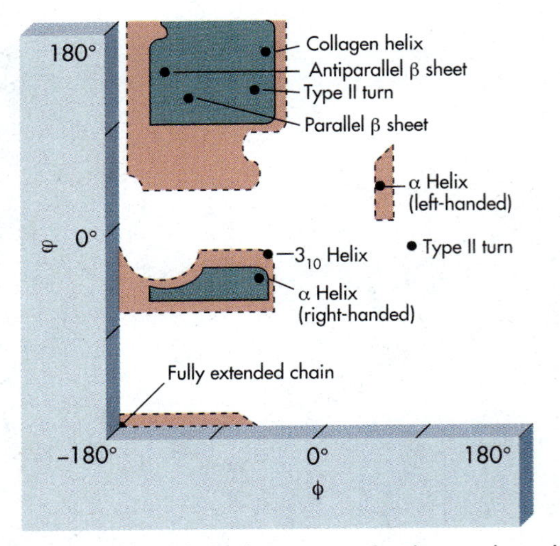

Ramachandran plot. Solid lines encircle the commonly observed combinations for φ and ψ. Dotted lines give the outer limits for an alanine residue. Large dots identify φ and ψ combinations that generate structural elements, such as the α helix and β sheet, commonly found in proteins (Chapter 4). The positions of the type II turn dots correspond to the second and third residues in the turn. White regions depict φ and ψ combinations that are seldom if ever found in proteins. (Based on Moran et al., 1994.)

geometrical arrangements are possible for a peptide chain. The values of ϕ and ψ depicted in Figure 3.5 b, for example, lead to an extended peptide chain in which all of the amide planes are in a common plane and neighboring side chains are on opposite sides of this common plane. Other values of ϕ and ψ can cause a peptide chain to coil and fold upon itself and can bring side chains into much closer proximity. This concept is revisited in Chapter 4 during the examination of the structural elements within proteins.

The importance of the trans, planar nature of peptide bonds can only be fully appreciated and clearly visualized by studying molecular models or, better yet, by building models. Colored miniature marshmallows and toothpicks can be used for some crude model building. A variety of commercial molecular model building kits are also available. When either building or studying models, be aware of bond angles. The bond angles about typical *cis*- and *trans*-peptide bonds are given on page 79. **[Try Problems 3.28 and 3.29. They provide a review of hybridization, bond angles, and molecular geometries.]**

Cyclosporin A (cyclosporine) and **FK506 (tacrolimus)** (Figure 3.7) are immunosuppressive drugs that are highly effective in preventing host rejection of transplanted organs and in treating some autoimmune diseases. Although the detailed mechanism of action of these compounds is uncertain, their immediate targets include enzymes known as **immunophilins** that catalyze the trans to cis isomerization of peptide bonds to specific proline residues within proteins. Such interconversions appear to be the **rate-limiting step** (the slowest step in a sequence of steps) in the folding of some proteins into biologically active conformations (Chapter 4).

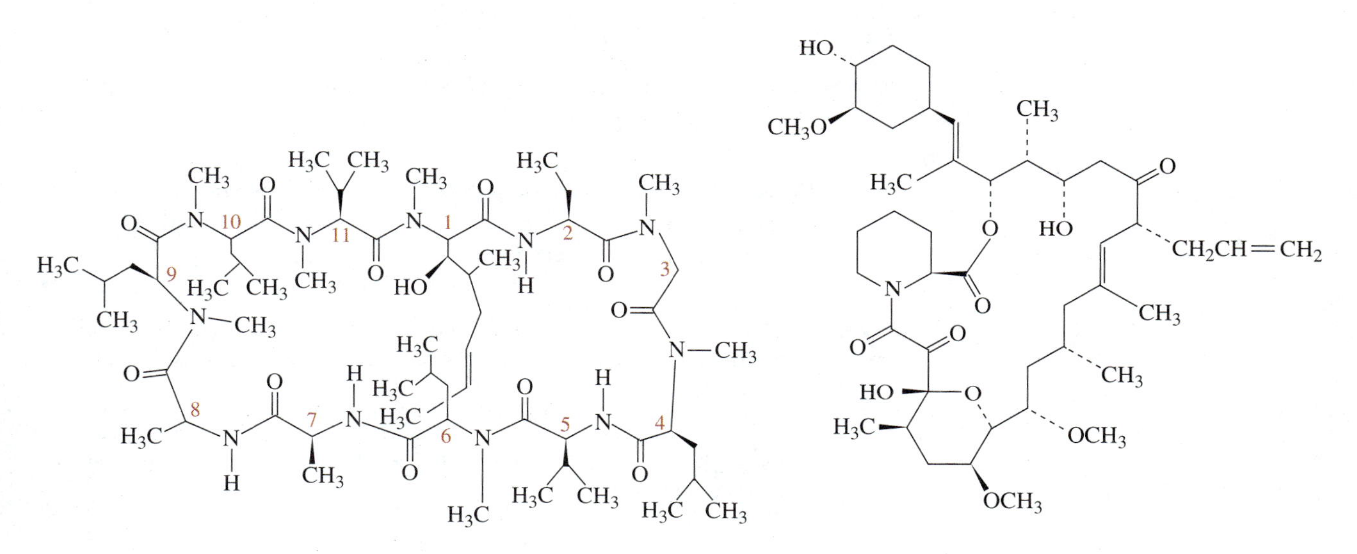

FIGURE 3.7
Both cyclosporin A (cyclosporine) and FK 506 (tacrolimus) are used to prevent host rejection of transplanted organs and to treat autoimmune diseases. Cyclosporin A is a derivative of a cyclic peptide. Its eleven amino acid residues are numbered.

Abbreviating and Naming Peptides

A peptide is abbreviated by listing, from left to right, the abbreviations for its amino acid residues starting with the N-terminal residue and proceeding in sequence to the C-terminal residue. Hyphens are placed between the amino acid abbreviations when three-letter abbreviations are utilized. A tripeptide containing phenylalanine (N-terminal), tyrosine, and valine (C-terminal) is abbreviated Phe-Tyr-Val or FYV. The pentapeptide in Figure 3.3 has the abbreviations Gly-Glu-Val-Ser-Lys and GEVSK. The amino acid residues in a peptide are numbered sequentially from the N-terminus to the C-terminus.

An official name for a peptide can be obtained by sequentially naming each amino acid from the N-terminus to the C-terminus, and then changing the *-ine* or *-ic* in each name, except the name of the C-terminal amino acid, to *-yl.* Glutaminyl, asparaginyl, cysteinyl, and tryptophyl, the modified names for glutamine, asparagine, cysteine, and tryptophan, are the only exceptions to this rule. In this book, hyphens are placed between the modified amino acid names. The peptide in Figure 3.3 is glycyl-glutamyl-valyl-seryl-lysine. According to the *Guinness Book of World Records,* the longest word in the English language is obtained by naming a protein using the procedure just described. In practice, proteins and polypeptides are always given alternative official names that usually contain one to three words. **[Try Problems 3.30 and 3.31]**

***PROBLEM 3.16** Give one example of an activated (high energy, unstable) derivative of a carboxylic acid.

***PROBLEM 3.17** What determines the numerical value of the equilibrium constant for a reaction?

PROBLEM 3.18 Create a hexapeptide by adding an amino acid residue to the pentapeptide in Figure 3.3. How many different hexapeptides could be created with this maneuver? Explain.

***PROBLEM 3.19** Which of the amino acid residues in Figure 3.3 contain side chains that could react with additional amino acids to form a branched peptide? Explain.

PROBLEM 3.20 Circle each atom in Figure 3.3 that could H bond with a water molecule.

PROBLEM 3.21 Write the structural formula for a peptide containing one residue of alanine, one residue of cysteine, and one residue of phenylalanine. How many different tripeptides can be produced that contain each of these three amino acids?

PROBLEM 3.22 Write the structural formula for the peptide in Figure 3.3 as it would predominantly exist in a solution at (a) pH 1.0, (b) pH 6.0, (c) pH 10.5, and (d) pH 14. Assume that the functional groups in the peptide have the same pK_a values as the equivalent functional groups in free amino acids (Table 3.5). The skills required to complete this problem are reviewed in Section 2.7.

PROBLEM 3.23 What would be the net charge to the nearest half of a charge unit on the peptide in Figure 3.3 in a solution at pH 1.0? pH 4.3? pH 8.0? pH 13? Assume that the functional groups in the peptide have the same pK_a values as the equivalent functional groups in free amino acids (Table 3.5). The skills required to complete this problem are reviewed in Section 2.7.

PROBLEM 3.24 Write the structural formula for a dipetide with a net charge (to the nearest half of a charge unit) of −2 at pH 7.0.

PROBLEM 3.25 If the pH of a solution is above the pI for a peptide in that solution, will the peptide have a net negative charge, a net positive charge, or a zero net charge? Explain.

***PROBLEM 3.26** Write expanded structural formulas for the cis and trans isomers of $CH_3—CH═CH—CH_3$ and 1,2-dimethylcyclopentane.

***PROBLEM 3.27** Use structural formulas for specific nonpeptides to provide additional examples of delocalized π bonding, resonance, and tautomerism.

***PROBLEM 3.28** What are the predicted bond angles about the carbon and the nitrogen in a peptide bond assuming that each of these atoms is sp^2 hybridized? Explain.

***PROBLEM 3.29** What are the predicted bond angles about the α-carbon in each amino acid residue within a peptide? Explain.

PROBLEM 3.30 Write the structural formula for Ser-Ala-Gly-Asn and then (a) number each amino acid residue, (b) label the C-terminal, (c) use an arrow to identify each peptide bond, and (d) give the official name for this peptide.

PROBLEM 3.31 Use one-letter abbreviations for the amino acids to write an alternative abbreviation for the tetrapeptide in problem 3.30.

3.6 EXAMPLES OF BIOLOGICALLY ACTIVE NONPROTEIN PEPTIDES

Now that we have noted the general properties of nonprotein peptides, we can examine some specific examples. **Vasopressin** (also called **antidiuretic hormone [ADH]**) is a modified nonapeptide produced within the pituitary gland.

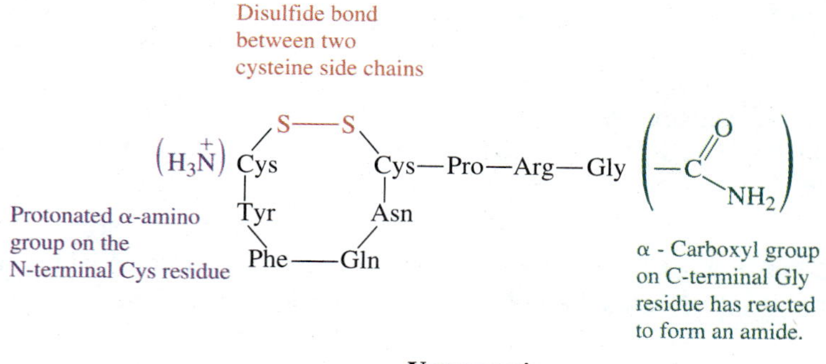

Vasopressin

Its C-terminal amino acid residue has reacted to form an amide and its two cysteine side chains have been oxidatively coupled to form a disulfide (—S—S—). It is well documented that vasopressin acts as a hormone to stimulate the reabsorption of water by the kidneys (an **antidiuretic action**) and to stimulate the contraction of smooth muscles. The latter activity contributes to the regulation of blood pressure, since the muscles in blood vessels are prime targets of this hormone. Vasopressin also helps regulate the release of certain other hormones and functions as a neurotransmitter. Preliminary evidence indicates that this peptide's interactions with the central nervous system may affect memory consolidation and sexual behavior. It is not uncommon for a hormone to trigger distinctive responses in different target tissues and to serve nonhormone functions as well. **[Try Problems 3.32 and 3.33]**

The replacement of the phenylalanine in vasopressin with isoleucine and the replacement of arginine with leucine leads to another pituitary hormone known as **oxytocin.**

* Amino acid residues that differ from corresponding residues in vasopressin

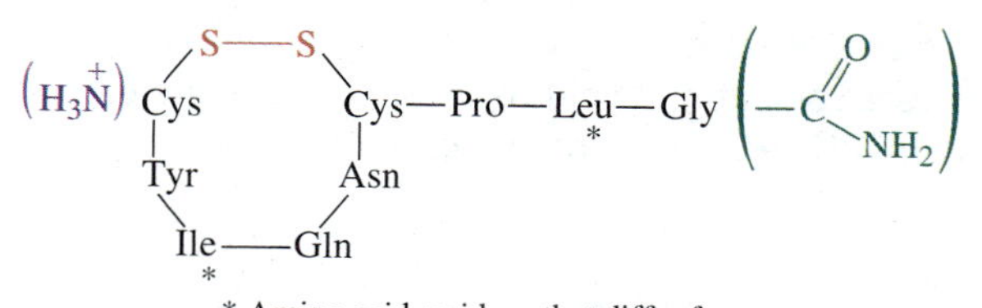

Oxytocin

Oxytocin stimulates **lactation** (milk production) and the contraction of uterine muscles in pregnant women. The latter activity explains why this nonprotein peptide is commonly employed to induce labor. Since oxytocin also impacts the central nervous system, it provides still another example of a neuropeptide.

In general, it is the amino acid sequence within a peptide that ultimately determines its chemical and physical properties and its biological activities. A change in a single amino acid side chain within a peptide can have a drastic impact on its net charge, solubility, stability, reactivity, receptor binding ability, and so on. Here, we are dealing directly with structure–function relationships. In the case of oxytocin and vasopressin, a change in two side chains certainly leads to marked differences in biological activities. The detailed molecular actions of these two hormones are still being explored. **[Try Problems 3.34 and 3.35]**

The central nervous system contains multiple categories of receptors that bind **morphine** (Figure 3.8) and other **opiates** (compounds related to morphine). These receptors largely account for the physiological effects of these **analgesics** (pain relievers). Several classes of endogenous (produced within the body) peptides, including the **enkephalins, endorphins,** and **dynorphins,** adhere to the same receptors, and their discovery has generated tremendous excitement in neurobiochemistry (biochemistry of the nervous system), psychiatry, and related areas of the health sciences. The amino acid sequence of some of these **opiate peptides** is given in Table 3.8. Each listed peptide has the amino acid sequence Tyr-Gly-Gly-Phe at its N-terminus. *When a structural feature is conserved within all members of a class of compounds, this usually indicates that the conserved feature is required for biological activity, and it provides a starting point for probing structure–function relationships.*

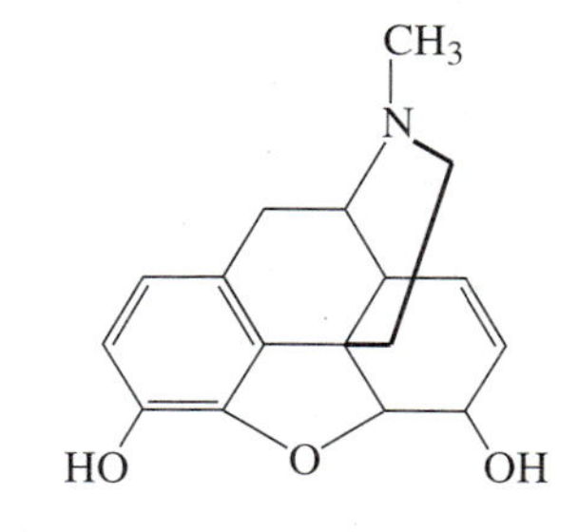

FIGURE 3.8
The structure of morphine.

Most of the opiate peptides are neurotransmitters that can function as natural pain relievers. These compounds may be at least partly responsible for the natural "highs" experienced by runners and other athletes. Acupuncture may also stimulate their release. Since these compounds appear to affect a variety of neurological and physio-

TABLE 3.8

The Structure of Some Opiate Peptides

Peptide	Structure
Met-Enkephalin	Tyr-Gly-Gly-Phe-Met
Leu-Enkephalin	Tyr-Gly-Gly-Phe-Leu
α-Endorphin	Tyr-Gly-Gly-Phe-Met-Thr-Ser-Glu-Lys-Ser-Gln-Thr-Pro-Leu-Val-Thr
γ-Endorphin	Tyr-Gly-Gly-Phe-Met-Thr-Ser-Glu-Lys-Ser-Gln-Thr-Pro-Leu-Val-Thr-Leu
β-Endorphin	Tyr-Gly-Gly-Phe-Met-Thr-Ser-Glu-Lys-Ser-Gln-Thr-Pro-Leu-Val-Thr-Leu-Phe-Lys-Asn-Ala-Ile-Ile-Lys-Asn-Ala-His-Lys-Lys-Gly-Gln
$Dynorphin_{1–8}$	Tyr-Gly-Gly-Phe-Leu-Arg-Arg-Ile
$Dynorphin_{1–17}$	Tyr-Gly-Gly-Phe-Leu-Arg-Arg-Ile-Arg-Pro-Lys-Leu-Lys-Trp-Asp-Asn-Gln

logical processes, the opiate peptide literature should make interesting reading for many years.

In the body, the opiate peptides are cut out of larger polypeptides by enzymes that catalyze the hydrolysis of specific peptide bonds. Oxytocin, vasopressin, and many other biologically active peptides are also generated from larger polypeptide precursors. The probable significance of this is examined in Chapter 5.

PROBLEM 3.32 Write the structural formula for vasopressin as it would primarily exist at pH 7.0. Use the pK_a values provided in Table 3.5.

PROBLEM 3.33 Study the structural formula drawn for Problem 3.32, and then circle each atom that could H bond with a water molecule.

PROBLEM 3.34 Which has more nonpolar character, oxytocin or vasopressin? Explain.

PROBLEM 3.35 To the nearest half of a charge unit, calculate the net charge on both oxytocin and vasopressin at pH 7.0. Assume that their functional groups have the same pK_a values as equivalent functional groups in free amino acids (see Table 3.5).

3.7 MONOSODIUM GLUTAMATE AND ASPARTAME: THE IMPORTANCE OF MODERATION

This section can be viewed as an extension of the previous section; it explores some of the biological activities of a specific amino acid and a derivatized dipeptide. **Monosodium glutamate (MSG)** is a sodium salt of glutamic acid, one of the 20 protein amino acids:

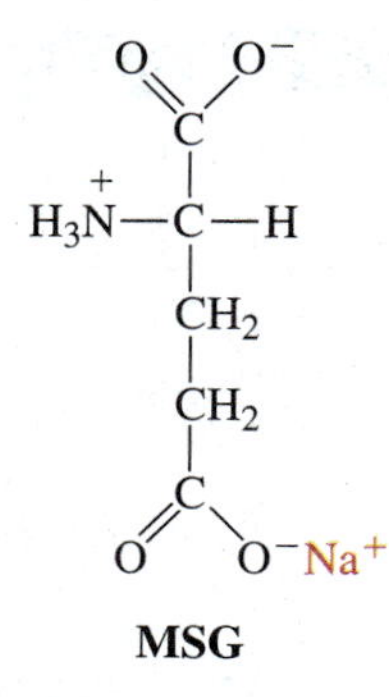

MSG

The term **glutamate,** in general, refers to any construct of glutamic acid in which a carboxyl group exists in its conjugate base (salt) form. Glutamic acid is a required building block for proteins, and it is also a key neurotransmitter in the mammalian central nervous system. It exists as a salt (glutamate) at physiological pH values.

MSG (sold as **Accent**) is a flavor enhancer used in many homes and restaurants. Overconsumption of this compound can lead to the **Chinese-restaurant syndrome,** an ailment with symptoms including dizziness, headaches, and fatigue. This syndrome was initially identified in individuals who frequently consumed large quantities of MSG while eating at Chinese restaurants; hence the name. Although the average person can consume relatively large amounts of this food additive without any serious side effects, a small fraction of the population (probably fewer than 2%) is extremely sensitive. People who are sensitive should avoid foods to which large amounts of MSG have been added, and they may need to limit their consumption of mushrooms, peas, and certain other vegetables that naturally contain significant quantities of this compound. The hydrolyzed vegetable protein (HVP) found in some grocery store products contains up to 40% MSG. Glutamate is one of many compounds that are needed by the body but that may be toxic when consumed in large amounts. There is something to be said for moderation.

Aspartame (NutraSweet) is an artificial sweetener and the methyl ester of L-Asp-L-Phe:

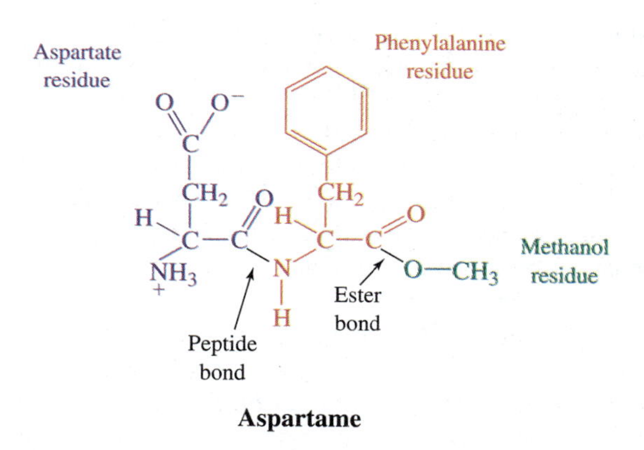

Aspartame

One gram of aspartame will provide 200 to 300 times the sweetness of one gram of **sucrose** (table sugar). Some researchers are questioning the safety of aspartame because it can be hydrolyzed in foods, beverages, and people to produce phenylalanine, aspartic acid, and methanol. Evidence indicates that both phenylalanine and aspartic acid, like glutamic acid, can lead to neurological symptoms in sensitive individuals, and it is well documented that a shot glass of methanol is enough to kill the average person. Overconsumption of phenylalanine by individuals with the genetic disease **phenylketonuria (PKU**; Section 19.7) can lead to mental retardation. The label on many aspartame-containing products bears a warning for PKU patients.The question becomes: "How much is too much?" Keep in mind that relatively little aspartame is required to satisfy even the biggest sweet tooth.

Aspartame provides another example of nature's stereospecificity. In the sweet form of aspartame, both the aspartic acid and the phenylalanine residues have an L configuration. If the configuration of either or both residues is changed, the resultant molecule is either bitter or tasteless.

EXHIBIT 3.1

Some Disorders That Lead to Abnormal Amino Acid Profiles in Physiological Fluids

- Arteriosclerosis
- Arthritis
- Endocrine disorders
- Infections, bacterial and viral
- Cancers
- Neurological disorders
- Kidney failure
- Vitamin deficiencies
- Head injuries

3.8 AMINO ACID ANALYSIS

To study any peptide in detail, we must determine the relative amounts of each of the 20 common protein amino acids within it, a process known as **amino acid analysis.** The amino acids within a peptide play a central role in determining its chemical and biological properties, a point emphasized during the discussion of structure–function relationships in Section 3.6. Once a pure sample of a peptide is in hand, there are three major steps involved in amino acid analysis: number one, hydrolysis of the peptide to yield free amino acids; number two, separation of the free amino acids; and number three, the quantitative detection of the separated amino acids (Table 3.9). Once a peptide has been completely hydrolyzed in step 1, steps 2 and 3 are usually carried out with an automated **amino acid analyzer.** Such an instrument can complete an analysis in less than an hour and some can quantitatively detect attomole (10^{-18} mole) levels of amino acids. A dedicated computer analyzes the experimental data, and final results are printed in a summary table. **[Try Problems 3.36 and 3.37]**

The value of amino acid analyzers extends well beyond the determination of the amino acid composition of peptides. These instruments are also used to determine the **amino acid profiles** of physiological fluids, including blood and urine. Such profiles are of significant diagnostic value, and they are often used for patient management as well. Exhibit 3.1 lists some of the disorders that lead to changes in amino acid concentrations in physiological fluids.

Most amino acid analyzers employ chromatography to separate the freed (by hydrolysis) amino acids, and many use spectrophotometric analysis to detect them quantitatively. These and several other techniques have proved invaluable in the purification and analysis of amino acids, peptides (including proteins), and most other biologically important compounds.

This chapter concludes with a Methods Section that provides an introduction to chromatography, spectrophotometric analysis, electrophoresis, and radioisotopes, four of the most powerful tools available to biochemists. Other tools will be encountered and examined in the chapters that follow. The methods of biochemistry are an essential and integral part of the discipline.

TABLE 3.9

Amino Acid Analysis of a Peptide

	What Is Accomplished	How Commonly Accomplished
Step 1	Peptide is hydrolyzed to generate free amino acids	Acid-catalyzed hydrolysis; base-catalyzed hydrolysis; enzymatic digestion
Step 2	Free amino acids are separated from one another	Chromatography[a]; electrophoresis[a]
Step 3	Separated amino acids are quantitatively detected	Spectrophotometric analysis[a]; reaction of amino acids with fluorescent reagents

[a]Subsequent sections of this chapter describe these tools of biochemistry.

PROBLEM 3.36 Why must a peptide be purified before amino acid analysis?

PROBLEM 3.37 Does amino acid analysis provide any information about the order in which amino acids are joined together within a peptide? Explain.

METHODS SECTION

3.9 CHROMATOGRAPHY

Chromatography is one of several common techniques used to separate the individual components within mixtures. Such techniques are vitally important in biochemistry because an amino acid, peptide, or any other compound, must be **purified** (separated from all other compounds) before either its chemical or its biological properties can be unambiguously examined. During chromatography, the compounds to be separated are carried through or over a solid or a liquid-coated solid (the **stationary phase**) by a flowing liquid or gas (the **mobile phase**). If the mobile phase is a liquid, we speak of **liquid chromatography. Gas chromatography** employs a gas as the mobile phase, so it can be used only to separate volatile compounds (a gas can only carry other gases). During liquid column chromatography, the only type of chromatography that will be examined in any detail in this text, a liquid mobile phase is continuously passed through a **column** (a hollow tube) containing a stationary phase. The liquid that emerges from the column is commonly collected in a series of sequentially numbered tubes. When this procedure is employed, the contents of each tube is said to represent a **column fraction.** If the mobile and stationary phases have been properly selected, the different compounds in a mixture added to the top of the column will emerge from the bottom of the column at different times and be collected in different column fractions. Figure 3.9 illustrates the basic features of a typical chromatographic system.

Those column fractions that contain compounds of interest can be immediately identified if the compounds are colored. Special procedures are required to locate amino acids and other noncolored substances. Amino acids are routinely located by treating part of each column fraction with **ninhydrin** to generate a colored compound known as **Ruheman's blue.**

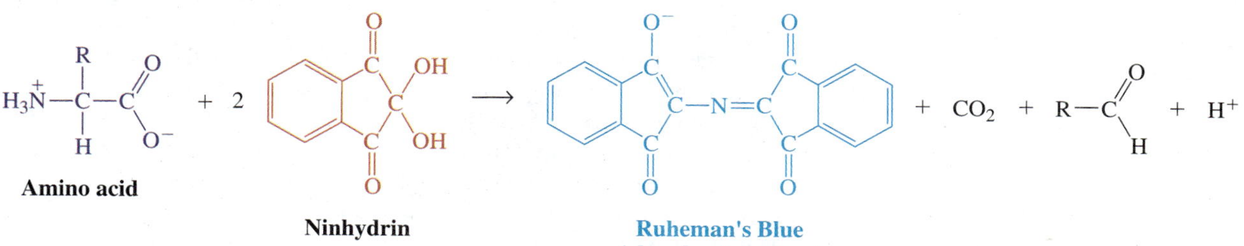

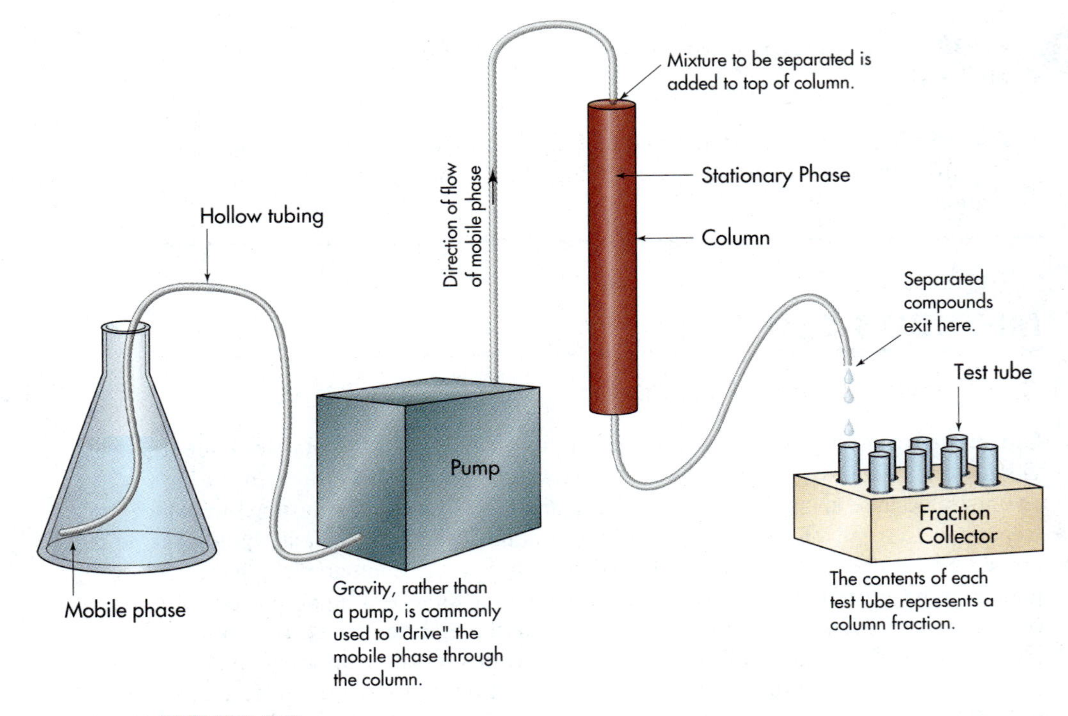

FIGURE 3.9

The basic features of a chromatographic system.

The concentration of amino acid in a sample is proportional to the amount of Ruheman's blue formed. The yield of Ruheman's blue can be estimated visually or it can be determined quantitatively by measuring the absorbance of the reaction mixture at 570 nm (Section 3.11). Figure 3.10 illustrates the separation of amino acids that can be achieved with the chromatography columns used in many automated amino acid analyzers (Section 3.8).

Ion exchange chromatography employs a stationary phase that is either positively or negatively charged. Compounds with a net positive charge **(cations)** bind to negative stationary phases **(cation exchange resins)** while compounds with a net negative charge **(anions)** bind to positively charged stationary phases **(anion exchange resins).** The greater the net charge on a compound, the more tightly it tends to bind to an oppositely charged stationary phase. Ions in the mobile phase compete with the compounds to be separated for binding to the stationary phase. The less tightly a compound binds to the stationary phase, the more readily it is displaced by the competing ions. After a compound is displaced from the stationary phase by a mobile phase ion, the continuously flowing mobile phase carries it farther down the column, where it binds to another charged functional group on the stationary phase. Each compound undergoes many rounds of binding and displacement before it finally emerges from the column. The least tightly bound compound emerges first; the most tightly bound compound last. **[Try Problem 3.38]**

Gel filtration chromatography, also known as **gel permeation chromatography, molecular sieve chromatography,** and **exclusion chromatography,** employs

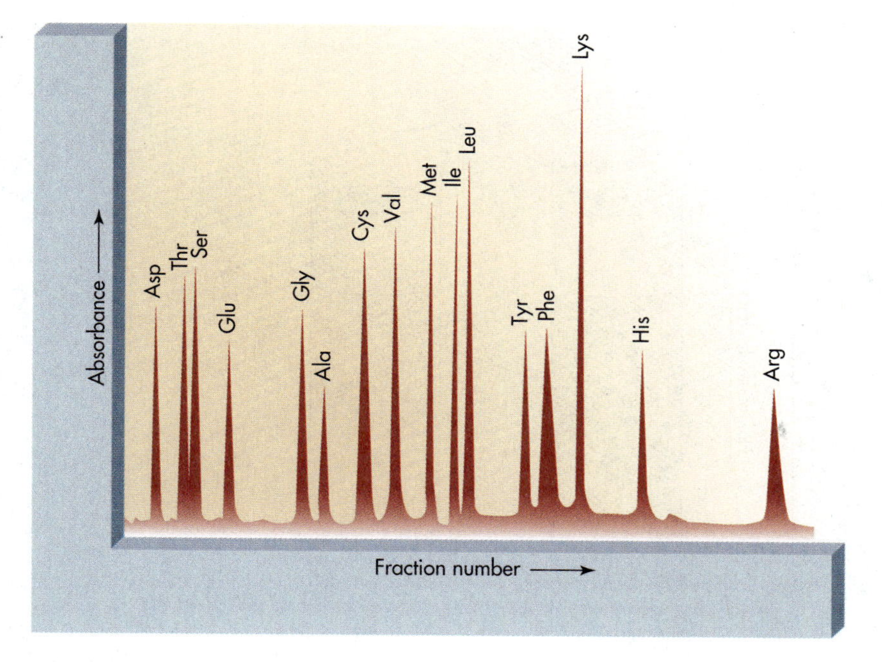

FIGURE 3.10
Separation of amino acids with ion exchange chromatography. Based on Brown & Rogers (1987).

a stationary phase consisting of many very tiny porous beads that are, ideally, chemically inert. Each bead contains multiple uniformly-sized pores. As sample molecules are carried over the beads by the mobile phase, small compounds that can pass into the pores diffuse into the beads. Once inside, a compound bounces around until its random movements bring it back into the mobile phase. It is carried further down the column before it once again diffuses into a bead of the stationary phase. Hence, the progress of small compounds through the column is retarded as they repeatedly enter and then exit the beads. In contrast, large compounds that cannot pass into the pores of the beads are rapidly carried over the surface of the beads and emerge from the column ahead of the smaller ones (Figure 3.11). Since gel filtration beads are available with a variety of pore sizes, a wide size range of compounds can be separated from one another with this chromatographic technique. **[Try Problems 3.39 through 3.41]**

Since any given compound will always emerge from a standardized chromatographic system at a fixed, predictable time after it has been added to a column, chromatography can be used to characterize compounds, as well as purify them. Since different compounds tend to emerge **(elute)** at different times, the coelution of a known and an unknown compound suggests that they may be one and the same. Such **coelution** does not constitute definitive identification of the unknown, because different compounds occasionally have the same chromatographic mobilities. If, however, two compounds coelute from multiple unique chromatographic systems, we can be fairly confident that they are indeed identical.

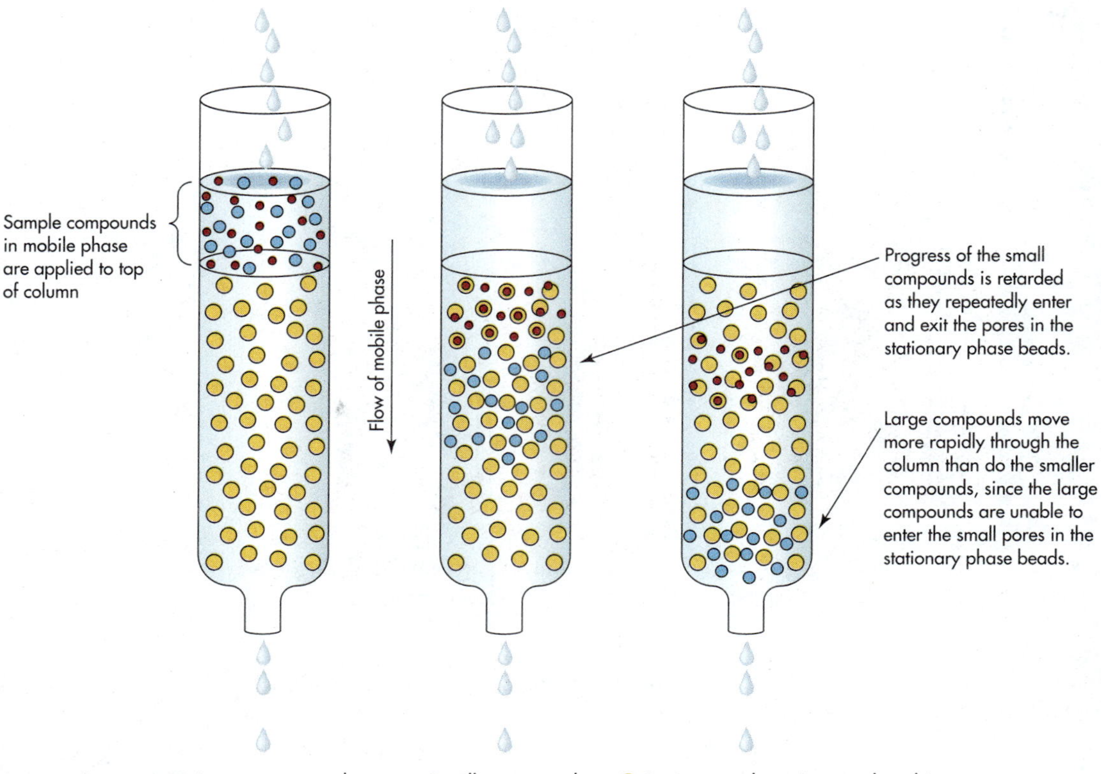

FIGURE 3.11
Gel filtration chromatography.

There are many kinds of liquid chromatography in addition to ion exchange chromatography and gel filtration chromatography. These include thin layer chromatography (TLC), adsorption chromatography, partition chromatography, reverse phase chromatography, high-pressure (or high-performance) liquid chromatography (HPLC), and affinity chromatography. The key point to remember is that all chromatographic systems have certain basic features in common, the ones described in the opening paragraph of this section.

PROBLEM 3.38 A mixture contains three dipeptides: Glu-Glu, Ala-Lys, and Lys-Lys.

a. Which dipeptide will be the first to emerge from a cation exchange column at pH 7.0? Explain.
b. Which dipeptide will be the first to emerge from an anion exchange column at pH 7.0? Explain.
c. Can these three peptides be readily separated from one another on a cation exchange column at pH 7.0? pH 1.0? pH 14? Explain.

PROBLEM 3.39 Will the shape of a molecule influence its rate of migration through a gel filtration column? Explain.

PROBLEM 3.40 True or false? The larger the molecular weight of an organic compound, the more rapidly it will tend to migrate through a gel filtration column. Explain.

PROBLEM 3.41 Which one of the following compounds will tend to emerge first from a gel filtration column? Which will tend to emerge last? Explain.

a. An amino acid
b. A tripeptide
c. A heptapeptide
d. Insulin (a small protein)
e. Oxytocin

3.10 ELECTROPHORESIS

Electrophoresis, like chromatography, is a technique that can be used to purify, characterize, and identify compounds. In most instances, the sample to be analyzed is applied to a moistened, buffered layer of paper, cellulose acetate, agarose, starch, polyacrylamide, or other support material positioned between two electrical poles **(electrodes).** When a voltage is applied across the electrodes, the resultant current carries compounds with net negative charges toward the positive pole (the **anode**) and compounds with net positive charges toward the negative pole (the **cathode**). Uncharged molecules and compounds with a zero net charge will remain at the **origin,** the site of sample application. Figure 3.12 illustrates the basic features of an electrophoresis system.

The rate of migration of a compound with a net charge toward an electrode is primarily determined by its precise net charge, its size, its shape, the support material, and the current. Similarly-sized compounds with the same charge to size ratio are difficult to separate electrophoretically. Most other charged compounds can be separated under properly selected electrophoretic conditions. During prolonged elec-

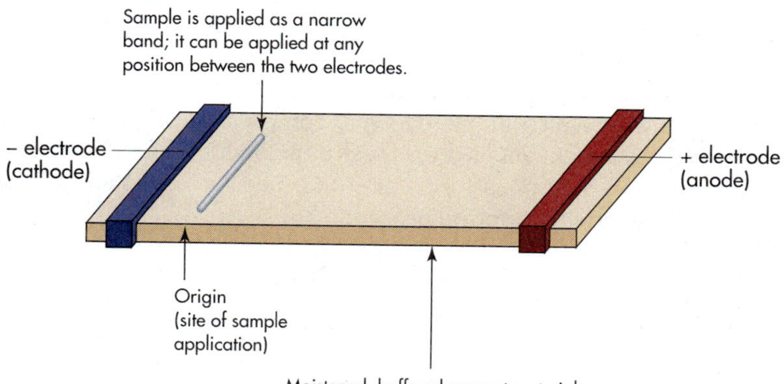

FIGURE 3.12 The basic features of an electrophoresis system. Compounds with net positive charges migrate from the origin toward the cathode, whereas those with net negative charges move toward the anode. Compounds with no net charge remain at the origin.

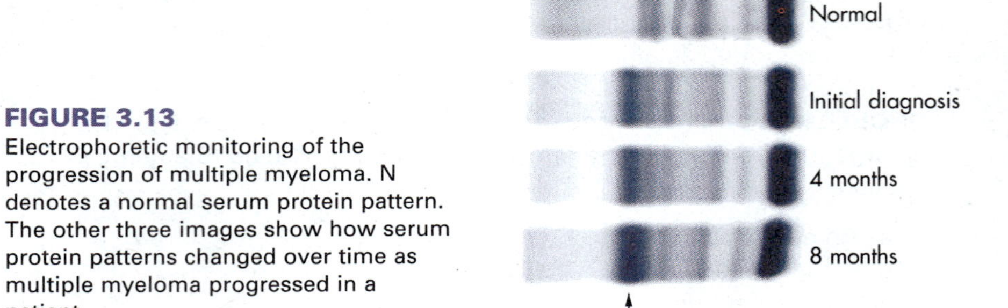

FIGURE 3.13
Electrophoretic monitoring of the progression of multiple myeloma. N denotes a normal serum protein pattern. The other three images show how serum protein patterns changed over time as multiple myeloma progressed in a patient.

trophoresis, molecules with net charges eventually migrate to the electrodes and react with them. In practice, an electrophoretic run is usually terminated before this happens. **[Try Problems 3.42 and 3.43]**

A majority of the compounds in the body, including most peptides, are colorless and cannot be directly visualized during or following electrophoresis, a problem already explored briefly during the discussion of chromatography (Section 3.9). Therefore, an agarose strip will look virtually the same before and after it has been used for the electrophoresis of serum proteins. If the same strip is treated with a **protein fixative** (a chemical that precipitates and fixes proteins on electrophoretic supports) and a protein-specific stain after electrophoresis, the strip will develop colored regions that correspond to the location of the proteins at the end of the electrophoretic run. The stained proteins appear as bands because the original serum sample was applied to a narrow rectangular area on the support material (see Figure 3.12). Electrophoretic patterns for serum proteins are used in the diagnosis of a variety of human diseases (Exhibit 3.2), including multiple myeloma (Figure 3.13). Diagnosis is often aided by the quantitative determination of the amount of protein in each band. This can be accomplished with an instrument, known as a **densitometer,** that scans the support material and measures the absorbance (Section 3.11) of each stained band.

EXHIBIT 3.2

Some Diseases That Are Diagnosed Using Electrophoretic Analysis of Serum Proteins

- Multiple myeloma
- Hodgkins disease
- Multiple sclerosis
- Viral hepatitis
- Nephrotic syndrome
- Diabetes
- Collagen diseases
- Rheumatic fever

The distance a compound migrates under standard electrophoretic conditions is a characteristic property of that compound. If two compounds migrate to different positions under the same electrophoretic conditions, we can confidently conclude that the two compounds are different. Molecules separated by electrophoresis can be washed out of the support material and collected for additional studies or further purification.

PROBLEM 3.42 Which of the following compounds will migrate toward the cathode and which will migrate toward the anode during electrophoresis at pH 7.0? pH 2.0? pH 10.5? See Tables 3.5 and 2.8 for pK_a's and some structural formulas.

a. Glu
b. Lys
c. Ala-Val-Cys-Asp
d. Lactic acid
e. Methyl amine

PROBLEM 3.43 Which of the following sets of compounds contain molecules that can readily be separated from one another with electrophoresis? For this problem, assume that the support material is inert and has no impact on the migration rate of the compounds involved.

Set a. Gly, acetic acid
Set b. Gly, Glu
Set c. Lys, Ala
Set d. Gly, Ala
Set e. Methanol, ethanol

3.11 SPECTROPHOTOMETRIC ANALYSIS

Electromagnetic radiation consists of oscillating electric and magnetic fields that carry energy and travel at the speed of light (1.86×10^5 mi/s [3.00×10^8 m/s] in a vacuum). In some experiments, electromagnetic radiation behaves as a stream of particles. In others, it appears to have wavelike properties. When electromagnetic radiation is treated as an energy wave, it can be diagrammatically represented as follows:

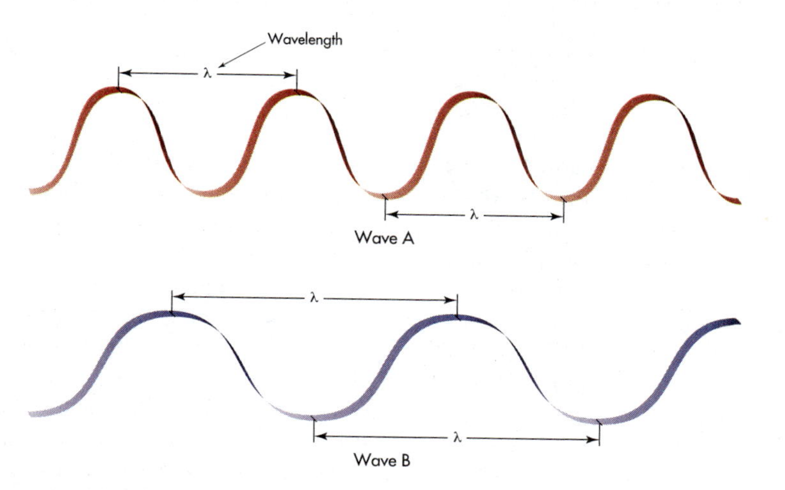

The distance between two adjacent peaks or two adjacent troughs in a wave is known as its **wavelength** (abbreviated with the Greek letter **lambda,** λ). Wavelength ranges are used to divide electromagnetic radiation into eight major classes (Table 3.10). Any two classes, **X-rays** and **visible light** for example, differ only in wavelength and energy content. The shorter the wavelength, the greater its energy content. **[Try Problems 3.44 and 3.45]**

Compounds, and functional groups within compounds, tend to **absorb** some wavelengths of electromagnetic radiation but not others, and each compound usually possesses distinctive absorption characteristics. When a compound or functional group does absorb an electromagnetic wave, it is elevated from a **ground state** energy level to an **excited state.** The total amount of electromagnetic radiation ab-

TABLE 3.10

Eight Major Classes of Electromagnetic Radiation

Class	Approximate Range of Wave Lengths (meters)[a]	Comments
Extremely low frequency electromagnetic fields	Over 1000	Produced by power lines and electrical appliances; possible health consequences highly controversial.
Radiowaves	1 to 1000	AM band, 190–560 m; FM band 2.8–3.4 m; television bands, 1.4–1.7, 3.4–4.0, 4.2–5.6 m.
Microwaves	10^{-4} to 1	Includes radar; used in microwave ovens.
Infrared (IR) light	7×10^{-7} to 10^{-4}	Felt as heat; helps maintain the temperature balance on Earth by escaping into outer space.
Visible light	4×10^{-7} to 7×10^{-7}	Contains all the colors of the rainbow; individual wavelengths correspond to specific colors.
Ultraviolet (UV) light	10^{-8} to 4×10^{-7}	Most UV light from the sun is absorbed by the ozone layer; leads to skin cancer and sunburns.
X-Rays	10^{-11} to 10^{-8}	Used for the diagnosis or treatment of multiple disorders; carcinogenic and mutagenic.[b]
Gamma- (γ-) rays	Less than 10^{-11}	One form of nuclear radiation (Section 3.12); used in the radiation therapy of cancer; carcinogenic and mutagenic.

[a]The meter (m) is the basic unit of distance in the SI System of Measurement. Nanometer (nm; 10^{-9} meter) is another unit of distance commonly used to express wavelength.
[b]Carcinogenic - leads to cancer; mutagenic - leads to mutations (changes in DNA).

sorbed by a sample is known as its **absorbance (A)** or **optical density (OD).** An instrument used to measure absorbance is called a **spectrophotometer. Spectrophotometric analysis** refers to any analysis that involves absorbance measurements.

In biochemistry, most absorbance measurements are made on dilute solutions. If a solvent is selected that does not absorb at the wavelengths of interest, the absorbance of the solution will result solely from the solute molecules. You should assume that this is the case for all of the discussions and problems that follow.

The absorbance of **monochromatic light** (electromagnetic radiation of a single wavelength) by a dilute solution of a compound is usually proportional to the concentration of that compound if there is no other substance in the solution that absorbs at the selected wavelength. This fact, known as **Beer's law** or the **Beer–Lambert law,** can be mathematically expressed as follows:

$$A = \epsilon bc$$

where
A = absorbance
ϵ = absorptivity (extinction coefficient)
b = pathlength
c = concentration of absorbing solute

The **absorptivity** is an experimentally determined number that is a constant for a given compound at a fixed wavelength. The absorptivity is usually altered if one changes either the compound that is being examined or the wavelength that is employed for the absorbance measurement. When solute concentration is expressed as **molarity** (moles of solute per liter of solution), the absorptivity is called the **molar absorptivity.** Molar absorptivity has units of liter per mole per centimeter ($L\ mol^{-1}cm^{-1}$). In this text, ϵ is defined as molar absorptivity unless clearly specified otherwise. The **pathlength** is the distance the monochromatic light travels through the sample solution. A 1-cm pathlength is standard, and you should assume that this pathlength is used for all absorbance measurement unless specified otherwise. The wavelength of light used for an absorbance measurement impacts absorbance through its effect on absorptivity; it does not appear directly in the Beer's law expression.

Given the value of any three of the four variables in the Beer's law expression, we can calculate the value of the fourth variable.

$$A = \epsilon bc \qquad c = \frac{A}{\epsilon b} \qquad \epsilon = \frac{A}{bc} \qquad b = \frac{A}{\epsilon c}$$

An absorbance measurement can be used to calculate c, for example, if the values of the other two variables (ϵ and b) are known. Absorbance measurements are commonly used in both clinical and research laboratories for the quantitative determination of the concentrations of biologically important compounds. The concentrations of amino acids in the column fractions of an amino acid analyzer are sometimes determined in this manner (Section 3.8). **[Try Problems 3.46 through 3.49]**

PROBLEM 3.44 What is the wavelength in centimeters of each of the two electromagnetic waves illustrated on page 95? Express each of these two wavelengths in meters.

PROBLEM 3.45 Which one of the two electromagnetic waves depicted on page 95 carries (contains) the greatest amount of energy? Explain.

PROBLEM 3.46 A 1×10^{-3} *M* solution of drug MX-2 has an absorbance at 552 nm (abbreviated A_{552}) of 0.37. Calculate the molar absorptivity at 552 nm (abbreviated ϵ_{552}) of this drug. Will the molar absorptivity be the same at 643 nm? 390 nm? Explain. Assume a 1-cm pathlength.

PROBLEM 3.47 The ϵ_{258} of phenylalanine is 195 L mol^{-1} cm^{-1}. Calculate the molarity of a phenylalanine solution that has an $A_{258} = 0.19$.

PROBLEM 3.48 To what major class of electromagnetic radiation does a wave belong if it has a wavelength of 552 nm? 258 nm? 10 nm? 10^5 nm?

PROBLEM 3.49 The ϵ_{218} of tryptophan is 33,500 L mol^{-1} cm^{-1}. Calculate the A_{218} of a 2.2×10^{-5} *M* tryptophan solution.

3.12 RADIOISOTOPES

"The value of radioisotopes to the biological sciences is undisputed. Without their availability and use we would be woefully ignorant of a vast range of physiological and biochemical processes. . . . The range of applications is vast and beyond the scope of any book." This quote from the preface to *Radioisotopes in Biology: A Practical Approach,* by R. J. Slater, accurately describes the importance of radioisotopes in biochemistry. Radioisotopes are equally important in medicine, where they are used for clinical analysis, diagnosis, and therapy. In this section, some biologically important aspects of radioisotopes are reviewed and then a glimpse is provided of some specific biomedical applications of these unique tools.

Isotopes and Nuclear Radiation

Isotopes are atoms with the same **atomic number** (number of protons) but different **mass numbers** (number of protons plus number of neutrons). Most samples of naturally occurring elements contain multiple isotopes; there are three isotopes of hydrogen, three of carbon, two of nitrogen, and so on. An isotope is abbreviated by writing its mass number as a superscript and its atomic number as a subscript to the left of its chemical symbol. For example, the three naturally occurring isotopes of carbon are abbreviated ${}^{12}_{6}C$, ${}^{13}_{6}C$, and ${}^{14}_{6}C$ and are designated carbon-12, carbon-13, and carbon-14, respectively. **[Try Problem 3.50]**

Some isotopes, called **radioisotopes,** have a built-in instability; their nuclei undergo *random, spontaneous* disintegrations. When a radioisotope disintegrates (decays), it emits from its nucleus one or more high-energy particles (waves, rays) collectively known as **nuclear radiation.** Samples of matter that contain radioisotopes and emit nuclear radiation are said to be **radioactive.** A **curie (Ci)** is that amount of radioactive material that undergoes 3.7×10^{10} disintegrations per second (dps). A **becquerel (Bq),** the official SI unit of radioactivity, is that amount of radioactive material that undergoes 1 dps. **[Try Problem 3.51]**

The three most common classes of nuclear radiation are **alpha (α) rays, beta (β) rays,** and **gamma (γ) rays** (Table 3.11). Some unstable atoms emit a single type of nuclear radiation, whereas others simultaneously emit multiple classes. The type of

TABLE 3.11

Properties of the Three Major Classes of Nuclear Radiation

Name	Symbols	Description	Charge	Mass (u)[a]	Velocity[b]	Penetrating Ability
Alpha	α, ${}^{4}_{2}He^{2+}$, ${}^{4}_{2}\alpha$	Helium nucleus	+2	4.0026	Up to 10% speed of light	Low
Beta	β, ${}^{0}_{-1}e$, ${}^{0}_{-1}\beta$	Electron	−1	0.000548	Up to 90% speed of light	Low to moderate
Gamma	γ, ${}^{0}_{0}\gamma$	Electromagnetic wave	0	0	Speed of light	High

[a]u = atomic mass unit.
[b]Speed of light = 3×10^{8} m/s = 1.86×10^{5} mi/s.

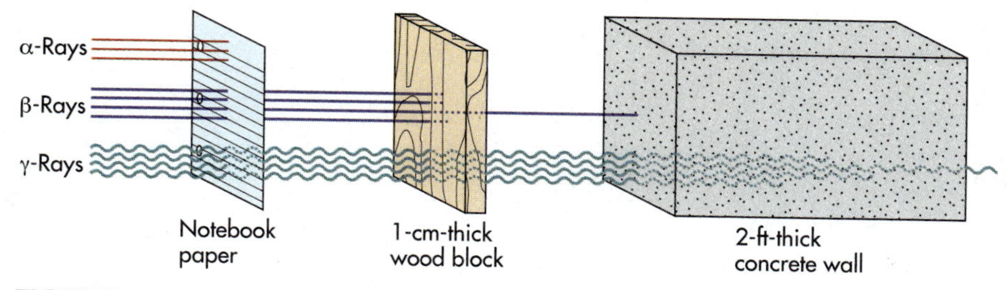

FIGURE 3.14
Penetrating ability of nuclear radiation. (Note: The concrete wall is not to scale. At true scale, it would be 61 times as thick as the wood block.)

radiation emitted, and its energy content, is determined by the specific radioistope involved. Carbon-14, for example, is a beta emitter, whereas radon-222 ($^{222}_{86}Rn$) emits both alpha and gamma rays.

Alpha (α) rays are emitted from the nuclei of disintegrating atoms at up to 10% the speed of light. They consist of two protons bundled together with two neutrons and are identical to the nucleus of the most common isotope of helium, $^{4}_{2}He$. Because α-rays are relatively slow and massive, they quickly transfer their energy to surrounding matter and have little penetrating ability (Figure 3.14). A typical α-ray travels only 4 to 5 cm in air and is unable to pass through human skin. External exposure poses no health risk.

Beta (β) rays are electrons that are ejected from the nuclei of disintegrating atoms at up to 90% the speed of light. Some of these negative, low-mass particles can travel over a meter in air and can penetrate more than a centimeter into living tissue. Although external β-rays cannot reach some internal organs, both internal and external exposure can lead to a significant health risk.

Gamma (γ) rays, like X-rays and visible light, are a class of electromagnetic radiation (Section 3.11) and travel at the speed of light. Because this third major class of nuclear radiation can pass completely through the human body, all tissues are at risk with either internal or external exposure. A thick wall of concrete or similar material is required to provide effective shielding. **[Try Problem 3.52]**

Half-Life

The random nature of the disintegration process makes it impossible to predict when a single atom of a radioisotope will disintegrate. However, if a sample contains a very large number of atoms of the same radioisotope, we can determine the time required for one half of these atoms to disintegrate. This time period, known as a **half-life,** is constant for each radioisotope. Distinctive radioisotopes normally have unique half-lives (Table 3.12). Since the passage of each half-life eliminates one half of the unstable atoms present at the beginning of that half-life, many half-lives must pass before all of the atoms in a sample of a radioisotope disintegrate (Figure 3.15). Because half-life is a statistical phenomenon based on the assumption that there are a very large number of unstable atoms, the concept of half-life becomes meaningless long before the last unstable atom disintegrates. At this point, the amount of radioisotope remaining is usually of no practical significance. **[Try Problem 3.53]**

TABLE 3.12

Half-Lives of Some Radioisotopes

Isotope	Half-Life
$^{235}_{92}U$	7.04×10^{8} years
$^{14}_{6}C$	5.72×10^{3} years
$^{137}_{55}Cs$	30.3 years
$^{90}_{38}Sr$	29.1 years
$^{3}_{1}H$	12.3 years
$^{60}_{27}Co$	5.27 years
$^{32}_{15}P$	14.3 days
$^{131}_{53}I$	8.04 days
$^{222}_{86}Rn$	3.82 days
$^{99m}_{43}Tc$	6.02 hours
$^{214}_{84}Po$	164 microseconds

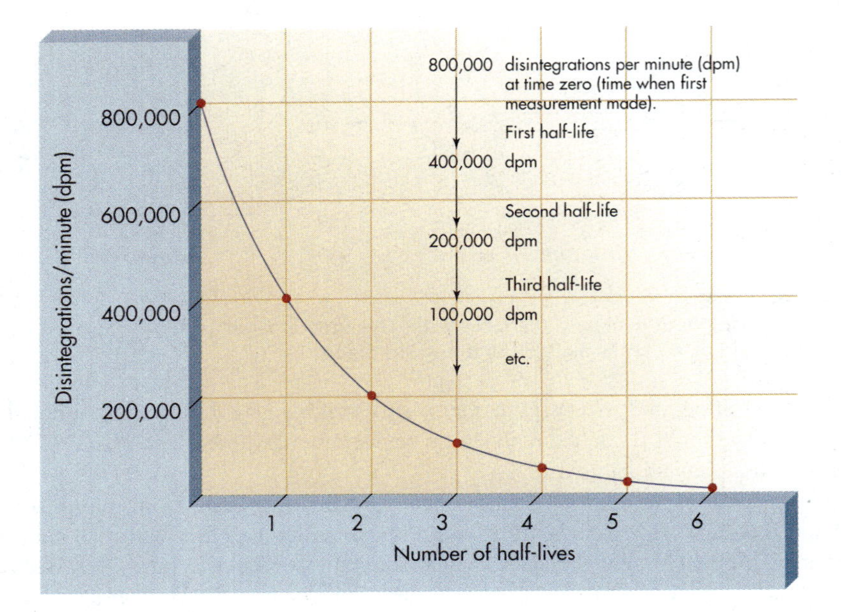

FIGURE 3.15
Impact of half-lives on the number of unstable atoms in a sample of a radioisotope.

Health Consequences of Radiation Exposure

When nuclear radiation passes through matter, including living tissue, it is eventually stopped as it transfers its kinetic energy to surrounding atoms and molecules. This transfer of energy can literally rip molecules apart. Since the transferred energy can also ionize (strip electrons from) atoms and molecules, nuclear radiation is described as **ionizing radiation**. X-rays are also ionizing rays. Some of the ions and **free radicals** (molecules with unpaired electrons) generated by nuclear radiation and X-rays are highly reactive; they lead to alterations in proteins, DNA, lipids, and other compounds. Such radiation-induced changes account for the health risks associated with radiation exposure.

A single, large dose of nuclear radiation can be lethal, a fact reaffirmed with over 30 acute fatalities following the Chernobyl nuclear power plant disaster in the Soviet Union in 1986 (Table 3.13). The probable consequences of continuous, low-level exposure include an increased cancer risk, an increased mutation rate, and symptoms that mimic some aspects of premature aging (Exhibit 3.3). Many health-related questions remain unanswered. What is the precise relationship between dose and risk? Does every increase in exposure lead to an increase in risk? What is an acceptable level of exposure and risk? **[Try Problem 3.54]**

All living organisms are constantly exposed to ionizing radiation, since they are partially constructed from radioisotopes, including ^{14}C. Small amounts of naturally occurring radioisotopes in air, water, soils, and building materials further contribute to the ever-present background radiation (Table 3.14). Medical procedures, the nuclear power industry, and a variety of consumer products make additional contributions to our radiation exposure.

EXHIBIT 3.3

Some Probable Consequences of Continuous Exposure to Low Levels of Nuclear Radiation

- Impaired immune function
- Increased cancer risk
- Increased mutation risk
- Increased risk of fetal damage
- Increased cataract risk
- Premature aging (?)[a]

[a] In experimental animals low doses of nuclear radiation can induce changes that closely mimic multiple age-related changes.

TABLE 3.13

Short-Term Consequences of High-Level, Short-Duration Nuclear Radiation Exposure

Dose (Sv)[a]	Short Term Health Consequences
0–0.5	No consistent symptoms.
0.5–2.0	Loss of white blood cells; nausea, vomiting; about 10% die within months at 2.0 Sv.
2.0–4.0	Loss of blood cells; fever, hemorrhage, hair loss, nausea, vomiting, diarrhea, fatigue, skin blotches; about 20% die within months.
4.0–5.0	Same symptoms as 2.0–4.0 Sv but more severe; increased risk of infections because of the lack of white blood cells; 50% die within months at 4.5 Sv.
5.0–10	Severe gastrointestinal damage, cardiovascular collapse, central nervous system damage; doses above 7.0 Sv are fatal within a few weeks.
100	Death within hours.
1000	Death within minutes.

[a]Sv = Sievert, an SI unit equal to the amount of radiation that gives a dose, in humans, equivalent to a dose of X-rays causing an energy absorption of 1 joule per kilogram of tissue.

TABLE 3.14

Average Annual Ionizing Radiation Exposure in the United States and United Kingdom[a]

	UK[b] (mSv)	USA[c] (mSv)
Natural sources		
Radioisotopes of radon and thoron	1.3	2.0
Terrestrial γ-rays	0.35	0.28
Cosmic rays	0.25	0.27
Ingested natural radioisotopes	0.30	0.39
Total	2.2	2.9
Artificial sources		
Medical procedures	0.30	0.53
Fallout (including Chernobyl)	0.01	0.0006
Miscellaneous[d]	0.02	0.06–0.14
Total	0.33	0.59–0.67

[a]The averages conceal a wide range of exposure; the exposure of single individuals may differ markedly from the average.
[b]Based on measurements made in 1988.
[c]Based on measurements made in 1987.
[d]Includes occupational exposure, luminous dial watches, smoke detectors, industrial waste, TV viewing, etc.

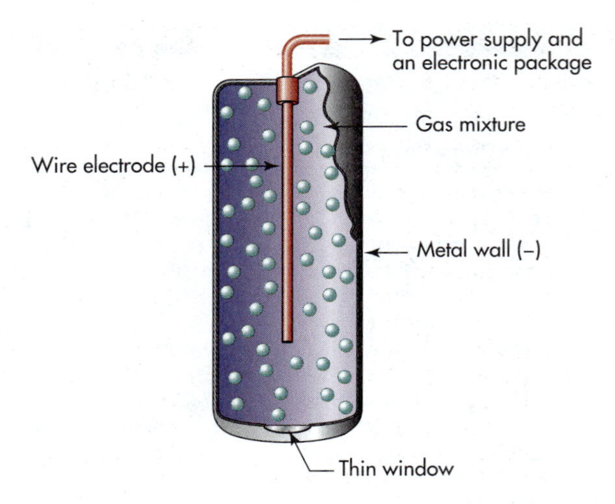

FIGURE 3.16
A Geiger tube. Radiation entering a Geiger tube through the thin window strips electrons from gas molecules within the tube. As the negative electrons rush toward the positive central electrode, they dislodge additional electrons from gas molecules. An electric pulse is created when the "avalanche of electrons" reaches the central electrode.

Detection of Nuclear Radiation

Nuclear radiation, like diagnostic X-rays, cannot be sensed by the body; we cannot see, hear, smell, taste, or feel such radiation. This fact raises the practical question of how nuclear radiation is detected and measured.

One of the most common devices used to detect nuclear radiation is the **Geiger–Müller counter.** At the heart of this detector is a hollow metal tube, called a **Geiger tube,** that is filled with a special gas mixture (Figure 3.16). A metal wire, called an **electrode** or **filament,** runs down the middle of the tube from one end. The opposite end is covered with a thin window, often made of mica. A high-voltage power supply generates a positive charge on the central electrode and a negative charge on the wall of the tube. When nuclear radiation enters the tube through the thin end-window, it ionizes (strips electrons from) part of the atoms within the gas mixture. The free, negatively charged electrons rush toward the positively charged central electrode and dislodge additonal electrons from other atoms of gas along the way. An electric pulse is generated when the **avalanche of electrons** reaches the central electrode. Such pulses can also be amplified, detected, and counted. The amplified pulses can also be used to create the audible clicks that are often associated with Geiger-Müller counters.

A Geiger–Müller counter is unable to detect all of the nuclear radiation emitted by a sample, because a significant fraction of that radiation never enters the Geiger tube and some of the radiation that does enter escapes undetected. In spite of these problems, it is possible to calibrate such counters and use them for the quantitative detection of nuclear radiation. **[Try Problem 3.55]**

Scintillation counters are a second major class of detectors. In these instruments, the energy from the radiation to be measured is absorbed by chemicals called **scin-**

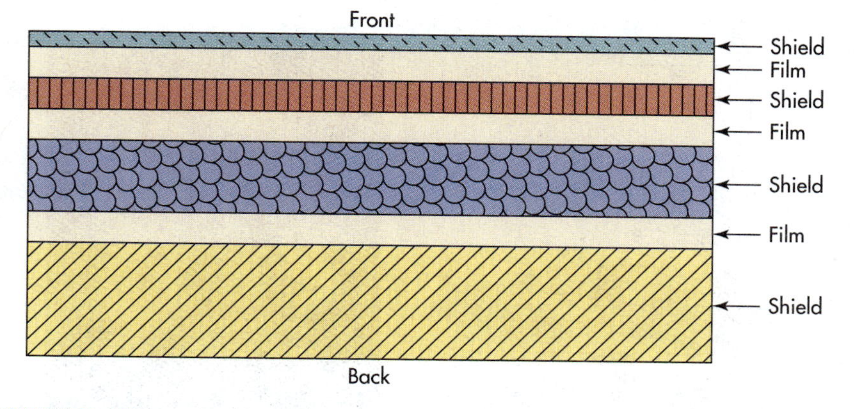

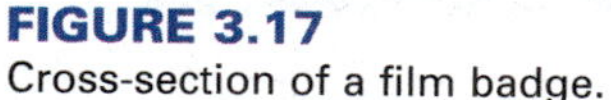

FIGURE 3.17
Cross-section of a film badge.

tillators or **fluors** that re-emit part of the energy as visible light. The flashes of visible light **(scintillations)** are a measure of the amount of radiation that has entered the scintillator, and it is these scintillations that are directly detected and counted.

Occupational radiation exposure is commonly monitored with film badges that are clipped to dresses or shirts. A **film badge** consists of multiple layers of X-ray film sandwiched between shields of increasing efficiency (Figure 3.17). Radiation striking the badge exposes one or more layers of film depending on its penetrating ability. The total amount of film exposure is a measure of the dose of radiation hitting the film badge, and the number of layers of film exposed provides some information about the type of radiation and its energy content. Film badges must be periodically collected and developed to assess exposure. X-ray film can also be used to detect radioisotopes in flat samples such as agarose gels, paper chromatograms, leaves, and cultured cells. The film is developed after it is placed over the sample of interest (in the dark) for an appropriate period of time. This technique, called **autoradiography,** allows location of radioisotopes with high sensitivity and high resolution.

Some Uses of Radioisotopes.

Radioisotopes are used routinely in biochemistry and medicine as tracers and as diagnostic, therapeutic, and analytical agents. Nuclear radiation is employed for many other purposes as well, including the sterilization of foods and other items. The ability to detect extremely small quantities of radioisotopes makes them ideal tools for following **(tracing)** the distribution and ultimate fate of virtually any substance within a living organism. Suppose, for example, that we inject a small amount of aspartame uniformly labeled with carbon-14 into the blood of rats. By isolating and counting the aspartame in various tissues from the rats at different times after injection, we can determine which tissues have taken up this artificial sweetner and how rapidly they have done so. By measuring the amount of carbon-14 that appears in other compounds over time, we can also determine the metabolic fate of the injected

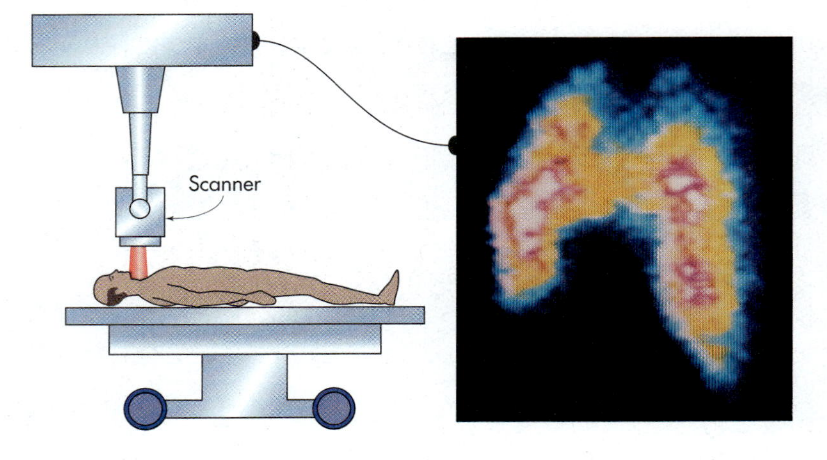

FIGURE 3.18
Thyroid scan following the administration of iodine-131. The plot (scan) to the right shows the location in the thyroid of each disintegrating ^{131}I atom whose emitted radiation has been detected by the scanner. Based on Brown & Rogers (1987).

aspartame. Many proteins, for example, become labeled with carbon-14 as phenylalanine produced from the hydrolysis of aspartame is incorporated into these polymers. The complete oxidation of part of the phenylalanine leads to the production of carbon-14 labeled CO_2. The rate of appearance of carbon-14 in urine reveals how rapidly aspartame and its metabolites are excreted. **[Try Problem 3.56]**

Iodine-131 provides a specific example of the diagnostic value of radioisotopes. External radiation detectors are employed to measure the rate of iodine-131 accumulation within the thyroid gland after the oral administration of an aqueous solution of KI or NaI. A normal thyroid will accumulate 7.5% to 25% of the oral dose in 6 hours and 12% to 46.5% in 24 hours. **Hyperthyroidism,** thyroid **cancer,** and other thyroid dysfunctions lead to characteristic changes in iodine-131 accumulation. A special scintillation detector called a **scintillation camera** can be used to produce a photographic image of the thyroid gland after it has accumulated enough radioisotope (Figure 3.18), a technique known as **imaging** or **scanning.** Any group of cells can be imaged if an appropriate radioisotope is selectively concentrated by these cells. Radioisotope-based imaging is used routinely to obtain information about the size, shape, and metabolic condition of many different tissues, organs, and tumors. **[Try Problem 3.57]**

The use of nuclear radiation to treat cancer and other ailments is known as **radiation therapy.** The use of iodine-131 to treat both hyperthyroidism and cancer is a specific example. Such treatment normally entails the oral administration of much larger doses of radioisotope than are used for diagnostic purposes. Because both normal and cancerous thyroid cells selectively concentrate iodine-131. they tend to be exposed to much higher levels of nuclear radiation than other types of cells. The duration of exposure is set by the 8.06-day half-life of iodine-131 and the rate at which iodine is eliminated from the body. With hyperthyroidism, an amount of iodine-131 is administered that will kill or damage just enough thyroid cells to reduce thyroid hormone production to a normal level. In cancer, the goal is to destroy all of the cancerous thyroid cells.

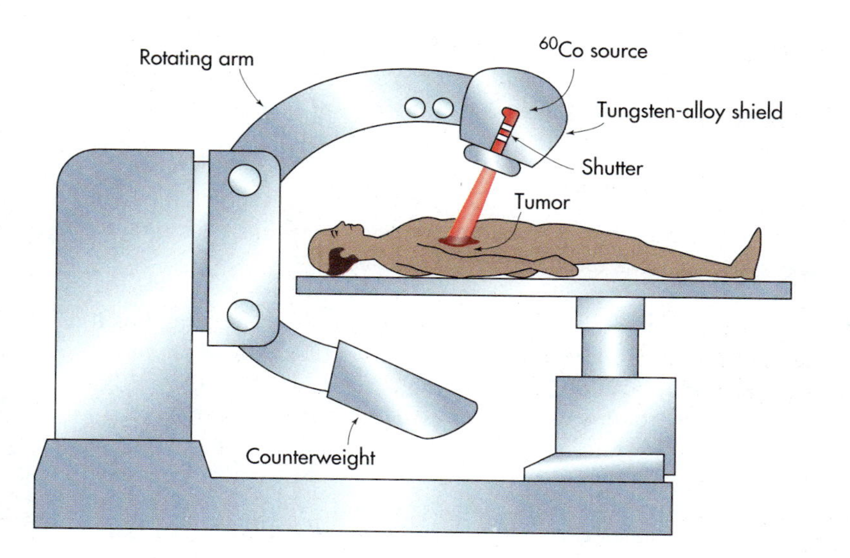

FIGURE 3.19
Cancer treatment with cobalt-60. The radiation source moves along a circular track, rotating the radioactive beam around the patient so that only the tumor receives continuous radiation. Based on Brown & Rogers (1987).

A beam of γ-rays provided by an external cobalt-60 or cesium-137 source can also be used for radiation therapy of cancer (Figure 3.19). In such cases, therapists attempt to minimize damage to normal cells by focusing the beam on the cancerous cells. Unfortunately, all radiation therapy leads to some damage to normal cells, and the extent of damage is substantial in some instances. Cells that divide rapidly, including bone marrow cells, intestine cells, and hair cells, are particularly sensitive. This explains why patients undergoing extensive radiation therapy commonly experience diarrhea, nausea, and a loss of white blood cells and hair cells. The relatively high sensitivity of rapidly dividing cancer cells to nuclear radiation accounts for any benefits associated with radiation therapy. It is ironic that a patient cured of cancer today with radiation therapy acquires an increased risk of future cancer as a consequence of that therapy. **[Try Problem 3.58]**

Radioimmunoassays (RIAs) illustrate the use of radioisotopes as analytical tools. RIAs are discussed along with the examination of enzyme-linked immunoassays (EIAs) and enzyme-linked immunosorbent assays (ELISAs) in Section 5.18. The analysis of immunoglobulins in Section 4.11 provides the groundwork for that discussion.

PROBLEM 3.50 Write abbreviations for four different isotopes of chlorine. There is no requirement that the isotopes you choose be stable or naturally occurring.

PROBLEM 3.51 Calculate the dpm within 10 μCi of carbon-14. Compare this to the dpm within 10 μCi of radon-222.

PROBLEM 3.52 Does a nurse increase his radiation exposure by sitting close to a patient who has been injected with (a) an alpha-emitting radioisotope, or (b) a gamma-emitting radioisotope? Explain.

PROBLEM 3.53 A general rule of thumb states that most samples of radioisotopes used for biomedical purposes will decay to a virtually harmless level within 10 half-lives. What fraction or percentage of the initial unstable atoms will remain after this period of time?

PROBLEM 3.54 List three practical, common sense steps that can be taken to reduce radiation exposure when working around a patient or a biological sample that is emitting nuclear radiation.

PROBLEM 3.55 The counting efficiency of a detector is defined as the percentage of the total disintegrations that are actually detected and recorded (counted). Calculate the counting efficiency if 29,000 cpm (counts per minute) are recorded when a sample undergoes 98,000 dpm (disintegrations per minute).

PROBLEM 3.56 All of the nonapeptides within a rat liver are isolated and separated 2 hours after the rat has been injected with ^{14}C-labeled glutamate, one of the amino acids used to assemble peptides. Will each of the separated nonapeptides contain the same amount of ^{14}C? Explain.

PROBLEM 3.57 A radioisotope used for diagnostic human imaging must be differentially concentrated by the cells to be imaged. What other properties or characteristics should the isotope possess?

PROBLEM 3.58 Will you increase your radiation exposure by sitting next to a friend who has recently been exposed to a massive dose of external γ-rays during cobalt-60 therapy? Explain.

SUMMARY

Twenty common amino acids serve as the basic building blocks, but not the sole building blocks, for the construction of the thousands of proteins and other peptides found within every living organism. Since the human body cannot synthesize adequate amounts of 9 of these amino acids, these 9 are classified as "essential" (from the standpoint of human nutrition); they must be obtained in the diet. With one exception, each of the 20 common amino acids is a carboxylic acid with a hydrogen, an amino group, and a side chain attached to its α-carbon (Figure 3.1). Proline, the only exception, is technically an imino acid because its side chain has reacted with its α-nitrogen to form a heterocyclic ring. The side chain of each amino acid distinguishes it from all other amino acids and accounts for its unique chemical and physical properties. Seven side chains possess weak acid functional groups (Tables 3.4 and 3.5), and four of these exist predominantly in a charged form at pH 7.0. At the same pH, 6 of the side chains are polar but primarily uncharged and the remaining 10 are nonpolar or borderline nonpolar (Table 3.6).

Each common protein amino acid, except glycine, is a chiral compound whose specific configuration can be designated by using either the *R–S* system or the D–L system. When the D–L system is employed, all of the chiral protein amino acids have an L configuration about their α-carbons. Their D enantiomers are neither produced nor utilized by most living organisms, including humans.

Peptides are created by joining the α-amino groups and α-carboxyl groups of neighboring amino acids through amide (peptide) bonds. Peptide bonds exhibit restricted rotation and have characteristic bond angles. Most possess a trans configuration. Peptide chains can be visualized as sequences of peptide (amide) planes whose relative orientations are determined by ϕ and ψ rotations about specific bonds to α-carbons. Naturally occurring peptides that contain 50 or more amino acid residues and with biological activities that depend on a specific folding pattern are classified as proteins.

A peptide has directional characteristics and, by convention, its amino acid residues are both named and numbered from the N-terminus to the C-terminus. It is the ordering of the amino acid side chains along a peptide's backbone that determines its physical, chemical, and biological properties. The net charge on a peptide is a function of pH, and that unique pH at which it has a zero net charge is called its isoelectric point (abbreviated pI).

The determination of the relative amount of each amino acid in a peptide is known as amino acid analysis. Such analysis normally involves the hydrolysis of the peptide followed by the separation (with chromatography) and quantitative detection (absorbance measurements often used) of each free amino acid in the hydrolysate. The last two steps in this operation are commonly performed by an automated amino acid analyzer.

Some of the common protein amino acids and some of their derivatives are neurotransmitters within the central nervous system. Part of these amino acids serve as precursors for the production of other compounds in addition to peptides. Monosodium glutamate is a common flavor enhancer. All amino acids are a potential source of energy; they can be used as fuel by most cells.

Peptides play an enormous number of roles within all living organisms. Many are chemical messengers that function as hormones, growth factors, or neurotransmitters. Oxytocin and vasopressin are specific examples. Together, these two nonapeptides help regulate a wide variety of physiologically important processes, including lactation, urine production, muscle contraction, and the operation of the central nervous system. Enkephalins, endorphins, and dynorphins, three classes of opiate peptides, provide additional examples of "messenger" peptides. The popular artificial sweetener known as aspartame or NutraSweet is a methyl ester of a dipeptide.

Chromatography is one of the four key biochemical tools introduced in this chapter. During chromatography, a gaseous or liquid mobile phase is used to carry the components of a mixture over a stationary phase. The more tightly a component "adheres" (using the term very loosely, particularly in the case of gel filtration chromatography) to the stationary phase, the more slowly it travels and the later it elutes from a chromatography column. Since a compound tends to elute at a unique and characteristic time under carefully controlled conditions, chromatography can be used to help identify and characterize compounds, as well as to purify them.

The migration of compounds in an electric field is called electrophoresis. Samples under study are placed on a moist support material between two electrical poles. Compounds with a net positive charge move toward the cathode and those with a net negative charge toward the anode. Any compounds that are uncharged or have a zero net charge remain at the point of application. In most cases, a compound's electrophoretic mobility is predominantly determined by the ratio of its net charge to its size. Compounds with different charge to size ratios tend to migrate at dissimilar rates and to separate from one another. Electrophoresis, like chromatography, can be used to characterize, identify, and purify compounds.

Spectrophotometric analysis refers to any analysis that involves absorbance measurements. Since different compounds tend to have unique absorption characteristics, a pure substance can often be characterized and identified with absorbance measurements. Absorbance measurements can also be used to detect a wide variety of compounds quantitatively. The absorbance of monochromatic light by a dilute solution of a compound is usually proportional to solute concentration (assuming that no other substance in the solution absorbs at the wavelength employed), a fact known as Beer's law:

$$A = \epsilon bc$$

Given a numerical value for any three of the four variables in Beer's law, we should be able to calculate the value of the remaining variable.

Radioisotopes are unstable atoms that undergo random spontaneous disintegrations and emit nuclear radiation, including α, β, and γ-rays. The emitted radiation allows us to locate these atoms sensitively and quantitatively. At high doses, the same radiation quickly kills individual cells and multicellular organisms. At low doses, nuclear radiation increases our cancer risk and contributes to several additional health problems.

Biochemists commonly tag compounds with radioisotopes in order to determine their distribution, chemical transformations, and ultimate fate within living organisms. In medicine, radioisotopes and compounds labeled with radioisotopes are routinely employed as analytical, diagnostic, and therapeutic agents.

EXERCISES

3.1 Write the structural formula for each amino acid whose structure you have been asked to learn. If a pH is not specified, it is usually acceptable to write a weak acid functional group in either its protonated or nonprotonated form.

3.2 Given the α-amino acids shown below and approximate pK_a values (Note: some of these are not protein amino acids), which of these amino acids:

a. Have a side chain with a −1 charge (to the nearest half charge unit) at pH 1? pH 7? pH 14?

b. Have a side chain that can H bond with a water molecule?

c. Have no functional group in their side chain?

d. Have a functional group in their side chain that can exist in both a weak acid and a conjugate base form?

e. Are alcohols? carboxylic acids? thiols? amines? phenols?

f. Have nonpolar side chains?

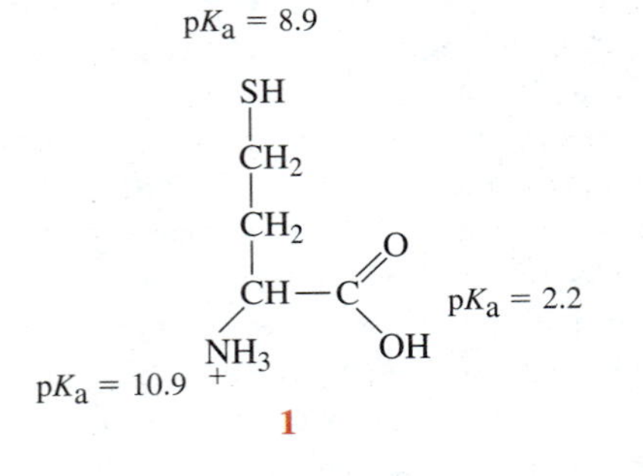

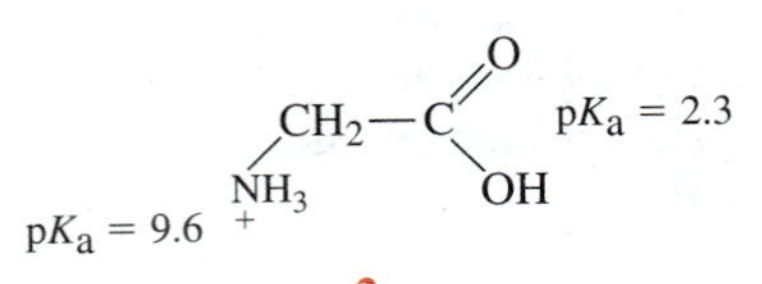

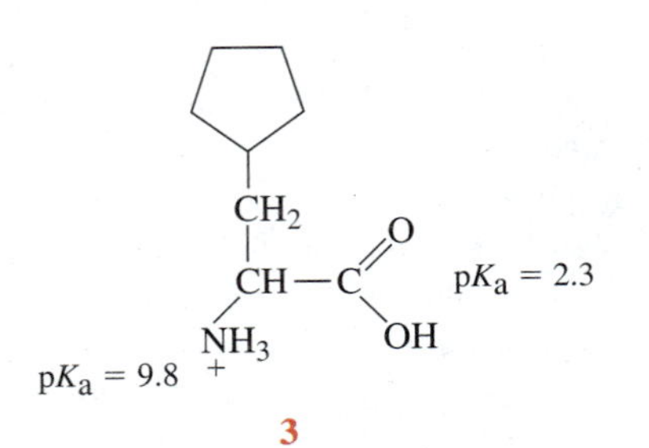

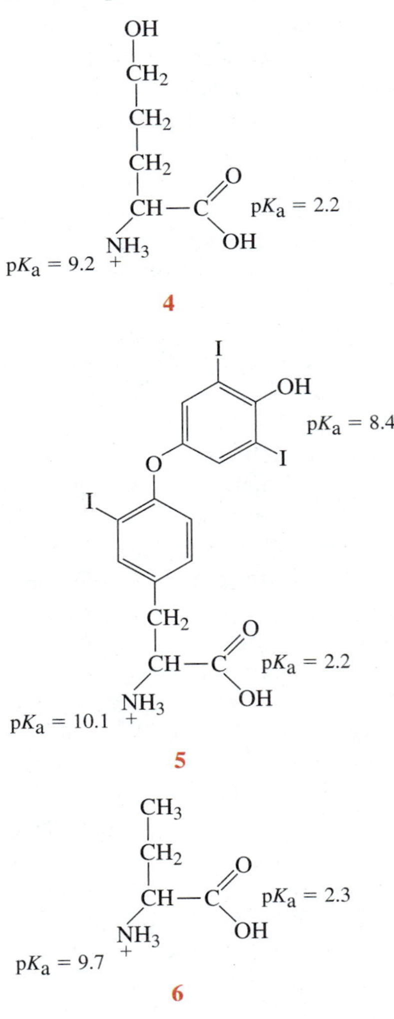

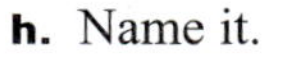

3.3 Write the structural formula for:

a. A β-amino acid that is a phenol.
b. A γ-amino acid that is an alcohol.
c. Any α-amino acid with three chiral centers.
d. D-Phenylalanine.
e. L-Valine.

3.4 Write a structural formula for the mirror image of each of the following compounds:

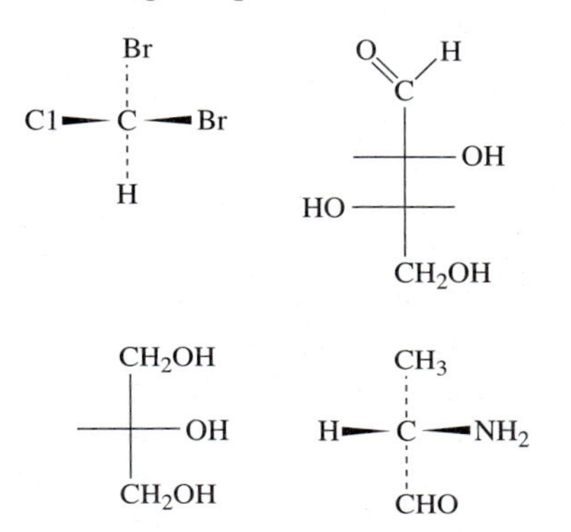

Which of the mirror images represent enantiomers of the given compound? Explain.

3.5 Would amino acid analysis of a peptide allow you to determine the molecular weight of that peptide? Explain.

3.6 Given the peptide at the bottom of this column:

a. Identify the C-terminal end.
b. Circle each amino acid residue.
c. Use an arrow to identify each bond that exhibits restricted rotation.
d. Put a square around each H atom that is capable of H bonding.
e. Put an asterisk (*) to the right of each nonpolar side chain.
f. Identify each amide plane.
g. Write its abbreviations.

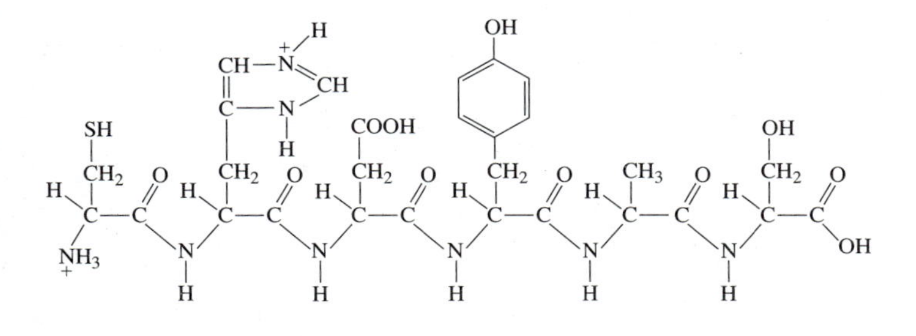

h. Name it.
i. Identify the terms in Table 3.1 that describe it.
j. What is the net charge on this peptide to the nearest one half charge unit at pH 1? pH 6? pH 9? pH 12? Assume that the weak acid functional groups in the amino acid residues have the same pK_a values as the corresponding functional groups in free amino acids (Table 3.5).
k. Will this peptide move toward the cathode (−) or the anode (+) during electrophoresis at pH 1? pH 6? pH 9? pH 12?
l. Will this peptide bind more tightly to an anion exchange column or to a cation exchange column during chromatography at pH 1? pH 6? pH 9? pH 12?

3.7 List the biological and commercial (if any) functions, activities, and uses of each of the following:

a. Oxytocin
b. Endorphins
c. Vasopressin
d. Glutamate

3.8 Amino acid analysis of an unknown peptide reveals that it contains three different amino acids in the following molar ratios; 2 Ala, 1 Thr, 2 Leu. Give the abbreviations for four different pentapeptides and one decapeptide that will all yield these same results if subjected to amino acid analysis. Can the four pentapeptides readily be separated from one another with electrophoresis at pH 7? Explain.

3.9 Write the structural formula for the tetrapeptide Pro-Met-His-Arg as it will primarily exist at (a) pH 1 (b) pH 5 (c) pH 8; and (d) pH 12. Assume that the pK_a's for the functional groups involved are the same as the pK_a's for the corresponding functional groups on free amino acids (Table 3.5). Will this tetrapeptide bind more tightly to an anion exchange resin or to a cation exchange resin during chromatography at pH 1? pH 5? pH 8? pH 12? Explain.

3.10 True or false? Any dipeptide can be readily separated from any polypeptide with gel filtration chromatography. Explain.

3.11 A 1×10^{-5} M solution of antibiotic Q has an $A_{220} = 1.08$. Calculate ϵ_{220} for antibiotic Q. What additional information is needed to calculate the ϵ_{345} of this antibiotic?

3.12 Drug Z has an ϵ_{260} of 2.3×10^3 L mol^{-1} cm^{-1}. Calculate the concentration of drug Z in a solution that yields an $A_{260} = 0.56$ when a 2-cm pathlength is used for the absorbance measurement.

3.13 Calculate the A_{340} and A_{560} of a solution that contains 2×10^{-5} M drug T ($\epsilon_{340} = 2 \times 10^{+4}$ L mol^{-1} cm^{-1}, $\epsilon_{560} = 0$ L mol^{-1} cm^{-1}) and 4×10^{-5} M drug P ($\epsilon_{340} = 1 \times 10^4$ L mol^{-1} cm^{-1}, $\epsilon_{560} = 3 \times 10^4$ L mol^{-1} cm^{-1}).

3.14 A 0.01 M solution of a compound with a molar absorptivity of 4×10^2 L mol^{-1} cm^{-1} has an absorbance of 1. What pathlength was used for this absorbance measurement?

3.15 Write a balanced equation for the production of disodium glutamate from monosodium glutamate and sodium hydroxide.

3.16 Methanol, a hydrolysis product of aspartame, is oxidized in the human body, and the oxidation products contribute to its toxicity. What products are formed when methanol is oxidized?

3.17 Write an accepted abbreviation for the isotope of phosphorus that contains 15 neutrons.

3.18 True or false? ${}^{14}_{6}C$ is an isotope of ${}^{14}_{7}N$. Explain.

3.19 The formula for calculating the half-life of a sample of a radioisotope is:

$$\text{Half-life} = \frac{0.301t}{\log (N_0/N)}$$

$$N_0 = \text{dpm at time zero}$$

$$N = \text{dpm at time } t$$

Calculate the half-life of a sample of a radioisotope that underwent 6200 dpm 6 hours ago and registers 5440 dpm now.

3.20 How many μCi of ${}^{99m}Tc$ (half-life 6.02 hours) will remain in a patient 24 hours after the injection of 1 μCi of this isotope? Assume that no Tc is eliminated from the body (in urine or feces, for example) during this 24-hour period.

3.21 For many years most health officials in the United States advocated routine chest X-rays for the detection of tuberculosis. Suggest why this is no longer the case. Note: Chest X-rays are still probably the best option when it comes to routine screening.

3.22 The detectors in both Geiger and scintillation counters directly detect specific "byproducts" from the interaction of nuclear radiation with matter, not the nuclear radiation itself. What "byproduct" is directly detected within a Geiger counter? A scintillation counter?

3.23 A scintillation counter with a counting efficiency of 30% records 37,000 counts per minute when used to count a blood sample containing iodine-131. Calculate the number of curies and the number of becquerels within this blood sample.

3.24 Describe an experiment that could be used to quickly and easily determine how rapidly drug X (a stable, nondigestable compound) enters the blood following oral administration. Assume that a ${}^{14}C$-labeled sample of drug X is available.

3.25 Thirty minutes after a ${}^{14}C$-labeled sample of compound F is injected into a cancer patient, virtually all of the radioactivity is found to be located within cancer cells; very little resides in normal cells. Can we confidently conclude that cancer cells selectively concentrate compound F? Explain.

3.26 List all the terms that appeared in *bold* print in this chapter. Make certain that you can describe or define each term. When appropriate, make certain that you can summarize the significance, use, or importance of the substance, technique, etc., represented by a term. Although this exercise will not be repeated in future chapters, you should repeat this process for bold terms in all chapters that follow.

SELECTED READINGS

Amato, I., Looking glass chemistry, *Science* 256, 964–966 (1992).

A research news article that stresses the biological importance of enantiomers and reviews efforts to synthesize individual enantiomers.

Balter, M., Filtering a river of cancer data, *Science* 267, 1084–1086 (1995).

Describes how a series of accidents in Russia is giving researchers an opportunity to study the health effects of long-term, low doses of nuclear radiation.

Bennett, W. R. Jr., Electromagnetic fields and power lines, *Sci. Am. Science and Medicine* 2 (4), 68–77 (1995).

Examines the claim that weak electromagnetic fields from power lines, transformers, and home appliances are a threat to health.

Choi, D. W., Bench to bedside: The glutamate connection, *Science* 258 241–243 (1992).

Reviews the role of glutamate as a neurotransmitter.

Cohen, J., Getting all turned around over the origins of life on earth, *Science* 267, 1265–1266 (1995).

Addresses the question "What is the origin of the stereospecificity exhibited by living organisms?"

Creighton, T. E., *Proteins: Structures and Molecular Properties*, 2nd ed. New York: W. H. Freeman, 1993.

Chapter 1 includes an excellent discussion of the chemistry of amino acids plus an introduction to the peptide bond. The detection of amino acids and peptides, gel filtration chromatography, and electrophoresis are also examined. Ample references are provided.

Dickerson, R. E., and I. Geis, *The Structure and Action of Proteins*. New York: Harper & Row, 1969.

A classic. Contains exceptional diagrams and a thorough discussion of peptide bonds, amide planes, and peptide structure.

Doll, R., H. J. Evans, and S. C. Darby, Parental exposure not to blame, *Nature* 367, 678–680 (1994).

Presents evidence that the "Gardner hypothesis" is wrong. The "Gardner hypothesis" states that an excess of childhood leukemia (a form of cancer) near a nuclear power plant in England is caused by parental exposure to ionizing radiation.

Eijgenraam, F., Chernobyl's cloud: A lighter shade of gray, *Science* 252, 1245–1246 (1991).

A study of the Chernobyl disaster finds much anxiety but no major impact on the health of the surrounding population as of 1990.

Mann, C. C., Radiation: balancing the record, *Science* 263, 470–473 (1994).

Examines the intentional exposure of people in the United States to radioisotopes and nuclear radiation during the World War II era. Both knowing and unknowing individuals were used as guinea pigs in experiments designed to assess some of the health consequences of radioisotopes and nuclear radiation.

Nutrition update: Who's afraid of MSG, *Consumers Reports on Health* 4(1), 6 (1992).

Summarizes the effects of MSG on health.

Robyt, J. F., and B. J. White, *Biochemical Techniques: Theory and Practice*. Prospect Heights, IL: Waveland Press, 1987.

Contains separate chapters on chromatography, electrophoresis, and absorption photometry. Literature references are provided.

Silverstein, R. M., G. C. Bassler, and T. C. Morrill, *Spectrometric Identification of Organic Compounds*, 5th ed. New York: John Wiley & Sons, 1991.

Examines the use of absorbance measurements to determine molecular structure.

Slater, R. J., *Radioisotopes in Biology: A Practical Approach*. New York: Oxford University Press, 1990.

A guide to working with radioisotopes in the biochemistry lab. Includes discussions of scintillation counters, autoradiography, tracer experiments, radioimmunoassays, and much more.

Swinbanks, D., and C. Anderson, Search for contaminant in EMS outbreak goes slowly, *Nature* 358, 96 (1992).

Examines the possible causes of the tryptophan-linked EMS outbreak in 1989.

4 Proteins

CHAPTER OUTLINE

Chapter 3 defined proteins as polypeptides that contain 50 (approximately) or more amino acid residues and that must coil or fold in a specific manner to be biologically active. Nonprotein peptides are generally smaller and architecturally less complex than proteins. Everything learned about the structure and properties of peptides in Chapter 3 applies directly to proteins.

The word *protein* is derived from a Greek root meaning "of first importance." Aptly so, since the biological significance of proteins cannot be overemphasized. The secret to life itself lies in these polymers and in the ways in which they are utilized by living organisms. Proteins play more roles and a greater variety of roles in humans and other organisms than any other class of compounds. Many function as catalysts, hormones, or transport agents. Some serve as structural building blocks for skin, tendons, cartilage, bone, and other body components. Still others are fundamental participants in muscle action, the visual process, the immune response and the operation of the nervous system. These and additional functions are summarized in Table 4.1.

This chapter examines some of the general properties and structural features of proteins. Specific proteins are then used to illustrate these properties and features and to further explore the theme of structure-function relationships, a topic introduced in Chapter 3 during the discussion of oxytocin, vasopressin, and other nonprotein peptides. Additional biochemical themes are encountered along the way. These include structural hierarchy, specificity, association of like with like, and evolution. Since information for the production of proteins is carried in DNA, the genetic material in all cells, proteins are also linked to the theme of information transfer (Exhibit 1.4). Chapter 5 takes a closer look at enzymes, one of the most important classes of proteins.

4.1 SOME GENERAL CHARACTERISTICS OF PROTEINS

All proteins contain one or more polypeptide chains. Some, in addition, encompass one or more nonpeptide components known as **prosthetic groups**. Proteins that lack nonpeptide components are described as **simple**, and those that contain a prosthetic group are said to be **complex** or **conjugated**. Complex proteins can be classified on the basis of their prosthetic groups (Table 4.2). **[Try Problems 4.1 and 4.2]**

Although all proteins are relatively large, some are over 100 times larger than others. Molecular weights range from about 5000 to over a million daltons. A **dalton** is a unit of mass that corresponds to 1/12th the mass of a ^{12}C atom. Proteins and other biological polymers are appropriately described as **macromolecules** (the prefix *macro-* means large, long, or excessive). Gel filtration chromatography (Section 3.9), polyacrylamide gel (PAGE) electrophoresis (Section 16.9), centrifugation (Section 16.7), and mass spectrometry are some of the techniques used for molecular weight determinations. **[Try Problems 4.3 and 4.4]**

At one point in history, each protein was assigned to one of six major classes based on its solubility in selected solvents. Although most of this solubility-based classification scheme has been abandoned, the terms *albumin* and *globulin* still rear their heads in the biomedical and biochemical literature. To understand this litera-

TABLE 4.1

Some Major Functions of Proteins

Function	Sample Protein(s)	Comments
To control the rates of reactions	Pepsin	The rate of virtually every reaction in an organism is controlled by an individual protein catalyst (an enzyme). Pepsin catalyzes the hydrolysis of dietary proteins in the human stomach.
	Nitrogenase	Nitrogenase is a bacterial enzyme responsible for the fixation of nitrogen in the root nodules of particular plants.
To coordinate and control reactions that occur in different parts of an organism	Insulin	Hormones are chemical messengers that move from initial sites of production to select target sites, where they help regulate specific reactions and physiological processes. Insulin is a hormone that helps regulate the utilization of fats, carbohydrates, and proteins by specific target cells within humans and some other animals.
To transport other molecules	Hemoglobin	Many substances are carried through the human body by proteins. Hemoglobin, a protein found only in red blood cells, transports O_2 from the lungs to outlying tissues and carries CO_2 from outlying tissues back to the lungs.
To defend organisms	Immunoglobulin G	Immunoglobulin G (IgG) is produced by the human immune system in response to foreign agents (virus, bacteria, etc.) known as immunogens. Once produced, IgG binds to these agents and helps the body neutralize, destroy, or eliminate them.
	B_T toxins	Spores of the bacterium *Bacillus thuringiensis* release toxic proteins when eaten by insects. This bacterium is being used to help make crops insect resistant.
To provide structure, support, and protective coverings	Collagen	Hair, skin, fingernails, connective tissues, cellular membranes, and other components of the human body are largely constructed from proteins. Collagen is the major protein component in skin, tendon, cartilage, bone, and blood vessels.
	Viral coat proteins	Most viruses consist of a nucleic acid-containing core enclosed in a protective protein coat.
To provide mobility	Actin, myosin	During the contraction of muscle, actin and myosin filaments slide past one another in an energy-requiring process. Human muscle is primarily composed of these two proteins.
To bind and store other compounds	Ferritin	Ferritin binds and stores iron in the human spleen.
To prevent excess bleeding	Fibrin	Fibrinogen is converted to fibrin during the final stage of blood clotting in humans. A blood clot is a clump of fibrin molecules.
To provide vision	Rhodopsin	Rhodopsin is a pigment (colored compound) in the human eye that, when excited by light, initiates the transmission of visual signals to the brain.
To regulate gene expression	Transcription factors	Transcription factors bind to specific segments of DNA and control the expression of genes. Since a vast majority of all genes carry information for the construction of proteins, protein production represents the final expression of most genes.
To regulate the flow of substances through membranes	Ion channels	The flow of ions through specific ion channels creates the electrical signals that play a central role in the propagation of nerve impulses and the functioning of nerve cells.

TABLE 4.2

Some Classes of Complex Proteins

Prosthetic Group	Class of Proteins
Nucleic acid	Nucleoprotein
Lipid	Lipoprotein
Carbohydrate	Glycoprotein
Metal ion	Metalloprotein
Heme	Hemoprotein
Phosphate	Phosphoprotein

ture, it helps to know that **albumins** are proteins that are soluble in pure water. **Globulins** are insoluble in pure water but are soluble in dilute salt solutions.

Proteins that are spherical, ellipsoidal, or generally compact in shape are called *globular proteins*. The term *globular proteins* must not be confused with the term *globulins*, a term associated with the solubility-based classification system described immediately above. Some, but not all, globular proteins are globulins. Proteins that are shaped like rods, strings, or fibers are categorized as **fibrous**. Globular proteins, including enzymes and some polypeptide hormones, tend to be more water soluble and more easily inactivated than fibrous ones. Fibrous proteins, such as collagen (in skin) and keratin (in hair), are usually tough and insoluble.

Globular proteins, in general, have been found to contain four distinguishable structural elements: primary structure, secondary structure, tertiary structure, and quaternary structure. This classification system was developed to emphasize the structural hierarchy that exists within them. Each of these structural elements is examined thoroughly in the sections that follow. Before we move into detailed discussions, it seems appropriate to provide a brief overview with the use of a simple analogy: a string of beads on a wire (primary structure) can be coiled and wound to form helices and sheets (secondary structure) that can, in turn, be folded together in a highly specific manner (tertiary structure). Multiple strings of coiled and folded beads can then be packed together in a precise geometry (quaternary structure) (Figure 4.1).

PROBLEM 4.1 What products are produced by the complete hydrolysis of a simple protein? A metalloprotein?

PROBLEM 4.2 What distinguishes a protein from a nonprotein peptide?

PROBLEM 4.3 An average-sized amino acid residue contains around 16 atoms. Approximately how many atoms are required to build a protein that contains 500 amino acid residues?

PROBLEM 4.4 Two simple proteins contain the same number of amino acid residues but different ratios of individual amino acids. Do these two proteins necessarily have the same, or even similar, molecular weights? Explain.

Primary structure

Secondary structure

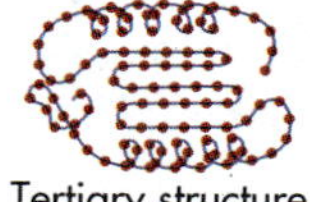
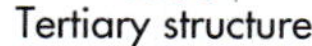

FIGURE 4.1
An overview of primary, secondary, tertiary, and quaternary structure.

4.2 PRIMARY STRUCTURE

The **primary structure** of a protein is its amino acid sequence, the order in which amino acids are joined together along its polypeptide chain(s). The primary structure determines the sequence of side chains along the polymer backbone. With 20 basic building blocks (the 20 common protein amino acids), the number of possible primary structures is enormous. To appreciate this fact, consider the question: "How many different polypeptides can be constructed that each contains 100 amino acid residues?" The answer is 20 raised to the hundredth power or roughly 10 followed by 130 zeros. Since a protein with 100 residues is relatively small, the potential protein diversity is indeed astronomical. The millions of distinct proteins within the human body, including millions of similar but unique immunoglobulins (Section 4.11), contain an infinitesimal fraction of the potential primary structures.

Protein Purification

The large number of distinct proteins within an organism often makes the purification of a specific protein a challenging endeavor. A protein, however, must be purified before its primary structure and its other structural elements can be analyzed. Chromatography (Section 3.9) normally plays a central role in this process. Many purification schemes also employ a variety of additional techniques, including electrophoresis (Section 3.10), centrifugation (Section 16.7), and salt-induced precipitation. Section 5.3 examines protein purification in more detail. Table 5.2 is a summary of a purification scheme for a hypothetical enzyme.

Amino Acid Sequence Determination

Once a sample of a pure protein is in hand, sequencing can be initiated. The amino acid sequences of relatively small polypeptides are commonly determined with an **automated peptide sequenator**. This computer-controlled device uses **Edman degradation** to cleave one amino acid residue at a time from the N-terminal end of the peptide (Figure 4.2). Once removed as a **phenylthiohydantoin (PTH) derivative,** an amino acid residue can be identified by comparing the chromatographic mobility of its PTH derivative to the chromatographic mobility of PTH derivatives of known structure. Other options also exist for amino acid identification.

Since the reactions involved in the stepwise removal of amino acid residues do not proceed to completion, a small amount of the original polypeptide will remain after a sequenator has attempted to remove the N-terminal amino acid from each molecule in a protein sample. Following the second cleavage cycle, most remaining peptides will be two amino acid residues shorter than the original one. However, a small quantity of peptide one amino acid residue shorter than the original peptide, and an even smaller amount of the original peptide, will have survived within the reaction chamber of the sequenator. After approximately 60 cleavage cycles, the accumulation of such byproducts makes the interpretation of further cleavage results difficult or impossible. This limits the size of a protein that can be directly sequenced with this approach.

To sequence polypeptide chains with over 60 amino acid residues, the chain is first cleaved into fragments containing 60, or fewer, amino acid residues. A variety of reactions, including the enzyme-catalyzed hydrolysis of specific peptide bonds, have been developed for this purpose. The cleavage fragments are purified by chromatography, and then sequenced separately. To determine the order in which the cleavage fragments were joined in the original polypeptide, we compare the sequences of two sets of fragments generated under two different fragmentation (cleavage) conditions. The type of comparison involved is best illustrated with a specific example.

Assume that peptides a, b, and c are the only products formed when peptide X is cleaved under a specific set of reaction conditions:

Ala–Trp–Tyr–Asp
Peptide a

Phe–Gly–His–Met–Ile–Thr–Cys–Lys
Peptide b

Glu–Val–Met–Ser–Val–Arg
Peptide c

Further assume that peptide d is among the peptides produced when peptide X is digested under a different set of cleavage conditions:

Ser–Val–Arg–Phe–Gly–His–Met
Peptide d

The amino acid sequence of peptide d allows us to determine that peptide b was joined to the C-terminal end of peptide c within the original peptide X. This determination is based on the observation that an amino acid sequence unique to the N-terminal end of peptide b is joined in peptide d to an amino acid sequence unique

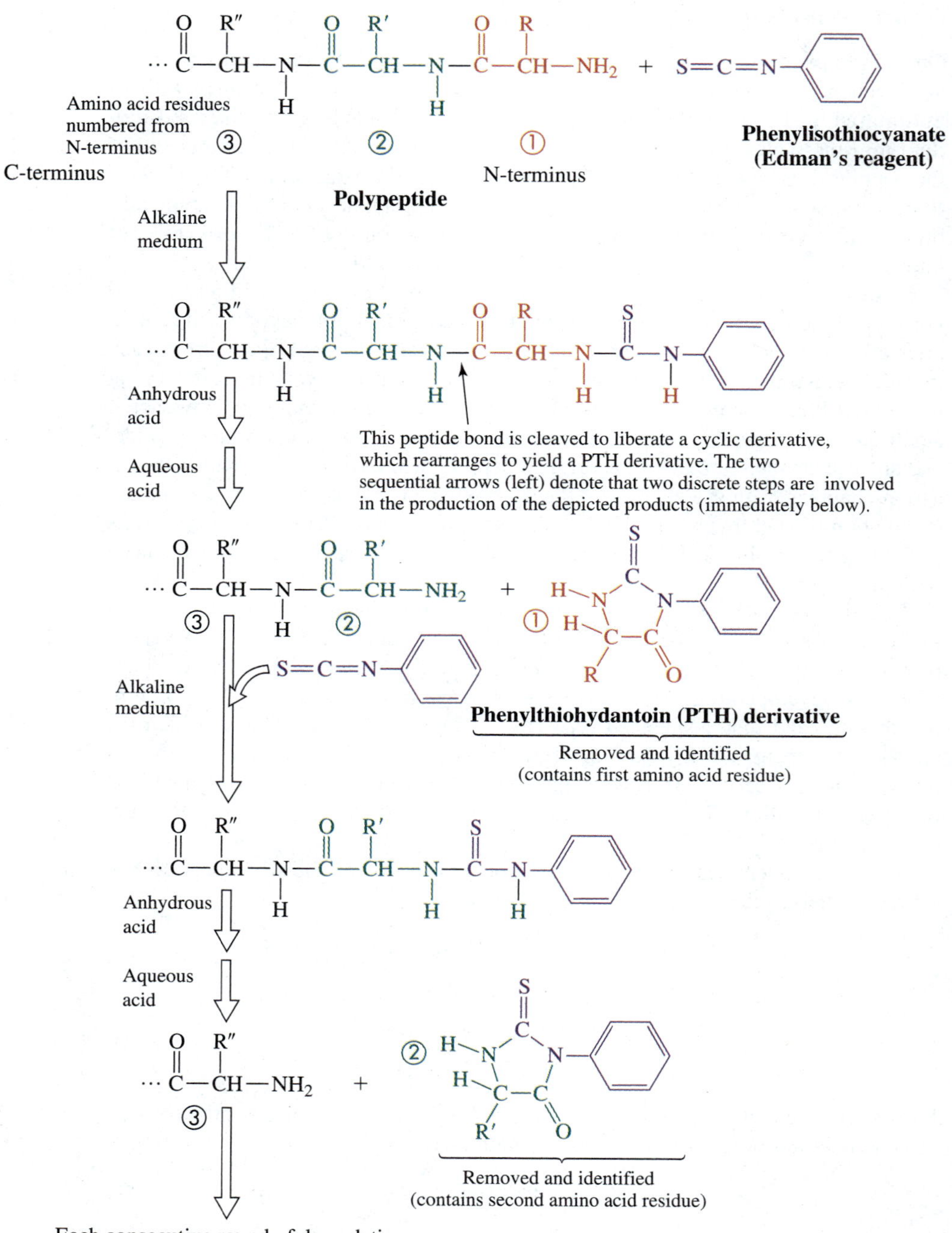

FIGURE 4.2
Edman degradation.

to the C-terminal end of peptide c. Similar comparisons of amino acid sequences in peptides produced under multiple digestion conditions allow us to determine unambiguously the location of peptide a in peptide X and the total amino acid sequence within any polypeptide. [Try Problem 4.5]

Other sequencing strategies also exist. Since the monomer sequence within a DNA molecule is more easily determined than the amino acid sequence of a polypeptide, polypeptide sequences are often read from those DNA sequences that code for their production. The details of this approach are examined during the discussion of the genetic code in Section 16.5. There is one major drawback: we cannot identify those amino acid residues, if any, whose side chains have been chemically modified after they have been incorporated into the polymer chain. Such chemical modifications can have a major impact on the structure and function of a protein (Sections 4.10, 5.13, and 18.11).

The primary structure has already been determined for thousands of proteins isolated from many species. The number of sequenced human proteins will increase extremely rapidly over the next several years as a consequence of the **Human Genome Project (Initiative)** (Section 16.9), which will ultimately yield the monomer sequence of the entire **human genome** (all human DNA). Rapid progress is also being made in the sequencing of several other genomes. The accumulating protein and DNA sequence information is being stored and studied at computerized data banks around the world. Such studies are providing valuable information on structure–function relationships in proteins, the storage and expression of genetic information, molecular evolution, and other topics. Since hemoglobins from different animals have been found to possess unique primary structures, electrophoretic mobilities, and crystalline forms, an analysis of hemoglobin on stone tools even allows archaeologists to determine what animals were killed or processed with these tools thousands of years ago.

PROBLEM 4.5 Given the following groups of peptides generated when peptide Y is digested under two different sets of cleavage conditions:

Group 1—From one cleavage reaction

Ile–Cys–Lys–Glu–Ser
Arg–Asp–Val–Leu–Val–Gly–Pro–Ser
Phe–Ala–Gly–Phe–Gly–Gly–Asp–Glu
Met–His–Tyr–Thr

Group 2—From a second cleavage reaction

Gly–Pro–Ser–Met
Gly–Asp–Glu
His–Tyr–Thr–Ile–Cys
Arg–Asp–Val–Leu–Val
Lys–Glu–Ser–Phe–Ala–Gly–Phe–Gly

What is the primary structure of peptide Y?

4.3 SECONDARY STRUCTURE

The amino acid chains within proteins do not remain in a fully extended state; they twist and coil into a variety of shapes. Ordered, local structural features known as **helices** and **β-sheets** (also called **β-pleated sheets**) are commonly generated. Collectively, these structural features make up the **secondary structure** of proteins. A protein lacking all such structural features is said to have no secondary structure. Fifty to 80% of the amino acid residues in most globular proteins are part of secondary structural elements, and virtually all of the amino acid residues in many fibrous proteins exist in helical or sheet regions.

Helices

The term ***helix*** refers to an ordered coiling of the amide planes in one or more polypeptide chains. If a single chain is involved, a structure is generated that looks much like a slinky or a metal spring. The major single-chained helices that have been observed in polypeptides and proteins are the **α-helix** (3.6 amino acid residues per turn), the **3_{10}-helix** (3 residues per turn) and the **π-helix** (4.4 residues per turn) (Figures 4.3). Each helix is associated with a unique and characteristic value for ϕ and ψ (Section 3.5). Each helix also possesses unique dimensions and a unique geometry (Table 4.3).

A novel helical structure, labeled a **parallel *β*-helix**, was discovered in a microbial enzyme in 1993. Its general significance, if any, remains to be determined. It will not be examined in this text.

All of the helices depicted in Figure 4.3 are said to be **right handed**. When viewed from either end, a right-handed helix spirals in a clockwise direction (as you follow it from the end nearest the observer to the more distant end) whereas a left-handed helix spirals in a counterclockwise direction. In a protein, right-handed helices are normally favored over left-handed ones, because left-handed coiling tends to generate considerable steric interference (hindrance). If the L-amino acid residues in a protein were replaced with D-amino acid residues, the left-handed helix would be favored. **[Try Problems 4.6 and 4.7]**

All of the single-chained helices observed within proteins are, to a large extent, maintained through hydrogen bonding between amide hydrogen atoms in one region of the helix and amide carboxyl groups in amide planes further along the helix. This is clearly illustrated in Figure 4.3. The α-helix is by far the most frequently observed

TABLE 4.3

Some Properties of the Major Polypeptide Helices

Helix	ϕ*	ψ*	Residues per Turn	Number of Atoms in Loops Created by H Bonding	Pitch (Rise [nm]/Turn)
α	−57	−47	3.6	13	0.54
3_{10}	−49	−26	3.0	10	0.60
π	−57	−70	4.4	16	0.52

*Values in degrees for right-handed helices.

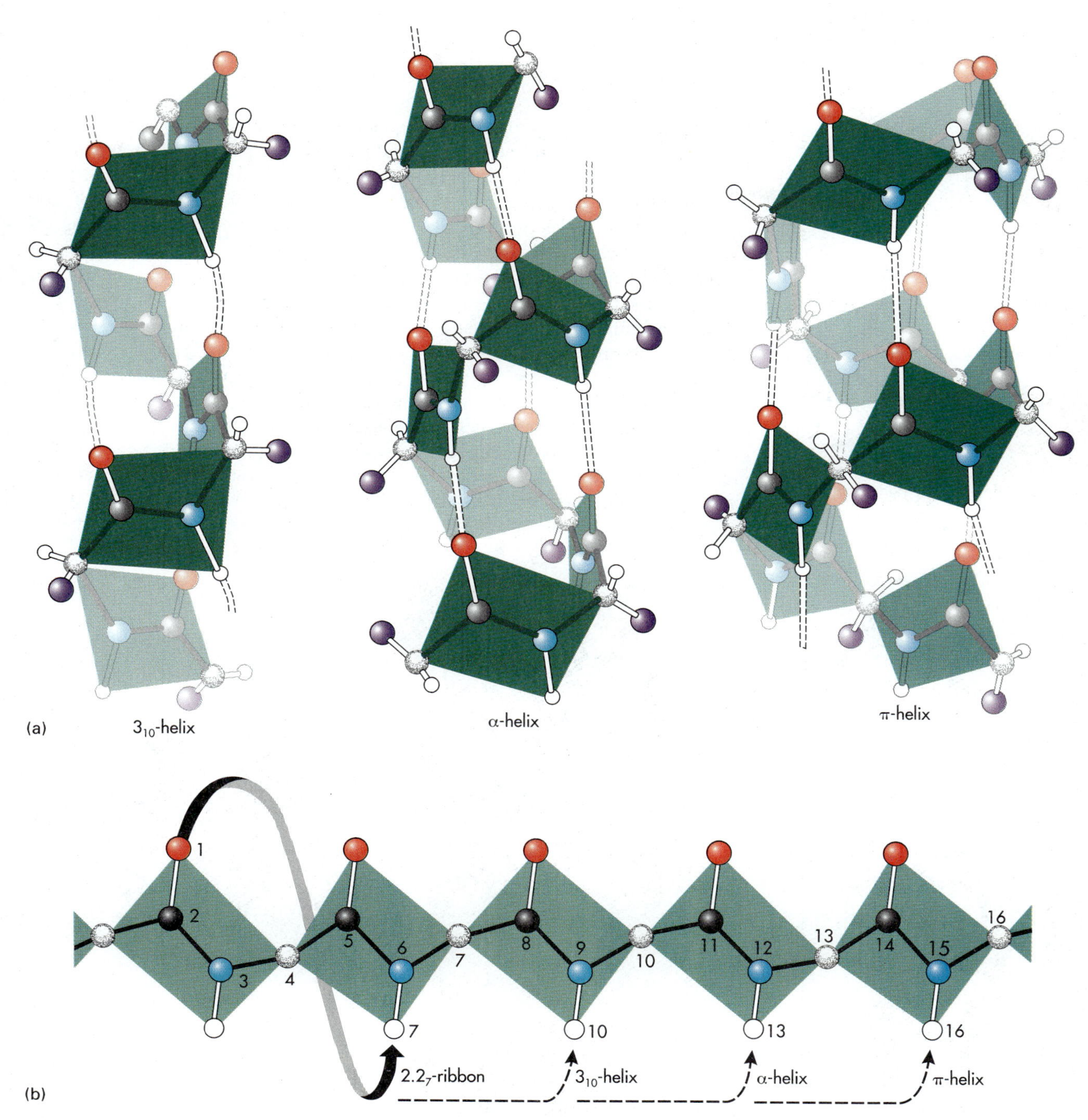

FIGURE 4.3

A 3_{10}-helix, an α-helix and a π-helix. (a) The 3^{10}-, α-, and π-helices differ in their patterns of hydrogen bonding, as shown in (b). Hydrogen bonds in the α-helix are particularly unstrained, making the α-helix especially stable. The α-carbons are stippled, with small attached spheres for hydrogens and larger spheres for side groups. (b) Hydrogen bonding pattern from one turn of the helix to the next, for four related helices. The principal number in the helix notation denotes the number of residues per turn and the subscript tells the number of atoms in the ring formed by closing the hydrogen bond. Thus the α-helix is called the 3.6_{13} helix (see numbering), and the π-helix, the 4.4_{16}-helix. © Irving Geis.

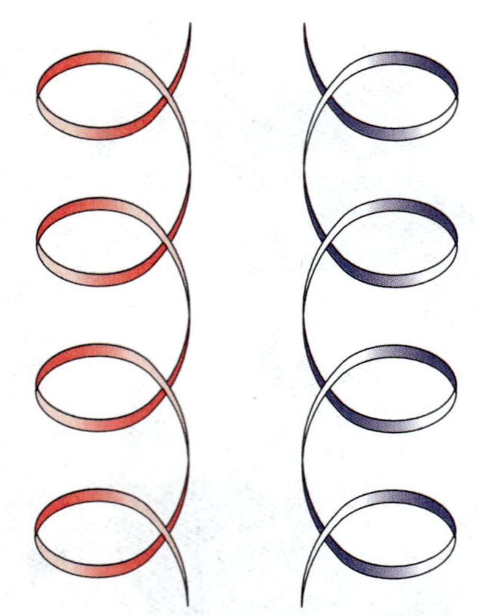

helix within globular proteins, because this coiling pattern leads to near-optimum H bonding and often leads to a stable environment for amino acid side chains. The H bonding is approximately parallel to the helix axis, and it encompasses all of the available amide hydrogens along the coiled region of the polypeptide backbone.

The side chains along an α-helix tilt toward the N-terminus and extend away from the cylinder created by the coiled amide planes. Certain sequences of side chains lead to steric interference, like-charge repulsions, or other destabilizing interactions that weaken an α-helix or prevent its formation initially. Proline, whose side chain is bonded to its α-nitrogen, is incompatible with an α-helix, unless it is one of the first three amino acid residues from the N-terminus of the helix.

The diagrams in Figure 4.3 and the **ball-and-stick models** commonly used to model the construction of molecules can create serious misconceptions about the structure of peptides. In both cases, the "balls" depict the approximate relative location of the nuclei of bonded atoms. They are not designed to portray the relative sizes of whole atoms. If this fact is not taken into account, the ball-and-stick representations give the impression that there is considerable empty space within peptides and most other molecules. The more expensive **space-filling models** avoid this problem by encompassing most of the volume occupied by a molecule and those atoms within it. A study of the space-filling model pictured in Figure 4.4 reveals that an

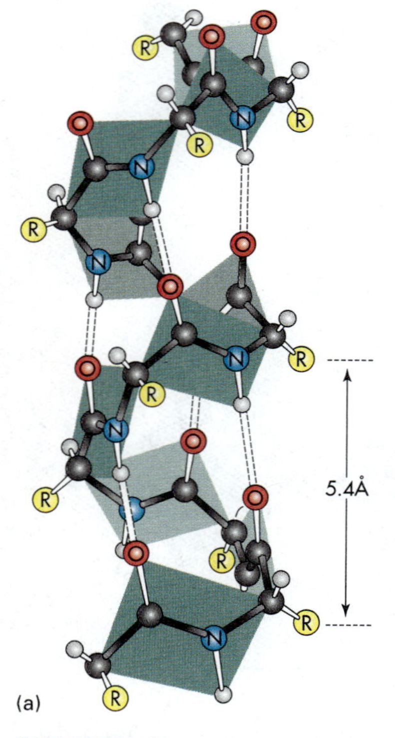

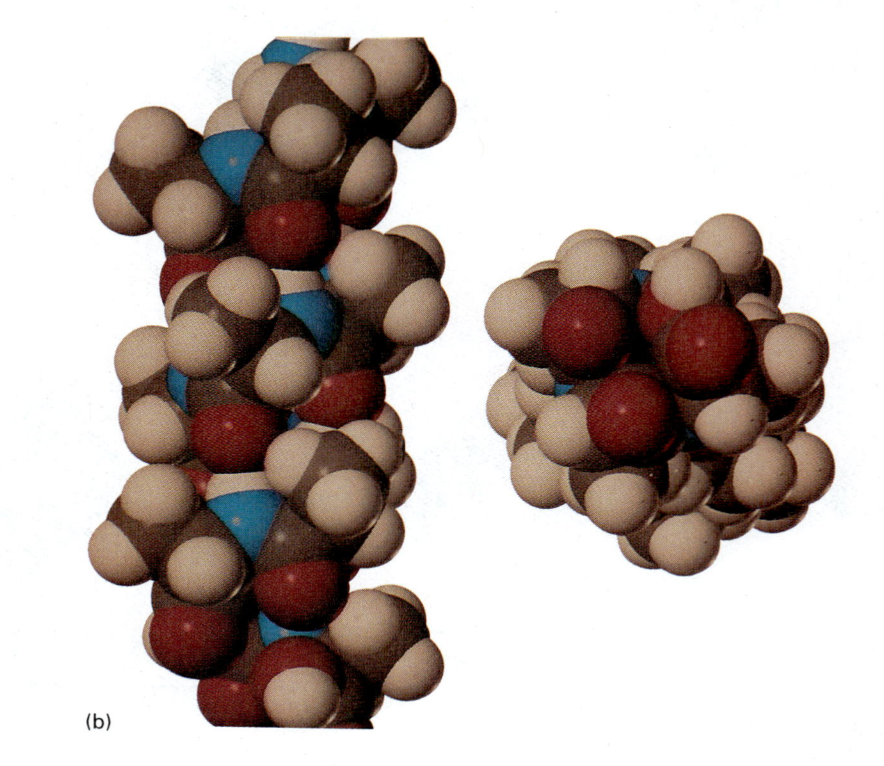

FIGURE 4.4
Space filling and ball-and-stick models of an α-helix. (a) Side view of a ball-and-stick model. R's represent side chains. Double dashed lines (====) depict H bonds. Amide planes are clearly identified. © Irving Geis. (b) Side view (left diagram) and top view (right diagram) of a space-filling model of polyalanine.

α-helix is a tightly packed, relatively rigid structural element. Three-dimensional visualization of α-helices on computer screens can provide similar structural information. The unoccupied volume in the core of an α-helix is much too small to accommodate even a water molecule. Tight yet nonhindered packing is a stabilizing factor within proteins. α-Helices are an important part of the structural backbone within those proteins in which they reside.

Sheets

Sheets represent the second half of the secondary structure story. The term *β-pleated sheet* refers to a sheetlike arrangement constructed by placing multiple segments of almost fully extended peptide chains, known as **β-strands**, side by side. In globular proteins, the β-strands are usually different segments of the same polypeptide, and most are five to ten amino acid residues in length.

Sheets can be either **parallel** or **antiparallel** (Figure 4.5). In parallel sheets, all β-strands run in the same direction. In an antiparallel sheet, immediately adjacent

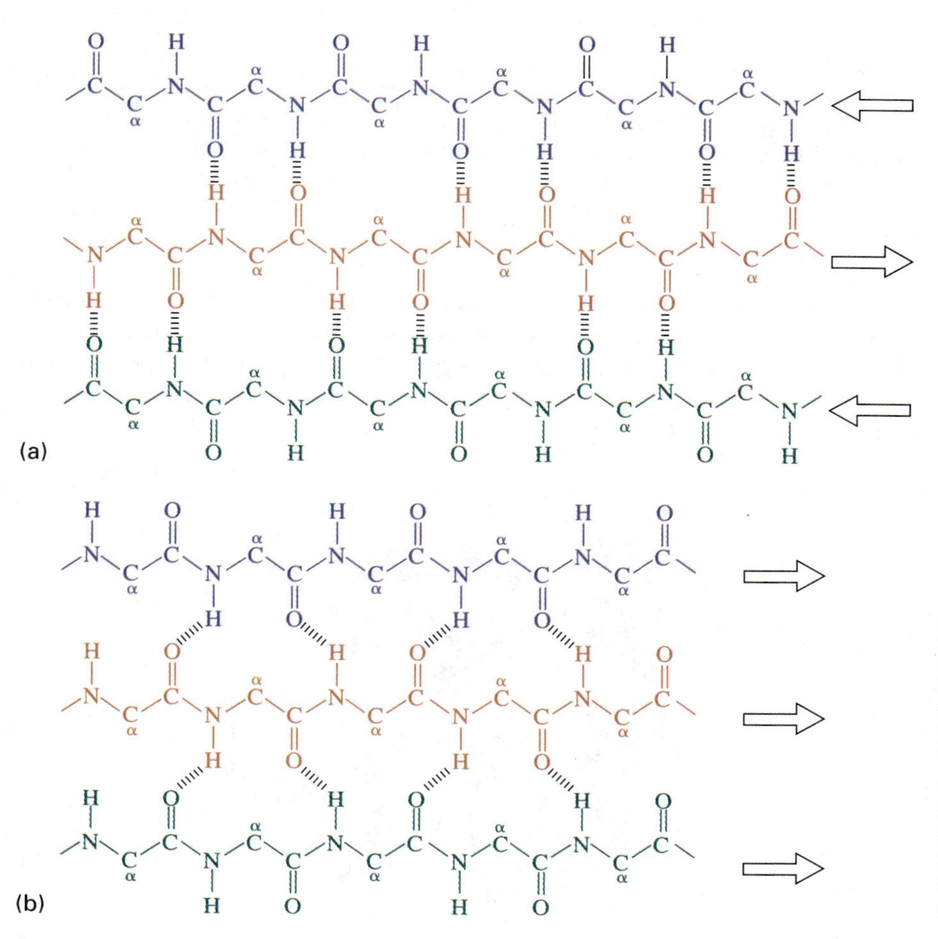

FIGURE 4.5
Parallel and antiparallel β-pleated sheets. (a) An antiparallel β-pleated sheet containing three β-strands. Each α-carbon is bonded to a hydrogen and a side chain that are not shown. Hashed lines (IIIIII) represent H bonds. (b) A parallel β-pleated sheet containing three β-strands. In both (a) and (b), the arrows to the right point towards the C-terminus of each β-strand.

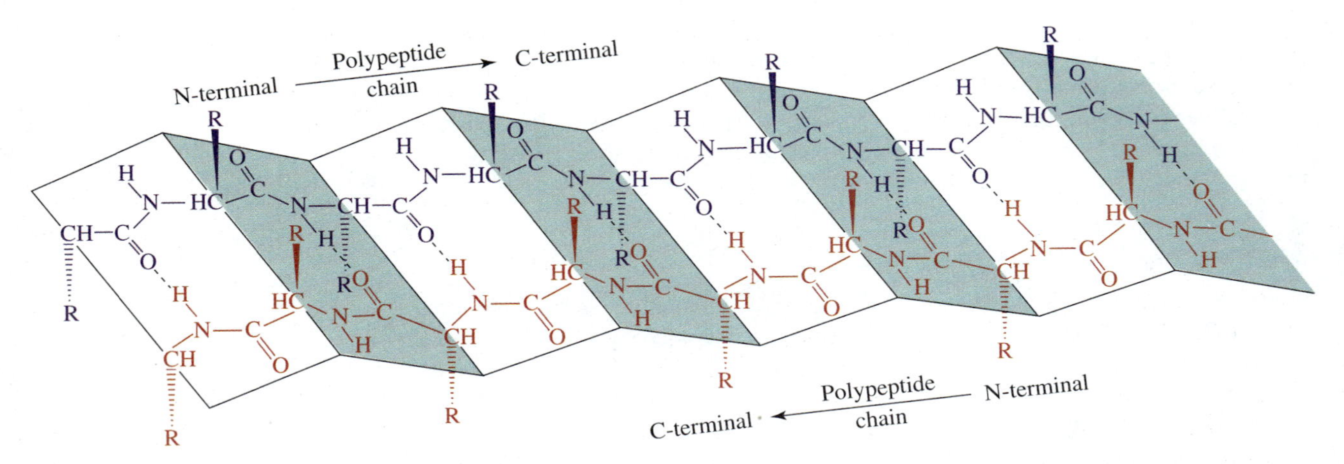

FIGURE 4.6
A three-dimensional diagram of an antiparallel β-pleated sheet. The dashed lines (----) are H bonds between β-strands. R's represent side chains. Along each β-strand, adjacent side chains are on opposite sides of the sheet. Based on Brown and Rogers (1987).

β-strands run in opposite directions, one N-terminal to C-terminal and the other C-terminal to N-terminal. An occasional protein contains **mixed β-sheets** in which some adjacent β-strands are parallel and others are antiparallel. In all β-sheets, the separate β-strands are held together through H bonds between amide hydrogen atoms and amide oxygens. Although a single H bond is fairly weak, large numbers of H bonds cooperate to make both sheets and helices relatively stable structural elements.

Although the two-dimensional drawings in Figure 4.5 depict β-sheets as smooth, flat surfaces, the three-dimensional diagram in Figure 4.6 clearly illustrates that this is not, in reality, the case. The backbone of β-sheets contains a regular pattern of pleats or ripples. In addition, virtually all of the β-sheets embedded in globular proteins are twisted or curled, not flat (Figure 4.7). In Figure 4.7, flat arrows, running

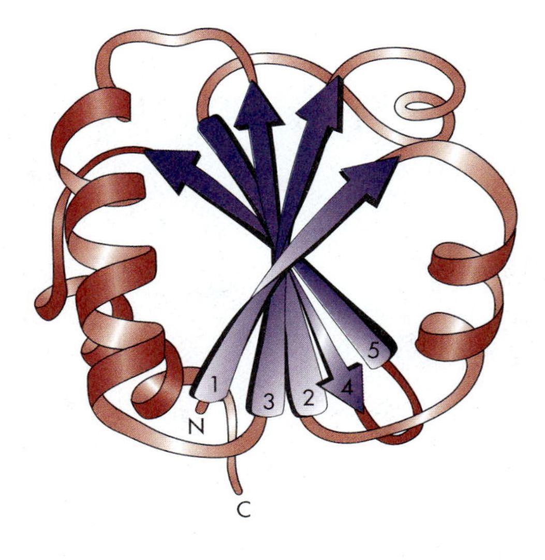

FIGURE 4.7
A twisted β-sheet in thioredoxin from *E. coli.* Thioredoxin contains a single polypeptide chain; the N marks its N-terminal and the C its C-terminal. Flat arrows identify β-strands, coiled ribbons identify α-helices, and ropes depict segments of the polymer that exist outside of secondary structural elements. The five-stranded β-sheet is markedly twisted. Based on Branden and Tooze (1991).

N-terminal to C-terminal, identify β-strands, coiled ribbons represent α-helices; and ropes depict segments of polymer that exist outside of secondary structural elements. Such **schematic diagrams** will be used throughout this text. The protein depicted in Figure 4.7 contains a single polypeptide strand whose N-terminal and C-terminal are labeled.

Along a single β-strand, each side chain is on the opposite side of the β-sheet from its two immediate neighbors. Although this arrangement eliminates side chain interactions along individual β-strands, it does not prevent steric interference or other interactions between side chains on adjacent β-strands. Steric interference between bulky side chains, repulsions between like-charged side chains, and other interactions can weaken a β-sheet or prevent its formation initially.

Loops and Bends

In order for a polypeptide chain to fold into a globular shape, a significant number of its amino acid residues must exist in bends and loops. A bend in which a polymer chain abruptly reverses its direction is described as a **reverse turn**, **β-bend**, or **tight turn**. Tight turns commonly involve four sequential amino acid residues and are often stabilized by hydrogen bonds (Figure 4.8). Some textbooks treat certain turns,

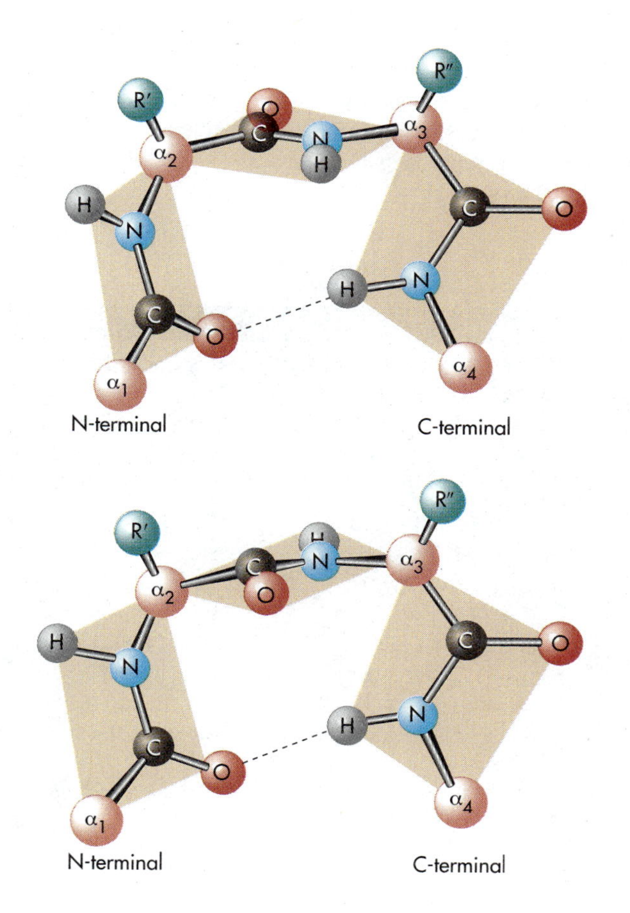

FIGURE 4.8
Examples of tight turns. R's represent side chain. Dashed lines (----) identify H bonds. Based on Garrett and Grisham (1995).

loops, and other geometrical arrangements as secondary structural elements. This practice, which is not universally accepted, is not followed here. **[Try Problems 4.8 and 4.9]**

Primary Structure Determines Secondary Structure

The amino acid sequence within a polypeptide determines what secondary structural elements will be formed along its polymer chain. Some sequences of side chains fold into α-helices, other sequences form β-sheets or tight turns, and still others lead to additional geometric arrangements. By comparing the primary and the secondary structures of many proteins, it is possible to calculate the frequency with which any given amino acid residue appears in each of the secondary structural elements. The results of such an analysis are summarized in Table 4.4.

Proline is one of the amino acids found infrequently in both helices and sheets. This is explained by the lack of a hydrogen on the nitrogen in an internal proline residue (limits H bonding possibilities) and by the restricted rotation about the α-carbon (because of the ring system in proline). The presence of proline commonly initiates a bend or a loop.

The preferences tabulated in Table 4.4 are not strong enough to allow us routinely and confidently to predict secondary structure from primary structure. This is partially explained by the impact of neighboring amino acid residues on each other's tendency to participate in the formation of secondary structural elements. We can more confidently predict the secondary structure associated with a novel primary structure if we can identify similar amino acid sequences within polypeptides of known structure. Such an approach is described as **modeling by homology**. **[Try Problems 4.10 through 4.13]**

TABLE 4.4

A Summary of the Amino Acid Residues Found Most Often and Least Often Within α-Helices, β-Sheets, and Tight Turns

Secondary Structural Element	Eight Most Common Amino Acid Residues[a]	Eight Least Common Amino Acid Residues[b]
α-Helix	Glu Met Ala Leu Lys Phe Gln Trp	Gly Pro Asn Tyr Cys Ser Thr Arg
β-Sheet	Val Ile Tyr Trp Phe Leu Cys Thr	Glu Asp Pro Ser Lys Gly Ala Asn
Tight turn	Gly Asn Pro Asp Ser Cys Tyr Lys	Ile Val Met Leu Phe Ala Glu Trp

[a]The amino acid found most often is listed first, followed by the amino acid that occurs with the second greatest frequency, etc.

[b]The amino acid found least often is listed first, followed by the amino acid that occurs with the next to the lowest frequency, etc.

PROBLEM 4.6 How many amino acid residues exist within the α-helix in Figure 4.3(a)? How many complete turns does this helix contain?

PROBLEM 4.7 Identify the N-terminal and the C-terminal end of each helix in Figure 4.3(a).

PROBLEM 4.8 Circle each amino acid residue in Figure 4.8.

PROBLEM 4.9 Identify the α-helices, β-sheets, and tight turns in Figure 4.7.

PROBLEM 4.10 List those amino acids, in addition to proline, that are seldom found in either sheets or helices but are very common in tight turns.

PROBLEM 4.11 Polyglutamic acid (a polypeptide containing only glutamic acid residues) spontaneously coils into an α-helix at pH 1 but not at pH 7. Explain.

PROBLEM 4.12 Will the following amino acid sequence more likely reside within a sheet region or a helical region within a protein (assuming that it exists in one or the other)? Explain.

Ala–His–Glu–Leu–Met–His–Glu–Leu–Lys–Phe–Ala

***PROBLEM 4.13** Define steric hindrance (interference; strain). Explain how steric hindrance can prevent a peptide fragment from becoming part of a secondary structural element.

4.4 TERTIARY STRUCTURE

The structural hierarchy of a protein moves from secondary structure to tertiary structure. **Tertiary structure** refers to the overall three-dimensional arrangement of a polypeptide chain, including the folding of secondary structural elements with respect to one another. Although there are an enormous number of ways in which a polypeptide chain can fold, there is usually only one folding pattern that leads to a biologically active (**native**) molecule. The native folding pattern may be the thermodynamically most stable one, or it may be a higher energy conformation more accessible kinetically (Figure 4.9). If the time required to reach the most stable folding pattern is long, a polypeptide will be unable to attain this conformation during its limited lifespan. Since it is the sequence of amino acids that ultimately directs and determines the folding, one should, in theory, be able to predict tertiary structure from primary structure. So far, however, biochemists have failed in their attempts to construct reliable predictive algorithms. Most of those attempting to do so operate on the assumption that the native structure is the thermodynamically most stable one.

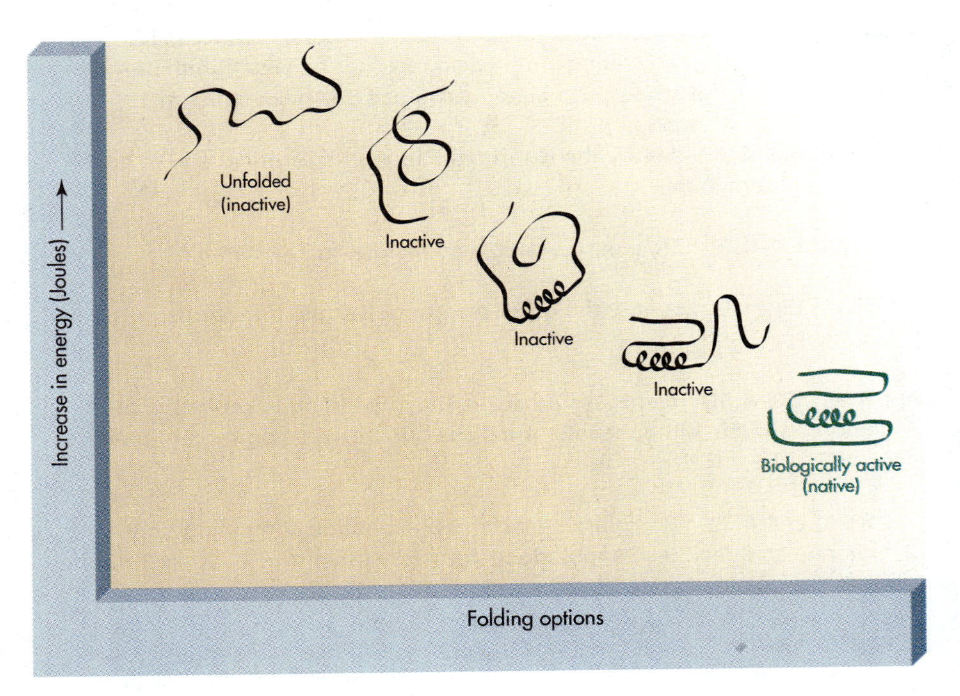

FIGURE 4.9
Energetics of protein folding. Generally, it is believed that the thermodynamically most stable folding pattern corresponds to the biologically active (native) form of a protein.

A Rule for Folding

In an aqueous environment, the biologically active form of a globular protein is usually one in which the majority of its nonpolar amino acid side chains have been folded inside and its polar side chains coat the surface. This general folding rule (summarized as **nonpolar in, polar out**) makes sense when we recall that like associates with like, a major biochemical theme (Exhibit 1.4). Table 4.5 divides amino acid residues into internal, external, and variable on the basis of the polarity of their side chains. The variable residues have side chains that are compatible with either the hydrophobic interior or the hydrophilic exterior. Unless indicated otherwise, it should be assumed that the proteins encountered in this text primarily exist and function in an aqueous environment. **[Try Problems 4.14 and 4.15]**

TABLE 4.5

The Classification of Amino Acids on the Basis of Their Usual Location Within Globular Proteins

Location	Amino Acid Residues
Internal	Ile Leu Met Phe Val
External	Arg Asn Asp Gln Glu His Lys
Variable	Ala Cys Gly Pro Ser Thr Trp Tyr

The Maintenance of Tertiary Structure

The tertiary structure of a globular protein is predominantly maintained through hydrophobic forces, H bonds, salt bridges, and disulfide bonds (Sections 2.1 and 2.2). One example of each of these types of interactions is provided in Figure 4.10. Although not shown in Figure 4.10, van der Waals interactions also help maintain tertiary structure.

Disulfide bonds are generated through the oxidative coupling of two cysteine side chains. Hydrophobic forces normally encompass most, if not all, of a protein's nonpolar side chains; they are largely responsible for the aggregation of nonpolar side chains within the core of a typical globular protein (the nonpolar in, polar out rule).

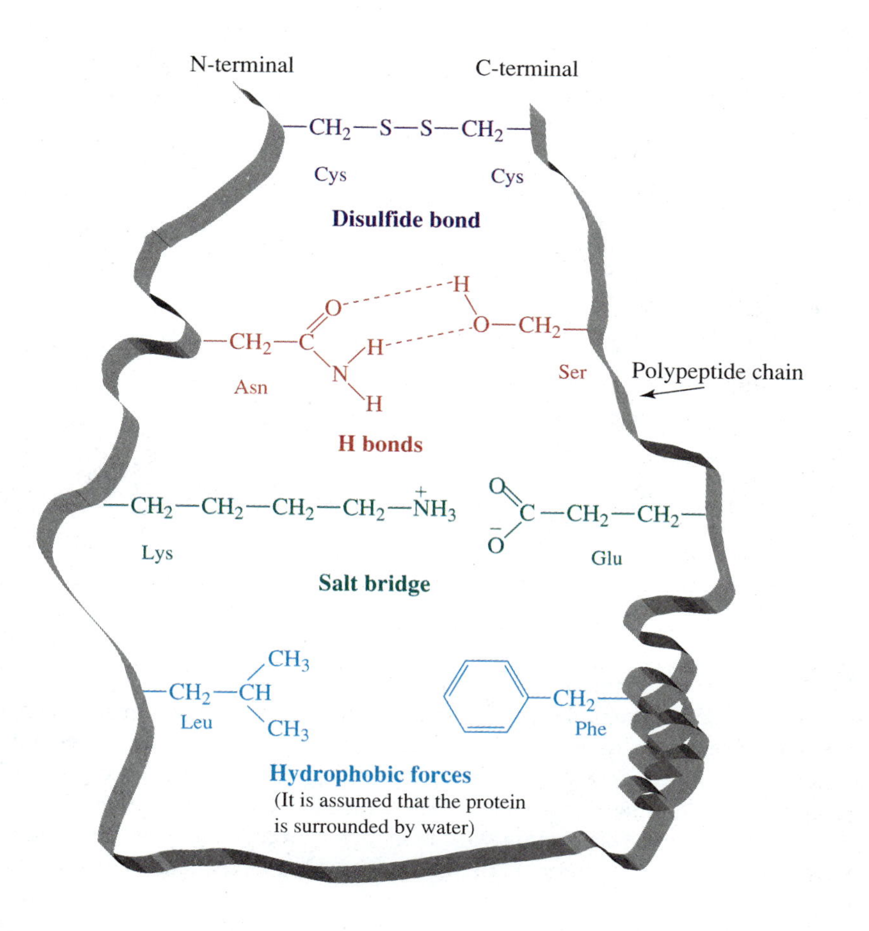

FIGURE 4.10
An illustration of the major classes of bonds that maintain the tertiary structure of proteins.

A salt bridge is formed when a positively charged side chain is attracted to a negatively charged side chain, a common occurrence in proteins. Most salt bridges are located at or near the surface, and they may involve N-terminal α-amino groups or C-terminal α-carboxyl groups as well as charged side chains. Each peptide bond and polar side chain contains hydrogen, oxygen, or nitrogen atoms that are able to participate in H bonding. Although H bonds are most prevalent on the surface of globular proteins, they can also be tolerated internally. Since H bonding tends to neutralize the partial charges that reside on the bonded atoms, an α-helix with its extensive network of hydrogen bonds can be completely buried within the nonpolar core of a globular protein. **[Try Problem 4.16]**

Within most proteins, **disulfide bonds** are the only covalent bonds that help maintain tertiary structure, and they are readily cleaved by mild reducing agents. Since the cytoplasm of a cell tends to house a reductive environment, disulfide bonds are less common in intracellular proteins than in extracellular ones. Because they are covalent, disulfide bonds are relatively strong, and they often play a key role in helping extracellular proteins maintain a native structure in potentially hostile environments. Copying nature, industrial biochemists commonly enhance the stability of proteins by increasing the number of disulfide bonds within them.

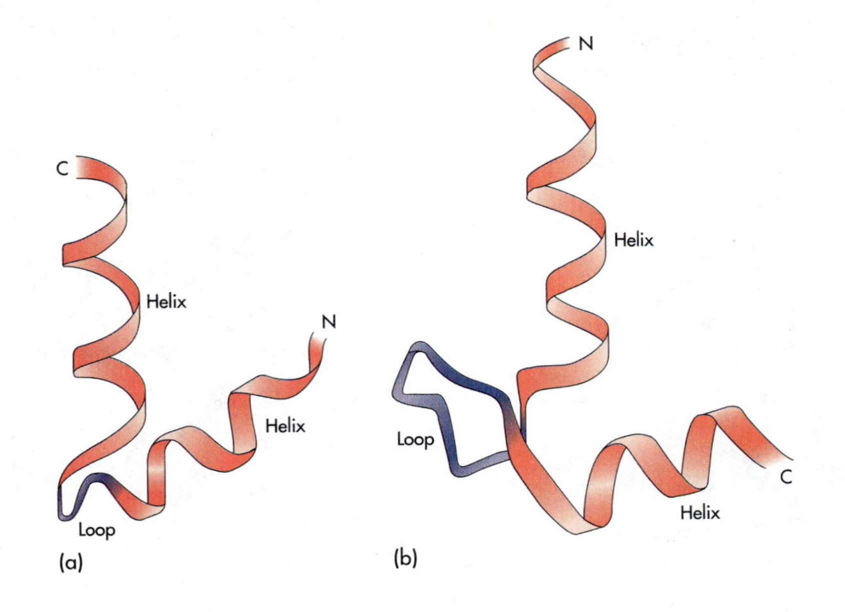

FIGURE 4.11
Two helix–turn–helix motifs. (a) A DNA-binding motif. (b) A calcium-binding motif. Based on Branden and Tooze (1991).

Motifs

Certain secondary structural elements fold together in a very similar manner in many different proteins. Such recurring geometric arrangements, called **supersecondary structures** or **motifs**, constitute part of the tertiary structure of those proteins in which they reside. A **helix–turn–helix** (also called **helix–loop–helix**) **motif** consists of two α-helices and a connecting short loop region arranged in a specific geometry. A **DNA-binding motif** (common in proteins that bind DNA) and a **calcium-binding motif** (found in multiple calcium-binding proteins) provide specific examples (Figure 4.11). When a β-sheet folds upon itself to create a cylinder, the resultant motif is called a **β-barrel**. β-barrels commonly possess six to eight β-strands. The association of a β-barrel with α-helices can generate an **α/β-barrel** (Figure 4.12). The α/β-barrel depicted in Figure 4.12 contains eight parallel β-strands, and it has been detected in well over a dozen different proteins. The α-helical regions that lie between the individual β-strands are packed around the outside of the barrel in a specific geometry. A wide variety of other motifs have also been characterized, including some with intriguing names such as **Greek keys, zinc fingers,** and **leucine zippers.** Part of these will be examined in future chapters. **[Try Problems 4.17 and 4.18]**

Domains

A **domain** is a relatively stable, independently folded region within the tertiary structure of a globular protein. Individual polypeptide chains possess from one to several dozen domains. Each domain may encompass one or more motifs. In a multidomain polypeptide, each domain contains a single segment of polypeptide that will usually maintain its unique folding pattern in the absence of the remainder of the polymer chain. Domains having more than 30% of their amino acid sequence in common normally adopt the same folding pattern. Distinctly folded domains nor-

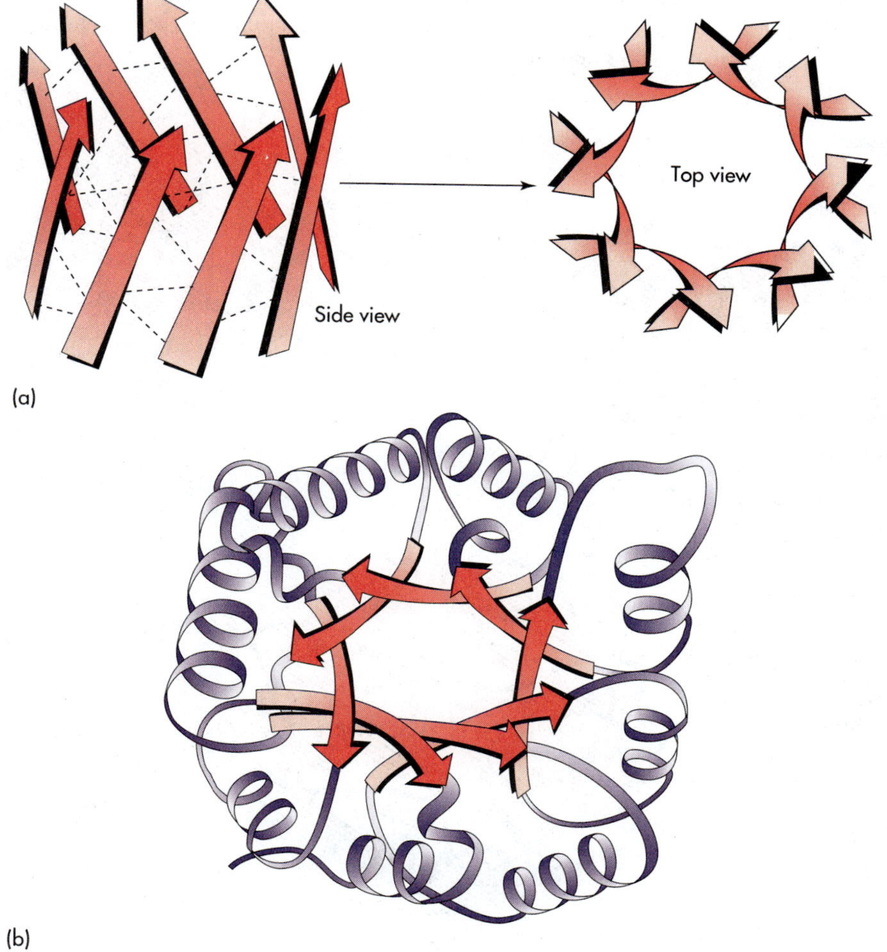

FIGURE 4.12
A β-barrel and α/β-barrel. (a) A β-barrel containing eight parallel β-strands. (b) The α/β-barrel in triose phosphate isomerase. Based on Hein et al. (1993).

mally serve different functions, whereas structurally related domains tend to serve similar functions, even when present in different proteins. Evolution commonly leads to novel proteins by selecting unique combinations of preexisting domains, a key ingredient in the major theme of evolution, diversity, and change. **[Try Problem 4.19]**

Domains are divided into multiple classes on the basis of their secondary structural elements: **α-domains** contain α-helices, **β-domains** possess β-sheets, and **α/β-domains** encompass both helices and sheets. The secondary structural elements within a domain commonly fold into one or more motifs that collectively constitute its core. The loop regions and complex coilings that connect secondary structural elements tend to surround the hydrophobic core and add details to the outer regions of tertiary structure. These loops and coils often contain those amino acid residues that make up the binding sites and catalytic sites commonly found within domains.

In 1968 the tertiary structure was known for only eight proteins. Today, that number is in the thousands. Figure 4.13 provides schematic diagrams for several proteins and protein domains. **[Try Problem 4.20]**

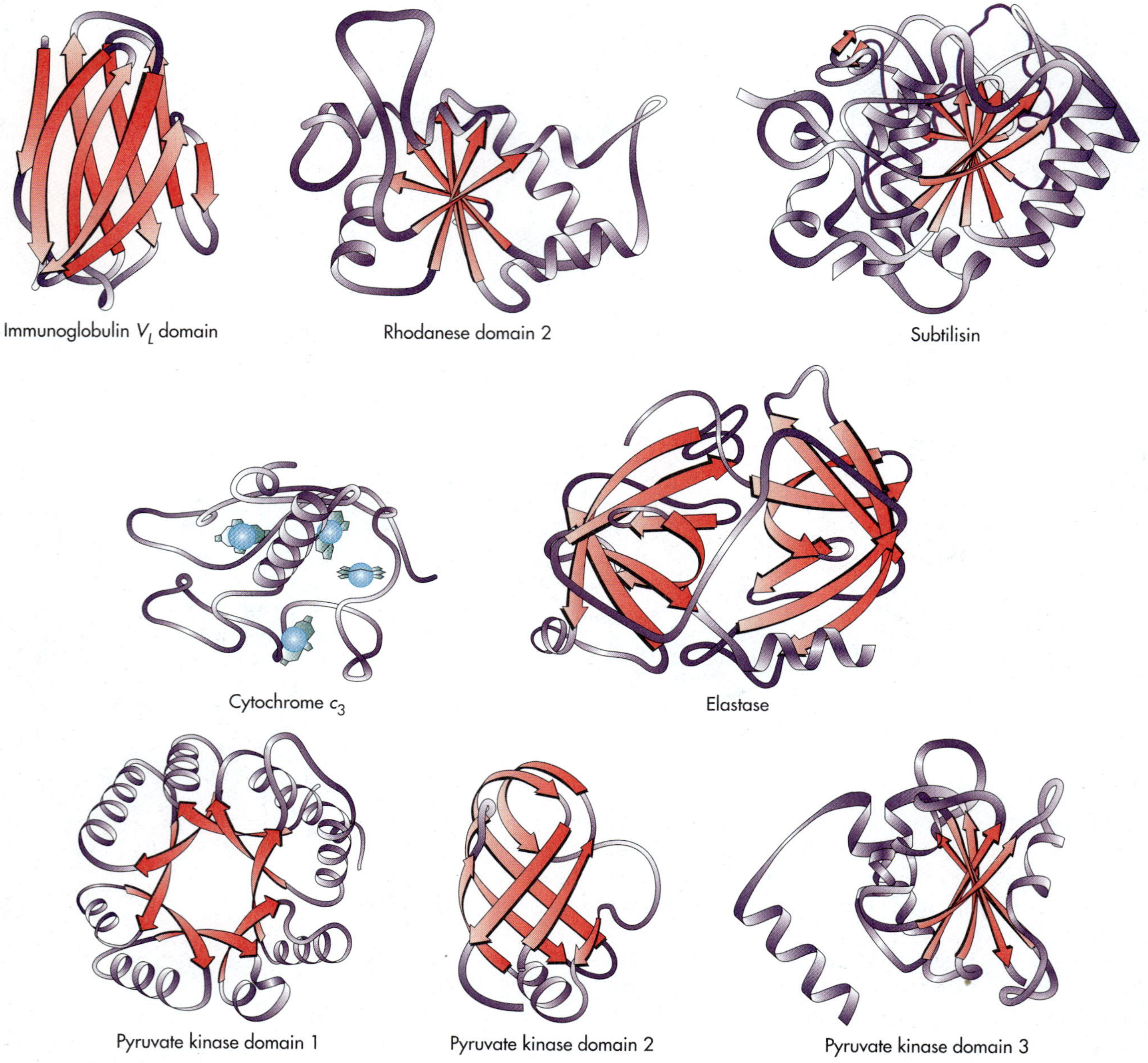

Immunoglobulin V_L domain

Rhodanese domain 2

Subtilisin

Cytochrome c_3

Elastase

Pyruvate kinase domain 1

Pyruvate kinase domain 2

Pyruvate kinase domain 3

FIGURE 4.13
Schematic diagrams of some proteins and protein domains. Each domain contains a single polypeptide chain. Coiled ribbons represent helical regions, flat arrows depict β-strands, and ropes identify segments of polymer that do not exist within secondary structural elements. The four satellite-like objects in cytochrome C_3 are prosthetic groups that help maintain tertiary structure. Based on Zubay (1983).

Chaperones

For many years it was believed that polypeptides spontaneously folded into their native forms during or immediately after their assembly from amino acids. It is now known, however, that protein molecules called **chaperones** sometimes participate in this process. Chaperones of the **Hsp70 family** (Hsp is an acronym for **heat-shock proteins**, proteins synthesized in above normal amounts after a cell is exposed to elevated temperatures or certain other stresses) function by binding to specific regions along polypeptide chains and, in the process, blocking off-pathway folding reactions that lead to inactive and thermodynamically less stable (than native form) tertiary structures or to inactive polypeptide aggregates (Figure 4.14). **Hsp60 chaperones** appear to enclose unfolded polypeptides in a protected environment and, in the process, promote the on-pathway folding that generates native tertiary structures. Hsp60 proteins are known as **chaperonins** and are most often found in bacteria, chloroplasts, and mitochondria. Since chaperones do not appear to contain any in-

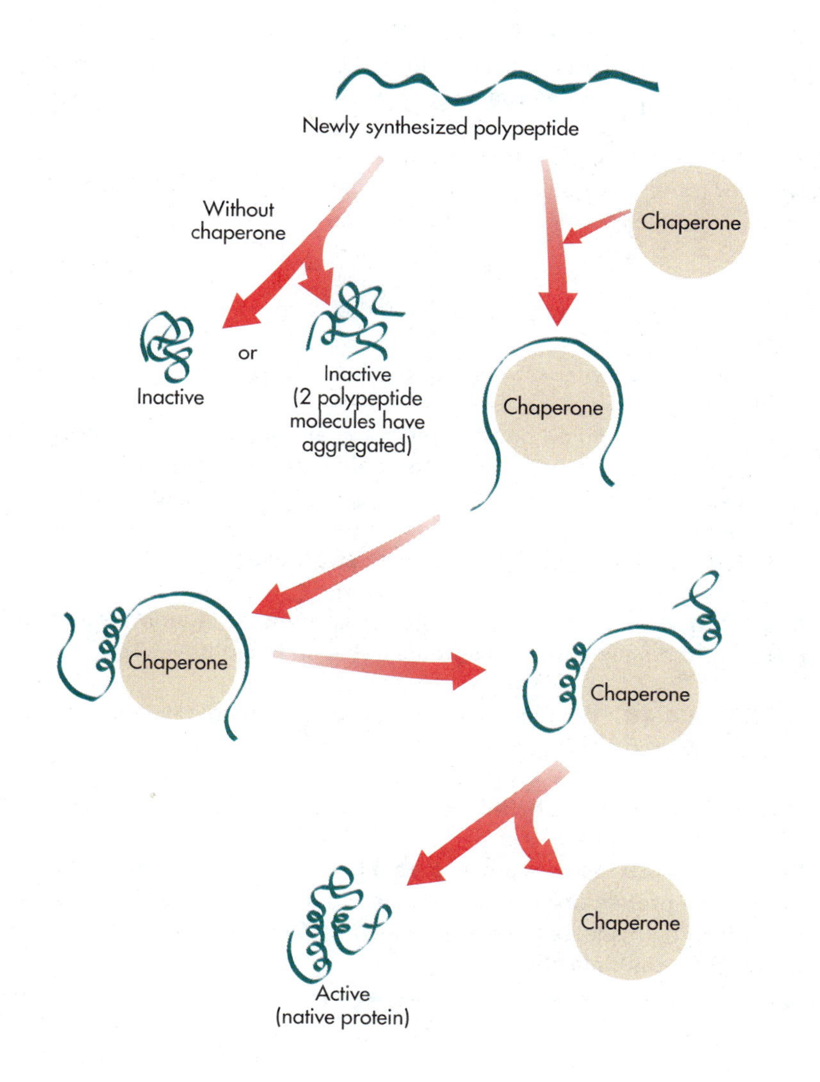

FIGURE 4.14
Chaperone blocking of off-pathway folding.

formation used to determine the ultimate conformation of a protein, they are said to assist in self-folding. Chaperones commonly require energy supplied by adenosine triphosphate (ATP), and there is growing evidence that some function while polypeptide chains are still in the process of elongation. In certain instances, additional proteins are also involved in the attainment of native folding patterns. These include enzymes responsible for disulfide bond formation and others that catalyze cis–trans isomerization within the amide planes of particular proline residues (Section 3.5).

A subset of chaperones plays a central role in the transduction of certain signals within cells. These chaperones appear to participate in the assembly and maintenance of "poised" receptors whose ligands function as signaling agents. The steroid hormones (Section 10.3) are some of the signaling molecules whose activities are linked to chaperone-associated receptors. Signal transduction is examined in detail in Chapter 10.

PROBLEM 4.14 The rule "nonpolar in, polar out," assumes that a protein exists in an aqueous environment. What is the folding rule for a protein in a nonpolar environment? Explain.

PROBLEM 4.15 In an aqueous environment, will the following peptide fragment more likely be buried inside a globular protein or located on its surface? Explain.

Leu–Ala–Gly–Pro–Phe–Ser–Val

Note: In reality, part could be buried and part could be on the surface.

PROBLEM 4.16 Draw structural formulas illustrating:

a. Hydrogen bonding between histidine and glutamic acid side chains.
b. Hydrogen bonding between threonine and serine side chains.
c. A salt bridge between an aspartic acid side chain and an arginine side chain.
d. Hydrophobic interactions between tryptophan and valine side chains.

PROBLEM 4.17 What is the approximate number of amino acid residues in each of the helix–turn–helix motifs depicted in Figure 4.11?

PROBLEM 4.18 What is the approximate bond angle between the two α-helices in each motif in Figure 4.11?

PROBLEM 4.19 List the major differences between domains and motifs.

PROBLEM 4.20 How many separate helices and sheets exist within each protein and protein domain depicted in Figure 4.13? When counting sheets, watch for single sheets folded or twisted in such a manner that they can be mistaken for multiple sheets.

4.5 QUATERNARY STRUCTURE

Quaternary structure is at the top of the ladder in the structural hierarchy of globular proteins. **Quaternary structure** refers to the manner in which separate polypeptide chains pack together in those proteins, called **oligomers**, that contain multiple noncovalently coupled polypeptide chains (Figure 4.15). Although most very large proteins are oligomers, there are well-documented examples of single-chain giants. In mammals, the synthesis of some fatty acids is performed by a single polypeptide chain containing around 2300 amino acid residues and seven domains. In *E. coli*, the same process involves seven distinct proteins, each corresponding to one of the seven mammalian domains. No one is certain why nature uses multiple domains within a single polypeptide chain in some instances, yet calls upon oligomers or separate proteins to perform similar functions in other cases. **[Try Problems 4.21 and 4.22]**

Quaternary structure is predominantly determined and maintained by some of the same forces that determine and maintain tertiary structure: hydrogen bonds, hydrophobic forces, and salt bridges. By convention, proteins in which distinct polypeptide chains are held together by disulfide bonds are said to have no quaternary structure since the polypeptide chains are covalently linked. In most cases, the native, biologically active quaternary structure of an oligomer probably represents the thermodynamically most stable packing for its **subunits**. Protein chaperones are commonly involved in the assembly process. Quaternary structure, like secondary and tertiary structure, is ultimately determined by primary structure.

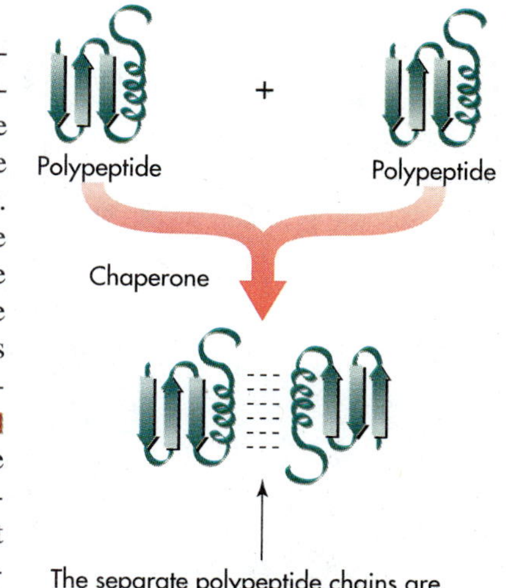

FIGURE 4.15 Quaternary structure is the manner in which separate polypeptide chains are packaged together in those proteins that contain multiple noncovalently coupled polypeptide chains.

PROBLEM 4.21 What term is used to describe an oligomer that contains two polypeptide chains? Four polypeptide chains?

PROBLEM 4.22 What is the minimum number of domains in an oligomer containing four subunits? Explain.

4.6 DENATURATION

Collectively, the secondary, tertiary, and quaternary structures of a globular protein constitute its **conformation**. The natural, biologically active form of a protein represents its **native conformation**. A globular protein that has been totally unfolded without breaking any covalent bonds is said to be **denatured**. A denatured protein is biologically inactive, and it is usually less ordered and less water soluble than its native counterpart. A protein that has been partly unfolded is said to be partially denatured. It may or may not retain some biological activity. **[Try Problem 4.23]**

Since the native conformation of a protein is largely maintained through hydrogen bonds, hydrophobic forces, and salt bridges, any treatment that weakens or ruptures such bonds can lead to **denaturation** (unfolding; Figure 4.16). These treatments include heat, extremes in pH, and organic solvents (Exhibit 4.1). Partial denaturation and an associated loss of biological activity can result from even slight changes in a protein's environment. To a large extent, this explains why relatively small shifts in blood pH can be lethal (Section 2.5) and why high fevers can lead to neurological damage and even death. The discussion of enzymes in Chapter 5

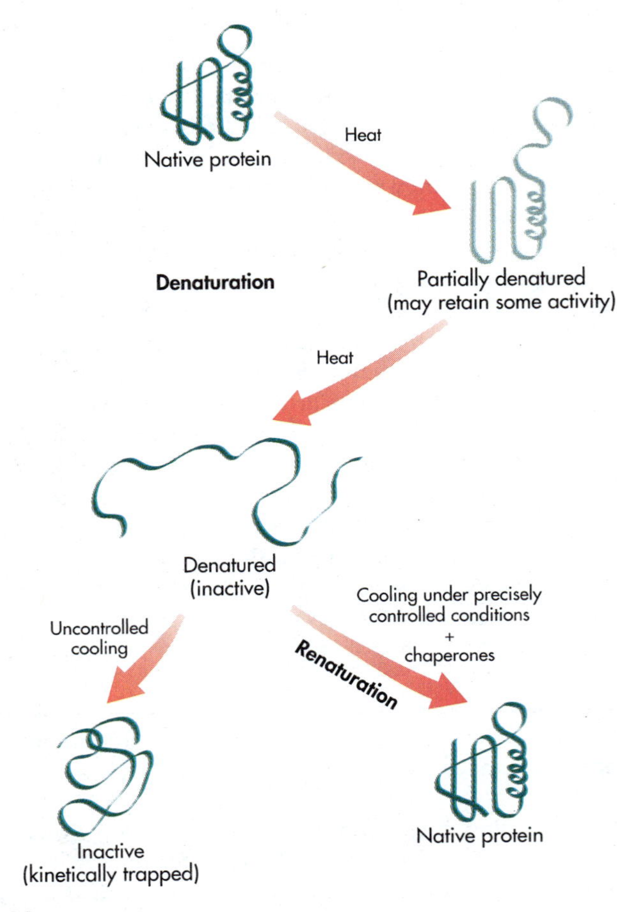

FIGURE 4.16
Denaturation and renaturation.

EXHIBIT 4.1
Some Treatments and Conditions That Denature Most Globular Proteins[a]

- Extremes in pH
- High temperature
- Organic solvents
- High salt concentrations
- Detergents
- Mechanical agitation

[a]Because different proteins commonly exhibit markedly different sensitivities to denaturation, it is not possible to generalize about the precise conditions under which a protein will become completely unfolded. Some proteins, for example, completely unfold at much lower temperatures than others. Some are much more sensitive to pH changes, etc.

emphasizes the ease with which most globular proteins can be inactivated through denaturation. [Try Problem 4.24]

In some instances, the human body takes advantage of the conformational sensitivity of proteins. Dietary proteins, for example, are exposed to rather harsh conditions in the stomach to unfold them and make their peptide bonds more vulnerable to attack by digestive enzymes. Many eukaryotic cells, including human cells, use relatively small differences in lysosomal and cytosolic pH to reduce the activity of lysosomal enzymes that may leak into the cytosol (Section 2.5).

Since the native conformation of a protein is programmed into its primary structure, denaturation (which does not modify amino acid sequence) should, in theory, be a reversible process. In practice, it is often difficult to find those conditions under which a denatured protein will refold into its native state, a process described as **renaturation.** Proteins tend to refold into inactive conformations that, even when less stable than the native conformation, are converted to the native form at a negligible rate (Figure 4.16). A protein in such an inactive state is said to be **kinetically trapped** (Note: kinetics and rates of reactions are reviewed in Chapter 5). The appropriate use of molecular chaperones may enable biochemists to overcome this problem.

PROBLEM 4.23 Why does the unfolding of a native protein tend to decrease its water solubility?

PROBLEM 4.24 Which of the classes of bonds that maintain the conformations of proteins are most sensitive to disruption by:

a. High pH values?
b. High concentrations of ethanol?
c. High temperatures?

Explain.

4.7 STRUCTURE–FUNCTION RELATIONSHIPS AND EVOLUTION

No doubt about it: proteins are architecturally complicated. The number of possible conformational variations is so enormous that it is difficult to comprehend. However, only a minute fraction of this potential structural diversity appears to be utilized by living organisms; a growing body of evidence indicates that it will be possible to place most individual proteins into one of 1000 to 2000 families on the basis of their folding patterns. Not all of the folding families are equally populated; over 30% of the roughly 3000 proteins in the protein structure database in 1994 adopt one of nine basic folding patterns. Although similar folding patterns do not necessarily imply that two proteins are related, those with similar folds usually have kindred sequences or functions as well. According to one hypothesis, most of the proteins within a folding family have diverged from a single ancestor or a small number of ancestors. This proposal is consistent with the widely held belief that all present-day organisms have evolved from the same ancient cell.

Over time, mutations and evolutionary selection shape and reshape a protein in response to biological needs and changing environments. In the process, a molecule emerges that is optimized to occupy a specific functional niche. The tight coupling between structure and function is documented by the hundreds of genetic diseases that are linked to changes in single amino acid side chains within specific proteins (Chapter 19). The next four sections explore the topic of structure–function relationships by examining some of the proteins that play central roles within the human body. In the process, three specific examples are used to illustrate the structural hierarchy within globular proteins.

4.8 MYOGLOBIN

Myoglobin is the conjugated protein responsible for the red color of steak. Its primary function is to bind and store O_2 in muscle, and its high concentration in the muscle of diving mammals accounts for their ability to remain under water for prolonged periods. Based on its biological role, we would expect myoglobin to have evolved into a container that selectively binds O_2 when it is present at high concentrations, and then releases it when surrounding O_2 levels drop and the bound oxygen is needed by a cell. How accurate is this prediction? This question will be answered by examining sperm whale myoglobin, rather than human myoglobin, since this abundant protein has been more thoroughly studied than its closely related human counterpart.

The Structure of Myoglobin

Sperm whale myoglobin contains a single polypeptide chain plus a single iron-containing **heme group** (Figure 4.17). Although the heme group is predominantly hydrophobic in nature, it does contain two polar, charged (at physiological pH values) propanoic (propionic) acid side chains ($-CH_2CH_2COO^-$). In this instance, the term *side chain* refers to any group attached to the planar, aromatic **porphyrin ring** at the heart of the heme group. **[Try Problem 4.25]**

The primary structure of myoglobin is the order in which its 153 amino acids are joined within its single polypeptide chain (Figure 4.18). The secondary structure of myoglobin consists of 8 helical regions that are almost perfect α-helices. These helical regions are labeled with the capital letters A through H, starting at the N-terminus. They range in length from 7 to 24 residues and account for 118 of myoglobin's 153 amino acids. Myoglobin contains no β-sheets.

The tertiary structure of myoglobin consists of a single domain, known as a **globin fold**, which is created by folding the eight helical regions together to form a pocket or box for the heme group (Figure 4.19). This native conformation encompasses multiple bends and loops. Hydrophilic amino acid side chains are distributed over the outer surface. Hydrophobic side chains are buried inside and line the heme pocket. The proposal that most globular proteins fold according to the rule "nonpolar in, polar out" was initially based on theoretical considerations plus a careful

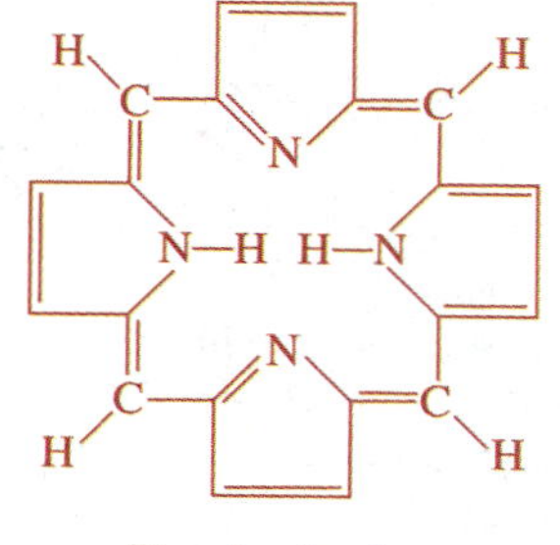

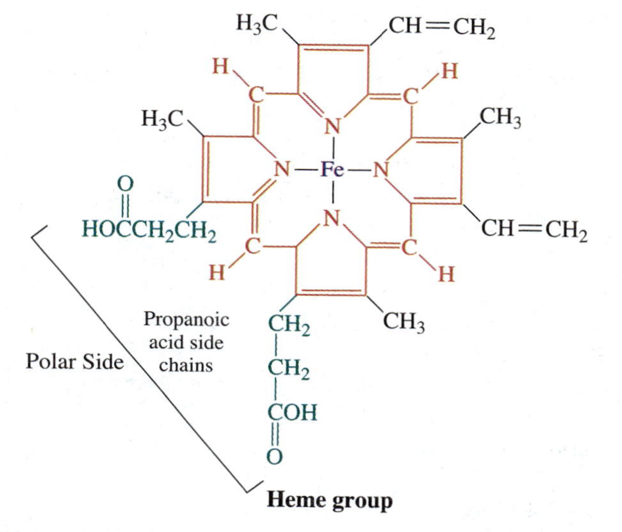

FIGURE 4.17
The porphyrin ring and the heme group. The propanoic acid side chains are predominantly dissociated at physiological pH values.

study of the structure of myoglobin and a limited number of other proteins. Although some H bonds and salt bridges do exist, the tertiary structure of myoglobin is maintained primarily through hydrophobic interactions. There are no disulfide bonds. Myoglobin's overall dimensions are approximately 4.4 by 4.4 by 2.5 nm. **[Try Problems 4.26 and 4.27]**

Myoglobin has no quaternary structure since it contains a single polypeptide chain. A protein must be an oligomer to have quaternary structure.

Oxygen Binding by Myoglobin

The heme group fits into the heme pocket in such a manner that its two charged carboxyl groups extend out of the pocket and interact with external, polar amino acid side chains and surrounding water molecules. A histidine side chain on one side of the porphyrin ring binds to the Fe^{2+} in the center of the heme group. In **oxymyo-**

Val–	Leu–	Ser–	Glu–	Gly–	Glu–	Trp–	Gln–	Leu–	Val–	10
NA1	NA2	A1	A2	A3	A4	A5	A6	A7	A8	
Leu–	His–	Val–	Trp–	Ala–	Lys–	Val–	Glu–	Ala–	Asp–	20
A9	A10	A11	A12	A13	A14	A15	A16	AB1	B1	
Val–	Ala–	Gly–	His–	Gly–	Gln–	Asp–	Ile–	Leu–	Ile–	30
B2	B3	B4	B5	B6	B7	B8	B9	B10	B11	
Arg–	Leu–	Phe–	Lys–	Ser–	His–	Pro–	Glu–	Thr–	Leu–	40
B12	B13	B14	B15	B16	C1	C2	C3	C4	C5	
Glu–	Lys–	Phe–	Asp–	Arg–	Phe–	Lys–	His–	Leu–	Lys	50
C6	C7	CD1	CD2	CD3	CD4	CD5	CD6	CD7	CD8	
Thr–	Glu–	Ala–	Glu–	Met–	Lys–	Ala–	Ser–	Glu–	Asp–	60
D1	D2	D3	D4	D5	D6	D7	E1	E2	E3	
Leu–	Lys–	Lys–	His–	Gly–	Val–	Thr–	Val–	Leu–	Thr–	70
E4	E5	E6	E7	E8	E9	E10	E11	E12	E13	
Ala–	Leu–	Gly–	Ala–	Ile–	Leu–	Lys–	Lys–	Lys–	Gly–	80
E14	E15	E16	E17	E18	E19	E20	EF1	EF2	EF3	
His–	His–	Glu–	Ala–	Glu–	Leu–	Lys–	Pro–	Leu–	Ala–	90
EF4	EF5	EF6	EF7	EF8	F1	F2	F3	F4	F5	
Gln–	Ser–	His–	Ala–	Thr–	Lys–	His–	Lys–	Ile–	Pro–	100
F6	F7	F8	F9	FG1	FG2	FG3	FG4	FG5	G1	
Ile–	Lys–	Tyr–	Leu–	Glu–	Phe–	Ile–	Ser–	Glu–	Ala–	110
G2	G3	G4	G5	G6	G7	G8	G9	G10	G11	
Ile–	Ile–	His–	Val–	Leu–	His–	Ser–	Arg–	His–	Pro–	120
G12	G13	G14	G15	G16	G17	G18	G19	GH1	GH2	
Gly–	Asn–	Phe–	Gly–	Ala–	Asp–	Ala–	Gln–	Gly–	Ala–	130
GH3	GH4	GH5	GH6	H1	H2	H3	H4	H5	H6	
Met–	Asn–	Lys–	Ala–	Leu–	Glu–	Leu–	Phe–	Arg–	Lys–	140
H7	H8	H9	H10	H11	H12	H13	H14	H15	H16	
Asp–	Ile–	Ala–	Ala–	Lys–	Tyr–	Lys–	Glu–	Leu–	Gly–	150
H17	H18	H19	H20	H21	H22	H23	H24	HC1	HC2	
Tyr–	Gln–	Gly								153
HC3	HC4	HC5								

FIGURE 4.18 The primary structure of sperm whale myoglobin. The amino acids are listed from the N-terminal residue to the C-terminal residue. The label below each residue refers to its position in an α-helical region or a nonhelical region. B4 is the fourth residue in the B helix, EF7 is the seventh residue in the nonhelical region between the E and F helices, etc.

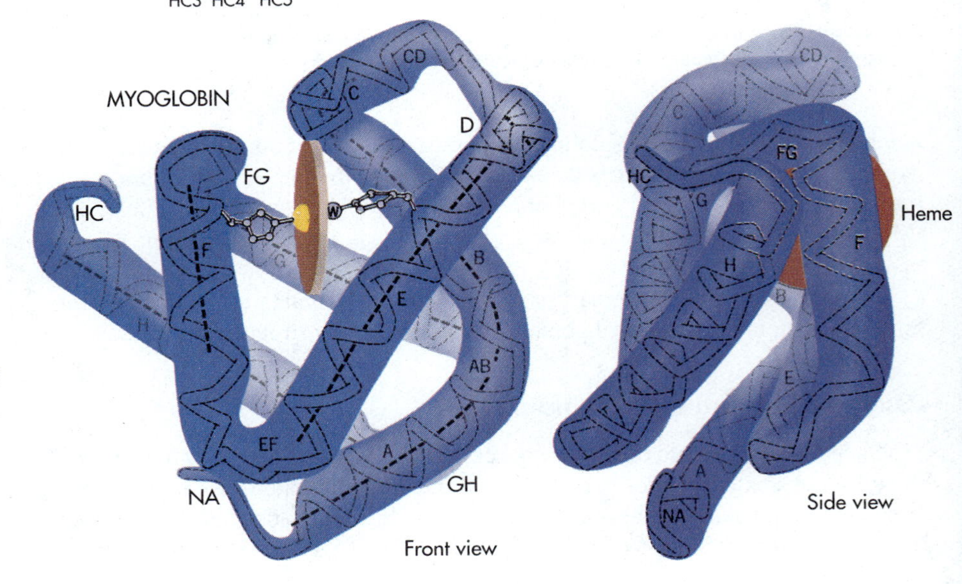

FIGURE 4.19 The secondary and tertiary structures of myoglobin. The myoglobin molecule is built up from eight stretches of α-helix that form a box for the heme group: helices E and F form the walls; helices B, G, and H are the floor; and the CD corner closes the open end. The rope outlines the course of the main chain. The front view clearly shows the interaction of the heme group with two histidine side chains, one to the left and one to the right. Ⓦ identifies the site where O_2 is bound within oxymyoglobin. © Irving Geis.

FIGURE 4.20
A space-filling model for myoglobin. The purple atoms are atoms in the heme group.

globin (myoglobin bound to oxygen), an O_2 molecule on the opposite side of the porphyrin ring binds to the same octahedrally coordinated Fe^{2+}. That O_2 is sandwiched between the Fe^{2+} and a second histidine side chain within the heme pocket (Figure 4.19). In **deoxymyoglobin** (myoglobin with no bound O_2), the O_2 binding site is vacant. The heme group, which is primarily bound by hydrophobic interactions, can be reversibly removed from its pocket. Figure 4.20 shows a space-filling model for myoglobin. **[Try Problem 4.28]**

Myoglobin's structure is definitely consistent with initial expectations. The polypeptide portion of this protein provides a box with an environment in which the heme group can selectively and reversibly bind O_2. The resultant binding curve (see Figure 4.24, described in the following section) leads to the acquisition of O_2 when it is present at high concentrations and to its release when O_2 levels are low. In the absence of the polypeptide chain, the binding of O_2 by the heme group would not be as selective (many other substances would also bind to the heme), the O_2 binding curve would be shifted from its physiologically optimal position, and the Fe^{2+} in the heme group would be rapidly oxidized to Fe^{3+}. An oxidized (Fe^{3+}) heme is unable to bind O_2. Even with a polypeptide wrap, the iron in the heme group of myoglobin is occasionally oxidized. The resultant nonfunctional molecule, known as **metmyoglobin**, can be reactivated by enzymes that catalyze the reduction of its iron to a 2+ oxidation state.

PROBLEM 4.25 What does it mean to say that the porphyrin ring system is planar? Is the heme group as a whole planar? Explain.

PROBLEM 4.26 How many bends or loops exist within a myoglobin molecule? How many residues exist within the smallest bend or loop? The largest? How many of the bends and loops are created, in part at least, by proline residues?

PROBLEM 4.27 Is there any folding pattern that would allow disulfide bonds to be formed within myoglobin? Explain.

***PROBLEM 4.28** Given the following equation for the binding of O_2 by myoglobin (abbreviated Mb):

$$Mb + O_2 \rightleftharpoons MbO_2$$

Write the equilibrium constant expression for this reaction. When myoglobin is chemically modified so that its affinity for O_2 increases, will the equilibrium constant for this reaction increase, decrease, or remain the same? Explain. If more O_2 is added to a reaction mixture at equilibrium, will the equilibrium constant increase, decrease, or remain the same? Explain.

4.9 HEMOGLOBIN

Hemoglobin is the second protein selected to illustrate protein structure and structure-function relationships. The hemoglobin story, like the myoglobin story, is tightly linked to oxygen and energy production. Each cell in the human body must have a constant supply of energy to survive. Most of this energy comes from the oxidation of glucose (Chapter 6) and other fuels. This oxidation requires O_2 and generates CO_2 as depicted in the following equation:

$$\text{Glucose} + 6O_2 \rightleftharpoons 6CO_2 + 6H_2O + \textbf{energy}$$

Most of the O_2 for energy production is carried from the lungs to cells throughout the body by hemoglobin, a transport protein found only within **red blood cells (erythrocytes)**. The CO_2 produced enters the blood stream, where a portion of it binds to hemoglobin for transport to the lungs. Every time you exhale, CO_2 is eliminated from your body. Every time you inhale, your lungs are filled with a fresh supply of O_2.

Based on its functions, we would predict that hemoglobin, like myoglobin, would have evolved into a selective container for O_2, a container that picks up O_2 at the relatively high O_2 concentrations present in the lungs, and then continuously releases this O_2 as blood moves further and further away from the lungs to regions of lower and lower partial pressures of O_2. It is essential that the O_2 be distributed throughout the body rather than deposited at a limited number of sites. This requires that hemoglobin's O_2 binding curve be shaped differently than the O_2 binding curve for myoglobin. Myoglobin is designed to dump most of its O_2 at one site in response to relatively slight changes in O_2 concentration.

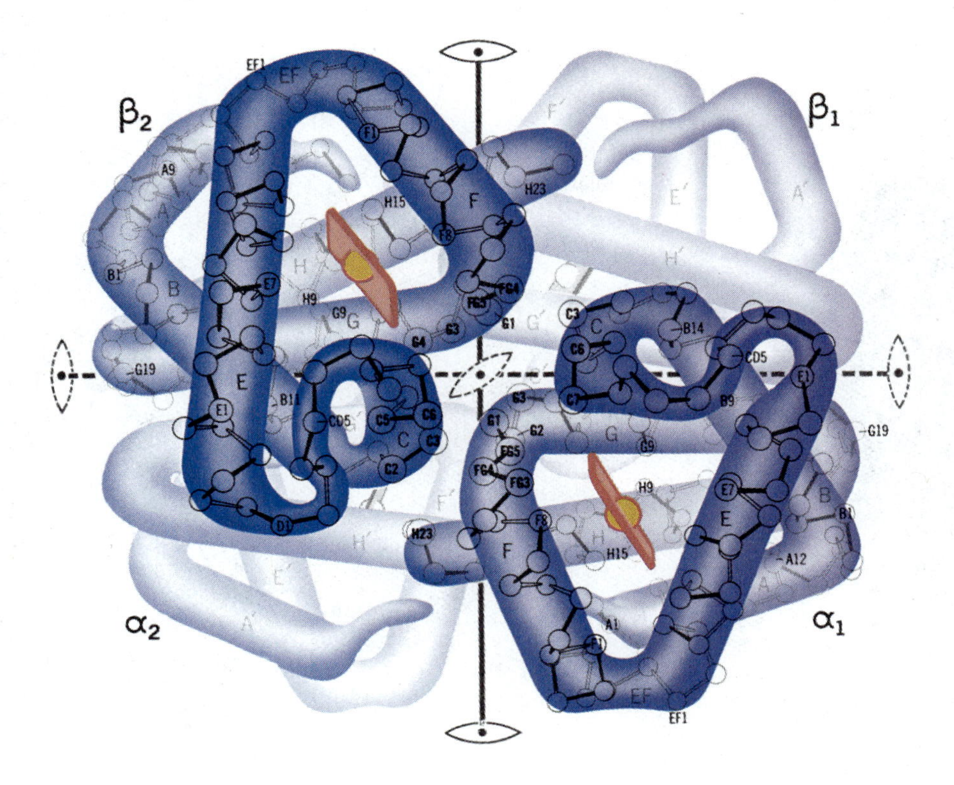

FIGURE 4.21
The secondary, tertiary, and quaternary structures of hemoglobin. The four subunits are labeled α_1, α_2, β_1, and β_2. Each amino acid residue is represented by a circle. Large capital letters identify helical regions. The single small capital letters associated with circles identify residues in helical regions. Pairs of small capital letters identify residues outside of helices. Residues in both helical and nonhelical regions are numbered from N-terminus to C-terminus. The rectangles with spheres in the middle depict heme groups. The solid vertical line is a true axis of symmetry. The dashed horizontal line is a pseudo-axis of symmetry. © Irving Geis.

The Structure of Hemoglobin

A single human hemoglobin molecule contains four separate polypeptide chains: two identical **α chains** and two identical **β chains** (Figure 4.21). The Greek letters α and β are used by biochemists for many purposes. In this instance, they are names given to two specific and unique polypeptide chains. Each α chain contains 141 amino acid residues that coil into 7 α-helical regions and then wrap into a globin fold domain. Similarly, the 146 amino acid residues in each β chain form 8 α-helical regions and a globin fold. The conformation and overall appearance of each α chain and each β chain are very similar to those of myoglobin. Like myoglobin, each α and β chain contains one heme group, multiple bends, no pleated sheets, and no disulfide bonds. Whereas myoglobin has hydrophilic amino acid side chains distributed more or less uniformly over its surface, each α and β chain of hemoglobin has a surface region that is rich in hydrophobic side chains. The quaternary structure of hemoglobin is generated when the hydrophobic regions in separate subunits pack together in a highly specific manner in order to escape from surrounding water molecules. Although hydrophobic forces primarily drive the packing of the subunits, hydrogen bonds and salt bridges account for approximately one third of the bonds that maintain quaternary structure. There are no disulfide bonds between subunits. The overall dimensions of a mammalian hemoglobin molecule are approximately 6.4 by 5.5 by 5.0 nm. **[Try Problem 4.29]**

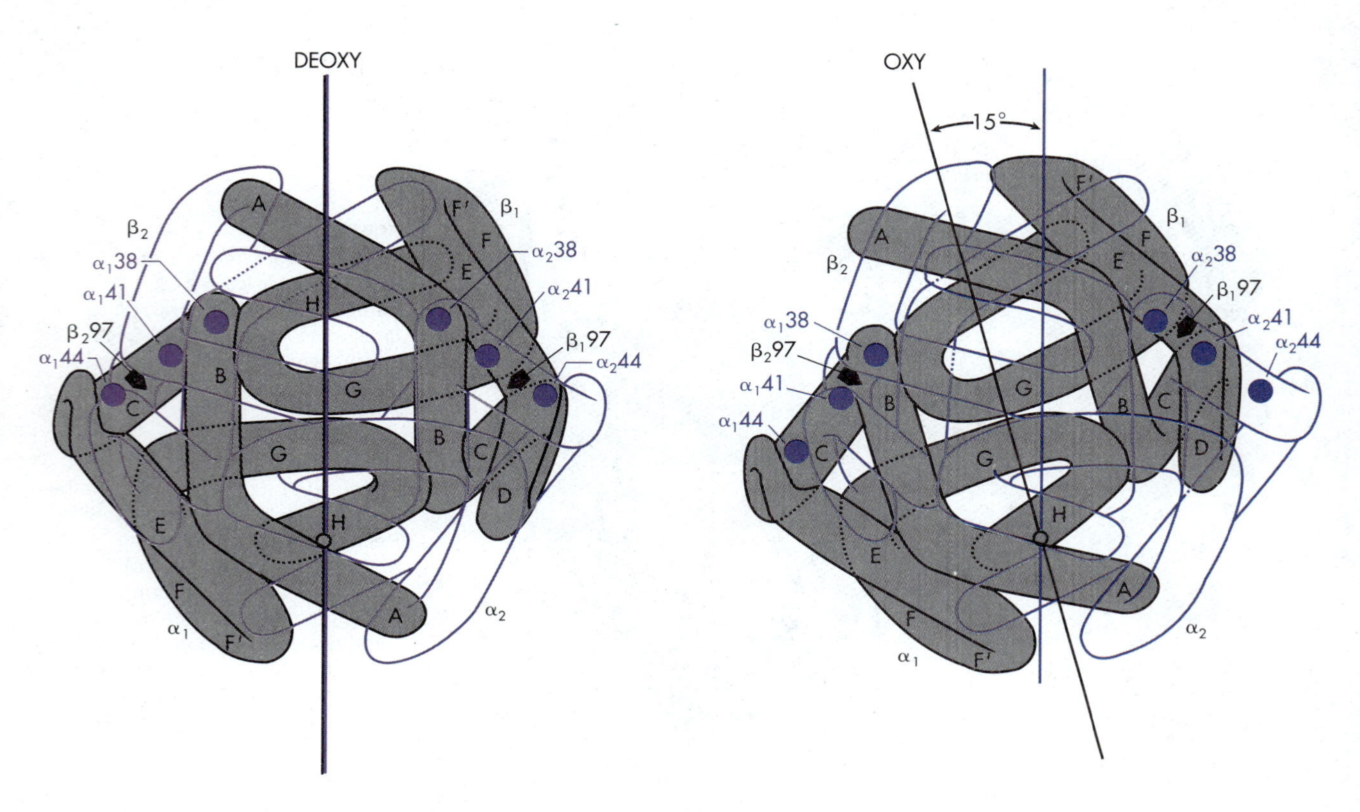

FIGURE 4.22
O_2-Induced shifts in hemoglobin conformation. The quaternary structure of hemoglobin and the tertiary structure of individual polypeptide subunits shift as one O_2 at a time is acquired during the oxygenation process. In oxyhemoglobin, the $\alpha_1\beta_1$ dimer is rotated 15° relative to the $\alpha_2\beta_2$ dimer. © Irving Geis.

Oxygen Binding by Hemoglobin

There are four O_2 binding sites on each hemoglobin molecule, since each heme–polypeptide subunit can bind one O_2 (in basically the same manner that myoglobin binds O_2). The binding of an O_2 to any one subunit in nonoxygenated hemoglobin (**deoxyhemoglobin**) leads to changes in the conformation of the other three subunits and, in the process, increases their O_2 affinity (Figure 4.22). The binding of a second O_2 increases the O_2 affinity of the remaining nonoxygenated subunits in a similar fashion. The binding of a third O_2 has a comparable effect. After the first O_2 has bound, each of the next three binds more and more readily (Figure 4.23). In molecules with multiple binding sites, binding is said to be **cooperative** if binding at one site impacts the binding at other sites. Cooperative binding usually entails the at-

$$\text{Hb} + 4\,\text{O}_2 \rightleftharpoons \text{HbO}_2 + 3\,\text{O}_2 \rightleftharpoons \text{Hb(O}_2)_2 + 2\,\text{O}_2 \rightleftharpoons \text{Hb(O}_2)_3 + \text{O}_2 \rightleftharpoons \text{Hb(O}_2)_4$$

FIGURE 4.23
Cooperative binding of O_2 by hemoglobin.

tachment of one or more ligands to each of the subunits within an oligomeric protein. It is the positive cooperative binding of O_2 by hemoglobin that accounts for the tendency of individual molecules of this protein to carry either four O_2 or no O_2; any given molecule remains partially oxygenated for only a fleeting period of time:

$$\text{Hb} + 4\,O_2 \rightleftharpoons \text{Hb}(O_2)_4$$

Deoxyhemoglobin Oxyhemoglobin

A collection of hemoglobin molecules, in contrast to individual molecules, is often partially oxygenated, some of the molecules carry four O_2 and the rest carry none. The percentage of oxygenation increases as the ratio of $\text{Hb}(O_2)_4$ molecules to Hb molecules (no O_2 bound) rises. The cooperative binding of O_2 by individual hemoglobin molecules is a consequence of the quaternary structure of hemoglobin and the communication that takes place between its four subunits. In this case, *communication* refers to changes in the folding of one subunit in response to the oxygenation of neighboring subunits. **[Try Problems 4.30 and 4.31—the answer to Problem 4.30 examines the molecular basis for the communication between hemoglobin subunits.]**

Oxygen Binding Curves

The graph obtained when the fraction or percent of myoglobin or hemoglobin molecules bound to O_2 (y-axis) is plotted against the concentration of O_2 (x-axis) is known as an **oxygen binding curve** or **oxygen dissociation curve**. Since O_2 binds as O_2 concentrations increase and dissociates when O_2 concentrations drop, the name chosen for the curve becomes a matter of perspective. Figure 4.24 presents the oxygen binding curve for both myoglobin and hemoglobin under physiological conditions. The concentration of O_2 is expressed as the partial pressure of O_2 in **torrs** [a torr is the pressure exerted by a column of mercury 1 mm high; 760 torr = 1 at-

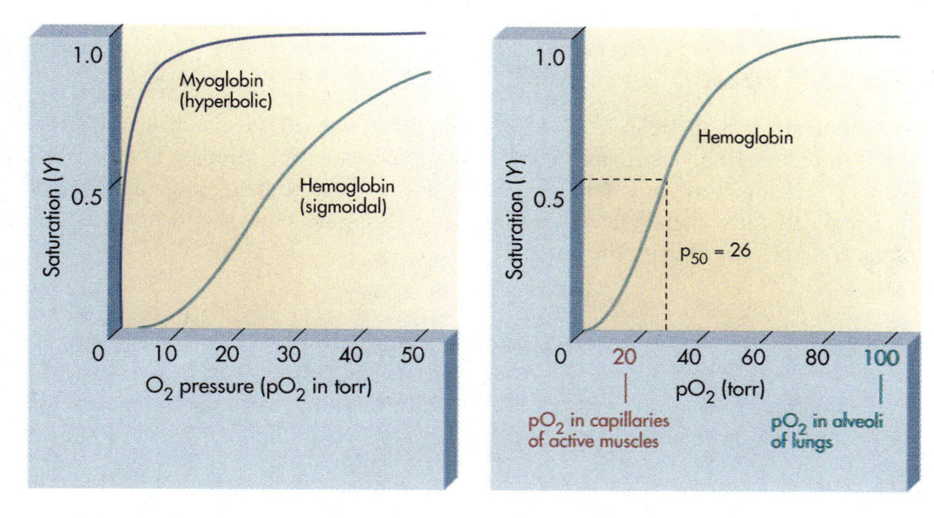

FIGURE 4.24
Oxygen binding curves for myoglobin and hemoglobin.

mosphere (atm) = 101 kilopascal (kPa)]. At any given O_2 concentration, myoglobin tends to be more completely oxygenated than hemoglobin. This facilitates the natural and desired flow of O_2 from hemoglobin in the blood to myoglobin within muscle cells.

The binding curve for myoglobin is **hyperbolic,** whereas that of hemoglobin is **sigmoidal**. The terms *sigmoidal* and *hyperbolic* refer to uniquely and specifically shaped lines that can be defined with standard mathematical equations. With hyperbolic binding curves, a relatively small drop in O_2 concentration leads to an extensive loss of O_2 once O_2 levels have reached a certain value (follow the O_2 binding curve for myoglobin in Figure 4.24 from high O_2 levels to low O_2 levels). Similar changes in O_2 concentration usually lead to the release of a much smaller fraction of the bound O_2 when the carrier is attached to O_2 in a sigmoidal fashion (study the hemoglobin O_2 binding curve in Figure 4.24). *Sigmoidal binding curves normally reflect cooperative binding*. It is the sigmoidal binding of O_2 by hemoglobin that allows this protein to distribute O_2 throughout the body. Hyperbolic binding would cause hemoglobin to deposit too much of its O_2 at a limited number of sites. Cells at other sites in the body would have difficulty acquiring adequate O_2. **[Try Problem 4.32]**

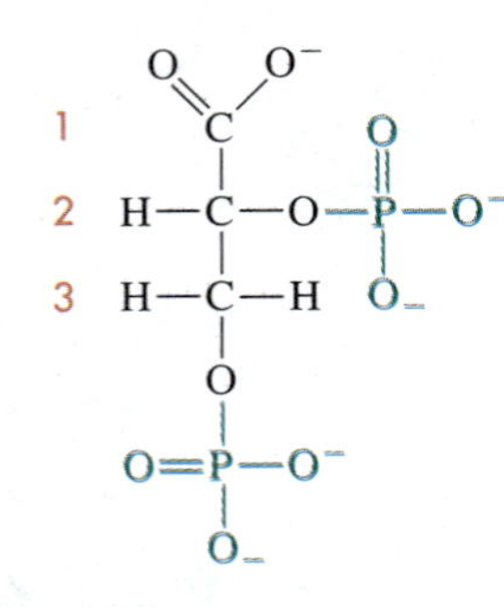

FIGURE 4.25
2,3-Bisphosphoglycerate.

There are multiple physiologically important substances, called **effectors**, that *reversibly* bind to hemoglobin and shift its O_2 binding curve. Each of these effectors, including $\mathbf{H^+}$ and **2,3-bisphosphoglycerate** (**2,3-BPG**; Figure 4.25), binds selectively to a unique site on deoxyhemoglobin and, in the process, reduces its ability to bind O_2. Although O_2 and effectors bind at separate sites, O_2 cannot bind (at least not as readily) when an effector is bound, and vice versa. Consequently, O_2 must compete with the effector for deoxyhemoglobin. Viewed from a different perspective, effectors facilitate the dissociation of oxyhemoglobin by pulling hemoglobin away from the O_2. The action of effectors is best understood in terms of Le Châtelier's principle and competing equilibria (see Problem 4.33). The net result: when an effector is present, it takes a higher O_2 concentration to oxygenate any given fraction of hemoglobin molecules. **[Try Problem 4.33]**

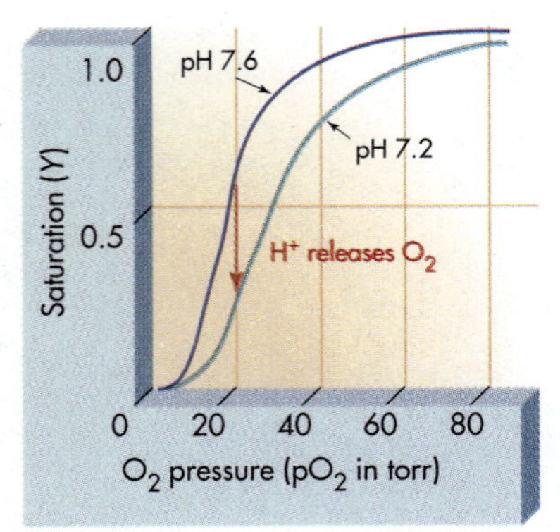

FIGURE 4.26
The Bohr effect. As the H^+ concentration increases (pH decreases), the fraction of hemoglobin molecules oxygenated at any specified O_2 pressure decreases. This leads to the dumping of O_2 by hemoglobin (see arrow on graph).

The Bohr Effect

The **Bohr effect** is the release of O_2 by hemoglobin as the H^+ (one of the effectors listed in the previous paragraph) concentration increases (pH drops). The Bohr effect is physiologically important because tissues that are oxidizing large amounts of fuel, and that consequently need large quantities of O_2, generate H^+ by dumping copious amounts of CO_2 into the blood:

$$H_2O + CO_2 \rightleftharpoons H_2CO_3$$

$$H_2CO_3 \rightleftharpoons \mathbf{H^+} + HCO_3^-$$

By facilitating the release of O_2, this H^+ helps ensure that extra O_2 is deposited in those active tissues that are consuming the most O_2 (study Figure 4.26 and Problem 4.33).

The molecular explanation for the Bohr effect lies in the modifications in subunit conformation brought about by the binding or release of O_2. The reshuffling of conformations changes the environment of particular weak acid functional groups and, in the process, alters the p*K*a's for these functional groups (Section 3.5). These changes in p*K*a's make deoxyhemoglobin a stronger base than oxyhemoglobin. Added H^+ reacts preferentially with the deoxyhemoglobin and removes it from the deoxyhemoglobin-oxyhemoglobin equilibrium. Oxyhemoglobin releases O_2 as the system moves to restore equilibrium (Le Châtelier's principle).

2,3-Bisphosphoglycerate

2,3-Bisphosphoglycerate (2,3-BPG), the second hemoglobin effector identified on page 146, is another key participant in the regulation of the body's O_2 delivery system. When O_2 pressure in the lungs is reduced, hemoglobin leaving the lungs is not as fully oxygenated as it is in the presence of normal O_2 concentrations, and the amount of O_2 delivered to tissues is reduced. **Hypoxia** (tissue O_2 deficiency) tends to result. The body responds to a prolonged decrease in O_2 concentration by increasing the concentration of 2,3-BPG within red blood cells. Since 2,3-BPG binds selectively to deoxyhemoglobin and pulls it out of the deoxyhemoglobin–oxyhemoglobin equilibrium, the additional 2,3-BPG reduces the amount of O_2 bound to hemoglobin at any given O_2 concentration. This leads to a shift in hemoglobin's oxygen binding curve and enhances the release of O_2 to tissues.

The 2,3-BPG effect is illustrated in Figure 4.27. In the presence of normal 2,3-BPG concentrations (left curve, Figure 4.27), hemoglobin releases about 30% of the

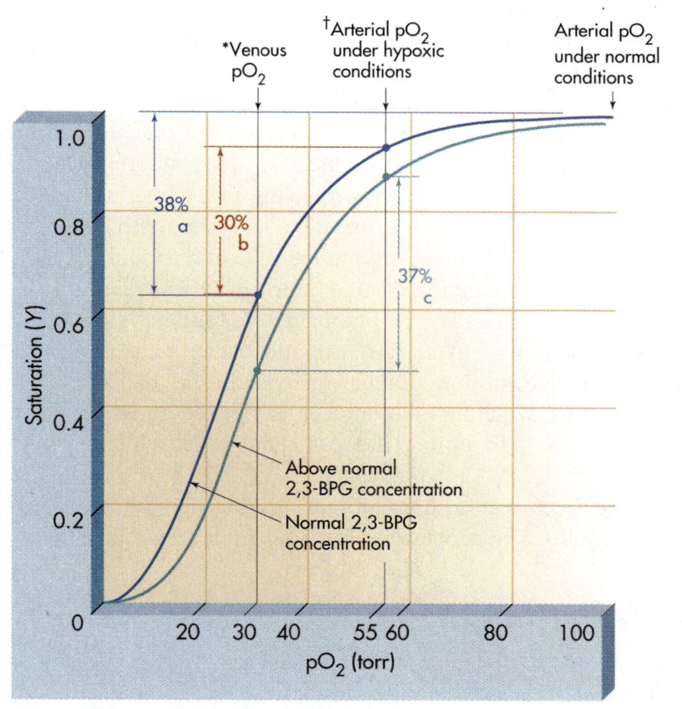

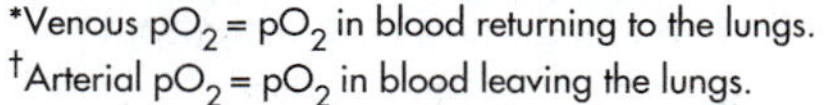

FIGURE 4.27
The effect of 2,3-BPG concentration on the oxygen binding curve of hemoglobin. (a) As the pO_2 shifts from 100 torr (normal pressure in lungs) to 30 torr (approximate concentration in blood returning to lungs) at normal 2,3-BPG concentrations, 38% of the total O_2 that hemoglobin is capable of carrying is released. (b) As the pO_2 shifts from 55 torr (lungs with below normal O_2 pressure) to 30 torr at normal 2,3-BPG concentrations, only 30% of the total O_2 that hemoglobin is capable of carrying is released. (c) At an increased 2,3-BPG concentration, 37% of the total O_2 that hemoglobin is capable of carrying is released as the pO_2 shifts from 55 torr to 30 torr. Based on Voet and Voet (1995).

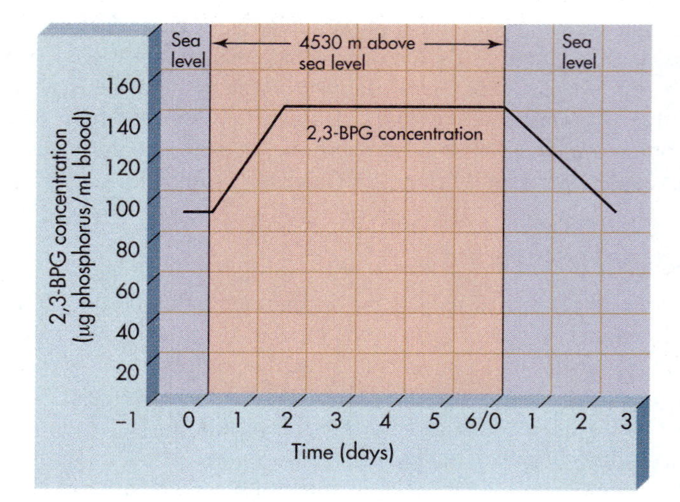

FIGURE 4.28
Altitude-induced shifts in blood 2,3-BPG concentration. The pO_2 of arterial blood is close to 100 torr at sea level and approximately 55 torr at 4530 m above sea level. Based on Voet and Voet (1995).

total oxygen it is capable of carrying as it moves from hypoxic lungs (with an assumed pO_2 of approximately 55 torr) to venous blood (which normally has a pO_2 of around 30 torr). At slightly higher 2,3-BPG levels (right curve in Figure 4.27), the same hemoglobin would release oxygen corresponding to about 37% of its total carrying capacity. This compares well with the 38% release observed when both lung O_2 concentrations and blood 2,3-BPG concentrations are normal. In this example, the 23% increase in O_2 delivery $\{[(37 - 30)/30] \times 100 = 23\%\}$ could be very significant from the standpoint of health and survival.

Numerous factors can lead to prolonged changes in lung O_2 levels and to shifts in 2,3-BPG concentration. One of the most common is a change in altitude, since the air at higher altitudes normally has a reduced O_2 concentration when compared to air at lower altitudes. Immediately after traveling to a higher altitude, a person will tend to be hypoxic, to be short of breath, and to have more difficulty exercising. Red blood cell 2,3-BPG concentrations start to increase almost immediately, but they do not reach a new plateau (assuming that the same low O_2 concentration is maintained) for 1 to 2 days (Figure 4.28). The O_2 binding curve for hemoglobin and the delivery or O_2 to tissues continue to shift during this adaptation period. High-altitude adaptation also includes a more gradual increase in the total number of red blood cells and the number of hemoglobin molecules per red blood cell. Upon returning to lower altitudes, the high altitude-induced changes gradually disappear. **[Try Problem 4.34]**

Athletes sometimes train at high altitudes to enhance the ability of the body to deliver O_2 to muscles. The benefits are lost soon after they return to lower altitudes.

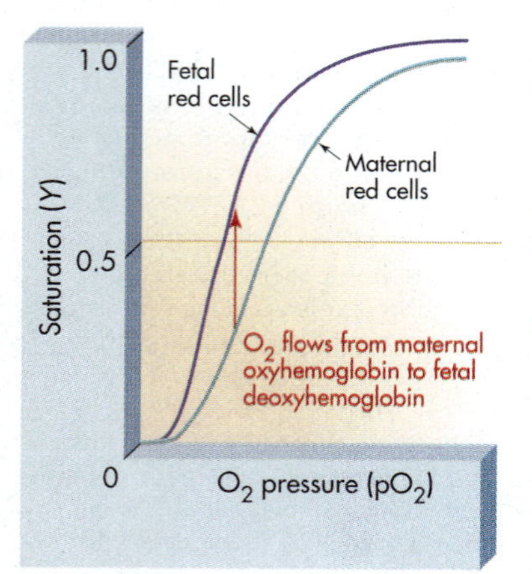

FIGURE 4.29
The oxygen binding curve for hemoglobin F.

Fetal Hemoglobin

The hemoglobin that has been described to this point in the text, hemoglobin A_1, is normally the most abundant hemoglobin molecule in adults. The blood of a fetus contains predominantly **hemoglobin F** and relatively little hemoglobin A_1. Hemoglobin F contains two α chains and two γ chains, rather than two α and two β chains. The γ and β chains have the same number of amino acids but unique (although similar) primary structures and slightly different conformations. Under physiological conditions, these differences cause 2,3-BPG to bind less tightly to hemoglobin F than to hemoglobin A_1. Consequently, the O_2 binding curve of hemoglobin F falls to the left of the binding curve for hemoglobin A_1 (Figure 4.29); at any given O_2 concentration the fetal hemoglobin is more highly oxygenated than its adult counterpart. This facilitates the natural and desired flow of O_2 from the blood of the mother to the blood of the fetus. During the first 6 months after birth, fetal hemoglobin concentrations drop drastically while the levels of the adult hemoglobins (there are multiple forms) increase. Most adults continue to produce small amounts of hemoglobin F. **[Try Problem 4.35]**

CO_2 Transport

Oxygen-consuming organisms, including humans, are faced with the need to transport CO_2 as well as O_2. A majority of the CO_2 that enters human blood at tissues is transported to the lungs as HCO_3^- (see the equation for the reaction of CO_2 with H_2O illustrated in the discussion of the Bohr effect). A small fraction is transported as an unreacted dissolved gas. Most of the remainder of the CO_2 is carried to the lungs in the form of **carbamates** after it has reacted with the N-terminal α-amino groups within hemoglobin (Table 4.6):

$$\text{Polypeptide—NH}_2 + \text{CO}_2 \rightleftharpoons \text{polypeptide—NHCOOH}$$

$$\text{Polypeptide—NHCOOH} \rightleftharpoons \text{polypeptide—NHCOO}^- \text{ (a carbamate)} + \text{H}^+$$

TABLE 4.6

Major Transport Forms of CO_2

Transport Form	Percentage Transported
HCO_3^-	78
Carbamino-hemoglobin	13
CO_2 (dissolved)	9

The H^+ generated in the process contributes to the Bohr effect and helps hemoglobin release extra O_2 at those tissues where a large amount of O_2 is being used to oxidize fuels and generate CO_2. When the HCO_3^- and carbamates reach the lungs, CO_2 is exhaled causing these CO_2 carriers to release their cargo (Le Châtelier's principle).

Structure, Function, and Evolution

In the hemoglobin story we find once again that structure is tightly coupled to function. This coupling is reflected in the cooperative binding of O_2, the presence of multiple effector binding sites, and the existence of multiple forms of hemoglobin. The initial structural predictions we made have proven accurate, even though they did not include some of the structural details encountered in this section. More information about the subtleties of hemoglobin function would have been required to make more detailed structural predictions.

A wide variety of organisms, including some invertebrates, produce their own unique myoglobins or hemoglobins. Collectively, these O_2-binding agents represent the globin family of proteins. Based on the similarities in their amino acid sequences and their folding patterns, it is probable that the genes for each present-day globin have evolved from a single ancestral gene. If this is true, gene duplication must have been involved in the creation of the myoglobin and hemoglobin genes found in humans.

PROBLEM 4.29 Study Figure 4.21. Which helices form the walls of the heme pocket in the α chains of hemoglobin? Which provide a floor?

PROBLEM 4.30 Suggest how the noncovalent binding of a molecule to a protein can change the protein's conformation. Can the primary structure of a protein be changed by such noncovalent binding?

PROBLEM 4.31 Each mature red blood cell contains roughly 280 million hemoglobin molecules. Approximately how many O_2 are carried by a single red blood cell when its hemoglobin is 80% saturated with O_2?

PROBLEM 4.32 Why is it impossible for myoglobin to bind O_2 cooperatively?

***PROBLEM 4.33** Assume that the following reaction is at equilibrium (the rate of the forward reaction equals the rate of the reverse reaction and the concentrations of reactants and products are not changing):

$$Hb + 4\ O_2 \rightleftharpoons Hb(O_2)_4$$

Will the concentration of $Hb(O_2)_4$ increase, decrease, or remain the same if the equilibrium is temporarily destroyed by removing some Hb? Explain. The answer lies in Le Châtelier's principle. Write equations (of the type given immediately above for the binding of O_2 to hemoglobin) for the reversible binding of H^+ and 2,3-BPG to deoxyhemoglobin (Hb). An effector, in binding Hb, removes it from the deoxyhemoglobin–oxyhemoglobin equilibrium.

PROBLEM 4.34 Which of the following conditions normally lead to elevated 2,3-BPG levels within red blood cells? Explain.

a. Lung cancer
b. Broken leg
c. Pneumonia
d. Sinus infection

PROBLEM 4.35 Predict whether the O_2 binding curve for myoglobin lies to the right or to the left of the O_2 binding curve for hemoglobin F? Explain.

4.10 COLLAGEN AND α-KERATIN

Our examination of protein structure and structure–function relationships continues with a glimpse at **collagen,** the most abundant protein in mammals. It represents from one fourth to one third of the total protein in the human body. It accounts for over 70% of the dry weight of skin, and it is the major protein component in tendon, cartilage, bone, blood vessels, and teeth. In each of these instances, collagen serves as an extracellular (outside of cells) supporting element. Because of its biological role, collagen needs to be physically strong, water insoluble, and relatively resistant to chemical change. What structural features would a protein need to acquire these properties? Multiple covalently linked polypeptide chains would probably be important from the standpoint of physical toughness. An abundance of nonpolar side chains on the surface would ensure water insolubility and create a relatively nonreactive coating. Tertiary structure may be nonexistent because tertiary structures tend to be rather easily disrupted. How accurate are these predictions?

The Structure of Collagen

The basic building blocks for all collagen molecules are polypeptides containing approximately 1000 amino acid residues. These polypeptides contain several nonstandard amino acids in addition to a majority of the 20 common protein amino acids. The uncommon amino acids, including 4-hydroxyproline (Hyp) and 5-hydroxylysine (Hyl), are produced through the enzyme-catalyzed modification of particular common amino acids after they have been incorporated into polymer chains.

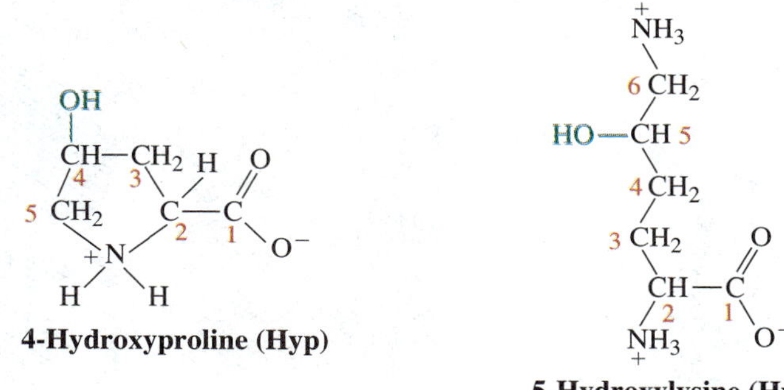

4-Hydroxyproline (Hyp)

5-Hydroxylysine (Hyl)

Glycine, hydroxyproline (mostly 4-hydroxy but some 3-hydroxy)/proline, alanine, glutamic acid, and arginine account for approximately 33%, 21%, 10%, 5%, and 5%, respectively, of all the amino acid residues in a typical collagen polypeptide. Every third amino acid residue is usually a glycine, and the sequence Gly–Pro–Hyp recurs many times. This unique primary structure causes most collagen polypeptides to twist into extended left-handed helices. Three such helices then wrap together to give a right-handed coil known as a **superhelix** or **triple helix** (Figure 4.30). In a similar manner, separate strands of twine can be wrapped to produce a rope. Initially, hydrophobic interactions and H bonds are the predominant forces that hold the three polypeptide chains together within a superhelix. Parallel superhelixes often pack together in **staggered arrays** to form collagen fibers (Figure 4.31). The positioning of proline, alanine, and other nonpolar amino acids on the surface of collagen accounts for its insolubility in water (the "nonpolar in, polar out" rule applies only to globular proteins in an aqueous environment). **[Try Problem 4.36]**

As collagen matures and ages, covalent cross-links accumulate between separate triple helices and between separate polypeptide chains within individual triple helices. Some of these cross-links involve the side chains of hydroxylysine residues. The covalent attachment of carbohydrates to specific side chains makes collagen a glycoprotein (Table 4.2). Individual collagen molecules differ in size, carbohydrate

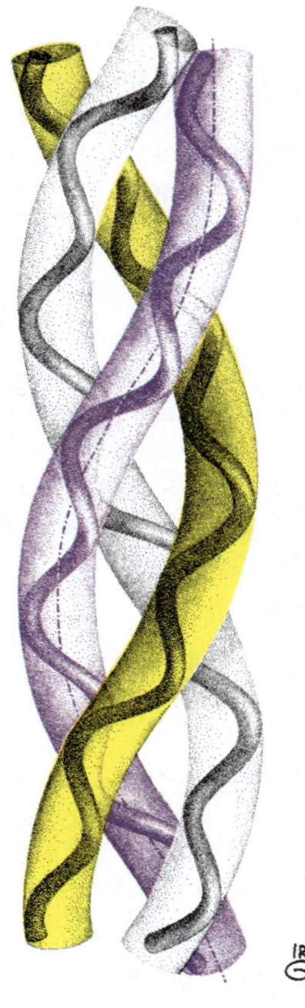

(a)

(b)

FIGURE 4.30
A collagen triple helix. (a) An artist's representation. © Irving Geis. (b) A space-filling model. Tripos, Inc.

Collagen triple helix

Staggered array

FIGURE 4.31
A staggered array of collagen triple helices.

content, the ratio of various amino acids, and the extent of covalent cross-linking. In some collagens, less ordered, globular-type domains separate superhelical regions. Simple variations on a common theme allow multiple forms of collagen to serve related yet unique roles at numerous locations within the body. As a consequence of its overall architecture, a collagen fiber tends to possess tremendous tensile strength; some fibers 1 mm in diameter can hold over 20 pounds. **[Try Problem 4.37]**

Abnormal Collagens

Abnormal collagens are responsible for multiple medical problems. **Scurvy,** a vitamin C deficiency disease (Section 10.1), leads to skin lesions, fragile blood vessels, and bleeding gums. These symptoms have been traced to hydroxyproline-deficient collagen, which is relatively weak and easily disrupted. The skeletal deformities in **homocystinuria,** a genetic disease (Chapter 19), are a consequence of abnormalities in the covalent cross-linking of collagen. The replacement of specific glycine residues with serine residues leads to **Ehlers–Danlos syndrome,** a genetic disease characterized by loose joints.

α-Keratins

The proteins in hair, fingernails, wool, claws, quills, horns, hooves, and feathers are similar in overall design to collagen. The basic building blocks for these proteins are polypeptides, known as **α-keratins,** that coil into extended (nonfolded) right-handed α-helices. In hair, four α-keratins wrap together into a left-handed (probably) superhelix called a **protofibril.** Protofibrils are bundled together in a specific geometry to form microfibrils that are held together, in part, by disulfide bonds (see Figure 4.32 on the next page). The permanent wave process for hair involves the reduction of disulfide bonds, the reshaping of strands of hair, and the oxidative regeneration of new disulfide bonds to lock the hair into its newly created shape.

In both collagen and hair, multiple coiled coils are linked through covalent bonds to create tough, fibrous molecules that encompass little if any ordered tertiary structure. Hydrophobic side chains contribute to water insolubility and tend to reduce chemical reactivity. Once again, structure is well matched to function and initial structural predictions are found to be validated. **[Try Problem 4.38]**

PROBLEM 4.36 What structural features of collagen polypeptides prevent them from coiling into α-helices? Explain.

PROBLEM 4.37 The collagen in the body accumulates covalent cross-links throughout life. As a consequence, the collagen in older people is more highly cross-linked than that in young people. What symptoms of old age can be at least partially explained by these facts?

PROBLEM 4.38 A hair fiber can be drawn to twice its original length, but a collagen fiber has little, if any, stretch. Explain these facts on the basis of the molecular structures of hair and collagen.

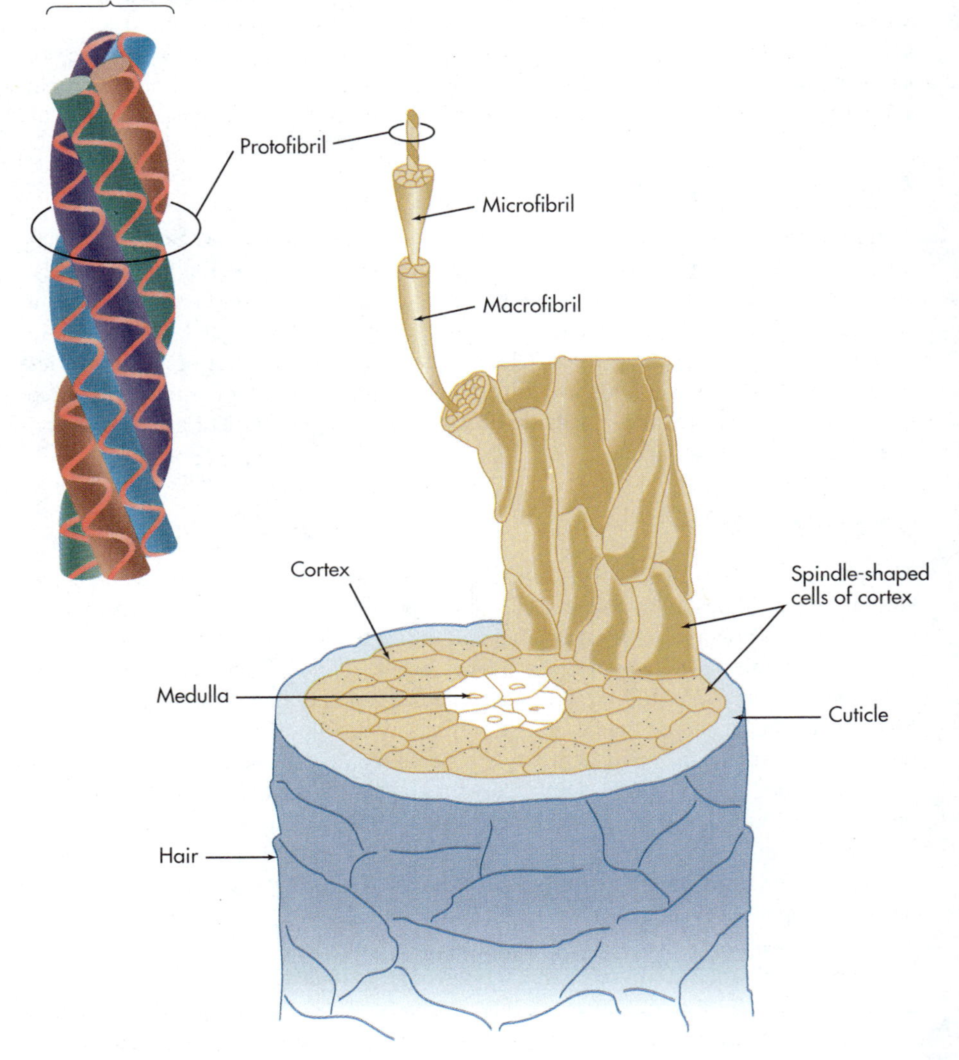

FIGURE 4.32
The structure of hair. From Moran et al. (1994).

4.11 IMMUNOGLOBULINS (ANTIBODIES)

Immunoglobulins (also called **antibodies**) are proteins secreted by the immune system after it has been challenged with foreign agents (viruses, bacteria, some cancer cells, and certain other materials), known as **immunogens** or **antigens** (Table 4.1). Immunoglobulins will be used to further explore the topic of structure–function relationships. Once produced, an immunoglobulin (Ig) binds to the antigen that has triggered its production and, in the process, neutralizes it, immobilizes it or tags it for destruction or elimination by other components of the immune system. In many instances, immunoglobulins function as suicide bombers; they are themselves destroyed as they participate in the destruction of antigens. The specific sites on foreign agents that stimulate antibody production are known as **antigenic determinants.** Most antigenic determinants are constructed from peptides, carbohydrates or nucleic acids. Although each antigen normally leads to the production of multiple distinct antibodies, an individual antibody normally recognizes and binds only that antigenic determinant which triggered its production. Based on its functions, an antibody should contain at least two structurally distinct binding sites, one that attaches to antigen and a second that communicates (probably through direct binding) with other immune system components.

The Structure of Immunoglobulins

There are five functional classes of antibodies, abbreviated **IgA, IgD, IgE, IgG,** and **IgM.** The basic structural unit within all classes is a tetramer that contains two identical **light chains** and two identical **heavy chains** (which are about twice the size of the light chains), all linked through disulfide bonds. Although the individual classes possess unique heavy chains, they all share common light chains. There are two different classes of light chains, designated λ (lambda) and κ (kappa). It is the structure of IgG, the most abundant immunoglobulin in normal human blood, that will be examined in this section. Since IgG is the only antibody capable of entering a fetus through the placenta, it plays a central role in protecting the unborn.

An IgG molecule consists of a single tetramer. Its two light chains are identical and each contains two domains, a **variable domain** (varies from one IgG to the next) and a **constant domain** (is the same for all immunoglobulins within a given class). A heavy chain possesses four domains, three constant and one variable. Each heavy chain is joined to one light chain and to an identical second heavy chain through disulfide bonds. The resultant Y-shaped molecule has two identical **antigen binding sites,** one at the end of each prong in the Y (Figure 4.33). The base or stem of the Y participates in the recruitment of other immune system components after an IgG binds an antigen. Since carbohydrate residues are attached to some amino acid side chains, IgG, like collagen, is technically a **glycoprotein** (Table 4.2).

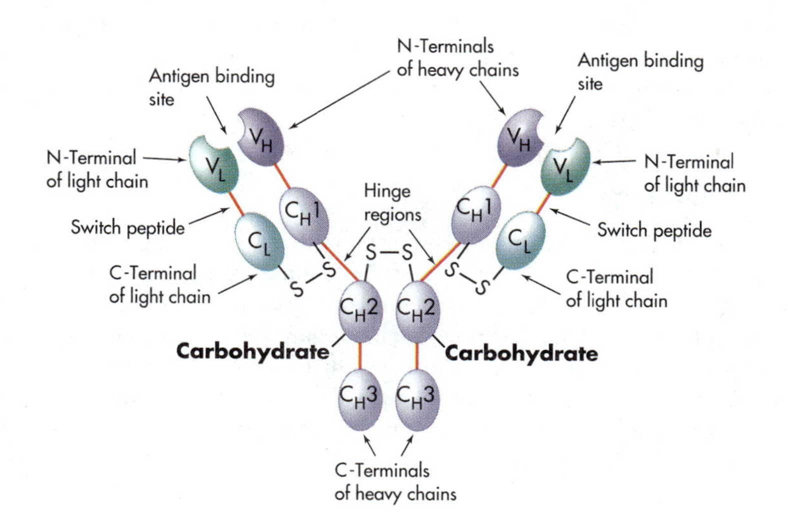

FIGURE 4.33
An IgG molecule. Each ellipsoid (⬬) represents a separate domain. V_L is the variable light-chain domain, C_L the constant light-chain domain, V_H the variable heavy-chain domain, and so on.

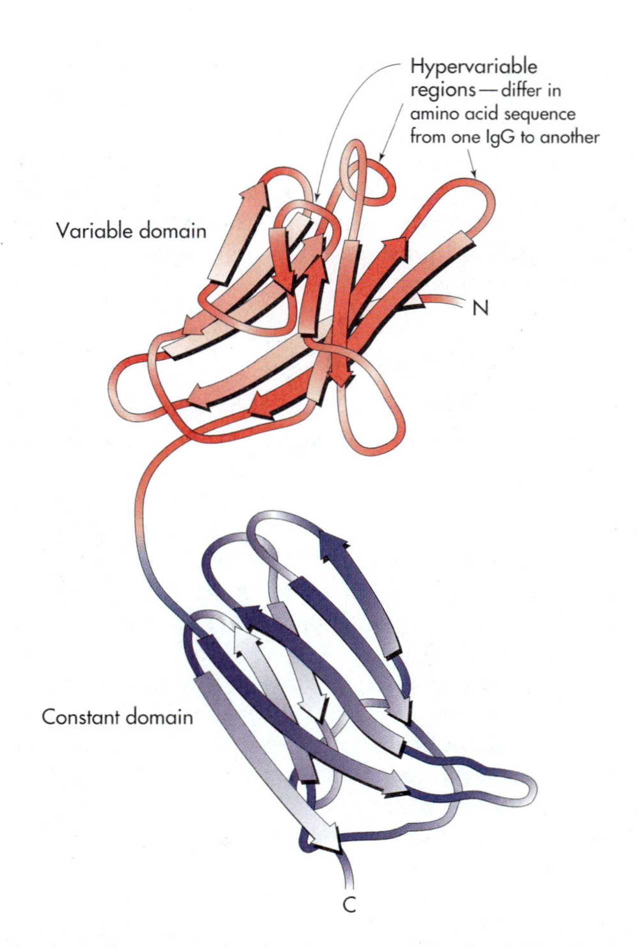

FIGURE 4.34
Immunoglobulin folds within an IgG light chain. Based on Branden and Tooze (1991).

The domains in IgG are described as **immunoglobulin folds.** Each contains two antiparallel β-sheets that are stacked one on top of the other and held together, in part, by a disulfide bond (Figure 4.34). All constant domains have three β-strands in one sheet and four in the second. The variable domains contain a four-strand sheet and a five-strand sheet. Since the loop regions are relatively short, most of the approximately 110 amino acid residues within each immunoglobulin fold reside within a β-strand. **[Try Problems 4.39 and 4.40]**

Antigen Binding

Distinct variable domains within IgG molecules differ in both length and amino acid sequence. An antigen binding site is created when the variable domain of one light chain attaches in a highly specific manner to a variable domain within a heavy chain. This pairing brings together six **hypervariable loops** (Figure 4.34), three from the heavy chain and three from the light chain, that contain the specific amino acid residues found at an antigen binding site (also called the **combining site**). A precise conformation at the combining site ensures an appropriate "fit" between antigen and antibody. An antibody binds an antigen through noncovalent interactions, predominantly H bonds, salt bridges and hydrophobic forces.

The human body has the ability to produce many billion unique IgG molecules, each designed to bind to a specific antigen or a select group of antigens. This **antibody diversity** enables the body's immune system to mount an attack against virtually any foreign agent that it encounters, even novel antigenic agents generated through evolutionary changes in viruses and bacteria. In spite of this antibody diversity, humans are victims of numerous pathogens, including the AIDS virus. Our immune systems may need some improvements if we are to remain among the fit that survive.

PROBLEM 4.39 Starting at the N-terminal end, sequentially number each β-strand in each domain in Figure 4.34. Each domain contains multiple β-strands that are adjacent in primary structure yet part of separate β-sheets. Give one example.

PROBLEM 4.40 Would you expect a globin fold to be more or less easily denatured than an immunoglobulin fold? Explain.

4.12 PROTEINS AND NUTRITION

The average American consumes about 1.4 g of protein/kg body weight/day. Adults probably need around 0.8 g/kg/day, children up to 2.4 g/kg/day. Pregnant women also have elevated requirements.

Dietary proteins are hydrolyzed in the stomach and intestine to generate free amino acids, dipeptides, and tripeptides. These products of digestion are actively carried (energy is required) into absorptive cells in the intestine where the peptides are further hydrolyzed to generate a final pool of protein-derived amino acids. Part of the amino acids are used by the absorptive cells themselves, but most are passed on to the blood stream, from which they are distributed throughout the body. Proteins are the major source of dietary amino acids. [Try Problem 4.41]

From a nutritional standpoint, proteins can vary greatly in quality. **High-quality proteins** are easily digested and contain the proper balance of essential amino acids (Section 3.4). A protein that contains all of the essential amino acids is called a **complete protein.** In general, animal protein is of higher nutritional quality than plant protein because plant proteins are often deficient in one or more of the essential amino acids. Since not all plants are deficient in the same amino acids, a vegetarian can satisfy amino acid requirements by consuming a carefully selected combination of plants. Contrary to popular belief, it is not necessary that all of the essential amino acids be consumed at the same meal. [Try Problem 4.42]

The protein in hair is a good source of amino acids for dietary supplements.

Dietary amino acid supplements are produced from a variety of plant and animal materials. India, for example, has constructed a factory designed to isolate amino acids from up to 1200 tons of human hair (Section 4.10) each year. Barber shops and Hindu temples supply the raw materials. The purified amino acids are marketed to the pharmaceutical and cosmetic industries as well as the food processing industry.

PROBLEM 4.41 Write a balanced equation for the complete hydrolysis of Asp–Ser–Gly as it would occur in the human intestine. Show structural formulas for all reactants and products.

PROBLEM 4.42 Does an adult who is lifting weights and building bulky muscles need more dietary amino acids than the average adult? Explain.

METHODS SECTION

4.13 X-RAY DIFFRACTION

No one has ever viewed a protein directly because the most powerful light microscopes are unable to provide even a fuzzy image of these macromolecules. How, then, have biochemists learned so much about their coilings, foldings, and packings? To a large extent, by X-ray diffraction.

X-Ray diffraction analysis of a protein involves the study of the diffraction (bending, scattering) of X-rays by electron clouds within proteins as X-rays pass through a protein crystal. Photographic film or other X-ray detectors are used to locate the scattered X-rays (Figure 4.35). Diffraction patterns, along with complex mathematical analysis, can be used to construct an electron density map for a cross

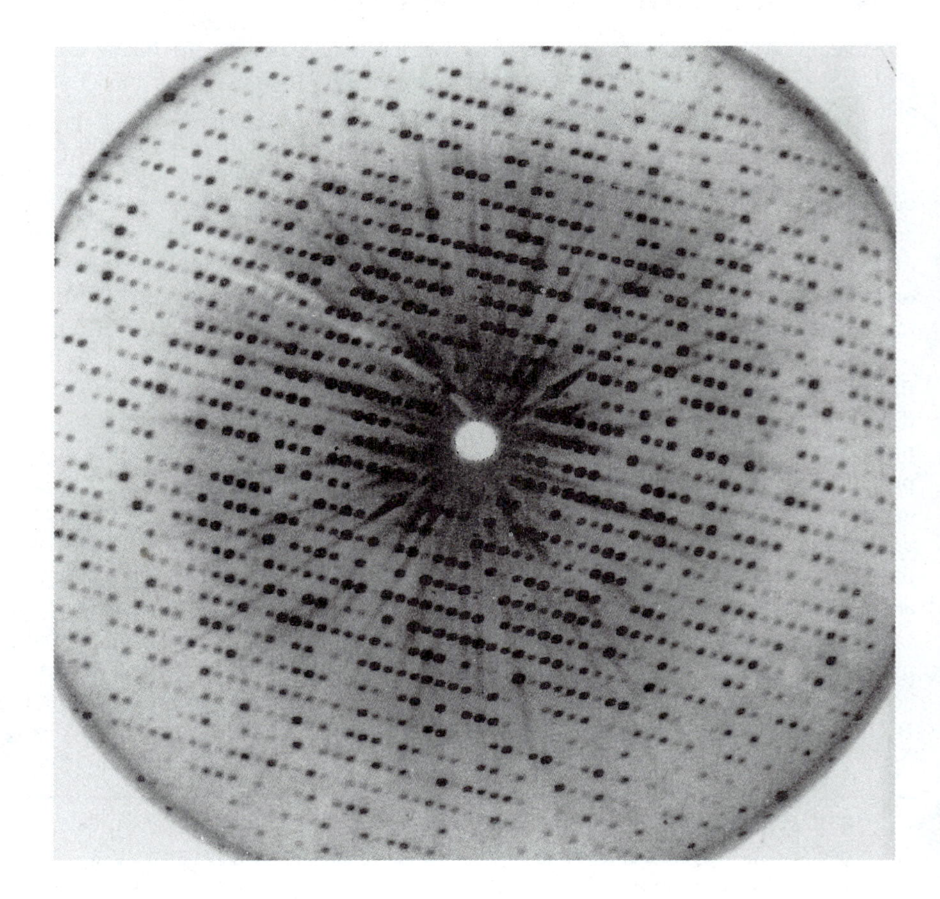

FIGURE 4.35
An X-ray diffraction pattern for a crystal of sperm whale myoglobin. Each dark spot represents a region where scattered electrons have exposed a photographic plate. Mathematical analysis allows determination of the relative locations of electron clouds within the crystal from the locations of the exposed spots. (Courtesy of John Kendrew, Cambridge University.)

section of the protein. Since the electron cloud of an atom is most dense around its nucleus, the regions of greatest electron density identify the locations of atomic nuclei. If electron density maps of neighboring cross sections are placed on transparent film and stacked one on top of another, one can create a three-dimensional image that depicts the relative location of each atomic nucleus within the protein (Figure 4.36). Alternatively, computers can be used to convert X-ray diffraction data into three-dimensional images on a computer screen. Various lines of evidence indicate that the structure of a protein within a crystal is often similar to its structure in solution.

X-Ray diffraction analysis can be used to determine the relative location of each atom in virtually any compound that can be crystallized. This technique has supplied much of the detailed structural knowledge regarding many complex compounds of biological interest, including DNA.

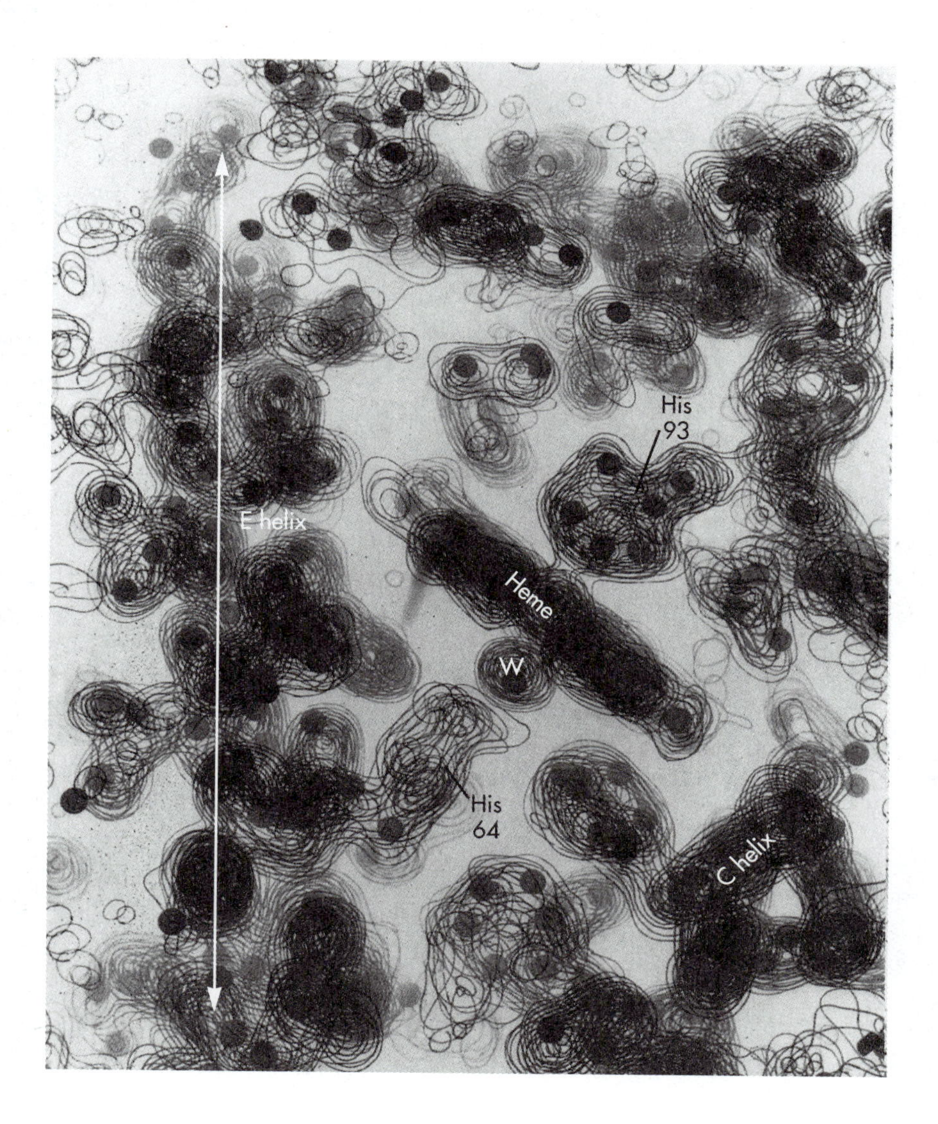

FIGURE 4.36
Stacked electron density maps for a small portion of myoglobin. Dots mark the location of nonhydrogen atoms. The heme group (side view), two histidine side chains, two α-helices and a water molecule (W) are labeled. The water molecule sits at the site occupied by O_2 in oxyhemoglobin. The C helix (lower right) extends into the plane of the paper, and its hollow core is readily visible.

4.14 NUCLEAR MAGNETIC RESONANCE SPECTROSCOPY

Nuclear magnetic resonance (NMR) spectroscopy has evolved into a second technique that can be used to determine the secondary and tertiary structure of small proteins (usually fewer than 100 amino acids), and, eventually, it may allow us to ascertain the conformations of large proteins as well. In contrast to X-ray diffraction studies, protein crystals are not required; analysis can be performed in aqueous solutions that more closely approximate physiological conditions.

NMR Theory

NMR spectroscopy is based on the fact that certain nuclei, such as those within ^{1}H, ^{13}C, ^{15}N, and ^{31}P atoms, behave as if they are spinning and act like tiny magnets. When placed in a powerful external magnetic field, such nuclei tend to align themselves so that their magnetic dipoles are oriented parallel to the field. A slight majority of the dipoles point in the same direction as the field and are said to exist in a low-energy spin state. The remainder point in the opposite direction and exist in a high-energy spin state (Figure 4.37). A spinning, low-energy nucleus will absorb a radiowave if the wave's energy content (determined by its wavelength, which, in

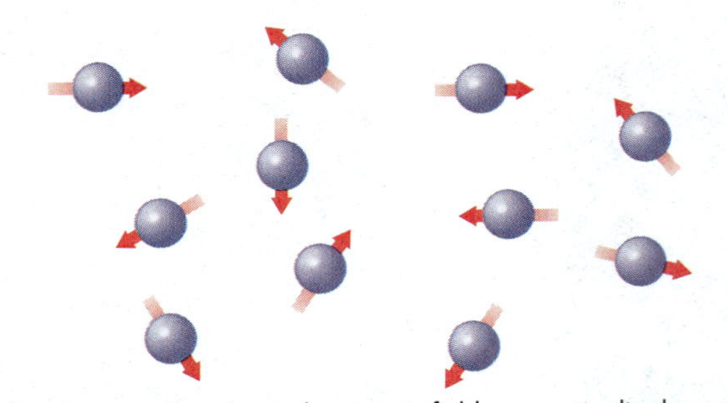

(a) In the absence of an external magnetic field, magnetic dipoles are oriented randomly.

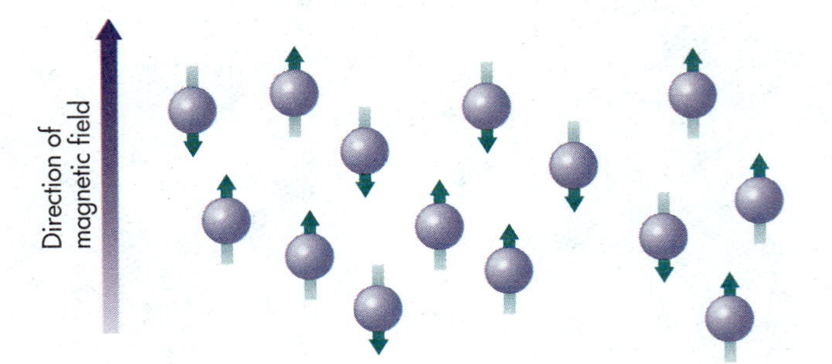

(b) In the presence of a powerful external magnetic field, a slight majority of the magnetic dipoles point in the same direction as the field (low-energy spin state), and the rest point in the opposite direction (high-energy spin state).

FIGURE 4.37
The orientation of magnetic dipoles in the presence and absence of a powerful external magnetic field.

turn, is determined by its frequency of oscillation) is exactly equal to the energy difference between the two spin states, a fact predicted by quantum theory. The energy difference between the spin states is determined by the strength of the external magnetic field, the specific isotope involved (^{1}H, ^{13}C, etc.), and the electron clouds in the vicinity of the absorbing nucleus. **[Try Problems 4.43 and 4.44—they provide a review of quantum theory]**

NMR Spectra

The graph obtained when we plot intensity of absorption (*y*-axis) against the frequencies of the radiowaves absorbed (usually expressed as a **chemical shift** relative to the frequency at which a standard absorbs) is called a **one-dimensional nuclear magnetic resonance spectrum (NMR spectrum),** because a magnetic nucleus is said to resonate (oscillate between its two spin states) when exposed to those wavelengths it absorbs. In a constant external magnetic field, individual atoms of the same isotope tend to absorb different, characteristic wavelengths of radiofrequency radiation when they exist in unique electron environments (Figure 4.38). By studying ^{1}H and ^{13}C NMR spectra, it is possible to determine the connectedness of the hydrogen and carbon atom network within relatively simple organic molecules, a task routinely performed by organic chemists. **[Try Problem 4.45]**

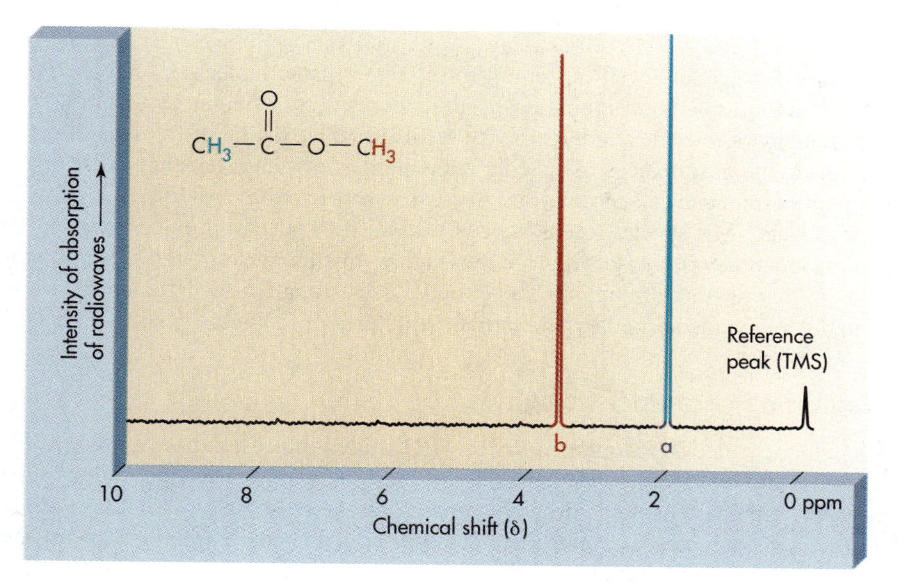

FIGURE 4.38
^{1}H NMR spectrum for methyl acetate. The three H's (colored blue green) in the acetate residue all reside in identical electron environments; they all absorb at the same frequency and collectively account for peak a. The three additional H's (colored red) are in the same electron environment, but one that is distinct from that of the acetate H's. Collectively the three additional H's account for peak b. In general, the number of peaks in a compound's ^{1}H NMR spectrum is equal to the number of chemically distinct classes of H's within the compound. TMS is trimethyl sulfoxide, a reference compound.

Magnetic Resonance Imaging (MRI)

A one-dimensional NMR spectrum for a large organic molecule, such as a protein, is difficult to interpret; unique absorbances resulting from nuclei in similar but different environments cannot readily be resolved with existing instrumentation. By measuring the absorbances of specific combinations of radiofrequency pulses having unique frequencies and durations, we can construct what are described as two-dimensional NMR spectra (see advanced texts). These spectra provide novel information about the relative locations of those atoms that interact through space as well as those that interact through covalent bonds. The through-space interactions allow us to determine the secondary and tertiary structures of relatively small proteins.

Magnetic Resonance Imaging

A technique called **MR tomography** or **MRI (magnetic resonance imaging)** uses ^{1}H NMR spectroscopy to "visualize" water molecules in humans and to determine the times required for the hydrogen atoms in the water to "relax" from an excited spin state into a nonexcited spin state. The spin–spin relaxation time varies from one tissue to the next and tends to be different in diseased and in normal tissues. For this reason, MR tomography can be used to locate tumors, vascular abnormalities, and certain other pathological conditions without exposing patients to X-rays (known to be harmful). So far, there is no evidence that the magnetic fields and radiofrequency pulses involved pose a health risk.

***PROBLEM 4.43** Summarize "quantum theory."

***PROBLEM 4.44** Describe what happens to a spinning, low-energy nucleus (in a magnetic field) when it absorbs a radiowave.

PROBLEM 4.45 How many peaks exist in an ^1H NMR spectrum for (a) propanone (acetone: CH_3COCH_3), (b) ethyl methyl ether ($CH_3CH_2OCH_3$), and (c) ethyl propanoate ($CH_3CH_2COOCH_2CH_3$)? Explain.

SUMMARY

Proteins are structurally and functionally diverse molecules that play more roles in humans and other organisms than any other class of compounds. Fibrous proteins, including collagen and α-keratin, tend to be tough and water insoluble. They are used to build a variety of materials that support and protect specific tissues and the human body as a whole. Skin, hair, fingernails, blood vessels, and connective tissues are largely constructed from these proteins. Globular proteins tend to be water soluble and relatively fragile. They include enzymes, hormones, antibodies, transport agents, storage compounds, regulatory agents, and visual proteins.

Most globular proteins possess a structural hierarchy that includes three or four structural elements: (1) primary structure—the order in which amino acids are connected along a polypeptide chain; (2) secondary structure—the helical and sheet regions within a polypeptide; (3) tertiary structure—the precise folding of secondary structural elements to generate a polypeptide chain with a specific three-dimensional shape and geometry; and (4) quaternary structure—the specific manner in which separate polypeptide chains pack together in those proteins that contain multiple noncovalently-linked polypeptides.

Primary structure is most often determined through the sequential removal of one amino acid at a time from the N-terminal end of a polypeptide. Edman degradation is normally employed, and the process has been automated. Secondary, tertiary, and quaternary structures are predominantly ascertained with X-ray diffraction and NMR spectroscopy. X-ray diffraction involves the study of the manner in which X-rays are diffracted (scattered) as they pass through a protein crystal. Subsequent mathematical analysis reveals the relative location of each atomic nucleus within the protein that has been crystallized. NMR spectroscopy measures the intensity of absorbance of radiowaves by specific atomic nuclei in proteins that have been placed in a powerful external magnetic field. Since the electron environment of an absorbing nucleus helps determine what frequencies of radiowave radiation it absorbs, an analysis of the specific frequencies that are absorbed by a protein provides information about the relative location of the numerous nuclei within it.

Tertiary structures encompass one or more relatively stable, independently folded regions called domains that commonly contain recurring folding patterns known as motifs or supersecondary structures. Helix–turn–helix motifs, β-barrels, and α/β-barrels are specific examples of supersecondary structures. The secondary, tertiary, and quaternary (if present) structures of a globular protein constitute its conformation; they determine its overall three-dimensional shape and geometry.

Most globular proteins must be folded into a highly specific conformation to be biologically active. In an aqueous environment, nonpolar side chains fold into the core of a globular protein whereas polar side chains, including charged side chains, remain on the surface; "nonpolar in, polar out." The native (biologically active) form of a protein normally represents its thermodynamically most stable folding pattern. Chaperones and enzymes sometimes assist in the folding process. Since the native conformation is primarily maintained (sometimes totally) by relatively weak noncovalent interactions, most globular proteins can be easily unfolded without breaking covalent bonds, a process described as denaturation. A denatured protein is normally water insoluble and inactive.

Over time, evolution has selected proteins whose structures are optimized to occupy specific functional niches. Both myoglobin and hemoglobin, for example, are containers constructed from α-helices. They are designed to hold O_2 and to release that O_2 only under specific conditions. Myoglobin resides in muscle, where it stores O_2 until the O_2 in its environment drops below a critical concentration. At that point, myoglobin's hyperbolic O_2 binding curve allows it to dump most of its cargo in response to slight additional decreases in O_2 concentration. Each myoglobin molecule is assembled from one polypeptide chain and a heme group, and it possesses a single O_2 binding site.

Hemoglobin is found only in red blood cells, where it continuously circulates throughout the body, picking up O_2

in the lungs and delivering it to every cell in the body. Hemoglobin also helps transport CO_2 from tissues back to the lungs. Each hemoglobin molecule encompasses four polypeptide chains, four heme groups, and four O_2 binding sites. Hemoglobin's quaternary structure and the communication that takes place between its subunits lead to the cooperative binding of O_2 and a sigmoidal binding curve. It is the sigmoidal binding behavior of hemoglobin that allows it to distribute O_2 throughout the body as blood moves away from the lungs to distant tissues containing lower and lower O_2 concentrations. The exact position of hemoglobin's O_2 binding curve is constantly shifting in response to changing concentrations of multiple effectors, including H^+ and 2,3-BPG. The effectors help make the body's O_2 delivery system responsive to both internal and external changes in O_2 concentration. Fetuses produce a special hemoglobin with an O_2 binding curve that lies to the left of the binding curve for adult hemoglobin. This facilitates the transfer of O_2 from the blood of the mother to the blood of the fetus.

Collagen and hair provide two other examples of the precision with which evolution has led to the coupling of structure with function. In both instances, fibrous helical polypeptides are wound together to generate coiled coils (superhelixes) that are partially held together by covalent bonds. Multiple superhelixes are subsequently packed together in a specific geometry, and then connected through additional covalent bonds. The final product is a tough, water-insoluble material. Collagen is employed as a structural element for the construction of skin, cartilage, bone, blood vessels, and a variety of other body components.

Immunoglobulin G molecules (IgG's) are Y-shaped glycoproteins designed to recognize and bind bacteria, viruses, and other "foreign" agents known as antigens. Each of the millions of different IgG's contains 4 covalently linked polypeptide chains, 12 domains, and 2 antigen combining sites, one located at the end of each prong of the Y. Each combining site is constructed from the hypervariable loops in one heavy-chain domain and one light-chain domain. Each domain (called an immunoglobulin fold) in an IgG contains two stacked β-sheets that are held together, in part, by a disulfide bond. Once an antigen is bound to IgG, the base of the Y interacts with other immune system components, which are ultimately responsible for the elimination or destruction of that antigen.

Nine of the 20 common protein amino acids cannot be synthesized in adequate amounts by the human body; they must be continuously supplied in the diet. Dietary proteins represent the primary source of these essential amino acids (Chapter 3). Since a "typical" animal protein has a better balance of the essential amino acids than a "typical" plant protein, animal protein tends to be of higher nutritional quality.

EXERCISES

4.1 A nucleoprotein is an albumin that binds ligands cooperatively. What does this description tell you about the properties and structure of this protein?

4.2 List five major functions of proteins in the body.

4.3 List five specific proteins encountered in this chapter; then list the functions of each.

4.4 True or false? All proteins contain one or more polypeptide chains. Explain.

4.5 Give the abbreviations for all possible tripeptides containing one copy each of aspartic acid, phenylalanine, and valine. How many tetrapeptides can be produced that contain one copy each of serine, methionine, arginine, and tyrosine? Take a guess at the number of different peptides that can be produced containing one copy of each of the 20 protein amino acids.

4.6 The two groups of peptides below are generated when peptide P is digested under two different sets of cleavage conditions.

Group 1 (from the first digestion):
Met–Phe–Glu–Met–His–Ile–Lys
Asp–Gly–Ser–Lys
Ala–Thr–Met–Pro–Tyr–Arg
Gln–Val

Group 2 (from the second digestion):
Phe–Glu–Met
Pro–Tyr–Arg–Gln–Val
Asp–Gly–Ser–Lys–Met
His–Ile–Lys–Ala–Thr–Met

What is the primary structure of peptide P?

4.7 When protein K and protein M are separately subjected to amino acid analysis (Chapter 3) they yield exactly the same results. Can you confidently conclude that the two proteins have the same primary structure? Explain.

4.8 Study Figures 4.18 and 4.19. How many amino acid residues exist within the F helix? The H helix? Which other helices appear to make contact with the F helix? Predict whether the helix–helix contacts in myoglobin primarily involve nonpolar side chains or polar side chains? Explain.

4.9 How many H bonds exist within an α-helix that contains a total of 36 amino acid residues? How many complete turns does this helix contain?

4.10 Given the following peptide fragment:

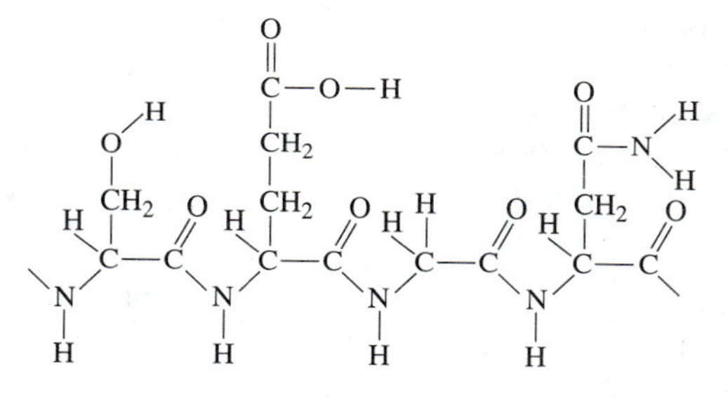

a. Assume that this fragment is located in the center of a long α-helical region, and then circle each atom that will be H bonded within the helix.

b. Put a square around each atom that will be H bonded if this fragment exists within the middle of an extended β-sheet.

c. Will this fragment more likely be folded into the interior of a globular protein or located on its surface? Explain.

4.11 Can a polypeptide chain coil into a helix that is not an α-, 3_{10}-, or π-helix? Explain.

4.12 Invent a globular protein that contains three α-helical regions and two β-sheets. Draw a schematic diagram (similar to the one in Figure 4.7) illustrating the secondary and tertiary structure of this imaginary protein.

4.13 What class or classes of bonds maintain the primary structure of proteins? The secondary structure of proteins? The tertiary structure of proteins? The quaternary structure of proteins? Which class listed represents the strongest class of bonds?

4.14 How many amino acid residues exist in the ball-and-stick model of the π-helix pictured in Figure 4.3? Compare this number to the number of amino acid residues in the 3_{10}-helix depicted in the same figure.

4.15 Which of the amino acids have side chains that can participate in salt bridge formation at pH 1? at pH 7? at pH 14?

4.16 Can a single amino acid residue simultaneously be part of both an α-helical region and a β-sheet region within a protein? Explain.

4.17 It is not uncommon for the amino acid residues on one surface of an α-helix to consist of internal and variable residues (Table 4.5) while the opposite surface contains external and variable residues. Suggest why this is the case.

4.18 The following polypeptide can be divided into two segments that would tend to be located internally within a globular protein and two segments that would tend to be located on the surface. Identify the two internal segments and the two external segments.

Arg–Cys–Pro–Gly–Ser–Glu–Asn–Gly–Ala–
Leu–Pro–Trp–Phe–Ser–Met–Cys–Gly–His–Tyr–
Glu–Asn–Thr–Lys–Tyr–Ala–Ile–Phe–Met–Pro–
Cys–Val

4.19 Exercise 4.18 might leave you with the false impression that internal amino acid residues are always clustered together along the primary structure of a globular protein, whereas external residues reside in separate clusters. In reality, internal and external residues may be mixed together within even short segments of primary structure. Explain how this is possible without violating the rule "nonpolar in, polar out."

4.20 Explain why extracellular proteins tend to contain more disulfide bonds than intracellular ones.

4.21 List, and then describe, two specific motifs.

4.22 List three different treatments that will destroy the quaternary structure of most proteins.

4.23 Arrange the following in their proper hierarchy within the structure of proteins: primary structure, secondary structure, tertiary structure, quaternary structure, domains, motifs, α-helices, and β-sheets. List the lowest ranked structural element first.

4.24 How do chaperones facilitate the production of a native protein. To what class of compounds do chaperones belong?

4.25 When a pure sample of protein Y is run through a gel filtration column (Section 3.9) at pH 7.0, a single peak elutes 10 minutes after the sample is added to the column. When the chromatography is repeated at pH 1.0, a single peak elutes after 27 minutes. Explain these observations. Assume that no covalent bonds are cleaved at pH 1.

4.26 Myoglobin is a nutritionally complete protein. What does this tell you about its structure?

4.27 Seventy percent ethanol solutions have antiseptic properties. What is the molecular explanation for this fact?

4.28 Assume that a water molecule is a box 0.1 by 0.2 by 0.3 nm; then calculate the approximate number of water molecules that can be packed into the volume occupied by a single myoglobin molecule. Assume that a myoglobin molecule is a rectangular box with the dimensions 4.4 by 4.4 by 2.5 nm.

4.29 What structural feature of myoglobin keeps four myoglobin molecules from coming together to form a tetramer similar to hemoglobin?

4.30 Will myoglobin bind and carry O_2 normally if one of the two histidine residues in its heme pocket is replaced with an alanine? Explain.

4.31 Explain why the bends and loops in myoglobin have a larger percentage of polar residues than is found in the helical regions.

4.32 We quickly die without O_2. What is the molecular explanation for this fact?

4.33 Which of the following pairs of compounds contain members that can be readily separated from one another with gel filtration chromatography?

PAIR 1 Myoglobin, hemoglobin
PAIR 2 HbA, HbF
PAIR 3 α chain of Hb, β chain of Hb
PAIR 4 Ala, myoglobin
PAIR 5 A decapeptide, myoglobin

4.34 If a nonpolar side chain in the heme pocket of myoglobin is replaced with an aspartic acid side chain, will the heme group bind more tightly or less tightly? Explain.

4.35 What effect will a decrease in the equilibrium constant for the binding of O_2 by hemoglobin have on the position of the sigmoidal O_2 binding curve depicted in Figure 4.24? Explain.

4.36 In the absence of α chains, the β chains of hemoglobin will aggregate into tetramers that can bind and carry O_2. The binding of O_2, however, is not cooperative. Suggest why this is the case.

4.37 Why does hemoglobin have a larger fraction of hydrophobic amino acid residues than myoglobin?

4.38 How many domains exist within a single myoglobin molecule? Hemoglobin molecule?

4.39 List the three major forms in which CO_2 is transported from tissues to the lungs.

4.40 Death caused by carbon monoxide (CO) poisoning is primarily due to suffocation. Carbon monoxide binds very tightly (over 100 times more tightly than O_2) but reversibly to the O_2 binding sites of hemoglobin. Hemoglobin bound to carbon monoxide cannot bind O_2 and vice versa. Suggest a treatment for carbon monoxide poisoning.

4.41 What effect will carbon monoxide have on the O_2 binding curve for hemoglobin? Explain.

4.42 The plasma membrane that surrounds cells has a nonpolar core. Given this fact, suggest why the intravenous administration (injection into a vein) of 2,3-BPG does not increase the concentration of 2,3-BPG within red blood cells.

4.43 If the concentration of 2,3-BPG in red blood cells is much greater than that corresponding to the curve to the right in Figure 4.27 will the amount of O_2 delivered to tissues be greater, less, or the same as that depicted by this curve? Explain.

4.44 Study Figure 4.26. What percentage of the total O_2 that hemoglobin is capable of binding will be released if the pO_2 shifts from 80 torr to 20 torr at pH 7.6? AT pH 7.2?

4.45 A study of Figure 4.26 might lead us to predict that at very high pH's (10 to 12) hemoglobin will bind O_2 so tightly that it will not release a significant amount of O_2, even at very low O_2 concentrations. Explain why this is not the case.

4.46 Give a molecular explanation for why steak from an old steer tends to be tougher than steak from a young steer.

4.47 Draw a graph similar to the one in Figure 3.9 clearly illustrating the type of results obtained when collagen is subjected to amino acid analysis.

4.48 List two unusual amino acid residues found in collagen. Are these residues present in a newly assembled collagen polypeptide?

4.49 Explain why hair protein and collagen are more difficult to digest than most globular proteins.

4.50 From a nutritional standpoint, does collagen represent a high-or a low-quality protein? Explain.

4.51 How many separate domains exist within a single IgG molecule? Are these domains α, β, or α/β?

4.52 If the two heavy strands of IgG are cleaved at the hinge region (Figure 4.33), will the light chain–containing fragments be able to bind antigen? Explain.

4.53 Describe the primary, secondary, and tertiary structures of IgG. Does IgG have a quaternary structure? Explain.

4.54 List the major advantages and disadvantages of NMR spectroscopy (compared with X-ray diffraction) when this technique is employed to determine protein conformations.

4.55 How many peaks exist in the ^{13}C NMR spectrum for each of the compounds in Problem 4.45?

SELECTED READINGS

Ang, D., K. Liberek, D. Skowyra, M. Zylicz, and C. Georgopoulos, Biological role and regulation of the universally conserved heat shock proteins, *J. Biol. Chem.* 266, 24233–24236 (1991).

A "minireview" dealing with the structure, function, and regulation of production of certain heat-shock proteins, including some molecular chaperones.

Bohen, S.P., A. Kralli, and K.R. Yamamoto, Hold 'em and fold 'em: Chaperones and signal transduction, *Science* 268, 1303–1304 (1995).

Discusses potential roles for chaperones in steroid receptor-linked signaling.

Branden, C., and J. Tooze, *Introduction to Protein Structure*. New York: Garland Publishing, 1991.

Examines the basic themes and principles of protein structure with emphasis on structure-function relationships. Contains an excellent introduction to structure determination with X-ray diffraction and NMR. Each chapter concludes with a substantial list of selected readings.

Bray, D., Protein molecules as computational elements in living cells, *Nature* 376, 307–312 (1995).

Proposes that the major function of many proteins is the transfer and processing of information. Such proteins may be linked in "circuits" that perform a variety of simple computational tasks including amplification, integration, and information storage.

Creighton, T. E., *Proteins: Structures and Molecular Properties*, 2nd ed. New York: W. H. Freeman and Company, 1993.

Topics include the chemical properties of polypeptides, protein biosynthesis, evolutionary origins of protein sequences, noncovalent interactions within proteins, conformational properties of polypeptide chains, the folding of globular proteins, enzyme catalysis, and protein degradation. There are ample references at the end of each section and a limited number of problems (most based on the research literature).

Dickerson, R.E., and I. Geis, *The Structure and Action of Proteins*. New York: Harper & Row, Publishers,1969.

A classic! Examines structure–function relationships in specific fibrous and globular proteins, including multiple enzymes.

Dickerson, R.E., and I. Geis, *Hemoglobin: Structure, Function, Evolution, and Pathology*. Menlo Park, California: Benjamin/Cummings Publishing Company, 1983.

Another classic! Everything you wanted to know about hemoglobin but were afraid to ask. Exceptional graphics.

Doolittle, R.F., and P. Bork. Evolutionarily Mobile Modules in Proteins, *Sci. Am.* 269(4), 50-56 (1993).

Examines protein domains and their role in the evolution of proteins.

Green, P., D. Lipman, L. Hillier, R. Waterston, D. States, and J-M. Claverie, Ancient conserved regions in new gene sequences and the protein databases, *Science* 259, 1711-1716 (1993).

Describes the search for ancient evolutionarily conserved regions in present-day proteins.

Hall, S.S., Protein images update natural history, *Science* 267, 620-624, (1995).

Reviews the techniques used to determine and image protein structure.

Kamtekar, S., J.M. Schiffer, H. Xiong, J.M. Babik, and M.H. Hecht, Protein design by binary patterning of polar and nonpolar amino acids, *Science* 262, 1680-1685 (1993).

Examines a strategy for studying the role of polar and nonpolar amino acids in determining the folding of proteins.

Olson, A.J., and D.S. Goodsell, Visualizing biological molecules, *Sci. Am.* 267(5), 76-81 (1992).

Reviews the technology used to visualize biological molecules. Emphasis is on computer graphics.

Orengo, C. A., D.T. Jones, and J.M. Thornton, Protein superfamilies and domain superfolds, *Nature* 372, 631-634 (1994).

Presents a classification system for proteins based on a comparison of both primary and tertiary structures.

Perutz, M., *Protein Structure: New Approaches to Disease and Therapy*. New York: W. H. Freeman and Company, 1992.

Designed to help readers "become familiar with, and put to practical use, the great body of new knowledge of molecular anatomy, physiology, and pathology (of proteins) that is now being created." The first chapter provides an excellent introduction to protein structure determination. The book, which is roughly 300 pages long, concludes with 23 pages of references.

Schneider, D., MRI goes back to the future, *Sci. Am.* 272(3), 42,(1995).

Examines new MRI designs that will lower imaging costs and make these machines more user friendly.

Smith, D.F., Steroid receptors and molecular chaperones, *Sci. Am. Science & Medicine* 2(4), 38–47 (1995).

Examines the role of chaperones in the functioning of steroid hormones. Chaperones appear to modulate the activity of steroid receptors.

5 Enzymes

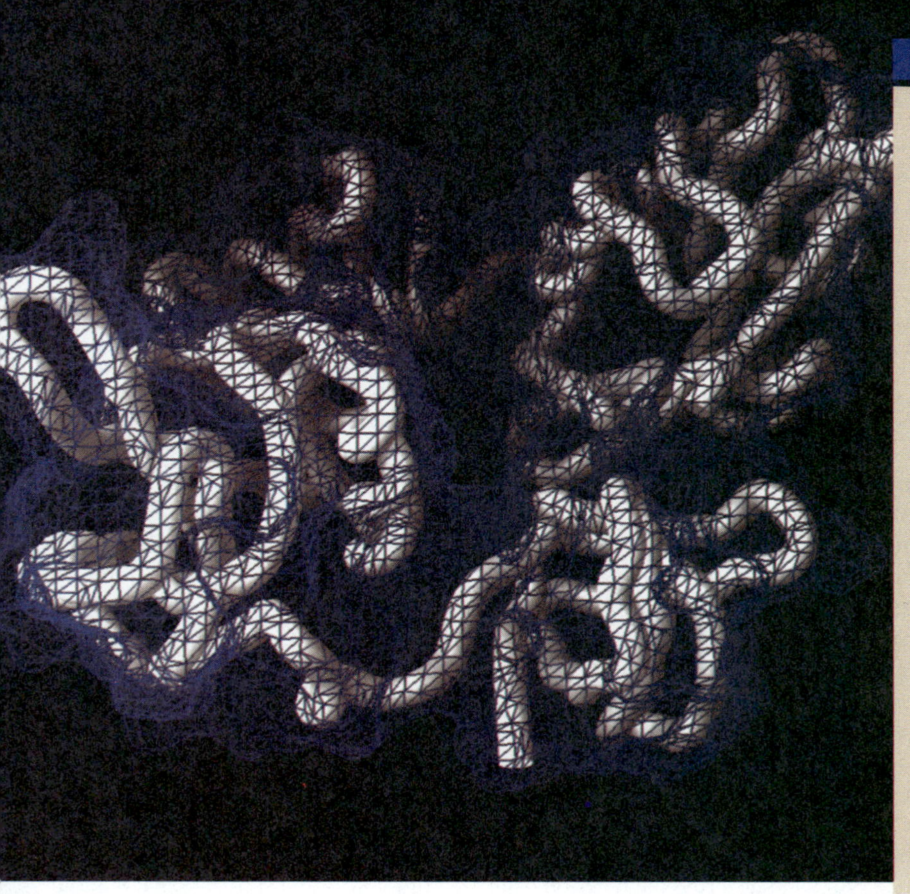

CHAPTER OUTLINE

Thousands of chemical reactions occur within the human body and within most other living organisms at all times. A vast majority of these reactions would not proceed, at least not at a biologically significant rate, in the absence of **catalysts** (substances that increase the rates of reactions without being permanently modified in the process). Thus, to a very large degree, the catalysts in an organism determine what reactions occur and how rapidly they occur. Collectively, these reactions maintain life.

Biological catalysts are called enzymes. They include a large number of globular proteins and a much smaller number of more recently discovered ribonucleic acids (RNAs). The RNA catalysts (Section 17.2) are known as **ribozymes.** In this text, the term *enzyme* refers to the classical protein catalysts unless specified otherwise. Protein enzymes possess the structural hierarchy and other characteristics described for globular proteins in Chapter 4.

This chapter examines the general properties of enzymes and the regulation of their catalytic activities. It also explores some of the applications of enzymes in medicine. Several of the major biochemical themes (Exhibit 1.4) are featured in the process, including experimentation and hypothesis construction, evolution, catalysis and the regulation of reaction rates, structure–function relationships, structural hierarchy and compartmentation, communication, and specificity. Because information for the production of both classes of enzymes is encoded in DNA, enzymes are linked to the theme of information transfer as well.

Chapter 6 moves from enzymes into a discussion of carbohydrates in preparation for a look at nucleotides, bioenergetics, metabolism, and other topics. The carbohydrate and enzyme stories are directly interrelated; amino acid side chains on many enzymes are coupled covalently to carbohydrate molecules.

5.1 NAMING AND CLASSIFYING ENZYMES

The Nomenclature Committee of the International Union of Biochemistry and Molecular Biology has adopted a systematic approach for naming and classifying enzymes. Each enzyme is placed in one of six major divisions or classes on the basis of the type of reaction that it catalyzes (Table 5.1). Each major division is further divided into subclasses and subsubclasses. Each class, subclass, and subsubclass is numbered so that any given enzyme can be unambiguously identified with a unique sequence of code numbers. **[Try Problem 5.1]**

Many enzymes were discovered, named, and categorized long before the adoption of a systematic classification system. For this reason, and because of the cumbersome nature of the systematic system, the Nomenclature Committee has suggested that a single trivial name be retained (or invented) for each enzyme and that this trivial name be used along with the systematic name. Guidelines exist for the construction of trivial names, now described as **recommended names.** Some recommended names, such as pepsin and chymotrypsin, give no clue to the nature of the reaction catalyzed by the enzyme. Many recommended names, however, have an ***-ase*** ending and describe the reaction catalyzed or identify the substrate that is modified during the reaction. DNA polymerases catalyze the polymerization of monomers into DNA. Asparaginase catalyzes the hydrolysis of asparagine. Lactate dehy-

TABLE 5.1

The Six Major Classes of Enzymes

Main Class	Type of Reaction Catalyzed	Specific Example
1. Oxidoreductases	Oxidation–reduction reactions.	Lactate (CH_3—CHOH—C(=O)—O^-) + NAD^+ —(Lactate dehydrogenase)→ Pyruvate (CH_3—C(=O)—C(=O)—O^-) + NADH + H^+
2. Transferases	Transfer of an intact group of atoms from a donor molecule to an acceptor molecule.	Glycogen phosphorylase (an enzyme) + ATP —(Glycogen phosphorylase kinase)→ glycogen phosphorylase—phosphate (a phosphorylated enzyme) + ADP
3. Hydrolases	Hydrolysis of bonds.	Glycogen synthase—phosphate (a phosphorylated enzyme) + H_2O —(Glycogen synthase phosphatase)→ glycogen synthase (an enzyme) + phosphate
4. Lyases	Cleavage of C—C, C—O, C—N, and other bonds by means other than hydrolysis or oxidation; includes reactions that eliminate water to leave double bonds or add water to a double bond.	H_2CO_3 —(Carbonic anhydrase)→ CO_2 + H_2O
5. Isomerases	Interconversion of various isomers, such as cis ⇌ trans, L ⇌ D, aldehyde ⇌ ketone.	D-Alanine (H—C($\overset{+}{N}H_3$)(CH_3)—C(=O)—O^-) —(Alanine racemase)→ L-Alanine ($H_3\overset{+}{N}$—C(H)(CH_3)—C(=O)—O^-)
6. Ligases	The joining of two molecules with the coupled hydrolysis of ATP or a similar triphosphate.	Acetate (CH_3—C(=O)—O^-) + CoASH + ATP —(Acetyl-SCoA synthetase)→ Acetyl-SCoA (CH_3—C(=O)—SCoA) + ADP + phosphate

drogenase (LDH) catalyzes the dehydrogenation (removal of hydrogen) of lactate to yield pyruvate. No doubt about it, enzyme nomenclature can be confusing; several thousand enzymes have already been identified and named. To get a start on this topic, learn the names of those enzymes that your instructors emphasize.

PROBLEM 5.1 To which one of the six major classes of enzymes does LDH belong? Place those enzymes that catalyze the conversion of *cis*-peptide bonds to *trans*-peptide bonds (Section 4.4) in the appropriate major class?

5.2 RATES OF REACTIONS

The measurement of reaction rates (velocities) lies at the heart of enzyme biochemistry. Such measurements normally involve the determination of the amount of product formed or the amount of reactant modified per unit of time. **Micromoles per second** (**μmol/s**) and **millimoles per minute (mmol/min)** are common units of rate. The **initial velocity** (**v_0**) for a reaction is its rate immediately after reactants have been mixed to initiate the reaction. At this time, there is very little product in the reaction mixture, and the reverse reaction can, for most purposes, be ignored. **[Try Problem 5.2]**

EXHIBIT 5.1
Ways to Increase Reaction Rates for Bimolecular Reactions

- Increasing Reactant Concentration:
    - Increases number of collisions per second.
- Increasing Temperature:
    - Increases number of collisions per second.
    - Increases the fraction of collisions that generate the necessary activation energy.
- Adding a Catalyst:
    - Lowers activation energy.
    - May orient reactants for productive interactions.
    - May operate in additional ways as well.

Collision Theory

What determines reaction rates? The currently accepted answer lies in **collision theory.** According to this theory, properly oriented reactants must collide, and that collision must lead to the acquisition (by the reactants) of a certain minimum amount of energy, known as the **activation energy** (**ΔG‡** or **E_a**), before a reaction can proceed. The activation energy is an energy barrier that separates reactants and products. An activation energy is associated with both bimolecular (two reactants) and unimolecular (a single reactant rearranges or dissociates) reactions and with both **endergonic** (energy is taken in and the products are of higher energy than the reactants) and **exergonic** (energy is released and the products are of lower energy than reactants) transformations. The activation energy for a reaction is commonly illustrated with a **progress of reaction diagram** in which energy (*y*-axis) is plotted against "progress of reaction" (*x*-axis) (Figure 5.1). The relatively unstable state in which reactants find themselves at the top of the activation energy barrier is known as the **transition state** for the reaction. **[Try Problems 5.3 and 5.4]**

Collision theory predicts that the rate of a bimolecular reaction can be increased in at least four ways: (1) by increasing the number of collisions per second, (2) by increasing the average energy of collision, (3) by lowering the activation energy, and (4) by optimizing the relative orientation of reactants at the time of collision (Exhibit 5.1). All of these predictions have been verified experimentally. One can boost collision frequencies and enhance the velocity of bimolecular reactions by either elevating the temperature of the reaction mixture or increasing reactant concentrations, or both. Higher temperatures additionally enhance reaction rates by increasing the fraction of productive collisions. The temperature effects are explained by the more rapid molecular motion that is associated with a rise in temperature. As

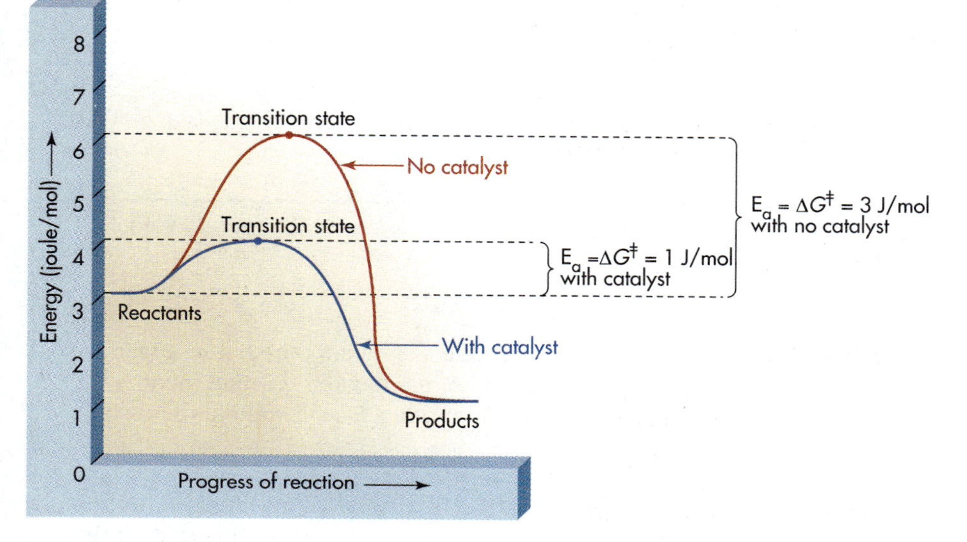

FIGURE 5.1
A progress of reaction diagram. Reactants must pass over an energy barrier, known as the activation energy ($\Delta G^{\ddagger}$) for the reaction, before they can be converted to products. Catalysts increase reaction rates by lowering the activation energy. Catalysts sometimes increase reaction rates through other mechanisms as well.

molecules travel more rapidly, they collide more frequently and more forcefully. Since more forceful collisions are more energetic collisions, they carry a larger fraction of the colliding reactants over the activation energy barrier. The fraction of productive collisions can also be enhanced by lowering the activation energy while holding the temperature constant, a feat that can be accomplished by catalysts. Catalysts can also enhance reaction rates by binding reactants and, in the process, orienting them for productive interactions. The proper positioning of colliding functional groups is an essential prerequisite for the making and breaking of chemical bonds. **[Try Problem 5.5]**

Equilibrium Versus Reaction Rates

The rate of a reaction is independent of the equilibrium constant for that reaction, a concept central to the understanding of life processes. At a constant temperature, equilibrium constants are determined by the free energy difference between reactants and products (Section 8.2), not by activation energies, collision frequencies, and relative orientations during collisions. It is possible for a reaction to have a very large equilibrium constant yet occur at a very slow rate, and vice versa. In the presence of a catalyst, equilibrium is reached more rapidly, but the equilibrium constant and the final equilibrium are unchanged. Problems 5.6 through 5.9 are designed to reinforce these concepts and to provide a review of equilibrium and equilibrium constants. Chapter 8 encompasses a more thorough review of these topics. **[Try Problems 5.6 through 5.9]**

Most of the reactions within a living organism never reach equilibrium regardless of how rapidly they occur, because the reactants or products in one reaction tend to be reactants or products in one or more additional chemical transformations. A reaction cannot reach equilibrium if reactants or products are constantly being added to or removed from a reaction mixture. There is some truth in the saying "Old biochemists never die, they just reach equilibrium."

PROBLEM 5.2 A reaction can be summarized with the following balanced equation:

$$B + C \rightleftharpoons D + 2\,E$$

At time zero, 1 mol of C and 1 mol of B are mixed together in a reaction mixture that contains no D or E. At 0.5 minutes the reaction mixture contains 5 mmol of product D, some product E, and some unreacted B and C.

a. Express the rate of the reaction as millimoles of D formed per minute and as micromoles of D formed per second. Does this rate represent the initial velocity for this reaction? Explain.
b. How much product E will exist in the reaction mixture at 0.5 minutes?
c. Calculate the rate of formation of E in micromoles per second.
d. Express the rate of the reaction as millimoles of B converted to products per minute.

PROBLEM 5.3 Place a curve on Figure 5.1 illustrating what type of results will be obtained with a catalyst that leads to half the decrease in activation energy observed with the catalyst depicted in this figure.

PROBLEM 5.4 Draw a progress of reaction diagram for a reaction with:

a. A relatively low activation energy.
b. A relatively high activation energy.
c. Products possessing more energy than reactants.

***PROBLEM 5.5** Define or describe "heat" and "temperature."

***PROBLEM 5.6** Write the equilibrium constant expression for each of the following reactions:

$$H_2CO_3 \rightleftharpoons HCO_3^- + H^+$$

$$CH_3OH + CH_3COOH \rightleftharpoons CH_3COOCH_3 + H_2O$$

$$3\,A + 2\,B \rightleftharpoons 2\,C + 2\,D$$

***PROBLEM 5.7** True or false? If the equilibrium constant for a reaction is very small, there will be more reactants than products in a reaction mixture at equilibrium. Explain.

***PROBLEM 5.8** True or false? If the equilibrium constant for a reaction is very small, we can be confident that the rate of the reaction will be very slow. Explain.

***PROBLEM 5.9** Explain why the equilibrium constant for a bimolecular reaction is independent of the initial concentration of reactants.

5.3 SOME ENZYME-RELATED UNITS AND ENZYME ASSAYS

Both researchers and clinicians are commonly faced with the task of determining and expressing both enzyme concentration and the catalytic capability of enzyme preparations. A variety of terms, units, and approaches are associated with this endeavor.

Katals and International Units

Units of enzyme activity are based on the measurement of reaction rates. The official unit of enzyme activity (recommended by the International Union of Biochemistry and Molecular Biology) is the **katal (kat),** the amount of enzyme that converts 1 mol of reactant to product in 1 second under standard reaction conditions (usually optimal). In the health sciences, one commonly encounters the **international unit (U),** that amount of enzyme that leads to the formation of 1 μmol of product per minute under standard conditions. Since 1 μmol/min equals 1.67×10^{-8} mol/s, 1 U equals 1.67×10^{-8} kat. In clinical chemistry, enzyme concentrations in serum and other biological specimens are commonly recorded as U/mL or U/L. [Try Problem 5.10]

Specific Activity and Enzyme Purification

The **specific activity** of an enzyme preparation is the number of international units or katals per milligram of protein. Specific activity is a measure of enzyme purity; its value reaches a maximum and then remains constant when all of the molecules in a sample are active enzyme molecules. The purification of an enzyme is usually monitored by determining the specific activity of the enzyme preparation after each purification step. A purification scheme for a hypothetical enzyme is depicted in Table 5.2. Like the purification scheme for most proteins, it entails multiple chromatographic procedures (Section 3.9). An enzyme is usually certified as pure when a variety of further efforts at purification lead to no additional increase in specific activity. Analytical electrophoresis and analytical chromatography can also be employed to assess the purity of an enzyme preparation; a pure sample of an enzyme will normally yield a single peak or band. A pure oligomeric enzyme may yield multiple peaks or bands if all of its subunits are not identical. [Try Problems 5.11 and 5.12]

Enzyme Assays

Under appropriate standard conditions, the amount of enzyme in a reaction mixture determines the initial velocity of the catalyzed reaction. This fact is the basis for **en-**

TABLE 5.2

A Purification Scheme for a Hypothetical Enzyme

Purification "Step"	Total Milligrams Protein	Total U[a]	Specific Activity (U/mg protein)
Isolate crude cellular extract	12,000	150,000	12.5
$(NH_4)_2SO_4$ fractionation[b]	4,000	140,000	35
Gel filtration chromatography	500	120,000	240
Ion exchange chromatography (pH 6.0)	75	95,000	1,260
Ion exchange chromatography (pH 7.8)	6	80,000	13,300

[a]U = international unit. Some enzyme molecules are denatured or "lost" at each step.
[b]Distinct proteins tend to precipitate at different $(NH_4)_2SO_4$ concentrations.

zyme assays, the experimental determination of the amount of enzyme in a sample. During an assay, a known amount of the sample whose enzyme content is to be analyzed is added to a standard reaction mixture and then initial velocity is measured. This initial velocity can be used to "read" the enzyme concentration in the reaction mixture from a graph prepared by plotting reaction rate (y-axis) versus known concentration of enzyme (x-axis). In the ideal case, initial velocity is proportional to enzyme concentration, and the graph (called a **standard curve**) is a straight line. **[Try Problem 5.13]**

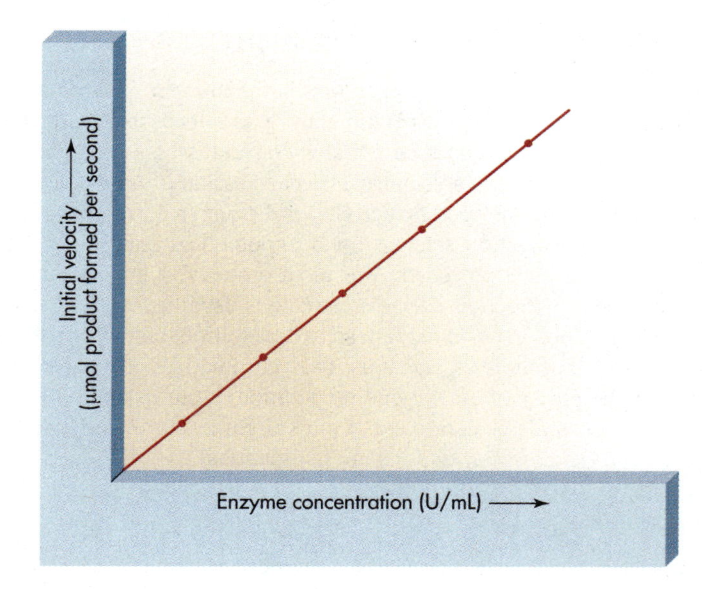

An idealized standard curve.

PROBLEM 5.10 A 0.1 mL sample of human serum contains enough enzyme X to convert 50 μmol of substrate to product per second under a standard set of reaction conditions. Calculate the international units of enzyme X per milliliter of serum. How many katals in 1.0 mL of serum?

PROBLEM 5.11 What fold purification (relative to the crude extract) is attained with the purification scheme depicted in Table 5.2?

PROBLEM 5.12 Calculate the number of international units in 2 g of an enzyme preparation whose specific activity is 10^5 U/mg protein.

PROBLEM 5.13 After a 10 μL aliquot of an enzyme preparation is added to a standard reaction mixture, the initial velocity is found to be 0.01 μmol/min. Calculate the number of U/mL within the original enzyme preparation.

5.4 SOME GENERAL PROPERTIES OF ENZYMES

Enzymes, like many other catalysts, increase reaction rates through a variety of mechanisms, including lowering of activation energies. Enzymes, however, tend to be substantially more efficient than their nonbiological counterparts; some enzymes can increase reaction rates (when compared with noncatalyzed reactions) close to 10^{20}-fold, and many are capable of boosting rates over 10^{10}-fold. Since enzymes are such phenomenal catalysts, most are required at extremely low concentrations by those organisms that produce them. The thousands of distinctive enzymes in a typical cell account for only a small fraction of the cell's total mass. **[Try Problem 5.14]**

Cofactors and Coenzymes

Enzymes, like other proteins, are classified as simple or complex (Section 4.1). The prosthetic groups (nonpolypeptide components) within a complex enzyme are called **cofactors.** Most organic cofactors, **coenzymes,** are derivatives of the water-soluble vitamins (Section 10.1, Table 10.6). Metal ions are common inorganic cofactors. Those enzymes whose active forms encompass one or more tightly bound metal ions (usually Fe^{2+}, Fe^{3+}, Cu^{2+}, Zn^{2+}, Mn^{2+}, Co^{3+}, or Mo^{2+}) are labeled **metalloenzymes.** They include roughly one third of all known enzymes. Enzymes that require loosely bound metal ions (often Na^+, K^+, Mg^{2+}, or Ca^{2+}) from solution are classified as **metal-activated enzymes.** **[Try Problem 5.15]**

Subtrate Specificity

The reactants in an enzyme-catalyzed reaction are said to be **substrates.** Most enzymes are extremely specific and will catalyze the reaction of a single substrate or a very limited number of substrates. Enzymes can even distinguish stereoisomers (configurational isomers). For this reason, they are said to be **stereospecific.** Hence, humans and most other organisms use only L-amino acids (no D-amino acids) to build proteins (Section 3.3). The remarkable substrate specificity of enzymes allows an organism to control individually the rates of most of the reactions that occur within it, an ability fundamental to life itself. [Try Problems 5.16 and 5.17]

PROBLEM 5.14 What is the minimum number of enzyme molecules required to convert 10^6 molecules of reactant to product? Assume that no product is formed in the absence of enzyme. Explain.

PROBLEM 5.15 List the chemical bonds or forces that are primarily responsible for maintaining:

a. The primary structure of an enzyme.
b. The secondary structure of an enzyme.
c. The tertiary structure of an enzyme.
d. The quaternary structure of an enzyme.

PROBLEM 5.16 Estimate the number of unique enzymes within a cell in which 2500 different reactions are occurring.

PROBLEM 5.17 Explain why you would expect to find some enzymes in brain cells that are not present in liver cells.

5.5 ENZYME–SUBSTRATE INTERACTIONS

The first step in an enzyme-catalyzed reaction is the formation of an **enzyme–substrate complex** (ES), which is subsequently converted to free enzyme (E) and product (P). In the simplest case, the overall process involves two steps:

$$E + S \rightleftharpoons ES \rightleftharpoons E + P$$

Active Sites

Each enzyme contains one or more **active sites,** which are composed of a **substrate binding site** and a **catalytic site** that commonly overlap. The binding site usually consists of a chiral pocket or groove where amino acid side chains and prosthetic groups (if present) initially (at least) attach to substrate molecules through noncovalent interactions. The chirality of this binding site accounts for the ability of enzymes to distinguish enantiomers and diastereomers; a chiral binding site is required to distinguish chiral compounds. By analogy, your chiral right hand can distinguish a right–handed glove from a left–handed glove, a feat that could not be accomplished by an achiral hand (Figure 5.2). [Try Problems 5.18 through 5.20]

The active sites on enzymes are created and maintained through the highly specific folding of one or more polypeptide chains, for it is this folding that brings together the amino acid side chains that constitute these sites. Any condition or treatment that modifies the secondary, tertiary, or quaternary structure of a protein tends

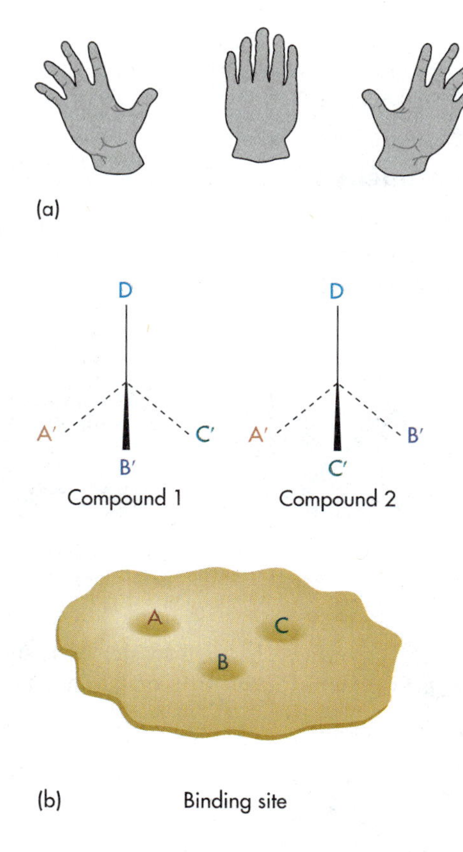

FIGURE 5.2
Stereospecific binding requires chiral binding sites. (a) An achiral hand can not distinguish a left-handed glove from a right-handed one. (b) This chiral binding site will accommodate compound 1 but not compound 2.

to change the binding and catalytic activities of an enzyme by changing the architecture of its active site. Since primary structure ultimately determines the other structural elements within a protein, its integrity is also essential for normal activity. **[Try Problems 5.21 through 5.23]**

The Nature of Catalysis

When binding a substrate, an enzyme tends to shift the structure of that substrate toward the structure of the transition state for the reaction catalyzed, and it positions the substrate for specific interactions with the amino acid side chains (and prosthetic groups, in some cases) at the catalytic site. In bimolecular reactions, the binding of the two substrates also locks them into relative orientations that facilitate the reaction, a process that partially accounts for the catalytic power of some enzymes. As the reaction proceeds, the transition state is usually stabilized through its interactions with the enzyme, and the activation energy is lowered in the process. The net result? The reaction occurs more rapidly than in the absence of the catalyst.

Amino acid side chains at the active sites of enzymes sometimes become covalently attached to substrates, intermediates, or products during the course of a reaction, a process called **covalent catalysis.** Even when catalysis is covalent, the enzyme is returned to its original state at the end of the reaction. Detailed catalytic mechanisms have been proposed for a variety of enzymes. Specific examples are examined in Sections 5.16 and 5.17.

The Lock and Key Model and the Induced Fit Model

Some enzymes interact with substrates through what approximates a **lock and key model.** Their active sites have unique, largely predetermined shapes that are designed to accept specifically shaped substrates in much the same manner that a lock accepts a key:

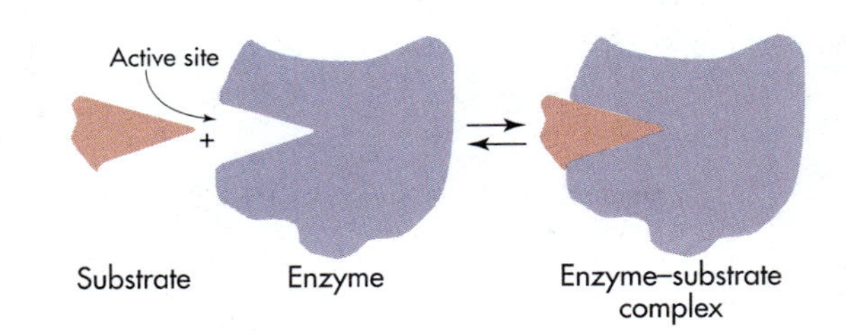

In most instances, however, active sites change their shapes as substrates bind. Such enzymes are said to act through an **induced fit model,** since the substrate induces the enzyme to fit around it:

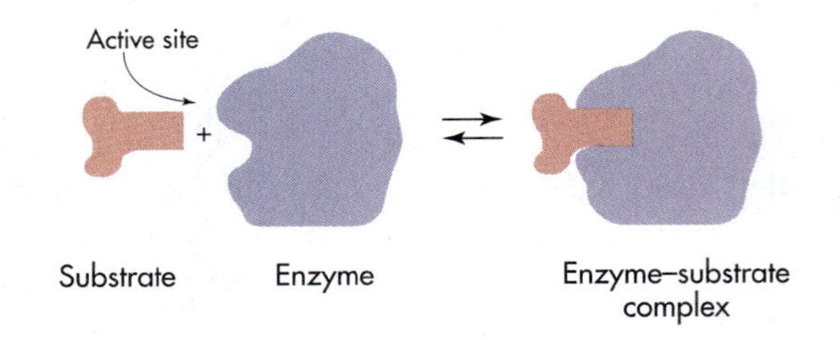

The conformation of the substrate is also modified or locked into a specific arrangement during the binding process.

When trying to visualize either an induced fit or a lock and key interaction between an enzyme and its substrate(s), keep in mind that enzymes and other proteins do not have totally rigid and inflexible structures. Individual atoms within proteins are constantly vibrating and rotating, and covalent bonds are stretching, contracting, and bending. At body temperature (37°C), most noncovalent bonds are continually breaking and reforming. A protein is sometimes said to "breathe" because of the atomic motions within it.

pH Optimum

Since the conformation of proteins is partially determined by pH, enzyme–substrate interactions and catalytic activity are pH dependent. Most human enzymes are po-

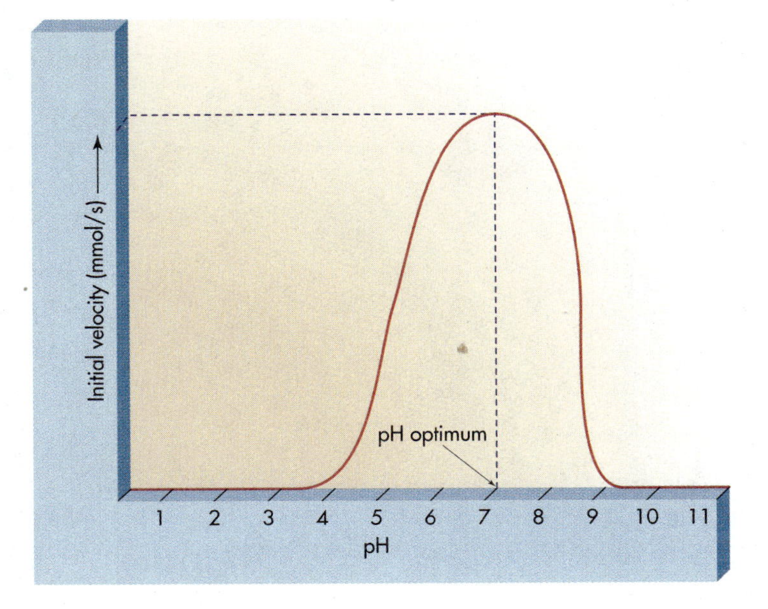

FIGURE 5.3
The effect of pH on the catalytic activity of a typical human enzyme. Although most human enzymes have a pH optimum between 6 and 8, each enzyme has a unique and characteristic sensitivity to pH. The precise shape of the pH sensitivity curve varies from one enzyme to the next; it is seldom bell shaped.

tent catalysts at pH values between approximately 6 and 8 (often described as the **physiological pH range** because most human body fluids have pH values within this range) but are denatured and inactive at extremes in pH (Figure 5.3). The pH at which an enzyme is most active is called its **pH optimum.** Some proteins are much more sensitive to pH changes than others, so one encounters both sharp and broad optima. The potentially lethal consequences of acidosis and alkalosis (shifts in blood pH; Section 2.5) are primarily explained by pH-induced changes in the conformation and activity of enzymes and other proteins. Occasionally, one finds an enzyme with a pH optimum far outside the physiological range. Since pepsin catalyzes the hydrolysis of dietary proteins in regions of the stomach where the pH is usually close to 1.5, evolution has selected an enzyme that exhibits peak activity at this pH (Table 5.3). **[Try Problems 5.24 and 5.25]**

The pH optimum of an enzyme sometimes plays a physiologically important role in determining when and where it is functional. Certain key enzymes in green plants, for example, are activated by shifts in pH associated with exposure to light (Section 15.6). The digestive enzymes in lysosomes (Section 1.3) are active at the relatively low pH values that exist within these organelles, but they are much less active at the pH of the cytosol. This property protects a cell from autodigestion if a lysosome should rupture or leak.

TABLE 5.3

The pH Optima for Select Enzymes

Enzyme	pH Optimum[a]
Pepsin	1.5
Phosphoglyceromutase	5.9
Urease	6.6
Carboxypeptidase	7.5
Succinate dehydrogenase	7.6
Trypsin	7.8
Alkaline phosphatase	9.5
Glycerol kinase	9.8

[a]Although few enzymes have a pH optimum of 7.0, most human enzymes exhibit peak activity at some pH between 6 and 8.

Temperature Optimum

Temperature, like pH, affects enzyme–substrate interactions and reaction rates. Section 5.2 explained why the rates of noncatalyzed, bimolecular reactions increase as the temperature rises. For similar reasons, the same trend is observed for most enzyme-catalyzed reactions at temperatures where the native folding of the enzyme

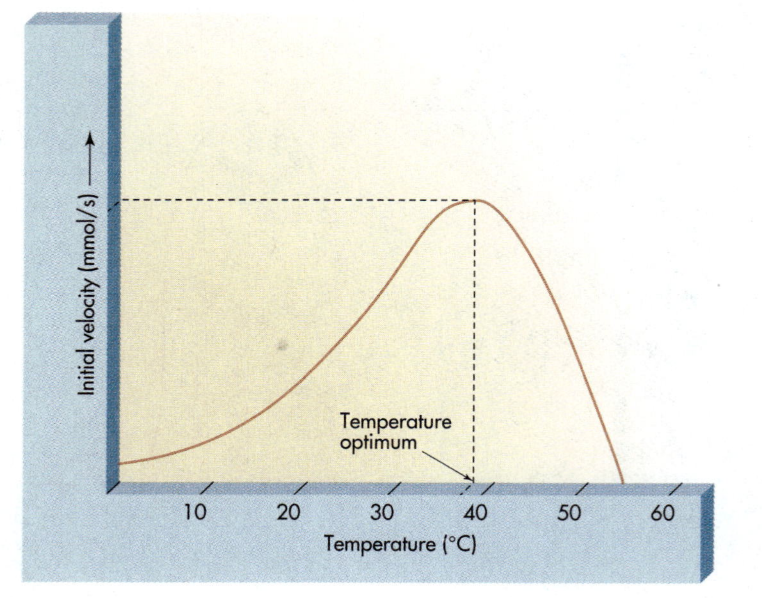

FIGURE 5.4
The effect of temperature on the catalytic activity of a typical human enzyme. Although most human enzymes exhibit maximum activity at temperatures close to body temperature (37°C), each enzyme tends to have a unique and characteristic sensitivity to temperature. The precise shape of the temperature sensitivity curve varies from one enzyme to the next.

is maintained. At higher temperatures, reaction rates usually drop sharply because of catalyst denaturation (Figure 5.4). Deaths resulting from high fevers are principally a result of changes in the conformation and activity of enzymes and other proteins.

Although most human enzymes have a temperature optimum near 37°C and are inactivated well before they reach 90°C, the enzymes in some of the bacteria living in Yellowstone Park's hot springs function well at temperatures close to the boiling point of water (100°C). One of these temperature–resistant enzymes plays a key role in the polymerase chain reaction (PCR), a patented process for amplifying the amount of DNA within biological samples (Section 17.6). Proteins are versatile polymers that can be constructed to function in a wide variety of environments. **[Try Problems 5.26 and 5.27]**

***PROBLEM 5.18** True or false? If an enzyme can distinguish geometric isomers (cis and trans isomers), we can confidently conclude that the enzyme contains a chiral binding site. Explain.

PROBLEM 5.19 What classes of noncovalent bonds are responsible for holding enzymes and substrates together within enzyme–substrate complexes?

PROBLEM 5.20 True or false? Enzymes bind substrates in basically the same manner that the polypeptide portion of myoglobin binds a heme group. Explain.

PROBLEM 5.21 List three treatments or conditions that will inactivate most enzymes.

PROBLEM 5.22 What term is used to describe an enzyme that has been unfolded and inactivated without breaking covalent bonds?

PROBLEM 5.23 Explain why a change in the primary structure of an enzyme frequently decreases its catalytic activity. Could a change in primary structure increase the catalytic activity of an enzyme? Explain.

PROBLEM 5.24 Provide a detailed molecular explanation for why the conformation of an enzyme changes as the pH of its environment is modified.

PROBLEM 5.25 Would you expect pepsin to have much, if any, catalytic activity at pH 7.0? Explain.

PROBLEM 5.26 Autoclaves, which are commonly used to sterilize medical instruments, produce superheated steam under pressure. Suggest what specific molecular changes are principally responsible for the autoclave-induced death of bacteria.

PROBLEM 5.27 Add a curve to Figure 5.4 illustrating the type of results obtained with an enzyme that is more sensitive to heat denaturation than the typical enzyme. Add a second curve illustrating the type of results obtained with a temperature-resistant enzyme.

5.6 ENZYME KINETICS

The determination of pH and temperature optima falls in the realm of enzyme kinetics. **Kinetics,** in general, is the study of motion. **Enzyme kinetics** is the study of how rapidly reactants "move" to products during enzyme-catalyzed reactions. Thus, enzyme kinetics is the study of reaction rates, a topic introduced in Section 5.2. From kinetic studies, we can learn how to regulate the activity of enzymes for medical purposes or other reasons. Kinetic studies also represent one of a limited number of tools that biochemists can use to probe the detailed mechanisms of enzyme catalysis. By carefully comparing reaction rates under a multitude of unique reaction conditions, we can gain considerable insight into the binding and catalytic processes that are at the heart of enzyme action.

The Michaelis–Menten Equation, K_m and V_{max}

Many enzyme-catalyzed reactions yield hyperbolic curves when one plots initial velocity, v_o, versus substrate concentration [S] (Figure 5.5). Such enzymes are said to follow **Michaelis–Menten kinetics** and the hyperbolic curves are mathematically described by the **Michaelis–Menten equation** (derived in Section 5.7):

$$v_o = \frac{V_{max}[S]}{K_m + [S]}$$

Knowing the value of any three of the four variables in this equation (v_o, K_m, V_{max}, and [S]), you should be able to calculate the remaining variable.

$\mathbf{V_{max}}$ is the **maximum initial velocity** for the reaction. It is attained at extremely high substrate concentrations, and it corresponds to the velocity at the plateau of the hyperbolic plot in Figure 5.5. The **Michaelis constant (K_m)** is defined as that substrate concentration which leads to an initial velocity equal to one half V_{max}. In many instances, the K_m for an enzyme is approximately equal to the **dissociation constant (K_d)** for its enzyme–substrate complex:

$$ES \rightleftharpoons S + E$$

$$K_m \approx K_d = \frac{[S][E]}{[ES]}$$

Although there are numerous exceptions to this generalization, you should assume that it is applicable when solving the problems in this text. When this relationship is valid, the smaller the K_m, the greater the affinity of an enzyme for its substrate and the lower the substrate concentration required to attain maximum initial velocity (or any given fraction of maximum initial velocity). The precise mathematical relationship between K_m and K_d is derived in Section 5.7. **[Try Problems 5.28 through 5.31]**

Figure 5.5 reveals that, when working with a Michaelis–Menten enzyme (one that exhibits Michaelis–Menten kinetics) at low substrate concentrations, initial velocity

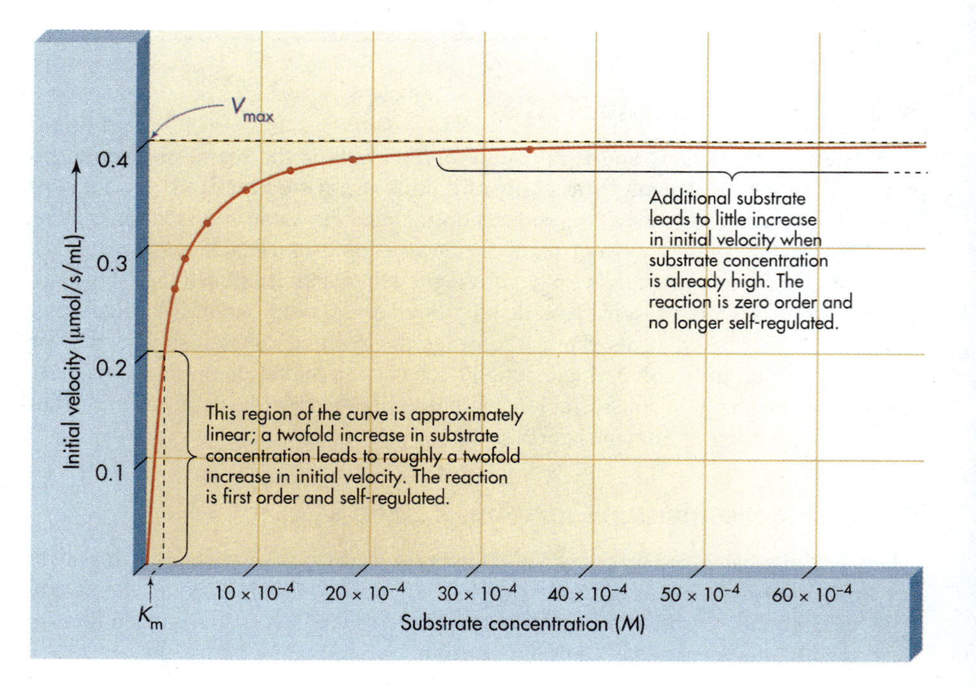

FIGURE 5.5
The effect of substrate concentration on the initial velocity of a reaction catalyzed by a Michaelis–Menten enzyme.

increases in an approximately linear fashion as more substrate is added to the reaction mixture. In contrast, a further increase in substrate concentration has no significant impact on initial velocity if substrate concentration is already very high. What is the explanation for this behavior? The answer lies in the fact that substrates must bind to enzymes before they can react.

Assume that the following reaction is initially at equilibrium:

$$E + S \rightleftharpoons ES$$

How will this reaction respond if we temporarily destroy the equilibrium by suddenly adding more substrate to the reaction mixture? Le Châtelier's principle predicts that there will be a net forward reaction until equilibrium is once again established. At the new equilibrium, a larger fraction of the enzyme molecules will be bound to substrate than was the case initially. If we continue to add substrate, one eventually reaches the point where all (from a practical standpoint) of the enzyme molecules are continuously bound to substrate and in the process of generating product. At this point, every enzyme molecule is "working" all the time, and the further addition of substrate is unable to enhance initial velocity; the reaction is at maximum initial velocity. The enzyme is said to be **saturated** with substrate, and the substrate concentration is said to be **saturating.** **[Try Problems 5.32 through 5.34]**

Initial Velocity Versus Enzyme Concentration

When we double the concentration of an enzyme in a reaction mixture, we double the number of "workers" and double the amount of "work" that can be accomplished during any fixed time period. For this reason, maximum initial velocity is proportional to enzyme concentration (Figure 5.6). In general, the amount of product formed during initial velocity measurements remains the same as long as the con-

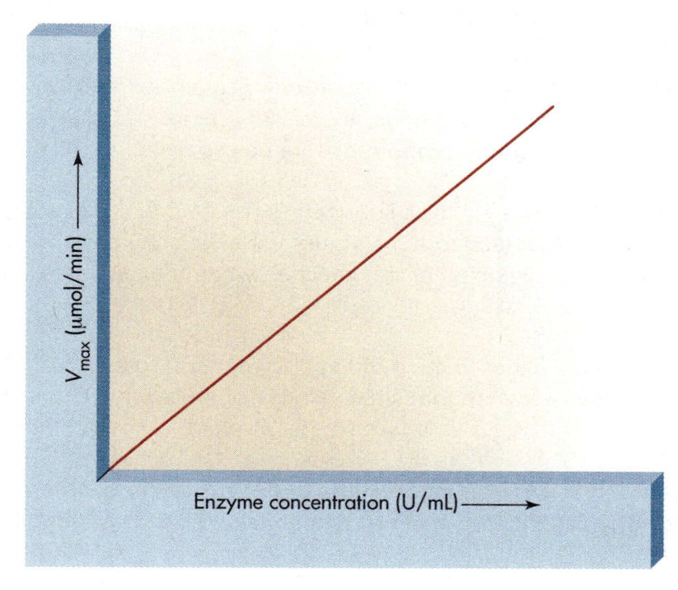

FIGURE 5.6
The effect of enzyme concentration on the maximum initial velocity for an enzyme-catalyzed reaction. Initial velocity is proportional to enzyme concentration when substrate concentration is saturating. Maximum initial velocity (V_{max}) is attained only at saturating substrate concentrations.

centration of enzyme multiplied by time of reaction is constant; a fixed amount of enzyme can do twice as much "work" if it is given twice as much time to do so.

Self-Regulating Reactions

At nonsaturating substrate concentrations, an enzyme-catalyzed reaction tends to be **self-regulating**: as substrate concentration rises, the rate of the reaction increases; and as substrate concentration drops, the rate of the reaction decreases (study Figure 5.5). The degree of self-regulation is maximum at low substrate concentrations, but it is still significant at substrate concentrations close to the K_m for an enzyme. Since, in numerous instances, evolution has led to the selection of enzymes with K_m's close to the substrate concentrations that they normally encounter, many enzyme-catalyzed reactions probably exhibit some degree of self-regulation in vivo (within living organisms). However, it is important to recognize that an enzyme may display markedly different kinetics within its in vivo environment than it does under standard assay conditions in the laboratory. In vivo, self-regulation helps prevent substrate concentrations from dropping too low or increasing to potentially toxic levels. [Try Problem 5.35]

PROBLEM 5.28 Explain why the K_m and V_{max} for an enzyme–substrate pair tends to be different at different pH values, even when all other reaction conditions are held constant.

PROBLEM 5.29 Enzyme R, which has a single substrate binding site, will catalyze the oxidation of each of three different substrates. Under standard conditions, the K_m is 2×10^{-3} *M* for substrate A, 7×10^{-2} *M* for substrate B, and 6×10^{-4} *M* for substrate C. Which substrate binds most tightly to enzyme R? Which enzyme–substrate complex has the smallest dissociation constant? Does this information allow you to determine which one of the three enzyme-catalyzed reactions has the greatest V_{max}? Explain.

PROBLEM 5.30 Assume that in an enzyme-catalyzed reaction under standard conditions $v_o = 5$ μmol/s, $V_{max} = 35$ μmol/s and [S] = 0.0002 *M*. Calculate the K_m for the enzyme–substrate pair involved.

PROBLEM 5.31 The K_m for a substrate is 8×10^{-5} *M* under standard assay conditions that lead to a maximum initial velocity of 2 μmol/min. Calculate the initial velocity for this reaction when the substrate concentration is 2×10^{-5} *M*.

PROBLEM 5.32 What fraction of the enzyme molecules is bound to substrate molecules when the initial velocity of an enzyme-catalyzed reaction is one-half V_{max}? One-tenth V_{max}? Explain.

PROBLEM 5.33 What fraction of the enzyme molecules is bound to substrate when substrate concentration equals K_m?

PROBLEM 5.34 The initial velocity for an enzyme-catalyzed reaction is 3 mmol/min in the presence of 4×10^{-3} *M* substrate. Describe a simple experiment that can be run to determine whether this initial velocity is the maximum initial velocity.

PROBLEM 5.35 Given the following graph obtained while studying a specific enzyme-catalyzed reaction:

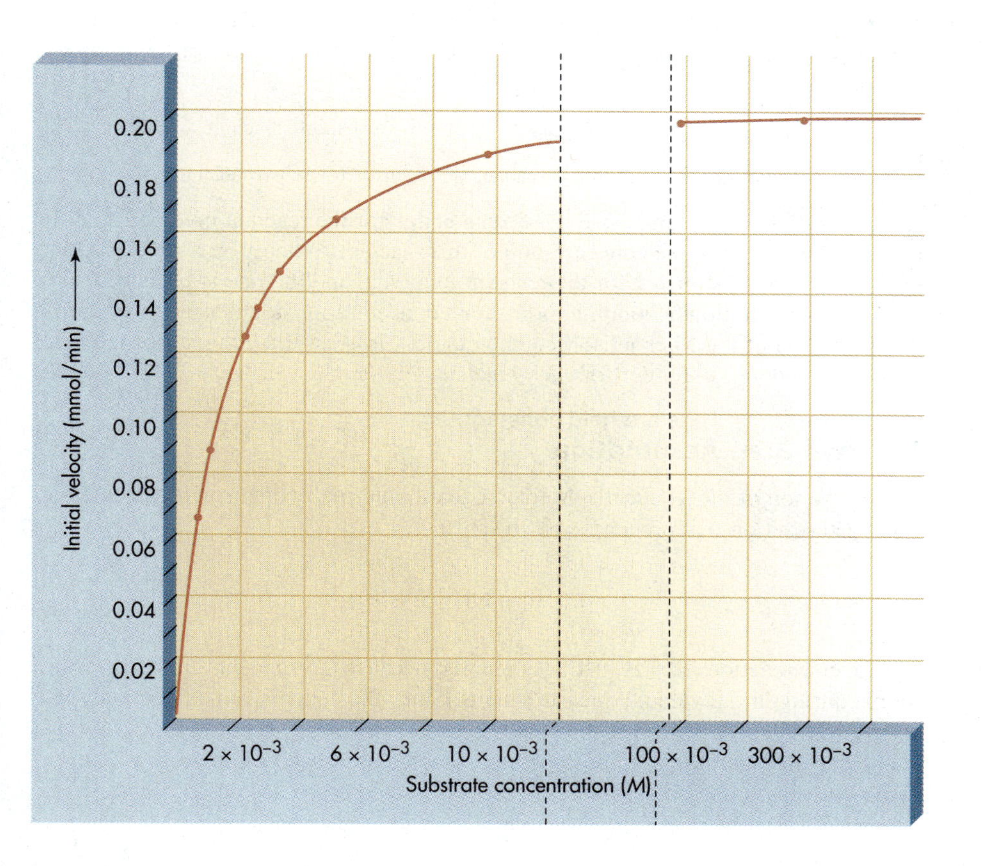

a. What is the approximate V_{max} for this reaction?
b. What is the approximate K_m for the enzyme–substrate pair involved?
c. What is the initial velocity when the substrate concentration is 2×10^{-1} *M*? What change occurs in the initial velocity when the substrate concentration is increased to 3×10^{-1} *M*?
d. At what substrate concentrations is there a significant degree of self-regulation?

5.7 DERIVATION OF THE MICHAELIS–MENTEN EQUATION

The Michaelis–Menten equation is central to modern enzyme kinetics, so it is appropriate to derive this equation, even in an introductory text. Since this derivation involves rate constants, a review of rate constants is in order.

Rate Constants and Reaction Order

The initial rate of a chemical reaction can normally be expressed as a constant, called a **rate constant (*k*),** multiplied by the concentration of reactant(s) raised to an appropriate exponent:

$$\text{Initial rate} = k[\text{reactant}]^x \qquad \text{if there is a single reactant}$$

$$\text{Initial rate} = k[\text{reactant}_1]^x[\text{reactant}_2]^y \qquad \text{if there are two reactants}$$

If x or $y = 0$, the reaction is said to be **zero order** for the reactant involved, and the rate is independent of the concentration of this reactant. When x or $y = 1$, the kinetics is said to be **first order** for the reactant involved, and the rate is proportional to reactant concentration; a doubling of reactant concentration doubles the rate. During the derivation of the Michaelis–Menten equation, it is assumed that all of the reactions involved are first order for each reactant. **[Try Problem 5.36]**

The Two Step Assumption

The derivation of the Michaelis–Menten equation normally begins with an enzyme-catalyzed reaction that proceeds in two steps:

$$E + S \underset{k_{-1}}{\overset{k_1}{\rightleftharpoons}} ES \overset{k_2}{\longrightarrow} E + P$$

The reverse reaction ($E + P \longrightarrow ES$) is ignored because the concentration of P is negligible during initial velocity measurements. Since the enzyme can exist in only two states, either free or bound to substrate, the total enzyme concentration ($[E_t]$) must equal the concentration of free enzyme ($[E_f]$) plus the concentration of the enzyme–substrate complex ($[ES]$):

$$[E_t] = [E_f] + [ES] \qquad \text{or, rearranged,} \qquad [E_f] = [E_t] - [ES]$$

The Rate-Limiting Step

It is next assumed that the rate limiting step (slowest step; Figure 5.7) during the conversion of substrate to product is the breakdown of the ES complex into E + P.

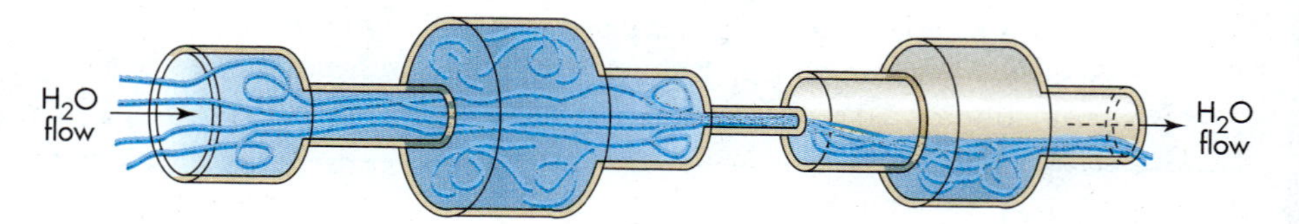

FIGURE 5.7
Rate-limiting step. In a pipe that varies in diameter along its length, water cannot flow any more rapidly than it can flow through the segment of pipe of smallest diameter. Thus, the flow of water through the smallest diameter segment is the rate-limiting step for the flow of water through the pipe.

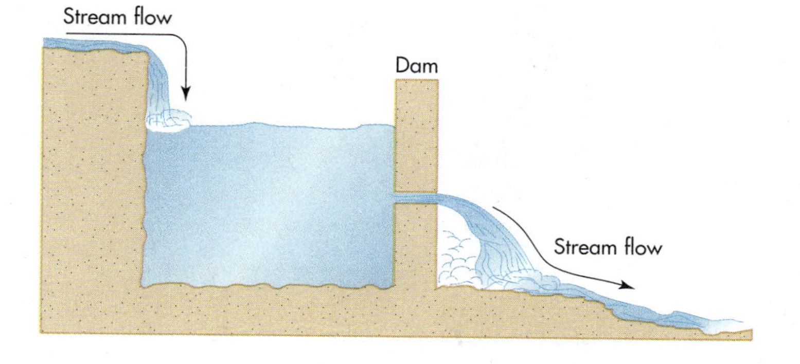

FIGURE 5.8
Steady state. The level of water behind the dam remains constant when the rate at which water enters the reservoir is the same as the rate at which water leaves the reservoir. When this requirement is met, the amount of water in the reservoir is at a steady state.

Under these conditions, the initial rate for the reaction is:

$$\text{Initial rate} = v_o = k_2[\text{ES}]$$

When all of the enzyme is bound to substrate, $[\text{E}_t] = [\text{ES}]$ and $v_o = V_{max}$:

$$v_o = V_{max} = k_2[\text{E}_t]$$

The Steady State Assumption

We proceed with the derivation by assuming that the concentration of ES instantly reaches a steady state and remains unchanged during initial velocity measurements (Figure 5.8). Although the ES concentration remains constant, the reactions involving this complex are not at equilibrium; the rate of ES production from E and S is simply equal to the rate of ES breakdown into E + P and E + S:

$$\text{Rate of ES formation} = k_1[\text{S}][\text{E}_f] = k_1[\text{S}]([\text{E}_t] - [\text{ES}])$$

$$\text{Rate of ES breakdown} = k_{-1}[\text{ES}] + k_2[\text{ES}]$$

$$k_1[\text{S}]([\text{E}_t] - [\text{ES}]) = k_{-1}[\text{ES}] + k_2[\text{ES}]$$

To solve for [ES], we first multiply $([\text{E}_t] - [\text{ES}])$ by $k_1[\text{S}]$ on the left side of the last equation (immediately above), and then factor out [ES] on the right:

$$k_1[\text{E}_t][\text{S}] - k_1[\text{ES}][\text{S}] = [\text{ES}](k_{-1} + k_2)$$

When the term $k_1[\text{ES}][\text{S}]$ is added to both sides of this equation we obtain:

$$k_1[\text{E}_t][\text{S}] = [\text{ES}](k_{-1} + k_2) + k_1[\text{ES}][\text{S}]$$

Factoring out [ES] on the right side of our new equation leads to:

$$k_1[\text{E}_t][\text{S}] = (k_{-1} + k_2 + k_1[\text{S}])[\text{ES}]$$

Dividing both sides of the factored equation by $k_{-1} + k_2 + k_1[\text{S}]$ yields:

$$[\text{ES}] = \frac{k_1[\text{E}_t][\text{S}]}{k_{-1} + k_2 + k_1[\text{S}]}$$

When both the numerator and denominator on the right side of this equation are divided by k_1, it is simplified to:

$$[\mathrm{ES}] = \frac{[\mathrm{E_t}][\mathrm{S}]}{(k_{-1} + k_2)/k_1 + [\mathrm{S}]}$$

Multiplying both sides by k_2 leads to:

$$k_2[\mathrm{ES}] = \frac{k_2[\mathrm{E_t}][\mathrm{S}]}{(k_{-1} + k_2)/k_1 + [\mathrm{S}]}$$

Since $k_2[\mathrm{ES}] = v_o$ and $k_2[\mathrm{E_t}] = V_{max}$ (see discussion of rate-limiting step), this equation can be rewritten:

$$v_o = \frac{V_{max}[\mathrm{S}]}{(k_{-1} + k_2)/k_1 + [\mathrm{S}]}$$

After **K_m is defined as $(k_{-1} + k_2)/k_1$** (ratio of substrate "off" rate constants to substrate "on" rate constant), we arrive at the Michaelis–Menten equation:

$$v_o = \frac{V_{max}[\mathrm{S}]}{K_m + [\mathrm{S}]}$$

It is this equation which mathematically describes the hyperbolic plot characteristic of enzymes exhibiting Michaelis–Menten kinetics (Figure 5.5).

A Closer Look at K_m

In Section 5.6 we noted that K_m equals the substrate concentration that yields an initial velocity equal to one-half V_{max}, and claimed that K_m often approximates the equilibrium constant for the dissociation of the enzyme–substrate complex. The derivation of the Michaelis–Menten equation allows us to examine these facts in more detail.

To explain the relationship between K_m and V_{max}, we assume that v_o equals one-half V_{max} and then substitute $1/2V_{max}$ into the Michaelis-Menten equation in place of v_o:

$$v_o = 1/2V_{max} = \frac{V_{max}[\mathrm{S}]}{K_m + [\mathrm{S}]}$$

Dividing both sides by $1/2V_{max}$:

$$1 = \frac{2[\mathrm{S}]}{K_m + [\mathrm{S}]}$$

Multiplying both sides by $K_m + [\mathrm{S}]$:

$$K_m + [\mathrm{S}] = 2[\mathrm{S}]$$

Subtracting [S] from both sides:

$$K_m = [\mathrm{S}]$$

Thus, K_m and [S] are numerically the same when the initial assumption ($v_o = 1/2V_{max}$) is satisfied. K_m is equal to that substrate concentration at which half the enzyme molecules are bound to substrate and working to catalyze the reaction. Under these conditions, half of the maximum amount of work is being accomplished. Maximum work (corresponds to V_{max}) is attained when all of the enzyme molecules are bound to substrate and in the process of converting substrate to product.

To understand the relationship between K_m and the dissociation constant (K_d) for ES, we must examine the relationship between rate constants and equilibrium constants. At equilibrium, the rate of the forward reaction equals the rate for the reverse reaction. In the reaction $ES \underset{k_1}{\overset{k_{-1}}{\rightleftharpoons}} E_f + S$ (initially encountered as the reverse reaction during the discussion of the two step assumption):

$$\text{Rate of the forward reaction} = k_{-1}[ES]$$

$$\text{Rate of the reverse reaction} = k_1[E_f][S]$$

At equilibrium:

$$k_{-1}[ES] = k_1[E_f][S]$$

Dividing both sides of this equation by k_1[ES]:

$$\frac{k_{-1}}{k_1} = \frac{[E_f][S]}{[ES]} = K_{eq} \text{ for the dissociation of ES} = K_d$$

After reexamining the theoretical definition of K_m [$K_m = (k_{-1} + k_2)/k_1$], it becomes apparent that:

$$K_m \approx \frac{k_{-1}}{k_1} = K_d \quad \text{when} \quad k_{-1} >> k_2$$

Restated, K_m is approximately equal to the dissociation constant for ES whenever k_2 is much smaller than k_{-1}. Under these conditions, $k_{-1} + k_2$ is approximately equal to k_{-1}. **[Try Problem 5.37]**

At this point, two definitions of K_m have been encountered:

a. K_m = substrate concentration when $v_o = 1/2V_{max}$ (known as **apparent K_m**) and

b. $K_m = \dfrac{k_{-1} + k_2}{k_1}$ (known as **theoretical K_m**)

In the absence of activators, inhibitors and other complicating factors, the apparent K_m is numerically equal to the theoretical K_m. Unless stated otherwise, any further reference to K_m in this text refers to the apparent K_m, which may or may not (depending on the specific reaction environment) equal the theoretical K_m.

More Complex Kinetics

It is common practice to describe any enzyme that yields a hyperbolic v_o versus [S] plot as a Michaelis–Menten enzyme. However, it has been discovered that some of these enzymes do not adhere to all of the assumptions involved in the derivation of the Michaelis–Menten equation. Certain reactions, for example, proceed in three steps rather than two:

$$E + S \rightleftharpoons ES \rightleftharpoons EP \rightleftharpoons E + P$$

Although the Michaelis–Menten equation still defines the hyperbolic v_o versus [S] curve, the theoretical interpretation of K_m and V_{max} differs in these cases. The theoretical interpretation of K_m and V_{max} may also differ in the presence of activators or inhibitors.

PROBLEM 5.36 Given the following equation:

$$v_o = k[\text{substrate}]^2$$

a. Is this a zero-,first-, second-, or third-order reaction? Explain.
b. How will v_o be impacted by a doubling of substrate concentration?

PROBLEM 5.37 Express the equilibrium constants for the reactions $E + S \underset{k_{-1}}{\overset{k_1}{\rightleftharpoons}} ES$ and $ES \underset{k_{-2}}{\overset{k_2}{\rightleftharpoons}} E + P$ in terms of rate constants.

5.8 TURNOVER NUMBERS AND k_{cat}

V_{max} and the derivation of the Michaelis–Menten equation tie in closely to the concept of turnover number and k_{cat}. The **turnover number** of an enzyme is the number of substrate molecules converted to products per unit time by a single enzyme molecule when substrate is saturating (enzyme operating at V_{max}). The turnover number, which has units of reciprocal time (usually s^{-1} [1/s] or min^{-1} [1/min]), is equal to the rate constant for the rate-limiting step of the catalytic process (assuming that the same step is rate limiting under all reaction conditions). This rate constant is called the $\boldsymbol{k_{cat}}$. If a reaction proceeds through the classical two-step process ($E + S \underset{k_{-1}}{\overset{k_1}{\rightleftharpoons}} ES \underset{k_{-2}}{\overset{k_2}{\rightleftharpoons}} E + P$) where the second step is rate limiting:

$$V_{max} = k_2[E_t]$$

$$k_2 = \frac{V_{max}}{[E_t]} = \text{turnover number} = k_{cat}$$

TABLE 5.4

The Turnover Number for Some Enzymes

Enzyme	Turnover Number (per minute)
Catalase	600,000,000
Carbonic Anhydrase	36,000,000
3-Ketosteroid isomerase	16,800,000
β-Amylase	1,100,000
Penicillinase	120,000
Lactate dehydrogenase	60,000
β-Galactosidase	12,000
Phosphoglucomutase	1,240
Lysozyme	30

Turnover numbers typically range from 10^3 to 10^6 per minute, although they fall considerably outside this range for some enzymes (Table 5.4). For a mental challenge, try to visualize a single enzyme molecule converting over 600 million reactants to products in less than a minute (see catalase in Table 5.4). Since a catalyst cannot modify a compound until contact has been made, the turnover numbers for enzymes are limited by collision frequencies. Evolution has selected some enzymes, including catalase, that approximate perfect catalysts; they catalyze a reaction almost every time they encounter a substrate molecule. **[Try Problem 5.38]**

PROBLEM 5.38 The turnover number for an enzyme is a rough measure of (circle the one best answer):

a. The amount of enzyme in a sample.
b. The stability of the enzyme.
c. The size of the enzyme.
d. The catalytic efficiency of the enzyme.
e. The specificity of the enzyme.

Assume that the rate of the reaction involved is insignificant in the absence of enzyme.

5.9 DETERMINATION OF K_m AND V_{max}

We have defined K_m and V_{max} and described their theoretical significance. How are these constants determined? The hyperbolic curve obtained when we plot initial velocity versus substrate concentration for a Michaelis–Menten enzyme can be used to estimate K_m and V_{max} as indicated in Figure 5.5. Since sophisticated curve fitting programs are required to fit a hyperbola to experimental points, this procedure is seldom used in practice.

Lineweaver-Burk Plots

The traditional method for determining K_m and V_{max} is based on the **Lineweaver–Burk equation,** a mathematical rearrangement of the reciprocal of the Michaelis–Menten equation:

$$\frac{1}{v_o} = \left(\frac{K_m}{V_{max}}\right)\frac{1}{[S]} + \frac{1}{V_{max}}$$

Lineweaver–Burk equation

This equation is of the general type $y = mx + b$. Such equations lead to a straight line with a slope of m and a y-intercept of b when y is plotted versus x. Thus, when the reciprocal of the initial velocity (1 divided by the initial velocity) is plotted against the reciprocal of the substrate concentration (1 divided by substrate concentration), a straight line is obtained with a slope equal to K_m divided by V_{max} and a

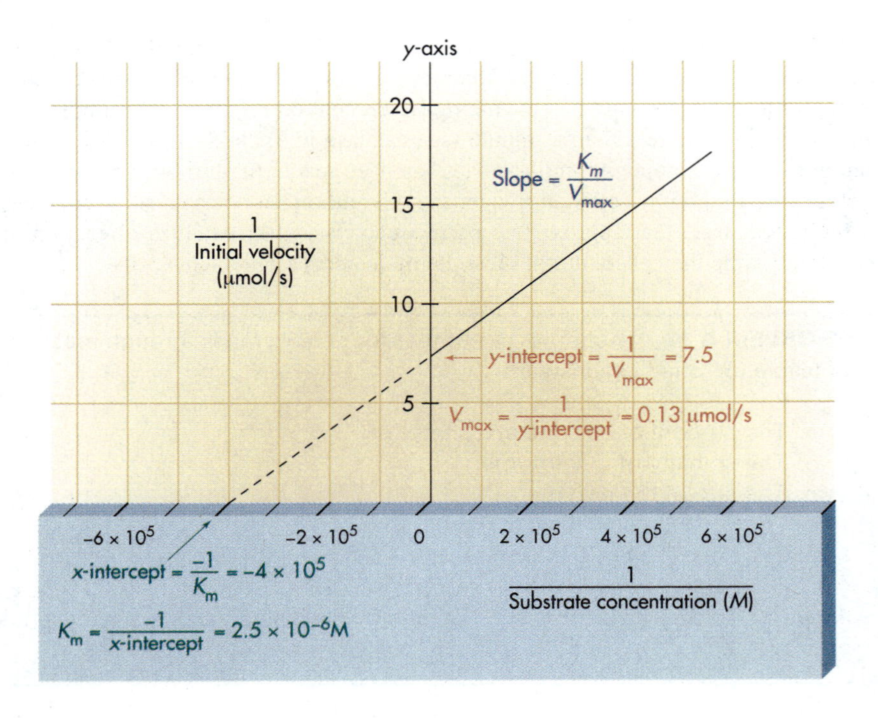

FIGURE 5.9
A Lineweaver–Burk plot.

y-intercept equal to 1 divided by V_{max} (Figure 5.9). If the straight line obtained is extended to the left of the *y*-axis until it crosses the *x*-axis, the intercept on the *x*-axis is equal to −1 divided by K_m. This extension is strictly a mathematical maneuver since the substrate concentration cannot, in reality, be negative, and there can be no experimental points to the left of the *y*-axis.

Lineweaver–Burk plots (also called **double–reciprocal plots**) are preferred over Michaelis–Menten plots for calculating K_m and V_{max}, since it is relatively easy to draw the best straight line through a set of experimental points. Several other graphical approaches exist for the determination of K_m and V_{max}. Some of these are described in more encyclopedic texts. It is important to recognize that the K_m calculated graphically corresponds to the apparent K_m.

Today, K_m and V_{max} calculations are often performed with the use of personal computers and appropriate software packages. Software is available that generates and "interprets" Lineweaver–Burk plots.

The Preparation of Lineweaver–Burk Plots

The first step in the experimental preparation of a Lineweaver–Burk plot is the measurement of the initial velocity of an enzyme-catalyzed reaction at several different substrate concentrations with all other reaction conditions held constant. One next completes a table of the following type:

v_o (mmol/min)	[S] (molar)	$1/v_o$	1/[S]
0.14	1.00×10^{-4}	7	10×10^3
0.17	1.25×10^{-4}	6	8×10^3
0.20	1.67×10^{-4}	5	6×10^3
0.25	2.50×10^{-4}	4	4×10^3
0.33	5.00×10^{-4}	3	2×10^3

The two columns to the left represent the tabulation of experimental results. The reciprocal values in the two columns to the right are calculated to provide the y and the x coordinates that locate the individual points on the graph to be prepared. The actual graphing of these points yields a Lineweaver–Burk plot. Problem 5.39 calls for the preparation and interpretation of a double-reciprocal plot based on the information in this table. [Try Problem 5.39]

Uses of K_m and V_{max}

Why calculate K_m's and V_{max}'s? Of what value are these constants? They are used to characterize enzymes in much the same way that melting points and boiling points are employed to characterize simpler organic compounds; each enzyme tends to have a unique K_m and V_{max} under any specific set of reaction conditions (pH, temperature, enzyme concentration, etc.). Thus, if a human enzyme and a plant enzyme both catalyze the same reaction but their K_m's and V_{max}'s differ under standard conditions, they are not identical molecules. If they have the same K_m's and V_{max}'s, they are, more likely than not, identical.

The study of K_m's and V_{max}'s provides some insight into enzyme–substrate interactions, including substrate specificity and substrate preferences (in those cases where an enzyme has alternative substrates). The catalytic mechanisms of enzymes can also be probed with a critical analysis of K_m's and V_{max}'s determined under a variety of reaction conditions. Although these claims will not be examined in detail, you should recognize that they relate to the derivation of the Michaelis–Menten equation and to the fact that K_m's and V_{max}'s can provide some information about rate constants, binding constants, and catalytic efficiencies.

PROBLEM 5.39 Prepare a Lineweaver-Burk plot from the data provided in the table above. Use the prepared graph to calculate the K_m and V_{max} for the enzyme–substrate pair involved.

5.10 ENZYME INHIBITORS AND SUBSTRATE INHIBITION

Enzyme inhibitors can be used along with K_m and V_{max} determinations to probe enzyme action. They can also be employed to regulate the rates of enzyme-catalyzed reactions, often with therapeutic benefits. An **enzyme inhibitor** is any compound that binds to an enzyme and inhibits its action without markedly disrupting its overall structure. Although acids, bases, organic solvents, and temperature can inactivate enzymes, they are not considered inhibitors. They are nonspecific denaturing agents that can lead to drastic changes in the folding of most globular proteins. A vast array of inhibitors has been divided into a variety of classes on the basis of differences in the manner in which they interact with enzymes or enzyme–substrate complexes. This section examines three classical categories of inhibitors: competitive, noncompetitive, and uncompetitive. Substrate inhibition, a concept closely related to competitive inhibition, is explored as well.

Competitive Inhibitors

A **competitive inhibitor** is one whose action can be completely reversed at high substrate concentrations. Most, but not all, competitive inhibitors compete with substrates for binding to the active sites on enzymes. Such inhibitors are usually similar in structure to a substrate and are mistaken for a substrate by an enzyme. When a competitive inhibitor is bound, a substrate cannot bind and no reaction occurs. Since the binding of both inhibitor and substrate is reversible, high concentrations of inhibitor can virtually eliminate the binding of substrate. Similarly, very high concentrations of substrate can block the binding of inhibitor. To understand why this is the case, we need to study the two competing reactions, where I stands for inhibitor, and think of Le Châtelier's principle.

$$E + S \rightleftharpoons ES$$

$$E + I \rightleftharpoons EI$$

Assume that both substrate and inhibitor are present along with a fixed amount of enzyme and that each of the competing reactions is initially at equilibrium. What changes occur if more inhibitor is suddenly added? The increased concentration of inhibitor temporarily destroys the equilibrium of the reaction of enzyme with inhibitor, which responds (Le Châtelier's principle) with a net forward movement that increases the concentration of the enzyme–inhibitor complex and lowers the concentration of free enzyme (enzyme not bound to either substrate or inhibitor). The drop in free enzyme concentration disrupts the equilibrium of the reaction of enzyme with substrate. The equilibrium for this reaction is reestablished with a net reverse reaction that leads to the release of some enzyme from the enzyme–substrate complex. The net result? The added inhibitor has taken some enzyme away from the substrate. At very high inhibitor concentrations, virtually all of the enzyme exists within an enzyme–inhibitor complex, and there is negligible enzyme bound to substrate. At this point, the reaction proceeds at its noncatalyzed rate. In the same manner, very high substrate concentrations can tie up virtually all of the enzyme and eliminate the inhibition. This explains why the V_{max} for an enzyme-catalyzed reaction is unchanged in the presence of a competitive inhibitor. **[Try Problem 5.40]**

How does a competitive inhibitor affect K_m? When the substrate must compete with an inhibitor for the active site, it takes a higher substrate concentration to force half of the enzyme into an enzyme–substrate complex (required for the initial velocity to equal one-half V_{max}). Consequently, K_m increases.

The **sulfonamide drugs, methotrexate,** and **3′-azido-2′,3′-dideoxythymidine triphosphate (AZT triphosphate)** are specific examples of competitive inhibitors (Table 5.5). All of these therapeutic agents block the action of one or more enzymes

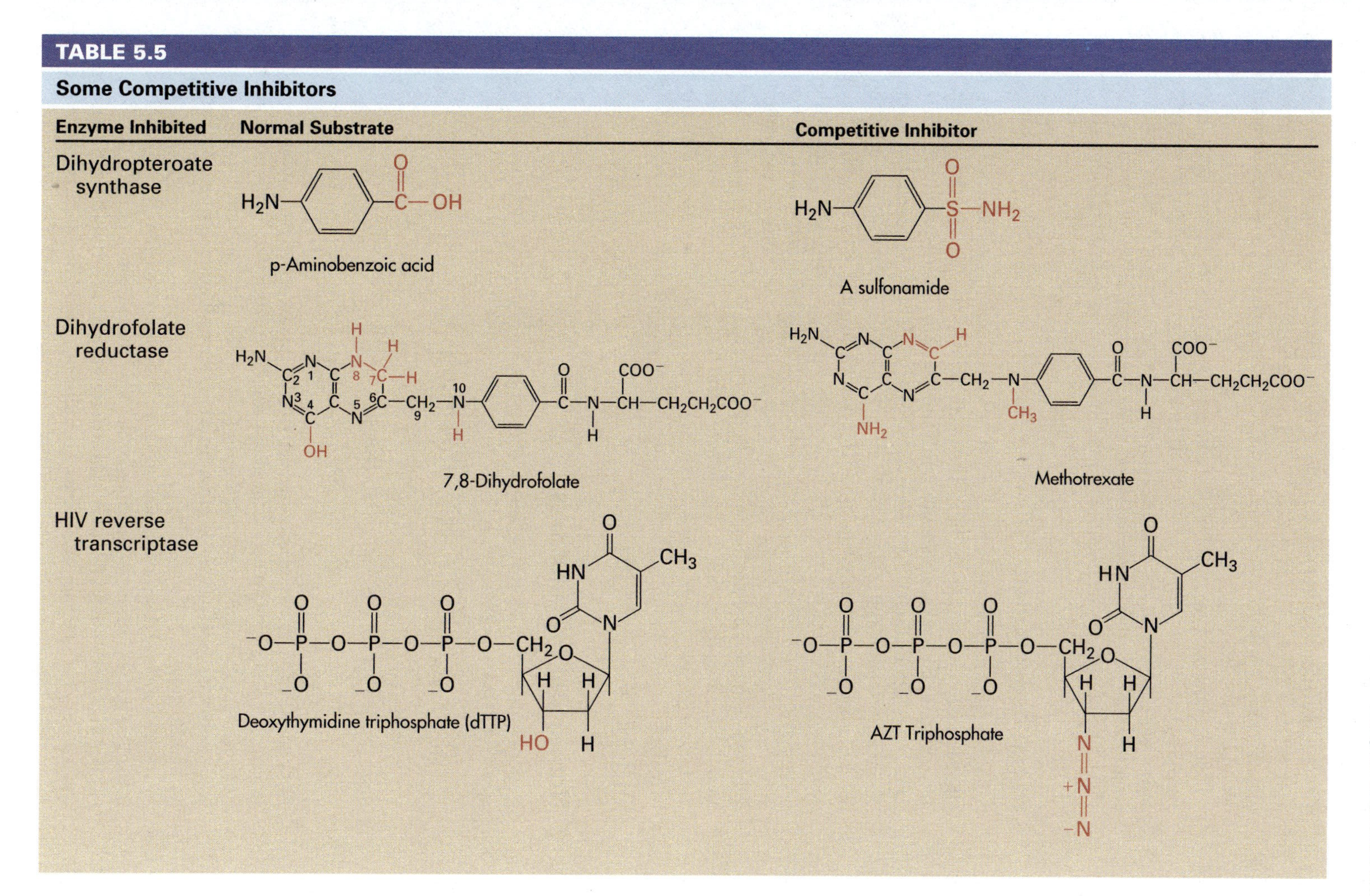

TABLE 5.5

Some Competitive Inhibitors

Enzyme Inhibited	Normal Substrate	Competitive Inhibitor
Dihydropteroate synthase	p-Aminobenzoic acid	A sulfonamide
Dihydrofolate reductase	7,8-Dihydrofolate	Methotrexate
HIV reverse transcriptase	Deoxythymidine triphosphate (dTTP)	AZT Triphosphate

that play a central role in the life cycle of tumor cells or a pathogen. AZT is used to treat AIDS, a disease linked to the HIV virus, whereas bacteria are the targets for the sulfonamides. Methotrexate is one of many enzyme inhibitors in the chemical arsenal against cancer.

Some organisms use home-made competitive inhibitors to regulate the rates of certain reactions that occur within them. Green plants, for example, produce a competive inhibitor of rubisco [the enzyme that catalyzes carbon fixation during photosynthesis (defined in Chapter 8)] when moved from the light into the dark (Section 15.6). The inhibitor is degraded when the plants are once again illuminated.

Substrate Inhibition

Alternate substrates compete with each other for enzyme active sites in the same manner that competitive inhibitors compete with substrates for active sites. This phenomenon, known as **substrate inhibition,** will be illustrated with a physiologically important example. Ethanol (the "active" ingredient in alcoholic beverages), methanol (an industrial solvent and common gas line antifreeze) and ethylene glycol (an antifreeze for automobile radiators) are alcohols commonly encountered in modern society. The toxicity of all three of these compounds is due primarily to their oxidation to aldehydes and carboxylic acids within the body, a multistep process that involves a liver enzyme known as alcohol dehydrogenase. Since all the alcohols are oxidized with the catalytic assistance of the same active site on alcohol dehydrogenase, the binding of one alcohol to an active site blocks the binding and oxidation of the other alcohols. For this reason, ethanol, which is much less toxic than the other two alcohols, is commonly employed in the treatment of methanol and ethylene glycol poisoning; it blocks their oxidation to more toxic substances.

Noncompetitive Inhibitors

A **noncompetitive inhibitor** binds reversibly to both free enzyme and the enzyme–substrate complex:

$$E + I \rightleftharpoons EI$$

$$ES + I \rightleftharpoons ESI$$

Substrate is also able to bind the EI complex:

$$EI + S \rightleftharpoons ESI$$

Both the EI and ESI complexes are catalytically inactive. Since the inhibitor and the substrate usually bind to separate, nonoverlapping sites, high substrate concentrations cannot prevent inhibitor binding and vice versa. The separate binding sites also account for the fact that, more often than not, the inhibitor is structurally unrelated to the substrate.

In the simplest cases of noncompetitive inhibition, the inhibitor binds free enzyme and the ES complex with equal ease (the binding constant is the same in both cases). Such inhibitors are called **pure noncompetitive inhibitors.** Since a given concentration of inhibitor inactivates a certain fraction of the total enzyme molecules (incapacitates part of the "workers"—Section 5.6), a noncompetitive inhibitor lowers V_{max}. The enzyme molecules not bound to inhibitor have a normal affinity for substrate and exhibit a normal K_m. The Michaelis constant is independent of active enzyme concentration. **[Try Problem 5.41]**

Certain heavy metal ions, including Pb^{2+} (lead II) and Hg^{2+} (mercury II), are examples of noncompetitive inhibitors. The toxicity of these common environmental agents is primarily a consequence of their interactions with sulfhydryl groups (—SH) and other susceptible functional groups in numerous proteins, including many enzymes. The chelating agents (bind heavy metal ions) employed therapeutically to remove lead and mercury ions from poisoned enzymes tend to tie up certain essential metal ions as well. Mg^{2+}- and Ca^{2+}-activated enzymes (Section 5.4) are often inactivated as a side effect of the therapy.

Uncompetitive Inhibitors

Uncompetitive inhibitors bind enzyme–substrate complexes but are unable to bind free enzyme:

$$E + I \rightleftharpoons \text{no reaction}$$

$$ES + I \rightleftharpoons ESI$$

The ESI complex is catalytically inactive, so the V_{max} decreases in the presence of inhibitor. K_m also declines since the $E + S \rightleftharpoons ES$ reaction shifts right as inhibitor removes ES (Le Châtelier's principle); in the presence of inhibitor, it takes less substrate to half saturate the enzyme.

Comparison of Some Common Classes of Inhibitors

Table 5.6 summarizes the differences between competitive, pure noncompetitive, and uncompetitive inhibitors. Figure 5.10 illustrates how each of these three classes of inhibitors impacts a Lineweaver–Burk plot. Such graphs are routinely used to characterize and distinguish these and other classes of inhibitors. **[Try Problems 5.42 and 5.43]**

TABLE 5.6

Some Differences Between Competitive, Pure Noncompetitive, and Uncompetitive Inhibitors

Class of Inhibitor	Effect on K_m	Effect on V_{max}	Binding Specificity
Competitive	Increases	None	Binds to free enzyme but not ES complex
Pure noncompetitive	None	Decreases	Binds to both free enzyme and ES complex
Uncompetitive	Decreases	Decreases	Binds to ES complex but not free enzyme

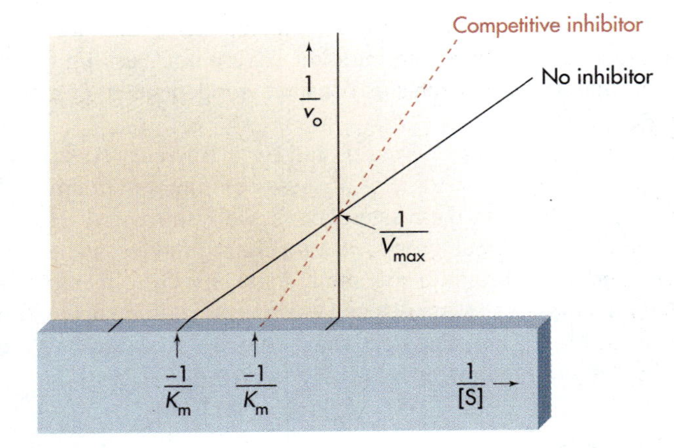

(a)

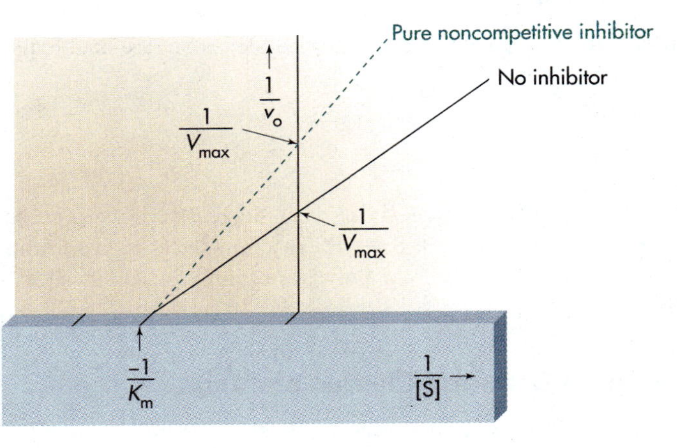

(b)

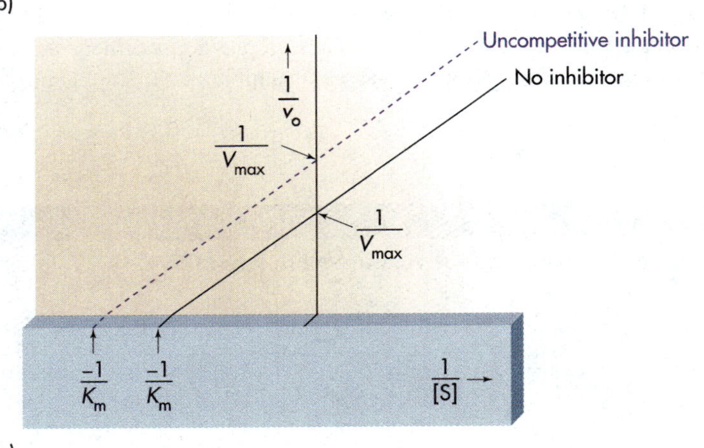

(c)

FIGURE 5.10
The impact of inhibitors on Lineweaver–Burk plots. (a) A competitive inhibitor increases apparent K_m and shifts the x-intercept toward the origin. The y-intercept and V_{max} are unchanged. (b) A pure noncompetitive inhibitor decreases V_{max} and shifts the y-intercept away from the origin. The x-intercept and K_m are unchanged. (c) An uncompetitive inhibitor decreases both apparent K_m and V_{max} and causes both the x- and y-intercept to shift away from the origin. The slope (K_m/V_{max}) is unchanged.

Other Categories of Inhibitors

Inhibitors have been divided into a variety of categories in addition to the three featured in this section. **Irreversible inhibitors,** for example, are those that bind so tightly (usually covalently) that, from a practical standpoint, the enzyme–substrate complex never dissociates after it is formed. A **suicide inhibitor** (also called a **suicide substrate** or **mechanism-based inactivator**) is a substrate analog that binds to the active site of an enzyme, and then becomes covalently attached to this site as the catalyst attempts to convert it to a normal product. Suicide inhibitors are a special class of irreversible inhibitors.

Inhibitors and the Study of Structure–Function Relationships

How can the study of enzyme inhibitors contribute to our understanding of the structure and the functioning of enzymes? A glimpse at the answer will be provided by examining a specific example. Assume that an inhibitor, which is known to react selectively with free sulfhydryl groups, inhibits a native enzyme. One can conclude that the inhibited enzyme contains one or more cysteine residues. Furthermore, the cysteine residue(s) must have been close to the surface of the enzyme to be accessible to the inhibitor (assuming that the reaction with the inhibitor occurred under conditions at which the enzyme maintains its native conformation). If the reactive cysteine residue(s) were at the active site, a high concentration of substrate might be able to protect the enzyme from the inhibitor by occupying the active site and, in the process, shielding the active site sulfhydryl group(s). If the inhibitor increased K_m but had no effect on V_{max}, this would suggest that one or more cysteine residues were involved in substrate binding but not in catalysis. Since we cannot directly view enzymes, biochemists must resort to a variety of indirect studies when attempting to develop a detailed understanding of these phenomenal protein catalysts.

PROBLEM 5.40 Is the concentration at which substrate becomes saturating greater, less, or the same in the presence of a competitive inhibitor? Explain.

PROBLEM 5.41 Explain why most competitive inhibitors are structurally similar to a substrate for the enzyme they inhibit, whereas most noncompetitive inhibitors are not.

PROBLEM 5.42 Add a line to each graph in Figure 5.10 clearly showing what results will be obtained in the absence of inhibitor with an enzyme concentration equal to one-half of that used to prepare the original graph. Which class of inhibitor is "mimicked" with the reduction in enzyme concentration? Explain.

PROBLEM 5.43 Add a line to each graph in Figure 5.10 showing what results will be obtained in the presence of inhibitor concentrations much greater than those used to prepare the graphs.

5.11 ISOZYMES

This section moves from enzyme inhibitors into one of a variety of enzyme-related topics discussed in the remainder of this chapter. Distinct enzymes that catalyze the same reaction are said to represent a group of **isoenzymes** or **isozymes.** Groups of isozymes commonly, but not always, consist of oligomeric proteins that differ from one another in subunit composition. A classical example is **lactate dehydrogenase (LDH),** which catalyzes the reduction of pyruvate to lactate. The coenzyme NADH, the reducing agent, is simultaneously oxidized to NAD^+:

$$\underset{\textbf{Pyruvate}}{CH_3-\overset{\overset{\displaystyle O}{\|}}{C}-\overset{\overset{\displaystyle O}{\|}}{C}-O^-} + NADH + H^+ \rightleftharpoons \underset{\textbf{Lactate}}{CH_3-\overset{\overset{\displaystyle OH}{|}}{CH}-\overset{\overset{\displaystyle O}{\|}}{C}-O^-} + NAD^+$$

Two different polypeptide chains, called M (for muscle) and H (for heart) chains, are employed to assemble five different LDH molecules, each of which contains a total of four noncovalently linked subunits:

H_4

H_3M

H_2M_2

HM_3

M_4

Each tissue has a unique ratio of these LDH isozymes. Heart muscle, for example, contains predominantly H_4, whereas skeletal muscle is rich in M_4 but contains little H_4. Section 5.18 describes how these tissue-specific differences are used for diagnostic purposes. **[Try Problem 5.44]**

Why has the human body evolved to produce different forms of the same enzyme? Probably because the same catalytic activity is sometimes needed in multiple types of cells that house markedly different environments or substrate concentrations. Thus, one isozyme may function well in the environment that exists in liver cells but possess little catalytic activity under conditions that prevail in heart cells. At any given concentration, each isozyme has a unique K_m and V_{max}, and distinct isozymes tend to respond differently to changes in pH, inhibitor concentrations, and other variables.

PROBLEM 5.44 How many isozymes can be produced from two different subunits if each isozyme molecule contains a total of three subunits? A total of five subunits? Explain.

5.12 ALLOSTERIC ENZYMES AND FEEDBACK INHIBITION

Allosteric enzymes, like isozymes, are a special category of enzymes. They play a central role in the continual coordination and regulation of the rates of the thousands of unique reactions that maintain life.

Allosteric Effectors

An **allosteric enzyme** is a protein that contains one or more binding sites, called **allosteric sites** (*allo-*, derived from Greek, means *other*), that are usually separate from its active site(s). The distinctive molecules that bind to allosteric sites are known as **effectors** or **modulators** because their binding alters (affects, modulates) the activity of the enzyme. When the effector is a substrate, it is described as **homotropic.** A **heterotropic effector** is one that differs from substrates.

Most effectors are classified as **ligands,** small substances that bind selectively to large molecules. Those that are heterotropic commonly have markedly different structures from the substrates of the enzymes they regulate. Many effectors are reactants, products, or intermediates in sequences of chemical reactions known as metabolic pathways. Adenosine triphosphate (ATP), glucose 6-phosphate, alanine, and citrate (Figure 5.11), for example, are some of the effectors that help regulate the activity of specific enzymes involved in the conversion of glucose to pyruvate, a process known as glycolysis. The metabolic logic behind the modulatory action of these effectors will be examined in Chapter 11.

Allosteric enzymes allow a cell to continuously fine tune the rates of key reactions and entire metabolic pathways as effector concentrations fluctuate in response to changing activities and needs. Effectors provide one mechanism through which distinct metabolic pathways, and even separate cells and organelles, communicate with one another. Some allosteric enzymes possess over a dozen unique allosteric sites with part of these designed to accommodate effectors from one metabolic pathway and others constructed to interact with distinctive effectors from one or more additional pathways.

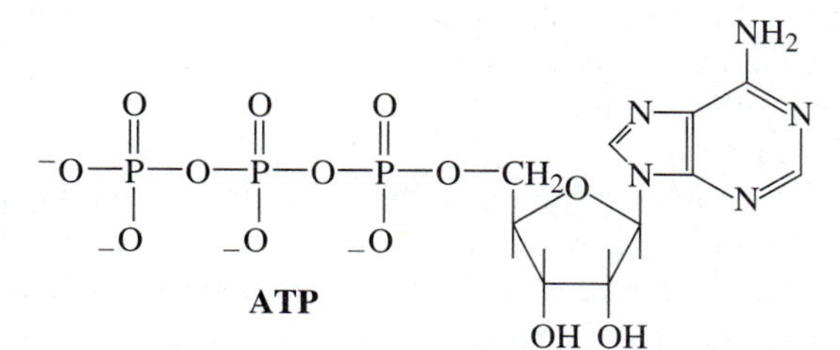

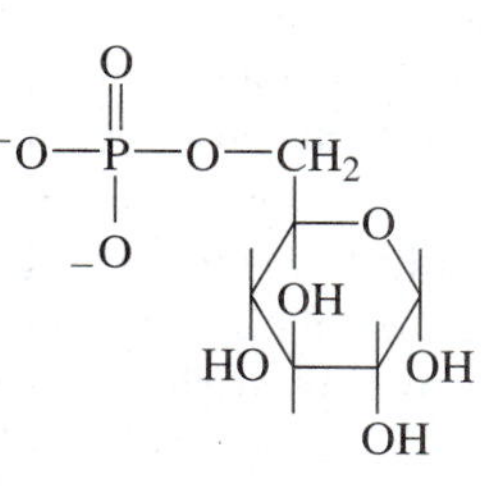

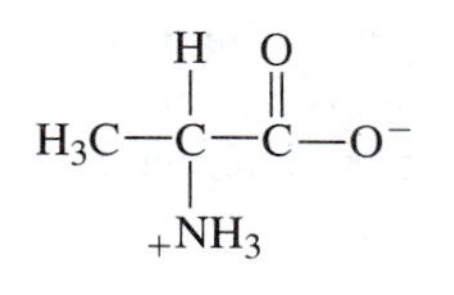

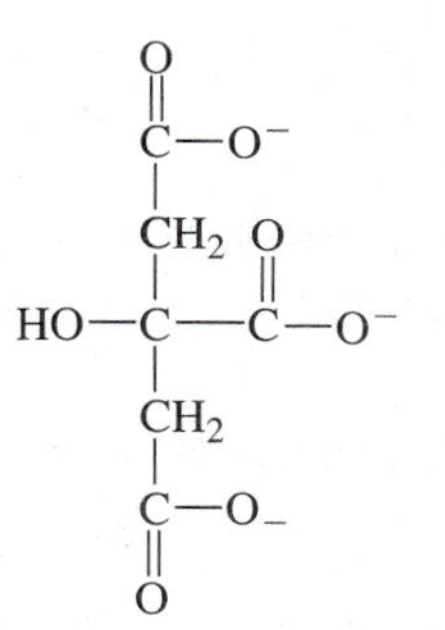

FIGURE 5.11
Some effectors that help regulate glycolysis.

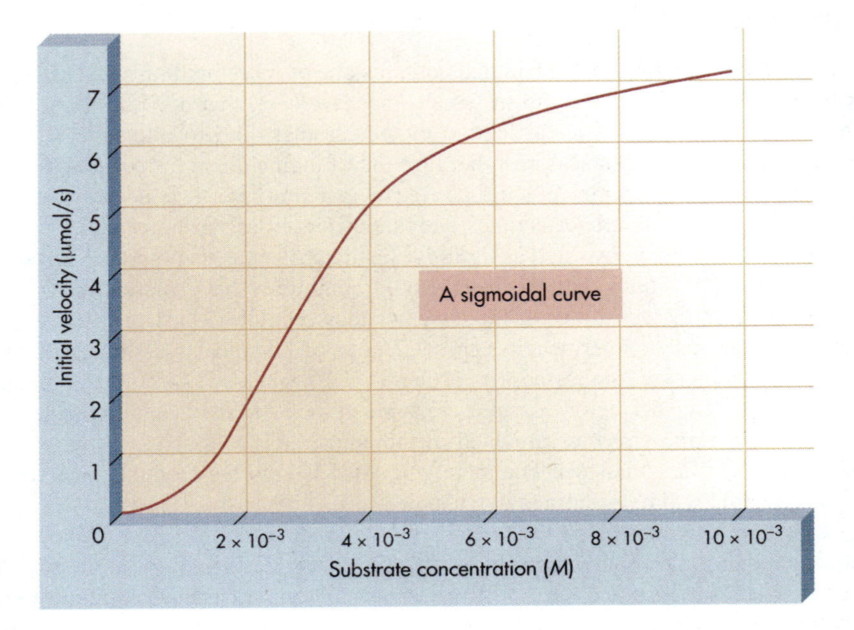

FIGURE 5.12
An initial velocity versus substrate concentration plot for a typical allosteric enzyme.

Cooperative Binding

Allosteric enzymes are normally symmetrical oligomeric proteins with an active site and one or more allosteric sites on each subunit. Since effector binding impacts substrate binding, effector–substrate interactions (usually indirect) are described as cooperative. The cooperativity is said to be **positive** when the effector enhances enzyme activity and **negative** when enzyme action is inhibited. Cooperative interactions are responsible for the sigmoidal initial velocity versus substrate concentration plots that are characteristic of most allosteric enzymes (Figure 5.12). The basic themes of specificity, regulation of reaction rates, structural hierarchy, structure–function relationships, and cellular communication all intertwine in the realm of allosteric enzymes. **[Try Problem 5.45]**

Protein effectors and cooperative binding were first introduced in Chapter 4 during the discussion of hemoglobin. You may want to revisit that discussion at this time. The sigmoidal O_2 binding curve for hemoglobin is a consequence of the cooperative binding of O_2. In this case, the cooperativity is described as positive and homotropic, since the binding of one O_2 enhances the binding of additional O_2's. Homotropic interactions, in general, are those that involve multiple copies of a single ligand. Carbon dioxide (CO_2), 2,3-bisphosphate, and H^+ are all negative heterotropic effectors of hemoglobin; they inhibit the binding of a substance (O_2) that differs from themselves. Although hemoglobin is not an enzyme, it can be classified as an allosteric protein on the basis of its mode of action. **[Try Problem 5.46]**

Models for Allosterism

There are two popular models that have been developed to explain the interactions of effectors with allosteric enzymes. The **sequential model** applies to those allosteric enzymes that initially exist in a single conformation. The simplest form of this model proposes that the reversible binding of an effector changes the conformation of the enzyme as a whole. The resulting shifts in amino acid side chains at the active sites stimulate or inhibit the activity of the enzyme.

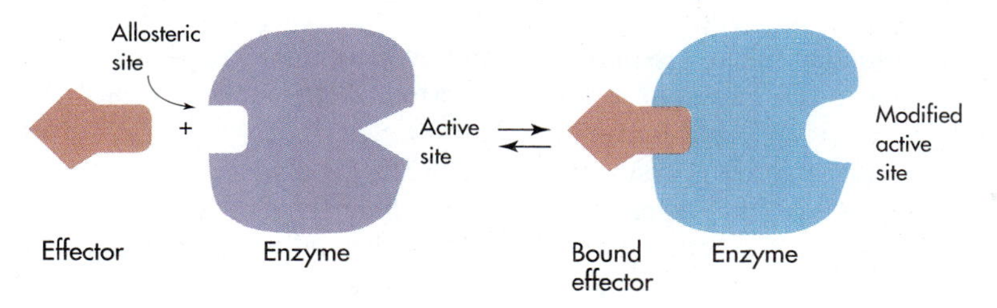

Although not shown, each allosteric enzyme contains multiple allosteric sites for each effector and multiple active sites. The binding of a single effector may not have the same impact on all of the active sites. According to the sequential model, *sequential* binding of effectors leads to the progressive activation or inactivation of the enzyme.

The simplest form of the **concerted (symmetry) model** proposes that an allosteric enzyme exists in interconvertible more active (relaxed, R) and less active (tight or tense, T) conformations (forms) that are at equilibrium in the absence of substrates or effectors:

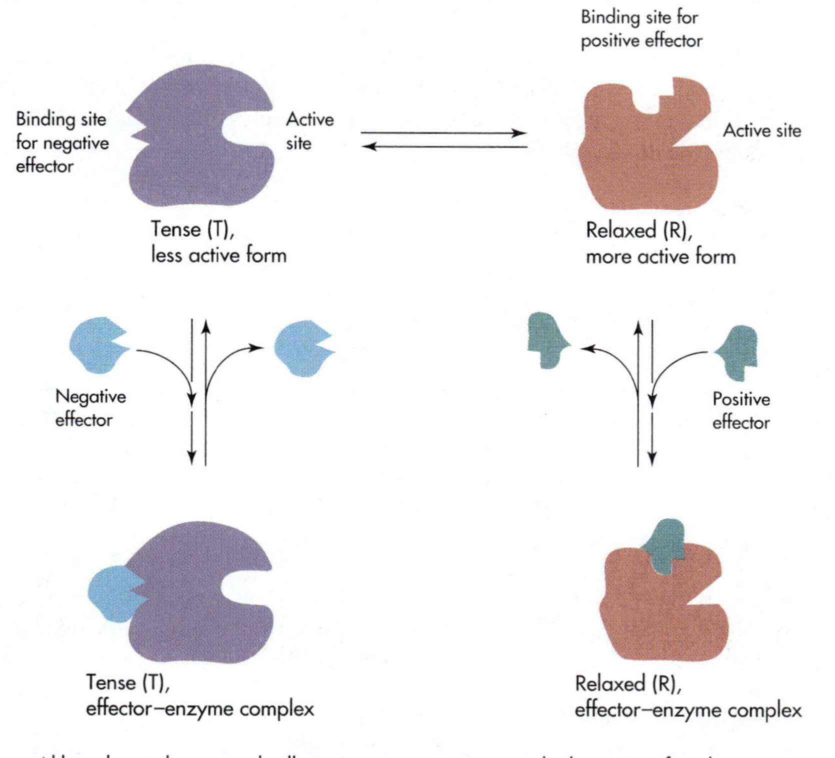

Although not shown, each allosteric enzyme contains multiple copies of each allosteric site and multiple active sites. According to the concerted model, all of the active sites are either relaxed (R) or tense (T); when one active site switches its conformation (from T or R or vice versa) all of the other active sites switch in a *concerted fashion*.

The R conformation contains binding sites for substrates and positive effectors, whereas the T arrangement contains binding sites for negative effectors. When a substrate or positive effector binds to the active arrangement of the enzyme, the concentration of free (unbound) R form decreases. This decrease in free R form destroys the equilibrium that initially existed between free R form and free T form. As the reaction spontaneously moves to restore equilibrium (Le Châtelier's principle), there is a net conversion of free T form to free R form. Since both bound (by substrate or positive effector) and free R conformations are catalytically competent, the presence of a substrate or a positive effector increases the fraction of the total enzyme that is capable of converting substrate to product. Thus, the addition of a positive effector to a reaction mixture leads to an increase in the rate of product formation as long as the enzyme is not already saturated with substrate or positive effector. Negative effectors bind to the free T form of the allosteric enzyme and shift the R–T equilibrium in favor of the T form. The rate of product formation decreases as the fraction of the total enzyme that exists in the T form increases. **[Try Problem 5.47]**

Feedback Inhibition

A product of an enzyme-catalyzed reaction or a sequence of enzyme-catalyzed reactions sometimes prevents its own overproduction by allosterically inactivating one or more of the enzymes responsible for its synthesis. This phenomenon is described as **feedback inhibition.** If the product is generated through a sequence of enzyme-catalyzed reactions, it will usually inhibit one of the early enzymes in the sequence and, in the process, prevent the buildup of intermediates as well as the excessive accumulation of final products. Pretty clever! Specific examples of feedback inhibition will be encountered in later chapters.

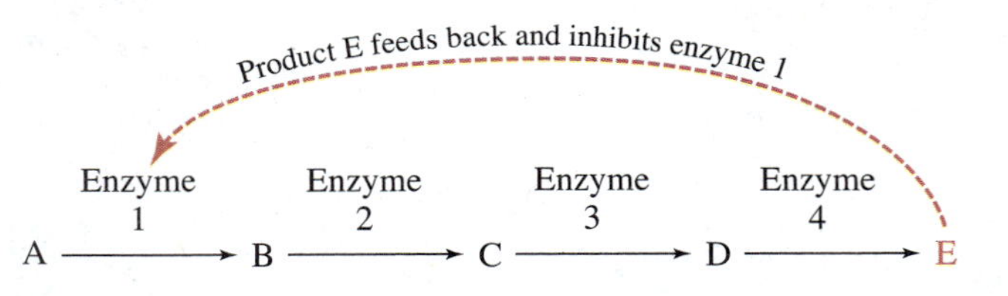

PROBLEM 5.45 An undergraduate research assistant reports that an allosteric enzyme functioning in the presence of a 1×10^{-4} *M* concentration of each of five different effectors is found to have exactly the same catalytic activity as it has when no effectors are present. Is this possible? Explain. If so, of what value are the effector binding sites in this instance?

PROBLEM 5.46 Describe the similarities and differences between the O_2 binding sites on hemoglobin and the active sites on allosteric enzymes. How many binding sites does a single hemoglobin molecule contain?

PROBLEM 5.47 From the plot of enzyme kinetic data in Figure 5.12 estimate the apparent K_m and V_{max} for the allosteric enzyme involved. Is the theoretical interpretation of K_m and V_{max} the same for allosteric enzymes and classical Michaelis–Menten enzymes? Explain.

5.13 PROTEIN KINASES AND PROTEIN PHOSPHATASES

The previous section has examined the role of allosteric effectors in the regulation of enzyme activity. This form of rate modulation falls under the general category of **ligand binding.** Cofactors, including coenzymes, can also be classified as regulatory ligands, since the activity of some enzymes is controlled by cofactor availability. A second general strategy for regulating the activity of existing enzymes is **covalent modification** (the enzyme-catalyzed making or breaking of covalent bonds). It is common for a covalent modification to either fire up an inactive form of an enzyme or shut down an active molecule. In such instances, the covalent change can be visualized as throwing a molecular switch that turns on (up) or off (down) the biological activity associated with the modified protein. Phosphorylation is a common example of covalent modification. Peptide bond cleavage (see Section 5.14) and the formation of disulfide bonds are additional examples.

Phosphorylation is a reversible process that is used to regulate the activity of a wide variety of proteins, not just enzymes. Hormones, growth factors and neurotransmitters are all targets for selective phosphorylation. The perceived general significance of this phenomenon is well documented; the 1992 Nobel Prize for Physiology and Medicine was awarded for pioneering work done in this area. Enzymes known as **protein kinases** are directly responsible for phosphorylation. Each protein kinase catalyzes the transfer of a phosphoryl group from ATP (Section 7.2) to a hydroxyl group in a particular serine, threonine, or tyrosine side chain within a targeted protein (a substrate for the kinase). ATP is converted to adenosine diphosphate (ADP) in the process. The hydrolytic removal of a phosphoryl group from a phosphorylated protein is catalyzed by an enzyme known as a **protein phosphatase.** A target protein may be repeatedly phosphorylated and dephosphorylated:

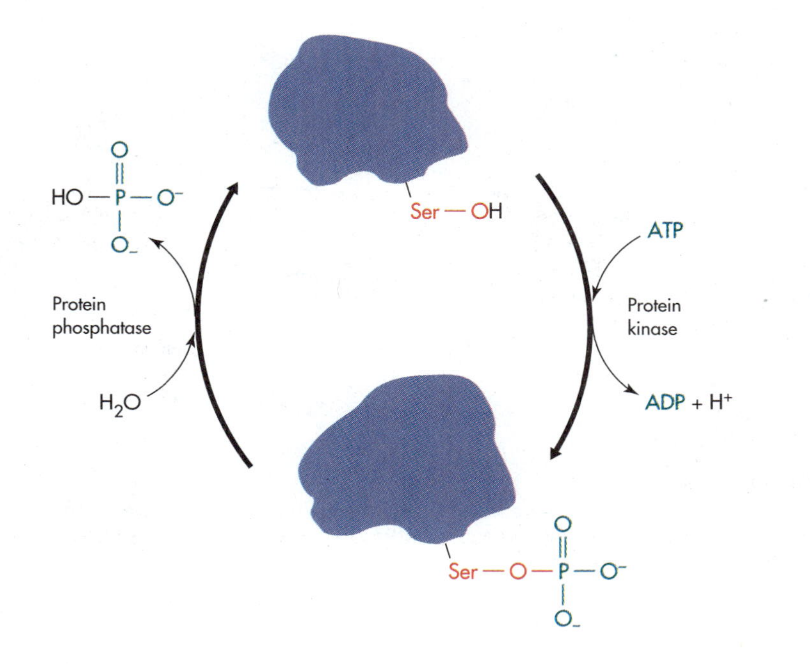

The extent to which a protein is phosphorylated at any instant is determined mainly by the relative concentrations and catalytic activities of its kinases and phosphatases. **[Try Problems 5.48 and 5.49]**

Protein kinases and protein phosphatases, like most enzymes, are highly selective catalysts, and a given kinase or phosphatase usually reacts with only one unique protein or a limited number of proteins. Distinctive kinases are normally involved in the phosphorylation of different side chains within the same protein. A typical human cell probably contains well over 100 different kinases and phosphatases. The exact number is unknown. **[Try Problems 5.50 and 5.51]**

Both the synthesis and the degradation of glycogen (Section 6.6), a polymer of glucose, provide specific examples of enzyme regulation through phosphorylation. The key enzyme involved in the breakdown of glycogen is **glycogen phosphorylase**, which catalyzes the removal of one glucose residue at a time from glycogen (Figure 5.13 and Section 12.1). During glycogen assembly, **glycogen synthase** transfers one activated glucose unit at a time to a growing glycogen chain (Section

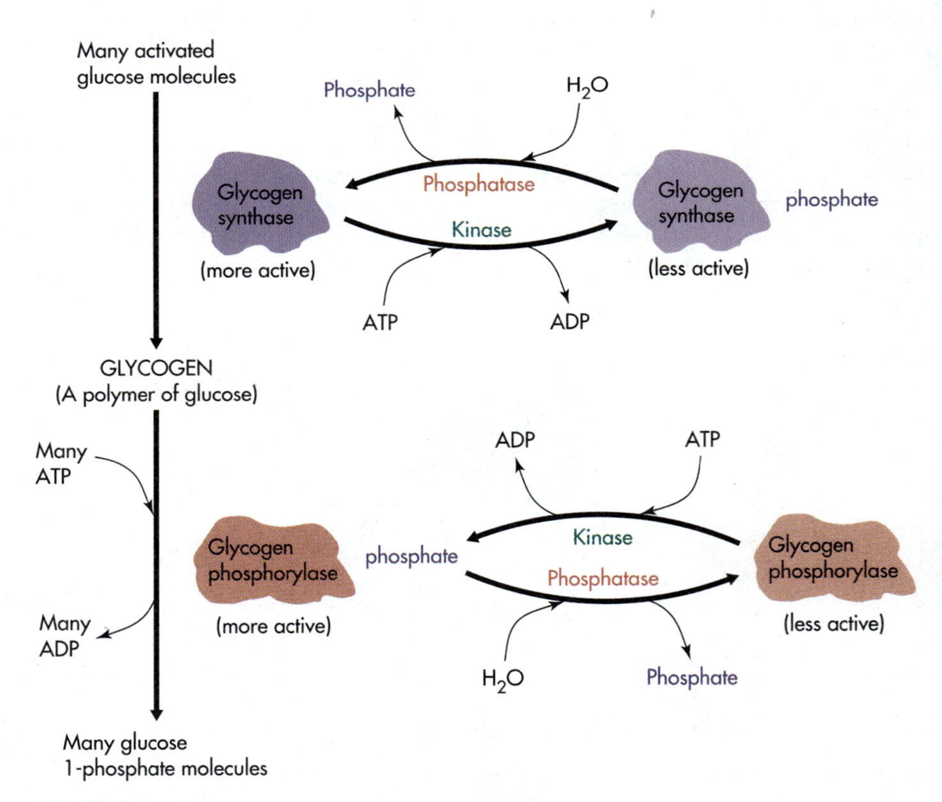

FIGURE 5.13
The regulation of glycogen metabolism through enzyme phosphorylation. The kinase and phosphatase that control glycogen synthase are distinct from those that regulate glycogen phosphorylase.

14.5). Each of these two enzymes exists in a phosphorylated and a nonphosphorylated state. The phosphorylated construct of the synthase is virtually inactive *in vivo,* but the nonphosphorylated species is a potent catalyst. The reverse is true for glycogen phosphorylase; the phosphorylated protein is more active. For each enzyme, the concentrations of catalytically active protein kinases and protein phosphatases primarily determine the ratio of the active and inactive forms. Kinase and phosphatase activities are, in turn, ultimately under the control of hormones that signal when it is advantageous for the body to either synthesize or degrade glycogen. These and other examples of regulation through phosphorylation will be examined in more detail in subsequent chapters.

PROBLEM 5.48 Write the structural formula for a phosphorylated serine residue, and then write an equation for the hydrolysis of this phosphate ester. Include structural formulas for all products.

PROBLEM 5.49 Explain in molecular detail how the phosphorylation of a single serine side chain can alter the catalytic activity of an enzyme.

5.14 ZYMOGENS AND THE DIGESTION OF PROTEINS

The digestion of dietary proteins consists of enzyme–catalyzed hydrolysis to generate small oligopeptides and free amino acids that are absorbed by intestine cells. The oligopeptides are further hydrolyzed in the intestine cells to yield a pool of protein-derived amino acids that enter the blood and are distributed throughout the body (Section 4.12). This section examines some of the enzymes involved and provides an additional example of enzyme regulation through covalent modification.

Proteinases

The enzymes in the stomach and intestine that catalyze the hydrolysis of dietary proteins are known as **proteinases** and **peptidases** (Table 5.7). **Endopeptidases** and **endoproteinases** target internal peptide bonds, whereas **exopeptidases** and **exoproteinases** catalyze the hydrolysis of only C-terminal or N-terminal bonds. Most proteinases and peptidases demonstrate some sequence specificity by preferentially catalyzing the hydrolysis of peptide bonds between particular amino acid residues. **[Try Problem 5.50]**

Protein biochemists commonly use proteinases and peptidases as molecular scissors to "cut" (technically, they catalyze hydrolysis) polypeptide chains at specific sites, a claim documented in Section 4.2 during the discussion of primary sequence determination. Some proteinases are also used in meat tenderizers, where they catalyze the partial hydrolysis of the proteins in tough meat before cooking. In addition, cells employ proteinases to help break down enzymes and other proteins that may be damaged or no longer needed. Even normally functioning proteins tend to be continuously hydrolyzed and replaced with newly synthesized ones, a process described

TABLE 5.7

Some Enzymes Involved in the Digestion of Dietary Proteins

Enzyme	Site of Production	Site of Action	Zymogen	Proteinase Specificity[a]
Pepsin	Stomach	Stomach	Pepsinogen	Not very specific. Some preference for peptide bonds on C-terminal side of Phe and Leu.
Trypsin	Pancreas	Intestine	Trypsinogen	Strong preference for peptide bonds on C-terminal side of Arg and Lys.
Chymotrypsin	Pancreas	Intestine	Chymotrypsinogen	Preference for peptide bonds on the C-terminal side of Phe, Trp, and Tyr.
Carboxypeptidase A	Pancreas	Intestine	Procarboxy-peptidase A	C-terminal peptide bond only. Inactive if C-terminal amino acid is Arg, Lys, or Pro.

[a]Each enzyme catalyzes the hydrolysis of select peptide bonds. Peptide bond preference is determined by a unique specificity pocket (binding pocket) that accommodates select amino acid side chains or select amino acid residues. The molecular basis for the specificity of chymotrypsin is examined in detail in Section 5.16.

as **protein turnover** (Section 14.1). By modifying catalyst concentration, the proteinase-catalyzed hydrolysis of enzymes can even be used to help regulate reaction rates.

Zymogens

What prevents those cells that synthesize digestive enzymes from being damaged by these potent catalysts? The answer lies in **zymogens,** inactive precursors to enzymes. Since the zymogens for digestive enzymes are activated only after they have entered the stomach or intestine, the molecules they encounter along the way are protected from digestion. The activation of zymogens normally involves the selective hydrolysis of one or more peptide bonds followed, in some cases, by the release of one or more peptide fragments:

$$\text{Zymogen} + H_2O \rightleftharpoons \text{Active enzyme} + \text{Peptide(s)}$$

The structural changes that result either create or uncover an active site. Figure 5.14 illustrates how the removal of a peptide fragment can provide access to a previously blocked active site. Table 5.7 identifies the zymogens of some of the enzymes involved in the digestion of dietary proteins.

The acidic fluids in the stomach are primarily responsible for the hydrolytic conversion of **pepsinogen** to active **pepsin.** These acidic fluids also denature (unfold) dietary proteins and, in the process, make their internal peptide bonds more susceptible to enzymatic attack and hydrolysis. Pepsin itself is one of the few proteins in the human body that has evolved to maintain its native structure and biological activity at the low pH's that exist in the stomach (Table 5.3).

Trypsinogen is activated through the catalytic action of a unique endopeptidase known as **enteropeptidase** (formerly **enterokinase**), which is assembled by cells in the small intestine. **Trypsin,** once produced through enteropeptidase action, is capa-

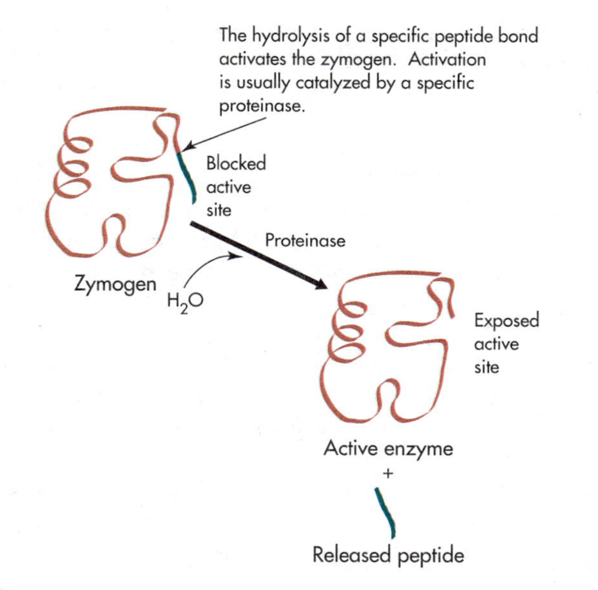

FIGURE 5.14
An illustration of zymogen activation.

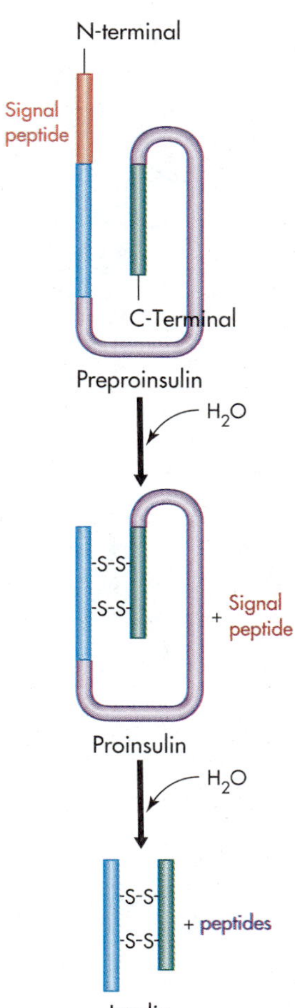

ble of catalytically activating trypsinogen, **chymotrypsinogen,** and **procarboxypeptidase.** The premature activation of trypsinogen and chymotrypsinogen within the pancreas leads to **pancreatitis,** a potentially lethal disease. **[Try Problems 5.51 and 5.52]**

Preproproteins and Proproteins

Zymogens were first discovered for the digestive enzymes. Since that time, a wide variety of other enzymes and nonenzyme proteins have been found to be formed from longer chained, inactive precursors. Such precursors, now generally described as **proproteins,** are activated by specific proteinases at locations that may be far removed from their sites of production. Proproteins represent one of several general mechanisms the body uses to regulate when and where active protein molecules are generated. Following nature's examples, chemists commonly attempt to synthesize pharmacologically active substances in inactive forms (called **prodrugs**) that are activated only at selective sites within the body.

Neuropeptides (Section 3.6) and insulin (Sections 10.12 and 12.4) are specific examples of nonenzyme peptides that are produced from proproteins. **Insulin,** a pancreatic hormone with 51 amino acid residues, is cut out of **preproinsulin** (109 amino acids) through a two-step process in which **proinsulin** (86 amino acids) is an intermediate. The *pre-* in the term *preproinsulin* and **preproprotein** (in general) indicates the presence of a terminal amino acid sequence, known as a **signal peptide.** A signal peptide often plays a central role in targeting a newly synthesized protein to a specific compartment within a cell. In the process, such peptides determine whether or not a protein will be secreted by a cell (Section 18.9).

PROBLEM 5.50 Which one of the enzymes listed in Table 5.7 is most specific? Which of the listed enzymes are endopeptidases and which are exopeptidases?

PROBLEM 5.51 The half-life for an enzyme within the human body is the time required for one-half of the molecules in a sample of the enzyme to be inactivated, destroyed, or eliminated. The proteinases listed in Table 5.7 tend to have limited half lives within the stomach and intestine. Explain this fact.

PROBLEM 5.52 An individual is found to produce trypsinogen molecules with an abnormal primary structure. However, the active proteinase produced from this zymogen is perfectly normal, both structurally and functionally. How is this possible?

5.15 MULTIENZYME SYSTEMS

Two or more enzymes that function together to bring about a sequence of reactions (a metabolic pathway) are said to be a **multienzyme system.** The multiple enzymes may be separate solutes in the same solution, or they may be packed together in a highly specific manner to generate a **multienzyme complex.** Alternatively, the unique catalytic activities in a metabolic pathway may reside within distinct domains along a single polypeptide chain called a **multifunctional enzyme.** When the enzymes in a multienzyme system are separated from one another in solution, the intermediate reactants and products must diffuse randomly from one enzyme to the next: a rather inefficient arrangement. In multienzyme complexes, the product of the first enzyme in the system is usually passed directly (with minimum diffusion) to the second enzyme. The product of the second enzyme is transferred directly to the third enzyme, and so on. Initial reactants enter at specific sites, final products exit at separate sites, and intermediates never exit the complex. Since the intermediates involved are not available to enzymes outside the complex, the process leads to **metabolic channeling.** A multienzyme complex, by minimizing diffusion and side reactions, can bring about a directed sequence of reactions in a highly organized and highly efficient manner. **[Try Problem 5.53]**

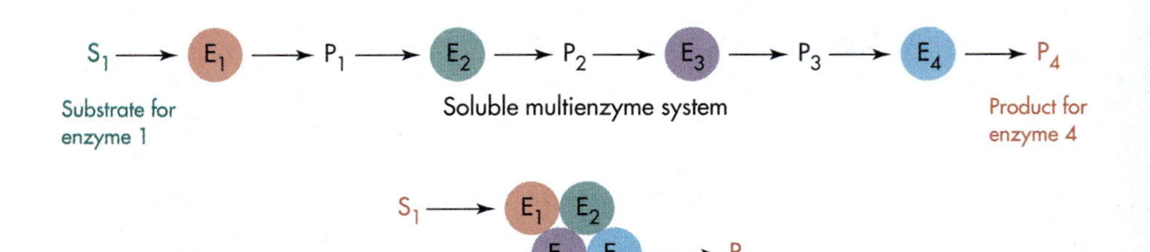

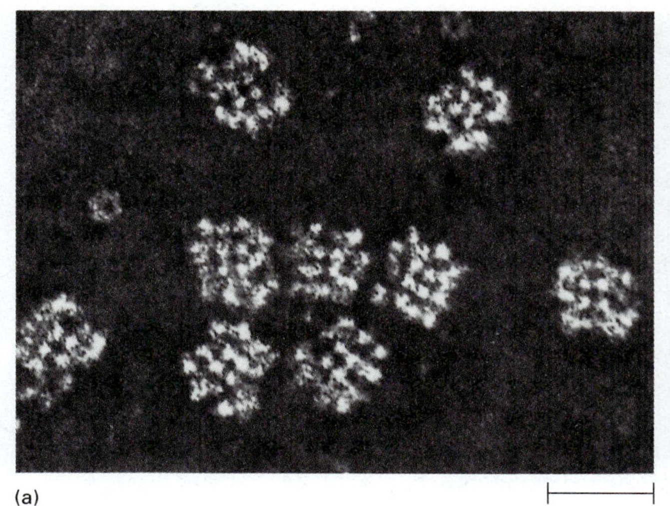

(a)

0.05 μm

(b)

FIGURE 5.15
Pyruvate dehydrogenase complex. (*a*) Electron micrograph of the pyruvate dehydrogenase complex isolated from *E. coli,* showing its subunit structure. (*b*) Interpretive model of the *E. coli* pyruvate dehydrogenase complex. The core of the complex consists of 24 identical subunits of dihydrolipoamide acetyltransferase (E_2) arranged as 8 trimers (white) in a cube-like structure. Twelve pyruvate dehydrogenase (E_1) dimers (blue) are bound along the 12 edges and 6 dihydrolipoamide dehydrogenase dimers (yellow) are bound on the 6 faces of the cube.

The pyruvate dehydrogenase complex (Figure 5.15) is one of several multienzyme systems that are examined in subsequent chapters. This complex possesses three distinct, noncovalently linked enzymes that collectively convert pyruvate to carbon dioxide (CO_2) and acetyl-coenzyme A (acetyl-SCoA):

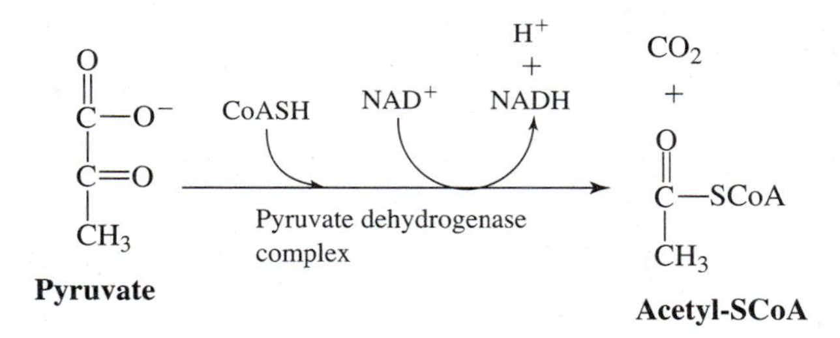

Nicotinamide adenine dinucleotide (NAD^+), an oxidative coenzyme, is reduced to NADH in the process (see Section 11.5 for more details). Each functional unit of the complex contains multiple copies of each enzyme, and all of these protein molecules are arranged in a symmetrical cluster. Within the complex, the pyruvate is decarboxylated and oxidized, and then the resultant acetyl group is attached to a coenzyme A molecule through a thioester link. Since all intermediates are covalently attached to enzyme-bound coenzymes, they are unable to escape from the complex.

Multienzyme complexes may be either free or membrane bound, and several membrane-bound complexes may be carefully positioned on the same membrane such that the ultimate product of one complex tends to become the initial reactant in a second. The tight physical association of distinct enzymes generates a structural hierarchy that leads to the compartmentation of reactions. Multienzyme complexes are used by organisms to help facilitate, control, and coordinate the thousands of reactions that occur within them. **[Try Problem 5.54]**

PROBLEM 5.53 True or false? Neither soluble multienzyme systems nor multifunctional enzymes can contribute to metabolic channeling. Explain.

PROBLEM 5.54 List the other strategies (in addition to the production of multienzyme complexes) encountered in this chapter that are used by organisms to help facilitate, control, and coordinate the reactions that occur within them.

5.16 A CATALYTIC MECHANISM FOR CHYMOTRYPSIN

The preceding sections have provided a general introduction to a variety of enzyme-related topics. It is now time to take a closer look at some specific enzymes. Chymotrypsin (Table 5.7) is one of two enzymes chosen to explore structure–function relationships in enzymes and to illustrate mechanisms of catalysis. The model presented is the one generally accepted today, and it is supported by a combination of physical, kinetic, and chemical studies. X-Ray diffraction (Section 4.13) has provided most of the structural details, and the kinetic and chemical studies have provided some insight into the dynamics of enzyme–substrate interactions. The kinetic analyses have included the determination of K_m's and V_{max}'s in the presence of a wide variety of substrates and inhibitors. Many of the chemical studies have involved the selective, covalent modification of chymotrypsin, followed by an analysis of the binding and catalytic activities of the modified protein.

The trypsin-catalyzed hydrolysis of chymotrypsinogen, a zymogen with a single polypeptide chain (Table 5.7), generates an active enzyme that contains three peptide chains (A, B, and C). Chymotrypsin's preferential hydrolysis of peptide bonds at the C-terminal side of aromatic amino acid residues is explained by a largely hydrophobic **specificity pocket** that is just the right size and shape to accommodate the side chain of phenylalanine (Phe), tryptophan (Trp), or tyrosine (Tyr) (Figure 5.16). Amino acid residues 215 through 219 (numbers based on the sequential numbering of chymotrypsinogen from N-terminus to C-terminus), which run along one side of the pocket, help the enzyme bind a peptide substrate by H bonding to it in an antiparallel β-pleated sheet arrangement. The specificity pocket, as well as the catalytic site, are created by the specific folding and packing of the A, B, and C chains. Chy-

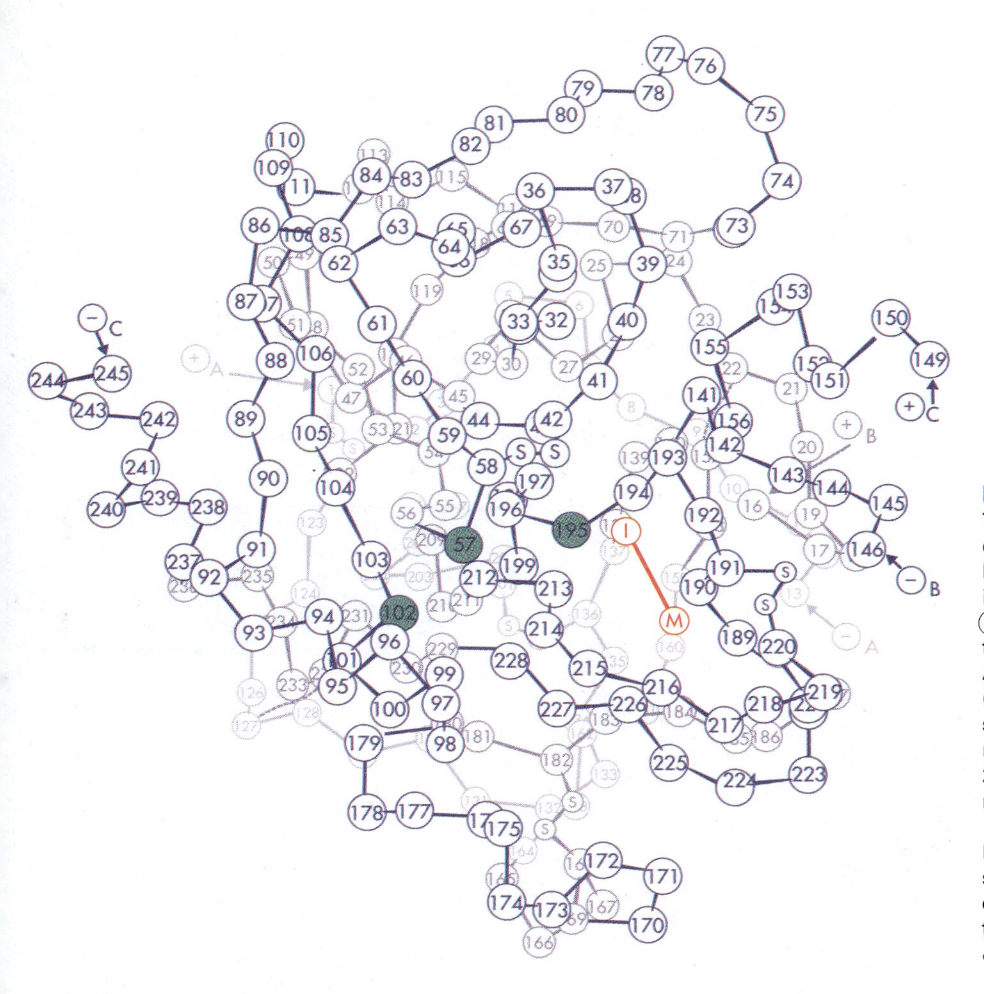

FIGURE 5.16
The structure of a chymotrypsin–inhibitor complex. Each amino acid residue is identified with a numbered circle. (+) and (−) mark the N- and C-terminal ends, respectively, of the A, B, and C chains. The inhibitor ((I)—(M)) (in red) occupies the specificity pocket formed by residues 184–191 and residues 214–227. The green colored residues, Asp 102, His 57, and Ser 195, make up the catalytic triad. Numbers are based on the sequential numbering of chymotrypsinogen from N-terminus to C-terminus. Based on Dickerson & Geis (1969). © Irving Geis.

motrypsin clearly illustrates the central role that tertiary structure plays in generating and maintaining the active sites of enzymes. **[Try Problems 5.55 through 5.57]**

Serine 195, histidine 57, and aspartic acid 102 constitute the core of the catalytic site of chymotrypsin; they are said to make up a **catalytic triad.** In the first step of the catalytic process (Figure 5.17), His 57 acts as a general base to remove a proton from the side chain of Ser 195. Asp 102 facilitates this reaction by stabilizing the protonated form of His 57 (the negative Asp tends to "neutralize" the positive charge that the His acquires). The removal of the proton from the Ser 195 side chain leaves the oxygen in the side chain with a negative charge and makes it a better **nucleophile** [nucleus- or positive (+) charge–seeking group]. Good nucleophiles, in general, possess a negative charge or a partial negative charge, and they usually contain a nonbonding pair of valence electrons. The negative oxygen in the Ser side chain mounts a nucleophilic attack on the carbon (has a partial positive charge) in the carbonyl group within the peptide bond to be hydrolyzed. The attacked carbon has a partial

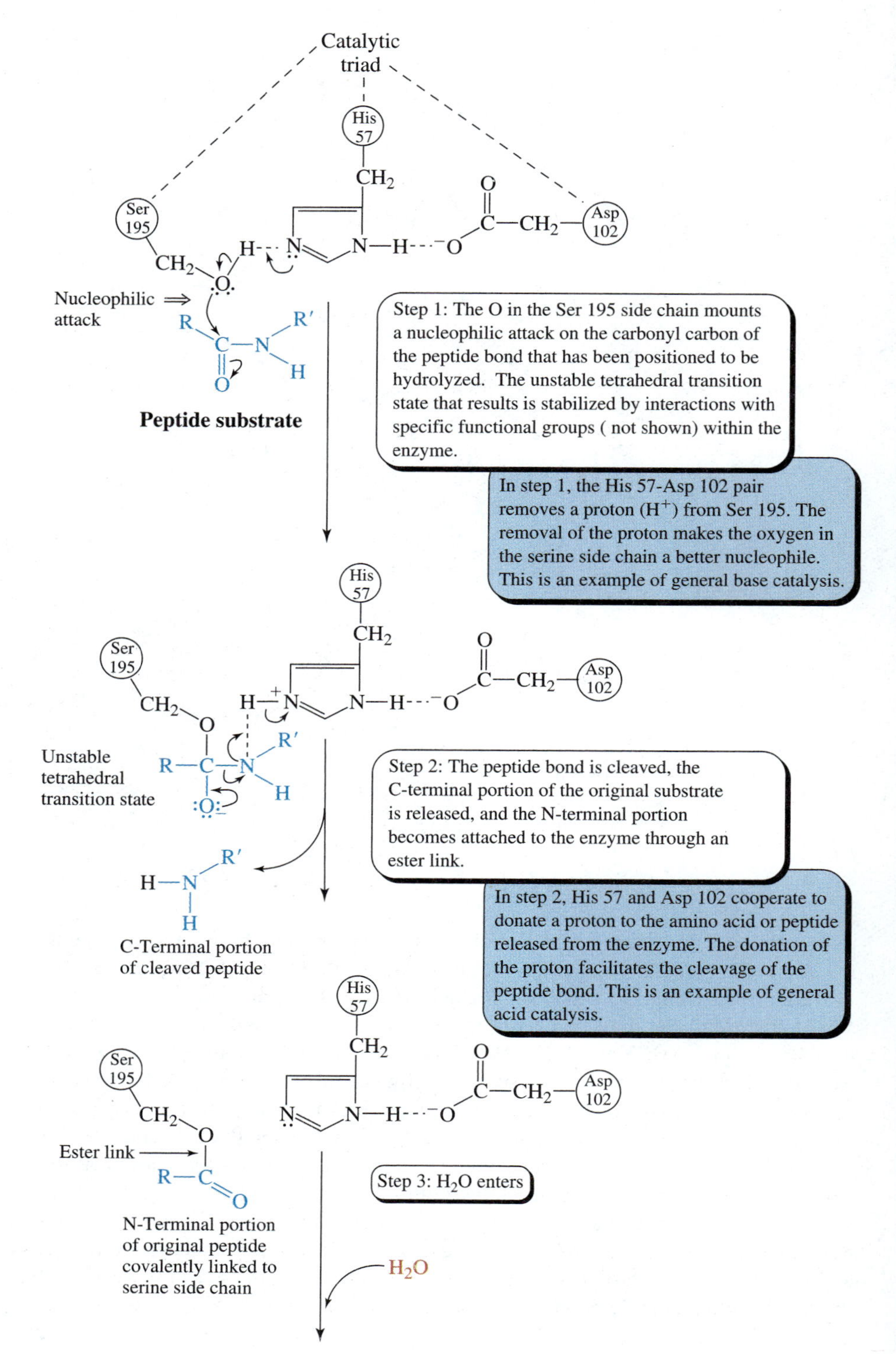

FIGURE 5.17
A proposed mechanism of action for chymotrypsin.

(Continued)

FIGURE 5.17
(Continued)

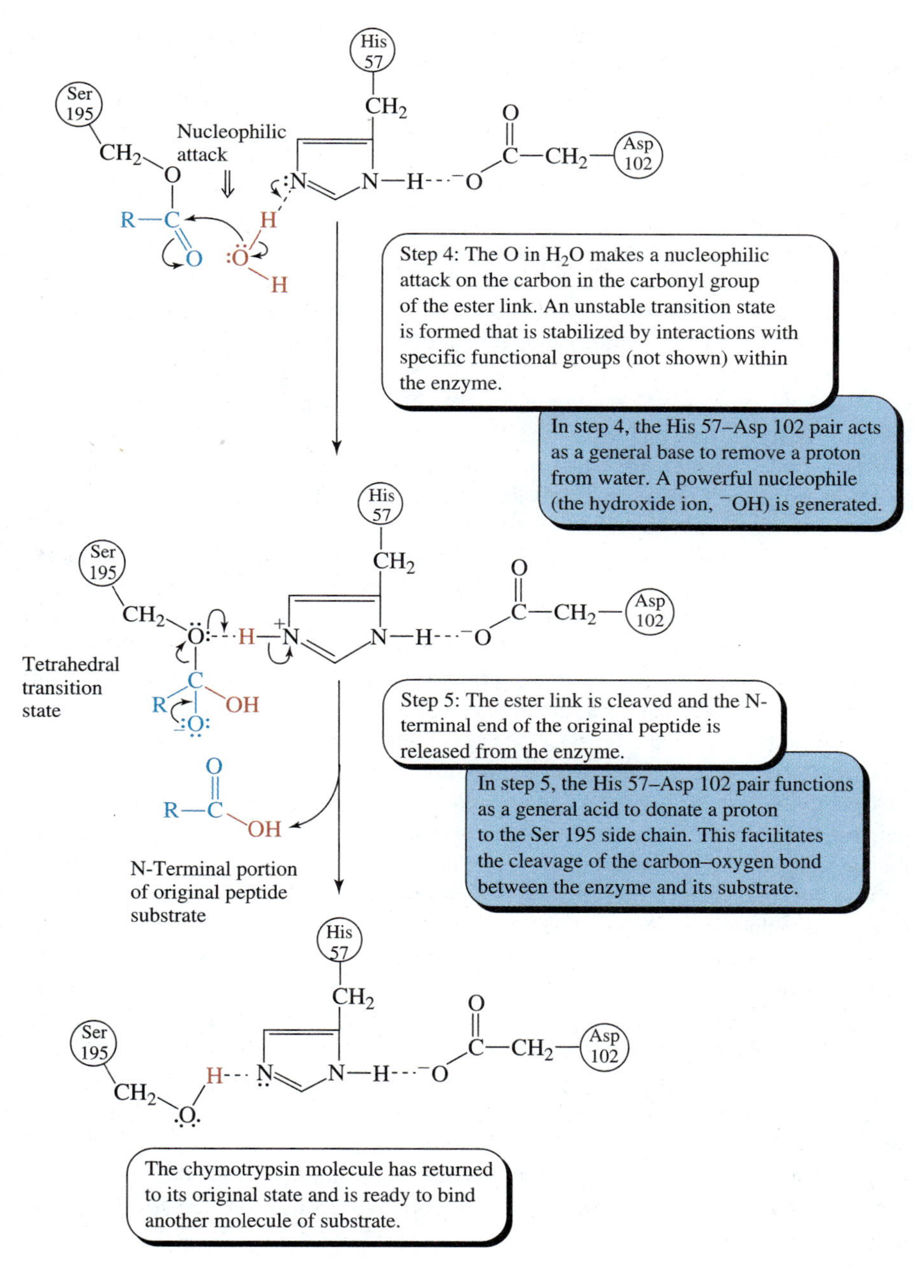

positive charge as a consequence of the electronegativity of the oxygen to which it is covalently bonded. The sequence of events described to this point provides an example of **general base catalysis,** catalysis that involves the removal of a proton by any base (a general base).

The attack on the carbonyl carbon leads to a sharing of a pair of electrons (constitutes the formation of a covalent bond) between the attacked carbon and the oxygen in the Ser side chain and to the formation of a tetrahedral transition state. The transition state is stabilized by interactions (not shown) with particular functional groups within the enzyme. This stabilization lowers the activation energy for the reaction and, to a large extent, accounts for the catalytic activity of chymotrypsin. **[Try Problem 5.58]**

Step 2 of the mechanism (Figure 5.17), entails the actual cleavage of the targeted peptide bond within the bound substrate (peptide). The His 57–Asp 102 pair facilitates bond cleavage by donating a proton to the amide nitrogen as the carbon–nitrogen bond is cleaved. This provides an example of **general acid catalysis,** catalysis that involves the donation of a proton by any acid (a general acid). Since the dissociated form of histidine is the conjugate base of its protonated form (a weak acid), histidine can function as either a general acid or a general base (depending on whether it is protonated or nonprotonated). The cleavage of the carbon–nitrogen bond in the transition state leaves the N-terminal portion of the original peptide (the substrate for chymotrypsin) attached to the Ser 195 through an ester link. This provides an example of **covalent catalysis,** catalysis that entails the covalent attachment of a substrate or a reaction intermediate to the enzyme.

After the C-terminal portion of the cleaved peptide dissociates from the active site and a water molecule enters (step 3), the His 57–Asp 102 pair once again participates in general base catalysis by removing a proton from water. The removal of the proton generates a hydroxide ion (a powerful nucleophile) that attacks the carbonyl carbon (has a partial positive charge) in the ester link that holds the N-terminal portion of the original substrate to Ser 195 at the active site. A second tetrahedral transition state is formed (step 4). Once again, the transition state is stabilized by interactions with specific functional groups (not shown) on the enzyme.

During the last step of the mechanism (step 5, Figure 5.17), the His 57–Asp 102 pair functions as a general acid; it donates a proton that facilitates the cleavage of the carbon–oxygen link between the enzyme and its substrate. The catalytic triad is returned to its original state, and the N-terminal portion of the original substrate is free to dissociate from the enzyme.

The mechanism of action of chymotrypsin encompasses several features that are frequently encountered in the mechanisms of action of other enzymes. These commonly shared features include general acid–base catalysis, covalent catalysis, and transition state stabilization. The addition of a proton to a reactant or the removal of a proton from a reactant often enhances its reactivity, even in the absence of enzymes; many nonenzymatic reactions are acid or base catalyzed. The covalent attachment of a substrate or intermediate to an enzyme can stabilize the substrate or intermediate and can position it for reaction with another substrate or particular functional groups on the enzyme. The stabilization of a transition state lowers the activation energy for a reaction.

PROBLEM 5.55 Study Figure 5.16. Use an arrow to identify each clearly visible disulfide bond. Identify (by number) those residues that make up the bottom of the hydrophobic specificity pocket.

PROBLEM 5.56 List three amino acids whose side chains can be used by an enzyme to line a hydrophobic pocket.

PROBLEM 5.57 Predict which peptide bond(s) would be cleaved preferentially if a negatively charged aspartate side chain existed at the bottom of the specificity pocket in chymotrypsin.

PROBLEM 5.58 The description of step 1 in Figure 5.17 states that "The unstable transition state that results is stabilized by interactions with specific functional groups (not shown) within the enzyme." What types of interactions would stabilize the transition state? Explain.

5.17 A CATALYTIC MECHANISM FOR TYROSYL-tRNA SYNTHETASE

Tyrosyl-tRNA synthetase from *Bacillus stearothermophilus* will be used to provide a second specific example of the mechanism of enzyme catalysis. This enzyme is a dimer containing two identical polypeptide chains, each with 418 amino acid residues. It catalyzes the following overall reaction:

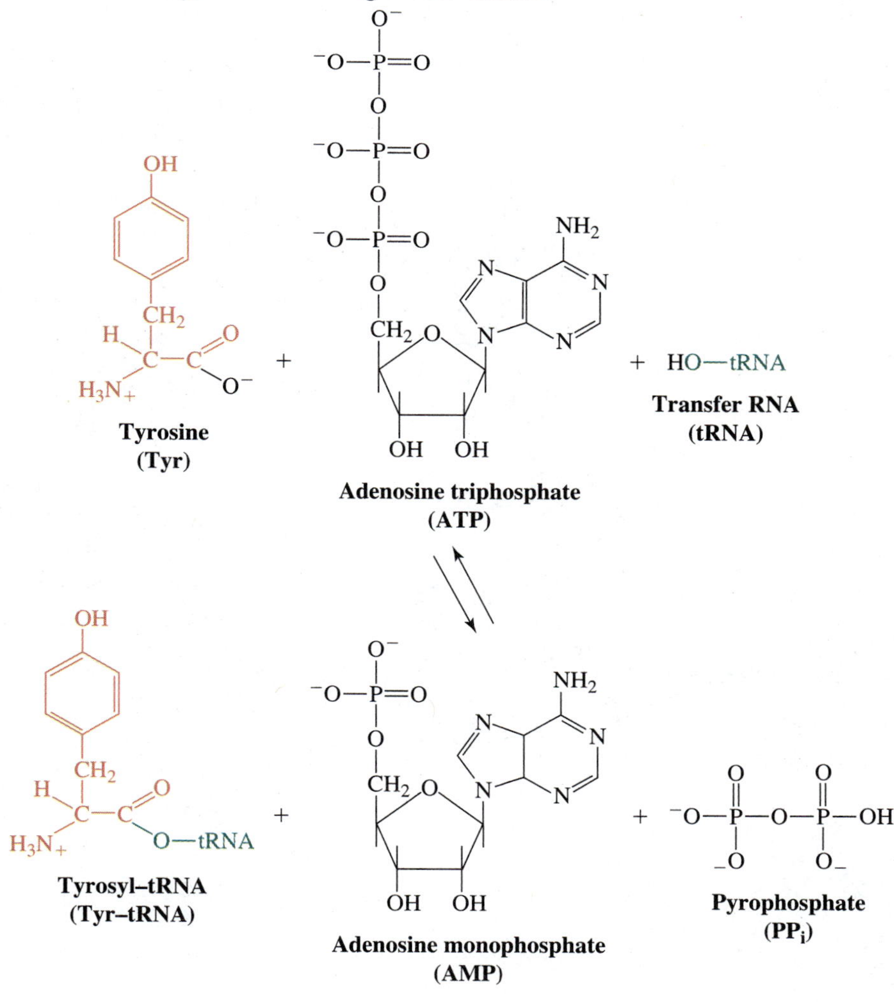

The attachment of the tyrosine to the tRNA activates this amino acid and prepares it for incorporation into a growing polypeptide chain during polypeptide biosynthesis (Chapter 18). Transfer RNAs (tRNAs) are polymers of nucleotides and a major class of nucleic acids. Their structures and functions are examined in Section 16.6. ATP is analyzed in Section 7.2.

The enzyme-catalyzed formation of Tyr–tRNA proceeds in two discrete steps (see below). The Tyr–AMP formed in step 1 is not considered a legitimate product; it is an unstable intermediate that remains attached to the enzyme and is stabilized through interactions with amino acid side chains at the active site. Much of what is

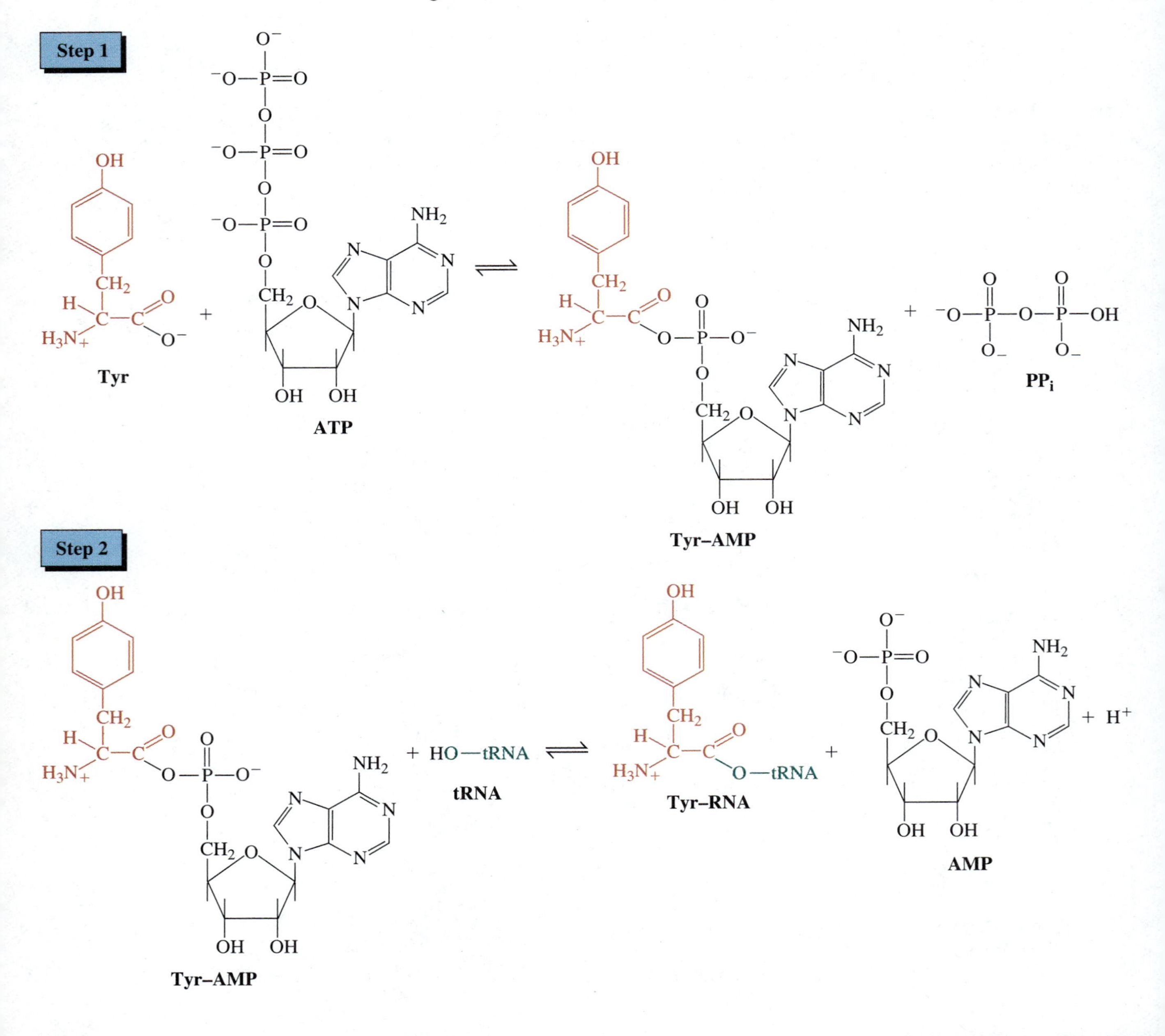

known about the mechanism of the Tyr–tRNA synthetase–catalyzed reaction has been learned by studying modified forms of the enzyme created through the replacement of particular amino acid residues with other amino acid residues of choice **(site-specific replacements).** If an amino acid replacement alters the binding or catalytic activity of the enzyme, the amino acid residue involved must participate in the binding or catalytic process or must play a central role in the creation or maintenance of the active site. Its specific role can sometimes be determined through a careful kinetic analysis (including the determination of K_m and V_{max}) of the modified enzyme.

The proposed mechanism for the first step (the only step whose mechanism is examined in this text) in the overall reaction catalyzed by Tyr–tRNA synthetase is presented in Figure 5.18. At the active site, over a dozen amino acid side chains interact through H bonding with substrates, the transition state, or the products of step 1. Site-specific amino acid replacement studies suggest the Asp 176 and Tyr 34 are associated with the tyrosine binding pocket and play a key role in determining the amino acid specificity of the enzyme. Along with Gln 173, Asp 78, and Tyr 169, these amino acid residues help secure the tyrosine and reaction intermediates to the active site. The formation of the transition state leads to a marked increase in the number of H bonds between the enzyme and its substrates. This increase in H bonding stabilizes the transition state on the enzyme and is hypothesized to account for a majority of the catalytic activity of the enzyme. In contrast to chymotrypsin, catalysis by Tyr–tRNA synthetase does not entail general acid–base catalysis or covalent catalysis. **[Try Problems 5.59 and 5.60]**

PROBLEM 5.59 Study Figure 5.18. Compare the number of H bonds between enzyme and substrates with the number of H bonds between the enzyme and the transition state complex. What is the proposed significance of the increase in H bonding during transition state formation?

***PROBLEM 5.60** What specific class of organic compound is created with the formation of the covalent bond between tyrosine and AMP? What are some of the chemical characteristics of this class of compound?

5.18 THE USES OF ENZYMES IN MEDICINE

It seems appropriate to conclude our introduction to enzymes with a glimpse at some of the medical uses of these amazing proteins. Enzymes are routinely employed as analytical tools, diagnostic agents, and therapeutic agents. This section examines one specific example from each of these three areas. In each case, the unique value of the enzyme lies in its specificity and in its enormous catalytic power.

Enzymes as Analytical Tools

Many biologically important molecules are present at extremely low concentrations in body fluids and are mixed with a large number of other molecules, some structurally very similar. Consequently, the detection and quantitative measurement of molecules of clinical interest commonly require the utilization of sensitive and

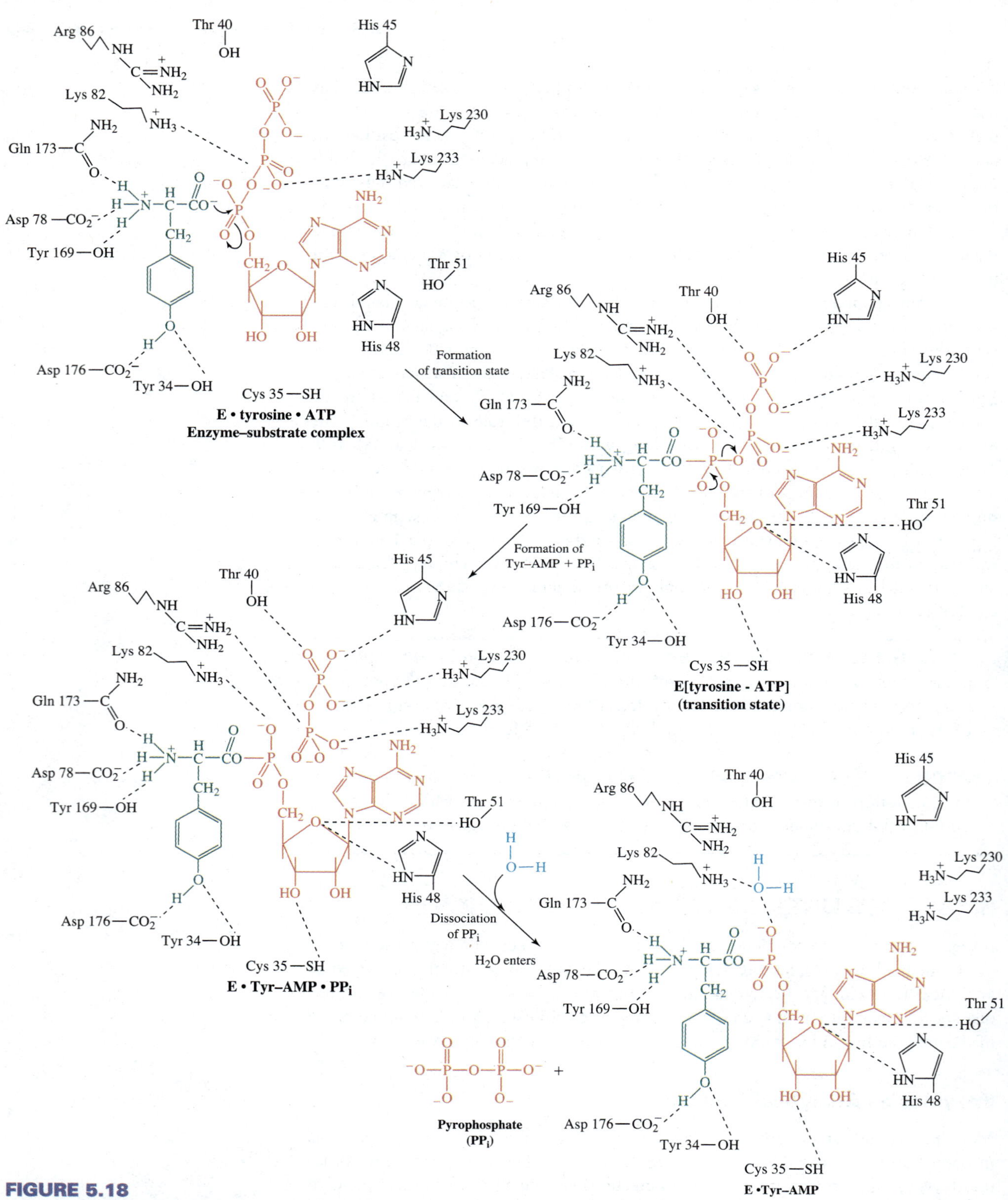

FIGURE 5.18

A proposed mechanism of action for tyrosyl–tRNA synthetase. Based on Creighton (1993).

highly selective procedures. **Immunoassays** are frequently the technique of choice. However, since all immunoassays require an antibody (immunoglobulin; Section 4.11) that reacts selectively with the compound to be analyzed, this methodology can only be employed to analyze substances, called antigens (immunogens), that stimulate antibody production.

The antibodies for immunoassays are normally obtained by injecting rabbits or other experimental animals with purified, known samples of the substance (antigen) to be analyzed. After an appropriate period of time, antigen-specific antibodies can be isolated from the blood of the injected animals. Alternatively, antibody-producing B-cells (specialized white blood cells isolated from the blood of the injected animals) can be fused with melanoma tumor cells to create antibody-producing hybrid cells known as **hybridomas.** Hybridomas, in contrast to B-cells, can be cloned (grown to produce many genetically identical cells), and each clone (set of genetically identical cells) functions as a factory to manufacture many identical copies of the same antibody, that single antibody which the fused B-cell was genetically programmed to synthesize. An antibody isolated from a hybridoma clone is known as a **monoclonal antibody.**

To carry out an immunoassay, one normally adds excess antibody to the sample whose antigen concentration is to be analyzed (Figure 5.19). The antigen–antibody complex that forms is separated from unreacted antibody and then measured quantitatively. An enzyme is commonly placed on an antigen-specific antibody to "tag" it for more sensitive detection. This leads to what is described as an **enzyme-linked immunoassay (EIA).** The tag on the antibody is usually an enzyme that catalyzes

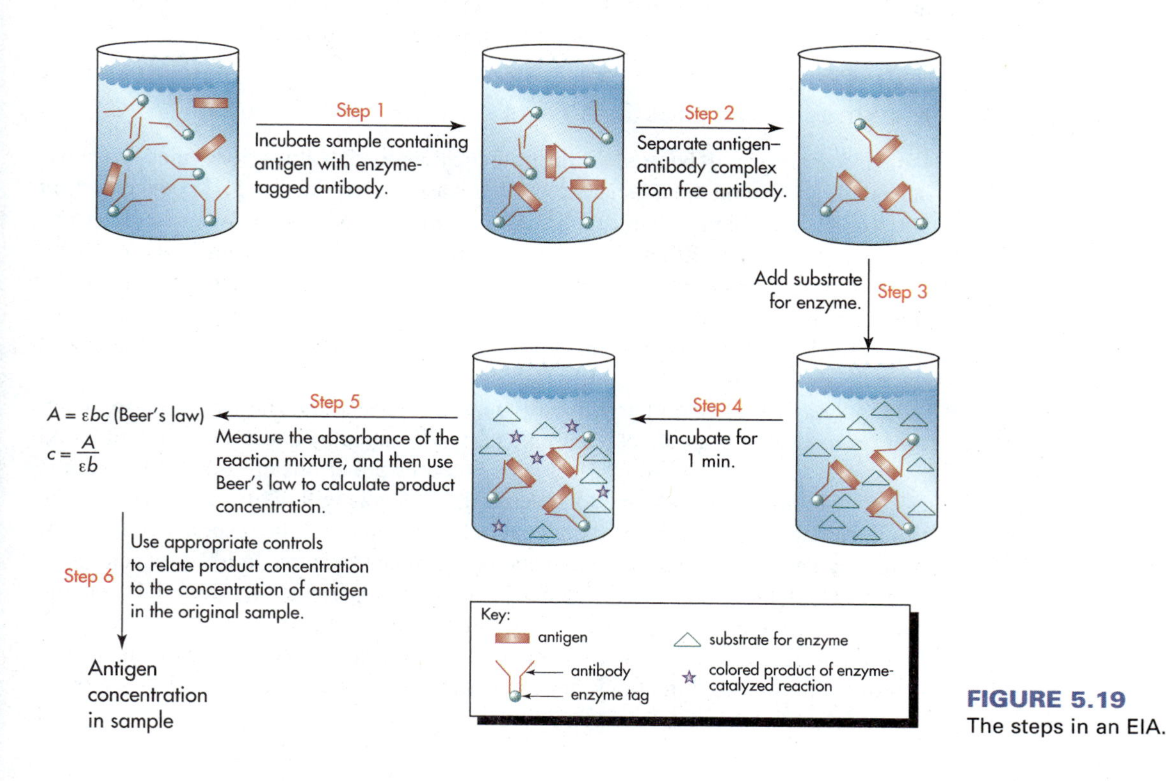

FIGURE 5.19
The steps in an EIA.

the conversion of a colorless reactant to a colored product that can be detected spectrophotometrically (Section 3.11). The amount of colored product formed per unit time is a measure of enzyme concentration, which is, in turn, a measure of the amount of antigen and antibody in the antigen-antibody complex. Samples containing known concentrations of antigen are used to prepare a standard curve [rate of colored product formation (*y*-axis) plotted against known concentrations of antigen (*x*-axis)] that can subsequently be used to calculate graphically the concentration of antigen in a sample of unknown concentration. **[Try Problems 5.61 and 5.62]**

An **enzyme-linked immunosorbant assay (ELISA)** is a special form of EIA. In this case, an untagged antibody is attached to an insoluble solid support and then reacted with the antigen in the sample being analyzed. Once the antigen is bound to the solid phase through its antibody, it is reacted with a second antibody (selected to bind to the antigen at a site that is not blocked by the first antibody), which is tagged with an enzyme. The amount of tagged antibody that adheres to the solid support through the bound antigen is a measure of the quantity of antigen originally present in the analyzed sample.

A **radioimmunoassay** is similar to an EIA. In a radioimmunoassay, a fixed, known amount of isotopically labeled antigen is incubated with a limiting amount of untagged antibody (not attached to an enzyme) and a specific amount of the sample whose antigen concentration is to be determined. The unlabeled antigen (from the sample of interest) and labeled antigen compete for binding sites on the antibody; the higher the concentration of unlabeled antigen, the smaller the amount of labeled antigen that binds. The amount of labeled antigen bound is ascertained by isolating and counting the antigen–antibody complex. A standard curve [amount of labeled antigen bound (*y*-axis) versus known concentration of unlabeled antigen (*x*-axis)] is used to determine the concentration of antigen in the experimental sample.

Enzymes as Diagnostic Agents

Isozyme analysis will be used to illustrate the diagnostic value of enzymes. Most tissues in the body contain a lactate dehydrogenase (LDH, Section 5.11) concentration about 500 times higher than that found in normal blood serum. When such tissues are damaged, some of their cells rupture and cause elevated serum LDH levels by dumping their contents into the blood. A determination of the relative amounts of the five LDH isozymes and the total concentration of LDH in a serum sample can provide valuable diagnostic information about which tissues have been damaged and the extent of the damage.

Suppose that a person has a chest pain and suspects that she has had a **myocardial infarction** (a heart attack; heart muscle is damaged by a restricted flow of blood to the heart). What can serum LDH studies tell us about this episode? If the pain is associated with tissue damage, total serum LDH levels will likely be elevated. If the tissue damage is indeed heart damage, the H_4 isozyme will be responsible for most of the increase, since this isozyme is the predominant isozyme in heart muscle (Section 5.11). If the pain is associated with damage to another tissue, other specific isozymes will tend to be selectively elevated because of the unique ratio of isozymes in different tissues. Serum isozyme patterns are routinely used to help diagnose myocardial infarctions, infectious hepatitis, muscle diseases, and a variety of other

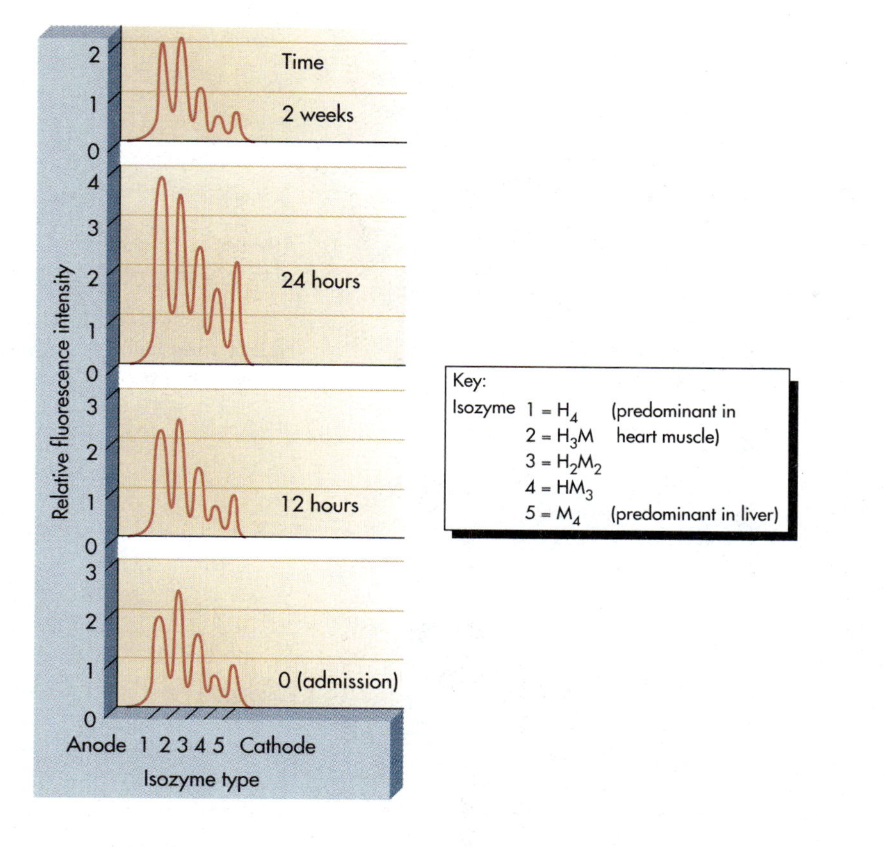

FIGURE 5.20
Changes in serum LDH isozyme levels following a myocardial infarction. The LDH isozymes were separated from one another with electrophoresis and then detected by their fluorescence. The isozyme pattern at time zero (the time of admission) is close to normal. Within 24 hours, the total LDH concentration has increased and the concentration of LDH_1 (H_4) has become greater than that of LDH_2 (H_3M), changes characteristic of a myocardial infarction. The increase in the relative concentration of LDH_5 (M_4) indicates that the liver, as well as the heart, has suffered some damage. Two weeks after the episode, LDH isozyme levels are back to normal.

pathological conditions. Figure 5.20 illustrates the serum LDH levels in one individual at select times after a myocardial infarction. **[Try Problem 5.63]**

Enzymes as Therapeutic Agents

The therapeutic uses of enzymes are as varied as the diagnostic uses. The ailments treated with enzymes include cancer, blood clotting disorders, genetic defects, inflammation, digestive problems, drug toxicities, and kidney failure. The treatment of one form of leukemia (a type of cancer) with asparaginase provides a specific example. **Asparaginase** catalyzes the hydrolysis of asparagine (Asn) to yield aspartic acid (Asp) and ammonia:

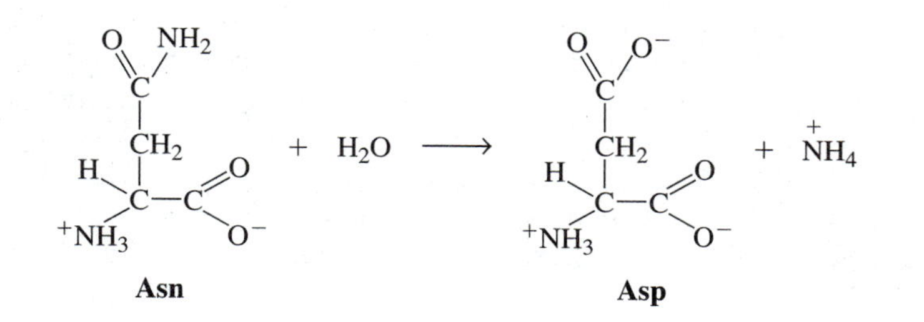

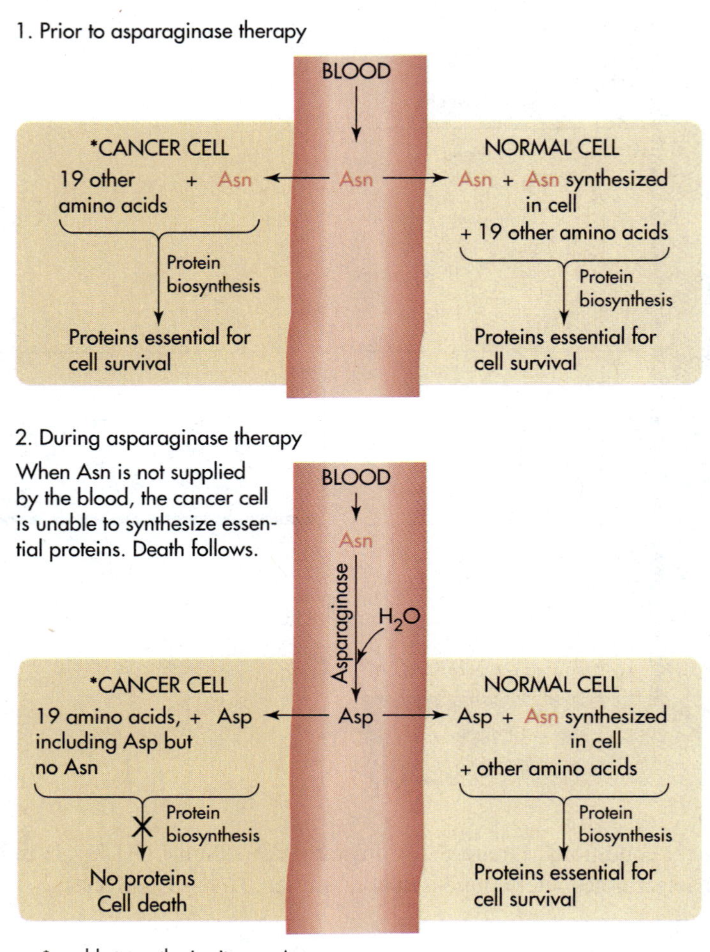

FIGURE 5.21
Asparaginase therapy for cancer.

Normal cells can synthesize an adequate amount of asparagine and do not require an external supply of this amino acid. Some leukemia cells, however, cannot synthesize enough asparagine to meet their needs, so their survival depends on the asparagine that is normally supplied in blood (Figure 5.21). Asparaginase has been injected into the blood of some leukemia patients in an attempt to destroy the asparagine in their blood and, in the process, deprive their cancer cells of this essential nutrient. Unfortunately, the asparaginases studied so far have short half lives in circulation, tend to stimulate the production of antibodies, or lead to other side effects. Persistent biochemists are gradually developing strategies for overcoming these problems. **[Try Problem 5.64]**

PROBLEM 5.61 Explain how a single antigen molecule can lead to the production of millions of colored product molecules during an EIA.

PROBLEM 5.62 Most EIAs are stereospecific and can readily distinquish antigens that are enantiomers. Is this stereospecificity due to the stereospecificity of the enzyme or to the stereospecificity of immunoglobulin? Explain.

PROBLEM 5.63 Redraw the graph in Figure 5.20, clearly illustrating what results would have been obtained if the myocardial infarction had been twice as severe.

PROBLEM 5.64 Would orally administered asparaginase be as effective as intravenously injected asparaginase in reducing serum asparagine levels? Explain.

SUMMARY

Present evidence indicates that thousands of distinct reactions are occurring within most living organisms at any given instant. The rate of the vast majority of these reactions is controlled by biological catalysts known as enzymes. Although some enzymes are ribonucleic acids (RNAs), the term *enzyme,* unless specified otherwise, normally refers to the more abundant protein catalysts. Since each biological reaction tends to be controlled by a separate enzyme, the total number of enzymes within the body is roughly equal to the number of reactions it sustains. Instructions for the assembly of enzymes are stored in DNA, the depository for all genetic information. Under guidelines adopted by the International Union of Biochemistry and Molecular Biology, each enzyme is given both a systematic name and a trivial name. Enzyme concentration is normally expressed as katals (kat)/mL or international units (U)/mL.

All protein enzymes contain one or more active sites at which substrates (reactants) bind and catalysis occurs. These sites are created by the specific folding of polypeptide chains, and they usually consist of chiral grooves or pockets of the proper size, shape, and polarity to accommodate the substrates. Precise engineering at the binding site accounts for the extreme substrate specificity, including the stereospecificity, exhibited by most enzymes. The catalytic power of enzymes is partly explained by their ability to bind substrates and to align them for optimum interactions with other substrates and with amino acid residues at the active site. The ability of enzymes to lower activation energies through the stabilization of transition states also plays a role in catalysis. Enzymes, like other catalysts, have no impact on reaction equilibrium; in the presence of a catalyst, equilibrium is attained more quickly, but the equilibrium constant remains the same. Although most enzymes change their shapes as substrates bind (the induced fit model for enzyme–substrate interaction), some substrates interact with active sites through what approximates a lock and key mechanism.

The catalytic capability of an enzyme fluctuates with pH, temperature, cofactor availability, the concentrations of activators or inhibitors, and other variables. Although different enzymes tend to have unique pH and temperature optima, most human enzymes exhibit peak activity between pH 6 and 8 and at temperatures close to body temperature (37°C). The term *cofactor* refers to the prosthetic groups for those enzymes which are complex proteins. Organic cofactors are called coenzymes. The most prevalent coenzymes are produced from the water-soluble vitamins. The most common inorganic cofactors are metal ions.

The study of the rates of enzyme-catalyzed reactions is described as enzyme kinetics, and such studies lie at the heart of enzyme biochemistry. Kinetic analysis can provide detailed information about pH and temperature optima, substrate specificity, catalytic efficiencies, enzyme inhibitors, reaction mechanisms, and additional topics.

Most kinetic studies involve the measurement of reaction rates immediately after reactants have been mixed and before a significant amount of product has accumulated. Under such conditions, the reverse reaction can be ignored, and the measured rates are known as initial velocities (v_o's).

Some years ago, Michaelis and Menten derived a mathematical equation that does an adequate job of expressing the relationship between the initial velocity for many enzyme-catalyzed reactions and substrate concentration:

$$v_o = \frac{V_{max}[S]}{K_m + [S]}$$

Maximum initial velocity (V_{max}) is attained at very high (saturating) substrate concentrations. The Michaelis constant (K_m) is that substrate concentration which leads to an initial velocity of one-half the V_{max}. Enzymes whose action can be described with this equation are said to follow Michaelis–Menten kinetics and to be Michaelis–Menten enzymes. Although the derivation of the Michaelis-Menten equation is based on the assumption that an enzyme-catalyzed reaction proceeds in two steps ($E + S \rightleftharpoons ES \rightleftharpoons E + P$), this equation also describes the activity of some enzymes with a disparate mode of operation. Graphically, the Michaelis–Menten equation defines the hyperbolic curve that is diagnostic of Michaelis–Menten kinetics:

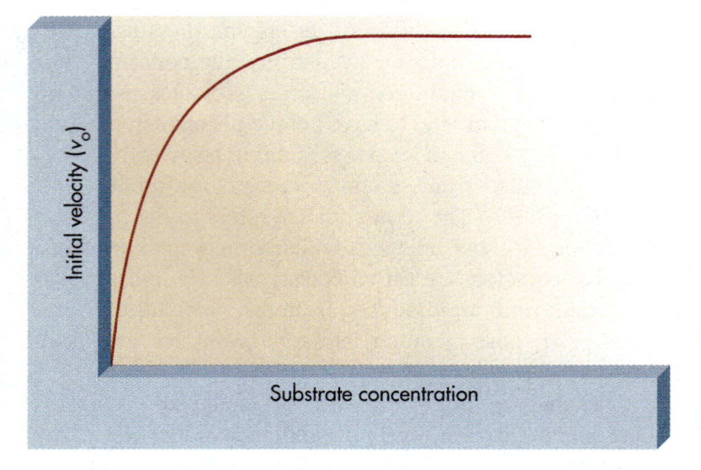

At nonsaturating substrate concentrations, Michaelis–Menten enzymes exhibit some degree of self-regulation: as substrate concentration rises, initial velocity follows suit; and when the substrate concentration is lowered, initial velocity also drops.

Several unique approaches have been developed for the graphical calculation of K_m's and V_{max}'s. One of the most commonly employed techniques, known as a Lineweaver–Burk plot or a double-reciprocal plot, is based on the following algebraic rearrangement of the Michaelis–Menten equation:

$$\frac{1}{v_o} = \left(\frac{K_m}{V_{max}}\right)\frac{1}{[S]} + \frac{1}{V_{max}}$$

When $1/v_o$ (*y*-axis) is plotted against $1/[S]$ (x-axis) for a Michaelis–Menten enzyme, the result is a straight line with a slope of K_m/V_{max}, a *y*-intercept equal to $1/V_{max}$, and an *x*-intercept equal to $-1/K_m$. The calculation of K_m and V_{max} is normally based on the values of the *x*- and *y*-intercepts.

Many substances in our pharmacies, environment, and diet bind to enzymes and inhibit their ability to convert substrates to products. These inhibitors include some heavy metal ions, certain pesticides, and a wide variety of drugs. Competitive inhibitors are usually structurally similar to substrates, and most compete with substrates for binding to active sites. They increase K_m's but have no impact on V_{max}'s. Noncompetitive inhibitors bind at sites distinct from the active sites, so it is possible for both substrate and inhibitor to bind separately or simultaneously. Pure noncompetitive inhibitors reduce the catalytic activity of enzymes without affecting their affinity for substrates; V_{max} decreases while K_m remains the same. Uncompetitive inhibitors, which bind to the enzyme–substrate complex but not to free enzyme, reduce both V_{max} and K_m. Inhibitors are commonly characterized and identified using Lineweaver–Burk plots. Inhibitors are used as tools to probe the mechanism of enzyme-catalyzed reactions and as therapeutic agents.

Isozymes (also called isoenzymes) are distinct enzymes that catalyze the same reaction. Some isozymes are oligomeric proteins that differ from one another in subunit composition. Since different tissues and organs tend to contain unique ratios of the members of a group of isozymes, serum isozyme levels can be used to help diagnose specific tissue damage. By constructing isozymes, an organism can optimize the rate of the same reaction in each of a variety of cellular environments.

The body employs multiple strategies to control when and where active enzyme molecules are generated, to regulate the activity of individual enzymes after they have been created, and to facilitate and coordinate sequences of enzyme-catalyzed reactions. These strategies include ligand binding, covalent modification, the construction of multienzyme complexes, and the utilization of zymogens.

Allosteric (regulatory) enzymes often set the pace of key reactions and entire metabolic pathways. Their catalytic power is controlled, at least in part, by ligands called effectors or modulators. These regulatory substances, which may be either stimulatory (positive) or inhibitory (negative), bind reversibly to enzymes at sites that are usually distinct from active sites. The activity of allosteric enzymes tends to be adjusted on a continuous basis, often by multiple effectors. Since these catalysts yield a nonhyperbolic curve (usually a sigmoidal one) when initial velocity is plotted against substrate concentration, they provide examples of non-Michaelis–Menten enzymes. Both a sequential and a concerted model have been developed to explain enzyme–effector interactions.

Covalent modification is frequently utilized to turn on (up) or turn off (down) the catalytic activity of an existing polypeptide. This process can be accomplished quickly and is often reversible. Phosphorylation is the most common example. In this instance, the concentration of the active form (may be either the phosphorylated or the nonphosphorylated construct) is primarily regulated by protein kinases (catalyze phosphoryl group transfer from ATP to a protein) and protein phosphatases (catalyze the hydrolytic removal of a phosphoryl group from a protein), whose own activities may, in turn, be controlled by other agents, including hormones and other kinases and phosphatases.

Multienzyme complexes consist of two or more enzymes packaged together in a specific geometry. Such assemblies may be free or membrane bound. They normally lead to metabolic channeling, and they provide a mechanism through which a cell can carry out a sequence of reactions in a highly efficient manner. The final product of a multienzyme complex commonly feeds back and regulates its own rate of synthesis by allosterically interacting with a key enzyme within the complex.

Proteinases and certain other classes of enzymes tend to be initially synthesized as inactive precursors, called zymogens, which are subsequently activated only when needed and often at sites far removed from their sites of production. This process helps prevent certain cells from being damaged by the potentially harmful enzymes that they assemble. From one perspective, zymogens can be visualized as rapidly mobilizable storage forms of enzymes.

A detailed mechanism of action has been hypothesized for dozens of enzymes, including chymotrypsin. The most active form of this proteolytic enzyme contains three peptide chains, which fold together in a highly specific manner to create an active site that contains a hydrophobic specificity pocket and a catalytic triad consisting of three amino acid side chains. Chymotrypsin appears to facilitate peptide bond hydrolysis by binding a peptide and, in the process, positioning a peptide bond for attack by a powerful nucleophile. Chymotrypsin also stabilizes the two unstable tetrahedral transition states that are encountered during the reaction mechanism. Chymotrypsin's action provides an example of both general acid–base catalysis and covalent catalysis.

Tyrosyl–tRNA synthetase is a second enzyme whose mechanism of action has been rather thoroughly examined. Its catalytic power lies principally in its ability to stabilize an unstable transition state. Its action does not appear to involve either acid–base catalysis or covalent catalysis.

Enzymes are used in medicine as analytical tools, diagnostic agents, and therapeutic agents. Their unique value lies in their enormous catalytic power and in their remarkable specificity. Specific medical applications include the measurement of antigen levels with enzyme-linked immunosorbent assays (ELISAs), the use of serum lactate dehydrogenase (LDH) isozyme levels to help diagnose myocardial infarctions, and the treatment of a unique category of leukemia with asparaginase.

EXERCISES

5.1 Draw a progress of reaction diagram for a reaction (reaction A) that will tend to occur relatively rapidly even though the products are of higher energy than the reactants. Draw a progress of reaction diagram for a second reaction (reaction B) that will tend to occur relatively slowly in spite of the fact that the reactants are of much higher energy than the products. Put a linear numerical scale on each energy axis and then estimate (from the graph) the activation energy for each reaction. Which reaction will benefit the most from an efficient catalyst? Explain.

5.2 Given the following balanced equation:

$$E + F \rightleftharpoons 2\,P$$

Assume that 4×10^{-3} mmol of P are formed in 10 seconds after 10 mmol of E are mixed with 10 mmol of F to start the reaction. Express the rate of disappearance of E in mmol/min and mmol/s. Calculate the rate of appearance of P in mmol/min and mmol/s. What is the initial velocity for this reaction? Explain.

5.3 A sample of enzyme has the ability to convert 0.02 μmol of substrate to product per minute under standard reaction conditions, which include a saturating substrate concentration. How many international units does this sample of enzyme contain? How many katals does this sample contain? What additional information is needed in order to calculate the turnover number for this enzyme?

5.4 Will 25 U of crude, nonpurified enzyme X contain more, less, or the same number of active enzyme molecules as 25 U of a highly purified sample of the same enzyme? Assume that the crude preparation contains no activators or inhibitors.

5.5 A crude cellular extract contains 10 g of protein and 5.0×10^{12} U of enzyme X. Gel filtration chromatography yields an enzyme solution containing 0.7 g of protein and 1.2×10^{12} U of enzyme X. Calculate the specific activity of the enzyme X preparations before and after gel filtration. What fold purification is achieved with the chromatography?

5.6 Given the following information:

$$G + C \rightleftharpoons X + B \qquad K_{eq} = 10^{15}$$

$$M + N \rightleftharpoons R + S \qquad K_{eq} = 10^{-8}$$

$$A + B \rightleftharpoons E + F \qquad K_{eq} = 10$$

What, if anything, do the equilibrium constants tell us about the relative rates of these three reactions? Explain.

5.7 Are hydrolysis reactions bimolecular reactions or unimolecular reactions? Explain.

***5.8** An increase in temperature increases the rates of bimolecular reactions. What effect, if any, does an increase in temperature have on the equilibrium constants for such reactions? Explain.

5.9 Are all enzymes catalysts? Are all catalysts enzymes? Explain.

5.10 The nonenzyme proteins within the human body account for a much larger percentage of total body weight than the enzyme proteins. However, the number of different enzyme proteins is much greater than the number of unique nonenzyme proteins. What is the explanation for this fact?

5.11 The needs and activities of a particular type of cell are found to be constantly and rapidly changing. Would you expect the half-life for most of the enzymes in such a cell to be short or long? Explain.

5.12 What products are formed when a simple enzyme is completely hydrolyzed?

5.13 Describe the relationship among cofactors, coenzymes, and prosthetic groups.

5.14 What feature of an enzyme allows it to distinguish the members of an enantiomer pair?

5.15 According to the lock and key model for enzyme–substrate interaction, as an enzyme binds its substrates(s) the enzyme (Circle all correct answers):

- **a.** Changes its primary structure.
- **b.** Changes its secondary structure.
- **c.** Changes its tertiary structure.
- **d.** Is hydrolyzed.
- **e.** Is denatured.
- **f.** Changes its pH optimum.
- **g.** None of the above.

Explain.

5.16 Put a square around the possible responses in Exercise 5.15 that accurately describe an enzyme that interacts with its substrate(s) through the induced fit model. Explain your choices.

5.17 Prepare a graph clearly showing the type of results expected if the initial velocity of an enzyme-catalyzed reaction (y-axis) is plotted against pH (x-axis) for an enzyme that has a sharp pH optimum at pH 5.2.

5.18 You will quickly die if the pH of your blood shifts from 7.4 (normal) to 8.4. Provide a detailed molecular explanation for this fact.

5.19 The pH optimum for an enzyme is usually close to the pH at which the enzyme normally functions. Suggest how the body benefits from this relationship.

5.20 A student claims that an enzyme with many weak acid functional groups is much more sensitive to pH changes than an enzyme with very few weak acid functional groups. Is this necessarily true? Explain.

5.21 Which one of the following amino acids would you least expect to find at the substrate binding site of an enzyme when the substrate for the enzyme is the oxalate ion ($^-$OOC—COO$^-$) and the pH optimum for the enzyme is 7.4? Explain.

a. Ala
b. Lys
c. Ser
d. Glu
e. Asn
f. Gly

5.22 Enzyme W has a broad pH optimum and has approximately the same catalytic activity at pH 6.5 and pH 7.6. Compound X is a potent competitive inhibitor of enzyme W at pH 6.5 but has little inhibitory activity at pH 7.6. What is the most likely explanation for these facts?

5.23 Will the initial velocity of an enzyme-catalyzed reaction increase, decrease, or remain the same when more substrate is added to a reaction mixture containing a saturating substrate concentration? Explain.

5.24 Assume that the reaction of an enzyme with its substrate to form an enzyme–substrate complex is at equilibrium. Will the concentration of the enzyme–substrate complex increase, decrease, or remain the same when more free enzyme is added? Explain.

5.25 The V_{max} for an enzyme-catalyzed reaction is 25 mmol of product formed per minute under a standard set of reaction conditions. What is the initial velocity of the enzyme-catalyzed reaction at a substrate concentration where 18% of the total enzyme molecules are bound to substrate? Explain.

5.26 Suppose an enzyme can catalyze the hydrolysis of both compound Z and compound W. The same active site is involved in both reactions, and only one substrate molecule can be bound to this active site at any time. The K_m for the hydrolysis of compound Z is 2×10^{-2} M whereas the K_m for the hydrolysis of compound W is 1×10^{-5} M. What, if anything, does this tell us about the relative V_{max}'s for the hydrolysis of compounds Z and W? When a reaction mixture contains equal concentrations of both compounds Z and W, which enzyme–substrate complex (EZ or EW) will be present at highest concentration in this mixture? Explain.

5.27 A colony of bacteria occasionally finds itself exposed to a highly toxic compound, CPD X. When exposed, the bacteria produce a highly specific enzyme that catalyzes the detoxification of CPD X by catalyzing its hydrolysis. Do you expect the K_m of the detoxifying enzyme to be very large or very small? Explain.

5.28 Explain why the K_m for an enzyme will usually be different for each of its substrates.

5.29 True or false? The K_m for an enzyme is independent of substrate concentration. Explain.

5.30 Under conditions where K_m is an accurate measure of the equilibrium constant for the reaction ES $\rightleftharpoons$ E + S, will K_m be a function of enzyme concentration? Explain.

5.31 Will the initial velocity of an enzyme-catalyzed reaction increase, decrease, or remain the same when the K_m for the enzyme is increased without modifying its V_{max}. Assume that the substrate concentration is constant and nonsaturating and that all other reaction conditions also remain unaltered. Explain.

5.32 Calculate the initial velocity for an enzyme-catalyzed reaction in which $K_m = 0.00034$ M, $V_{max} = 22.2$ μmol/s, and [S] = 0.00012 M.

5.33 The following data are collected when an enzyme-catalyzed reaction is studied under a standard set of reaction conditions:

Substrate Concentration (M)	Initial Velocity (mmol/s)
1.0×10^{-4}	0.024
1.7×10^{-4}	0.031
2.5×10^{-4}	0.036
5.0×10^{-4}	0.045

Use these data to prepare a Lineweaver–Burk plot, and then use this plot to determine the V_{max} for the reaction and the K_m for the enzyme–substrate pair involved. Place a line on the Lineweaver–Burk plot clearly showing the results expected in the presence of twice as much enzyme, with all other conditions unaltered.

5.34 Given the following data for an enzyme-catalyzed reaction under standard reaction conditions in the presence and absence of inhibitor:

Substrate Concentration (M)	No Inhibitor: Initial Velocity (mmol/min)	With Inhibitor: Initial Velocity (mmol/min)
2.0×10^{-3}	0.123	0.057
2.5×10^{-3}	0.133	0.067
3.3×10^{-3}	0.145	0.081
5.0×10^{-3}	0.161	0.100
10×10^{-3}	0.175	0.135

Calculate the K_m and V_{max} for the enzyme–substrate pair involved in the presence and in the absence of inhibitor. Is the inhibitor competitive, noncompetitive, or uncompetitive? Explain.

5.35 Based on the derivation of the Michaelis–Menten equation, $V_{max} = k_2[E_t]$ and $K_m = (k_{-1} + k_2)/k_1$. Identify the reaction associated with each rate constant encountered in these definitions.

5.36 Express the equilibrium constant for the following reaction as a ratio of rate constants:

$$E + P \rightleftharpoons ES$$

5.37 The amount of enzyme in a sample is seldom expressed in grams or moles. Suggest why this is the case.

5.38 Turnover numbers can be calculated only for enzymes that have been purified and whose molecular weights have been determined. Explain why this is the case.

5.39 An enzyme has a single active site at which it can bind and catalyze the hydrolysis of either ester A or ester B. The enzyme cannot bind ester A and ester B simultaneously. Will the K_m for the hydrolysis of ester A increase, decrease, or remain the same in the presence of ester B? Will the V_{max} for the hydrolysis of ester A increase, decrease, or remain the same in the presence of ester B? Explain.

5.40 Suggest how the binding of an inhibitor at a site far removed from the active site on an enzyme can lead to competitive inhibition.

5.41 Two competitive inhibitors, I and I′, each inhibit enzyme W by competing with substrate for the single active site on the enzyme. Given the following information:

$$S + E \rightleftharpoons ES \qquad K_{eq} = 30$$

$$I + E \rightleftharpoons EI \qquad K_{eq} = 0.1$$

$$I' + E \rightleftharpoons EI' \qquad K_{eq} = 200$$

a. Which inhibitor, I or I′, is the most potent inhibitor? Explain.

b. In a reaction mixture that contains equal concentrations of S, I, and I′, which enzyme complex, ES, EI, or EI′, is present in the highest concentration? Explain.

c. Assume that E and I are mixed in a ratio at which 90% of the total enzyme exists in an EI complex and 10% of the enzyme is in a free uncombined state. Further assume that the mixture contains no substrate and that I′ (the other inhibitor) is gradually added after the reaction $E + I \rightleftharpoons EI$ has reached equilibrium. Place a line or curve on a graph (plot percentage of total enzyme in an EI complex on the Y-axis and the amount of I′ added on the X-axis) clearly illustrating how the percentage of total enzyme in an EI complex will change as I′ is added. Will the concentration of free I increase, decrease, or remain the same as I′ is added? Explain.

5.42 Prepare graphs clearly illustrating the type of results expected for a typical human enzyme if you:

a. Plotted V_{max} (y-axis) against the concentration of a pure noncompetitive inhibitor (x-axis).

b. Plotted V_{max} (y-axis) against the concentration of a competitive inhibitor (x-axis).

c. Plotted V_{max} (y-axis) against pH (x-axis) from pH 1.0 to pH 12.0.

d. Plotted v_o (y-axis) against competitive inhibitor concentration (x-axis). Assume that substrate concentration and all other reaction conditions are held constant.

5.43 Will the y-intercept on a Lineweaver–Burk plot increase, decrease, or remain the same in the presence of a competitive inhibitor? A pure noncompetitive inhibitor? An uncompetitive inhibitor?

5.44 An enzyme catalyzes the hydrolysis of the following compound:

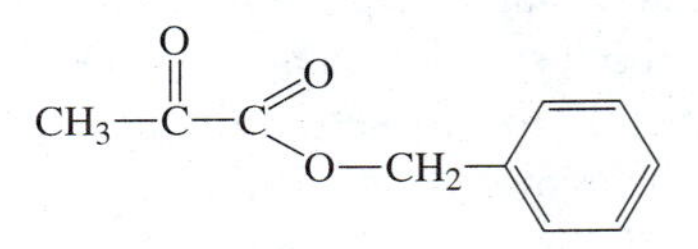

Write the structural formula of two compounds that might reasonably be tested when searching for competitive inhibitors of the enzyme involved. Explain how you arrived at your answer.

5.45 Three isozymes, I, II, and III, exist for enzyme R. These isozymes have K_m values of 0.02, 0.002, and 0.0001 *M*, respectively, with the single substrate that exists for enzyme R. Which isozyme will most likely be present at highest concentrations within a cell when the substrate concentration within the cell is normally around 0.00009 *M*? Explain. Which, if any, of the three isozymes will be saturated with substrate at this substrate concentration? Explain.

5.46 True or false? In the concerted model of allosterism, negative effectors bind to the inactive (T) form of the allosteric enzyme and positive effectors bind to the active (R) form. Explain.

5.47 Describe the similarities and differences between active sites and allosteric sites.

5.48 Assume that compound X can be converted to F, K, L, or I through the following enzyme-catalyzed reactions:

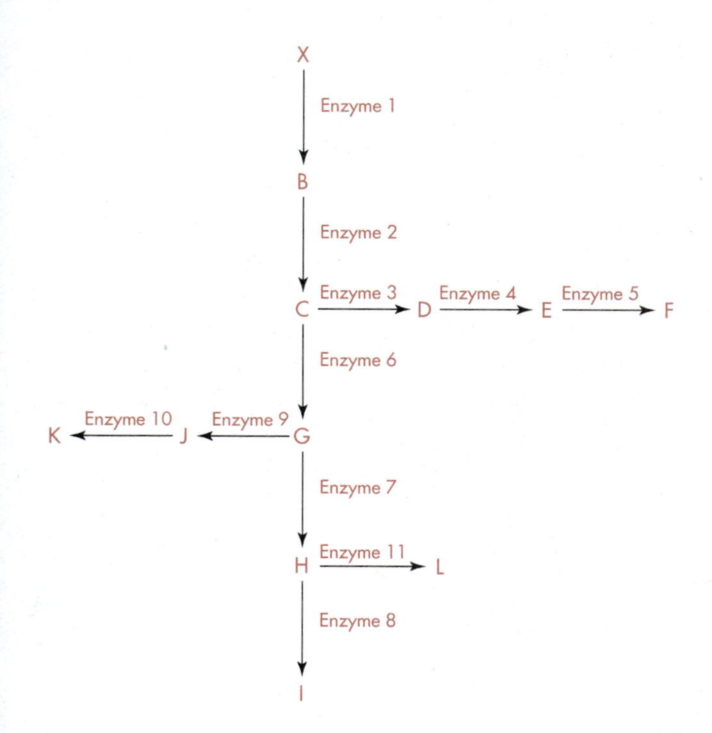

a. Compound F feeds back and allosterically inhibits its own production without inhibiting the production of K, L, or I. Which enzyme is most likely inhibited by F? Explain.

b. Compound K feeds back and allosterically inhibits its own production as well as the production of L and I. In contrast, the production of F is not inhibited by high concentrations of K. What enzyme is inhibited by K? Explain.

5.49 True or false? Any allosteric inhibitor that functions according to the concerted model is a competitive inhibitor. Explain.

5.50 Describe how a single protein kinase with a single protein substrate can lead to specific changes in the catalytic activities of many enzymes.

5.51 To which one of the six major international classes of enzymes do the protein phosphatases belong?

***5.52** Which of the following changes represent covalent modifications within an enzyme?

a. Oxidation of a serine side chain to form an aldehyde.

b. Methylation (leads to ether formation) of a serine side chain.

c. Hydrogen bonding of a serine side chain to aspirin.

d. Heat denaturation.

e. Removal of the N-terminal amino acid residue.

Explain.

5.53 Is a serine side chain primarily uncharged, positively charged, or negatively charged at pH 7? Is a phosphorylated serine side chain primarily uncharged, positively charged, or negatively charged at pH 7? Explain. How do your answers relate to the fact that the phosphorylation of a protein often acts as a molecular switch that either activates or inactivates the protein?

5.54 A multienzyme complex contains eight enzymes, each catalyzing a separate essential step in the conversion of compound Q to compound W. Will the total inhibition of one enzyme in the complex necessarily reduce the rate of conversion of compound Q to compound W? Will the partial inhibition of one enzyme in the complex necessarily reduce the rate of conversion of compound Q to compound W? Explain.

5.55 Many diabetics regularly receive injections of insulin (a protein hormone) that help them to control their blood sugar levels. Why is the insulin injected rather than administered orally?

5.56 Write the structural formula for a dipeptide, and then write an equation for the hydrolysis of this peptide. Describe the role that this type of reaction plays in the digestion of dietary proteins and the activation of proproteins.

5.57 True or false? Zymogens are proproteins. Explain.

5.58 The zymogen for an enzyme contains 107 amino acid residues. The active enzyme contains three noncovalently linked peptide chains, one with 56 amino acid residues, one with 40 amino acid residues, and the third with 3 amino acid residues. What is the minimum number of peptide bonds that would need to be hydrolyzed to convert the zymogen to the active form of the enzyme? Explain.

5.59 Each of the chemical reactions that occurs within the human body falls into one of six major classes or categories. List those six classes.

5.60 Which one of the six major international classes of enzymes catalyzes the conversion of D-alanine to L-alanine?

5.61 When an ester, rather than an amide, is bound to the active site of chymotrypsin, is there any chance that this enzyme will catalyze the hydrolysis of the ester? Explain.

5.62 Chymotrypsin would not be as effective a catalyst at pH 1 as at pH 8 (approximate pH optimum), even if the conformation of chymotrypsin were exactly the same at both pH values. Provide a specific and detailed molecular explanation for this fact.

5.63 Within a single polypeptide chain that contains multiple tyrosine residues, chymotrypsin may preferentially catalyze the hydrolysis of the peptide bond adjacent to a particular tyrosine. What is the probable explanation for this fact?

5.64 Chymotrypsin preferentially catalyzes the hydrolysis of peptide bonds at the C-terminal side of aromatic amino acid residues. However, this enzyme will catalyze the hydrolysis of peptide bonds at other amino acid residues at lower efficiencies. For each of the following pairs of amino acid residues, circle that member of the pair whose C-terminal peptide bond would probably be most susceptible to chymotrypsin-catalyzed hydrolysis:

Pair 1 Met : Glu
Pair 2 Arg : Val
Pair 3 Gly : Leu

Explain your choices.

5.65 In enzyme immunoassays (EIAs), the sensitivity of the assay is increased by the presence of the enzyme tag. Can the sensitivity of an immunoassay be increased by tagging an antibody with a radioisotope (Section 3.12) rather than an enzyme? Explain.

5.66 Explain why it is often difficult to attach an enzyme to an antibody without inactivating the enzyme.

5.67 Heart damage leads to an increase in lactate dehydrogenase (LDH) activity in blood serum. Could tissue damage ever lead to a sudden reduction in the concentration of a specific serum enzyme? Explain.

5.68 True or false? If liver cells are the only cells in the human body that produce enzyme X, an increase in serum levels of enzyme X following a virus infection probably indicates that the infection has led to some liver damage. Explain.

5.69 Most of the asparagine in both normal and leukemia cells is normally utilized for one purpose. What is that purpose?

5.70 Explain how asparaginase treatment kills certain leukemia cells while having no impact on most normal cells.

5.71 Do you expect the activity of Tyr–tRNA synthetase to be very sensitive to changes in pH? Explain.

5.72 Which sections of this chapter deal with structure–function relationships?

5.73 Describe the structural hierarchy within a multienzyme complex.

5.74 Give one example (from this chapter) of experimentation and hypothesis construction. Note: In most cases, the experimentation was mentioned but not described in detail.

SELECTED READINGS

Branden, C., and J. Tooze, *Introduction to Protein Structure*. New York: Garland Publishing, 1991.

Three of 17 chapters deal specifically with enzyme structure and function. Includes a very readable section on chymotrypsin.

Chan, M.K., S. Mukund, A. Kletzin, M.W.W. Adams, and D.C. Rees, Structure of a hyperthermophilic tungstopterin enzyme, aldehyde ferredoxin oxidoreductase, *Science* 267, 1463–1469, 1995.

Presents the crystal structure of an enzyme from Pyrococcus furiosus, *an archaeon that grows optimally at 100° C. A relatively small solvent-exposed surface area, and a relatively large number of both ion pairs and buried atoms may contribute to the extreme thermostability of this enzyme.*

Creighton, T.E., *Proteins: Structures and Molecular Properties*, 2nd ed. New York: W. H. Freeman and Company, 1993.

Chapter 9 titled "Enzyme Catalysis," includes four major sections: "The Kinetics of Enzyme Action," "Theories of Enzyme Catalysis," "Examples of Enzyme Mechanisms," and "Regulation of Enzyme Activity." There are ample references at the end of each section and a limited number of problems (most based on the research literature).

Dickerson, R.E., and I. Geis, *The Structure and Action of Proteins*. New York: Harper & Row, 1969.

A classic! Examines structure–function relationships in specific fibrous and globular proteins, including multiple enzymes.

Dugas, H., *Bioorganic Chemistry: A Chemical Approach to Enzyme Action*, 3rd ed. New York: Springer-Verlag, 1995.

Clearly documents the importance of organic chemistry in biochemistry. A comprehensive text with an extensive reference section.

Efiok, B.J.S., *Basic Calculations for Chemical and Biological Analysis*. AOAC International, 1993.

A brief problems-oriented supplement. Chapter 4 is titled "Enzyme Assays and Activity."

Enzyme Nomenclature: Recommendations (1992) of the Nomenclature Committee of the International Union of Biochemistry and Molecular Biology. San Diego: Academic Press, 1992.

Describes the rules of enzyme nomenclature, and then lists the recommended names, the systematic names, and the classification for 3196 enzymes.

Moran, L.A., K.G. Scrimgeour, H.R. Horton, R.S. Ochs, and J.D. Rawn, *Biochemistry*, 2nd ed. Englewood Cliffs, NJ: Neil Patterson/Prentice-Hall, 1994.

An encyclopedia-type biochemistry text with excellent coverage of enzymes.

Pawson, T., Getting down to specifics, *Nature* 373, 477–478, 1995.

Analyzes a technique for rapidly determining the specificities of protein kinases.

Perutz, M., *Protein Structure: New Approaches to Disease and Therapy*. New York: W. H. Freeman and Company, 1992.

Contains an interesting chapter on how knowledge of the structure of enzymes is employed to design drugs. Includes many references to the research literature.

Saier, M.H., Jr., *Enzymes in Metabolic Pathways: A Comparative Study of Mechanism, Structure, Evolution, and Control*. New York: Harper & Row, 1987.

A brief introduction to enzymes. A strong section on coenzymes.

Taubes, G., X-ray movies start to capture enzyme molecules in action, *Science* 266, 364–365, 1994.

Describes efforts to get "as close as you can imagine to a moving picture or video of molecules and enzymes changing conformations" in real time.

Voet, D., and J.G. Voet, *Biochemistry*, 2nd ed. New York: John Wiley & Sons, 1995.

A comprehensive biochemistry text. Part III is titled Mechanisms of Enzyme Action. *Contains references and some problems.*

6 Carbohydrates

CHAPTER OUTLINE

This chapter moves from enzymes into carbohydrates. The direct link between the two lies in glycoenzymes, enzymes with a carbohydrate component. Enzymes are indirectly associated with carbohydrates through metabolism. Each reaction involved in the metabolism of carbohydrates is catalyzed by a separate enzyme.

The term "**carbohydrate**" is based on a general formula that characterizes most members of this major class of biochemicals: $C_n(H_2O)_n$ [carbo(carbon)hydrate(water)]. Although carbohydrates (also called **saccharides:** from the Greek word *sakcharon,* meaning sugar) now include compounds that do not accommodate this simple formula, most of the carbohydrates encountered in this text fit the classical definition.

Carbohydrates that cannot be hydrolyzed to yield smaller saccharides are called **monosaccharides.** The joining of monosaccharides through hydrolyzable acetal bonds generates **disaccharides, trisaccharides,** other **oligosaccharides,** and **polysaccharides** (Table 6.1). A similar prefix-based classification system was encountered during the discussion of peptides in Chapter 3 (Table 3.1). Water soluble, sweet-tasting carbohydrates are called **sugars.** Although most sugars are monosaccharides, the most common sugar, **table sugar** or **sucrose,** is a disaccharide.

Carbohydrates are versatile compounds that play an enormous number of roles within most living organisms, including humans. They are, for example, widely utilized as both an immediate fuel and a storage form of energy. Particular polysaccharides serve as structural or protective agents in countless plants, bacteria, and animals. Other polysaccharides function as lubricants around skeletal joints and as adhesives between cells. A variety of saccharides, either alone or attached to proteins, lipids, or nucleic acids, operate as signaling molecules, recognition agents or hormones. Both glycoproteins, including glycoenzymes, and glycolipids (the prefix *glyco-* designates a carbohydrate component) help fill a particularly large number of functional niches.

On the commercial side, carbohydrates are used as materials for the manufacture of paper, cardboard, and clothing. They are also employed as thickening agents in the food and textile industries, as inert fillers for medical pills, as supports for chromatography and electrophoresis, and as ingredients in numerous other products.

This chapter is primarily devoted to a discussion of the structure, properties, general functions, and uses of some common carbohydrates. A number of the major biochemical themes crop up along the way, and some weave their way throughout the chapter. These themes include specificity, structure–function relationships, catalysis, evolution, and communication. Much of the material covered in this chapter is essential in Chapter 7, since nucleotides and related compounds contain one or more carbohydrate residues.

TABLE 6.1

A Classification System for Carbohydrates

Class	Number of Monosaccharide Residues
Disaccharide	2
Trisaccharide	3
Tetrasaccharide	4
Pentasaccharide	5
Oligosaccharide	2 to 11
Polysaccharide	Over 11

6.1 MONOSACCHARIDES

Monosaccharides are the simplest carbohydrates. They serve multiple functions within all living organisms: most can be oxidized to produce energy; some are used to build oligosaccharides and polysaccharides; and some serve as a source of carbon for the production of fats, nucleic acids, amino acids, and other classes of compounds.

TABLE 6.2

A Classification System for Monosaccharides

Class	Number of Carbon Atoms
Triose	3
Tetrose	4
Pentose	5
Hexose	6
Heptose	7
Octose	8
Nonose	9

Classification

The simplest monosaccharides are nonbranched, polyhydroxy (although the prefix *poly-* usually designates "many," in this case, it denotes "more than one") aldehydes or ketones that contain from three to nine carbon atoms. One carbon resides within a carbonyl group (C═O) while each of the remaining carbons normally carries a hydroxyl group (—OH). Monosaccharides that are ketones are classified as **ketoses,** whereas those that are aldehydes are called **aldoses.** Monosaccharides are also categorized on the basis of the number of carbon atoms they possess (Table 6.2). The two classification systems are commonly merged to generate the terms "**aldohexoses,**" "**ketopentoses,**" and so on. The names of most carbohydrates terminate in "ose." Similarly, an *-ase* ending is characteristic of most enzyme names (Section 5.1). **[Try Problem 6.1]**

Chirality and Stereoisomers

Almost all monosaccharides contain one or more chiral centers. When all of the carbons in a monosaccharide are placed in a vertical column with the carbonyl carbon toward the top, a hydroxyl group written to the right of a chiral carbon denotes one configuration, and a hydroxyl group drawn to the left depicts a second configuration. After the carbons are numbered from the end closest to the carbonyl, the configuration about the highest numbered chiral center is used to assign a D (—OH right) or an L (—OH left) to the molecule as a whole.

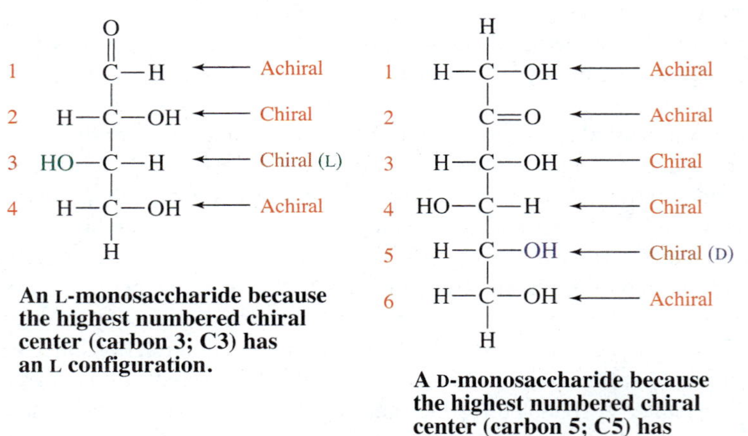

An L-monosaccharide because the highest numbered chiral center (carbon 3; C3) has an L configuration.

A D-monosaccharide because the highest numbered chiral center (carbon 5; C5) has a D configuration.

Since D and L configurations are not possible about an achiral carbon, attached atoms and functional groups can be drawn in any desired orientation. Similar conventions have been employed in Section 3.3 to designate and illustrate the configurations about chiral centers within amino acids. **[Try Problem 6.2]**

The names and structures of all D-aldoses containing from 3 to 6 carbon atoms are given in Figure 6.1. Sugars that differ only in the configuration about a single carbon atom (D-glucose and D-galactose, for example) are **epimers** of one another.

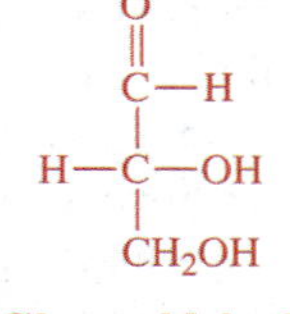

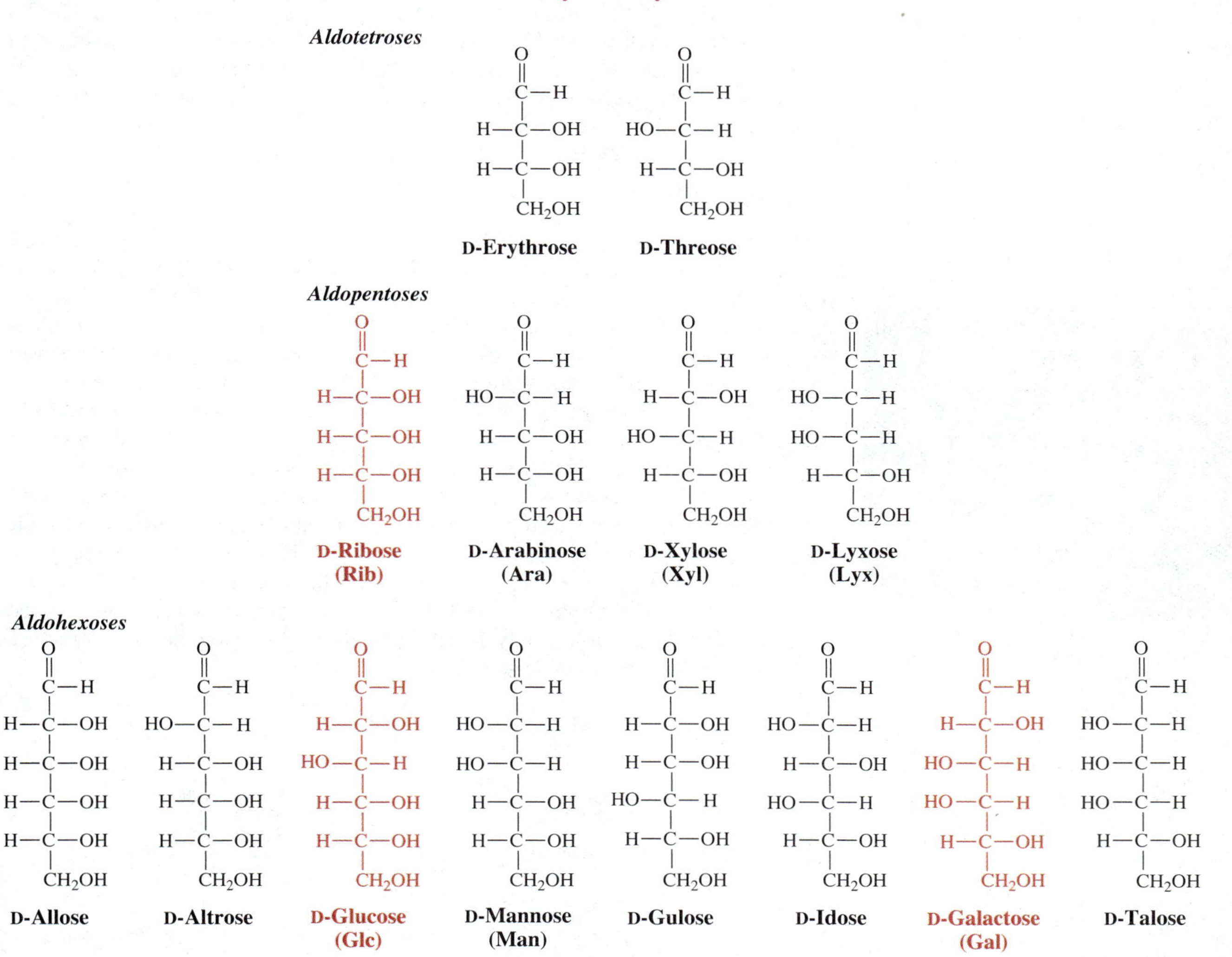

FIGURE 6.1
All D-aldoses containing from three to six carbons. The sugars encountered most often in this text are shown in red.

The enantiomer (nonsuperimposable mirror image isomer) of each D-aldose is an L-monosaccharide that is assigned the same name. The enantiomer of D-glucose, for example, is L-glucose. All of the aldoses in Figure 6.1 that possess the same number of carbon atoms are **diastereomers** (stereoisomers that are not enantiomers). The total number of stereoisomers (enantiomers plus diastereomers) for a monosaccharide is equal to $2^n - 1$, where n equals the number of chiral centers present. Since D-glucose contains 4 chiral centers, this monosaccharide is one of a group of 16 (2^4) stereoisomers. Seven of the 15 stereoisomers ($2^4 - 1$) of D-glucose are shown in Figure 6.1. The rest are L-monosaccharides. Problems 3.10 through 3.13 (Chapter 3) provide a review of stereoisomerism and some associated terminology. **[Try Problems 6.3 and 6.4]**

Stereospecificity

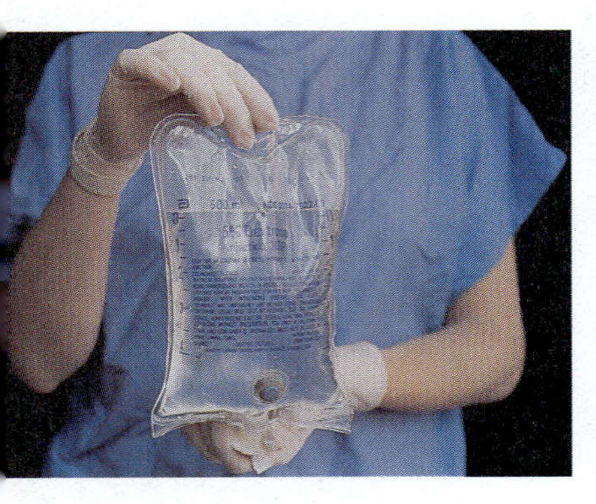

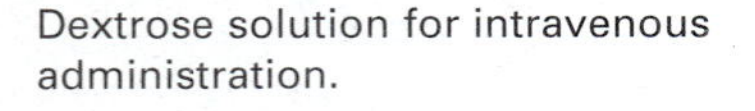
Dextrose solution for intravenous administration.

Most living organisms can synthesize and utilize only one member (if either) of a pair of enantiomeric carbohydrates, still another example of stereospecificity. Specificity is one of the most frequently occurring themes in biochemistry. The human body is usually D-specific when it comes to monosaccharides. **D-Glyceraldehyde, D-ribose, D-galactose,** and **D-glucose** (also called **blood sugar** and **dextrose**) are the monosaccharides in Figure 6.1 that will be encountered most often in this text. D-Glucose is the most abundant monosaccharide in the body, and it is routinely used for intravenous (within a vein) feeding in hospitals. Your health depends on the maintenance of a relatively constant blood glucose concentration (Section 10.12).

A series of D-ketoses exists that is closely related to the D-aldose series illustrated in Figure 6.1. **Dihydroxyacetone** and **D-fructose** (also called **levulose** and **fruit sugar**) are two frequently encountered members of this series. Dihydroxyacetone has no D or L in its name because it possesses no chiral centers. Fructose is present at relatively high concentrations in some fruits, and it is primarily responsible for the rich taste of honey (approximately 40% fructose). On a gram to gram basis, fructose is about twice as sweet as table sugar. **[Try Problem 6.5]**

CH_2OH
|
C=O
|
CH_2OH

Dihydroxyacetone

CH_2OH
|
C=O
|
HO—C—H
|
H—C—OH
|
H—C—OH
|
CH_2OH

D-Fructose

Colored atoms identify the ketone functional group.

Deoxyribose

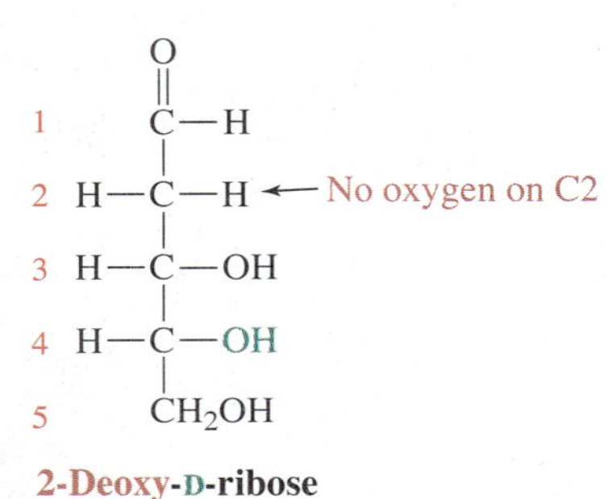

The complete hydrolysis of DNA yields, along with other products, a rather unique monosaccharide known as **2-deoxy-D-ribose.** In this sugar, the C2 hydroxyl group has been reduced, a process that leads to the removal of the oxygen (as implied by the "deoxy" in the name). Chapter 15 describes the position of 2-deoxy-D-ribose within the architecture of DNA. **[Try Problems 6.6 and 6.7]**

PROBLEM 6.1 Given the following monosaccharides:

a. Number each carbon atom in each monosaccharide.
b. Which of these monosaccharides are aldoses? Ketoses? Trioses? Aldotetroses? Pentoses? Aldohexoses? Heptoses?

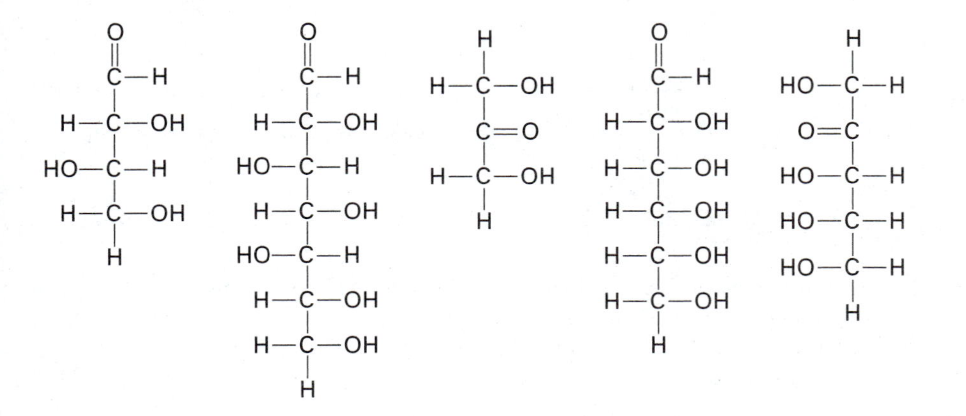

PROBLEM 6.2 Define "configuration." Identify each chiral center in each monosaccharide in Problem 6.1, and then determine the configuration about the highest numbered chiral center in each monosaccharide. Which are D-monosaccharides and which are L-monosaccharides?

PROBLEM 6.3 Write the Fischer projection formula (reviewed in Section 3.3) for L-glucose.

PROBLEM 6.4 How many enantiomers and how many diastereomers exist for D-ribose?

PROBLEM 6.5 Are there any stereoisomers for dihydroxyacetone? Are there any stereoisomers for dihydroxyacetone phosphate (dihydroxyacetone attached to phosphate through an ester link), an intermediate in the biological oxidation of D-glucose? Explain.

PROBLEM 6.6 Which is chemically more reactive, ribose or 2-deoxyribose? Explain.

PROBLEM 6.7 Write the Fischer projection formula for 3-deoxy-L-ribose.

6.2 ANOMERS

Most monosaccharides can reside in multiple forms that tend to exist in equilibrium. Since this phenomenon is related to the chemistry of hemiacetals, a review of some organic chemistry is in order.

Hemiacetals

Alcohols react with aldehydes and ketones to form **hemiacetals:**

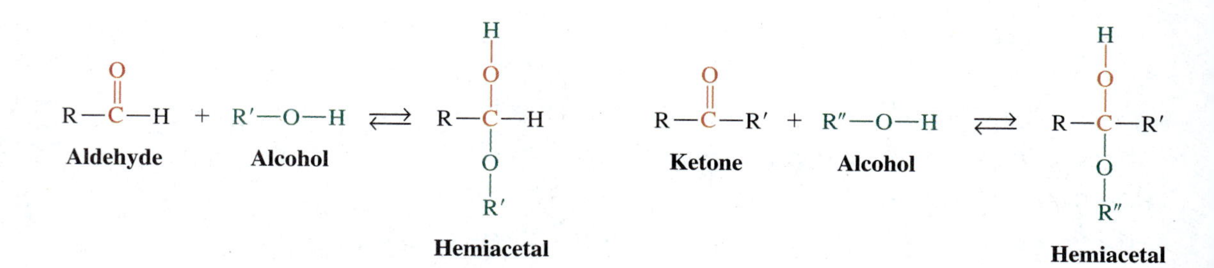

Hemiacetal formation is reversible under physiological conditions. The product of the reaction between an alcohol and a ketone was previously called a hemiketal, instead of a hemiacetal. Although this older terminology still lingers within the scientific literature, it is not utilized in this text.

Cyclic Hemiacetals of Monosaccharides

Monosaccharides with five or more carbon atoms tend to react internally to form cyclic hemiacetals that are often more stable than the corresponding open chain forms. In the case of glucose, the C4, C5, or C6 hydroxyl group can add across the carbonyl group to form a heterocyclic ring with five, six, or seven atoms, respectively. The addition of the C5 hydroxyl group is favored under physiological conditions.

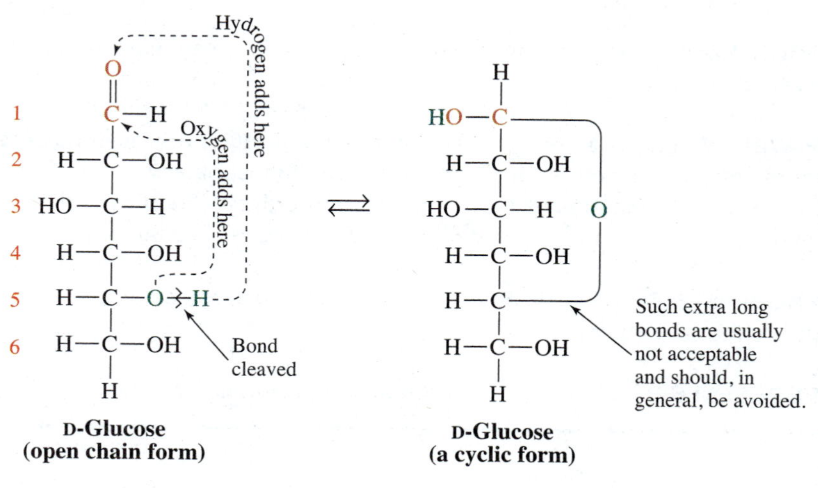

To understand how this addition is possible, we must recognize that the carbon chain in glucose is not really linear. The carbonyl carbon is sp^2 hybridized and has approximately 120° bond angles. Each of the other carbon atoms is sp^3 hybridized with roughly 109° bond angles. Consequently, the carbon chain in glucose can twist in such a manner that the C5 hydroxyl group is very close to the carbonyl group with which it reacts:

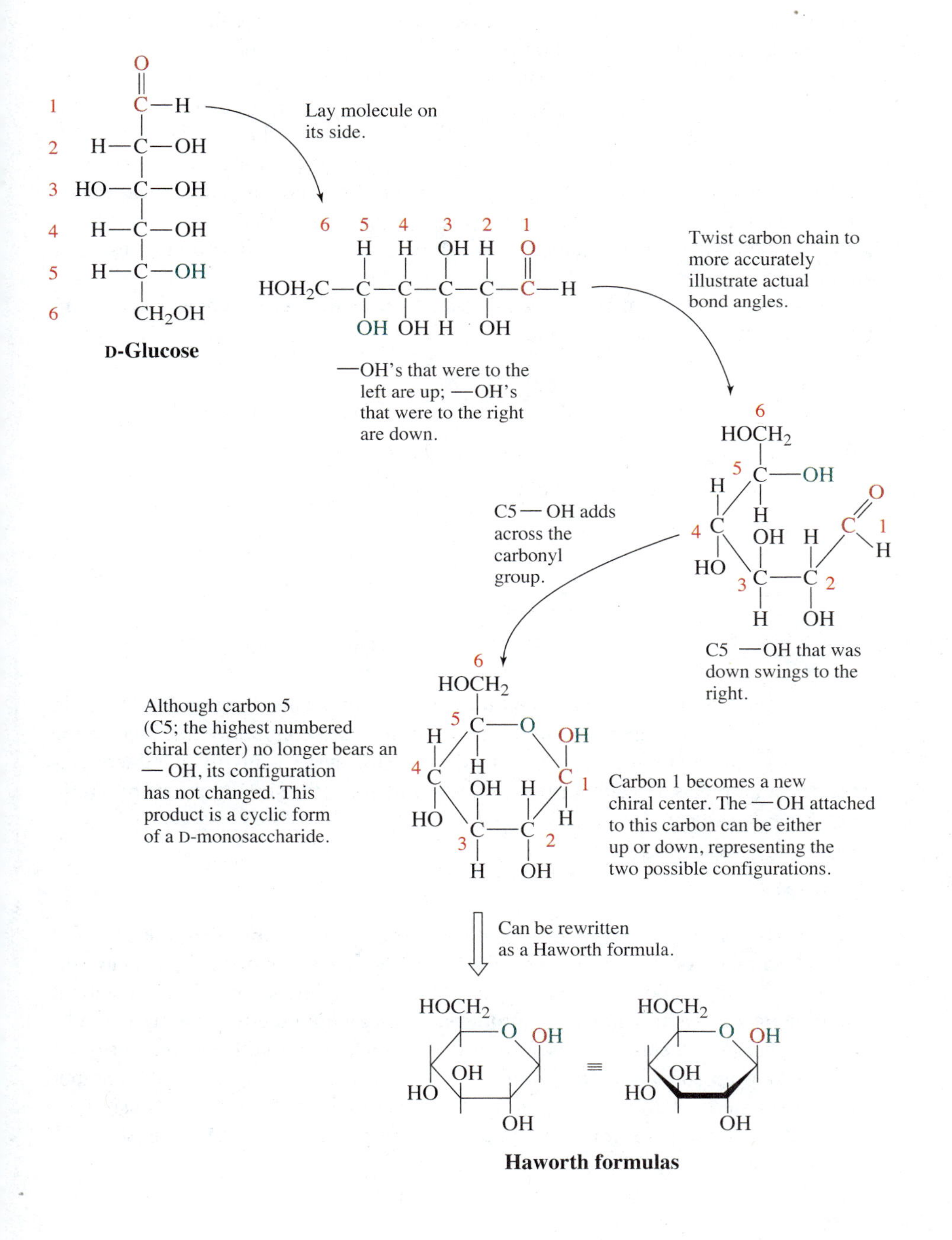

To appreciate these and other structural features of glucose you should build or study molecular models.

The cyclic hemiacetal form of glucose is depicted in the structures on the previous page with three different formulas, including two representations of a **Haworth formula.** In Haworth formulas, the ring is intended to be perpendicular to the plane of the paper and the ring oxygen is normally in the rear. Light and heavy lines are sometimes added (as shown on the previous page) to provide a three-dimensional perspective. The hemiacetal carbon is normally positioned to the right. When a hydroxyl group attached to a chiral ring carbon is placed above the ring, it designates one configuration. A hydroxyl group positioned below the ring identifies the other possible configuration. C5 is still the highest numbered chiral center and still has a D configuration, even though it lacks a hydroxyl. In the corresponding hemiacetal form of *L-glucose,* the $—CH_2OH$ group attached to C5 is positioned below the ring, rather than above it. The hydroxyl group associated with the C5 in the open chain form of glucose adds across the carbonyl group during the cyclization process. **[Try Problem 6.8]**

A six-membered sugar ring is known as a **pyranose ring,** whereas a five-membered sugar ring is a **furanose ring.**

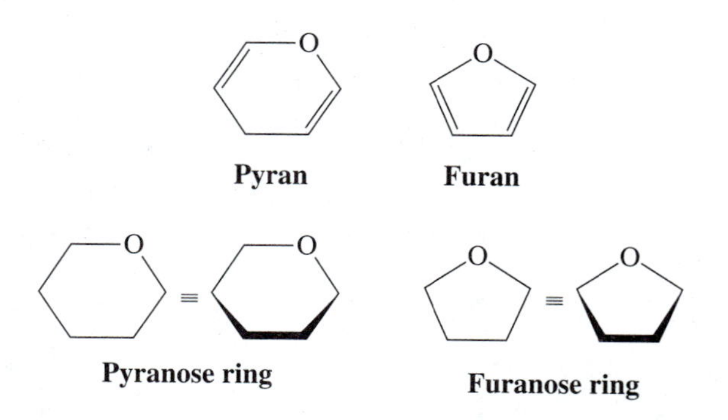

These are usually the most stable and the most common ring systems found within monosaccharides. The terms *pyranose* and *furanose* are commonly included in the names of cyclic monosaccharides to identify the ring systems involved. Glucopyranose, for example, is a cyclic form of glucose containing the six-membered pyranose ring. **[Try Problem 6.9]**

Anomers

When a monosaccharide reacts with itself to form a cyclic hemiacetal, the carbonyl carbon becomes a new chiral center, called an **anomeric carbon,** which can have either of two configurations. Stereoisomers that differ only in their configuration about an anomeric carbon are known as **anomers.** When naming the two anomers of a D-monosaccharide, a beta (β) is used to designate the anomer with the anomeric hydroxyl group up (in a Haworth formula), while an alpha (α) specifies the anomer with the anomeric hydroxyl group down. Once again, we find biochemists using the Greek letters α and β for multiple, independent purposes. The α and β used to distinguish anomers have absolutely nothing to do with protein sheets, protein helices, or hemoglobin subunits. If neither an α or a β appears in the name of a monosaccharide, you should assume that the name refers to a noncyclic, open-chain form.

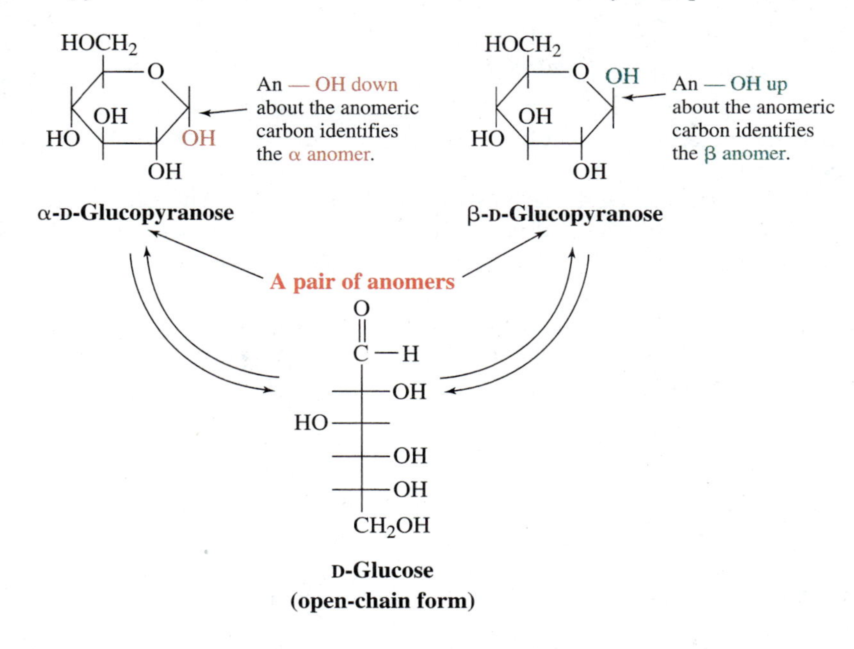

In aqueous systems, including body fluids, a trace amount of the open-chain form of a monosaccharide exists in equilibrium with multiple cyclic forms, including some that differ in ring size (Table 6.3). Although the ring size should ideally be designated in the name (by using furanose or pyranose, for example), it is commonly omitted. The names "α-D-glucose" or "β-D-glucose" imply a pyranose form. In contrast, "α-D-ribose" and "β-D-ribose" ordinarily refer to furanose constructs, because β-D-ribofuranose residues appear in nucleotides, nucleic acids, and a variety of other compounds. Similarly, the names α-D-fructose and β-D-fructose carry an implied "furanose," because the furanose forms of these ketoses tend to be selectively utilized as substrates for enzymes. It is the structural and stereospecificity of enzyme binding sites, not the abundance or relative stability of potential reactants, that de-

TABLE 6.3

Relative Amounts of the Predominant Anomeric Forms for Some Monosaccharides in Water at 40°C

	Relative Amount (percent)			
	α-Pyranose	β-Pyranose	α-Furanose	β-Furanose
D-Ribose	20	56	6	18
D-Glucose	36	64	$<<1$	$<<1$
D-Fructose	3	57	9	31

termines what compounds are metabolized by living organisms. This key concept is central to life itself. **[Try Problems 6.10 and 6.11]**

Chair and Boat Conformations

The Haworth formulas for the pyranose and furanose rings may make it appear that these ring systems are flat and planar. This is approximately true for the five-membered furanose ring system, but it is definitely not the case for pyranose rings. The pyranose ring exists in puckered conformations most of the time, since puckering reduces the ring strain that is associated with the planar arrangement. **Chair** and **boat forms** represent two conformational extremes. The chair conformation is usually favored because it normally leads to less steric hindrance between the substituents attached to a sugar ring.

Chair conformation

Boat conformation

***PROBLEM 6.8** Explain why the C2 and C3 hydroxyl groups in glucose have very little tendency to add across the carbonyl.

PROBLEM 6.9 Write a Haworth formula for α-D-glucofuranose.

PROBLEM 6.10 Write equations showing the two steps involved in the conversion of α-D-fructofuranose to its anomer.

PROBLEM 6.11 Write Haworth formulas for the two pyranose anomers of D-galactose (see Figure 6.1 for its open chain form). Name each anomer.

6.3 REACTIONS OF MONOSACCHARIDES

Monosaccharides undergo numerous reactions in addition to the formation of anomers. These reactions are the ones characteristic of alcohols, aldehydes, and ketones. Alcohols can be oxidized, dehydrated, and alkylated. They also react with carboxylic acids, phosphoric acid, nitric acid, sulfuric acid, and certain other acids to form esters. Phosphate esters are particularly important in the metabolism chapters, because the phosphorylated forms of monosaccharides are usually the metabolically active forms. Aldehydes and some hydroxy ketones are readily oxidized, and all aldehydes and ketones can be reduced. Hemiacetal formation was reviewed in the preceding section. Acetal formation is reviewed in the following section. Biologically important examples of most of these reactions will be encountered as you proceed through this text. Problems 6.12 through 6.16 can be used to review some relevant organic chemistry. **[Try Problems 6.12 through 6.16]**

One additional reaction of monosaccharides is worthy of special comment. This is their nonenzymatic reaction with amino groups (—NH_2) on proteins to form **Schiff bases** (also called **imines**) and, subsequently, **Amadori products** (Figure 6.2). These reactions are responsible for some of the problems associated with diabetes (Section 12.4), and it is hypothesized that, over time, they lead to an accumulation of damaged proteins that contributes to the decline in cell and tissue function associated with biological aging (Section 10.12).

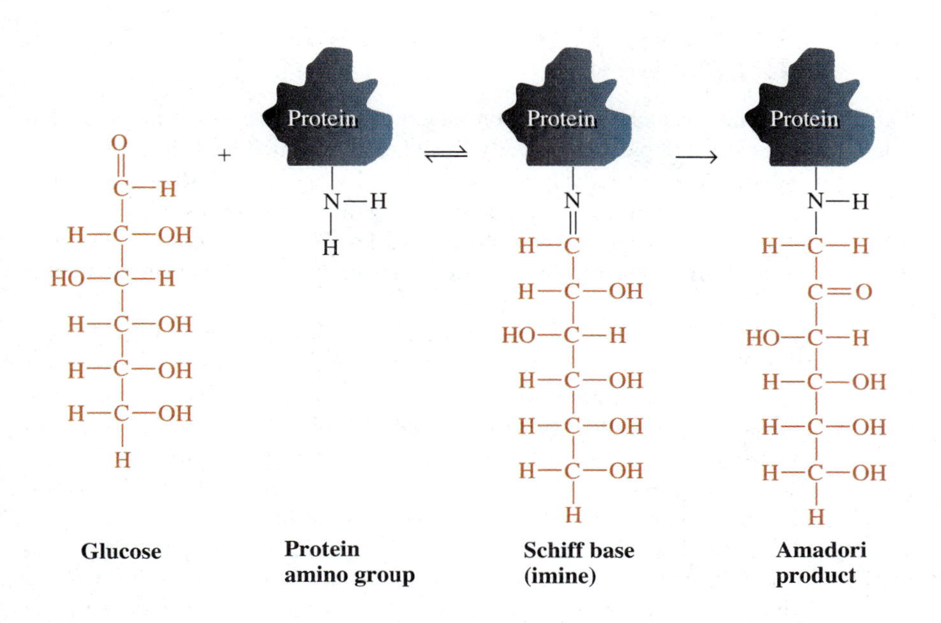

FIGURE 6.2
Reaction of glucose with a protein amino group to form a Schiff base and Amadori product.

Sugars that can exist in a form that contains a free carbonyl group are very easily oxidized and are called **reducing sugars,** because any compound that is oxidized reduces the compound responsible for its oxidation (Section 8.7). Most monosaccharides and disaccharides are reducing sugars. **[Try Problem 6.17]**

***PROBLEM 6.12** Define oxidation, reduction, dehydration, and alkylation.

***PROBLEM 6.13** What class or classes of organic compounds are produced when an alcohol is oxidized? Dehydrated? Alkylated?

***PROBLEM 6.14** What class or classes of compounds are produced when an aldehyde is oxidized? Reduced?

***PROBLEM 6.15** Use structural formulas to write a balanced equation for the reaction of the C2 hydroxyl group of β-D-glucose with hydrogen phosphate to form an ester (glucose 2-phosphate). Note: Since the equilibrium constant for this reaction is small, this is not a good way to produce this ester.

***PROBLEM 6.16** How many ester bonds can simultaneously exist to the same phosphate residue? Explain.

PROBLEM 6.17 Which, if any, of the specific carbohydrates whose formulas have been encountered to this point in the chapter are nonreducing sugars? Explain.

6.4 GLYCOSYLATED HEMOGLOBIN AND DIABETES

Glycosylated Hemoglobin in Diabetics

Subject	Percentage of Total Hemoglobin Glycosylated
Normal	6.9
Typical diabetic	15.0
Well-controlled diabetic	11.2
Poorly controlled diabetic	21.4

This section examines a clinically important example of a reaction of monosaccharides. **Diabetes** is a complex, genetically linked disease which leads to a characteristic **hyperglycemia** (elevated blood glucose level) [Section 12.4]. Since hyperglycemia can cause serious health problems, most diabetics use insulin injections, diet, or other measures to control their blood sugar levels. Their success can be monitored by periodically measuring the blood concentration of **glycosylated hemoglobin.**

Glycosylated hemoglobin refers to hemoglobin with covalently attached sugar residues. It is formed through the spontaneous (nonenzyme-catalyzed) reaction of free amino groups on hemoglobin with the sugars (primarily glucose) normally encountered in blood. The resultant Amadori products (Figure 6.2) can be separated from nonglycosylated hemoglobin with electrophoresis or chromatography. Since glycosylation is a bimolecular reaction, its rate increases as blood glucose levels rise (Section 5.2). As a consequence, the blood concentration of glycosylated hemoglobin reflects the "average" blood glucose level over the 120-day lifespan of a typical red blood cell. Glycosylated hemoglobin levels will be above normal if a diabetic has been hyperglycemic at any time during the 120 days before blood collection; the higher the levels, the less success a diabetic has had in controlling hyperglycemia. **[Try Problems 6.18 and 6.19]**

PROBLEM 6.18 What property of glycosylated hemoglobin allows it to be electrophoretically separated from normal hemoglobin?

PROBLEM 6.19 What blood proteins, in addition to hemoglobin, are susceptible to Amadori product formation? Explain.

6.5 DISACCHARIDES

Biologically, one of the most important reactions of monosaccharides is the formation of acetals, including oligosaccharides and polysaccharides. An acetal and water are the products of the reaction of a hemiacetal and an alcohol:

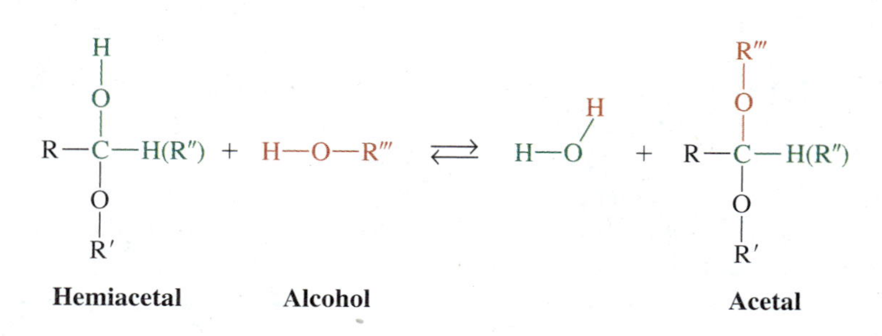

Glycosides

When the hemiacetal that reacts to form an acetal is a cyclic form of a monosaccharide and the alcohol is a second monosaccharide, the product is a **disaccharide.**

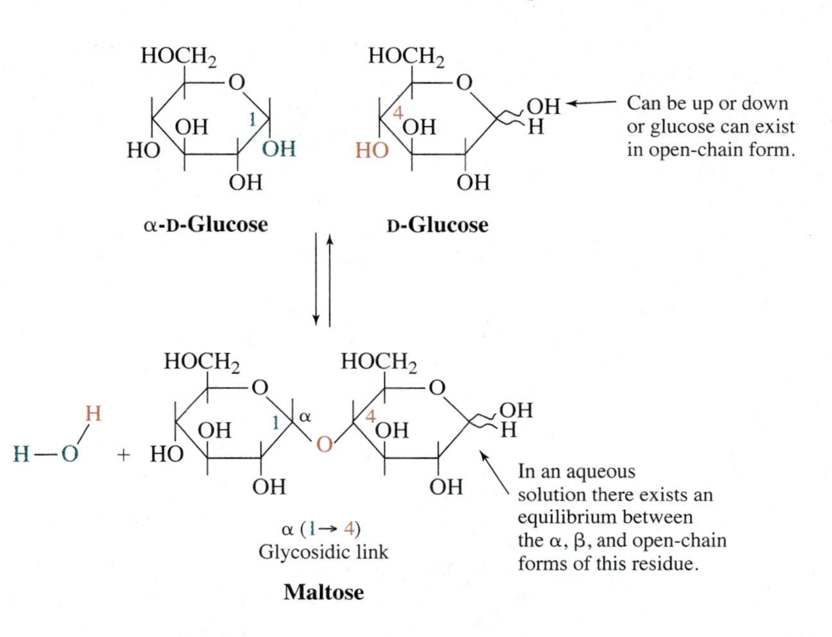

Since the reverse process constitutes a hydrolysis reaction, disaccharides can be hydrolyzed to generate two monosaccharides. **[Try Problem 6.20]**

An acetal link involving the cyclic form of a monosaccharide is known as a **glycosidic link** or, more specifically, an ***O*-glycosidic link** (bond) (Figure 6.3). The replacement of the oxygen in an *O*-glycosidic link with a nitrogen creates an ***N*-glycosidic link.** *N*-glycosidic links are found within all nucleotides and nucleic acids (Chapters 7 and 16). Compounds, such as disaccharides, that contain a glycosidic link are known as **glycosides.**

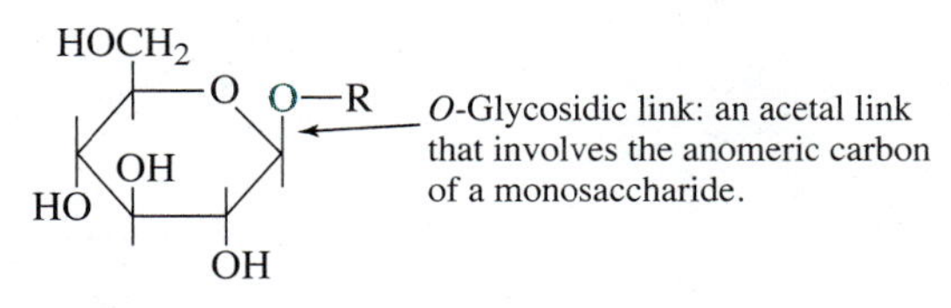

***O*-Glycoside (glycoside)**

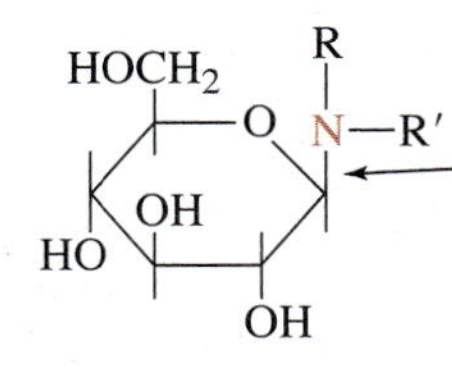

N-Glycosidic link: an *N*-acetal link (N in place of O) that involves the anomeric carbon of a monosaccharide.

***N*-Glycoside**

FIGURE 6.3
O-Glycosidic and *N*-glycosidic links. Although the monosaccharide shown is glucose, a glycoside or *N*-glycoside may contain any monosaccharide residue.

Maltose and Cellobiose

Maltose is the first of four specific disaccharides that are examined in this section. Since the formation of **maltose** (equation given above) entails the reaction of the hemiacetal group associated with C1 of an α-D-glucose molecule with the C4 hydroxyl group of a second glucose, the bond that results is labeled an α(1 ⟶ 4) glycosidic link. Combining this terminology with some that has been previously employed, maltose can be named *O*-α-D-glucopyranosyl-(1 ⟶ 4)-D-glucose. Acetal formation involving one α-D-glucose and the C3 hydroxyl group of a second glucose leads to an α(1 ⟶ 3) glycosidic link and to a disaccharide that is a structural isomer of maltose.

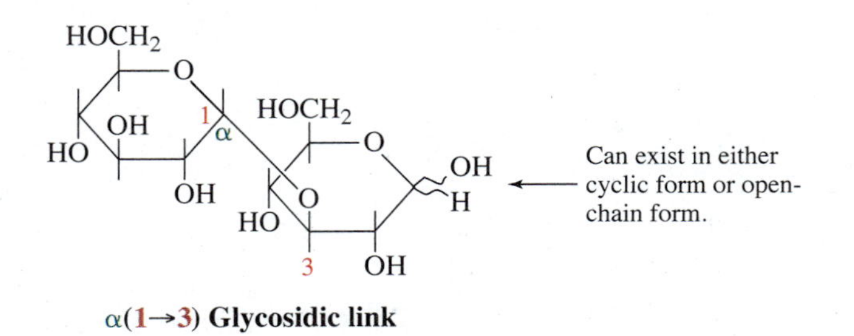

α(1→3) **Glycosidic link**

The reaction of the hemiacetal carbon (anomeric carbon) of β-D-glucose with the C4 hydroxyl group of a second glucose molecule leads to **cellobiose** [*O*-β-D-glucopyranosyl-(1 ⟶ 4)-D-glucose], a disaccharide that differs from maltose in the configuration about its single acetal carbon.

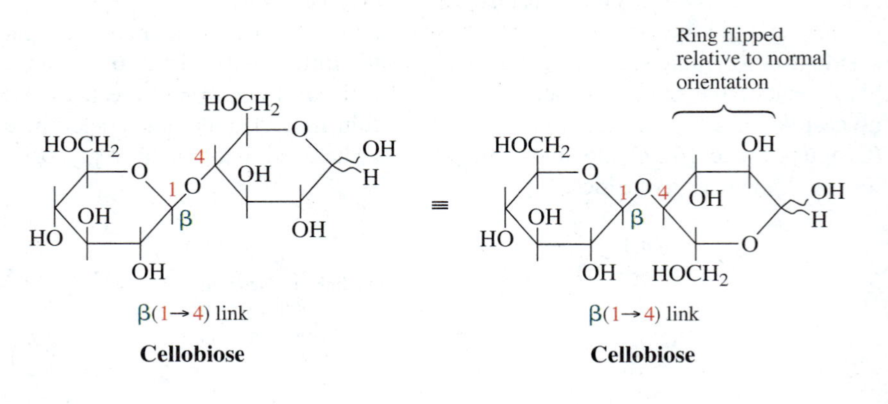

Maltose and cellobiose are intermediates in the synthesis and breakdown of starch and cellulose, respectively. **[Try Problem 6.21]**

Lactose

Lactose [*O*-β-D-galactopyranosyl-(1 ⟶ 4)-D-glucose] contains a galactose joined to glucose through a β(1 ⟶ 4) glycosidic link as shown at the top of the next page. Because mammalian milk is the only known natural source of this disaccharide, it is often called **milk sugar.** The lactose in human milk (approximately 5% by weight) is an important energy and carbon source for nursing infants. **[Try Problem 6.22]**

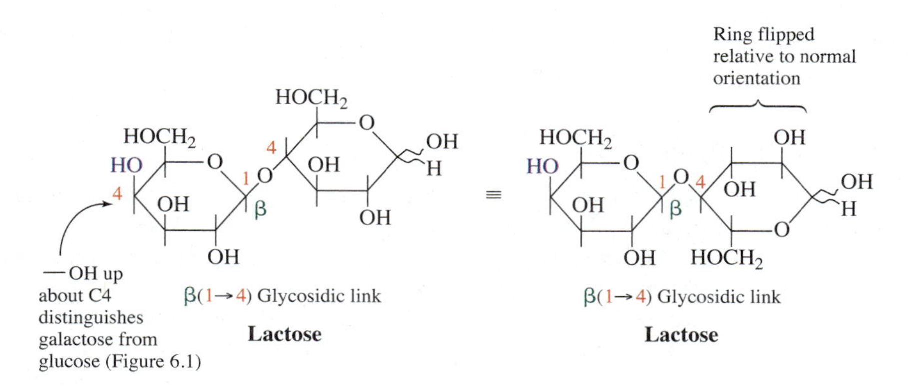

Sucrose

When the hemiacetal carbon (anomeric carbon; C1) of α-D-glucose reacts with the hemiacetal carbon of β-D-fructose (C2), an α,β(1 ⟶ 2) glycosidic link is formed and the product is known as **sucrose, invert sugar,** or **table sugar** [*O*-α-D-glucopyranosyl-(1 ⟶ 2)-β-D-fructofuranoside].

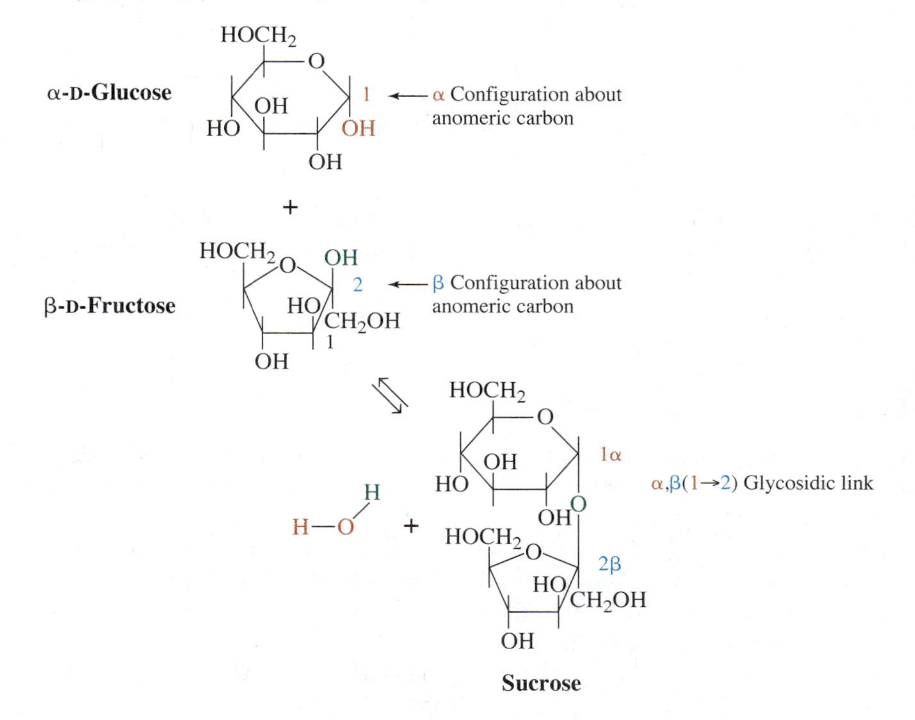

Each of the anomeric carbons in this disaccharide is tied up in an acetal link. Since there are no open chain forms of this sugar and, consequently, no free carbonyl groups, sucrose is relatively resistant to oxidation. It is a **nonreducing sugar.** In gen-

Sugarcane and sugar beets are the two major sources of commercial sucrose. In the United States, sugarcane (shown above) is grown primarily in Florida, Louisiana, Hawaii, and Texas.

Relative Sweetness of Some Sugars

Sugar	Relative Sweetness[a]
Fructose	170
Sucrose	100
Glucose	70
Maltose	40
Galactose	30
Lactose	15

[a]Sucrose (Table Sugar) is arbitrarily assigned a value of 100.

eral, an *-ide* ending within a carbohydrate name (for example, furanoside) denotes a sugar that is nonreducing. Sucrose serves as a storage form of energy in many plants, including sugar cane and sugar beets. Dietary sucrose is primarily used by the human body as an energy and carbon source (Section 6.9). Humans do not synthesize or store this oligosaccharide. **[Try Problem 6.23]**

Other Oligosaccharides

Many additional disaccharides and larger oligosaccharides have been isolated from a variety of living organisms, including humans. Some are intermediates in the synthesis or breakdown of other carbohydrates, but part are biologically active on their own. The **oligosaccharins,** for example, are plant oligosaccharides that function as hormones. Oligosaccharide residues also play key functional roles within many glycoproteins and glycolipids.

PROBLEM 6.20 There are three major forms of maltose that exist in equilibrium with one another in aqueous solutions. Write the structural formula for each of these forms.

PROBLEM 6.21 How many different disaccharides can be produced by joining two β-D-glucopyranose molecules through a glycosidic link? One of these is cellobiose. Write the structural formula for each of the other disaccharides, and then name each glycosidic link.

PROBLEM 6.22 What specific compounds are produced when lactose is hydrolyzed?

PROBLEM 6.23 True or false? A nonreducing sugar is formed any time two monosaccharides are joined through a glycosidic link between their anomeric carbons. Explain.

6.6 POLYSACCHARIDES

When 12 or more monosaccharides are joined through acetal links, the product is classified as a polysaccharide rather than an oligosaccharide (Table 6.1). Polysaccharides (also called **glycans**) serve as structural elements and as a storage form of energy and carbon in both plants and animals. A few are designed to occupy other functional niches as well. **Homopolysaccharides** (Table 6.4) contain only a single type of monomeric unit, (the prefix "homo" means "same") whereas **heteropolysaccharides** ("hetero" means "different") are constructed from two (most often) or more different monosaccharides, usually arranged in a repeating sequence. Polysaccharides, in contrast to proteins (Chapter 4) and nucleic acids (Chapter 16), are often branched. Most are constructed from D-glucose, D-mannose, D-fructose, D-galactose, D-xylose, D-arabinose, and derivatives of these sugars, including **D-glucuronic acid** and ***N*-acetyl-D-glucosamine.** **[Try Problem 6.24]**

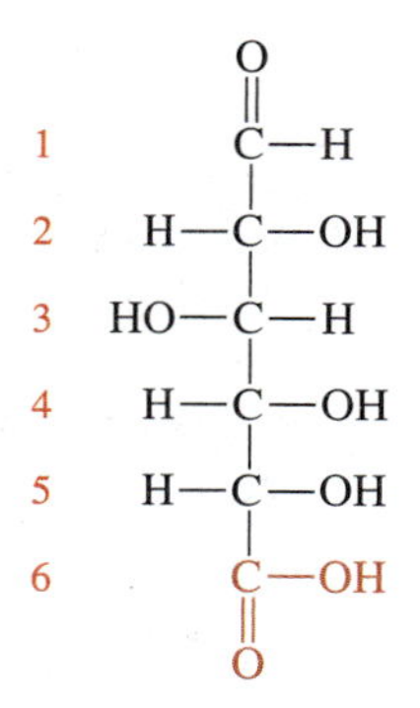

D-Glucuronic acid

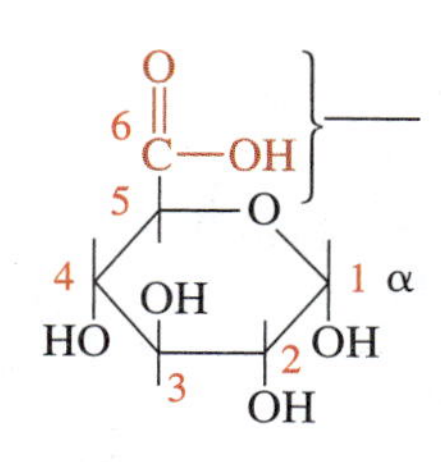

The primary hydroxyl group on C6 of glucose has been oxidized to yield a carboxylic acid. The carboxyl group will exist predominantly in a dissociated form at pH 7.0.

α-D-Glucuronic acid

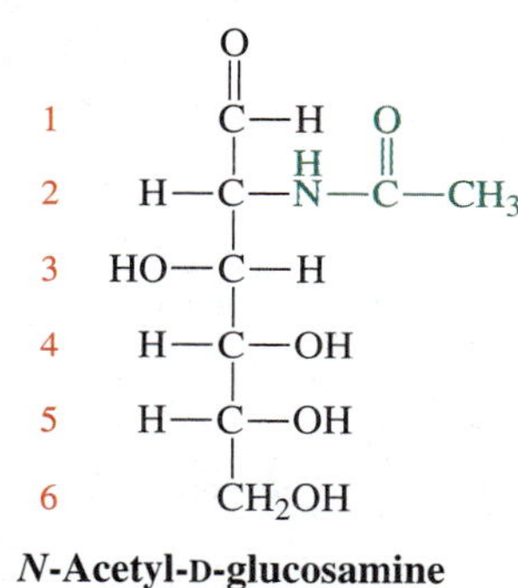

N-Acetyl-D-glucosamine

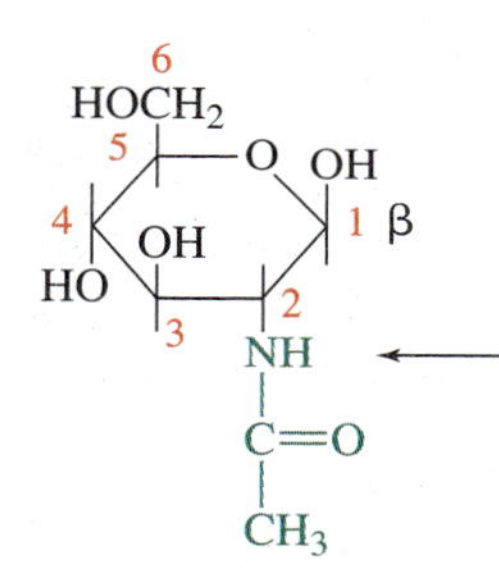

The C2 hydroxyl group on glucose has been replaced with an amino group that has been acetylated.

N-Acetyl-β-D-glucosamine

TABLE 6.4

Role and Chemical Nature of Some Homopolysaccharides

Homopolysaccharide	Repeating Monosaccharide	Link(s) Between Monosaccharide Units	Number of Monosaccharides Per Molecule	Role or Function
Starch				
Amylose	D-Glucose	α(1 → 4)	Few thousand to around 500,000	Store energy and carbon in plants
Amylopectin	D-Glucose	α(1 → 4) α(1 → 6) branches every 24–30 residues	Up to roughly 10^6	Store energy and carbon in plants
Glycogen	D-Glucose	α(1 → 4) α(1 → 6) branches every 8–12 residues	Up to several million	Store energy and carbon in bacteria and animals
Chitin	*N*-Acetyl-D-glucosamine	β(1 → 4)	Very large	Structural element in exoskeletons of insects, spiders, and crustaceans
Cellulose	D-Glucose	β(1 → 4)	Up to roughly 15,000	Structural element in plant cell walls

Starch

Starch is a storage homopolysaccharide of glucose that is deposited as granules within individual plant cells. A potato is a plant tuber composed primarily of this glycan. When a plant needs either energy or carbon, it can hydrolyze some of its stored starch and use the released glucose as a fuel or as a source of carbon for the construction of other compounds.

Starch molecules are divided into two classes: **amylose** and **amylopectin.** Amylose molecules are linear polymers containing up to 500,000 glucose molecules joined in one continuous chain through $\alpha(1 \longrightarrow 4)$ glycosidic links.

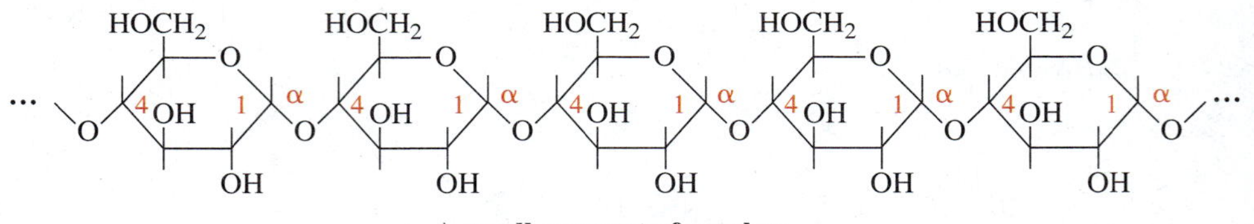

A small segment of amylose

Amylopectin molecules are branched polymers of glucose in which $\alpha(1 \longrightarrow 4)$-bonded chains are attached through $\alpha(1 \longrightarrow 6)$ glycosidic links to other $\alpha(1 \longrightarrow 4)$-bonded chains.

An amylopectin molecule with branched branches

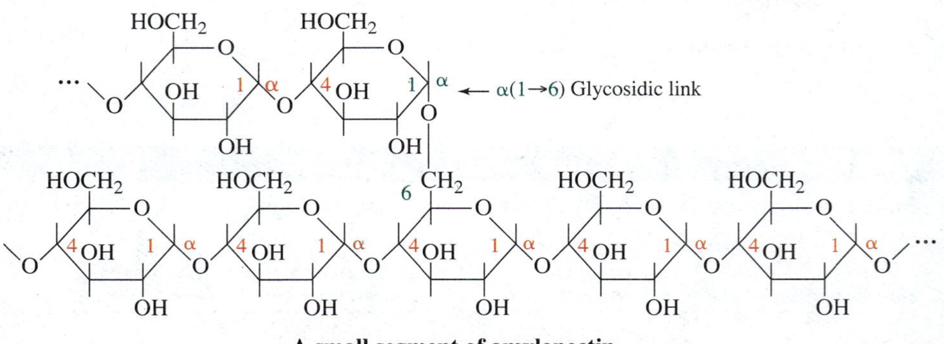

A small segment of amylopectin

On the average, there is one branch for every 25 to 30 glucose residues, and branches may, themselves, be branched. Amylopectin molecules tend to be extremely large; some contain around one million glucose residues. **[Try Problems 6.25 through 6.27]**

Glycogen

Glycogen is a storage polysaccharide found in at least small amounts within many animal cells. In the human body, it is most abundant in skeletal muscle (1 to 2% wet weight) and the liver (up to 10% wet weight). It is chemically identical to amylopectin, except more highly branched (one branch for every 8 to 12 residues). Since glycogen would be called starch if it resided within a plant, it is sometimes referred to as **animal starch.**

Liver glycogen plays a central role in the maintenance of blood sugar levels (Section 10.12). Muscle glycogen regularly provides fuel for muscle action, and it can be particularly important during athletic competition. Since there is a limit to the amount of glycogen that can be stored in muscle, marathon runners and other endurance athletes burn virtually all of their muscle glycogen long before they complete an event. The finish is predominantly fat-powered. To extend the time that glycogen is available, endurance athletes commonly use exercise and dietary regimens in attempts to increase the amount of muscle glycogen before competition. Starvation, as well as exercise, rapidly depletes glycogen reserves.

Marathon runners burn virtually all of their muscle glycogen before they reach the finish line. The finish is predominantly fat-powered.

Chitin

Chitin is the major structural element in the exoskeleton of insects, spiders, and crustaceans, and it is a component of the cell walls of many algae and fungi. This glycan is a linear homopolymer with *N*-acetyl-D-glucosamines connected through $\beta(1 \longrightarrow 4)$ glycosidic links. It is the most abundant animal polysaccharide on Earth. Chitin has been found to promote wound healing, and efforts are being made to construct suture material from it. The removal of the acetyl groups from chitin yields **chitosan,** a compound used in bandages, burn dressings, food additives, drug capsules, and cosmetics. **[Try Problem 6.28]**

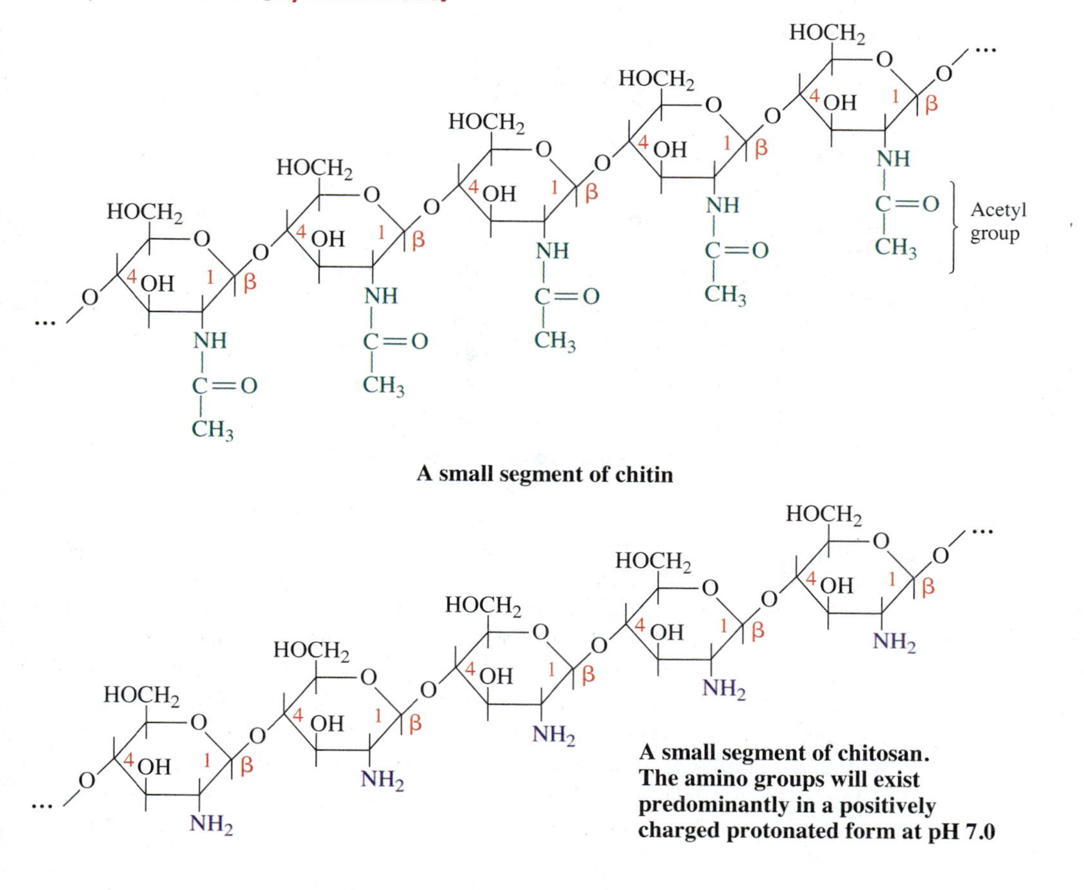

A small segment of chitin

A small segment of chitosan. The amino groups will exist predominantly in a positively charged protonated form at pH 7.0

Cellulose

Cellulose is a structural polysaccharide that accounts for over half of the total carbon in the biosphere. Its tough, water-insoluble nature allows it to provide both support and a protective barrier for a variety of organisms, including most plants and some marine invertebrates. As a major component of plant cell walls, it is present in particularly large amounts in stems, stalks, and woody elements. Wood is about 50% cellulose and paper roughly 90%. Cotton contains over 90% cellulose, so a person wearing cotton is, basically, clad in a polysaccharide. Although the human body does not synthesize this glycan, the body does assemble other polysaccharides that are used for structural purposes.

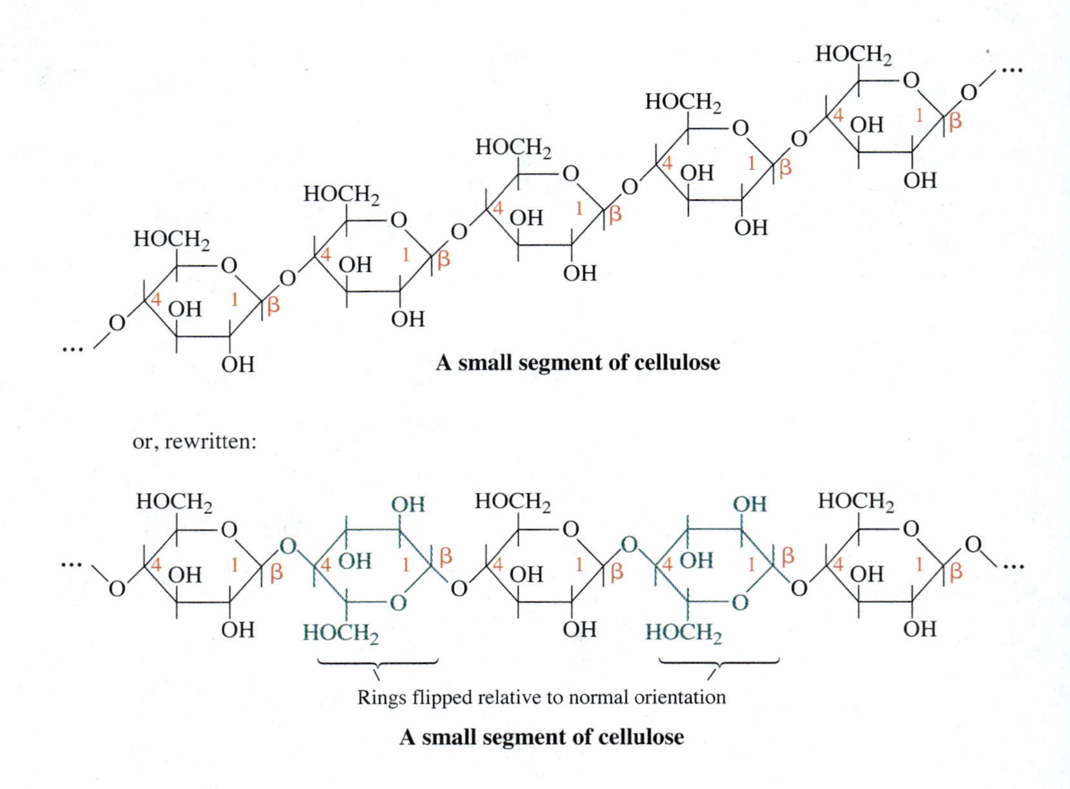

Cellulose is a linear homopolymer containing β(1 ⟶ 4)-linked glucose residues. The only structural difference between cellulose and amylose is the configuration about their anomeric carbons. Whereas the α configurations in amylose and amylopectin cause starch molecules to coil into dense storage granules, the β configurations in cellulose allow these macromolecules to stretch out and hydrogen bond with neighboring, parallel polysaccharide strands to form tough fibrils of the type found in plant cell walls—good examples of structure–function relationships (Figure 6.4). Similarly, some sequences of amino acids cause polypeptide chains to coil into α-helices, whereas other amino acid sequences cause peptides to stretch into β-strands or related, more-extended arrangements (Section 4.3). **[Try Problems 6.29 and 6.30]**

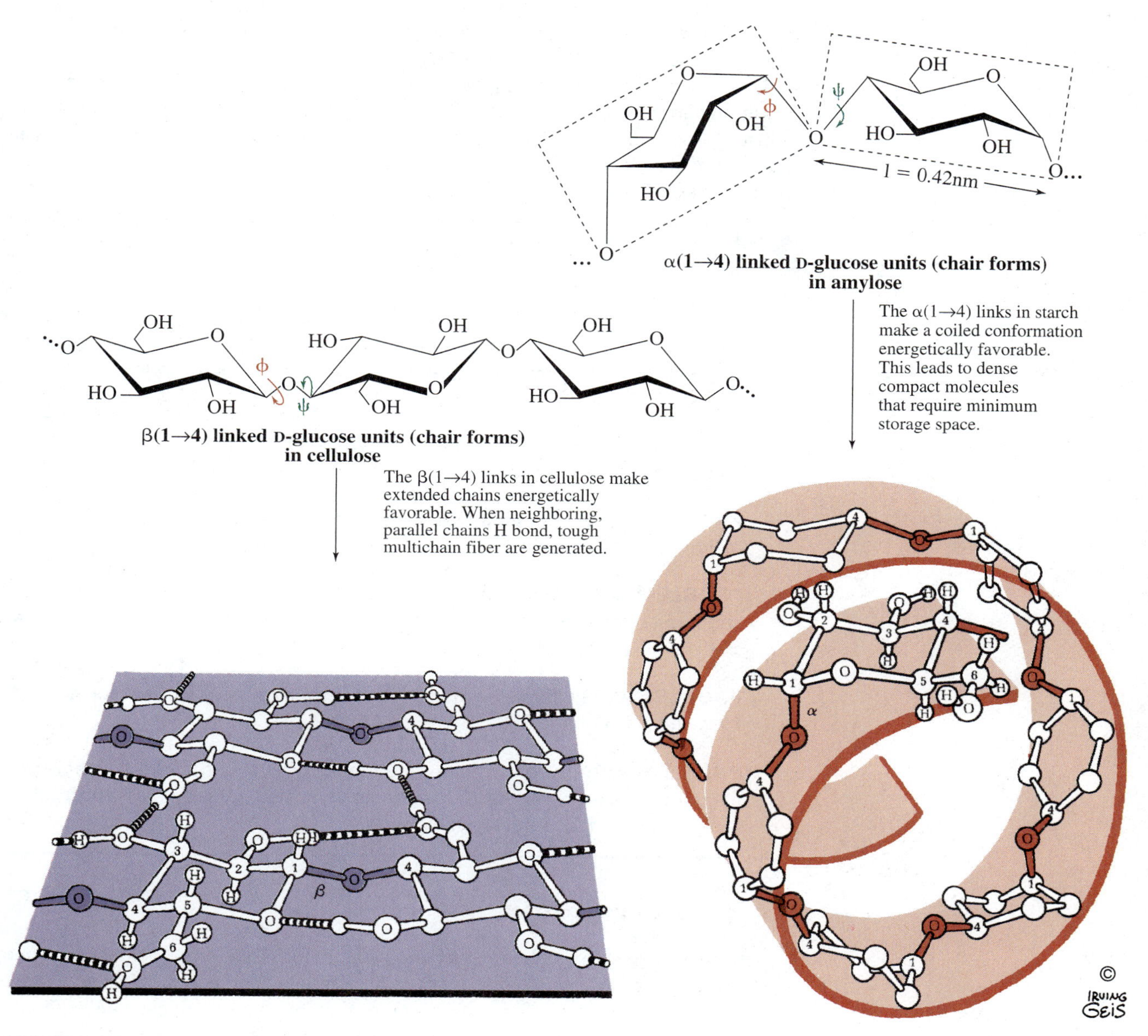

FIGURE 6.4
Energetically favored conformations for glucose residues in amylose and cellulose.

Other Polysaccharides

Although the list of biologically and biomedically important polysaccharides is very long, only a few additional examples are mentioned in this chapter. **Agar** and **agarose,** two structural polysaccharides in plants, are both commonly encountered in biochemistry laboratories. Agar preparations are used to grow cells, and agarose is used as a support for both chromatography and electrophoresis. **Dextrans** are storage polysaccharides in plants, and they are employed as blood extenders and as stationary phases for chromatography. The dextrans formed by bacteria growing on teeth are an important component of dental plaque. Heparin (Section 6.8) is a human heteropolysaccharide and an **endogenous** (produced within your body) **anticoagulant** (inhibits blood clotting). Hyaluronic acid and chondroitin sulfates serve as lubricants and structural elements in humans (Section 6.8).

***PROBLEM 6.24** What class of organic compounds is formed when an amino group is acetylated? Can members of this class of compounds be hydrolyzed?

PROBLEM 6.25 Circle each glucose residue in the segments of amylose and amylopectin depicted on page 254. Identify each anomeric carbon.

PROBLEM 6.26 Branches other than $\alpha(1 \longrightarrow 6)$ branches are theoretically possible along an $\alpha(1 \longrightarrow 4)$-bonded amylopectin chain. Identify or describe these other possible branches.

PROBLEM 6.27 Create a branched branch by adding a single glucose residue to the segment of amylopectin depicted on page 254.

PROBLEM 6.28 What chemical reaction is involved in the conversion of chitin to chitosan? How does the net charge on chitosan differ from that on chitin at physiological pH values?

PROBLEM 6.29 Circle each anomeric carbon in the structural formulas for cellulose on page 256.

***PROBLEM 6.30** What is the predominant product when a cotton shirt is completely hydrolyzed? Explain why concentrated solutions of strong acids rapidly "burn" holes in cotton clothing.

6.7 GLYCOPROTEINS

Although some carbohydrates, including starch and glycogen, function in a free state, many are covalently linked to other molecules in their biologically active form. The covalent attachment of carbohydrates to proteins leads to **glycoproteins** (Table 4.2). The carbohydrate content of such compounds ranges from under 1% to over 90% by weight. Multiple carbohydrate components, each a mono-, oligo-, or polysaccharide, may be attached to the same protein, usually through *O*-glycosodic links to the side chains of threonine or serine or through *N*-glycosidic links to asparagine side chains. D-Glucose, D-galactose, D-mannose, *N*-acetyl-D-glucosamine, *N*-acetyl-D-galactosamine and ***N*-acetylneuraminic acid** (**sialic acid,** see next page) residues are common. The presence of carbohydrate residues can affect the pI, hydrophobicity, size, conformation, and rigidity of a protein. Glycoproteins in which the carbohydrate components predominate tend to form viscous solutions and are known as **mucoproteins** or **proteoglycans.** **[Try Problems 6.31 and 6.32]**

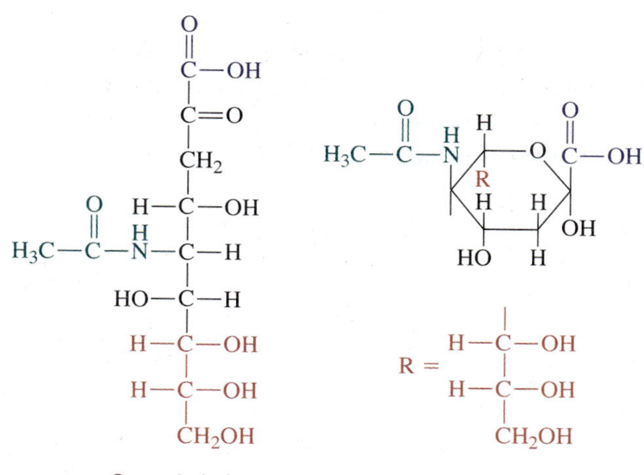

Open-chain form α-Pyranose form

***N*-Acetylneuraminic acid. The —COOH group is dissociated and negatively charged at physiological pH values.**

Growing evidence indicates that a majority of the proteins within the body may well be glycoproteins. Given this fact, it is not surprising that glycoproteins have been found to serve a particularly large number of diverse roles. Immunoglobulins (Chapter 4) and **fibrinogen** (plays a central role in blood clotting) are glycoproteins. Many protein hormones, enzymes, transport proteins, receptors, and structural proteins possess carbohydrate residues. Some of the functions of the carbohydrate components of glycoproteins, including the stabilization of active conformations, protection against proteolytic degradation, and service as antigenic determinants, are summarized in Exhibit 6.1. The role of carbohydrates as antigenic determinants is reflected in the existence of different **blood groups** (A, B, O, and AB), a phenomenon associated with carbohydrates attached to proteins on the surface of red blood cells.

EXHIBIT 6.1

Some Major Functions of the Carbohydrate Components of Glycoproteins

- Recognition of specific cellular receptors.
- Regulation of protein clearance from the body.
- Stabilization of active protein conformations.
- Protection against catalytic degradation by proteinases.
- Participation in intracellular adhesion.
- Regulation of cell–cell contact.
- Creation of antigenic determinants.

PROBLEM 6.31 Write the structural formula for a glucose joined to serine through a β-glycosidic link.

PROBLEM 6.32 Write the structural formula for the compound produced when the β-pyranose form of *N*-acetylneuraminic acid is deacylated.

6.8 MUCOPOLYSACCHARIDES

Modifications on common structural themes allow carbohydrates to serve many diverse functions. **Mucopolysaccharides** (*muco* meaning viscous), which dissolve or disperse in water to give syrupy, gel-like mixtures, provide additional examples. They serve as lubricants and shock absorbers around bone joints, they are structural elements within animal cell coats, and they are a major component of the gelatinous **ground substance** that occupies many of the extracellular spaces within the human body. Connective tissues are particularly rich in this class of glycan.

Hyaluronic Acid

Hyaluronic acid, the most abundant mucopolysaccharide in the body, is a major component of **synovial fluid** (the lubricating fluid around bone joints) and connective tissues. It is a linear polymer in which D-glucuronic acid and *N*-acetyl-D-glucosamine residues alternate along the polymer chain.

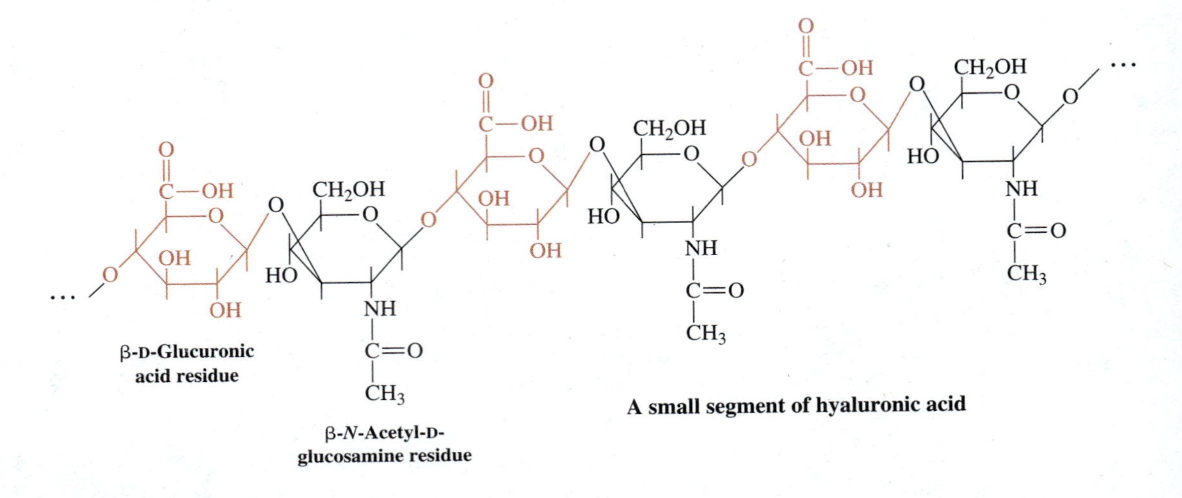

A small segment of hyaluronic acid

A single hyaluronic acid molecule may contain as many as 50,000 monomers. **[Try Problems 6.33 and 6.34]**

Chondroitin Sulfates

Chondroitin sulfates provide additional examples of mucopolysaccharides. They are linear polymers containing alternating D-glucuronic acid and *N*-acetyl-D-galactosamine residues in which some of the hydroxyl groups have reacted to form sulfate esters.

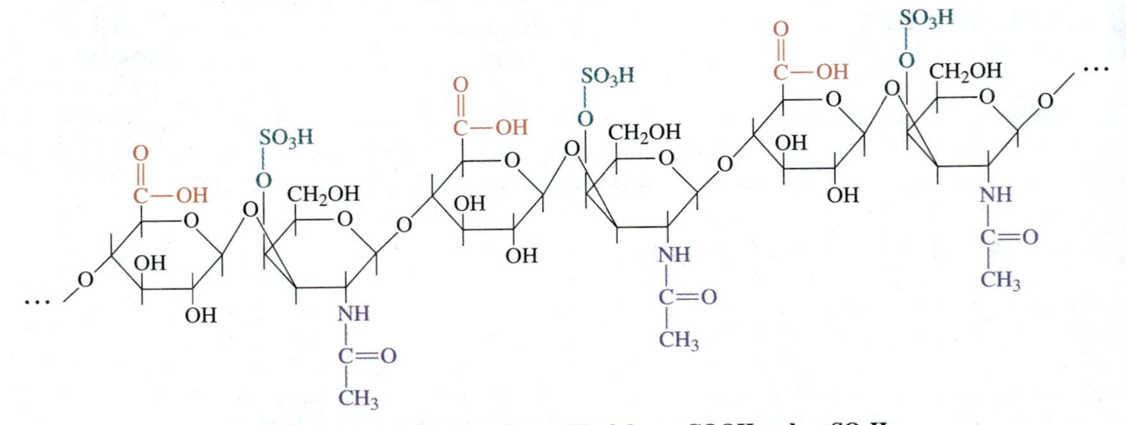

A segment of chondroitin sulfate. All of the —COOH and —SO_3H groups are dissociated and negatively charged at physiological pH values

These glycans are common in the ground substance of connective tissues and, along with the protein collagen (Section 4.10), are used to make an **artificial skin** which is sometimes used in the treatment of burns. **[Try Problems 6.35 and 6.36]**

Heparin

Heparin, another mucopolysaccharide, is used as an anticoagulant both by the body and in medicine. It is produced by the **mast cells** that line arterial walls and is thought to prevent runaway clotting following an injury. Although the exact structure of heparin has not yet been determined, it is known to contain sulfated glucuronic acid residues and glucosamine residues.

PROBLEM 6.33 What types of *O*-glycosidic bonds exist between the monomers in hyaluronic acid?

PROBLEM 6.34 What is the net charge on a hyaluronic acid molecule at pH 7.0 if the molecule contains 25,000 sugar residues? Explain.

PROBLEM 6.35 What types of *O*-glycosidic linkages exist between the monomers in chondroitin sulfates?

PROBLEM 6.36 What compounds are produced when chondroitin sulfate is completely hydrolyzed?

6.9 SOME DIETARY CONSIDERATIONS

Carbohydrates make up as much as 90% of the diet of some individuals, and they account for about 50% of the caloric intake of a typical American. In the United States, the average college-age male consumes approximately 400 g of carbohydrate per day, the vast majority in the form of starch and sucrose.

Digestion of Carbohydrates

Most dietary oligo- and polysaccharides are digested by hydrolytic enzymes known as **glycosidases** (because their substrates are glycosides) to yield monosaccharides, the only carbohydrates that can be absorbed from the human digestive system. **Sucrase,** which catalyzes the hydrolysis of sucrose, and the **α-amylases,** which catalyze the hydrolysis of the $\alpha(1 \longrightarrow 4)$ links in amylose, amylopectin, and glycogen, are specific glycosidases. Sucrase, also known as **invertase,** operates within the upper part of the small intestine (the **duodenum**). Salivary glands and the pancreas each manufacture a different α-amylase, one found in saliva and the second in the digestive fluids of the duodenum. Although the digestion of starch and glycogen begins as food is mixed with saliva in the mouth, the enzymatic digestion of most oligo- and polysaccharides does not begin until food reaches the small intestine. **[Try Problem 6.37]**

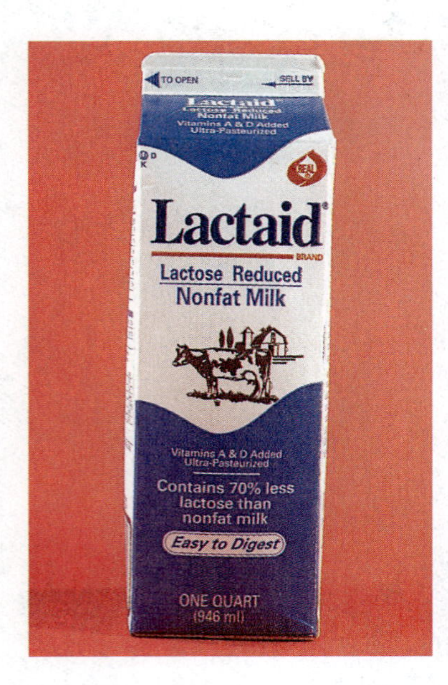

A product for the lactose intolerant.

Beano has been marketed as "a scientific and social breakthrough."

Carbohydrate Intolerances

Roughly 70% of the world's population begins losing the ability to synthesize **lactase,** the glycosidase that catalyzes the digestion of the lactose in milk, immediately after they have been weaned. As a consequence, many adults, and even some children, develop a **lactose intolerance.** After drinking too much milk or consuming too many other dairy products, they tend to experience abdominal pain, bloating, gas, diarrhea, and occasionally other symptoms that result from the CO_2, H_2, and irritating organic acids produced by intestinal bacteria as they utilize the nondigested lactose. Affected individuals can cope with their intolerance by avoiding milk and related products, by consuming commercially available low-lactose products, or by taking lactase tablets along with their meals and snacks. In most instances, lactose intolerance is genetically determined. It is particularly common among Asians, many West African peoples, Native Americans, certain Mediterranean peoples, and their descendants. Peoples who have depended historically on herding and dairy products, including northern Europeans and the Masai in Africa, exhibit a low incidence of lactose intolerance as a consequence of evolution. Evolution and change is one of the central themes in biochemistry (Exhibit 1.4). **[Try Problem 6.38]**

Beano, which is marketed as a scientific and social breakthrough, contains **α-galactosidase.** This enzyme catalyzes the hydrolysis of certain indigestible (in humans) oligosaccharides found in beans. It is these oligosaccharides that, when utilized by intestinal bacteria, are responsible for the gas attacks experienced by many bean lovers. By stimulating the hydrolysis of these carbohydrates, Beano, it is claimed, shuts down gas production. Preliminary "clinical" trials indicate that there is some validity to the claims.

Intestinal Absorption

Monosaccharides move from the intestines into the blood stream through a variety of mechanisms including **active transport, facilitated diffusion,** and **free diffusion** (Figure 6.5; see also Sections 9.10 and 9.11). **Diffusion,** in general, refers to the spontaneous movement of a compound from a region of higher concentration to one of lower concentration. It consumes no energy. Facilitated diffusion is brought about by a protein carrier, whereas free diffusion is a carrier-free process. Active transport involves a specific carrier protein, requires energy, and can occur against a concentration gradient (that is, a compound can be transported from a region of relatively low concentration to one of higher concentration). Most glucose is absorbed through active transport, but fructose tends to be absorbed through facilitated diffusion. Certain other monosaccharides are absorbed only through free diffusion. Once monosaccharides have entered the blood, they are carried to cells throughout the body, where they are utilized primarily as a fuel and as a source of carbon for anabolism. If too much starch or sugar is consumed, part of the carbon ends up in fats. **[Try Problem 6.39]**

Dietary Fiber

Dietary **fiber** (formerly called **roughage**) is defined as plant material that is resistant to digestion. It is classified as either soluble or insoluble, depending on its solubility

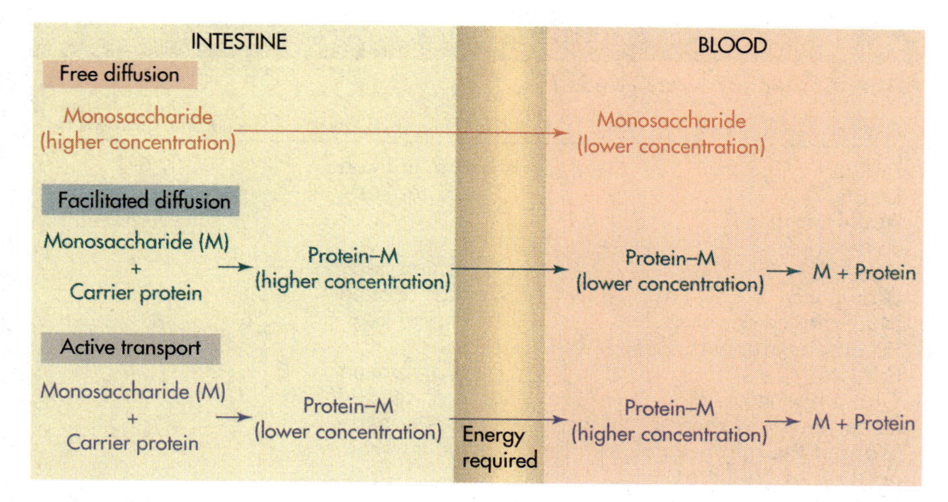

FIGURE 6.5
Intestinal absorption of monosaccharides. Energy, when required, is usually provided by ATP, a high-energy compound examined in Chapters 7 and 8. Carrier proteins are normally imbedded in membranes; they pick up monosaccharides on one side and release them on the opposite side.

in hot water. Cellulose, the major component of most insoluble dietary fiber, is indigestible because the human body is not genetically programmed to produce enzymes, called **cellulases,** capable of catalyzing the hydrolysis of its $\beta(1 \longrightarrow 4)$ glycosidic links. Amylases, which catalyze the hydrolysis of the $\alpha(1 \longrightarrow 4)$ glycosidic links in glycogen and starch, are $\alpha(1 \longrightarrow 4)$ specific. Enzyme specificity (a central theme of biochemistry) has many consequences. Although a small amount of cellulose is hydrolyzed with the catalytic assistance of cellulases synthesized by bacteria in the large intestine, most passes right on through the body. The cellulases assembled by bacteria in the rumen of cattle allow these organisms to utilize the cellulose in grains and grasses. In humans, both grains and vegetables are common dietary sources of cellulose and other fiber (see Table 6.5 at the top of the next page). **[Try Problem 6.40]**

There has been considerable publicity in both the public and scientific press about the possible benefits and risks of dietary fiber. Insoluble fiber, by softening and adding bulk to the contents of the intestine, helps alleviate constipation and increases the rate at which fecal material is transported through the intestinal tract. The more rapid movement of fecal material, coupled with the tendency of certain fiber to bind particular substances, reduces the absorption of many chemicals, including carcinogens (cancer-causing agents; some produced by bacteria in the intestine) and certain beneficial nutrients. The claim that dietary fiber can reduce the risk of heart disease and specific types of cancer, including colon cancer, is still under investigation. Further research is also needed to verify the proposed benefits of fiber for those with diabetes (Section 12.4) and **diverticulosis** (an intestinal disorder characterized by the presence of diverticula (small sacs) that bulge outward from the inner lining of the intestine).

TABLE 6.5

Fiber Content of Some Foods

Food	Quantity or Amount	Fiber Content (g)
Beans	1/2 cup, cooked	5–8
Lentils	1 cup, cooked	7
Apple or pear	1	4
Banana	1	3
Barley	1 cup, cooked	6
Brown rice	2/3 cup, cooked	3
Macaroni or spaghetti	1 cup, cooked	2
Potato, baked, with skin	1	4
Carrots	1/2 cup, cooked	3
Green beans	1/2 cup, cooked	2
Asparagus or broccoli	1/2 cup, cooked	2
Wonder 9-Grain Bread	2 slices	6
Tortilla, whole wheat	1	3
Pita or tortilla, white flour	1	1
White or French bread	2 slices	1
Wasa Fiber Plus crispbread	1 ounce	9
Health Valley fruit bars	1 ounce	4
Wheaties	1 cup	3
Post Raisin Bran	1 cup	8

Dietary Sugars

The typical American diet contains a large amount of refined sucrose plus an appreciable amount of other sugars. Our high sugar consumption has been claimed to contribute to an enormous number of health problems, ranging from tooth decay to heart disease and diabetes. The present status of these claims is examined in Section 6.11.

PROBLEM 6.37 What two monosaccharides are produced in largest amounts during the digestion of the carbohydrates in a typical United States diet?

PROBLEM 6.38 What is the simplest way to convert normal milk to low lactose milk?

PROBLEM 6.39 What mechanism is involved in the intestinal absorption of dietary amino acids? Hint: Review Section 4.12.

PROBLEM 6.40 Why are green leafy vegetables, which are rich in carbohydrate, recommended for someone wanting to lose weight?

6.10 THE ENTEROINSULAR AXIS

The dietary carbohydrate story provides an interesting example of the chemical communication that constantly occurs throughout the human body, one of the major

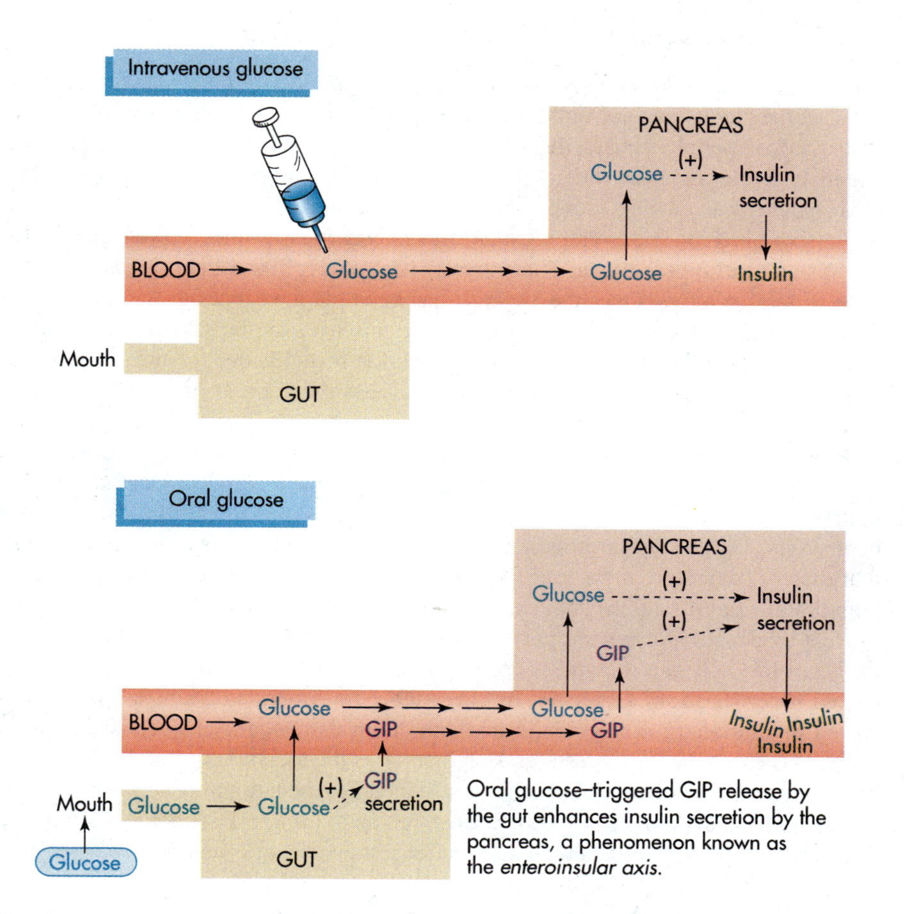

FIGURE 6.6
Effect of intravenous and oral glucose on blood insulin concentration. GIP is gastric inhibitory polypeptide.

themes of biochemistry (Exhibit 1.4). **Insulin** is a protein hormone that, among other things, stimulates the uptake of blood glucose by muscle cells. An increase in blood glucose concentration usually leads to an increased release of insulin from the pancreas and to an increase in blood insulin levels. However, the elevation in blood glucose concentration through the ingestion of glucose leads to a greater increase in blood insulin concentration than is observed when the same change in blood glucose concentration is achieved with the intravenous feeding of glucose. The effect of oral glucose on insulin, known as the **enteroinsular axis,** is thought to be at least partly explained by the production of **gastric inhibitory polypeptide** (**GIP,** a hormone) within the gut in response to the presence of glucose. Gut-produced GIP moves through the blood to the pancreas, where it stimulates the release of insulin when glucose is also present in the blood. In the absence of glucose, GIP has no effect on blood insulin levels (Figure 6.6).

GIP is one of several gut hormones that have been found to help the gut communicate with the rest of the body. Such hormones act to prime the body so that it can rapidly and efficiently utilize ingested foods. In the body, as in business, industry, politics, and many other endeavors, communication is a key to efficient operation and survival.

6.11 SUGARS: FACTS AND FICTION

The dietary carbohydrate story would not be complete without an examination of some of the claims that surround the sugar-rich diet that is so common in the United States. Most people equate the term "sugar" with sucrose, or table sugar. Biochemists define sugars as water-soluble, sweet-tasting carbohydrates. Sugars include most naturally occurring mono- and disaccharides, plus some larger oligosaccharides (Exhibit 6.2). The average United States citizen consumes approximately 133 lbs of sugar (all types) a year. That corresponds to over 500 cal per day and close to 25% of total calories. Over the years, it has been claimed that our national "sweet tooth" contributes to diabetes, obesity, heart disease, **hypoglycemia** (abnormally low blood sugar levels), hyperactivity in children, tooth decay, and other health problems. How does the scientific community view these claims today? **[Try Problem 6.41]**

Prolonged **hyperglycemia** (abnormally high blood sugar levels) can damage the eyes, kidneys, nervous system, and circulatory system, so diabetics (lack the normal ability to return high blood sugar concentrations to usual values) have traditionally been advised to strictly limit their sugar intake. This advice was based on the belief that dietary sugars are more rapidly absorbed from the intestine than complex carbohydrates (mainly starch) and, consequently, generate or aggravate hyperglycemia. Present evidence does not support this belief; dietary starch is almost instantly hydrolyzed to yield glucose and it appears to be as rapidly absorbed (hydrolysis required for absorption) as sugars. In May 1994, the American Diabetes Association issued revolutionary new guidelines which claim that sugar-rich diets are no more dangerous for diabetics than starch-rich diets. It is the total amount of digestible carbohydrate that must be regulated. To safely increase the amount of sugar in their diets, diabetics must reduce the amount of starch and vice versa. Although dietary carbohydrates can lead to diabetic complications, they do not cause diabetes (see Section 12.4).

On a gram to gram basis, sugars are no more fattening than proteins. Dietary fat, rather than sugar, is linked to most cases of obesity. Although sugars may increase the heart disease risk for a small number of "carbohydrate-sensitive" individuals, sugar consumption is not considered a significant risk factor for the general public. What about hypoglycemia? Hypoglycemia is much less common than once believed, and dietary sugar does not appear to play a significant role in most instances; dietary sugar-linked hypoglycemia, known as **reactive hypoglycemia,** is very rare. In spite of the continuing controversy over the impact of diet on mood and behav-

EXHIBIT 6.2
Some Ways That Sugars Are Listed on Food Labels

Sucrose (table sugar)	Maple syrup
Glucose (dextrose)	Honey
Fructose	Corn syrup
Maltose	Molasses
Lactose	Brown sugar

ior, most recent studies fail to support the claim that dietary sugar induces hyperactivity in children. In fact, several clinical trials suggest that meals rich in carbohydrates, including sugars, tend to make people feel relaxed or even sluggish. What, then, are the risks of a "sweet tooth"?

For most people, the major direct risk is probably tooth decay. Dietary sugars are converted by bacteria in the mouth to the acids that cause cavities. The risk can be reduced by confining sweets to mealtimes. That way, teeth are exposed to a barrage of acids a limited number of times each day. Meals also increase saliva output which helps wash away sugars and neutralizes acids.

Refined sucrose, which accounts for most of the sugar in a typical United States diet, is often described as "**empty calories.**" This terminology comes from the fact that normal table sugar contains no vitamins, no amino acids, no minerals, and no other nutrients, only calories. From a practical standpoint, honey (primarily glucose and fructose) and brown sugar (usually sucrose with a little molasses added for color and flavor) deserve a similar classification. If too many of the total calories in a diet are empty, the remaining calories may not supply an adequate amount of the essential nutrients, particularly if the remaining calories are supplied by some of the common junk foods. This means that high sugar consumption, coupled with other poor dietary habits, can lead to potentially serious health problems, but the sugar is not the primary culprit. [Try Problem 6.42]

Many of the health problems once attributed to sugar now appear to be linked to the fat that is so often found in sugary foods. Excess dietary fat is definitely associated with multiple health problems, including obesity and heart disease (Chapter 9).

PROBLEM 6.41 Corn syrup is produced by partially hydrolyzing starch. Which specific monosaccharides and disaccharides are present in corn syrup?

PROBLEM 6.42 Are the calories in purified starch as "empty" as those in refined table sugar? Explain.

SUMMARY

Monosaccharides are polyhydroxy aldehydes or ketones and derivatives of these compounds. They all contain three to nine carbons, and most are chiral. D-Glyceraldehyde, D-ribose, D-galactose, and D-glucose are the most commonly encountered aldoses. Dihydroxy acetone and D-fructose are representative ketoses. Monosaccharides containing five or more carbons tend to react internally to form cyclic hemiacetals that are ordinarily depicted with Haworth formulas. The cyclization process creates a new chiral center, the anomeric carbon, and leads to alpha (α) and beta (β) anomers. The most stable cyclic form of a monosaccharide usually contains a pyranose or furanose ring, and it exists in equilibrium with an open-chain form and one or more other cyclic forms. The most abundant forms of D-glucose (Table 6.3) are shown at the top of the next page. The starred compounds are present in relatively small amounts.

Monosaccharides readily undergo a variety of chemical reactions, including oxidation, reduction, dehydration, esterification, alkylation, and Schiff base (imine) formation. The cyclic forms of monosaccharides frequently react with alcohols to form special classes of acetals known as glycosides. The acetal link in a glycoside is called a glycosidic bond or link.

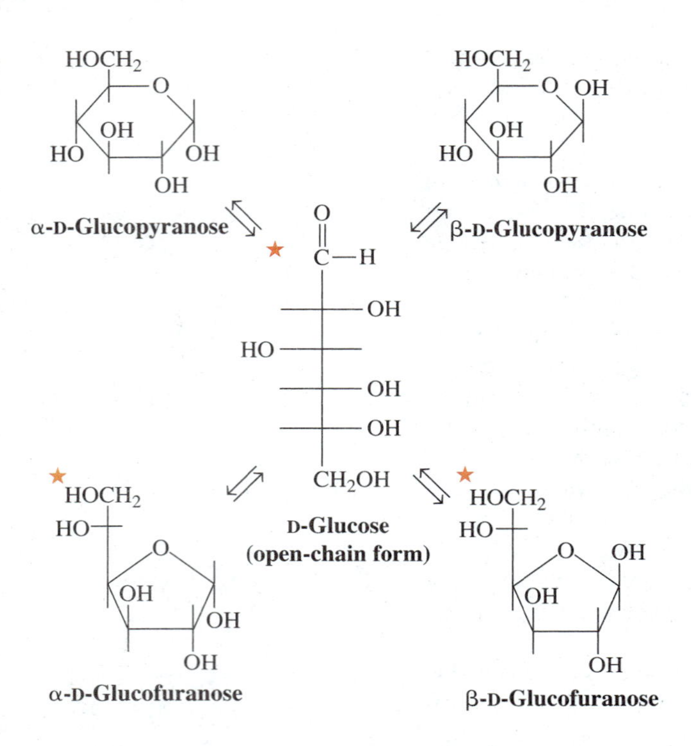

Glucose and certain other saccharides tend to react spontaneously with amino groups on proteins to generate Amadori products. Amadori products may play a role in normal biological aging, and their concentration in blood can be used to help monitor blood glucose levels in diabetics. Because glycosylated hemoglobin (an Amadori product) concentration remains elevated for some time after a bout of hyperglycemia, the periodic measurement of the abundance of this derivatized red blood cell protein reveals how successful a diabetic has been in maintaining normal blood glucose levels.

The joining of monosaccharides through glycosidic links creates disaccharides, other oligosaccharides, and polysaccharides. Maltose and cellobiose are disaccharides and intermediates in the synthesis and breakdown of glycogen and cellulose, respectively. Lactose (milk sugar) and sucrose (table sugar), two common dietary disaccharides, serve as fuels and carbon sources for those organisms that ingest or produce them. Some oligosaccharides serve as biological messengers whereas others play key structural and functional roles within glycoproteins and glycolipids.

Both starch (in plants) and glycogen (in animals) are homopolymers of glucose that function as readily mobilizable storage forms of fuel and carbon. In amylose, the linear form of starch, all monomers are joined through $\alpha(1 \rightarrow 4)$ glycosidic links. Amylopectin contains $\alpha(1 \rightarrow 4)$-linked chains with $\alpha(1 \rightarrow 6)$ branches. Glycogen is chemically identical to amylopectin, except it tends to be more highly branched and of greater molecular weight. Liver glycogen plays a central role in maintaining normal blood sugar levels, whereas muscle glycogen provides fuel for muscle action. Most other human cells contain lower concentrations of glycogen.

Cellulose (in most plants) and chitin (in insects, spiders, crustaceans, some algae, and certain other organisms) are linear homopolymers that function as structural glycans. Cellulose, a major component of stems, stalks, and woody elements in plants, is assembled by joining numerous glucose through $\beta(1 \rightarrow 4)$ glycosidic links. In chitin, the monomer is *N*-acetyl-D-glucosamine rather than glucose. Other polysaccharides include agarose and the dextrans (used as supports for chromatography and electrophoresis), heparin (a structurally complex anticoagulant), hyaluronic acid, and chondroitin sulfates. Hyaluronic acid is a heteropolysaccharide and functions as a lubricant and shock absorber around bone joints. Chondroitin sulfates are structurally similar to hyaluronic acid, but, in contrast to hyaluronic acid, they tend to be highly sulfated. Both hyaluronic acid and chondroitin sulfates are mucopolysaccharides and major structural elements in connective tissue.

A majority of the proteins in the human body may possess a carbohydrate component. Glycoproteins are known to include immunoglobulins and fibrinogen plus a variety of hormones, enzymes, transport agents, receptors, and structural elements. The carbohydrate moieties, which range from under 1% to over 90% by weight, serve a wide variety of roles.

Carbohydrates make up well over 50% of the substances in a typical American diet. Although starch and sucrose normally account for a vast majority of these carbohydrates, glycogen, cellulose, lactose, and fructose are also consumed on a regular basis. Starch, glycogen, sucrose, and lactose are hydrolyzed in the digestive system with the catalytic assistance of specific glycosidases. Released monosaccharides, along with any free dietary monosaccharides, move from the intestines into the blood stream through a variety of mechanisms, including active transport, facilitated diffusion, and free diffusion. Once in the blood, the monosaccharides are delivered to cells throughout the body, where they can be oxidized to produce energy or utilized to synthesize glycogen, hyaluronic acid, certain amino acids, fats, and numerous other substances.

The human body is not genetically programmed to assemble enzymes that catalyze the hydrolysis of cellulose and certain other oligo- and polysaccharides. These digestion-resistant carbohydrates, which are primarily of plant origin, make a major contribution to dietary fiber. The possible perils and benefits of dietary fiber, including the lowering of heart disease and cancer risks, are under intense investigation.

In the past, dietary sugar, primarily sucrose, has been claimed to contribute to numerous health problems, including obesity, heart disease, diabetes, hyperactivity in children, and tooth decay. Ongoing studies fail to support many of these claims. For most individuals, the major risk of a "sweet tooth" is probably tooth decay. However, we must make certain that purified sugars (empty calories) do not replace other foods to the point where a diet becomes deficient in essential nutrients.

The human gut communicates with the rest of the body in order to prepare the body to promptly and efficiently utilize those foods that it does ingest. The enteroinsular axis provides a specific example. In the presence of dietary glucose, gut-derived gastric inhibitory polypeptide (GIP) enhances the release of insulin into the blood by the pancreas. Insulin is required for the proper utilization of dietary glucose.

EXERCISES

6.1 Given the following monosaccharides:

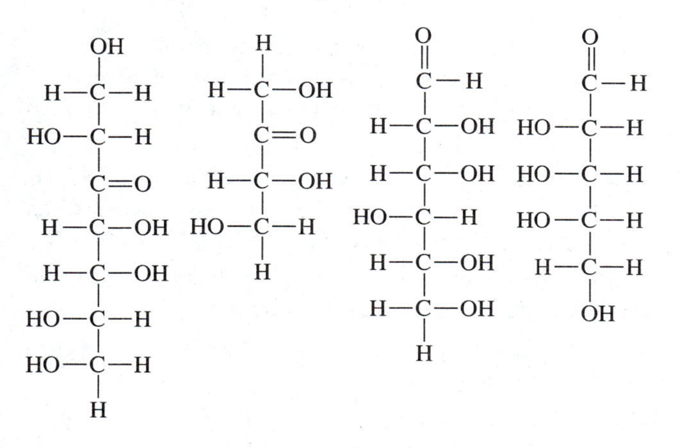

a. Number each carbon chain.
b. Label each monosaccharide as D or L.
c. Circle each chiral center.
d. Write the structural formula for the enantiomer of each compound.
e. Which, if any, of these monosaccharides are aldoses? Heptoses? Ketoses? Oligosaccharides? Pentoses?

6.2 Are a pair of anomers enantiomers? Diastereomers? Stereoisomers? Structural isomers? Explain.

6.3 Write the structural formulas for 3-deoxy-D-glucose, α-D-galactopyranose, 2-deoxy-β-D-ribofuranose, and β-D-fructopyranose.

6.4 In an aqueous solution, most glucose molecules exist in a pyranose form. A very small fraction, however, will exist in other forms. Describe these other forms.

6.5 Write Fischer projection formulas for four different L-aldoheptoses.

6.6 Write the structural formula for a hemiacetal form of D-glucose that contains a seven-membered ring system.

6.7 Write the structural formulas for the compounds formed when the C2 hydroxyl group in β-D-ribose is (a) methylated, (b) phosphorylated, and (c) acetylated.

6.8 What functional group distinguishes reducing sugars from nonreducing sugars?

6.9 What monosaccharide has the common names "dextrose" and "blood sugar?"

6.10 Write the structural formula for the compound formed when the carbonyl group in D-glucose is reduced.

6.11 Write the structural formula for the compound formed when the aldehyde group in D-ribose is oxidized.

6.12 Write the structural formula for the compound formed when a single phosphoric acid molecule reacts to form two ester links, one with the C3 hydroxyl group of β-D-ribose and a second with the C6 hydroxyl group in β-D-fructose.

6.13 Explain how hyperglycemia increases the blood concentration of glycosylated hemoglobin.

6.14 Does glycosylated hemoglobin bind and carry O_2 and CO_2 normally? Explain.

6.15 Write the structural formulas for two ribose residues joined through (a) an $\alpha(1 \longrightarrow 2)$ glycosidic link, (b) an $\alpha(1 \longrightarrow 5)$ glycosidic link, and (c) an $\alpha,\beta(1 \longrightarrow 1)$ glycosidic link. Which of these disaccharides are reducing sugars? Explain.

6.16 What products are formed with the complete hydrolysis of starch? Fructose? Sucrose? Glucose? Glycogen? Chitin? Cellulose? Lactose? If any compound cannot be hydrolyzed, put "no hydrolysis."

6.17 What products are formed with the complete digestion (in the intestine) of starch? Glycogen? Lactose?

6.18 Can a linear homopolymer be made from ribose? Fructose? Mannose? Galactose? Can a branched homopolymer be made from ribose? Fructose? Mannose? Galactose? Explain.

6.19 The human body is unable to synthesize cellulose, whereas certain other organisms are unable to build glycogen. What determines which carbohydrates can be produced by any given organism?

6.20 List two specific glycoproteins or classes of glycoproteins, and then list the major roles or functions of each.

6.21 Give an example of a polysaccharide that is a polyanion at pH 7.0.

6.22 What classes of noncovalent bonds hold chondroitin sulfates and proteins together within connective tissues?

6.23 What are the major functions of hyaluronic acid?

6.24 Why is the human body unable to digest cellulose and chitin?

6.25 List three mechanisms through which compounds are absorbed from the intestine into the blood stream. Which of these transport mechanisms require energy?

6.26 Explain why a cow that has been treated with a large dose of an antibiotic (a compound produced by microorganisms that specifically inhibits bacterial growth) is often unable to digest grass (mostly cellulose).

6.27 Compound X is present at high concentrations in blood but present at low concentrations in the diet. Can compound X be absorbed from the intestines through facilitated diffusion? Explain.

6.28 It has been suggested that many cases of colon cancer are due to carcinogens produced by microorganisms that grow in the large intestine. Assuming this to be true, how does dietary fiber reduce the risk of colon cancer?

6.29 True or false? All dietary carbohydrates are equally fattening. Explain.

6.30 Which one of the following statements best describes the biological activity of gastric inhibitory polypeptide (GIP)?

- **a.** It stimulates the digestion of starch.
- **b.** It stimulates the absorption of glucose from the intestine.
- **c.** It stimulates the release of insulin by the pancreas.
- **d.** It stimulates the production of intestinal amylase.
- **e.** It increases the rate of movement of food through the intestine.

6.31 Which will more likely lead to prolonged hyperglycemia in a normal person: intravenous glucose or an equal amount of oral glucose? Explain.

6.32 From a nutritional standpoint, is the sugar in fruit any better than refined sucrose? Explain.

6.33 Suggest why glucose is sweet tasting and starch, a polymer of glucose, is not.

6.34 What specific bonds hold: (a) sugars and polypeptides together within glycoproteins; (b) separate cellulose molecules together within a plant cell wall; (c) monomers together within glycogen; and (d) fructose and water molecules together in an aqueous solution of fructose?

SELECTED READINGS

Cerami, A., H. Vlassara, and M. Brownlee, Glucose and aging, *Sci Am*. 256(5), 90–97 (1987).

Examines the chemistry of protein glycosylation and its possible biological consequences.

Dwek, R. A., Glycobiology: More functions for oligosaccharides, *Science* 269, 1234–1235 (1995).

Provides examples of oligosaccharides that have a structural role within glycoproteins and emphasizes the diverse functions and properties of oligosaccharide prosthetic groups.

Fiber bounces back, *Consumer Reports on Health* 7(3), 25–28 (1995).

Examines the evidence that dietary fiber helps fight diseases.

Moran, L. A., K. G. Scrimgeour, H. R. Horton, R. S. Ochs, and J. D. Rawn, *Biochemistry,* 2nd ed. Englewood Cliffs, NJ, Neil Patterson/Prentice Hall, 1994.

An encyclopedic biochemistry text with additional details regarding the structure and properties of carbohydrates. Contains an excellent section on glycoproteins and an adequate bibliography.

Pennisi, E., Chitin craze, *Sci News* 144, 72–74 (1993).

Discusses the properties of chitin and some of its practical uses.

Roehrig, K.L., *Carbohydrate Biochemistry and Metabolism.* Westport, Connecticut: AVI Publishing Company, 1984.

A very readable introduction to carbohydrates. A brief bibliography follows each chapter. Topics include: carbohydrate chemistry and nomenclature; digestion and absorption of carbohydrates; and industrial uses of carbohydrates.

Voet D., and J. G. Voet, *Biochemistry,* 2nd ed. New York: John Wiley & Sons, 1995.

A comprehensive biochemistry text with references and some problems. Chapter 10 is titled Sugars and Polysaccharides.

What's wrong with sugar? *Consumer Reports on Health* 6(10), 114–115, 1994.

Reviews the possible health risks of dietary sugar.

Wurtman, R. J., and J. J. Wurtman, Carbohydrates and depression, *Sci Am* 260(1), 68–75 (1989).

Explores the link between mood-related disorders, including seasonal affective disorder (SAD), and carbohydrate craving.

7 Nucleotides and Some Related Compounds

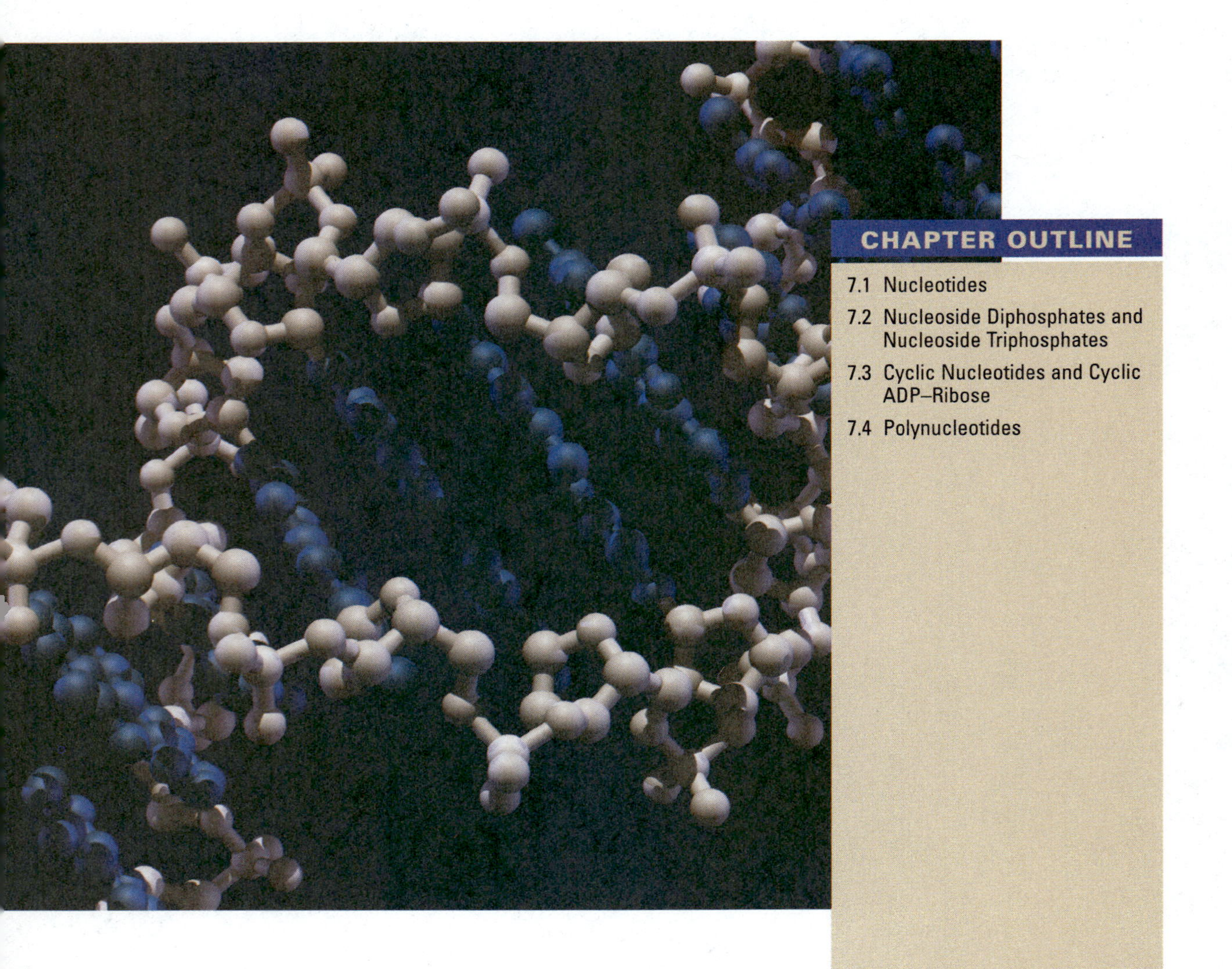

CHAPTER OUTLINE

Nucleotides are the building blocks from which nucleic acids, both RNA and DNA, are constructed. Nucleotide residues are also found within a wide variety of other compounds, including ATP, several important biochemical messengers, and some coenzymes. These compounds play central roles in a multitude of biochemical processes. Since all nucleotides and related compounds contain one or more sugar residues, the knowledge acquired in Chapter 6 will be invaluable in understanding the discussions that follow.

This chapter provides an introduction to the structure, chemical properties, and biological functions of nucleotides, oligonucleotides, polynucleotides, nucleoside triphosphates, cyclic nucleotides, and cyclic ADP–ribose. In the process it touches on several major biochemical themes: specificity, catalysis and the regulation of reaction rates, cellular communication, structure-function relationships, and the transformation and use of energy. Some of the compounds examined in this chapter play key roles in bioenergetics, the topic of Chapter 8.

7.1 NUCLEOTIDES

A nucleotide contains a phosphoric acid residue, a sugar residue, and an organic base residue. In those nucleotides that are used to assemble nucleic acids, the organic base is always a purine or a pyrimidine. The building blocks and the overall architecture of nucleotides are examined in this section.

Phosphoric Acid

Phosphoric acid (Table 2.8) is a triprotic acid (contains three dissociable hydrogens) that can form anhydrides and can react with carbohydrates and other alcohols to form esters (Section 6.3). Its dissociated or salt forms are known as **phosphates.** The three acidic hydrogens in phosphoric acid have pK_a's of approximately 2.2, 7.2, and 12.4.

O
‖
HO—P—OH
|
OH

$pK_{a_1} = 2.2$
$pK_{a_2} = 7.2$
$pK_{a_3} = 12.4$

Phosphoric acid

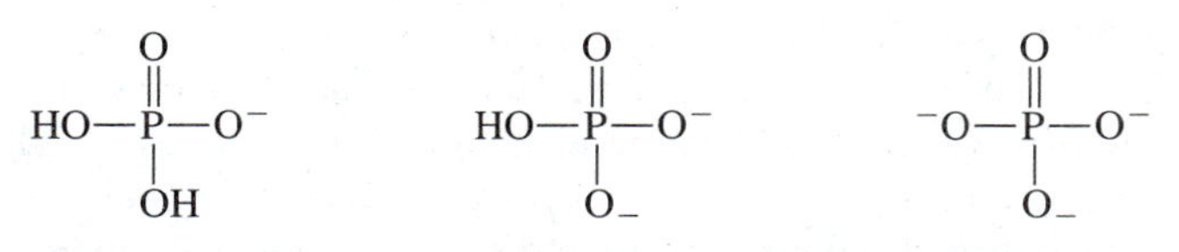

Dihydrogen phosphate **Hydrogen phosphate** **Phosphate**

The precise pK_a's for phosphoric acid esters and anhydrides depend on the nature of any neighboring functional groups, the cations present in the environment, and other variables. The two dissociable hydrogens in a phosphoric acid monoester commonly have pK_a's around 1 and 6, whereas the single dissociable hydrogen in a diester tends to have a pK_a close to 1.

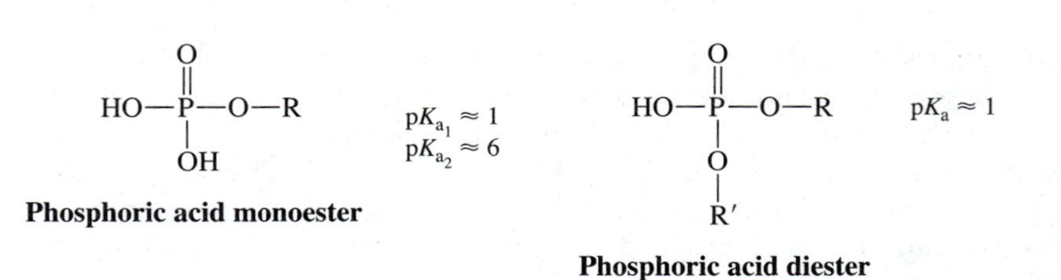

The pK_a's for **pyrophosphoric acid** (PP_i), the simplest phosphoric acid anhydride, are roughly 1.0, 2.5, 6.1, and 8.5. At physiological pH's, phosphoric acid residues usually contribute one or more negative charges to those compounds in which they reside, because the dissociated form of a weak acid functional group is favored at pH's above its pK_a (Section 2.7). **[Try Problem 7.1]**

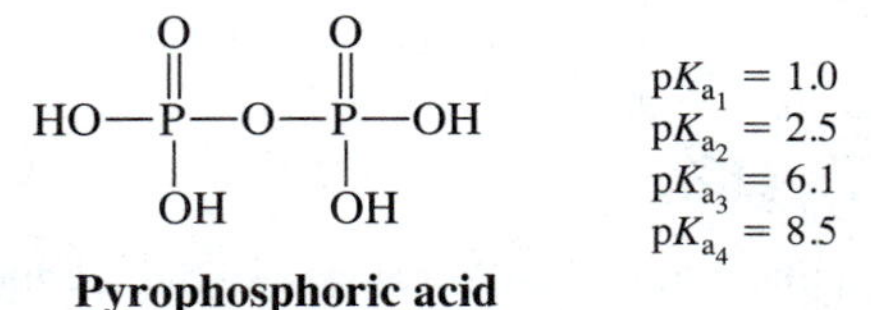

Ribose and Deoxyribose

The sugar residue in a nucleotide is either **β-D-ribofuranose** or **β-D-2-deoxyribofuranose:**

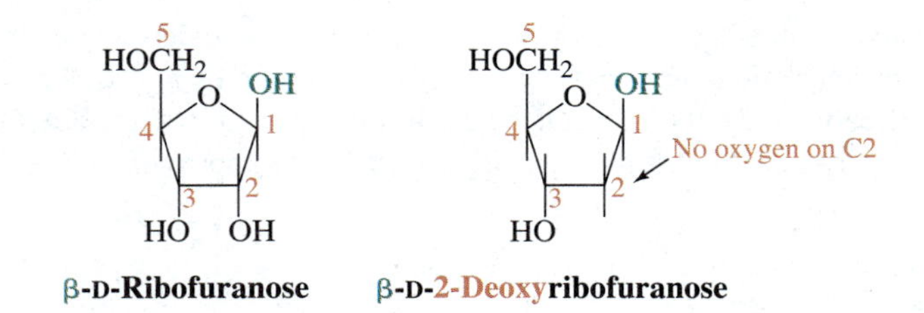

The β in the names indicates that the hydroxyl group is up about the anomeric carbon (the hemiacetal carbon) and the D designates the configuration about the highest numbered chiral center. In aqueous solutions, these cyclic aldopentoses exist in equilibrium with open-chain forms and α anomers. The C1 hydroxyl group can react with alcohols and amines to form *O*-acetals and *N*-acetals, respectively. All of the hydroxyl groups are capable of reacting with phosphoric acid, carboxylic acids, and certain other acids to form esters. You may want to reread Sections 6.1 through 6.3 if you are uncomfortable with any of the terms or concepts in this paragraph. **[Try Problem 7.2]**

Purines and Pyrimidines

Purine and pyrimidine are specific heterocyclic aromatic amines. Any compound that contains the heterocyclic ring found in pyrimidine is said to be a **pyrimidine.** A

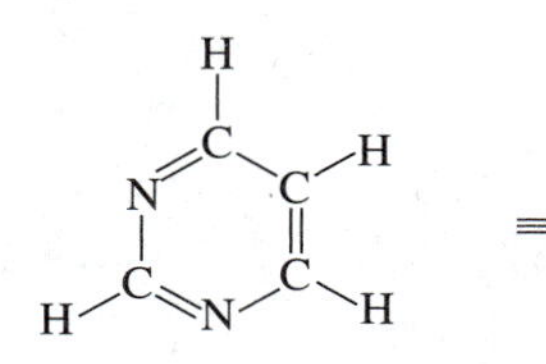

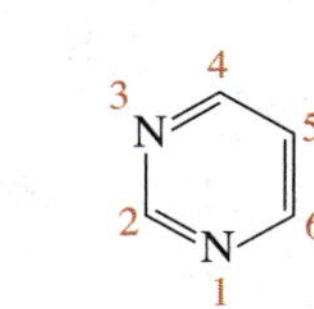

Pyrimidine

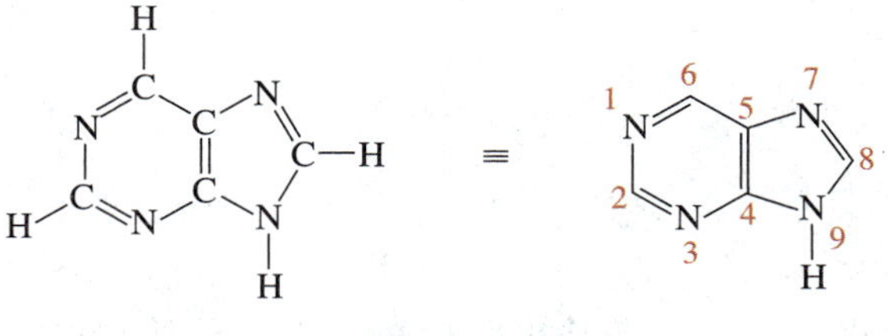

Purine

purine is characterized by the fused ring system that exists in purine. Such relationships are common in organic chemistry. Phenols, for example, are derivatives of the specific compound named phenol.

Three pyrimidines and two purines are found within those nucleotides, which are normally used to assemble nucleic acids:

Uracil (U)

Cytosine (C)

Thymine (T)

Pyrimidines

Adenine (A)

Guanine (G)

Purines

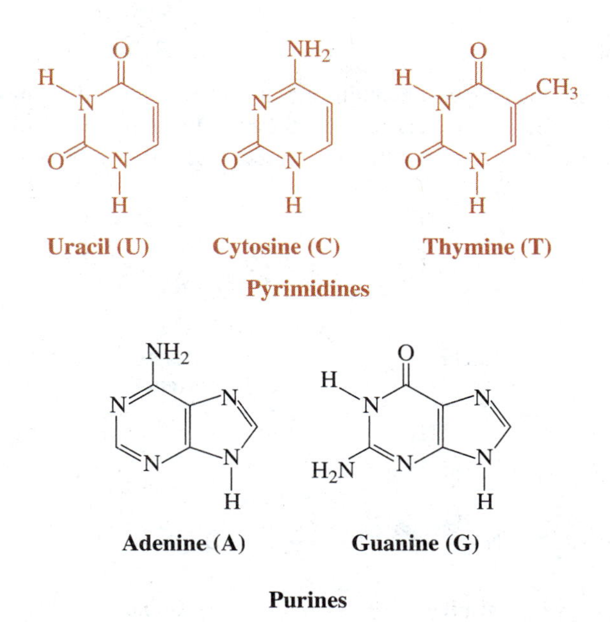

Although these **major** or **common bases** (amines are bases) can exist in both charged and uncharged forms, the uncharged forms (illustrated above) predominate at pH values between 5 and 9. Over 50 other bases occur much less frequently in nu-

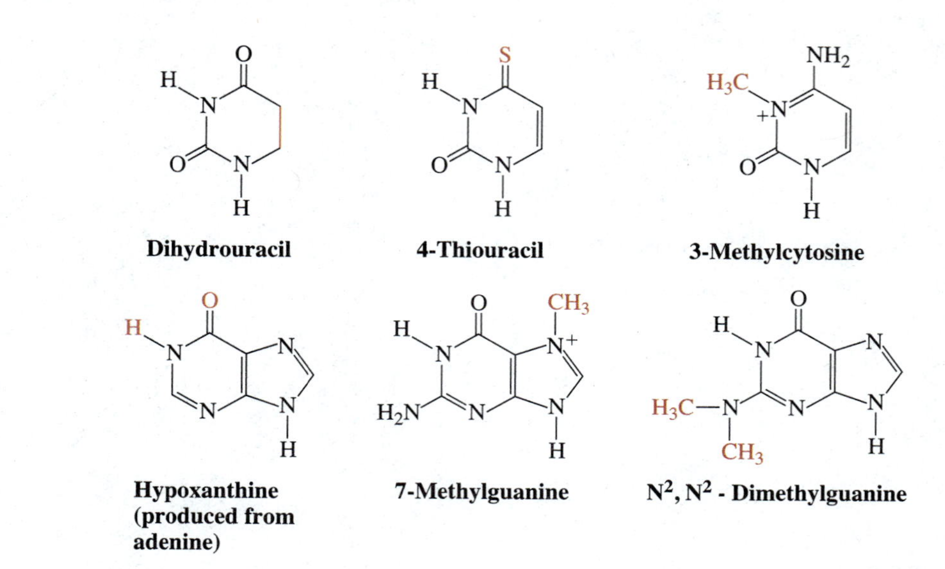

FIGURE 7.1
Some minor bases.

cleic acids (Figure 7.1). These **minor bases** are constructed through the enzyme-catalyzed modification of the five major bases after they have been incorporated into polymer strands.

Tautomerism

Because all of the major bases contain one or more functional groups that exhibit tautomerism (Figure 7.2 and Section 3.5), they exist in multiple **tautomeric forms** (forms that exist in equilibrium with one another and differ in the location of a sin-

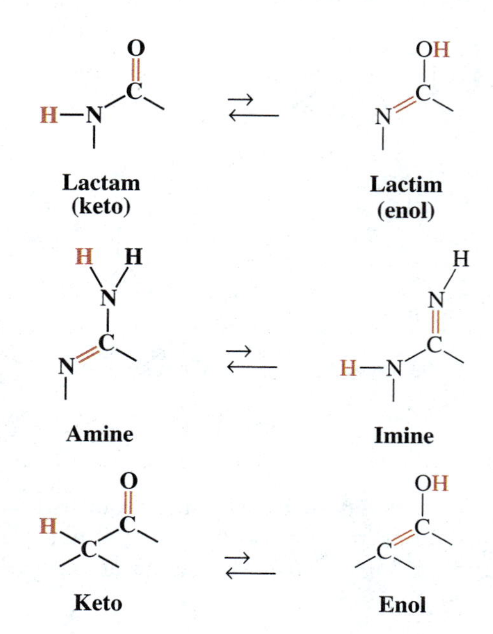

FIGURE 7.2
The tautomeric forms for some biologically important functional groups. In each case, the more stable form is in bold print.

gle H atom and a single double bond). The constructs in bold print in Figure 7.2 represent the most stable **tautomers** under physiological conditions. At equilibrium, the concentration of the most stable tautomeric form of a purine or pyrimidine is normally 10,000 to 100,000 times the concentration of the less stable tautomers. Chapters 17, 18, and 19 explain how tautomerism contributes to errors during DNA replication and how the resultant mutations (abnormal changes in DNA) can lead to genetic diseases, cancer, and other health problems. **[Try Problem 7.3]**

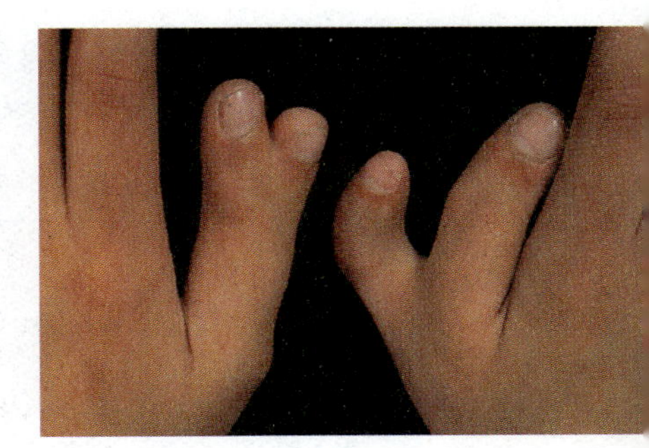

Tautomerism-linked mutations can lead to genetic diseases, including polydactyly (more than the normal number of fingers or toes).

Nucleosides

The common **nucleosides** are compounds in which the anomeric hydroxyl group of β-D-ribofuranose or β-D-2-deoxyribofuranose is bonded to N1 of a pyrimidine ring or N9 of a purine ring through an *N*-glycosidic link.

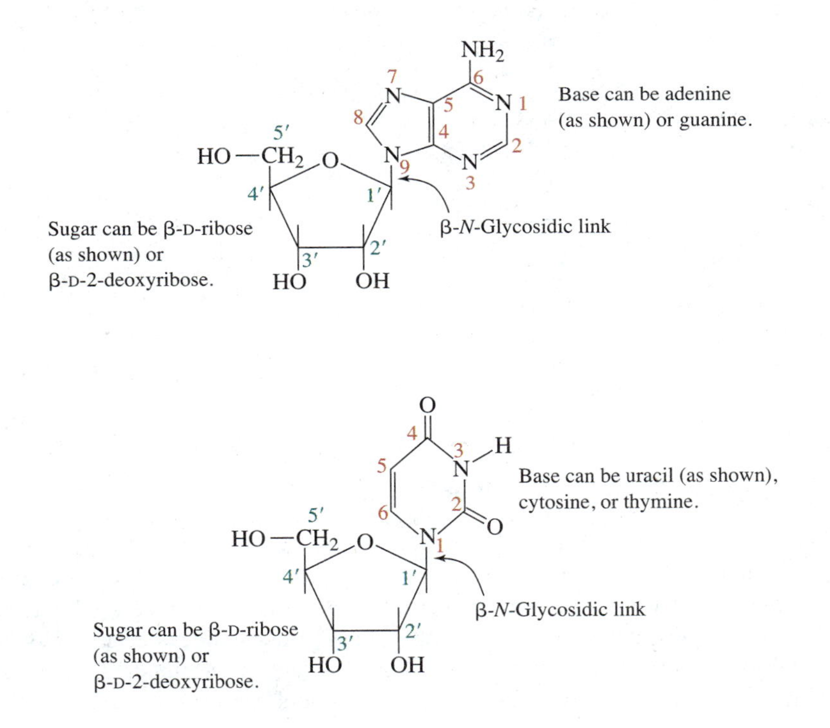

Although the sugars and bases can, in theory, be attached at other positions and through alternative classes of covalent bonds, such molecular combinations are relatively uncommon in living organisms. When the *N*-glycosidic link in a nucleoside is cleaved through hydrolysis, a sugar and a base are released as products. **[Try Problems 7.4 and 7.5]**

Nucleosides containing D-ribose are called **ribonucleosides,** and those containing D-2-deoxyribose are described as **deoxyribonucleosides.** The names of the major nucleosides are summarized in Table 7.1. The atoms in the purine or pyrimidine ring of a nucleoside are identified with regular numbers and the atoms in sugar rings are labeled with primed numbers (see the nucleoside structures just given).

TABLE 7.1

Summary of the Common Nucleosides and Nucleotides

Base	Sugar	Nucleoside	Nucleotide	Abbreviation of Nucleotide[a]
Adenine (A)	D-Ribose	Adenosine	Adenosine 5′-monophosphate or adenylate	5′-AMP or 5′-rAMP
	D-2-Deoxyribose	Deoxyadenosine	Deoxyadenosine 5′-monophosphate or deoxyadenylate	5′-dAMP
Guanine (G)	D-Ribose	Guanosine	Guanosine 5′-monophosphate or guanylate	5′-GMP or 5′-rGMP
	D-2-Deoxyribose	Deoxyguanosine	Deoxyguanosine 5′-monophosphate or deoxyguanylate	5′-dGMP
Cytosine (C)	D-Ribose	Cytidine	Cytidine 5′-monophosphate or cytidylate	5′-CMP or 5′-rCMP
	D-2-Deoxyribose	Deoxycytidine	Deoxycytidine 5′-monophosphate or deoxycytidylate	5′-dCMP
Thymine (T)	D-Ribose	——— [b]	———	———
	D-2-Deoxyribose	Deoxythymidine[c] or thymidine	Deoxythymidine 5′-monophosphate or deoxythymidylate	5′-dTMP
Uracil (U)	D-Ribose	Uridine	Uridine 5′-monophosphate or uridylate	5′-UMP or 5′-rUMP
	D-2-Deoxyribose	——— [d]	———	———

[a] If no number precedes an abbreviation, a 5′ is implied.
[b] It is uncommon for D-ribose to be bonded to thymine.
[c] The deoxy- prefix is often omitted because thymine is rarely bonded to ribose.
[d] It is uncommon for deoxyribose to be bonded to uracil.

Nucleotides

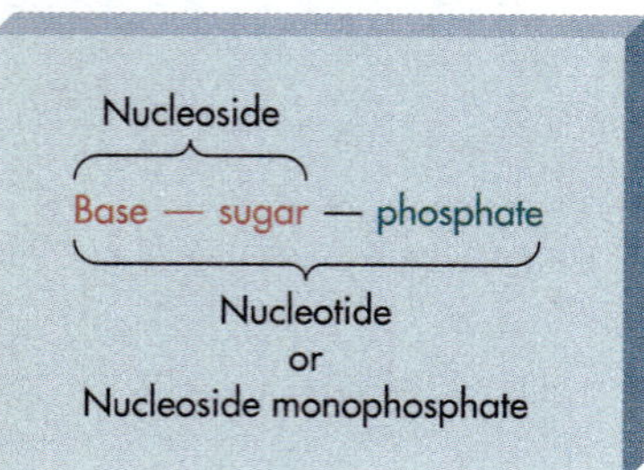

When a phosphoric acid residue is esterified to one of the hydroxyl groups on the sugar residue within a nucleoside, a **nucleotide** (also called a **nucleoside monophosphate**) is formed. Ribonucleosides can react to form 2′-, 3′- and 5′-nucleotides. Deoxynucleosides yield only 3′- and 5′-nucleotides because they lack a 2′-hydroxyl. Since the 5′-nucleotides are most frequently encountered, a 5′ is usually implied if a number is not specified in a nucleotide name or abbreviation.

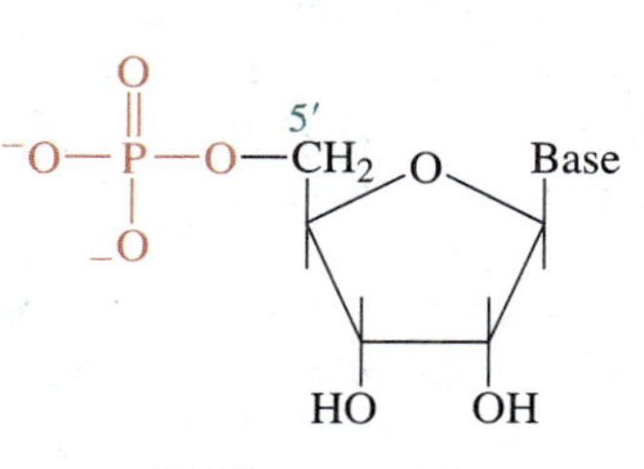

5′-Ribonucleotide

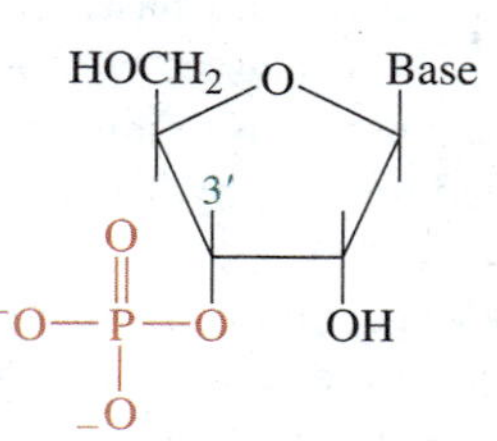

3′-Ribonucleotide

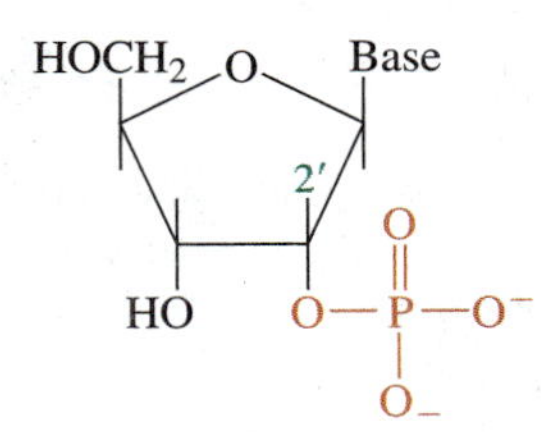

2′-Ribonucleotide

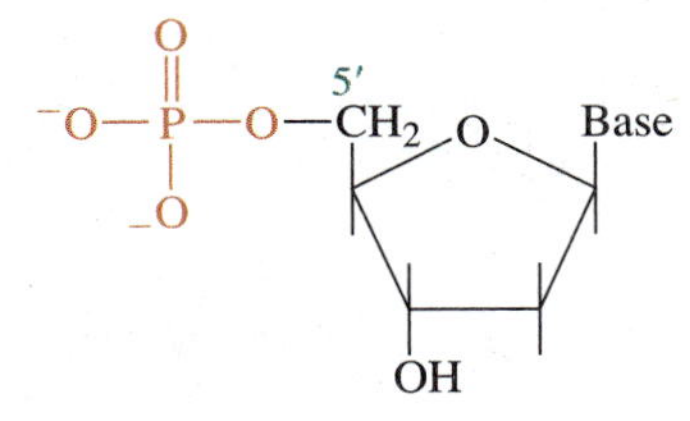

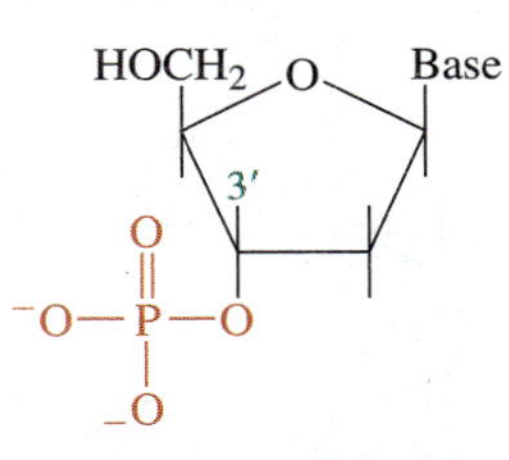

5′-Deoxyribonucleotide **3′-Deoxyribonucleotide**

Table 7.1 provides the names and abbreviations for the major nucleotides. **[Try Problems 7.6 and 7.7]**

PROBLEM 7.1 Calculate the net charges (to the nearest half charge unit) on phosphate, methyl phosphate, and dimethyl phosphate at both pH 5 and pH 7.2. Assume that these molecules have the pK_a's specified at the top of page 274. Hint: Review Section 2.7.

PROBLEM 7.2 Write the structural formula for the acetal formed by reacting α-D-ribofuranose with (a) ethanol, and (b) ethyl amine ($CH_3CH_2NH_2$). What term is used to describe an acetal created through the reaction of the cyclic form of a monosaccharide with an alcohol or an amine? Hint: Review Section 6.5.

PROBLEM 7.3 Write the structural formulas for three tautomers of uracil and three tautomers of guanine.

PROBLEM 7.4 What determines which compounds are produced within a living organism?

PROBLEM 7.5 Write the structural formula for the nucleoside formed when β-D-2-deoxyribose is joined to (a) uracil, and (b) adenine.

PROBLEM 7.6 Write the structural formulas for (a) 2′-CMP, (b) 5′-dAMP, and (c) 3′-GMP.

PROBLEM 7.7 Calculate the net charge (to the nearest half charge unit) on a nucleotide at pH 6.0. Repeat these calculations at pH 7.0 and pH 8.0. Assume that the phosphoric acid residue has pK_a's of 1 and 6.

7.2 NUCLEOSIDE DIPHOSPHATES AND NUCLEOSIDE TRIPHOSPHATES

A nucleotide is a nucleoside attached to one phosphate through an ester link. The attachment of a second phosphate through an ester link to a second sugar hydroxyl group generates a **nucleoside bisphosphate (NBP)**. If the second phosphate is attached to the existing phosphate residue through an anhydride link, the product is a **nucleoside diphosphate (NDP).** The addition of a third phosphate to the terminal phosphate in a nucleoside diphosphate leads to a **nucleoside triphosphate (NTP).**

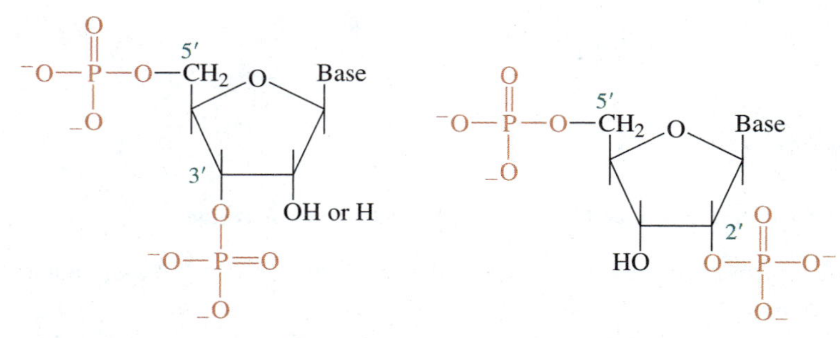

3′,5′-Nucleoside bisphosphate **2′,5′-Nucleoside bisphosphate**

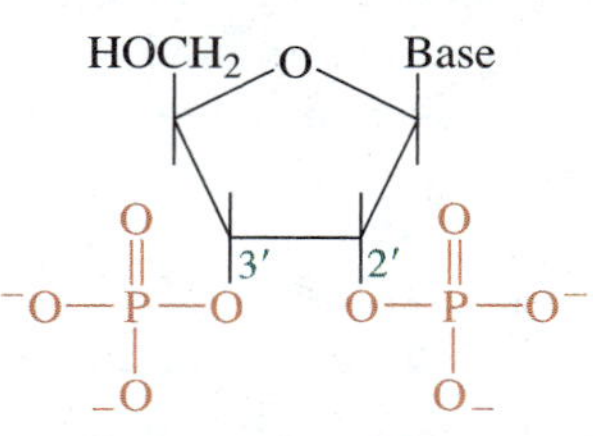

2′,3′-Nucleoside bisphosphate

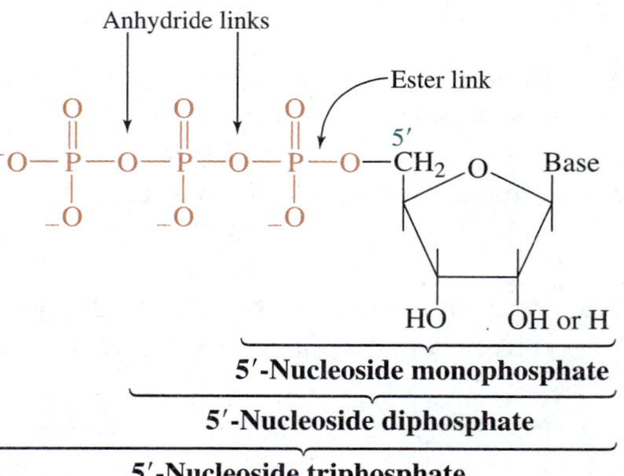

5′-Nucleoside monophosphate

5′-Nucleoside diphosphate

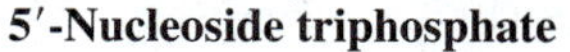

5′-Nucleoside triphosphate

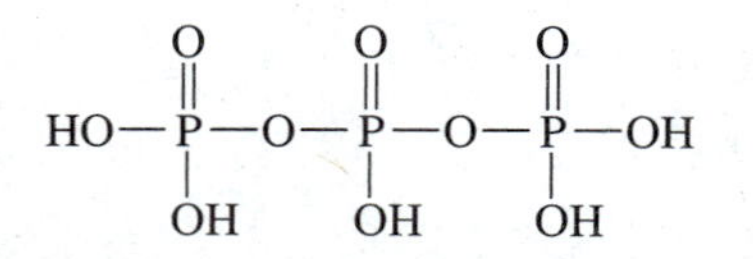

Tripolyphosphoric acid (PPP_i)

From a slightly different perspective, nucleoside diphosphates and triphosphates are esters of pyrophosphoric acid (PP_i; Section 7.1) and **tripolyphosphoric acid (PPP_i)**, respectively.

In general, the prefix **bis** indicates that separate copies of the same substituent are attached at each of two unique sites on a molecule. For a previous example, see 2,3-bisphosphoglycerate (2,3-BPG) in Section 4.9. The prefix **tris** denotes three identical groups with each attached at a unique site. The prefix **di** implies that two copies of a group have been joined and that the resultant "dimer" has been incorporated into a compound. Similarly, **tri** indicates that a "trimer" has been linked to a compound.

The most common nucleoside diphosphates and nucleoside triphosphates are summarized in Table 7.2. Both 2′- and 3′-diphosphates and triphosphates also exist. The abbreviations in Tables 7.1 and 7.2 should be memorized since they will be encountered on a regular basis throughout the remainder of this text. Although it is best to include the 5′ in these abbreviations, in practice, it is often deleted. If an abbreviation lacks a number, a 5′ is normally implied. **[Try Problem 7.8]**

Both nucleoside diphosphates and nucleoside triphosphates are energy-rich compounds that release a considerable amount of energy when hydrolyzed. The source of this energy is examined in Section 8.4. Nucleoside triphosphates are the immediate source of energy for a majority of the energy-requiring processes within most organisms, including humans. ATP is the most widely distributed and highly publicized compound in this regard; it is commonly described as the energy currency of cells (Section 8.4). Nucleoside triphosphates play an assortment of other roles as well. They are, for example, precursors for the synthesis of most nucleotide-containing compounds, including nucleic acids (Chapter 17) and cyclic nucleotides (Section 7.3). ATP serves as a neurotransmitter in both the peripheral and the central nervous system, whereas GTP plays a central role in the operation of numerous molecular switches that will be encountered in future chapters.

TABLE 7.2

A Summary of the Most Common Nucleoside Diphosphates and Nucleoside Triphosphates

Nucleoside Diphosphates[a]		Nucleoside Triphosphates[a]	
Ribo-	**Deoxyribo-**	**Ribo-**	**Deoxyribo-**
Adenosine 5′-diphosphate (5′-ADP)	Deoxyadenosine 5′-diphosphate (5′-dADP)	Adenosine 5′-triphosphate (5′-ATP)	Deoxyadenosine 5′-triphosphate (5′-dATP)
Guanosine 5′-diphosphate (5′-GDP)	Deoxyguanosine 5′-diphosphate (5′-dGDP)	Guanosine 5′-triphosphate (5′-GTP)	Deoxyguanosine 5′-triphosphate (5′-dGTP)
Uridine 5′-diphosphate (5′-UDP)	———	Uridine 5′-triphosphate (5′-UTP)	———
———	Deoxythymidine 5′-diphosphate (5′-dTDP)	———	Deoxythymidine 5′-triphosphate (5′-dTTP)
Cytidine 5′-diphosphate (5′-CDP)	Deoxycytidine 5′-diphosphate (5′-dCDP)	Cytidine 5′-triphosphate (5′-CTP)	Deoxycytidine 5′-triphosphate (5′-dCTP)

[a]If no number precedes an abbreviation, a 5′ is implied.

PROBLEM 7.8 Write the structural formulas for 2′-UTP, 3′-CDP, 2′,3′-GBP, 5′-dATP, and 3′,5′-dTBP.

7.3 CYCLIC NUCLEOTIDES AND CYCLIC ADP–RIBOSE

The ability of phosphate to form multiple ester links allows a phosphate residue in a nucleotide to simultaneously attach to two sites on the sugar residue. The product is called a **cyclic nucleotide** or a **cyclic nucleoside monophosphate. Adenosine 3′,5′-cyclic monophosphate (cyclic AMP or cAMP)** and **guanosine 3′,5′-cyclic monophosphate** (cyclic GMP or cGMP) are biologically important examples.

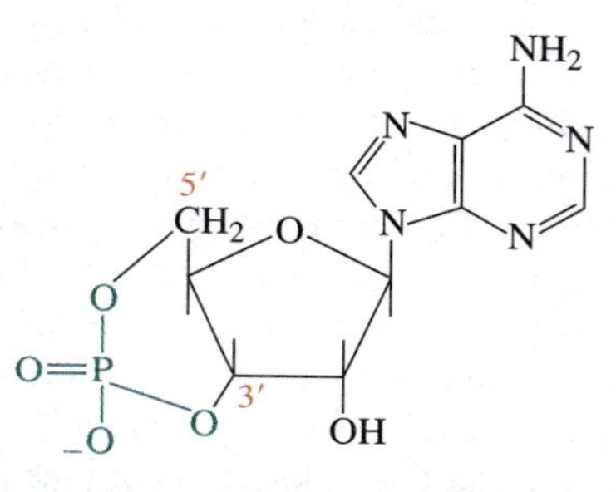

Adenosine 3′,5′-cyclic monophosphate
(3′,5′-cAMP or cAMP)

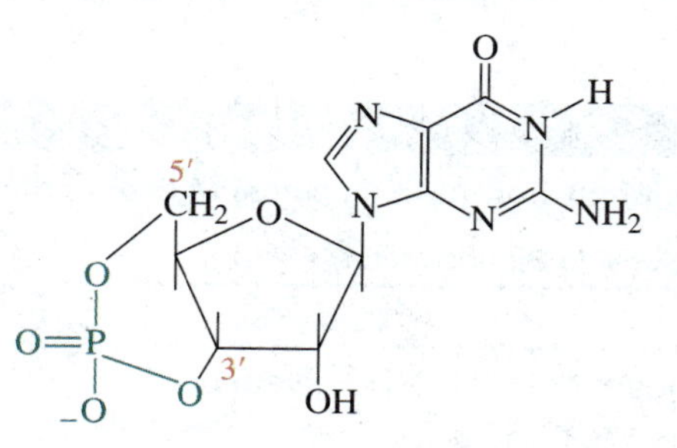

Guanosine 3′,5′-cyclic monophosphate
(3′,5′-cGMP or cGMP)

If no numbers appear in the abbreviation for a cyclic nucleotide, 3′,5′- is implied.

Cyclic nucleotides help regulate a large number of biological processes by controlling the activities of the specific proteins (usually allosteric enzymes) to which they bind. They normally function as key participants in signal transduction pathways. The propagation of a signal through cyclic nucleotide-linked pathways commonly turns on or turns off the expression of specific genes, modulates the activity of particular existing enzymes, or alters ion channels in membranes. The intracellular synthesis and degradation of cyclic nucleotides are modulated by a wide range of external "signals." [Try Problem 7.9]

The cyclic nucleotides are classified as second messengers because they only modulate cellular processes after a first messenger provides a signal that alters their rate of production or degradation. During the visual process, for example, a photon of light is the first messenger. It stimulates the hydrolytic breakdown of cGMP, a second messenger whose drop in concentration leads to the closure of ion channels and the transmission of nerve signals. Fluctuations in cAMP levels mediate the action of a wide variety of hormones.

Cyclic nucleotides are produced from nucleoside triphosphates through reactions catalyzed by enzymes called **cyclases.** The inactivation of these second messengers usually entails the **phosphodiesterase**-catalyzed hydrolysis of an internal phosphate ester link. Catalysis and the regulation of reaction rates, one of the central themes of biochemistry (Exhibit 1.4), is encountered time and time again. In this instance, the catalysis theme is connected to signal transduction and the cellular communication theme. Several cyclic nucleotide-linked signal transduction pathways are examined in detail in Chapter 10 and the metabolism chapters that follow.

Cyclic adenosine diphosphate ribose (cADPR) is another nucleotide-containing second messenger. cADPR is hypothesized to play a central role in **calcium mobilization** (the release of Ca^{2+} from organelles into the cytosol or the movement of Ca^{2+} into a cell) within numerous types of cells. Calcium mobilization is involved in muscle contraction, cellular secretion, and sundry other processes. The glucose-induced release of insulin from the pancreas, for example, is mediated by cADPR and changes in Ca^{2+} concentrations. Specific cyclases catalyze the production of cADPR from NAD^+, a coenzyme whose structure is provided in Figure 11.1.

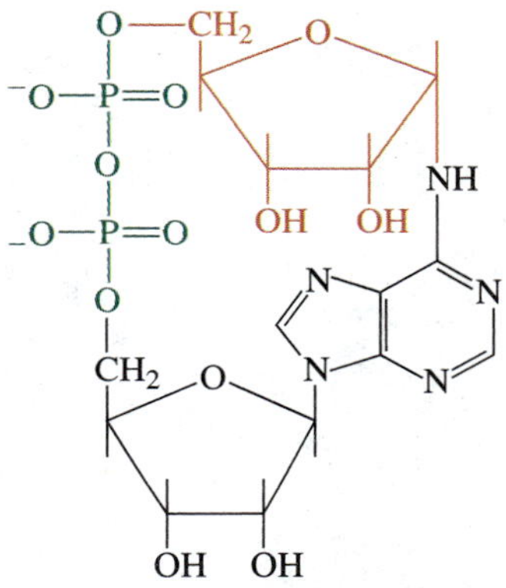

Cyclic adenosine diphosphate ribose (cyclic ADP-ribose or cADPR)

PROBLEM 7.9 The binding of a second messenger to its allosteric site on a regulatory enzyme:

a. Denatures the enzyme.
b. Modulates the catalytic activity of the enzyme.
c. Alters the primary structure of the enzyme.
d. Catalyzes the hydrolysis of the enzyme.
e. Leads to the phosphorylation of the enzyme.
f. None of the above.

Circle the one best answer. Hint: Review Section 5.12.

7.4 POLYNUCLEOTIDES

The ability of a single phosphate residue to form multiple ester links not only accommodates the formation of cyclic nucleotides, it also allows separate nucleotides to be joined to form **oligo-** and **polynucleotides**. Adjacent nucleotide residues are held together through 3′,5′-**phosphodiester bonds**; the 3′-hydroxyl group in one nucleotide residue and the 5′-hydroxyl group in a second are bonded through separate ester links to the same phosphate residue.

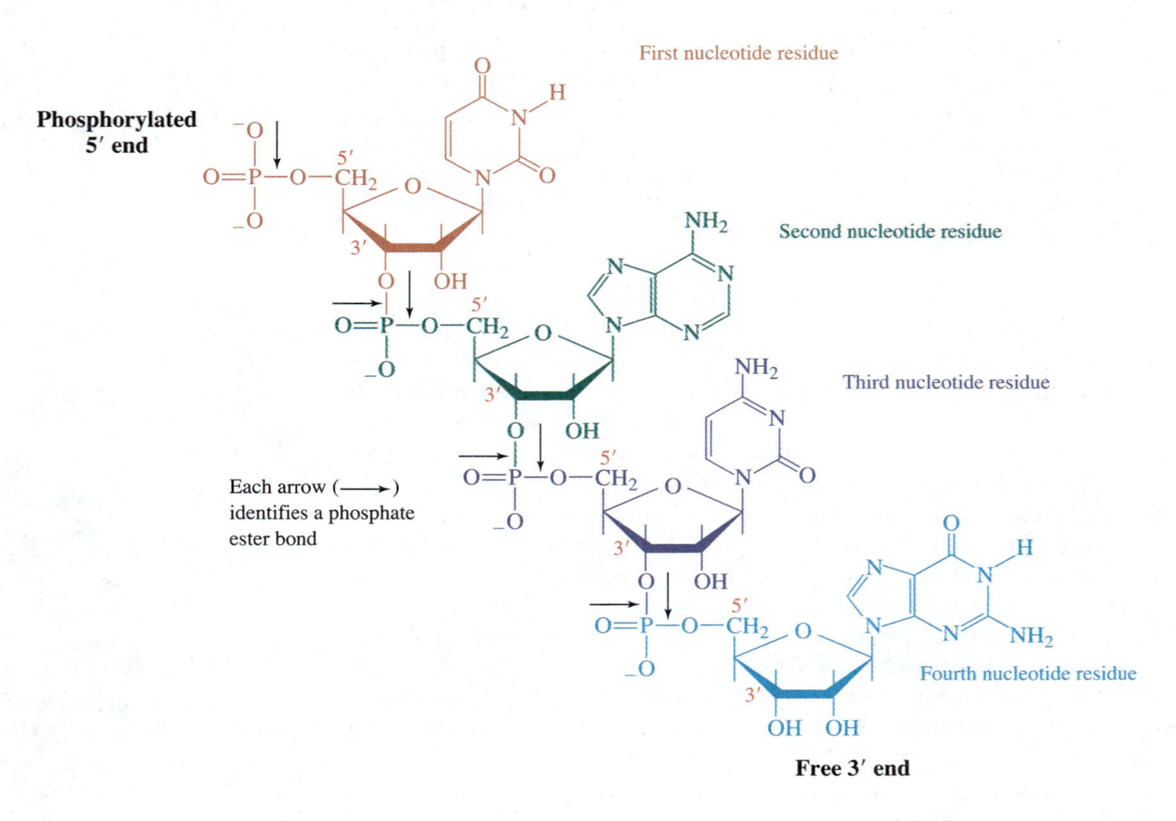

That end of an oligo- or polynucleotide with a 5′-hydroxyl group that is not bonded to a neighboring nucleotide residue is known as its **5′ end.** The opposite end, the **3′ end,** has a 3′-hydroxyl group that is not attached to an adjacent monomer. Both the 3′- and 5′-hydroxyl groups of internal nucleotide residues are esterified to neighboring residues. The hydroxyl groups at the 3′ and 5′ ends may be either free or phosphorylated. Nucleotide residues are numbered consecutively from the 5′ to the 3′ end.

There are a variety of accepted abbreviations for the tetranucleotide depicted at the top of this page:

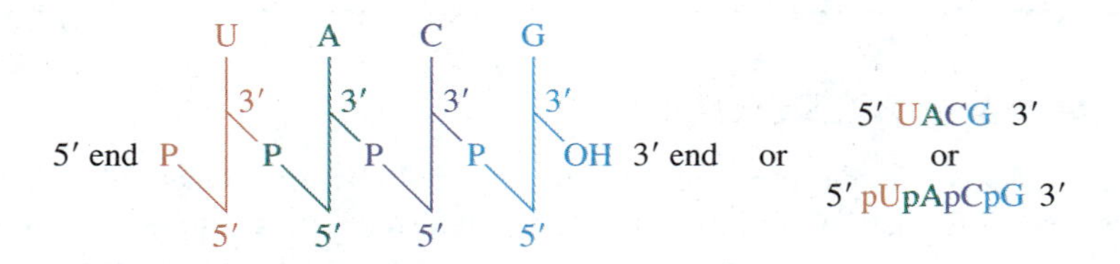

Since all nucleotide residues are identical except for their bases, the abbreviation for a base is also used as an abbreviation for the nucleotide that contains it. If it is not

TABLE 7.3

RNA and DNA

Nucleic Acid	Sugar Residue	Common Bases	Usual Number of Nucleotide Residues Per Molecule	Usual Number of Polymer Strands per Molecule
Ribonucleic Acid (RNA)	Ribose	A C G U	60 to over 10,000	1
Deoxyribonucleic Acid (DNA)	Deoxyribose	A C G T	Thousands to millions	2

specified, you should assume that in an abbreviation for an oligo- or polynucleotide the 5′ end is to the left and the 3′ end is to the right. A "d" before an abbreviation indicates that the monomers involved are deoxyribonucleotides rather than ribonucleotides.

Because each phosphate residue has a dissociable hydrogen with a pK_a of approximately 1, polynucleotides exist as polyanions at physiological pH. Mg^{2+} and basic proteins (positively charged at physiological pH) are common counter ions for polynucleotides; they bind phosphate residues and neutralize their charges. Since isolated charges are inherently unstable, negative ions tend to be associated continuously with positive ions (Section 2.7). **[Try Problem 7.10]**

Ribonucleic acids (RNAs) are polynucleotides in which the monomers are ribonucleotides. Their common bases are adenine, guanine, cytosine, and uracil. Deoxyribonucleic acids (DNAs) contain deoxyribonucleotide residues and the common bases adenine, guanine, cytosine, and thymine. Some characteristics of RNA and DNA are summarized in Table 7.3. Both classes of nucleic acids are examined in detail in Chapter 16.

PROBLEM 7.10 Write the structural formula for dpApTpGp and then:

a. Use an arrow to identify each ester bond.
b. Clearly identify the 3′ end and the 5′ end.
c. Put a circle around each atom that can participate in H bonding.
d. Calculate the approximate net charge (to the nearest half charge unit) on this compound at pH 7.0. Use the pK_a's given in Section 7.1.

SUMMARY

Nucleotides (nucleoside monophosphates) and related compounds are ubiquitously distributed within all living organisms. The nucleotides themselves are the monomers from which oligo- and polynucleotides, including nucleic acids, are constructed. Each nucleotide consists of a phosphate esterified to a nucleoside. The nucleoside encompasses a sugar residue joined to a nitrogenous base residue through a β-*N*-glycosidic link. The sugar is either β-D-ribofuranose or β-D-2-deoxyribofuranose. The base is normally a pyrimidine (usually C, U, or T) or a purine (usually A or G).

Nucleoside triphosphates, including ATP, are energy-rich compounds that are the immediate source of energy for most of the energy-requiring process within living organisms. They are also precursors for the production of nucleic acids, cyclic nucleotides, and most other nucleotide-containing compounds. ATP, in addition, serves as a neurotransmitter, whereas GTP is a central participant in the operation of many molecular switches.

Cyclic AMP and cGMP are second messengers in numerous signal transduction pathways. Changes in intracellular cyclic nucleotide concentrations in response to an external first messenger lead to alterations in the activities of specific proteins. The ultimate consequence of fluctuations in cyclic nucleotide concentration is usually the selective adjustment of the catalytic activities within the host cell. Cyclic adenosine diphosphate ribose (cADPR), another nucleotide-containing second messenger, is involved in the regulation of Ca^{2+} mobilization.

Oligo- and polynucleotides, including RNA and DNA, are constructed by joining nucleotides through 3′,5′-phosphodiester links. The common bases in RNA are A, G, C, and U, whereas the common bases in DNA are A, G, C, and T. Minor bases are synthesized from common bases after the common bases have been incorporated into nucleic acid chains. Oligo- and polynucleotides have directional properties because they possess a 3′ end that is distinguishable from a 5′ end. Polynucleotides are polyanions at physiological pH. Mg^{2+} and basic proteins are common counter ions.

EXERCISES

7.1 Write the structural formulas for:

a. Guanosine
b. 2-Methyladenine
c. 3′-dGDP
d. 2′,5′-cUMP
e. Two tautomers of AMP
f. dpTpApCp
g. ppUpC
h. Adenosine 2′,3′,5′-trisphosphate
i. 5′-CTP
j. Ribose joined to adenine through an α-*N*-glycosidic link

7.2 Identify each ester bond, each anhydride bond, each glycosidic bond, and each hydrolyzable bond in each compound in Exercise 7.1.

7.3 Write a balanced equation for the complete hydrolysis of:

a. UMP
b. GTP
c. pCpU
d. dTDP
e. cADPR

Provide structural formulas for all reactants and products.

7.4 Calculate the net charge (to the nearest half charge unit) on each of the following compounds at pH 7.0. Use the pK_a's given in Section 7.1.

a. AMP
b. pCpUpGp
c. Adenosine
d. pCpApGpUpApApApC

7.5 Cyclic AMP commonly modulates the activity of protein kinases. What type of reaction is catalyzed by protein kinases? Hint: Review Section 5.13.

7.6 Write two additional abbreviations for pApCpCpGp.

7.7 Write a balanced equation for the cyclase-catalyzed formation of cAMP from ATP. This is a unimolecular reaction that yields two products. Provide structural formulas for all reactants and products.

SELECTED READINGS

Alberts, B., D. Bray, J. Lewis, M. Raff, K. Roberts, and J. D. Watson, *Molecular Biology of the Cell*, 3rd ed. New York: Garland Publishing, 1994.

An up-to-date and encyclopedic textbook that examines nucleotide structure and function. A chapter titled "Cell Signaling" provides good coverage of cAMP-linked signaling pathways. Numerous references are provided.

Blackburn, G. M., and M. J. Gait, Editors, *Nucleic Acids in Chemistry and Biology,* New York: Oxford University Press, 1990.

Contains some details about the structure and properties of nucleotides that are not presented in most introductory level biochemistry texts.

Galione, A., Cyclic ADP–ribose: A new way to control calcium, *Science* 259, 325-326 (1993).

A brief review of cADPR biochemistry.

Linder, M. E., and A. G. Gilman, G proteins, Sci. Am. 267 (2), 56-65 (1992).

Examines the role of GTP in molecular switches and provides some examples of cAMP-linked signal transduction.

Moran, L. A., K. G. Scrimgeour, H. R. Horton, R. S. Ochs, and J. D. Rawn, *Biochemistry,* 2nd ed. Englewood Cliffs, NJ: Neil Patterson/Prentice-Hall, 1994.

An encyclopedia-type biochemistry text with a separate chapter titled "Nucleotides."

Stryer, L., Visual excitation and recovery, *J. Biol. Chem.* 266, 10711-10714 (1991).

A minireview that describes the role of cGMP in the visual process.

Voet, D., and J.G. Voet, *Biochemistry*, 2nd ed. New York: John Wiley & Sons, 1995.

An encyclopedic text with references and problems.

8 Bioenergetics

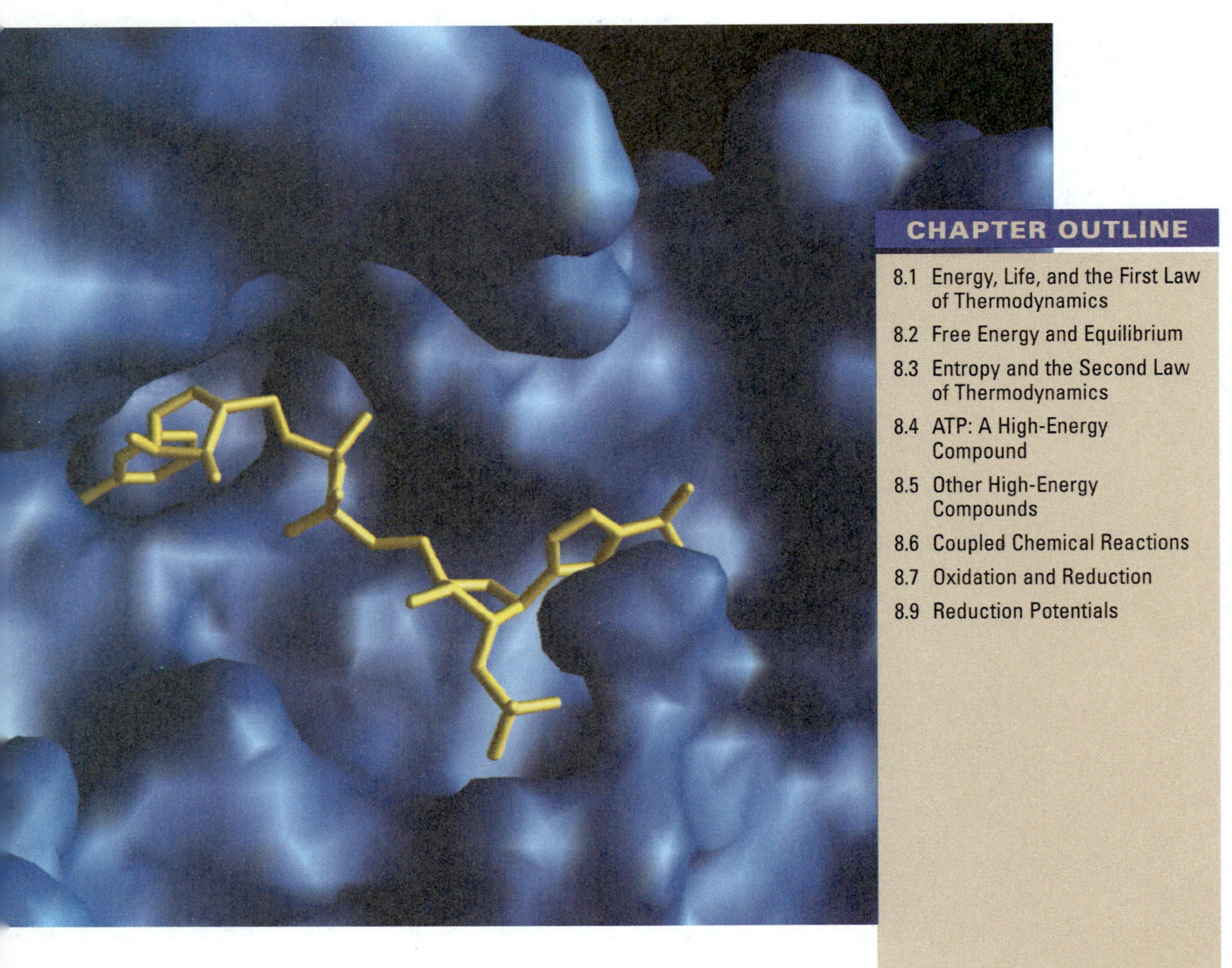

CHAPTER OUTLINE

The nucleoside triphosphates examined in Chapter 7 play a central role in **bioenergetics,** the study of the energy transformations that occur within living organisms and the exchange of energy between organisms and their environment. Bioenergetics is a branch of thermodynamics, the study of energy and natural change. Since living organisms, and those reactions that occur within them, are subject to the same thermodynamic laws as inanimate objects, some knowledge of thermodynamics is an essential prerequisite to the study of metabolism.

All organisms require a constant supply of energy to survive. In humans, this energy is used for growth, muscle action, brain action, and a variety of other processes (Exhibit 8.1). Most of the energy used by humans and many other organisms is provided through the oxidation of organic fuel molecules acquired from their environment. Molecular oxygen (O_2) is the oxidizing agent in this stepwise process whose net reaction is:

$$\text{Organic substance} + O_2 \rightleftharpoons H_2O + CO_2 + \textbf{ENERGY}$$

There are additional inorganic products when the organic fuel molecule contains chemical elements in addition to hydrogen, oxygen, and carbon. **[Try Problems 8.1 and 8.2]**

Organisms that acquire their energy from organic substances are called **heterotrophs. Autotrophs** employ light energy or chemical energy from a source other than organic compounds to synthesize organic compounds from CO_2. Autotrophs provide the organic fuel molecules used by heterotrophs; they are the primary producers in all food chains. Most of the energy used by the human body can be traced to solar energy (sunlight) that was transformed into chemical energy by photosynthetic autotrophs.

Photosynthesis is the process through which plants and other photosynthetic organisms convert light energy into chemical energy within carbohydrates (Chapter 15). Molecular oxygen (O_2) is a byproduct. Photosynthesis, coupled with the oxidation of organic fuel molecules by heterotrophs, leads to a cycling of carbon and oxygen between autotrophs, heterotrophs and the environment (Figure 8.1).

This chapter reviews the first and second laws of thermodynamics and provides an introduction to bioenergetics. In the process, it lays a foundation for an examination of membrane transport in Chapter 9 and the discussion of metabolism in Chapters 10 through 18. The transformation, storage, and use of energy is the featured biochemical theme. Supporting themes include structure–function relationships, specificity, and the regulation of reaction rates. **[Try Problems 8.1 and 8.2]**

EXHIBIT 8.1

Some Major Uses of Energy Within the Human Body

- Signal transduction and amplification
- Muscle contraction and other movement
- Anabolism (involved in growth and development plus additional processes, including tissue repair and turnover)
- Active transport across membranes (basis for nerve action)
- Maintenance of normal body temperature

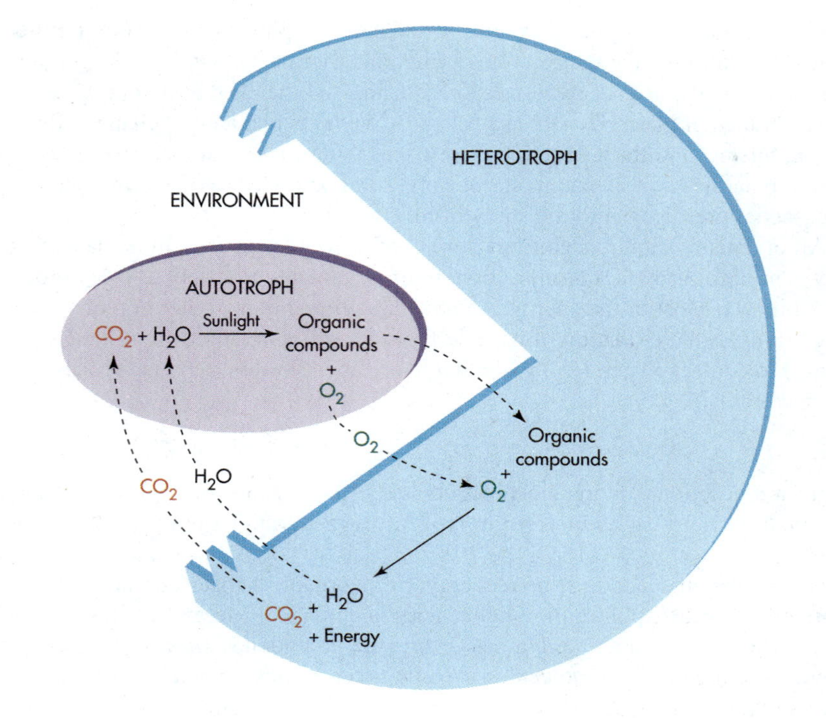

FIGURE 8.1
The carbon and oxygen cycles.

PROBLEM 8.1 Provide a molecular explanation for why a person will usually die if deprived of O_2 for more than a few minutes.

***PROBLEM 8.2** Write a balanced equation for the oxidation (by O_2) of each of the following compounds to CO_2 and H_2O: (a) ethanol (C_2H_6O) and (b) glucose ($C_6H_{12}O_6$).

8.1 ENERGY, LIFE, AND THE FIRST LAW OF THERMODYNAMICS

Energy is the capacity or ability to do work; anything that has the potential to perform work possesses energy. Most of the work performed by living organisms falls into one of three major classes: mechanical work that moves a cell, organelle, or other object against a gravitational or frictional force; gradient work that creates concentration gradients and electrical gradients across membranes; and synthetic work that entails the making and breaking of bonds and the production of compounds.

Energy is divided into two broad categories: energy associated with motion **(kinetic energy),** and energy attributed to the position or condition of an "object" **(potential energy).** Wind energy (movement of air molecules) and solar energy (trav-

A rock balanced on a natural spire posseses considerable potential energy.

eling electromagnetic waves) are examples of kinetic energy. Chemical bond energy is potential energy because it is associated with the condition or position of atoms and electrons within matter. Gravitational energy and the energy associated with concentration gradients are other forms of potential energy (Table 8.1). **[Try Problem 8.3]**

During chemical and physical changes, energy is neither created nor destroyed; the total amount of energy remains constant. Energy can, however, be converted from one form to another. These experimental facts are known as the **first law of thermodynamics.** During a chemical reaction whose products contain less energy than reactants, some chemical energy is transformed into another form, typically heat. A chemical reaction leading to products that possess more energy than the reactants is said to consume energy. The consumed energy, which is acquired from the "environment," is not lost. It is converted to chemical energy within products. **[Try Problem 8.4]**

Life is maintained through energy transformations and the exchange of energy between an organism and its environment. Living organisms do not create or destroy energy. Thus, the total amount of energy in a cell is the same immediately before and immediately after it oxidizes a fuel molecule. Oxidation simply converts a portion

TABLE 8.1

Some Forms of Kinetic and Potential Energy

Some Forms of Kinetic Energy	Some Forms of Potential Energy
Solar energy (sunlight)	Chemical bond energy
Wind energy	Gravitational energy
Heat (thermal) energy	Concentration gradient energy
Electrical energy	Nuclear energy
Tide energy	

of the chemical energy in the fuel molecule into a form that can be utilized to perform biologically useful work. Part of the energy of oxidation is also released as heat. In humans, this heat helps maintain normal body temperature.

***PROBLEM 8.3** Describe the relationship between heat and temperature.

***PROBLEM 8.4** Describe the major energy transformations involved in the operation of a hydroelectric dam.

8.2 FREE ENERGY AND EQUILIBRIUM

The **free energy** (abbreviated ***G*** for **Gibbs free energy**) of any pure element in its most stable form at a temperature of 298 K (25° C) and a pressure of 101.3 kPa (1 atm) is assigned a value of zero. The free energy of a compound is defined as the change in free energy that is associated with the formation of that compound from pure elements. The **free energy change (Δ*G*)** for a chemical reaction is equal to the free energy of products ($G_{products}$) minus the free energy of reactants ($G_{reactants}$):

$$\Delta G = G_{products} - G_{reactants}$$

The Δ*G* for a reaction is a measure of the theoretical maximum amount of work that the reaction can accomplish at constant temperature and pressure. Most of the mechanical work, gradient work, and synthetic work performed by living organisms (previous section) is driven by chemical reactions in which products possess substantially lower free energy than reactants and Δ*G* is relatively large and negative.

The detonation of fireworks initiates exergonic chemical reactions. Part of the chemical energy is converted to visible light.

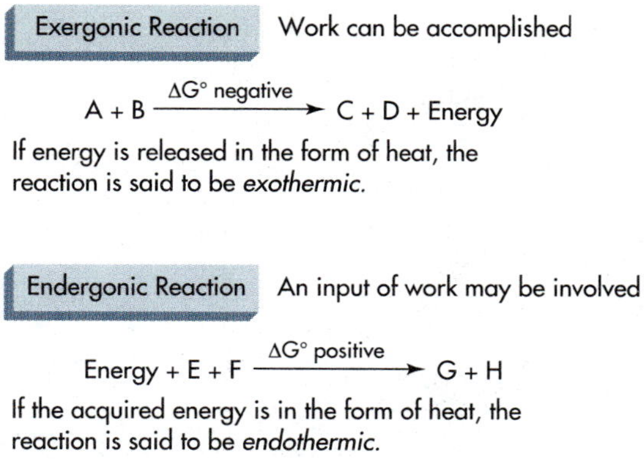

Exergonic Reaction   Work can be accomplished

$$A + B \xrightarrow{\Delta G^\circ \text{ negative}} C + D + \text{Energy}$$

If energy is released in the form of heat, the reaction is said to be *exothermic.*

Endergonic Reaction   An input of work may be involved

$$\text{Energy} + E + F \xrightarrow{\Delta G^\circ \text{ positive}} G + H$$

If the acquired energy is in the form of heat, the reaction is said to be *endothermic.*

FIGURE 8.2
Exergonic and endergonic reactions.

Standard Free Energy Change

The ΔG for a reaction at 298 K when all reactants and products are present at 1 *M* concentration or at partial pressure of 101.3 kPa, if gases, is called its **standard free energy change** and is abbreviated $\mathbf{\Delta G^\circ}$**.** Although 298 K is most often selected as the standard temperature, ΔG° can be calculated at any selected "standard" temperature. All ΔG° values given in this text have been calculated at 298 K. ΔG° is primarily used by chemists and physicists.

Biochemists tend to work with **biological standard free energy change (**$\mathbf{\Delta G^{\circ\prime}}$**)** which is defined as the free energy change for a reaction under standard biological conditions: 298 K (normally), a partial pressure of 101.3 kPa for gaseous reactants and products, 10^{-7} *M* H^+ (pH 7), 55.5 *M* H_2O and a 1 *M* concentration of all other nongaseous reactants and products. Standard biological conditions are internationally agreed upon reference points. They do not represent conditions that are "standard" within living organisms. A reaction is described as **exergonic** if free energy is released ($\Delta G^{\circ\prime}$ is negative and work can be accomplished) and as **endergonic** when free energy is taken up ($\Delta G^{\circ\prime}$ is positive and an input of work may be involved; no work can be accomplished) (Figure 8.2). **[Try Problems 8.5 and 8.6]**

The Equilibrium Concept

Before we examine the relationship between ΔG and chemical equilibrium, a brief review of equilibrium is in order. Consider the generic reaction:

$$aA + bB \rightleftharpoons cC + dD$$

The capital letters correspond to reactants or products and the lower case letters represent coefficients in a balanced equation. Products C and D begin to accumulate in a reaction mixture immediately after A and B are mixed together (assuming that the rate of the reaction is greater than zero). As C and D accumulate, they react together to regenerate A and B (the reverse reaction) (Figure 8.3). Given enough time, the rate of the reverse reaction will eventually equal the rate of the forward reaction. At this point, the reaction is **at equilibrium,** and the reactants and products will remain at their equilibrium concentrations unless the equilibrium is disturbed. The equilibrium

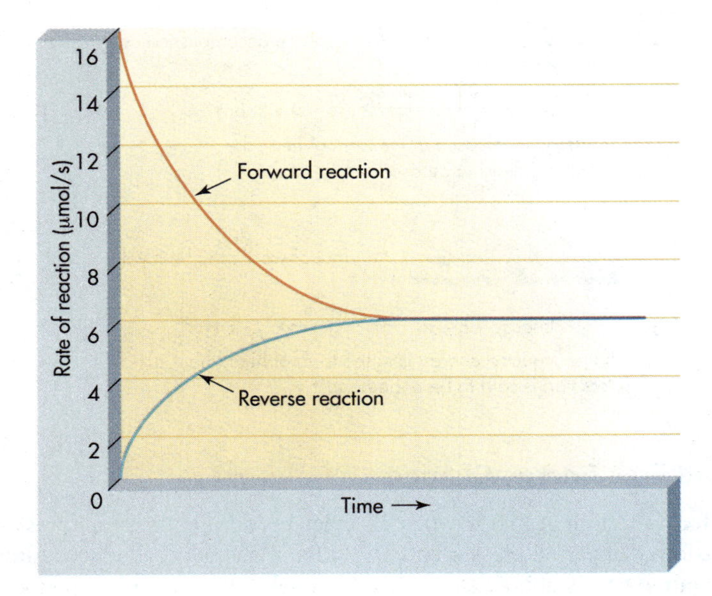

FIGURE 8.3
Changes in reaction rates during the establishment of equilibrium.

concentration of reactants is not necessarily equal to the equilibrium concentration of products, and although there is no net reaction, both the forward and reverse reactions are still occurring.

All of the reactions in a living organism are constantly moving toward equilibrium. Few ever reach this thermodynamic goal, since the reactants and products in one reaction are usually reactants or products in one or more additional reactions. As C and D are produced from A and B, for example, they may be immediately transformed into E and F, and the reaction of A with B may not be the only reaction that is generating some of compound C. There is some truth to the old saying, "Old biochemists never die; they just reach equilibrium."

Equilibrium Constants and $\Delta G°$

The equilibrium constant for a reaction can be calculated by using equilibrium concentrations (in molarity) for reactants and products and the equilibrium constant expression:

$$\textbf{Equilibrium constant} = K_{\textbf{eq}} = \frac{[\mathrm{C}]^{\mathrm{c}}[\mathrm{D}]^{\mathrm{d}}}{[\mathrm{A}]^{\mathrm{a}}[\mathrm{B}]^{\mathrm{b}}}$$

The larger the equilibrium constant, the greater the concentration of products relative to reactants at equilibrium. If a K_{eq} is very large, a reaction is commonly said to go to completion because an insignificant amount of reactants remains at equilibrium. A very small K_{eq} denotes a very low concentration of products relative to reactants at equilibrium and identifies a reaction that does not spontaneously proceed (in the forward direction) to a significant extent. **[Try Problem 8.7]**

EXHIBIT 8.2
Some Useful Relationships

1 cal = 4.184 J
R = 8.315 J/mol · K
8.315 J/mol · K = 1.987 cal/mol · K
K = °C + 273
25°C = 298 K
At 298 K, RT = 2.479 kJ/mol
2.479 kJ/mol = 0.592 kcal/mol
ln X = 2.303 log X
1 atm = 101.3 kPa

The $\Delta G°$ for a reaction determines its equilibrium constant:

$$\Delta G° = -RT \ln K_{eq}$$

where R is the gas constant = 8.315 J/mol · K; T is the absolute temperature (K); and K_{eq} is the equilibrium constant. If we choose to work with common logarithms (base 10), rather than natural logarithms (base e, an irrational number equal to approximately 2.7183), this equation becomes:

$$\Delta G° = -2.3RT \log K_{eq}$$

The natural log to common log relationship is ln X = 2.303 log X. Exhibit 8.2 is a tabulation of this and other useful conversion factors. In this text, *you should assume that the temperature for a reaction is 298 K (25°C) unless specified otherwise.*

A negative $\Delta G°$ indicates that under standard conditions the products are more stable (possess less free energy) than the reactants and that the equilibrium constant for the reaction is greater than one. The reaction is said to be **spontaneous** because there is a spontaneous net conversion of reactants to products under standard conditions. The natural tendency for the reaction to move toward equilibrium provides a **positive driving force** for the reaction. A positive $\Delta G°$ denotes that products have more free energy (are less stable) than reactants and that the equilibrium constant is less than one. There is a spontaneous net conversion of products to reactants (under standard conditions), and the driving force (for the forward reaction) is said to be negative. If $\Delta G° = 0$, a reaction is at equilibrium under standard conditions and $K_{eq} = 1$. **[Try Problem 8.8]**

The $\Delta G°$ for a reaction plays absolutely no role in determining its rate, a key point to remember. The rate of a reaction is determined by activation energy, reactant concentrations, temperature, and other variables (Section 5.2). Table 8.2 at the top of the next page summarizes the interpretation of $\Delta G°$ values. **[Try Problem 8.9]**

The relationship between $\Delta G°'$ and K_{eq}' is exactly the same as that between $\Delta G°$ and K_{eq}:

$$\Delta G°' = -\text{RT} \ln K_{eq}'$$

TABLE 8.2

Interpretation of $\Delta G°$

			Under Standard Conditions[a]				
$\Delta G°$	K_{eq}	Rate of Reaction	Positive Driving Force for Forward Reaction	Spontaneous Forward Reaction	Net Forward Reaction to Reach Equilibrium	Net Reverse Reaction to Reach Equilibrium	At Equilibrium
+	<1	Unpredictable	No	No	No	Yes	No
0	1	Unpredictable	No	No	No	No	Yes
−	>1	Unpredictable	Yes	Yes	Yes	No	No

[a] 298 K (usually), 1 *M* concentration of each nongaseous reactant and product, and a partial pressure of 101.3 kPa for each gaseous reactant or product.

In calculating a K_{eq}', the concentrations of H^+ and H_2O are assumed to be constant at 10^{-7} *M* and 55.5 *M*, respectively, and the concentrations of these substances are normally incorporated into the numerical value of K_{eq}'. (Technically, the activity of 10^{-7} *M* H^+ and the activity of 55.5 *M* H_2O are both defined as one.) For this reason, $[H^+]$ and $[H_2O]$ do not usually appear in K_{eq}' expressions even when they are reactants or products in a reaction. The K_{eq}' expression for the reaction ATP + $H_2O \rightleftharpoons$ ADP + P_i, for example, is written:

$$K_{eq}' = \frac{[\text{ADP}][\text{P}_i]}{[\text{ATP}]} \left(\text{instead of } \frac{[\text{ADP}][\text{P}_i]}{[\text{ATP}][\text{H}_2\text{O}]} \right)$$

Unless specified otherwise, any future mention of an equilibrium constant in this text refers to K_{eq}'. **[Try Problem 8.10]**

The $\Delta G°'$ for a reaction is a measure of its K_{eq}' in much the same way that pH is a measure of $[H^+]$. Since the relationship is logarithmic, a relatively small change in $\Delta G°'$ corresponds to a relatively large change in K_{eq}' (Table 8.3).

TABLE 8.3

The Relationship Between K_{eq}' and $\Delta G°'$ at 298 *K*

K_{eq}'	$\Delta G°'$ (kJ/mol)	$\Delta G°'$ (kJ/mol)	K_{eq}'
0.0001	22.8	40	9.6×10^{-8}
0.001	17.1	30	5.4×10^{-6}
0.01	11.4	20	3.1×10^{-4}
0.1	5.7	10	1.8×10^{-2}
1.0	0.0	0.0	1.0
10	−5.7	−10	5.7×10^{1}
100	−11.4	−20	3.2×10^{3}
1,000	−17.1	−30	1.8×10^{5}
10,000	−22.8	−40	1.0×10^{7}

$$\Delta G°' = -RT \ln K_{eq}'$$

Free Energy Change Under Physiological Conditions

Actual physiological conditions normally differ from the standard biological conditions described on page 293. How do we calculate the free energy change for a reaction under physiological conditions, and what is its significance? Once again, consider the generic reaction:

$$aA + bB \rightleftharpoons cC + dD$$

When the concentrations of A, B, C, and D are those that exist under physiological conditions, the free energy change ($\Delta G_{physiological}$, abbreviated ΔG_p) is:

$$\Delta G_p = \Delta G^{\circ\prime} + RT \ln \frac{[C]^c[D]^d}{[A]^a[B]^b}$$

In this case, $[C]^c[D]^d/[A]^a[B]^b = K_{eq}'$ if, and only if, the reaction is at equilibrium under physiological conditions. It is important to remember that the concentrations of reactions and products in the equilibrium constant expression are equilibrium concentrations.

When a reaction is at equilibrium under physiological conditions, ΔG_p equals zero and $\Delta G^{\circ\prime} = -RT \ln ([C]^c[D]^d/[A]^a[B]^b)$. There is no driving force for the reaction, and no work can be extracted from the reaction. A negative ΔG_p means that, under physiological conditions, the reaction must move from left to right (corresponds to a net forward reaction) to reach equilibrium. The more negative the ΔG_p, the greater the driving force for this movement, the further the reaction is from equilibrium, and the greater the amount of work that can be performed as the reaction moves to equilibrium. A positive ΔG_p means that, under physiological conditions, the driving force is in the reverse direction; there must be a net conversion of C and D into A and B to reach equilibrium, and a net forward reaction requires an input of energy. **[Try Problems 8.11 and 8.12]**

Most biochemistry texts, including this one, present the $\Delta G^{\circ\prime}$ for a reaction rather than the more meaningful (from a physiological standpoint) ΔG_p. The reason is straightforward. It is extremely difficult to determine ΔG_p because it is extremely difficult to ascertain the usual reactant and product concentrations within the intracellular environment where biochemical reactions take place. Reactant and product concentrations are variable, even at a given site within a single cell, so ΔG_p values are not constant; they tend to fluctuate within certain normal ranges. Whereas $\Delta G^{\circ\prime}$ values can be calculated with certainty, *all ΔG_p values should be considered estimates.*

***PROBLEM 8.5** Calculate the molar concentration of water in a pure sample of water. Water has a density of 1 g/mL.

PROBLEM 8.6 Explain why $\Delta G^\circ = \Delta G^{\circ\prime}$ for reactions that involve no water and no H^+.

PROBLEM 8.7 Ethanol and acetic acid react to generate ethyl acetate and water.

$$\text{Ethanol} + \text{acetic acid} \rightleftharpoons \text{ethyl acetate} + \text{water}$$

Assume that in a reaction mixture at equilibrium, [ethanol] = 2 *M*, [acetic acid] = 2 *M*, [ethyl acetate] = 0.4 *M*, and [water] = 5 *M*. Calculate the K_{eq} (not K_{eq}') for the formation of ethyl acetate under the conditions that exist in this reaction mixture.

PROBLEM 8.8 Will the addition of more product to a reaction mixture at equilibrium increase, decrease, or have no effect on its K_{eq}? Its $\Delta G°$? Explain.

PROBLEM 8.9 Does a catalyst increase, decrease, or have no effect on the $\Delta G°$ for a reaction? Explain. Hint: Review Section 5.2.

PROBLEM 8.10 Explain why $[H^+]$ and $[H_2O]$ within many regions of a cell tend to remain constant even when reactions are occurring that produce or consume these substances.

PROBLEM 8.11 True or false? ΔG_p is always negative when $\Delta G°'$ is negative. Explain.

PROBLEM 8.12 Calculate the ΔG_p for the reaction ATP + $H_2O \rightleftharpoons$ ADP + P_i. Assume: $\Delta G°' = -31.0$ kJ/mol; 25° C; pH = 7; and intracellular concentrations of ATP, ADP, and P_i are 2.35 m*M*, 0.200 m*M*, and 1.60 m*M*, respectively. Will a net forward reaction or a net reverse reaction be required to reach equilibrium? Explain.

8.3 ENTROPY AND THE SECOND LAW OF THERMODYNAMICS

The ΔG for a reaction is determined by the enthalpy and the entropy content of reactants and products. **Enthalpy** (abbreviated ***H***) is a measure of heat content, whereas entropy (abbreviated *S*) is a quantitative measure of the randomness or disorder in a system. Any chemical or physical process that leads to a more random distribution of particles or to an increase in temperature at constant pressure increases entropy. Elevated temperatures lead to "thermal disorder" by boosting atoms and molecules to higher energy levels, where there are more options for the distribution of the energy.

At constant temperature and pressure, the relationship among free energy, enthalpy, and entropy can be expressed as:

$$\Delta G = \Delta H - T\Delta S$$

where T is the absolute temperature (K), $\Delta H = H_{products} - H_{reactants}$, and $\Delta S = S_{products} - S_{reactants}$. ΔH normally has units of J/mol, and ΔS is expressed as J/mol · K. A negative ΔH means that products contain less enthalpy than reactants and the reaction is **exothermic** (heat given off) (Table 8.4). A positive ΔH is associated with **endothermic** (heat taken in) reactions. Similarly, a negative ΔS denotes less random product, whereas a positive ΔS denotes an increase in randomness. ΔG shifts in a

TABLE 8.4

Interpretation of ΔH and ΔS

Variable	Value	Interpretation
ΔH	Positive	Reaction is endothermic
	Negative[a]	Reaction is exothermic
	Zero	Heat content of reactant and products is the same
ΔS	Positive[a]	Products are more random (disordered) than reactants
	Negative	Reactants are more random (disordered) than products
	Zero	Extent of randomness (disorder) is the same in reactants and products

[a]Favors the forward reaction because it makes a negative contribution to ΔG.

negative direction (it may still be positive if originally positive) as ΔH becomes more negative or ΔS becomes more positive; a loss of enthalpy or an increase in entropy favors the forward reaction. When all reactants and products are in their standard state, $\Delta G° = \Delta H° - T\Delta S°$. **[Try Problems 8.13 and 8.14]**

The **second law of thermodynamics** states that the entropy of the universe increases during all chemical and physical changes. This does not necessarily mean that ΔS is positive for all chemical reactions, since part of the universe can become less random (lose entropy) as long the entropy of the entire universe increases. During a chemical reaction, the universe is the reaction plus its surroundings. Although life appears to involve the creation of order out of disorder, there is a net boost in randomness when we consider the universe as a whole. A living organism loses entropy at the expense of its environment.

Although the human body as a whole is more ordered than its environment, numerous life-sustaining processes are partly driven by an associated increase in entropy. Many of these processes, including protein folding, the binding of ligands to some proteins, lipid bilayer formation in membranes, and double-helix formation in nucleic acids, involve the aggregation of nonpolar groups (like associates with like) and the creation of hydrophobic forces. Hydrophobic forces are a consequence of the increase in entropy acquired as nonpolar groups are excluded from a polar aqueous environment (Section 2.2). When nonpolar groups are mixed with water molecules, the randomness with which water molecules interact with each other is reduced. **[Try Problem 8.15]**

The decrease in entropy associated with the assembly of a living organism is brought about through an input of energy from the environment. For most organisms, including humans, the ultimate source of this energy is sunlight (Chapter 15). The immediate source of the energy used in most entropy-reducing processes is ATP, an energy-rich compound produced during photosynthesis and the oxidation of organic compounds. The next section examines ATP and the molecular basis for its relatively high free energy content. Subsequent chapters explore ATP production and some of the many life-supporting processes that rely upon energy from ATP to provide order (reduce entropy).

PROBLEM 8.13 Under what enthalpy and entropy conditions does $\Delta G^\circ = 0$? What is the value of K_{eq} when $\Delta G^\circ = 0$?

PROBLEM 8.14 True or false? At a high enough temperature, all reactions with a positive ΔS (increase in entropy) will become spontaneous. Assume that ΔS and ΔH are independent of temperature. Explain.

PROBLEM 8.15 Will the ΔS for a dissociation reaction at constant temperature and pressure be positive, negative, or zero? Explain.

8.4 ATP: A HIGH-ENERGY COMPOUND

Previous sections have examined ΔG° and the relationship between ΔG° and K_{eq}. In the process, they have emphasized that the thermodynamic driving force for a reaction is determined by its K_{eq} and how far the reaction is from equilibrium. Organic chemists use this knowledge when designing synthetic procedures. Consider, for example, the synthesis of a carboxylic acid ester. Although such an ester can be produced by reacting an alcohol with a carboxylic acid, the driving force for this reaction is relatively small and the percentage yield is relatively low. Organic chemists normally overcome these problems by using an acid chloride or acid anhydride in place of the carboxylic acid. Since acid chlorides and anhydrides contain more free energy than carboxylic acids, these "activated acid derivatives" lead to a more neg-

TABLE 8.5

Effect of High-Energy Reactants on ΔG° and K_{eq}

Reaction[a]	ΔG°	K_{eq}
$R{-}C(=O){-}OH + HO{-}R' \rightleftharpoons R{-}C(=O){-}O{-}R' + HOH$ Carboxylic acid + alcohol ⇌ ester + H_2O	Usually close to zero	Usually close to one
* $R{-}C(=O){-}O{-}C(=O){-}R + HO{-}R' \rightleftharpoons R{-}C(=O){-}O{-}R' + HO{-}C(=O){-}R$ Anhydride + alcohol ⇌ ester + carboxylic acid	Significantly negative	Large
* $R{-}C(=O){-}Cl + HO{-}R' \longrightarrow R{-}C(=O){-}O{-}R' + HCl$ Acid chloride + alcohol → ester + HCl	Even more negative	Larger

[a]An * indicates a high-energy compound.

ative $\Delta G°$ (larger K_{eq}) for ester formation (Table 8.5). Acid chlorides and acid anhydrides are examples of high-energy compounds. Living organisms employ the same strategy as organic chemists; they use high-energy compounds to drive reactions. **[Try Problem 8.16]**

High-Energy Compounds

In biochemistry, a **high-energy compound (energy-rich compound)** is a compound containing one or more bonds whose $\Delta G°'$ for hydrolysis is more negative than −25 kJ/mol. A bond whose hydrolysis leads to the release of such a large amount of energy is commonly called a **"high-energy bond."** Because the bond itself does not possess the energy released, *this terminology is misleading and should be abandoned.* The source of the released energy can usually be traced to some destabilizing structural feature within reactants or to the stabilization of products through resonance, ionization, isomerization, or solvation. Several of the most common biologically important high-energy compounds are examined in this section and the section that follows.

Nucleoside Triphosphates

ATP is the most widely distributed and widely utilized high-energy compound within the human body. Since ATP is coupled to so many energy-requiring processes, including muscle action and brain action, it is often described as the "energy currency" of cells and organisms. In heterotrophs, a majority of ATP is produced from ADP and P_i using energy released during the oxidation of organic fuel molecules (Sections 11.8 and 11.9). Like all nucleoside triphosphates (NTP's), ATP contains a phosphate ester bond and two phosphate anhydride bonds (Figure 8.4). The approximate $\Delta G°'$ for the hydrolysis of each of these follows:

$$\text{NTP} + H_2O \rightleftharpoons \text{NDP} + P_i \qquad \Delta G°' = -31 \text{ kJ/mol}$$

$$\text{NTP} + H_2O \rightleftharpoons \text{NMP} + PP_i \qquad \Delta G°' = -32 \text{ kJ/mol}$$

$$\text{NDP} + H_2O \rightleftharpoons \text{NMP} + P_i \qquad \Delta G°' = -31 \text{ kJ/mol}$$

$$\text{NMP} + H_2O \rightleftharpoons \text{N} + P_i \qquad \Delta G°' = -14 \text{ kJ/mol}$$

The $\Delta G°'$ for hydrolysis is considerably less negative for a typical phosphate ester bond than for a phosphate anhydride bond. The $\Delta G°'$ for the hydrolysis of most phosphate anhydride bonds is roughly the same, around −31 kJ/mol.

ATP is the immediate source of energy for muscle action, brain action, and numerous other life-sustaining processes.

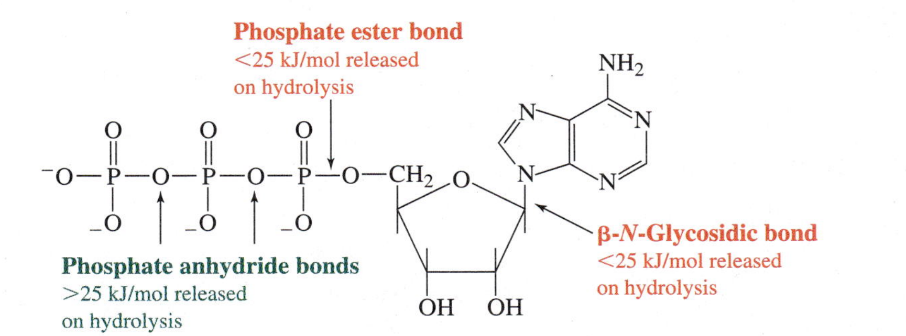

FIGURE 8.4
The structure of ATP.

Under physiological conditions, the free energy of hydrolysis for ATP is usually closer to −50 kJ/mol than the −31 kJ/mol observed under standard biological conditions. The difference is largely explained by the reactant and product concentrations that exist within living organisms. $\Delta G^{\circ\prime}$ calculations assume that all reactant and product concentrations are 1 *M* (except $[H^+]$ and $[H_2O]$), whereas ΔG_p calculations are based on actual intracellular concentrations (Section 8.2). The consequences are illustrated in Problem 8.12 where the concentrations provided for ATP, ADP, and P_i are similar to those found within many cells. The precise value for ΔG_p tends to vary from cell to cell, and it even fluctuates within individual cells as reactant and product concentrations shift in response to changes in cellular activity and metabolic needs. **[Try Problem 8.17]**

The ΔG_p for ATP hydrolysis is also a function of pH, Mg^{2+} concentration, and other variables. The pH effect is explained by shifts in the extent of dissociation of the hydroxyl groups in phosphate residues. As the pH rises, net negative charge increases and leads to an increase in like-charge repulsion in phosphate anhydrides. This destabilizes anhydrides relative to their hydrolysis products, and ΔG_p becomes more negative. The importance of like-charge repulsion is examined in more detail in the paragraphs that follow. Mg^{2+}, by complexing with phosphate anhydrides and shielding negative charges from one another, stabilizes these anhydrides relative to their hydrolysis products (Figure 8.5). Consequently, the free energy of hydrolysis of the Mg^{2+}:ATP complex is less negative than that for the hydrolysis of free ATP.

The Structural Basis for Free Energy Differences

The large negative $\Delta G^{\circ\prime}$ for the hydrolysis of phosphate anhydride bonds indicates that the reactants contain substantially more free energy than products. This energy

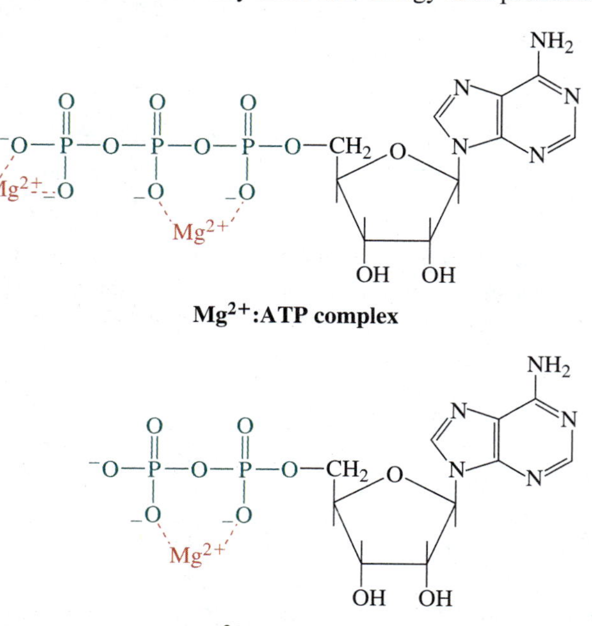

FIGURE 8.5
Mg^{2+}:phosphate anhydride complexes.

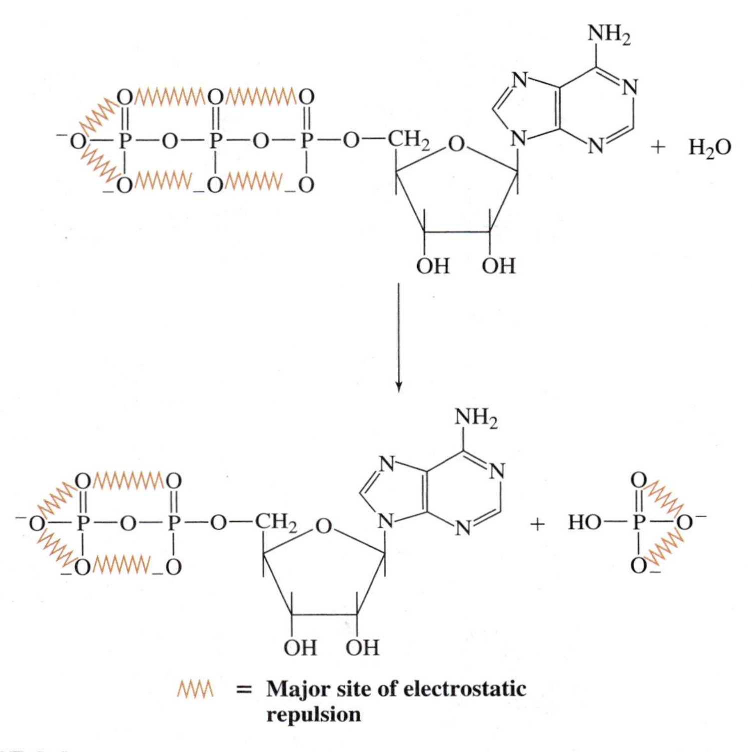

FIGURE 8.6
Some of the major sites of electrostatic repulsion in ATP and its hydrolysis products.

difference is explained by a greater like-charge repulsion in reactants (relative to products) and the stabilization of products resulting from resonance, ionization, and solvation. ATP hydrolysis will be used to examine these phenomena.

Three of the four hydroxyl groups in the triphosphate moiety of ATP are almost completely dissociated at pH 7. The fourth hydroxyl group exists predominantly in a dissociated form as well. The clustering of negative charges that results leads to like-charge repulsion that destabilizes ATP and makes a significant contribution to its free energy content. A reduction in the number of clustered negative charges from roughly −4 (in ATP) to −3 (in ADP) helps account for the release of energy as ATP is hydrolyzed to ADP plus P_i (Figure 8.6). Similar reductions in charge clustering are associated with the hydrolysis of ATP to yield AMP + PP_i. **[Try Problem 8.18]**

The resonance stabilization of the products of ATP hydrolysis also contributes to the large free energy difference between reactants and products. Before this claim is examined in detail, it is appropriate to review the general concept of resonance. When two or more valid Lewis (electron dot) structures can be drawn that differ only in the placement of electrons, a compound will have some characteristics of all these structures. The individual structures are called **resonance forms (structures)** and the compound is said to exist as a **resonance hybrid** of all its resonance forms (Figure 8.7). The resonance hybrid possesses less free energy (is more stable) than its resonance forms. The precise free energy difference constitutes the **resonance en-**

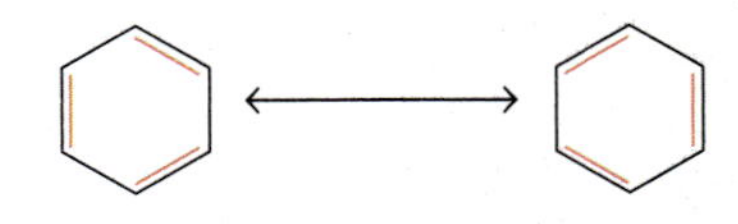

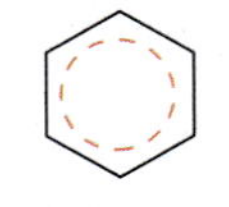

FIGURE 8.7
Resonance in benzene. The benzene ring does not contain alternating double and single bonds. Each bond in the ring has partial double-bond and partial single-bond character. Benzene is more stable than either of its resonance forms.

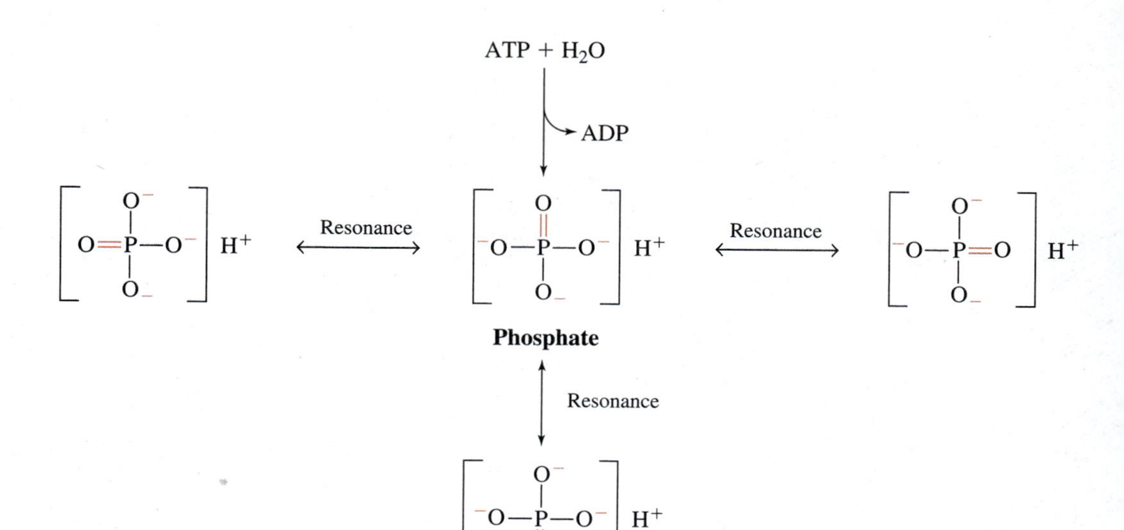

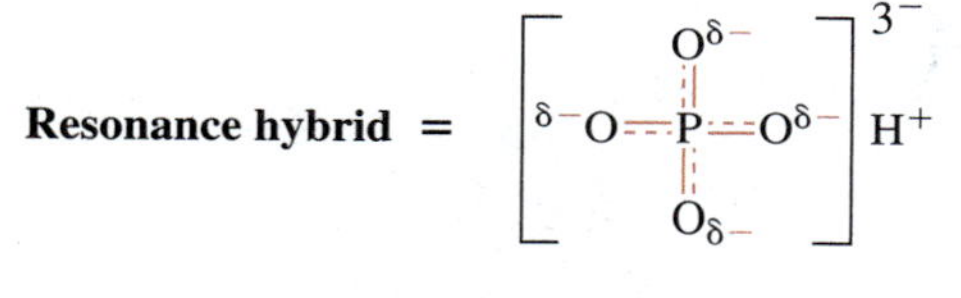

Note: δ^- denotes a partial negative charge (Section 2.1)

FIGURE 8.8
The resonance stabilization of phosphate.

ergy for the compound, and the compound is said to be **resonance stabilized.** In Chapter 3, resonance was used to help explain the restricted rotation that exists about peptide bonds (Section 3.5). Problems 8.19 and 3.27 provide a further review of resonance and related topics. **[Try Problem 8.19]**

At pH 7, a free phosphate group can exist in more resonance forms than a phosphate group tied up in an ester or anhydride link (Figure 8.8). As a consequence, the release of a phosphate through the hydrolysis of a phosphate ester or anhydride leads to an increase in resonance energy and to a release of free energy. The increase in resonance possibilities is associated with the dissociation of H^+ from the hydroxyl group within the phosphate created during hydrolysis. Although the dissociation constant is very small ($pK_a = 12$), the dissociation still opens the door for additional resonance forms.

The $\Delta G^{\circ\prime}$ for the hydrolysis of ATP is also impacted by the dissociation of an H^+ from the ADP produced. This dissociation, which leads to an increase in both resonance energy and entropy, is normally considered part of the overall hydrolysis process:

$$ATP^{4-} + H_2O \rightleftharpoons ADP^{3-} + P_i^{2-} + H^+ \qquad \Delta G^{\circ\prime} = -31 \text{ kJ/mol}$$

Under physiological conditions, buffers tend to pull the reaction toward completion by removing part of the H^+ (LeChâtelier's principle). **[Try Problem 8.20]**

An additional variable influencing free energies of hydrolysis is the interaction between solvent and solutes. The **solvation energy** for a compound is the energy released (stability acquired) as a consequence of noncovalent interactions between the compound and solvent molecules (usually water in biological systems). The solvation energy is lower for ATP than it is for the hydrolysis products of ATP. Since solvation energy is difficult to measure, particularly under physiological conditions, the precise contribution of solvation energy to the $\Delta G°'$ for the hydrolysis of ATP is uncertain.

***PROBLEM 8.16** Write the equation for the formation of methyl acetate from methanol and (a) acetic acid, (b) acetic anhydride, and (c) acetyl chloride. In each case, show structural formulas for all reactants and products. Which reaction has the smallest K_{eq}?

PROBLEM 8.17 Will the ΔG_p for ATP hydrolysis (to ADP + P_i) increase, decrease, or remain the same under conditions where the ADP to ATP ratio increases as a consequence of rapid ATP hydrolysis? Explain.

PROBLEM 8.18 At very low pH, the ΔG for ATP hydrolysis becomes positive. Explain this fact.

PROBLEM 8.19 The acetate ion (CH_3COO^-) exhibits resonance. Write structural formulas depicting its two resonance forms and its resonance hybrid.

PROBLEM 8.20 Write formulas for the resonance forms of the terminal phosphate residue in the fully dissociated form of ATP.

8.5 OTHER HIGH-ENERGY COMPOUNDS

Although ATP is the "energy currency" of cells, a variety of other high-energy compounds participate in the energy transformations that maintain life. Nucleoside triphosphates containing bases other than adenine provide specific examples. This section examines additional examples.

Mixed Anhydrides

Carboxylate-phosphate mixed anhydrides, like phosphate anhydrides, are biologically important, high-energy compounds. Although a reduction in like-charge repulsion plays a relatively minor role in driving their hydrolysis, the resonance stabilization of products makes a significant contribution to the $\Delta G°'$ for hydrolysis (see Figure 8.9 at the top of the next page). Once again, the dissociation of H^+ from prod-

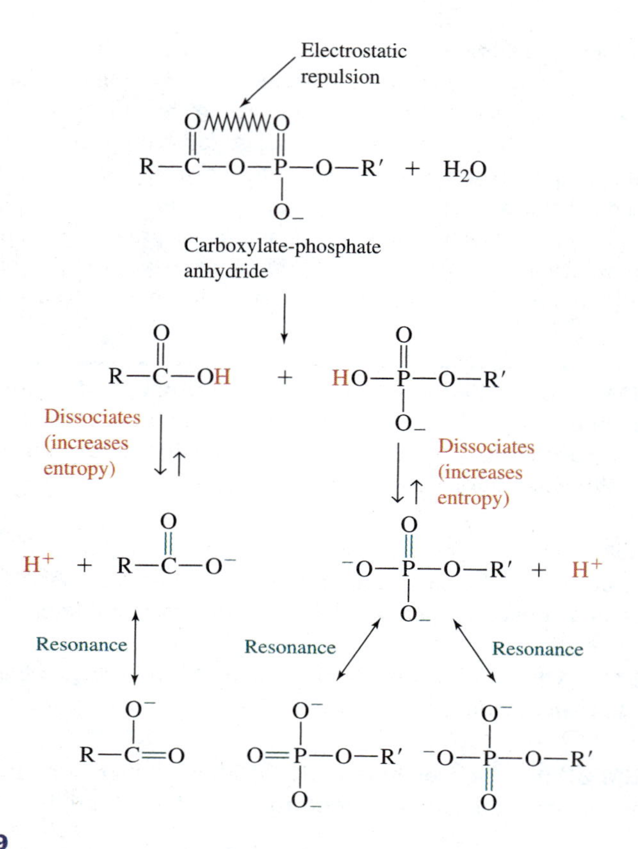

FIGURE 8.9

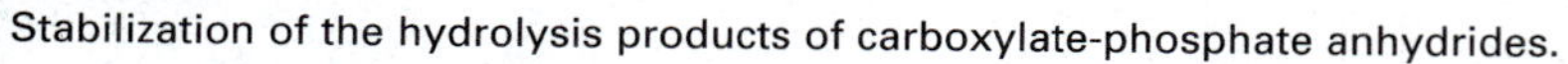

Stabilization of the hydrolysis products of carboxylate-phosphate anhydrides.

ucts (increases entropy) is considered part of the hydrolysis process. Solvation energies may also contribute to the observed free energy differences.

The most thoroughly studied mixed anhydride is probably 1,3-bisphosphoglycerate, an intermediate in the breakdown of glucose into two pyruvates (a process called glycolysis). During glycolysis, part of the free energy within 1,3-bisphosphoglycer-

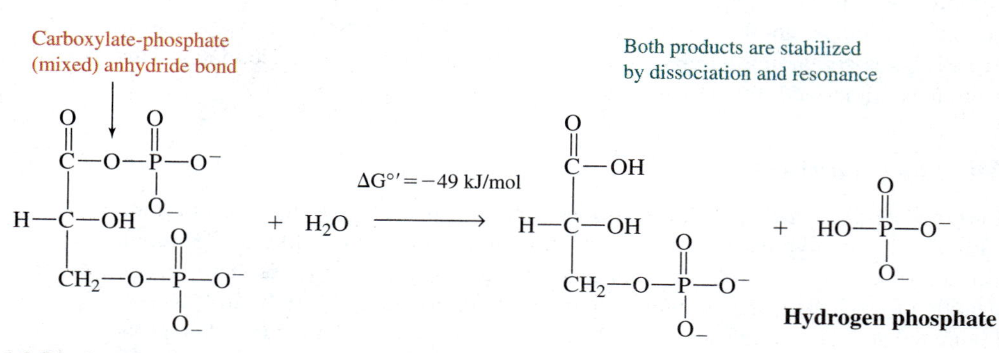

ate is transformed into free energy within ATP as a phosphoryl group is transferred to ADP (Section 14.2):

$$\text{1,3-Bisphosphoglycerate} + \text{ADP} \rightleftharpoons \text{3-phosphoglycerate} + \text{ATP}$$
$$\Delta G^{\circ\prime} = -19 \text{ kJ/mol}$$

The free energy in the 1,3-bisphosphoglycerate provides a thermodynamic driving force for ATP production.

Enoyl Phosphates

Phosphoenolpyruvate (PEP), another intermediate in glycolysis, is an example of an **enoyl phosphate,** a third class of high-energy compound. The large negative $\Delta G^{\circ\prime}$ for the hydrolysis of this unusual phosphate ester is, for the most part, explained by the resonance stabilization of the released phosphate plus the spontaneous conversion of the enol form of a ketone (the basis for the term "enoyl phosphate") to a more stable keto form (Figure 8.10). This tautomerization, which is considered part of the overall reaction, makes a negative contribution to the $\Delta G^{\circ\prime}$.

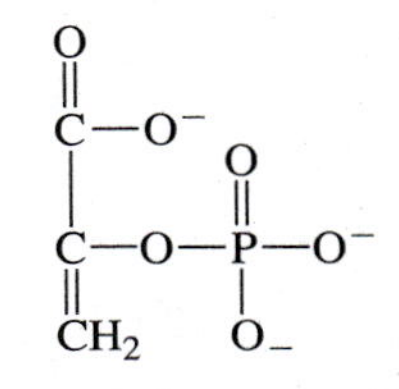

Phosphoenolpyruvate (PEP)

Tautomers are compounds that exist in equilibrium and differ in the location of a single H and a single double bond. Tautomerism was used in Section 3.5 (along with resonance and delocalized π bonding) to help explain the planar nature of peptide bonds. It was also encountered in Section 7.1 during the discussion of the purines and pyrimidines found within nucleotides and nucleic acids. Figure 7.2 summarizes some of the biologically important functional groups that exhibit tautomerism. **[Try Problem 8.21]**

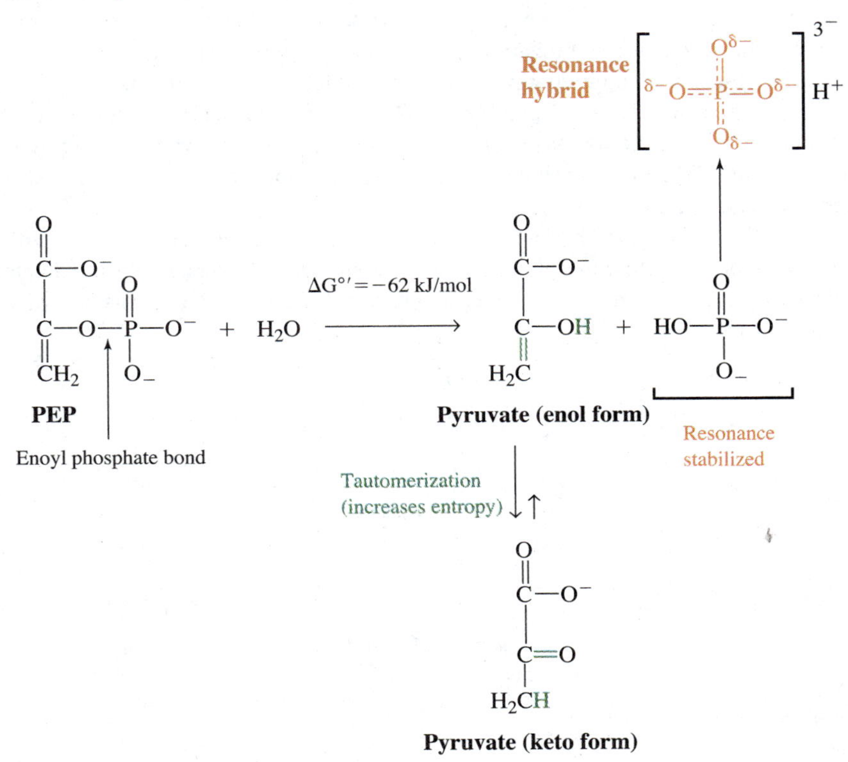

FIGURE 8.10
The hydrolysis of PEP.

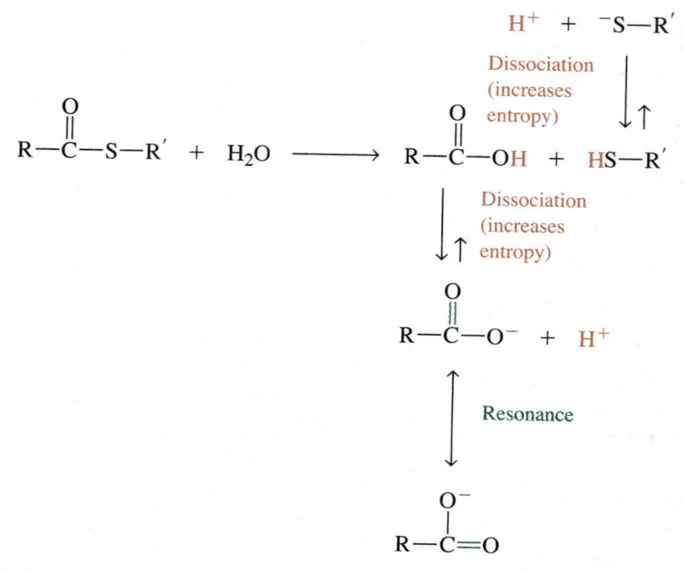

FIGURE 8.11
Stabilization of the hydrolysis products of thioesters.

Thioesters

Thioesters are a fourth class of physiologically important high-energy compounds. Their large, negative $\Delta G^{\circ\prime}$ for hydrolysis is primarily explained by product dissociation and the resonance stabilization of a product (Figure 8.11). Oxygen esters do not usually qualify as high energy compounds, since the esters themselves exhibit some resonance stabilization (Figure 8.12). This resonance stabilization reduces the free energy difference between oxygen esters and their hydrolysis products. Thioesters fail to display similar resonance stabilization because the larger sulfur atom less readily participates in resonance.

Acetyl-SCoA is one of the most common and widely distributed thioesters within living organisms, including humans. It will be encountered repeatedly in Chapters 11 through 14, four of the chapters devoted to metabolism. Its hydrolysis is depicted in Figure 8.13.

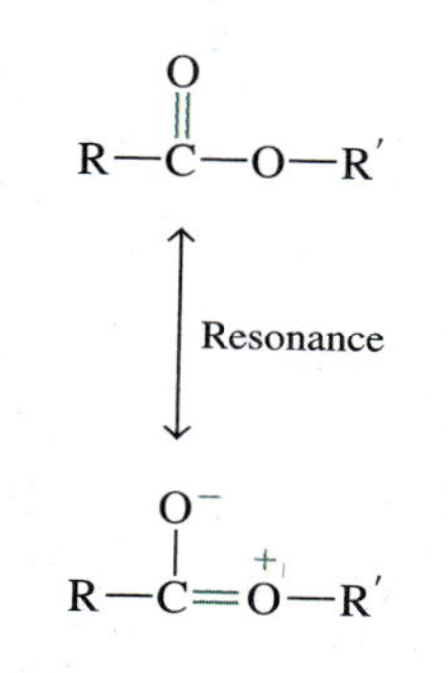

FIGURE 8.12
Resonance in oxygen esters.

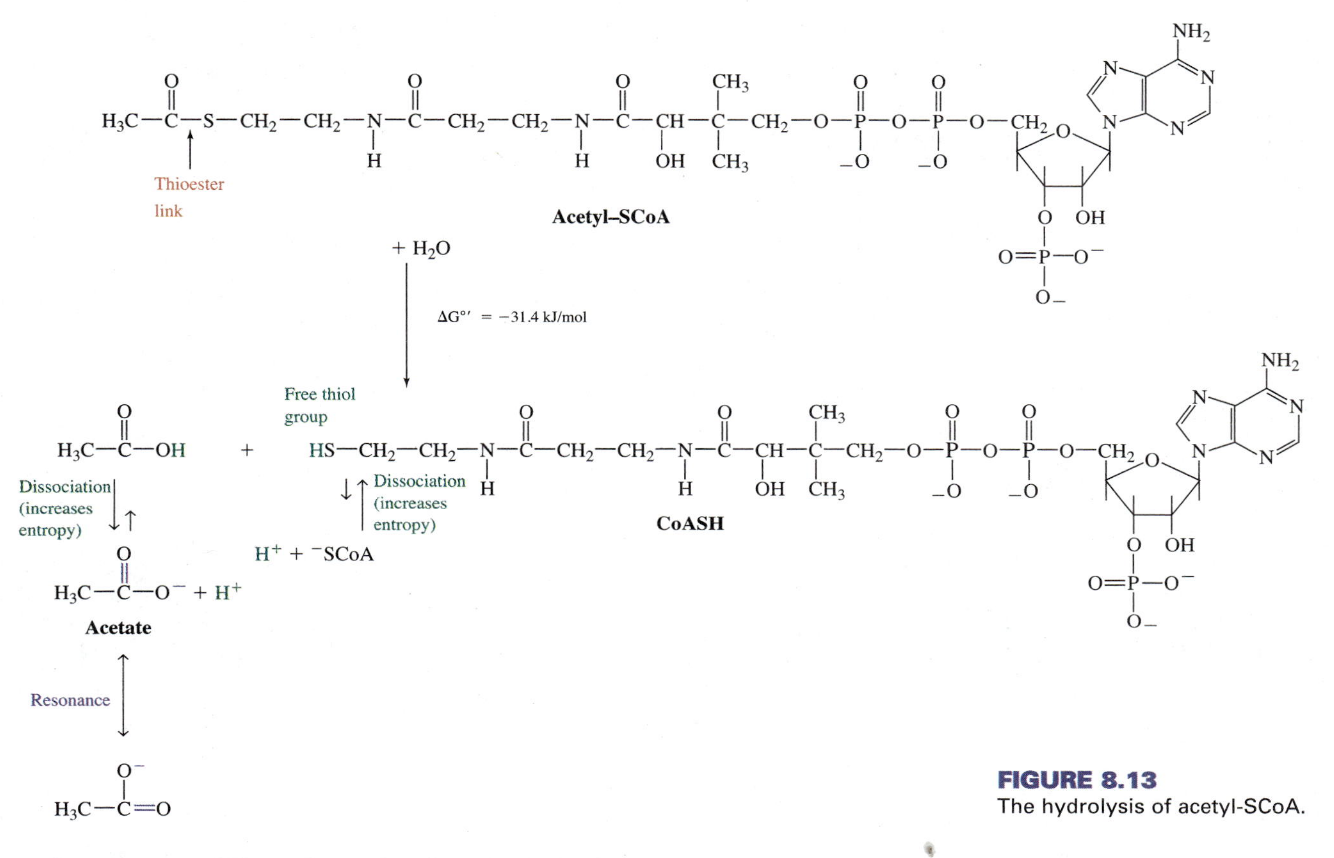

FIGURE 8.13
The hydrolysis of acetyl-SCoA.

A Summary of Free Energies for Hydrolysis

Table 8.6 summarizes the approximate $\Delta G^{\circ\prime}$ for those classes of high-energy compounds encountered most often in this text, whereas Table 8.7 provides the $\Delta G^{\circ\prime}$ for the hydrolysis of a variety of specific metabolites.

TABLE 8.6

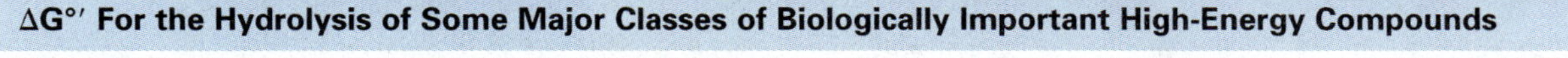

$\Delta G^{\circ\prime}$ For the Hydrolysis of Some Major Classes of Biologically Important High-Energy Compounds

Class of Compound	Hydrolysis Reaction[a]	Approximate $\Delta G^{\circ\prime}$ (kJ/mol)
Phosphate anhydrides	$R{-}O{-}PO(O^-){-}O{-}PO(O^-){-}O{-}R' + H_2O \longrightarrow R{-}O{-}PO(O^-){-}O^- + H^+ + R'{-}O{-}PO(O^-){-}O^- + H^+$	−30

(Continued)

TABLE 8.6 *(Continued)*

$\Delta G°'$ For the Hydrolysis of Some Major Classes of Biologically Important High-Energy Compounds

Class of Compound	Hydrolysis Reaction[a]	Approximate $\Delta G°'$ (kJ/mol)
Carboxylate-phosphate anhydrides (acyl phosphates)	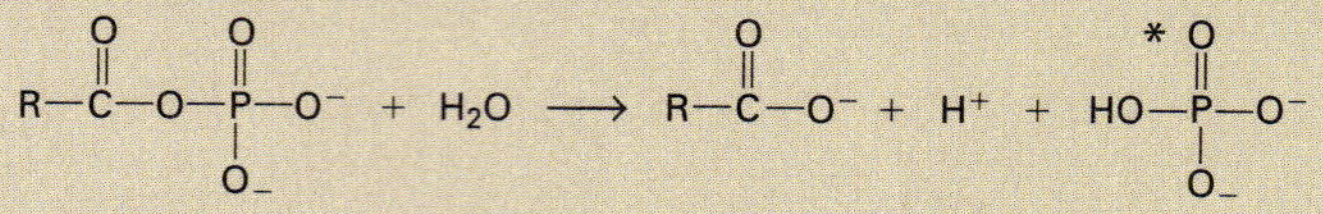 $R{-}C(=O){-}O{-}P(=O)(O^-){-}O^- + H_2O \longrightarrow R{-}C(=O){-}O^- + H^+ + {}^{*}HO{-}P(=O)(O^-){-}O^-$	–50
Enoyl phosphates	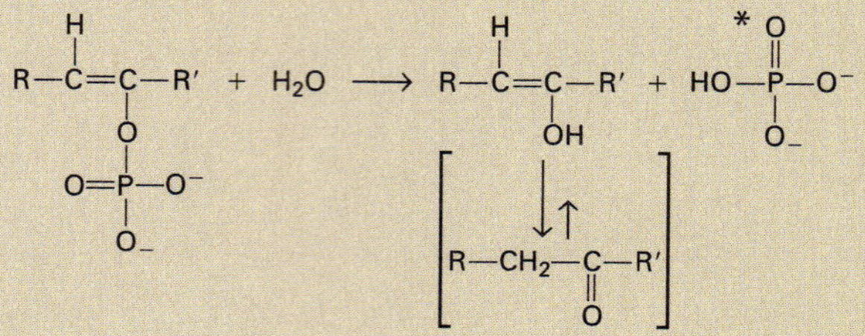 $R{-}C(H){=}C(O{-}P(=O)(O^-){-}O^-){-}R' + H_2O \longrightarrow R{-}C(H){=}C(OH){-}R' + {}^{*}HO{-}P(=O)(O^-){-}O^-$; $\left[\downarrow\uparrow\ R{-}CH_2{-}C(=O){-}R'\right]$	–60[b]
Thioesters	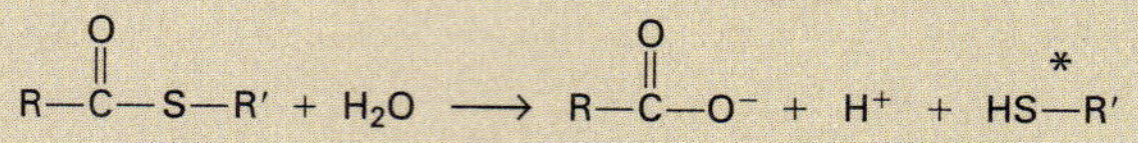 $R{-}C(=O){-}S{-}R' + H_2O \longrightarrow R{-}C(=O){-}O^- + H^+ + {}^{*}HS{-}R'$	–30

[a]The dissociation of products is considered part of the hydrolysis reactions. Although the products identified with a * dissociate, they are depicted in their nondissociated form since this is their predominant form at physiological pH values.
[b]This value includes the tautomerism of the enol product.

TABLE 8.7

$\Delta G°'$ for the Hydrolysis of Selected Metabolites

Compound	Class of Bond Hydrolyzed	$\Delta G°'$ (kJ/mol)
Phosphoenolpyruvate (PEP)	Enoyl phosphate	–62
1,3-Bisphosphoglycerate	Carboxylate-phosphate (mixed) anhydride	–49
Acetyl Phosphate	Carboxylate-phosphate (mixed) anhydride	–43
Pyrophosphate (PP_i)	Phosphate anhydride	–33
ATP (to ADP + P_i)	Phosphate anhydride	–31
ATP (to AMP + PP_i)	Phosphate anhydride	–32
AMP (to Adenosine + P_i)	Phosphate ester	–14
Acetyl-SCoA	Thioester	–31
Glucose 1-phosphate	Phosphate ester	–21
Glucose 6-phosphate	Phosphate ester	–14
Glycerol 3-phosphate	Phosphate ester	–9

Some Kinetic Considerations

How can ATP and other high-energy compounds survive within a cell when the thermodynamic driving force for their hydrolysis is so large? Why isn't ATP hydrolyzed as rapidly as it is formed? The answer lies in a fact emphasized in Section 5.2 and again in Section 8.2; the $\Delta G^{\circ\prime}$ for a reaction does not determine its rate. Since most biologically important, high-energy compounds have a large activation energy for hydrolysis, they are not hydrolyzed at a significant rate under physiological conditions in the absence of catalysts. Similarly, a high activation energy for the oxidation of organic compounds by O_2 protects firewood from spontaneous combustion. Enzymes determine when and, by their cellular location, where high-energy metabolites are utilized. The importance of specificity and catalysis, two of the central themes in biochemistry, cannot be overemphasized. **[Try Problem 8.22]**

***PROBLEM 8.21** Study Figure 7.2 and then write a structural formula for a tautomer of each of the following compounds.

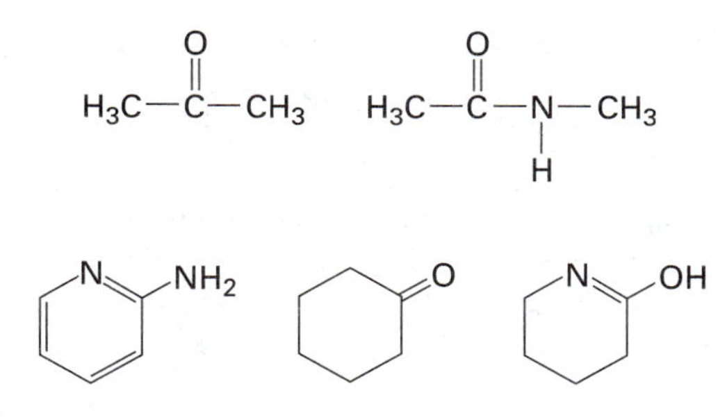

Identify the most stable member of each tautomer pair.

PROBLEM 8.22 Draw a progress of reaction diagram for (a) an endergonic reaction with a relatively low activation energy, and (b) an exergonic reaction with a relatively high activation energy. Which reaction has the greatest thermodynamic driving force? Which reaction will tend to proceed most rapidly in the absence of a catalyst? Explain. Hint: Review Section 5.2.

8.6 COUPLED CHEMICAL REACTIONS

High-energy compounds are used by all organisms to provide a driving force for thermodynamically unfavorable reactions. This is made possible through the coupling of reactions. Two reactions are **coupled** when one or more of the reactants or products in one reaction are reactants or products in the second reaction. In cells, coupled sequences of reactions constitute metabolic pathways. Many of the intermediate reactants and products within a pathway may not show up within the net reaction for the pathway.

Consider the following generic reactions:

$$A + B \rightleftharpoons C + D \qquad \Delta G^{\circ\prime} = +18 \text{ kJ/mol}$$
$$C \rightleftharpoons X \qquad \Delta G^{\circ\prime} = +7 \text{ kJ/mol}$$
$$X + Y \rightleftharpoons Z \qquad \Delta G^{\circ\prime} = -46 \text{ kJ/mol}$$

The net reaction for the three coupled reactions is:

$$A + B + Y \rightleftharpoons D + Z \qquad \Delta G^{\circ\prime} = -21 \text{ kJ/mol}$$

The $\Delta G^{\circ\prime}$ for the net reaction is equal to the sum of the $\Delta G^{\circ\prime}$ values for the three reactions in the sequence. In this example, the net reaction is highly favorable (has a large negative $\Delta G^{\circ\prime}$ and a large $K_{eq}{}'$) in spite of the unfavorable nature of the two initial reactions within the pathway. By coupling the third reaction with the first two reactions, the driving force for the conversion of A and B to D is greatly enhanced; the free energy change becomes large and negative. Living organisms routinely use this strategy to drive and direct metabolic processes.

Turning to a biologically important example:

$$\text{Glucose} + P_i \rightleftharpoons \text{glucose 6-phosphate} + H_2O \qquad \Delta G^{\circ\prime} = +14 \text{ kJ/mol}$$
$$\text{ATP} + H_2O \rightleftharpoons \text{ADP} + P_i \qquad \Delta G^{\circ\prime} = -31 \text{ kJ/mol}$$
$$\text{Glucose} + \text{ATP} \rightleftharpoons \text{glucose 6-phosphate} + \text{ADP} \qquad \Delta G^{\circ\prime} = -17 \text{ kJ/mol}$$

The direct phosphorylation of glucose (first reaction) is not spontaneous ($K_{eq} < 1$) because its free energy change is positive. If, however, the phosphorylation of glucose is coupled to the hydrolysis of ATP (the second reaction), this phosphorylation becomes spontaneous; the free energy change for the net reaction is negative. **[Try Problems 8.23 and 8.24]**

The ATP-driven phosphorylation of glucose within cells does not, in reality, entail the hydrolysis of ATP. A phosphoryl group is transferred from ATP to glucose during a reaction catalyzed by an enzyme that is classified as a transferase, not a hydrolase (Section 5.1). From a thermodynamic standpoint, however, the result is the same as if a phosphoryl group were added and an ATP then hydrolyzed. For this reason, ATP hydrolysis is commonly said (inaccurately) to drive the phosphorylation. Many of the other reactions that are often said to be driven by ATP hydrolysis also proceed through group transfer reactions rather than hydrolysis.

PROBLEM 8.23 Given the following information:

$$X \rightleftharpoons P_i + Y$$
$$P_i + \text{ADP} \rightleftharpoons \text{ATP}$$

What can you conclude about the $\Delta G^{\circ\prime}$ for the first reaction if the coupled reactions have a positive thermodynamic driving force under standard biological conditions? Explain.

PROBLEM 8.24 True or false? If the $\Delta G^{\circ\prime}$ for a metabolic pathway is negative, we can be confident that the pathway will proceed spontaneously within a cell. Explain.

8.7 OXIDATION AND REDUCTION

In humans and other heterotrophs, most of the energy in ATP and other high-energy compounds is acquired through the oxidation of organic compounds. **Oxidation** is the loss of electrons, whereas **reduction** is the gain of electrons. The two processes are always coupled; a donor can only give up electrons when an acceptor is available to receive them. Since most oxidation–reduction reactions **(redox reactions)** are reversible, once a donor has given up an electron (is oxidized) it becomes an electron acceptor:

$$\text{Electron donor} \rightleftharpoons e^- + \text{electron acceptor}$$

Similarly, an acceptor becomes an electron donor after it has received an electron (is reduced):

$$\text{Electron acceptor} + e^- \rightleftharpoons \text{electron donor}$$

Hence, each reactant in a redox reaction can exist in an **oxidized form** (accepts electrons) and a **reduced form** (donates electrons):

$$\text{A oxidized form} + \text{B reduced form} \rightleftharpoons \text{A reduced form} + \text{B oxidized form}$$

The equilibrium constant for the reaction is determined by which oxidized form has the greatest affinity for electrons (see the next section). Because the electron donor reduces the electron acceptor, it is said to be the **reducing agent** or **reductant** (Table 8.8). The electron acceptor is the **oxidizing agent** or **oxidant.**

During redox reactions, the transferred electrons can be passed directly from donors to acceptors:

$$Fe^{2+} + Cu^{2+} \rightleftharpoons Fe^{3+} + Cu^{+}$$

Alternatively, the transferred electrons are often delivered to the acceptor by an electron carrier:

$$XH_2 + Y \rightleftharpoons X + YH_2$$

In this generic example, hydrogens carry electrons from the electron donor (XH_2) to the electron acceptor (Y). This is a common phenomenon within living organisms.

A neutral hydrogen atom carries one electron:

$$H \rightleftharpoons e^- + H^+$$

A negative **hydride ion** (H^-) can deliver two electrons:

$$H^- \rightleftharpoons 2\,e^- + H^+$$

TABLE 8.8

Redox Terminology

Electron Donor (Reduced Form)	Electron Acceptor (Oxidized Form)
Is oxidized as it donates one or more electrons	Is reduced as it accepts one or more electrons
Is the reducing agent (reductant)	Is the oxidizing agent (oxidant)
Is converted to an electron acceptor with the loss of one or more electrons	Is converted to an electron donor with the gain of one or more electrons

The hydrogen frequently becomes covalently attached to the compound that it reduces (the one to which it delivers its electron cargo). Although such a hydrogen still shares the electrons that it has delivered, the electrons are usually assigned (from the standpoint of oxidation–reduction) to the atom to which the hydrogen has become bonded, since this atom normally has a greater electron affinity than hydrogen. Similarly, the electrons delivered by the hydrogen are originally (with rare exceptions) assigned to the atom covalently bonded to the hydrogen within the electron donor. In general, the shared electrons in a covalent bond are assigned (from the standpoint of oxidation–reduction) to the atom with the greater electron affinity. If covalently bonded atoms possess the same electron affinity, each is assigned the electron(s) it contributed for sharing. **[Try Problem 8.25]**

In some instances, the carrier-linked transfer of electrons leads to the production of a single product from the combination of the oxidizing agent with the reducing agent:

$$H_2 + CH_2{=}CH{-}CH_3 \rightleftharpoons CH_3{-}CH_2{-}CH_3$$

$$CH_3CH_3 + O_2 \rightleftharpoons CH_2OHCH_2OH$$

Molecular hydrogen (H_2) and ethane (CH_3CH_3) are the electron donors in the above reactions. The atoms delivering the electrons (H in the first example, H and C in the second example) become covalently attached to the atoms that accept the electrons (C in the first example; O in the second example). Since the atoms that accept the electrons have a greater electron affinity than the atoms that deliver the electrons, the acceptor atoms are assigned all of the shared electrons. Part of these shared electrons were originally assigned to atoms within the electron donor. The oxidation of organic compounds frequently involves the addition of oxygen or the removal of hydrogen. Reduction often entails the addition of hydrogen or the removal of oxygen. **[Try Problem 8.26]**

The complete biological oxidation of organic fuel molecules involves the flow of electrons from these molecules to O_2 through a series of intermediate electron carriers that function as an electron bucket brigade:

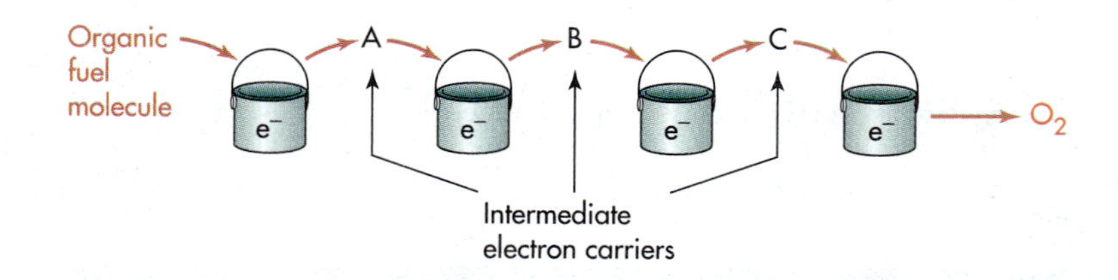

A detailed analysis of this electron transport system is provided in Section 11.8. Each step in the system is an exergonic redox reaction in which electrons are spontaneously transferred from a carrier with a lower electron affinity to one with a higher electron affinity. The net reaction is:

$$\text{Organic fuel} + O_2 \rightleftharpoons CO_2 + H_2O + \text{energy}$$

Other inorganic products are formed if the fuel molecule contains elements in addition to C, O, and H. Part of the energy released is used to create a concentration gradient that is employed to join ADP and P_i to make ATP. In humans, breathing allows the body to acquire O_2 from the air and to eliminate CO_2.

PROBLEM 8.25 In each of the following redox reactions, identify the oxidizing agent, the reducing agent, the electron donor, the electron acceptor, the compound oxidized and the compound reduced:

$$2\ CH_3SH + CH_3CH_2S{-}SCH_2CH_3 \rightleftharpoons CH_3S{-}SCH_3 + 2\ CH_3CH_2SH$$

$$CH_3CHO + 2\ CuO \rightleftharpoons CH_3COOH + Cu_2O$$

$$H_2O_2 + CH_3OH \rightleftharpoons 2\ H_2O + H_2C{=}O$$

***PROBLEM 8.26** Rank the following classes of organic compounds from least highly oxidized to most highly oxidized: alcohols, alkanes, aldehydes, carboxylic acids, and ketones.

8.8 REDUCTION POTENTIALS

Since oxidation and reduction are always coupled, an electron acceptor normally acquires an electron as the electron is being released from a donor. However, the process can be visualized as proceeding in two distinct steps, the loss of an electron by a donor followed by the gain of an electron by an acceptor:

$$\text{donor(1)} \rightleftharpoons e^- + \text{acceptor(1)}$$

$$e^- + \text{acceptor(2)} \rightleftharpoons \text{donor(2)}$$

$$\text{Net reaction} = \text{donor(1)} + \text{acceptor(2)} \rightleftharpoons \text{acceptor(1)} + \text{donor(2)}$$

Each step involves a donor:acceptor pair and is said to represent a **half-reaction.** As pointed out in the previous section, the K_{eq} for the reaction is determined by the relative electron affinities of the two acceptors. The forward reaction is favored when acceptor(2) has a greater electron affinity than acceptor(1).

The **standard reduction potential** (*E*°) is a quantitative measure of the electron affinity of the acceptor in a half-reaction. By international agreement, the half-reaction $H^+ + e^- \rightleftharpoons \frac{1}{2}H_2$ is assigned a standard reduction potential of exactly zero when $[H^+] = 1\ M$ and H_2 is at 101.3 kPa. This hydrogen half-reaction is used as the reference half-reaction for all other *E*° measurements. The standard reduction potential for any other half-reaction can be determined by placing one end of a wire (called an electrode) into a beaker containing 1 *M* H^+ plus H_2 at 101.3 kPa (the reference half-cell) and the other end of the wire into a separate beaker containing 1 *M*

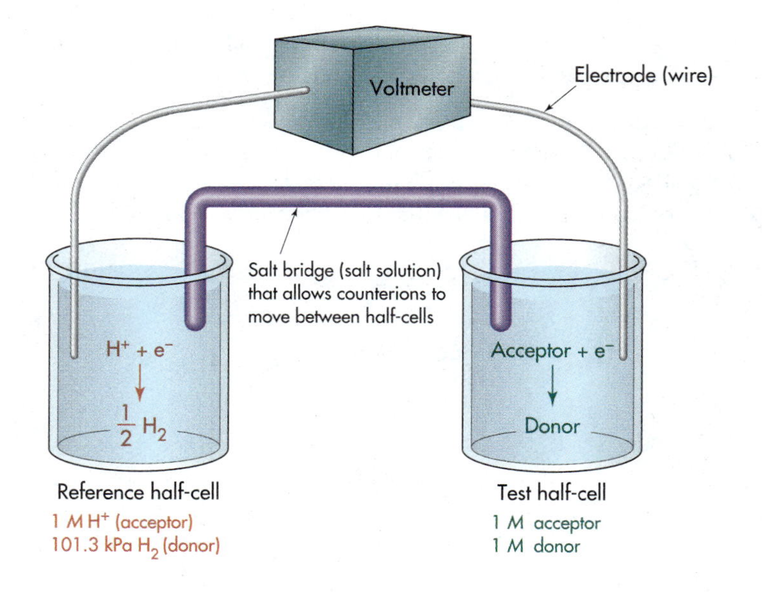

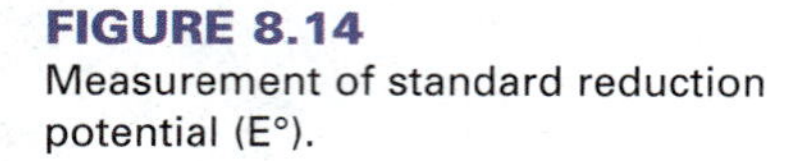

FIGURE 8.14
Measurement of standard reduction potential (E°).

acceptor and 1 *M* donor for a second half-reaction (the test half-cell) (Figure 8.14). If the half-reaction in the test half-cell has a different reduction potential than the reference half-reaction, a current will flow through the wire from one half-cell to the other. By convention, a test half-reaction is said to have a positive reduction potential if electrons flow from the reference half-cell to the test half-cell. Such a flow indicates that the acceptor in the test half-reaction has a greater affinity for an electron than does the acceptor (H^+) in the hydrogen half-reaction. The electromotive force (in volts, V) between the two half-cells is equal to the standard reduction potential for the test half-reaction. Recall that the reference half-reaction has been arbitrarily assigned an $E°$ of zero.

The standard reduction potential is the reduction potential for a test half-reaction when its acceptor and donor are both at 1 *M* (101.3 kPa if a gas). The reduction potential at other acceptor and donor concentrations (in the test half-reaction) can be calculated from $E°$ with the following equation:

$$E = E° + \frac{RT}{n\mathfrak{F}} \ln \frac{[\text{electron acceptor}]}{[\text{electron donor}]}$$

where R is the gas constant = 8.315 J/mol · K; T is the absolute temperature (K); n is the number of electrons transferred per molecule; and $\mathfrak{F}$ is the **Faraday** constant = 96.48 kJ/V · mol. At 298 K (25°C) this equation becomes:

$$E = E° + \frac{0.026\ \text{V}}{n} \ln \frac{[\text{electron acceptor}]}{[\text{electron donor}]}$$

The **standard biological reduction potential** (**E°′**) for a half-reaction is the reduction potential when $[H^+] = 10^{-7}$ *M* (pH = 7) and $[H_2O] = 55.5$ *M* within the test half-cell. The acceptor and donor concentrations (other than H^+ and H_2O) are still 1 M in the test half-cell and the reference half-cell still contains 1 M H^+ and H_2 at 101.3 kPa. The reduction potential for the test half-reaction at other concentrations of its acceptor and donor can be calculated as above:

$$E' = E°' + \frac{RT}{n\mathfrak{F}} \ln \frac{[\text{electron acceptor}]}{[\text{electron donor}]}$$

Table 8.9 provides the E°′ for some biologically important half-reactions at 298 K. Electrons will (under standard conditions) spontaneously flow from the electron donor in one half-reaction to the electron acceptor in any other half-reaction with a more positive (it may be negative, but less negative) reduction potential. Thus, the following reactions will proceed spontaneously under standard conditions:

$$\text{Lactate} + \text{oxaloacetate} \rightleftharpoons \text{malate} + \text{pyruvate}$$

$$\text{Fumarate} + \text{malate} \rightleftharpoons \text{succinate} + \text{oxaloacetate}$$

TABLE 8.9

Standard Biological Reduction Potentials

Half-Reaction[a]	$E°'$ (Volts)
$1/2 O_2 + 2\ H^+ + 2\ e^- \rightleftharpoons H_2O$	0.816
$Fe^{3+} + 1\ e^- \rightleftharpoons Fe^{2+}$	0.771
Cytochrome a_3 (Fe^{3+}) + e^- ⇌ cytochrome a_3 (Fe^{2+})	0.350
Cytochrome *a* (Fe^{3+}) + e^- ⇌ cytochrome *a* (Fe^{2+})	0.29
Cytochrome *c* (Fe^{3+}) + e^- ⇌ cytochrome *c* (Fe^{2+})	0.254
Cytochrome c_1 (Fe^{3+}) + e^- ⇌ cytochrome c_1 (Fe^{2+})	0.22
Cytochrome *b* (Fe^{3+}) + e^- ⇌ cytochrome *b* (Fe^{2+})	0.077
Ubiquinone + 2 H^+ + 2 e^- ⇌ Ubiquinol	0.045
$Fumarate^-$ + 2 H^+ + 2 e^- ⇌ $succinate^-$	0.031
FAD + 2 H^+ + 2 e^- ⇌ $FADH_2$ (in flavoproteins)[b]	~zero
$Oxaloacetate^-$ + 2 H^+ + 2 e^- ⇌ $malate^-$	−0.166
$Pyruvate^-$ + 2 H^+ + 2 e^- ⇌ $lactate^-$	−0.185
Acetaldehyde + 2 H^+ + 2 e^- ⇌ ethanol	−0.197
FAD + 2 H^+ + 2 e^- ⇌ $FADH_2$ (free coenzyme)	−0.219
Glutathione + 2 H^+ + 2 e^- ⇌ 2 reduced glutathiones	−0.23
Lipoic acid + 2 H^+ + 2 e^- ⇌ dihydrolipoic acid	−0.29
NAD^+ + H^+ + 2 e^- ⇌ NADH	−0.320
$NADP^+$ + H^+ + 2 e^- ⇌ NADPH	−0.320
Cystine + 2 H^+ + 2 e^- ⇌ 2 cysteines	−0.340
$Acetoacetate^-$ + 2 H^+ + 2 e^- ⇌ β-$hydroxybutyrate^-$	−0.346
α-Ketoglutarate + CO_2 + 2 H^+ + 2 e^- ⇌ isocitrate	−0.38

[a]In each half reaction, the oxidized form (the electron acceptor and oxidizing agent) is to the left and the reduced form (the electron donor and reducing agent) is to the right.
[b]FAD is the oxidized form of flavine adenine dinucleotide and $FADH_2$ is the reduced form.

A glance at Table 8.9 provides an explanation for why O_2 is the ultimate electron acceptor during the oxidation of fuel molecules in the human body; O_2 is the strongest oxidizing agent (has the greatest electron affinity) among those normally found within the body. **[Try Problems 8.27 through 8.29]**

The $\Delta G^{\circ\prime}$ for a redox reaction can be calculated from the reduction potentials for its two half-reactions:

$$\Delta G^{\circ\prime} = -n\mathfrak{F}\Delta E^{\circ\prime}$$

where n is the number of electrons transferred, $\mathfrak{F}$ is the Faraday constant and $\Delta E^{\circ\prime}$ is the difference in reduction potential for the two half-reactions. $\Delta E^{\circ\prime}$ is equal to the $E^{\circ\prime}$ for the half-reaction containing the oxidizing agent minus the $E^{\circ\prime}$ for the half-reaction involving the reducing agent.

Consider the reaction:

$$\text{NAD}^+ + \text{ethanol} \rightleftharpoons \text{acetaldehyde} + \text{NADH}$$

Since NAD^+ is the oxidizing agent:

$$\Delta E^{\circ\prime} = -0.320 - (-0.197) = -0.123 \text{ V} \qquad \text{(see Table 8.9)}$$

$$\Delta G^{\circ\prime} = -(2)\,(96.48 \text{ kJ/V} \cdot \text{mol})\,(-0.123 \text{ V}) = +23.7 \text{ kJ/mol}$$

This reaction will not proceed spontaneously under standard conditions; a net reverse reaction is required to reach equilibrium. The ΔG_p can be calculated for a reaction once $\Delta G^{\circ\prime}$ and the physiological concentrations of reactants and products have been determined (Section 8.2). The $K_{eq}{}'$ for a redox reaction can also be computed once $\Delta G^{\circ\prime}$ has been ascertained. **[Try Problem 8.30]**

PROBLEM 8.27 Study Table 8.9. Within each of the following sets of oxidizing agents, circle the one agent with the greatest electron affinity. This corresponds to the strongest oxidizing agent. Within each set, put a rectangle around the agent most readily reduced.

Set A: NAD^+; cystine; pyruvate; cytochrome *c* (Fe^{3+})
Set B: NAD^+; FAD (in flavoproteins); FAD (free); ubiquinone.

PROBLEM 8.28 Calculate the E' for the NAD^+ half-reaction when $[NAD^+] = 0.10$ m*M* and $[NADH] = 5.0$ m*M*.

PROBLEM 8.29 List two oxidizing agents from Table 8.9 that will be spontaneously reduced by ubiquinol under standard biological conditions.

PROBLEM 8.30 Calculate the $\Delta G^{\circ\prime}$ and the $K_{eq}{}'$ for the following reactions at 298 K:

a. Lactate + oxaloacetate $\rightleftharpoons$ malate + pyruvate
b. Fumarate + malate $\rightleftharpoons$ succinate + oxaloacetate

---

SUMMARY

Living organisms must obey the same physical and chemical laws as inanimate objects. Energy can be transformed from one form to another and can be exchanged between an organism and its environment, but it cannot be created or destroyed (first law of thermodynamics). Although organisms are more ordered (contain less entropy) than their environment, the entropy of the universe as a whole increases during the creation and maintenance of life (second law of thermodynamics).

Enthalpy (H) is a measure of heat content, whereas entropy (S) is a measure of randomness or disorder. The free energy difference between reactants and products in a chemical reaction ($\Delta G = G_{\text{products}} - G_{\text{reactants}}$) is determined by their enthalpy difference ($\Delta H = H_{\text{products}} - H_{\text{reactants}}$), their entropy difference ($\Delta S = S_{\text{products}} - S_{\text{reactants}}$), and the temperature of the reaction mixture: $\Delta G = \Delta H - T\Delta S$. The ΔG for a reaction at 298 K (normally) with a 1 M concentration of all reactants and products (101.3 kPa if gases) is defined as the standard free energy difference and abbreviated $\Delta G°$. The ΔG under standard biological conditions (298 K, pH = 7, $[H_2O]$ = 55.5 M, and 1 M concentration [101.3 kPa if gases] of all other reactants and products) is designated $\Delta G°'$.

The $\Delta G°$ for a reaction determines its equilibrium constant: $\Delta G° = -RT \ln K_{eq}$. For the generic reaction aA + bB $\rightleftharpoons$ cC + dD, $K_{eq} = [C]^c[D]^d/[A]^a[B]^b$ where all concentrations are equilibrium concentrations. Similarly, $\Delta G°' = -RT \ln K_{eq}'$. Since the $[H^+]$ and $[H_2O]$ are normally incorporated into the numerical value of K_{eq}', they do not usually appear in K_{eq}' expressions. A negative $\Delta G°'$ means that, under standard biological conditions, a reaction has a positive driving force (is spontaneous) and a $K_{eq} > 1$. A positive $\Delta G°'$ identifies a reaction with a negative driving force (under standard conditions) and a $K_{eq} < 1$. When $\Delta G°' = 0$, a reaction is at equilibrium under standard conditions. A decrease in enthalpy or increase in entropy makes a negative contribution to ΔG and favors the forward reaction. Many biologically important processes are driven, in part at least, by an increase in entropy.

Under physiological conditions:

$$\Delta G = \Delta G_p = \Delta G°' + RT \ln \frac{[C]^c[D]^d}{[A]^a[B]^b}$$

where all concentrations are actual intracellular concentrations, which may or may not correspond to equilibrium concentrations. The more negative the ΔG_p, the greater the driving force for the reaction and the further it is from equilibrium under physiological conditions. By operating at reactant and product concentrations far from equilibrium, a cell can provide a strong driving force for key reactions. The continual production and removal of reactants and products prevent most intracellular reactions from reaching equilibrium while an organism is alive.

$\Delta G°'$ plays no role in determining the rate of a reaction. A reaction may have an extremely large driving force yet proceed very slowly and vice versa. The rate of most *in vivo* reactions is determined by the catalytic activity of enzymes that function, in part, by lowering activation energies.

High-energy compounds are the immediate source of energy for most energy-requiring processes within the human body. The large negative $\Delta G°'$ for the hydrolysis of these compounds ($\Delta G°' < -25$ kJ/mol) is usually a consequence of the destabilization of reactants by like-charge repulsion, the stabilization of products through resonance, tautomerization, dissociation, or solvation, or a combination of these factors. The major classes of high-energy compounds include phosphate anhydrides, carboxylate-phosphate anhydrides, enoyl phosphates, and thioesters. ATP, a phosphate anhydride, is the most widely distributed and commonly employed high-energy compound within living organisms. It is frequently described as the "energy currency" for cells.

Metabolic pathways are sequences of coupled chemical reactions, in which the ΔG_p for the net reaction of the pathway equals the sum of the ΔG_p's for the individual reactions involved. Unfavorable steps along the pathway can be made favorable by coupling them with subsequent steps that possess a large negative ΔG_p. ATP is commonly cleaved during those steps that provide a positive driving force for a metabolic pathway.

Most of the energy used by the human body is obtained through the complete oxidation of organic molecules acquired in the diet, an exergonic process with a large negative $\Delta G°'$:

$$\text{Organic fuel} + O_2 \rightleftharpoons CO_2 + H_2O + \text{energy}$$

The oxidation involves the flow of electrons from the fuel molecule to O_2 through various intermediates that, collectively, function as an electron bucket brigade. Each transfer of electrons represents a redox reaction in which the electron donor is oxidized by the electron acceptor (reduced in the process). The electrons are often transferred while bound to atoms that function as electron carriers. The carrier atoms frequently become covalently bonded to the compounds that are reduced (that accept the electrons). The path of electron flow is determined by the relative electron affinities of those substances that participate in the electron bucket brigade. Electrons are always transferred from an acceptor with a lower electron affinity to one with a higher electron affinity.

The reduction potential for the half-reaction involving an electron acceptor is a quantitative measure of the electron affinity of that acceptor. Under standard conditions, the hydrogen half-reaction (H^+ is electron acceptor, H_2 is electron donor) has been assigned a standard reduction potential ($E°$) of zero. The standard reduction potentials for other half-reactions are determined by measuring the voltage between a standard hydrogen half-cell (the reference half-cell) and a second half-cell (the test half-cell) in which [electron acceptor] = [electron donor] = 1 *M* (101.3 kPa for gases) for the test half-reaction.

The standard biological reduction potential ($E°'$) is the reduction potential when the H^+ concentration in the test half-cell is 10^{-7} *M* rather than 1 *M* and $[H_2O] = 55.5$ *M*. The E' for a half-reaction at any given concentration of electron acceptor and electron donor can be calculated with the following equation:

$$E' = E°' + \frac{RT}{n\mathfrak{F}} \ln \frac{[\text{electron acceptor}]}{[\text{electron donor}]}$$

The $\Delta G°'$ for a redox reaction can be calculated from the difference in the standard biological reduction potentials for the two half reactions involved:

$$\Delta G°' = -n\mathfrak{F}\Delta E°'$$

ΔG_p and K_{eq}' can be calculated from $\Delta G°'$. Intracellular reactant and product concentrations are required for the ΔG_p computations.

EXERCISES

8.1 Indirectly, the sun provides most of the energy used by the human body. It is also the source of the majority of the energy consumed by our industrialized society. Explain these facts.

8.2 Describe the energy transformations that occur during the starting and running of an automobile.

8.3 How is the CO_2 produced from the oxidation of organic matter within the human body eliminated from the body? What role does hemoglobin play in this process?

8.4 Most animals continuously eliminate more water than they consume, yet they maintain a relatively constant percentage of water. How is this possible?

8.5 True or false? There are only two ways a cell can change its energy content: It can take up energy from its surroundings, or it can release energy to its surroundings. Explain.

8.6 True or false? At 298 K, all reactions are at equilibrium when the concentration of each nongaseous reactant and product is 1 M and the partial pressure of each gaseous reactant and product is 101.3 kPa. Explain.

8.7 The equilibrium constant for the conversion of X to Y ($X \rightleftharpoons Y$) is 3.0 at 298 K. If we start with 0.10 *M* X, what is the ratio of Y to X at equilibrium? Will the equilibrium ratio increase, decrease, or remain the same if the initial concentration of X is doubled? Calculate the concentration of Y in equilibrium with X when [X] = 0.064 *M*.

8.8 Calculate the $\Delta G°$ for the reaction featured in Exercise 8.7.

8.9 True or false? It is impossible for $\Delta G_p'$ to be negative when $\Delta G°'$ is large and positive. Explain.

8.10 True or false? For a reaction to have a large driving force, that reaction must exist far from equilibrium. Explain.

8.11 True or false? A large negative ΔG_p always indicates that the equilibrium constant for a reaction is large. Explain.

8.12 Calculate the ΔG_p for the following reaction:

$$\text{Glucose 1-phosphate} + H_2O \rightleftharpoons \text{glucose} + P_i$$

$$\Delta G°' = -20.9 \text{ kJ/mol}$$

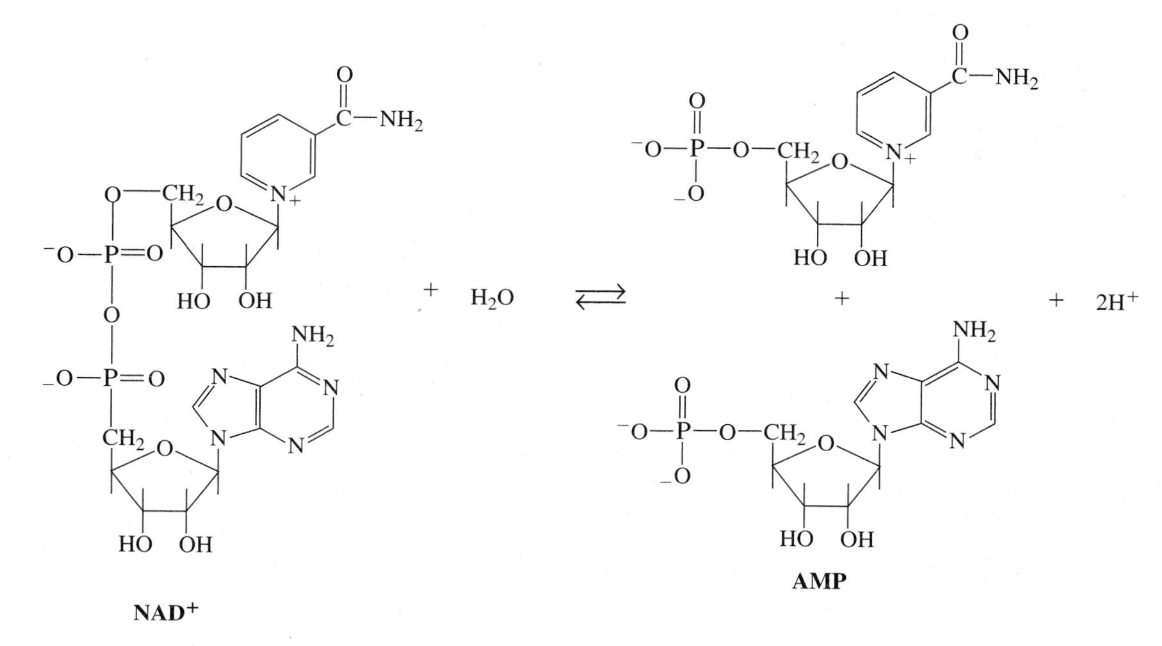

Part **b** of exercise 8.14.

Assume that the intracellular concentrations of glucose 1-phosphate, glucose, and P_i are 0.0010 m*M*, 10 m*M*, and 2.0 m*M*, respectively, at 25° C and pH 7. Will a net forward or a net reverse reaction be required to reach equilibrium? Explain.

8.13 Given the following information:

Hydrolysis Reaction Number	$\Delta G^{\circ\prime}$ (kJ/mol)
1	+40
2	−8
3	0
4	−48
5	+28
6	−27

a. Which reactions are spontaneous under standard biological conditions?
b. Which reactions have an equilibrium constant that is less than 1 under standard biological conditions?
c. Which reactions represent the hydrolysis of a high-energy compound?
d. Which reaction proceeds most rapidly?

8.14 What is the approximately $\Delta G^{\circ\prime}$ for each of the following reactions?

a. ATP + 2 $H_2O \rightleftharpoons$ AMP + 2 P_i
b. See above.

8.15 Does K_{eq} ever have a negative value? Explain.

8.16 True or false? It is possible for a "high-energy compound," as defined in Section 8.4, to possess less total free energy than some "low-energy compounds." Explain.

8.17 ATP can be hydrolyzed to yield AMP plus PP_i or hydrolyzed to yield ADP plus P_i. Compare the ability of these reactions to provide a driving force for a coupled chemical reaction under standard biological conditions.

8.18 The overall $\Delta G^{\circ\prime}$ for a synthetic process is +150 kJ/mol. If ATP hydrolysis is coupled to this process, how many moles of ATP must be hydrolyzed to provide a negative $\Delta G^{\circ\prime}$ for the process?

8.19 ATP is the phosphoryl donor for protein kinase–catalyzed phosphorylation reactions (Section 5.13). Using names or abbreviations of reactants and products, write an equation for the protein kinase–catalyzed phosphorylation of protein X.

8.20 Does 5′-adenosine tetraphosphate contain more, less, or the same amount of free energy as ATP? Explain.

8.21 Does 3′,5′-ABP (adenosine with one phosphate at the 3′ position and one phosphate at the 5′ position [Chapter 7]) contain more, less, or the same amount of free energy as 5′-ADP? Explain.

8.22 Write structural formulas for two resonance forms of PP_i.

8.23 Given the following reaction:

$$\text{3-Phosphoglycerate} \rightleftharpoons \text{2-phosphoglycerate}$$

$$\Delta G^{\circ\prime} = +4.40 \text{ kJ/mol}$$

Calculate the K_{eq}' for this reaction at 298 K. If the ΔG_p is +25, will a net forward or a net reverse reaction be required (under physiological conditions) to reach equilibrium? Explain.

8.24 Under what intracellular conditions does $\Delta G_p = \Delta G^{\circ\prime}$?

8.25 Study Table 8.7. What does this table tell you about the relative rates of hydrolysis of PEP, ATP, and glucose 1-phosphate? Explain. What does this table tell you about the ΔG_p's for the hydrolysis of PEP, ATP, and glucose 1-phosphate? Explain.

8.26 Given the following reaction:

$$ATP + H_2O \rightleftharpoons AMP + PP_i \qquad \Delta G^{\circ\prime} = -32 \text{ kJ/mol}$$

Will the ΔG_p for this reaction tend to increase, decrease, or remain the same if PP_i concentration drops because of its enzyme-catalyzed hydrolysis? Explain.

8.27 True or false? The ΔG for an exothermic reaction is always negative? Explain.

8.28 True or false? An increase in entropy contributes to the thermodynamic driving force for diffusion. Explain. Diffusion is the movement of a substance from higher to lower concentration.

8.29 Does entropy increase, decrease, or remain the same as water evaporates in a sealed container at constant temperature? Explain.

8.30 Is CO_2 oxidized or reduced as it is incorporated into glucose ($C_6H_{12}O_6$) during photosynthesis-linked processes? Explain.

8.31 Identify the reactions below that are redox reactions. For each redox reaction, identify the oxidizing agent, the reducing agent, the reactant oxidized, the reactant reduced, the electron donor, and the electron acceptor.

- **a.** $4\ Fe + 3\ O_2 \rightleftharpoons 2\ Fe_2O_3$
- **b.** ATP + GDP $\rightleftharpoons$ ADP + GTP
- **c.** $Zn + Fe^{2+} \rightleftharpoons Zn^{2+} + Fe$
- **d.** ATP + Glucose $\rightleftharpoons$ glucose 6-phosphate + ADP
- **e.** Glucose 1-phosphate $\rightleftharpoons$ glucose 6-phosphate
- **f.** $CH_3CHO + H_2 \rightleftharpoons CH_3CH_2OH$

8.32 In each case, record whether the given reactant is oxidized or reduced during the indicated transformation.

- **a.** Alcohol $\Rightarrow$ ketone
- **b.** Thiol $\Rightarrow$ disulfide
- **c.** Aldehyde $\Rightarrow$ alcohol
- **d.** Aldehyde $\Rightarrow$ carboxylic acid
- **e.** Benzene $\Rightarrow$ cyclohexane
- **f.** NADH $\Rightarrow$ NAD^+

8.33 Define and distinguish E, E', E°, and $E^{\circ\prime}$.

8.34 True or false? The E for a half-reaction increases as the ratio of electron acceptor to electron donor increases. Explain.

8.35 Which of the following reactions are spontaneous under standard biological conditions?

- **a.** NADH + $NADP^+ \rightleftharpoons NAD^+$ + NADPH
- **b.** $FADH_2 + NAD^+ \rightleftharpoons$ FAD + NADH + H^+
- **c.** Lactate + glutathione $\rightleftharpoons$ pyruvate + 2 reduced glutathiones
- **d.** Ubiquinol + pyruvate $\rightleftharpoons$ ubiquinone + lactate

In each reaction, identify the oxidizing agent and the reducing agent.

8.36 Calculate E' for the pyruvate half-reaction when [pyruvate] = 10.0 mM and [lactate] = 2.00 mM. Assume 298 K and 101.3 kPa.

8.37 List three reducing agents from Table 8.9 that are more powerful reducing agents than $FADH_2$ (in flavoproteins). List three oxidizing agents that are more powerful oxidizing agents than FAD (in flavoproteins).

8.38 Calculate the $\Delta G^{\circ\prime}$ and the K_{eq}' for the following reaction at 298 K:

$$2\ Fe^{3+} + \text{succinate} \rightleftharpoons 2\ Fe^{2+} + \text{fumarate} + 2\ H^+$$

8.39 Calculate the $\Delta E^{\circ\prime}$ for the following redox reaction at 298 K:

$$A + B \rightleftharpoons C + D \qquad \Delta G^{\circ\prime} = +23.5 \text{ kJ/mol}$$

Assume that a single electron is transferred.

SELECTED READINGS

Lehninger, A.L., D.L. Nelson, and M.M. Cox, *Principles of Biochemistry,* 2nd ed. New York: Worth Publishers, 1993.

Chapter 13, titled "Principles of Bioenergetics," includes two dozen references and 23 problems.

Moran, L.A., K. G. Scrimgeour, H.R. Horton, R.S. Ochs, and J.D. Rawn, *Biochemistry,* 2nd ed. Englewood Cliffs, NJ: Neil Patterson/Prentice-Hall, 1994.

An encyclopedic biochemistry text that discusses bioenergetics in a chapter titled "Introduction to Metabolism." Includes a limited number of problems and some selected readings.

Smith, C.A., and E.J. Wood, *Energy in Biological Systems,* Hong Kong: Chapman & Hall, 1991.

A 171-page paperback that discusses metabolism with an emphasis on bioenergetics. Contains both questions and references.

Stenesh, J., *Core Topics in Biochemistry,* Kalamazoo, MI: Cogno Press, 1993.

Devotes over 150 pages to bioenergetics. Provides a more thorough analysis of this topic than the other selected readings in this text. Includes some good problems.

9 Lipids and Biological Membranes

CHAPTER OUTLINE

All of the previous chapters, including Chapter 8, Bioenergetics, provide part of the framework required for a detailed discussion of metabolism, the collection of the enzyme-catalyzed reactions within living organisms. Lipids, membranes, vitamins, and hormones are additional components of this framework.

Lipids are those compounds produced by living organisms that are virtually insoluble in water but are soluble in relatively nonpolar solvents, including chloroform, carbon tetrachloride, and diethyl ether.

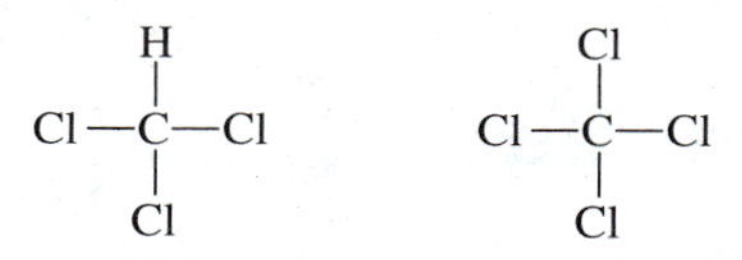

Chloroform (trichloromethane) **Carbon tetrachloride (tetrachloromethane)**

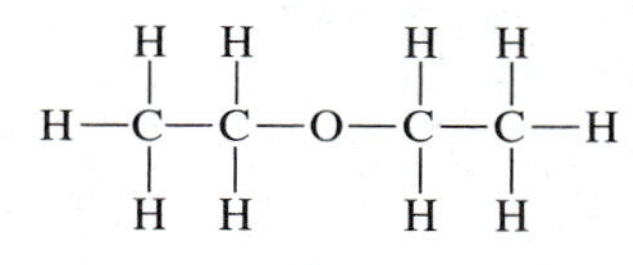

Diethyl ether (ethoxyethane)

Such solvents are called **fat solvents,** since they dissolve fats. Although some lipids are strictly nonpolar in character, many others possess a polar "head" and a nonpolar "tail" and are, therefore, amphipathic compounds (Section 2.2). Since lipids encompass an extremely wide variety of substances with diverse structures and functions, the only property many have in common is a predominantly nonpolar character. Lipids include fats, waxes, phospholipids, sphingolipids, glycolipids, eicosanoids, steroids, lipoproteins, and the fat-soluble vitamins (Figure 9.1). **[Try Problems 9.1 and 9.2]**

One of the major roles of lipids is to provide a nonpolar core for the membranes that surround all cells and envelop the organelles within eukaryotic cells. Membranes provide a protective barrier that is tough yet flexible. Separate membranes can fuse together and segments of membranes sometimes "pinch off" to create membrane-enclosed bodies. The lipids, proteins, and other components within membranes determine what substances enter and leave both cells and organelles, and they play a central role in regulating the rates at which this flow of material occurs. Enzymes associated with membranes participate in a broad range of physiologically important processes, including ATP production (Section 11.9) and nerve action (Section 10.9).

The outer surface of a plasma membrane contains a multitude of binding sites that collectively accommodate a wide range of ligands, including hormones, viruses, growth factors, neurotransmitters, and drugs. Ligands were defined in Chapter 2 as relatively small substances that bind specifically to larger molecules or particles. The ligand binding sites on membranes are called **receptors.** Receptor–ligand interactions play a central role in intercellular communication and in the response of cells to changes in their environment.

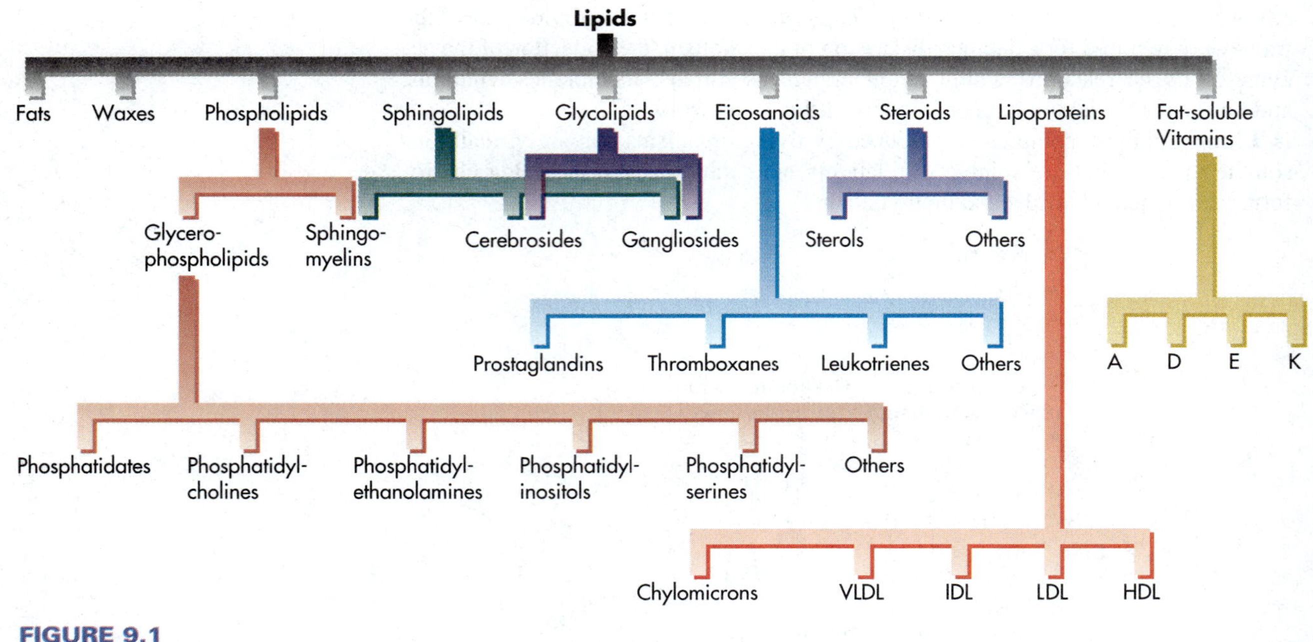

FIGURE 9.1
Classification of lipids.

This chapter adds to the framework required for a discussion of metabolism by providing an initial look at some of the common classes of lipids and by examining the chemical nature of membranes and membrane transport. Major biochemical themes include experimentation and hypothesis construction, transformation and use of energy, structural hierarchy and compartmentation, catalysis and the regulation of reaction rates, association of like with like, structure–function relationships, and specificity. Chapter 10 completes the framework for the exploration of metabolism by examining vitamins, hormones, and signal transduction across membranes.

Problem 9.1 Like dissolves like! Interpret this general rule for predicting solubility.

***Problem 9.2** Write the structural formulas for two relatively nonpolar solvents in addition to chloroform, carbon tetrachloride, and diethyl ether.

LIPIDS: STRUCTURE AND FUNCTION

9.1 FATS AND FATTY ACIDS

Fats, a major category of lipids, are naturally occurring carboxylic acid esters of glycerol. The esterification of one, two, or all three of the hydroxyl groups in glycerol leads to **monoacylglycerols (monoglycerides), diacylglycerols (diglycerides),**

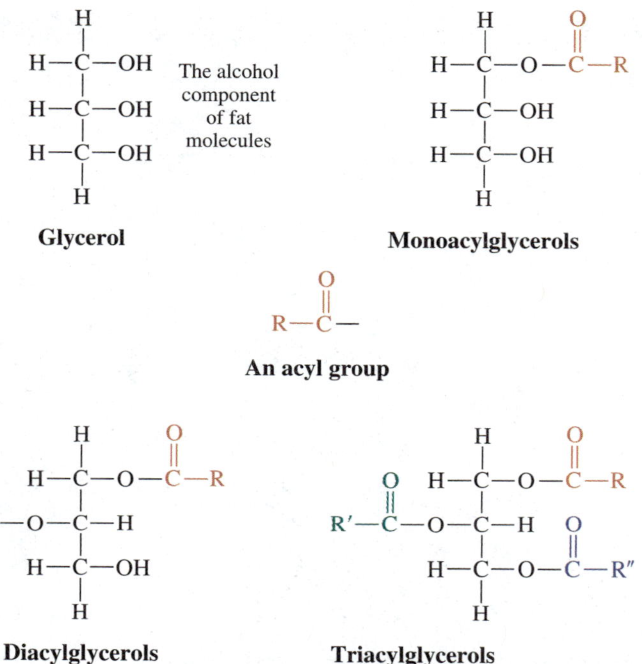

FIGURE 9.2
The structure of fats. R, R′, and R″ are usually hydrocarbon chains containing 11 to 19 carbons. Table 9.1 provides specific examples.

and **triacylglycerols (triglycerides),** respectively (Figure 9.2). A sample of natural fat is predominantly composed of triacylglycerols with much smaller amounts of mono and diacyl compounds. Those carboxylic acids that are used by living organisms to build fat molecules are known as **fatty acids.** **[Try Problems 9.3 and 9.4]**

Structure and Classification of Fatty Acids

Most fatty acids are linear molecules containing 12, 14, 16, 18, or 20 carbon atoms. A smaller number have been discovered that contain a different number of carbons, possess branches, or encompass ring systems. Fatty acids are classified on the basis of the number of carbon-to-carbon double bonds they contain: **saturated,** zero; **monounsaturated,** one; and **polyunsaturated,** more than one. The carbon-to-carbon double bonds usually possess a cis configuration and those in polyunsaturated molecules are normally separated by a single methylene [—CH_2—] group. Table 9.1 on the following page provides the names, structures, symbols, and melting points for some commonly encountered fatty acids. **[Try Problem 9.5]**

The Greek letter *omega* (ω) is frequently used in nutrition and the biomedical sciences to designate that carbon in a fatty acid which is furthest removed from the carboxyl group. When this practice is employed, the location of double bonds is identified by numbering the carbon chain from the ω-carbon rather than the carboxyl carbon. An **ω-3 fatty acid,** for example, is one in which there is a double bond between the third and fourth carbons from the ω-end. Two of the most frequently

TABLE 9.1

Some Common Fatty Acids

Systematic Name	Common Name	Symbol[a]	Structure[b]	Melting Point (°C)
SATURATED				
Dodecanoic acid	Lauric acid	12:0	$CH_3(CH_2)_{10}COOH$	44.2
Tetradecanoic acid	Myristic acid	14:0	$CH_3(CH_2)_{12}COOH$	52
Hexadecanoic acid	Palmitic acid	16:0	$CH_3(CH_2)_{14}COOH$	63.1
Octadecanoic acid	Stearic acid	18:0	$CH_3(CH_2)_{16}COOH$	69.6
Eicosanoic acid	Arachidic acid	20:0	$CH_3(CH_2)_{18}COOH$	75.4
UNSATURATED				
9-Hexadecenoic acid	Palmitoleic acid	$16{:}1(\Delta^{9})$	$CH_3(CH_2)_5CH{=}CH(CH_2)_7COOH$	−0.5
9-Octadecenoic acid	Oleic acid	$18{:}1(\Delta^{9})$	$CH_3(CH_2)_7CH{=}CH(CH_2)_7COOH$	13.4
9,12-Octadecadienoic acid	Linoleic acid	$18{:}2(\Delta^{9,12})$	$CH_3(CH_2)_4(CH{=}CHCH_2)_2(CH_2)_6COOH$	−9
9,12,15-Octadecatrienoic acid	α-Linolenic acid	$18{:}3(\Delta^{9,12,15})$	$CH_3CH_2(CH{=}CHCH_2)_3(CH_2)_6COOH$	−17
6,9,12-Octadecatrienoic acid	γ-Linolenic acid	$18{:}3(\Delta^{6,9,12})$	$CH_3(CH_2)_4(CH{=}CHCH_2)_3(CH_2)_3COOH$	
5,8,11,14-Eicosatetraenoic acid	Arachidonic acid	$20{:}4(\Delta^{5,8,11,14})$	$CH_3(CH_2)_4(CH{=}CHCH_2)_4)(CH_2)_2COOH$	−49.5
5,8,11,14,17-Eicosapentaenoic acid	EPA	$20{:}5(\Delta^{5,8,11,14,17})$	$CH_3CH_2(CH{=}CHCH_2)_5(CH_2)_2COOH$	−54

[a] 12:0, for example, is number of carbon atoms:number of double bonds. The superscripts to a Δ are the numbers of the first carbons in each double bond. Carbon chains are numbered beginning with the carboxyl carbon.
[b] All double bonds have a cis configuration. The carboxyl groups exist in a dissociated form at physiological pH values.

encountered ω-3 fatty acids are **eicosapentaenoic acid** (EPA) ($20{:}5\Delta^{5,8,11,14,17}$) and **α-linolenic acid** ($18{:}3\Delta^{9,12,15}$). Interpretation of fatty acid symbols, such as $20{:}5\Delta^{5,8,11,14,17}$ and $18{:}3\Delta^{9,12,15}$, is explained in footnote *a* of Table 9.1. **[Try Problem 9.6]**

Functions of Fatty Acids

The biological functions of fatty acids extend well beyond their role as building blocks for fats. Most phospholipids and glycolipids, for example, contain fatty acid residues. **Arachidonic acid** ($20{:}4\Delta^{5,8,11,14}$, Table 9.1) serves as a precursor for the synthesis of the eicosanoids, and it appears to help regulate the release of glutamate by nerve cells, a process that is hypothesized to play a central role in learning and memory. Glutamate, one of the 20 common protein amino acids, is the major excitatory neurotransmitter in the central nervous system. **Anandamide** (from a Sanskrit word meaning internal bliss) contains an arachidonic acid residue joined to ethanol amine through an amide link (Figure 9.3). This lipid is the putative (reputed, supposed) endogenous ligand for those receptors on nerve cell membranes to which

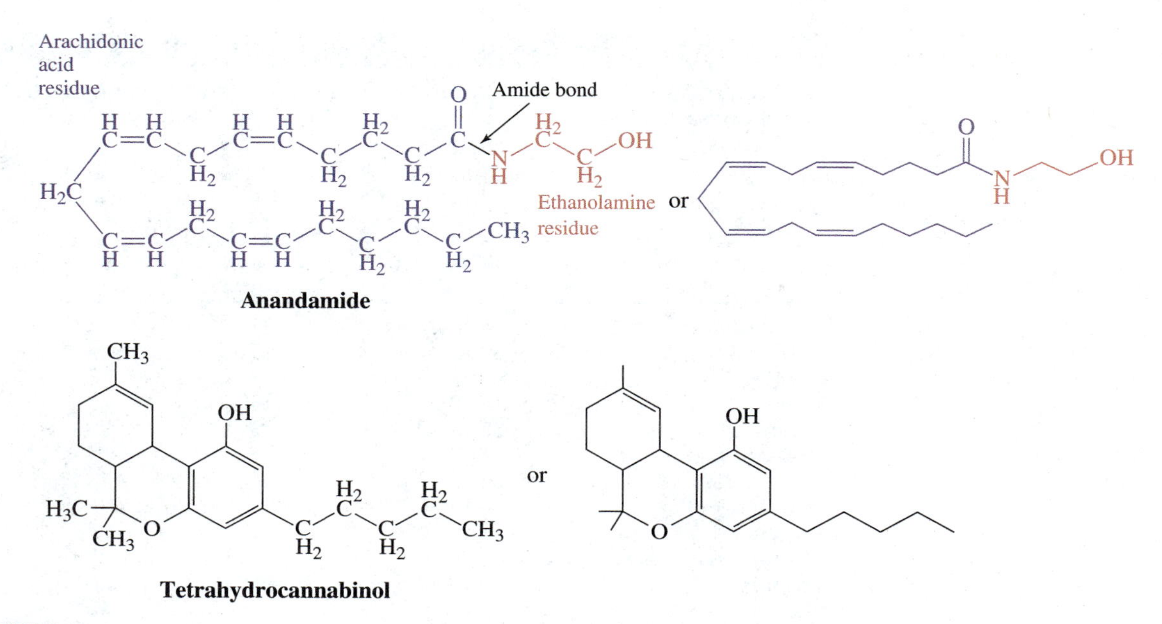

FIGURE 9.3
Anandamide and tetrahydrocannabinol.

tetrahydrocannabinol (the active ingredient in marijuana) attaches. An **endogenous** substance is one produced within the organism in which it resides. The primary amide of *cis*-9-octadecenoic acid [18:1 (Δ^9); oleic acid], named ***cis*-9-octadecenoamide,** appears to be a signaling molecule involved in the induction of physiological sleep. **Docosahexaenoic acid** [**DHA**; 22:6 ($\Delta^{4,7,10,13,16,19}$)], a nutrient in human milk, is essential for normal brain and eyesight development in fetuses and in premature infants. In addition, cells sometimes attach specific fatty acids to particular proteins, and, in most of these instances, the "tagged" proteins are unable to function normally without the fatty acid moiety. The fatty acid residues normally anchor the "tagged" proteins to membranes (Section 9.9). **[Try Problem 9.7]**

Animal Fats Versus Plant Fats

Fats differ markedly in fatty acid composition. Most animal fats, including butter (milk fat) and the fat in red meats, are primarily constructed from saturated fatty acids and are said to be saturated. Most plant fats possess predominantly unsaturated fatty acid residues and are categorized as unsaturated (Table 9.2). Because the tropical oils, including palm oil and coconut oil, are rich in saturated fats, they are rather atypical plant fats.

The cis configuration about the carbon–carbon double bonds found within most plant fats leads to a kinking and a slight shortening of the hydrocarbon chains within fatty acid residues (Figure 9.4). As a consequence, there is a reduction in the number of packing interactions between neighboring fat molecules when unsaturated fats are compared with saturated fats containing the same number of carbons. Since the conversion of a solid to a liquid involves a partial disruption of intermolecular (be-

TABLE 9.2

Approximate Percentage of Saturated and Unsaturated Fatty Acids in Some Dietary Fats

	Solidification Point (°C)	[a]Percentage Saturated Fatty Acids	[a]Percentage Unsaturated Fatty Acids: Mono	Poly
Coconut oil	14 to 22	92	6	2
Corn oil	−20 to −10	13	25	62
Cottonseed oil	−13 to +12	27	18	55
Olive oil	−6	14	77	9
Palm oil	35 to 42	53	38	9
Peanut oil	3	18	48	34
Safflower oil	−18 to −13	9	13	78
Sesame oil	−6 to −4	14	42	44
Soybean oil	−16 to −10	15	25	60
Sunflower-seed oil	−17	11	21	68
[b]Butter	20 to 23	68	28	4
[b]Lard	27 to 30	41	43	16
Margarine (soft)	near 22	19	53	28
Shortening		25	68	7

[a]The values given are representative values. Specific samples of the fats and oils may contain different percentages than those listed.
[b]Animal fats. The other fats and oils are of plant origin.

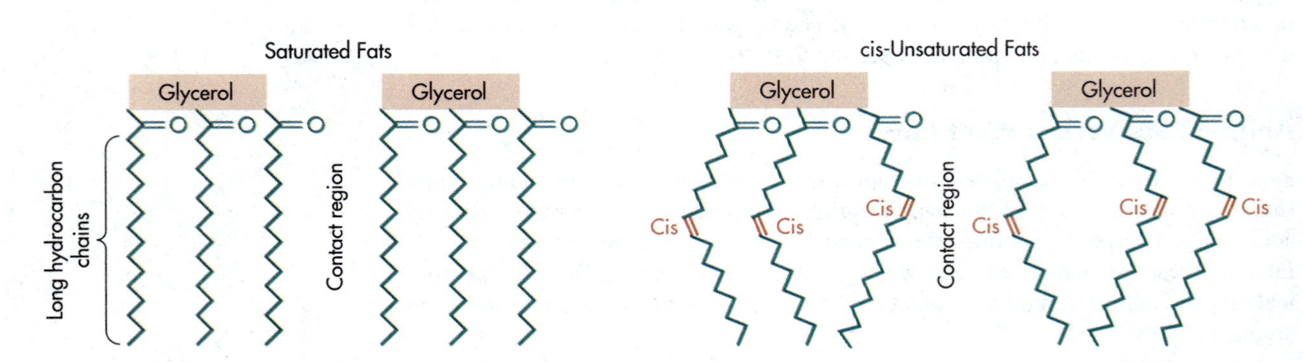

FIGURE 9.4
The impact of cis double bonds on intermolecular contacts. The presence of cis double bonds reduces the efficiency with which neighboring fat molecules pack together; there are fewer neighbor–neighbor interactions (contacts). These diagrams exaggerate the effect of cis double bonds to more clearly illustrate this general phenomenon.

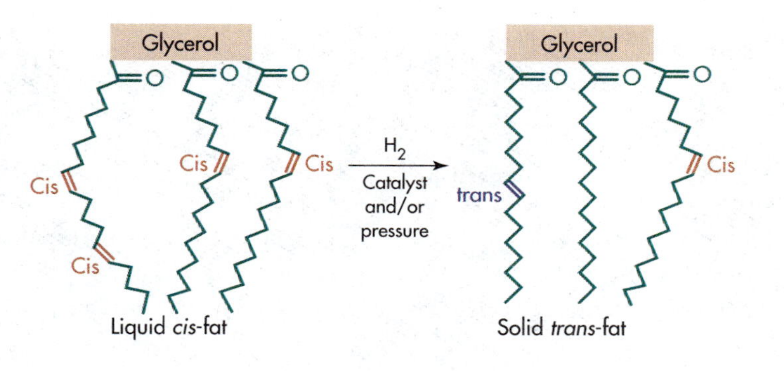

FIGURE 9.5
The hydrogenation of vegetable oils and formation of *trans*-fats.

tween separate molecules) interactions, the reduction in neighbor–neighbor contacts favors the solid to liquid transition. Hence, unsaturated plant fats tend to have lower melting points than more saturated animal fats. If, however, the fatty acid residues in a saturated triacylglycerol are small, this saturated fat molecule may have a lower melting point (solidification point) then a triacylglycerol containing larger, unsaturated fatty acid residues. **Vegetable oils** are plant fats that, in general, melt below room temperature.

The hydrogenation (addition of H_2) of vegetable oils eliminates carbon-to-carbon double bonds. This raises the melting point of the oils and leads to the formation of solid fats, which are commonly marketed as margarines and shortenings. If the hydrogenation process is carefully controlled, hydrogenation is incomplete and the **hydrogenated vegetable oils** retain more carbon-to-carbon double bonds than are found in most animal fats, including butter and lard (Table 9.2). The hydrogenation process, however, does convert a portion of the nonhydrogenated cis double bonds to their trans isomers (Figure 9.5). The possible health consequences of this chemical transformation are under investigation. The results of some studies indicate that dietary ***trans*-fats** may increase heart disease risk, whereas other studies have found a correlation between the amount of *trans*-oleic acid stored in the body and a reduction in heart attack risk. **[Try Problem 9.8]**

Functions of Fats

Most of the triacylglycerol molecules within the human body reside within oily droplets in the cytoplasm of **adipocytes** (Figure 9.6), the specialized fat storage cells found in **adipose** (fat) **tissue. Hepatocytes** (liver cells) and intestine cells typically hold smaller amounts of stored triacylglycerols. Although fat-laden adipose tissue provides insulation, pads vital organs, and stores fat-soluble vitamins and other lipids, the major biological role of triacylglycerols is the storage of energy. Since the complete oxidation (by O_2) of 1 g of fat releases more than twice as much energy as the oxidation of 1 g of either carbohydrate or protein (Table 9.3), fats are well matched to this biological mission. Particular fat molecules do play additional roles as well. Diacylglycerols, for example, sometimes serve as second messengers during the transduction of signals across membranes (Section 10.7). **[Try Problem 9.9]**

TABLE 9.3

Energy of Oxidation for Fats, Carbohydrates, and Proteins

	Energy of Oxidation	
Fuel	**kcal/g**	**kJ/g**
Fat	9	38
Carbohydrate	4	17
Protein	4	17

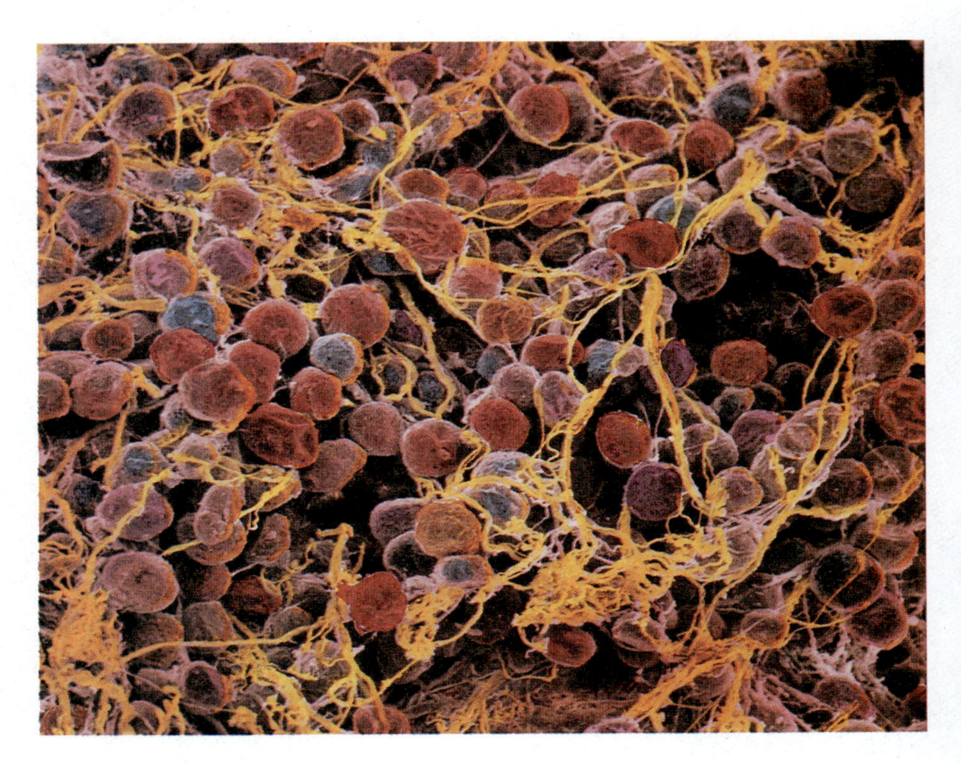

FIGURE 9.6
Scanning electron micrograph of adipocytes (brown, red). Each contains a fat globule that occupies nearly the entire cell.

Some Dietary Considerations

Dietary fats (mainly triacylglyercols) are digested primarily in the small intestine where an enzyme, **pancreatic lipase,** catalyzes the hydrolysis of part of their ester links. To be functional, pancreatic lipase must be complexed with a second pancreatic protein, **colipase.** The fatty acids and monoacylglycerols formed during digestion are absorbed by cells lining the small intestine (the intestinal mucosa) and converted back into triacylglycerols (triglycerides) before they are released into the bloodstream for transport throughout the body. A hypothesized link between blood triglyceride levels and heart disease risk has been difficult to tie down.

Dietary fats promote satiety, are carriers of fat-soluble vitamins, are the major source of essential fatty acids, and help maintain healthy skin and hair. The **essential fatty acids,** including linoleic acid and linolenic acid (Table 9.1), are those required fatty acids that cannot be synthesized by the human body. Essential amino acids are defined in a similar manner (Section 3.4). The essential fatty acids are precursors for many of those lipids known as eicosanoids (Section 9.5).

Although some dietary fat is essential for health, the average American consumes far too much fat, 35 to 40% of total calories. Most nutritionists recommend that no more than 30% of consumed calories come from fat, and many experts argue that even this level of fat consumption is excessive (Figure 9.7a). Since there is a convincing correlation between the total amount of saturated fat in the diet and heart disease risk, fats rich in saturated fatty acid residues should be reduced to less than 10% of total calories. Red meat and dairy products are the primary sources of saturated

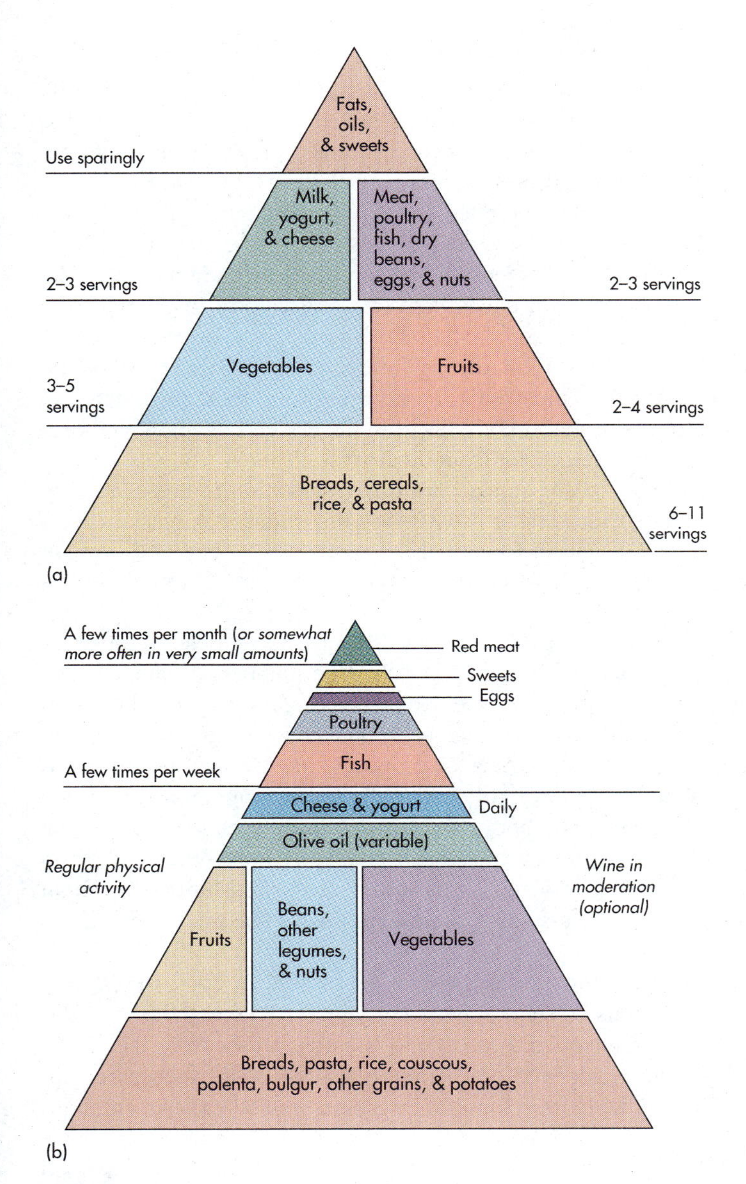

The Official Food Guide Pyramid, a descendant of the four food groups concept promoted widely in the United States in the 1950s and 1960s, was endorsed by the United States Department of Agriculture in 1992. According to this pyramid, your diet should consist mainly of breads, cereals, rice, pasta, vegetables, and fruits. Fats, oils, and sweets should be used sparingly. Red meats, poultry, eggs, and dairy products normally provide most of the dietary fat.

The Mediterranean Diet Pyramid, based on the traditional Mediterranean diet that is associated with low rates of chronic disease and high life expectancies, was endorsed in 1994 by Harvard School of Public Health, Oldways Preservation & Exchange Trust, and the World Health Organization European Regional Office. Although the base of this pyramid is similar to the base of the Official Food Guide Pyramid, the upper portion is unique. Beans, other legumes, and nuts are given a more favorable ranking. Moderate amounts of dietary fat are judged to be accepted as long as the fat comes from olive oil and other oils rich in monounsaturated fatty acids and as long as the fat does not lead to excess body weight. Dairy products, fish, and poultry should be consumed in smaller amounts. Eggs, sweets, and red meat fill out the top of the pyramid. Red meat should be consumed no more than a few times per month. Regular exercise is promoted along with the pyramid. Moderate consumption of wine is categorized as optional, because this practice is linked to reduced heart disease risk. However, wine consumption may elevate other risks, including cancer risk and the risk of becoming an alcoholic.

FIGURE 9.7
Two different views on the ideal food pyramid. (a) Source: U.S. Department of Agriculture/U.S. Department of Health & Human Services. (b) Copyright 1994 Oldways Preservation & Exchange Trust.

fat in a typical United States diet. The saturated fat story is complicated by evidence that at least one saturated fatty acid, stearic acid (Table 9.1), may pose little heart risk. However, most saturated dietary fats contain numerous saturated fatty acids in addition to stearic acid. **[Try Problem 9.10]**

Diets relatively high in fats containing monounsaturated fatty acid residues, such as the traditional Mediterranean diet (Figure 9.7b), may be heart safe. Fats possessing certain ω-3 fatty acids, including EPA ($20{:}5\Delta^{5,8,11,14,17}$) and α-linolenic acid ($18{:}3\Delta^{9,12,15}$), may even lower heart disease risk. This does not necessarily mean that we can safely consume all of the unsaturated fat desired. Such fat has been found to increase cancer risk in experimental animals, and some epidemiological studies suggest that a cancer–dietary fat link may also exist in humans.

Dietary fat, both saturated and unsaturated, is also a major contributor to obesity, another serious health problem in the United States. The human body does not "view" all calories to be the same. There is a much stronger correlation between fat intake and percentage body fat than exists between total calorie intake and percentage body fat. A calorie of fat is more "fattening" than a calorie of protein or carbohydrate for at least two reasons: First, dietary fat causes the body to conserve fat that would, in the absence of the dietary fat, be exported from fat depots; and second, more energy is required to build body fat from dietary carbohydrate and protein than is needed to construct body fat from dietary fat. Whereas roughly 23% of the calories in carbohydrate are consumed during its conversion to body fat, only 3% of the calories in dietary fat are lost during the same process. A partial explanation for this difference is found in Chapter 13. The dietary fat and body weight connection is complex; additional aspects of this relationship are still under investigation.

Weight-loss programs abound, and billions of dollars are spent each year in attempts to shed excess pounds. Success requires that the number of calories consumed be less than the number of calories the body burns or eliminates in some other manner. Growing evidence indicates that each individual is genetically programmed to have a body weight within a certain range, a proposal known as **set point theory.** Those who insist on maintaining a weight outside the lower programmed limit are usually facing an uphill battle. The body's **basal metabolic rate** (determines the amount of energy consumed by the body at rest), which tends to be relatively low in individuals programmed to be overweight, seems to decline even further as weight drops below this limit. The body may also begin to oxidize some muscle protein to conserve fat.

Researchers are still searching for the specific genes that set body weight (assuming that set point theory is correct). One potential candidate is the ***obese (ob)* gene.** The protein hormone encoded in this gene is called ***leptin*** from the Greek word for *thin.* Leptin, which appears to act directly on the brain, suppresses appetite, reduces food intake, and alters some basic metabolism in experimental animals. The injection of leptin into leptin-deficient obese mice leads to around a 30% loss in body weight, almost exclusively from the loss of fat. When injected into normal mice, leptin leads to the loss of almost all body fat. A second obesity-linked gene encodes an enzyme named **carboxypeptidase E** which appears to be involved in converting certain inactive proproteins into active proteins, including some hormones. According to one hypothesis, abnormalities in genes for carboxypeptidase E lead to metabolic imbalances that lead to obesity. The precise nature of these metabolic imbalances is under investigation. Regardless of one's genetics, exercise should be at the heart of any weight-loss program. Exercise burns calories, boosts basal metabolic rates, relieves stress, and benefits the body in additional ways.

Problem 9.3 Mono- and diacylglycerols exist that differ in structure from those shown in Figure 9.2. Write generalized structural formulas (of the type shown in Figure 9.2) for these compounds. The mono- and diacylglycerols depicted in Figure 9.2 are the ones found in living organisms and the ones produced during the biosynthesis of triacylglycerols (Section 14.9).

Problem 9.4 Circle each chiral center in each compound in Figure 9.2. Explain why these chiral centers have a unique configuration.

Problem 9.5 Write the structural formula for a polyunsaturated carboxylic acid containing 14 carbons and a trans configuration about each carbon-to-carbon double bond. Experimentally, how can you determine whether or not this carboxylic acid is a component of human fat?

Problem 9.6 Which of the compounds in Table 9.1 are ω-6 fatty acids? ω-9 fatty acids?

Problem 9.7 Name two distinct, hydrolyzable chemical bonds that can be used to covalently attach a fatty acid to a protein.

Problem 9.8 True or false? Completely hydrogenated vegetable oils are saturated fats. Explain.

Problem 9.9 What structural feature of fats accounts for the fact that 1 g of fat contains over twice the calories as 1 g of protein or carbohydrate? Hint: The oxidation of organic compounds usually involves the addition of oxygen and/or the removal of hydrogen.

Problem 9.10 A cookie contains 50 g of digestible carbohydrate, 25 g of fat, and 5 g of protein. Calculate the percentage of the total calories contributed by the fat.

9.2 WAXES

Natural **waxes,** a second major class of lipids, are fatty acid esters of high molecular weight alcohols. Both the fatty acid and the alcohol are usually saturated. Myricyl palmitate, the major component of beeswax, provides a specific example (Figure 9.8). Waxes serve as protective coatings and water barriers for many organisms. Feathers, animal skins, fur, and leaves are commonly coated with these nonpolar compounds. Some organisms use waxes for other purposes as well. Sperm whales, for example, employ waxes to control buoyancy and to produce sonic booms. **[Try Problem 9.11]**

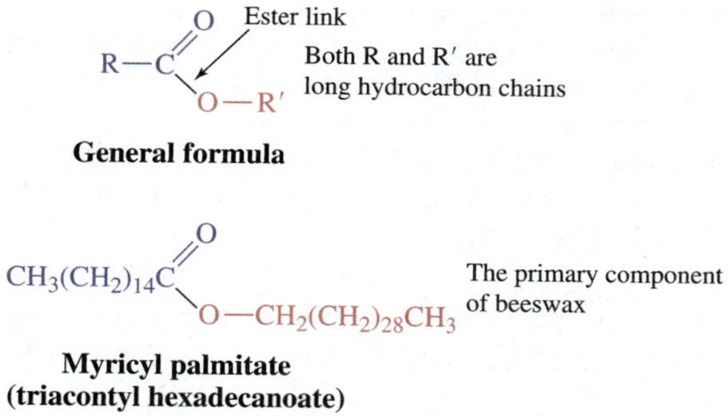

$$R{-}C(=O){-}O{-}R'$$

Ester link

Both R and R′ are long hydrocarbon chains

General formula

$$CH_3(CH_2)_{14}C(=O){-}O{-}CH_2(CH_2)_{28}CH_3$$

The primary component of beeswax

Myricyl palmitate (triacontyl hexadecanoate)

FIGURE 9.8
Structure of waxes.

Although natural waxes have been used to make candles, lubricants, cosmetics, adhesives, and varnishes, synthetic substances are now used in place of lipids in many of these products. The "paraffin wax" commonly employed as a sealant during home canning operations is not a natural wax or even an ester. It is a mixture of high molecular weight alkanes.

Problem 9.11 True or false? Fats are substantially more reactive than natural waxes. Explain.

9.3 PHOSPHOLIPIDS

Phospholipids, a third major class of lipids, are nonpolar (predominantly) substances containing one or more phosphate residues. Many also possess one or more fatty acid residues. The **glycerophospholipids (phosphoglycerides)** are the most abundant members of this class. The simplest glycerophospholipids are the **phosphatidic acids,** diacylglycerols joined to phosphoric acid through an ester link. Salts of phosphatidic acids, known as **phosphatidates,** predominate at physiological pH. The many unique phosphatidates that have been characterized differ only in their fatty acid residues. The other glycerophospholipids are derivatives of phosphatidates in which an alcohol has been esterified to the phosphate residue. These subclasses of glycerophospholids include the **phosphatidylcholines, phosphatidylethanolamines, phosphatidylinositols,** and **phosphatidylserines** (Table 9.4). **[Try Problems 9.12 and 9.13]**

All phospholipids are amphipathic compounds containing polar "heads" that tend to dissolve in water and nonpolar "tails" that tend to aggregate in an aqueous environment. For this reason, phospholipids spontaneously form micelles or **lipid bilayers** (Section 2.2 and Figure 9.9) in the presence of water. A lipid bilayer constitutes the core of all biological membranes, and phospholipids, including those in Table 9.4, are normally the most abundant lipids in this bilayer. When a lipid bilayer is induced to fold into a spherical vesicle containing an aqueous core, the product is called a **liposome** (Figure 9.9). Since these artificially produced compartments readily fuse with plasma membranes, they are used as drug delivery systems and are em-

TABLE 9.4

Some Glycerophospholipids

Subclass	Structure	Alcohol Esterified to Phosphatidate
Phosphatidates	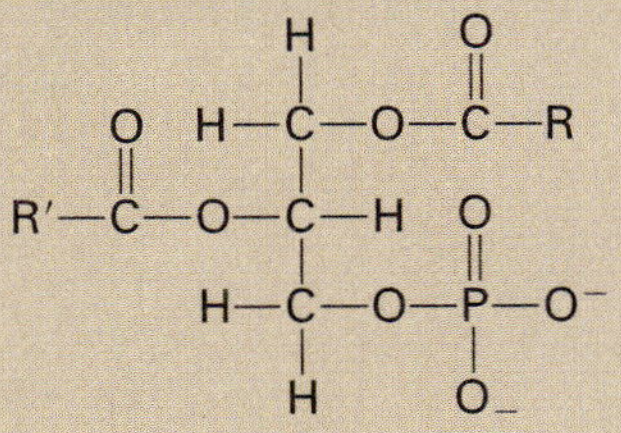 H O O H—C—O—C—R R′—C—O—C—H O H—C—O—P—O^- H O^-	—
Phosphatidylcholines (lecithins)	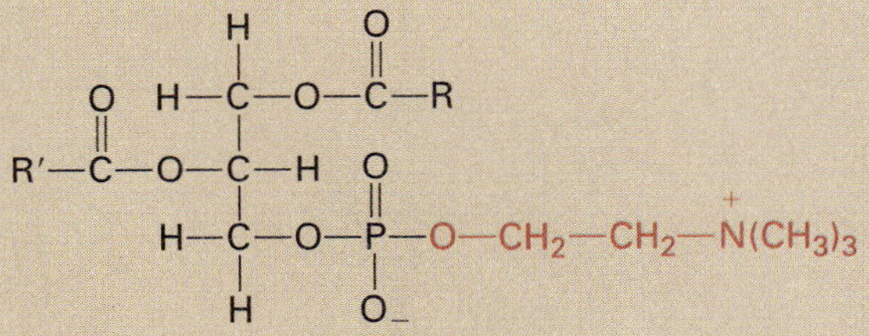 H O O H—C—O—C—R R′—C—O—C—H O H—C—O—P—O—CH_2—CH_2—$\overset{+}{N}(CH_3)_3$ H O^-	HO—CH_2—CH_2—$\overset{+}{N}(CH_3)_3$ (choline)
Phosphatidylethanolamines (cephalins)	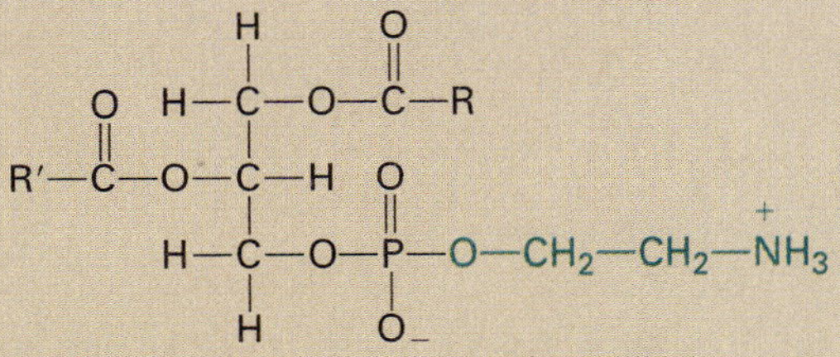 H O O H—C—O—C—R R′—C—O—C—H O H—C—O—P—O—CH_2—CH_2—$\overset{+}{N}H_3$ H O^-	HO—CH_2—CH_2—$\overset{+}{N}H_3$ (ethanolamine)
Phosphatidylinositols	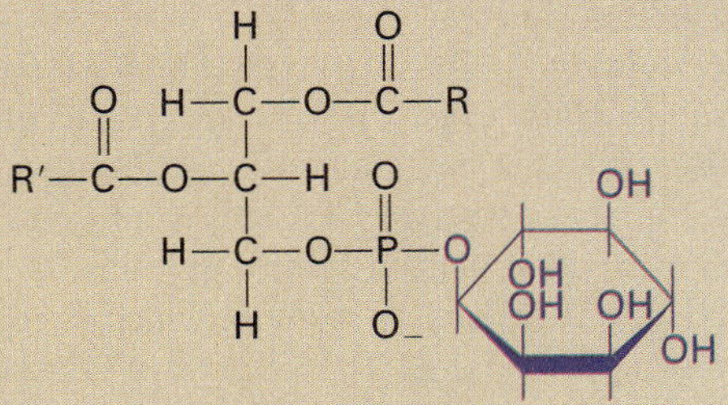 H O O H—C—O—C—R R′—C—O—C—H O OH H—C—O—P—O OH H O^- OH OH OH	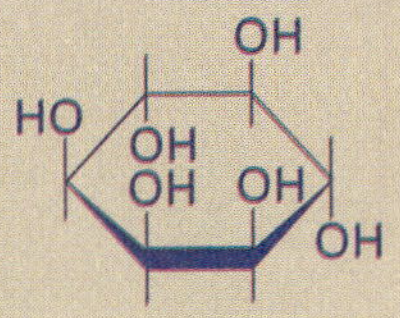 OH HO OH OH OH OH (myo-inositol)
Phosphatidylserines	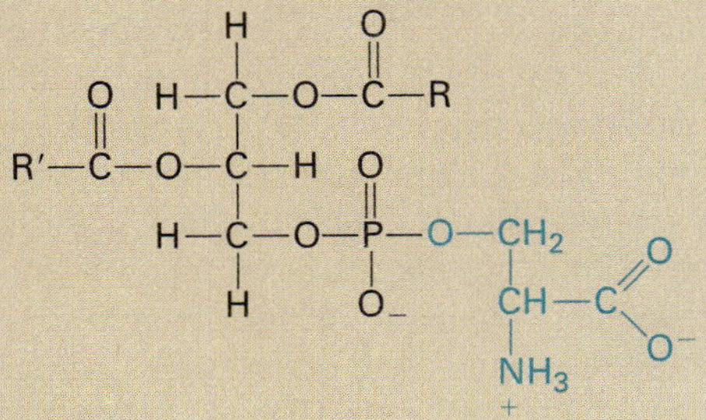 H O O H—C—O—C—R R′—C—O—C—H O H—C—O—P—O—CH_2 O H O^- CH—C $\overset{+}{N}H_3$ O^-	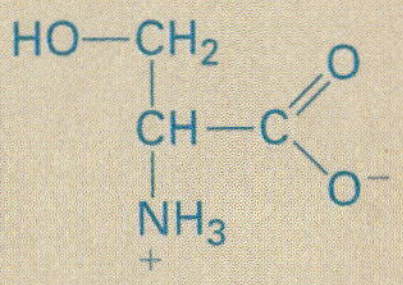 HO—CH_2 O CH—C $\overset{+}{N}H_3$ O^- (serine)

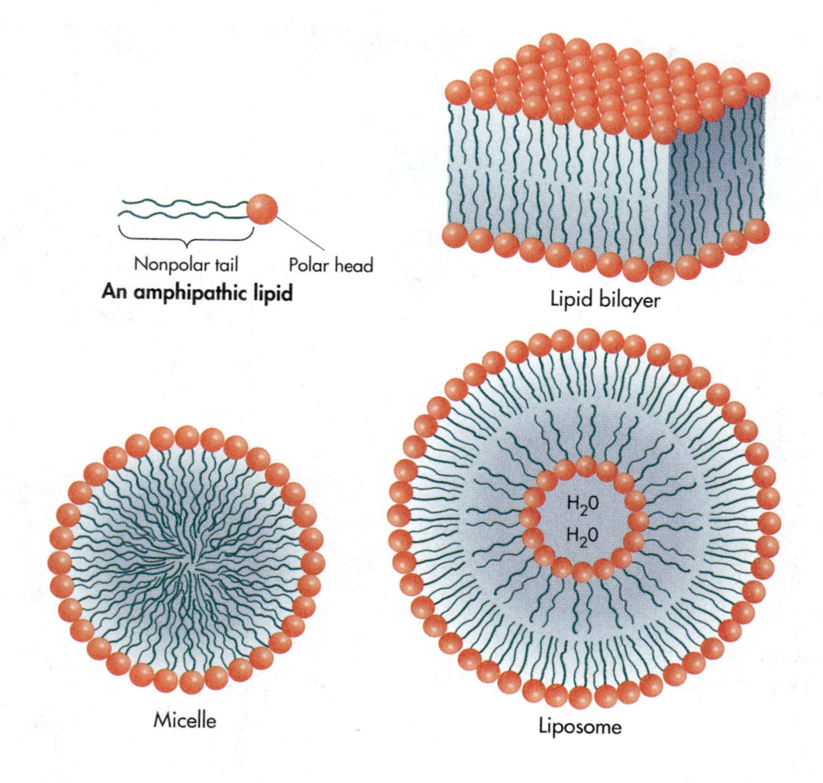

FIGURE 9.9
A micelle, a lipid bilayer, and a liposome.

ployed to introduce recombinant DNAs into host cells during gene therapy (Table 19.5). **[Try Problems 9.14 and 9.15]**

The **sphingomyelins** are the second most abundant and widely distributed phospholipids (Figure 9.10). They make up one of the three major classes of **sphingolipids,** lipids that contain a **sphingosine** residue. All sphingolipids possess a **ceramide** moiety comprised of a fatty acid joined to the amino group of shingosine through an amide bond. In sphingomyelins, the C1-hydroxyl group of sphingosine is attached to a phosphate residue, which is joined through an ester link to either choline or ethanolamine. The C3-hydroxyl group of sphingosine is normally free. Sphingomyelins, like the glycerophospholipids, are amphipathic compounds and components of the lipid bilayer in a wide variety of cellular membranes. The lipid portion of the **myelin sheath** (the membrane that surrounds nerve fibers), for example, contains approximately 5% sphingomyelins. **[Try Problem 9.16]**

A growing number of phospholipids are now known to have unique and physiologically important roles outside of their contributions to the construction and maintenance of lipid bilayers. Particular phosphatidylcholines and phosphatidylinositols (Table 9.4) are coupled to membrane receptors for specific hormones, growth factors, neurotransmitters, and other ligands. These phospholipids play a central role in signal transduction following ligand binding, a fact examined in detail in Chapter 10. **Platelet-activating factor (PAF)** is a hormone and an ether lipid described as an **alkylacylglycerophospholipid** (Figure 9.11). After it has been secreted by a white blood cell or another producer, it binds to specific receptors on a variety of target

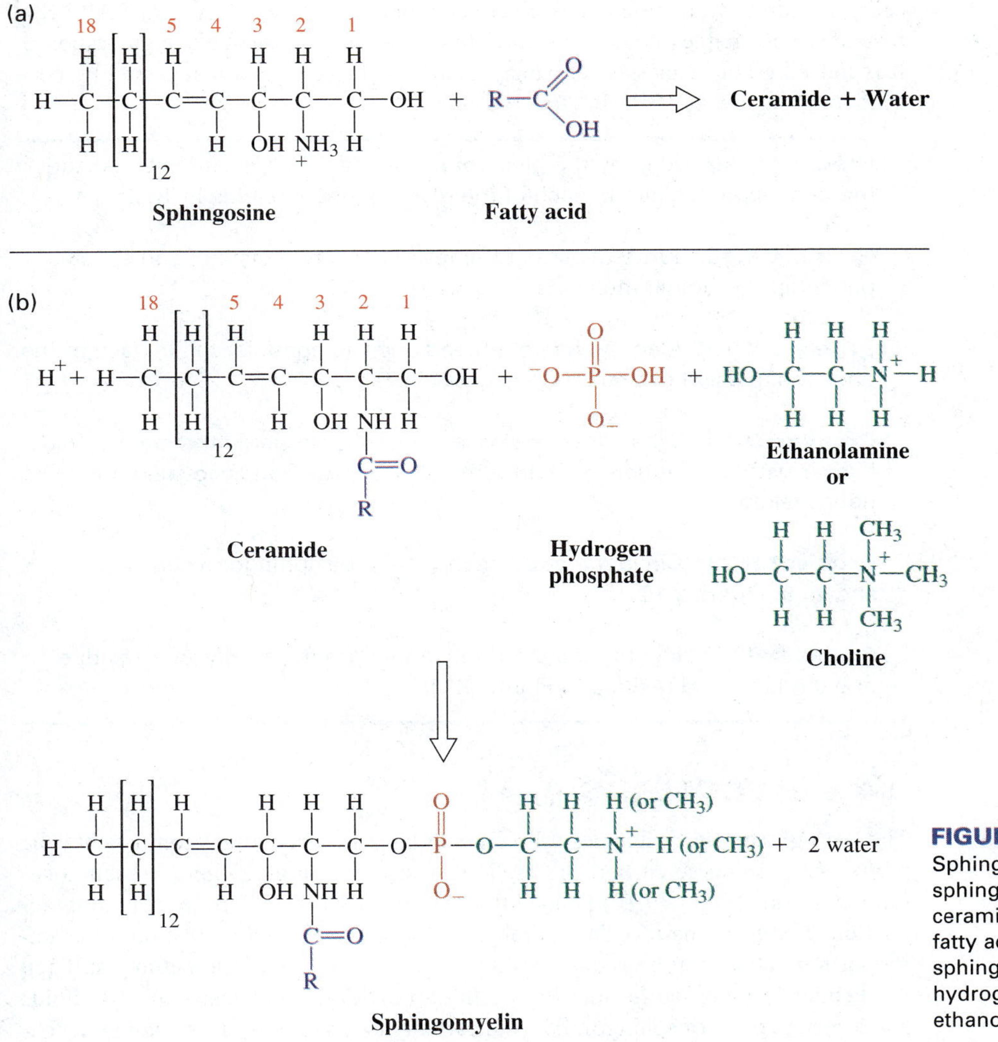

FIGURE 9.10
Sphingosine, ceramide, and sphingomyelin. (a) The formation of ceramide from sphingosine and a fatty acid. (b) The formation of sphingomyelin from ceramide, hydrogen phosphate, and ethanolamine or choline.

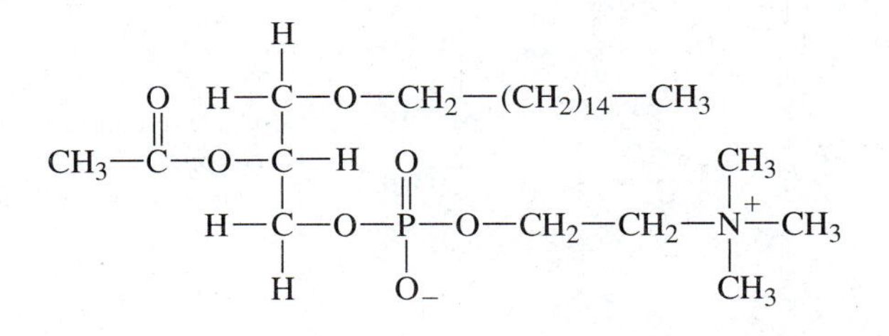

FIGURE 9.11
Platelet-activating factor (PAF).

cells, including white blood cells, liver cells, lung cells, and brain cells. PAF helps regulate a wide range of processes, including inflammation and the allergic response. It is indeed a potent messenger, since some responsive cells are activated by concentrations as low as 10^{-12} *M*. [Try Problem 9.17]

Problem 9.12 Identify the glycerol residue, the fatty acid residues, and the phosphate residue in each of the glycerophospholipids in Table 9.4.

Problem 9.13 Write the structural formula for two distinct and specific phosphatidylcholine molecules.

Problem 9.14 Identify the polar head and the nonpolar tail in each of the phospholipids whose structure is given in Table 9.4.

Problem 9.15 What class or classes of bonds primarily maintain a lipid bilayer within an aqueous environment? Are these bonds covalent or noncovalent?

Problem 9.16 Circle the polar head within the sphingomyelin structure shown in Figure 9.10.

Problem 9.17 Identify the ether functional group, the glycerol residue, and the fatty acid residue in Figure 9.11.

9.4 GLYCOLIPIDS

The term **glycolipid** designates a lipid containing a carbohydrate component. The prefix *glyco-* refers to carbohydrates. The most common glycolipids contain a ceramide residue (Figure 9.10) bonded through its C1-hydroxyl group to a monosaccharide or oligosaccharide. The carbohydrate component, which constitutes the major portion of the polar head of these amphipathic substances, is sometimes sulfated or chemically modified in other ways. In **cerebrosides,** a subclass of glycolipids and a second class of sphingolipids, the carbohydrate component is a monosaccharide (frequently glucose or galactose) (Figure 9.12). Some membranes are particularly rich in cerebrosides. Galactose-containing cerebrosides, for example, account

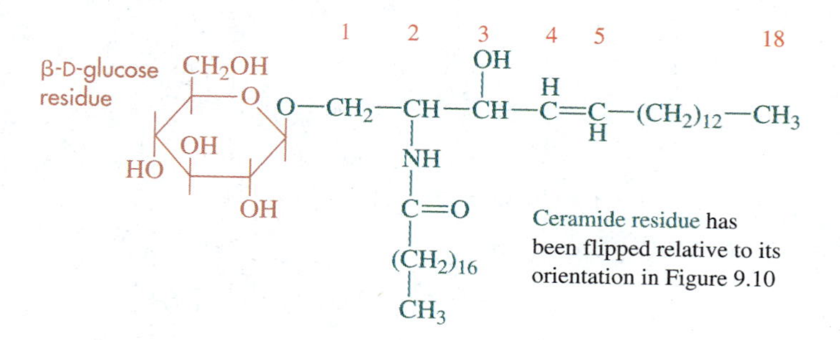

FIGURE 9.12
A cerebroside.

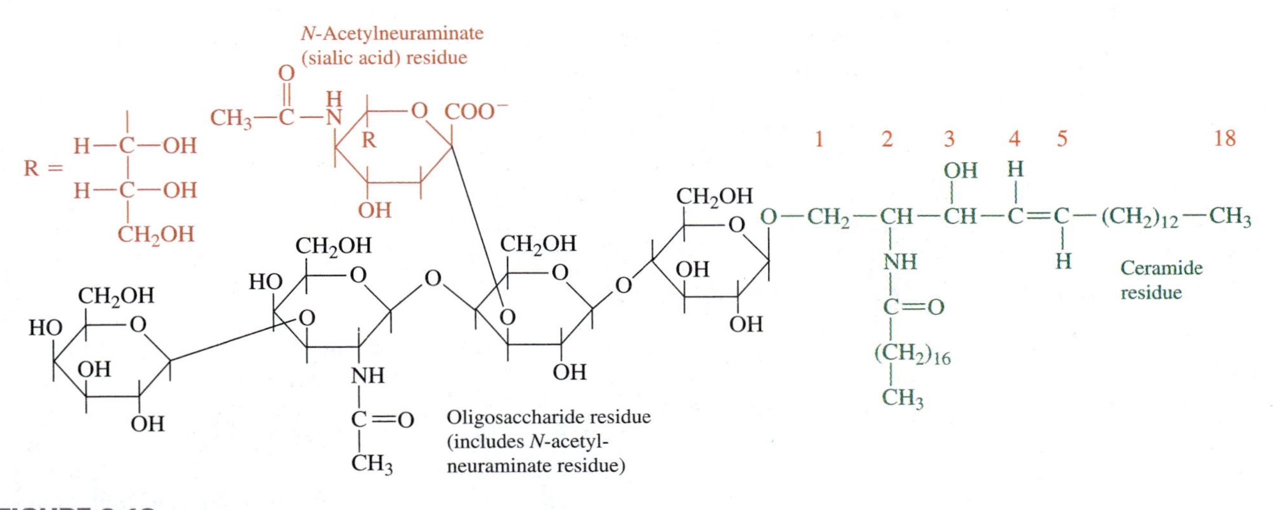

FIGURE 9.13
A ganglioside.

for approximately 15% of the total lipids in the myelin sheath. **[Try Problems 9.18 through 9.20]**

Gangliosides, like cerebrosides, are both glycolipids and sphingolipids. In these amphipathic compounds, an oligosaccharide component containing at least one *N*-acetylneuraminate residue (the salt form of *N*-acetylneuraminic acid) replaces the monosaccharide moiety that characterizes cerebrosides (Figure 9.13). Since *N*-acetylneuraminic acid is negatively charged at physiological pH, the head of these membrane lipids is more polar than the uncharged head associated with cerebrosides.

Glycolipids, like many phospholipids, are more than just building blocks for a lipid bilayer. Glycolipid "markers" on cell membranes are involved in such key processes as cell–cell interactions, the control of cell division and differentiation, the transformation of normal cells into cancer cells, infection by viruses, and the binding of hormones and other substances by cells. Blood group antigens, which determine blood type (A, B, AB, or O), are specific examples of glycolipid markers. In general, it is the carbohydrate component of glycolipids that is specifically "recognized" during binding events. Likewise, the carbohydrate moieties in glycoproteins play a key role when these compounds serve as cellular markers (Section 6.7).

The central importance of glycolipids in life processes is documented by the existence of Tay–Sachs disease, a common recessively inherited ailment (Chapter 19). An enzyme deficiency in affected individuals leads to the accumulation of a partially-degraded ganglioside that contributes to nervous system degeneration. A typical victim dies between 3 and 4 years of age after gradually becoming blind, paralyzed, and unaware of surroundings.

Problem 9.18 Use common sense to define "glycosphingolipids". Are sphingomyelins glycosphingolipids? Explain.

Problem 9.19 What specific chemical link joins the carbohydrate and ceramide residues in Figure 9.12?

***Problem 9.20** To what classes of organic compounds does the cerebroside in Figure 9.12 belong? Use an arrow to identify each hydrolyzable bond in this lipid.

9.5 EICOSANOIDS

The **eicosanoids** (from the Greek word *eikosi,* meaning twenty) are an additional major class of lipids. All contain 20 carbon atoms, and most are produced from arachidonic acid ($20{:}4\Delta^{5,8,11,14}$; Table 9.1). They encompass multiple subclasses including **prostaglandins (PG), thromboxanes (TX)** and **leukotrienes (LT)** (Figure 9.14). The names for these three subclasses are based on those cells in which they were first detected: prostaglandins in the prostate gland; thromboxanes in thrombocytes; and leukotrienes in leukocytes.

The prostaglandins are characterized by a five-membered carbon ring, while the thromboxanes possess a six-membered ether (oxane) ring. Although some leukotrienes lack a ring system, they all contain three conjugated double bonds. The individual members of the subclasses of eicosanoids are identified with a capital let-

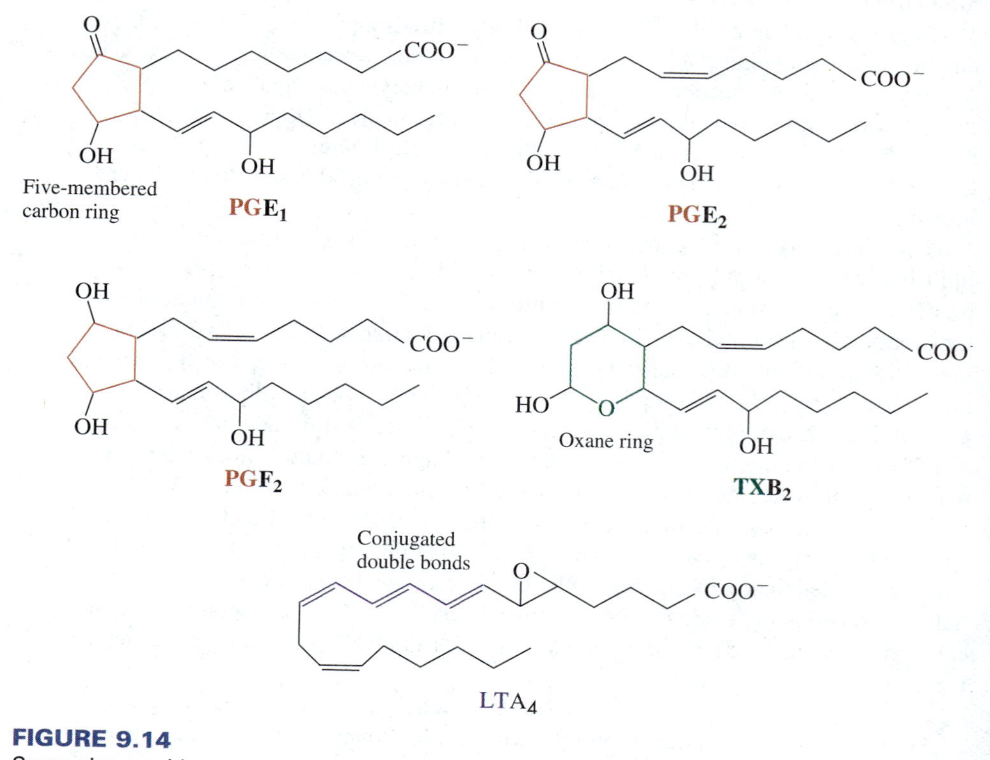

FIGURE 9.14
Some eicosanoids.

ter followed by a subscript that indicates the number of double bonds outside of ring systems. PGE_2, for example, is the abbreviation for a prostaglandin E that contains two carbon to carbon double bonds outside of its five-membered ring. All prostaglandins assigned the same capital letter have the same substituents on their five-membered ring. Similar considerations are used to assign letters to unique categories of thromboxanes and leukotrienes. **[Try Problems 9.21 and 9.22]**

Most eicosanoids are now viewed as "local hormones" by many biochemists, and they are classified as hormones (chemical messengers; examined in detail in Section 10.3) in this text. Most are produced in minute amounts, have brief half-lives, and travel only short distances to reach their target cells. Some also impact the metabolism of the cells that produce them. They commonly bind to specific receptors on target cells and, in the process, trigger the production of a second messenger, often cyclic AMP (cAMP) or cyclic GMP (cGMP) (Sections 7.3 and 10.5). One or more eicosanoids are produced by most human cells, and they help regulate a broad range of physiological processes including smooth muscle contraction, nerve transmission, blood clotting, electrolyte balance, inflammation, immune response, and tissue repair. Similar eicosanoids may bring about opposite physiological responses. PGF_2, for example, stimulates venous muscle contraction whereas PGE_1 relaxes the same muscles.

A variety of drugs, including some being developed and tested as anticancer agents, antiviral agents, or sleeping pills, are eicosanoids or eicosanoid derivatives. Certain other drugs act by modifying eicosanoid metabolism. The anti-inflammatory properties of **aspirin** and several other nonsteroidal drugs are attributed to their inhibition of one or more enzymes involved in prostaglandin production. The ability of small doses of aspirin (a baby aspirin, or possibly less, a day) to lower the risk of heart attacks and stroke is hypothesized to stem from its eicosanoid-linked inhibition of platelet aggregation, an event involved in the formation of blood clots.

RU486 (mifepristone), the highly publicized abortion pill, is normally administered along with a prostaglandin that helps expel the embryo by strengthening uterine contractions. RU486 (see Figure 9.16) is, itself, a steroid and a progesterone (see Figure 9.16) antagonist; it blocks progesterone action by occupying progesterone receptors without stimulating a response. RU486, progesterone and other steroids are examined in the next section. The medical uses of the known eicosanoids and eicosanoid-linked drugs are expanding rapidly, and continuing research is steadily uncovering other physiologically important members of this complex and heterogeneous group of lipids.

Problem 9.21 Interpret the abbreviation TXB_3

Problem 9.22 To what major classes of organic compounds does PGE_1 belong?

9.6 STEROIDS

Steroids, another major category of lipids, are a diverse group of compounds that play a variety of roles within most living organisms, including humans. They are

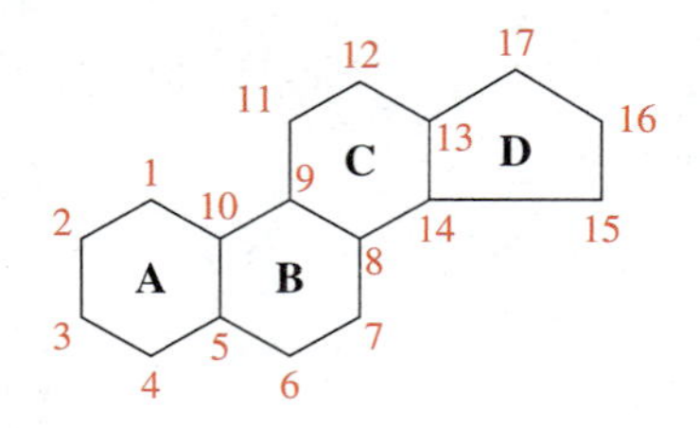

FIGURE 9.15
Perhydrocyclopentanophenanthrene.

characterized by the presence of a perhydrocyclopentanophenanthrene ring system in which the four rings are identified with the capital letters A, B, C, and D (Figure 9.15). Individual steroids differ in the location and the number of double bonds within the rings or in the nature or location of ring substituents or both. **[Try Problem 9.23]**

Sterols are steroids in which a hydroxyl group is attached to C3 and a hydrocarbon chain is attached to C17 of the steroid nucleus. **Cholesterol** (Figure 9.16), a sterol synthesized mainly in the liver, is the most abundant steroid in the human body. It is a normal component of many animal cell membranes and a precursor to the synthesis of bile acids, steroid hormones, and one form of vitamin D. Cholesterol is commonly viewed as a "bad guy" in spite of the essential and normal roles that it plays within the human body. This image stems from a highly significant correlation between blood cholesterol levels and risk of heart disease. Cholesterol is rarely found in plants or prokaryotes.

The **corticosteroids,** including **cortisol** (Figure 9.16), are steroid hormones. These products of the adrenal cortex help regulate the metabolism of electrolytes (compounds that dissolve in water to yield solvated ions and solutions that conduct electricity), proteins, carbohydrates, and a variety of other substances. Cortisol **(hydrocortisone)** and **cortisone,** the 11-keto derivative of cortisol, are well known in medicine for their anti-inflammatory properties. **[Try Problem 9.24]**

Human **sex hormones** and their derivatives are also steroids. They include **testosterone,** a male sex hormone, and the two female sex hormones **estradiol** (an **estrogen**) and **progesterone** (Figure 9.16). The sex hormones regulate the development and maintenance of secondary sexual characteristics and sexual organs. Commer-

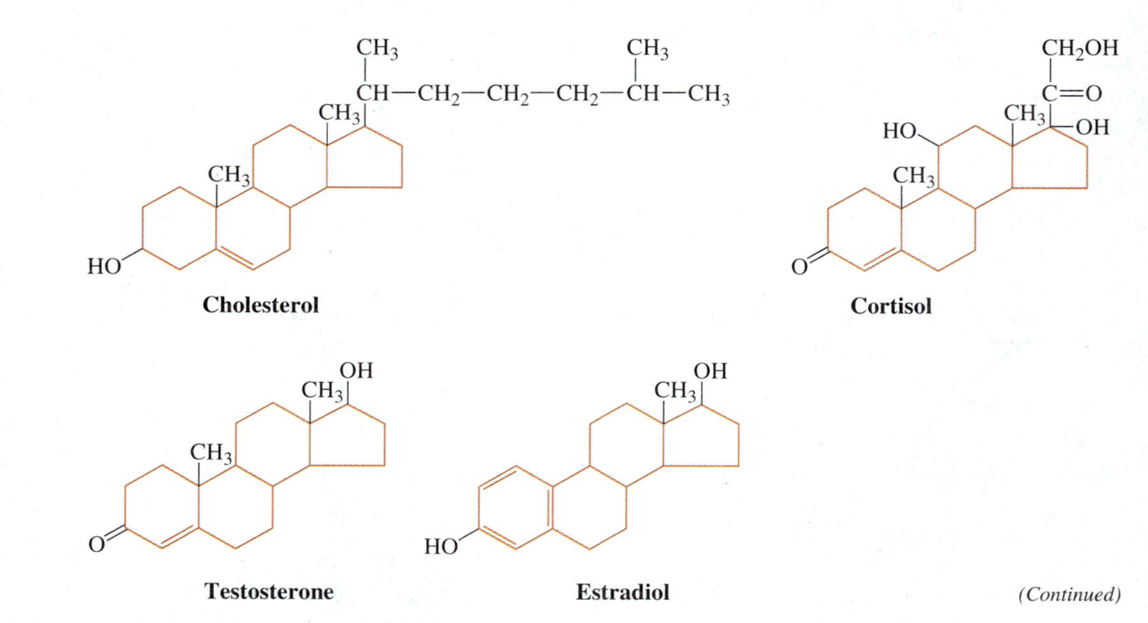

FIGURE 9.16
Some steroids.

(Continued)

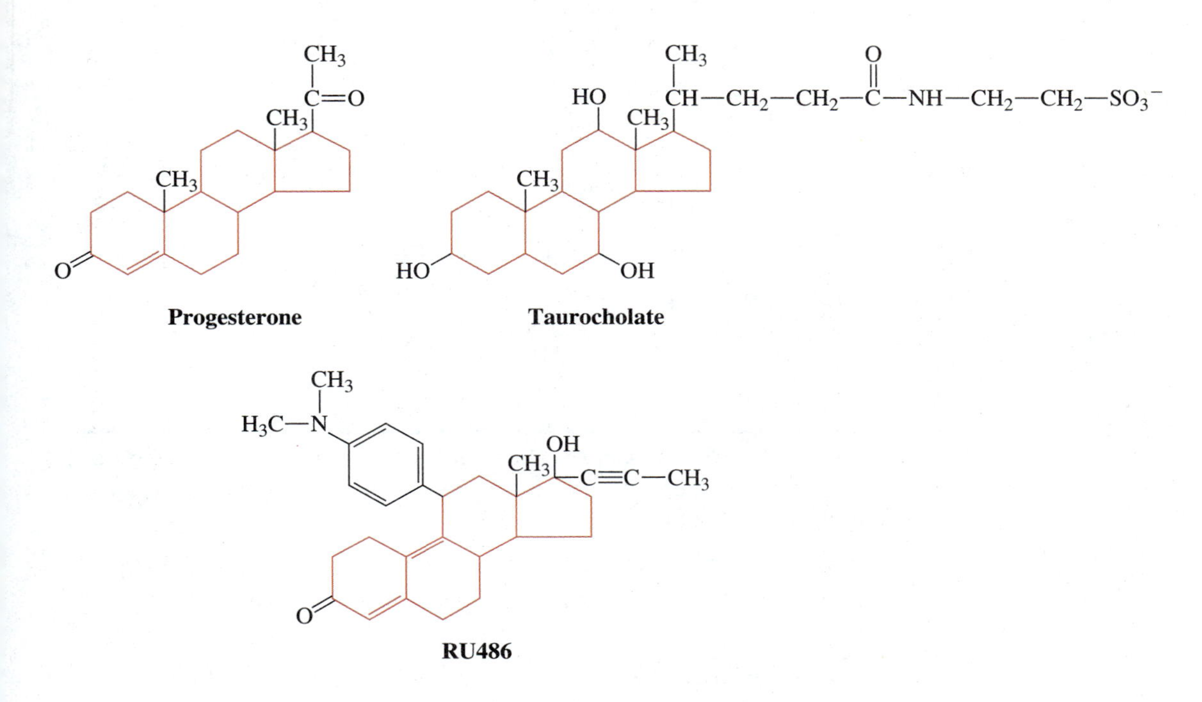

cially, derivatives of these steroids are used in birth control pills and in multiple therapeutic agents, including some designed to treat postmenopausal symptoms. Male sex hormones and some of their artificially synthesized derivatives are known as **anabolic steroids,** because they stimulate specific anabolic (synthetic) processes, including muscle production. These substances are sometimes abused by athletes who are attempting to improve performance. This practice can lead to abnormal aggressiveness, liver damage, and heart disease. It can also cause impotence and trigger changes in secondary sexual characteristics. Hormone imbalances, in general, can lead to serious health problems. **[Try Problem 9.25]**

Bile is a mixture of steroids, other organic compounds, and certain inorganic substances. It is produced by the liver, stored in the **gall bladder,** and delivered to the upper region of the small intestine (the **duodenum**) by the **common bile duct. Bile salts** and phosphatidylcholines are the most abundant components of bile. The bile salts, including **taurocholate** (Figure 9.16), are amphipathic steroids with polar and nonpolar parts. They aid digestion in the small intestine by helping nonpolar lipids mix with the more polar digestive enzymes. In a similar manner, amphipathic, surface-active detergents help nonpolar dirt mix with wash water in a washing machine. **[Try Problem 9.26]**

Several antimicrobial steroids have been identified in plants, and one, named **squalamine** (Figure 9.17), was isolated from the dogfish shark *(Squalus acanthias)* in 1993. This antibiotic kills a broad range of microorganisms, including ***Candida,*** a fungus that commonly infects individuals with suppressed immune systems.

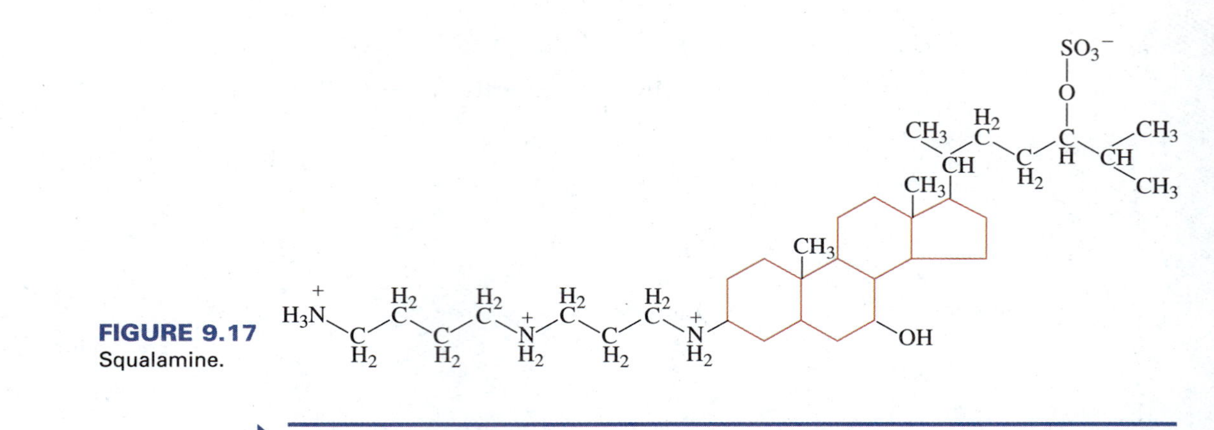

FIGURE 9.17
Squalamine.

Problem 9.23 Circle each chiral center in the steroids shown in Figure 9.16. Explain why these steroids have specific configurations (not shown) about these chiral centers.

Problem 9.24 Study Figures 9.15 and the structure of cortisol in Figure 9.16, and then write the structural formula for cortisone, the 11-keto derivative of cortisol.

Problem 9.25 Use past knowledge or common sense to describe two secondary sexual characteristics.

Problem 9.26 Circle each polar functional group in the formula for taurocholate in Figure 9.16. Which one of the circled groups is most polar?

9.7 LIPOPROTEINS

Since lipids, including steroids, do not dissolve in aqueous fluids, blood lipids are usually bound to specific carrier proteins. Although unesterified fatty acids tend to be transported by serum albumin, many other lipids bind to one or more of over a dozen unique proteins to form lipid–protein complexes known as **lipoproteins.** The protein components of these complexes are called **apolipoproteins** or **apoproteins.** The prefix *apo-* designates a complex protein minus its normal prosthetic group. The liver and the intestine are the primary sites of both apolipoprotein production and lipoprotein assembly. Differences in density are used to divide lipoproteins into five major classes: **chylomicrons, very low-density lipoproteins (VLDL), intermediate-density lipoproteins (IDL), low-density lipoproteins (LDL),** and **high-density lipoproteins (HDL)** (Table 9.5). One additional lipoprotein, **lipoprotein(a),** has also been identified. It is sometimes classified as an LDL variant because the removal of apolipoprotein(a) from the lipoprotein(a) complex leaves an LDL particle. All lipoproteins contain a hydrophobic core (primarily triacylglycerols or cholesterol esters) surrounded by amphipathic lipids (predominantly phospholipids and free cholesterol) and apolipoproteins (Figure 9.18). The apolipoproteins and the polar heads of the amphipathic lipids account for the ability of these lipid complexes to mix with blood. **[Try Problems 9.27 and 9.28]**

TABLE 9.5

Some Properties of the Five Major Classes of Lipoproteins

	Chylomicron	VLDL	IDL	LDL	HDL
Density (g/mL)	<0.95	0.95–1.006	1.006–1.019	1.019–1.063	1.063–1.210
Diameter (nm)	75–1200	30–80	25–35	18–25	5–12
Percentage protein (dry weight)	1–2	10	18	25	33
Major core lipids	Triacylglycerols	Triacylglycerols	Triacylglycerols, cholesterol esters	Cholesterol esters	Cholesterol esters
Major apolipoproteins	A-I, A-II B-48 C-I, C-II, C-III	B-100 C-I, C-II, C-III E	B-100 C-I, C-II, C-III E	B-100	A-I, A-II C-I, C-II, C-III D E

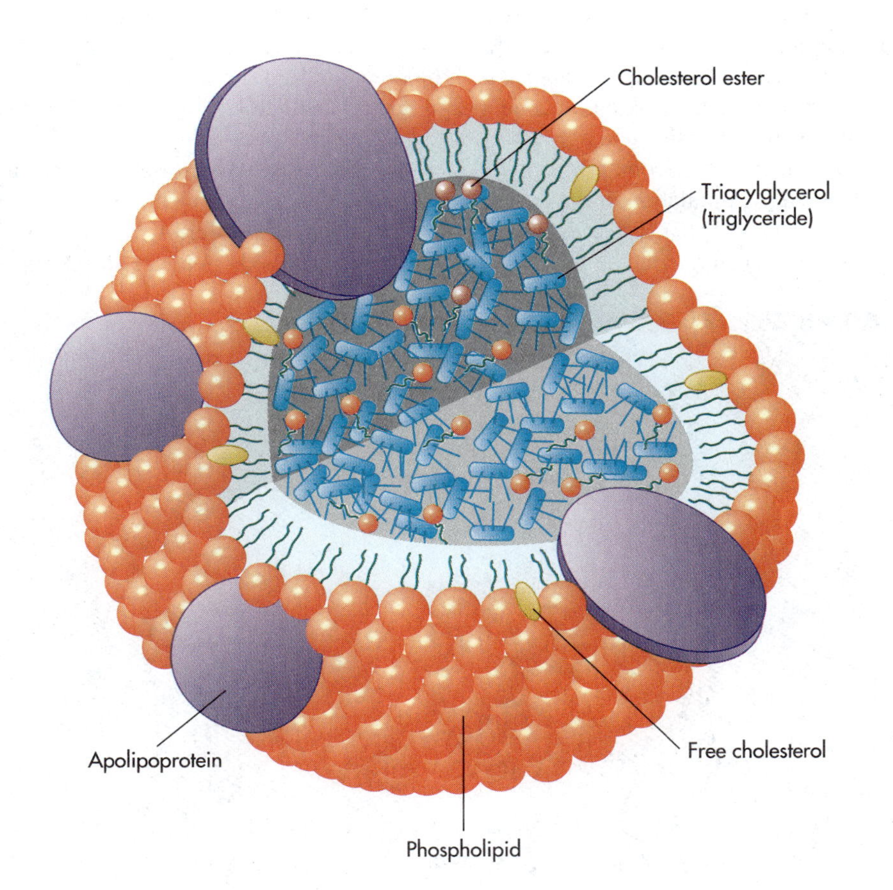

FIGURE 9.18
Generalized structure of a lipoprotein, a spherical particle consisting of a core of triglycerols and cholesterol esters with a shell ~ 3 nm thick containing apolipoproteins, phospholipids, and unesterified cholesterol. Apolipoproteins are embedded with their hydrophobic edges oriented toward the core and their hydrophilic edges toward the outside. Modified from R. B. Weinberg. Lipoprotein metabolism: hormonal regulation. From Devlin, 1992.

The lipoproteins are more than inert transport agents. Most play key roles in determining how the lipids they carry are utilized. The role of lipoproteins in cholesterol metabolism and heart disease is examined in the following section. Lipoprotein metabolism is coupled to other disease processes, as well. A particular variant *(E4)* of the gene that encodes apolipoprotein E is, for example, linked to an elevated risk of **Alzheimer's disease,** a degenerative brain disorder afflicting roughly four million individuals in the United States. Having one copy of the variant gene elevates risk three- to fourfold, and the presence of a second copy leads to a substantial additional risk. The molecular explanation for this correlation is under investigation.

Problem 9.27 The hydrolysis of a specific cholesterol ester yields cholesterol plus stearic acid. Write the structural formula for the cholesterol ester.

Problem 9.28 Which one of the properties of the lipoproteins listed in Table 9.5 correlates best with their density?

9.8 CHOLESTEROL AND HEART DISEASE

Heart disease, which is responsible for approximately 2000 deaths per day in the United States, is the number one killer in the United States. Primary risk factors include genetics (one's gene content), high blood cholesterol levels, high blood pressure, and cigarette smoking. Secondary risk factors include lack of exercise, stress, poor nutrition, obesity, and age.

One of the most common forms of heart disease is **atherosclerosis,** which involves the accumulation of cholesterol-rich deposits along the inner walls of blood vessels (Figure 9.19). These deposits (plaques) lead to the **hardening (sclerosis) of arteries** and can restrict the flow of blood throughout the body. A **heart attack (my-**

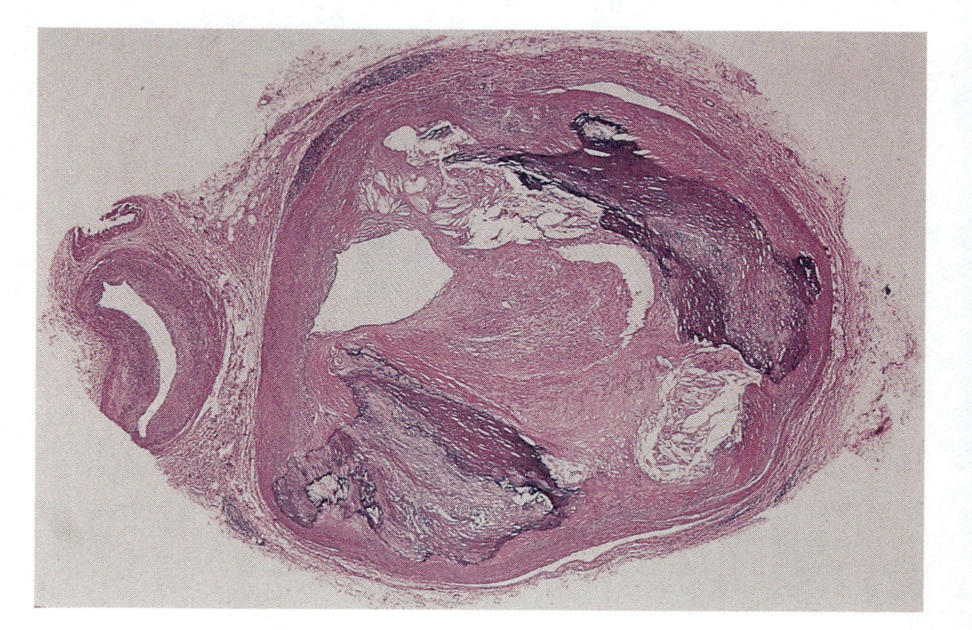

FIGURE 9.19
Cross-section of a plaque-laden artery.

ocardial infarction) results when an artery that carries blood to the heart becomes severely clogged and a blood clot further restricts the delivery of an adequate supply of O_2 to heart muscle. Heart muscle cells are damaged, heart function is impaired, and death may follow. A **stroke** results when a blood clot restricts the flow of blood to the brain. Atherosclerosis is not a new disease; the 5000-year-old body of the "Ice Man" discovered under the ice in the Tyrolean Alps in 1991 has partially clogged arteries. [Try Problem 9.29]

Although much remains to be learned about the formation of the cholesterol-rich atherosclerotic plaques, there is convincing evidence that plaque formation is more likely when blood cholesterol levels are high. For this reason, the American Heart Association and many health professionals have been urging individuals to monitor their blood cholesterol levels and to take corrective action if these levels are too high. Although there is some controversy over what level is ideal, most experts agree that blood cholesterol concentrations should be kept below 200 mg/dL (the prefix deci [d] means 0.1). Diets low in cholesterol and saturated fat but high in soluble fiber are therapeutic. Exercise is also beneficial. However, because one's blood cholesterol level is largely determined by genetics, there are definite limitations on what can be accomplished through life style changes. Drug therapy is often required to lower cholesterol levels in genetically susceptible individuals. Although some individuals are unable to lower their cholesterol levels to the extent desired, they need to recognize that cholesterol is just one of several risk factors, as noted at the beginning of this section. By controlling other risk factors, it may be possible to lower overall risk to an acceptable range (Figure 9.20).

Dietary fat normally has a greater impact on blood cholesterol levels than does dietary cholesterol itself, a fact that is partially explained by the tendency of the body to produce considerably more cholesterol than is found in a typical diet. Thus, a diet rich in saturated fat but low in cholesterol may lead to a greater heart disease risk than one that is relatively high in cholesterol but low in saturated fat. Since many people are unaware of the poorly understood fat–cholesterol connection, manufacturers sometimes mislead trusting consumers to believe that their saturated fat–laden products are heart-safe by removing cholesterol, and then proudly advertising this accomplishment. Although the price may differ, there is probably no significant difference between the health risks of cholesterol-free butter and normal butter. However, dietary cholesterol does elevate the risk of death from coronary heart disease independently of its effect on serum cholesterol levels. The total intake of both cholesterol and saturated fat should be limited.

The cholesterol–heart disease connection is complicated by the fact that risk is not a direct function of total blood cholesterol concentration. The distribution of cholesterol between high-density lipoproteins (HDL) and low-density lipoproteins (LDL) (Table 9.5), the primary transporters of blood cholesterol, is extremely important. Since HDL escorts most of its cholesterol to the liver for disposal, HDL-cholesterol is often described as "good" cholesterol. It is the LDL-cholesterol that is the primary source of the steroid found in plaque. The picture is further clouded by the presence of multiple subclasses of both HDL and LDL, with individual subclasses possessing distinctive biological activities. Although further research is definitely needed to fully sort out the "good," the "bad," and the "neutral," the absolute and relative concentrations of HDL-cholesterol and LDL-cholesterol are presently employed to help assess heart disease risk (Table 9.6). [Try Problem 9.30]

WHAT'S YOUR HEART-ATTACK RISK?

This test allows people who have no history of coronary disease to calculate their odds of having a heart attack within 1, 5, or 10 years. The test, based mainly on data from the four largest or longest American studies of coronary risk, was designed for Consumer Reports on Health by Ted Pass, Ph.D., president of StrateCision, Inc. (a Wellesley, Mass., software-development firm specializing in medical-risk assessment), with the help of Harvard researcher JoAnn Manson, M.D.

ADD OR SUBTRACT POINTS FOR EVERY "YES" ANSWER TO ITEMS 1-9:	WOMEN	MEN
1. EXERCISE		
Do you get little or no regular exercise?	Plus 6	Plus 2
2. WEIGHT		
Calculate your body mass index (BMI) with the following formula: Multiply your weight in pounds by 705. Divide the result by your height in inches. Then divide by your height in inches again.		
Is your BMI between 21 and 25?	Plus 2	Plus 0
Is your BMI between 25 and 29?	Plus 3	Plus 2
Is your BMI more than 29?	Plus 6	Plus 4
3. DIABETES		
Do you have diabetes?	Plus 11	Plus 8
4. SMOKING		
If you're an ex-smoker, did you quit in the past 5 years?	Plus 4	Plus 1
If you smoke, do you smoke fewer than 15 cigarettes a day?	Plus 8	Plus 2
Do you smoke 15 to 24 cigarettes a day?	Plus 15	Plus 4
Do you smoke more than 24 cigarettes a day?	Plus 18	Plus 6
5. FAMILY HISTORY		
Did either of your parents have a heart attack before age 60? [1]	Plus 9	Plus 9
6. ANTIHYPERTENSIVE MEDICATION		
Do you take medicine to control your blood pressure? [2]	Plus 1	Plus 1
7. ESTROGEN		
If postmenopausal, are you currently taking estrogen alone?	Minus 5	
Are you currently taking estrogen plus progestin?	Minus 3	
If you don't currently take estrogen, did you previously take it (with or without progestin)?	Minus 2	
8. ASPIRIN		
Do you take low doses of aspirin at least every other day?	Minus 4	Minus 4
9. ALCOHOL		
Do you drink alcohol in moderation (2 to 15 drinks per week?)	Minus 4	Minus 4
ADD UP YOUR POINTS SO FAR	Subtotal:	Subtotal:
NOW DO THESE CALCULATIONS FOR ITEMS 10-12, ROUNDING TO THE NEAREST WHOLE NUMBER:		
10. BLOOD PRESSURE		
Multiply your systolic pressure (the higher number) by 0.15 if you're a woman, 0.14 if you're a man.	Plus	Plus
11. AGE		
Multiply your age by 0.8 if you're a woman, 0.51 if you're a man.	Plus	Plus
12. CHOLESTEROL		
Multiply your total-cholesterol level by 0.06 if you're a woman, 0.07 if you're a man.	Plus	Plus
Multiply your HDL level by 0.3 if you're a woman, 0.25 if you're a man.	Minus	Minus
ADD UP YOUR POINTS FOR ITEMS 10-12.	Subtotal:	Subtotal:
ADD THE TWO SUBTOTALS TO GET YOUR TOTAL SCORE.	Total:	Total:
CONSULT THE TABLES BELOW.		

WOMEN
Probability of a heart attack within:

Score	1 year	5 years	10 years
0-60	<0.1%	<0.4%	<1%
61-70	0.1-0.2%	0.4-1%	1-3%
71-80	0.2-0.5%	1-3%	3-7%
81-85	0.5-1%	3-5%	7-12%
86-90	1%	5-8%	12-19%
91-95	1-2%	8-13%	19-29%
96-100	2-4%	13-20%	29-43%

MEN
Probability of a heart attack within:

Score	1 year	5 years	10 years
0-35	<0.1%	<0.4%	<1%
36-45	0.1-0.2%	0.4-1%	1-3%
46-55	0.2-0.6%	1-3%	3-7%
56-65	0.6-2%	3-8%	7-17%
66-70	2%	8-13%	17-27%
71-75	2-4%	13-20%	27-40%
76-80	4-6%	20-30%	40-56%

[1] An early heart attack is one that occurs before age 55 in men, 65 in women. However, this test is based on studies that used age 60 for both sexes.

[2] Taking antihypertensive medication is a sign that your blood pressure was once elevated.

FIGURE 9.20
Heart-attack risk test. Courtesy of Ted Pass, Ph.D. and StrateCision, Inc.

TABLE 9.6

Recommended Levels of Blood HDL- and LDL-Cholesterol[a]

	LDL-cholesterol (mg/dL)	HDL-cholesterol (mg/dL)
Coronary patients	<100	>35
Middle-aged individuals with no coronary disease	<130	>35

[a]The benefits of cutting cholesterol in healthy people wane by age 65 to 75 and may disappear entirely by age 80.

The blood concentration of LDL-cholesterol is determined, in part, by how rapidly LDL-cholesterol is removed from blood by various cells, including adipocytes and hepatocytes (liver cells). Before LDL-cholesterol can enter a cell, the LDL particle containing the cholesterol must bind to an **LDL receptor,** a protein located on the outer surface of the plasma membrane. Once bound, both the LDL and its receptor enter the cell through a process known as **receptor-mediated endocytosis.** During this event, a **clathrin-coated pit** (in the plasma membrane) bearing occupied LDL-receptors folds into the cell and is "pinched" off to form a **coated vesicle** (Figure 9.21). **Clathrin,** a protein, plays a central role in the overall process.

A cell's ability to take up cholesterol is largely determined by the inherited nature of its LDL receptors. This claim is supported by the existence of **familial hypercholesterolemia (FH),** a genetic disease associated with the production of nonfunctional LDL receptors from information encoded in abnormal LDL receptor genes. Roughly 1 in every 500 people in the United States inherits one normal LDL receptor gene along with one copy of the abnormal gene responsible for familial hypercholesterolemia. These heterozygous individuals usually have normal triacylglycerol (triglyceride) and HDL levels, but LDL concentrations typically range from 320 to 500 mg/dL. Most start to develop heart disease in their thirties and forties. Individuals homozygous for familial hypercholesterolemia possess two copies of the abnormal LDL receptor gene and no normal copies of this gene. They have extremely high levels of blood LDL-cholesterol and tend to develop serious atherosclerotic plaques by their early teens. Few live past the age of 20. **[Try Problem 9.31]**

Attempts to assess heart disease risk are further complicated by the presence of lipoprotein(a), which appears to be an independently inherited risk factor for at least some individuals. You are genetically programmed to possess a relatively fixed blood lipoprotein(a) level throughout your lifetime, and this concentration is unaffected by LDL, HDL, exercise, smoking, and many other variables. However, your lipoprotein(a) concentration varies by as much as 1000-fold from some other individuals. For certain categories of individuals, the higher the concentration, the greater the heart disease risk. The molecular basis for this correlation is still uncertain.

When concentrating on the cholesterol–heart disease link, it is easy to lose sight of the normal and essential roles of cholesterol within cells. The positive side of the cholesterol story is brought back into focus as the following section explores the role of cholesterol and other lipids in membrane structure and function.

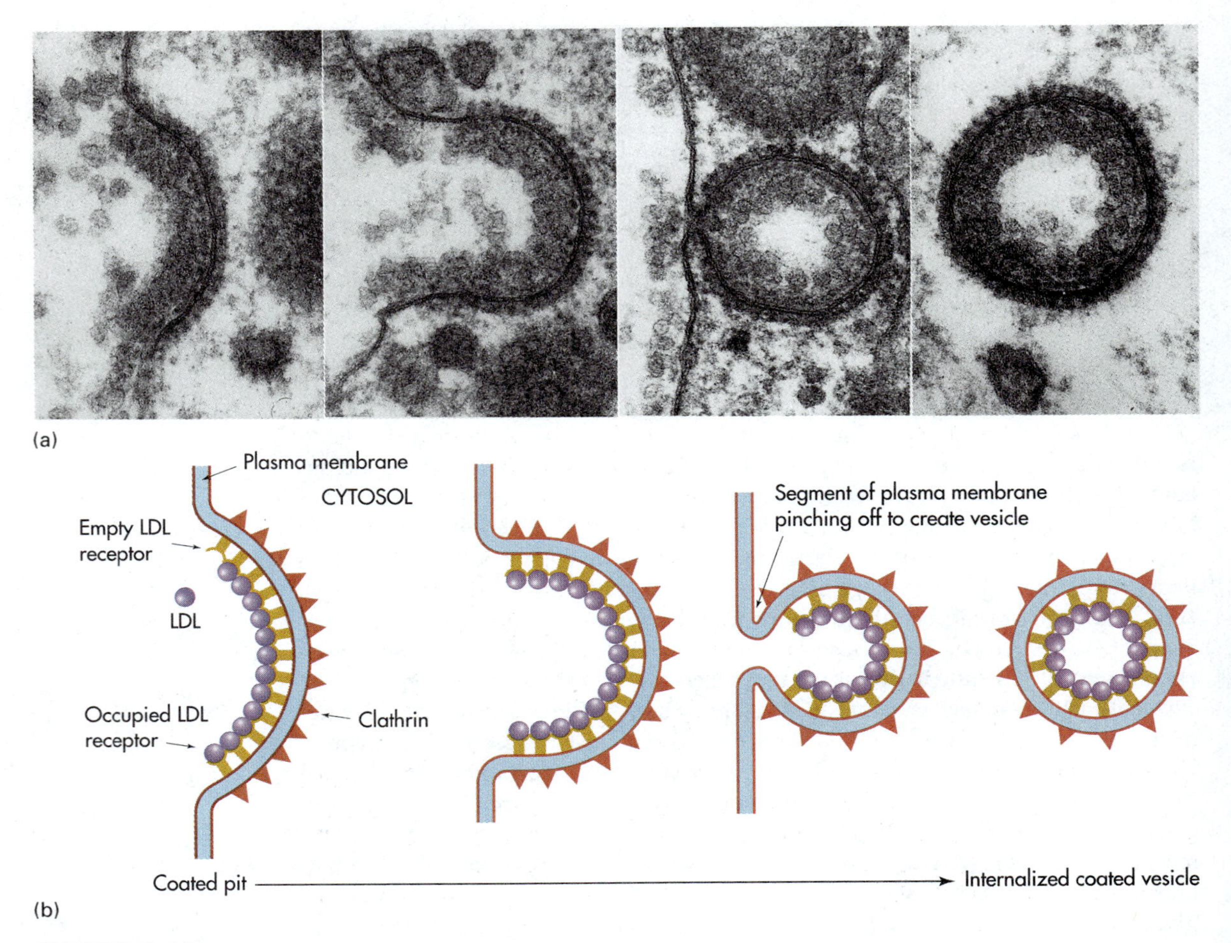

FIGURE 9.21
Receptor-mediated uptake of LDL by a cell. (a) Electron micrograph of receptor-mediated endocytosis. (b) Interpretation of the electron micrograph.

Problem 9.29 When cells run out of O_2 they quickly run out of another essential commodity as well. What is this other commodity? It is the loss of this commodity that directly leads to cell death in the absence of O_2.

Problem 9.30 Explain why total blood cholesterol concentration, alone, is a poor measure of heart disease risk.

Problem 9.31 Heart disease is not usually considered a genetic (inherited) disease. Should it be, since blood cholesterol levels are partly determined by the genes one inherits? Explain.

MEMBRANES AND MEMBRANE TRANSPORT

9.9 THE FLUID–MOSAIC MODEL

Each cell is surrounded by a lipid-rich membrane known as its plasma membrane. Eukaryotic cells contain a multitude of additional membranes including those associated with mitochondria, Golgi bodies, the nucleus, lysosomes, and the endoplasmic reticulum (Section 1.3). Membranes are much more than just protective physical barriers. They regulate what substances enter and leave cells and organelles, and they possess a wide range of biologically active components, including enzymes. Membranes play a central role in cell–cell communication and in the communication of cells with their external environment.

The core of a biological membrane consists of a lipid bilayer about 5 nm thick (Figure 9.9 and Figure 9.22) that is primarily constructed from phospholipids, glycolipids and cholesterol. This bilayer is predominantly maintained through noncovalent interactions between its lipids. The lipid bilayer is sandwiched between two layers of **peripheral** or **surface proteins,** most attached to the membrane through noncovalent interactions with other membrane proteins or with the polar heads of the amphipathic lipids within the bilayer. A few peripheral proteins are bound more tightly by covalent links to lipids associated with the bilayer. Proteins containing a covalently attached fatty acid (Section 9.1), for example, become securely anchored to a membrane when the nonpolar tail of the fatty acid inserts itself into the bilayer. **Integral proteins** extend into and, in some cases, completely span the bilayer. Their interactions with peripheral proteins often help the peripheral proteins maintain an appropriate position within the overall architecture of a membrane. Many of the membrane-spanning proteins **(transmembrane proteins)** are involved in the transport of substances across membranes. Specific examples are examined in Section

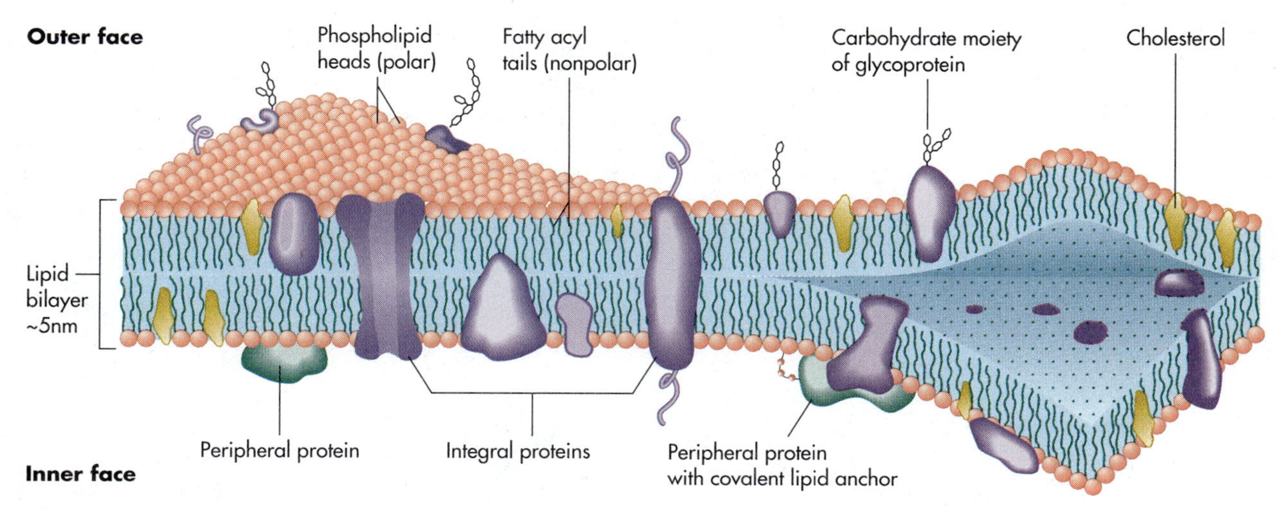

FIGURE 9.22
Fluid–mosaic model for membranes. Based on Lehninger et al., 1993.

TABLE 9.7

Molecular Composition of Some Membranes[a]

	% Protein (by dry weight)	% Lipid (by dry weight)
Plasma Membranes		
Human red blood cell	60	40
Rat liver	62	38
E. coli	75	25
Organelle Membranes		
Rat liver nuclear	71	29
Rat liver mitochondrial (inner)	80	20
Rat liver mitochondrial (outer)	60	40
Rat liver Golgi	40	60
Spinach chloroplast	48	52

[a]The percentages in this table are approximate. Percentages vary over time as both proteins and lipids come and go from membranes. Distinct regions of the same membrane may also differ in composition. The glycolipids and glycoproteins in membranes contain carbohydrate residues that typically account for 1 to 9% of the total dry weight of a membrane.

9.11. This description of a membrane is known as the **fluid-mosaic model.** A typical membrane has an average thickness close to 8 nm and contains roughly 40% lipid and 60% protein. **[Try Problem 9.32]**

Table 9.7 summarizes the molecular composition of some representative membranes. In addition to lipids and proteins, most membranes contain a small fraction of carbohydrate that is usually bonded covalently to the lipids and proteins. The relative amounts of many specific lipids and proteins fluctuates, sometimes markedly, from membrane to membrane and even from region to region within the same membrane. The inside and outside of a membrane also differ both structurally and functionally. Most of the lipids and proteins within membranes are capable of lateral movement, and particular membrane components can shift from surface to surface as well (Figure 9.23). Some proteins on the inside surface of plasma membranes are at least partially immobilized through their covalent attachment to the **cytoskeleton,** an internal network of protein scaffolding (Section 1.3). Collectively, the lipids

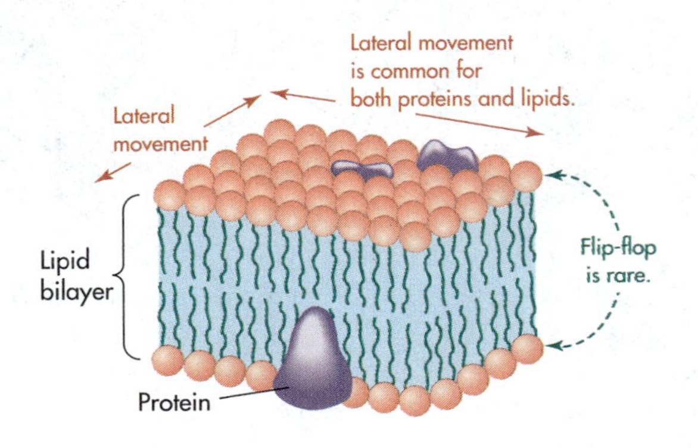

FIGURE 9.23
Mobility of membrane components.

within a bilayer play the central role in determining the fluidity and permeability of a membrane. Cholesterol and other steroids are particularly important in this regard; the rigid steroid ring interacts with, and partially immobilizes, neighboring lipids in the bilayer. Most membrane components have limited half-lives and some, including cholesterol, readily dissociate from membranes. Membranes are dynamic rather than static systems; they undergo constant changes as they participate in numerous cellular processes. **[Try Problems 9.33 and 9.34]**

Problem 9.32 Predict what specific classes of noncovalent bonds are involved in holding peripheral proteins and lipid bilayers together within a membrane.

Problem 9.33 What structural feature of membranes makes the lateral movement of membrane components possible?

Problem 9.34 What does it mean to say that the lipids and proteins within a membrane have limited half-lives? How does a cell benefit from this turnover?

9.10 MEMBRANE TRANSPORT: AN OVERVIEW

Any membrane through which some substances can pass and others cannot is said to be **semipermeable.** The semipermeability of biological membranes enables cells to regulate the concentrations of most of the substances within them, an ability central to life itself. Although most small nonpolar molecules, including O_2 and CO_2, can move freely into and out of cells and organelles (Figure 9.24), the lipid bilayer in biological membranes resists the passage of ions, polar molecules, and polymers. The ability of polar water molecules to traverse biological membranes freely is a notable exception that has not been fully explained. **[Try Problem 9.35]**

Many small, lipid bilayer–excluded substances are transported across membranes attached to specific transmembrane proteins called **transporters, translocators, translocases,** or **permeases.** Certain other excluded substances traverse membranes by passing through selective channels or less selective pores constructed from transmembrane proteins that do not always bind to the translocated molecules during their journey. Polymers and macromolecular complexes commonly enter and leave cells through membrane exchange processes known as endocytosis and exocytosis, respectively. Certain proteins are transported through membranes as they are being synthesized, a transport mechanism that will be examined during the discussion of translation (RNA-directed polypeptide synthesis) in Chapter 18.

Transporter Kinetics

Although many transporters do not catalyze chemical transformations and are technically not enzymes, they do speed up the flow of substances across membranes and do share many characteristics with enzymes. Like enzymes, transporters normally

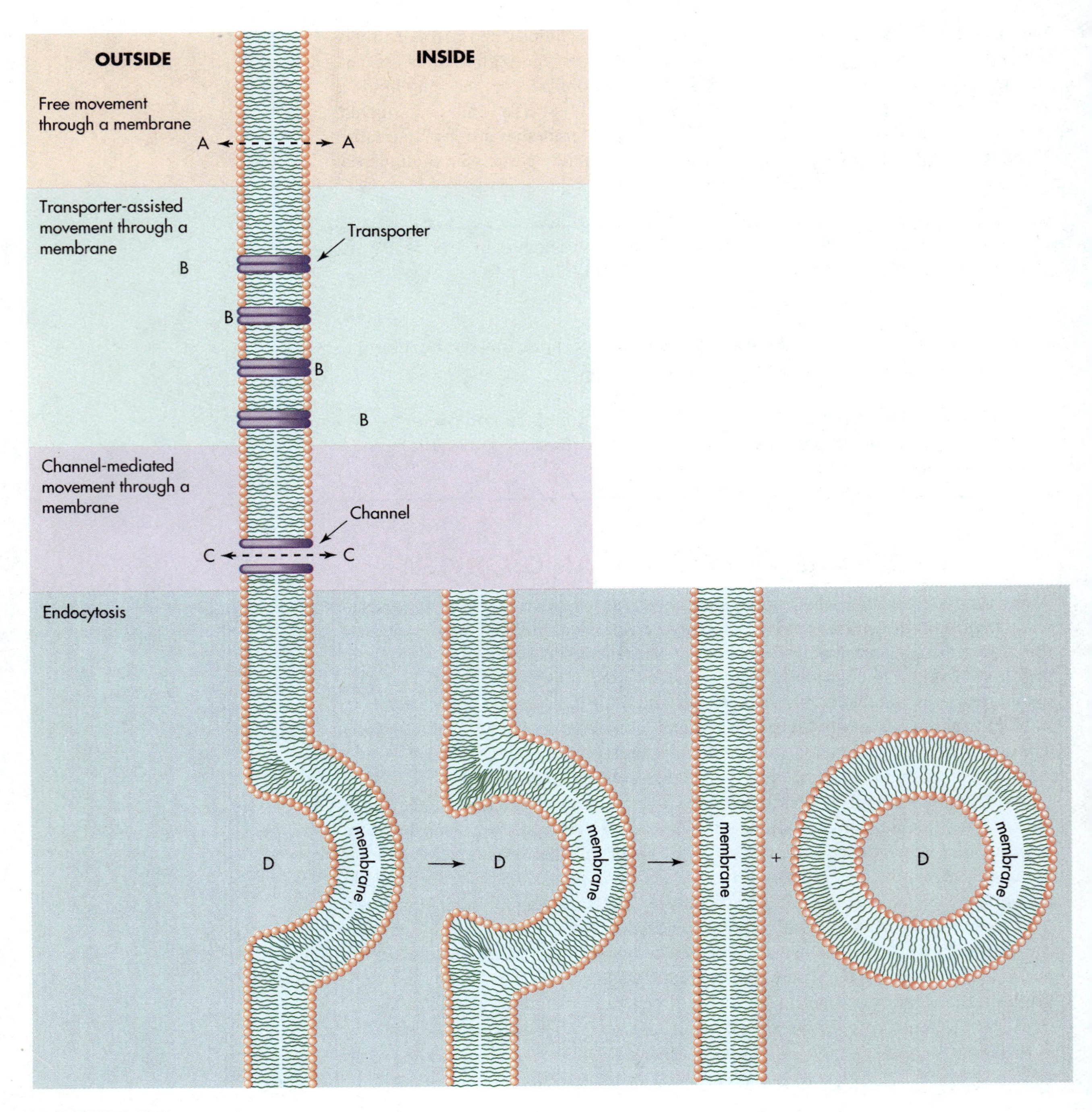

FIGURE 9.24
Some mechanisms by which substances move across membranes. A, B, C, and D are distinct substances that are translocated across membranes by different mechanisms.

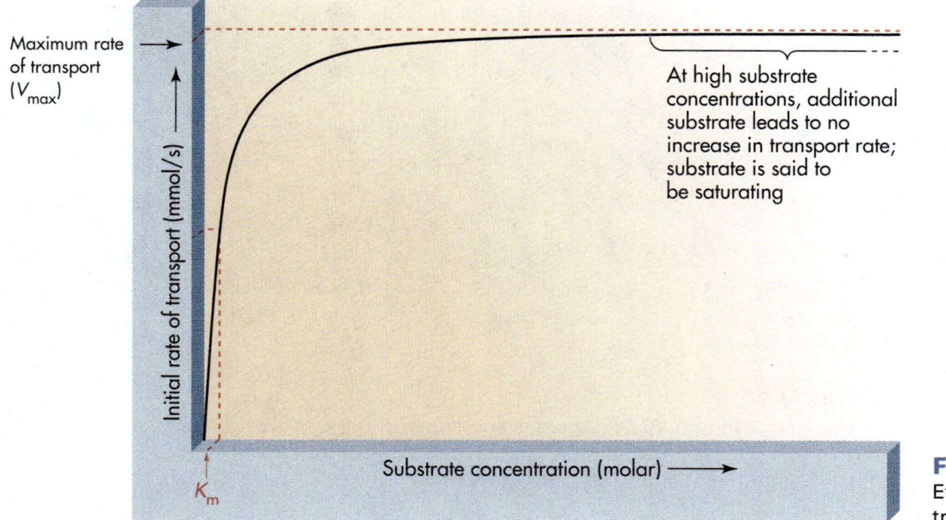

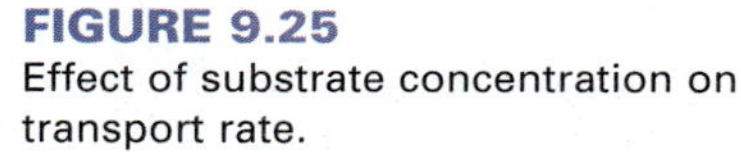

FIGURE 9.25 Effect of substrate concentration on transport rate.

possess highly selective, stereospecific binding sites for their *substrates* (those substances they transport), and they adhere to their substrates through noncovalent interactions.

When substrate concentration (x-axis) is plotted against initial rate of transport (y-axis), a hyperbolic plot (Figure 9.25) is frequently obtained that is analogous to the Michaelis–Menten plot (Section 5.6) and that can be mathematically described with a Michaelis–Menten-type equation:

$$\text{Initial rate of transport} = \frac{V_{max}[S]}{K_m + [S]}$$

The initial rate of transport is the rate of transport when the concentration of substrate on the opposite side of the membrane is zero. At high substrate concentrations, all of the binding sites on Michaelis–Menten transporters are, from a practical standpoint, occupied at all times; each transporter is working at full capacity and transport is proceeding at maximum initial velocity (V_{max}). Additional substrate cannot increase the transport rate. The existing substrate concentration is said to be saturating and the transporter is saturated with substrate. Transporters that behave in this fashion are said to exhibit **saturation kinetics.** The K_m for a substrate–transporter pair is that substrate concentration that leads to an initial rate of transport equal to one half the V_{max}. Like enzyme-catalyzed reactions, transporter-linked translocation systems are susceptible to both competitive and noncompetitive inhibition. **[Try Problem 9.36]**

Active Versus Passive Transport

The transport of a substance across a membrane is said to be **active transport** if energy is consumed during the process, and it is described as **passive transport** if there

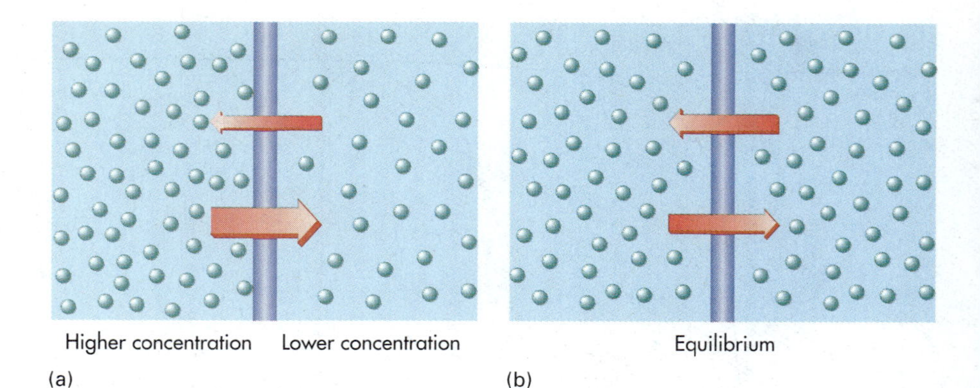

FIGURE 9.26
Diffusion. (a) During diffusion, there is spontaneous net movement of a substance from a region of higher concentration to one of lower concentration. (b) At equilibrium there is no concentration gradient and no net movement.

is no energy requirement. Passive transport always involves the spontaneous net movement of a substance from a region of higher concentration to one of lower concentration, a process known as **diffusion** (Figure 9.26). Diffusion is thermodynamically driven by the increase in entropy associated with the reduction of a concentration gradient. All other factors being equal, a system will spontaneously move to its most random (highest entropy) state (Section 8.3). When a substance must bind to a protein transporter before it diffuses, the process is described as **facilitated diffusion.** **[Try Problem 12.37]**

The transport of a substance across a biological membrane from a region of lower concentration to a region of higher concentration is always an active process that requires a transporter. The transporter involved is commonly called a **pump,** since energy is used to "pump" a substance up a concentration gradient. Active transport is said to be **primary** when the consumed energy is provided directly by a primary energy source such as the hydrolysis of ATP, a photon of light, or a redox reaction. **Secondary active transport** utilizes the energy associated with one concentration gradient to produce a second concentration gradient; as the energy provider flows from a region of high concentration to one of lower concentration, a second substance simultaneously flows to (is pumped to) a region of higher concentration. Because the concentration gradient that drives secondary active transport is generated through primary active transport, the energy for all active transport can be traced ultimately to one or more primary energy sources, most often the hydrolysis of ATP. Active transport processes account for over 25% of the total energy consumed by the human body at rest. Much of this energy is used by nerve cells to create the Na^+ and K^+ concentration gradients that play a central role in the transmission of nerve impulses (Section 10.9). The properties of several membrane translocation systems, including the transporter-linked systems, are summarized in Table 9.8. **[Try Problems 9.38 and 9.39]**

TABLE 9.8

Properties of Some Membrane Translocation Systems

Translocation Process	Substance Bound to Protein During Translocation?	Energy Required?	Saturation Kinetics?	Can Produce a Concentration Gradient?	Examples of Translocated Substances[a]
Simple (free) diffusion	No	No	No	No	H_2O, O_2, NO
Facilitated diffusion (passive transport)	Yes	No	Yes	No	Glucose, Cl^-, HCO_3^-
Primary active transport	Yes	Yes	Yes	Yes	Na^+, K^+, Ca^{2+}, H^+
Secondary active transport	Yes	Yes	Yes	Yes	Glucose, amino acids, Ca^{2+}
Channels	No	No	No	No	Na^+, K^+, Ca^{2+}, Cl^-
Nuclear pores	No (small substances) Yes (large substances)	No (small substances) Sometimes (large substances)	No (small substances) Yes (large substances)	No (small substances) Yes (large substances)	Most small substances, specific proteins and nucleic acids

[a]The same substance may be translocated by multiple mechanisms. The mechanism employed will depend upon the membrane involved and the metabolic state of a cell.

Uniport and Cotransport

When a transporter moves a single substance through a membrane the process is described as **uniport. Cotransport** involves the simultaneous passage of two or more substances through a membrane. The cotransport of two substances in the same direction is labeled **symport.** During **antiport,** two substrates pass one another moving in opposite directions (Figure 9.27). **[Try Problem 9.40]**

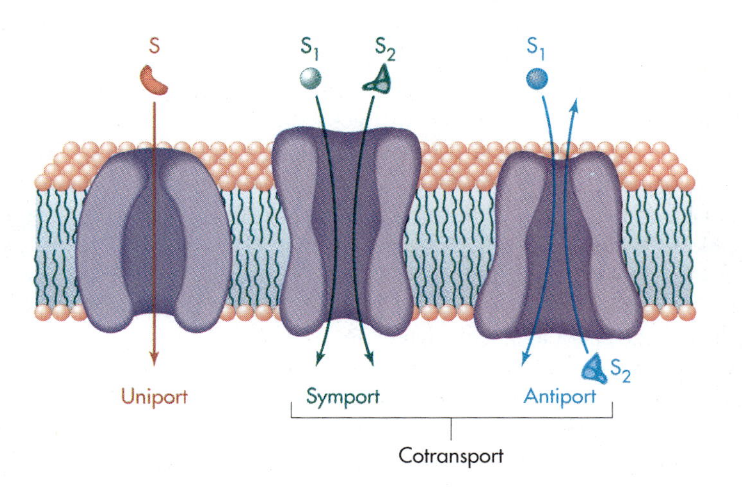

FIGURE 9.27
Uniport and cotransport. During uniport, a single substance is transported in one direction. In symport, two substances are cotransported in the same direction. In antiport, two substances are cotransported in opposite directions. Based on Lehninger et al., 1993.

Problem 9.35 Why do membranes tend to be permeable to small nonpolar substances but impermeable to most ions and polar substances?

Problem 9.36 What effect does a competitive inhibitor have on the K_m and V_{max} for a transporter-linked membrane translocation system exhibiting Michaelis–Menten kinetics? Hint: Review Section 5.10.

Problem 9.37 List three factors that help determine the rate of free (nonfacilitated) diffusion of a solute across a membrane.

Problem 9.38 How much ATP must be hydrolyzed (to ADP + P_i) to move one molecule of a substance through a membrane if 134 kJ/mol (32 kcal/mol) are consumed during the transport process? Assume that transport proceeds under standard conditions. Hint: Review Section 8.4.

Problem 9.39 What factor primarily determines the minimum amount of energy required to move a substance across a membrane against a concentration gradient?

Problem 9.40 List the major biochemical themes (Exhibit 1.4) encountered in this section.

9.11 SPECIFIC MEMBRANE TRANSPORTERS

This section examines in more detail two transporters that are associated with the plasma membranes of virtually all mammalian cells. Their general modes of action are similar to those of many other transporters.

Glucose Transporters (GluTs)

Glucose is the major fuel and a prime carbon source for many human cells, and the vast majority of the cells in the human body possess one or more **glucose transporters (GluTs).** Since individual tissues tend to have unique glucose needs, they often possess distinctive GluTs, or distinct combinations of GluTs, each with a unique K_m and V_{max}. Intestinal cells, for example, contain at least two distinctive GluTs, one that employs secondary active transport to absorb trace amounts of glucose from the contents of the intestine and a second that moves the absorbed glucose into the bloodstream through facilitated diffusion.

Glucose transporters that function through facilitated diffusion are numbered in the order in which they have been discovered. GluT1 is abundant in the plasma membranes of the endothelial cells which line blood vessels. GluT2 is a major transporter in cells that release glucose into the blood, including intestine cells, liver cells, and kidney cells. GluT3 is the dominant transporter in certain brain cells. GluT4 is present at particularly high levels in muscle and fat cells, and its role in the insulin–diabetes story is examined in Section 12.4. At the moment, no one is certain how many family members exist. GluT1 is the most thoroughly studied member of this family.

GluT1 is a single polypeptide chain containing 492 amino acids (Figure 9.28). Its primary structure (sequence of amino acids) consists of 12 predominantly hydrophobic segments that alternate with 13 primarily hydrophilic regions. Each of the hydrophobic segments coils into an α-helix, and the entire chain then weaves back and forth across the membrane in such a manner that each of the α-helices spans the

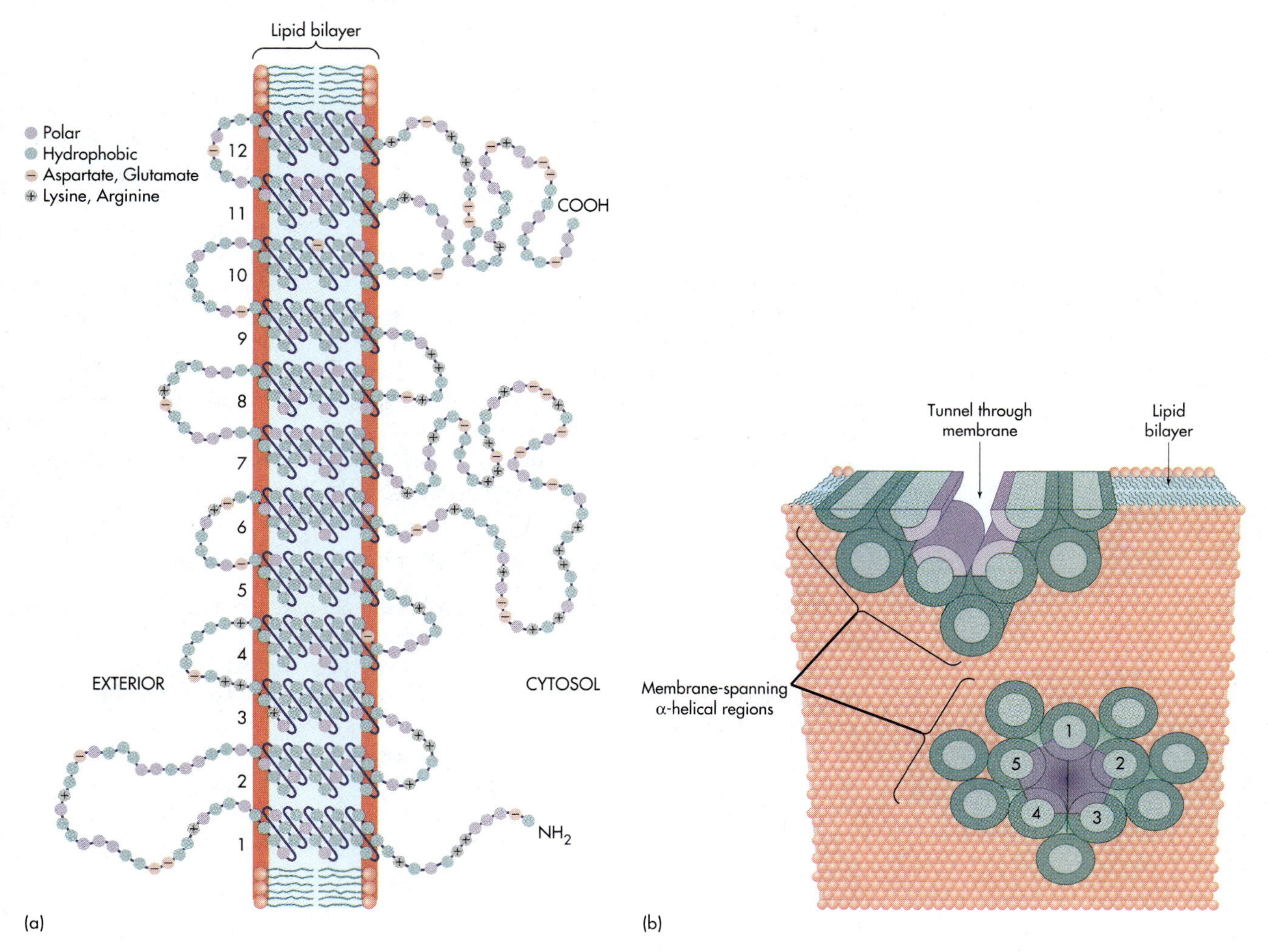

FIGURE 9.28
Structure of glucose transporter 1 (GluT1). (*a*) The proposed structure of GluT1 is a folded chain with twelve α helical regions that each traverse the lipid bilayer. Amino acids having charged groups appear as (+) or (−). Because such groups are more stable in water than in lipid, most are located outside the membrane rather than in it. (*b*) Membrane-spanning tunnel. The five numbered helices directly create the tunnel within GluT1. The numbers here do not correspond to the numbers assigned to helices in part a of this figure. From Lienhard et al., 1992.

lipid bilayer. In the process, the 12 transmembrane helices are brought together to generate a tunnel through which a glucose molecule can be passed. Hydrophilic side chains on the five helices that line the tunnel account for the hydrophilic nature of the tunnel. **[Try Problem 9.41]**

The actual movement of glucose into a cell through GluT1 is thought to involve a four-step cycle: in step 1, glucose attaches to a specific outward- facing binding site in or near the mouth of the transporter tunnel; in step 2, the transporter changes its conformation and, in the process, shifts the glucose to an inward facing position; in step 3, the glucose is released into the cell; and in step 4, the transporter returns to its original conformation (Figure 9.29). In the GluT1 of erythrocytes, the net result is roughly a 50,000-fold increase in the rate of movement of glucose into the cell. Since the transport process is readily reversible, internal glucose can exit the cell through a reversal of the steps involved in glucose import. There is no net flow of glucose when the internal and external glucose concentrations are equal; diffusion cannot generate a concentration gradient.

GluTs are not totally specific for D-glucose, because the monosaccharide binding sites on these transporters also accommodate certain other aldohexoses, including D-mannose and D-galactose. However, these alternative substrates have a much lower affinity than glucose for transporter binding sites. **[Try Problem 9.42]**

The uptake of glucose by intestine cells, as noted at the beginning of this section, is an example of secondary active transport. The symport system involved extracts energy from a Na^+ gradient. One glucose molecule moves into a cell along with each Na^+ that enters through facilitated diffusion. Thus, a glucose gradient is created at the expense of a Na^+ gradient. The Na^+ gradient is maintained through a primary active transport system that employs the Na^+K^+transporter.

The Na^+K^+Transporter (Na^+K^+ATPase; Na^+K^+Pump)

Inorganic ion concentration gradients play a key role in a variety of membrane-related processes, including secondary active transport (see previous paragraph), the regulation of pH, signal transduction across membranes, and ATP production. These concentration gradients, which represent a form of entropy-linked potential energy, are maintained through selective active transport systems (pumps) that commonly are powered by energy released during the cleavage of ATP. Table 9.9 summarizes some of the ion concentration gradients that exist across the plasma membrane of a typical mammalian cell.

The plasma membrane of all mammalian cells contains an antiport **Na^+K^+transporter** that is also called the **Na^+K^+ATPase** or the **Na^+K^+pump.** This protein exports three Na^+ and imports two K^+ for each ATP that is cleaved, and it is primarily responsible for the maintenance of the Na^+ and K^+ gradients depicted in Table 9.9. Since three positive charges are pumped out of a cell while only two positive charges are pulled in, the Na^+K^+transporter generates an electrical gradient (electrical potential) as well as concentration gradients. The electrical potential is established as the inside of the cell becomes negative relative to the outside and a voltage develops across the membrane. The overall gradient generated is said to be an **electrochemical gradient,** because concentration gradients are superimposed on the

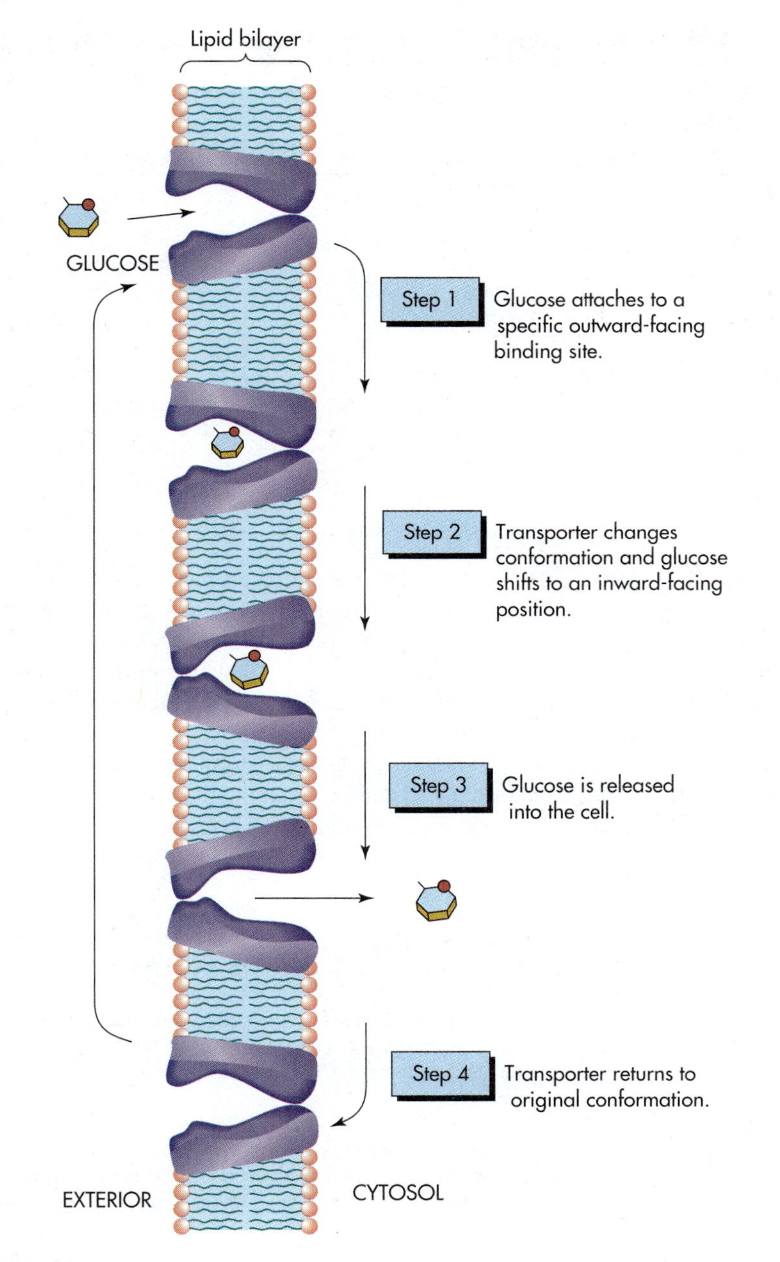

FIGURE 9.29
The transport of glucose by GluT1. Adapted from Lienhard et al., 1992.

TABLE 9.9

Some Ion Concentration Gradients Associated with a Typical Mammalian Cell

Ion	Intracellular Concentration (m*M*)	Extracellular Concentration (m*M*)
Na^+	5–15	145
K^+	140	5
Mg^{2+} [a]	0.5	1–2
Ca^{2+} [a]	10^{-4}	1–2
H^+	7×10^{-5}	4×10^{-5}
Cl^-	5–15	110

[a]The concentrations for Ca^{2+} and Mg^{2+} are those for the free ions. The total intracellular concentration for Mg^{2+} is approximately 20 m*M* and that for Ca^{2+} roughly 1.5 m*M*. Most of the Mg^{2+} and Ca^{2+} is bound to proteins, nucleic acids, and other substances. Ca^{2+} is also stored in certain organelles.

electrical one (Figure 9.30). The gradients generated by the Na^+K^+ transporter are involved in the transduction of signals across membranes; in the secondary active transport of amino acids, glucose, and certain other substances through membranes; and in the regulation of osmotic pressure and cell volume. **Osmotic pressure** is the pressure created by the flow of water into a cell or organelle that contains a higher concentration of solutes than exists externally.

Ouabain (Figure 9.31), a naturally occurring steroid derivative, inhibits the Na^+K^+ transporter and reduces the Na^+ and K^+ gradients that normally exist across membranes. It interferes with nerve action and can even cause a cell to swell and

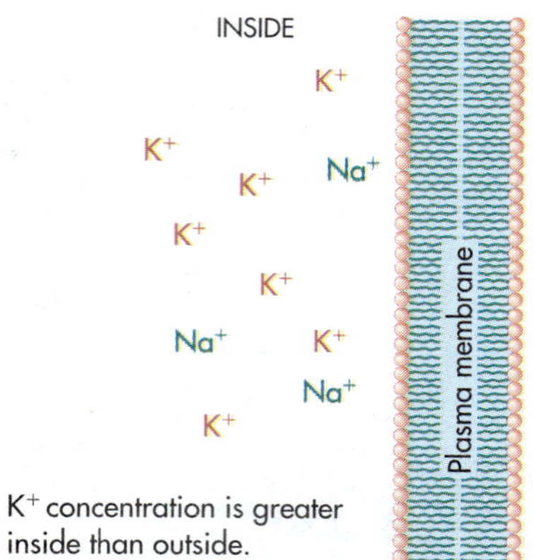

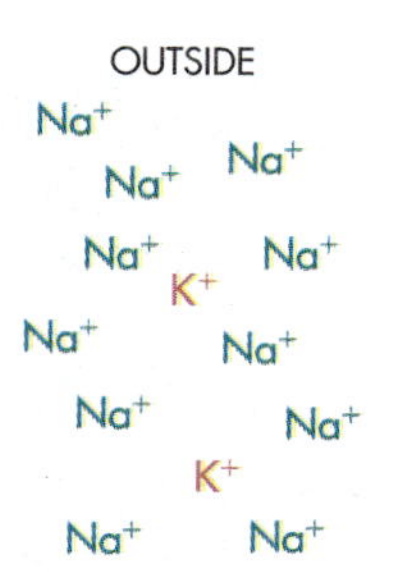

Na^+ concentration is greater outside than inside.

Outside is positive relative to inside.

If the membrane becomes permeable to K^+:
1. The K^+ concentration gradient provides a thermodynamic driving force for the movement of K^+ out of the cell.
2. The electrical gradient (more + outside than inside) provides a driving force for the movement of K^+ into the cell.
3. The net movement of K^+ is determined by the magnitude of the two gradients.

If the membrane becomes permeable to Na^+:
Both the Na^+ concentration gradient and the electrical gradient provide a thermodynamic driving force for the net movement of Na^+ into the cell.

FIGURE 9.30
An electrochemical gradient.

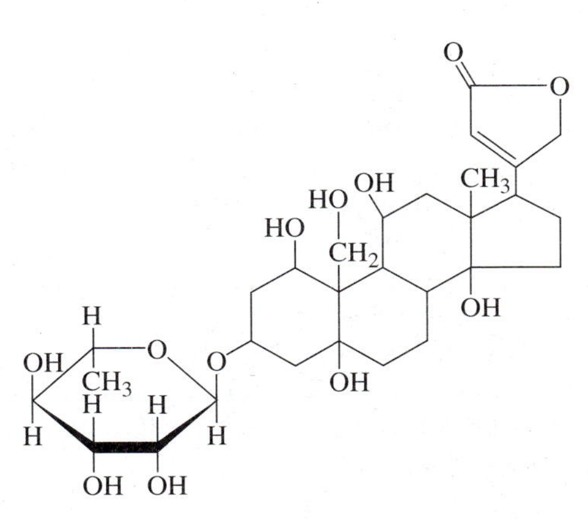

FIGURE 9.31
Ouabain.

burst because of changes in osmotic pressure. Native African hunters regularly use ouabain-containing juices from an African shrub to poison the tips of their hunting arrows. **[Try Problem 9.43]**

The Na^+K^+ transporter spans the plasma membrane and is constructed from two α subunits (polypeptides) and two β subunits. These subunits are totally unrelated to the α and β subunits found in hemoglobin. The α and β, as used here, are also unrelated to α-helices and β-pleated sheets. According to the current working model, the antiport of Na^+ and K^+ by the Na^+K^+ transporter is driven by a four-step cycle (Figure 9.32): step 1, three Na^+ bind to the transporter protein on the cytosolic side;

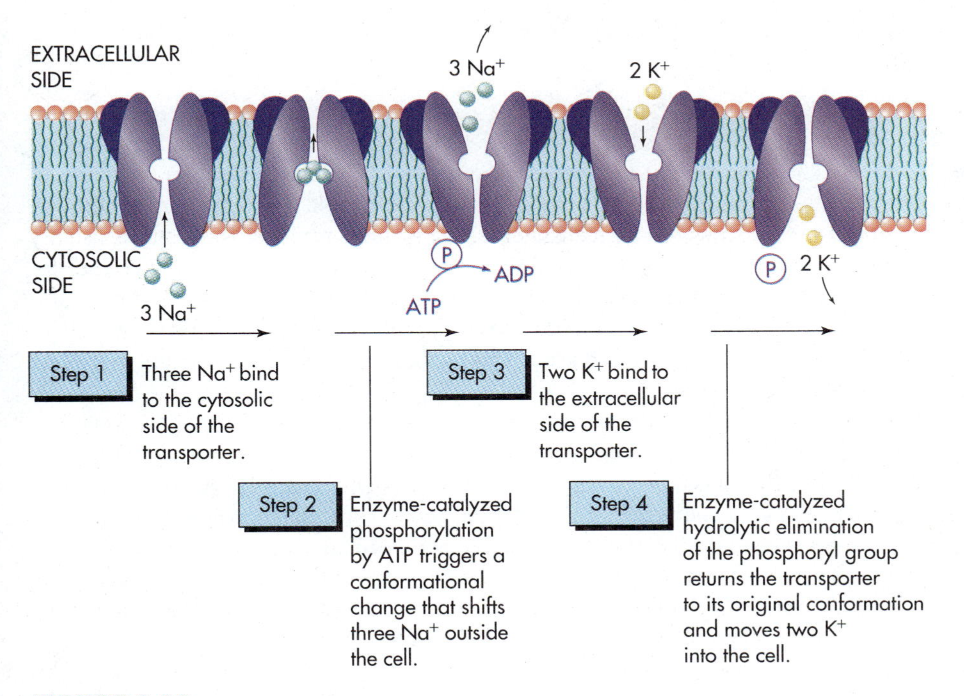

FIGURE 9.32
Ion transport by the Na^+K^+ transporter. Based on Devlin, 1992.

step 2, a protein kinase activity associated with the transporter catalyzes the ATP-linked phosphorylation of the transporter and triggers a conformational change that shifts the three Na^+ toward the outside of the membrane and facilitates their release; step 3, two K^+ bind to the outward-facing side of the phosphorylated transporter; and step 4, a hydrolase activity associated with the transporter catalyzes the hydrolytic elimination of the phosphate residue, causing the protein to return to its original conformation, an event that shifts the K^+ into the cell. The Na^+K^+transporter, like most enzymes that catalyze the hydrolysis of ATP, has an absolute requirement for Mg^{2+}. **[Try Problem 9.44]**

Problem 9.41 Explain why the 12 α-helical regions in GluT1 are all approximately the same length.

Problem 9.42 Explain how D-mannose inhibits the transport of D-glucose by GluT1. Explain why L-mannose is not an inhibitor.

Problem 9.43 Explain how the inactivation of the Na^+K^+transporter leads to changes in osmotic pressure.

Problem 9.44 Would you expect the Na^+K^+transporter to function in reverse as does GluT1? Explain.

9.12 MEMBRANE CHANNELS AND NUCLEAR PORES

Transporters facilitate the movement of polar and ionic substances through membranes by selectively attaching to these substances and providing a hydrophilic tunnel through which they can be passed. Transporter-linked translocation exhibits saturation kinetics and can be either active or passive (Table 9.8). In contrast, **channel proteins** provide a water-filled channel through which particular substances, usually inorganic ions, travel without attaching to the protein itself. Channel-linked translocation always involves simple diffusion, and it does not usually exhibit saturation kinetics, because there are no binding sites to be saturated. The specificity of a channel is determined by its size and shape and by the nature of the functional groups that line it. **[Try Problem 9.45]**

Channels are usually gated and the gates open and close rapidly in response to specific stimuli. Neurotransmitters, for example, open the ion channels that participate in the transmission of nerve impulses, a phenomenon discussed in Section 10.9. The defective gene responsible for the genetic disease known as cystic fibrosis (Chapter 19) encodes the **cystic fibrosis transmembrane conductance regulator,** a gated Cl^- channel that requires both cAMP-dependent phosphorylation and ATP hydrolysis to open. Certain other channels are opened by changes in a transmembrane electrical potential and are said to be **voltage gated.**

The nuclear envelope (Section 1.3) that separates the cytoplasm from the nucleoplasm in eukaryotic cells contains numerous water-filled pores that share several properties with membrane channels. Each **nuclear pore** is part of a **nuclear pore complex** that spans the double membrane of the nuclear envelope (Figure 9.33). Nu-

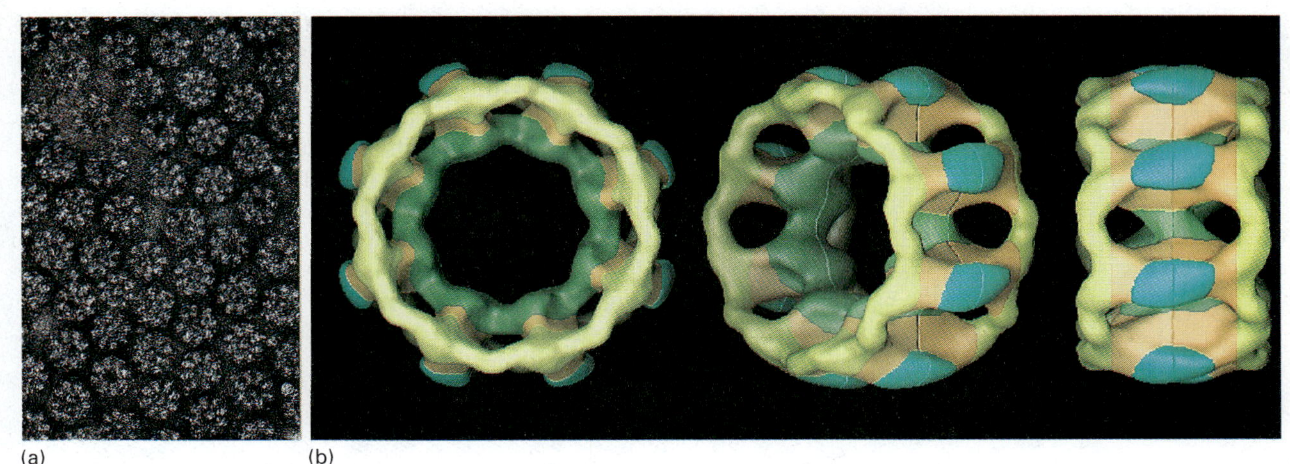

FIGURE 9.33
Nuclear pore complex. (a) Negatively stained electron micrograph of nuclear pore complexes released from the nuclear envelope by treatment with detergent. (b) Top, tilted, and side views of a computer reconstruction of a nuclear pore complex.

clear pores, which have a diameter of roughly 9 nm, are larger than channels. They allow most molecules with masses less than about 5000 daltons to pass freely between the nucleus and the cytoplasm. Nuclear pores, like membrane channels, do not bind the small molecules being transported and normally exhibit nonsaturation kinetics. In contrast to the smaller membrane channels, nuclear pores are not gated, they lack specificity (when transporting small molecules), and they participate in the transport of certain large molecules as well as small ones. Large-molecule transport is usually a selective, active (ATP is cleaved) process that entails the attachment of the transported substance to specific receptor proteins within the nuclear pore complex. Ribosomal subunits, RNAs, and specific enzymes are some of the many large particles that undergo transport through the nuclear pore complex.

Problem 9.45 Would a Na^+ channel more likely be lined with (a) hydrophobic amino acid side chains; (b) hydrophilic, uncharged amino acid side chains; or (c) positively charged amino acid side chains? Explain.

9.13 ENDOCYTOSIS AND EXOCYTOSIS

Transporters and channels are responsible for enhancing the movement of small polar and small charged substances through plasma membranes. Neither mechanism, however, is amenable to the transfer of macromolecules or large particles. Large substances and molecular aggregates commonly enter and leave cells by processes called endocytosis and exocytosis, respectively.

Endocytosis was introduced in Section 9.7 during the discussion of LDL and heart disease (Figure 9.21). The uptake of LDL by a cell involves the invagination and "pinching off" of LDL-containing clathrin-coated pits (on the plasma membrane) to create internalized, membrane-enclosed coated vesicles. The process is LDL specific because the receptors associated with the coated pits select LDL. Endocytosis, in general, proceeds similarly (Figure 9.34a). Once inside a eukaryotic cell, a membrane-enclosed vesicle normally fuses with the membrane of a specific organelle. In the process, the vesicle releases its contents into the organelle.

Exocytosis is basically the reverse of endocytosis. Internal membrane-enclosed vesicles released from the Golgi apparatus (Section 1.3) fuse with the plasma membrane, releasing their contents into the external environment (Figure 9.34b). Similar internal events allow cells to use membrane-enclosed vesicles to shuttle substances back and forth between organelles. Additional research in cell biology is needed in order to understand how the trafficking of vesicle-enclosed substances is regulated.

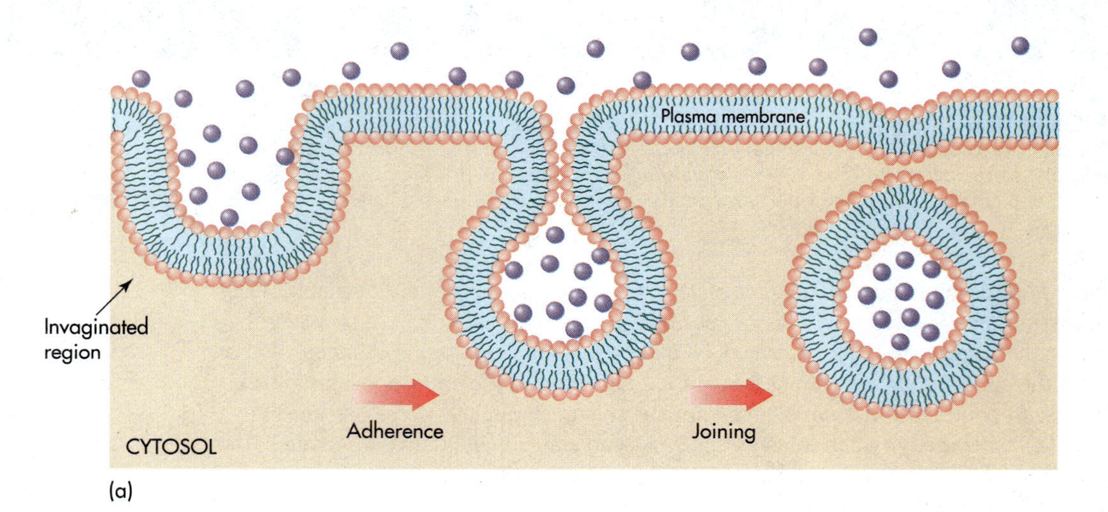

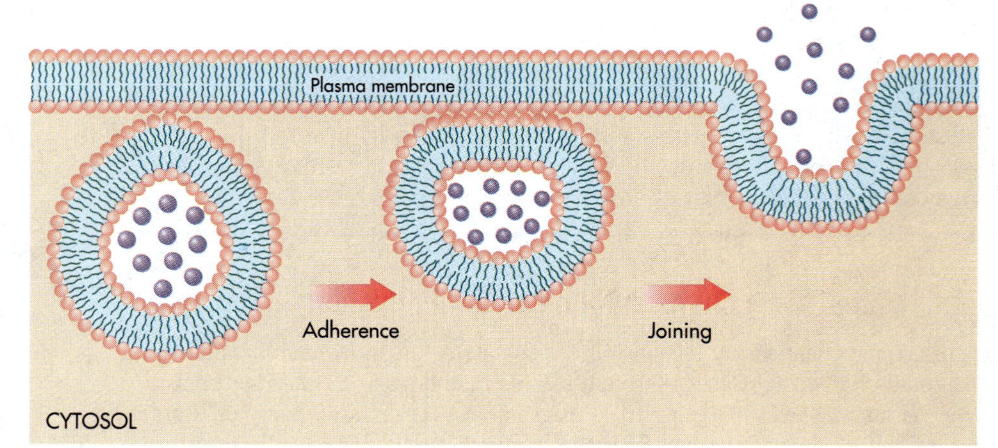

FIGURE 9.34
(a) Endocytosis. (b) Exocytosis.

9.14 CAVEOLAE

The surface of plasma membranes is coated with numerous ridges, bumps, pits, and indentations of various sizes and shapes. Although the role of the clathrin-coated pits in endocytosis has been extensively studied, the biological role, if any, of many of the other topological features remains uncertain. There is, however, growing evidence that a class of indentations known as caveolae is also involved in the importation of substances by cells.

Caveolae ("tiny caves") are small membrane-lined caves around 50 nm in diameter that are fused to the plasma membrane (Figure 9.35). They possess a protein marker, analogous to the clathrin associated with clathrin-coated pits, called **cave-**

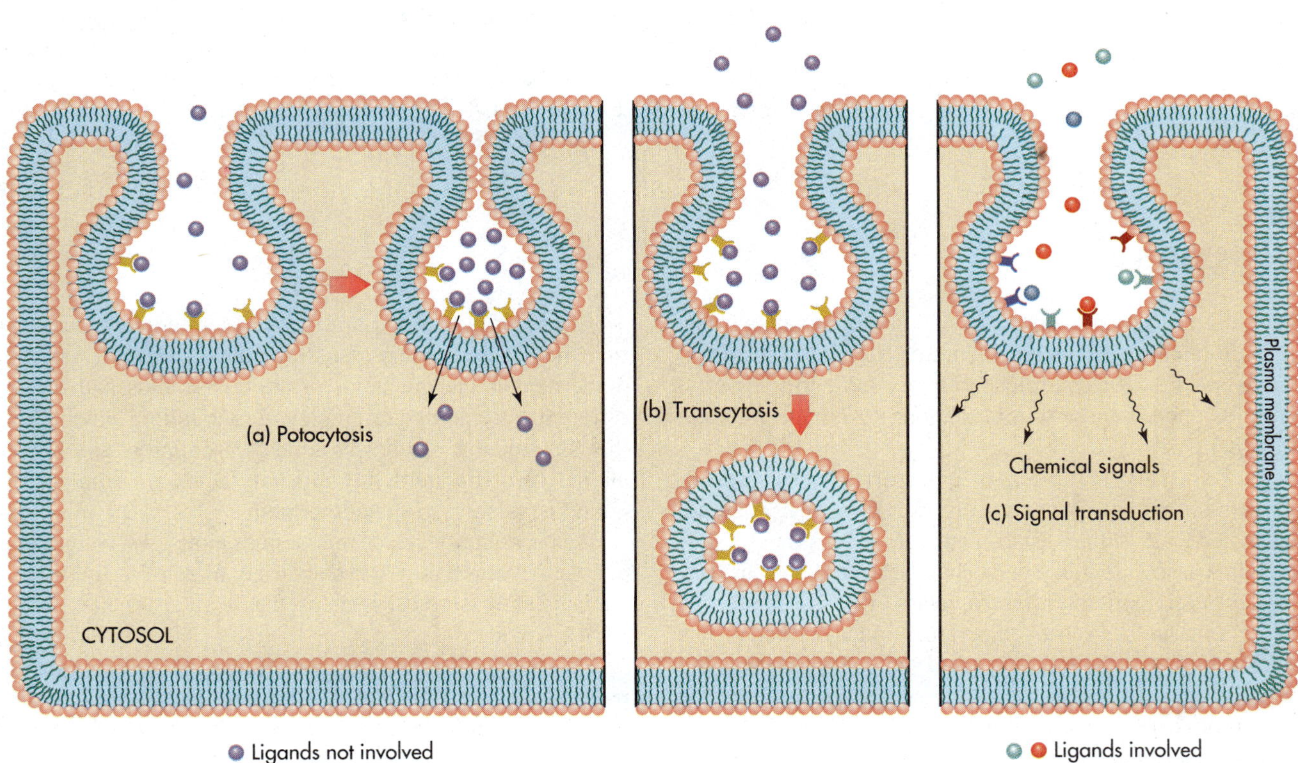

FIGURE 9.35
Potocytosis, transcytosis, and signal transduction by caveolae. (a) In potocytosis, after ligands bind to specific receptors in caveolae, the "caves" are sealed until the trapped ligands move into the parent cell. (b) During transcytosis, ligand-containing caveolae "pinch off," move across the cell, and reattach to the plasma membrane. The caveolae do not open and the contents of the caveolae are not released into the cell. (c) With signal transduction, the binding of regulatory agents to receptors in caveolae triggers the release of chemical signals into the cell.

olin (not shown in Figure 9.35). They also contain receptors for those substances that they sequester for a cell. After the receptors have accumulated the substrate targeted for transport, the openings of some caveolae close, creating a compartment in which a high concentration of a substrate facilitates the movement of that substrate through the caveolae membrane into the parent cell. The process is known as **potocytosis.** The prefix *poto-* comes from a Latin verb meaning *to drink.* In contrast to the vesicles created during endocytosis, the sealed caveolae remain attached to the plasma membrane during potocytosis. The vitamin folic acid (Section 10.1) was one of the first substances found to be taken up by cells through potocytosis.

It is hypothesized that caveolae participate in other processes as well. During **transcytosis,** closed substrate-containing caves detach from the plasma membrane on one side of the cell and then reattach to the plasma membrane on the opposite side after traversing the cytoplasm of the cell. This provides a mechanism for transporting substances through cells and releasing them on the other side. Caveolae are also called "**signaling organelles**" by some who propose that they play a key role in the propagation of signals across the plasma membrane. This proposal is based on preliminary evidence that some of the tiny caves harbor many substances involved in signal transduction, including receptors for hormones, growth factors, and other regulatory agents.

SUMMARY

A lipid is any compound produced by living organisms that is predominantly nonpolar in nature. The major classes of lipids are fats, waxes, phospholipids, sphingolipids, glycolipids, eicosanoids, steroids, lipoproteins, and fat-soluble vitamins.

Fats are fatty acid esters of glycerol, primarily triacylglycerols, that are stored mainly in adipocytes. They are classified as either saturated or unsaturated. The carbon–carbon double bonds in unsaturated fats usually possess a cis configuration. Vegetable oils are plant fats. Most are liquids at room temperature and unsaturated. The hydrogenation of vegetable oils yields margarines and shortenings that contain *trans*-fats and a reduced number of carbon–carbon double bonds. Fats are a major storage form of energy and carbon.

Dietary fats promote satiety, carry fat-soluble vitamins, are a source of essential fatty acids, and help maintain healthy skin and hair. Excess dietary fat contributes to obesity and may increase cancer risk. A strong correlation exists between saturated dietary fat and heart disease risk. Preliminary evidence indicates that dietary *trans*-fat may be an independent heart risk factor. Dietary fat is digested by lipase-catalyzed hydrolysis in the small intestine.

Although fatty acids are the major building blocks of fats, they have a variety of additional roles as well. Arachidonic acid is the precursor to the eicosanoids and many different fatty acids are employed as building blocks for the construction of phospholipids, glycolipids, and other classes of compounds. Certain fatty acids are commonly bound covalently to specific proteins.

Waxes are fatty acid esters of high molecular weight alcohols. Although they serve a variety of roles in living organisms, they are most often used as a water-repellent protective barrier.

Phospholipids are lipids that contain one or more phosphate residues. In addition, most glycerophospholipids contain a glycerol residue, a residue of a second alcohol, and one or more fatty acid residues. The sphingomyelins, a second category of phospholipids, are constructed from sphingosine, fatty acids, phosphate, and choline or ethanolamine. All phospholipids are amphipathic compounds. Some are structural and functional components of membranes, and others are involved in signal transduction across membranes. Still others function as hormones.

Glycolipids contain a carbohydrate component that may be either a monosaccharide (in cerebrosides) or an oligosaccharide (in gangliosides). The most common glycolipids also contain a sphingosine and a fatty acid residue. Glycolipids are frequently involved in "recognition" processes.

Steroids are characterized by a distinctive fused ring system. In sterols, including cholesterol, a hydroxyl group and other substituents are attached to the steroid ring. Some steroids are hormones, others aid in digestion, and still others are components of membranes. Steroids play additional roles as well.

Cholesterol is a normal membrane component and a precursor to a variety of compounds, including bile acids and steroid hormones. Despite this fact, cholesterol is often considered a "bad guy" because of its highly publicized correlation with heart disease risk. Cholesterol is a major component of the artery-clogging plaques that accumulate during atherosclerosis. Although blood cholesterol levels are largely determined by genetics, they can be modified to some extent by diet, exercise, and medication.

The eicosanoids include prostaglandins, thromboxanes, and leukotrienes. All contain 20 carbon atoms and most function as "local hormones." One or more eicosanoids are produced by most cells in the human body, and they help regulate numerous processes including muscle contraction, nerve transmission, blood clotting, electrolyte balance, and the immune response. The action of a variety of drugs has been traced to their impact on eicosanoid metabolism.

Lipoproteins are lipid–protein complexes that transport lipids in blood and play a central role in determining where and how these transported lipids are utilized. Lipoproteins are divided into five major subclasses on the basis of their density. HDL and LDL are the lipoproteins primarily responsible for transporting cholesterol, and the distribution of this sterol between these particles is a significant heart risk indicator. It is primarily the cholesterol in LDL that is deposited in arteries. Blood LDL-cholesterol levels are largely determined by the efficiency with which liver cells and other cells remove LDL from the blood by receptor-mediated endocytosis. Familial hypercholesterolemia is a genetic disease associated with defective LDL receptors. In some individuals, blood lipoprotein(a) concentration is an independently inherited heart risk factor.

Biological membranes are semipermeable barriers that help regulate what substances enter and leave cells and organelles. They normally possess a large number of enzymes and receptors, and they participate in a wide range of physiologically important events. They consist of a lipid bilayer associated with peripheral and integral proteins. The lipid bilayer and membrane-spanning proteins determine permeability. In general, only small nonpolar substances can pass directly through the bilayer. Polar, ionic, and polymeric substances rely upon protein transporters, channels, and pores to traverse membranes. Some substances also enter cells through processes called endocytosis and potocytosis.

Specific protein transporters are involved in both facilitated diffusion and active transport. Active transport, in contrast to diffusion, requires energy and can move a substance against a concentration gradient. Active transport is classified as primary or secondary, depending on the source of the energy used during the transport process. Many transporters exhibit Michaelis–Menten-type kinetics and distinctive transporters have unique K_m values and V_{max} values for the substrates they transport.

A family of uniport transporters exists for the translocation of glucose across plasma membranes by facilitated diffusion. Each family member, including GluT1, possesses multiple, membrane-spanning, α-helical regions that are brought together to produce a hydrophilic tunnel through which glucose can be passed. The transporters oscillate between two conformations, one with the glucose binding site positioned on the outside of the membrane and the second with the glucose binding site exposed to the interior of the cell.

The Na^+K^+ transporter is an antiport membrane pump that employs energy from ATP to create Na^+ and K^+ gradients. This transporter, like GluT1, is a membrane-spanning protein that functions by switching between two conformations. One conformation is phosphorylated and the second is nonphosphorylated. Phosphorylation by ATP is the energy-consuming step in the transport cycle. Na^+ and K^+ gradients are involved in secondary active transport, signal transduction across membranes, and the control of osmotic pressure.

Channels, in contrast to transporters, do not possess specific binding sites for the substances they shuttle through membranes. Net movement is by diffusion and the kinetics involved is usually nonsaturating. Channels are normally gated and substrate specific. Although nuclear pores function as large, nongated, nonspecific openings for the transport of small molecules and ions, they function as specific transporters for a variety of large substances, including proteins and nucleic acids.

Caveolae are hypothesized to be involved in a number of events, including potocytosis, transcytosis, and the transduction of signals across membranes. During potocytosis, caveolae remain attached to the plasma membrane. In contrast, the membrane-enclosed vesicles associated with endocytosis and exocytosis are separated from the plasma membrane during certain stages of these processes.

EXERCISES

9.1 Write the structural formula for all the different triacylglycerols that can be constructed using glycerol, stearic acid, and palmitoleic acid as building blocks.

9.2 Vegetable oils tend to become rancid more rapidly than animal fats. Provide a chemical explanation.

9.3 Write the structural formula for a triacylglycerol that will most likely exist as a liquid at room temperature and pressure. Write the structural formula for a second triacylglycerol that will most likely exist as a solid under similar conditions.

9.4 Write the structural formula for an ω-3, continuous-chain, unsaturated carboxylic acid containing 12 carbons.

***9.5** What is the chemical difference between vegetable oils and car oils? Provide a chemical explanation for why it would be disastrous to try to use a vegetable oil in your car.

9.6 What products are formed when fats are completely oxidized by O_2 within the body?

9.7 Explain why nonpolar toxins tend to accumulate within fatty tissues.

9.8 Calculate the number of calories in (a) a cookie that contains 10 g of digestible carbohydrates, 2 g of protein, and 2 g of fat, and (b) a cookie that contains 5 g of digestible carbohydrates, 5 g of protein, and 5 g of fat. For each cookie, calculate the percentage of the total calories that comes from fat.

9.9 List three normal and beneficial roles or functions of fats within the human body. List three potential risks associated with the overconsumption of saturated fats.

9.10 Is there any health risk associated with a diet that contains absolutely no fat? Explain.

9.11 List those amino acids with side chains that can be covalently attached to a fatty acid through an ester link.

9.12 Identify each hydrolyzable bond in each compound in Table 9.4.

9.13 Fats are not usually considered amphipathic compounds. Explain.

9.14 Which of the following terms or statements accurately describe a sphingomyelin? A lipid, a fat, a phospholipid, a glycolipid, an amphipathic compound, a steroid, a common membrane component, a chiral compound, an eicosanoid, a protein.

9.15 Identify each term or statement in Exercise 9.14 that accurately describes:

a. a cerebroside
b. cholesterol
c. a monoacylglycerol
d. PGE_2.

9.16 Will the addition of a carbohydrate residue to a lipid increase or decrease its polarity? Explain.

9.17 Write the structural formula for *trans*-oleic acid and its cis isomer. Explain why the *trans*-fatty acid has a higher melting point.

9.18 Write the structural formula for two different glucose-containing cerebrosides.

9.19 What products are generated when the cerebroside in Figure 9.12 is completely hydrolyzed?

9.20 Which, if any, of the steroids in Figure 9.16 are saponifiable (can be hydrolyzed by basic solutions)?

9.21 List three normal functions of steroids within living organisms. For each function, name one specific steroid that serves this function.

9.22 What property must a compound possess before it can exhibit geometric isomerism? Is geometric isomerism possible in cholesterol? Explain.

9.23 If the pK_a for the carboxyl group in PGF_2 is approximately 4, will PGF_2 primarily exist in its salt or its nonsalt form at pH 7? Will the salt form be more or less soluble in water than the nonsalt form? Explain.

9.24 Which fatty acid is the precursor for most eicosanoids?

9.25 Explain why the apolipoproteins found within lipoproteins tend to have a polar side and a nonpolar side. List three amino acids you might expect to find on the nonpolar side of an apolipoprotein.

9.26 Can the apolipoprotein content of a lipoprotein particle be used to determine the major class to which the lipoprotein particle belongs? Explain.

9.27 Most cholesterol esters are products of the reaction of cholesterol with fatty acids. Explain why cholesterol esters are part of the hydrophobic cores of lipoproteins, whereas free cholesterol is found among the outer amphipathic lipids within lipoproteins.

9.28 What property of waxes allows them to be used as waterproofing agents by various organisms?

9.29 What property of peripheral proteins tends to keep such proteins from spontaneously moving across a membrane from one side to the other?

***9.30** Researchers must use detergents to separate membranes into their component molecules. Explain. Are covalent bonds broken in the process?

9.31 Receptors normally interact in a noncovalent manner with their ligands. Predict what specific classes of noncovalent bonds are involved in binding cholesterol to a protein receptor.

9.32 Explain why carbohydrate residues are not usually buried inside the lipid bilayer of a membrane.

9.33 What effect does the doubling of the number of transporters have on the V_{max} for the transport of a substrate across a membrane? What effect does this doubling have on K_m? Explain.

9.34 If a substrate that is destined to be transported across a membrane is normally present at very low concentrations, would you expect its transporter to have a large or small K_m? Explain.

9.35 Describe the similarities and differences between facilitated diffusion and active transport and between primary and secondary active transport.

9.36 Although transporters A and B both transport the same substrate, transporter A consumes energy and transporter B does not. Can you conclude that transporter A moves the transported substance more rapidly across a membrane? Explain.

9.37 An α-helix rises 0.56 nm per turn. Use this information to calculate the approximate thickness of the lipid bilayer depicted in Figure 9.28.

9.38 List four amino acids that could be used to line the tunnel in GluT1 (Figure 9.28).

9.39 The net transport of glucose from liver to blood by GluT2 automatically slows down following the consumption of a high-carbohydrate meal. Assume that the concentration of glucose in liver is approximately the same before and after the meal, and then explain the change in net rate of glucose transport.

9.40 Since mannose has a lower affinity for GluT1 than glucose, would the K_m for the transport of mannose by GluT1 be greater, less, or the same as the K_m for the transport of glucose? Explain.

9.41 True or false? Facilitated diffusion is unable to increase the magnitude of a concentration gradient. Explain.

9.42 Which step in the transport of Na^+ and K^+ by the Na^+K^+ transporter (Figure 9.32) is the energy-consuming step in the transport process?

9.43 Under what conditions is the rate of transport of Na^+ by the Na^+K^+ transporter independent of Na^+ concentration? Explain.

9.44 What does it mean to say that "A concentration gradient possesses entropy-linked potential energy"?

9.45 Summarize the similiarities and differences between channels and membrane transporters.

9.46 Transport through a channel does not usually exhibit saturation kinetics. Describe one set of conditions that leads to an exception to this generalization.

9.47 Define and distinguish *endocytosis, exocytosis, potocytosis,* and *transcytosis.*

SELECTED READINGS

Alberts, B., D. Bray, J. Lewis, M. Raff, K. Roberts, and J. D. Watson, *Molecular Biology of the Cell,* 3rd ed. New York: Garland Publishing, 1994.

An up-to-date and encyclopedic textbook on the structure and function of cells. It contains numerous references and multiple chapters dealing with membrane structure and function.

Barinaga, M., "Obese" protein slims mice, *Science* 269, 475–476 (1995).

A brief review of impact of leptin (ob protein, the protein encoded in ob *gene) on weight loss in mice.*

Barinaga, M., Pot, heroin unlock new areas for neuroscience, *Science* 258, 1882–1884 (1992).

Summarizes research on anandamide.

Cravatt, B.F., O. Prospero-Garcia, G. Siuzdak, N.B. Gilula, S.J. Henriksen, D.L. Boger, and R.A. Lerner, Chemical characterization of a family of brain lipids that induce sleep, *Science* 268, 1506–1509 (1995).

Article provides evidence that fatty acid primary amides may represent a previously unrecognized class of biological signaling molecules.

Cutting cholesterol: more vital than ever, *Consumer Reports on Health* 7(2), 13–14 (1995).

Describes a "landmark trial" showing that cutting blood cholesterol levels can indeed reverse atherosclerosis and save lives.

Fat for the Fit, *Consumer Reports on Health* 4(6), 41–43 (1992).

Examines the link between dietary fats and heart disease.

Goedert, M., W. J. Strittmatter, and A. D. Roses, Risky apolipoprotein in brain, *Nature* 372, 45–46 (1994).

Reviews efforts to determine the molecular basis for the correlation between Alzheimer's disease risk and the E4 variant of the gene for apolipoprotein E.

Hoberman, J. M., and C. E. Yesalis, The history of synthetic testosterone, *Sci. Am.* 272(2), 76–81 (1995).

Examines the past and present uses of anabolic steroids and some possible future applications of these potent physiological agents.

Lasic, D. D., and D. Papahadjopoulos, Liposomes revisited, *Science* 267, 1275–1276 (1995).

Reviews the properties of different classes of liposomes and the use of liposomes in cancer chemotherapy and gene therapy.

Lawn, R. M., Lipoprotein(a), in heart disease, *Sci. Am.* 266(6), 54–60 (1992).

Examines the link between lipoprotein(a) and heart disease.

Lienhard, G. E., J. W. Slot, D. E. James, and M. M. Mueckler, How cells absorb glucose, *Sci. Am.* 266(1), 34–39 (1992).

Describes the biochemistry of some glucose transporters.

Moran, L. A., K. G. Scrimgeour, H. R. Horton, R. S. Ochs, and J. D. Rawn, *Biochemistry,* 2nd ed. Englewood Cliffs, NJ: Neil Patterson/Prentice Hall, 1994.

An encyclopedia-type biochemistry text with selected readings and problems. Chapter 7 is titled "Lipids and Membranes."

New *trans* fat studies muddy the waters, *Science News* 147(8), 127 (1995).

Describes studies which reveal that high body stores of trans-oleic acid correlate with increased protection against heart attacks.

Noyori, R., and M. Suzuki, Organic synthesis of prostaglandins: advancing biology, *Science* 259, 44–45 (1993).

Reviews the use of organic synthesis to develop therapeutic prostaglandins.

Rink, T. J., In search of a satiety factor, *Nature* 372, 406–407 (1994).

*Describes the isolation of a mouse obese (*ob*) gene that encodes a protein (ob protein) hypothesized to function as a "satiety factor" by suppressing food intake and appetite. The mouse protein is 84% the same as its human counterpart.*

Roses, A.D., Apolipoprotein E and Alzheimer disease, *Sci. Am. Sci. Med.* 2(5), 16–25 (1995).

Reviews evidence that the common, late-onset form of Alzheimer disease is associated with abnormalities in a gene that encodes apolipoprotein E.

Sadava, D. E., *Cell Biology: Organelle Structure and Function.* Boston: Jones and Bartlett Publishers, 1993.

The book begins with two chapters on the general nature of membranes and then examines the unique membrane associated with each organelle.

Stamler, J., and J. D. Neaton, Benefits of lower cholesterol, *Sci. Am. Sci. Med.* 1(2), 28–37 (1994).

Explores the link between cholesterol and heart disease.

Travis, J., Cell biologists explore "tiny caves," *Science* 262, 1208–1209 (1993).

A brief review on caveolae.

Wolfe, S. L., *Molecular and Cellular Biology.* Belmont, California: Wadsworth Publishing Company, 1993.

Provides a thorough coverage of membrane structure and function. Contains references and review questions.

10 Vitamins, Hormones and an Introduction to Metabolism

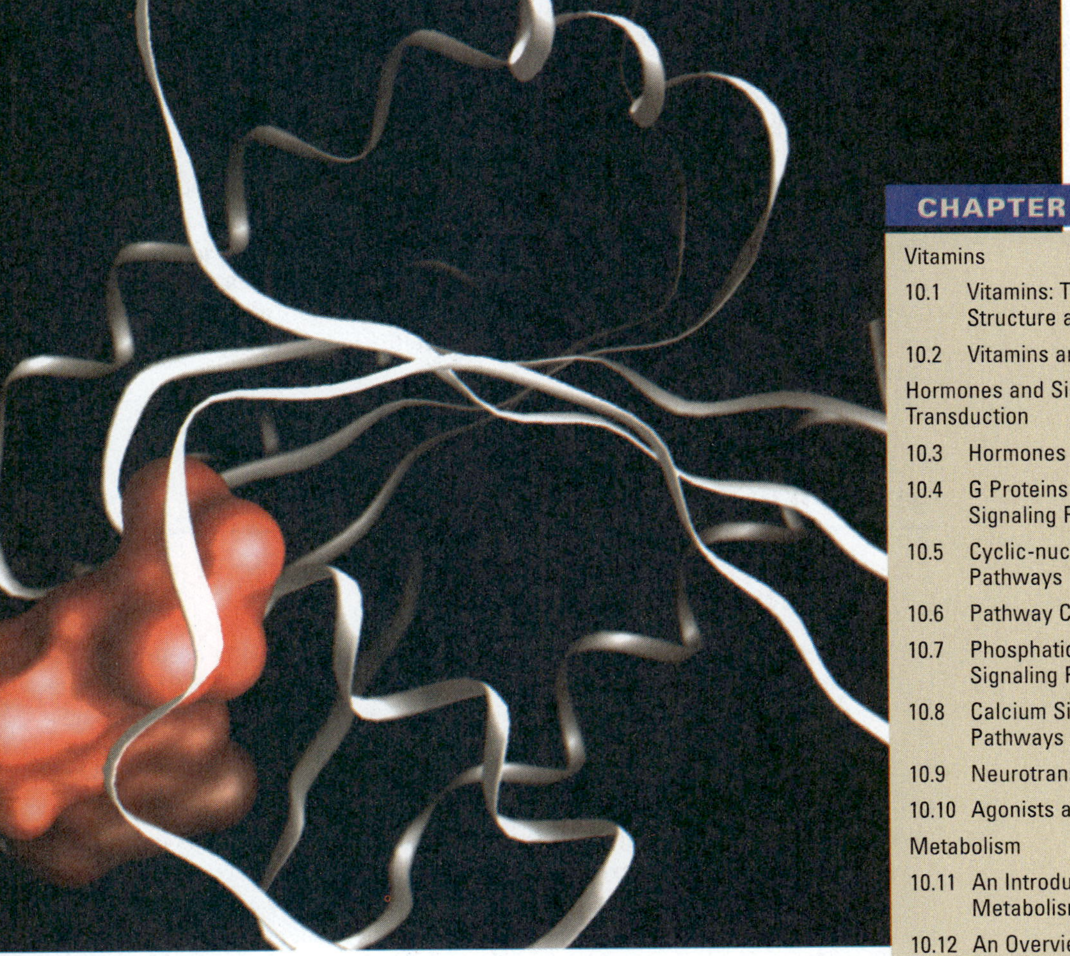

CHAPTER OUTLINE

Chapter 9, "Lipids and Biological Membranes," completed the process of introducing DNA, RNA, proteins, carbohydrates, and lipids, those classes of compounds listed in Table 1.1 titled "The Major Classes of Biochemicals". It also added to the framework required for the study of metabolism. A discussion of vitamins and hormones is needed to complete that framework.

Human life requires roughly two dozen chemical elements and involves 50,000 to 100,000 distinct chemical reactions. Virtually all of these reactions are catalyzed by enzymes, and many must be continuously regulated and coordinated. Metabolism is the sum of all the enzyme-catalyzed reactions that occur within a living organism. It is divided into two major categories: anabolism, the collection of all synthetic reactions; and catabolism, the sum of the degradative reactions (Section 1.2). Most of the chemical transformations that occur within living organisms clearly fall into one of these two categories.

Vitamins are essential nutrients that are converted by the human body into a wide variety of substances, including part of the hormones and most of the coenzymes crucial to metabolism. Some also function as antioxidants, compounds that protect other compounds from oxidation by serving, themselves, as targets for oxidizing agents. Hormones are chemical messengers that help coordinate and regulate both catabolism and anabolism. Most human cells contain specific receptors for multiple hormones.

This chapter begins with an examination of vitamins followed by an investigation of hormones and their mode of action. A formal introduction to metabolism draws the chapter to a close. Featured biochemical themes are experimentation and hypothesis construction, transformation and use of energy, structural hierarchy and compartmentation, specificity, communication, and catalysis and the regulation of reaction rates. Chapters 11 through 18 introduce the metabolism of carbohydrates, lipids, proteins, and nucleic acids.

VITAMINS

10.1 VITAMINS: THEIR NATURE, STRUCTURE, AND FUNCTIONS

The human body can produce most of the organic substances that it needs from the carbohydrates (Chapter 6), lipids (Chapter 9), and other organic compounds present in abundant amounts in virtually every diet. However, there are a few required substances the body cannot produce at all or cannot produce in sufficient amounts. These **essential nutrients** include vitamins, essential amino acids (Section 3.4), essential fatty acids (Section 9.1), and particular chemical elements (Table 10.1). **Vitamins** are those essential organic nutrients that are required by the body in trace amounts. They are **micronutrients.** The essential amino acids and essential fatty acids are not usually considered vitamins, because the body needs considerably larger quantities of these substances. There is a chance that additional compounds will be added to the list of essential nutrients as biochemists learn more about human metabolism. **[Try Problem 10.1]**

TABLE 10.1

Nutrients Required by the Human Body

	Amino Acids	Fatty Acids	Vitamins	Elements[g]
ESTABLISHED AS ESSENTIAL	Arginine[a]	Linoleic[b]	Ascorbic acid	Calcium
	Histidine	Linolenic	Biotin[c]	Chlorine
	Isoleucine		Folic acid	Chromium
	Leucine		Niacin[d]	Copper
	Lysine		Pantothenic acid	Fluorine
	Methionine		Pyridoxine	Iodine
	Phenylalanine		Riboflavin	Iron
	Threonine		Thiamine	Magnesium
	Tryptophan		Vitamin B_{12}	Manganese
	Valine		Vitamin A	Molybdenum
			Vitamin D[e]	Phosphorus
			Vitamin E	Potassium
			Vitamin K[c]	Selenium
				Silicon
				Sodium
				Sulfur
				Zinc
POSSIBLY ESSENTIAL[f]				Arsenic
				Cobalt
				Nickel
				Tin
				Vanadium

[a]Unnecessary for adults but required for normal growth and development in children.
[b]Linoleic acid is a precursor for arachidonic acid which, in turn, is the precursor for many eicosanoids.
[c]Requirement is largely met by the biosynthetic activities of the intestinal flora.
[d]Requirement can be met with adequate dietary tryptophan.
[e]Requirement can be met by exposure to sunlight.
[f]The five elements listed are necessary for normal growth of young rats fed only with highly purified diets.
[g]See Problem 10.1

Classification

Vitamins are divided into two major categories, **fat-soluble vitamins** (**A, D, E,** and **K**) and **water-soluble vitamins** (**B-complex** and **vitamin C** [**ascorbic acid,** ascorbate]). **Biotin, cobalamin (B_{12}), folic acid** (folate, **folacin**), **niacin, pantothenic acid** (pantothenate), **pyridoxine (B_6), riboflavin (B_2),** and **thiamine (B_1)** make up the B complex. The abbreviations B_1, B_2, B_3, B_4, and so on, were assigned to components within the B complex as they were isolated and characterized. Some of the abbreviations were subsequently abandoned when the components involved were found to be different forms of the same vitamin or to be identical to previously named and identified compounds. Inositol, choline, and **para-aminobenzoic** acid are vitamin-like substances sometimes classified as part of the B complex, but no convincing evidence establishes a dietary requirement in humans (Figure 10.1). Inositol and choline, both building blocks for some of the lipids examined in Chapter 9, are known to be essential nutrients for some nonhuman species.

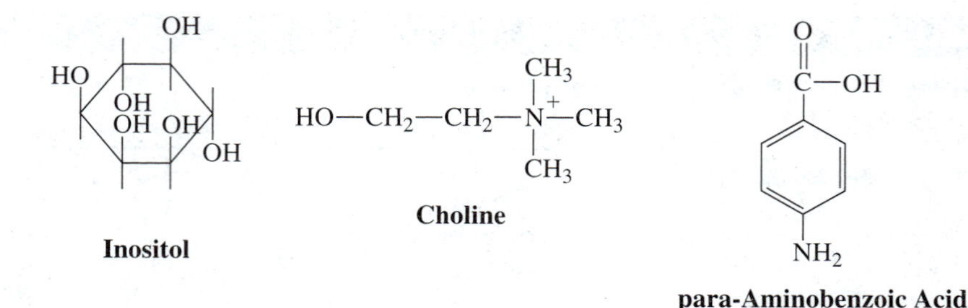

FIGURE 10.1
Inositol, choline, and para-aminobenzoic acid.

All of the fat-soluble vitamins and some of those in the B complex exist in multiple forms; they are families of compounds. The term "vitamin D," for example, refers to ergocalciferol (D_2) and cholecalciferol (D_3) and, sometimes, to select mono- and dihydroxy derivatives of these compounds. The active forms of vitamin A are retinol, retinal, and retinoic acid. These compounds, along with derivatives of vitamin A, are known as **retinoids.** The vitamin E family includes four **tocopherols** and four **tocotrienols** with **α-tocopherol** being the most abundant and most active member. The multiple forms of a vitamin are often interconvertible, and some are functionally interchangeable as well.

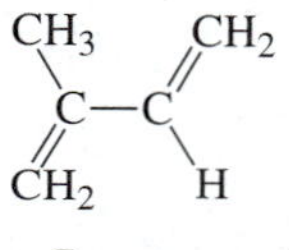

The fat-soluble vitamins, being nonpolar and soluble in fat and other nonpolar solvents, are a class of lipids (Figure 9.1). All are synthesized (in those organisms that produce them) from isoprene and are, for that reason, said to be **isoprenoids.** Excess quantities of these vitamins tend to be stored in fat-containing cells within the liver. These reserves can, potentially, eliminate the dietary requirements for vitamins A and D for several months. The tendency to store fat-soluble vitamins provides a partial explanation for the toxicity sometimes observed with vitamins A and D. In the United States, both **hypervitaminosis A** and **hypervitaminosis D** have become significant health problems because some foods are fortified with these micronutrients and more and more people are turning to vitamin tablets to supplement their diets. Some cases of vitamin D "poisoning," for example, have been traced to fortified milk containing an excess of this compound. More is not necessarily better, even when it comes to essential nutrients. **[Try Problem 10.2]**

In contrast to the fat-soluble vitamins, most water-soluble vitamins are inactivated or eliminated fairly rapidly and are not stored in significant amounts within the human body. Hence, they must be consumed on a regular basis. The inactivation and elimination of the water-soluble vitamins helps explain why it is more difficult to overdose on these vitamins than on some of the fat-soluble ones. However, pyridoxine and niacin are definitely toxic at high doses, and there is preliminary evidence that large enough quantities of several other water-soluble vitamins may also lead to adverse side effects. **[Try Problem 10.3]**

Deficiency Symptoms, Structures, and Dietary Sources

Inadequate vitamin consumption or the inadequate absorption of vitamins from the intestine can lead to deficiency diseases, including scurvy, rickets, beriberi and pellagra (Table 10.2). These diseases emphasize the importance of the vitamins in metabolism. Some therapeutic drugs and various illnesses can also lead to vitamin deficiency symptoms by disrupting normal vitamin metabolism. Tables 10.3 and 10.4 provide the structural formulas, a few of the nutritional sources, and some of the deficiency symptoms for the vitamins. These tables should be studied thoroughly. **[Try Problems 10.4 through 10.6]**

TABLE 10.2

Some Vitamin Deficiency Diseases

Disease[a]	Deficient Vitamin
Scurvy	Vitamin C (ascorbic acid)
Rickets	Vitamin D
Beriberi	Vitamin B_1 (thiamine)
Pellagra	Niacin
Pernicious anemia	Vitamin B_{12} (cobalamin)
Megaloblastic anemia	Folic acid (folacin)
Night blindness	Vitamin A

[a]See Tables 10.3 and 10.4 for deficiency symptoms.

Recommended Daily Allowances

Nutritionists are still struggling to determine what level of vitamin consumption is optimum. The question is complex since the daily requirement for any given vitamin is a function of age, gender, general health, pregnancy, other dietary components, and additional variables. Smoking boosts the need for vitamin C, and the use of oral contraceptives may increase the requirement for riboflavin. The **intestinal flora,** the collection of microorganisms that live in the intestine, produces particular vitamins and, in certain instances, may produce a large fraction of the necessary supply. Some vitamins exist in multiple forms that are absorbed into the blood stream and metabolized at different rates and may exhibit distinctive biological activities. One form of vitamin D is produced when a precursor in the skin, 7-dehydrocholesterol, is irradiated with UV light.

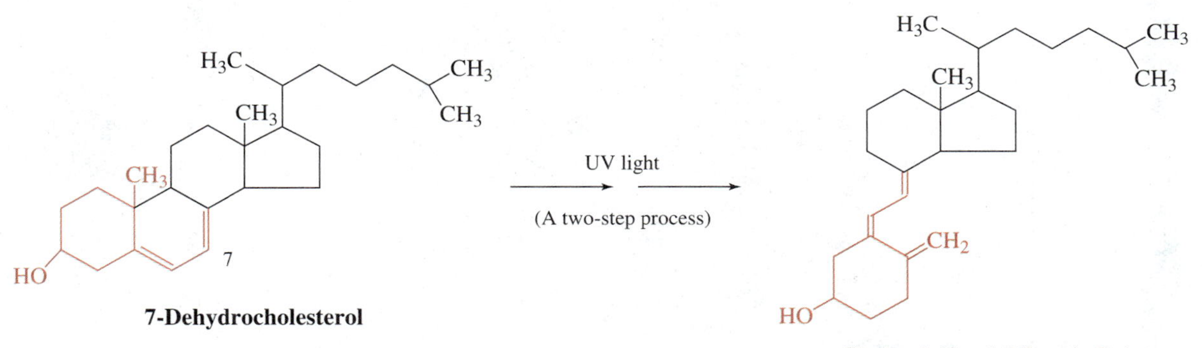

Cholecalciferol (Vitamin D_3)

When UV exposure is high, there is no dietary requirement, and vitamin D is technically no longer a vitamin. Since UV exposure increases the risk of skin cancer, it is probably inadvisable to replace dietary vitamin D with tanning parlors or days on the beach.

The **recommended daily allowance (RDA)** for a vitamin is the amount that prevents deficiency symptoms in most normal individuals without producing any toxic effects. The RDAs are set by the Food and Nutrition Board of the National Research Council and are reexamined on a periodic basis. They are summarized in Table 10.5. Certain RDAs are controversial, and additional research is definitely needed to refine the presently accepted values. Vitamin E, for example, was not assigned an RDA until 1968, and its RDA has been reduced twice since then. Attempts to induce vitamin E deficiency symptoms in normal human adults have failed.

TABLE 10.3

Water-Soluble Vitamins

Vitamin	Structure	Some Common Dietary Sources	A Partial Listing of Deficiency Symptoms in Humans
1. Thiamine (vitamin B_1)	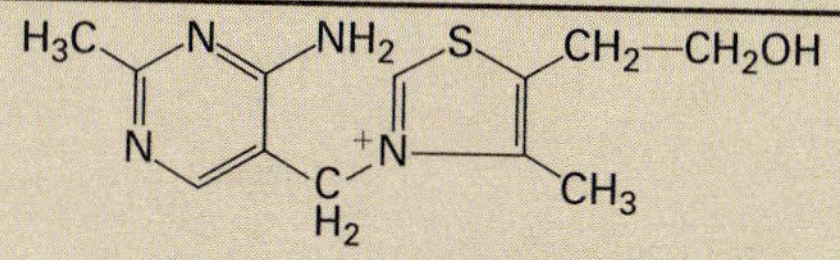 H_3C, N, N, NH_2, S, $CH_2—CH_2OH$, ^+N, CH_3, $C H_2$	Liver, meat, milk, vegetables, whole grains, nuts	Weight loss, muscle wasting, sensory changes, mental confusion, enlargement of the heart, constipation
2. Riboflavin (vitamin B_2)	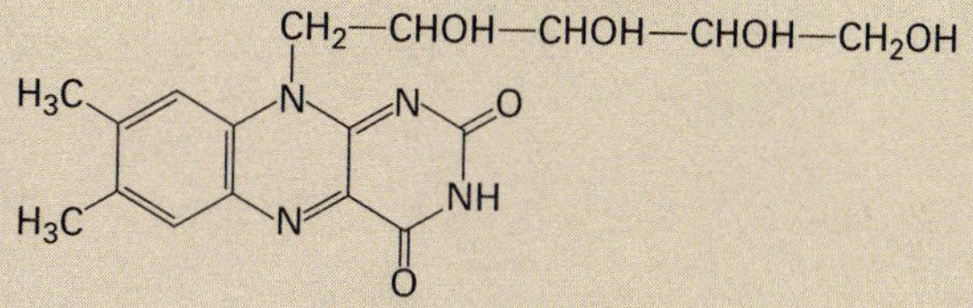 $CH_2—CHOH—CHOH—CHOH—CH_2OH$; H_3C, H_3C, N, N, N, O, NH, O	Liver, wheat germ, eggs, milk, green leafy vegetables, meat	Magenta-colored tongue, fissuring at the corners of mouth and lips, dermatitis, behavioral changes
3. Niacin (nicotinic acid)	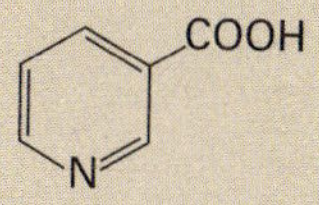 COOH, N	Meat, liver, cereals, legumes	Dermatitis when exposed to sunlight, weakness, insomnia, impaired digestion, diarrhea, dementia, irritability, memory loss, headaches
4. Pyridoxine (vitamin B_6 or pyridoxol)	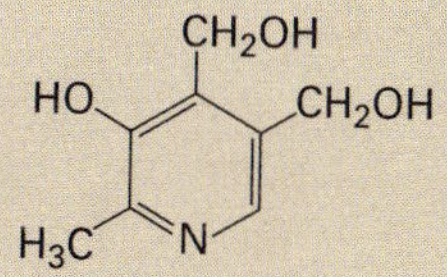 CH_2OH, HO, CH_2OH, H_3C, N	Egg yolk, fish, meat, lentils, nuts, fruits, vegetables	In infants, convulsions, dermatitis, weight loss, apathy, irritability, weakness
5. Pantothenic acid	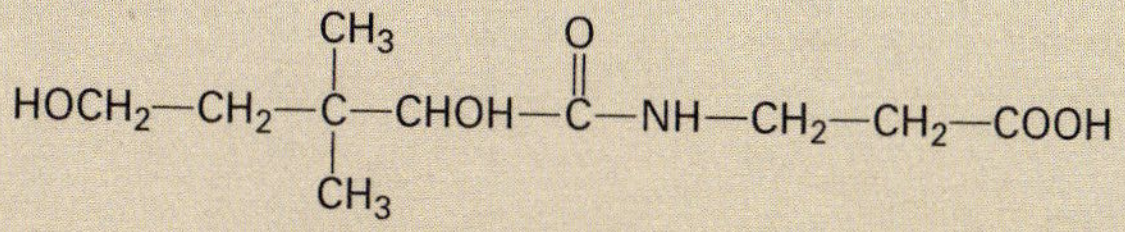 $HOCH_2—CH_2—C(CH_3)_2—CHOH—C(=O)—NH—CH_2—CH_2—COOH$	Eggs, peanuts, liver, meat, milk, cereals, vegetables	Vomiting, malaise, abdominal distress, cramps, fatigue, insomnia
6. Biotin	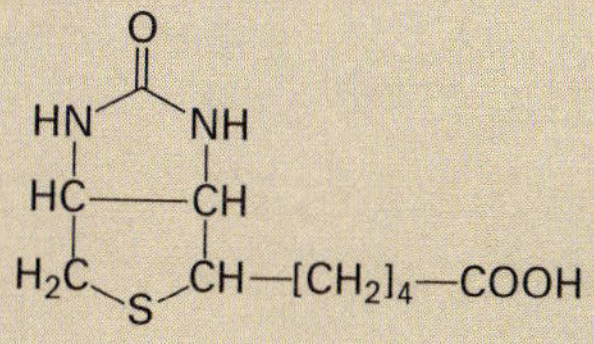 O, HN, NH, HC—CH, H_2C, S, $CH—[CH_2]_4—COOH$	Liver, yeast, meat, peanuts, eggs, chocolate, dairy products, grains, fruits, vegetables	Dermatitis, skin dryness, depression, muscle pain, nausea, anorexia (appetite loss)

(Continued)

TABLE 10.3 ***(Continued)***

Water-Soluble Vitamins

Vitamin	Structure	Some Common Dietary Sources	A Partial Listing of Deficiency Symptoms in Humans
7. Folic acid (folacin)	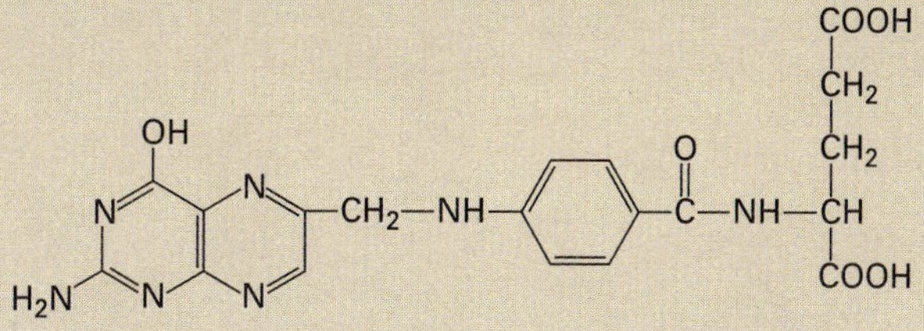 OH N N H_2N N N $-CH_2-NH-$ (benzene ring) $-\overset{O}{\overset{\Vert}{C}}-NH-CH$ COOH CH_2 CH_2 COOH	Yeast, liver, green vegetables, some fruits	Possibly hostility and paranoid behavior, anemia leading to weakness, tiredness, sore tongue, diarrhea, irritability, headache, heart palpitations
8. Ascorbic acid (vitamin C)	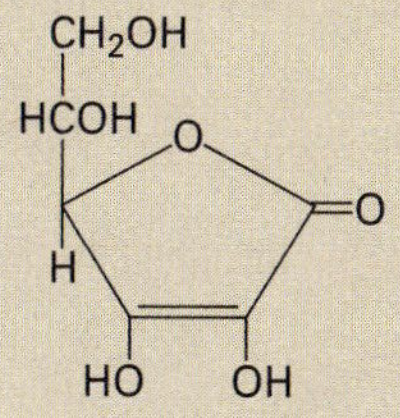 CH_2OH HCOH O =O H HO OH	Vegetables and fruit	Sore gums, loose teeth, joint pain, edema, anemia, fatigue, depression, impaired iron absorption, impaired wound healing
9. Cobalamin (vitamin B_{12}) Note: Multiple forms exist.	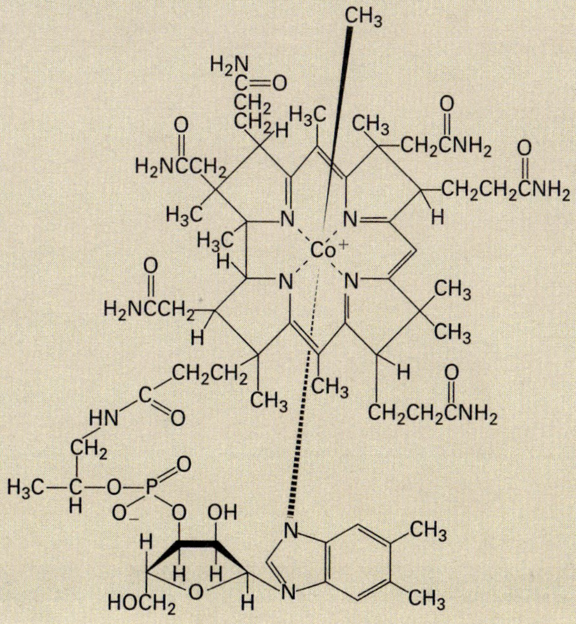 CH_3 H_2N C=O CH_2 CH_2 H H_3C CH_3 $-CH_2\overset{O}{\overset{\Vert}{C}}NH_2$ $H_2N\overset{O}{\overset{\Vert}{C}}CH_2$ $-CH_2CH_2\overset{O}{\overset{\Vert}{C}}NH_2$ H H_3C N N H_3C H Co^+ $H_2N\overset{O}{\overset{\Vert}{C}}CH_2$ H N N CH_3 CH_3 CH_2CH_2 CH_3 CH_3 H $CH_2CH_2\overset{O}{\overset{\Vert}{C}}NH_2$ HN C=O CH_2 $H_3C-\underset{H}{C}-O-\overset{O}{\overset{\Vert}{P}}-O$ O^- O OH H H H $HOCH_2$ O H N N CH_3 CH_3	Meat, shellfish, fish, milk, eggs	Neurological disorders, anemia leading to tiredness, sore tongue, constipation, headache, heart palpitations

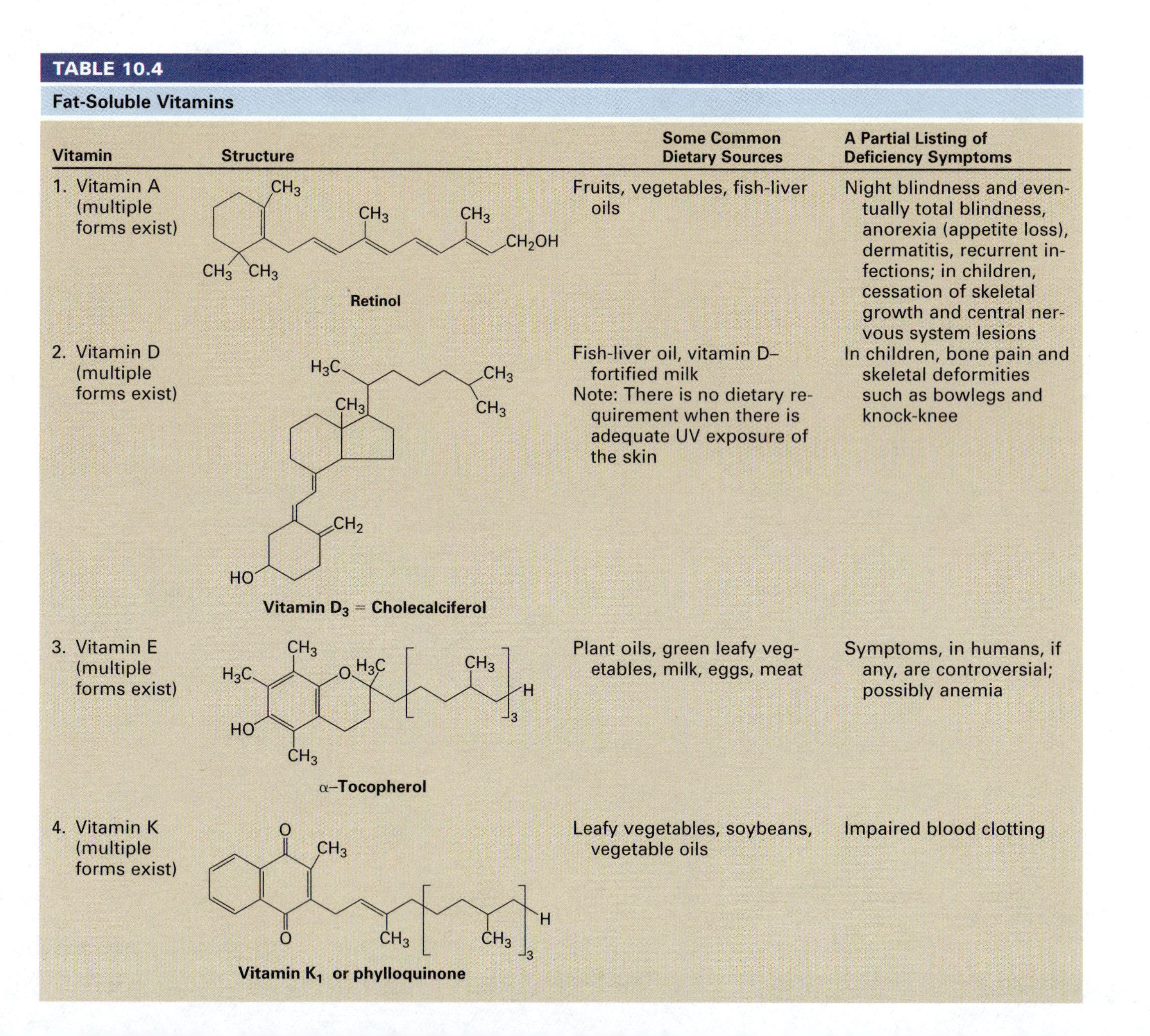

TABLE 10.4

Fat-Soluble Vitamins

Vitamin	Structure	Some Common Dietary Sources	A Partial Listing of Deficiency Symptoms
1. Vitamin A (multiple forms exist)	**Retinol**	Fruits, vegetables, fish-liver oils	Night blindness and eventually total blindness, anorexia (appetite loss), dermatitis, recurrent infections; in children, cessation of skeletal growth and central nervous system lesions
2. Vitamin D (multiple forms exist)	**Vitamin D_3 = Cholecalciferol**	Fish-liver oil, vitamin D–fortified milk Note: There is no dietary requirement when there is adequate UV exposure of the skin	In children, bone pain and skeletal deformities such as bowlegs and knock-knee
3. Vitamin E (multiple forms exist)	**α–Tocopherol**	Plant oils, green leafy vegetables, milk, eggs, meat	Symptoms, in humans, if any, are controversial; possibly anemia
4. Vitamin K (multiple forms exist)	**Vitamin K_1 or phylloquinone**	Leafy vegetables, soybeans, vegetable oils	Impaired blood clotting

TABLE 10.5

Recommended Daily Allowances for the Vitamins[a,b]

Category	Age (yr) or Condition	Weight[c] (kg)	Weight[c] (lb)	Height[c] (cm)	Height[c] (in)	Protein (g)	Fat-Soluble Vitamins: A (μg RE)[d]	D (μg)[e]	E (mg α-TE)[f]	K (μg)	Water-Soluble Vitamins[h]: C (mg)	Thiamine (mg)	Riboflavin (mg)	Niacin (mg NE)[g]	B_6 (mg)	Folate (μg)	B_{12} (μg)
Infants	0.0–0.5	6	13	60	24	13	375	7.5	3	5	30	0.3	0.4	5	0.3	25	0.3
	0.5–1.0	9	20	71	28	14	375	10	4	10	35	0.4	0.5	6	0.6	35	0.5
Children	1–3	13	29	90	35	16	400	10	6	15	40	0.7	0.8	9	1.0	50	0.7
	4–6	20	44	112	44	24	500	10	7	20	45	0.9	1.1	12	1.1	75	1.0
	7–10	28	62	132	52	28	700	10	7	30	45	1.0	1.2	13	1.4	100	1.4
Males	11–14	45	99	157	62	45	1,000	10	10	45	50	1.3	1.5	17	1.7	150	2.0
	15–18	66	145	176	69	59	1,000	10	10	65	60	1.5	1.8	20	2.0	200	2.0
	19–24	72	160	177	70	58	1,000	10	10	70	60	1.5	1.7	19	2.0	200	2.0
	25–50	79	174	176	70	63	1,000	5	10	80	60	1.5	1.7	19	2.0	200	2.0
	51+	77	170	173	68	63	1,000	5	10	80	60	1.2	1.4	15	2.0	200	2.0
Females	11–14	46	101	157	62	46	800	10	8	45	50	1.1	1.3	15	1.4	150	2.0
	15–18	55	120	163	64	44	800	10	8	55	60	1.1	1.3	15	1.5	180	2.0
	19–24	58	128	164	65	46	800	10	8	60	60	1.1	1.3	15	1.6	180	2.0
	25–50	63	138	163	64	50	800	5	8	65	60	1.1	1.3	15	1.6	180	2.0
	51+	65	143	160	63	50	800	5	8	65	60	1.0	1.2	13	1.6	180	2.0
Pregnant						60	800	10	10	65	70	1.5	1.6	17	2.2	400	2.2
Lactating	1st 6 months					65	1,300	10	12	65	95	1.6	1.8	20	2.1	280	2.6
	2nd 6 months					62	1,200	10	11	65	90	1.6	1.7	20	2.1	260	2.6

[a]From: The Food and Nutrition Board, 1989.

[b]The allowances, expressed as average daily intakes over time, are intended to provide for individual variations among most normal persons as they live in the United States under usual environmental stresses. Diets should be based on a variety of common foods in order to provide other nutrients for which human requirements have been less well defined.

[c]Weights and heights of reference adults are actual medians for the U.S. population of the designated age, as reported by NHANES II. The median weights and heights of those under 19 years of age were taken from Hamill et al., 1979. The use of these figures does not imply that the height-to-weight ratios are ideal.

[d]Retinol equivalents. 1 retinol equivalent = 1 μg of retinal or 6 μg of β-carotene.

[e]As cholecalciferol. 10 μg of cholecalciferol = 400 IU of vitamin D.

[f]α-Tocopherol equivalents. 1 mg of D-α-tocopherol = 1 α-TE.

[g]1 NE (niacin equivalent) is equal to 1 mg of niacin or 60 mg of dietary tryptophan.

[h]No RDA has been established for biotin or pantothenic acid.

This failure may be linked to the ability of the body to recycle vitamin E; the oxidized form of this compound can be reduced through a sequence of reactions hypothesized to involve vitamin C. Although no RDA has been established for pantothenic acid or biotin, 5 to 10 mg per day is probably adequate for adults. Section 10.2 further examines the controversy surrounding RDAs.

There is growing evidence that aging may have an even greater impact on dietary vitamin requirements than is indicated in Table 10.5. The elderly usually have a reduced ability to absorb and utilize pyridoxine (B_6) and cobalamin (B_{12}), and a diminished ability to convert 7-dehydrocholesterol to vitamin D upon exposure of skin to sunlight. The absorption of vitamin A increases, rather than decreases, with age, and this leads to an enhanced risk of hypervitaminosis. Dietary folic acid requirements also drop because of a rise in folate-producing bacteria in the intestines. Since medication use tends to increase with age, the age–vitamin connection is often complicated by drug–nutrient interactions. **[Try Problems 10.7 and 10.8]**

Molecular Functions

How does the body utilize vitamins? Biochemists have only a partial answer to this question. Most water-soluble vitamins are precursors to coenzymes (Table 10.6), and some of these have additional functions as well. Coenzymes are organic prosthetic groups within complex enzymes (Section 5.4). They are essential for the catalytic activity of those enzymes that contain them, and they are encountered repeatedly in Chapters 11 through 15. Table 10.7 provides a partial listing of the processes in which the fat-soluble vitamins and vitamin C, and compounds derived from them, are known to participate. Much remains to be learned about the precise nature of their participation. It will probably be many years before all of the deficiency symptoms of any vitamin can be adequately explained in terms of its detailed metabolic interactions.

TABLE 10.6

Vitamins Known To Be Precursors to Coenzyme

Vitamin Precursor	Coenzyme	Major Type of Reaction in Which the Coenzyme Participates
Niacin	Nicotinamide adenine dinucleotide (NAD^+); nicotinamide adenine dinucleotide phosphate ($NADP^+$)	Oxidation–reduction
Riboflavin (B_2)	Flavin adenine dinucleotide (FAD); flavin mononucleotide (FMN)	Oxidation–reduction
Pantothenic acid	Coenzyme A (CoASH)	Acyl group transfer
Thiamine (B_1)	Thiamine pyrophosphate (TPP)	Aldehyde group transfer
Biotin	Biocytin	CO_2 fixation (carboxylation)
Pyridoxine (B_6)	Pyridoxal phosphate	Amino group transfer
Folic acid (folate, folacin)	Tetrahydrofolic acid (tetrahydrofolate)	Transfer of one-carbon fragments
Cobalamin (B_{12})	Cobalamin coenzymes—including 5′-deoxyadenosylcobalamin	Alkylation

TABLE 10.7

Some Metabolic Processes Involving Fat-Soluble Vitamins or Vitamin C

Vitamin	Some Metabolic Processes in Which it or its Derivatives Participate
A	Vision; production and maintenance of skeletal tissue; production of sperm; placenta development and maintenance; immune response
D	Regulation of calcium and phosphorus metabolism, including bone growth and maintenance; cell growth and differentiation; immune response
E	Membrane production and maintenance; scavenging of free radicals
K	Calcium metabolism; blood clotting
C	Iron metabolism; collagen, connective tissue, and ground substance production; protein metabolism; scavenging of free radicals

PROBLEM 10.1 There are a number of chemical elements essential for human life that are not listed in Table 10.1. These are the elements that are present in adequate amounts in the carbohydrates and other organic nutrients present in virtually every diet. List these additional elements.

PROBLEM 10.2 What property of fat-containing cells allows them to serve as the major depots for the storage of fat-soluble vitamins?

PROBLEM 10.3 Are the water-soluble vitamins predominantly polar or nonpolar? Explain. Note: No structural information is required to answer this question.

***PROBLEM 10.4** Which, if any, of the vitamins (as depicted in Tables 10.3 and 10.4) are alcohols? Carboxylic acids? Amines? Aromatic compounds? Sulfides? Ethers? Esters? Thiols? Phenols?

***PROBLEM 10.5** Which water-soluble vitamin (as depicted in Table 10.3) has the least number of polar functional groups? Consider a charged functional group a polar one.

PROBLEM 10.6 List those vitamins for which fruit is a common dietary source.

PROBLEM 10.7 True or false? Genetics has a significant impact on a healthy individual's vitamin requirements. Explain.

PROBLEM 10.8 Define the term *half-life.* Suggest why different vitamins and different forms of the same vitamin usually have unique half-lives within your body. Does the half-life of a vitamin have an impact on dietary requirements? Explain.

10.2 VITAMINS AND HEALTH

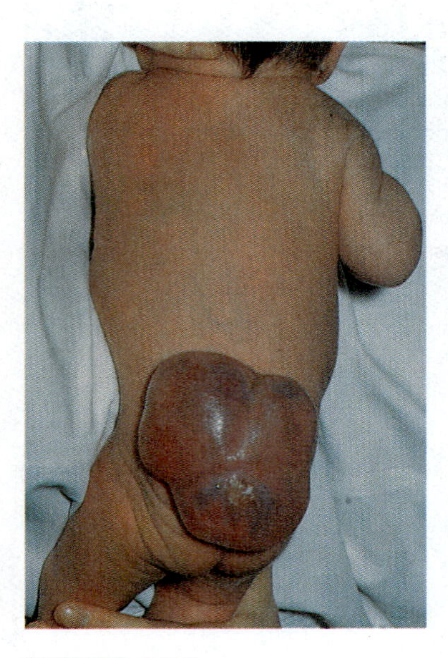

FIGURE 10.2
A folate deficiency can lead to spina bifida, a congenital defect in which one or more vertabrae fail to develop completely. In some cases, an infant is born with a raw swelling over the spine and is destined to be severely handicapped as a consequence of spinal cord damage.

Vitamins A, C, E, folic acid, pyridoxine (B_6), and cobalamin (B_{12}) have all been topics of controversies in the scientific literature and the public press for a number of years. In each case, the controversies have centered primarily around the potential health benefits of the vitamin: Does vitamin C prevent or cure the common cold? Do vitamins A, E, and folic acid reduce cancer risk? Does vitamin E slow biological aging and protect against heart disease? Does folic acid reduce the incidence of spina bifida and other neural tube defects in fetuses? Do folic acid, pyridoxine, and cobalamin provide protection against heart attacks? Do pyridoxine and folic acid strengthen the immune system? The fact that these and related questions are taken seriously indicates that some vitamins may have health benefits well beyond the avoidance of deficiency diseases, and many experts now recommend that the focus of the RDAs shift from deficiency symptoms to a reduction in the risk of heart disease, cancer, and other health problems.

A consensus is gradually being reached on the folic acid–birth defect connection. In September 1992, the U.S. Public Health Service advised that "all women capable of becoming pregnant" consume 0.4 mg of folic acid a day to reduce the incidence of neural tube defects (Figure 10.2). The vitamin needs to be ingested before conception because the defects usually occur during the first 2 weeks of pregnancy, and it can take several days for a prospective mother to build up a protective level of this nutrient. In October 1993, the U.S. Food and Drug Administration (FDA) agreed to allow the labels on foods and supplements that contain folic acid to publicize the vitamin's ability to reduce the risk of birth defects. The same agency may soon require that the flour in bread and other wheat products be "enriched" in folic acid. Similarly, most milk is enriched in vitamin D to reduce the incidence of rickets, and most dietary salt is iodized to reduce the incidence of **goiter,** an iodine deficiency disease.

In the judgment of many experts, there is convincing evidence that vitamin E also provides some benefits beyond the avoidance of deficiency symptoms. However, several major questions about this antioxidant remain unanswered: What are all of its benefits and how significant are they? What is the optimum dose? Can high doses be toxic? What specific reactions or processes are responsible for its benefits? Vitamin E appears to provide some protection against both heart disease and some forms of cancer. At the same time, this antioxidant may preferentially protect cancer cells (when compared with normal cells) once they have formed. Vitamin E may provide relief for those with arthritis, may reduce cataract risk, and may enhance immune function. **Cataracts** are an abnormal clouding of the lens of the eye. Although vitamin E does extend the maximum life span of certain experimental animals, its impact on aging in humans is uncertain.

In virtually all instances, vitamin E is hypothesized to function by protecting cellular components from oxidizing agents. It eliminates oxidizing agents by reducing them; it is oxidized in the process (Section 8.7). In the case of heart disease, vitamin E may reduce risk by decreasing the rate of oxidation of low-density lipoproteins (Sections 9.7 and 9.8). Vitamin E may slow biological aging by protecting numerous body components from **free radicals** (highly reactive oxidizing agents with unshared electrons). The **superoxide** free radical and its **superoxide dismutase**–linked metabolism are depicted in Figure 10.3. The **free radical theory of aging** proposes

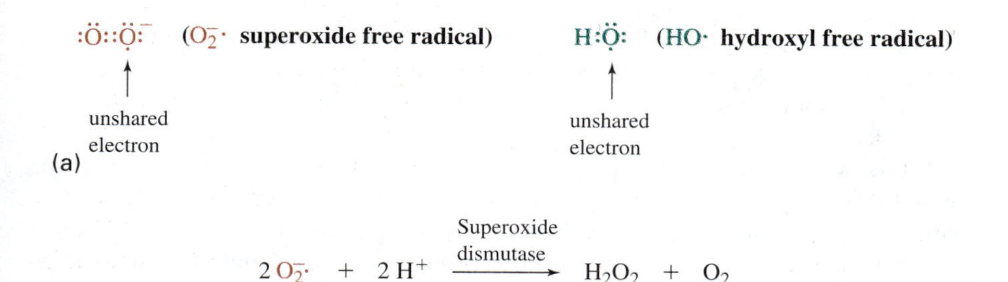

FIGURE 10.3
(a) Two free radicals thought to be involved in biological aging. (b) The $O_2^-\cdot$ free radical is so common and so toxic that enzymes have evolved to catalyze its conversion to less reactive and less toxic substances.

that biological aging is partially a consequence of the accumulation of oxidative damage induced by free radicals.

Although some retinoids are employed in cancer chemotherapy, the potential health benefits of vitamin A, vitamin C, pyridoxine and cobalamin remain topics of debate and research. The vitamin A story is linked to **β-carotene,** a yellow pigment and antioxidant with a number of proposed health benefits. β-Carotene and related compounds are known as **carotenoids.** They are **provitamins,** substances that can be cleaved to release vitamins (Figure 10.4). β-Carotene is the major source of the vitamin A in a typical United States diet. It is abundant in dark green and yellow vegetables, including broccoli and carrots. There is so much β-carotene in carrots that

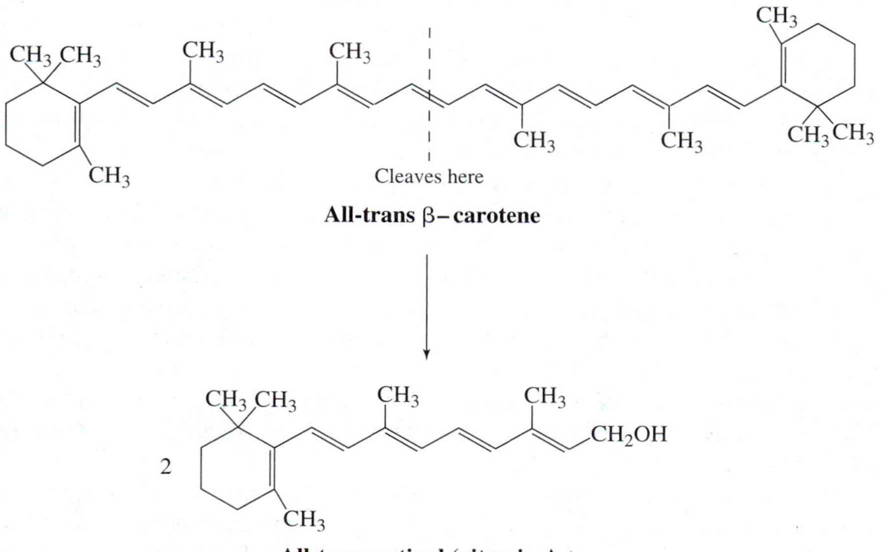

FIGURE 10.4
Conversion of β-carotene to vitamin A.

the ingestion of large quantities of this vegetable can cause the skin to turn yellow-orange as a consequence of β-carotene accumulation. High doses of carotenoids do not lead to vitamin A toxicity because the body regulates their conversion to this micronutrient.

While we wait for more definitive answers to vitamin-related questions, we should recognize that many nutritionists still question the benefits of vitamin supplements for normal individuals on a well-balance diet (see food pyramids Figure 9.7). Vitamins are part of the **functional components** in the **functional foods** in a well-balanced diet. The term *functional foods* was coined by the Food and Nutrition Board of the Institute of Medicine in 1994 to describe "foods with ingredients thought to prevent disease." The ingredients that prevent disease are the functional components. Present evidence indicates that many functional foods, including fruits and vegetables, contain unidentified functional components that are unavailable in capsule form. This helps explain why existing dietary supplements are no substitute for a well-balanced diet and why good nutrition, including an abundance of fruits and vegetables, is central to good health.

HORMONES AND SIGNAL TRANSDUCTION

10.3 HORMONES

Although certain metabolites of vitamins A and D are hormones, the link between vitamins and hormones is a relatively weak one. About all that most vitamins and hormones share in common is having one or another central role in metabolism.

A Definition

EXHIBIT 10.1
Some Endocrine Glands

Adrenals
Ovaries
Pancreatic islets
Parathyroids
Pituitary
Testes
Thyroid

Endocrinology is the study of hormones. **Hormones** (a term derived from the Greek word *hormaein* meaning *to excite*) were originally defined as those compounds that stimulated specific metabolic processes at distant **target sites** after being secreted into the blood by ductless glands known as the **endocrine glands** (Exhibit 10.1). The original definition has subsequently been expanded to include compounds produced at sites other than endocrine glands, to include compounds that inhibit metabolic processes, to include compounds that do not enter the blood stream, and to include compounds that travel only short distances to their target sites. The term *hormone* is still in a state of evolution and, at the moment, the distinction among hormones, growth factors, cytokines, neurotransmitters, and certain other classes of compounds is somewhat fuzzy. In this text, **growth factors** are considered hormones that cause resting cells to undergo division and, in some instances, differentiation. Similarly, **cytokines** are treated as hormones that help regulate the immune response and certain other processes. Some cytokines are growth factors. Hormones that participate in the communication between leukocytes are called **interleukins,** one of several subclasses of cytokines.

All hormones have one property in common: They are chemical messengers that help regulate and coordinate metabolic processes. In many instances, the secretion of the hormone itself is also subject to direct or indirect feedback regulation from its target site(s). Such feedback helps determine the duration and magnitude of a hormone-induced metabolic response. The detailed nature of the feedback regulation varies from hormone to hormone and is rather complex in some instances. A specific example of feedback regulation is examined in Section 10.5. **[Try Problem 10.9]**

Hormones regulate most physiological processes, including growth and development. Dwarfism is sometimes a consequence of a deficiency in human growth hormone, whereas an excess of the same protein produces giants.

Chemical Nature

Several hundred human hormones have now been identified and the list is growing steadily. A large fraction of the presently characterized hormones are either nonprotein peptides (Section 3.6) or steroids (Section 9.6). Proteins (Chapter 4), eicosanoids (Section 9.5), and nonpeptide amines are other prominent classes. The nonpeptide amine hormones include **dopamine, norepinephrine,** and **epinephrine (adrenaline),** compounds synthesized from the amino acid tyrosine and known as **catecholamines** (Figure 10.5). Certain metabolites of vitamin A and vitamin D are also classified as hormones by most investigators.

Some hormones are produced by a single type of cell, have a single target site, and primarily regulate a single metabolic process. Others are produced by multiple cell types, have a number of target sites, and regulate more than one metabolic

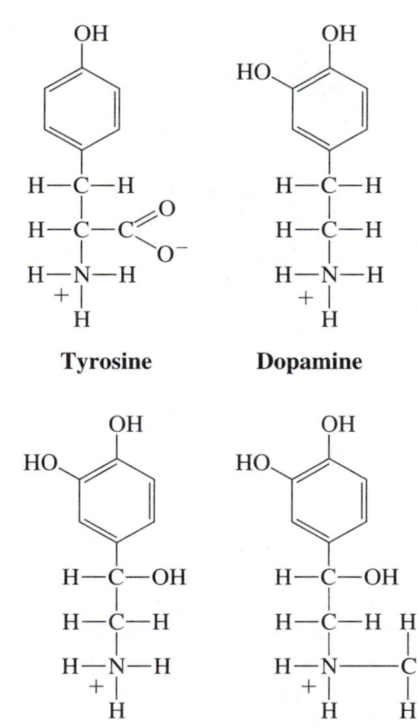

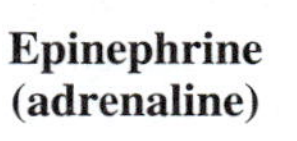

FIGURE 10.5
Tyrosine and the catecholamines.

TABLE 10.8

Some Human Hormones

Hormone or Class of Hormones	Site of Production	Chemical Nature	Some Processes Regulated
Vasopressin (also called antidiuretic hormone [ADH])	Pituitary	Oligopeptide (9 residues)	Blood pressure, water balance
Insulin	Pancreas	Polypeptide (51 residues)	Blood sugar levels; carbohydrate, protein, and fat metabolism
Glucagon	Pancreas	Polypeptide (29 residues)	Blood sugar levels; carbohydrate and fat metabolism
Epinephrine (a catecholamine, also called adrenaline)	Adrenal medulla	Tyrosine derivative	Smooth muscle contraction; heart function; carbohydrate and fat metabolism
Estrogens (estradiol, estrone)	Ovaries	Steroids	Maturation and function of sex organs
Androgens (testosterone)	Testes	Steroids	Maturation and function of sex organs
Thyroxine	Thyroid	Iodinated tyrosine derivative	Stimulates many metabolic processes; regulates growth and development
Parathyroid hormone	Parathyroid	Polypeptide (84 residues)	Calcium and phosphate metabolism
1,25-Dihydroxycholecalciferol	Kidney	Steroid derivative produced from vitamin D_3	Calcium metabolism, bone formation

process. A variety of hormones have important biological roles that are distinguishable from their hormonal functions. The characteristics of several hormones are summarized in Table 10.8. Three of these, insulin, glucagon and epinephrine, are featured in Section 10.12, an overview of fuel metabolism.

Mode of Action

Since hormones commonly have an impact on gene expression, it is appropriate to begin a discussion of the mode of hormone action with a review of the role of DNA in life processes. DNA is the inherited material within cells and organisms. It contains information for the synthesis of all of a cell's proteins, including its enzymes. The enzymes within a cell determine what reactions occur within it. Information for the production of proteins (contain polypeptides) is contained in segments of DNA known as genes. Most genes encode a specific polypeptide, and the final expression of a gene usually consists of the formation of an active polypeptide. Gene expression entails two major events: **transcription,** the use of information in DNA to produce a messenger RNA (mRNA) (a nuclear event in eukaryotes); and **translation,** the mRNA-directed assembly of a polypeptide (a cytoplasmic event). **Transcription factors,** proteins that control transcription, play a central role in the regulation of gene expression. Gene expression, in turn, plays a central role in regulating a cell's metabolism by determining what enzymes are present in a cell and their rates of production. The details of gene expression are examined in Chapters 16 through 18.

The biochemical action of all hormones depends upon their binding to specific **receptor proteins** on or within target cells. A typical human cell contains a multitude of such receptors that are designed to keep it in touch with what is going on in neighboring cells and more distant parts of the body. Such communication is central to survival since many life-sustaining processes depend upon the continual coordination of the activities of an enormous number of cells.

Since most water-soluble hormones, including peptide hormones, protein hormones, and catecholamines, are unable to pass freely through the lipid bilayer within cell membranes, their receptors are associated with plasma membranes. The attachment of such a hormone to a binding site on the exterior surface of its membrane-spanning receptor modulates the activity of a receptor-associated enzyme. This change in enzyme activity leads directly or indirectly to alterations in one or more proteins (usually enzymes) within the receptor-bearing cell, an event that constitutes the transduction of a signal across the membrane (Figure 10.6). The membrane-associated enzyme is frequently involved in the production, mobilization, or acquisition of one or more **second messengers** that help bring about the changes signaled by the *first messenger,* the hormone. Cyclic-AMP (cAMP), cyclic-GMP (cGMP), Ca^{2+}, diacylglycerols, and **inositol 1,4,5-trisphosphate (IP_3)** are common second messengers (Figure 10.7). Cyclic ADP-ribose (cADPR; Section 7.3) is an additional example. **[Try Problems 10.10 through 10.12]**

Once produced, a second messenger usually alters the catalytic activity of one or more protein kinases, enzymes that catalyze the ATP-linked phosphorylation of a protein (Section 5.13). Alterations in a protein kinase commonly lead to changes in additional enzymes that may also be protein kinases. The domino effect continues as long as the change in activity of a newly altered enzyme leads to changes in the activity of one or more other enzymes. Such a sequence of events is described as a **reaction cascade.** Since catalysts alter catalysts within a cascade, the signal is amplified at each step. Some cascades terminate in the cell nucleus where they lead to modifications in protein transcription factors and to changes in gene expression. Other cascades terminate with the activation or inactivation of other classes of proteins, frequently specific enzymes. The reactions that occur between the binding of a hormone to its plasma membrane-associated receptor and the termination of its associated reaction cascade(s) constitute its **signaling pathway(s).** Collectively, the changes in enzyme activity brought about by a hormone account for the overall effects of the hormone. Specific examples of cascades and signaling pathways are examined in the sections that follow. Because of the central role of kinases in signaling pathways, drugs are being developed that regulate hormone action by controlling kinase activity. **[Try Problem 10.13]**

Relatively nonpolar hormones, including steroid hormones, thyroid hormones, and certain derivatives of vitamins D and A, appear to pass freely through cell membranes (Figure 10.8). The receptors for these hormones are usually located inside target cells rather than on their outer surfaces. Once formed, the hormone–receptor complexes normally bind to DNA or DNA–protein complexes within the nucleus and modify the rate of expression of specific genes. The genes involved typically encode specific enzymes. Thus *both water-soluble and nonpolar hormones regulate metabolic processes by indirectly modifying the catalytic activities within a cell.* **[Try Problems 10.14 and 10.15]**

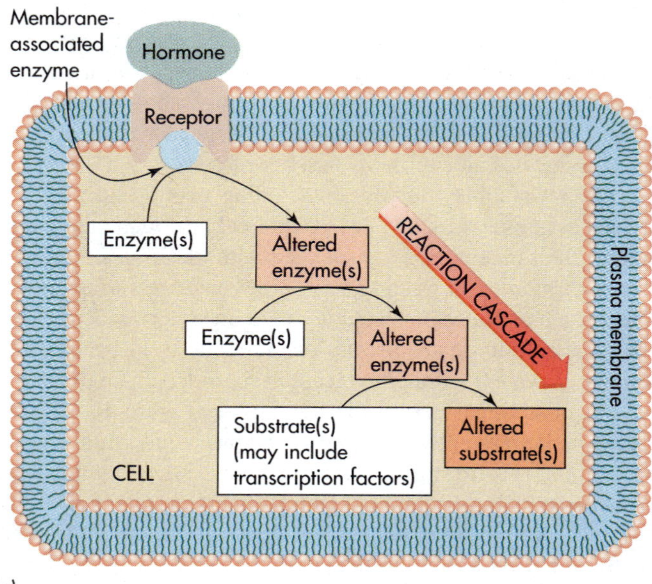

(a)

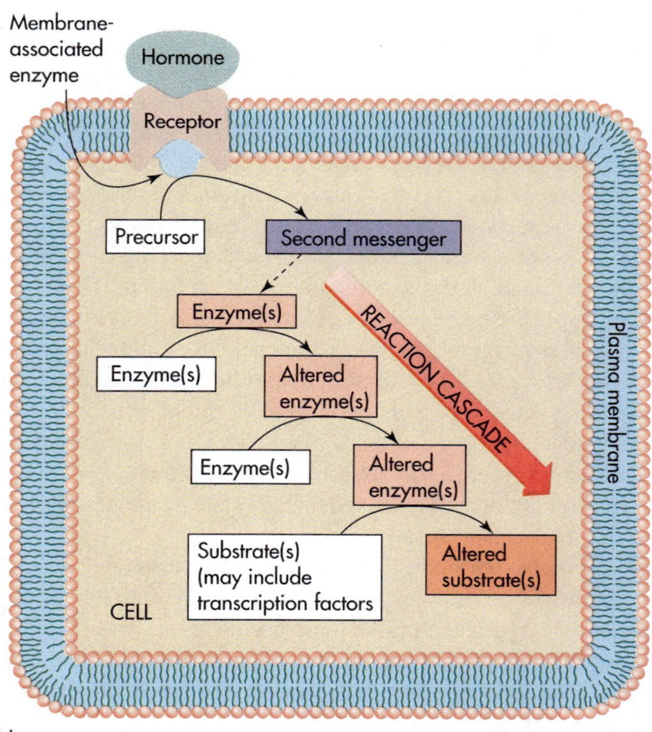

(b)

FIGURE 10.6
General mode of action for water-soluble hormones. (a) The binding of a hormone to a plasma membrane receptor alters a membrane-associated enzyme, whose catalytic activity modulates the activity of one or more cytoplasmic enzymes. The cytoplasmic enzymes may, in turn, lead to changes in the biological activity of additional enzymes or other compounds, including transcription factors. (b) An alternative mode of hormone action involves the production of a second messenger with the catalytic assistance of the hormone-linked, membrane-associated enzyme. The second messenger initiates a reaction cascade by binding to one or more enzymes and, in the process, changing their catalytic capabilities. In all cases, a hormone changes the metabolism in a target cell by indirectly modifying the catalytic activities within it.

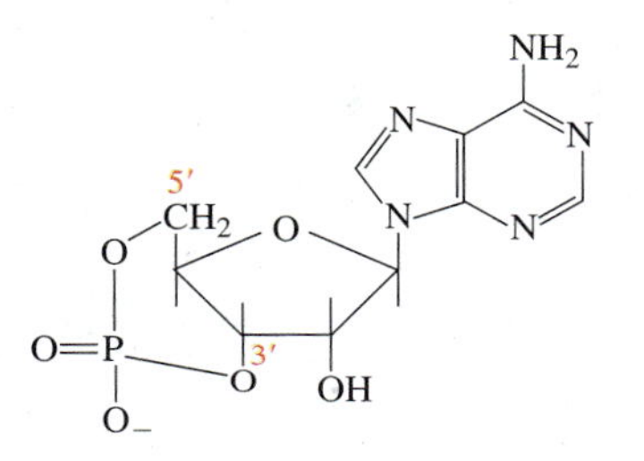

Adenosine 3′,5′-cyclic monophosphate (cAMP)

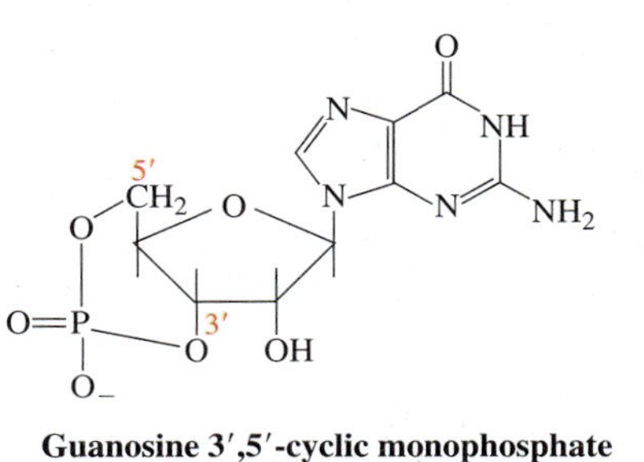

Guanosine 3′,5′-cyclic monophosphate (cGMP)

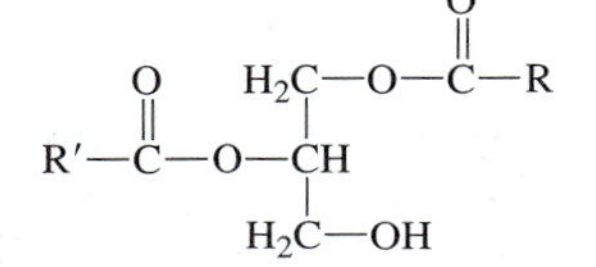

Diacylglycerol (DAG)

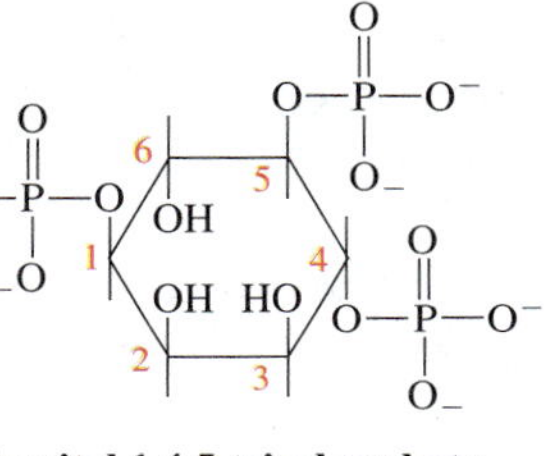

Inositol 1,4,5-trisphosphate (IP_3)

Ca^{2+}

Divalent calcium ion

FIGURE 10.7
Some second messengers.

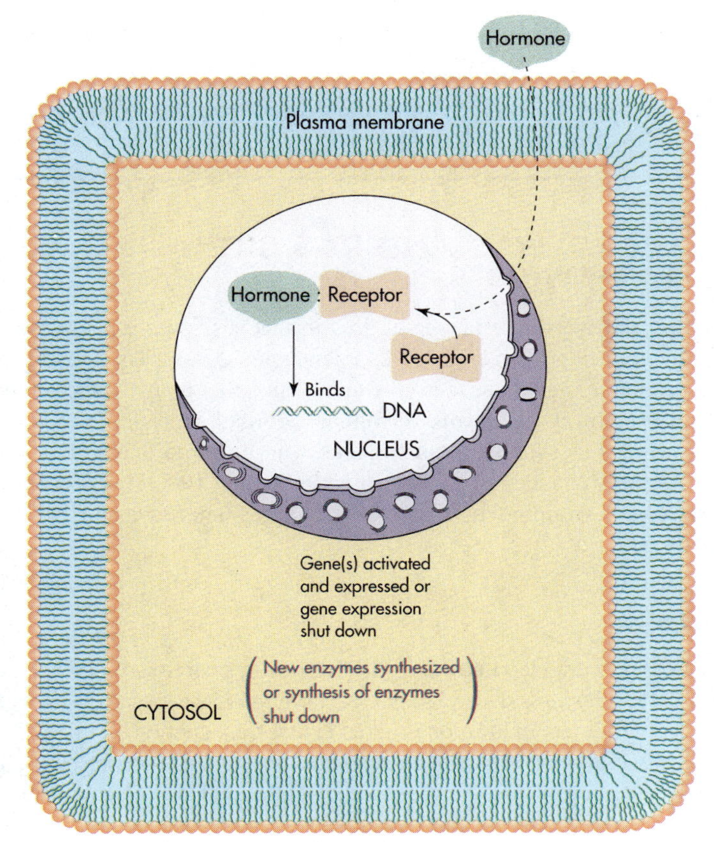

FIGURE 10.8
General mode of action for nonpolar hormones. After passing freely through the plasma membrane and nuclear membrane, a nonpolar hormone attaches to a protein receptor within the nucleus. The hormone:receptor complex alters gene expression by binding to DNA or a DNA:protein complex.

PROBLEM 10.9 It is common for a single process to be stimulated by one hormone and inhibited by a second. Suggest how an organism benefits from such a relationship.

PROBLEM 10.10 Which of the compounds in Figure 10.7 are water soluble and which are lipids? Which bear a net charge at physiological pH? Explain.

PROBLEM 10.11 The binding of a hormone to a receptor is normally a reversible process involving noncovalent interactions. List the bonds or forces most likely involved.

PROBLEM 10.12 Would you expect the binding of a hormone to its receptor to be stereospecific? Explain.

PROBLEM 10.13 Suggest how the binding of a second messenger to an intracellular enzyme modifies the catalytic activity of that enzyme.

PROBLEM 10.14 Gene expression involves two major processes, transcription and translation. Does the binding of a eukaryotic hormone–receptor complex to DNA directly alter the rate of transcription, the rate of translation, or both? Explain.

PROBLEM 10.15 What kinetic phenomenon makes it possible for the catalytic activities in a cell to determine what reactions occur within that cell? Hint: Review Section 5.2.

10.4 G PROTEINS AND RAS SIGNALING PATHWAYS

Ras signaling pathways are universal growth factor signaling pathways that contain a **Ras protein** (a protein encoded in a gene named *Ras*). These pathways are of more than academic interest because the genes encoding many of the proteins associated with the pathways are **proto-oncogenes,** normal genes that can be converted to cancer-causing genes known as **oncogenes.** Cancer, proto-oncogenes, oncogenes, and the Ras–cancer connection are discussed in Section 19.8. Since Ras proteins are G proteins, we will examine the nature of G proteins before we analyze Ras signaling pathways.

G Proteins

GTPases are enzymes, also known as **GTP-binding proteins,** that catalyze the hydrolysis of GTP, a nucleoside triphosphate examined in Section 7.2. **G proteins** are GTPases that participate in hormone-linked signal transduction. GTPases, including G proteins, possess other activities in addition to their GTPase activities. The non-GTPase activities depend on the presence of bound GTP. GTPases cycle between

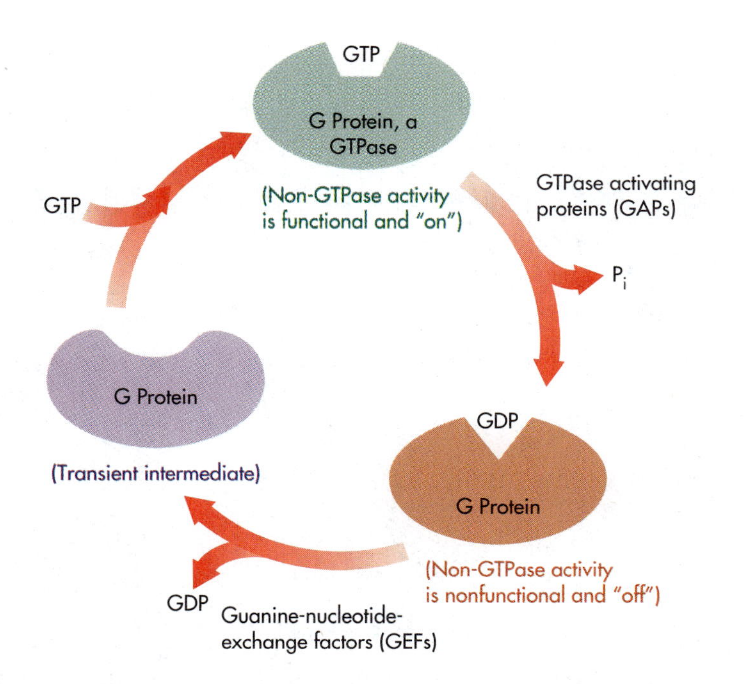

FIGURE 10.9
GTPase cycling. GTP is guanosine triphosphate, GDP is guanosine diphosphate, and P_i is phosphate. A GTPase possesses a non-GTPase activity as well as a GTPase activity. The non-GTPase activity is only functional ("on") when the GTPase is bound to GTP. GTPase activating proteins (GAPs) activate the GTPase domain within a GTPase and trigger the hydrolysis of the bound GTP, an event that turns "off" (inactivates) the non-GTPase activity. Guanine-nucleotide-exchange factors (GEFs) return the GTPase to its "on" (from the standpoint of its non-GTPase activity) state by stimulating the replacement of GDP with GTP.

three states: GTP-bound, "empty," and GDP-bound (Figure 10.9). The cycling is always in the same direction since GTP hydrolysis is, from a practical standpoint, irreversible.

GTPase cycling creates a **molecular switch** with the GTP-bound form representing the "on" or active (from the standpoint of the non-GTPase activity) position and the GDP-bound state being the "off" or inactive position. The "empty" state is a short-lived transient intermediate. Proteins that throw the switch from "on" to "off" are dubbed **GTPase activating proteins (GAPs)**; they activate the GTPase-catalyzed hydrolysis of GTP. Proteins that mediate the replacement of GDP with GTP and throw the switch from "off" to "on" are called **guanine–nucleotide-exchange factors (GEFs).** An enormous number of GTPases have already been discovered, and the list is growing constantly. Ras proteins are specific examples. The 1994 Nobel Prize in Medicine was awarded to Alfred Gilman and Martin Rodbell for their discovery of G proteins and their mechanism of action. **[Try Problem 10.16]**

The Regulation of Ras Proteins

Growth factor receptors sit at the head of Ras signaling pathways. These integral membrane proteins are known as **tyrosine kinase-associated receptors** because they possess a tyrosine kinase activity as well as a growth factor binding site. A **tyrosine kinase,** in general, is a protein kinase that catalyzes the phosphorylation of particular tyrosine residues on specific target proteins. Once a growth factor attaches to the outside of a tyrosine kinase-associated receptor, the receptor autophosphorylates (Figure 10.10). This phosphorylation makes the cytosolic side of the receptor attractive to **adaptor proteins** that recruit other proteins known as **Ras exchange factors,** specific examples of guanine–nucleotide-exchange factors (see previous

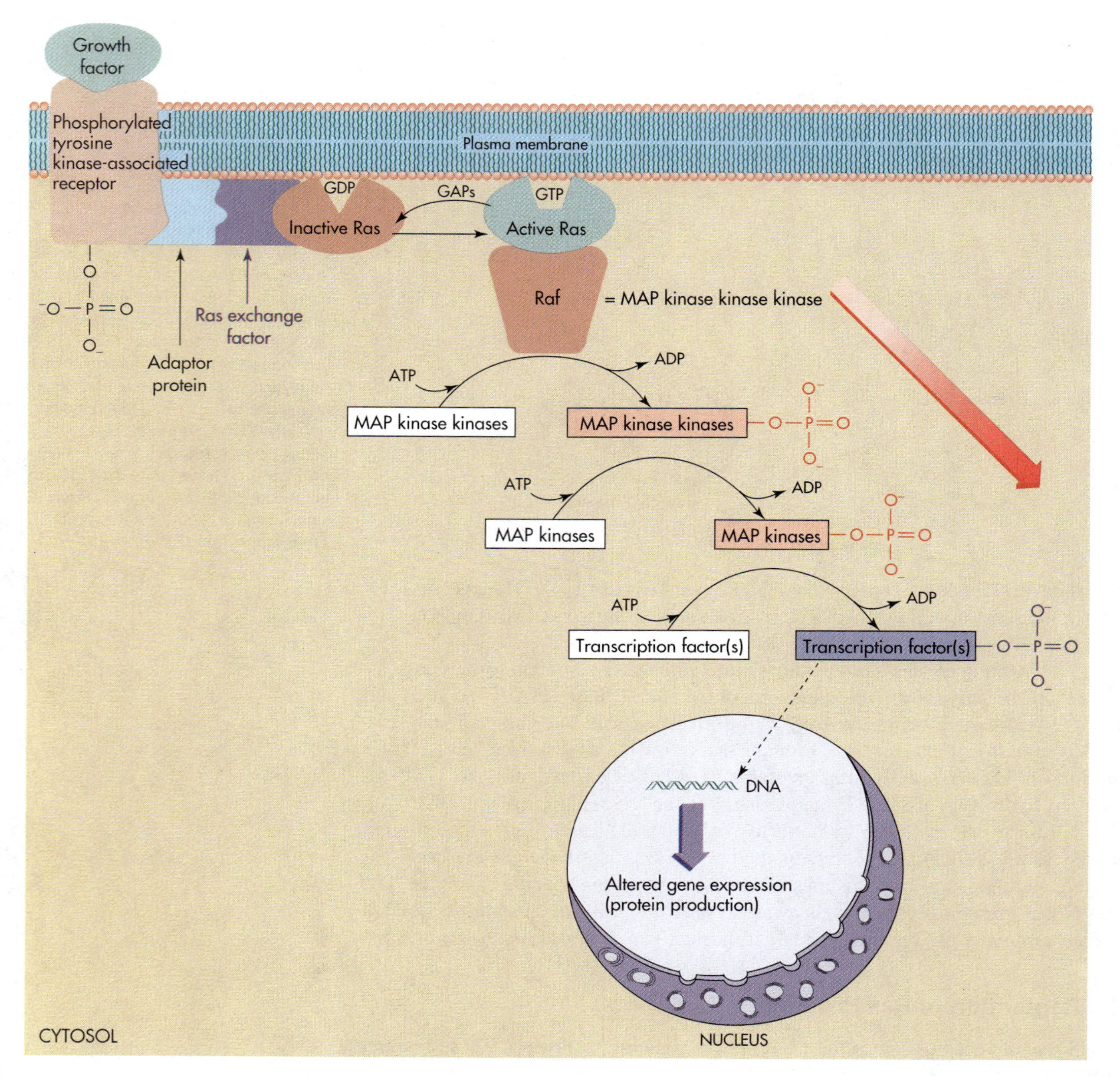

FIGURE 10.10
A Ras signaling pathway. Growth factor binding leads to the autophosphorylation of a tyrosine kinase-associated receptor. An adaptor protein attaches to the phosphorylated receptor and then recruits a Ras exchange factor. The exchange factor turns "on" the G protein known as Ras by stimulating the exchange of GDP for GTP. The length of time Ras protein remains active is regulated, in part, by GTPase activating proteins (GAPs), which enhance the GTPase activity of Ras protein. Once turned "on", Ras protein anchors Raf protein to the plasma membrane and leads to its activation. The activation of Raf (a MAP [mitogen-activated protein] kinase kinase kinase) initiates a kinase cascade that leads to the phosphorylation of transcription factors and the activation of banks of genes whose products are involved in cell division or differentiation or both.

paragraph). The term *adaptor proteins* comes from the fact that these polypeptides "plug" into phosphorylated receptors and then into exchange factors. The exchange factors bind to a Ras protein and throw its molecular switch; they turn "on" Ras protein (activate its non-GTPase activity) by stimulating the replacement of GDP with GTP. Specific GTPase-activating proteins also have their hands on the switch. After binding to a Ras protein, they stimulate its GTPase activity, that activity which catalyzes the hydrolysis of GTP to GDP and P_i and turns the switch from the "on" to the "off" position. The relative concentrations of Ras exchange factors and GTPase activating proteins help determine the amount of time that a Ras protein resides in its active (GTP-bound; "on") state.

Kinase Cascades

Once a Ras exchange factor has turned "on" a Ras protein, the Ras protein binds to a **Raf protein** (product of a gene named *Raf*) and secures it to the plasma membrane where it is activated by mechanisms still under investigation. Thus, Ras proteins serve as regulable devices for anchoring other proteins to membranes. Activated Raf protein, a protein kinase known as a **MAP** (mitogen-activated protein) **kinase kinase kinase** or **MAPKK kinase,** catalyzes the phosphorylation of kinases known as **MAP kinase kinases (MAPKKs). Mitogens** are substances, including growth factors, that trigger cell division. The MAPKKs catalyze the phosphorylation of MAP kinases (MAPKs) which catalyze the phosphorylation of transcription factors or other proteins. There are families of MAPKKs and MAPKs, and distinctive cells possess unique members of these families or unique combinations of family members. Transcription factors are the final substrates for most of the kinase cascades. The targeted transcription factors regulate the expression of banks of genes that encode proteins involved in cell division or differentiation or both. **[Try Problems 10.17 and 10.18]**

The Ras proteins so far discovered fall into three main families. Collectively, they help control an extraordinarily broad range of cellular processes. Although some of the details of Ras signaling are still uncertain and may vary from pathway to pathway, the basic design and the essential features of Ras signaling pathways appear to have been determined.

PROBLEM 10.16 Write the equation for the hydrolysis of GTP to yield GDP plus P_i. Provide structural formulas for all reactants and products. Why is this reaction irreversible?

PROBLEM 10.17 Will a change in a *Ras* gene (encodes Ras protein) that increases the affinity of the encoded Ras protein for GDP increase, decrease, or have no effect on a cell's response to a growth factor whose action is coupled to the abnormal protein? Explain. Assume that all of the other activities of the abnormal Ras protein are normal.

PROBLEM 10.18 Describe how the protein kinase-catalyzed phosphorylation of an enzyme can alter its catalytic activity.

10.5 CYCLIC NUCLEOTIDE SIGNALING PATHWAYS

The action of a large number of polar hormones is linked to signaling pathways in which cAMP (Figure 10.7; Section 7.3) serves as a second messenger. Some of these hormones are listed in Exhibit 10.2.

cAMP Signaling Pathways

Cyclic-AMP signaling pathways, like most other signaling pathways linked to water-soluble hormones, begin when a hormone binds to the outside of a membrane-spanning receptor on the plasma membrane. This binding "turns on" a membrane-associated G protein by triggering the replacement of GDP with GTP. The activated G protein interacts with an enzyme known as **adenylate cyclase,** an integral membrane protein with its active site positioned on the cytosolic surface of the plasma membrane. Adenylate cyclase catalyzes the conversion of ATP to cAMP (Figure 10.11). The cAMP produced enters the cytosol where it activates a **cAMP-dependent kinase (protein kinase A)**, which catalyzes the phosphorylation of specific target proteins that may themselves be kinases. The net result is a reaction cascade that modifies the activity of one or more ultimate target proteins, usually enzymes. Distinctive cells produce unique cAMP-dependent kinases that modulate the activity of different intracellular enzymes. Some of the enzymes targeted by cAMP-linked kinase cascades will be identified as the metabolism story unfolds. **[Try Problem 10.19]**

The G protein in a cAMP signaling pathway may either stimulate or inhibit the adenylate cyclase to which it is coupled. A hormone coupled to a stimulatory G pro-

EXHIBIT 10.2

Some cAMP-Linked Hormones

- Adrenocorticotropic hormone (ACTH)
- Beta-adrenergic catecholamines (includes epinephrine)
- Calcitonin
- Follicle-stimulating hormone (FSH)
- Glucagon
- Histamine
- Human chorionic gonadotropin (hCG)
- Luteinizing hormone (LH)
- Luteinizing hormone–releasing hormone (LHRH)
- Melanocyte-stimulating hormone (MSH)
- Parathyroid hormone (PTH)
- Prostaglandin E_1 (PGE_1)
- Serotonin
- Thyrotropin-releasing hormone (TRH)
- Thyroid-stimulating hormone (TSH)
- Vasopressin

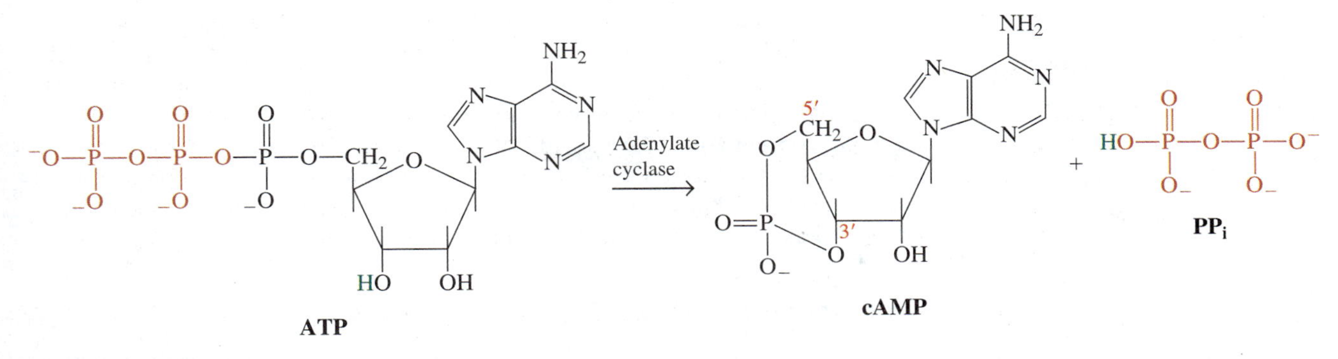

FIGURE 10.11
Production of cAMP.

tein (**G_s protein**) leads to an increase in the intracellular concentration of cAMP, and one coupled to an inhibitory G protein (G_i protein) has the opposite effect. In some instances, the same hormone is coupled to a G_i protein in one tissue but associated with a G_s protein in a second distinctive tissue. Epinephrine and prostaglandin E_1 (PGE_1) fall into this category.

Cyclic nucleotide phosphodiesterases and **protein phosphatases** are additional actors in the events that follow the binding of a hormone to an adenylate cyclase–coupled receptor (Figure 10.12). The former enzymes catalyze the hydrolysis of cAMP to yield 5′-AMP, a compound inactive as a second messenger. The latter enzymes catalyze the hydrolytic removal of phosphoryl groups from the substrates of the protein kinases (Section 5.13). These two classes of enzymes largely determine the half-lives of cAMP and the altered enzymes generated during the kinase cascades. These half-lives, in turn, help determine the magnitude of the overall

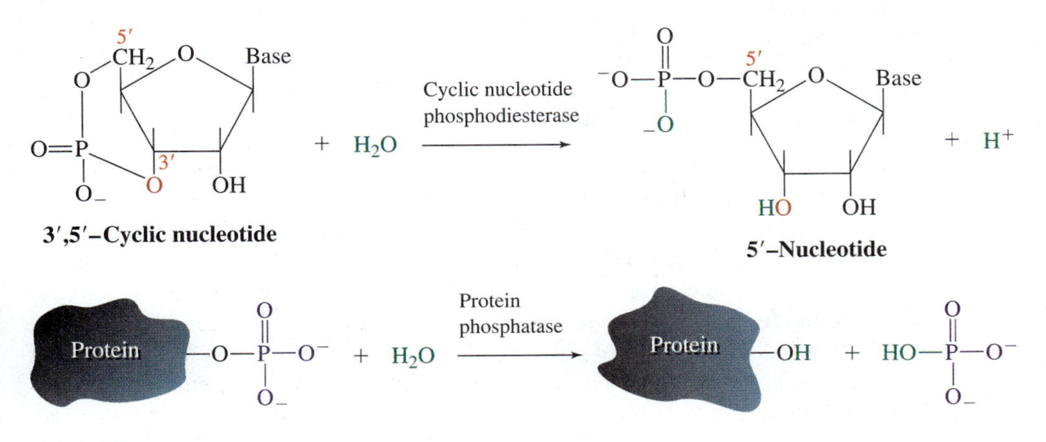

FIGURE 10.12
Cyclic nucleotide phosphodiesterase and protein phosphatase.

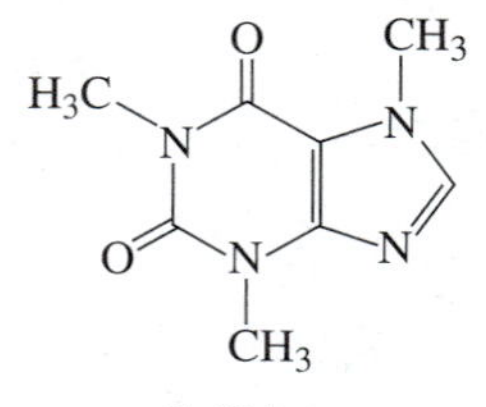

Caffeine

response that is triggered by the binding of a hormone to its receptor. Thus, the effects of **caffeine** are partially explained by its inhibition of cAMP phosphodiesterase, an event that prolongs and intensifies the response elicited by certain hormones. Other factors that help regulate the response elicited by a hormone include hormone concentration, the half-life of the hormone, receptor concentration, the affinity of the receptor for the hormone, the nature of the coupled G protein, GTP concentration, the kinetic properties of the adenylate cyclase, and the kinetic properties of the cell's cAMP-dependent kinase(s). **[Try Problems 10.20 and 10.21]**

An Epinephrine-linked Signaling Pathway

The effect of epinephrine on liver glycogen metabolism is brought about by a classical cAMP signaling pathway. Epinephrine is a catecholamine (Figure 10.5) released into the blood by the adrenal medulla in response to low blood glucose levels (Section 10.12). The hormone is carried to the liver, where it attaches to a receptor associated with the plasma membrane of a liver cell. Epinephrine binding leads to a change in the conformation of the receptor. This change brings about the replacement of GDP with GTP on the G_s protein associated with the receptor, an event that turns on this G_s protein (Figure 10.13). The activated G_s protein stimulates a membrane-bound adenylate cyclase that catalyzes the production of cAMP. At this point, the membrane-associated phase of the signaling pathway is complete.

The cAMP produced moves into the cytoplasm, where it binds to and, in the process, activates a cAMP-dependent kinase. This kinase activates a second kinase known as **phosphorylase b kinase** by catalyzing its phosphorylation. Activated phosphorylase b kinase catalyzes the phosphorylation of **glycogen phosphorylase b,** an enzyme whose phosphorylated form catalyzes the breakdown of glycogen (a polymer of glucose; Section 6.6) into many glucose 1-phosphates. After glucose 1-phosphate is hydrolyzed with the catalytic assistance of a cytosolic enzyme, the free glucose enters the blood, where it elevates the low blood sugar concentration that initially triggered the release of epinephrine. As blood sugar levels rise, the glucose "feeds back" to shut down the release of epinephrine from the adrenal medulla. The production of many hormones is partially regulated by feedback inhibition. **[Try Problem 10.22]**

cGMP as a Second Messenger

Cyclic-GMP, like cAMP, is a common second messenger. It modulates **cGMP-dependent protein kinases (protein kinase Gs)** and, in the process, regulates distinctive reaction cascades. Cyclic-GMP production is catalyzed by both membrane-bound and cytosolic **guanylate cyclases** that are activated by a variety of stimuli. Some of the cytosolic guanylate cyclases are turned on by **carbon monoxide (CO)** or **nitric oxide (NO)**, two gaseous messengers. At low concentrations, CO and NO help control a large number of physiologically important processes including blood flow, penile erection, blood clotting, skeletal muscle contraction, and immune response. Both messengers function as neurotransmitters (Section 10.9), and they may be involved in memory and learning.

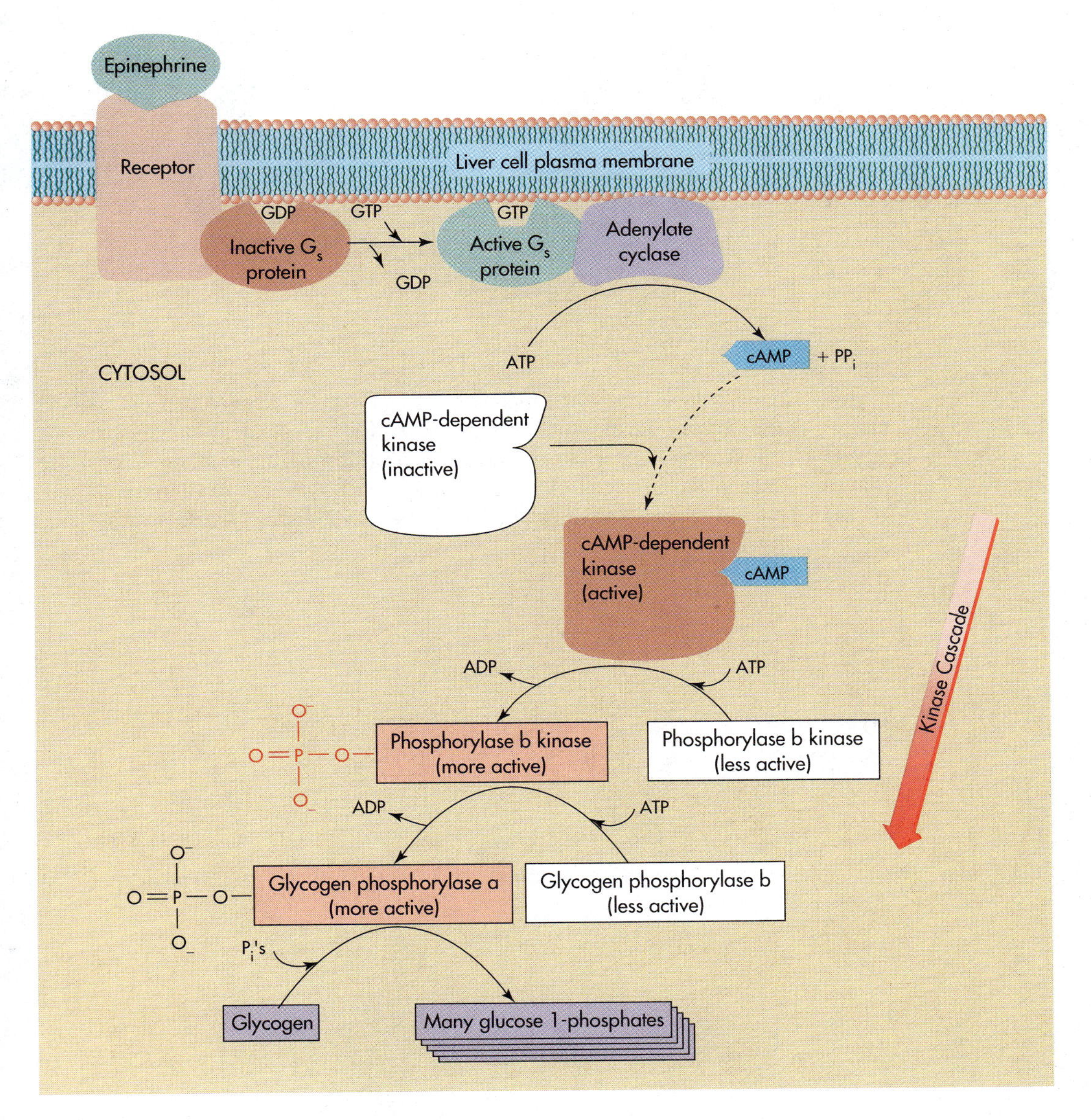

FIGURE 10.13
An epinephrine-linked signaling pathway. The binding of epinephrine to its plasma membrane receptor activates a G_s protein that stimulates the adenylate cyclase–catalyzed production of cAMP, a second messenger. Cyclic-AMP binds to and activates a cAMP-dependent kinase that activates phosphorylase b kinase by catalyzing its phosphorylation. The activated phosphorylase b kinase catalyzes the phosphorylation of glycogen phosphorylase b. The glycogen phosphorylase a that is generated catalyzes the breakdown of glycogen (a polymer of glucose) into many glucose 1-phosphates.

Nitric oxide, which was voted molecule of the year by *Science* in 1992, is normally generated from the amino acid arginine by Ca^{2+}-dependent enzymes known as **NO-synthases (NOSs)**, whereas CO is produced during the breakdown of heme (the prosthetic group in myoglobin and hemoglobin) by **heme oxygenase** (Figure 10.14). There are several isozymes of NO-synthase with distinct tissue distributions.

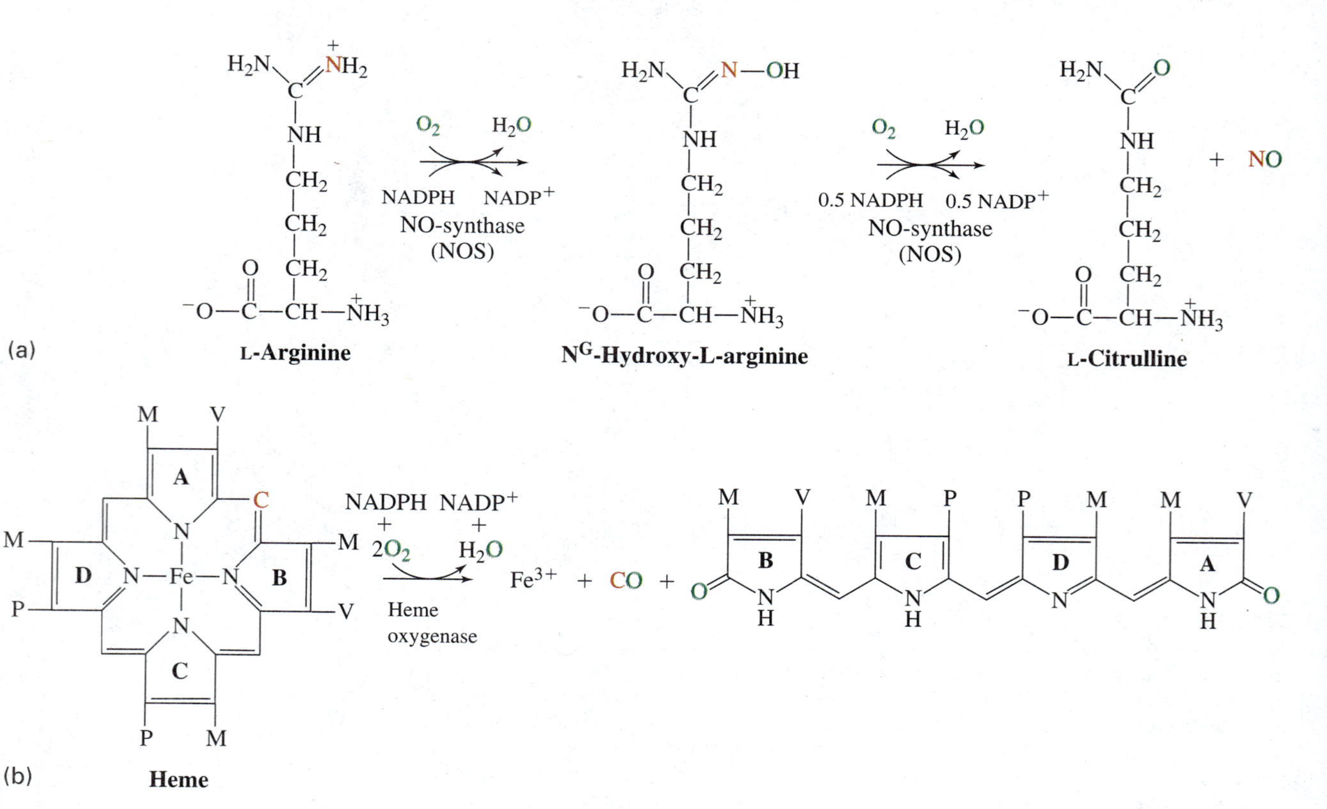

FIGURE 10.14
(a) Production of nitric oxide. NADPH is a coenzyme and a biological reducing agent (Tables 8.8 and 10.6). (b) Production of carbon monoxide. P, propionyl ($—CH_2—CH_2—COO^-$); M, methyl ($—CH_3$); V, vinyl ($—CH{=}CH_2$).

It is the breakdown of **nitroglycerin** and certain other organic nitrates into NO that accounts for their ability to relax blood vessels and help restore the flow of blood to the heart during and following a heart attack. **[Try Problem 10.23]**

PROBLEM 10.19 Hormone concentration and GTP concentration both influence the rate of production of cAMP by a target cell. List four other compounds whose concentration will impact the rate of cAMP production.

PROBLEM 10.20 Use structural formulas to write the equation for the phosphodiesterase-catalyzed hydrolysis of cAMP to yield 5′-AMP.

PROBLEM 10.21 Use structural formulas to write (a) the equation for the hydrolytic dephosphorylation of a phosphorylated serine side chain and (b) the equation for the phosphorylation of a serine side chain assuming that ATP is the phosphoryl donor.

PROBLEM 10.22 Identify the participants in the epinephrine-linked signaling pathway depicted in Figure 10.13 that are proteins.

***PROBLEM 10.23** Although carbon monoxide is a natural biological messenger, it is better known as a toxic gas that binds to hemoglobin and blocks its oxygen binding sites. What is the major source of the carbon monoxide in the air we breathe?

10.6 PATHWAY CROSS TALK

Multiple hormone signaling pathways often operate simultaneously within a single cell, and, in many instances, it is essential that the activities of these separate pathways be coordinated. This requires that the pathways communicate with one another, a process described as **cross talk.**

Part of the cross talk between Ras pathways and cAMP pathways involves cAMP itself. This second messenger leads to modifications in specific components within Ras-linked kinase cascades. What molecules are modified and the chemical nature of the modifications are under investigation. Such modifications could, in theory, either stimulate or inhibit a Ras pathway, and cAMP may have opposite effects in different cells. cAMP signaling pathways and calcium signaling pathways (Section 10.8) are linked in many cells through the inhibition of adenylate cyclases by Ca^{2+}. Cross talk is a "hot" topic, one in which rapid progress is being made.

10.7 PHOSPHATIDYLINOSITOL SIGNALING PATHWAYS

Some hormone receptors are coupled, through a G protein, to a membrane-bound **phospholipase C.** Hormone binding turns on the G protein by stimulating the replacement of GDP with GTP. Once "turned on", the G protein fires up the

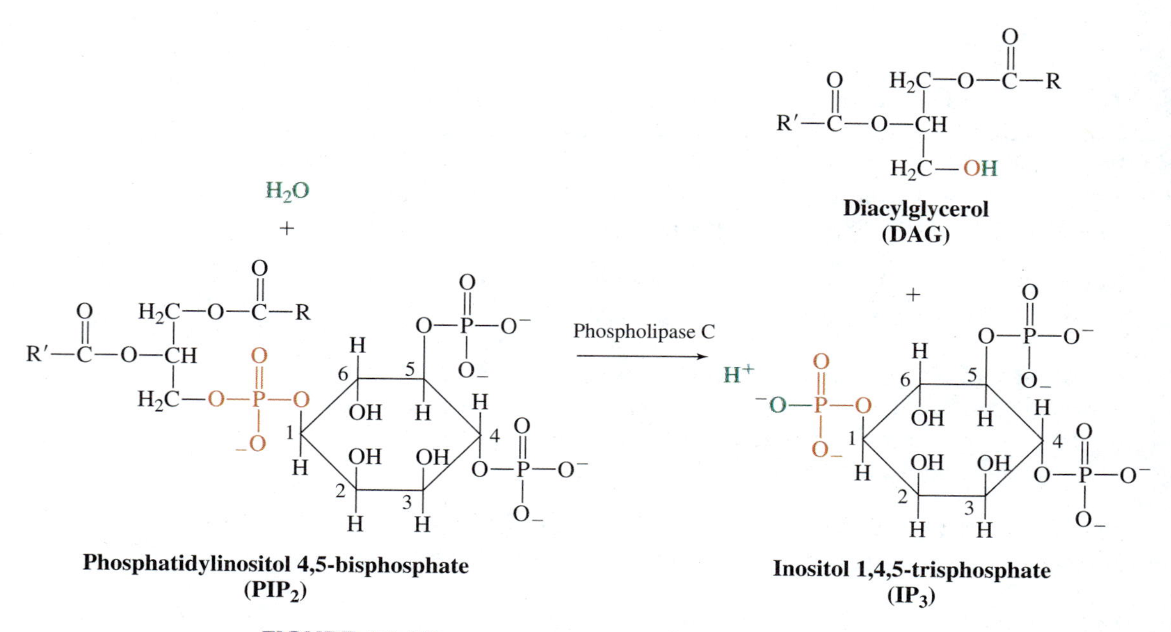

FIGURE 10.15
Phosphatidylinositol 4,5-bisphosphate (PIP_2) hydrolysis catalyzed by phospholipase C.

phospholipase C, which catalyzes the hydrolysis of phosphatidylinositol 4,5-bisphosphate (PIP_2) to yield a diacylglycerol (DAG) plus inositol 1,4,5-trisphosphate (IP_3), both of which function as second messengers (Figure 10.15).

Diacylglycerol as a Second Messenger

The diacylglycerol normally activates a membrane-associated, Ca^{2+}-dependent **protein kinase C** that catalyzes the phosphorylation of additional proteins and may initiate a kinase cascade (Figure 10.16). Distinctive types of cells possess unique protein kinase C isozymes or unique combinations of isozymes. The role of diacylglycerols as second messengers may extend well beyond the activation of protein kinase C, since the hydrolysis of some diacylglycerols yields arachidonic acid which can be converted to numerous metabolically active eicosanoids. In some instances, diacylglycerols from the enzyme-catalyzed hydrolysis of phosphatidylcholines (rather than PIP_2) are also involved in the modulation of protein kinase C.

IP_3 as a Second Messenger

The IP_3 produced by the catalytic action of the hormone-activated phospholipase C migrates to the endoplasmic reticulum, where its binding to specific receptors triggers the release of stored Ca^{2+}, causing its influx into the cytosol. Depending on the cell involved, the rise in cytosolic Ca^{2+} may modulate the activity of a broad range

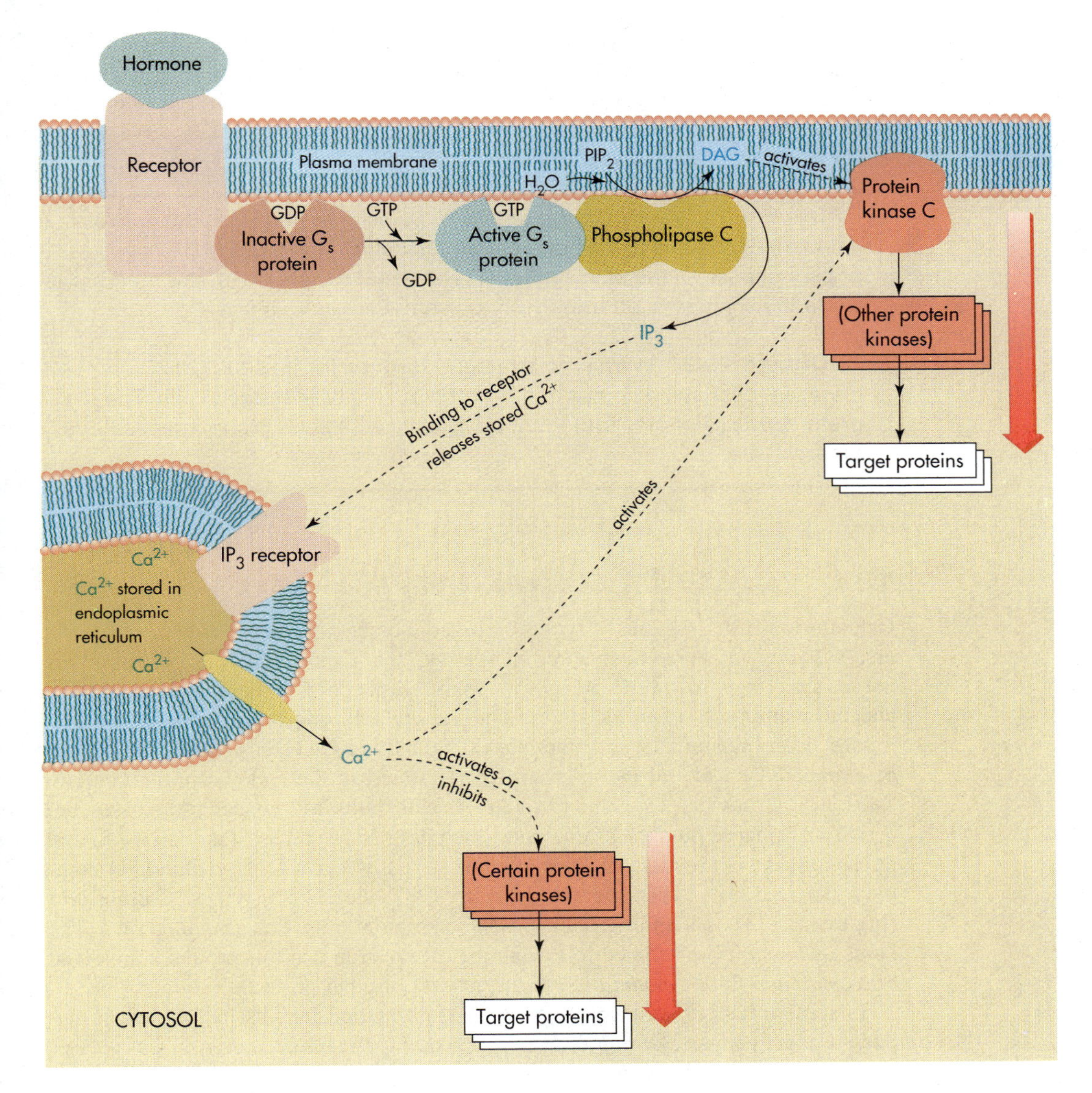

FIGURE 10.16
Diacylglycerol and inositol 1,4,5-trisphosphate as second messengers. The binding of a hormone to a plasma membrane receptor activates a G_s protein that stimulates the phospholipase C–catalyzed hydrolysis of phosphatidylinositol 4,5-bisphosphate (PIP_2) to yield diacylglycerol (DAG) plus inositol 1,4,5-trisphosphate (IP_3). The DAG activates a membrane-associated protein kinase C that leads to the initiation of a kinase cascade. The IP_3 binds to IP_3 receptors on the endoplasmic reticulum and triggers the release of stored Ca^{2+} into the cytosol. The resultant increase in Ca^{2+} concentration modulates the activity of multiple enzymes, including certain protein kinases that are capable of initiating kinase cascades.

of proteins, including adenylate cyclase and multiple protein kinases (Section 10.8). The net result is once again the initiation of one or more reaction cascades that modulate the catalytic activities within a cell. Well over a dozen different inositol phosphates have been identified in various cells. Some, in addition to IP_3, serve as second messengers. **[Try Problems 10.24 and 10.25]**

PROBLEM 10.24 When a membrane-associated phosphatidylinositol 4,5-bisphosphate is hydrolyzed with the catalytic assitance of phospholipase C, the diacylglycerol remains associated with the membrane while the IP_3 moves off into the cytoplasm. Explain this fact.

PROBLEM 10.25 Write the structural formula for inositol 1,3,4,5-tetrakisphosphate, a compound found in a variety of human cells. The prefix *tetrakis* denotes four identical groups with each group attached at a different site.

10.8 CALCIUM SIGNALING PATHWAYS

Calcium ion (Ca^{2+}) is one of several inorganic biochemical messengers. Two others, NO and CO, were encountered in Section 10.5. Cytosolic Ca^{2+} concentrations are normally kept below $10^{-7} M$ by Ca^{2+} pumps imbedded in the plasma membrane and the membranes of mitochondria and the endoplasmic reticulum. Certain hormones, neurotransmitters, and other physiologically active agents stimulate an influx of extracellular Ca^{2+} or the release of Ca^{2+} stored in the endoplasmic reticulum. Section 10.7 provides a specific example. As cytosolic Ca^{2+} concentration rises, the activity of a large number of enzymes, including NO-synthase (Section 10.5) and phospholipase C (Section 10.7), is affected. Usually it is a Ca^{2+}–calmodulin complex, not free Ca^{2+}, that interacts with the Ca^{2+}-sensitive enzymes. **Calmodulin** (Figure 10.17) is an acidic protein with four high-affinity Ca^{2+} binding sites. The name *calmodulin* was coined following the observation that this protein is involved in the *calcium*-linked *modulation* of the activity of other proteins.

In some instances, an increase in Ca^{2+} concentration inhibits, rather than stimulates, a target enzyme. During the visual process, for example, a drop in Ca^{2+} levels (indirectly triggered by a photon of light) activates a guanylate cyclase and increases the production of cGMP, another second messenger in the **phototransduction cascade.** Calcium's mode of action is commonly linked to other second messengers. Ca^{2+} inhibits some adenylate cyclases as well as some guanylate cyclases (Section 10.6).

10.9 NEUROTRANSMITTERS

The transmission of nerve impulses further illustrates the importance of chemical messengers. In this case, the messengers involved are known as **neurotransmitters.** Although a case can be made for classifying neurotransmitters as hormones, they are usually classified separately.

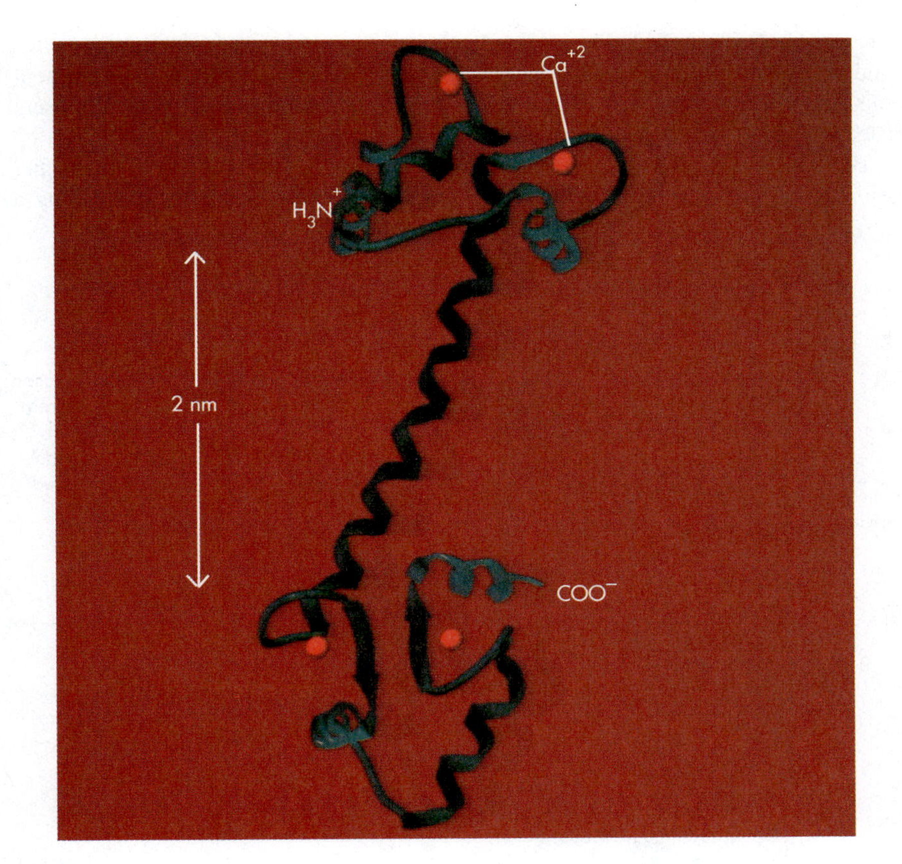

FIGURE 10.17
Calmodulin contains two Ca^{2+}-binding domains separated by a seven-turn α-helix. Each domain has two Ca^{2+} binding sites, where negatively charged aspartate and glutamate side chains form ionic bonds to Ca^{2+}.

Those cells of the human nervous system, including brain cells, that are involved in the transmission of nerve impulses are known as **neurons.** Each neuron contains a cell body with **dendrites** and an **axon** extending from it (Figure 10.18). The dendrites are antenna-like projections that receive signals from neighboring cells, whereas the axon is involved in the transmission of signals to other cells. Most neurons are closely associated with other neurons and, in some instances, muscle cells as well. The junctions between neighboring neurons and between neurons and muscle cells are called **synapses.**

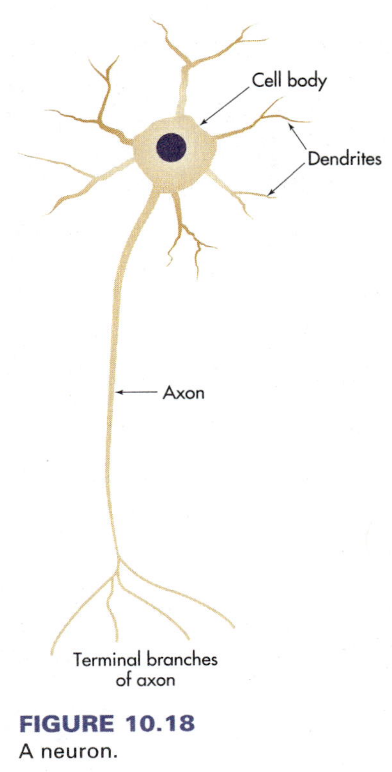

FIGURE 10.18
A neuron.

Signal Transduction

Once a neuron is stimulated, a nerve impulse is transmitted along its axon by an electrical signal that, when it reaches the end of the axon, leads to the release of neurotransmitters into synapses (Figure 10.19). The neurotransmitters are housed in storage vesicles, which are emptied into synapses through exocytosis, a process described in Section 9.13. The released neurotransmitters traverse the synapses and bind to specific receptors associated with the plasma membrane of postsynaptic neurons (neurons receiving a signal across a synapse) or muscle cells. For the present discussion, it is assumed that the receptors are on neurons.

Before it binds a neurotransmitter, a postsynaptic neuron, like most of the other cells in the human body, possesses an electrochemical gradient maintained by a

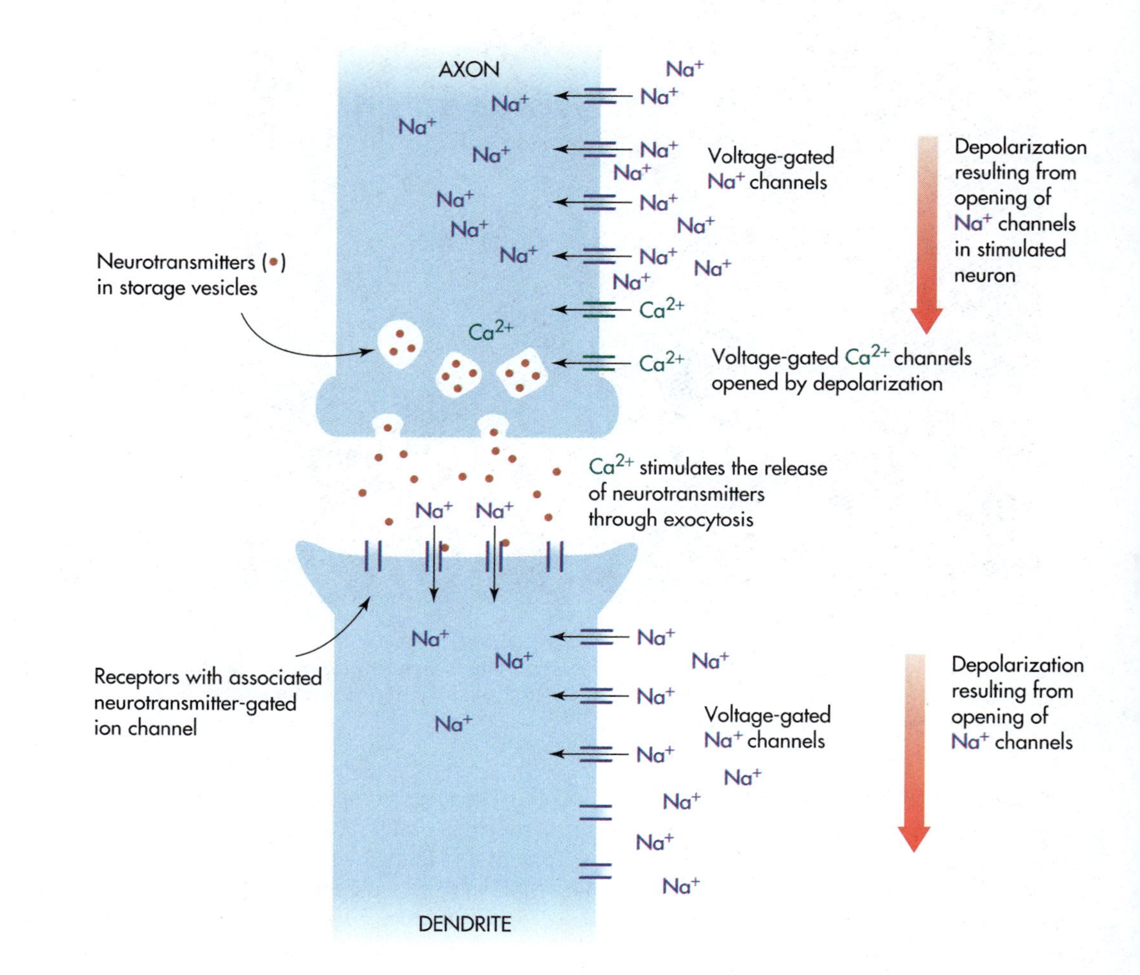

FIGURE 10.19
Signal transduction by neurotransmitters. Following the stimulation of a neuron, a nerve impulse (a progressive, unidirectional depolarization of the plasma membrane brought about by an influx of Na^+) travels to the ends of its axon, where the opening of voltage-gated Ca^{2+} channels leads to the release of neurotransmitters from storage vesicles into synapses. The neurotransmitters diffuse across the synapses, where they bind to receptors on dendrites of postsynaptic neurons and open Na^+ channels. The influx of Na^+ initiates another nerve impulse, which travels to the ends of the axon of the postsynaptic neuron. As the impulse opens voltage-gated Ca^{2+} channels, the postsynaptic neuron becomes a presynaptic neuron.

Na^+K^+transporter (Section 9.11). The docking of a stimulatory neurotransmitter to its receptor opens an ion channel and allows Na^+ ions to diffuse into the postsynaptic neuron. This flow decreases the electrical potential (gradient) across the membrane, a process described as **depolarization.** The depolarization of the membrane at the neurotransmitter receptor site opens neighboring voltage-gated Na^+channels. A **voltage-gated ion channel** is one that opens in response to changes in the electrical potential across the membrane (Section 9.12). As Na^+ flows through the opened channels, voltage-gated channels are opened further along the membrane and a nerve impulse is transmitted along the neuron (Figure 10.19). Select K^+ channels also open during this process. Since each ion channel is opened for only a brief period and cannot be reopened immediately, signal propagation is unidirectional.

When the electrical signal associated with the progressive depolarization of a neuron membrane reaches the ends of an axon, it opens voltage-gated Ca^{2+} channels and leads to an influx of Ca^{2+}. At this point, the postsynaptic neuron becomes a presynaptic neuron as the rise in Ca^{2+} concentration triggers the release of neurotransmitters into synapses. The process repeats itself as a nerve impulse "flows" from neuron to neuron. **[Try Problems 10.26 and 10.27]**

EXHIBIT 10.3

Some Representative Neurotransmitters

EXCITATORY
- Acetylcholine
- Aspartate
- ATP
- Dopamine
- Epinephrine
- Glutamate
- Histamine
- 5-Hydroxytryptamine (serotonin)
- Neuropeptides
- Norepinephrine

INHIBITORY
- 4-Aminobutyrate (γ-aminobutyrate, GABA)
- Glycine

Chemical Nature

Neurotransmitters are, structurally, very diverse compounds. Roughly 50 have been identified, and many of these also function as hormones. Several of those listed in Exhibit 10.3 have already been encountered in this text. Aspartate, glutamate, and glycine are three of the 20 amino acids normally used to assemble proteins (Section 3.2). ATP is the immediate source of energy for most of the energy-requiring processes in the human body (Section 8.4), and it is a substrate for the adenylate cyclase–catalyzed production of cAMP. Dopamine, epinephrine, and norepinephrine are catecholamines and hormones (Figure 10.5, Section 10.3). Neuropeptides are examined in Section 3.6 as examples of biologically active peptides. The structures of the other neurotransmitters listed in Exhibit 10.3 are depicted in Figure 10.20. **Serotonin** and **histamine** are produced from tryptophan and histidine (two protein amino acids), respectively. Histamine, which plays a central role in allergic responses, is

$CH_3-C(=O)-O-CH_2-CH_2-\overset{+}{N}(CH_3)_3$

Acetylcholine

(imidazole ring)$-CH_2-CH_2-\overset{+}{N}H_3$

Histamine

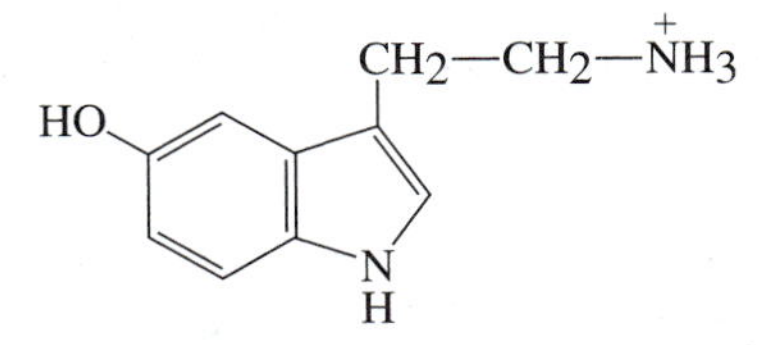

5-Hydroxytryptamine (serotonin)

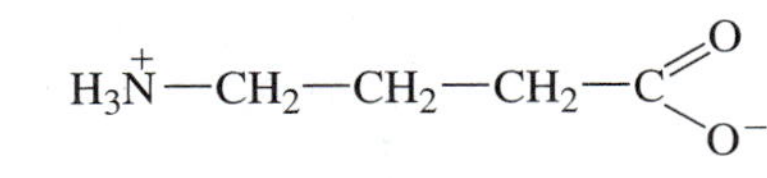

4-Aminobutyrate (gamma-aminobutyrate [GABA])

FIGURE 10.20
Structural formulas for some neurotransmitters.

also a hormone. **Antihistamines** are drugs that interfere with the synthesis or action of histamine. Glutamate is the precursor of **4-aminobutyrate (γ-aminobutyric acid,** or **GABA)**. Functionally distinct neurons produce different neurotransmitters or unique combinations of neurotransmitters. **[Try Problem 10.28]**

Both CO and NO (Section 10.5) play roles in signal transduction in neurons, and some biochemists are still trying to decide whether they should be classified as neurotransmitters (a practice followed in this text). They are definitely novel messengers, because they are gases and they lack the storage vesicles and plasma membrane receptors characteristic of classical neurotransmitters. Both compounds can diffuse freely through biological membranes. The traditional definition of a neurotransmitter must be modified if NO and CO are ranked among this category of compounds.

The Regulation of Nerve Impulses

Not all neurotransmitters are stimulatory. Some, as indicated in Exhibit 10.3, are inhibitory. Upon docking onto their receptors, the inhibitory transmitters open Cl^- channels and lead to an influx of Cl^- into postsynaptic neurons. These negative ions lead to hyperpolarization and make it more difficult for stimulatory neurotransmitters to depolarize a neuron. Inhibitory neurotransmitters, including GABA, help prevent the overstimulation of neurons. Clinically, the underproduction of GABA is associated with some epileptic seizures.

Most neurotransmitters have a very short life span within a synapse. Some are rapidly degraded by specific enzymes, but others are removed by reuptake or diffusion. These processes play a central role in controlling signal transduction between neurons.

A wide variety of substances, known as **neurotoxins,** can modulate the transmission of nerve impulses, with potentially lethal consequences. Some block Na^+ channels, some block neurotransmitter receptors, others inactivate enzymes that degrade neurotransmitters, and so on. Most insecticides, including **parathion,** are neurotoxins. **Tabun** and **sarin,** two other neurotoxins, are employed as nerve gases, while **succinylcholine** is used as a muscle relaxant during surgery. The active ingredient in **curare,** an Amazon arrow poison, is the neurotoxin ***d*-tubocurarine.** Protein neurotoxins are found in scorpion venom, in black widow spider venom, in snake venom, and in *Clostridium botulinum,* the bacterium responsible for **botulism.** The structure of some of these representative neurotoxins is illustrated in Figure 10.21.

PROBLEM 10.26 Ca^{2+} functions as a second messenger during the transmission of a nerve impulse. Is Na^+ also a second messenger during this process? Explain.

PROBLEM 10.27 What specific aspect of the transmission of nerve impulses accounts for most of the large amount of energy consumed by the human nervous system?

PROBLEM 10.28 Which, if any, of the neurotransmitters in Exhibit 10.3 are lipids? Carboxylic acids? Esters? Amines? Amides? Alcohols?

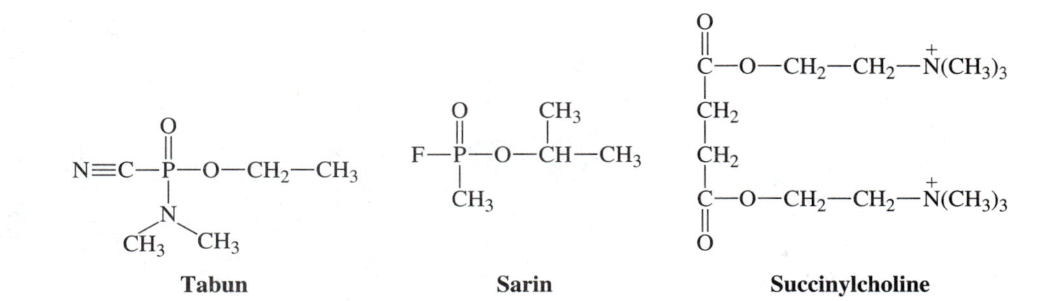

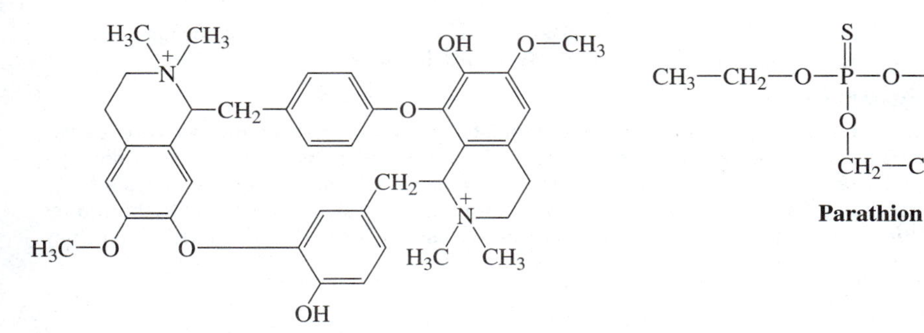

FIGURE 10.21
Structural formulas for some representative neurotoxins.

10.10 AGONISTS AND ANTAGONISTS

Hormone action, like the transmission of nerve impulses, can be modulated by specific chemicals. In some instances, this modulation is of clinical significance.

A wide variety of health problems involve abnormal hormones, insufficient hormone production, excessive hormone production, or an abnormal responsiveness of hormone target sites. The administration of a hormone or a hormone agonist (Figure 10.22) is a therapeutic option for the treatment of problems associated with inactive or partially active hormones and problems associated with insufficient hormone production. The use of insulin (a protein hormone; Table 10.8) to treat diabetes is a classic example (Section 12.4). A hormone **agonist,** in general, is a substance that binds to a receptor for a hormone and triggers the same response as the hormone itself.

Health problems that involve excess hormone production can sometimes be treated with **antagonists,** compounds that bind to receptors without eliciting the normal response. Neither a hormone nor an agonist can trigger a response if the receptor for a hormone is occupied by an antagonist. However, since hormones, agonists, and antagonists normally bind receptors in a reversible fashion, high concentrations of an agonist or hormone can overcome the effects of an antagonist. Some antihistamines (Section 10.9) and RU486 (mifepristone; Sections 9.5 and 9.6) are examples of antagonists.

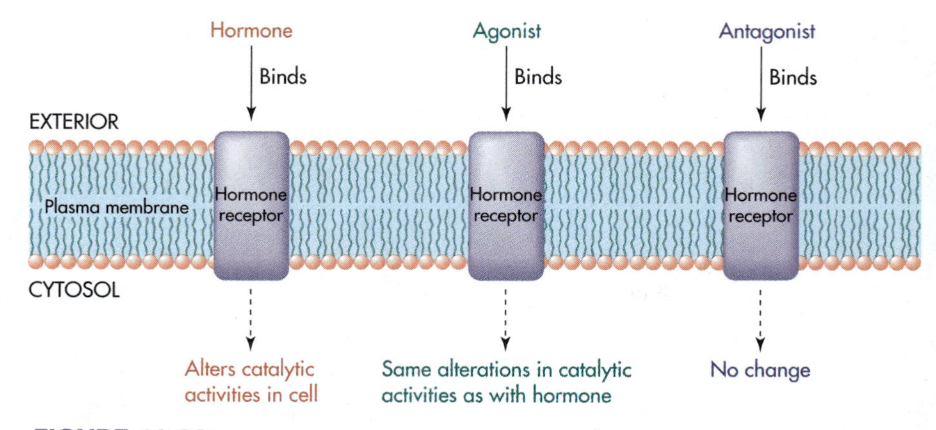

FIGURE 10.22
Agonists and antagonists. Hormones, agonists, and antagonists compete for the same binding sites. When an antagonist is bound to a receptor, neither hormone nor the agonist can bind. When part or all of the hormone receptors on a cell are occupied by antagonists, the cell is unable to respond normally to messages delivered by hormones. In the case of nonpolar hormone, receptors are usually located inside a cell rather than on the plasma membrane.

In some instances, hormones, agonists, or antagonists may also be used to treat diseases linked to abnormal target site responsiveness. Agonists or increased hormone levels produced by hormone administration can, for example, compensate for a receptor with a reduced hormone affinity. Advances in endocrinology have had, and will continue to have, a major impact on the health sciences. **[Try Problems 10.29 and 10.30]**

PROBLEM 10.29 List two specific molecular defects that would lead to a reduction in the responsiveness of a target cell to a normal water-soluble hormone.

PROBLEM 10.30 Many antagonists and agonists are structurally similar to those hormones whose action they modulate. Explain.

METABOLISM

10.11 AN INTRODUCTION TO METABOLISM

The preceding discussion of vitamins and hormones completes the framework for the discussion of metabolism. The metabolism of an organism is the sum of its enzyme-catalyzed reactions. Metabolism is divided into two broad categories, catabolism (degradative reactions) and anabolism (synthetic reactions). Catabolism is normally an exergonic process (energy given off), which involves the oxidation of the compound being degraded. Anabolism is normally an endergonic process (energy taken in) in which precursors are reduced to generate a final product. In humans and many other organisms, catabolism provides most of the energy, intermediates, and reducing power for anabolism. Hence, catabolism and anabolism are tightly coupled. A metabolite is a substance that is a reactant, product, or intermediate in metabolism.

The topics of energy transformation, oxidation–reduction, and reducing power are central to metabolism, and they are woven together in the preceding paragraph. Although these topics are examined in detail in Chapter 8, "Bioenergetics," a brief review is in order.

Oxidation is the loss of electrons and reduction the gain of electrons. Although the two processes are always coupled and normally proceed simultaneously, a redox reaction can be visualized as proceeding through two half reactions. The standard reduction potential ($E^{\circ\prime}$) is a quantitative measure of the electron affinity of the electron acceptor in a half-reaction under standard biological conditions. The higher the reduction potential for a half-reaction, the more powerful the oxidizing agent (the electron acceptor) within the half-reaction. The lower the standard reduction potential, the more powerful the reducing agent (the electron donor) within a half-reaction. The $\Delta G^{\circ\prime}$ and equilibrium constant for a redox reaction are determined by the difference in the standard reduction potentials for its two half reactions. **[Try Problem 10.31]**

FAD, NAD^+, and $NADP^+$ are three of the most important oxidizing agents involved in catabolism. They are all coenzymes derived from water-soluble vitamins (Table 10.6). Their reduction potentials are given in Table 8.9 and their structures will be examined at the point where they are first encountered during the detailed discussion of metabolism. Their roles are summarized in Table 10.9. As FAD, NAD^+, and $NADP^+$ oxidize specific metabolites they are reduced to $FADH_2$, NADH, and NADPH, respectively, three biological reducing agents. The electrons acquired by $FADH_2$ and NADH tend to be transferred to O_2 in a stepwise process coupled to the production of ATP from ADP and P_i. ATP is the immediate source of energy for many life-sustaining processes. Molecular O_2, with a standard reduction

TABLE 10.9

Some Key Participants in Metabolism

Participants	Major Role or Function
FAD NAD^+ $NADP^+$	Oxidizing agents for catabolism
$FADH_2$ NADH	Provide reducing power for ATP production
NADPH	Provides reducing power for anabolism
ATP	Provides energy for many energy-requiring processes, including muscle action, brain action, and anabolism

EXHIBIT 10.4

Some Factors That Determine the Rates of Enzyme-Catalyzed Reactions

Enzyme concentration
Substrate concentration
Coenzyme concentration
Allosteric effector concentration
Substrate inhibition
pH
Temperature
Ionic strength
Covalent modification of enzyme

potential of 0.816 V, is the ultimate oxidizing agent during catabolism. The electrons acquired by NADPH are used primarily for anabolism. Thus, NADPH provides most of the reducing power for anabolism. **[Try Problem 10.32]**

Both degradative and synthetic processes commonly involve sequences of coupled reactions that convert initial reactant(s) to some final product(s). Many intermediates are sometimes generated along the way. Such sequences of reactions are called **metabolic pathways.** The $\Delta G^{\circ\prime}$ for a metabolic pathway is equal to the sum of the $\Delta G^{\circ\prime}$ values for each step in the pathway (Section 8.6). Thus, a thermodynamically unfavorable reaction ($\Delta G^{\circ\prime} > 0$) can be made thermodynamically favorable by coupling it to one or more energy-releasing reactions ($\Delta G^{\circ\prime} < 0$) such as the hydrolysis of ATP. **[Try Problem 10.33]**

Since each metabolic reaction is catalyzed by an enzyme, the study of metabolism is, to a large extent, the study of enzymes. The study of enzymes is, in turn, tied directly to the expression of those genes that encode them. The rate of gene expression and the rate of enzyme degradation, together, determine the concentration of an enzyme. The concentration of an enzyme sets the upper limit for the rate of the reaction that it catalyzes. A reaction may proceed at a rate below its maximum rate because of a nonsaturating substrate concentration, inhibitors, nonoptimum pH, and other variables (Exhibit 10.4). You may want to review Chapter 5, "Enzymes," thoroughly before you continue your study of metabolism.

The key to life itself lies in the integration, coordination, and control of the thousands of reactions that are necessary to sustain life. Cells and organisms must continuously fine-tune the rates of individual reactions and entire metabolic pathways in response to constantly changing needs, stimuli, or environments. This regulation is normally accomplished by modifying the concentration or activity of an enzyme. In some cases, a change in the rate of a single reaction has lethal consequences. Many genetic diseases document this fact since they are linked to changes in the genes for single enzymes (Chapter 19).

Biochemists are gradually beginning to understand how different metabolic pathways within a single cell communicate with one another and how different cells and distant tissues exchange messages. This communication, which has emerged during evolution, is central to many of the delicate checks and balances that help maintain the thread of life. Some of these checks and balances have already been described. Others will be analyzed in this chapter and the chapters that follow. **[Try Problem 10.34]**

Within eukaryotic cells, the enzymes for some metabolic pathways are found only within the nucleus, the enzymes for other pathways are localized in mitochondria, and so on. This leads to an orderly and efficient division of labor within a cell. At the same time, this compartmentation facilitates or complicates, depending on one's point of view, the control and coordination of metabolic processes. If the product of one pathway is to become a reactant in a second pathway, that product must sometimes be carried through a membrane barrier. In one sense, this makes it more difficult for a cell to couple metabolic pathways; at the same time, however, it provides an additional point at which that coupling can be regulated. Numerous examples of compartmentation will be encountered and emphasized in the chapters that lie ahead. **[Try Problem 10.35]**

PROBLEM 10.31 Give a specific example of an oxidation–reduction (redox) reaction. Clearly identify the oxidizing agent and the reducing agent. Give a specific example of a reaction that is not a redox reaction.

PROBLEM 10.32 Study Table 8.9. Which is the most powerful reducing agent, NADH, NADPH, or $FADH_2$? Which is the strongest oxidizing agent, NAD^+, $NADP^+$, or FAD? Explain.

PROBLEM 10.33 All of the enzymes for a specific metabolic pathway often reside in a common multienzyme complex. How does an organism benefit from this relationship?

PROBLEM 10.34 Chapter 5 describes some of the strategies that have evolved for controlling when and where specific reactions occur and how rapidly these reactions occur. List those strategies.

PROBLEM 10.35 List the major organelles found within animal cells.

10.12 AN OVERVIEW OF FUEL METABOLISM

Each cell in the human body requires a constant supply of energy to survive. Virtually all of this energy is provided by the complete oxidation (by O_2) of organic fuels, a catabolic process that proceeds through numerous discrete reactions (Figure 10.23; Chapter 11). The need for a constant supply of energy generates the need for a continuous supply of fuel and O_2. Different types of cells commonly rely on different fuels, and even the same cell may preferentially use dissimilar fuels under different circumstances (Table 10.10). **[Try Problem 10.36]**

Under normal circumstances, brain cells rely primarily on glucose (also known as **blood sugar** and **dextrose**) as a fuel, because the unique capillaries that supply blood to the brain create a **blood–brain barrier** that prevents most fatty acids and a variety of other potential fuels from entering this organ. Since brain cells do not store fuel, elaborate regulatory mechanisms exist that are designed to ensure that blood glucose concentrations (glucose plus ketone bodies during starvation) are continuously high enough to meet the energy demands of the brain. This makes common

TABLE 10.10
Preferred Fuels for Selected Tissues

Tissue	Preferred Fuel Under Normal, Between-meals Circumstances
Brain	Glucose
Active muscle[a]	Glucose
Resting muscle	Fatty acids
Liver	Dietary amino acids
Adipose (fat)	Glucose

[a]During prolonged activity muscles gradually shift to fatty acids as their primary fuel.

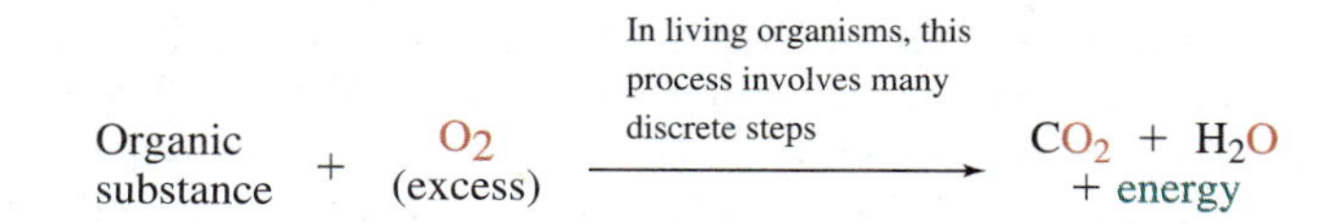

FIGURE 10.23
The complete oxidation of an organic fuel. Additional inorganic products are formed if the fuel contains elements in addition to C, H, and O.

sense from the standpoint of survival, and it must be kept in mind to appreciate the "biological logic" behind a major portion of human metabolism. **[Try Problem 10.37]**

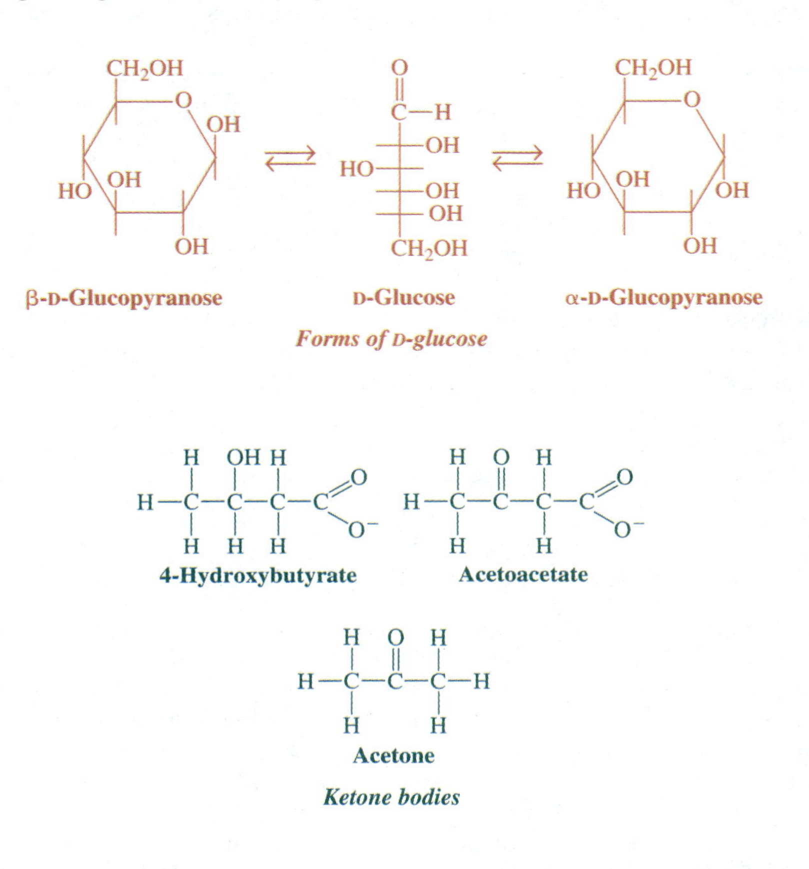

Forms of D-glucose

Ketone bodies

In the international system of measurement, blood solute concentrations are normally expressed as millimoles per liter of blood (m*M*). At the present time, however, most clinical labs express blood glucose levels in mg/dL, where 1 dL = 0.1 L. For glucose, 1 mg/dL = 0.055 mmol/L = 0.055 m*M*.

Blood glucose levels normally range from 80 mg/dL to 120 mg/dL. A person with a blood glucose level below 80 mg/dL is considered **hypoglycemic,** whereas blood glucose levels above 120 mg/dL lead to **hyperglycemia.** A person who is mildly hypoglycemic may feel sluggish or dizzy, and even brief periods of severe hypoglycemia can lead to a coma or death as the brain runs dangerously low on energy and fuel. Brief bouts of hyperglycemia probably pose little health risk, at least in the short run. Prolonged or severe hyperglycemia leads to potentially serious health problems as glucose becomes covalently cross-linked with proteins and other molecules within the body (Sections 6.4 and 12.4). These cross-linking reactions tend to alter or destroy the biological activities of the molecules involved. Such reactions occur at a slower rate at normal glucose concentrations, and gerontologists have suggested that they may contribute to normal biological aging. The body attempts to minimize the risks of both hypoglycemia and hyperglycemia by using multiple checks and balances to keep blood glucose levels within the "normal range" the majority of the time. **[Try Problems 10.38 and 10.39]**

Muscle cells, in contrast to brain cells, have a significant amount of stored energy in the form of glycogen, a branched polymer of glucose:

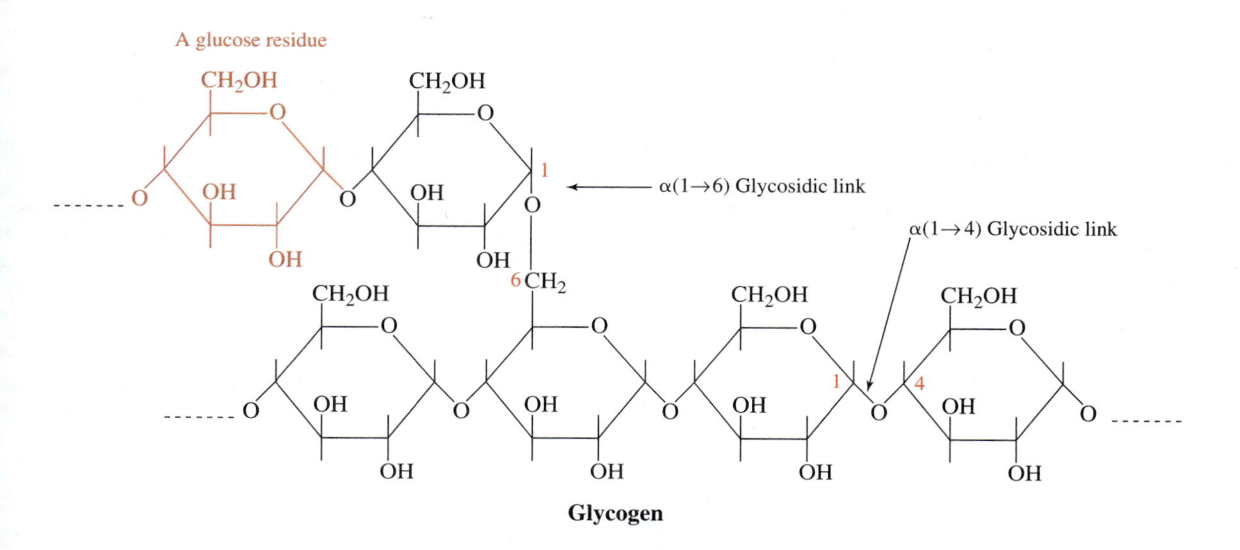

Glycogen

Glucose, from both the blood and stored glycogen, tends to be the fuel of choice for active muscles (Table 10.10). During vigorous exercise such as sprinting, muscles become **anaerobic** (oxygen deficient) as most of the available O_2 is rapidly consumed to produce energy (oxidize fuel). Under such conditions, a reduced amount of energy is produced through the partial oxidation of glucose, an O_2-independent process that is limited in duration because a drop in pH is associated with the accumulation of lactic acid (Figure 10.24). Resting muscle cells favor fatty acids and ketone bodies as fuels, and they tend to incorporate glucose into glycogen instead of burning it. This glycogen helps prepare the cells for the next round of activity. When blood glucose levels are low, muscle cells utilize less blood glucose as the body attempts to ensure that brain cells have a constant, adequate supply.

The major fuel of fat cells (adipocytes) is usually glucose. When blood glucose levels are high, much of the glucose that enters adipocytes is used to produce glycerol for triacylglycerol biosynthesis (Figure 10.25). As blood glucose levels drop, glucose concentrations within adipocytes also decrease, an event that triggers the hydrolysis of stored fats and the release of fatty acids into the blood. The released fatty acids serve as an alternate fuel to glucose for various cells, including muscle cells. The use of alternate fuels by other cells helps the body conserve blood glucose for brain cells. **[Try Problem 10.40]**

The liver plays a key role in regulating both blood glucose levels and fat metabolism. While doing so, it primarily oxidizes dietary proteins and amino acids to meet its own energy needs. The liver also uses some glucose and fatty acids as fuel. When blood glucose levels are low, the liver releases glucose into the blood and uses less

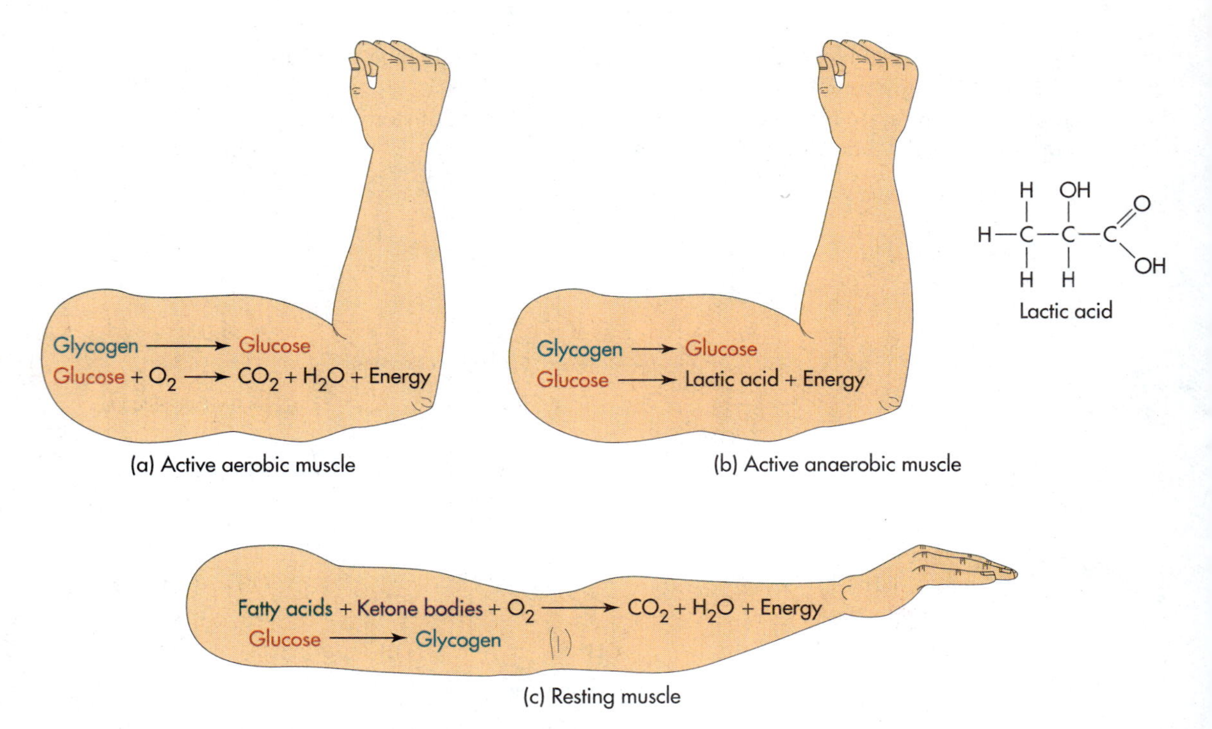

FIGURE 10.24
Fuel metabolism in muscle. (a) The complete oxidation of glucose (predominantly from glycogen) provides most of the energy used by muscles during short-duration aerobic activity. Glycogen reserves are depleted. (b) The partial oxidation of glucose to lactic acid provides most of the energy used by muscles during anaerobic activity. Glycogen reserves are depleted. (c) The complete oxidation of fatty acids and ketone bodies provides most of the energy used by resting muscles. Glucose is incorporated into glycogen to build up glycogen reserves.

glucose itself. The released glucose is obtained from glycogen stored within the liver and from **gluconeogenesis,** the synthesis of glucose from lactate, glycerol, amino acids, and other noncarbohydrate precursors (Figure 10.26). Low blood glucose levels also prompt the liver to produce and release ketone bodies that can be used as a fuel by many different cells, including brain cells. When blood glucose levels are high, liver cells (hepatocytes) take up glucose and convert it to glycogen and fatty acids. The fatty acids are secreted into the blood where they are transported by serum albumin (Section 9.7) to adipocytes for the assembly of triacylglycerols. Most of the fatty acids produced within the body are synthesized within the liver.

Most eukaryotic cells take up glucose through facilitated diffusion, a process that employs membrane-embedded glucose transporters (Section 9.11). In some cells, transporter availability is controlled partially by hormones, including insulin. Once glucose enters a cell it tends to be quickly phosphorylated with the catalytic assistance of a glucose kinase with a small K_m (substrate concentration that yields an initial velocity of $\frac{1}{2}$ V_{max}) for glucose. The small K_m allows the enzyme to efficiently

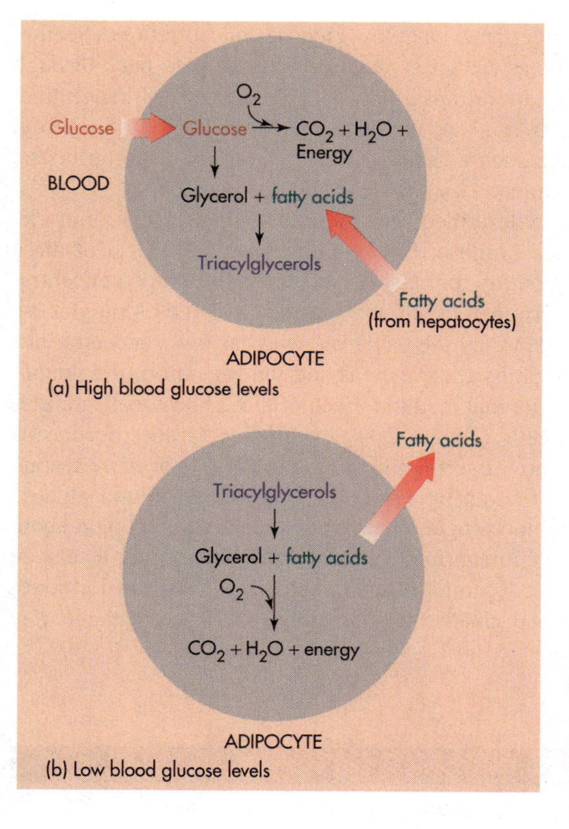

FIGURE 10.25
Fuel metabolism in adipocytes. (a) When blood glucose levels are high, imported glucose is the primary fuel and some glucose is converted to glycerol. Fatty acids are imported and joined to glycerol to produce triacylglycerols to beef up fat reserves. (b) When blood glucose levels are low, stored triacylglycerols are hydrolyzed. Glycerol and fatty acids are used as fuels and fatty acids are exported to provide an alternate fuel (one other than glucose) for many cells. The use of fatty acids as fuels by nonbrain cells conserves blood glucose for use by brain cells.

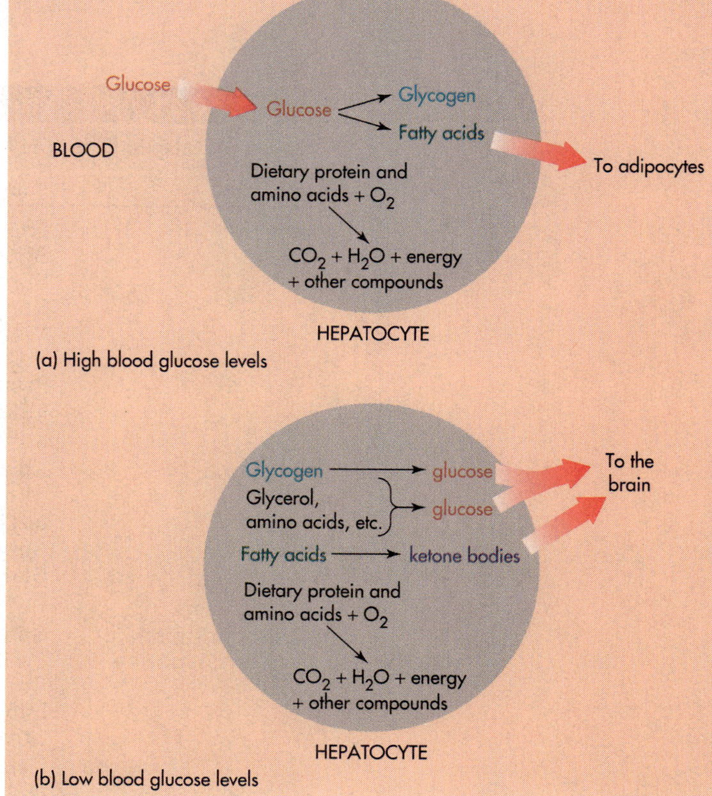

FIGURE 10.26
Fuel metabolism in hepatocytes. (a) When blood glucose levels are high, glucose is imported. Some of the glucose is incorporated into glycogen while most of the rest is converted to fatty acids. The fatty acids are exported to adipocytes where they are used to assemble triacylglycerols to build up fat reserves. Dietary proteins and amino acids are primary fuels. (b) When blood glucose levels are low, both glucose from glycogen and newly synthesized glucose are exported to ensure that the brain has an adequate amount of blood-supplied fuel. Fatty acids are converted to ketone bodies (alternate fuels for brain cells) which are also exported to the brain. Dietary protein and amino acids continue to be the primary fuels for hepatocytes themselves.

catalyze the phosphorylation of glucose, even at low glucose concentrations (Section 5.7). Since the charged phosphate ester that is generated cannot pass back through the plasma membrane, phosphorylation traps glucose inside the cell. Although liver cells possess an enzyme that can catalyze the removal of the added phosphoryl group, muscle and brain cells lack such a catalyst. Hence, liver can export glucose, whereas brain and muscle cells cannot. **[Try Problem 10.41]**

The preceding paragraphs have described how the metabolism of liver, muscle, and fat cells changes in response to high or low blood sugar levels. Most of these metabolic changes are brought about by four key hormones whose blood concentrations are, themselves, largely determined by blood sugar levels. High blood glucose concentrations enhance insulin release by the pancreas, whereas low concentrations stimulate the discharge of glucagon by the pancreas and the liberation of epinephrine and norepinephrine from the adrenal medulla. Each of these hormones modifies the metabolism of select target cells (Table 10.11) by binding to specific plasma membrane-associated receptors and triggering the production of one or more second messengers. The changes brought about by insulin (released when blood glucose levels are too high) tend to lower blood sugar levels while the changes brought about by glucagon, epinephrine, and norepinephrine (released when blood sugar levels are too low) tend to raise blood sugar concentrations. Collectively, these four hormones play a central role in keeping blood glucose levels within the "normal" range. **[Try Problem 10.42]**

TABLE 10.11

Some Metabolic Changes Brought About by Four Key Hormones

Hormone	Some Metabolic Changes It Brings About
Insulin	Enhances glycogen synthesis in muscle and liver Suppresses gluconeogenesis in liver and muscle Stimulates oxidation of glucose in liver and muscle Stimulates production of fatty acids by liver Promotes uptake of glucose by muscle and adipose tissue Promotes protein production in muscle Inhibits breakdown of protein in muscle Enhances fat production by adipocytes
Glucagon	Stimulates breakdown of glycogen (to yield glucose) in liver Inhibits glycogen synthesis in liver Suppresses fatty acid production in liver Stimulates gluconeogenesis in liver Stimulates release of fatty acids by adipocytes
Epinephrine and/or norepinephrine	Stimulates breakdown of glycogen in liver and muscle Stimules release of fatty acids by adipocytes Inhibits uptake of glucose by muscle Stimulates secretion of glucagon Suppresses secretion of insulin

TABLE 10.12

A Summary of Glycogen and Fatty Acid Metabolism in Selected Tissues

Tissue	Low Blood Glucose Levels[a]	High Blood Glucose Levels[a]
Active muscle	Glycogen degraded and glucose oxidized Fatty acids imported and oxidized	Glycogen degraded and glucose oxidized
Resting muscle	Fatty acids imported and oxidized	Glycogen synthesized from imported glucose Fatty acids imported and oxidized
Liver	Glycogen degraded and glucose exported Fatty acids imported and converted to ketone bodies for export	Glycogen synthesized from imported glucose Fatty acids synthesized from imported glucose and then exported
Adipose	Fats hydrolyzed and fatty acids exported	Triacylglycerols synthesized from imported fatty acids and glucose-derived glycerol

[a]A substance listed as oxidized is one utilized as a fuel.

Table 10.12 summarizes the key elements of glycogen and fatty acid metabolism in selected tissues. Table 10.11 makes it clear that this metabolism is largely regulated by hormones. We must keep in mind that active muscle uses a large amount of energy and that brain cell survival depends upon a constant supply of blood glucose or ketone bodies. Hepatocytes and adipocytes are constantly adjusting their metabolism to help meet the needs of both muscle and brain. The fuel and energy needs of most other cells are accommodated in the process. In the long run, it will pay to ponder the figures and tables in this section. It is helpful to acquire a broad and consolidated picture of fuel metabolism before you proceed to an analysis of the individual reactions and metabolic pathways involved.

PROBLEM 10.36 Use structural formulas to write a balanced equation for the total oxidation of glucose by O_2. Write a second balanced equation for the complete oxidation of stearic acid (a fatty acid) by O_2. Which reaction leads to the greater release of energy per gram of fuel? Why?

PROBLEM 10.37 Suggest why a blood barrier has evolved for the brain although a similar barrier does not exist for other organs and tissues.

PROBLEM 10.38 True or false? You are hyperglycemic for a short period after a typical American meal. Explain.

PROBLEM 10.39 Explain how an increase in blood glucose concentration leads to an increase in the rate of glucose-associated protein cross-linking reactions.

PROBLEM 10.40 Write the structural formula for a triacylglycerol. Circle the glycerol residue in your formula. What is the relationship between fat and triacylglycerol?

PROBLEM 10.41 Use structural formulas to write a balanced equation for the reaction of glucose with inorganic phosphate to yield glucose 6-phosphate. Cells normally use ATP as the source of a phosphoryl group during the phosphorylation of glucose. Rewrite the equation for the phosphorylation reaction taking this fact into account. Which one of the two equations has the larger equilibrium constant? Explain.

PROBLEM 10.42 The human body can alter the blood concentration of insulin, or any other hormone, by modifying the rate at which active hormone molecules are released into the blood. List one other strategy the body can employ to adjust the blood concentration of a hormone.

SUMMARY

Vitamins are organic compounds that the human body requires in small amounts but cannot produce. They must be obtained in the diet. Inadequate consumption can lead to deficiency diseases. An excess of certain vitamins, including vitamin A, vitamin D, pyridoxine, and niacin, can lead to severe toxicity. The recommended daily allowance (RDA) for a vitamin is the amount that prevents deficiency symptoms in most normal individuals without any toxic effects.

There is growing evidence that certain vitamins have health benefits beyond the avoidance of deficiency symptoms. Folic acid can reduce the risk of neural tube defects in fetuses. Vitamins A and E appear to reduce cancer risk, and they may have a positive impact on other health problems as well. The possible benefits of other vitamins, including vitamin C, pyridoxine, and cobalamin, are under investigation.

The fat-soluble vitamins (A, D, E, and K) are isoprenoids and lipids. Each of these vitamins and some of the water-soluble ones are families of compounds. Vitamins A and D are both precursors to hormones. Vitamin A is involved in vision, immune response, differentiation, and other processes. The primary role of vitamin D is the regulation of calcium and phosphorous metabolism, including the development and maintenance of bone. Vitamin K helps regulate calcium metabolism and blood clotting. Vitamin E is an antioxidant that protects other substances from oxidative damage.

The water-soluble vitamins are the B complex [biotin, cobalamin (B_{12}), folic acid, niacin, pantothenic acid, pyridoxine (B_6), riboflavin (B_2) and thiamine (B_1)] and vitamin C. The B-complex vitamins are precursors to coenzymes, whereas vitamin C, an antioxidant, is involved in iron metabolism, collagen production, and other processes.

Hormones are chemical messengers that play a central role in the constant communication that occurs between cells and tissues within multicellular organisms. The term *hormone* now encompasses growth factors, cytokines, eicosanoids, and certain other messengers, in addition to the traditional hormones produced by the endocrine glands. Structurally, hormones are very diverse; they include peptides, proteins, nonpeptide amines, steroids, eicosanoids, and other classes of compounds. Some hormones also function as neurotransmitters.

Most water-soluble hormones attach to membrane-spanning receptors associated with the plasma membrane of target cells. These receptors are usually coupled to a G protein that, following hormone binding, modifies the activity of a membrane-associated enzyme. This enzyme, in turn, directly modulates the activity of other proteins or catalyzes the production of one or more second messengers. Once produced, second messengers activate select target enzymes, normally protein kinases. The activated kinases phosphorylate specific proteins that may include other protein kinases. The net result is a reaction cascade that modulates the catalytic ability of particular target enzymes or alters gene expression by modifying transcription factors. The ultimate result is the modification of the catalytic activities within the target cells.

The receptors for fat-soluble hormones are located inside a target cell rather than on its surface. The receptor–hormone complex binds directly or indirectly to DNA and, in the process, alters the expression of specific genes. Since these genes normally encode enzymes or enzyme subunits, the net result is once again a change in the catalytic activities within target cells.

Second messengers include cAMP, cGMP, Ca^{2+}, diacylglycerols, inositol 1,4,5-trisphosphate, and cADPR. Carbon monoxide and nitric oxide, two gases, also function as second messengers. The Ras signaling pathway and the epinephrine-linked cAMP signaling pathway provide specific examples of signal transduction and reaction cascades. The Ras pathway is of considerable medical interest, because several of its participants are encoded in proto-oncogenes.

Neurotransmitters are the basis for communication at the junctions between neighboring neurons and the synapses between neurons and muscle cells. The traditional neurotransmitters are stored in membrane-enclosed vesicles in axons and are released into synapses in response to a stimulus-induced depolarization of axons. They include amino acids, peptides, ATP, catecholamines, serotonin, histamine, 4-aminobutyrate (GABA), and other compounds. Most are stimulatory but some, including 4-aminobutyrate and glycine, are inhibitory. When transported between neurons, stimulatory neurotransmitters open ion channels and initiate the propagation of a wave of depolarization (a nerve impulse) along the axon of the postsynaptic neuron. This event causes the postsynaptic neuron to become a presynaptic neuron as it triggers the release of stored neurotransmitters into its synapses. Inhibitory neurotransmitters open Cl^- channels and induce a hyperpolarization.

The transmission of nerve impulses can be altered by neurotoxins, agonists, and antagonists. Such neuroregulators are employed in medicine, agriculture, biological warfare, and research.

Metabolism is the total collection of enzyme-catalyzed reactions occurring within organisms. It is divided into catabolism (degradative reactions) and anabolism (synthetic reactions). Catabolism, in general, provides the energy, intermediates, and reducing power for anabolism. To a large extent, metabolism is the study of enzymes and the regulation of their activities. Enzymes are commonly clustered in metabolic pathways and, in eukaryotic cells, they are frequently compartmentalized within organelles.

A significant fraction of human metabolism is involved in the oxidation of fuels to produce energy, a commodity required continuously by each cell in the body. Under normal circumstances, glucose is the primary fuel for many cells. Ketone bodies, fatty acids, and amino acids are alternative fuels. The blood–brain barrier prevents most fuels other than glucose and ketone bodies from entering the brain. Normally, most of the glucose utilized by the human body is supplied in the diet, but under certain circumstances gluconeogenesis can make a substantial contribution. Although most cells store some fuel, usually glycogen or fat, brain cells lack a significant fuel reserve; they depend mainly upon a constant supply of blood glucose. Normally, multiple checks and balances keep blood glucose levels within a narrow range most of the time.

The metabolism of hepatocytes and adipocytes is largely designed to ensure that brain cells and muscle cells are provided with a constant and adequate supply of fuel. Much of this metabolism is regulated by insulin, glucagon, epinephrine, and norepinephrine, hormones whose release is controlled by blood glucose concentrations. Collectively, these hormones play a central role in regulating the production, uptake, release, and utilization of glucose and fatty acids by a variety of cells.

EXERCISES

10.1 Many Americans could eliminate the fat-soluble vitamins from their diets for several months without developing deficiency symptoms. In contrast, deficiency symptoms would be likely to surface within a week or two if you were completely deprived of dietary water-soluble vitamins. What is the explanation for these facts?

10.2 List the hydrolyzable functional groups found in vitamins.

10.3 Explain how boiling vegetables reduces their vitamin content.

10.4 List those vitamins in Tables 10.3 and 10.4 that are carboxylic acids. These vitamins bear a negative charge at pH 7.0. Explain.

10.5 List those vitamins that are alcohols, and then list three reactions that are characteristic of alcohols.

10.6 Would a pure sample of chemically synthesized vitamin C be as biologically active as an equal amount of pure vitamin C isolated from rose hips? Explain. Assume that the synthetic vitamin C has the same chirality as natural vitamin C.

10.7 List those vitamins that are known to be precursors to coenzymes. Name the coenzyme produced from each listed vitamin, and then describe the type of reaction in which each named coenzyme participates.

10.8 True or false? Many of the reactions that occur within the human body depend on a virtually constant supply of vitamins. Explain.

10.9 What distinguishes vitamins from essential trace elements (chemical elements required in very small amounts for normal growth and development)?

10.10 Why are the recommended daily allowances for some vitamins so controversial?

10.11 List three factors that help determine your vitamin requirements at any given time.

10.12 True or false? It is inadvisable for someone to consume large quantities of a multivitamin supplement, even when it contains an optimum ratio of all the vitamins. Explain.

10.13 Both hormones and enzymes help your body regulate rates of reactions. What distinguishes hormones and enzymes?

10.14 What distinguishes vitamins from hormones?

10.15 True or false? The receptor binding sites for most polar hormones are located on the internal surface of the plasma membrane rather than on the external surface. Explain.

10.16 What feature of a target cell allows a hormone to distinguish it from nontarget cells?

10.17 Describe what is meant by a *reaction cascade.*

10.18 Cyclic-AMP is a second messenger within liver cells. Does cAMP injected into the blood elicit the same response as that produced in vivo (within liver cells)? Explain.

10.19 True or false? The equilibrium constant for the binding of a hormone to its receptor is a function of pH. Explain.

10.20 True or false? The reversible binding of an antagonist to a hormone receptor is analogous to the reversible binding of a competitive inhibitor to the active site of an enzyme. Explain.

10.21 To what major classes of biochemicals do most hormones belong?

10.22 List two physiological processes that are regulated, in part, by nitric oxide or carbon monoxide. What second messenger is commonly coupled to nitric oxide action?

10.23 What second messengers are produced from phosphatidylinositol 4,5-bisphosphate? What specific substances bind to these messengers during their modulation of metabolism?

10.24 Hormone signal transduction consumes considerable energy. Identify those steps in both the Ras signaling pathway (Figure 10.10) and the epinephrine-linked cAMP signaling pathway (Figure 10.13) where a high-energy compound is cleaved and energy is consumed.

10.25 Give an example of a compound that functions as both a neurotransmitter and a hormone.

10.26 ATP is the phosphoryl donor for protein kinase–catalyzed phosphorylation reactions. Use names or abbreviations of reactants and products to write an equation for the protein kinase–catalyzed phosphorylation of protein X.

10.27 Describe the immediate role of GTP in the production of cAMP.

10.28 True or false? We can be confident that the phosphorylation of an amino acid residue some distance from the active site of an enzyme will have no impact on the catalytic activity of the enzyme. Explain.

10.29 Overstimulation or overinhibition of a process by a hormone can have lethal consequences for a cell or organism. List two enzymes that help prevent such overstimulation or overinhibition following the binding of a cAMP-linked hormone to a receptor. Clearly describe the catalytic activity of each listed enzyme.

10.30 Explain how high hormone concentrations can overcome the effects of an antagonist.

10.31 True or false? A hormone agonist enhances the affinity of a receptor for its hormone. Explain.

10.32 Would you expect different neurotransmitters to compete for the same receptors on postsynaptic neurons? Explain.

10.33 True or false? During the transmission of nerve impulses from neuron to neuron, ligand-gated ion channels trigger the opening of voltage-gated ion channels. Explain.

10.34 Explain how hyperpolarization results from the influx of Cl^- into a neuron.

10.35 Identify the neurotransmitters in Exhibit 10.3 that bear charged functional groups at pH 7. Explain.

10.36 Classify each of the following processes as either anabolic or catabolic;

- **a.** Transcription
- **b.** Conversion of glycogen to glucose 1-phosphate
- **c.** Digestion
- **d.** Building muscle

10.37 True or false? Multienzyme complexes represent one form of compartmentation within cells. Explain.

10.38 How does a cell benefit from the relatively short half-lives of most second messengers?

10.39 How does liver metabolism change as blood glucose drops from a high level to a low level? How does the brain benefit from these changes?

10.40 Under what conditions do muscle cells incorporate glucose into glycogen? What is the primary use or function of muscle glycogen?

10.41 Under what conditions do adipocytes release fatty acids into the blood? What is the usual fate of most of these fatty acids?

10.42 What four hormones play a central role in maintaining normal blood sugar levels? What are the primary target cells for each of these four hormones?

10.43 Explain why brain cells, but not muscle cells, are quickly damaged by low blood sugar levels.

10.44 Suggest a simple, short-run treatment for hypoglycemia.

SELECTED READINGS

Almers, W., How fast can you get? *Nature* 367, 682–683 (1994).
A brief review of synaptic transmission.

Barinaga, M., Carbon monoxide: Killer to brain messenger in one step, *Science* 259, 309 (1993).
Explores the possible roles of CO in nerve action.

Bhagavan, N. V., *Medical Biochemistry.* Boston: Jones and Bartlett Publishers, 1992.
A biochemistry text with medical emphasis. Provides a particularly thorough coverage of vitamins and hormones. Contains selected readings.

Bikle, D. D., A bright future for the sunshine hormone, *Sci. Am. Science and Medicine* 2(2), 58–67 (1995).
Reviews the biological roles of vitamin D and describes some of the possible therapeutic uses of vitamin D and its analogs.

Can B vitamins prevent heart disease?, *University of California at Berkeley Wellness Letter* 11(6), 1–2 (1995).
Examines evidence that folacin (folic acid), vitamin B_6 (pyridoxine), and vitamin B_{12} (cobalamin) may play a key role in averting heart attacks.

Can vitamin C save your life? *Consumer Reports on Health* 6(3), 25–27 (1994).
Reviews the evidence that vitamin C may help fight cardiovascular disease and cancer.

Changeux, J-P., Chemical signaling in the brain, *Sci. Am.* 269(5), 58–62 (1993).
Discusses neurotransmitters, with emphasis on the mode of action of acetylcholine.

Cooper, D. M. F., N. Mons, and J. W. Karpen, Adenylyl cyclases and the interaction between calcium and cAMP signalling, *Nature* 374, 421–424 (1995).
Explores part of the cross-talk that occurs between Ca^{2+} signaling pathways and cAMP signaling pathways.

De Luca, L. M., N. Darwiche, C. S. Jones, and G. Scita, Retinoids in differentiation and neoplasia, *Sci. Am. Science and Medicine* 2(4), 28–37 (1995).
Reviews the biological roles and clinical uses of retinoids.

Devlin, T. M., *Textbook of Biochemistry with Clinical Correlations,* 3rd ed. New York: Wiley–Liss, 1992.
An encyclopedia-type text with good coverage of vitamins, hormones, and basic metabolism. Contains both problems and selected readings.

Egan, S. E., and R. A. Weinberg, The pathway to signal achievement, *Nature* 365, 781–783 (1993).
A brief review of the Ras signaling pathway.

Fieg, L. A., and B. Schaffhausen, The hunt for Ras targets, *Nature* 370, 508–509 (1994).
Looks at some of the molecules regulated by Ras.

Hall, A., A biochemical function for Ras—at last, *Science* 264, 1413–1414 (1994).

Explores how Ras regulates protein kinase cascades.

Jones, A. M., Surprising signals in plant cells, *Science* 263, 183–184 (1994).

A review of some hormone signaling pathways in plants.

Levitzki, A., and A. Gazit, Tyrosine kinase inhibition: an approach to drug development, *Science* 267, 1782–1788 (1995).

Summarizes progress in the development of inhibitors of protein tyrosine kinases (key participants in numerous signal transduction pathways) and demonstrates the potential use of these inhibitors in the treatment of disease.

Napier, K., Green revolution, *Harvard Health Letter* 20(6), 9–12 (1995).

Examines some cancer-fighting foods and their "active ingredients."

Nishizuka, Y., Intracellular signaling by hydrolysis of phospholipids and activation of protein kinase C, *Science* 258, 607–614 (1992).

The title accurately describes this brief review.

Nowak, R., Beta-carotene: Helpful or harmful? *Science* 264, 500–501 (1994).

Discusses reports that β-carotene may increase cancer risk.

Packer, L., Vitamin E is nature's master antioxidant, *Sci. Am. Science and Medicine* 1(1), 54–63 (1994).

A very readable review of vitamin E and its biological functions.

Pawson, T., Protein modules and signalling networks, *Nature* 373, 573–580 (1995).

Analyzes some of the conserved protein domains (modules) shared by many of the proteins that participate in signal transduction following the binding of ligands to membrane receptors.

Snyder, S. H., More jobs for that molecule, *Nature* 372, 504–505 (1994).

Reviews the tissue distribution of the three distinct forms of NO synthase and the possible roles of nitric oxide (NO) in the relaxation of skeletal muscle.

Snyder, S. H., No endothelial NO, *Nature* 377, 196–197 (1995).

Reviews some of the functions of nitric oxide and describes the consequences of knocking out the mouse gene encoding nitric oxide synthase (NOS) within mouse endothelial cells.

Snyder, S. H., and D. S. Bredt, Biological roles of nitric oxide, *Sci. Am.* 266(5), 68–77 (1992).

Reviews the biological roles of NO.

Südhof, T. C., The synaptic vesicle cycle: a cascade of protein–protein interactions, *Nature* 375, 645–653 (1995).

Describes molecular models for key steps in the formation and functioning of synaptic vesicles, which play a central role in the transmission of nerve impulses.

Willett, W. C., Diet and health: What should we eat? *Science* 264, 532–537 (1994).

Examines the roles of dietary factors in disease and disease prevention.

11 The Oxidation of Glucose

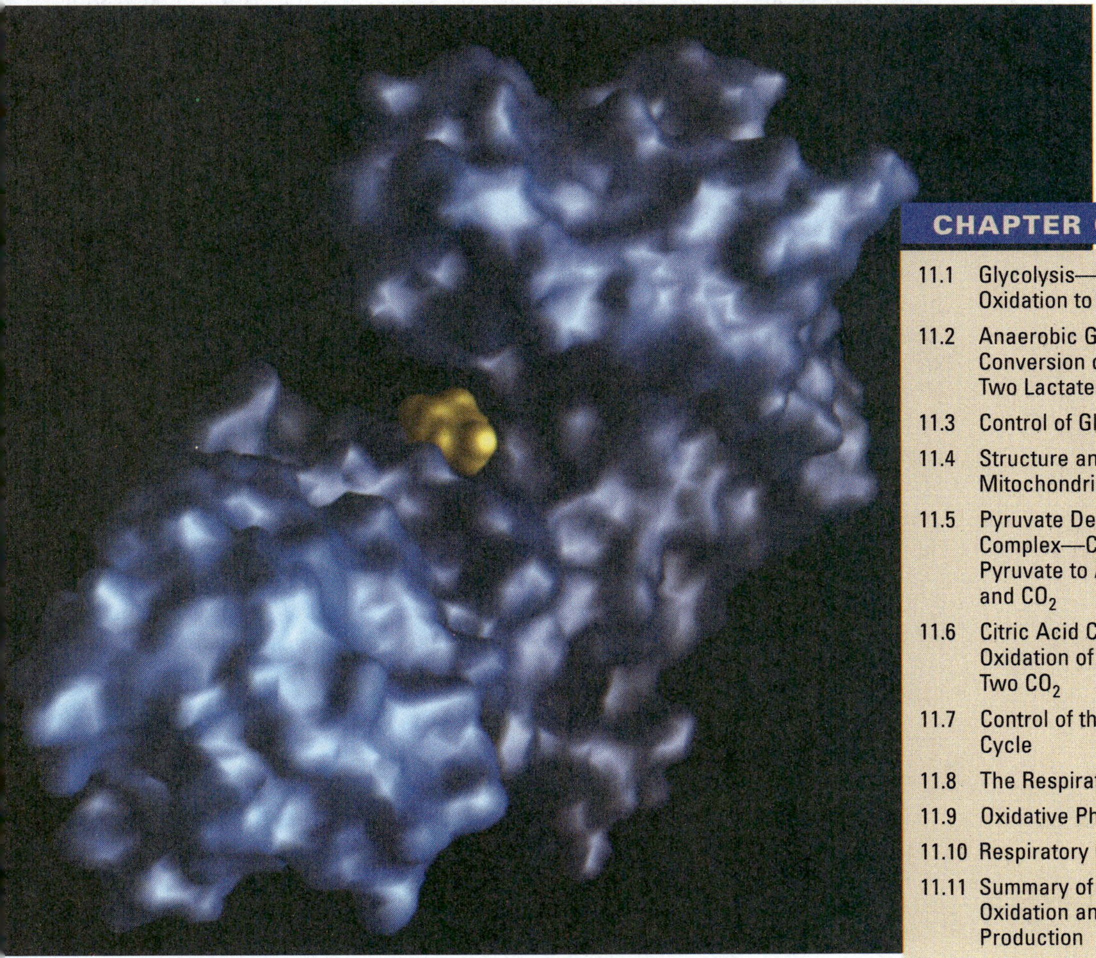

CHAPTER OUTLINE

Chapter 10 completed the framework required for the discussion of metabolism and provided an overview of fuel metabolism in humans. Humans and other heterotrophs rely upon the oxidation of carbohydrates and other dietary organic compounds to meet their energy needs. Starch, glycogen, cellulose, and sugars account for most of the dietary carbohydrate. Starch, glycogen, and cellulose are all polymers of glucose, whereas most sugars are monosaccharides or disaccharides (Chapter 6). Since the human body is not genetically programmed to digest and absorb cellulose, only the starch, glycogen, and sugars are metabolized. Cellulose contributes to our insoluble dietary fiber.

The catabolism of dietary carbohydrates begins in the mouth where the hydrolysis of starch and glycogen is initiated with the catalytic assistance of salivary amylases (Section 6.9). Catalytic activities associated with enzymes in the intestine lead to further hydrolysis of these polysaccharides and contribute to the hydrolysis of sucrose, lactose, and certain other oligo- and polysaccharides. Most of the digestible carbohydrates are ultimately hydrolyzed to yield monosaccharides, predominantly glucose. These monosaccharides enter the bloodstream and are distributed throughout the body. A variety of factors determine how much monosaccharide, if any, is removed from the blood by individual cells. Some of these factors are examined in Section 10.12. **[Try Problem 11.1]**

A cell normally has multiple options for utilizing the glucose and other monosaccharides that it does acquire. Since most other monosaccharides can be converted to glucose within a cell, they can, for many purposes, be considered metabolically equivalent to glucose. Glucose can be stored in glycogen, it can be used to generate intermediates for the production of other compounds, or it can be completely oxidized. The immediate fate of intracellular glucose is determined by the metabolic state of the cell and the state of the body as a whole. **[Try Problem 11.2]**

This chapter examines the reactions a cell employs to oxidize glucose completely. Most of these reactions are an integral part of one of three metabolic pathways: glycolysis, the citric acid cycle, and the respiratory chain. If a cell completely oxidizes glucose, it does so because it needs the energy released in the process (Section 10.12). For this reason, energy and energy transformations are emphasized throughout this chapter. Experimentation and hypothesis construction, structural hierarchy and compartmentation, structure–function relationships, catalysis, specificity, and communication stand out among the other major biochemical themes encountered. After Chapter 12 explores diabetes and some additional topics in carbohydrate catabolism, Chapter 13 provides an introduction to the catabolism of fats, proteins, and amino acids.

PROBLEM 11.1 What specific monosaccharides are produced during the complete hydrolysis of starch? Sucrose (table sugar)? Lactose (milk sugar)?

PROBLEM 11.2 Write the structural formulas for D-glucose and D-fructose. Which one of the six major international classes of enzymes (Table 5.1) is required to directly interconvert these two monosaccharides? Hint: Review Chapter 6.

11.1 GLYCOLYSIS—GLUCOSE OXIDATION TO TWO PYRUVATES

The complete oxidation of glucose begins with **glycolysis,** the sequence of coupled reactions that converts a glucose molecule containing six carbons into two pyruvate molecules each containing three carbons. The term *glycolysis* comes from the Greek *glykys,* meaning *sweet,* and *lysis,* meaning *to split.* During glycolysis, a sweet six-carbon sugar is split into two three-carbon fragments. **[Try Problem 11.3]**

Glycolysis is similar to a river that splits and rejoins; the aldolase-catalyzed reaction leads to two separate products that both enter the mainstream of this metabolic pathway.

An Overview

Glycolysis, also known as the **Embden–Meyerhof pathway,** is coupled to a net production of two ATPs and two NADHs (Figure 11.1). The net reaction is an oxidation–reduction (redox) reaction in which NAD^+ is the oxidizing agent and glucose, the reducing agent, is partially oxidized. The electrons acquired by each NAD^+ (carried by the hydrogen in NADH) can be passed into mitochondria and then to O_2 through the respiratory chain, a sequence of redox reactions coupled to the production of additional ATPs. In both glycolysis and the respiratory chain, redox reactions are the ultimate source of the energy for ATP production.

Figure 11.2 provides an overview of the individual steps in glycolysis. All of the sugar derivatives involved are D isomers. Although each step is catalyzed by a separate cytosolic enzyme, there is growing evidence that some of these enzymes are part of a highly organized multienzyme complex that exhibits metabolic channeling (Section 5.15). During metabolic channeling, reactants are passed directly from enzyme to enzyme and are, therefore, unable to equilibrate with free pools of the same metabolites.

The Conversion of Glucose to Two Glyceraldehyde 3-Phosphates

Glycolysis begins with the **hexokinase**-catalyzed phosphorylation of glucose to yield **glucose 6-phosphate.**

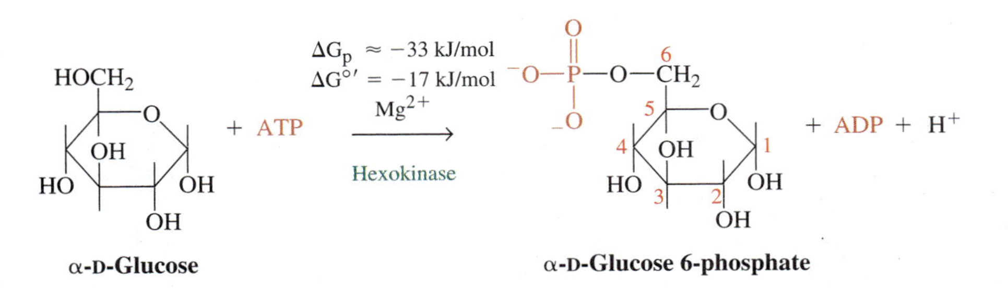

Hexokinase, like other kinases, requires Mg^{2+} and catalyzes a reaction in which ATP serves as a phosphoryl donor. Although ATP is commonly said to be a phosphate donor, it is actually a phosphoryl group $\left(-\overset{O}{\overset{\|}{P}}(O^-)-O^-\right)$ that is donated. The

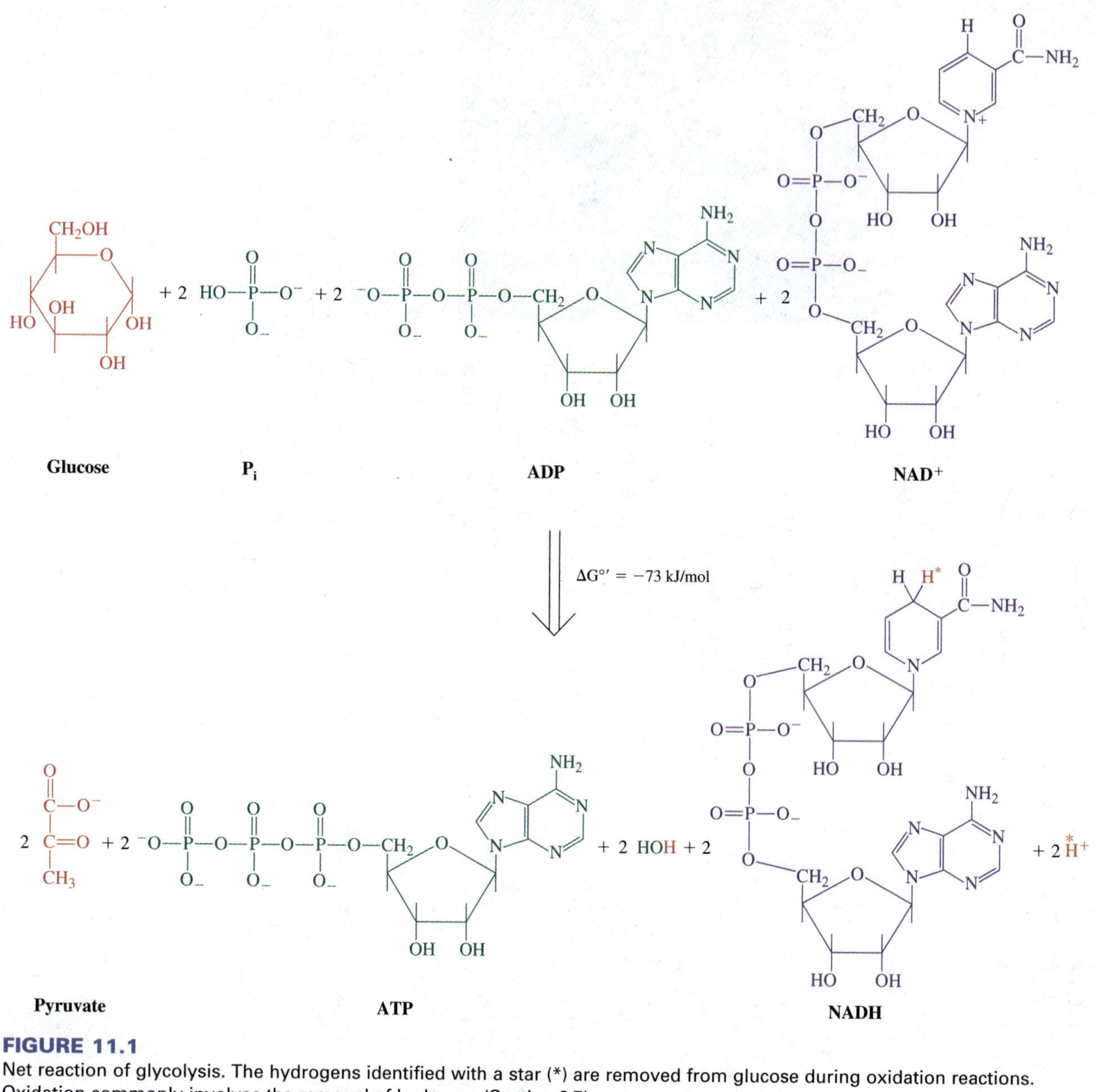

FIGURE 11.1
Net reaction of glycolysis. The hydrogens identified with a star (*) are removed from glucose during oxidation reactions. Oxidation commonly involves the removal of hydrogen (Section 8.7).

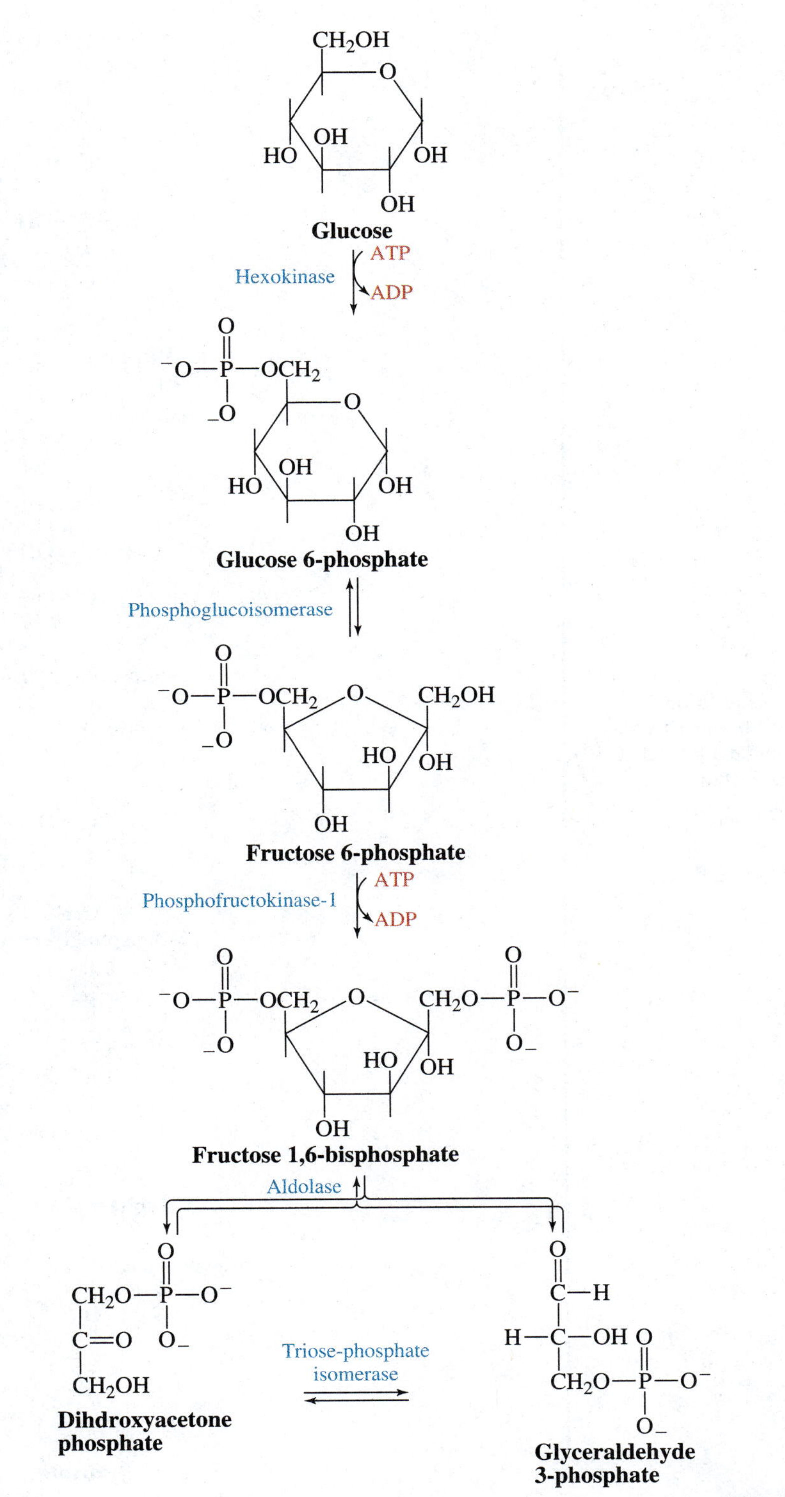

FIGURE 11.2 Glycolysis.

(Continued)

FIGURE 11.2
(Continued)

Each of these reactions occurs twice because two glyceraldehyde 3-phosphates are produced from one glucose.

Glyceraldehyde-3-phosphate dehydrogenase

$NAD^+ + P_i$

$NADH + H^+$

O O
‖ ‖
C—O—P—O⁻
| |
H—C—OH O₋
| O
| ‖
CH₂O—P—O⁻
|
O₋

1,3-Bisphosphoglycerate

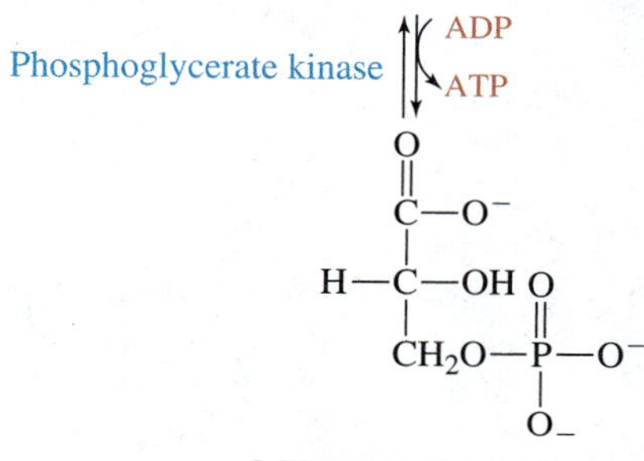

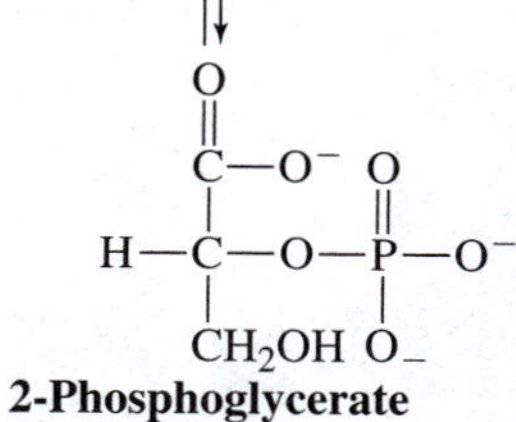

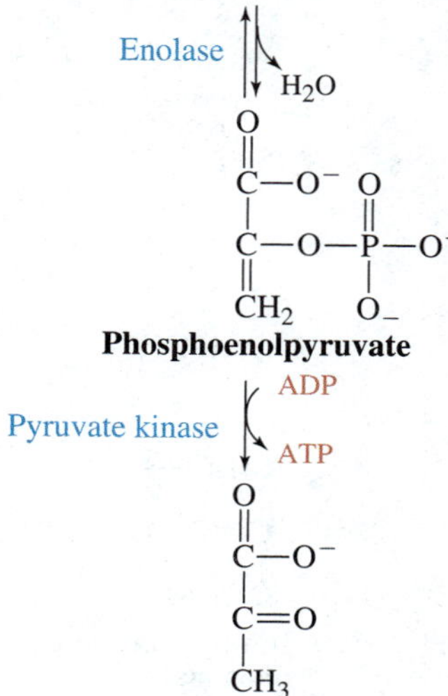

phosphoryl group ends up in a phosphate group $\left(\text{—O—}\overset{\text{O}}{\overset{\|}{\text{P}}}\text{—O}^-\right)$ with O^- below P, within the acceptor. Since the cleavage of an anhydride bond in ATP is coupled to the phosphorylation of glucose, the overall reaction has a $\Delta G^{\circ\prime} = -17$ kJ/mol and is virtually irreversible (Section 8.2). The estimated ΔG under physiological conditions (abbreviated ΔG_p) is −33 kJ/mol. The phosphorylation "activates" glucose for further metabolism and traps it inside the cell. **[Try Problems 11.4 and 11.5]**

Since thermodynamics is such an important component of the study of metabolism, the review of thermodynamics initiated in Section 10.11 is extended at this point. The $\Delta G^{\circ\prime}$ for a reaction is the free energy of products minus the free energy of reactants under standard biological conditions. The $\Delta G^{\circ\prime}$ of a reaction determines its equilibrium constant: $\Delta G^{\circ\prime} = -RT \ln K_{eq}'$. The more negative the $\Delta G^{\circ\prime}$ for a reaction, the larger its K_{eq}' and the more irreversible the reaction under standard biological conditions (Table 8.3).

$\Delta G^{\circ\prime}$ sometimes differs markedly from ΔG_p, because ΔG_p is a measure of the driving force for a reaction under physiological conditions. The further the value of ΔG_p is from zero, the further the reaction is from equilibrium and the greater the driving force for a net reaction. A negative ΔG_p means that under physiological conditions a net conversion of reactants to products is required to reach equilibrium and that there is a driving force for this reaction. A positive ΔG_p indicates that under physiological conditions a driving force exists for the net conversion of products to reactants. Since ΔG_p values change as reactant and product concentrations change, the ΔG_p value for a reaction fluctuates over time within each cell, and it may differ significantly in two different cells. Although biochemists would prefer to work with ΔG_p values, these values are difficult to determine; good estimates of ΔG_p are available for a limited number of reactions. In most cases, biochemists must rely on values for $\Delta G^{\circ\prime}$. You may wish to review Section 8.2 at this time.

The glucose 6-phosphate produced with the catalytic assistance of hexokinase is converted to **fructose 6-phosphate** during the **phosphoglucoisomerase**-catalyzed reaction.

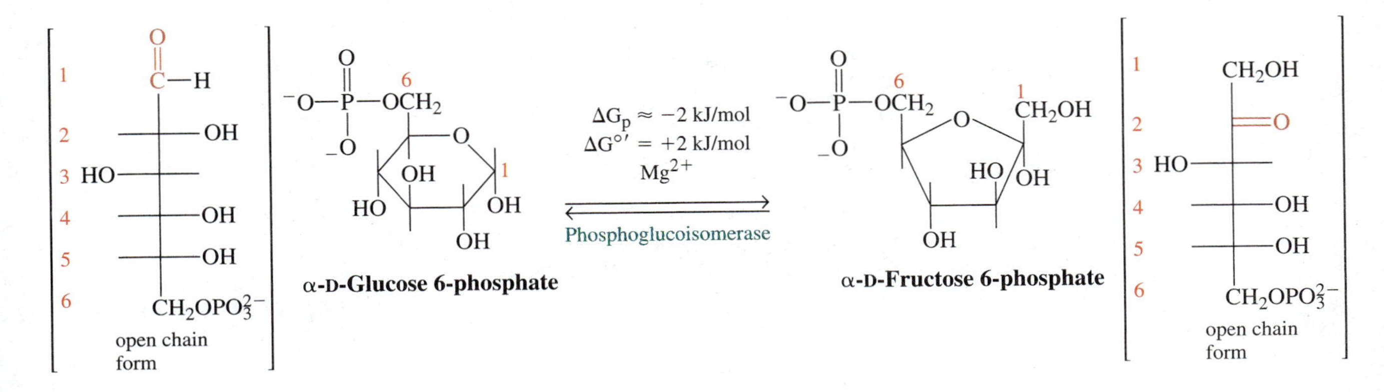

This isomerization reaction entails the shift of a carbonyl group from carbon number 1 (C1) in glucose 6-phosphate to C2. The reaction is readily reversible since no significant amount of free energy is consumed or released in the process. **[Try Problem 11.6]**

Fructose 6-phosphate is a substrate for **phosphofructokinase-1,** an enzyme that catalyzes the formation of **fructose 1,6-bisphosphate.** The *kinase* in the enzyme name once again indicates that ATP is a substrate and that the catalyzed reaction is a phosphorylation reaction.

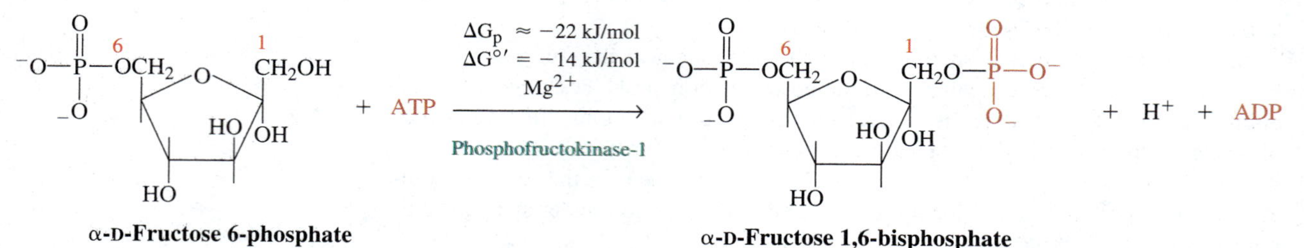

With a $\Delta G_p \approx -22$ kJ/mol, this is the second reaction in glycolysis that is considered irreversible. The two ATPs that are consumed through this point in the metabolic pathway are more than replaced as the process continues. **[Try Problem 11.7]**

The production of fructose 1,6-bisphosphate sets the stage for the splitting of the six-carbon chain of glucose into two three-carbon fragments. This splitting is catalyzed by the enzyme **aldolase,** which acquired its name from the reverse reaction, an example of an aldol condensation. **[Try Problem 11.8]**

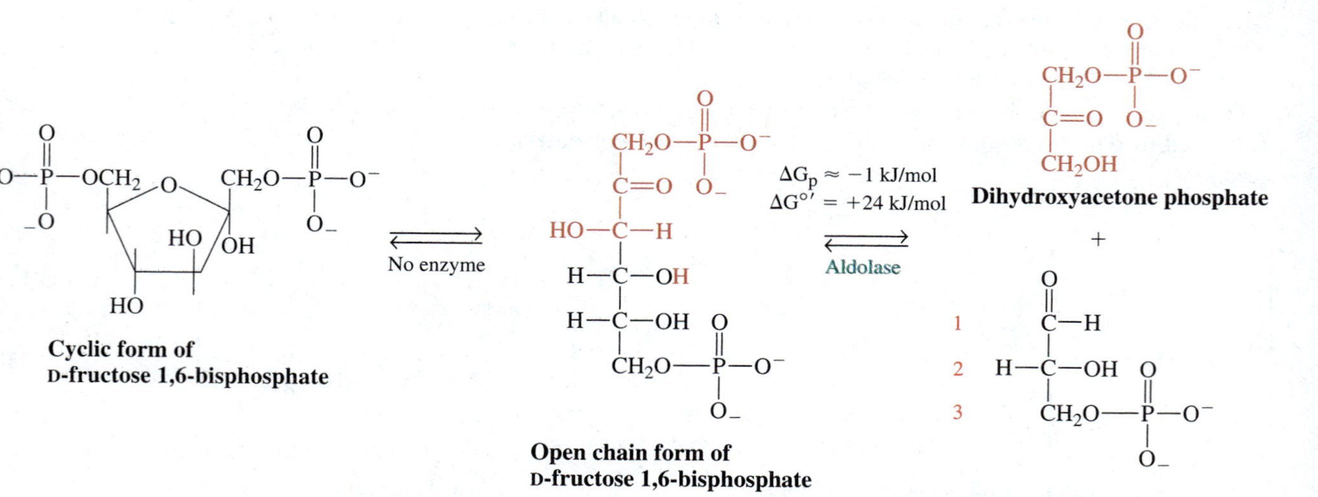

Although the aldolase-catalyzed reaction has a $\Delta G^{\circ\prime} = +24$ kJ/mol, ΔG_p is close to -1, and the reaction is readily reversible under physiological conditions. The **dihydroxyacetone phosphate** produced is converted to **glyceraldehyde 3-phosphate** through a reversible reaction catalyzed by **triose–phosphate isomerase.**

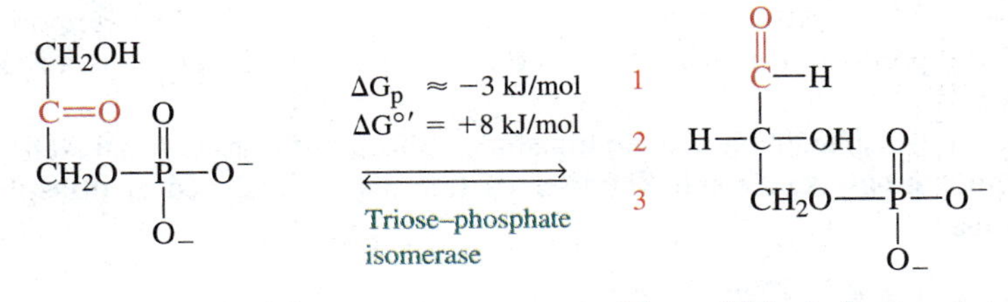

To this point in glycolysis, two molecules of glyceraldehyde 3-phosphate have been generated from the single molecule of glucose that originally entered this metabolic pathway. One glyceraldehyde 3-phosphate is an immediate product of the aldolase-catalyzed reaction and the second is the product of the triose–phosphate isomerase–catalyzed reaction. Since each of the glyceraldehyde 3-phosphates proceeds separately through the remaining payoff steps in glycolysis, these terminal steps occur twice during the total oxidation of one molecule of glucose to two pyruvates. The terminal steps in glycolysis are called payoff steps, because they account for the ATP and NADH produced by this metabolic pathway. **[Try Problem 11.9]**

The Payoff Steps in Glycolysis

Glyceraldehyde-3-phosphate dehydrogenase catalyzes the oxidation and phosphorylation of glyceraldehyde 3-phosphate during the next step toward two pyruvates.

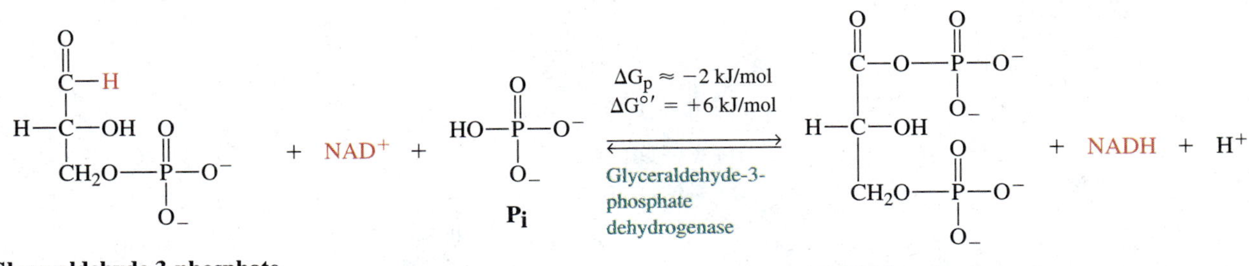

Dehydrogenases, in general, are oxidoreductases that catalyze the removal of hydrogen atoms from the substrates being oxidized. The oxidation and phosphorylation catalyzed by glyceraldehyde-3-phosphate dehydrogenase are said to be coupled because the energy released during the oxidation of the aldehyde to the carboxylic acid is used to drive the phosphorylation. NAD^+ is the oxidizing agent and $\Delta G_p \approx -2$ kJ/mol. The **1,3-bisphosphoglycerate** formed is a mixed (carboxylate–phosphate) anhydride and a high-energy compound (Section 8.5). The NADH generated is a potential source of approximately 2.5 ATP (based on 1991 estimates; see selected readings) when glycolysis is coupled to the respiratory chain. This is an important point, since ATP production is what the biological oxidation of glucose is all about. **[Try Problem 11.10]**

Each 1,3-bisphosphoglycerate transfers a phosphoryl group to an ADP to generate **3-phosphoglycerate** and ATP in a reaction that is catalyzed by **phosphoglycerate kinase.**

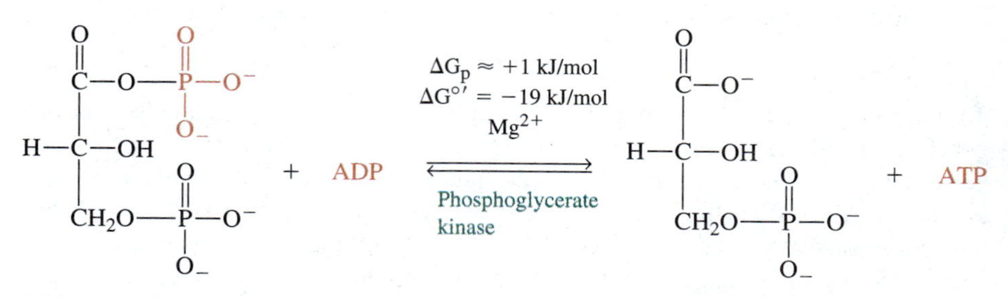

The transfer of a phosphoryl group from a substrate to ADP to generate ATP is known as **substrate-level phosphorylation.** In this instance, the cleavage of one anhydride bond provides the thermodynamic driving force for the formation of a second. Although the $\Delta G^{\circ\prime} = -19$ kJ/mol, the $\Delta G_p \approx +1$ kJ/mol. The two ATPs produced at this point in glycolysis (one from each of the two 1,3-bisphosphoglycerates generated from a single glucose) replace the ATPs used in the hexokinase- and phosphofructokinase-1–catalyzed reactions.

Each 3-phosphoglycerate is converted to **2-phosphoglycerate** with the catalytic assistance of **phosphoglycerate mutase,** another enzyme found in the cytosol.

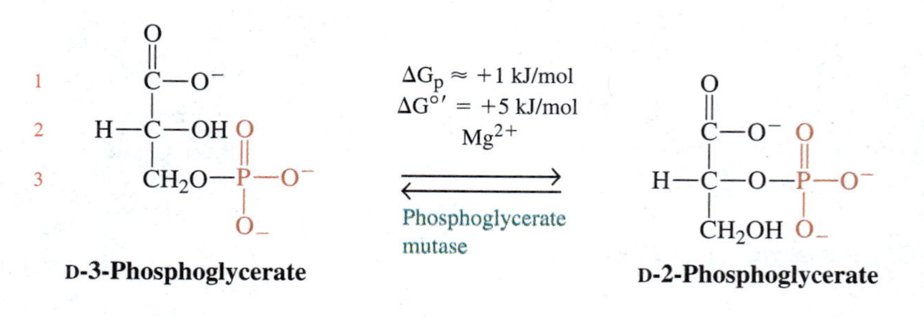

A **mutase,** in general, is an enzyme that catalyzes the transfer of a single functional group from one site to another on the same molecule. Mutases are a subclass of isomerases. The net result of the reversible phosphoglycerate mutase–catalyzed reaction is the shift of a phosphoryl group from C3 in glycerate to C2.

Glycolysis continues with a dehydration reaction catalyzed by **enolase.** One of the products is water. The other product, **phosphoenolpyruvate (PEP),** is a high-energy compound that is the phosphorylated **enol** form of pyruvate (Section 8.5), hence the name *enolase.*

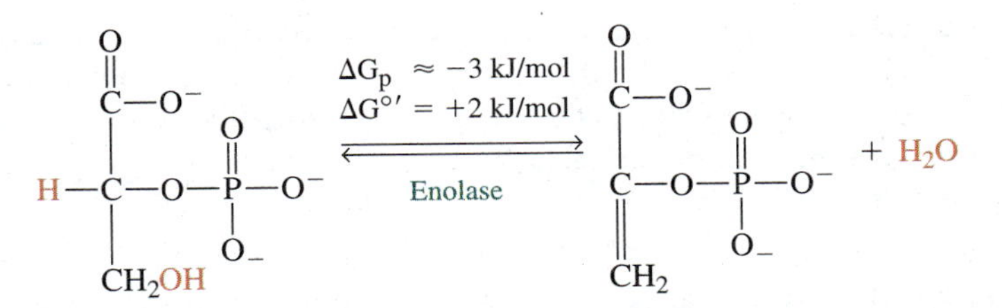

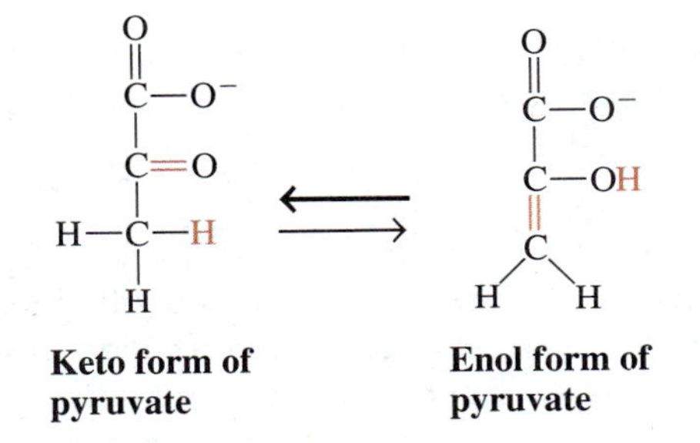

Phosphoenolpyruvate is an enoyl phosphate (Tables 8.6 and 8.7).

Pyruvate kinase catalyzes the transfer of a phosphoryl group from phosphoenolpyruvate to ADP during the final step in glycolysis, a second example of substrate-level phosphorylation.

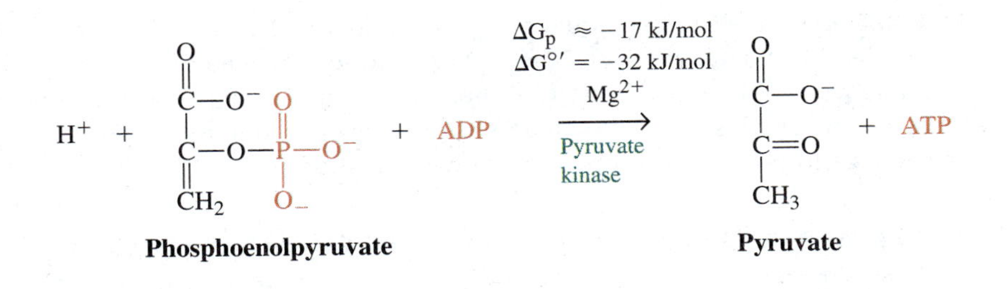

Although pyruvate is initially generated in its enol form, most of it quickly shifts to its more stable keto form. Once again, the energy released during the cleavage of one high-energy compound provides the thermodynamic driving force for the production of a second high-energy compound. An estimated ΔG_p of −17 kJ/mol makes the pyruvate kinase–catalyzed reaction the third step in glycolysis that is considered irreversible under physiological conditions. Since each molecule of glucose yields two phosphoenolpyruvates, the pyruvate kinase–catalyzed reaction accounts for two of the four ATPs produced directly during glycolysis. **[Try Problem 11.11]**

When a glucose is destined to be completely oxidized, the two pyruvates from glycolysis are further oxidized with the catalytic assistance of the pyruvate dehydrogenase complex, the citric acid cycle, and the respiratory chain. Complete oxidation depends upon an adequate supply of molecular oxygen, the terminal oxidizing agent in the respiratory chain (Section 11.8).

PROBLEM 11.3 Describe the relationship between *pyruvate* (an end product of glycolysis) and *pyruvic acid,* two terms that tend to be used interchangeably by biochemists. Which exists in largest amounts at physiological pH?

PROBLEM 11.4 What type of bond joins the phosphoryl group to glucose in glucose 6-phosphate? Is glucose 6-phosphate uncharged, positively charged, or negatively charged at pH 7.0? Explain.

PROBLEM 11.5 List the six major international classes of enzymes. To which one of these six classes does hexokinase belong? Explain. Hint: Review Table 5.1.

PROBLEM 11.6 If more fructose 6-phosphate is added to a phosphoglucoisomerase-catalyzed reaction at equilibrium, will the concentration of glucose 6-phosphate in the reaction mixture increase, decrease, remain the same, or change in an unpredictable manner? Will the equilibrium constant for this isomerization reaction increase, decrease, remain the same, or change in an upredictable manner? Explain.

PROBLEM 11.7 Is the K_{eq}' for the phosphofructokinase-1–catalyzed reaction >100, <100, or 100? Explain. Hint: Review Table 8.3.

***PROBLEM 11.8** Write an equation for the aldol condensation of one molecule of acetaldehyde (CH_3COH) with a second molecule of acetaldehyde. Write the equation for the reversal of the glycolytic reaction catalyzed by aldolase. In both equations, provide structural formulas for all reactants and products. Compare these two equations.

PROBLEM 11.9 Write structural formulas for glucose 6-phosphate and fructose 6-phosphate in their open chain forms. Describe the similarities between the phosphoglucoisomerase-catalyzed reaction and the triose–phosphate isomerase–catalyzed reaction.

PROBLEM 11.10 Use an arrow to identify the anhydride link in 1,3-bisphosphoglycerate.

PROBLEM 11.11 How many separate steps are involved in glycolysis? How many separate enzymes are involved? What is the minimum number of genes that must be expressed in order for a cell to carry out this metabolic pathway? The actual number of genes required is larger than this minimum number. Explain.

Although anaerobic glycolysis allows human cells to produce ATP for a brief period in the absence of oxygen, oxygen is required for sustained ATP production.

11.2 ANAEROBIC GLYCOLYSIS—CONVERSION OF GLUCOSE TO TWO LACTATES

Anaerobic glycolysis refers to glycolysis as it proceeds in the absence of O_2. Aerobic (with O_2) and anaerobic (without O_2) glycolysis differ only in the manner in which the NAD^+ reduced during glycolysis is regenerated to keep the pathway operational. In the presence of O_2, NAD^+ is regenerated as NADH passes two electrons to O_2 through the respiratory chain (Section 11.8). In the absence of O_2, the NAD^+ is regenerated as NADH reduces pyruvate (a ketone) to lactate (a secondary alcohol) in a reaction catalyzed by lactate dehydrogenase (LDH) (Figure 11.3). Unfortunately, the accumulation of lactic acid leads to a drop in pH that inhibits glycolytic enzymes at a time when the ATP from glycolysis is most needed. Phosphofructokinase-1, the key regulatory enzyme in glycolysis (Section 11.3), is particularly sensitive to inhibition by H^+. Although the lactate dehydrogenase–catalyzed reaction may allow a cell to survive for a short period in the absence of O_2, cells run low on energy and die if deprived of O_2 for more than a few minutes. **[Try Problem 11.12]**

The lactate that builds up in muscle cells during anaerobic glycolysis leads to pain and fatigue. Since training can enhance the ability of the heart and blood vessels to deliver O_2 to muscle, trained athletes accumulate less lactate at any given level of exercise. The lactate that accumulates in both trained and untrained individuals is converted back to pyruvate as normal O_2 levels are restored. The liver and the Cori cycle (Section 14.4) play a central role in this process. This pyruvate can be used for anabolism or oxidized further to provide additional energy for the production of ATP. **[Try Problem 11.13.]**

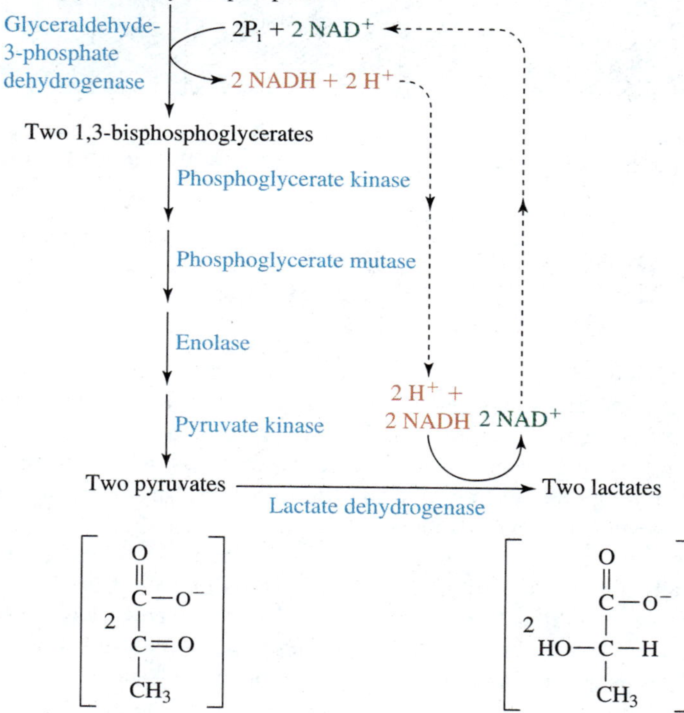

FIGURE 11.3
NAD$^+$ regeneration during anaerobic glycolysis.

PROBLEM 11.12 Lactate dehydrogenase is present in muscle cells at all times, yet little pyruvate is converted to lactate in the presence of O_2. What triggers the production of lactate as O_2 concentrations drop?

PROBLEM 11.13 Lactate dehydrogenase catalyzes the conversion of lactate back to pyruvate as a cell shifts from anaerobic to aerobic conditions. Will the rate of glycolysis increase, decrease, or remain the same during this process? Explain.

11.3 CONTROL OF GLYCOLYSIS

Glycolysis serves two major functions within a cell: (1) it generates intermediates for anabolism; and (2) it leads to a net production of ATP. ATP is often described as the energy currency of a cell, because it is the immediate source of energy for most energy-requiring reactions. The rate of glycolysis tends to rise and fall as the need for anabolic intermediates or energy increases and decreases, respectively. The ebb and flow of the entire metabolic pathway is controlled mainly by three key allosteric (regulatory) enzymes: phosphofructokinase-1, hexokinase, and pyruvate kinase. Allosteric enzymes possess distinct binding sites where regulatory ligands, called effectors or modulators, dock to modulate catalytic activity (Section 5.12).

Regulation of Phosphofructokinase-1

Phosphofructokinase-1 normally serves as the main floodgate for controlling the flow of glucose through glycolysis. It is activated by AMP, ADP, and fructose 2,6-bisphosphate, while it is inhibited by ATP and citrate. Although closely related to fructose 1,6-bisphosphate (an intermediate in glycolysis), fructose 2,6-bisphosphate is not a reactant in any metabolic pathway; it functions solely as a regulatory agent.

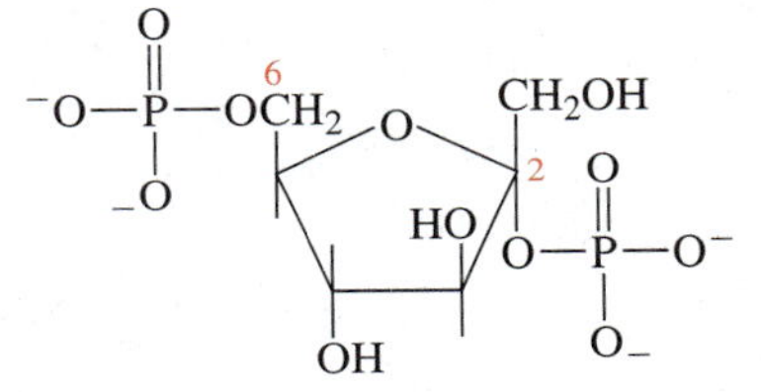

α-D-Fructose 2,6-bisphosphate

What is the metabolic logic behind the phosphofructokinase-1 effectors? Citrate can be produced from pyruvate, the end product of aerobic glycolysis. A high concentration of citrate signals an abundant concentration of intermediates for anabolism. Under aerobic conditions, it also signals a high energy charge since high citrate concentrations are linked to high ATP concentrations (Section 11.6). When a cell has an adequate energy charge and an adequate concentration of anabolic intermediates, it normally benefits by slowing down glycolysis to conserve glucose. The feedback inhibition of phosphofructokinase-1 by citrate accomplishes this.

The **energy charge** of a cell is the extent to which the ATP–ADP–AMP system is filled with ATP, and secondarily, ADP (Figure 11.4). ATP is the direct source of energy for primary active transport, muscle action, most anabolism, and a wide variety of additional processes (Section 8.4). Most ATP is produced by adding

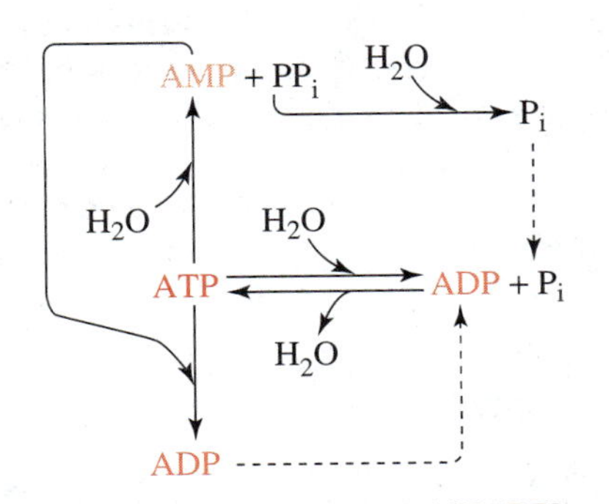

$$\text{Energy charge} = \frac{[\text{ATP}] + 1/2[\text{ADP}]}{[\text{ATP}] + [\text{ADP}] + [\text{AMP}]}$$

Energy charge = 1 when [ADP] = [AMP] = 0

Energy charge = 0 when [ATP] = [ADP] = 0

FIGURE 11.4 The ATP–ADP–AMP system and energy charge.

phosphoryl groups to ADP or AMP, and most AMP and ADP come from the dephosphorylation of ATP. Consequently, when ATP levels are high, AMP and ADP levels are low and vice versa. Low ratios of ATP to AMP signal a low energy charge, whereas high ratios denote a high energy charge. A high NADH to NAD^+ ratio is usually associated with a high ATP to AMP ratio, so high NADH concentrations also signal a high energy charge.

The observed modulation of phosphofructokinase-1 by AMP, ADP, and ATP, like the modulation by citrate, makes sense from an energy charge standpoint. The furnace (fueled by glucose and driven, in part, by glycolysis) should be turned up when the immediately accessible energy supply (ATP) within a cell is depleted, and it should be turned down when an adequate supply of ATP already exists. Hence, AMP and ADP are positive effectors and ATP is inhibitory.

Fructose 2,6-bisphosphate has a particularly strong affinity for phosphofructokinase-1, and it is frequently the dominant effector of this regulatory enzyme. The production and degradation of fructose 2,6-bisphosphate are controlled mainly by water-soluble hormones that bind to specific receptors that are integral proteins within the plasma membrane of target cells. Hormone binding triggers the production of second messengers that initiate reaction cascades (Section 10.3). These cascades regulate specific enzymes, including **phosphofructokinase-2,** the enzyme that catalyzes the transfer of a phosphoryl group from ATP to the 2-hydroxyl group (the basis for the *2* in the name *phosphofructokinase-2*) on fructose 6-phosphate. Thus, the concentration of fructose 2,6-bisphosphate fluctuates in response to signals from outside the cell. *Hormones commonly call upon individual cells to change the rates of their metabolic pathways in order to meet the needs of the body as a whole.* Chapter 12 reinforces this important general principle and examines in more detail the metabolic logic behind fructose 2,6-bisphosphate's activation of glycolysis. **[Try Problem 11.14]**

Since phosphofructokinase-1 is an allosteric enzyme, a sigmoidal curve is obtained when initial velocity (y axis) is plotted against fructose 6-phosphate concentration (x axis) (Section 5.12). Each effector shifts this curve in a characteristic and reversible manner (Figure 11.5). Positive effectors, such as ADP, increase the rate of the reaction at nonsaturating fructose 6-phosphate concentrations, whereas negative effectors, such as ATP, lead to a drop in initial velocity under the same conditions. Certain combinations of effector concentrations lead to the same catalytic activity that is observed in the absence of all regulators. When this is the case, an increase in the concentration of an inhibitor or a decrease in the concentration of an activator will lead to net inhibition; an increase in activator concentration or a decrease in inhibitor concentration will lead to a net enhancement of catalytic activity. **[Try Problem 11.15]**

Regulation of Hexokinase

Hexokinase, the first enzyme encountered in glycolysis, is allosterically inhibited by glucose 6-phosphate, a product of the reaction that it catalyzes. During glycolysis, glucose 6-phosphate accumulates only after fructose 6-phosphate builds up and leads to the reversal of the phosphoglucokinase-catalyzed reaction (see Problem 11.6). Fructose 6-phosphate only builds up when phosphofructokinase-1 is relatively inactive (the main floodgate for glycolysis is nearly closed) and glucose 6-phosphate

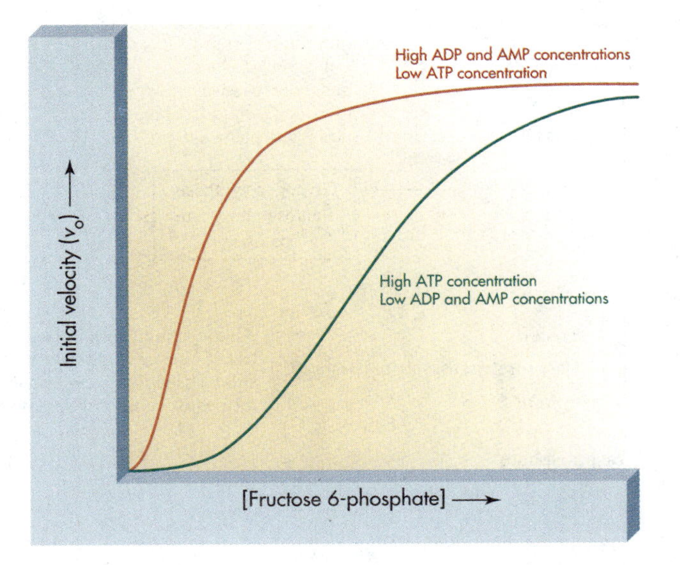

FIGURE 11.5
Impact of ADP concentration and ATP concentration on phosphofructokinase-1 activity. As [ATP] rises and [ADP] drops, the curve shifts right and it takes a higher fructose 6-phosphate concentration to attain any specific initial velocity. An increase in [ADP], along with an associated decrease in [ATP], shifts the sigmoidal curve to the left and reduces the fructose 6-phosphate concentration required to attain any particular initial velocity.

and fructose 6-phosphate are not being used in other metabolic pathways. Under these conditions, it makes no sense to rapidly convert glucose to glucose 6-phosphate. Feedback inhibition of hexokinase by glucose 6-phosphate ensures that this does not happen.

One might wonder why hexokinase is not the key regulatory enzyme in glycolysis, because it catalyzes the first step in this metabolic pathway. If glycolysis is not needed, wouldn't it make sense to shut it down before it even gets started? Why wait until the third step (catalyzed by phosphofructokinase-1) to install the main floodgate? The answers: Glucose 6-phosphate is involved in glycogen synthesis (Section 14.5) and the pentose phosphate pathway (Section 12.2), as well as glycolysis (Figure 11.6). When ATP and anabolic intermediates are in abundant supply, a cell benefits by shutting down glycolysis, increasing the rate of glycogen synthesis (to store glucose for future use), and increasing the rate of the pentose phosphate pathway (to generate reducing power for anabolism). Glycogen synthesis and the pentose phosphate pathway could not occur in the absence of glycolysis if glycolysis were completely shut down at step 1 (the hexokinase reaction), but they can continue in the absence of glycolysis when glycolysis is shut off at step 3 (the phosphofructokinase-1 reaction).

In liver, there is a separate enzyme, **glucokinase,** that converts glucose to glucose 6-phosphate when glucose is present in high concentration. Under these conditions, the body benefits from the conversion of this glucose to glucose 6-phosphate and then to glycogen (Section 10.12). Glucokinase is not inhibited by glucose 6-phosphate. Since it has a relatively large K_m, glucokinase is only of significance when glucose concentrations are above normal. **[Try Problem 11.16]**

Regulation of Pyruvate Kinase

Pyruvate kinase, the last enzyme in glycolysis, is regulated by several effectors, including ATP, alanine, and fructose 1,6-bisphosphate. ATP inhibits its own production

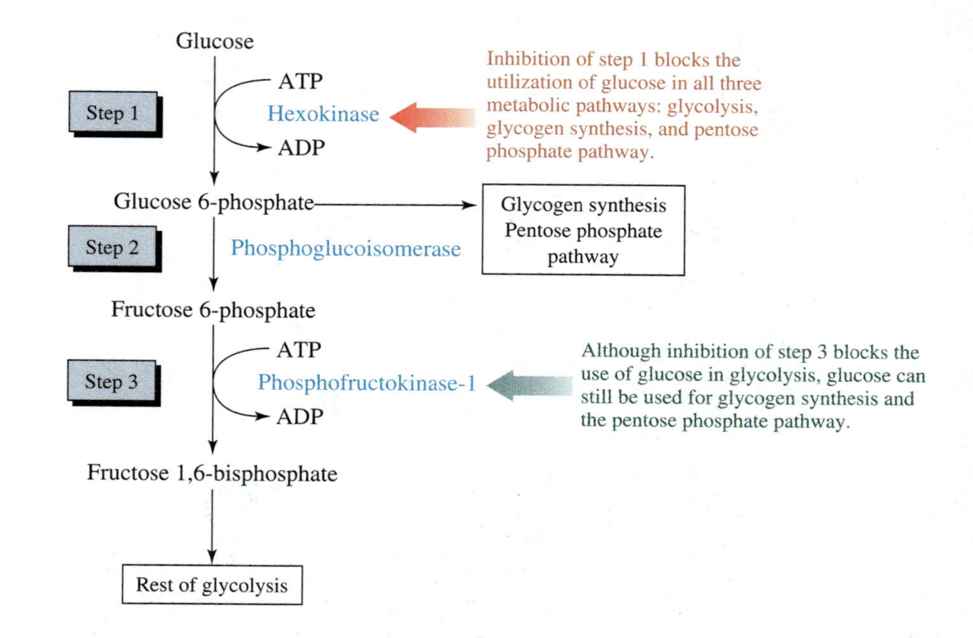

FIGURE 11.6
The reason glycolysis is primarily regulated at step 3 rather than at step 1.

as well as the production of pyruvate and the continued oxidation of glucose. When ATP is already present in a cell in adequate amounts, why produce more? The same metabolic logic applies to the inhibition of phosphofructokinase-1 by ATP.

The metabolic logic behind alanine's modulation of glycolysis is the same as that for citrate. Alanine, a second negative effector of pyruvate kinase and a protein amino acid, is synthesized from pyruvate. Since it is a widely-employed anabolic intermediate, high alanine concentrations signal an abundant supply of material for synthetic processes. By inhibiting pyruvate kinase, high alanine concentrations prevent a cell from sacrificing glucose to make more anabolic intermediates when such intermediates are already present in adequate amounts.

Fructose 1,6-bisphosphate, a product of the phosphofructokinase-1–catalyzed reaction, feeds forward to activate pyruvate kinase, because a high concentration of fructose 1,6-bisphosphate indicates that the main floodgate for glycolysis is wide open and that a high overall rate of glycolysis is desirable. To control the flux through an entire metabolic pathway, a regulatory enzyme must catalyze the rate-limiting step in that pathway. The feed-forward activation by fructose 1,6-bisphosphate ensures that the pyruvate kinase reaction does not become rate-limiting. In general, there is some metabolic logic behind the action of each effector for every regulatory enzyme, a consequence of so many years of evolution.

Regulation Summary

The regulation of glycolysis is summarized in Figure 11.7. The three regulatory enzymes involved, like most regulatory enzymes, catalyze reactions that are virtually irreversible. Since these enzymes exist in multiple forms and different cells sometimes contain unique isozymes, the regulatory details sometimes differ from one

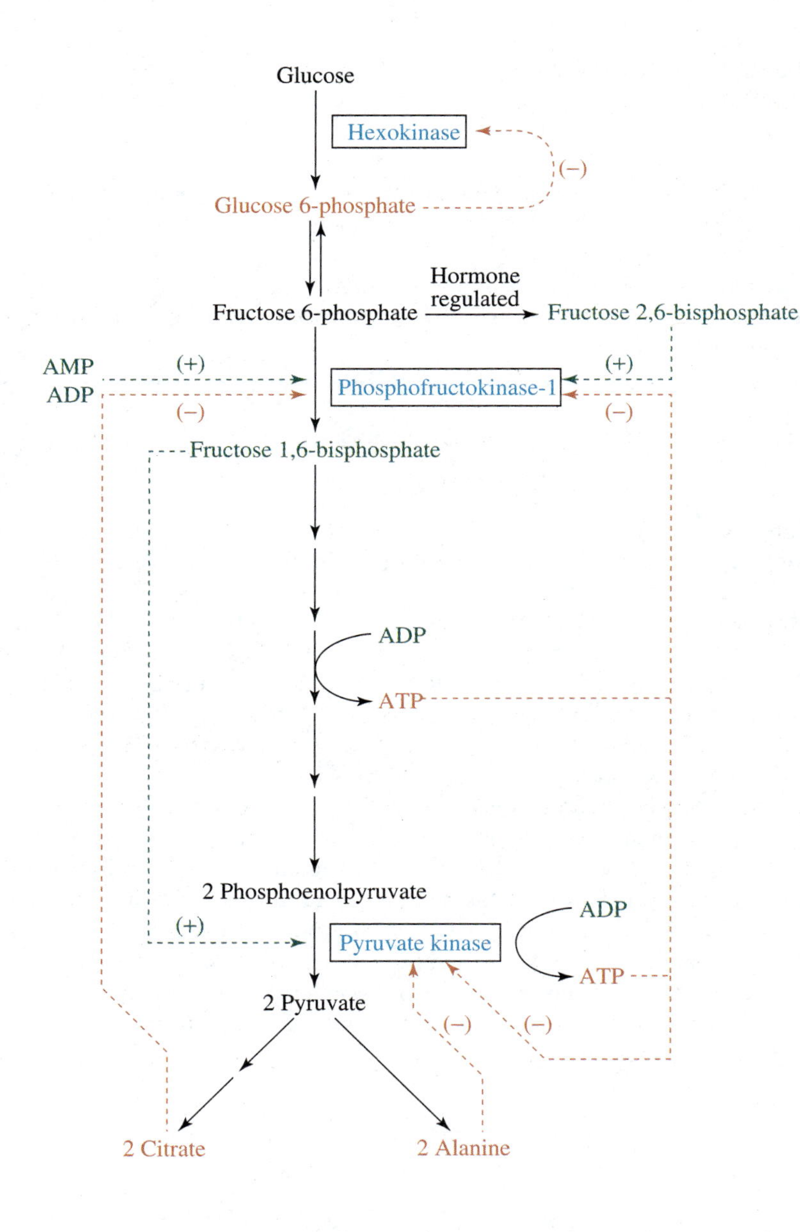

FIGURE 11.7
Regulation of glycolysis.

type of cell to another. The pyruvate kinase isozyme present in some cells, for example, is regulated by covalent modification (phosphorylation in this case) as well as allosteric effectors. Isozymes are distinctive enzymes that catalyze the same reaction. Each normally has a unique K_m, V_{max}, and sensitivity to modulation (Section 5.11). **[Try Problem 11.17]**

PROBLEM 11.14 If NAD^+ were an allosteric effector of phosphofructokinase-1, would you expect it to be a positive or a negative effector? Explain. Note: A high $[NAD^+]$ signals a low [ATP].

PROBLEM 11.15 True or false? All of the positive effectors of phosphofructokinase-1 bind to the same allosteric site on this enzyme. Explain.

PROBLEM 11.16 Explain why glucokinase has little impact on glucose 6-phosphate production at low glucose concentrations.

PROBLEM 11.17 If a large amount of pyruvate suddenly enters the cytosol of a cell, will the rate of glycolysis tend to increase, decrease, or remain the same? Explain.

11.4 STRUCTURE AND FUNCTION OF MITOCHONDRIA

After glucose has been oxidized to two pyruvates through glycolysis, the complete oxidation of glucose in eukaryotic cells shifts from the cytosol to mitochondria because of the compartmentation of enzymes. Mitochondria are described as the powerhouses of eukaryotic cells, since they produce most of a cell's ATP. These organelles possess all of the enzymes necessary to catalyze the oxidation of pyruvate to CO_2 and H_2O. The enzymes involved are distributed among three multienzyme systems: the pyruvate dehydrogenase complex; the citric acid cycle; and the respiratory chain. The respiratory chain, which is the primary source of ATP, also oxidizes the NADH produced during glycolysis to NAD^+.

Mitochondria are cylindroid organelles that contain an outer membrane separated from an inner membrane by an intermembrane space. The inner membrane repeatedly folds inward to from cristae (Figure 11.8). The fluid-filled volume enclosed by the inner membrane is known as the matrix. The number of mitochondria per cell ranges from zero for mature erythrocytes to over 500 for a typical liver or kidney cell. The pyruvate dehydrogenase complex and the citric acid cycle enzymes, with the exception of the succinate dehydrogenase complex, reside in the matrix. The succinate dehydrogenase complex and the respiratory chain are integral parts of the inner membrane. **[Try Problem 11.18]**

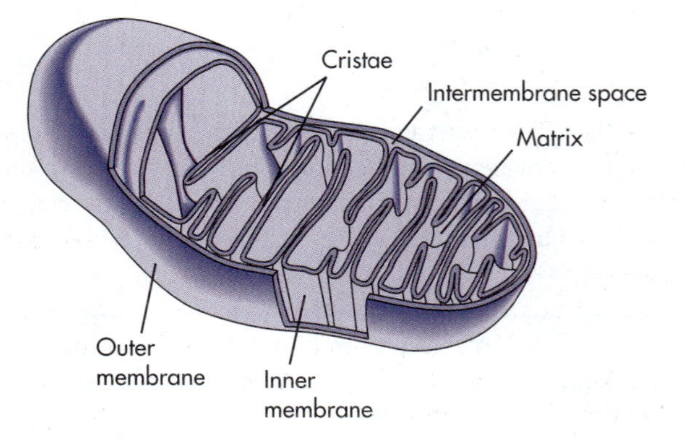

FIGURE 11.8
Diagram of a mitochondrion.

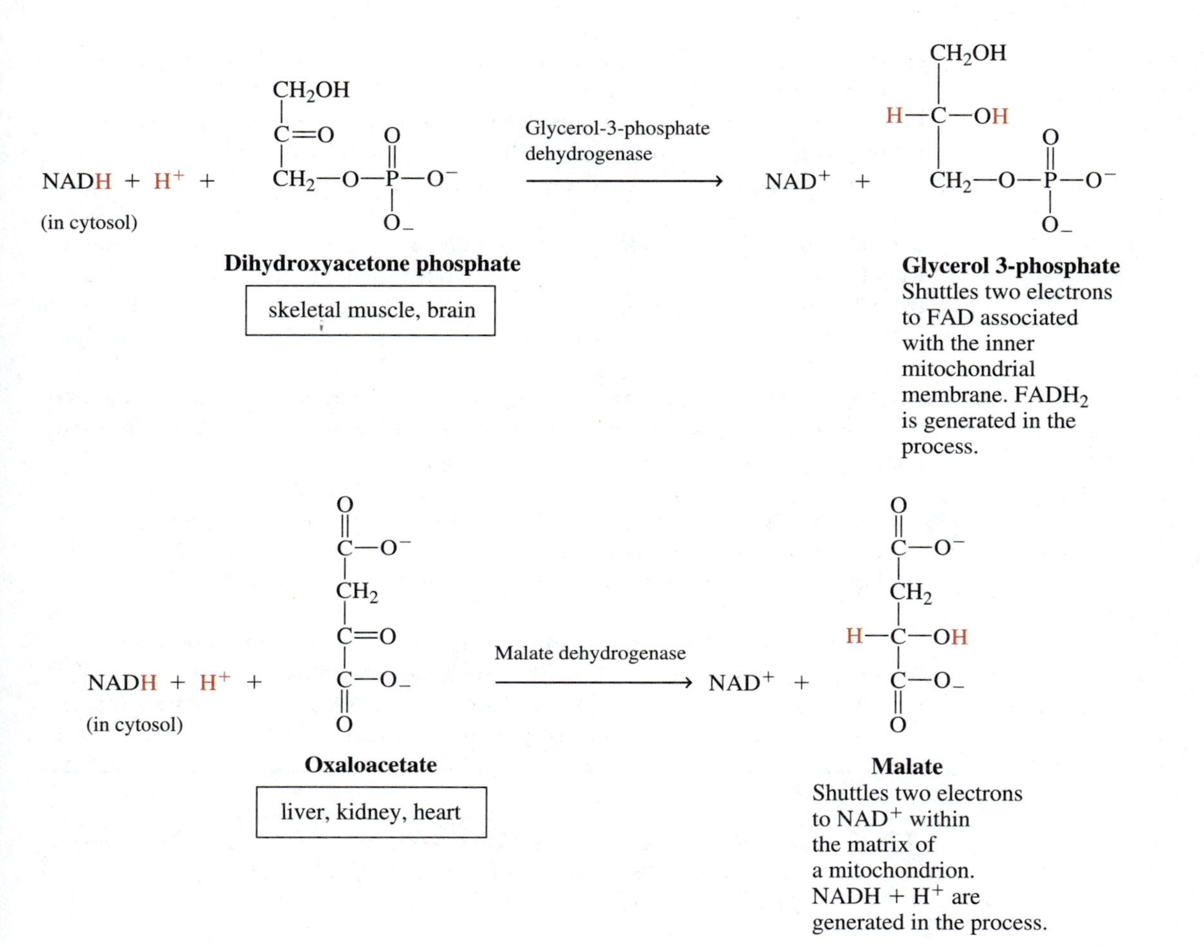

FIGURE 11.9
Compounds that shuttle electrons from NADH in the cytosol into mitochondria.

The NADH produced in the cytosol during glycolysis is oxidized as shuttle systems pass the electrons it acquired from glucose to the respiratory chain within mitochondria. The NADH is unable to deliver its electrons to the respiratory chain directly because it cannot pass through the inner mitochondrial membrane, and there is no transport protein. In skeletal muscle and brain cells, the two electrons carried by cytosolic NADH are most often picked up (along with associated hydrogens) by dihydroxyacetone phosphate, which is converted to **glycerol 3-phosphate** in the process (Figure 11.9). $FADH_2$ is generated as glycerol 3-phosphate shuttles two electrons to a flavin adenine dinucleotide (FAD; Table 10.6, Section 10.11) associated with the inner mitochondrial membrane. Approximately 1.5 ATP can be produced as the two electrons, originally extracted from glucose during glycolysis, flow from $FADH_2$ to O_2 through the respiratory chain. **[Try Problem 11.19]**

In liver, kidney, and heart cells, the two electrons carried by glycolysis-derived NADH in the cytosol tend to be passed to oxaloacetate to form **malate.** The malate enters the matrix of a mitochondrion through a specific membrane transporter and then passes the two electrons it acquired from NADH in the cytosol to NAD^+ within the mitochondrion. NADH is generated in the process. Roughly 2.5 ATP can be formed as these electrons are passed from the NADH in the mitochondrion to O_2 through the respiratory chain. Thus, the maximum number of ATPs that can be generated using the two electrons carried by a NADH in the cytosol is either 1.5 or 2.5, depending on which transport system shuttles the electrons into a mitochondrion. This fact will be revisited when the energy transformations associated with the oxidation of glucose are summarized in Section 11.11. During the previous discussion of glycolysis, it was assumed that 2.5 ATP are produced from each glycolysis-derived NADH.

The next five sections examine the pyruvate dehydrogenase complex, the citric acid cycle, and the respiratory chain. Keep in mind that, in eukaryotes, all of these metabolic processes are mitochondria-specific.

PROBLEM 11.18 Suggest why mature erythrocytes (red blood cells) contain no mitochondria. Hint: Red blood cells have a limited life span and their primary function is to transport O_2 and CO_2 (Section 4.9).

PROBLEM 11.19 From an organic chemistry standpoint, describe the reaction that occurs when NADH transfers its electrons to either dihydroxyacetone phosphate or oxaloacetate. Which one of the six major international classes of enzymes catalyzes these reactions?

11.5 PYRUVATE DEHYDROGENASE COMPLEX—CONVERSION OF PYRUVATE TO ACETYL-SCoA AND CO_2

The **pyruvate dehydrogenase complex** is located in the matrix of mitochondria, where it catalyzes the **oxidative decarboxylation** of pyruvate, an α-keto acid. This irreversible reaction ($\Delta G^{\circ\prime} = -33.4$ kJ/mol; ΔG_p is also large and negative) couples glycolysis to the citric acid cycle. The rapid diffusion of CO_2 out of mitochondria contributes to the irreversibility of this reaction (Le Châtelier's principle). Since two pyruvates are produced from one glucose, the pyruvate dehydrogenase complex–catalyzed reaction accounts for two of the six CO_2 produced during the total oxidation of one glucose molecule within the human body. The acetyl-SCoA produced is a high-energy compound (Section 8.5). The acetyl group in acetyl-SCoA is said to be *activated,* because it is chemically more reactive than a free acetate ion. Since the NADH from the pyruvate dehydrogenase complex is generated within a mitochondrion, its electrons can be delivered directly to the respiratory chain; no shuttle is required. **[Try Problem 11.20]**

Enzymes and Coenzymes

The pyruvate dehydrogenase complex is a multienzyme system that contains three different enzymes [**pyruvate dehydrogenase(E_1), dihydrolipoyl transacetylase(E_2),**

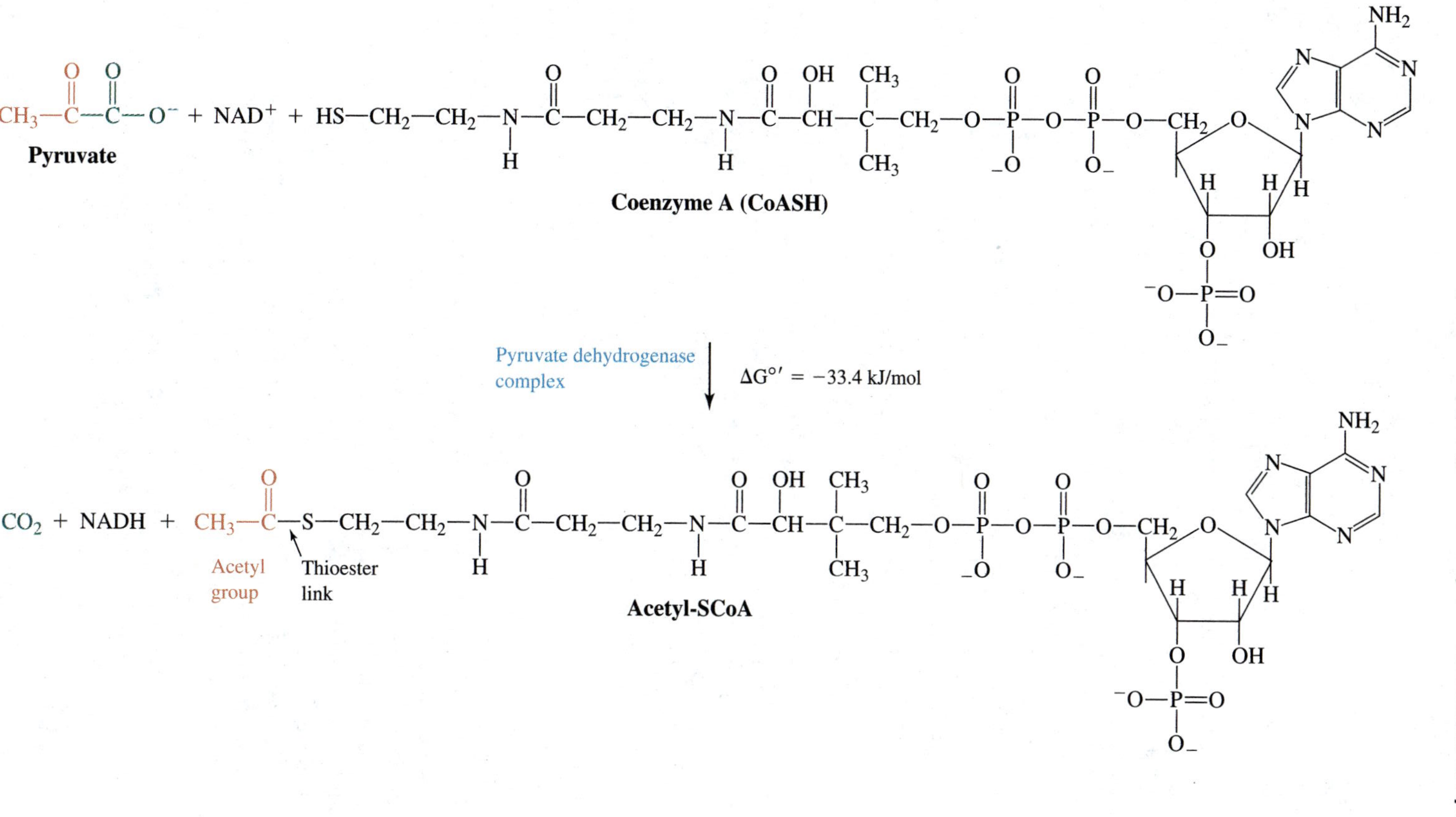
Pyruvate
Coenzyme A (CoASH)
Pyruvate dehydrogenase complex
ΔG°′ = −33.4 kJ/mol
Acetyl group
Thioester link
Acetyl-SCoA

and **dihydrolipoyl dehydrogenase (E_3)**] and five different coenzymes [thiamine pyrophosphate (TPP), lipoic acid, CoASH, FAD, and NAD^+]. In *E. coli,* the complete pyruvate dehydrogenase complex contains 12 copies of pyruvate dehydrogenase (two polypeptide chains per copy), 24 copies of dihydrolipoyl transacetylase, and 6 copies of dihydrolipoyl dehydrogenase (two polypeptide chains per copy), all noncovalently packaged together in a highly specific geometry (Figure 5.15). Eurkaryotic complexes tend to contain more copies of each of the three enzymes and to exhibit a unique geometry. Some also encompass multiple copies of a protein kinase and protein phosphatase that are involved in the regulation of catalytic activity through covalent modification (Section 5.13).

The structural formulas for NAD^+ and CoASH, two of the five coenzymes within the pyruvate dehydrogenase complex, have already been presented in this chapter (Figure 11.1, introduction to this section). The structures for the other three coenzymes are given in Figure 11.10. All of these coenzymes, except lipoic acid, are synthesized from water-soluble vitamins (Section 10.1). Since TPP, lipoic acid, and FAD function catalytically, they do not appear in the net reaction for the complex. **[Try Problem 11.21]**

Reaction Sequence

The pyruvate dehydrogenase complex catalyzes five sequential reactions that are efficiently coupled through the selective packaging of its enzymes (Figure 11.11). The process begins when pyruvate dehydrogenase (E_1) catalyzes the decarboxylation of pyruvate and the attachment of the acetaldehyde produced to TPP to generate hydroxyethyl–TPP (reaction 1). The same enzyme, E_1, employing the lipoic acid attached to dihydrolipoyl transacetylase (E_2) as an oxidizing agent, catalyzes the oxidation of the hydroxyethyl group in hydroxyethyl–TPP to an acetyl group and the attachment of this acetyl group to the reduced lipoic acid through a thioester link (reaction 2). The energy of oxidation provides the thermodynamic driving force for the formation of the high-energy thioester. E_2 next catalyzes the transfer of the acetyl group from lipoic acid to CoASH (reaction 3), a process in which the cleavage of one thioester link provides the free energy to create a second. The deacylated lipoic acid residue, which is covalently attached to a lysine side chain in E_2 through an amide link, physically swings from E_1, where it picked up the acetyl group, to dihydrolipoyl dehydrogenase (E_3). E_3 provides the catalytic assistance that couples the oxidation of the reduced lipoic acid on E_2 to the reduction of FAD (reaction 4). In a final step (reaction 5), E_3 catalyzes the oxidation of the $FADH_2$ formed in step 4. NAD^+ is the oxidizing agent; NADH plus H^+ and FAD are final products. Although TPP, lipoic acid, and FAD are all intimately involved in the overall process, they find themselves in their original state when all is said and done. Initial reactants enter at particular sites, and final products exit at other sites. Since the intermediates are passed from one enzyme to the next within this complex, they are unable to serve as substrates for other enzymes, an example of metabolic channeling (Section 5.15).

Regulation

The pyruvate dehydrogenase complex is a major control point for carbohydrate catabolism. In many organisms, it is allosterically inhibited by ATP, acetyl-SCoA, and

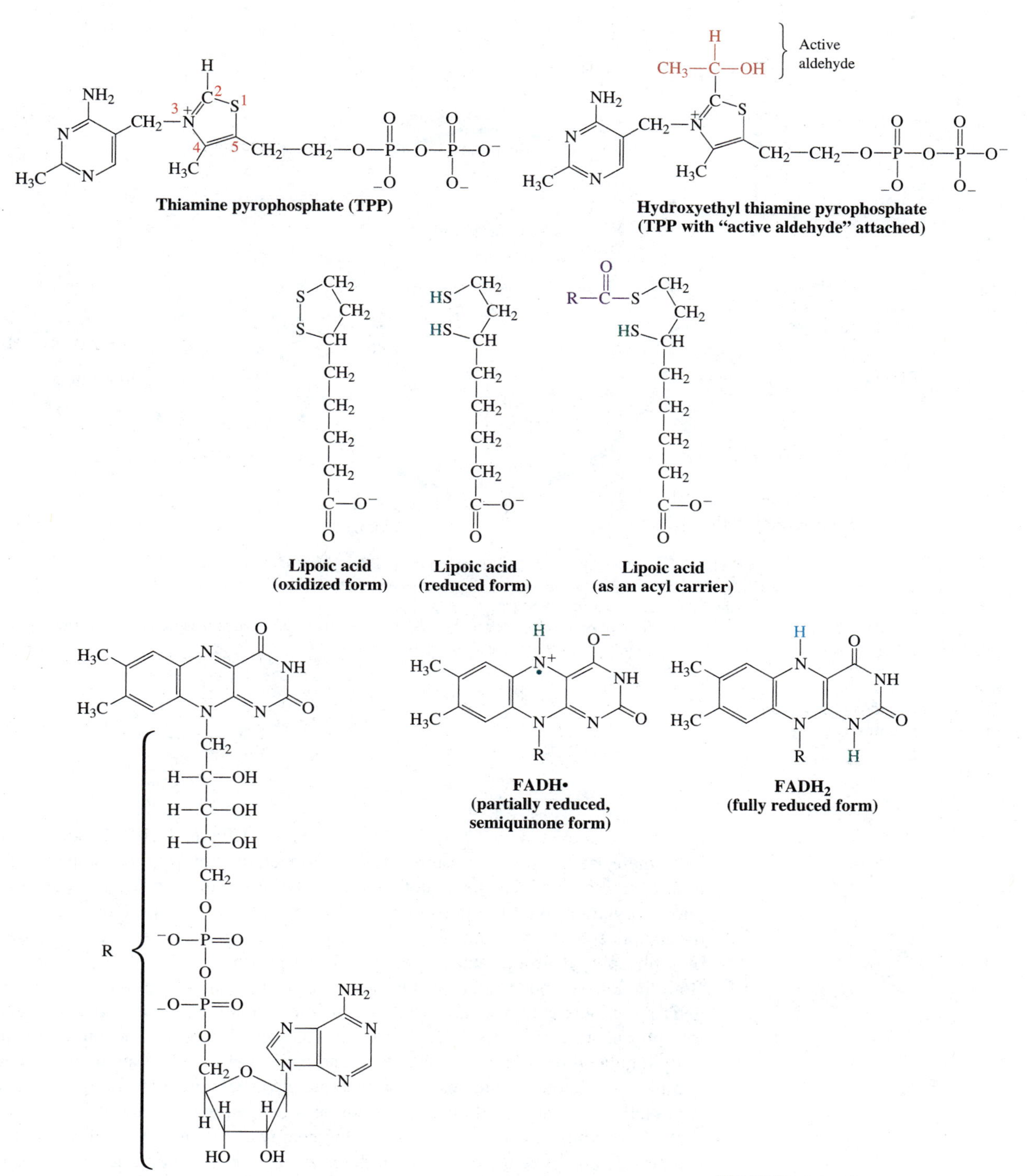

FIGURE 11.10
Structures of TPP, lipoic acid, and FAD.

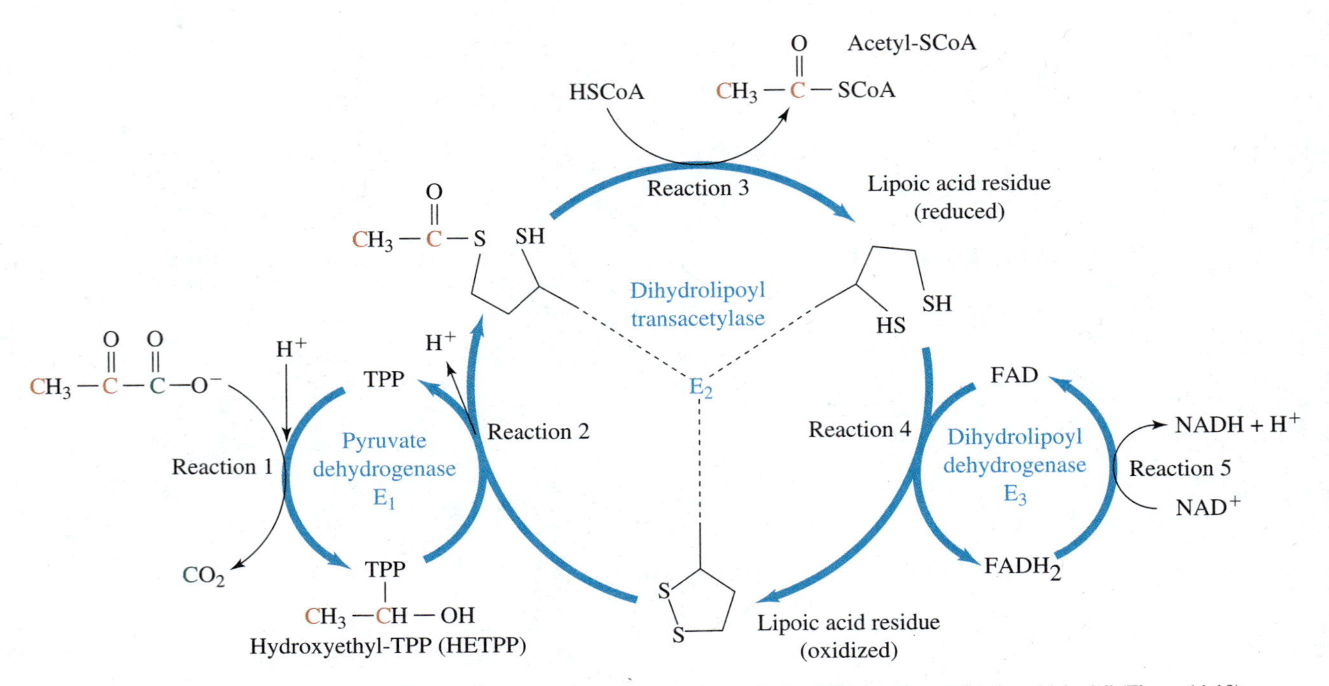

Reaction 1: Pyruvate is decarboxylated and the acetaldehyde generated is attached to TPP to create an "active aldehyde" (Figure 11.10).
Reaction 2: Lipoic acid oxidizes the two-carbon fragment to an acetyl group and becomes attached to this group through a thioester link.
Reaction 3: The acetyl group is transferred from lipoic acid to HSCoA.
Reaction 4: FAD converts the reduced (dithiol) form of lipoic acid back to the oxidized (disulfide) form and is reduced to $FADH_2$ in the process.
Reaction 5: NAD^+ oxidizes $FADH_2$ to FAD. NADH + H^+ are additional products.

FIGURE 11.11
Reactions catalyzed by the pyruvate dehydrogenase complex.

NADH and activated by AMP, CoASH, and NAD^+ (Figure 11.12). In some organisms, there are additional allosteric effectors as well, and in vertebrates, the complex is also regulated by covalent modification. The covalent modification consists of phosphorylation, with the phosphorylated form of the complex being inactive and the nonphosphorylated form active. A unique protein kinase catalyzes the addition of a phosphoryl group, and a specific protein phosphatase catalyzes its removal. Both the kinase and the phosphatase are part of the mammalian pyruvate dehydrogenase complex, and both are regulated by allosteric effectors. One of the stimulatory phosphatase effectors is Ca^{2+}, a second messenger whose concentration is partially regulated by hormones. The combined actions of the pyruvate dehydrogenase complex regulators help to ensure that the rate of carbohydrate catabolism meets the ever-changing energy and anabolic needs of a cell. In general, the pyruvate dehydrogenase complex is most active when the energy charge in a cell is low and there is a need to complete the oxidation of the pyruvate generated during glycolysis. **[Try Problem 11.22]**

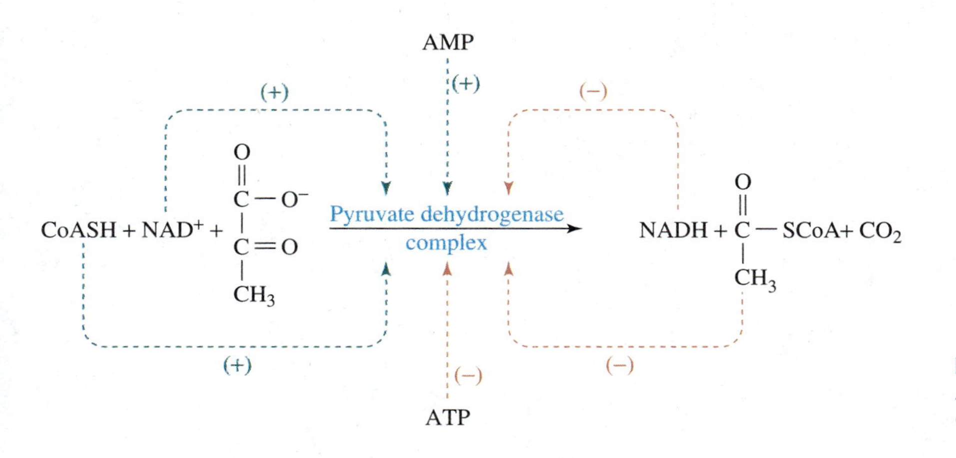

FIGURE 11.12
Allosteric regulation of the pyruvate dehydrogenase complex.

***PROBLEM 11.20** What organic product is formed when an α-keto acid is decarboxylated? What class of organic compound is generated when this decarboxylation product is oxidized?

PROBLEM 11.21 Define catalyst. What does it mean to say that TPP, lipoic acid, and FAD function catalytically within the pyruvate dehydrogenase complex?

PROBLEM 11.22 Explain the metabolic logic behind the inhibition of the pyruvate dehydrogenase complex by acetyl-SCoA and the activation of this complex by NAD^+.

11.6 CITRIC ACID CYCLE—OXIDATION OF ACETATE TO TWO CO_2

During the complete oxidation of glucose, the **citric acid cycle** [also called the **Krebs cycle** or **tricarboxylic acid (TCA) cycle**] picks up where the pyruvate dehydrogenase complex leaves off; it converts the acetyl groups in acetyl-SCoAs to CO_2 molecules. NADH, $FADH_2$, and GTP are additional products. The net result is a redox reaction in which both NAD^+ and FAD serve as oxidizing agents:

$$\underset{\textbf{Acetyl-SCoA}}{CH_3{-}\overset{O}{\overset{\|}{C}}{-}SCoA} + 3\,NAD^+ + FAD + GDP + P_i + 2\,H_2O \xrightarrow{\Delta G^{\circ\prime} = -58\text{ kJ/mol}} 2\,CO_2 + 3\,NADH + 2\,H^+ + FADH_2 + GTP + CoASH$$

The entire pathway is illustrated in Figure 11.13. Each of the eight steps is catalyzed by a separate enzyme or multienzyme complex. Since the oxaloacetate consumed in step 1 is regenerated in step 8, the pathway functions catalytically and is normally depicted as a cycle. Two loops through the pathway are required to complete the oxidation of one glucose molecule, because each glucose yields two acetyl-SCoAs. When fed into the respiratory chain, the pair of electrons from each NADH can lead

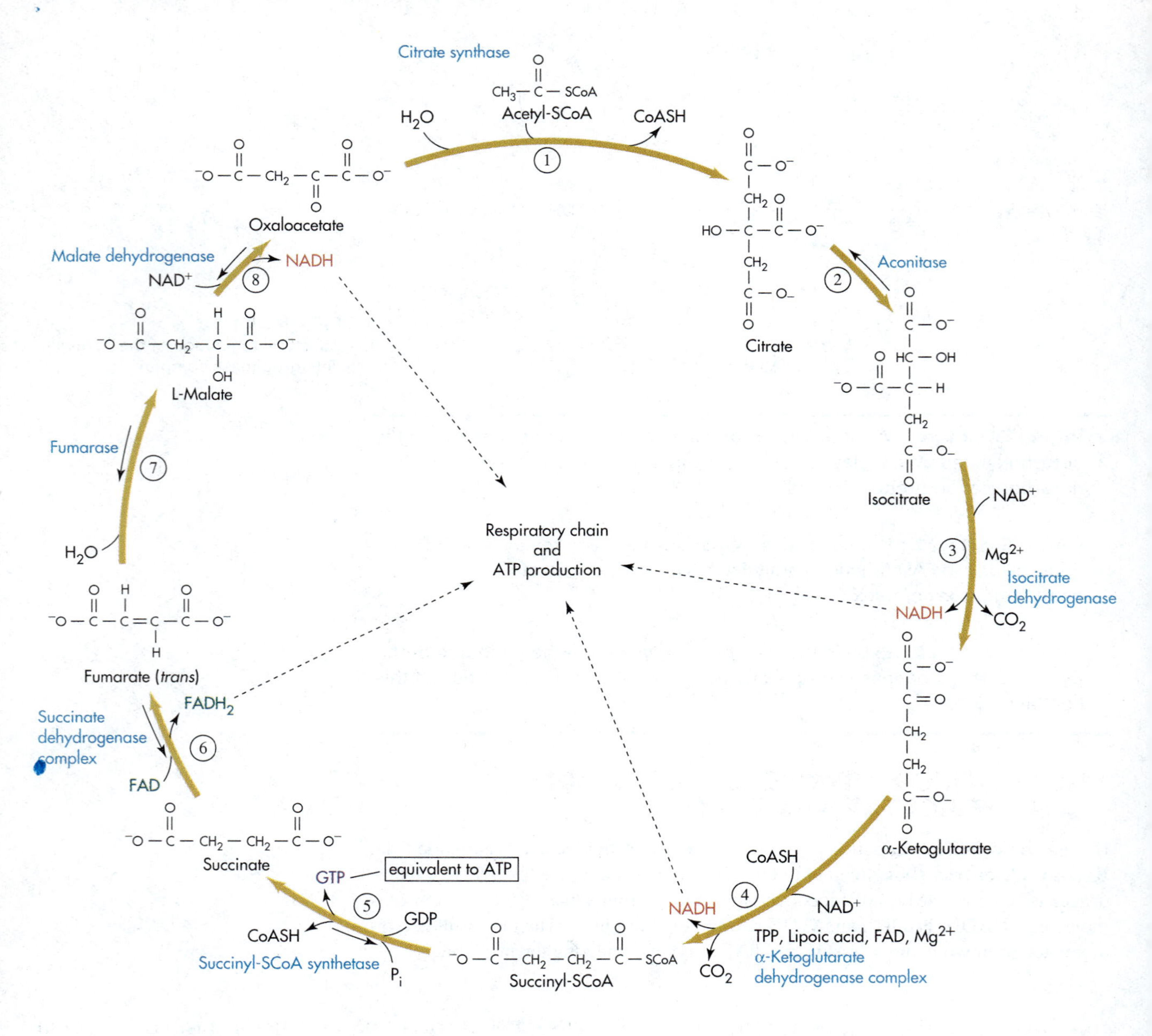

FIGURE 11.13
The citric acid cycle. Each NADH can lead to the production of approximately 2.5 ATPs and each $FADH_2$ to the production of approximately 1.5 ATPs.

to the production of roughly 2.5 ATP. Close to 1.5 ATP can be produced each time a $FADH_2$ passes a pair of electrons through the respiratory chain. From an energy standpoint, GTP is equivalent to ATP; it can even phosphorylate ADP to produce ATP:

$$\text{GTP} + \text{ADP} \rightleftharpoons \text{GDP} + \text{ATP} \qquad \Delta G^{\circ\prime} = 0 \text{ kJ/mol}$$

Step 1 of the citric acid cycle involves two reactions: an aldol condensation between acetyl-SCoA and **oxaloacetate** to give **citryl-SCoA,** and the hydrolysis of the citryl-SCoA to yield **citrate.**

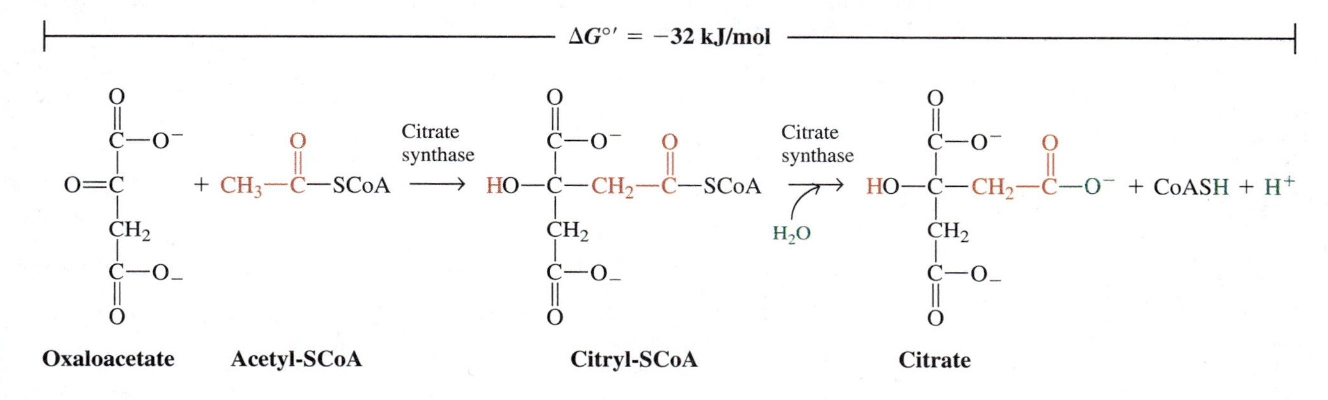

The hydrolysis of the thioester link in citryl-SCoA provides a thermodynamic driving force for citrate production and makes this step irreversible from a practical standpoint ($\Delta G^{\circ\prime} = -32$ kJ/mol). Since both the aldol condensation and the hydrolysis are catalyzed by **citrate synthase (condensing enzyme),** they are usually treated as a single step in this metabolic pathway. **[Try Problem 11.23]**

Step 2 is also a two phase process; dehydration followed by rehydration leads to isomerization. **Isocitrate** is the final product and **aconitase** is the catalyst for this reversible reaction ($\Delta G^{\circ\prime} = +6$ kJ/mol).

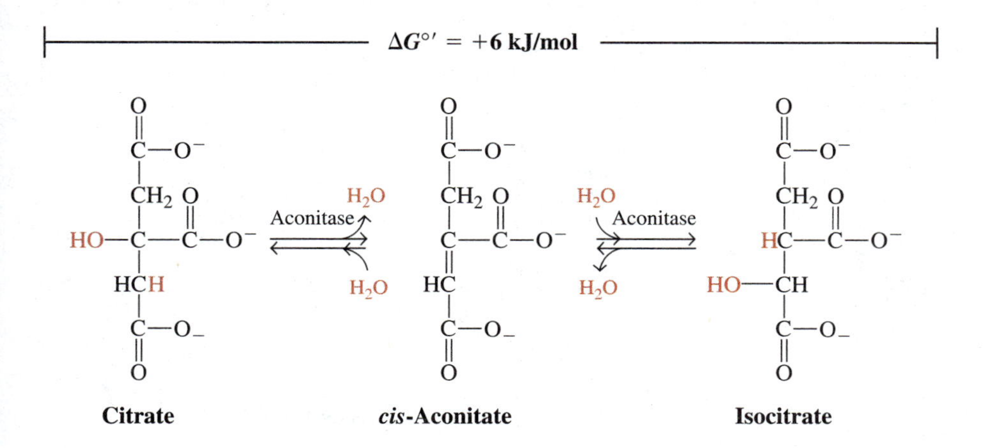

Aconitase is one of the most thoroughly studied examples of an **iron–sulfur protein (nonheme iron protein).** It contains four iron atoms that are complexed with sulfur atoms to form an **iron–sulfur cluster** (also called an **iron-sulfur center** or **iron-sulfur complex**). Some of the sulfur atoms exist as inorganic sulfides and others reside within cysteine side chains. In this instance, the cluster plays a central role in the binding and positioning of citrate at the enzyme's active site. **[Try Problem 11.24]**

Step 3 consists of the oxidative decarboxylation of isocitrate to yield **α-ketoglutarate,** NADH plus H^+, and CO_2. NAD^+ is the oxidizing agent and **oxalosuccinate** is an intermediate in this irreversible reaction ($\Delta G^{\circ\prime} = -21$ kJ/mol) catalyzed by **isocitrate dehydrogenase.**

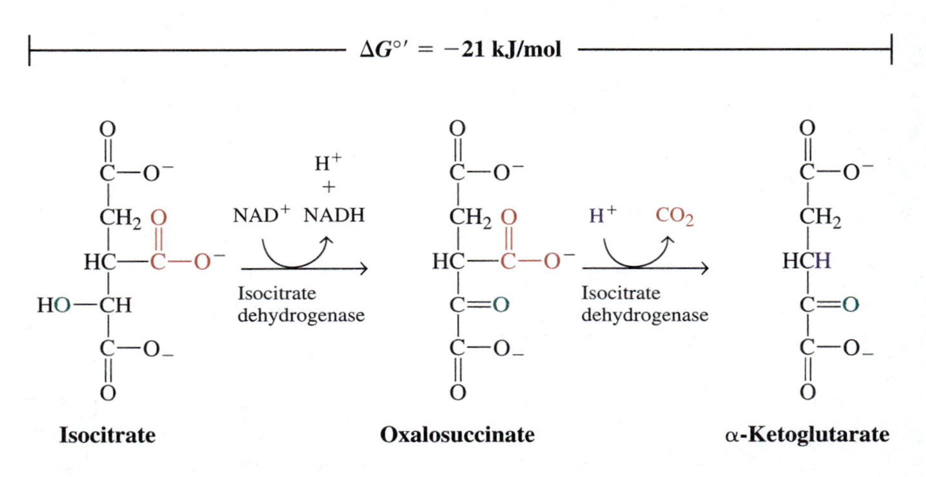

Two of the six CO_2 and two of the NADH produced during the total oxidation of glucose are generated by this step in the citric acid cycle. Each of the NADHs can be used to synthesize approximately 2.5 ATP.

Step 4 (below) is another irreversible oxidative decarboxylation reaction ($\Delta G^{\circ\prime} = -34$ kJ/mol). NAD^+ and CoASH react with α-ketoglutarate to yield **succinyl-SCoA,** CO_2, and NADH. This step is catalyzed by the **α-ketoglutarate dehydroge-**

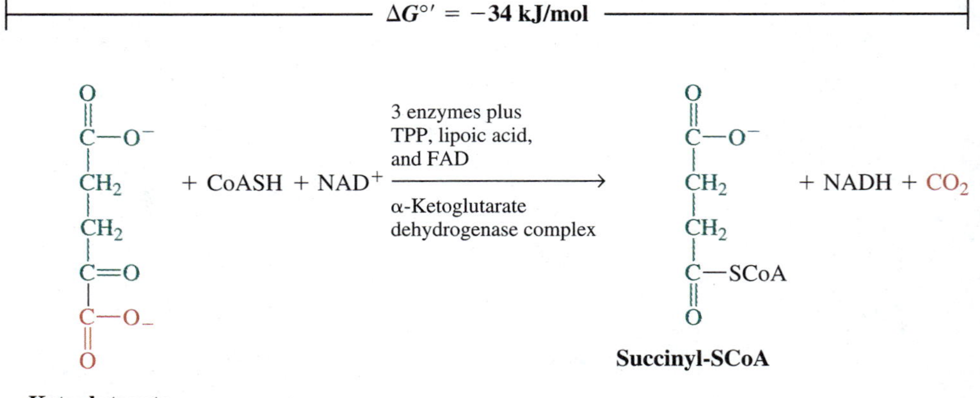

nase complex, a multienzyme system constructed from multiple copies of each of three separate enzymes and five distinct coenzymes. Since TPP, lipoic acid, and FAD function catalytically, they do not appear in the net reaction. Both the net reaction and the individual steps (not shown) catalyzed by the α-ketoglutarate dehydrogenase complex are very similar to those catalyzed by the pyruvate dehydrogenase complex. The reactions catalyzed by the α-ketoglutarate dehydrogenase complex account for the last CO_2 produced during the complete oxidation of glucose. Once again, each NADH generated is a potential source of roughly 2.5 ATP. The remainder of the cycle is involved mainly in the regeneration of the oxaloacetate that initiated the pathway and is required to keep the cycle functional.

In **step 5,** the cleavage of the thioester link in succinyl-SCoA drives the phosphorylation of GDP, a reversible process ($\Delta G°' = -3$ kJ/mol) and a third example of substrate-level phosphorylation (two examples were encountered in glycolysis). The final products of this **succinyl-SCoA synthetase**–catalyzed reaction are **succinate,** CoASH, and GTP.

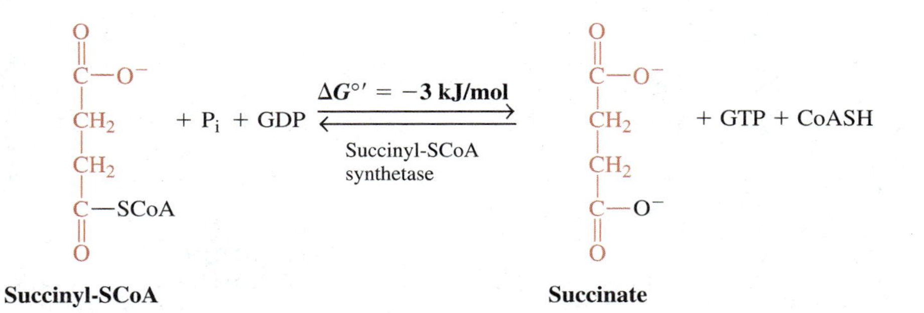

Step 6 is a reversible, stereospecific oxidation reaction ($\Delta G°' = 0$ kJ/mol) in which the **succinate dehydrogenase complex,** the only membrane-bound component of the citric acid cycle, catalyzes the FAD-linked removal of two hydrogens from succinate to generate **fumarate** (a trans-dicarboxylic acid) and $FADH_2$.

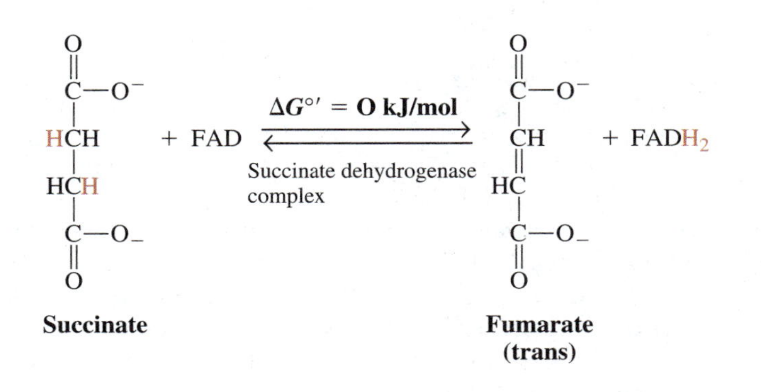

The electrons removed from succinate flow through at least one iron–sulfur cluster on the way to FAD and, ultimately, the respiratory chain. $FADH_2$ is a potential source of around 1.5 ATPs.

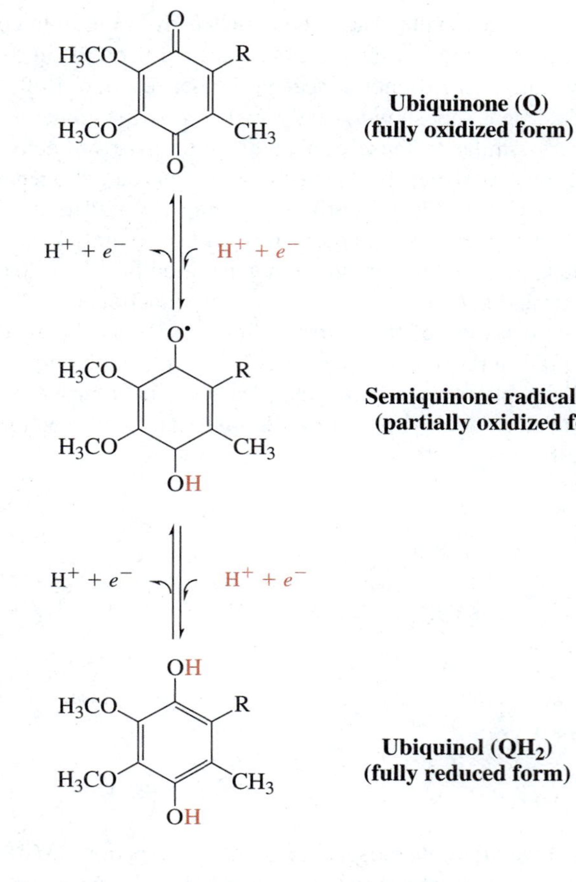

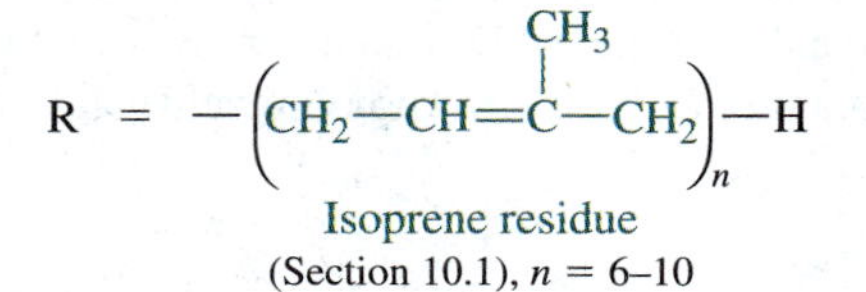

FIGURE 11.14
Ubiquinone and ubiquinol.

Because the $FADH_2$ generated in the succinate dehydrogenase complex is oxidized by **ubiquinone** (**coenzyme Q,** abbreviated **Q**; Figure 11.14) while within the complex, the totally reduced form of Q (**QH_2**; known as **ubiquinol**) is technically the final product of this multienzyme system. QH_2, in contrast to the membrane-bound $FADH_2$, is a mobile electron carrier. The succinate dehydrogenase complex is also known as **complex II** of the respiratory chain. It is closely associated with the other membrane-bound complexes that are components of this metabolic pathway (Section 11.8).

Step 7 consists of the reversible, stereospecific hydration of fumarate to give **L-malate,** a reaction catalyzed by **fumarase** ($\Delta G°' = -4$ kJ/mol):

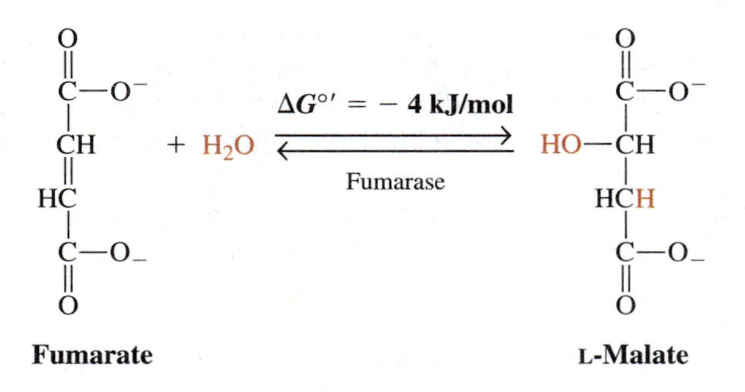

Water is added to an alkene to produce a secondary alcohol.

Step 8, the final step in the citric acid cycle, is the oxidation of the hydroxyl group in L-malate by NAD^+ to give oxaloacetate (a ketone) and NADH plus H^+.

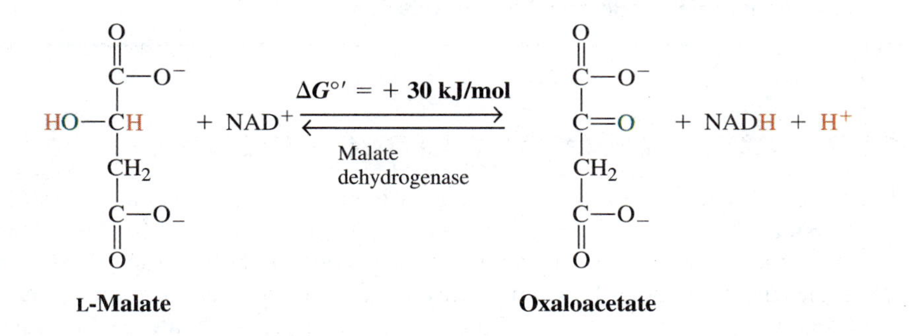

This reaction ($\Delta G°' = +30$ kJ/mol) is catalyzed by another stereospecific enzyme, **malate dehydrogenase.** The oxaloacetate formed is ready to condense with another molecule of acetyl-SCoA and initiate an additional round of oxidation through the citric acid cycle. Oxaloacetate must be regenerated to keep the citric acid cycle operational. The citric acid cycle, along with the respiratory chain, accounts for the majority of the ATP used by most human cells under aerobic conditions. **[Try Problems 11.25 and 11.26]**

The precise reaction conditions that exist within mitochondria during the operation of the citric acid cycle have not been determined. Hence, the ΔG_p values for the individual steps in this pathway are uncertain. The ΔG_p values for steps 1, 3, and 4, the irreversible steps in the cycle, are relatively large and negative, while the ΔG_p values for the remaining steps are close to zero.

Although the citric acid cycle is said to oxidize the acetyl group that initiates the cycle at step 1, the carbon atoms in the two CO_2 molecules released during the cycle are derived from carbons in the oxaloacetate that condensed with the acetyl group. This is explained by the stereospecific reactions and interactions that occur at multiple steps within the citric acid cycle. It is of no significance in the net reaction. As well, the carbons in the oxaloacetate used in one cycle can be traced to acetyl groups that entered previous cycles.

PROBLEM 11.23 Using CoASH as the formula for coenzyme A, write the structural formula for the product formed when dichloroacetyl-SCoA condenses with oxaloacetate. Identify the thioester link within this formula.

PROBLEM 11.24 True or false? Although citrate and isocitrate are isomers, aconitase, which catalyzes their interconversion, is a lyase. Explain.

PROBLEM 11.25 What is the maximum number of ATPs that can be produced from the complete oxidation of one acetyl group within the citric acid cycle? Assume that this cycle is coupled to a functional respiratory chain with O_2 as the terminal electron acceptor.

PROBLEM 11.26 The citric acid cycle, unlike glycolysis, shuts down under anaerobic conditions. Explain this fact.

11.7 CONTROL OF THE CITRIC ACID CYCLE

The citric acid cycle, like glycolysis, serves two major functions within a cell: (1) it helps oxidize fuel to produce energy; and (2) it provides intermediates for anabolism. Since the anabolic removal of any one member of the cycle depletes the entire cycle, the cycle must be continuously replenished by reactions that generate cycle intermediates. A key replenishment reaction is the **pyruvate carboxylase**–catalyzed conversion of pyruvate to oxaloacetate.

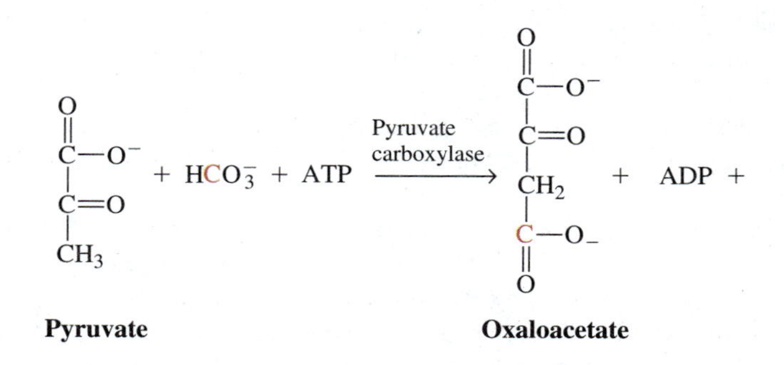

Certain amino acids, including aspartate and glutamate, and some fatty acids can also be transformed into cycle intermediates. Other replenishment reactions exist as well. Although the constant depletion and replenishment of the cycle may have an impact on the rate of cycling, it does not alter the only net reaction of the cycle, the oxidation of the acetyl group.

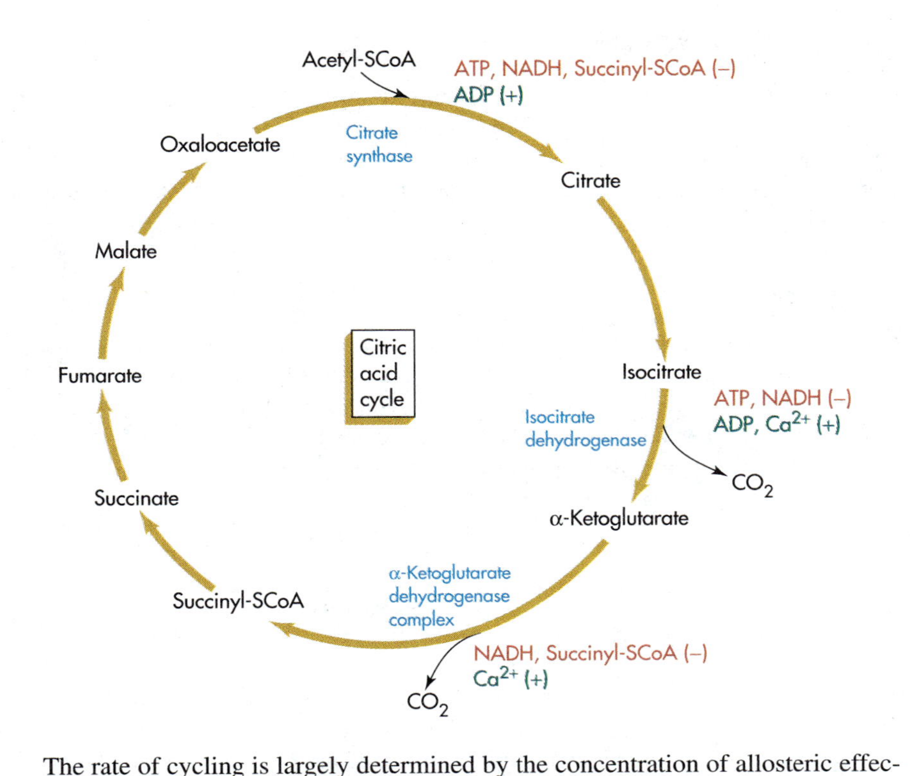

FIGURE 11.15
Some key regulators of the citric acid cycle.

The rate of cycling is largely determined by the concentration of allosteric effectors and the availability of oxaloacetate and acetyl-SCoA, the substrates for step 1. The regulatory enzymes are citrate synthase, isocitrate dehydrogenase, and α-ketoglutarate dehydrogenase complex. Like their counterparts in glycolysis, they catalyze the most irreversible reactions within the pathway. Although a wide variety of compounds have been found to modulate the activity of one or more of these enzymes *in vitro,* there is considerable controversy over which of these effectors are of functional significance under physiological conditions. The negative effectors include ATP (citrate synthase and isocitrate dehydrogenase), NADH (citrate synthase, isocitrate dehydrogenase, and α-ketoglutarate dehydrogenase complex), and succinyl-SCoA (citrate synthase and α-ketoglutarate dehydrogenase complex). The activators include ADP (citrate synthase and isocitrate dehydrogenase) and Ca^{2+} (isocitrate dehydrogenase and α-ketoglutarate dehydrogenase complex) (Figure 11.15). The citric acid cycle is not subject to covalent modification in mammals, but this regulatory option is employed by *E. coli* and certain other organisms. In general, the oxidation of the acetyl group within the citric acid cycle is most rapid when the energy charge of a cell is below normal as signaled by a low $NADH/NAD^+$ ratio and a low ATP/ADP ratio. At high $NADH/NAD^+$ ratios, the rate of cycling drops as a result of both the allosteric inhibition by NADH and the reduced concentration of NAD^+, a required reactant at three steps within the cycle. **[Try Problem 11.27]**

PROBLEM 11.27 Would you expect FAD to be a positive or a negative modulator if it were a modulator of the citric acid cycle? Explain.

11.8 THE RESPIRATORY CHAIN

FAD and NAD^+ are converted to $FADH_2$ and NADH as they pick up a pair of electrons during redox reactions in glycolysis, the citric acid cycle, and other metabolic pathways. Within mitochondria, the electrons are passed in a bucket brigade fashion from one compound to another along a series of electron carriers that make up a metabolic pathway known as the **electron transport system** or **respiratory chain.**

An Overview

Molecular oxygen (O_2) is the terminal electron acceptor within the respiratory chain, and two water molecules are formed as each O_2 acquires two pairs of electrons. The energy released within the respiratory chain provides a thermodynamic driving force for the coupled phosphorylation of ADPs to generate ATPs. The net reactions of the respiratory chain, including the coupled production of ATP, are:

$$\text{NADH} + 3.5\,\text{H}^+ + .5\,\text{O}_2 + 2.5\,\text{ADP} + 2.5\,\text{P}_i \xrightarrow{\Delta G^{\circ\prime} = -142\ \text{kJ/mol}} \text{NAD}^+ + 3.5\,\text{H}_2\text{O} + 2.5\,\text{ATP}$$

$$\text{FADH}_2 + .5\,\text{O}_2 + 1.5\,\text{ADP} + 1.5\,\text{P}_i + 1.5\,\text{H}^+ \xrightarrow{\Delta G^{\circ\prime} = -111\ \text{kJ/mol}} \text{FAD} + 2.5\,\text{H}_2\text{O} + 1.5\,\text{ATP}$$

These net reactions assume that 2.5 ATP are produced during the oxidation of NADH and 1.5 ATP are produced during the oxidation of $FADH_2$. A majority of the ATP produced within the human body, and most of the O_2 consumed by the body, are accounted for by these reactions. **[Try Problems 11.28 and 11.29]**

The amount of energy released at each step of the respiratory chain is determined by the difference in the reduction potential for the two half-reactions involved in the step:

$$\Delta E^{\circ\prime} = E^{\circ\prime}{}_{\text{acceptor}} - E^{\circ\prime}{}_{\text{donor}}$$

$$\Delta G^{\circ\prime} = -n\mathfrak{F}\Delta E^{\circ\prime}$$

$E^{\circ\prime}{}_{\text{acceptor}}$ is the standard biological reduction potential for the half reaction containing the electron carrier accepting electrons (the oxidizing agent), $E^{\circ\prime}{}_{\text{donor}}$ is the standard biological reduction potential for the half reaction containing the electron carrier that donates electrons (the reducing agent), n is the number of electrons transferred, and $\mathfrak{F}$ is the Faraday constant. You may wish to review Section 8.8 at this time. The flow of electrons is always from a reducing agent in a half reaction with a lower reduction potential to an oxidizing agent in a half reaction with a higher reduction potential. The total amount of energy released within the respiratory chain is substantial; $\Delta G^{\circ\prime} = -219$ kJ/mol when electrons are passed from NADH to O_2 and equals -157 kJ/mol when the electrons are delivered to the respiratory chain by $FADH_2$ (rather than NADH). It is this energy that is used to synthesize ATP. **[Try Problem 11.30]**

The respiratory chain is an integral part of the inner mitochondrial membrane. It is composed of an ATP synthase, the enzyme responsible for catalyzing the production of ATP, plus two mobile electron carriers and four electron-transporting protein complexes, designated **I, II, III,** and **IV** (Table 11.1). The protein complexes, each

TABLE 11.1

The Respiratory Chain Complexes

Respiratory Chain Complexes	Alternate Names
Complex I	NADH dehydrogenase complex NADH-ubiquinone oxidoreductase
Complex II	Succinate dehydrogenase complex Succinate-ubiquinone oxidoreductase
Complex III	Cytochrome reductase complex Ubiquinone-cytochrome *c* oxidoreductase
Complex IV	Cytochrome oxidase complex Cytochrome *c* oxidase

containing multiple tightly bound electron carriers, are able to migrate within the membrane, since they are not anchored to the membrane at specific sites. This section describes that portion of the respiratory chain involved in the transport of electrons, and Section 11.9 examines how the energy released during the transport of electrons is utilized to produce ATP.

The Transport of Electrons to O_2

A pair of electrons delivered to the respiratory chain by NADH and destined for O_2 is initially passed to ubiquinone (Q; Figure 11.14), one of the two mobile electron carriers, with the catalytic assistance of complex I (Figure 11.16). Complex II catalyzes the transfer of a pair of electrons to Q when $FADH_2$ delivers the electrons. Ubiquinol (QH_2), the reduced form of Q, passes the electrons it acquires from NADH or $FADH_2$ through complex III to **cytochrome *c* (cyt *c*),** the second mobile electron carrier along the respiratory chain. Since cyt *c* can accept only one electron, only one electron at a time exits complex III. Cyt *c* transfers the electron it receives to molecular oxygen through **complex IV.** The redox reaction catalyzed by complex IV accounts for a significant portion of the water that is eliminated from the human body each day. The same reaction allows some desert animals to survive for long periods in the absence of an external supply of water. Collectively, the reactions catalyzed by complexes I, III, and IV generate all of the energy used by the respiratory chain to synthesize ATP (Section 11.9).

The electron carriers within complex II (FAD and iron–sulfur clusters) are included in the discussion of the citric acid cycle, since this complex (also called the succinate dehydrogenase complex) is part of that cycle (Section 11.6). Table 11.2 identifies the individual electron carriers within complexes I, III, and IV. Each of these carriers falls into one of five major classes: flavins, quinones, iron–sulfur clusters, cytochromes, and copper atoms.

Complex I, which contains two electron carriers, catalyzes three reactions: the transfer of a pair of electrons from NADH to FMN; the transfer of a pair of electrons from $FMNH_2$ to iron–sulfur clusters; and the transfer of two electrons from iron–sulfur clusters to Q. During each transfer of electrons, the compound losing electrons is oxidized and the compound gaining electrons is reduced. The oxidized and re-

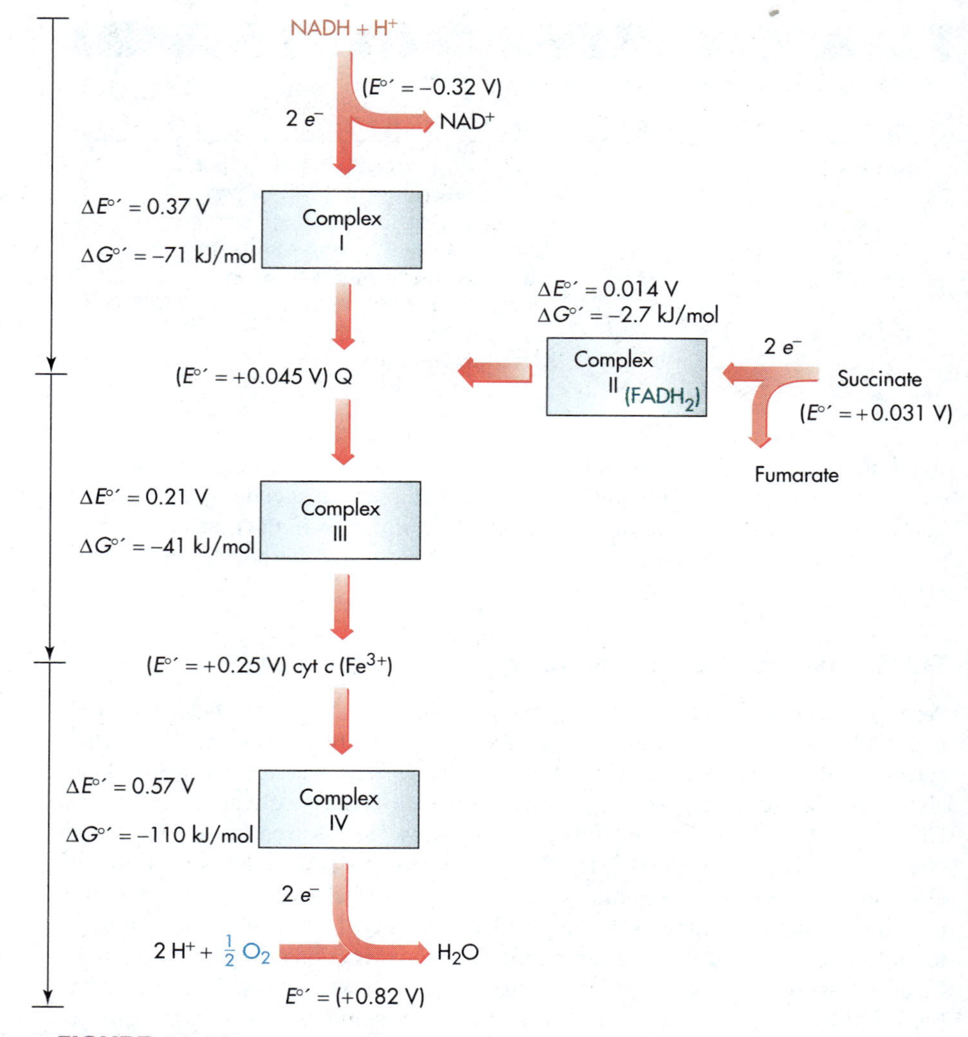

FIGURE 11.16
Electron transport within the respiratory chain.

duced forms of **flavin mononucleotide** (FMN) and Q (ubiquinone) are given in Figure 11.17 and 11.14, respectively. The iron–sulfur clusters contain iron ions, inorganic sulfide ions, and sulfur atoms within cysteine side chains (Figure 11.18). The iron atoms in such clusters are the actual electron carriers; they oscillate between a Fe^{3+} and a Fe^{2+} oxidation state as one electron at a time passes through on the way to oxygen.

Complex III encompasses two classes of electron carriers, iron–sulfur clusters and **cytochrome c_1.** Cytochromes are **heme proteins,** proteins containing a heme prosthetic group. Heme groups are classified on the basis of the side chains joined to

Reactions Within the Major Respiratory Complexes

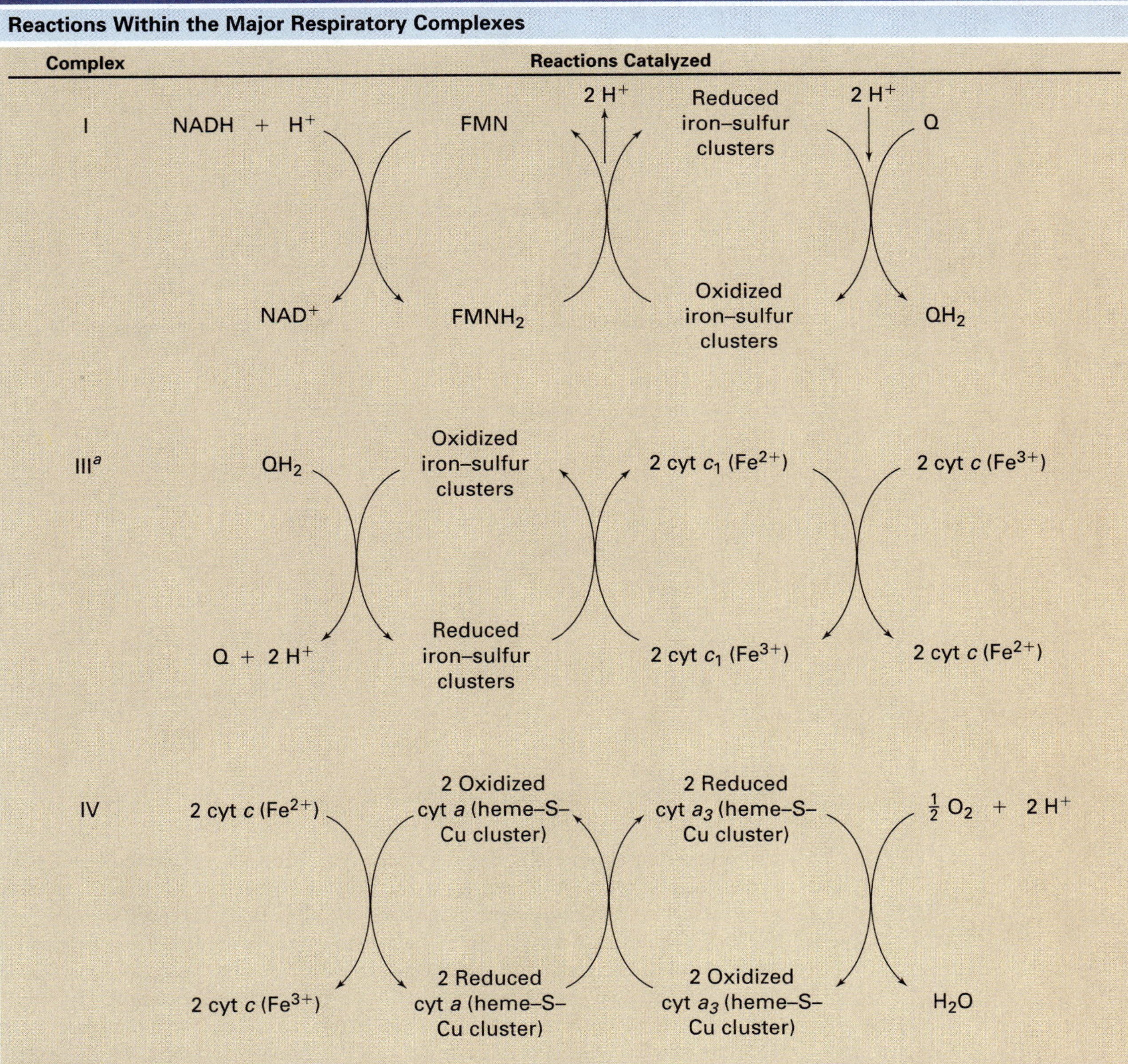

[a]Cyt *b* is also part of this complex. It stabilizes the semiquinone radical (Figure 11.14) produced from ubiquinone during the transfer of a single electron from QH_2 (a two-electron carrier) to an iron–sulfur cluster (a single-electron carrier).

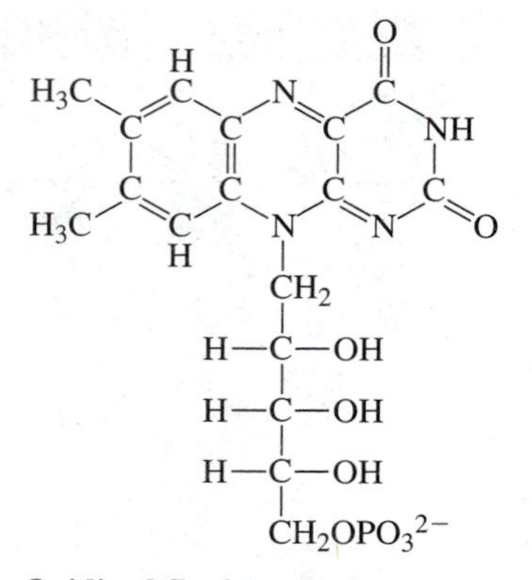

Oxidized flavin mononucleotide (FMN)

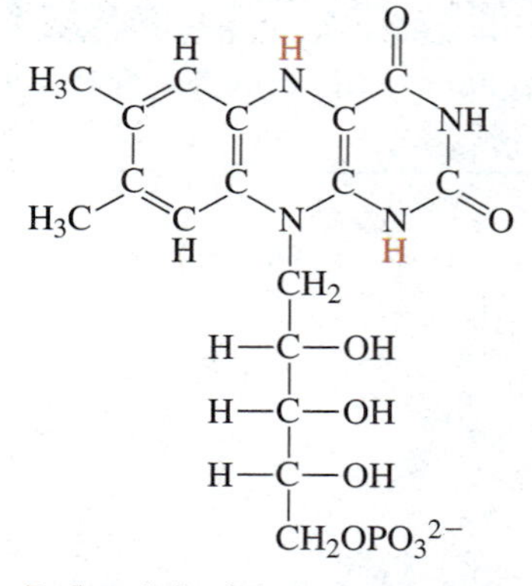

Reduced flavin mononucleotide ($FMNH_2$)

FIGURE 11.17
Structure of FMN and $FMNH_2$.

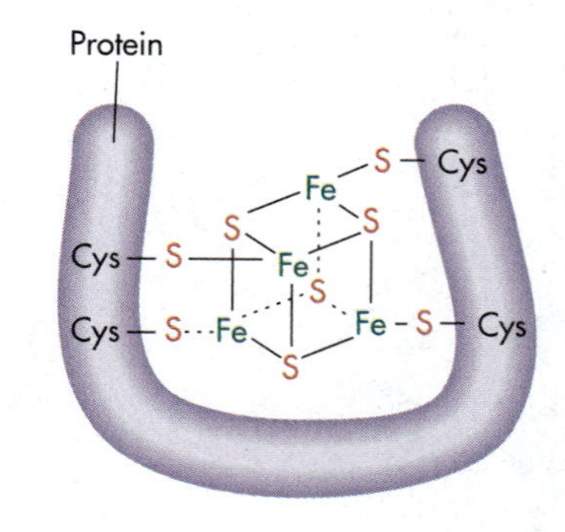

FIGURE 11.18
Proposed structures of two iron–sulfur clusters.

their ring system (Figure 11.19). The cytochromes, in turn, are classified on the basis of their heme groups. Although the *a* and *b* cytochromes possess a noncovalently bound heme, the heme prosthetic group is covalently attached to specific amino acid side chains within the *c* cytochromes. Each cytochrome is able to accept and carry a single electron. The heme iron cycles between a Fe^{3+} and Fe^{2+} oxidation state as an electron is accepted and then transferred by a cytochrome. In complex III, iron–sulfur clusters accept two electrons (one at a time) from QH_2 and then pass these electrons (one at a time) to cytochrome c_1. Cytochrome c_1, in turn, transfers electrons (one at a time) to cytochrome *c*. **[Try Problem 14.31]**

Complex IV employs cytochrome *a*, cytochrome a_3, and two copper atoms as electron carriers as it catalyzes the transfer of electrons from cytochrome *c* to O_2 (Table 11.2). One copper atom is associated with the heme iron in cyt *a* and the second is associated with the heme iron in cyt a_3. Both copper atoms cycle between the Cu^{2+} and Cu^{+} oxidation states as they participate in the transfer of electrons to O_2. Although complex IV is designed to keep partially reduced O_2 from escaping, some oxygen-containing free radicals, including some superoxide radicals ($O_2^{\cdot-}$), do break

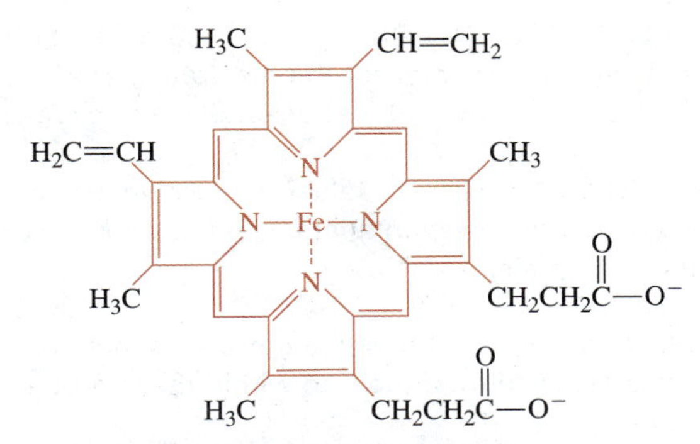

Iron protoporphyrin *IX*; the heme found in hemoglobin, myoglobin, and *b* cytochromes.

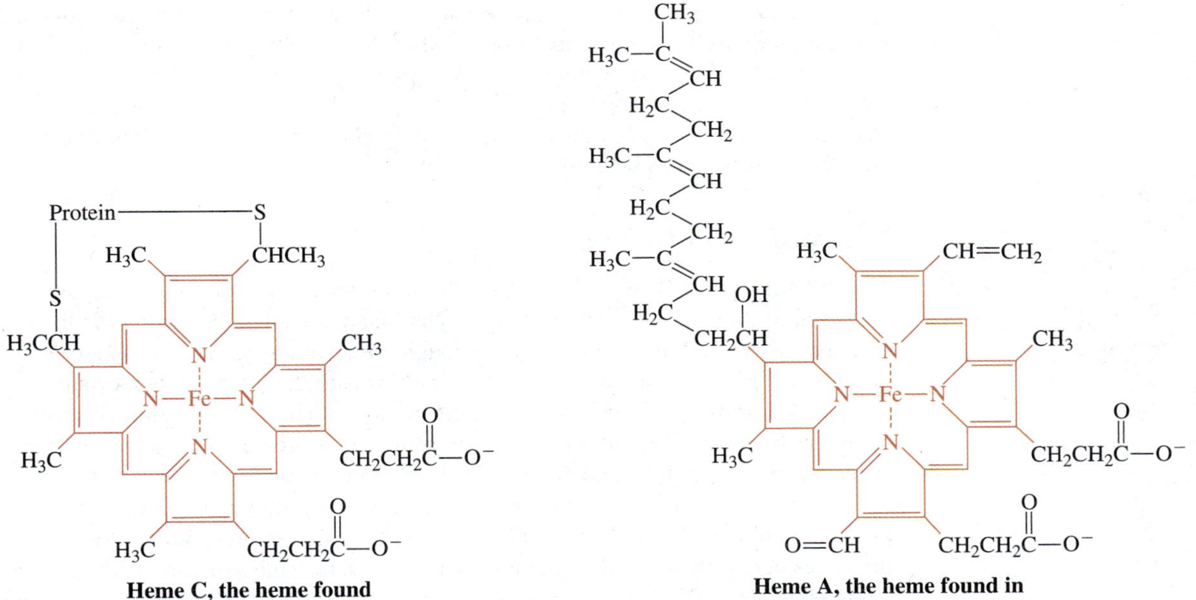

Heme C, the heme found in *c* cytochromes.

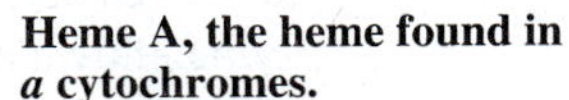

Heme A, the heme found in *a* cytochromes.

FIGURE 11.19
Heme groups.

away. Antioxidants, including vitamin E, and enzymes, including superoxide dismutase, play a key role in protecting the body from these reactive agents (Section 10.2). Under aerobic conditions, O_2 ultimately accepts all 24 of the electrons removed from glucose during its complete oxidation.

PROBLEM 11.28 Only one of the water molecules in each of the net reactions for the respiratory chain is produced from the transfer of electrons to molecular oxygen. What is the source of the additional water molecules?

PROBLEM 11.29 Not all of the O_2 used by the human body serves as a terminal electron acceptor for the respiratory chain. List the two enzymes encountered in Chapter 10 that use molecular oxygen (O_2) as a substrate.

PROBLEM 11.30 True or false? A negative $\Delta E°'$ is characteristic of a redox reaction that is spontaneous and has a $K_{eq} > 1$ under standard conditions. Explain.

PROBLEM 11.31 List two heme proteins that are not cytochromes. What is the role or function of the iron within these proteins? Hint: Review Chapter 4.

11.9 OXIDATIVE PHOSPHORYLATION

The respiratory chain–coupled production of ATP is known as **oxidative phosphorylation,** because the phosphorylation of ADP to make ATP is driven by energy released during oxidation reactions. For many years, it was assumed that the energy of oxidation was used to create a high-energy, phosphate-containing compound that transferred a phosphoryl group to ADP. This hypothesis has been abandoned because no such compound has been identified.

Chemiosmotic Theory

The **chemiosmotic theory** (Figure 11-20, a) provides the currently accepted explanation for how oxidation and phosphorylation are coupled. According to this theory, energy from the flow of electrons through the respiratory chain is used to pump protons (H^+) from the matrix of a mitochondrion into the intermembrane space, an example of primary active transport (Section 9.10). This translocation of protons generates both a **pH gradient** (the H^+ concentration is greater in the intermembrane space than in the matrix) and an **electrical gradient** (the matrix is negative relative to the intermembrane space) that possess a considerable amount of **electrochemical potential energy.** This energy is described as a **proton-motive force.** As protons flow back across the inner mitochondrion membrane through a membrane-bound enzyme known as ATP synthase, ADP and inorganic phosphate (P_i) are joined to make ATP (Figure 11.20, b). Roughly three H^+ must flow through ATP synthase for each ATP synthesized. **[Try Problems 11.32 through 11.34]**

ATP synthase consists of two major components, F_0 and F_1, that each contain multiple polypeptide chains. F_1 possesses 3 α, 3 β, 1 γ, 1 δ, and 1 ε chain. F_0 contains four or five unique polypeptides and multiple copies of at least three of these, a, b, and c. The three identical catalytic sites on each ATP synthase molecule are associated with F_1 which protrudes into the mitochondrial matrix. F_0 is membrane bound.

ATP synthesis is hypothesized to proceed through a process described as the binding-change mechanism (Figure 11.21). Each of the three catalytic sites on ATP synthase cycles through three conformations: "open," "loose," and "tight." Changes in conformation are driven by the flow of H^+ through the membrane-spanning

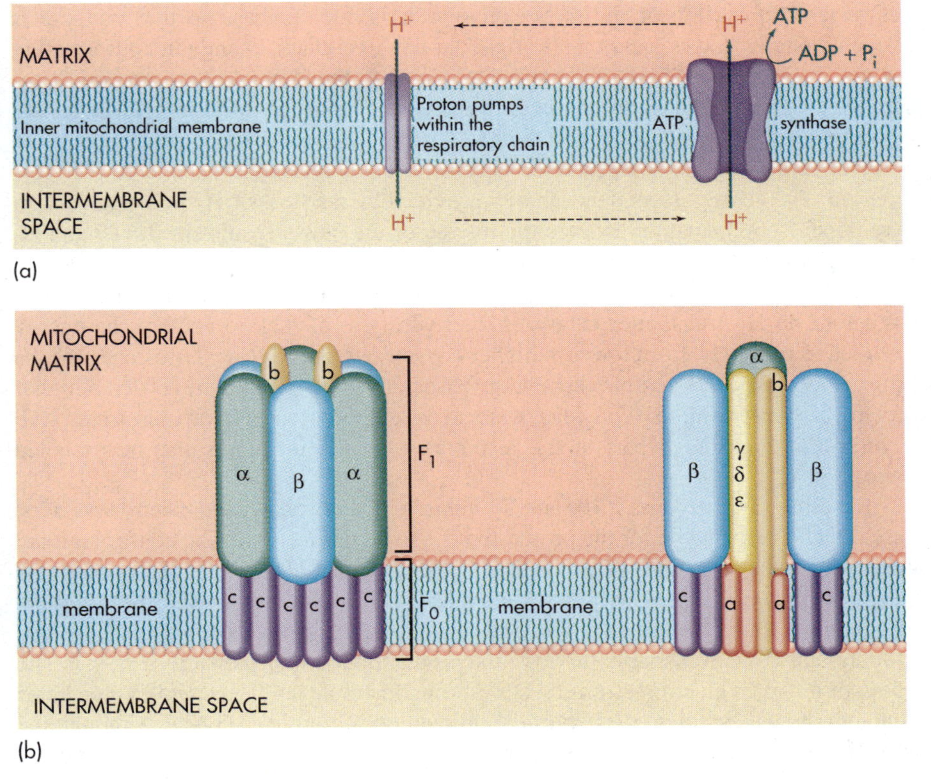

FIGURE 11.20
(a) Chemiosmotic theory. Pumps associated with the respiratory chain move H^+ from the mitochondrial matrix into the intermembrane space and create a proton gradient that possesses substantial electrochemical energy. The flow of H^+ back into the matrix through ATP synthase is coupled to the joining of ADP and P_i to generate ATP. (b) ATP synthase. *(Left)* The complete ATP synthase complex. The F_1 component contains nine polypeptide chains: 3α, 3β, 1γ, 1δ, and 1ε. F_0 contains four or five distinct polypeptides with multiple copies of some chains. The F_0 polypeptides shown are labeled a, b, and c. *(Right)* The ATP synthase complex cut in half lengthwise. (From Cross, 1994.)

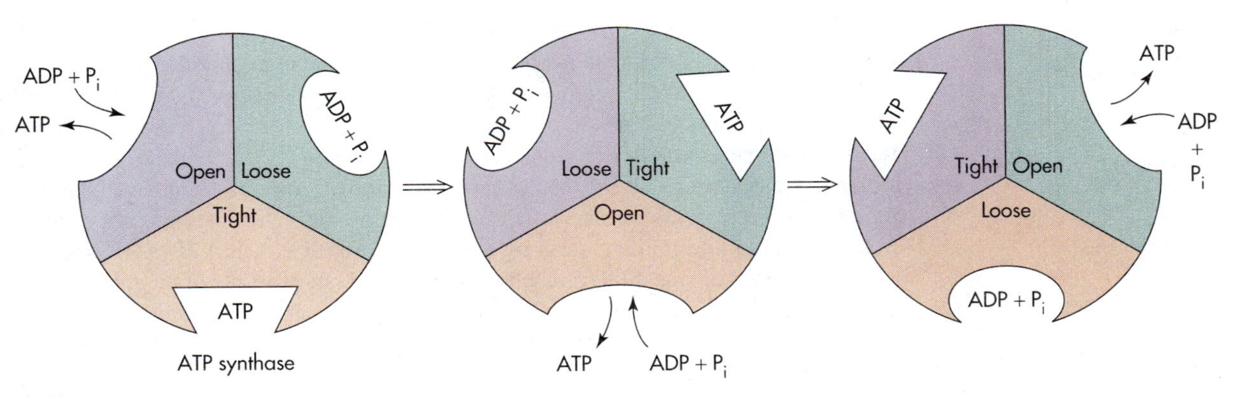

FIGURE 11.21
Binding-change mechanism for ATP synthesis. ATP synthase contains three catalytic sites that cycle among "open," "loose," and "tight" conformations. The cycling is driven by the flow of H^+ (not shown) through the membrane-bound synthase complex. The cycling of the separate catalytic sites is out of phase, so one site is "open," one "loose," and one "tight" at all times. When a site opens it releases an ATP that has been formed and binds ADP + P_i. The site subsequently shifts to a "loose" and then to a "tight" conformation. The shift to the tight conformation, so called because it binds ATP tightly, promotes ATP formation. The energy-dependent shift (energy provided by H^+ flow) back to the "open" conformation triggers

enzyme. The cycling of the three active sites is out of phase so that one site is "open," one is "loose," and one is "tight" at any time that a change in conformation is not in progress. When a catalytic site cycles to the "open" conformation, it releases an ATP that has already been formed and binds an ADP plus P_i. It subsequently changes to a "loose" and then to a "tight" conformation. The change to the "tight" conformation, so designated because it binds ATP tightly, leads to ATP formation. The energy-dependent (energy provided by the flow of H^+) change back to the "open" conformation triggers the release of the newly synthesized ATP and allows another ADP and P_i to bind.

ATP synthase is basically an H^+ transporter (ATPase) operating in reverse. Whereas an H^+ transporter catalyzes the hydrolysis of ATP to acquire the energy needed to move H^+ from lower to higher concentrations, ATP synthase catalyzes the joining of ADP and P_i employing energy associated with the flow of H^+ from higher to lower concentration. You may wish to reread Section 9.10 at this time. ATP synthase is considered part of the respiratory chain, and it is sometimes labeled **complex V.**

Respiratory complexes I, III, and IV, along with their associated electron carriers, are directly responsible for the proton-motive force used to drive the conformational changes associated with ATP synthesis; they pump H^+ into the intermembrane space. Complex II lacks a H^+ pump and makes no contribution to the proton-motive force. Each time a pair of electrons flows from NADH through complexes I, III, and IV, enough H^+ is transported to drive the production of approximately 2.5 ATP. The flow of a pair of electrons from $FADH_2$ through complexes II, III, and IV generates enough of a H^+ gradient to support the production of roughly 1.5 ATP. Although the details of H^+ transport are still under investigation, each complex appears to use a unique transport strategy.

Regulation

The rate of ATP production is primarily determined by the availability of ADP, one of the substrates for ATP synthase. The rate of ATP production increases as ADP concentration rises and drops as ADP levels decrease, a fact predictable from Michaelis–Menten kinetics (Section 5.6). Since oxidation and phosphorylation are normally tightly coupled (one cannot proceed in the absence of the other), ADP has a similar impact on the rate at which electrons flow through the respiratory chain. ADP is brought into mitochondria by a transport protein called **ATP-ADP translocase** which exports one ATP for each ADP imported, an example of an antiport system (Section 9.10). The transporter is powered by the same H^+ gradient that is involved in oxidative phosphorylation. Most of the ATP produced in mitochondria is exported for use in other parts of the cell. [Try Problem 11.35]

A variety of substances have been discovered that allow H^+ to "leak" back into the matrix without passing through an ATP synthase. Such substances, called **uncouplers,** reduce the H^+ gradient and allow electrons to flow through the respiratory chain in the absence of phosphorylation. In the presence of an uncoupler, the rate of electron flow increases, and most of the redox energy is immediately converted to heat (rather than H^+ gradients and ATP). A natural uncoupler called **thermogenin** or **uncoupling protein** keeps some animals from freezing during hibernation. Human infants also use the same uncoupler to help maintain normal body temperature. In both instances, the uncoupler is produced by specialized adipose tissue, known as **brown fat,** and the electrons flowing through the respiratory chain come primarily from the oxidation of fat rather than the oxidation of glucose.

PROBLEM 11.32 Write a balanced equation for the production of ATP from ADP and P_i. Provide structural formulas for all reactants and products. Is the K_{eq}' for this reaction >1000, <1000, or of intermediate value? Explain. Hint: Review Sections 8.2 and 8.4.

PROBLEM 11.33 True or false? Energy is needed to generate a concentration gradient. Explain.

PROBLEM 11.34 When the respiratory chain is transporting electrons, will the pH within the matrix of mitochondria be greater than, less than, or the same as the pH of the intermembrane space? Will the OH^- concentration be greater, less or the same in the matrix relative to that in the intermembrane space? Explain.

PROBLEM 11.35 Draw a graph illustrating the results obtained when initial velocity is plotted against substrate concentration for a Michaelis–Menten enzyme. Circle that region of the Michaelis–Menten plot where an enzyme-catalyzed reaction exhibits the greatest degree of self-regulation (rate increases as substrate concentrations increase and rate declines as substrate concentrations drop). Hint: Review Section 5.6.

11.10 RESPIRATORY CHAIN INHIBITORS

A number of substances have been identified that either block the flow of electrons through the respiratory chain or inhibit ATP synthase. Several of these have been

Carbon monoxide (CO) from auto exhaust and other sources inhibits oxidative phosphorylation by binding to Fe^{2+} in complex IV of the respiratory chain.

TABLE 11.3

Respiratory Chain Inhibitors

Respiratory Chain Inhibitors	Site/Mode of Action
Rotenone (a fish poison isolated from plants)	Binds to complex I and blocks the transfer of electrons from Fe–S clusters to ubiquinone (Q)
Carboxin	Binds to complex II and blocks the transfer of electrons from $FADH_2$ to ubiquinone (Q)
Antimycin A (an antibiotic)	Binds to complex III and blocks the transfer of electrons from ubiquinol (QH_2) to Fe–S clusters
Cyanide (one of the most potent human toxins)	Blocks electron flow by binding to Fe^{3+} within cytochromes of complex IV
Carbon monoxide (a common air pollutant found in auto exhaust)	Blocks electron flow by binding to Fe^{2+} within cytochromes of complex IV
Oligomycin (an antibiotic)	Blocks the flow of H^+ through ATP synthase

successfully employed as tools to probe the chemistry of oxidative phosphorylation. Representative respiratory chain inhibitors are listed in Table 11.3 and their structural formulas are given in Figure 11.22. **Rotenone, carboxin, antimycin A, cyanide,** and carbon monoxide each bind to and inactivate one or more electron carriers. **Oligomycin,** however, binds to the H^+ channel in ATP synthase and blocks the reentry of H^+ into the mitochondrial matrix.

11.11 SUMMARY OF GLUCOSE OXIDATION AND ENERGY PRODUCTION

When glucose is completely oxidized, CO_2, H_2O, and energy are the only products:

$$C_6H_{12}O_6 + 6\ O_2 \rightleftharpoons 6\ CO_2 + 6\ H_2O \qquad \Delta G^{\circ\prime} = -2823\ \text{kJ/mol}$$

The same amount of free energy is released regardless of where the reaction occurs, as long as it occurs under standard conditions. In a test tube, virtually all of the energy is lost as heat. In the body, part of the energy is employed to synthesize approximately 32 mol of ATP per mol of glucose oxidized (Table 11.4). Since, under standard biological conditions, the hydrolysis of each mole of ATP to ADP and inorganic phosphate releases 31 kJ (Section 8.4), the production of 32 mol of ATP from ADP and P_i traps (stores) 992 kJ (32 × 31). Hence, under standard conditions, approximately 35% of the total energy released when glucose is oxidized is trapped in ATP (992/2823 × 100 = 35%), an impressive feat once we recognize that man-

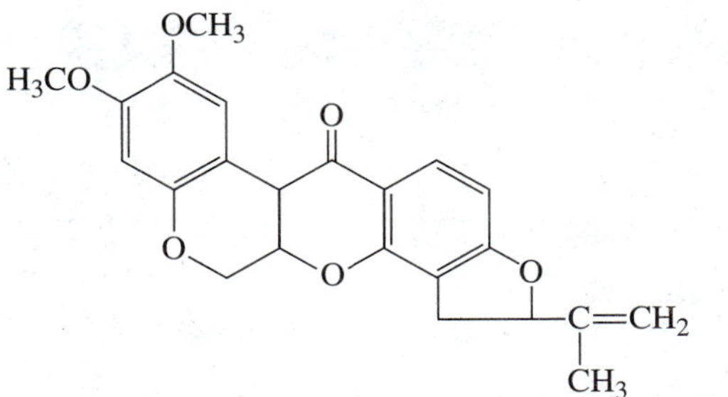

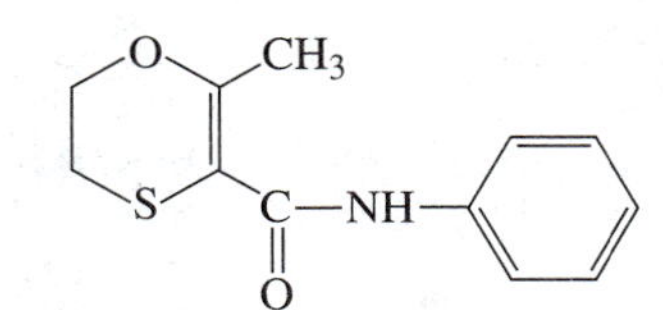

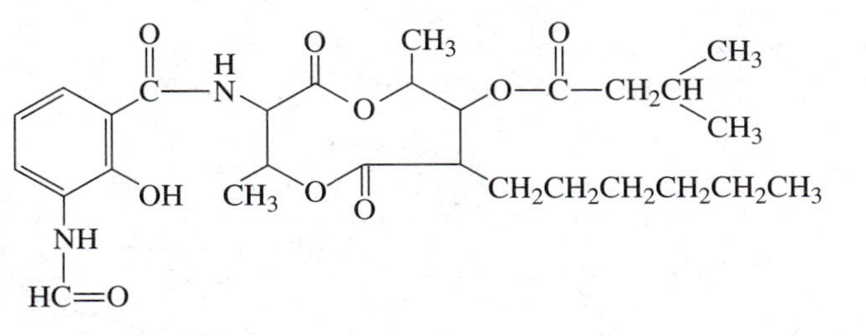

CN^-

Cyanide

CO

Carbon monoxide

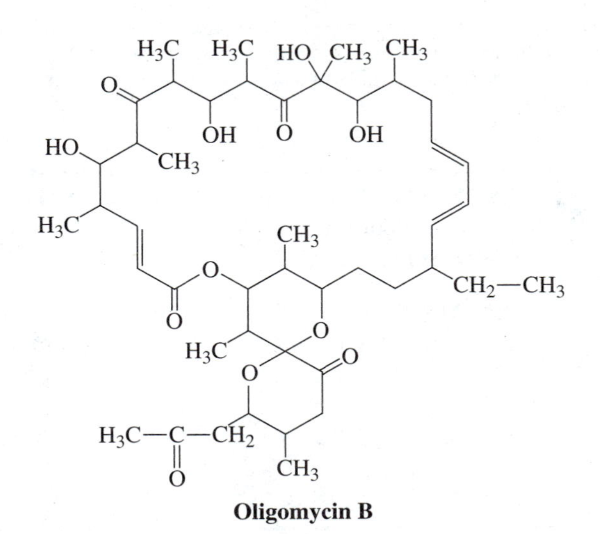

FIGURE 11.22
Respiratory chain inhibitors.

made machines typically operate at efficiencies below 30%. When pondering these figures, we must remember that ATP may store considerably more energy under physiological conditions than under standard conditions. Within most cells, the ΔG_p for the hydrolysis of ATP is probably between −50 and −60 kJ/mol (Section 8.4). **[Try Problems 14.36 through 14.38]**

TABLE 11.4

Approximate Net ATP Production During the Complete Oxidation of Glucose

	Reactions	ATP Production per Glucose[a]
Glycolysis (in cytosol)	Phosphorylation of glucose	−1
	Phosphorylation of fructose 6-phosphate	−1
	Transfer of phosphoryl groups from two 1,3-bisphosphoglycerates to two ADP	+2
	Transfer of phosphoryl groups from two phosphoenolpyruvates to two ADP	+2
Citric Acid Cycle (in mitochondria)	GTP production from the reaction of two succinyl-SCoA	+2
Respiratory Chain (in mitochondria)	Flow of electrons from two NADH produced in glycolysis to O_2	+5 or +3[b]
	Flow of electrons from two NADH produced by the pyruvate dehydrogenase complex to O_2	+5
	Flow of electrons from six NADH produced in the citric acid cycle to O_2	+15
	Flow of electrons from two $FADH_2$ produced in the citric acid cycle to O_2	+3
Approximate Net ATP Production		+32 or +30[b]

[a]The number of ATP produced in the respiratory chain during the oxidation of NADH and $FADH_2$ are estimates.
[b]The actual value depends on which shuttle is used to carry electrons from cytosolic NADH into mitochondria (see Figure 11.9). The shuttle that produces NADH inside mitochondria yields two more ATP per glucose than the shuttle that generates $FADH_2$.

PROBLEM 11.36 How many of the total ATPs synthesized (assume 32) during the complete biological oxidation of glucose are produced within the respiratory chain?

PROBLEM 11.37 If $\Delta G_p = -50$ kJ/mol for the hydrolysis of ATP to ADP and P_i, how much energy is trapped during the production of 32 mol of ATP from ADP and P_i?

PROBLEM 11.38 What is the approximate $\Delta G^{\circ\prime}$ for the following reaction?

$$ATP + 2\ H_2O = AMP + 2\ P_i$$

SUMMARY

Much of the energy used by the human body comes from the oxidation of glucose and other carbohydrates that can be converted to glucose. Some of the intermediates in the oxidation of glucose are used for anabolism. Complete oxidation involves the pyruvate dehydrogenase complex plus three metabolic pathways: glycolysis, the citric acid cycle, and the respiratory chain. This oxidation is coupled to the production of ATP, the immediate source of the energy for most energy-requiring processes in the body.

Glycolysis employs 10 cytosolic enzymes to convert one molecule of glucose to two molecules of pyruvate. The glucose is phosphorylated and then split into two three-carbon fragments, each containing a phosphate. After each three-carbon fragment is oxidized by NAD^+, it is converted to pyruvate through a sequence of reactions in which two ATPs are produced through substrate-level phosphorylation. Collectively, the two three-carbon fragments yield four ATPs. Since two ATPs are consumed by kinases during the early stages of glycolysis, the net production of ATP is 2. Under aerobic conditions, the NADH produced during the oxidation of each three-carbon fragment is reconverted to NAD^+ as it passes a pair of electrons into a mitochondrion and, ultimately, the respiratory chain. Under anaerobic conditions, the NADH from glycolysis is reoxidized to NAD^+ as lactate dehydrogenase converts pyruvate to lactate. The lactate dehydrogenase reaction allows a cell to produce some ATP in the absence of O_2.

Although each of the three irreversible steps in glycolysis is catalyzed by a regulatory enzyme, phosphofructokinase-1 is the prime target for allosteric regulators. Positive effectors of glycolysis include AMP, ADP, and fructose 2,6-bisphosphate; ATP, citrate, and alanine are some of the modulators that inhibit the conversion of glucose into two pyruvates. Fructose 2,6-bisphosphate, whose concentration is determined mainly by hormones, is a product of the phosphofructokinase-2–catalyzed reaction. In general, glycolysis occurs most rapidly when a cell needs energy or needs intermediates for anabolism. Low $NADH/NAD^+$ ratios and low ATP/ADP ratios signal a low energy charge. High citrate or alanine concentrations are indicative of an adequate supply of intermediates for anabolism.

The pyruvate dehydrogenase complex, which is confined to the matrix of mitochondria, employs three enzymes, five coenzymes, and five distinct reactions to convert the two pyruvates from one glucose into two acetyl-SCoAs and two CO_2 molecules. Two NAD^+ are reduced to two NADH in the process. The pyruvate dehydrogenase complex in humans is regulated by both allosteric effectors and covalent modification. The effectors include AMP (+), CoASH (+), NAD^+ (+), ATP (−), acetyl-SCoA (−), and NADH (−). A unique protein kinase and protein phosphatase are directly responsible for regulation through covalent modification. Both of these enzymes are themselves under allosteric control. The pyruvate dehydrogenase complex tends to be most active when the energy charge of a cell is low and there is a need to complete the oxidation of glucose that is initiated by glycolysis.

During the complete oxidation of glucose, the two acetyl-SCoAs (one from each of the two pyruvates generated from a single glucose) released by the pyruvate dehydrogenase complex condense with separate oxaloacetates to create two citrates and initiate the citric acid cycle. Seven additional reactions convert each acetyl group to two CO_2 and regenerate oxaloacetate. Three NADH, one $FADH_2$, and one GTP are also produced for each acetyl group that enters the cycle. The pathway is controlled by three regulatory enzymes, including isocitrate dehydrogenase. Effectors, including ADP (+), Ca^{2+} (+), ATP (−), NADH (−) and succinyl-SCoA (−), ensure that the cycle is most active when a cell is most in need of energy.

For each glucose that is run through glycolysis, the pyruvate dehydrogenase complex and the citric acid cycle, there is a net production of 2 ATP, 10 NADH, 2 $FADH_2$, 2 GTP, and 6 CO_2. Each NADH can lead to the production of approximately 2.5 ATPs when it passes the pair of electrons it has acquired from glucose through the respiratory chain to O_2, the terminal electron acceptor. Each $FADH_2$, when coupled to the respiratory chain, is a source of roughly 1.5 ATPs. Each GTP is equivalent to an ATP. Consequently, around 32 mol of ATP can be produced for each mole of glucose that is completely oxidized within the human body. ATP production is what glucose oxidation is all about.

The respiratory chain consists of four respiratory complexes, ATP synthase, and two mobile electron carriers, all associated with the inner mitochondrial membrane. All of the electron carriers fall into one of five classes: flavins, quinones, iron–sulfur clusters, cytochromes, and copper atoms. As electrons are passed from carrier to carrier, part of the energy released is used to pump H^+ from the matrix into the intermembrane space creating a H^+ gradient whose electrochemical energy is known as a proton-motive force. As the H^+ flows back into the matrix through ATP synthase, the proton-motive force provides the energy for ATP production (from ADP + P_i) through the binding-change mechanism. The overall process is described as oxidative phosphorylation, since the energy for the phosphorylation of ADP comes ultimately from the oxidation reactions associated with the transfer of electrons along the respiratory chain.

Only three of the four respiratory complexes pump H^+ into the intermembrane space. The electrons from NADH pass through all of these pump-containing complexes, whereas the electrons from $FADH_2$ traverse only two of the three.

In the respiratory chain, oxidation and phosphorylation are normally tightly coupled; one process does not occur without the other. Some cells, however, produce a protein that allows H^+ to leak back into the matrix without passing through ATP synthase. This leakage of H^+ uncouples oxidation and phosphorylation and allows oxidation to proceed in the absence of phosphorylation. In the presence of an uncoupler, most of the energy of fat or carbohydrate oxidation is converted to heat; very little ends up in ATP. Some organisms, including human infants, use uncouplers to produce part of the heat required to maintain an appropriate body temperature.

A variety of substances have been discovered that inhibit oxidative phosphorylation by either blocking the flow of electrons through the respiratory chain or inhibiting ATP synthase. Although such substances are toxic, several have been employed as tools to study the respiratory chain.

EXERCISES

11.1 List the major uses of energy by the human body. Hint: Review Chapter 8.

11.2 Sunlight is the ultimate source of most of the energy used by the human body. Explain this fact. Hint: Review Chapter 8.

11.3 What is the maximum net production of ATP from 1 mol of glucose when aerobic glycolysis is coupled to a functional respiratory chain? What is the maximum net production of ATP from 1 mol of glucose during anaerobic glycolysis? Explain.

11.4 Define "oxidation" and "reduction." Which of the following compounds are more highly oxidized than ethanol? Explain.

Ethanal (acetaldehyde)
Ethane
1,2-Ethanediol
Ethanoic acid (acetic acid)

11.5 Define "kinase." Are phosphoglycerate kinase and pyruvate kinase exceptions to this definition? Explain.

11.6 Identify the two steps in glycolysis that are furthest from equilibrium (are most irreversible) under physiological conditions. Are these two steps also furthest from equilibrium under standard biological conditions? Explain.

11.7 Identify those steps in glycolysis where the products in Figure 11.1 are generated.

11.8 True or false? The cytosolic pH of muscle cells drops as active muscle shifts from aerobic to anaerobic conditions. Explain.

11.9 Explain why the activity of enzymes is sensitive to changes in pH.

11.10 ATP inhibits glycolysis by allosterically inhibiting both phosphofructokinase-1 and pyruvate kinase. What problems would be created for a muscle cell if ATP also inhibited hexokinase, a third allosteric enzyme in glycolysis?

11.11 The accumulation of fructose 6-phosphate leads to the accumulation of glucose 6-phosphate as well. Would the accumulation of fructose 1,6-bisphosphate lead to the accumulation of both fructose 6-phosphate and glucose 6-phosphate? Explain.

11.12 True or false? The rate of glycolysis increases in muscle cells during anaerobic exercise. Explain.

11.13 True or false? Allosteric inhibitors are always structurally similar to the normal substrates for their target enzymes. Explain.

11.14 How many net ATPs can be produced from 1 mol of glucose during glycolysis in a mutant cell unable to assemble an active triose phosphate isomerase? How many net ATPs can be produced when glycolysis in the mutant cell is coupled to an active respiratory chain?

11.15 What are the immediate oxidizing agents for the complete oxidation of glucose within the body? These are the compounds that directly remove electrons from glucose.

11.16 What specific reactions account for most of the CO_2 that is eliminated from your body each time you exhale?

11.17 List those steps in glycolysis that are isomerization reactions.

11.18 List those enzymes in glycolysis that are

a. Oxidoreductases
b. Lyases
c. Transferases

11.19 The equilibrium constant for the pyruvate kinase–catalyzed reaction is large since $\Delta G^{\circ\prime} = -31.4$ kJ/mol. What does this fact reveal about the relative free energy of hydrolysis of enoyl phosphates and phosphate anhydrides?

11.20 A cell with a pair of abnormal genes is unable to carry out glycolysis under anaerobic conditions, but aerobic glycolysis proceeds normally. What polypeptide is encoded in the abnormal genes?

11.21 What distinguishes allosteric enzymes from nonallosteric enzymes?

11.22 Study the cellular regulation of glucose oxidation. Give one specific example of feedback inhibition and one specific example of feed-forward activation. Explain the metabolic logic behind the examples you have selected.

11.23 Would a drop in ATP concentration lead to an increase in the rate of glycolysis in the absence of an increase in AMP and ADP concentrations? Explain.

11.24 True or false? A eukaryotic cell that regularly uses large amounts of energy needs more mitochondria than a cell that uses much less energy. Explain.

11.25 Which of the reactions involved in the complete oxidation of glucose occur in the cytosol and which occur within mitochondria?

11.26 Based on the information given in this chapter, what is the minimum number of unique allosteric sites that you would expect to exist within the pyruvate dehydrogenase complex? Explain.

11.27 List those coenzymes that function as oxidizing agents during the complete oxidation of glucose within cells.

11.28 List the five coenzymes found within the pyruvate dehydrogenase complex. For each coenzyme, list its vitamin precursor (if any) and its role during the conversion of pyruvate to acetyl-SCoA.

11.29 List the two classes of enzymes that are involved in determining the extent of phosphorylation for those enzymes that are regulated by phosphorylation. What are the substrates for each of these two classes of enzymes?

11.30 Identify those steps in the citric acid cycle that represent redox reactions.

11.31 True or false? Organelles are responsible for some metabolic channeling. Explain.

11.32 List the two allosteric effectors that modulate *both* glycolysis and the citric acid cycle. How does a cell benefit from the shared effectors?

11.33 List the five major classes of electron carriers within the respiratory chain. Which of these are capable of transporting two electrons at a time?

11.34 What is the maximum number of ATPs that can be produced when a pair of electrons is passed through complex IV of the respiratory chain to oxygen? Explain. Assume that complex IV pumps twice as many H^+ across the intermitochondrial membrane as does complex III when a pair of electrons passes through.

11.35 True or false? The respiratory chain cannot operate in the absence of oxygen. Explain.

11.36 What specific amino acid residue is an integral part of the iron–sulfur clusters within iron–sulfur proteins?

11.37 What specific atom is reversibly oxidized as an electron flows through a cytochrome molecule? An iron–sulfur cluster? In each case, identify the usual oxidation states for this atom.

11.38 What distinguishes *a* cytochromes from *c* cytochromes?

11.39 Explain how iron deficiency directly reduces the body's ability to produce energy.

11.40 What specific enzyme is directly responsible for the production of ATP during oxidative phosphorylation? What is the direct (immediate) source of the energy used to produce this ATP?

11.41 From an energy transformation standpoint, what is the efficiency of ATP production when 25 ATPs are produced for each molecule of glucose oxidized under standard biological conditions?

11.42 Will a respiratory chain uncoupler increase, decrease, or have no effect on (a) the pH gradient associated with the inner mitochondrial membrane and (b) the amount of ATP produced per mole of glucose oxidized? Explain.

11.43 What impact does cyanide (a respiratory chain inhibitor) have on the rate of glycolysis and the rate of cycling of the citric acid cycle? Explain.

11.44 Identify the energy-requiring steps or events involved in the complete oxidation of glucose within the human body. In each case, identify the immediate source of the required energy.

SELECTED READINGS

Cross, R. L., Our primary source of ATP, *Nature* 370, 594–595 (1994).

A brief review of the structure and functioning of ATP synthase.

Gennis, R., and S. F. Ferguson-Miller, Structure of cytochrome *c* oxidase, energy generator of aerobic life, *Science* 269, 1063–1064 (1995).

Reviews the structure and functioning of cytochrome c oxidase (complex IV of the respiratory chain).

Hinkle, P. C., M. A. Kumar, A. Resetar, and D. L. Harris, Mechanistic stoichiometry of mitochondrial oxidative phosphorylation, *Biochemistry* 30, 3576–3582 (1991).

Provides estimates of the stoichiometry of ATP formation during oxidative phosphorylation.

Lehninger, A. L., D. L. Nelson, and M. M. Cox, *Principles of Biochemistry,* 2nd ed. New York: Worth Publishers, 1993.

An encyclopedia-type text, with a thorough coverage of metabolism.

Moran, L. A., K. G. Scrimgeour, H. R. Horton, R. S. Ochs, and J.D. Rawn, *Biochemistry,* 2nd ed. Englewood Cliffs, NJ, Neil Patterson/Prentice-Hall, 1994.

An encyclopedic biochemistry text containing a detailed discussion of metabolism. Includes a limited number of problems and some selected readings.

Smith, C. A., and E. J. Wood, *Energy in Biological Systems.* Hong Kong: Chapman & Hall, 1991.

A 171-page paperback that summarizes the basic metabolic pathways. Contains both questions and references.

Stryer L., *Biochemistry,* 4th ed. New York: W. H. Freeman and Company, 1995.

A comprehensive biochemistry text with good coverage of carbohydrate metabolism.

Werner, R., *Essential Biochemistry and Molecular Biology,* 2nd ed. New York: Elsevier Publishing Co., 1992.

A 463-page paperback that provides a concise and readable review of basic metabolism.

12 Additional Topics in Carbohydrate Catabolism

CHAPTER OUTLINE

Chapter 10 described the central role of glucose in fuel metabolism in humans. This monosaccharide, which is delivered to cells throughout the body by the blood system, is the major fuel for many cells, including brain cells. Since both hypoglycemia (too little blood glucose) and hyperglycemia (too much blood glucose) can lead to serious health problems, blood glucose levels are normally held within a rather narrow range most of the time. A number of diseases, including diabetes, are associated with abnormalities in glucose metabolism.

In humans, the metabolism of glucose is tightly coupled to the metabolism of glycogen (Figure 12.1), a branched polymer of glucose and a storage form of this sugar. Although most human cells contain at least small amounts of glycogen, the vast majority of glycogen resides in liver and muscle. The glycogen in liver plays a key role in buffering blood glucose levels; glucose from liver glycogen is released into blood when blood glucose levels are too low, and glucose is removed from blood to assemble liver glycogen when blood glucose levels are elevated (Table 10.12). Glucose from muscle glycogen is used as a fuel during periods of muscular activity. Starch and other polymers of glucose are storage forms of this monosaccharide in plants and microorganisms. **[Try Problems 12.1 and 12.2]**

Chapter 11 examines the roles of glycolysis, the pyruvate dehydrogenase complex, the citric acid cycle, and the respiratory chain in the oxidation of glucose and the coupled production of ATP. The pentose phosphate pathway provides an alternate route for oxidizing glucose. This route is coupled to the formation of NADPH rather than the production of ATP. NADPH from the pentose phosphate pathway accounts for most of the reducing power utilized during anabolism (Chapter 14).

The catabolism of glucose in most organisms involves the same metabolic pathways employed to oxidize glucose in humans. Additional catabolic pathways also exist in some instances. One of these accounts for alcoholic fermentation, the conversion of glucose to ethanol by yeast and certain other microorganisms under anaerobic conditions.

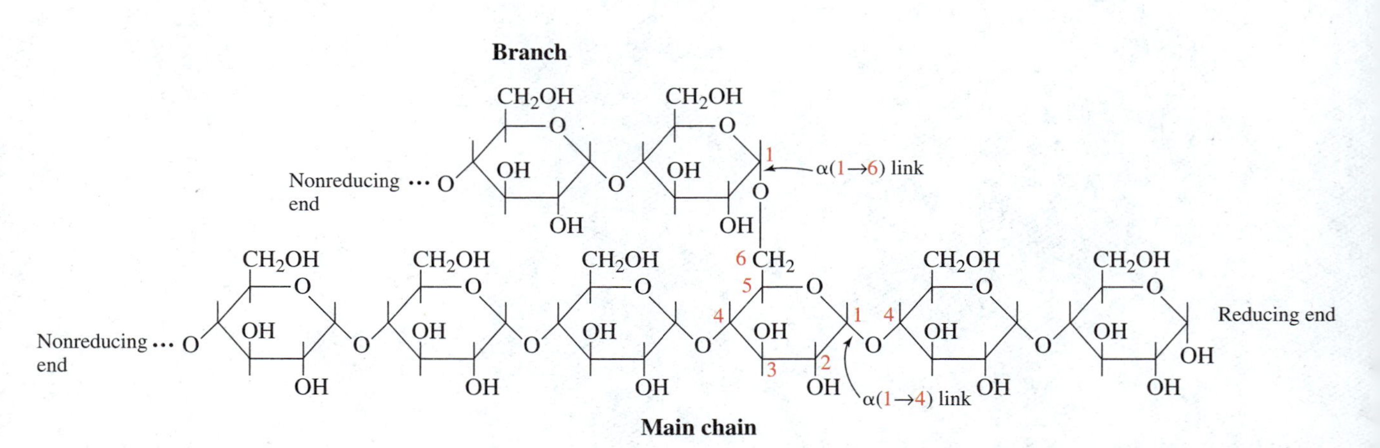

FIGURE 12.1
Glycogen. Because a typical glycogen molecule contains numerous branches, there are many nonreducing ends but only a single reducing end.

This chapter examines diabetes plus three glucose-linked catabolic processes, glycogenolysis, the pentose phosphate pathway, and alcoholic fermentation. Featured biochemical themes include energy transformation, communication, regulation of reaction rates, specificity, and hypothesis construction. Chapter 13 concludes the discussion of catabolism by examining the biological breakdown of fats and amino acids.

PROBLEM 12.1 To what classes of organic compounds does glycogen belong? Identify each hydrolyzable bond in Figure 12.1. Identify each anomeric carbon in Figure 12.1

PROBLEM 12.2 Define "buffer" as this term relates to the control of pH. Does "buffer" have a similar meaning when used to describe the role of the liver in regulating blood glucose levels? Explain.

12.1 GLYCOGEN CATABOLISM

Since glycogen is stored in the cytosol of eukaryotic cells, the enzymes involved in its metabolism are cytosolic. The breakdown of glycogen to release glucose (primarily as glucose 1-phosphate) is known as **glycogenolysis.** It requires two enzymes, **glycogen phosphorylase** and **debranching enzyme.**

Glycogenolysis

Glycogenolysis begins when glycogen phosphorylase catalyzes the cleavage of the terminal glucose residue from a nonreducing end of a glycogen molecule (Figure 12.2). As the $\alpha(1 \rightarrow 4)$ glycosidic link is severed, a phosphate residue is added to

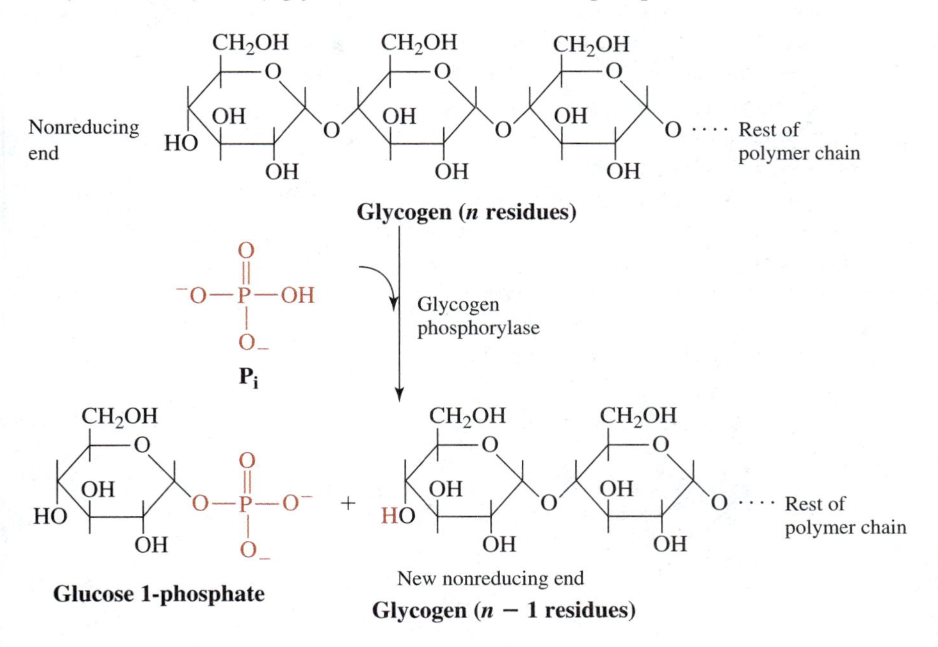

FIGURE 12.2 Phosphorolysis catalyzed by glycogen phosphorylase.

the released glucose and a hydrogen atom is added to the residual glycogen, a process called **phosphorolysis.** Phosphorolysis is the same as hydrolysis except a phosphate group, rather than a hydroxyl group, is added during the process of bond cleavage. The products of glycogen phosphorylase-catalyzed phosphorolysis are glucose 1-phosphate and a glycogen molecule one glucose residue shorter than the initial glycogen. The shortened glycogen can immediately be used as a substrate for another round of phosphorolysis. The stepwise removal of glucose residues by phosphorolysis continues to within four residues of an $\alpha(1 \longrightarrow 6)$ glycosidic link at a branch point.

After both a branch and its main chain have been cleaved to within four residues of the branch point, debranching enzyme catalyzes the transfer of a trisaccharide unit from the branch to the nonreducing end of the main chain (Figure 12.3). The same enzyme then catalyzes the hydrolysis of the $\alpha(1 \longrightarrow 6)$ glycosidic bond between the main chain and the solitary glucose residue remaining in the branch. Debranching enzyme's transferase activity is called **4-α-glucanotransferase,** and its hydrolase activity is named **amylo-1,6-glucosidase.** After the amylo-1,6-glucosidase-catalyzed removal of the side chain, glycogen phosphorylase returns to the job of catalyzing the degradation of the main chain. When the next branch is encountered, debranching enzyme once again steps in to help catalyze the elimination of the branch. Ultimately, a glycogen molecule can be broken down into many glucose 1-phosphates and a few free glucose (from the catalytic activity of the amylo-1,6-glucosidase domain of the debranching enzyme). The metabolic picture in a cell or organism normally changes before this occurs, and a cell begins to add glucose residues to the partially degraded glycogen to build up its glucose reserves.

A typical glycogen molecule contains an abundance of nonreducing ends, because its main chain has many branches and some branches are themselves branched. When glycogen needs to be mobilized, multiple nonreducing ends are simultaneously degraded with the catalytic assistance of separate glycogen phosphorylase molecules, an event leading to a burst of glucose 1-phosphate production. When more fuel is needed, it is often needed in a hurry. Since most glucose is released in a phosphorylated form, it is trapped within the glycogen-containing cell unless enzymes are available to catalyze the removal of the phosphoryl group. The necessary enzymes are present in liver but not in muscle. **[Try Problem 12.3]**

If the glucose from glycogen is destined to be completely oxidized, **phosphoglucomutase** catalyzes the conversion of glucose 1-phosphate to glucose 6-phosphate. The glucose 6-phosphate, an intermediate in glycolysis, is converted to two pyruvate with the catalytic assistance of glycolytic enzymes. Pyruvate is further oxidized with the catalytic assistance of the pyruvate dehydrogenase complex, the citric acid cycle and the respiratory chain (Chapter 11).

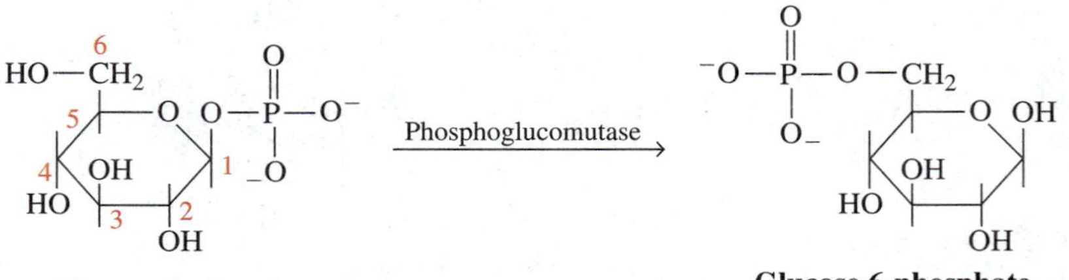

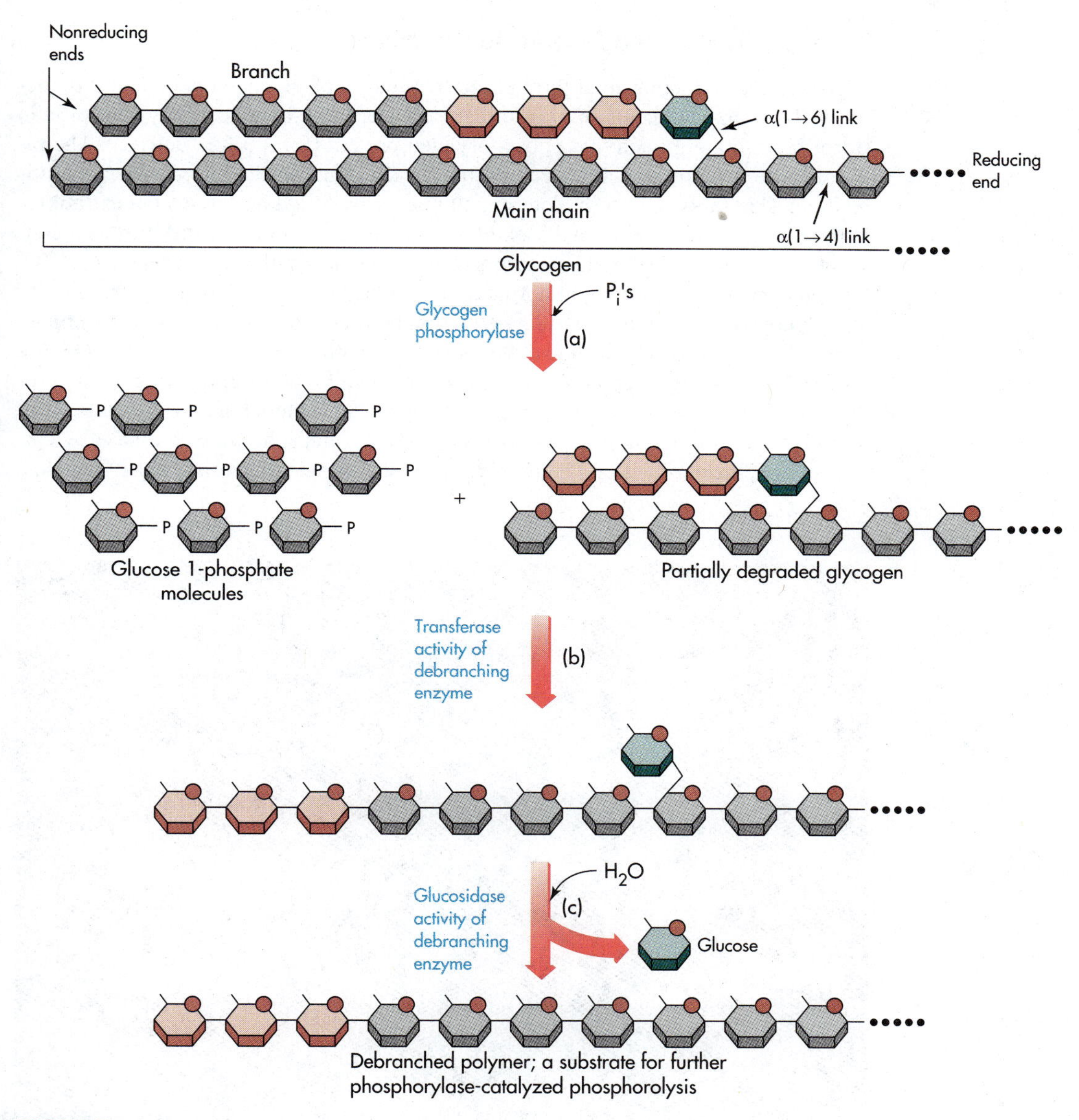

FIGURE 12.3
Debranching enzyme action. (a) Phosphorolytic stepwise removal of glucose residues from nonreducing ends. Process stops when phosphorylase comes within four residues of an $\alpha(1 \longrightarrow 6)$ glycosidic link. b) Debranching enzyme catalyzes the transfer of a trisaccharide from the four-residue branch to the nonreducing end of the main chain. (c) Debranching enzyme catalyzes hydrolysis of the $\alpha(1 \longrightarrow 6)$ acetal link that binds the remaining glucose residue to the main chain.

Epinephrine-Linked Regulation in Muscle

Glycogenolysis is controlled through the regulation of glycogen phosphorylase activity. This enzyme exists in two forms: a low-activity, nonphosphorylated protein (form **b**); and a high-activity, phosphorylated protein (form **a** for "active"). The ratio of these two forms plays a central role in determining the rate of glycogen utilization. This ratio is, in turn, determined mainly by blood hormone concentrations. Different types of cells normally exhibit distinctive hormone sensitivities as a consequence of unique hormone receptors or unique combinations of receptors.

Epinephrine is the principal hormone regulating glycogen phosphorylase activity and glycogen utilization *in muscle cells.* Epinephrine's mode of action is examined in the discussion of cyclic nucleotide signaling pathways in Section 10.5 (Figure 10.13). The binding of epinephrine to a plasma membrane receptor triggers the production of cAMP which activates a cAMP-dependent protein kinase (also called protein kinase A; Figure 12.4 a). This kinase activates phosphorylase b kinase by catalyzing its phosphorylation (Figure 12.4 b). Although not shown in these figures, a

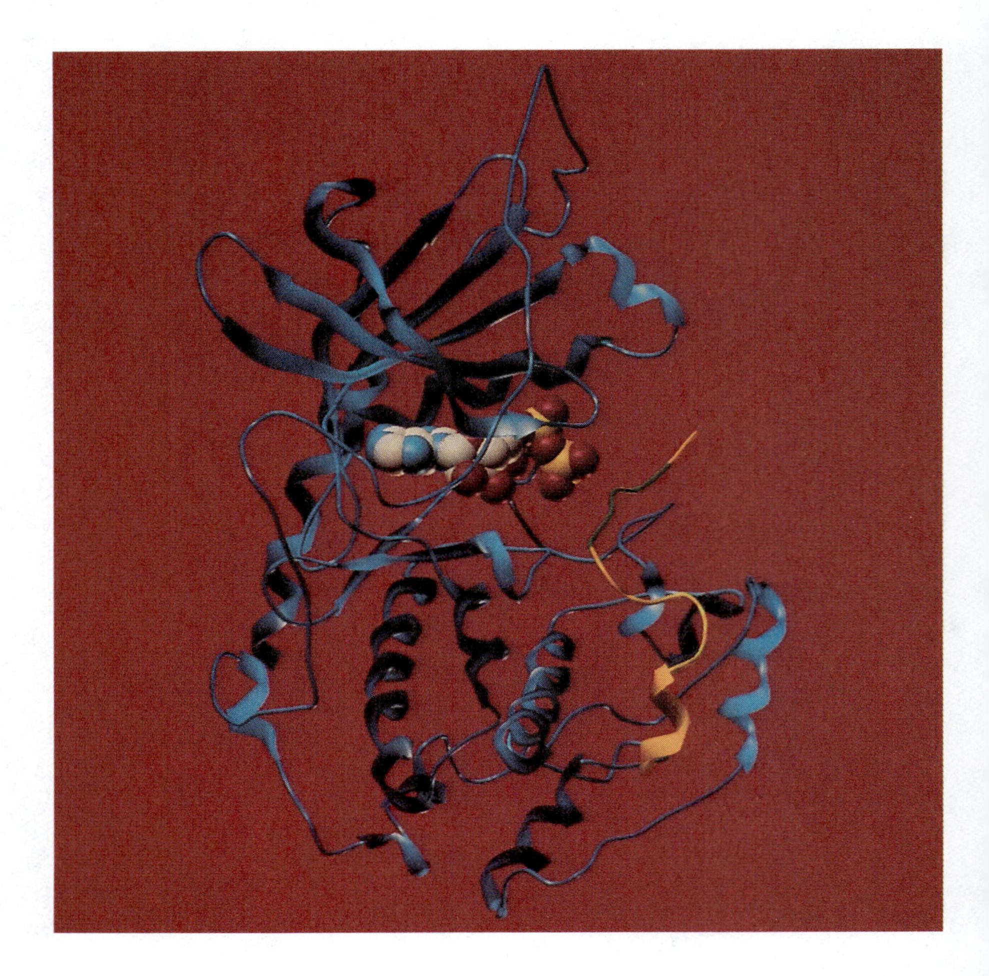

FIGURE 12.4 a
Cyclic AMP-dependent protein kinase complexed with ATP and a 20-residue segment of a naturally occurring protein kinase inhibitor.

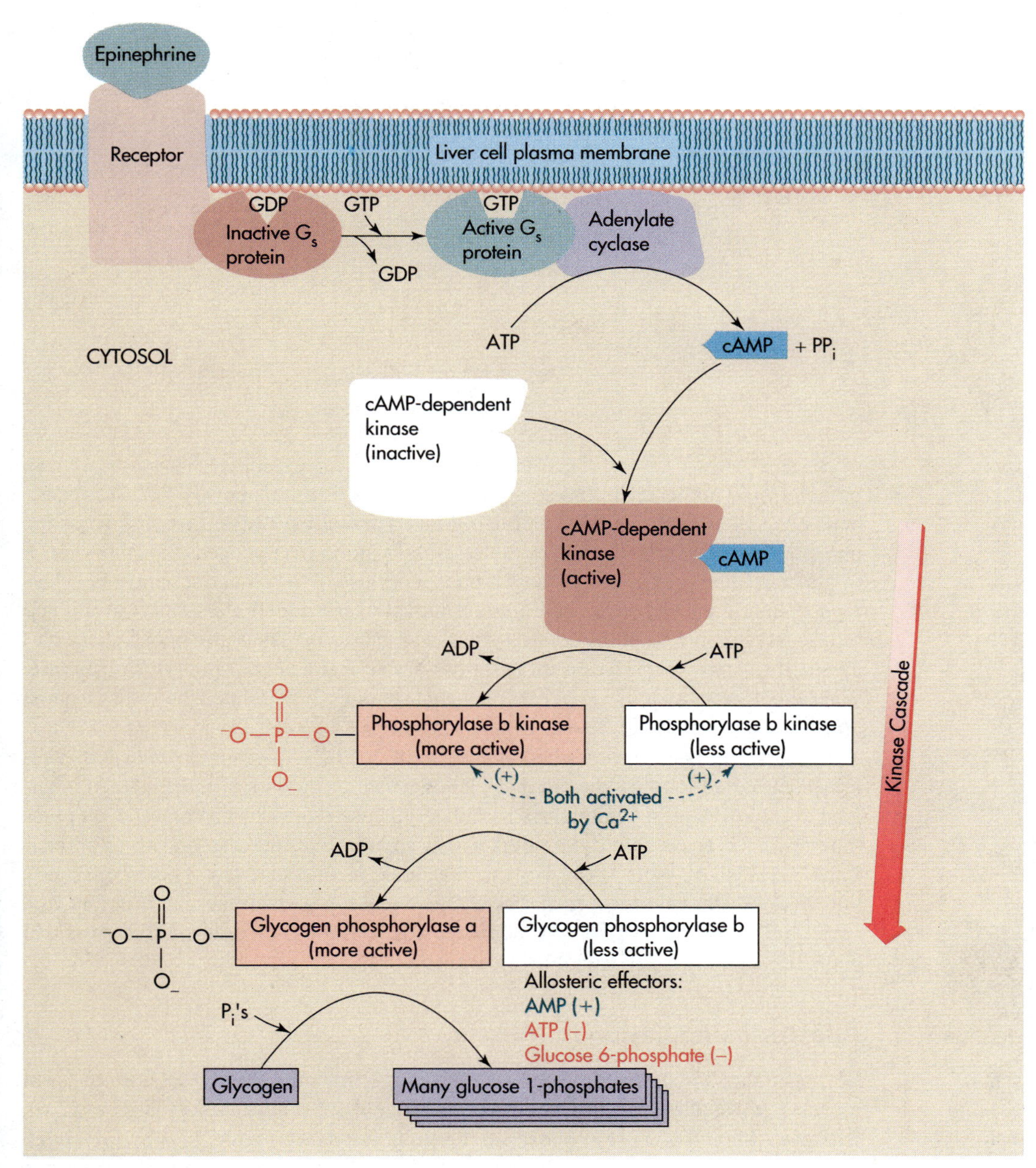

FIGURE 12.4 b
Regulation of glycogen phosphorylase activity in muscle cells. The binding of epinephrine to its plasma membrane receptor activates a G_s protein that stimulates the adenylate cyclase–catalyzed production of cAMP, a second messenger. Cyclic AMP binds to and activates a cAMP-dependent kinase that activates phosphorylase b kinase by catalyzing its phosphorylation. The activated phosphorylase b kinase catalyzes the phosphorylation and activation of glycogen phosphorylase b. The glycogen phosphorylase a formed catalyzes the breakdown of glycogen into many glucose 1-phosphates.

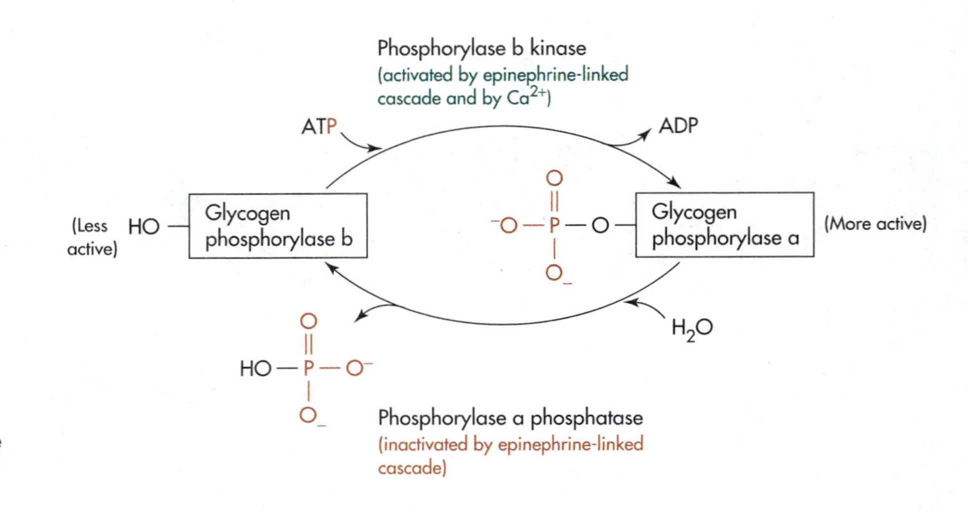

FIGURE 12.5
Regulation of the ratio of phosphorylase a to phosphorylase b in muscle cells.

branch of the epinephrine-linked signaling pathway simultaneously leads to the *inactivation* of **phosphorylase a phosphatase (protein phosphatase 1).** Activated phosphorylase b kinase catalyzes the phosphorylation of glycogen phosphorylase b to generate the more active glycogen phosphorylase a. When active, phosphorylase a phosphatase catalyzes the hydrolytic removal of the added phosphoryl group and returns glycogen phosphorylase to its less active b form. The same phosphatase catalyzes the dephosphorylation of the more active form of phosphorylase b kinase. The ratio of active kinase to active phosphatase directly regulates the ratio of phosphorylase a to phosphorylase b (Figure 12.5). **[Try Problems 12.4 and 12.5]**

The release of epinephrine from the adrenal medulla helps prepare humans and many animals for fight or flight.

Epinephrine is often described as the "fight or flight" hormone, because it is released into the blood by the adrenal glands when an individual is frightened, challenged, or stressed. By helping to mobilize glucose stored in glycogen, it prepares muscle for vigorous activity. Epinephrine is a catecholamine also known as adrenaline (Figure 10.5, Table 10.8). For this reason, epinephrine receptors are commonly known as **adrenergic receptors.** There are several types of adrenergic receptors, and each type is linked to a unique reaction cascade. It is the **β-adrenergic receptor** that helps regulate glycogen catabolism in muscle.

Allosteric Regulation in Muscle

Glycogen catabolism is controlled by regulatory enzymes, as well as hormones. In muscle, AMP allosterically stimulates glycogen phosphorylase b (Figure 12.6), whereas ATP and glucose 6-phosphate are inhibitory (see Figure 12.4 b). Glycogen phosphorylase a is unaffected. The regulatory activities of AMP and ATP make sense from an energy charge standpoint. When AMP levels are elevated, signaling a low energy charge, fuel needs to be mobilized. An abundance of energy, as signaled by a high ATP concentration, leads a muscle cell to conserve its fuel reserves. Glycogen is also conserved when a cell already has a high concentration of glucose in the form of glucose 6-phosphate.

Phosphorylase b kinase, which catalyzes the transformation of glycogen phosphorylase b to glycogen phosphorylase a (the more active form), is a second regula-

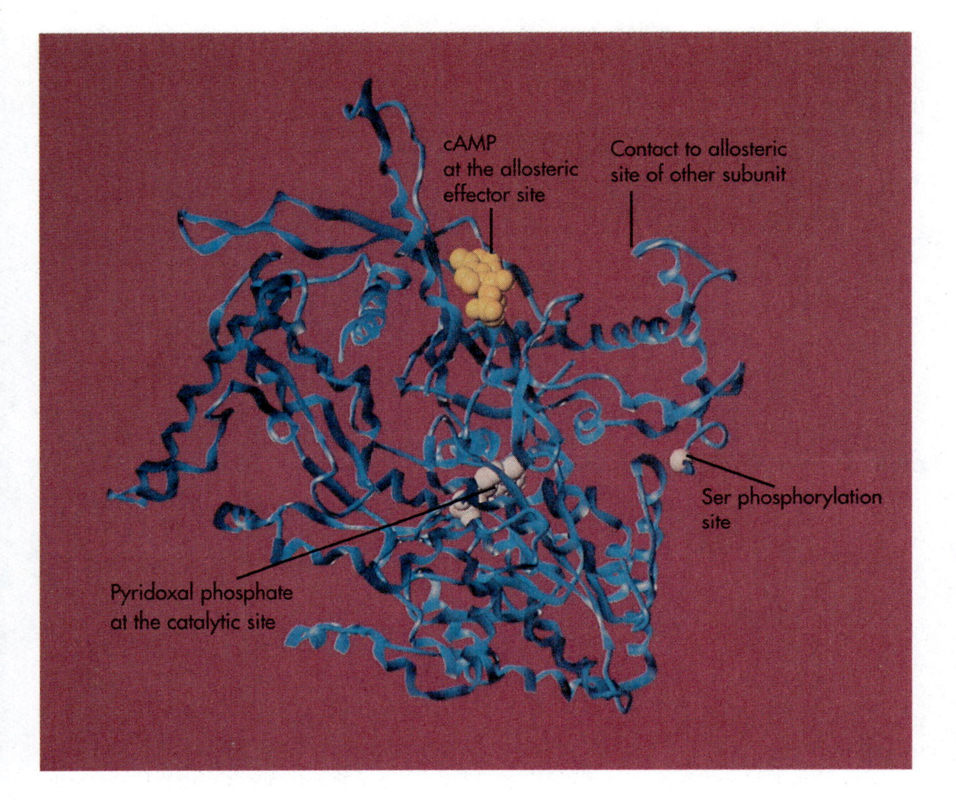

FIGURE 12.6
Phosphorylase b phosphatase subunit. One of the two identical subunits within phosphorylase b phosphatase. Several structural and functional features are identified, including the allosteric effector site, the catalytic site, and the phosphorylation site.

tory enzyme. Both the phosphorylated and nonphosphorylated forms of this enzyme contain a calmodulin (a Ca^{2+} binding polypeptide; Section 10.8) subunit and are activated by Ca^{2+}. The kinase is most active when both phosphorylated and bound to Ca^{2+} (see Figure 12.4 b). The molecular logic behind Ca^{2+}'s stimulation of glycogen mobilization lies in the stimulation of muscle contraction (an energy- and fuel-dependent process) by transient increases in Ca^{2+} concentration. When muscle is contracting, it needs to be rallying glucose to support anticipated future contractions. While hormone-induced changes tend to be relatively slow (seconds to minutes), allosteric effectors allow cells to respond almost immediately (within milliseconds) to changes in energy and fuel requirements.

Glucagon-Linked Regulation in Liver

Although epinephrine receptors do exist on hepatocytes, glucagon normally plays the central role in the hormonal regulation of glycogen mobilization *in liver.* Glucagon is a polypeptide hormone containing 29 amino acid residues (Table 10.8). Hepatocytes contain considerably more glucagon receptors than any other human cells. After glucagon is released into the blood by the pancreas in response to low blood sugar levels (Section 10.12), its binding to a liver cell plasma membrane receptor triggers a cAMP-linked reaction cascade similar to that of epinephrine. The glucose 1-phosphate generated from liver glycogen in response to glucagon or epinephrine is converted to glucose 6-phosphate by phosphoglucomutase. The subsequent **glucose-6-phosphatase**–catalyzed hydrolysis of glucose 6-phosphate frees the glucose for export into the blood stream. This elevates the low blood sugar level initially responsible for the release of glucagon. **[Try Problem 12.6]**

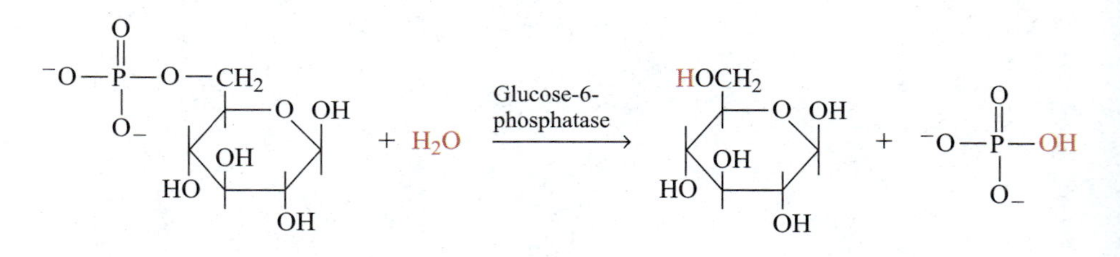

Allosteric Regulation in Liver

The glycogen phosphorylase in liver cells differs from that found in muscle cells. *In liver,* it is glycogen phosphorylase a, rather than glycogen phosphorylase b, that is allosterically regulated, and glucose is the effector (Figure 12.7). Glucose binding changes the conformation of glycogen phosphorylase a, making it more susceptible to dephosphorylation catalyzed by phosphorylase a phosphatase. The hydrolytic removal of the phosphoryl group reduces the total glycogen phosphorylase activity by converting phosphorylase a (the more active form) to phosphorylase b (the less active form). Since the concentration of glucose in the cytosol of liver cells is determined by the level of glucose in blood, glycogen phosphorylase a functions as a sensor to monitor blood sugar levels. When blood sugar levels are high and liver glucose no longer needs to be exported to avoid hypoglycemia, phosphorylase a detects this fact by binding glucose. The subsequent dephosphorylation of phosphorylase a slows the breakdown of glycogen and the release of glucose into blood. This is one

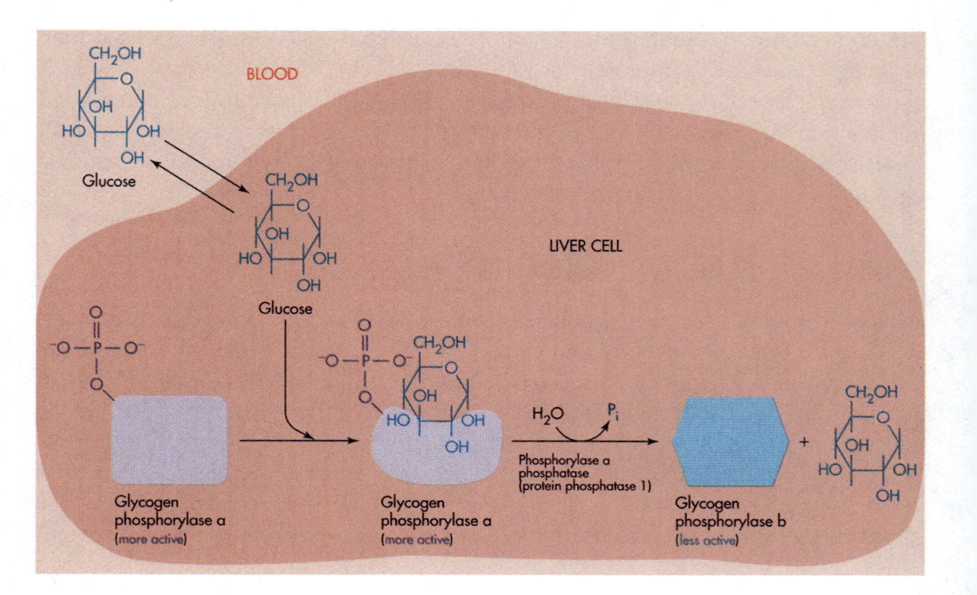

FIGURE 12.7 Glucose-linked regulation of glycogen phosphorylase in liver. As blood glucose levels rise, the associated increase in liver glucose concentration enhances the binding of glucose to glycogen phosphorylase a (Le Châtelier's principle). Glucose-induced conformational changes make glycogen phosphorylase a more susceptible to hydrolysis catalyzed by phosphorylase a phosphatase (protein phosphatase 1). The conversion of phosphorylase a to phosphorylase b leads to a drop in phosphorylase activity. Thus, glycogen mobilization is inhibited when there is already an abundance of glucose in blood.

more example of the many ingenious mechanisms that have evolved to coordinate and regulate glucose catabolism and a multitude of other metabolic processes.

PROBLEM 12.3 True or false? During glycogenolysis, glucose residues can be phosphorolyzed from a glycogen molecule containing 10 nonreducing ends 10 times as rapidly as they can be phosphorolyzed from a linear glycogen molecule. Explain.

PROBLEM 12.4 Which amino acids contain side chains that are common targets for protein kinase–catalyzed phosphorylation reactions? Explain how the phosphorylation of a single side chain in an enzyme can alter its catalytic activity.

PROBLEM 12.5 For each of the following changes indicate whether it will increase, decrease, or have no effect on the ratio of glycogen phosphorylase b to glycogen phosphorylase a in muscle cells under conditions where epinephrine is present in blood. Explain.

a. Inject an epinephrine antagonist into the blood.
b. Increase the number of epinephrine receptors per muscle cell.
c. Increase the intracellular concentration of Ca^{2+}.
d. Inhibit glycogen phosphorylase a phosphatase.
e. Activate phosphorylase b kinase.

PROBLEM 12.6 Why is it important that epinephrine and glucagon have relatively short half-lives in blood?

12.2 PENTOSE PHOSPHATE PATHWAY

The **pentose phosphate pathway** (also called the **pentose shunt** and the **hexose monophosphate shunt**) is another component of the glucose catabolism story. It is a versatile cytosolic pathway that accomplishes a variety of tasks. Its major function is usually the generation of pentoses and reducing power for anabolic processes. Ribose 5-phosphate, for example, can be removed from the pathway for nucleic acid biosynthesis, and erythrose 4-phosphate is a precursor for the synthesis of some protein amino acids in plants and microorganisms. When coupled with auxiliary enzymes, the pentose phosphate pathway also provides an alternative set of reactions for the oxidation of glucose to CO_2. The role of pentose phosphate pathway enzymes in photosynthesis is examined in Chapter 15.

NADPH

The reducing power generated by the pentose phosphate pathway comes in the reduced form of **nicotinamide adenine dinucleotide phosphate (NADPH),** a coenzyme derived from niacin (a B vitamin whose amide form is known as **nicotinamide**) and a close relative of NADH (Table 10.6). Although NADPH and NADH differ by only a single phosphate group (Figure 12.8), NADPH is selectively em-

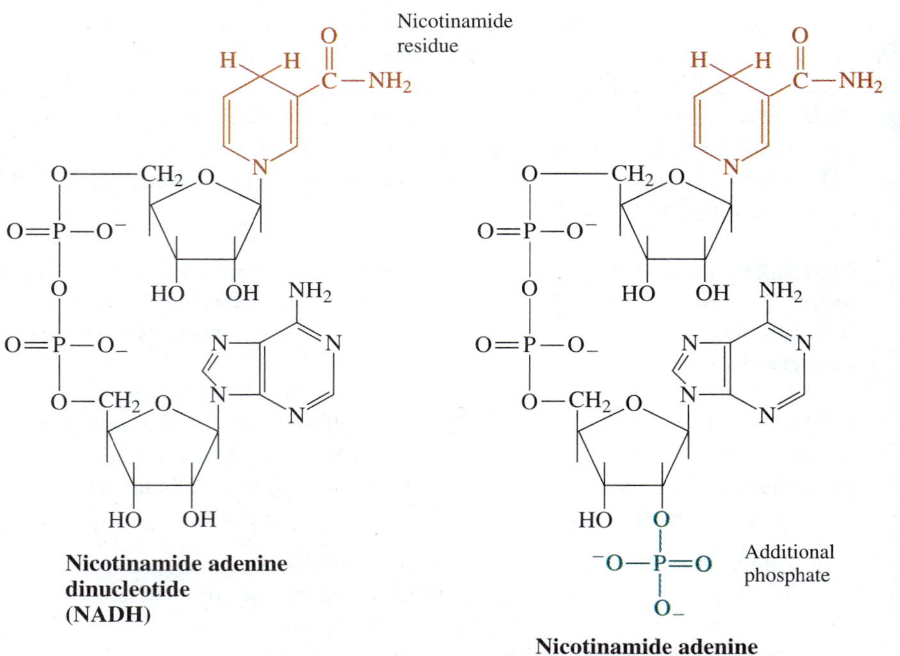

FIGURE 12.8
Structure of NADH and NADPH.

ployed as a reducing agent by many anabolic enzymes, including those involved in the synthesis of fatty acids and steroids from smaller precursors. The pentose phosphate pathway is a major source of this NADPH. **[Try Problem 12.7]**

Oxidative Phase

The pentose phosphate pathway begins with three reactions that collectively convert glucose 6-phosphate to **ribulose 5-phosphate** plus CO_2:

$$2\,\text{NADP}^+ + \text{H}_2\text{O} + \underset{\textbf{(6 carbons)}}{\text{glucose 6-phosphate}} \xrightarrow{\text{3 steps}} 2\,\text{NADPH} + 2\,\text{H}^+ + \underset{\textbf{(1 carbon)}}{\text{CO}_2} + \underset{\textbf{(5 carbons)}}{\text{ribulose 5-phosphate}}$$

The first and third reactions are redox reactions that generate NADPH, the reducing power emphasized in previous paragraphs (Figure 12.9). The second reaction is the hydrolysis of a lactone (cyclic ester) to yield a carboxylic acid. These three reactions represent the **oxidative phase** of the pentose phosphate pathway. The first step, catalyzed by **glucose-6-phosphate dehydrogenase,** is the rate-limiting step, and its rate is primarily determined by $NADP^+$ to NADPH ratio. The oxidative phase of the pentose phosphate pathway is most active when NADPH is in short supply and the

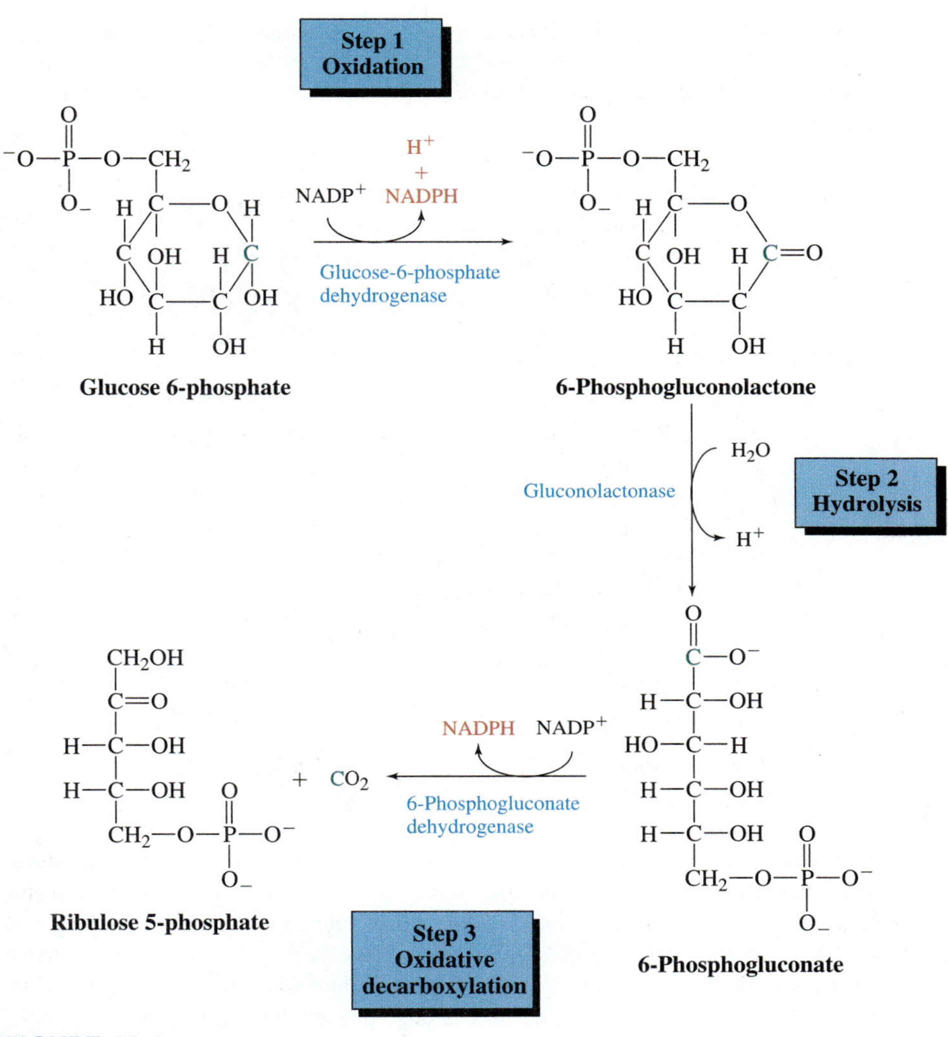

FIGURE 12.9
Oxidative phase of the pentose phosphate pathway.

$NADP^+$ to NADPH ratio is high. NADPH inhibits oxidation by competing with $NADP^+$ for its binding site on glucose-6-phosphate dehydrogenase. **[Try Problems 12.8 and 12.9]**

Nonoxidative Phase

The nonoxidative phase of the pentose phosphate pathway only operates at a significant rate when the oxidative phase generates more ribulose 5-phosphate than is needed for anabolism, because the rate of the nonoxidative phase is determined mainly by substrate availability. The nonoxidative phase converts excess ribulose 5-phosphate to intermediates in glycolysis that can be converted to glucose

6-phosphate, the substrate for the oxidative phase. The nonoxidative phase begins when ribulose 5-phosphates, final products of the oxidative phase, are converted to **ribose 5-phosphate** and **xylulose 5-phosphate** with the catalytic assistance of specific isomerases:

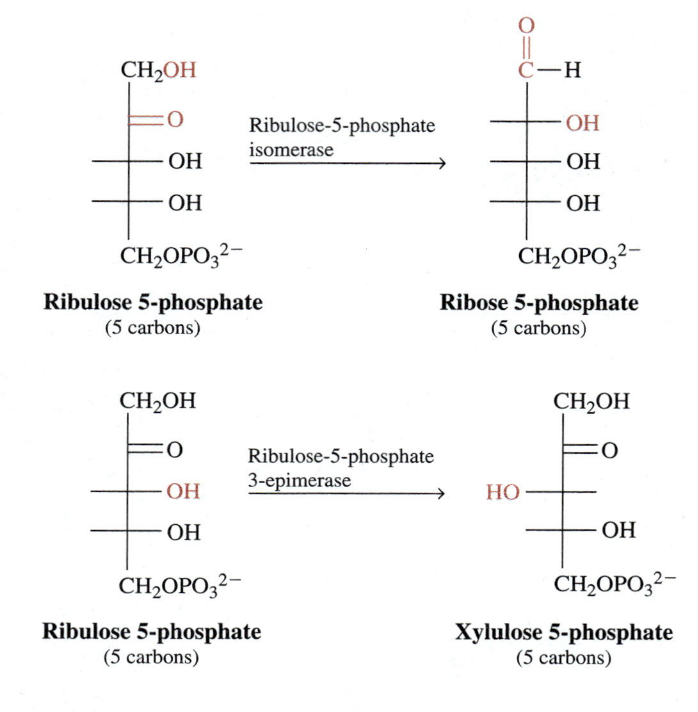

Epimerases are isomerases that catalyze the interconversion of epimers, diastereomers (non–mirror image stereoisomers) that differ in the configuration about a single chiral center. Stereoisomers are reviewed in Section 3.3 and revisited in Section 6.1. After the appropriate pentose phosphates have been produced, **transketolase** and **transaldolase** catalyze the transfers of two-carbon and three-carbon fragments, respectively, from ketose phosphates to aldose phosphates (see top of page 494). During three transfer reactions, monosaccharides are generated that range in size from glyceraldehyde 3-phosphate (contains three carbons) to **sedoheptulose 7-phosphate** (contains 7 carbons).

The net result of the three reactions is the conversion of three pentose phosphate molecules (15 carbons) to two hexose phosphate molecules (12 carbons) and one triose phosphate molecule (3 carbons, for a total of 15 product-linked carbons) that are intermediates in glycolysis. The details of the **nonoxidative phase** of the pentose phosphate pathway are provided in Figure 12.10. **[Try Problems 12.10 and 12.11]**

Complete Oxidation of Glucose 6-Phosphate

The pentose phosphate pathway–linked oxidation of glucose 6-phosphate involves auxiliary enzymes that convert the fructose 6-phosphate and glyceraldehyde 3-phos-

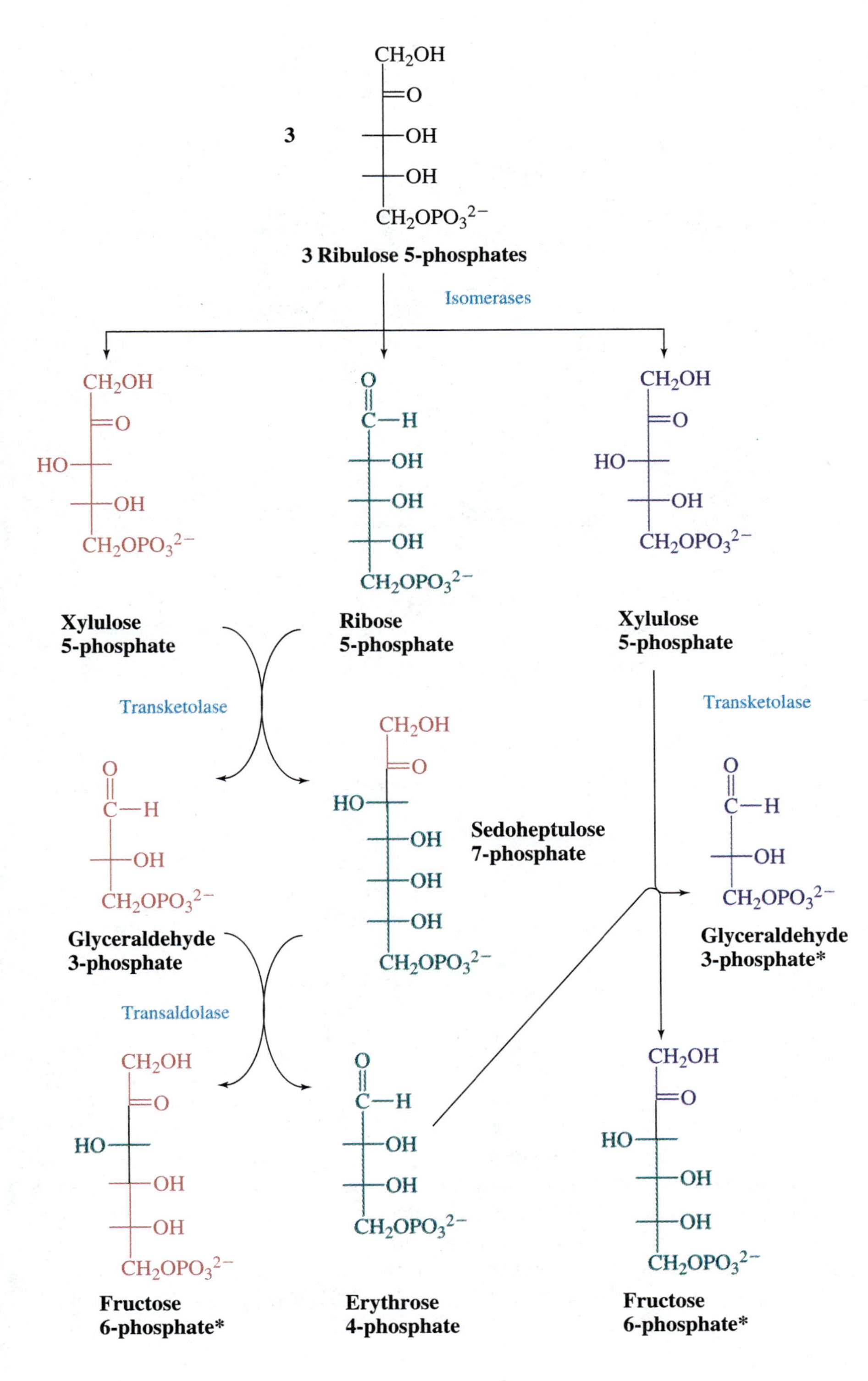

FIGURE 12.10
Nonoxidative phase of the pentose phosphate pathway. Compounds marked with asterisks are final products.

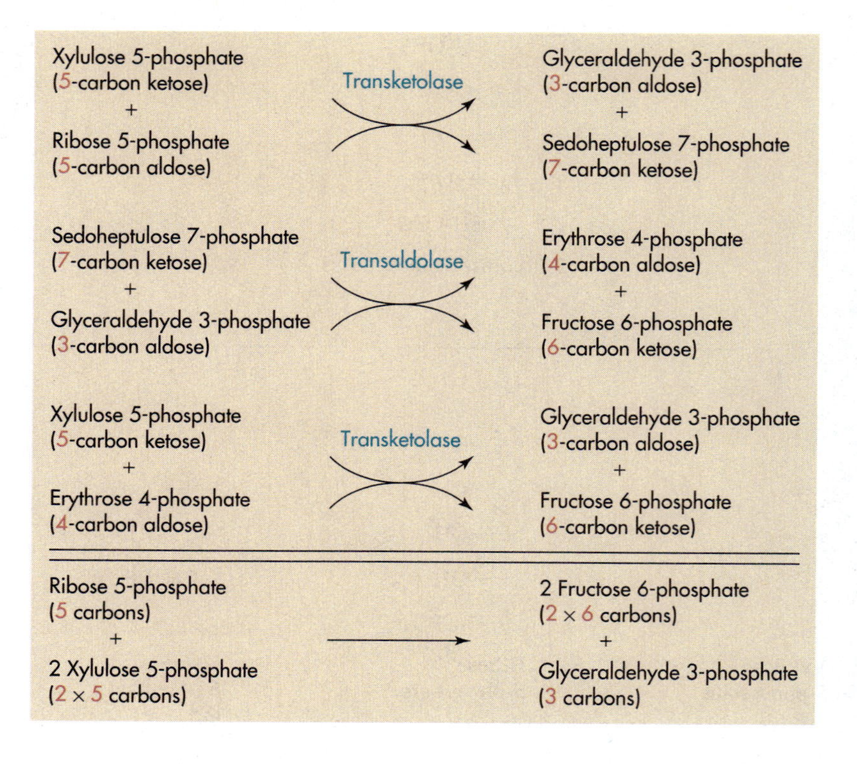

phate produced during the nonoxidative phase of the pentose phosphate pathway to glucose 6-phosphate:

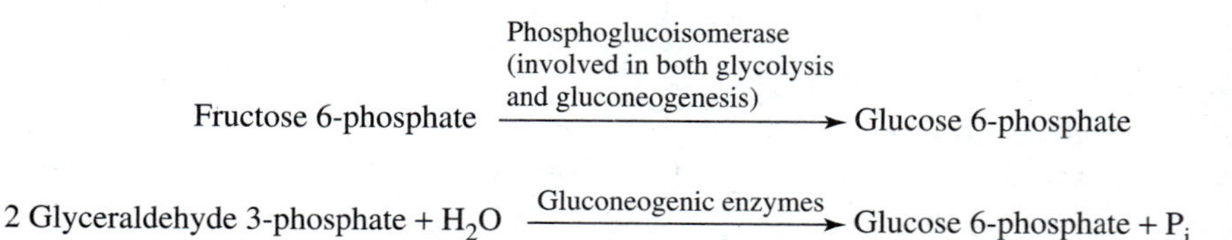

All of the auxiliary enzymes are participants in gluconeogenesis (the synthesis of glucose from noncarbohydrate precursors; Section 14.2). The complete oxidation of glucose 6-phosphate to CO_2 is summarized in Table 12.1. The process is cyclic since it begins with glucose 6-phosphates and ends with glucose 6-phosphates. One out of every six glucose 6-phosphates that enter the cycle (called the **pentose phosphate cycle**) is completely oxidized during the cycle. The five glucose 6-phosphates that are products of the cycle can, along with another glucose 6-phosphate, participate in another round of oxidation. In general, the pentose phosphate cycle only operates when a cell needs much more NADPH than pentoses for anabolism. The pentose phosphate pathway is the major source of NADPH. **[Try Problems 12.12 and 12.13]**

TABLE 12.1

Role of the Pentose Phosphate Pathway in the Oxidation of Glucose 6-Phosphate

Oxidative phase	12 $NADP^+$ + 6 H_2O + 6 glucose 6-phosphate → 12 NADPH + 12 H^+ + 6 CO_2 + 6 ribulose 5-phosphate
Isomerization reactions	6 Ribulose 5-phosphate → 2 ribose 5-phosphate + 4 xylulose 5-phosphate
Transaldolase plus transketolase reactions	2 Ribose 5-phosphate + 4 xylulose 5-phosphate → 4 fructose 6-phosphate + 2 glyceraldehyde 3-phosphate
Part of gluconeogenesis	4 Fructose 6-phosphate + 2 glyceraldehyde 3-phosphate + H_2O → 5 glucose 6-phosphate + P_i
Summary	12 $NADP^+$ + 7 H_2O + 6 glucose 6-phosphate → 5 glucose 6-phosphate + 12 NADPH + 12 H^+ + 6 CO_2 + P_i
Net reaction[a]	12 $NADP^+$ + 7 H_2O + glucose 6-phosphate → 12 NADPH + 12 H^+ + 6 CO_2 + P_i

[a]In actuality, each molecule of CO_2 produced comes from a separate glucose 6-phosphate.

PROBLEM 12.7 Write the structural formula for the oxidized form of nicotinamide adenine dinucleotide phosphate.

PROBLEM 12.8 From an organic chemistry standpoint, what class of compound is oxidized and what class of compound is produced during each oxidation reaction in Figure 12.9?

PROBLEM 12.9 When a lactone is hydrolyzed, what class of compound is formed in addition to a carboxylic acid? Circle each of the new functional groups created during the hydrolysis of the lactone in step 2 of Figure 12.9.

PROBLEM 12.10 List the three pentoses that are intermediates in the pentose phosphate pathway. Which of these compounds are structural isomers? Which are stereoisomers?

PROBLEM 12.11 Study the three reactions catalyzed by transaldolase or transketolase in Figure 12.10. Write a fourth reaction that you might reasonably expect to be catalyzed by transketolase if this enzyme were relatively nonspecific.

PROBLEM 12.12 During the respiratory chain–linked oxidation of glucose to CO_2, molecular oxygen (O_2) is the ultimate acceptor of the electrons removed (lost) from glucose (oxidation is the loss of electrons). An equal number of electrons are removed from glucose during its pentose phosphate pathway–linked oxidation to CO_2. How many electrons are removed from glucose during its oxidation to CO_2, and what is the ultimate fate of these electrons during those metabolic processes normally linked to the pentose phosphate pathway?

PROBLEM 12.13 Check the carbon balance in each equation in Table 12.1. Ignore the carbons in $NADP^+$ and NADPH, since their carbons cancel one another in those equations where these coenzymes appear.

12.3 ALCOHOLIC FERMENTATION

This section adds yet another topic to the glucose catabolism story. **Fermentation,** in general, is the anaerobic catabolism of glucose and certain other organic substances to yield products that support continued catabolism and the coupled generation of ATP. The anaerobic conversion of glucose to lactate (Section 11.2) is sometimes called **lactate fermentation. Alcoholic fermentation** is the conversion of glucose to **ethanol.** This process, which is a characteristic of yeast and a limited number of other microorganisms, entails glycolysis plus two auxiliary cytosolic reactions. The first auxiliary reaction, the conversion of pyruvate to **acetaldehyde** (ethanal) plus CO_2, is a decarboxylation reaction catalyzed by **pyruvate decarboxylase** (Figure 12.11). **Alcohol dehydrogenase** subsequently catalyzes the reduction of the acetaldehyde to ethanol. Ethanol production accomplishes for yeast what lactate production accomplishes for mammalian cells; it regenerates NAD^+ that can be used to keep glycolysis and ATP production going in the absence of oxygen. **[Try Problems 12.14 and 12.15]**

PROBLEM 12.14 What prevents pyruvate from being converted to acetyl-SCoA and CO_2 under anaerobic conditions?

PROBLEM 12.15 Write the structural formulas for acetate (ethanoate), acetaldehyde (ethanal), and ethanol. What compound is produced when acetaldehyde is oxidized? Reduced?

12.4 DIABETES

Because **diabetes** disrupts the normal metabolism of glucose, it too ties into the glucose catabolism story. Diabetes is a complex disease with many variations and causes. Early symptoms include excessive thirst and frequent urination. It is one of the most common metabolic diseases in the world, and in the United States it kills more people than any other disease except heart disease and cancer. Some forms of diabetes, like some forms of cancer and heart disease, have a strong genetic link.

EXHIBIT 12.1

Some Changes Associated with Untreated Insulin-Dependent Diabetes

- ↓ Concentration of insulin in blood
- ↑ Concentration of glucose in blood (hyperglycemia)
- ↑ Concentration of glucose in urine (glucosuria)
- ↑ Concentration of ketone bodies in blood (ketosis; acidosis)
- ↑ Rate of protein glycosylation (tissue damage)

Insulin-Dependent Diabetes Mellitus

Insulin-dependent diabetes mellitus (**IDDM;** also called **juvenile-onset diabetes,** and **type I diabetes**) is most often triggered by the autoimmune destruction of part of the pancreatic islet β-cells, those cells responsible for the production of insulin. The low insulin to glucagon ratio that results leads to largely predictable (from the information presented in Table 10.11) changes in carbohydrate, fat, and protein metabolism (Exhibit 12.1). The diagnostic hyperglycemia (elevated blood glucose levels) that results is caused by the increased production of glucose by the liver, the

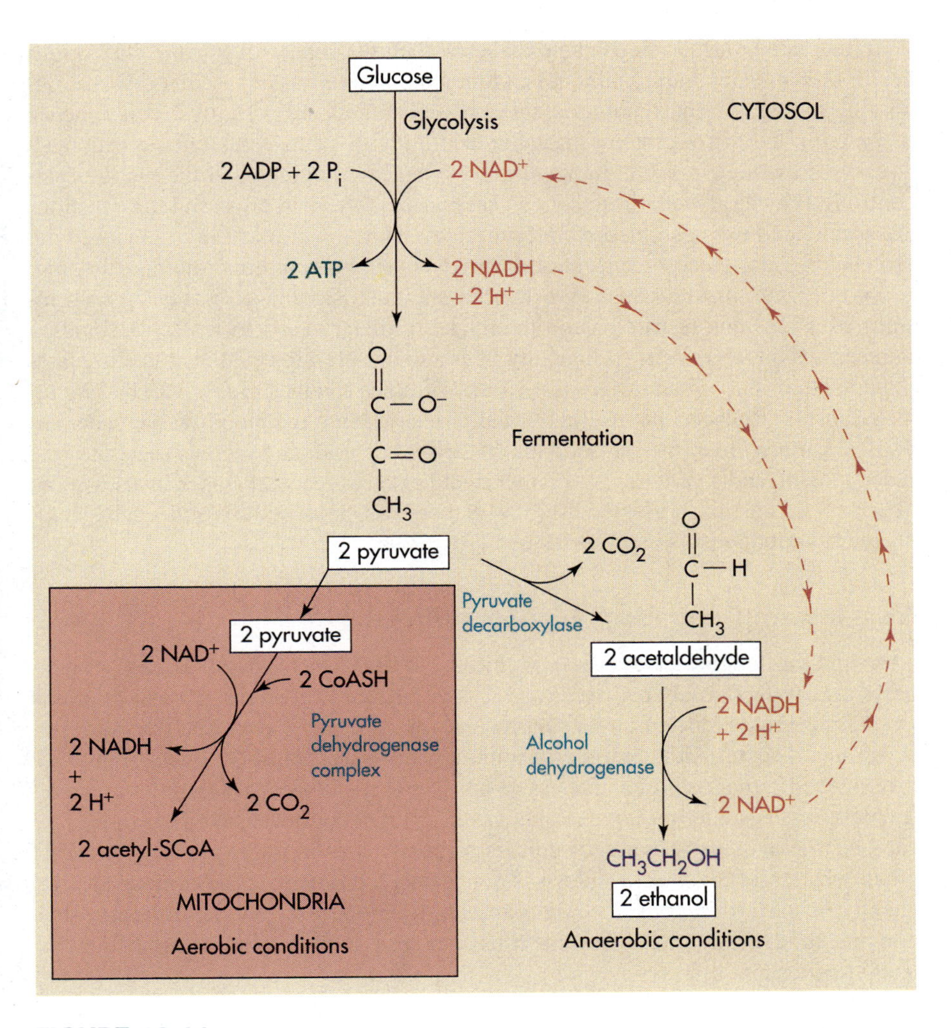

FIGURE 12.11
Alcoholic fermentation.

decreased utilization of glucose by the liver, and the decreased ability of insulin-dependent tissues to remove glucose from the blood. Hyperglycemia leads to **glucosuria** (an abnormally high concentration of glucose in urine), which is also of diagnostic value. In fact, *diabetes mellitus* means *excess excretion of sweet urine*. Fats are mobilized, and fatty acids become the fuel of choice for the liver and certain other cells. The increased combustion of fatty acids coupled with the decreased combustion of glucose accounts for a characteristic **ketosis** (an abnormally high blood concentration of ketone bodies) that can lead to acidosis. Acidosis, a below normal blood pH, is discussed in Section 2.6, and ketone bodies are described during the discussion of fuel metabolism in Section 10.12. The production of ketone bodies and their connection to glucose and fat catabolism are examined in Section 13.4. Untreated **ketoacidosis** (acidosis due to ketosis) can lead to a diabetic coma and even death. **[Try Problem 12.16]**

Insulin injections play a central role in the management of insulin-dependent diabetes mellitus.

If untreated, insulin-dependent diabetes mellitus leads, over time, to serious health problems, including blindness, kidney failure and cardiovascular disease. According to one current hypothesis, these complications are partially a consequence of the cross-linking of proteins through reactions with glucose. A similar reaction accounts for the high concentrations of glycosylated hemoglobins in the blood of uncontrolled (hyperglycemic) diabetics (Section 6.4). Protein cross-linking reactions are accelerated by high glucose concentrations. The slower reaction of glucose with proteins at usual glucose concentrations may contribute to normal biological aging.

At this time, insulin-dependent diabetes mellitus is managed primarily with insulin injections and restrictions on the intake of dietary carbohydrates. By binding to receptors on target cells (principally liver and muscle), injected insulin stimulates the uptake of glucose and triggers a variety of other changes (Table 10.11). The injected insulin replaces the insulin normally released into the blood by the pancreas. Pancreas transplants, the introduction of "bags" of normal pancreas cells into the blood stream, and a wide range of other treatment strategies are under investigation. Much of the insulin presently used by diabetics is produced by genetically engineered microorganisms (Section 19.11). **[Try Problem 12.17]**

Non–Insulin-Dependent Diabetes Mellitus

Non–insulin-dependent diabetes mellitus (**NIDDM;** also called **late-onset diabetes** and **type II diabetes**) typically strikes later in life, and it is most common among the obese. A person malnourished as a fetus runs an increased risk of developing this disease. Although non–insulin-dependent diabetics continue to produce insulin, their target cells become less responsive to the hormone. The decrease in responsiveness is sometimes due to changes in the number or the nature of insulin receptors. The diagnostic hyperglycemia is primarily a result of the reduced uptake of glucose by target cells. Ketoacidosis is not usually a problem, and diet alone can frequently be used to treat the disease (particularly in obese individuals). If untreated, non–insulin-dependent diabetes mellitus can lead to the same complications as insulin-dependent diabetes.

PROBLEM 12.16 True or false? A defective insulin gene can lead to insulin-dependent diabetes. Explain.

PROBLEM 12.17 The injection of too much insulin can cause a diabetic to go into a coma. What is the molecular explanation for this fact?

SUMMARY

All living organisms need a constant supply of energy to survive. Humans and other heterotrophs acquire this energy by oxidizing carbohydrates and other organic compounds. Blood glucose is a primary fuel for many human cells, including brain cells. Its complete oxidation usually involves glycolysis, the pyruvate dehydrogenase complex, the citric acid cycle, and the respiratory chain (Chapter 11).

In animals, glucose is stored as glycogen, a polysaccharide found in particularly large amounts in liver and muscle. Glucose from muscle glycogen tends to be oxidized within muscle cells where the released energy is used to support muscle contraction. Glucose from liver glycogen is exported to blood whenever blood sugar levels begin to drop below normal values. The mobilization of the glucose stored in glycogen involves two enzymes, glycogen phosphorylase and debranching enzyme. The former enzyme catalyzes the stepwise phosphorolysis of glucose residues joined through $\alpha(1 \longrightarrow 4)$ glycosidic links, whereas the latter enzyme catalyzes the removal of $\alpha(1 \longrightarrow 6)$ branches after these branches have been degraded (by glycogen phosphorylase) to four residues in length.

Glycogen utilization is controlled through the phosphorylation of glycogen phosphorylase and the binding of allosteric effectors to this regulatory enzyme. Phosphorylation is brought about by cAMP-linked reaction cascades that are initiated by the binding of hormones to plasma membrane receptors. Epinephrine is mainly responsible for the hormonal regulation of glycogen metabolism in muscle, whereas both glucagon (primarily) and epinephrine exercise some control in liver cells. In muscle, glycogen phosphorylase b is allosterically activated by AMP and inhibited by ATP and glucose 6-phosphate. Glucose is the key modulator of liver glycogen phosphorylase. In muscle, phosphorylase b kinase, the enzyme that catalyzes the activation of glycogen phosphorylase during hormone-induced reaction cascades, is also subject to allosteric regulation. Both phosphorylase b kinase and muscle contraction are stimulated by increases in Ca^{2+} concentration. Ca^{2+}'s stimulation of glycogenolysis helps ensure that contracting muscles have an adequate supply of fuel and energy.

Some glucose is catabolized by the pentose phosphate pathway rather than by glycolysis and coupled pathways. The major importance of the pentose phosphate pathway lies in the generation of pentoses and NADPH for anabolism.

When the pyruvate dehydrogenase complex, the citric acid cycle, and the respiratory chain shut down under anaerobic conditions, many organisms are capable of fermentation, the conversion of glucose and certain other compounds to products that support continued catabolism and the coupled production of ATP. In humans, pyruvate from glycolysis can be converted to lactate, whereas in yeast, pyruvate is transformed into ethanol. Alcoholic fermentation, like lactate fermentation, converts NADH into NAD^+ and allows glycolysis, with its associated production of ATP, to remain operational in the absence of O_2.

Diabetes, a common and complex human disease with many causes, disrupts the normal metabolism of glucose, fats, and amino acids. Susceptibility to diabetes is at least partially determined by genetics. Most insulin-dependent diabetes mellitus is a consequence of the autoimmune destruction of insulin-producing pancreatic islet β-cells. This form of diabetes is normally managed with diet and insulin administration. In non–insulin dependent diabetes mellitus, insulin production is normal but target cells fail to respond normally. Non–insulin dependent diabetes can often be managed with diet alone. Untreated diabetes leads to hyperglycemia, glucosuria, and, in insulin-dependent diabetes, ketosis as a consequence of alterations in insulin-linked metabolism. Untreated diabetes ultimately leads to blindness, kidney failure, cardiovascular disease and death.

EXERCISES

12.1 Explain why brain cells, but not muscle cells, are quickly damaged by low blood sugar levels. Hint: Review Section 10.12.

12.2 Suggest a simple, short-run treatment for hypoglycemia.

12.3 Describe the role of the liver in maintaining normal blood sugar levels.

12.4 Summarize the major differences between glycogen catabolism in liver and in muscle. What is the usual fate of the glycogen-derived glucose produced in each of these two tissues?

12.5 Liver cells can export glycogen-derived glucose but muscle cells cannot. Explain.

12.6 Explain why some athletes attempt to build up glycogen reserves prior to long endurance events.

12.7 Describe the role of protein kinases in the regulation of glycogen metabolism.

12.8 Describe the role of cAMP in the glucagon-induced mobilization of the glucose stored in glycogen.

12.9 Prepare a graph illustrating the type of results expected when cAMP concentration in a muscle cell (y-axis) is plotted against blood epinephrine concentration (x-axis). Epinephrine binds rapidly and reversibly to its receptors.

12.10 Write a balanced equation for the ATP-dependent phosphorylation of a tyrosine side chain. Provide structural formulas for all reactants and products.

12.11 Write the structural formula for a disaccharide containing two glucose residues joined through an $\alpha(1 \longrightarrow 4)$ glycosidic link. Write a balanced equation for the hydrolysis of this compound and a separate balanced equation for its phosphorolysis. What enzyme catalyzes the phosphorolysis of glycogen during glycogen mobilization in liver and muscle?

12.12 Explain why muscle cells in untreated insulin-dependent diabetics tend to burn more fat than glucose even when these individuals consume large amounts of carbohydrates.

12.13 List the major differences between insulin-dependent and non–insulin-dependent diabetes.

12.14 There are two distinct reasons why diabetics sometimes go into a coma. What are those two reasons? In which instance would it be inadvisable to administer insulin? Explain.

12.15 True or false? The oxidation of glucose by the pentose phosphate pathway is a major source of ATP. Explain.

12.16 List the two major roles of the pentose phosphate pathway and the two major roles of glycolysis.

12.17 What is the metabolic logic behind NADPH's inhibition of the pentose phosphate pathway?

12.18 How do the organisms that are capable of alcoholic fermentation benefit from alcohol production?

12.19 Human liver cells contain an alcohol dehydrogenase that catalyzes the oxidation of dietary alcohol to acetaldehyde, a reversal of the last step in alcoholic fermentation. Explain why this reaction usually proceeds in one direction in liver cells but in the opposite direction in fermenting yeast cells.

12.20 Both pyruvate decarboxylase (involved in alcoholic fermentation) and the pyruvate dehydrogenase complex (involved in the aerobic oxidation of glucose) catalyze the decarboxylation of pyruvate. In each case, what is the immediate fate of the acetaldehyde formed?

12.21 How many net ATP's are produced for each molecule of glucose catabolized through alcoholic fermentation? Hint: Study Figure 12.11.

12.22 Provide a molecular explanation of how you benefit from the production of adrenaline when in physical danger.

SELECTED READINGS

A new and better treatment for diabetes, *Consumers Reports on Health* 6(4), 44–46 (1994).

Discusses the results of recent studies indicating that the "tight control" of blood glucose levels can greatly reduce the risk of complications from insulin-dependent diabetes mellitus.

Barinaga, M., Shedding light on blindness, *Science* 267, 452–453 (1995).

Describes evidence that the growth factor VEGF may be "factor X," a protein thought to be responsible for vision loss in diabetics and premature infants.

Davies, J. L., Y. Kawaguchi, S. T. Bennett, J. B. Copeman, H. J. Cordell, L. E. Pritchard, P. W. Reed, S. C. L. Gough, S. C. Jenkins, S. M. Palmer, K. M. Balfour, B. R. Rowe, M. Farrall, A. H. Barnett, S. C. Bain, and J. A. Todd, A genome-wide search for human type I diabetes susceptibility genes, *Nature* 371, 130–136 (1994).

During a search of the human genome for genes that predispose people to type I diabetes, 20 different chromosome regions were detected that "showed some positive evidence for linkage to disease."

Hales, C. N., Fetal nutrition and adult diabetes, *Sci. Am. Science and Medicine* 1(3), 54–63 (1994).

Explores the role of nutrition and other nongenetic factors in non–insulin-dependent diabetes mellitus.

Lacy, P. E., Treating diabetes with transplanted cells, *Sci. Am.* 273(1), 50–58 (1995).

Reviews the progress that has been made in using pancreatic islet cell implants to treat diabetes.

Lehninger, A. L., D. L. Nelson, and M. M. Cox, *Principles of Biochemistry,* 2nd ed. New York: Worth Publishers, 1993.

An encyclopedia-type text with a thorough coverage of metabolism.

Moran, L. A., K. G. Scrimgeour, H. R. Horton, R. S. Ochs, and J. D. Rawn, *Biochemistry,* 2nd ed. Englewood Cliffs, NJ: Neil Patterson/Prentice-Hall, 1994.

An encyclopedia-type biochemistry text containing a detailed discussion of metabolism. Includes a limited number of problems and some selected readings.

Smith C. A., and E. J. Wood, *Energy in Biological Systems.* Hong Kong: Chapman & Hall, 1991.

A 171-page paperback that summarizes the basic metabolic pathways. Contains both questions and references.

Stryer, L., *Biochemistry,* 4th ed. New York: W. H. Freeman and Company, 1995.

A comprehensive text with good coverage of carbohydrate metabolism.

Werner R., *Essential Biochemistry and Molecular Biology,* 2nd ed. New York: Elsevier Publishing Co., 1992.

A 463-page paperback that provides a concise and readable review of basic metabolism.

13 Catabolism of Fats and Proteins

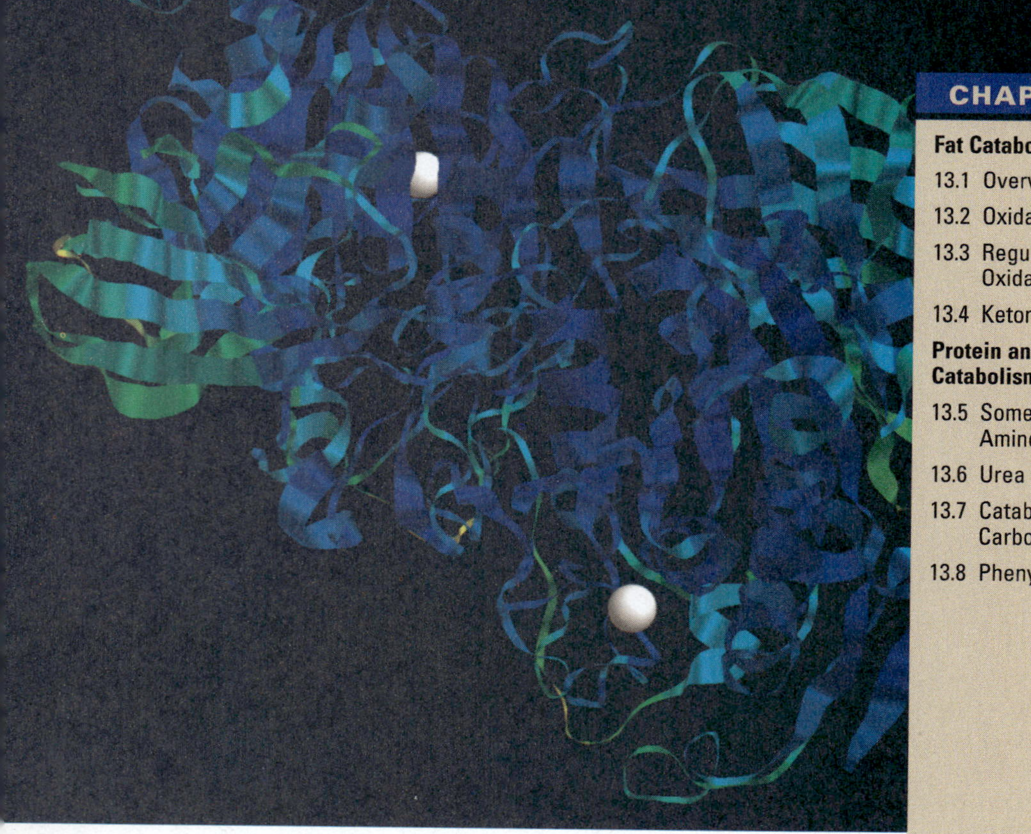

CHAPTER OUTLINE

Chapters 11 and 12 provide an introduction to carbohydrate catabolism. In the process, three metabolic pathways are encountered that play a central role in a large fraction of all catabolism, including the catabolism of fats and proteins. These three pathways are glycolysis, the citric acid cycle, and the respiratory chain.

In humans, fats and proteins, along with carbohydrates, account for the vast majority of the material in a typical diet. Although all three of these classes of compounds can be oxidized, the human body produces no protein whose principal role is to serve as a fuel. In contrast, both glycogen and fats (predominantly triacylglycerols) are synthesized to function as fuel molecules; they are storage forms of energy. Fats are particularly well suited for this mission since their calorie content per gram is over twice that of carbohydrates and proteins (Table 9.3). When compared to carbohydrates and fats, the biological oxidation of proteins is a relatively complex process because of the structural diversity of protein amino acids. The presence of sulfur and nitrogen links protein catabolism with sulfur and nitrogen metabolism, including the urea cycle.

This chapter examines fat and protein catabolism with emphasis placed on those reactions involved in the complete oxidation of these biological molecules. Energy transformation, structural hierarchy and compartmentation, regulation of reaction rates, specificity, and communication are prominent biochemical themes. Chapter 14 explores anabolism and the reciprocal regulation of catabolism and anabolism.

FAT CATABOLISM

13.1 OVERVIEW OF FAT METABOLISM

Dietary fats are hydrolyzed mainly in the small intestine with the catalytic assistance of pancreatic lipase (Figure 13.1). The fatty acids and monoacylglycerols produced are absorbed by cells lining the intestine and then reassembled into triacylglycerols by these same cells. The triacylglycerols are released into the bloodstream after they are packaged into lipoprotein particles known as chylomicrons. Blood does contain some free fatty acids, but liver and adipose tissue are the major sources of these compounds. The triacylglycerols carried by the chylomicrons are hydrolyzed with the catalytic assistance of **lipoprotein lipases** as blood circulates through tissues. The free fatty acids and glycerol produced are utilized by the lipase-producing tissues as fuels or as building blocks for phospholipids, triacylglycerols and other classes of compounds. **[Try Problem 13.1]**

When excess calories are consumed, dietary fats tend to be used by the human body to build up its own fat reserves. Most of these reserves are housed in specialized fat storage cells known as adipocytes. In times of fuel need, the body produces epinephrine, glucagon, and other hormones that lead to the mobilization of this stored fuel (Section 10.12). The mobilizing hormones bind to specific receptors on the plasma membranes of adipocytes and initiate cAMP-linked reaction cascades (Section 10.5) that activate intracellular lipases capable of catalyzing the hydrolysis of stored fats. The free fatty acids and glycerol produced are released into the blood and transported to those tissues in need of fuel. **Blood albumin,** the most abundant

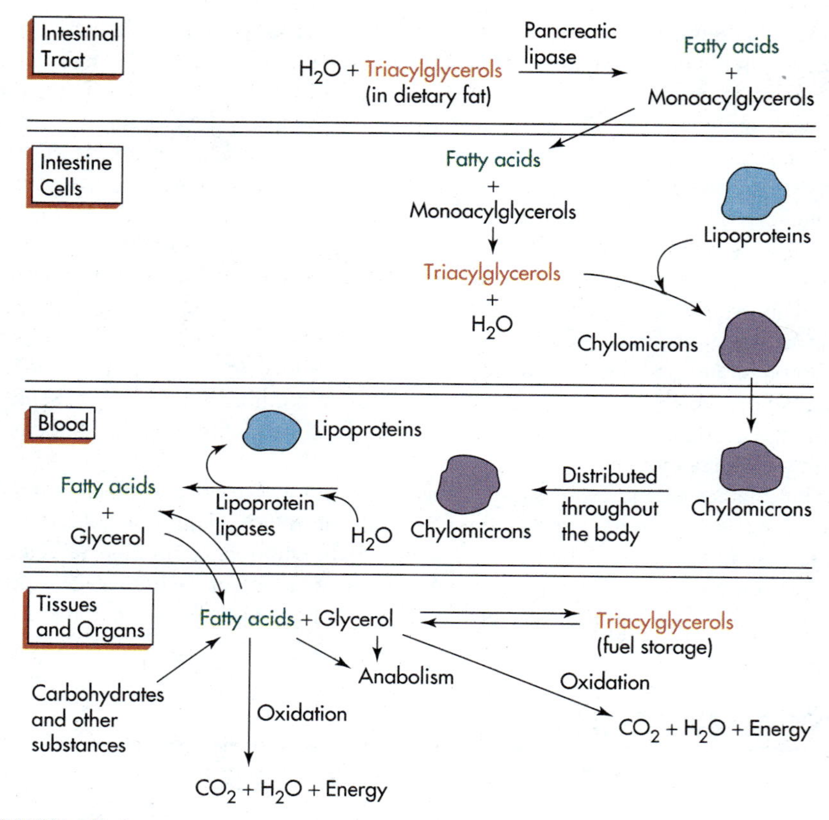

FIGURE 13.1
Overview of dietary fat metabolism.

protein in blood plasma, binds and carries the free fatty acids. The next section examines those reactions involved in the biological oxidation of fatty acids and glycerol. **[Try Problem 13.2]**

PROBLEM 13.1 Write an equation for the complete hydrolysis of a specific triacylglycerol. Provide structural formulas for all reactants and products.

PROBLEM 13.2 Write a balanced equation for the complete oxidation of stearic acid [$CH_3(CH_2)_{16}COOH$] by molecular oxygen.

13.2 OXIDATION OF FATS

Glycerol from stored fat tends to be removed from the blood by the liver and used for gluconeogenesis, the production of glucose from noncarbohydrate precursors

(Section 14.2). Glycerol can, however, be directly oxidized. Fatty acids from mobilized fats can be completely oxidized by those cells that remove them from blood.

Oxidation of Glycerol

The oxidation of glycerol begins in the cytosol where phosphorylation is followed by oxidation:

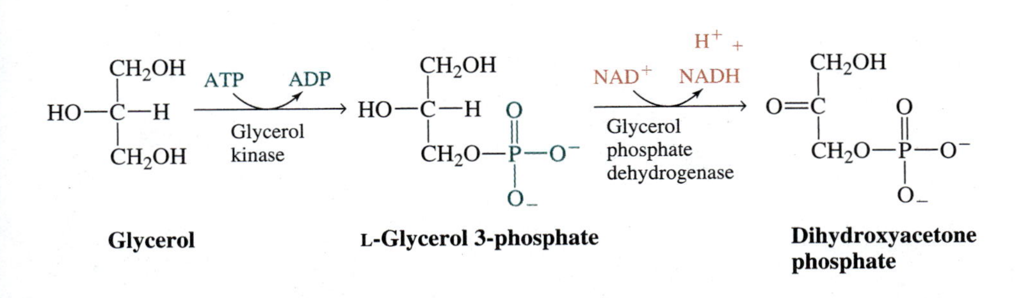

The dihydroxyacetone phosphate generated is converted to pyruvate with the catalytic assistance of part of the enzymes involved in glycolysis. After pyruvate is transformed into CO_2 and acetyl-CoA within the pyruvate dehydrogenase complex, the acetyl group is further oxidized within the citric acid cycle (Figures 11.13 and 13.2). The $FADH_2$ and NADH generated during these processes are used to synthesize ATP within the respiratory chain. One GTP (an ATP equivalent) is produced within the citric acid cycle itself. **[Try Problem 13.3]**

Fatty Acid Activation

Fatty acid catabolism begins on the cytosolic side of the inner mitochondrial membrane with the ATP-dependent **acyl-SCoA synthetase (fatty acid thiokinase)**–catalyzed formation of a thioester:

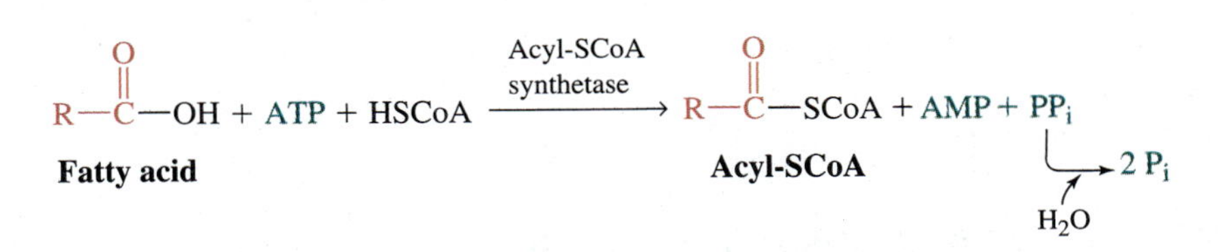

The coupled hydrolysis of ATP provides a thermodynamic driving force for this reaction, and the subsequent hydrolysis of the pyrophosphate (PP_i) produced helps "pull" the reaction to completion (Le Châtelier's principle). The fatty acid residue within an acyl-SCoA is said to be activated since thioesters are high-energy compounds (Section 8.5). Be alert to the distinction between an "acyl" group (RCO—) and an "acetyl" group (CH_3CH—); an acetyl group is a specific example of an acyl group.

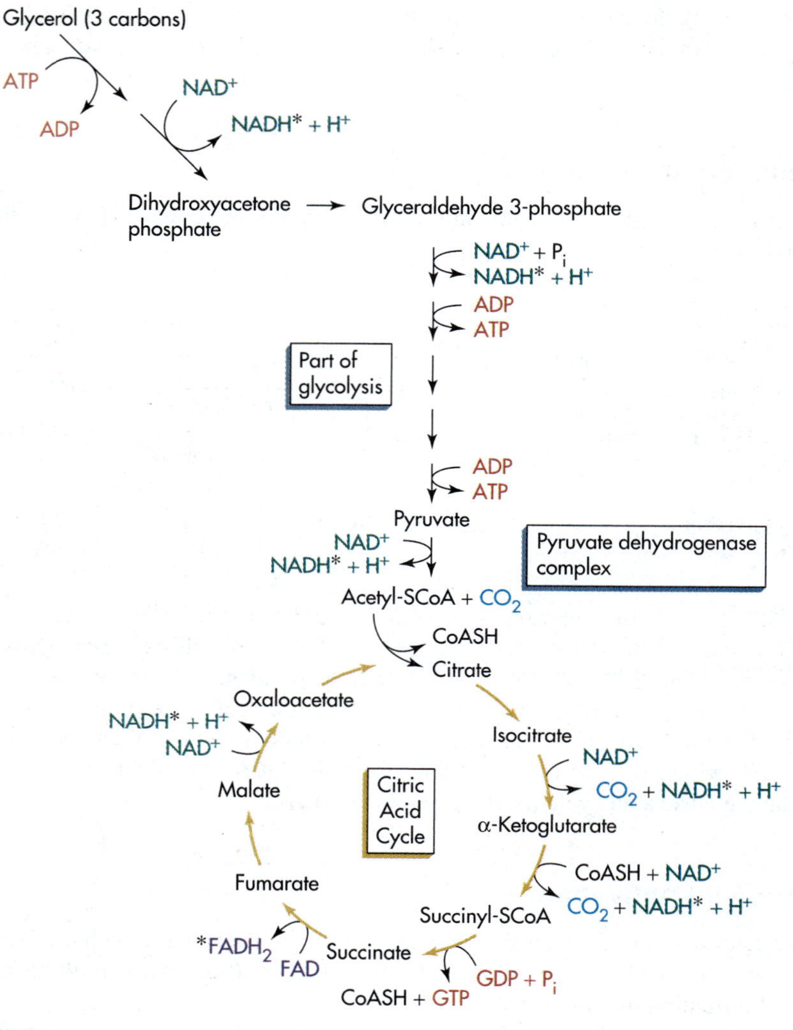

FIGURE 13.2
Oxidation of glycerol. Compounds marked with asterisks shuttle electrons to the respiratory chain.

Transport of Fatty Acids into Mitochrondria

An activated fatty acid must enter the matrix of a mitochondrion for further catabolism, since the enzymes involved are localized within this organelle (another example of compartmentation). Although some short-chain fatty acyl-SCoA molecules can pass directly through both the outer and the inner mitochondrial membranes, most fatty acyl-SCoAs are unable to penetrate the inner membrane. In these instances, the acyl group is transferred from CoASH to **carnitine** with the catalytic assistance of **carnitine acyltransferase I,** an enzyme located on the cytosolic side of the inner membrane:

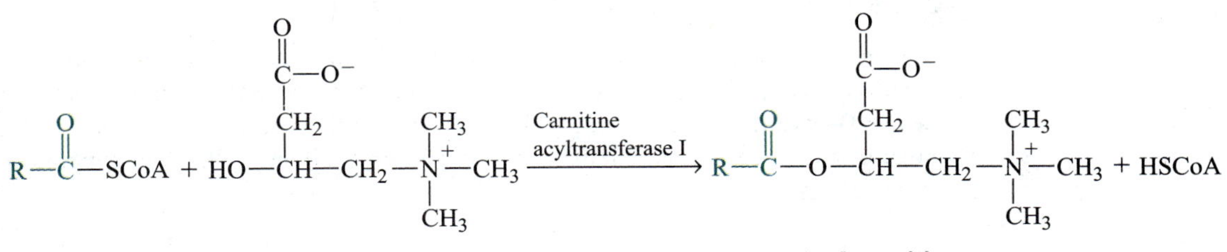

After the acyl-carnitine is carried into a mitochrondrion by a selective transporter (a membrane-spanning protein; Section 9.10), the acyl group is transferred to an internal CoASH molecule during a reaction catalyzed by **carnitine acyltransferase II,** an isozyme of carnitine acyltransferase I located on the inside of the inner membrane:

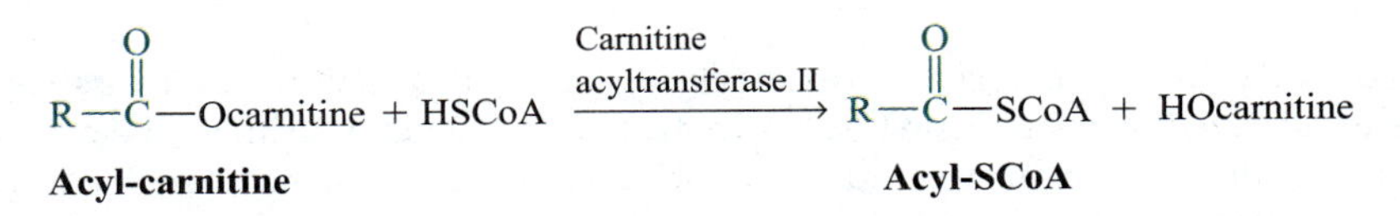

Thus, carnitine accepts an acyl group from a CoASH in the intermembrane space and delivers it to another CoASH which is confined to the matrix of a mitochondrion. The freed carnitine moves back into the intermembrane space where it picks up another acyl group for delivery to the mitochondrial matrix (Figure 13.3). **[Try Problem 13.4]**

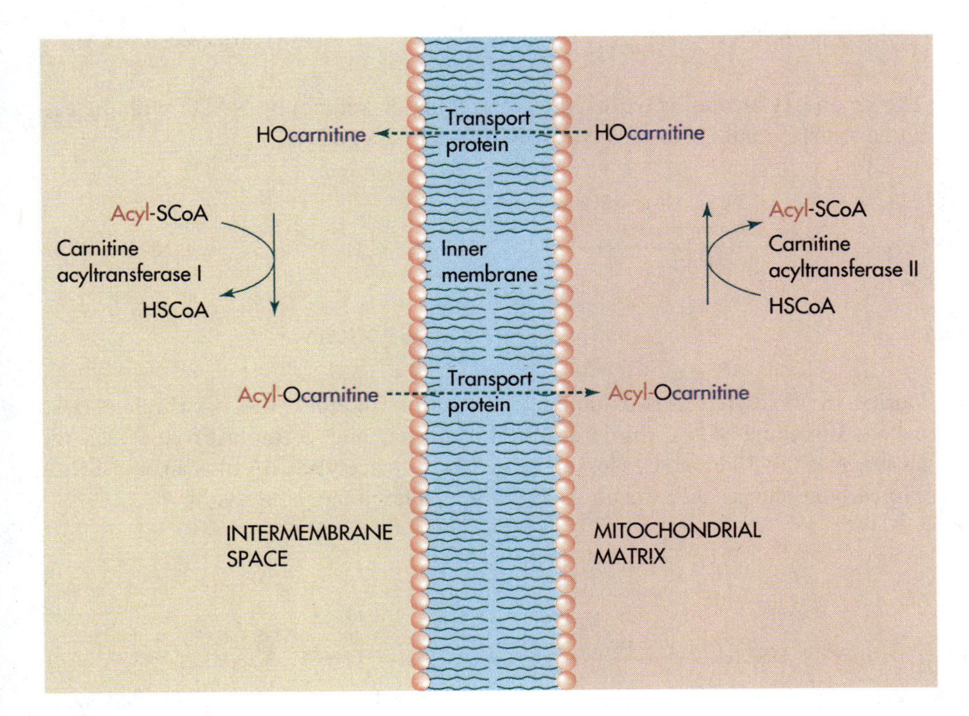

FIGURE 13.3
Acyl group transport into mitochondria.

β-Oxidation

Once inside a mitochondrion, an activated fatty acid that is destined to be completely oxidized is broken down into multiple acetyl-SCoA molecules through a process known as **β-oxidation** (Figure 13.4). The first step in this metabolic pathway is the removal of two specific hydrogens from the acyl group in acyl-SCoA to generate a 2,3-*trans*-alkene:

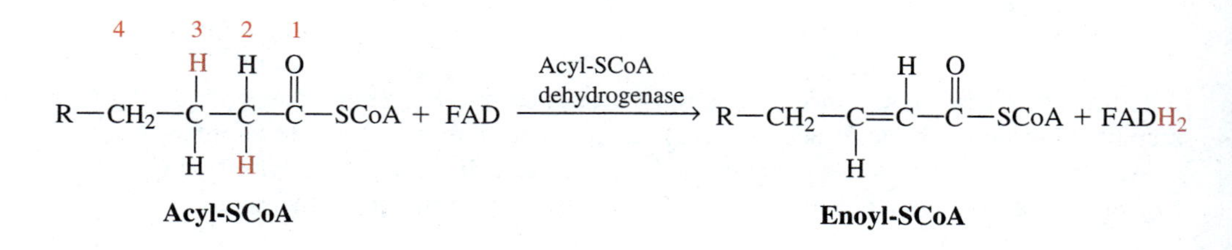

FAD is the hydrogen acceptor (oxidizing agent) in this **acyl-SCoA dehydrogenase**–catalyzed reaction. **Enoyl-SCoA hydratase,** a second mitochondrial enzyme, catalyzes the addition of water (a hydration reaction; hence the name hydratase) to the *trans*-alkene to generate a secondary alcohol.

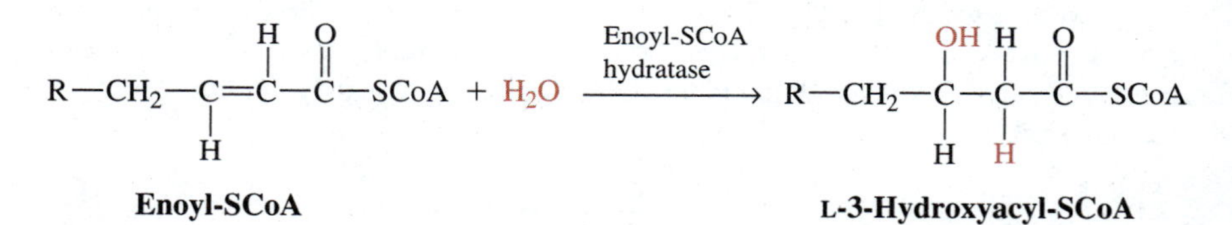

The secondary alcohol is oxidized to a 3-ketone (β-ketone) by NAD^+ with the catalytic assistance of **L-3-hydroxyacyl-SCoA dehydrogenase.**

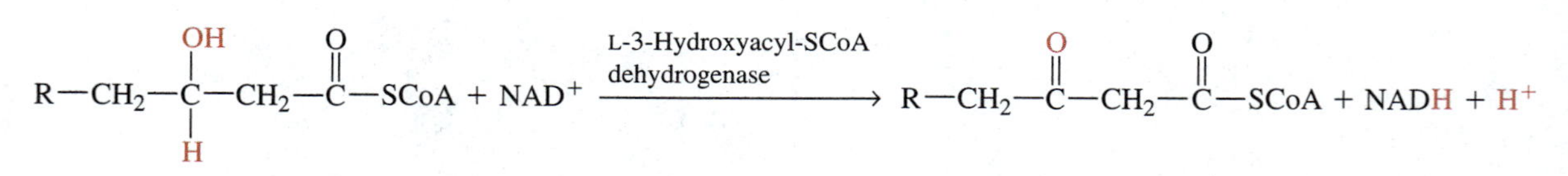

During the final step in β-oxidation, the enzyme **3-ketoacyl-SCoA thiolase** (also called **3-oxoacyl-SCoA thiolase, β-ketothiolase,** and **β-thiolase**) catalyzes the cleavage of the 3-ketone by CoASH to generate an acetyl-SCoA plus an acyl-SCoA two carbons shorter than the one that initially entered the pathway.

$$R{-}CH_2{-}\overset{O}{\overset{\|}{C}}{-}CH_2{-}\overset{O}{\overset{\|}{C}}{-}SCoA + HSCoA \xrightarrow[\text{thiolase)}]{\text{3-Ketoacyl-SCoA thiolase (3-oxoacyl-SCoA}} R{-}CH_2{-}\overset{O}{\overset{\|}{C}}{-}SCoA + HCH_2{-}\overset{O}{\overset{\|}{C}}{-}SCoA$$

3-Ketoacyl-SCoA (3-Oxoacyl-SCoA) **Acyl-SCoA** **Acetyl-SCoA**

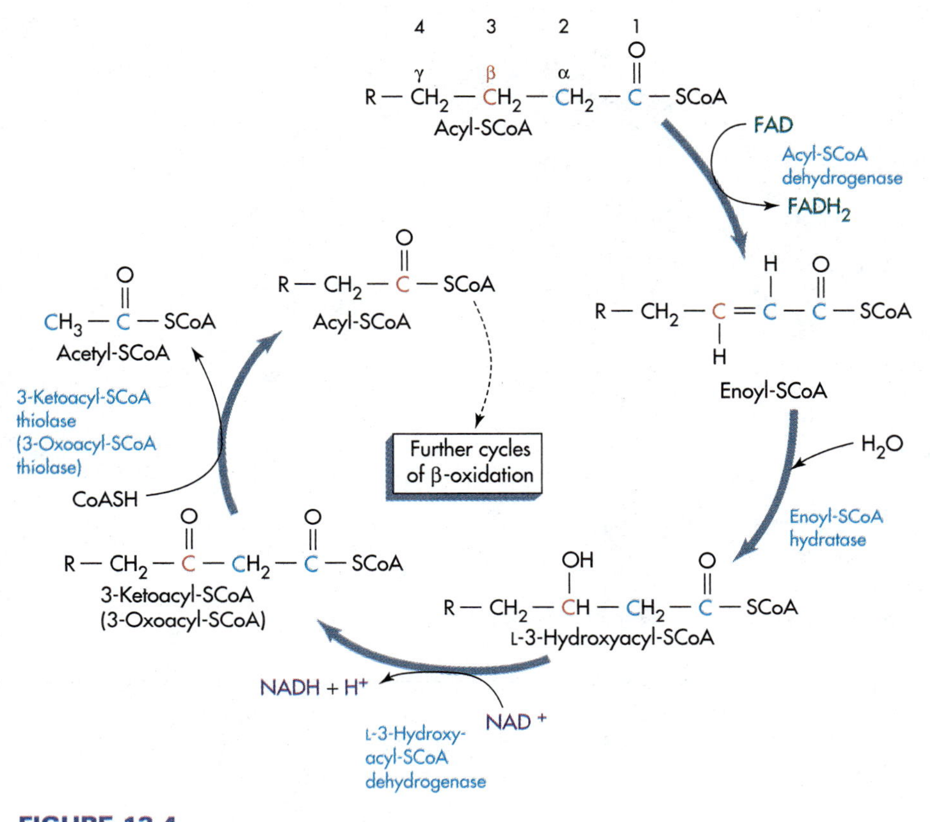

FIGURE 13.4
β-Oxidation.

This **thiolysis** reaction (cleavage of a bond through reaction with a thiol) is similar to hydrolysis and phosphorolysis. The β-carbon (3-carbon) of the original fatty acyl-SCoA is oxidized during the operation of the four-step catabolic pathway (Figure 13.4), hence the name β-oxidation. **[Try Problems 13.5 and 13.6]**

The acyl-SCoA produced during the first round of β-oxidation is shortened by two more carbons during a second trip through this metabolic pathway, and another acetyl-SCoA is generated in the process. Further rounds of β-oxidation release additional acetyl-SCoAs (Figure 13.5). If we start with a fatty acid with an even number of carbons, an acetoacetyl-SCoA is split into two acetyl-SCoAs during the last step in the final round of β-oxidation:

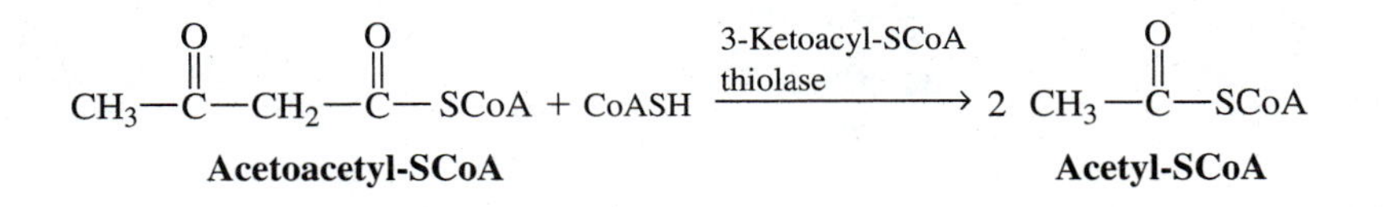

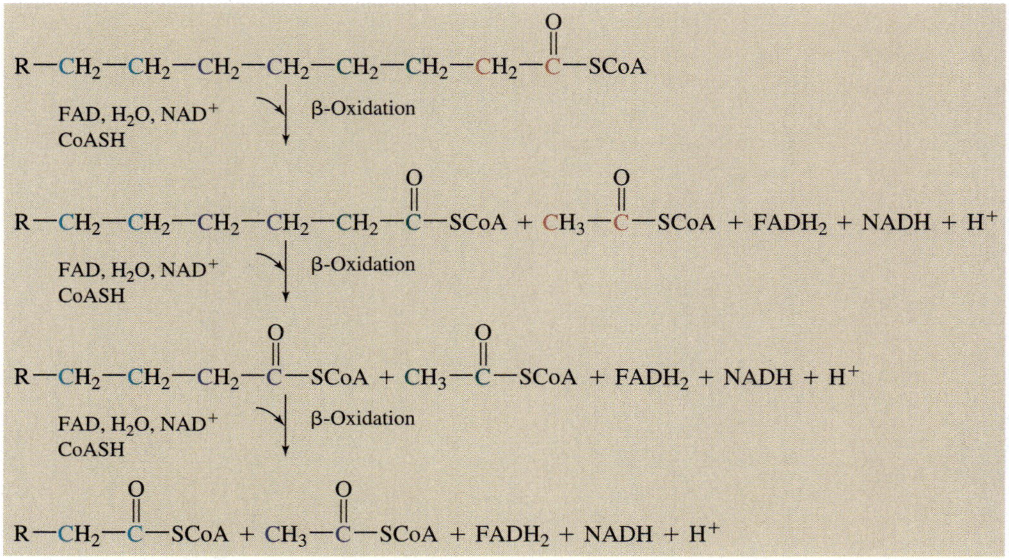

$$R{-}CH_2{-}CH_2{-}CH_2{-}CH_2{-}CH_2{-}CH_2{-}CH_2{-}\overset{\overset{\large O}{\|}}{C}{-}SCoA$$

$$\xrightarrow[\text{FAD, } H_2O\text{, NAD}^+\text{, CoASH}]{\beta\text{-Oxidation}}$$

$$R{-}CH_2{-}CH_2{-}CH_2{-}CH_2{-}CH_2{-}\overset{\overset{\large O}{\|}}{C}{-}SCoA + CH_3{-}\overset{\overset{\large O}{\|}}{C}{-}SCoA + FADH_2 + NADH + H^+$$

$$\xrightarrow[\text{FAD, } H_2O\text{, NAD}^+\text{, CoASH}]{\beta\text{-Oxidation}}$$

$$R{-}CH_2{-}CH_2{-}CH_2{-}\overset{\overset{\large O}{\|}}{C}{-}SCoA + CH_3{-}\overset{\overset{\large O}{\|}}{C}{-}SCoA + FADH_2 + NADH + H^+$$

$$\xrightarrow[\text{FAD, } H_2O\text{, NAD}^+\text{, CoASH}]{\beta\text{-Oxidation}}$$

$$R{-}CH_2{-}\overset{\overset{\large O}{\|}}{C}{-}SCoA + CH_3{-}\overset{\overset{\large O}{\|}}{C}{-}SCoA + FADH_2 + NADH + H^+$$

FIGURE 13.5
Release of acetyl-SCoA's through consecutive rounds of β-oxidation.

Thus, seven rounds of β-oxidation convert palmitoyl-SCoA, which contains a saturated fatty acid residue with 16 carbons, into eight acetyl-SCoAs (Figure 13.6). Propionyl-SCoA and acetyl-SCoA are the products of the last round of β-oxidation when an odd-numbered fatty acid is oxidized. **[Try Problem 13.7]**

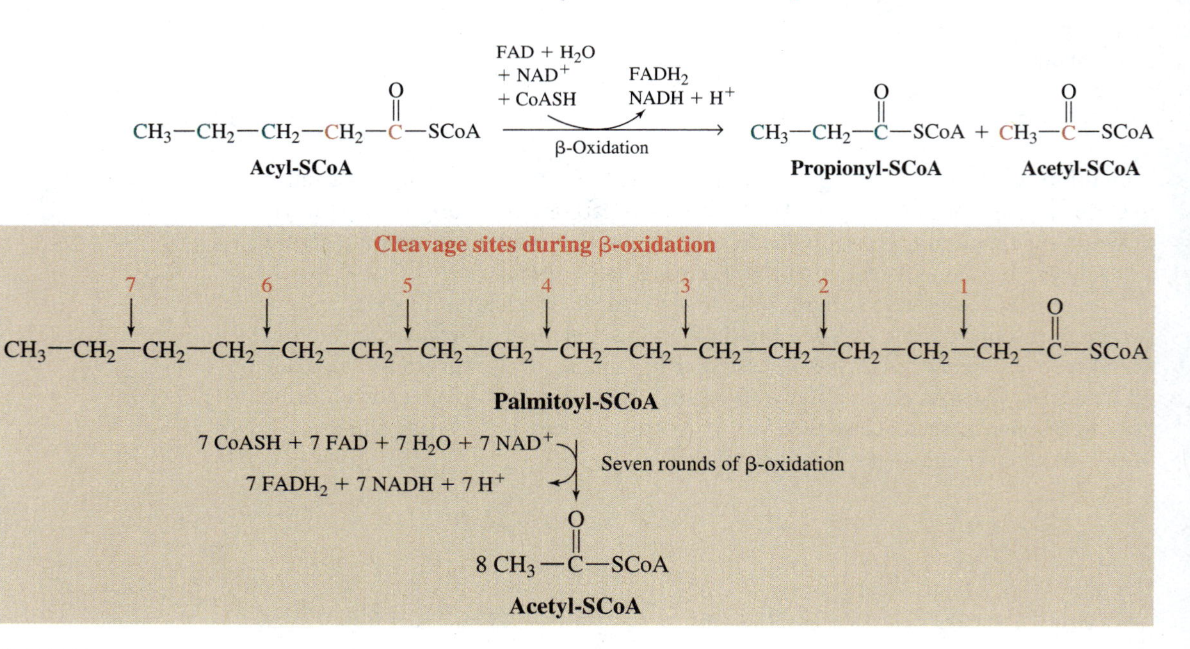

FIGURE 13.6
Conversion of palmitoyl-SCoA to eight acetyl-SCoA's.

Net Reaction for Palmitate Oxidation

The $FADH_2$ and NADH produced during β-oxidation lead to the generation of ATP when they pass electrons acquired from fatty acids to O_2 through the respiratory chain. Each NADH directly transfers its two acquired electrons to complex I of the respiratory chain, while electrons from $FADH_2$ flow to ubiquinone (Q) through multiple electron carriers (Section 11.8). When a fatty acid is to be completely oxidized, the acetyl-SCoAs from β-oxidation enter the citric acid cycle which, along with the respiratory chain, completes this catabolic process (see Figure 13.7 on the next page). Part of the energy released is utilized to synthesize additional ATP and some GTP (an ATP equivalent).

The net reaction for the biological oxidation of palmitate (the salt form of palmitic acid) is:

$$CH_3(CH_2)_{14}COO^- + H^+ + 23\ O_2 + 106\ P_i + 106\ ADP \xrightleftharpoons{\Delta G^{\circ\prime} = -9800\ kJ/mol} 16\ CO_2 + 122\ H_2O + 106\ ATP$$

All of the CO_2 is produced within the citric acid cycle, whereas the water and most of the ATP (Table 13.1) are produced within the respiratory chain. For each mole of palmitate oxidized under standard conditions, approximately 31 kJ are trapped in each of the 106 net mol of ATP generated. Since the total combustion of 1 mole of palmitate releases around 9800 kJ under standard conditions, close to 34% of the energy of oxidation is trapped in a usable form:

$$\frac{31\ \text{kJ/mol ATP} \times 106\ \text{mol ATP}}{9800\ \text{kJ}} \times 100 = 34\%$$

Although this is an impressive feat, the efficiency of the process may be even greater under physiological conditions. A similar efficiency is associated with the oxidation of glucose (Section 11.11). **[Try Problem 13.8]**

TABLE 13.1

Approximate Net Production of ATP During the Total Oxidation of Palmitate

Process	ATP
Palmitate + ATP + HSCoA → palmitoyl-SCoA + AMP + PP_i	−1
PP_i[a] (from palmitate activation) + H_2O → 2 P_i	−1
7 $FADH_2$[b] from 7 rounds of β-oxidation	+10.5
7 NADH[b] from 7 rounds of β-oxidation	+17.5
8 Acetyl-SCoA[b] oxidized through the citric acid cycle and the respiratory chain	+80
Total (Net)	+106

[a]The hydrolysis of the anhydride bond in PP_i is equivalent to the hydrolysis of an anhydride bond in ATP.

[b]The amount of ATP produced is an estimate based on the assumption that the oxidation of each $FADH_2$ and each NADH leads to the production of roughly 1.5 ATP and 2.5 ATP, respectively, within the respiratory chain (Section 11.8).

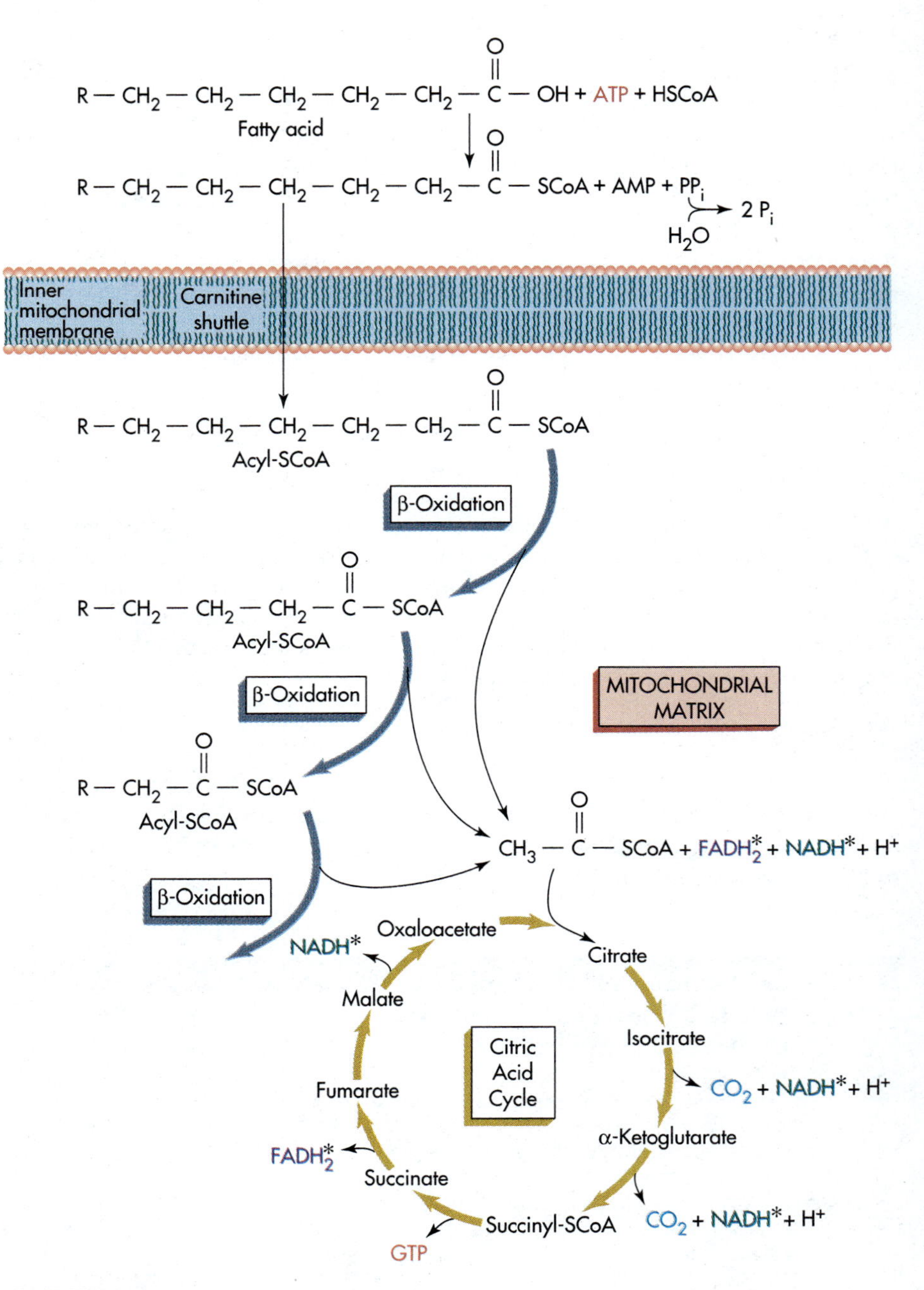

FIGURE 13.7
Summary of fatty acid oxidation. Compounds marked with asterisks deliver electrons to the respiratory chain.

Oxidation of Odd-Numbered Fatty Acids

The propionyl-SCoA produced during the final round of β-oxidation for an odd numbered fatty acid is converted, in three steps, to succinyl-SCoA, an intermediate in the citric acid cycle (Figure 13.8). **Propionyl-SCoA carboxylase,** a biocytin-containing enzyme, first catalyzes the covalent joining of a carbonate molecule to propionyl-SCoA to generate D-methylmalonyl-SCoA. The coupled hydrolysis of ATP provides a thermodynamic driving force for this reaction. **Biocytin (biotinyl-lysine)** is a biotin molecule joined through an amide link to a lysine side chain within an enzyme (Figure 13.9). Biotin is a B vitamin that participates in a variety of carboxylation reactions (Table 10.6). During the second step in propionyl-SCoA metabolism, an isomerase named **methylmalonyl-SCoA epimerase** (also called **methylmalonyl-SCoA racemase**) catalyzes the transformation of D-methylmalonyl-SCoA into L-methylmalonyl-SCoA. The L-methylmalonyl-SCoA is subsequently con-

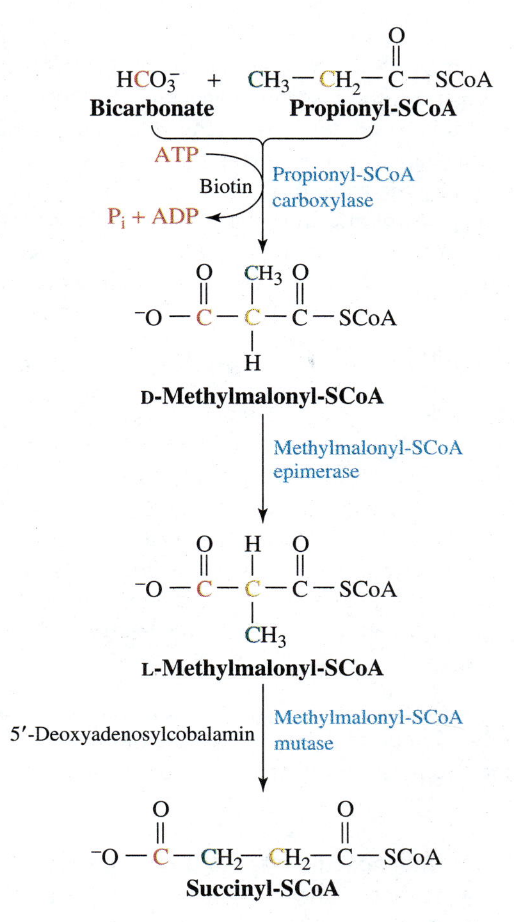

FIGURE 13.8
Propionyl-SCoA metabolism.

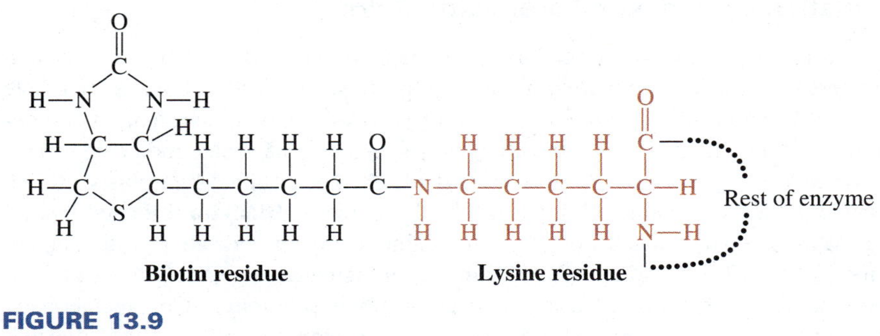

FIGURE 13.9
Biocytin.

verted to succinyl-SCoA with the assistance of **methylmalonyl-SCoA mutase,** a vitamin B_{12}-dependent enzyme. The vitamin B_{12} resides within **5′-deoxyadenosylcobalamin (coenzyme B_{12})**, one of several coenzymes produced from this cobalt-containing vitamin (Table 10.6). Enzymes containing 5′-deoxyadenosylcobalamin normally catalyze isomerization reactions in which a hydrogen atom and a substituent on an adjacent carbon exchange places. The succinyl-SCoA from the mutase-catalyzed reaction enters the citric acid cycle, where it can be converted to other intermediates in the cycle. Like these intermediates, succinyl-SCoA can be used directly or indirectly for any of a wide variety of metabolic processes, including gluconeogenesis. To be oxidized, succinyl-SCoA must be pulled out of the cycle, because the citric acid cycle can only oxidize entering acetyl groups (Section 11.6). The details of succinate oxidation can be found in more encyclopedic texts.

Oxidation of Unsaturated and Branched Fatty Acids

The preceding discussions assume that the fatty acid being oxidized is a linear, saturated one. Although additional enzymes are required and the details of some rounds of β-oxidation differ, unsaturated fatty acids are oxidized in a similar manner. The oxidation of branched fatty acids (relatively rare) sometimes involves α-oxidation as well as β-oxidation. During α-oxidation, it is the α-carbon (immediately adjacent to the carbonyl) that is oxidized. Certain organisms are also genetically programmed to oxidize the carbon furthest removed from the carboxyl group in a fatty acid, a process called ω-oxidation. The details of α- and ω-oxidation are not examined in this text.

PROBLEM 13.3 Calculate the approximate net number of ATP molecules produced during the total biological oxidation of one molecule of glycerol by O_2.

PROBLEM 13.4 What specific class of bonds joins a fatty acid to carnitine within an acyl-carnitine molecule? Is the equilibrium constant for the carnitine acyltransferase I–catalyzed reaction most likely large, small, or of intermediate value? Explain.

PROBLEM 13.5 When enoyl-SCoA hydratase catalyzes the addition of water to the *trans*-alkene in Figure 13.4, a specific secondary alcohol is produced. Additional alcohols can be synthesized through the addition of water to the same alkene. What are these other alcohols? Suggest why these other alcohols are not generated during the β-oxidation of saturated fatty acids.

PROBLEM 13.6 Identify the alpha (α)-, beta (β)- and gamma (γ)-carbons in the L-3-hydroxyacyl-SCoA molecule depicted in Figure 13.4.

PROBLEM 13.7 How many rounds of β-oxidation are required to completely oxidize a saturated fatty acid containing 19 carbons? How many acetyl-SCoAs are generated in the process?

PROBLEM 13.8 Calculate approximate net ATP production during the complete biological oxidation of stearic acid [$CH_3(CH_2)_{16}COOH$] by O_2.

13.3 REGULATION OF FATTY ACID OXIDATION

The regulation of human fatty acid oxidation begins with those hormones that control the mobilization and release (into the blood) of fatty acids stored within the fat reserves of adipocytes. Both epinephrine (produced by the adrenal medulla) and glucagon (a product of the pancreas) stimulate the mobilization process. Their secretion into blood is triggered by low blood fuel levels. Since the same two hormones mobilize the glucose in liver and muscle glycogen (Section 12.1), the human body simultaneously calls upon both of its two major fuel reserves in time of need. The action of insulin, which signals a fed and high-energy state when present at high concentration, opposes that of epinephrine and glucagon (Section 10.12).

Once acquired by a cell from the blood, the transport of fatty acids into mitochondria is allosterically modulated. This is normally the rate-limiting step in the oxidation process. Malonyl-SCoA, which accumulates only when there is an abundant supply of energy and carbohydrates, inhibits carnitine acyltransferase I (catalyzes the transfer of acyl groups to carnitine for transport into mitochondria). This inhibition helps a cell conserve its fatty acid reserves when there is already an adequate supply of energy and fuel. Once a fatty acid has been transported into a mitochondrion, its oxidation is subject to still further modulation. Both L-3-hydroxyacyl-SCoA dehydrogenase and 3-ketoacyl-SCoA thiolase, two enzymes involved in β-oxidation, are regulatory proteins. The former is inhibited by NADH and the latter is susceptible to feedback inhibition by acetyl-SCoA (Figure 13.10). **[Try Problem 13.9]**

PROBLEM 13.9 What is the metabolic logic behind the inhibition of β-oxidation by NADH?

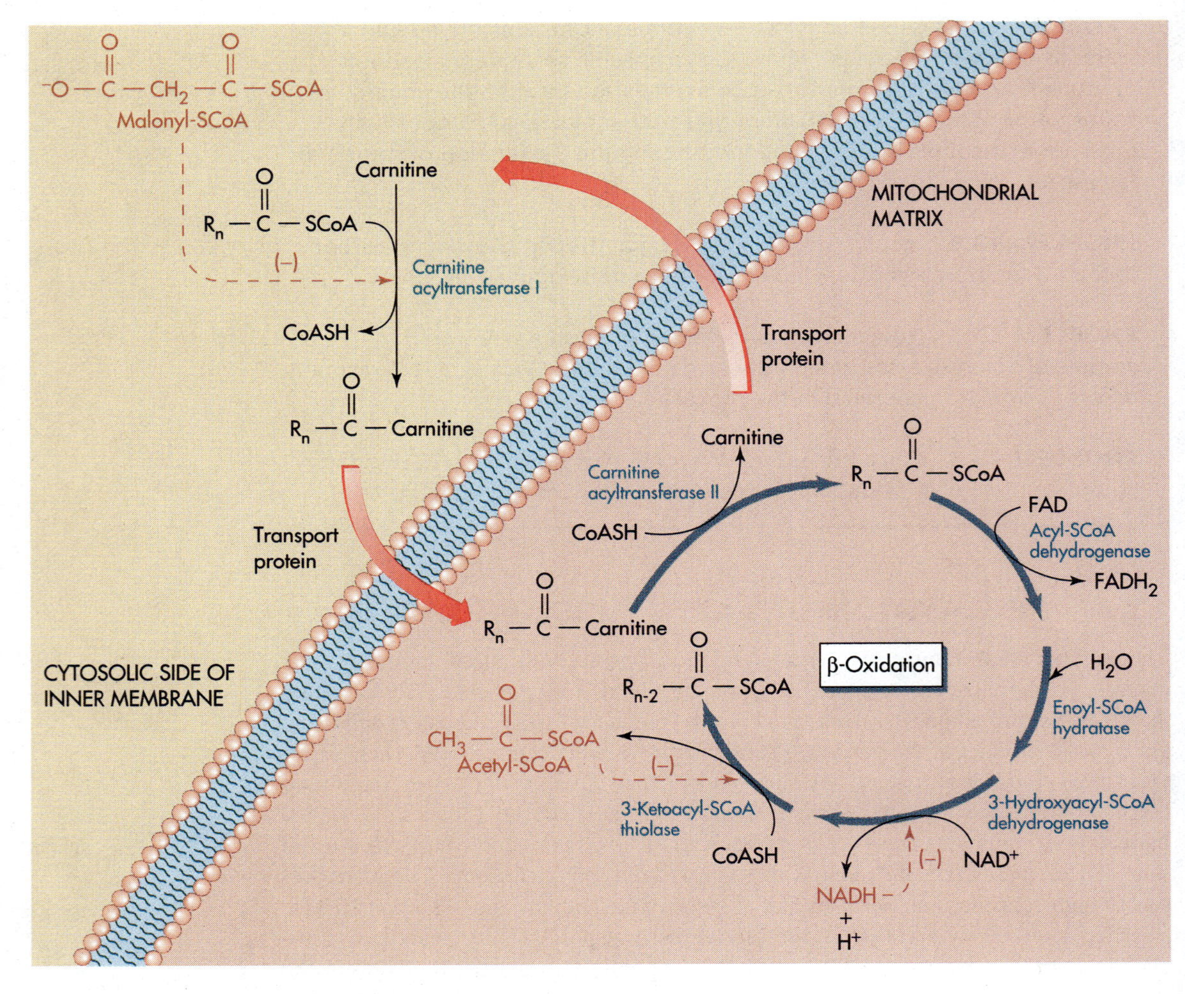

FIGURE 13.10
Allosteric regulation of fatty acid oxidation.

13.4 KETONE BODIES AND KETOSIS

Liver mitochondria, and to a lesser extent other mitochondria, use part of the acetyl-SCoA from the β-oxidation of fatty acids to build **acetoacetate, acetone,** and **3-hydroxybutyrate** (Figure 13.11). Known as **ketone bodies,** these substances are exported from the liver and carried by blood to other tissues, where they can be utilized as fuel molecules (Section 10.12). Since ketone bodies can breach the blood–brain barrier, they, along with glucose, account for most of the fuel consumed by the brain. Heart muscle and the renal cortex oxidize acetoacetate in preference to glucose.

Under normal circumstances, most of the acetyl-SCoA generated during β-oxidation and other metabolic processes condenses with oxaloacetate and enters the cit-

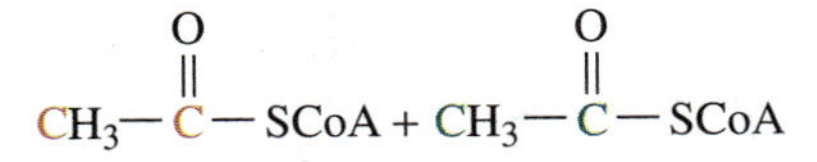

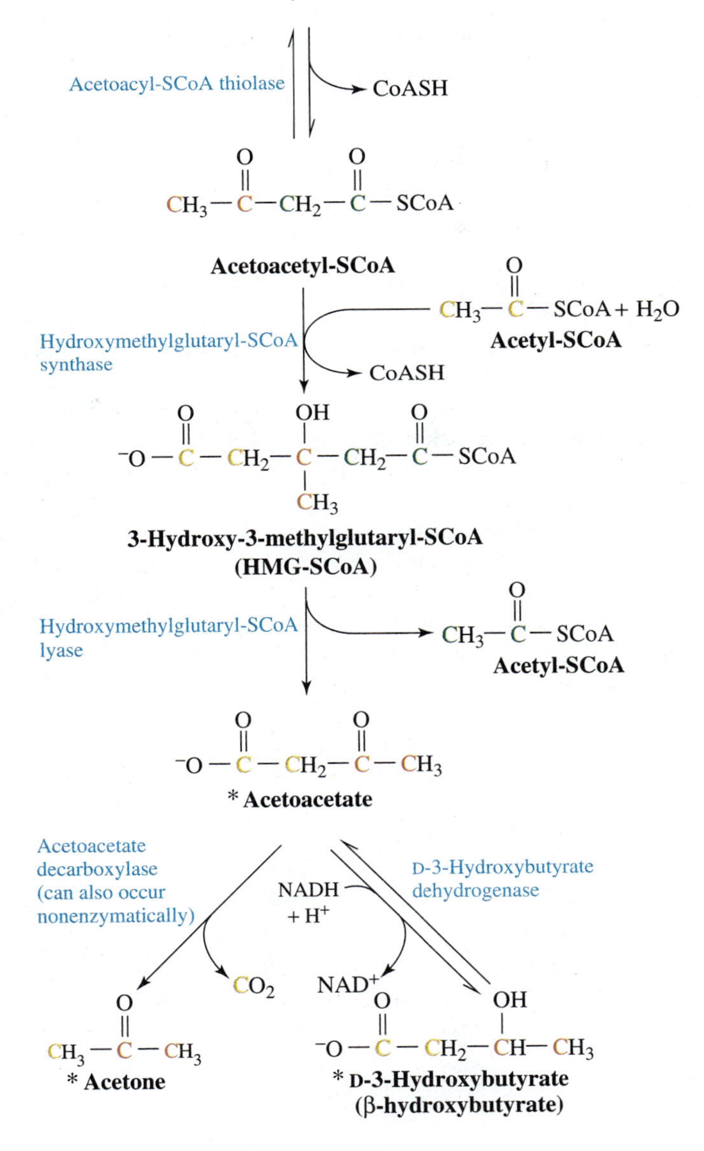

FIGURE 13.11
Formation of ketone bodies. Ketone bodies are identified with asterisks.

ric acid cycle to be oxidized. This process tends to keep ketone body concentrations relatively low. A shortage of oxaloacetate, however, can lead to an accumulation of acetyl-SCoA and an increase in the rate of production of ketone bodies. Such a shortage can result from either a drop in the rate of oxaloacetate synthesis or an increase in the rate at which it is consumed, or both. Oxaloacetate is primarily produced from pyruvate, the end product of aerobic glycolysis. This citric acid cycle replenishment reaction, which is catalyzed by pyruvate carboxylase, was examined during the discussion of citric acid cycle regulation in Section 11.7. Since pyruvate carboxylase is allosterically activated by acetyl-SCoA, an accumulation of acetyl-SCoA favors oxaloacetate formation when an adequate supply of pyruvate is available. Oxaloacetate is consumed during a variety of anabolic processes, including gluconeogenesis (Figure 13.12 and Section 14.2).

Ketone bodies tend to accumulate whenever the body is catabolizing relatively large amounts of fat while utilizing little, if any, carbohydrate. Under these circumstances, β-oxidation generates large amounts of acetyl-SCoA, but there is little pyruvate for oxaloacetate production. The oxaloacetate that is present tends to be converted to glucose to combat hypoglycemia and to ensure that the brain has an adequate supply of fuel. Starvation and diabetes (Section 12.4) are two common conditions that cause the body to switch from burning carbohydrates to burning fats. The resultant oxaloacetate shortage can lead to **ketosis,** an abnormally high concentration of ketone bodies in blood and urine. Affected individuals may produce enough acetone for one to smell it on their breath. The accumulation of acidic ketone bodies leads to **metabolic acidosis,** a metabolism-linked drop in blood pH. If severe, the acidosis leads to a coma and even death (Section 2.6). **[Try Problems 13.10 through 13.12]**

Ketosis tends to be self-regulating since acetoacetate inhibits fat oxidation by shutting down the release of fatty acids from adipocytes. Although the brain normally prefers glucose as a fuel, it gradually increases its use of ketone bodies as these metabolites become the predominant fuel in blood during starvation (Figure 13.13). When starvation is prolonged, up to 75% of the energy used by the brain is provided by acetoacetate.

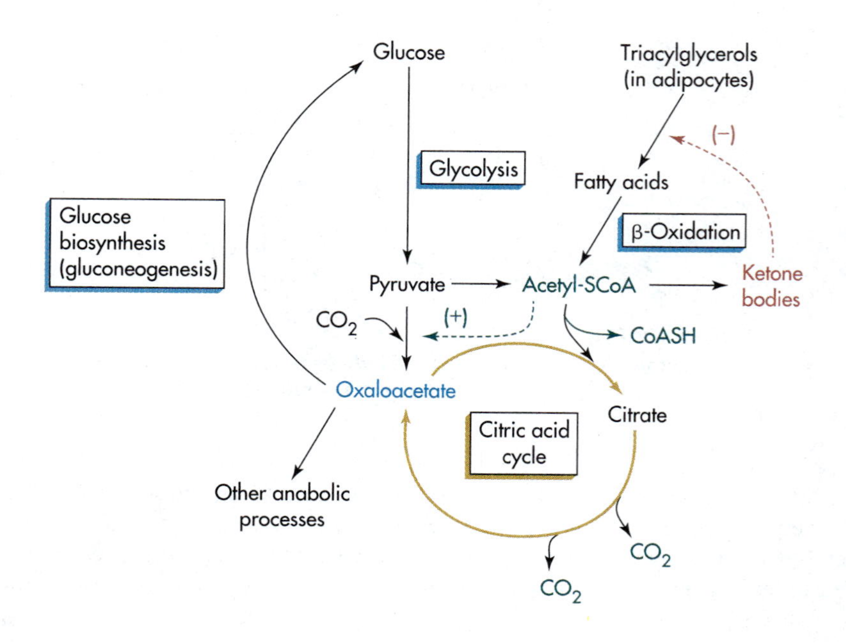

FIGURE 13.12
Production and use of oxaloacetate and acetyl-SCoA. An acetyl group that enters the citric acid cycle is not converted to oxaloacetate; it is oxidized to two CO_2 molecules before oxaloacetate is regenerated from citrate.

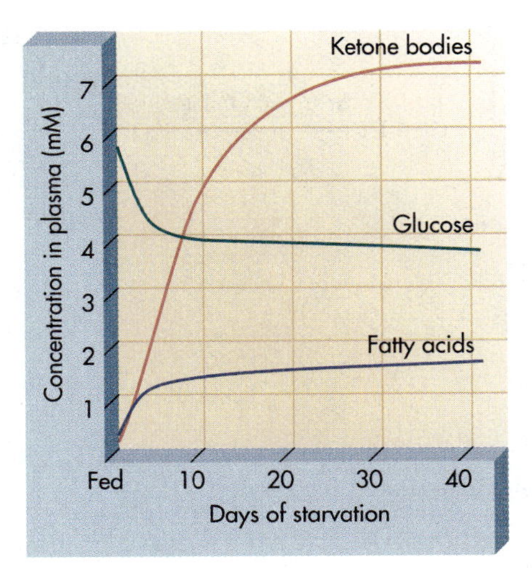

FIGURE 13.13
Effect of starvation on blood fuel levels. Plasma is the fluid part of blood, the part that remains when blood cells are removed.

The rest of this chapter concludes our introduction to catabolism by examining some of those reactions that participate in the breakdown of proteins and amino acids. The metabolic pathways involved are linked to some of the pathways that contribute to the oxidation of fats and carbohydrates. The conversion of certain amino acids to pyruvate or acetyl-SCoA, for example, connects protein catabolism to the topic of ketone bodies and ketosis.

PROBLEM 13.10 Which ketone bodies are acids? Are they weak or strong acids?

PROBLEM 13.11 Explain why the human body tends to burn more fat than carbohydrate during prolonged starvation.

PROBLEM 13.12 Calculate the approximate pH of a 0.100 M acetoacetic acid (pK_a = 3.62) solution. Hint: review Section 2.8.

PROTEIN AND AMINO ACID CATABOLISM

13.5 SOME NITROGEN-LINKED AMINO ACID CATABOLISM

Dietary proteins are digested in the stomach and small intestine to yield free amino acids (Section 4.12). The amino acids not used to assemble peptides and proteins tend to be utilized as fuel or converted to other classes of compounds, including fats and carbohydrates. Since one gram of protein or amino acid contains the same number of calories as one gram of carbohydrate (Table 9.3), proteins represent a legitimate fuel that releases considerable energy when oxidized. This section explores some of the nitrogen-linked reactions associated with amino acid catabolism.

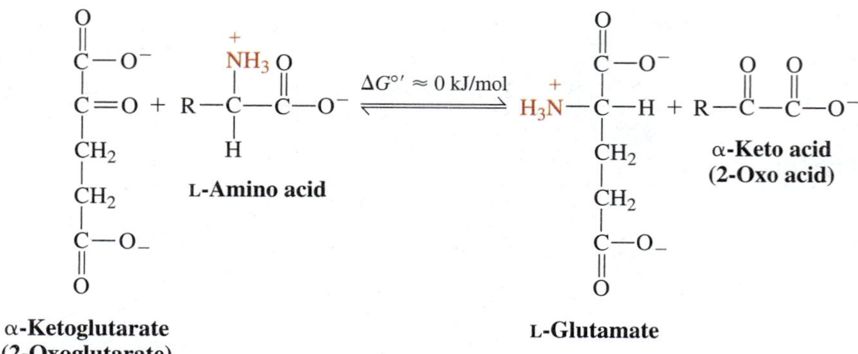

FIGURE 13.14
Aminotransferase-catalyzed reaction.

Aminotransferases

The first step in the catabolism of an amino acid in any organism is normally the removal of the α-amino group. This is most often accomplished with the assistance of enzymes known as **aminotransferases** or **transaminases.** Each of these proteins catalyzes the transfer of the α-amino group from an amino acid to α-ketoglutarate (usually) to form glutamate, itself a protein amino acid. The amino group donor is converted to an α-keto acid during this **transamination reaction** (Figure 13.14). A pool of aminotransferases participates in amino acid catabolism with each enzyme exhibiting some degree of substrate specificity. **[Try Problem 13.13]**

All of the aminotransferases share a common mechanism and require **pyridoxal phosphate,** a coenzyme produced from pyridoxine (vitamin B_6; Table 10.6). During an initial step, pyridoxal phosphate is transformed to pyridoxamine phosphate as it accepts an amino group from an amino acid (substrate 1, Figure 13.15). The amino acid is simultaneously converted to an α-keto acid (product 1). After this α-keto acid dissociates from the enzyme, α-ketoglutarate (substrate 2) docks at the active site. Glutamate (product 2) is formed during a second step as the amino group acquired by pyridoxamine phosphate is transferred to α-ketoglutarate. Pyridoxal phosphate is regenerated in the process. Collectively, the aminotransferases feed the α-amino group from most amino acids to α-ketoglutarate to form glutamate.

Excretory Forms for Nitrogen

The ultimate fate of the amino group carried by glutamate depends upon the organism involved and that organism's nutritional and metabolic state. Although amino groups can be used anabolically to construct a variety of nitrogen-containing compounds, excess amino groups tend to be excreted. Since most organisms do not store nitrogen, they must retain enough nitrogen to meet their short-term metabolic needs. The excretory form for the nitrogen in excess amino groups is determined by the organism in which it resides. The excretory form is $^+NH_4$ in most aquatic vertebrates, urea in many terrestrial vertebrates (including humans), and uric acid in birds and reptiles (Figure 13.16). Since plants do not accumulate excess nitrogen, they are not faced with a nitrogen elimination problem.

The bird droppings that coat roosting and nesting areas are a rich source of uric acid.

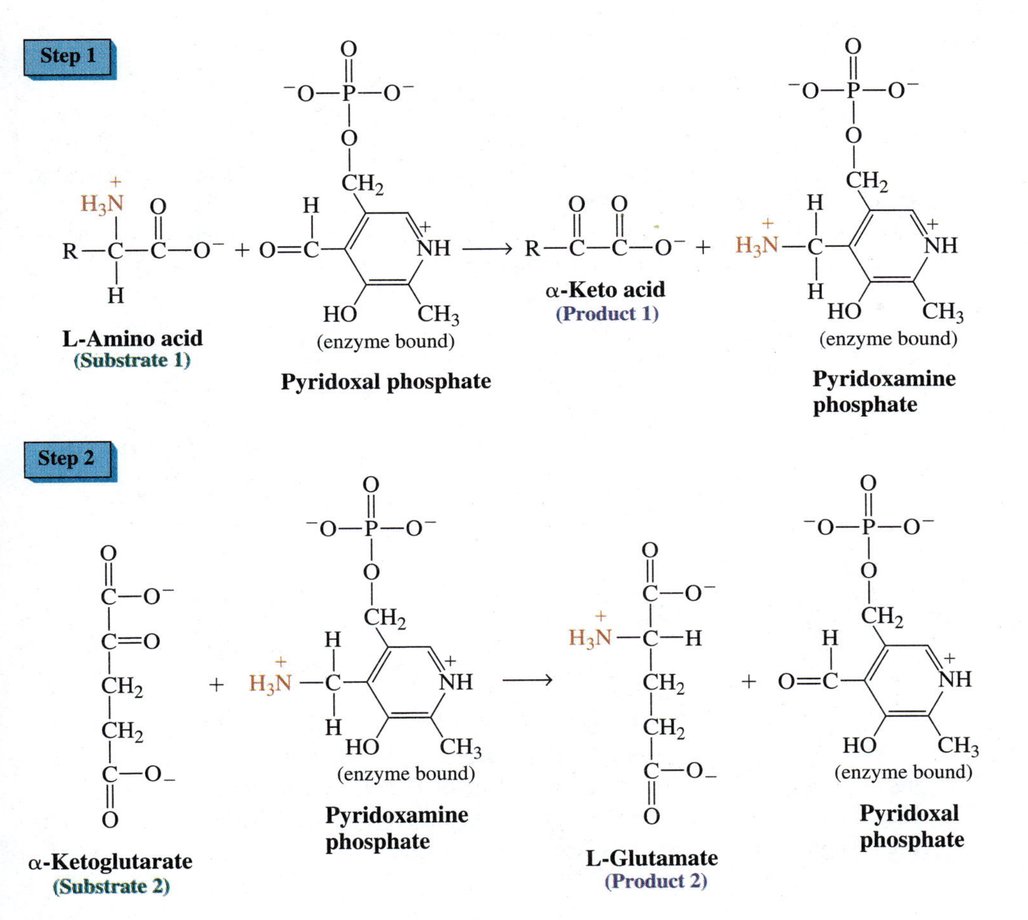

FIGURE 13.15
Role of pyridoxal phosphate in transamination. Between step 1 and step 2, α-ketoglutarate replaces the α-keto acid at the active site of the enzyme.

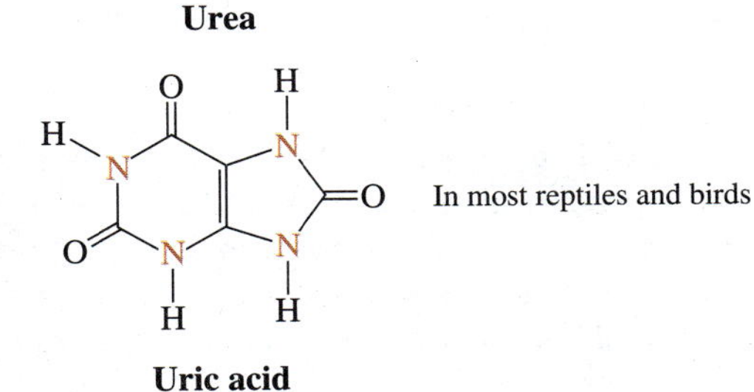

FIGURE 13.16
Excretory forms for the nitrogen in amino groups.

Amino Group Transport to the Liver

In humans, amino groups that are destined to be excreted collect in glutamate throughout the body and are then attached to a carrier for transport to hepatocytes (liver cells, the major site of urea production) (Figure 13.17). In most extrahepatic tissues, amino groups in glutamates are released as $^{+}NH_4$ before they are linked to a carrier. This reaction, catalyzed by mitochondrial **L-glutamate dehydrogenase,** is described as **oxidative deamination** since the amino group is removed as the carbon skeleton is oxidized to α-ketoglutarate by NAD^{+}. **[Try Problem 13.14]**

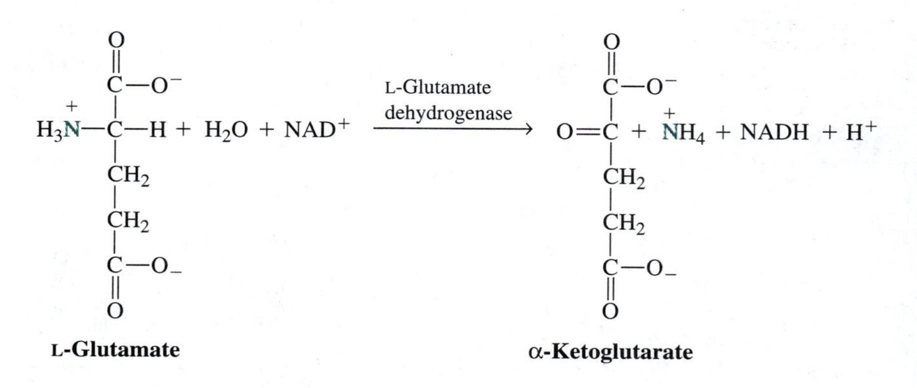

The $^{+}NH_4$ to be excreted, which has been produced by the L-glutamate dehydrogenase–catalyzed reaction, reacts with glutamate to form ***glutamine,*** a nontoxic carrier of amino groups that can pass through cell membranes. ATP is hydrolyzed during this **glutamine synthetase**–catalyzed reaction.

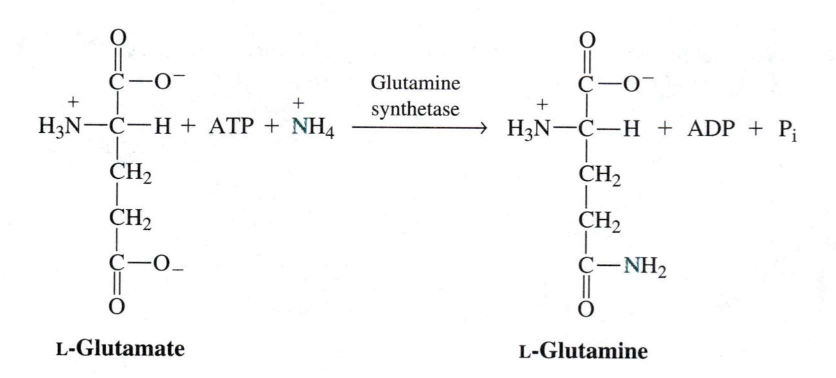

The glutamine enters the blood and is delivered to the liver. Once inside liver mitochondria, the amino group attached to the amide side chain of glutamine is converted back to $^{+}NH_4$ with the catalytic assistance of **glutaminase. [Try Problem 13.15]**

In muscle, *alanine,* rather than glutamine, often serves as the major agent for the transport of amino groups to the liver. The alanine is produced as amino groups col-

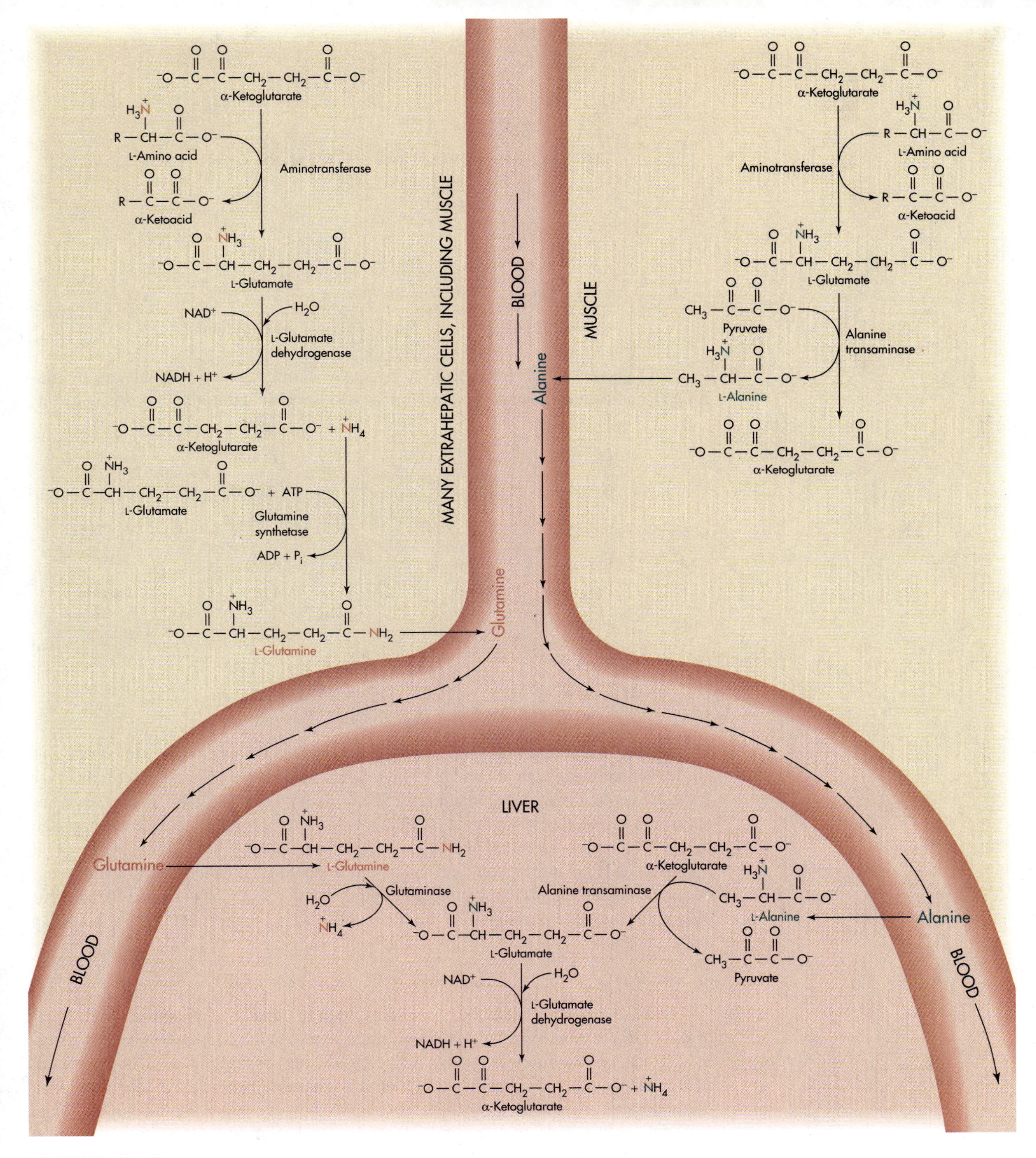

FIGURE 13.17
Transport of amino groups from extrahepatic amino acids to the liver.

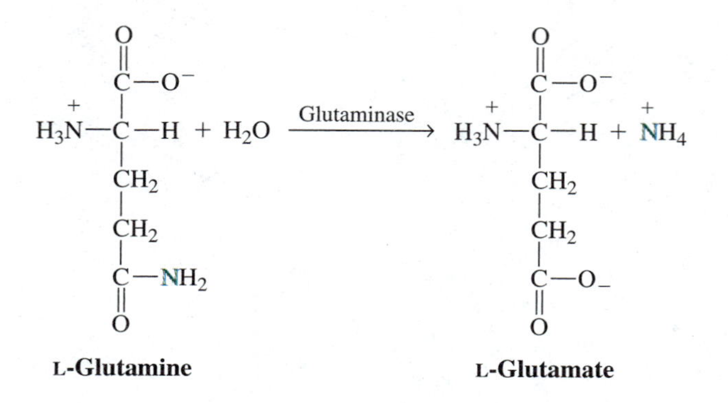

lected in glutamate are transferred to pyruvate during a reaction catalyzed by **alanine transaminase,** one of several amino acid–specific aminotransferases.

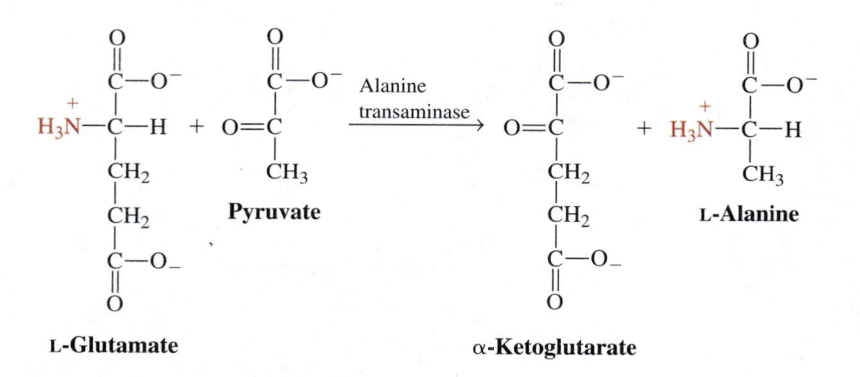

In the liver, the transported amino group is transferred from alanine to α-ketoglutarate, generating glutamate and pyruvate. Since pyruvate is often needed in the liver for gluconeogenesis, an organism kills two birds with one stone by using alanine as an amino group transport agent. Oxidative deamination of glutamate releases the transported amino group as $^+NH_4$.

Since $^+NH_4$ is quite toxic, particularly to brain cells, it cannot be allowed to accumulate within the human body. Although the toxicity of $^+NH_4$ has not been completely explained, a partial explanation lies in its ability to shift pH; $^+NH_4$ is a weak acid:

$$^+NH_4 \rightleftharpoons NH_3 + H^+ \qquad pKa = 9.5$$

$^+NH_4$ disrupts several metabolic processes through other mechanisms as well. Most of the $^+NH_4$ produced in the mitochondria of extrahepatic cells is attached to glutamate (forms glutamine) or pyruvate (yields alanine) as rapidly as it is generated, while most of the $^+NH_4$ produced in liver mitochondria is quickly incorporated into nontoxic urea. **[Try Problem 13.16]**

Some amino acids contain nitrogen atoms in addition to the one in the α-amino group. During catabolism, most of these nitrogens are also transported to the liver, converted to ammonium, and then incorporated into urea. Thus, most of the nitrogen from the catabolism of amino acids is ultimately excreted in the form of urea.

PROBLEM 13.13 What does the R represent in Figure 13.14 if the amino donor is serine? Phenylalanine?

PROBLEM 13.14 True or false? Since glutamate is oxidized as it is converted to α-ketoglutarate plus $^{+}NH_4$, α-ketoglutarate must be reduced as glutamate is regenerated during an aminotransferase-catalyzed reaction. Explain.

PROBLEM 13.15 Write the equation for the net reaction when the glutamine synthetase–catalyzed reaction is followed by the glutaminase-catalyzed reaction. Why doesn't this two-reaction sequence represent a waste of ATP?

PROBLEM 13.16 What is the relationship between "ammonia" and "ammonium?" What determines the relative amount of these two substances in a solution?

13.6 UREA PRODUCTION

The incorporation of $^{+}NH_4$ into urea entails carbamoyl phosphate synthesis plus the urea cycle. The urea produced in the liver is transported by blood to the kidneys for excretion. The kidneys also produce and excrete smaller amounts of free $^{+}NH_4$ that are involved in maintaining acid–base balance within the body.

Carbamoyl Phosphate Formation

The first step in urea production is the **carbamoyl-phosphate synthase I**–catalyzed formation of **carbamoyl phosphate.**

$$CO_2 + \overset{+}{N}H_4 + H_2O + 2\,ATP \xrightarrow[\text{synthase I}]{\text{Carbamoyl-phosphate}} H_2N{-}\overset{\overset{\large O}{\|}}{C}{-}O{-}\underset{\underset{\large O^-}{|}}{\overset{\overset{\large O}{\|}}{P}}{-}O^- + 2\,ADP + P_i + 3\,H^+$$

Carbamoyl phosphate

This ATP-dependent process occurs within the matrix of hepatic (liver) mitochondria and leads to the covalent coupling of an $^{+}NH_4$ nitrogen to a carbon atom. The $^{+}NH_4$ is normally from amino groups delivered to the liver by glutamine or alanine. **Carbamoyl-phosphate synthase II** resides in the cytosol where it is involved in the biosynthesis of pyrimidines (Chapter 14).

The Urea Cycle

Carbamoyl phosphate initiates a metabolic pathway known as the urea cycle by condensing with **L-ornithine** to produce **L-citrulline**. This transformation is catalyzed by a mitochondrial enzyme named **ornithine transcarbamoylase** (Figure 13.18). The L-citrulline is transported to the cytosol, where **argininosuccinate synthase** catalyzes its condensation with aspartate (derived mainly from the catabolism of other amino acids in the liver) to form **argininosuccinate.** The coupled hydrolysis of ATP to AMP plus PP_i provides a thermodynamic driving force for this process. The subsequent hydrolysis of PP_i "pulls" the reaction further toward completion by removing a product (Le Chatêlier's principle). Argininosuccinate is split into arginine and fumarate with the catalytic assistance of **argininosuccinate lyase (argininosuccinase)**, another cytosolic enzyme. In the final step of the cycle, cytosolic **arginase** catalyzes the hydrolysis of arginine to generate urea and L-ornithine. The L-ornithine is transported into a mitochondrion where it can condense with a carbamoyl phosphate to initiate another round through the urea cycle. Metabolic channeling (Section 5.15) appears to exist in spite of the involvement of both mitochondrial and cytosolic enzymes. [Try Problems 13.17 and 13.18]

Net Reaction and Some Energy Considerations

The net reaction of carbamoyl phosphate formation plus the urea cycle is:

$$CO_2 + {}^{+}NH_4 + 3\ ATP + {}^{-}O{-}\overset{O}{\overset{\|}{C}}{-}CH_2{-}\underset{{}^{+}NH_3}{\underset{|}{CH}}{-}\overset{O}{\overset{\|}{C}}{-}O^{-} + 2\ H_2O$$

Aspartate

$$\downarrow$$

$$H_2N{-}\overset{O}{\overset{\|}{C}}{-}NH_2 + 2\ ADP + 2\ P_i + PP_i + AMP + {}^{-}O{-}\overset{O}{\overset{\|}{C}}{-}\overset{H}{C}{=}\underset{H}{C}{-}\overset{O}{\overset{\|}{C}}{-}O^{-} + 7\ H^{+}$$

Urea — Fumarate

Excretion of urea and hydrolysis of the PP_i "pull" the reaction toward completion. One of the nitrogens in urea comes from ${}^{+}NH_4$ whereas the second nitrogen is contributed by aspartate. Under normal circumstances, the furmarate formed from aspartate is predominantly used for gluconeogenesis (glucose biosynthesis; Section 14.2). However, some fumarate is used to regenerate aspartate in a three-step process (occurs in mitochondria) that involves citric acid cycle enzymes plus a transaminase (Figure 13.19). The NADH produced during the recycling of part of the fumarate back to aspartate provides some of the energy (around 2.5 ATPs when a pair of electrons flows from NADH to O_2 through the respiratory chain) necessary to sustain urea production. Since the PP_i in the net reaction is hydrolyzed, four phosphate anhydride bonds are cleaved during the production of a single molecule of urea. The detoxification of ${}^{+}NH_4$ is definitely an energy-requiring process.

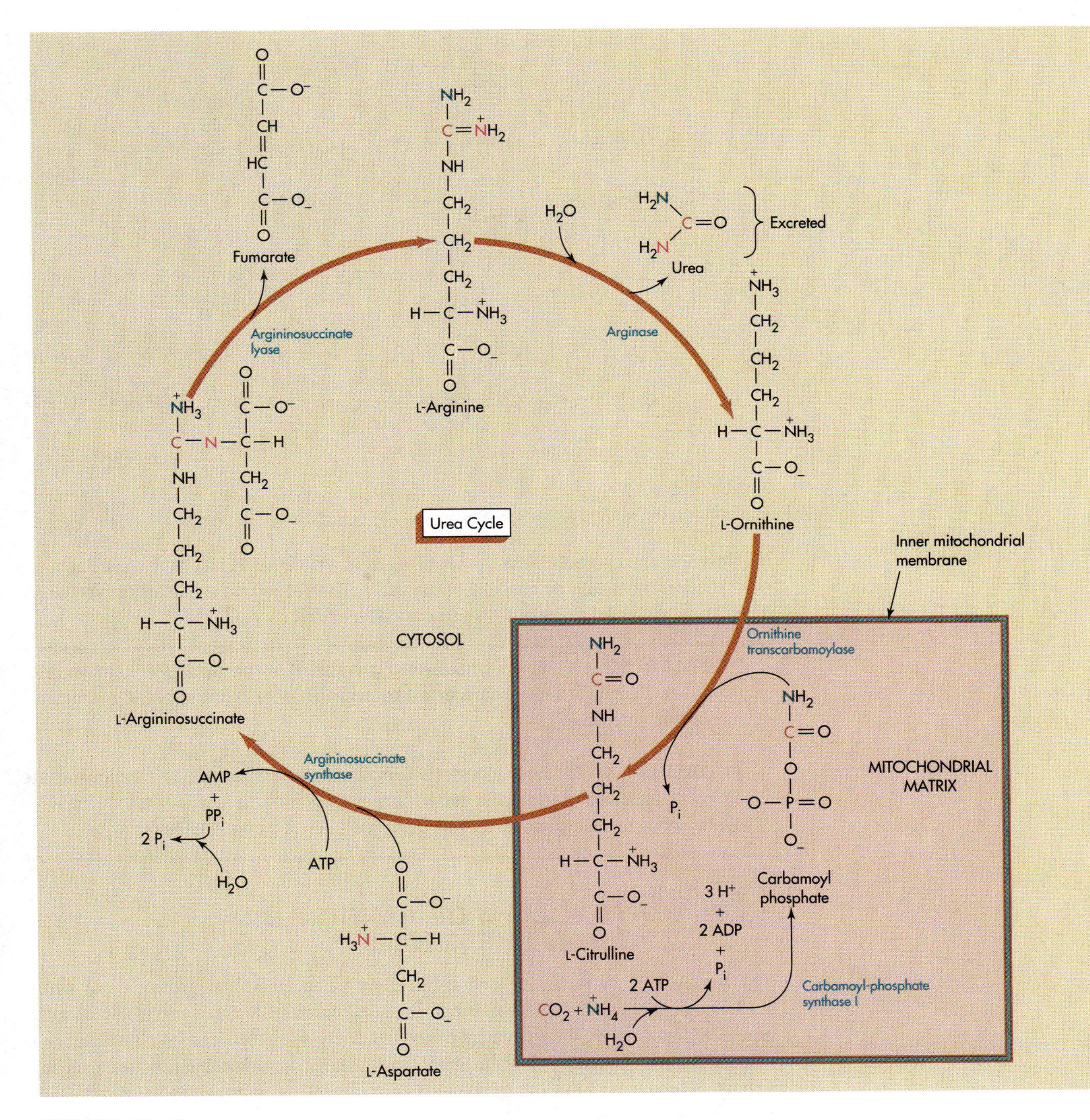

FIGURE 13.18
The urea cycle. The PP_i that is a product of the argininosuccinate synthase-catalyzed reaction is hydrolyzed to "pull" the reaction toward products (Le Châtelier's principle). The excretion of urea also helps "pull" the cycle.

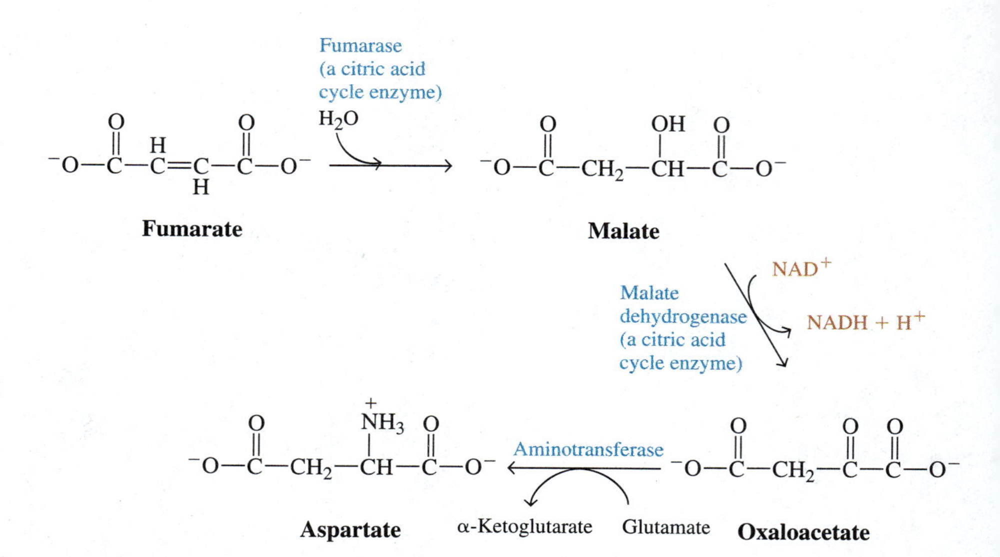

FIGURE 13.19
Conversion of fumarate to aspartate.

Now that the catabolic fate of the nitrogen in amino acids has been examined, the next section turns to a discussion of the catabolism of amino acid carbon skeletons. The ultimate fate of the sulfur in proteins is revealed as well.

PROBLEM 13.17 The alpha amino groups of some amino acids can end up in urea without being converted to ammonium. What specific reactions make this possible?

PROBLEM 13.18 Some genetic defects lead to a tendency to accumulate $^+NH_4$ as a consequence of a reduction in the catalytic activity for a urea cycle enzyme. Suggest one treatment for such a genetic disease.

13.7 CATABOLISM OF AMINO ACID CARBON SKELETONS

The removal of all nitrogen- and sulfur-containing functional groups from amino acids leaves carbon skeletons that can be metabolized through a multitude of interconnected pathways. When energy is needed, these skeletons can be completely oxidized. Whereas most of the nitrogens from nitrogen-containing functional groups end up in urea, the sulfur in amino acids is oxidatively catabolized to SO_2, a gas that is eliminated during respiration.

Carbon Skeleton Oxidation

The carbon skeletons in all 20 of the common protein amino acids are converted into compounds that are components of central metabolic pathways. All of these

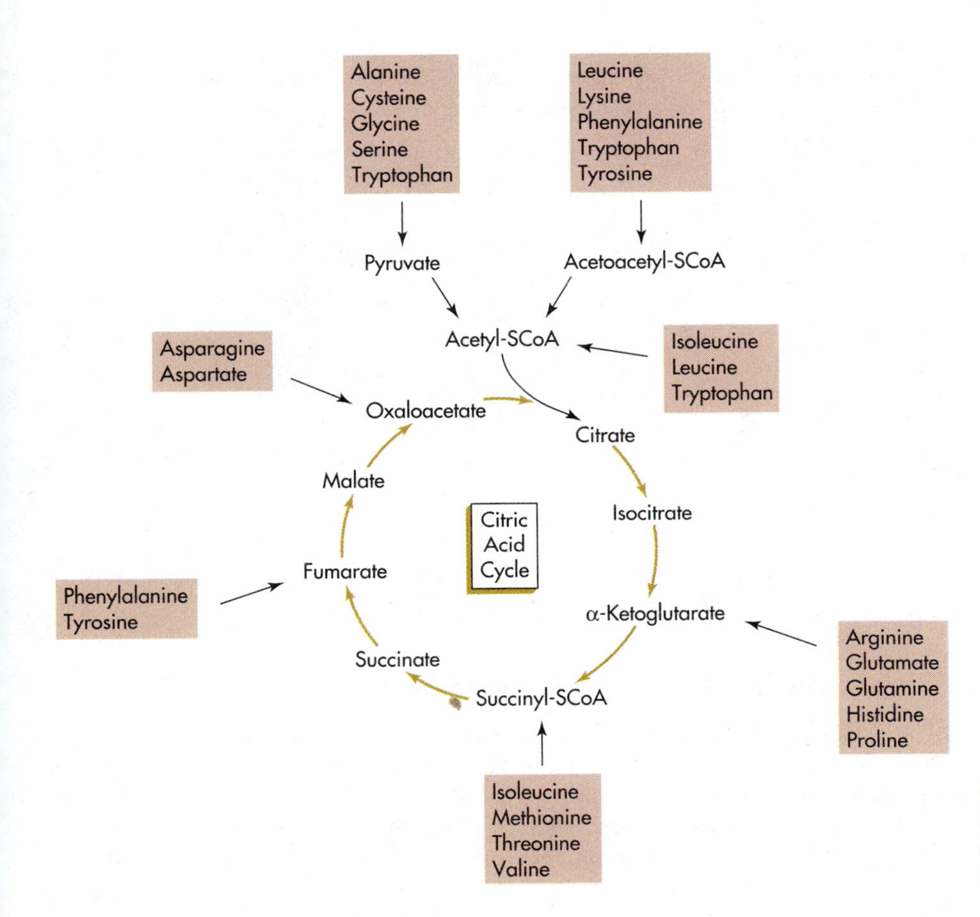

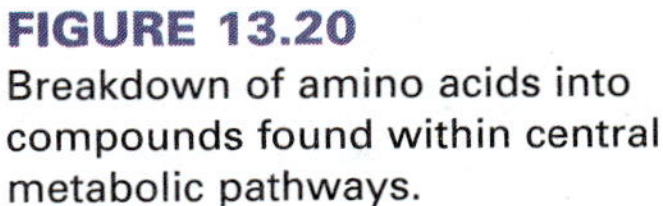
FIGURE 13.20
Breakdown of amino acids into compounds found within central metabolic pathways.

compounds can be fed into the citric acid cycle (Figure 13.20). Some conversions involve a single step, but most involve a sequence of reactions. The transformation of alanine into pyruvate illustrates the one-step movement of an amino acid into the mainstream of metabolism. In Section 13.8, a discussion of the catabolism of phenylalanine provides an example of a more complex degradative pathway. Arginine, glutamate, glutamine, histidine and proline are catabolized to α-ketoglutarate. Isoleucine, methionine, threonine, and valine can be converted to succinyl-SCoA. Fumarate is produced from phenylalanine and tyrosine, and asparagine and aspartate yield oxaloacetate. Alanine, cysteine, glycine, serine, and tryptophan are degraded to pyruvate. Leucine, lysine, phenylalanine, tryptophan, and tyrosine are catabolized to acetoacetyl-SCoA. Isoleucine, leucine, and tryptophan are converted to acetyl-SCoA without passing through either a pyruvate or an acetoacetyl-SCoA intermediate.

Some amino acids appear multiple times in Figure 13.20 because their carbon skeletons are severed to create distinctive fragments that are catabolized along different pathways. In some instances, a carbon skeleton can be catabolized along alternative pathways. The body uses each of the seven central metabolites produced from amino acids for either anabolism or energy production. During the complete

TABLE 13.2

Glucogenic and Ketogenic Amino Acids

Only Glucogenic	Only Ketogenic	Both Glucogenic and Ketogenic
Arginine	Leucine	Alanine
Asparagine	Lysine	Cysteine
Aspartate		Glycine
Glutamate		Isoleucine
Glutamine		Phenylalanine
Histidine		Serine
Methionine		Tryptophan
Proline		Tyrosine
Threonine		
Valine		

oxidation of amino acids, CO_2, H_2O, and energy are the final products generated from the carbon skeletons. Part of the energy released is employed to make ATP—the ultimate goal of fuel utilization. **[Try Problem 13.19]**

Glucogenic Versus Ketogenic Amino Acids

Those amino acids that are catabolized to pyruvate, α-ketoglutarate, succinyl-SCoA, fumarate, or oxaloacetate are often described as **glucogenic** because these degradation products can be readily utilized by the body to build glucose (Table 13.2). The **ketogenic** amino acids are those whose carbon skeletons are readily converted to acetyl-SCoA or acetoacetyl-SCoA, precursors to ketone bodies. The human body lacks the enzymes necessary to use acetate and acetoacetate as substrates for gluconeogenesis. It is the absence of these enzymes that prevents the body from converting fatty acids to carbohydrates, a point that will be reexamined in Chapter 14. The carbon skeletons of some amino acids can be converted to both glucose and ketone bodies. Such amino acids are classified as both glucogenic and ketogenic. This rather archaic classification system is not straightforward because of the many metabolic interconversions possible within the body. This paragraph would not even appear in this text if the terms "glucogenic" and "ketogenic" did not appear so frequently within the biochemical and biomedical literature.

PROBLEM 13.19 Calculate the approximate net production of ATP during the complete biological oxidation (by O_2) of the carbon skeleton derived from the aminotransferase-catalyzed removal of the amino groups in 1 mol of alanine.

13.8 PHENYLALANINE CATABOLISM

The catabolism of phenylalanine is examined in this section to provide an example of a relatively complex amino acid degradative pathway. Phenylalanine was selected for this purpose since inherited defects in phenylalanine metabolism are responsible

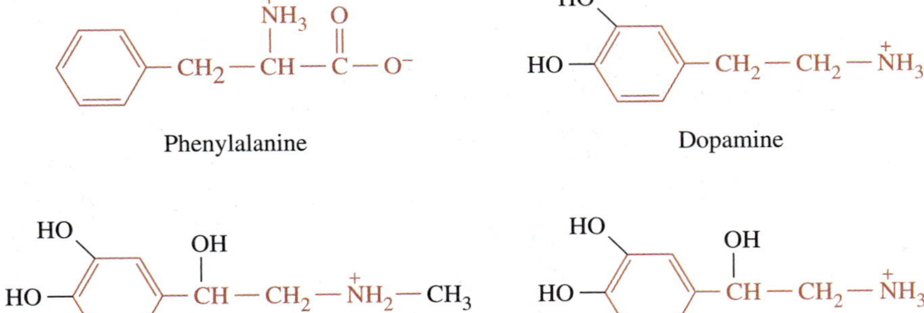

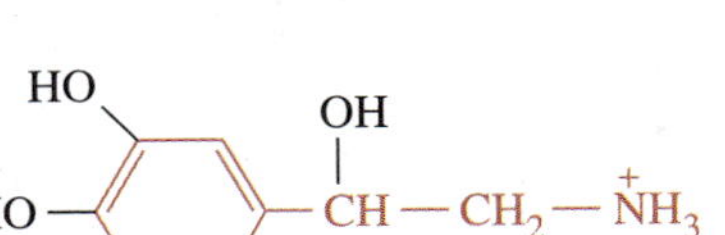

Epinephrine (adrenaline)

Norepinephrine (noradrenaline)

Melanins are irregular polymers of undetermined structure.

FIGURE 13.21
Some important metabolites produced from phenylalanine.

for several genetic diseases, including **phenylketonuria (PKU)** which is featured in Chapter 19. Phenylalanine is a normal precursor for the synthesis of the hormones epinephrine and norepinephrine, the neurotransmitter dopamine, and the pigment melanin (primarily responsible for skin color) (Figure 13.21).

Phenylalanine catabolism begins with the hydroxylation of the benzene ring to generate tyrosine, another protein amino acid (Figure 13.22). In this **phenylalanine hydroxylase**–catalyzed reaction, one of the oxygen atoms in an O_2 molecule adds to the benzene ring in phenylalanine while the second oxygen atom is reduced to water. NADH provides the reducing power. Because a pair of electrons from NADH flows to oxygen through dihydrobiopterin, the reduced form of the latter coenzyme **(tetrahydrobiopterin)** is the immediate reducing agent. The reaction between dihydrobiopterin and NADH is catalyzed by a separate enzyme named **dihydropteridine reductase.**

Phenylalanine hydroxylase is defective in most individuals with PKU. Since phenylalanine cannot be catabolized in a normal fashion, it accumulates and must be degraded by alternate pathways. Alanine and phenylpyruvate are produced when an aminotransferase catalyzes the transfer of phenylalanine's α-amino group to pyruvate (Figure 13.23). As phenylpyruvate accumulates along with phenylalanine, part of the phenylpyruvate is catabolized to phenylacetate and phenyllactate. All of these compounds are found at abnormally high levels in phenylketonurics on a normal diet. Their accumulation in blood is the basis for PKU screening of newborn infants. When present at elevated concentrations in early life, these metabolites impact brain development and lead to mental retardation. PKU can be managed by restricting the dietary intake of phenylalanine (Section 19.7). This disease is the basis for the warning label on foods containing Aspartame (Nutrasweet), a derivatized dipeptide containing phenylalanine (Section 3.7). **[Try Problem 13.20]**

Phenylalanine hydroxylase is a key enzyme in phenylalanine metabolism because carbon atoms in phenylalanine flow through tyrosine on the way to epinephrine, norepinephrine, dopamine, melanin, and citric acid cycle intermediates (see Figure

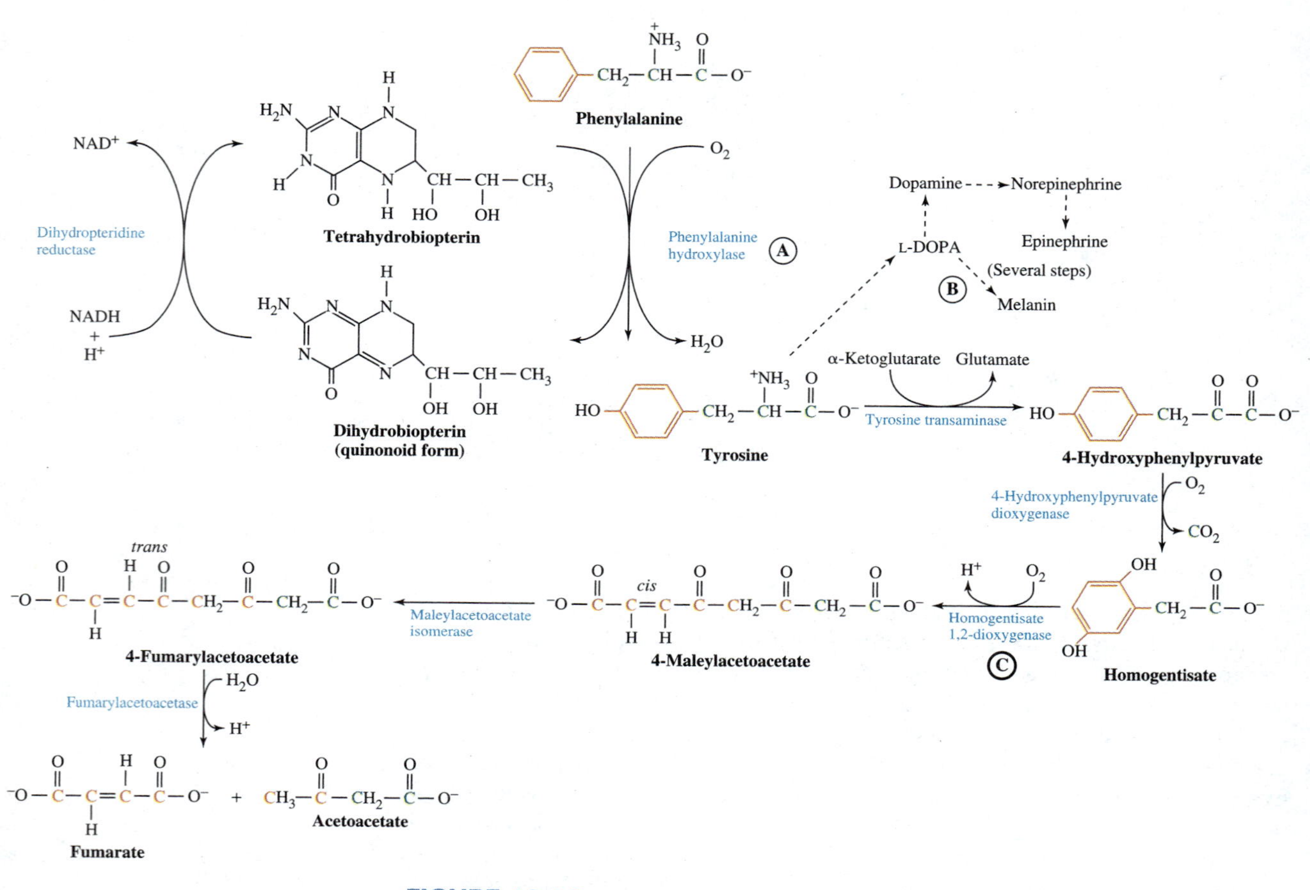

FIGURE 13.22
Phenylalanine catabolism. A defect in phenylalanine hydroxylase (A) leads to PKU. A defect in the pathway to melanin (B) leads to albinism. A defect in homogentisate 1,2-dioxygenase (C) leads to alcaptonuria.

13.22). A defect in the gene for **tyrosine 3-monoxygenase** (also called **tyrosinase** and **tyrosine 3-hydroxylase**), one of the enzymes (not shown in Figure 13.22) involved in the conversion of tyrosine to melanin, is responsible for a common form of **albinism.** Affected individuals **(albinos)** have white hair and pink skin (Figure 13.24). The pink skin is a consequence of hemoglobin and other pigments in blood. Albinos suffer visual problems and have an increased risk of skin cancer because they lack the protection (from sunlight) provided by melanin.

The catabolism of tyrosine, like the catabolism of most other amino acids, is initiated with the assistance of an aminotransferase (transaminase), which transfers tyrosine's α-amino group to α-ketoglutarate. The enzyme **4-hydroxyphenylpyruvate dioxygenase** catalyzes an oxidative decarboxylation reaction between molecular oxygen and the **4-hydroxyphenylpyruvate** that is formed (see Figure 13.22). The benzene ring in **homogentisate** is next severed oxidatively during a reaction

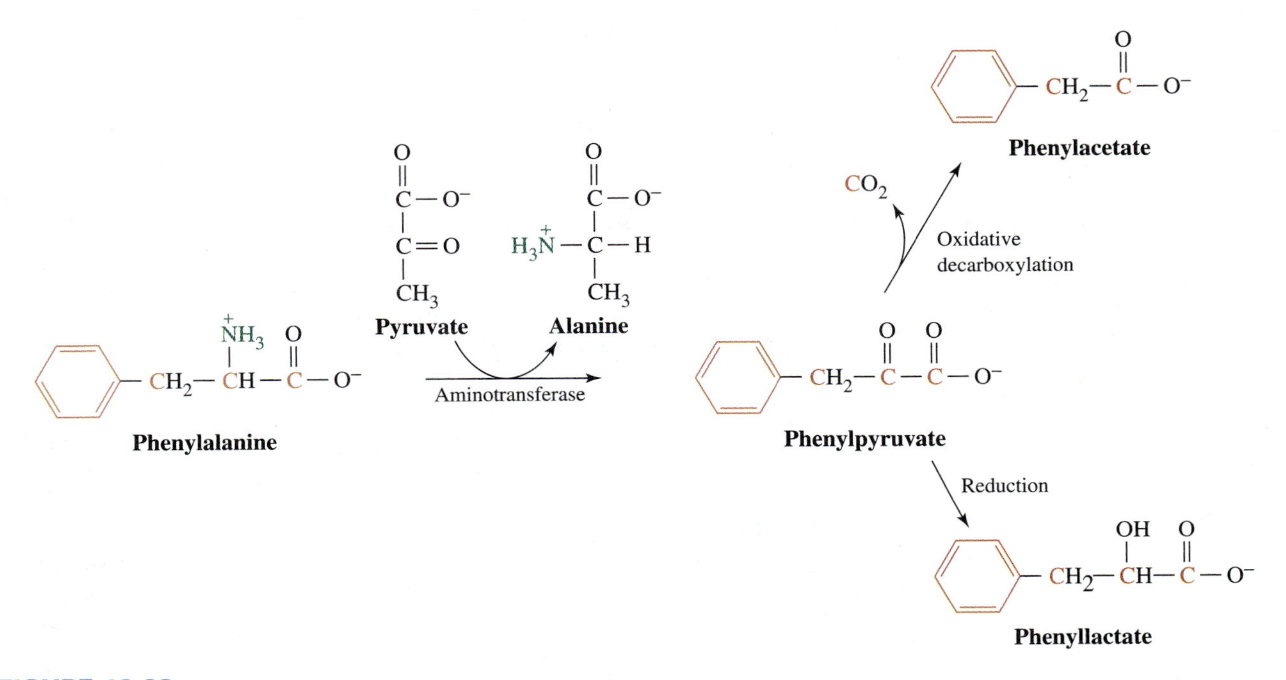

FIGURE 13.23
Alternate pathways for phenylalanine catabolism.

FIGURE 13.24
A child with albinism.

catalyzed by **homogentisate 1,2-dioxygenase.** A defect in the gene for this enzyme leads to **alcaptonuria,** a rare genetic defect making affected individuals prone to late-onset arthritis. The **maleylacetoacetate** produced by the homogentisate 1,2-dioxygenase–catalyzed reaction is transformed into **fumarylacetoacetate** in a reaction catalyzed by **maleylacetoacetate isomerase. Fumarylacetoacetase** catalyzes the cleavage of fumarylacetoacetate to yield fumarate and acetoacetate, two compounds that can feed into the citric acid cycle (see Figure 13.20). **[Try Problem 13.21]**

PROBLEM 13.20 Although PKU normally results from the production of a defective phenylalanine hydroxylase or the absence of this enzyme, the inactivation or absence of a second enzyme (shown in Figure 13.22) has an identical impact on phenylalanine metabolism. Identify this second enzyme. Explain. Some cases of PKU are linked to a defect in the gene for this second enzyme.

PROBLEM 13.21 Based on the information in Figure 13.22, what is the minimum number of genes needed to encode the enzymes required to catalyze the catabolism of phenylalanine to fumarate plus acetoacetate? The actual number of genes involved may be greater than this minimum. Explain.

SUMMARY

Dietary fat and protein can both be utilized as fuels. Although the human body does not produce any protein whose primary function is to serve as a fuel, it synthesizes triacylglycerols mainly for this purpose. Triacylglycerol reserves are stored predominantly in adipocytes.

The first step in the biological oxidation of fat is hydrolysis to release free fatty acids and glycerol. Oxidation followed by phosphorylation converts the glycerol to dihydroxyacetone phosphate. The dihydroxyacetone phosphate, like that produced from glucose during glycolysis, is catabolized to pyruvate, which can be totally oxidized with the assistance of the pyruvate dehydrogenase complex, the citric acid cycle, and the respiratory chain. The fatty acids from fats are activated and then transported into mitochondria by carnitine before oxidation. Oxidation employs three metabolic pathways: β-oxidation, the citric acid cycle, and the respiratory chain. Additional enzymes participate in the oxidation of certain fatty acids, including most that are unsaturated, odd numbered, or branched. Under standard conditions, close to 34% of the energy of fat oxidation is transformed into chemical energy in ATPs.

Fatty acid oxidation is controlled by regulating the rate of three processes: the release of fatty acids from adipocytes (a hormone-linked process), the transport of fatty acids into mitochondria, and β-oxidation.

Consecutive rounds of β-oxidation convert fatty acyl-SCoAs containing an even number of carbons in their acyl groups into multiple acetyl-SCoAs. Odd-numbered fatty acids yield a propionyl-SCoA and one acetyl-SCoA during the final round of β-oxidation. The acetyl groups in acetyl-SCoAs are completely oxidized with the catalytic assistance of the citric acid cycle and the respiratory chain. The oxidation of the propionyl group in a propionyl-SCoA requires additional enzymes.

Starvation, diabetes, and certain other "conditions" cause the human body to oxidize fats while oxidizing little carbohydrate. Oxaloacetate levels drop in mitochondria as the concentration of carbohydrate-derived pyruvate decreases. With a shortage in oxaloacetate, the acetyl-SCoA produced during β-oxidation is unable to enter the citric acid cycle as rapidly as it is formed. Part of the acetyl-SCoA that accumulates is converted to ketone bodies. Ketosis results when these bodies start to appear at elevated levels in blood and urine. Severe ketosis leads to acidosis, coma, and even death.

The catabolism of proteins begins with hydrolysis to release amino acids. In humans, most of the nitrogen atoms from the catabolism of amino acids throughout the body are collected in amino groups in glutamate, and then delivered to the liver within glutamine and alanine (transport forms of amino groups and toxic $^{+}NH_4$). A majority of this nitrogen is incorporated into urea, the major excretory form of nitrogen in most terrestrial vertebrates.

The energy-dependent production of urea from amino acid–derived $^{+}NH_4$ requires carbamoyl phosphate synthase I plus the urea cycle. Since the first step in the urea cycle occurs in mitochondria while the remaining three steps are catalyzed by cytosolic enzymes, two of the participants in the cycle must be continuously transported across the inner mitochondrial membrane. One of the nitrogens in urea is contributed by carbamoyl phosphate, and the second is delivered to the urea cycle by aspartate. Aspartate is converted to fumarate as it donates an amino group for urea production. Fumarate can be used for gluconeogenesis, the regeneration of aspartate, and other metabolic processes.

The sulfur in amino acids is oxidatively incorporated into SO_2. The carbon skeletons of the 20 common protein amino acids are catabolized in distinct but often converging pathways that generate a limited number of central metabolites. These metabolites can feed into the citric acid cycle or be siphoned off into other metabolic pathways. To be completely oxidized, these metabolites must be converted to acetyl-SCoA. Amino acids containing carbon skeletons that can readily be converted to glucose are said to be glucogenic. Ketogenic amino acids are those whose carbon skeletons can be transformed into ketone bodies. Some amino acids are both glucogenic and ketogenic.

Phenylalanine is catabolized to tyrosine and then to fumarate plus acetoacetate in a pathway that employs six enzymes. Tyrosine is a precursor for the synthesis of epinephrine, norepinephrine, dopamine, melanin, and other metabolically active substances. Defects in the genes for some of the enzymes involved in phenylalanine and tyrosine metabolism are linked to a variety of genetic diseases, including PKU, albinism, and alcaptonuria.

EXERCISES

13.1 Although a cell can produce some ATP during the partial oxidation of glucose under anaerobic conditions, it is unable to produce ATP from the partial oxidation of fatty acids under similar conditions. Explain this fact.

13.2 What enzymes are ultimately activated by the cAMP-linked cascade responsible for the hormone-induced mobilization of stored fats?

13.3 Explain how cAMP brings about different cascades in distinctive types of cells. This second messenger, for example, initiates a fat-mobilizing reaction cascade in adipocytes but a glycogen-mobilizing reaction cascade in muscle.

13.4 List the metabolic pathways involved in the total oxidation of saturated, even-numbered fatty acids within the human body.

13.5 Describe the relationship among the terms "acyl-SCoA," "acetyl-SCoA," and "acetoacetyl-SCoA."

13.6 How many rounds of β-oxidation are required to convert one molecule of lauric acid (a saturated fatty acid with 12 carbons) to 6 acetyl-SCoAs? How many ATP (give an estimate of the maximum number) can be produced from the complete oxidation of one molecule of lauric acid? What percentage of the ATP is generated within the respiratory chain?

13.7 What is the metabolic logic behind the inhibition of β-oxidation by acetyl-SCoA?

13.8 How many O_2 are consumed during the total oxidation of hexanoic acid, a 6-carbon, saturated fatty acid? What is the approximate net moles of ATP produced during the total oxidation of 1 g of this fatty acid? How many O_2 are consumed during the total oxidation of glucose (also contains 6 carbons)? What is the approximate net moles of ATP produced during the total oxidation of 1 g of glucose? On a gram to gram basis, which yields more ATP when oxidized, the fatty acid or glucose? How much more?

13.9 Could dietary amino acids provide some relief from ketosis? Explain.

13.10 Predict what specific reactions are involved in the total oxidation (by O_2) of 3-hydroxybutyrate (a ketone body) by a brain cell. Estimate maximum net ATP production per mole of 3-hydroxybutyrate oxidized. Ketone bodies are among a limited number of fuel molecules that can penetrate the blood–brain barrier and enter brain cells.

13.11 Explain why a diet rich in protein (relative to carbohydrate and fat) places an extra burden on the kidneys.

13.12 List all of the chemical elements (different types of atoms) found within the common protein amino acids. What specific compounds contain the amino acid–derived atoms following amino acid catabolism?

13.13 List those reactions (discussed in this chapter) in which $^{+}NH_4$ is a product or a reactant. Describe the biological role or significance of each of these reactions.

13.14 Which two amino acids can be converted to a citric acid cycle intermediate through a one-step reaction catalyzed by an aminotransferase?

13.15 The rate of synthesis of the enzymes involved in urea production increases when you are on a high-protein diet. What is the metabolic logic behind this fact? Similar changes in the rate of enzyme synthesis occur during prolonged starvation. Explain.

13.16 When PP_i is a product of a reaction, it is normally hydrolyzed to "pull" that reaction toward completion. Give two examples from this chapter that support this claim. Explain how the hydrolysis of PP_i "pulls" a reaction.

13.17 Use Figure 13.22 to explain how phenylalanine can be both glucogenic and ketogenic.

13.18 The absence of functional phenylalanine hydroxylase does not lead to albinism, even though this enzyme is required to convert phenylalanine to melanin. Explain.

SELECTED READINGS

Gurr, M. I., and J. L. Harwood, *Lipid Biochemistry: An Introduction,* 4th ed. London: Chapman & Hall, 1991.

The topics discussed include fat catabolism.

Lehninger, A. L., D. L. Nelson, and M. M. Cox, *Principles of Biochemistry,* 2nd ed. New York: Worth Publishers, 1993.

A comprehensive text with a thorough coverage of metabolism.

Moran, L. A., K. G. Scrimgeour, H. R. Horton, R. S. Ochs, and J. D. Rawn, *Biochemistry,* 2nd ed. Englewood Cliffs, NJ: Neil Patterson/Prentice Hall, 1994.

An encyclopedia-type biochemistry text containing a detailed discussion of metabolism. Includes a limited number of problems and some selected readings.

Smith, C. A., and E. J. Wood, *Energy in Biological Systems,* New York: Chapman & Hall, 1991.

A 171-page paperback that summarizes the basic metabolic pathways. Contains both questions and references.

Stryer, L., *Biochemistry,* 4th ed. New York: W. H. Freeman and Company, 1995.

Provides good coverage of metabolism.

Werner, R., *Essential Biochemistry and Molecular Biology,* 2nd ed. New York: Elsevier Publishing Co., 1992.

A 463-page paperback that provides a concise and readable review of basic metabolism.

14 Anabolism

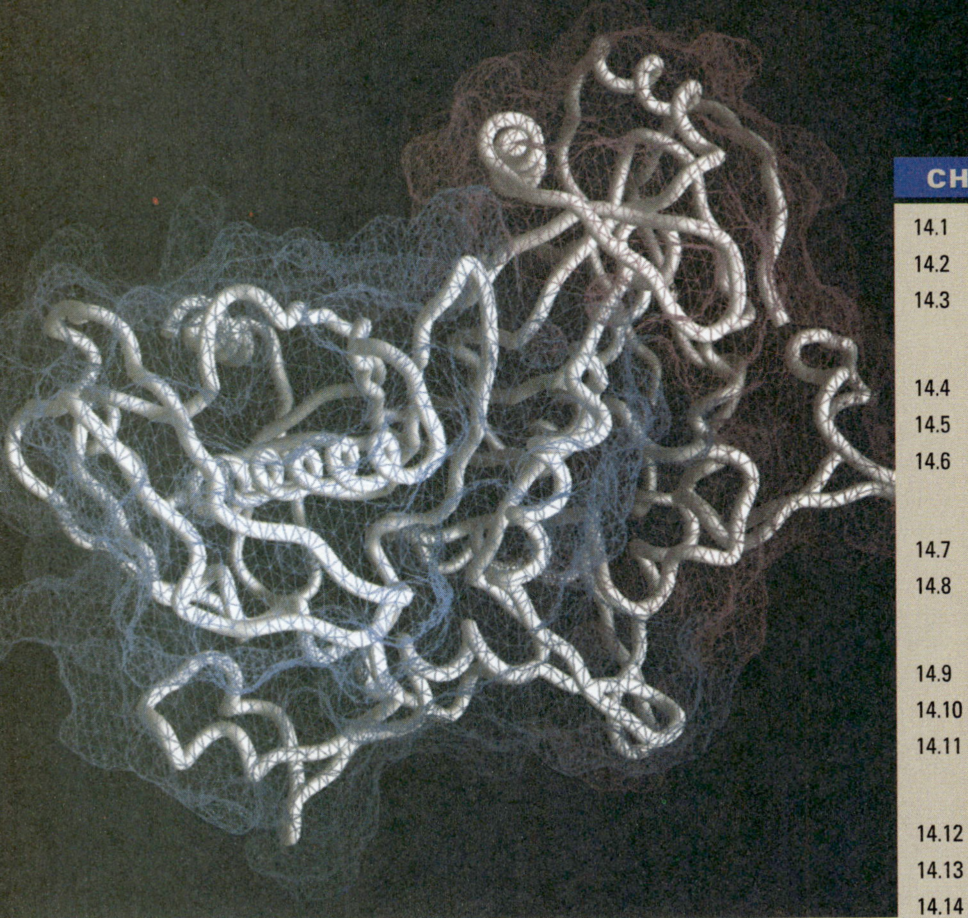

CHAPTER OUTLINE

Chapters 11 through 13 have provided an introduction to the catabolism of carbohydrates, fats, and proteins. A second part of the metabolism story entails anabolism, that collection of reactions responsible for the synthesis of the wide variety of substances required by a living organism. Most anabolism requires both a carbon source and an energy source. In plants and some microorganisms, the predominant carbon source is CO_2 and the primary energy source is sunlight. In humans, the catabolism of dietary organic compounds, including carbohydrates, fats, and amino acids, provides both the carbon-containing compounds and the energy used for anabolism. Catabolism also generates the reducing power essential for many synthetic processes.

Both the anabolic and catabolic capabilities of an organism are determined by the genetic information it inherits. This information, which is carried by DNA, encodes those enzymes that the organism is able to synthesize. Since most of the reactions that occur within an organism do not proceed at a biologically significant rate in the absence of catalysts, the enzymes produced by an organism determine what reactions occur within it. This, in turn, dictates the nature of the organism.

This chapter examines the **biosynthesis** (synthesis as it occurs within living organisms) of glucose, glycogen, fatty acids, triacylglycerols, amino acids, and certain other organic compounds. It also explores the coordination and regulation of anabolism and catabolism. This exploration helps provide an integrated picture of metabolism and biochemistry as a whole. The featured biochemical themes are energy transformation, structural hierarchy and compartmentation, communication, specificity, and the regulation of reaction rates. Chapter 15 examines photosynthesis and nitrogen fixation, two additional components of the anabolism story.

14.1 METABOLIC TURNOVER

Since most of the compounds in the human body and other living organisms are constantly being broken down and resynthesized, they are said to **turn over.** Proteins, RNAs, glycogen, glucose, triacylglycerols, and some amino acids are specific examples of compounds that do turn over. The time required for one half of the molecules in a pool of a metabolite to be eliminated or degraded is known as its half-life.

TABLE 14.1

Half-Lives of Some Membrane Components

Component	Half-Life (days) Myelin Membrane	Half-Life (days) Mitochondrial Membrane
Phosphatidylcholine	41	12
Phosphatidylserine	120	17
Sphingomyelin	>200	33
Cholesterol	>200	39
Proteins	35	21

Different compounds often have markedly different half-lives (Table 14.1), and even the same compound may have dissimilar half-lives under different metabolic conditions. The half-lives of proteins, for example, typically range from a few minutes to several months. In humans, individual proteins are broken down by one or more of the proteinases that constitute the body's protein degradation system.

Much of the metabolic turnover that occurs within the human body is related to the constantly changing needs of individual cells, specific organs, and the body as a whole. Cells commonly require a different combination of enzymes or different concentrations of enzymes at distinct points in time, and the optimal level for other substances is also subject to change. Under some circumstances, for example, a muscle cell needs to mobilize glycogen to obtain energy for contraction, whereas under other circumstances it benefits by storing glucose in glycogen. Some distinct enzymes are involved in the two processes. Molecules not providing an immediate benefit tend to be degraded. Even those enzymes and other compounds that are utilized on a regular basis are continuously catabolized and resynthesized (Table 14.2). **[Try Problem 14.1]**

TABLE 14.2

Half-Lives of Some Rat Liver Enzymes

Enzyme	Half-Life (hours)
Ornithine decarboxylase	0.2
RNA polymerase	1.3
Tyrosine aminotransferase	2.0
PEP carboxylase	5.0
Aldolase	118
Cytochrome *b*	130
Lactate dehydrogenase (LDH)	130

How does the body profit from the turnover of a compound, particularly one that is regularly or continuously utilized? Isn't it wasteful to degrade and resynthesize molecules? Evidence indicates that the cost of metabolic turnover is normally outweighed by its benefits. The benefits include the conservation of space (a cell can only store a limited number of molecules) and the regulation of cellular activities (a degraded molecule is inactive). Turnover also helps prevent the buildup of aged molecules that tend to become altered by the accumulation of damage over time.

DNA, one of the few molecules in a cell that it is not programmed to turn over completely, does indeed amass chemical alterations (mutations) in spite of a variety of repair systems that lead to the turnover of damaged regions in DNA. The **somatic mutation hypothesis** of aging proposes that biologic aging is partially a consequence of this cumulative process. This hypothesis is supported by a correlation between certain DNA repair capabilities and maximum life span in mammals. Since the DNA repair systems preserve genetic information, guard against cancer, and unite the fields of basic cell biology, cancer research, and toxicology, the enzymes of the DNA repair systems were voted molecules of the year by *Science* in 1994. **[Try Problem 14.2]**

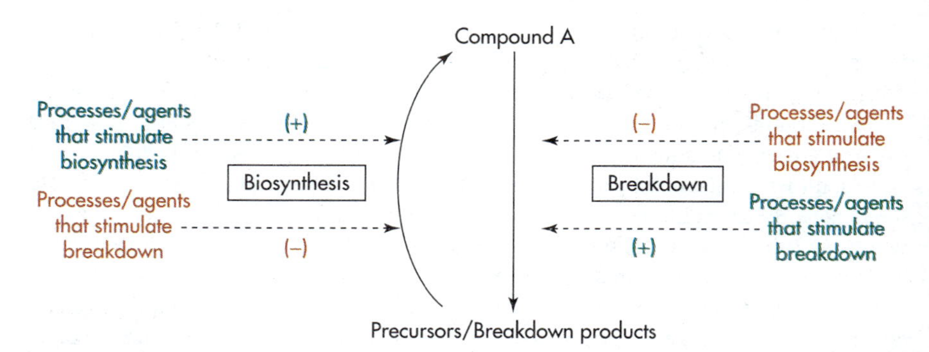

FIGURE 14.1
Reciprocal regulation of anabolism and catabolism.

The accumulation of molecular damage over time is responsible for the changes in facial features associated with aging.

Because molecular damage can inhibit metabolic turnover, even those molecules that normally do turn over may pile up in modified forms in older cells and organisms. There is, for example, a strong correlation between a person's age and the amount of cross-linked collagen (Section 4.11) that he or she possesses. The **cross-linkage hypothesis** of aging suggests that biological aging is partially the consequence of a gradual build up of cross-linked macromolecules, including collagen. **[Try Problem 14.3]**

When considering anabolism, catabolism, and metabolic turnover, there are three general principles worth remembering: (1) the catabolism of both dietary and endogenous substances provides energy, reducing power, and building blocks for anabolism; (2) the anabolic pathway involved in synthesizing a compound is not simply the reverse of the catabolic pathway involved in its degradation, although the two pathways frequently share some reactions and intermediates; and (3) the pathway involved in the production of a compound is normally repressed when there is a need to catabolize the compound, and vice versa. On the basis of this last principle, anabolism and catabolism are said to be **reciprocally regulated** (Figure 14.1). In the sections that follow, a small fraction of the anabolic pathways that have been analyzed are described along with several examples of reciprocal regulation.

PROBLEM 14.1 Would you expect a protein hormone such as insulin or glucagon to have a short or a long half-life in blood? Explain.

PROBLEM 14.2 Biological aging is more than simply the passage of time. It can be defined as a time-related decrease in an organism's ability to cope with its environment. Describe how the molecular changes associated with an accumulation of altered DNA (mutations) could lead to a decrease in your coping ability.

PROBLEM 14.3 Review Section 4.10, which examines the functions of collagen, and then explain how the cross-linking of collagen molecules could decrease a person's ability to cope with environmental stresses.

14.2 GLUCONEOGENESIS

Gluconeogenesis is the synthesis of glucose from noncarbohydrate precursors. This metabolic process is most active in liver, and the glucose produced by this organ plays a central role in the buffering of blood sugar levels (Sections 10.12 and 12.1). Kidneys assemble a smaller quantity of glucose through gluconeogenesis whereas the brain, skeletal muscles, and heart muscles tend to catabolize glucose rather than synthesize it.

Precursors for Gluconeogenesis

Under normal conditions, the primary precursors for gluconeogenesis are *glycerol* from the hydrolysis of fats, *lactate* from anaerobic glycolysis, and *alanine* from dietary proteins. Lactate and alanine tend to be quantitatively most important. During gluconeogenesis, lactate is converted to pyruvate through the reversal of the last step in anaerobic glycolysis, while alanine is converted to the same metabolite through a reversible transamination reaction (Figure 14.2). Hence, pyruvate is a precursor for most gluconeogenesis. Mammals are unable to convert fatty acids to glucose, because they are not genetically programmed to build the enzymes required to catalyze the synthesis of pyruvate or other precursors for gluconeogenesis from the acetyl-SCoA or other intermediates generated during fatty acid catabolism (review Figure 13.12). **[Try Problem 14.4]**

Relationship Between Glycolysis and Gluconeogenesis

Since there are three steps in glycolysis that are virtually irreversible, it is not thermodynamically feasible for a cell to convert pyruvate to glucose by reversing this

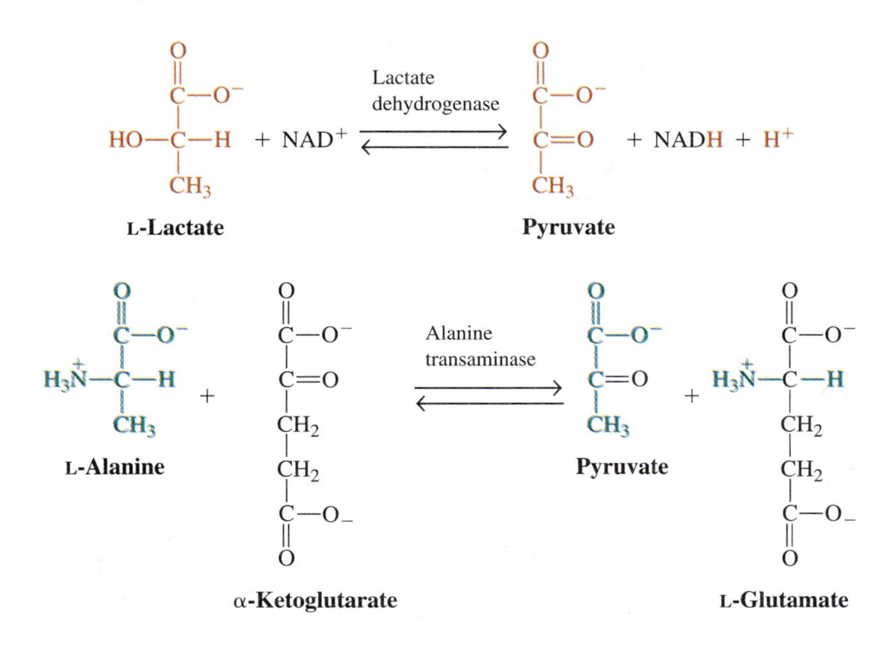

FIGURE 14.2
Conversion of lactate and alanine to pyruvate.

metabolic pathway. Gluconeogenesis couples the seven more readily reversible steps in glycolysis with four unique reactions that bypass the irreversible steps (Figure 14.3). The first bypassed step is the conversion of phosphoenolpyruvate (PEP) to pyruvate:

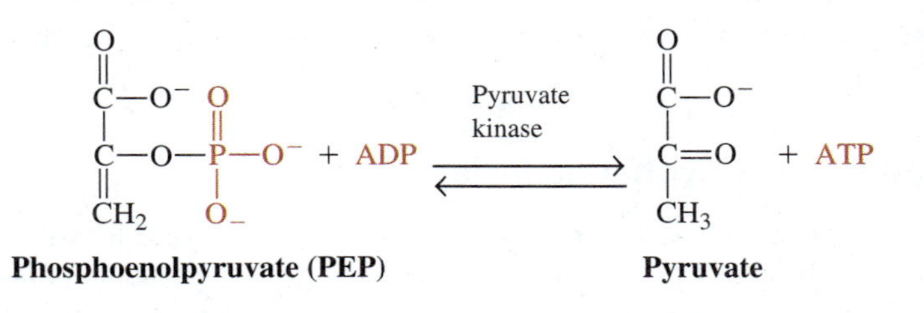

Since the reverse reaction has a highly unfavorable $\Delta G^{\circ\prime}$ of +31 kJ/mol, cells use two unique reactions to convert pyruvate to PEP and, in the process, reduce the $\Delta G^{\circ\prime}$ for PEP production to +0.84 kJ/mol.

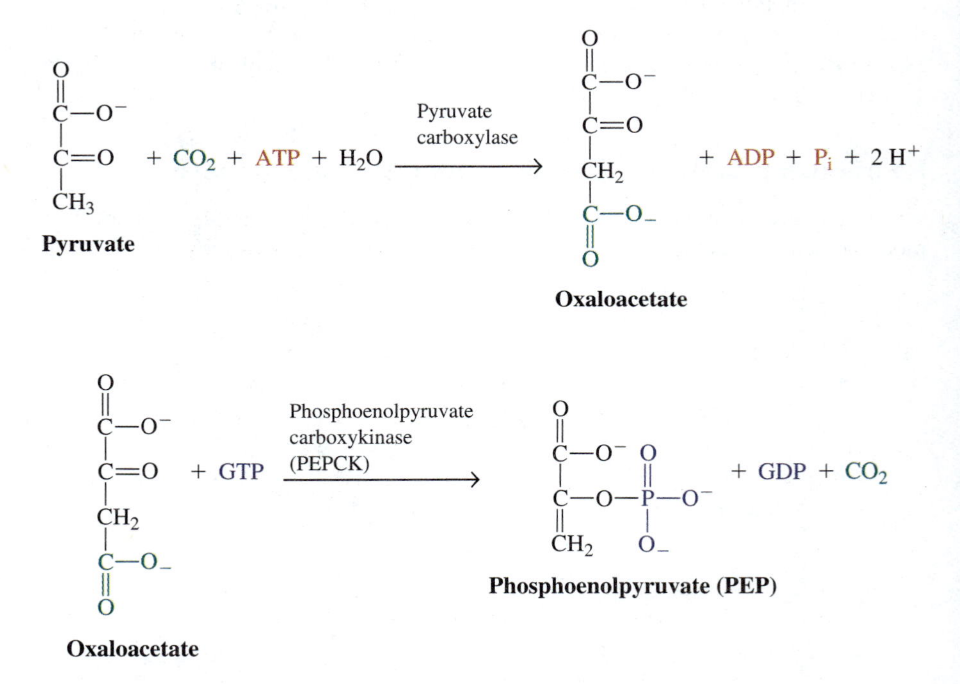

$\Delta G^{\circ\prime}$ must be negative for a reaction to be spontaneous under standard biological conditions (Section 8.2). The first reaction in PEP formation occurs in mitochondria where pyruvate carboxylase catalyzes the carboxylation of pyruvate to form oxaloacetate. Pyruvate carboxylase encompasses biocytin, a biotin-containing coenzyme involved in a variety of carboxylation reactions (Table 10.6). Pyruvate car-

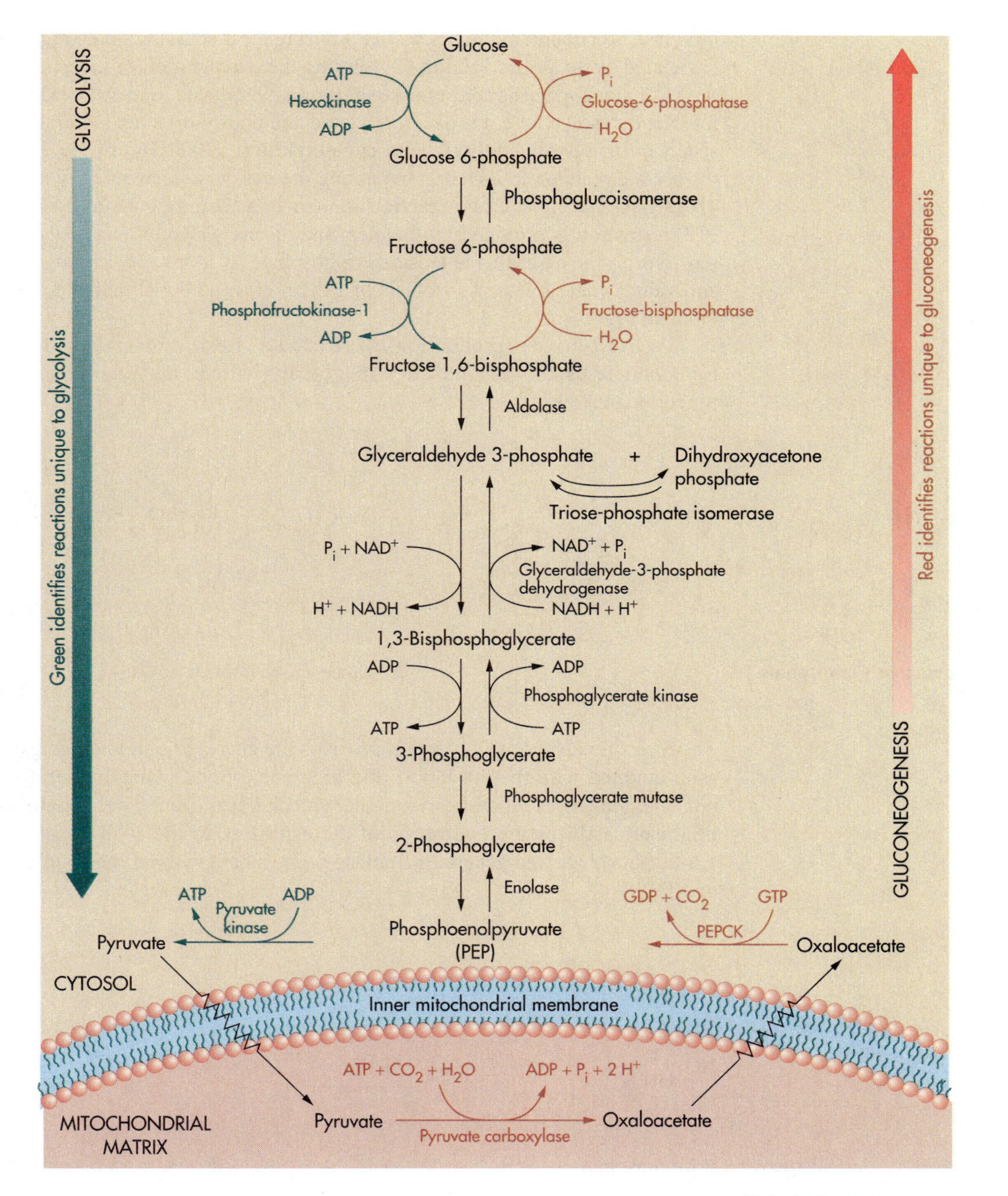

FIGURE 14.3
Glycolysis and gluconeogenesis.

boxylase is encountered in Section 11.7 during the discussion of the regulation of the citric acid cycle and in Section 13.4 during the examination of ketosis.

As gluconeogenesis continues, oxaloacetate produced in mitochondria is shuttled into the cytosol, where it is decarboxylated and phosphorylated with the catalytic assistance of **phosphoenolpyruvate carboxykinase (PEPCK)** to yield PEP. This is the second reaction involved in bypassing the pyruvate kinase–catalyzed reaction of glycolysis. During the two-reaction sequence, a thermodynamic driving force for PEP formation is provided by the cleavage of one ATP and one GTP; the $\Delta G^{\circ\prime}$ for the hydrolysis of each nucleoside triphosphate is about 30 kJ/mol, whereas the equivalent value for the hydrolysis of PEP is close to 60 kJ/mol (Section 8.5). **[Try Problem 14.5]**

The ATP-driven production of fructose 1,6-bisphosphate from fructose 6-phosphate is the second step in glycolysis that must be bypassed during gluconeogenesis.

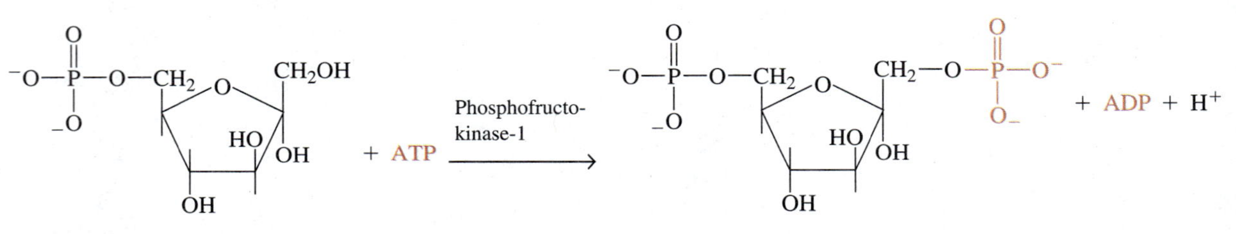

The direct reversal of this reaction involves the cleavage of a low-energy phosphate ester coupled with the formation of a high-energy phosphate anhydride, a process not energetically favorable. A unique gluconeogenic enzyme, **fructose-bisphosphatase,** catalyzes the hydrolysis of the ester link to the 1-phosphate in fructose 1,6-bisphosphate and generates fructose 6-phosphate without the coupled formation of ATP.

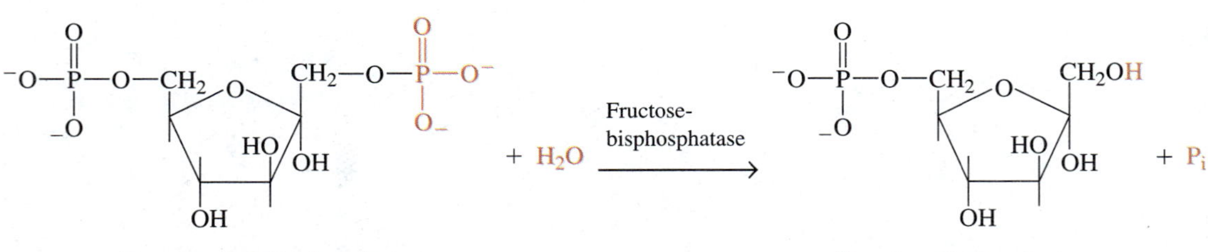

The $\Delta G^{\circ\prime}$ for this reaction is around −9 kJ/mol. Fructose-bisphosphatase, along with all of the enzymes for glycolysis, is found in the cytosol. **[Try Problem 14.6]**

The final bypassed step in glycolysis is the ATP-driven phosphorylation of glucose:

From a thermodynamic standpoint this reaction is analogous to the ATP-driven phosphorylation of fructose 6-phosphate discussed in the previous paragraph.

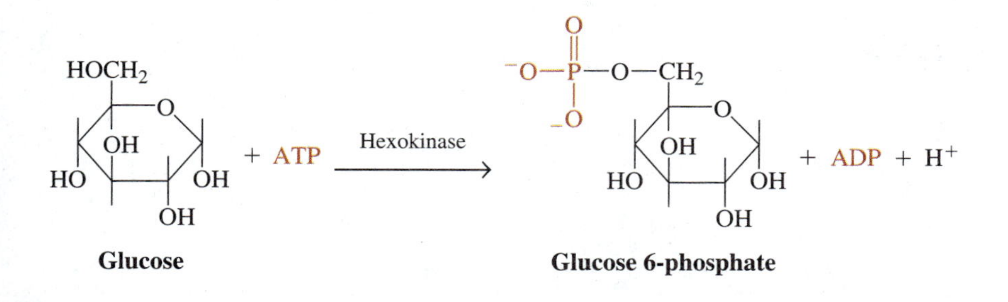

Glucose-6-phosphatase, the fourth and last gluconeogenesis-specific enzyme, generates glucose by catalyzing the hydrolysis of glucose 6-phosphate without the coupled formation of ATP.

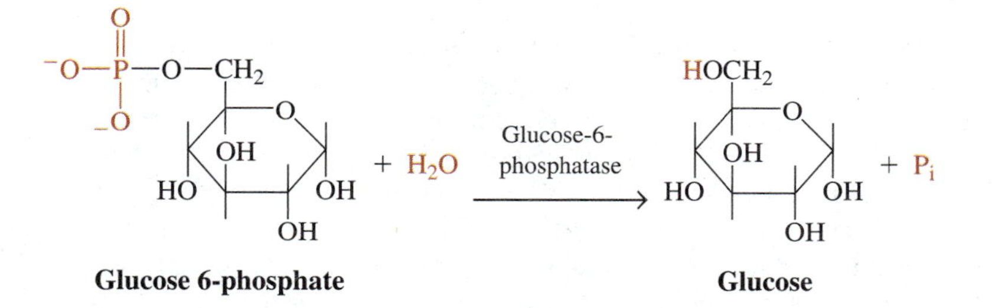

This enzyme is primarily housed within the endoplasmic reticulum in liver cells, and the free glucose generated is normally exported to the bloodstream. The export of glucose is one of multiple ways in which the liver responds to hypoglycemia (Sections 12.1 and 14.4).

Cells, including brain cells and muscle cells, that tend to catabolize rather than synthesize glucose lack a glucose-6-phosphatase. Any glucose 6-phosphate produced by these cells becomes trapped inside, since the cells do not possess transporters for sugar phosphates (charged molecules unable to pass unaided through the lipid bilayer of membranes). This trapping of glucose makes sense when one recognizes that brain and muscle cells normally need all of their glucose to meet their own metabolic requirements.

Summary of Gluconeogenesis and Reversal of Glycolysis

Exhibit 14.1 provides a summary reaction for both gluconeogenesis from pyruvate and the reversal of glycolysis. When compared to the reversal of glycolysis, gluconeogenesis entails the hydrolysis of four additional nucleoside triphosphates, two ATP, and two GTP. As a consequence, the overall $\Delta G^{\circ\prime}$ for glucose production shifts from a highly unfavorable +84 kJ/mol during the reversal of glycolysis to a favorable −38 kJ/mol during gluconeogenesis. **[Try Problem 14.7]**

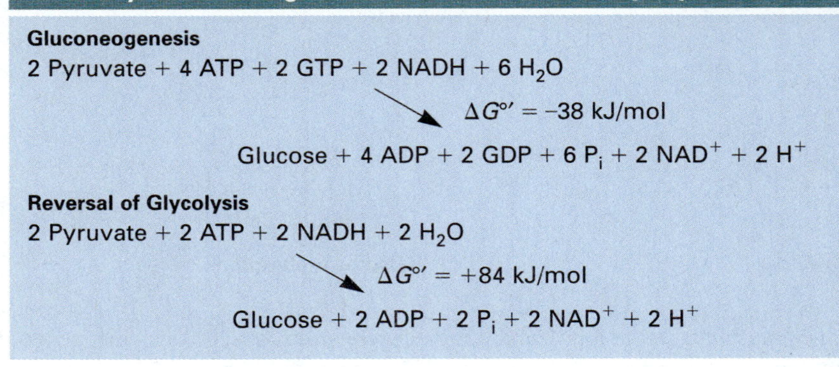

EXHIBIT 14.1

Summary of Gluconeogenesis and the Reversal of Glycolysis

Gluconeogenesis

$$2 \text{ Pyruvate} + 4 \text{ ATP} + 2 \text{ GTP} + 2 \text{ NADH} + 6 \text{ H}_2\text{O} \xrightarrow{\Delta G^{\circ\prime} = -38 \text{ kJ/mol}} \text{Glucose} + 4 \text{ ADP} + 2 \text{ GDP} + 6 \text{ P}_i + 2 \text{ NAD}^+ + 2 \text{ H}^+$$

Reversal of Glycolysis

$$2 \text{ Pyruvate} + 2 \text{ ATP} + 2 \text{ NADH} + 2 \text{ H}_2\text{O} \xrightarrow{\Delta G^{\circ\prime} = +84 \text{ kJ/mol}} \text{Glucose} + 2 \text{ ADP} + 2 \text{ P}_i + 2 \text{ NAD}^+ + 2 \text{ H}^+$$

Gluconeogenesis from Glycerol and Amino Acids

Gluconeogenesis from glycerol begins with an ATP-dependent phosphorylation reaction followed by an oxidation reaction (Figure 14.4). Because the dihydroxyacetone phosphate produced is an intermediate in both glycolysis and gluconeogenesis from pyruvate, it can be anabolized to glucose by following the flow of gluconeogenesis or be catabolized to pyruvate following the flow of glycolysis (see Figure 14.3). Its actual fate is determined by the immediate needs of a cell or organism at the time it is produced from glycerol. The dihydroxyacetone phosphate–linked oxidation of glycerol is examined in Section 13.2. **[Try Problem 14.8]**

Although alanine tends to be quantitatively the most important amino acid from the standpoint of gluconeogenesis, 18 of the 20 common protein amino acids can be employed as precursors for the production of glucose. These glucogenic amino acids are discussed in Chapter 13, and they are summarized in Table 13.2. The carbon skeletons from all of these amino acids can be converted directly or indirectly to oxaloacetate. Under certain conditions, including prolonged starvation, gluconeogenesis supported by protein-derived amino acids is a major source of blood glucose.

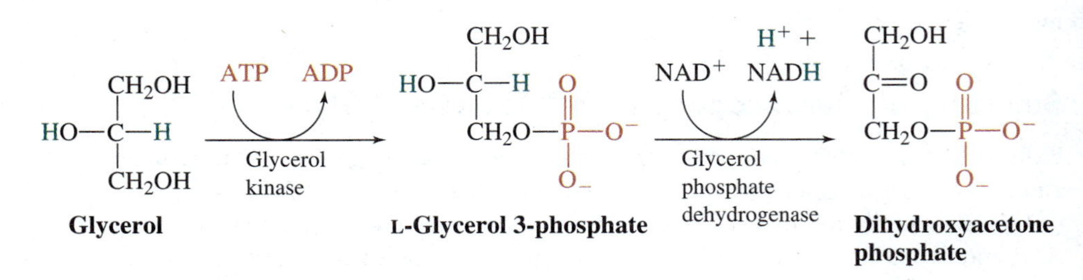

FIGURE 14.4
Conversion of glycerol to dihydroxyacetone phosphate.

PROBLEM 14.4 Does a high NAD^+/NADH ratio favor pyruvate production or lactate production? Lactate is not usually a quantitatively important precursor for gluconeogenesis under aerobic conditions. Explain.

PROBLEM 14.5 To which one of the six major international classes of enzymes does pyruvate carboxylase belong? Phosphoenolpyruvate carboxykinase?

PROBLEM 14.6 To which one of the six major international classes of enzymes does fructose-bisphosphatase belong?

PROBLEM 14.7 True or false? The more negative the $\Delta G^{\circ\prime}$ for a reaction, the smaller the equilibrium constant for that reaction. Explain. Hint: Review Section 8.2.

PROBLEM 14.8 Calculate the approximate net yield of ATP during the aerobic conversion of two glycerol to one glucose. Calculate the approximate net production of ATP during the aerobic conversion of two glycerol to two pyruvate. Assume that the reactions involved are coupled to a functional respiratory chain with O_2 as the terminal electron acceptor.

14.3 RECIPROCAL REGULATION OF GLYCOLYSIS AND GLUCONEOGENESIS

When gluconeogensis and glycolysis proceed simultaneously, the net reaction is:

$$2\ \text{ATP} + 2\ \text{GTP} + 4\ \text{H}_2\text{O} \rightleftharpoons 2\ \text{ADP} + 2\ \text{GDP} + 4\ \text{P}_\text{i} + 4\ \text{H}^+ + \text{energy}$$

This is a waste of energy, because an organism reaps no benefit from the hydrolysis of the nucleoside triphosphates. Since energy is too valuable a commodity to waste, glycolysis tends to slow down under conditions where a cell or organism benefits from rapid glucose production, and gluconeogenesis tends to slow down when a high rate of glycolysis is needed. This relationship is a specific example of reciprocal regulation.

Allosteric Effectors

The reciprocal regulation of glucose metabolism is primarily brought about by allosteric effectors, including several that modulate the enzyme-catalyzed interconversion of fructose 6-phosphate and fructose 1,6-bisphosphate. Phosphofructokinase-1, the enzyme that catalyzes the virtually irreversible formation of fructose 1,6-bisphosphate from fructose 6-phosphate and ATP during glycolysis, is activated by AMP and fructose 2,6-bisphosphate and inhibited by citrate and ATP (Section 11.3). In contrast, fructose-bisphosphatase, the enzyme that catalyzes the hydrolysis of fructose 1,6-bisphosphate during gluconeogenesis, is inhibited by AMP and fructose 2,6-bisphosphate and activated by citrate (Figure 14.5). **[Try Problem 14.9]**

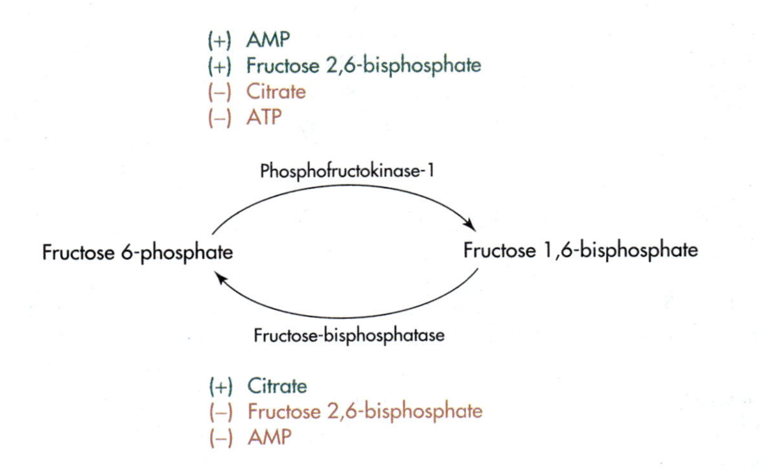

FIGURE 14.5
Reciprocal regulation of the interconversion of fructose 6-phosphate and fructose 1,6-bisphosphate.

The metabolic logic behind the regulatory activities of AMP, ATP, and citrate can be partially explained by an energy charge argument. Energy charge was defined in Chapter 11 as the extent to which the ATP–ADP–AMP system is filled with the higher energy compounds (ATP > ADP > AMP). AMP accumulates within a cell and stimulates glycolysis when the energy charge is low and there is a need to oxidize more glucose. AMP simultaneously slows down gluconeogenesis because this pathway, if allowed to function at a significant rate, would consume too much of the energy needed to "recharge" the cell. Gluconeogenesis, in the presence of active glycolysis, would also create the nonproductive cycle described at the beginning of this section and, in the process, consume pyruvate that could otherwise be shuttled into the citric acid cycle for energy production. An accumulation of citrate or ATP signals that the energy charge within a cell is high and that a cell should turn down glycolysis since additional energy is not required. In the process of suppressing glycolysis, the effectors shift the balance between glycolysis and gluconeogenesis in favor of gluconeogenesis. At the same time, citrate directly stimulates gluconeogenesis. **[Try Problem 14.10]**

Acetyl-SCoA is a fifth allosteric effector, and its modulation of the relative rates of glycolysis and gluconeogenesis also makes sense from the standpoint of energy charge considerations (Figure 14.6). This metabolite allosterically activates pyruvate carboxylase and, in the process, gluconeogenesis. At the same time, it allosterically inhibits the pyruvate dehydrogenase complex, which converts pyruvate to CO_2 and acetyl-SCoA. Thus, when acetyl-SCoA levels are high, normally signaling an adequate energy charge, pyruvate from glycolysis and other sources is conserved (by inhibition of its catabolism) for gluconeogenesis, and the first step in gluconeogenesis (from pyruvate) is simultaneously activated. Oxaloacetate from the pyruvate carboxylase–catalyzed reaction is also used to replenish the citric acid cycle when the need arises (Section 11.7). A steady supply of oxaloacetate is required to avoid ketosis when fat is catabolized (Section 13.4).

Hormone-Linked Regulation

The metabolic logic behind the allosteric activity of fructose 2,6-bisphosphate has been traced to glucagon and epinephrine, hormones released into the blood in response to low blood sugar levels (Section 10.12). These chemical messengers bind

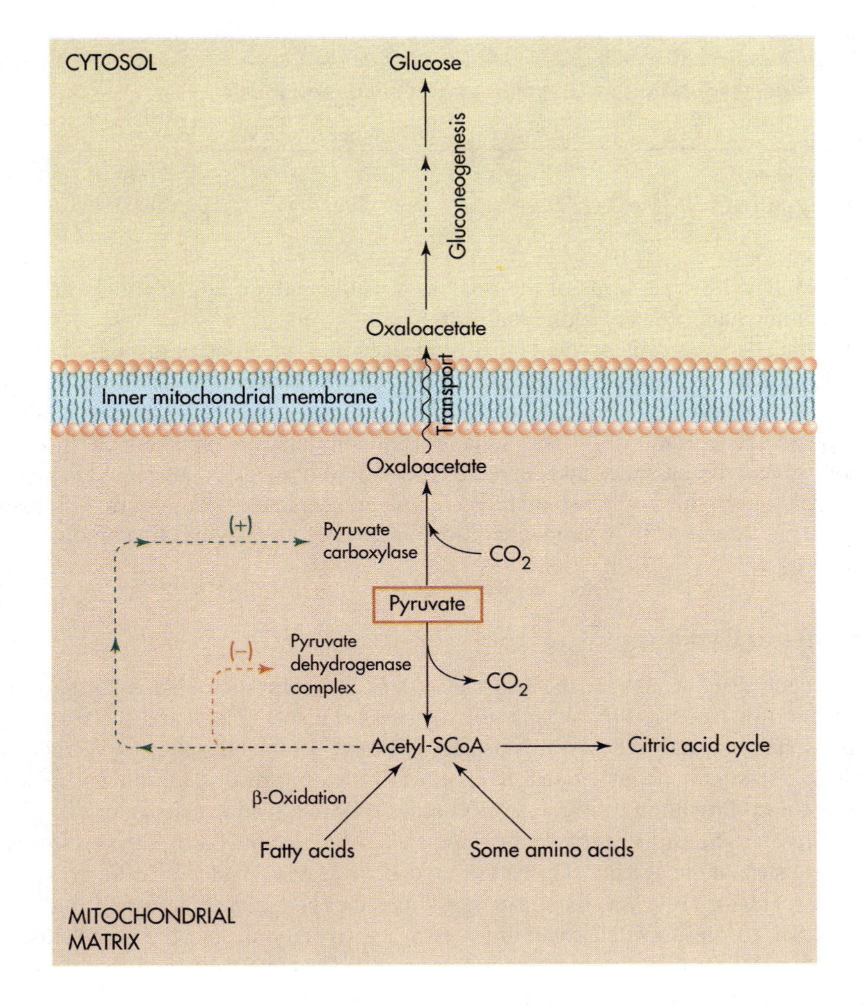

FIGURE 14.6
Modulation of gluconeogenesis through the regulation of pyruvate metabolism. Acetyl-SCoA, when present at high concentration, normally signals a high energy charge.

to specific receptors associated with the plasma membrane of liver cells and stimulate both gluconeogenesis and the export of glucose. The exported glucose enters the bloodstream, where it tends to counter the hypoglycemia that originally triggered the release of the hormones.

When glucagon binds to a membrane receptor on a hepatocyte, it triggers the internal production of cyclic AMP (cAMP), a second messenger that initiates a reaction cascade by binding a cAMP-dependent protein kinase (Section 10.5). Because the kinase that produces fructose 2,6-bisphosphate is inhibited by the cAMP-linked cascade and the phosphatase that catalyzes fructose 2,6-bisphosphate hydrolysis is stimulated by this cascade, glucagon-induced production of cAMP leads to a drop in fructose 2,6-bisphosphate concentration (Section 11.3). Thus, when the body calls upon the liver to increase blood sugar levels (by releasing glucagon into the blood), fructose 2,6-bisphosphate levels drop, the rate of glycolysis decreases (to conserve glucose for export), and the rate of gluconeogenesis increases (to produce glucose for export; Table 14.3). Epinephrine-linked reaction cascades stimulate gluconeogenesis in liver and glycolysis in muscle. The rates of metabolic pathways are par-

TABLE 14.3

Hormonal Regulation of Glycolysis and Gluconeogenesis

	Glucagon	Epinephrine	Insulin
Glycolysis	↓ (liver)	↑ (muscle)	↑ (liver, muscle)
Gluconeogenesis	↑ (liver)	↑ (liver)	↓ (liver, muscle)

tially regulated by the needs of the body as a whole and are not solely determined by the immediate needs of individual cells.

Insulin is a pancreatic hormone that is released into the blood in response to high blood sugar levels (Section 10.12). Insulin targets both liver and muscle cells, where insulin-linked cascades lead to an increase rather than a decrease in fructose 2,6-bisphosphate concentrations. In liver cells, insulin additionally stimulates the production of glycolytic enzymes and represses the assembly of gluconeogenic enzymes. Fatty acid biosynthesis and β-oxidation are also subject to such "long-term" (it takes a relatively long time to assemble additional copies of an enzyme) regulation (Section 14.8).

Substrate Cycling

It has been discovered that the ATP-dependent, phosphofructokinase-1–catalyzed conversion of fructose 6-phosphate to fructose 1,6-bisphosphate and the fructose-bisphosphatase–catalyzed hydrolysis of fructose 1,6-bisphosphate to fructose 6-phosphate simultaneously occur at similar rates within resting muscle cells. This **substrate cycling** allows the cell to shift rapidly from a resting state to one of high level activity. The ability to make the rapid switch is apparently worth the energy required to sustain the cycling. This substrate cycle was described as a **futile cycle** before its regulatory role was recognized. A detailed explanation of how substrate cycling leads to rapid switch capabilities is provided in some more comprehensive texts. Bumblebees and certain other insects use substrate cycling to generate the heat that is sometimes needed to maintain body temperature. **[Try Problem 14.11]**

PROBLEM 14.9 Describe the functions of both active and allosteric sites within a regulatory enzyme.

PROBLEM 14.10 *If* ATP were an allosteric effector for fructose-bisphosphatase, would you expect this metabolite to be an activator or an inhibitor? *If* NADH were an effector of the same enzyme, would it most likely be a positive or a negative one? Explain.

PROBLEM 14.11 Less than half of the energy released during the oxidation of glucose within a cell is trapped within ATP. An even smaller fraction of the energy "consumed" within most anabolic pathways ends up within final products. What is the immediate fate of most of the rest of the energy?

14.4 THE CORI CYCLE

Gluconeogenesis and its regulation are associated with a metabolic process in humans known as the **Cori cycle.** The Cori cycle is of major importance under a number of metabolic conditions, including those that exist during anaerobic exercise. During anaerobic exercise, pyruvate and NADH from glycolysis cannot be oxidized in mitochondria, since the citric acid cycle and respiratory chain shut down as O_2 levels drop. Muscle lactate dehydrogenase (LDH) isozymes catalyze the conversion of pyruvate and NADH to lactate and NAD^+ in order to provide the NAD^+ necessary for continued glycolysis and the associated production of ATP (Section 11.2).

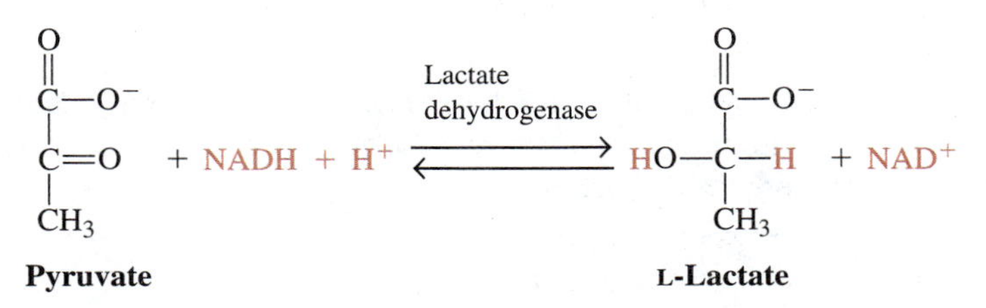

While the NAD^+ is being used to support glycolysis, part of the accumulating lactate moves into the blood and is carried to the liver where distinctive LDH isozymes and a favorable NAD^+/NADH ratio lead to the oxidation of lactate back to pyruvate. The favorable NAD^+/NADH ratio in liver is maintained with an active respiratory chain supported by an adequate supply of O_2. A drop in O_2 concentration in active muscle (because of rapid consumption of O_2) does not lead to a corresponding drop in O_2 level in the liver. The liver-generated pyruvate is converted to glucose within the liver and then exported to muscles for glycolysis. The anaerobic glycolysis of this glucose generates more lactate, which is carried back to the liver to generate more glucose for further muscle glycolysis; the Cori cycle is underway (Figure 14.7). **[Try Problems 14.12 and 14.13]**

The Cori cycle indirectly shifts part of the metabolic burden for ATP production from muscle to liver during anaerobic exercise, a physiologically important accomplishment. A constant supply of ATP is essential for muscle action and, although all

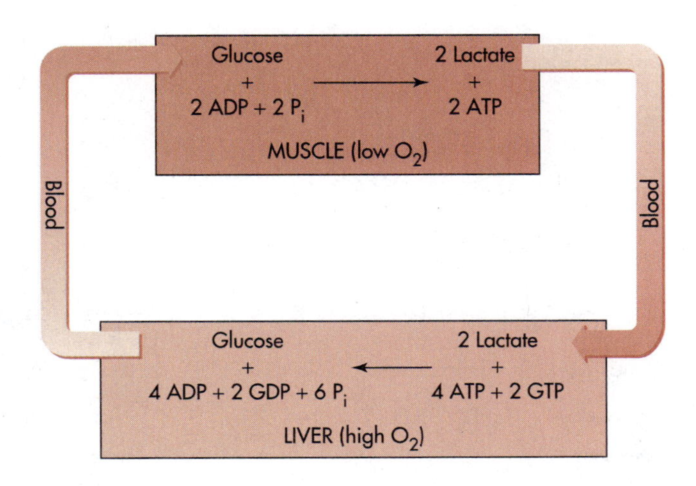

FIGURE 14.7
The Cori cycle.

of the utilized ATP is directly produced within muscle cells, part of this ATP is generated during the glycolysis of glucose assembled from muscle-derived lactate within the liver. The transport of lactate from muscle to liver also reduces the drop in muscle pH caused by lactic acid accumulation. Lactate is considered a dead-end metabolite because it is produced only from pyruvate and must be converted back to pyruvate to be metabolized.

PROBLEM 14.12 Although high NADH concentrations signal a high energy charge in aerobic muscle cells, this is not the case in anaerobic muscle. Explain.

PROBLEM 14.13 What keeps the NADH/NAD^+ ratio high in anaerobic muscle cells in spite of an increased rate for the lactate dehydrogenase–catalyzed reaction?

14.5 GLYCOGENESIS

Like gluconeogenesis and the Cori cycle, **glycogenesis** (the synthesis of glycogen from glucose) is an important component of the glucose metabolism story. Glycogen (Sections 6.6 and 12.1) is a branched polymer of glucose that serves as a readily mobilizable storage form of this monosaccharide. Although most cells in the body contain at least small amounts of glycogen, it is stored primarily in the cytosol of liver and muscle cells as insoluble granules. Glucose from muscle glycogen is used mainly as a fuel by muscle cells, whereas glucose from liver glycogen tends to be exported to help elevate blood glucose levels when blood glucose concentrations drop below normal. The phosphorolysis of glycogen to release glucose 1-phosphates, a process known as glycogenolysis, was examined in Section 12.1. **[Try Problem 14.14]**

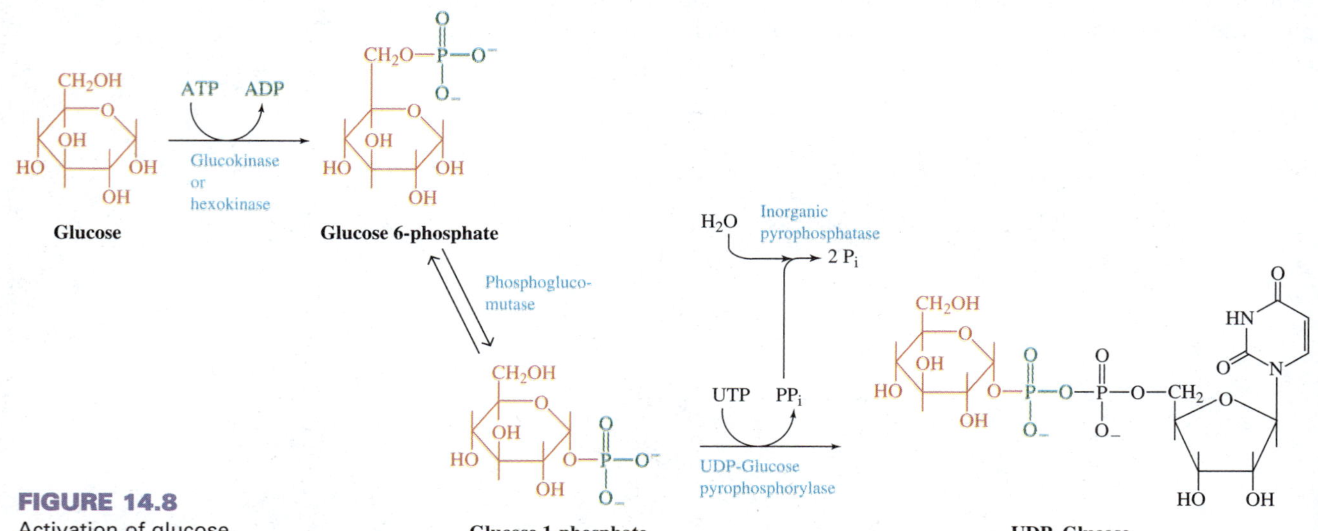

FIGURE 14.8
Activation of glucose.

Glucose must be activated before it can be added to a growing polymer chain. The first step in the activation process is a glucokinase- or hexokinase-catalyzed phosphorylation yielding glucose 6-phosphate (Figure 14.8). This ATP-dependent reaction is also the first step in glycolysis. After **phosphoglucomutase** catalyzes the transformation of glucose 6-phosphate to glucose 1-phosphate, **UDP-glucose pyrophosphorylase** catalyzes the reaction of glucose 1-phosphate with uridine triphosphate (UTP) to generate UDP-glucose and pyrophosphate (PP_i). The subsequent hydrolysis of the pyrophosphate "pulls" this reaction toward completion (Le Châtelier's principle). The glucose residue in UDP-glucose is activated because the cleavage of the pyrophosphate ester bond to glucose provides a thermodynamic driving force for the transfer of a glucosyl group from UDP to other compounds. Viewed from a slightly different perspective, the UDP residue in UDP–glucose is a good "leaving group."

The assembly of glycogen from a pool of UDP-glucose molecules begins when **UDPglucosyltransferase** catalyzes the transfer of a glucosyl group from UDP-glucose to the hydroxyl group of a specific tyrosine side chain in an enzyme named **glycogenin.**

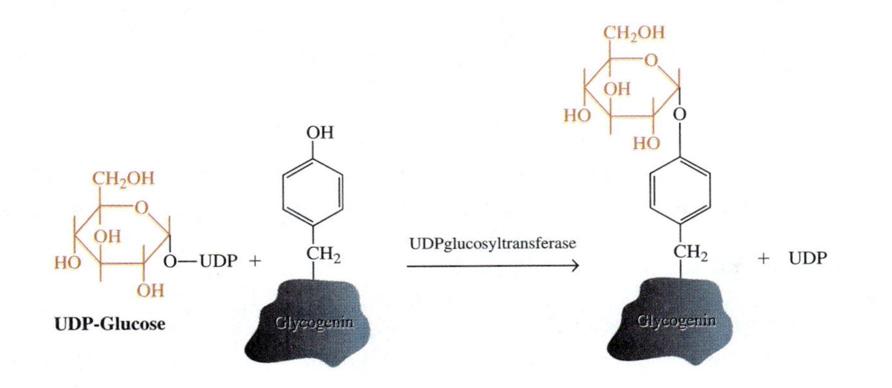

Glycogenin then begins the polymerization process by catalyzing the addition of a second glucosyl group to its attached glucose residue, an example of **autocatalysis**. An $\alpha(1 \longrightarrow 4)$ glycosidic link is formed.

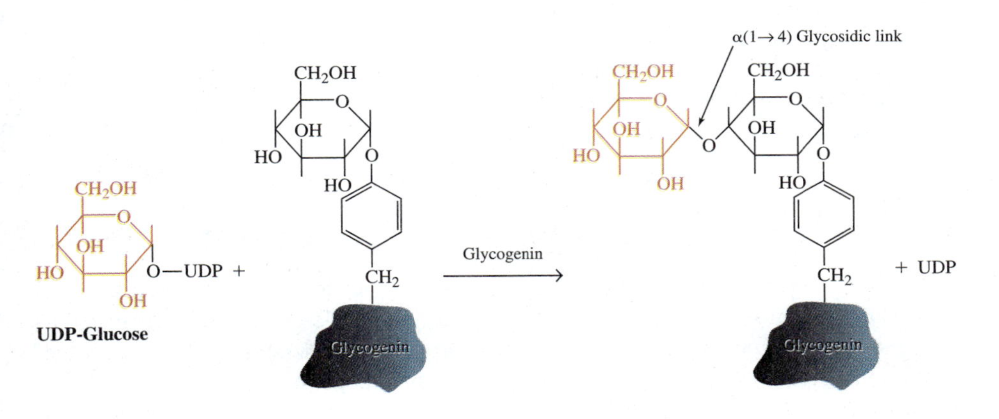

Additional glucosyl groups are added one at a time until a linear oligomer containing roughly seven monomer residues has been assembled. At this point, **glycogen synthase** takes over the elongation process; it continues to catalyze the stepwise addition of glucosyl groups (from UDP-glucose) to the growing polymer chain. After a dozen or so glucose residues have been joined in a continuous chain, **amylo-(1,4–1,6)-transglycosylase [branching enzyme]** catalyzes the transfer of an oligosaccharide fragment (usually six or seven glucose residues in length) from the growing end of the polymer to the 6-hydroxyl of a glucose residue further back along the polymer; an $\alpha(1 \longrightarrow 6)$ branch is created (see Figure 14.9). Separate glycogen synthase molecules can simultaneously catalyze the elongation of the branch and the original chain, each of which can subsequently undergo further branching. Branching, by creating multiple sites on a single molecule where glucosyl groups can be simultaneously added or removed, allows a cell to store glucose more rapidly when there is an abundant nutritional supply and to mobilize glucose more rapidly when it is needed. A glycogen molecule, which may contain up to 60,000 glucose residues, normally remains attached to the glycogenin that initiated its synthesis. **[Try Problems 14.15 through 14.17]**

PROBLEM 14.14 Write the structural formula for a segment of a glycogen molecule that contains an $\alpha(1 \longrightarrow 6)$ branch. Identify each $\alpha(1 \longrightarrow 4)$ and $\alpha(1 \longrightarrow 6)$ link within your structural formula.

PROBLEM 14.15 Section 12.1 (glycogen catabolism) noted that a glycogen molecule contains many nonreducing ends but only one reducing end. This statement assumes that a glycogen molecule is free (not bound to glycogenin). How many reducing ends does a glycogenin-bound glycogen molecule contain? Explain.

PROBLEM 14.16 True or false? The number of $\alpha(1 \longrightarrow 6)$ glycosidic links in a glycogen molecule is equal to the number of branches in the molecule. Explain.

PROBLEM 14.17 How many net phosphoric acid anhydride links are cleaved during the production of a linear glycogen molecule (contains no branches) from free glucose molecules? Assume that the process begins with a preformed octomer of glucose and that the glycogen molecule synthesized contains a total of 100 glucose residues. How many enzymes are directly involved in this synthesis?

14.6 RECIPROCAL REGULATION OF GLYCOGENESIS AND GLYCOGENOLYSIS

The simultaneous breakdown and synthesis of glycogen, like the simultaneous breakdown and synthesis of glucose itself, is a futile cycle that wastes energy. Consequently, glycogenesis and glycogenolysis are reciprocally regulated. Epinephrine and glucagon, hormones the body releases into the bloodstream when blood glucose levels are low and stored glucose needs to be mobilized, stimulate glycogenolysis

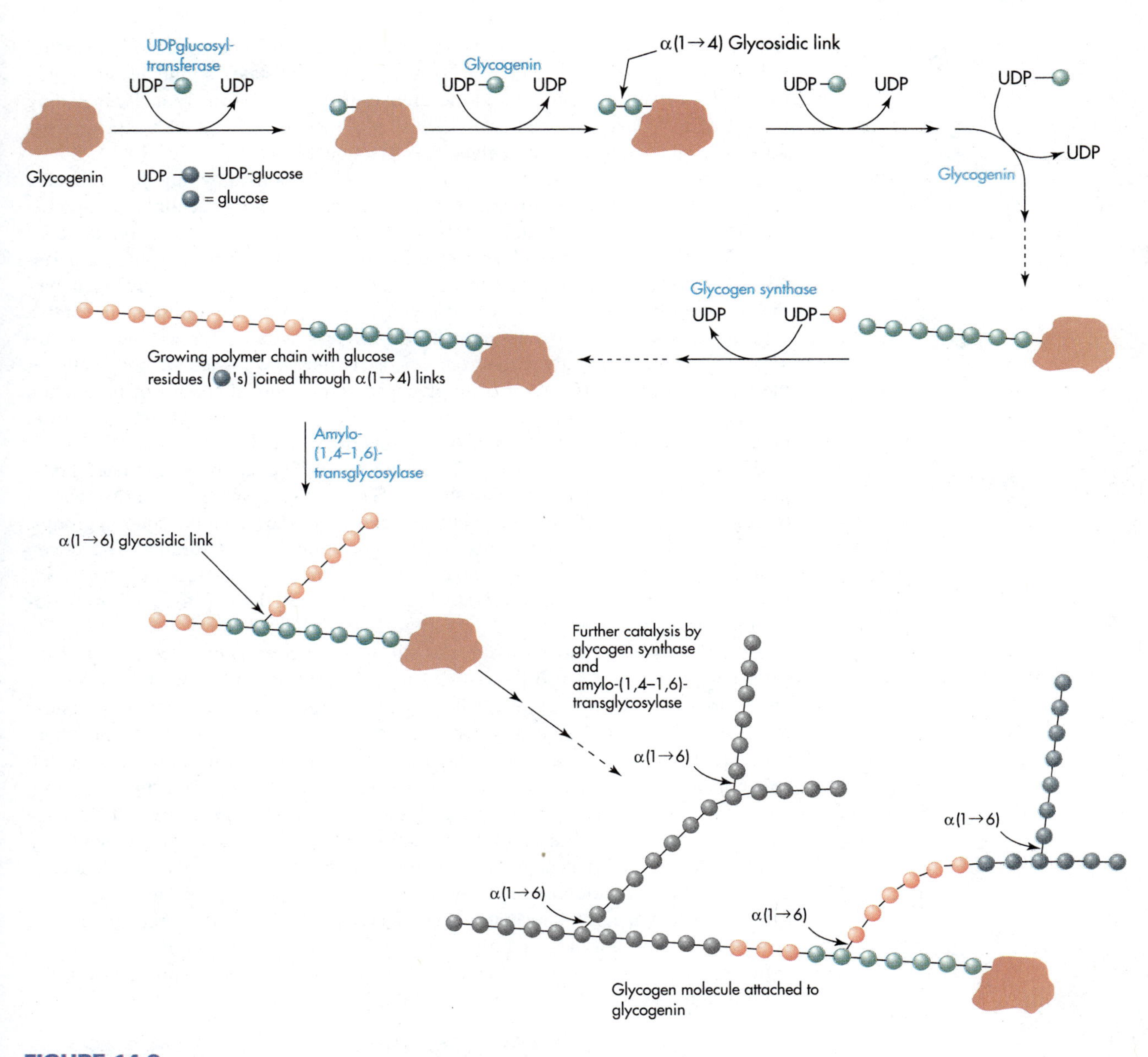

FIGURE 14.9
Glycogenesis.

and inhibit glycogenesis in liver and muscle. Insulin, which is produced in response to high blood glucose concentrations (signals a well-fed state), stimulates glycogenesis in both liver and muscle and the uptake of glucose by many cells, including muscle cells. The same three hormones also play a central role in the reciprocal regulation of gluconeogenesis and glycolysis (Section 14.3).

Epinephrine and glucagon modulate glycogen metabolism by binding to specific receptor sites on the outside of target cells and triggering the intracellular production of cAMP. The cAMP activates a protein kinase, which initiates reaction cascades that ultimately lead to the phosphorylation of both glycogen phosphorylase (involved in glycogen breakdown) and glycogen synthase (involved in glycogen synthesis). The phosphorylations convert glycogen phosphorylase b to the more active a form and convert glycogen synthase a to the less active b form (Figure 14.10). Thus, the production of glycogen from glucose is shut down when the body calls for the release of glucose residues from glycogen. This coupled control of production and degradation ensures that a cell does not continue to use energy and glucose to produce glycogen under conditions of high energy demand.

The changes brought about by the glucagon- and epinephrine-stimulated phosphorylation of glycogen phosphorylase b and glycogen synthase a are reversed through the protein phosphatase–catalyzed removal of the added phosphoryl groups. The dephosphorylation inhibits glycogen degradation while it stimulates glycogen synthesis. These changes come about as the consequence of "signals" (partly involving changes in hormone levels) that indicate to a cell that it is time to start building up glucose reserves.

Although insulin's mode of action is still under investigation, the activity of insulin receptors is known to be partially regulated by epinephrine. The binding of epinephrine to unique receptors, called **α_1-adrenergic receptors,** activates a phosphatidylinositol-linked signaling pathway (Section 10.7) that leads to the phosphorylation of insulin receptors and an associated reduction in the affinity of the receptors for insulin. The epinephrine-triggered phosphorylation of insulin receptors insures that insulin action is suppressed under conditions where epinephrine concentrations are high. A branch of the phosphatidylinositol-linked signaling pathway that inactivates insulin receptors leads to an increase in Ca^{2+} concentration that contributes to the activation of glycogen phosphorylase and inactivation of glycogen synthase. The epinephrine receptors that are coupled to the cAMP-linked regulation of glycogen phosphorylase and glycogen synthase activities are known as β-adrenergic receptors (Section 12.1). Table 14.4 summarizes the role of hormones in the reciprocal regulation of glycogenolysis and glycogenesis.

TABLE 14.4

Role of Hormones in the Reciprocal Regulation of Glycogenolysis and Glycogenesis

	Hormone		
	Glucagon	Epinephrine	Insulin
Glycogenesis	↓ (liver)	↓ (liver, muscle)	↑ (liver, muscle)
Glycogenolysis	↑ (liver)	↑ (liver, muscle)	↓ (liver, muscle)

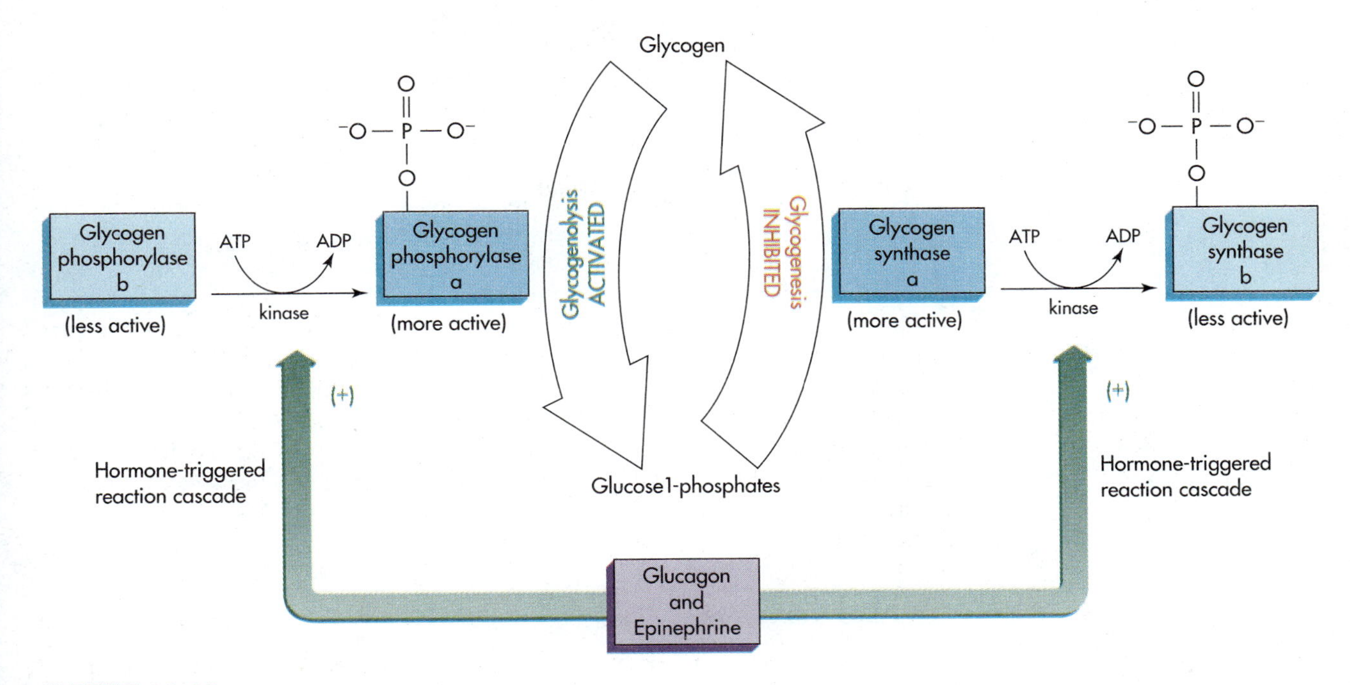

FIGURE 14.10
Hormonal regulation of glycogen phosphorylase and glycogen synthase.

Although hormones play the central role in the reciprocal regulation of glycogenolysis and glycogenesis, allosteric effectors are also involved. Section 12.1 examines the inhibition of muscle glycogen phosphorylase b by ATP and glucose 6-phosphate and the activation of this enzyme by AMP. The catalytic activity of liver glycogen phosphorylase is regulated by glucose concentration.

Because insulin, glucagon, and epinephrine also play major roles in the reciprocal regulation of glycolysis and gluconeogenesis, these hormones ensure that, when blood glucose levels are low, liver cells enhance both glycogenolysis and gluconeogenesis while suppressing glycolysis and glycogenesis. When blood glucose levels are high, glycogenolysis and gluconeogenesis are shut down, glycolysis is enhanced, and part of the abundant glucose is incorporated into glycogen to build up glucose reserves. The same three hormones play central roles in regulating the metabolism of fatty acids and fats, other classes of fuel molecules. The precise role of these hormones in fatty acid and triacylglycerol anabolism is explored in the next three sections. **[Try Problem 14.18]**

PROBLEM 14.18 Would it be possible for a cell to use allosteric effectors to slow down glycogen degradation while increasing the rate of glycogen synthesis if a single, reversible metabolic pathway were involved in both processes? Explain.

14.7 FATTY ACID BIOSYNTHESIS

In mammals, most fatty acids are synthesized in the cytosol of hepatocytes (liver cells) from acetyl groups generated within mitochondria. Since these acetyl groups can be derived from glucose, it is possible to convert carbohydrate to fat, as most weight watchers can testify. Carbohydrates, however, cannot be produced from fatty acids by the human body. Although many organisms are capable of such interconversions, humans are not genetically programmed to assemble the enzymes necessary to convert acetyl-SCoA from the β-oxidation of fatty acids (or any other source) to gluconeogenic precursors (Section 14.2). Mitochondrial acetyl groups destined for fatty acid biosynthesis are shuttled into the cytosol by an ATP-dependent transport system in which citrate serves as the acetyl carrier (Figure 14.11). The shuttled acetyl groups are transferred from citrate to CoASH within the cytosol. The NADPH that is a by-product of this transport process provides up to 57% of the reducing power required for fatty acid production. The remainder of the required reducing power is supplied by NADPH generated within the cytosol by the pentose phosphate pathway (Section 12.2). In photosynthetic plant cells, fatty acid biosynthesis normally occurs in chloroplasts rather than the cytosol.

NADH Versus NADPH

Since NADH and NADPH have identical reduction potentials (Table 8.9), it should be possible, from a thermodynamic standpoint, to use the two coenzymes interchangeably. In practice, however, nature tends to select NAD^+ as an oxidizing agent for catabolic pathways and to employ NADPH as a reducing agent for anabolic pathways (Sections 10.11 and 12.2). Although this division of labor is maintained primarily with coenzyme-specific enzymes, compartmental differences in coenzyme concentrations also help determine coenzyme utilization in eukaryotes.

Most of the NAD^+ reduced during catabolism is regenerated as NADH transfers the electrons it carries to O_2 through the respiratory chain. The NADPH oxidized during anabolism is primarily regenerated through the conversion of $NADP^+$ to NADPH within the pentose phosphate pathway. Thus, the reductive power of NADH is utilized within the respiratory chain to make ATP, whereas the reductive power of NADPH is devoted to biosynthesis. The oxidative power of $NADP^+$ is used to support the pentose phosphate pathway, whereas the oxidative power of NAD^+ supports other (the pentose phosphate pathway is a catabolic pathway) catabolic pathways. A summary of the production and use of NADH and NADPH appears in Table 14.5. **[Try Problem 14.19]**

TABLE 14.5

Production and Use of NADH and NADPH

	Primary Source	Primary Use
NADH	Catabolism (except the pentose phosphate pathway)	ATP production within the respiratory chain
NADPH	Pentose phosphate pathway	Reducing power for anabolism

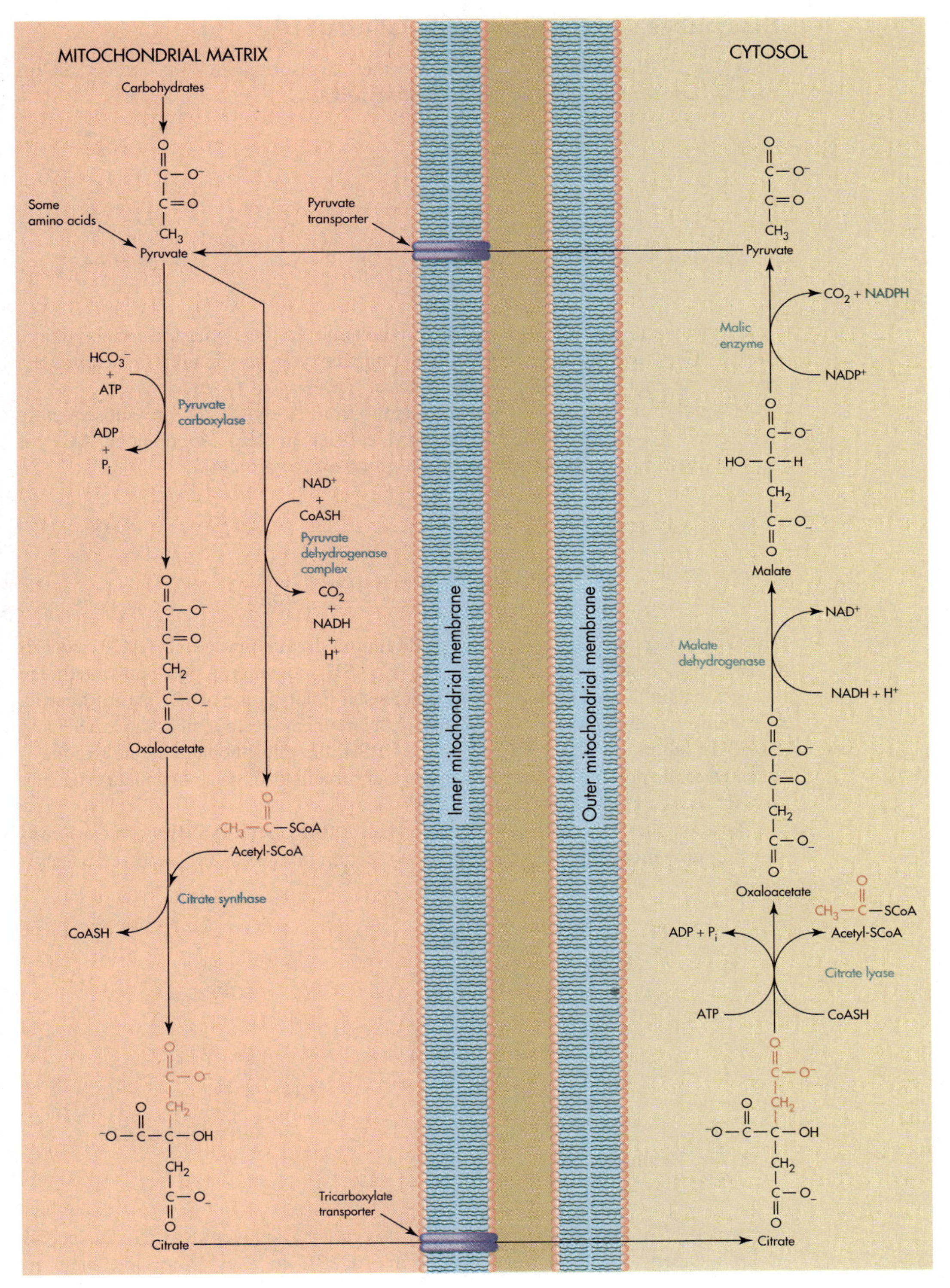

FIGURE 14.11 Acetyl group transport from the mitochondrial matrix to the cytosol.

Biosynthesis of Palmitic Acid (Palmitate)

Fatty acid biosynthesis begins in the cytosol with the **acetyl-SCoA carboxylase**–catalyzed formation of **malonyl-SCoA.**

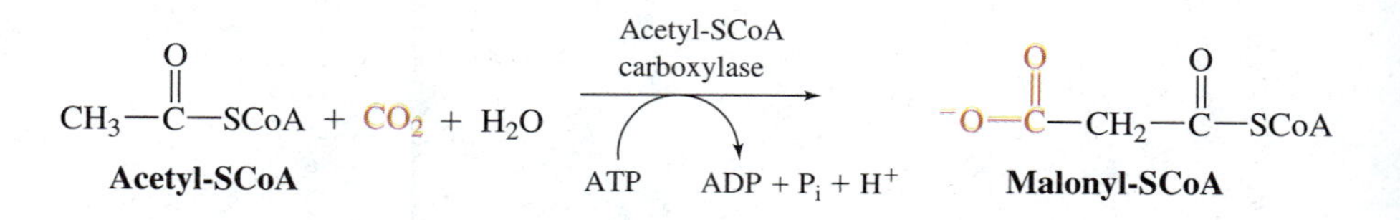

In this reaction, ATP provides the thermodynamic driving force for carboxylation. Acetyl-SCoA carboxylase is a biotin-containing enzyme very similar to the pyruvate carboxylase encountered in gluconeogenesis. **[Try Problem 14.20]**

Malonyl from malonyl-SCoA and acetyl from acetyl-SCoA are subsequently transferred to separate molecules of **acyl carrier protein** (abbreviated **ACP** or **ACPSH,** where the SH represents a thiol group on the protein).

$$\text{Malonyl-SCoA} + \text{ACPSH} \xrightarrow{\text{[ACP] malonyltransferase}} \text{Malonyl-SACP} + \text{CoASH}$$

$$\text{Acetyl-SCoA} + \text{ACPSH} \xrightarrow{\text{[ACP] acetyltransferase}} \text{Acetyl-SACP} + \text{CoASH}$$

These reactions are catalyzed by **[ACP] malonyltransferase** and **[ACP] acetyltransferase,** respectively. ACPSH, like CoASH, contains a phosphopantetheine residue that binds acyl groups through a thioester link (Figure 14.12). **Pantothenate,** a B vitamin, is the precursor for the phosphopantetheine residue in both CoASH and ACPSH (Tables 10.3 and 10.6). Within ACPSH, the phosphopantetheine residue is attached to the polypeptide chain of the carrier protein through a phosphate ester link to a serine side chain. **[Try Problem 14.21]**

The elongation of fatty acids begins with a **3-oxoacyl-[ACP] synthase (condensing enzyme)**–catalyzed condensation reaction that yields CO_2 and acetoacetyl-SACP.

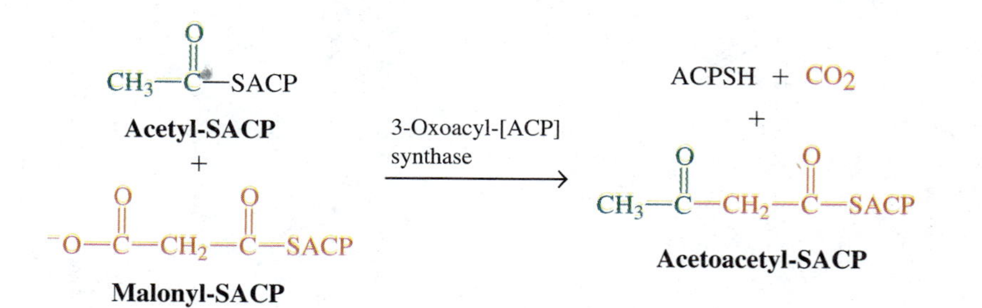

Acetoacetyl-SACP is an oxoacyl-SACP (also called a ketoacyl-SACP). The acetyl group in acetyl-SACP is transferred from ACPSH to a cysteine side chain in

3-oxoacyl-[ACP] synthase before its reaction within malonyl-SACP.

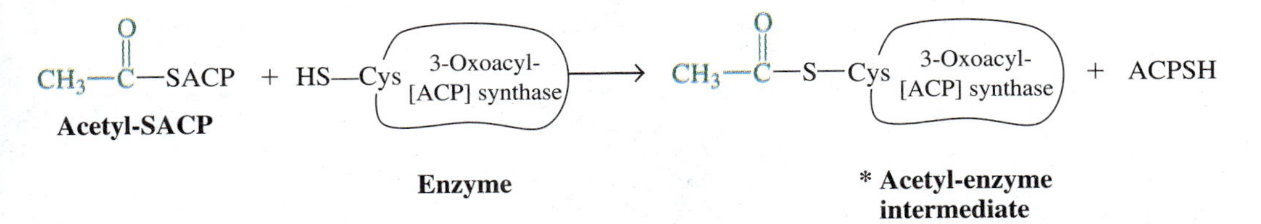

The asterisk identifies the immediate acetyl donor during acetoacetyl-SACP formation from acetyl-SACP and malonyl-SACP. The cleavage of a thioester link and the release of CO_2 from the malonyl group make the overall synthase-catalyzed reaction irreversible. Since the CO_2 released during the 3-oxoacyl-[ACP] synthase–catalyzed reaction is the one that was linked to acetyl-SCoA (to form malonyl-SCoA) during the ATP-driven reaction catalyzed by acetyl-SCoA carboxylase, ATP indirectly provides the thermodynamic driving force for the condensation reaction. The condensation of a three-carbon fragment with a two-carbon fragment is one of several fea-

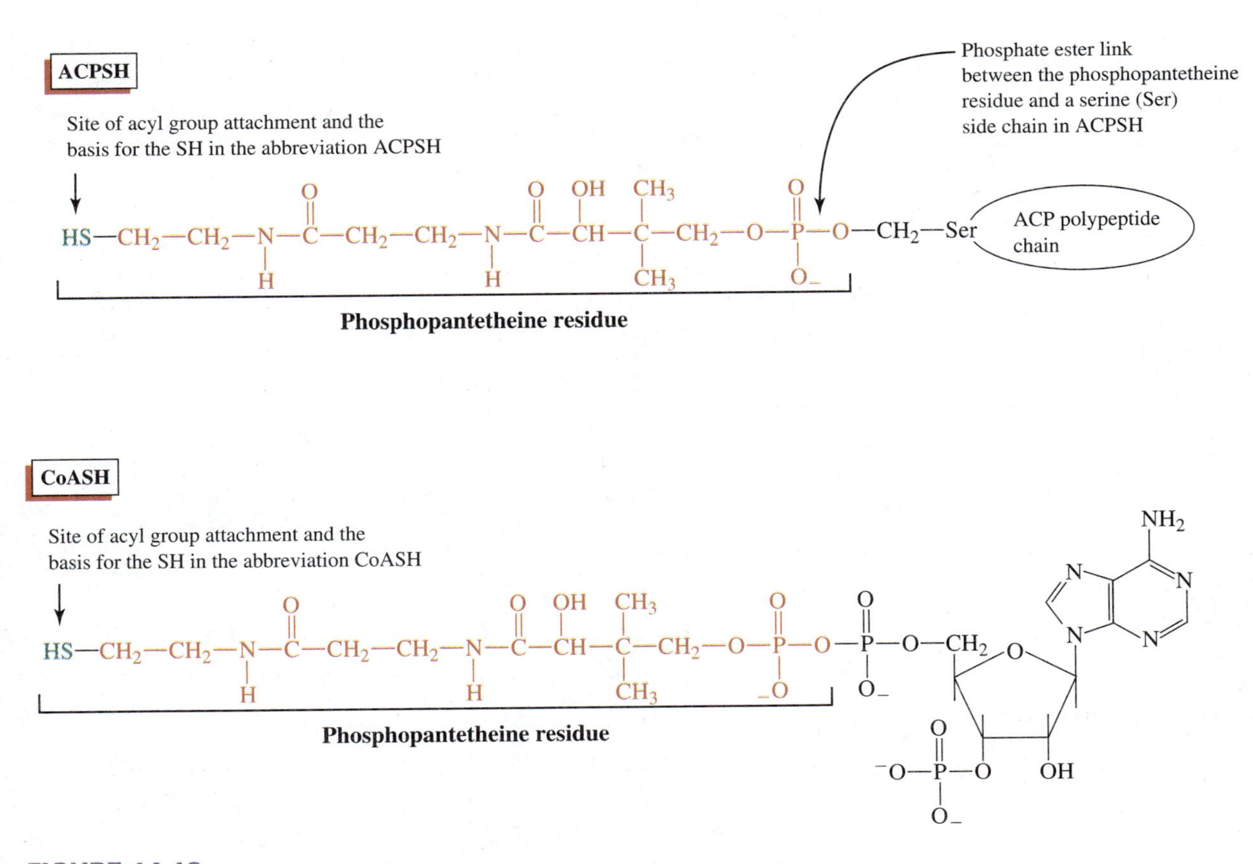

FIGURE 14.12
Phosphopantetheine residue within ACPSH and CoASH.

TABLE 14.6

Comparison of β-Oxidation and Fatty Acid Biosynthesis in Mammals

	β-Oxidation	Fatty Acid Biosynthesis
Location	Mitochondria	Cytosol
Acyl carrier	CoASH	ACPSH
Redox agent(s)	NAD^+, FAD	NADPH + H^+
Enzymes	Unique	Unique
Enzyme organization	Separate enzymes	Multifunctional enzyme
Alcohol intermediates	L-Configuration	D-Configuration
Energy balance	Exergonic	Endergonic
Reaction cycle	Two-carbon fragment released	Adds two-carbon fragment derived from three-carbon fragment (malonyl group)

tures that distinguish the pathway for fatty acid biosynthesis from the reversal of β-oxidation (Table 14.6).

NADPH generated within the acetyl group transport system (Figure 14.11) or the pentose phosphate pathway supplies the reducing power for the **3-oxoacyl-[ACP] reductase**–catalyzed conversion of acetoacetyl-SACP (called a β-ketone, 3-ketone, and 3-oxoacyl-SACP) to a secondary alcohol called D-3-hydroxybutyryl-SACP.

$$\underset{\textbf{Acetoacetyl-SACP}}{CH_3-\overset{O}{\overset{\|}{C}}-CH_2-\overset{O}{\overset{\|}{C}}-SACP} \xrightarrow[\text{NADPH + H}^+ \;\rightarrow\; \text{NADP}^+]{\text{3-Oxoacyl-[ACP] reductase}} \underset{\textbf{D-3-Hydroxybutyryl-SACP}}{CH_3-\overset{OH}{\overset{|}{CH}}-CH_2-\overset{O}{\overset{\|}{C}}-SACP}$$

At first glance, this process may be mistaken for the reversal of one step in β-oxidation (Figure 13.4). Closer examination reveals that β-oxidation involves: NAD^+, not $NADP^+$; an L-alcohol, not a D-alcohol; and CoASH, not ACPSH. The D-alcohol from β-reduction is dehydrated to a *trans*-alkene that is subsequently reduced to butyryl-SACP.

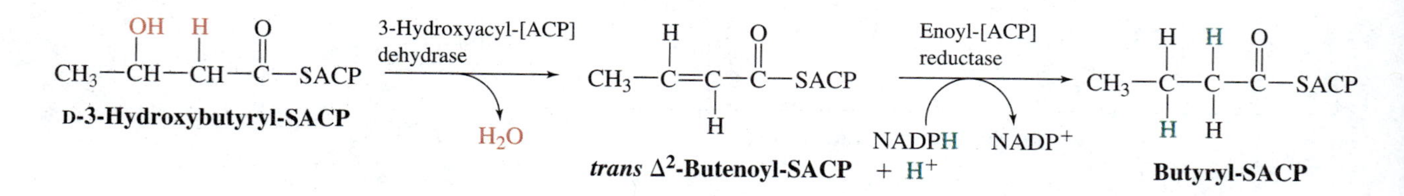

These reactions are catalyzed by **3-hydroxyacyl-[ACP] dehydrase** and **enoyl-[ACP] reductase,** respectively. The reducing power is once again provided by NADPH rather than NADH, $FADH_2$, or some other reducing agent.

The butyryl-SACP produced during the first round of elongation condenses with another malonyl-SACP as fatty acid biosynthesis continues.

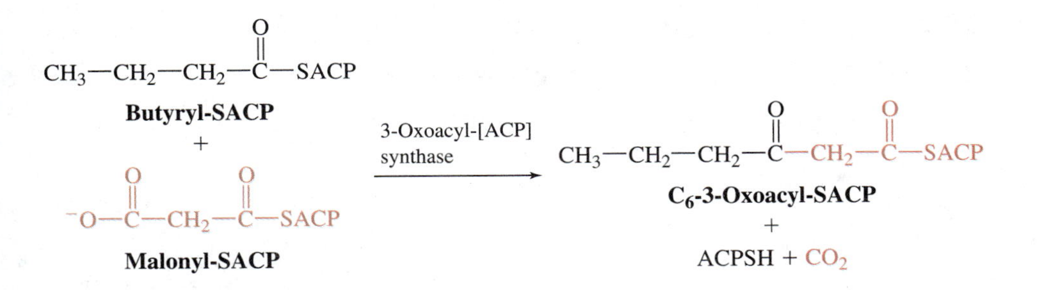

A second round of reduction–dehydration–reduction generates a C_6-acyl-SACP.

$$CH_3—CH_2—CH_2—CH_2—CH_2—\overset{O}{\overset{\|}{C}}—SACP$$

C_6-Acyl-SACP

Additional elongation cycles generate even-numbered acyl groups with up to 16 carbons.

$$CH_3—CH_2—CH_2—CH_2—CH_2—CH_2—CH_2—CH_2—CH_2—CH_2—CH_2—CH_2—CH_2—CH_2—CH_2—\overset{O}{\overset{\|}{C}}—SACP$$

C_{16}-Acyl-SACP = Palmitoyl-SACP

Elongation usually stops at this point since C_{16}-acyl-SACP (palmitoyl-SACP) is a poor substrate for 3-oxoacyl-[ACP] synthase. Palmitate is released from palmitoyl-SACP through enzyme-catalyzed hydrolysis. **[Try Problem 14.22]**

Figure 14.13 summarizes the biosynthesis of palmitate from eight cytosolic acetyl-SCoAs, while Figure 14.14 presents the net reaction for palmitate production. The required energy (provided by ATP) and reducing power (provided by NADPH) are derived from catabolic processes. In the human body, catabolism normally drives anabolism. **[Try Problem 14.23]**

Fatty Acid Synthase

In mammalian systems, the biosynthesis of palmitate from acetyl-SCoA and malonyl-SCoA takes place within a multienzyme complex named **fatty acid synthase.** This complex contains two identical polypeptide chains that each possess all of the participating catalytic activities plus the ACPSH sequence. Each polypeptide chain is an example of a **multifunctional enzyme,** a single protein with more than one catalytic activity. Acetyl-SCoA carboxylase, which catalyzes the formation of malonyl-SCoA, is a separate polypeptide and is not part of this complex. Fatty acid synthase, like most multienzyme complexes, tends to promote metabolic channeling (Section

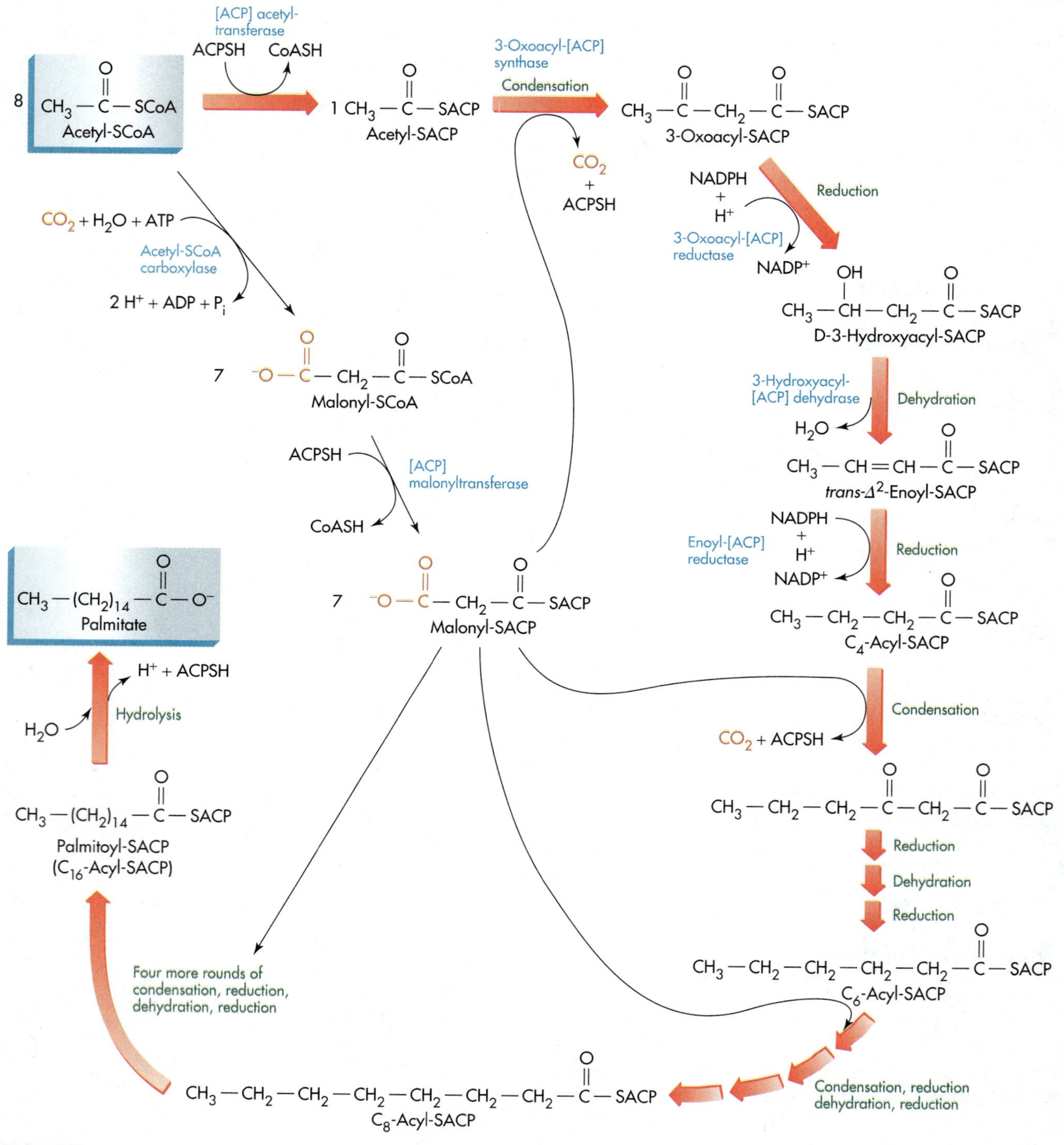

FIGURE 14.13
Summary of palmitate biosynthesis.

$$8\ \text{Acetyl-SCoA} + 7\ \text{ATP} + 14\ \text{NADPH} + 6\ \text{H}^+ \rightarrow \text{Palmitate} + 14\ \text{NADP}^+ + 8\ \text{CoASH} + 6\ \text{H}_2\text{O} + 7\ \text{ADP} + 7\ \text{P}_i$$

FIGURE 14.14
Net reaction for palmitate production from eight cytosolic acetyl-SCoAs.

5.15) and the direct, efficient coupling of a specific sequence of reactions. The multifunctional protein approach to complex construction leads to enhanced complex stability and ensures that all of the component enzymes are produced in exactly equal amounts at approximately the same time. **[Try Problems 14.24 and 14.25]**

Biosynthesis of Other Fatty Acids

Fatty acid synthase normally produces even-numbered, saturated fatty acids with no more than 16 carbons. Longer chained fatty acids and certain unsaturated fatty acids are synthesized from palmitate with the catalytic assistance of enzymes known as **elongases** and **desaturases** that are associated with mitochondria or the endoplasmic reticulum. Because mammals require linoleate [18:2($\Delta^{9,12}$)] and α-linolenate [18:3($\Delta^{9,12,15}$)] (Figure 14.15) but lack the enzymes necessary to introduce double bonds beyond the ninth carbon in a fatty acid, these polyunsaturated fatty acids are required in the diet and are classified as essential fatty acids (Section 9.1). At least part of the dietary linoleate is used to synthesize arachidonate [20:4($\Delta^{5,8,11,14}$)] which, in turn, is used by the human body to produce prostaglandins and other eicosanoids (Section 9.5). The eicosanoids, which function as "local hormones," help regulate a large number of processes including muscle contraction, electrolyte balance, inflammation, and tissue repair. Odd-numbered fatty acids, which are synthesized mainly by certain microorganisms, are produced when propionyl-SACP,

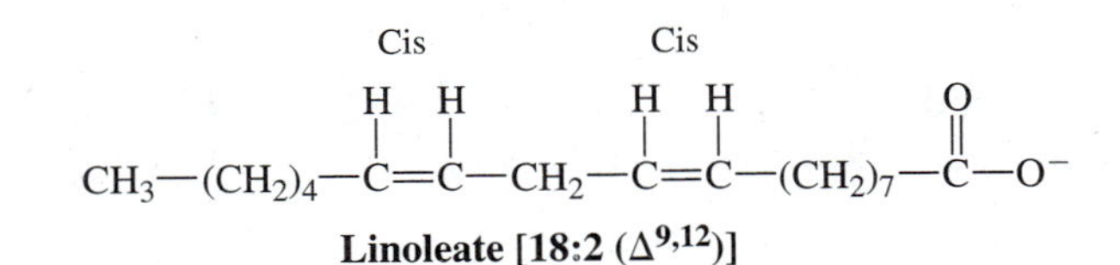

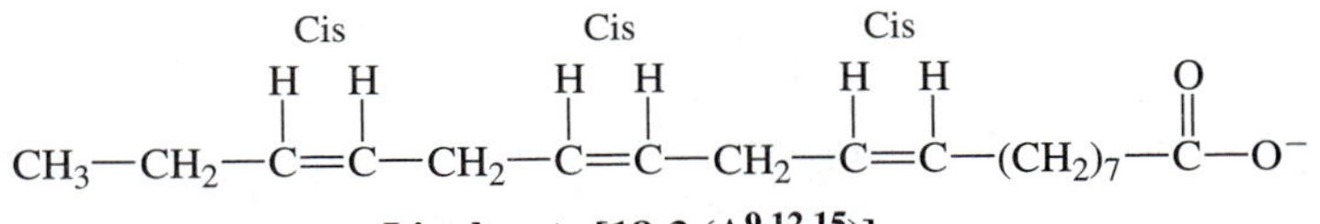

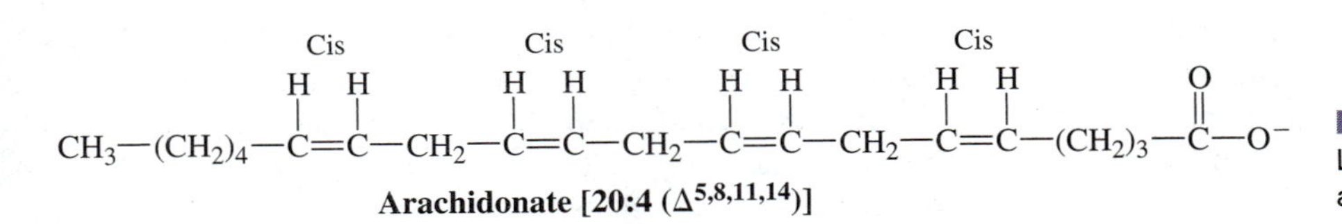

FIGURE 14.15
Linoleate, linolenate, and arachidonate.

rather than acetyl-SACP, reacts with malonyl-SACP during the initial condensation step leading to fatty acid biosynthesis. [Try Problem 14.26]

PROBLEM 14.19 Although NADH does not provide the immediate reducing power used for anabolism, it indirectly plays a central role in anabolism. Describe NADH's indirect contribution.

PROBLEM 14.20 To which one of the six major classes of enzymes does acetyl-SCoA carboxylase belong?

PROBLEM 14.21 Write the formula for an acetyl-SACP molecule by adding an acetyl group to the ACPSH depicted in Figure 14.12.

PROBLEM 14.22 Write the structural formula for the acyl group that exists after four elongation cycles during fatty acid biosynthesis.

PROBLEM 14.23 How many malonyl-SCoAs are involved in the biosynthesis of one palmitate? Explain.

PROBLEM 14.24 How many different enzymes participate in the synthesis of palmitoyl-SACP from acetyl-SCoA?

PROBLEM 14.25 What prevents the acyl-SACP produced as an intermediate during palmitate biosynthesis in humans from being immediately used as a substrate for β-oxidation?

PROBLEM 14.26 Write the structural formula for the oxoacyl (ketoacyl) group that is formed when propionyl-SACP condenses with malonyl-SACP.

14.8 RECIPROCAL REGULATION OF β-OXIDATION AND FATTY ACID BIOSYNTHESIS

The human body uses multiple strategies to regulate the relative rates of β-oxidation and fatty acid biosynthesis. In the process, both enzyme concentration and enzyme activity are modulated.

Long-Term Versus Short-Term Regulation

The regulation of enzyme concentrations is accomplished mainly through hormone-induced changes in the rate of enzyme production and degradation. It is often described as **long-term regulation,** since hours or even days may pass before there are significant shifts in enzyme concentration (Table 14.7). The shifts in hormone levels brought about by starvation or regular exercise lead to gradual increases in the concentration of the enzymes involved in fatty acid oxidation and to gradual decreases in the concentration of the enzymes involved in fatty acid biosynthesis. Hence, regular exercise is a key weapon in the fight against obesity. Sporadic exercise will not

TABLE 14.7

Long-Term Versus Short-Term Regulation of Reaction Rates

	Usual Time Required to Make a Significant Change	How Accomplished	Effect on Enzyme Concentration	Effect on Activity of Existing Enzyme
Short-Term Regulation	Seconds to minutes	Allosteric effectors Covalent modification of enzymes Control of substrate or coenzyme availability Control of pH and other variables	None	Increase or decrease
Long-Term Regulation	Hours to days	Alter rate of gene expression (enzyme synthesis) Alter rate of enzyme degradation	Increase or decrease	None, unless enzyme degraded

induce and maintain the long-term shifts in enzyme concentration that favor fat consumption over fat production.

Short-term regulation of fatty acid metabolism primarily involves the modulation of the catalytic activity of existing enzymes, a process brought about by allosteric effectors and hormone-induced covalent modifications of the enzymes themselves. Some short-term regulation is also achieved through the control of substrate or coenzyme availability. **[Try Problem 14.27]**

Regulation by Allosteric Effectors

Several of the enzymes involved in fatty acid metabolism in mammals are targets for specific allosteric effectors. The regulatory activities of most of these effectors make metabolic sense from the standpoint of energy charge considerations. An abundance of mitochondrial NADH and acetyl-SCoA, which signals a high energy charge, suppresses β-oxidation (a mitochondria-specific process) by inhibiting L-3-hydroxyacyl-SCoA dehydrogenase and 3-ketoacyl-SCoA thiolase (3-oxoacyl-SCoA thiolase) respectively (Section 13.3). Although high cytosolic citrate levels, which also signal a high energy charge, activate acetyl-SCoA carboxylase and fatty acid biosynthesis (a cytosolic event) in vitro, the in vivo significance of this activation is presently being questioned. High cytosolic citrate concentrations are associated with a high acetyl-SCoA concentration because citrate, which carries mitochondrial acetyl groups to the cytosol, is the major source of cytosolic acetyl-SCoA (Figure 14.11).

Collectively, allosteric effectors help ensure that when an abundant supply of energy exists, the oxidation of fatty acids is turned down to conserve existing energy stores. The production of more fatty acids is simultaneously stimulated to convert some of the excess energy to additional energy stores. If high concentrations of palmitoyl-SCoA eventually accumulate, this activated fatty acid feeds back and blocks further fatty acid synthesis by inhibiting acetyl-SCoA carboxylase, the enzyme that catalyzes the rate limiting step in fatty acid biosynthesis. Some aspects of the allosteric regulation of fatty acid metabolism are summarized in Figure 14.16.

When AMP concentrations are high, signaling a low energy charge, AMP (not cAMP) allosterically activates a protein kinase that catalyzes the phosphorylation

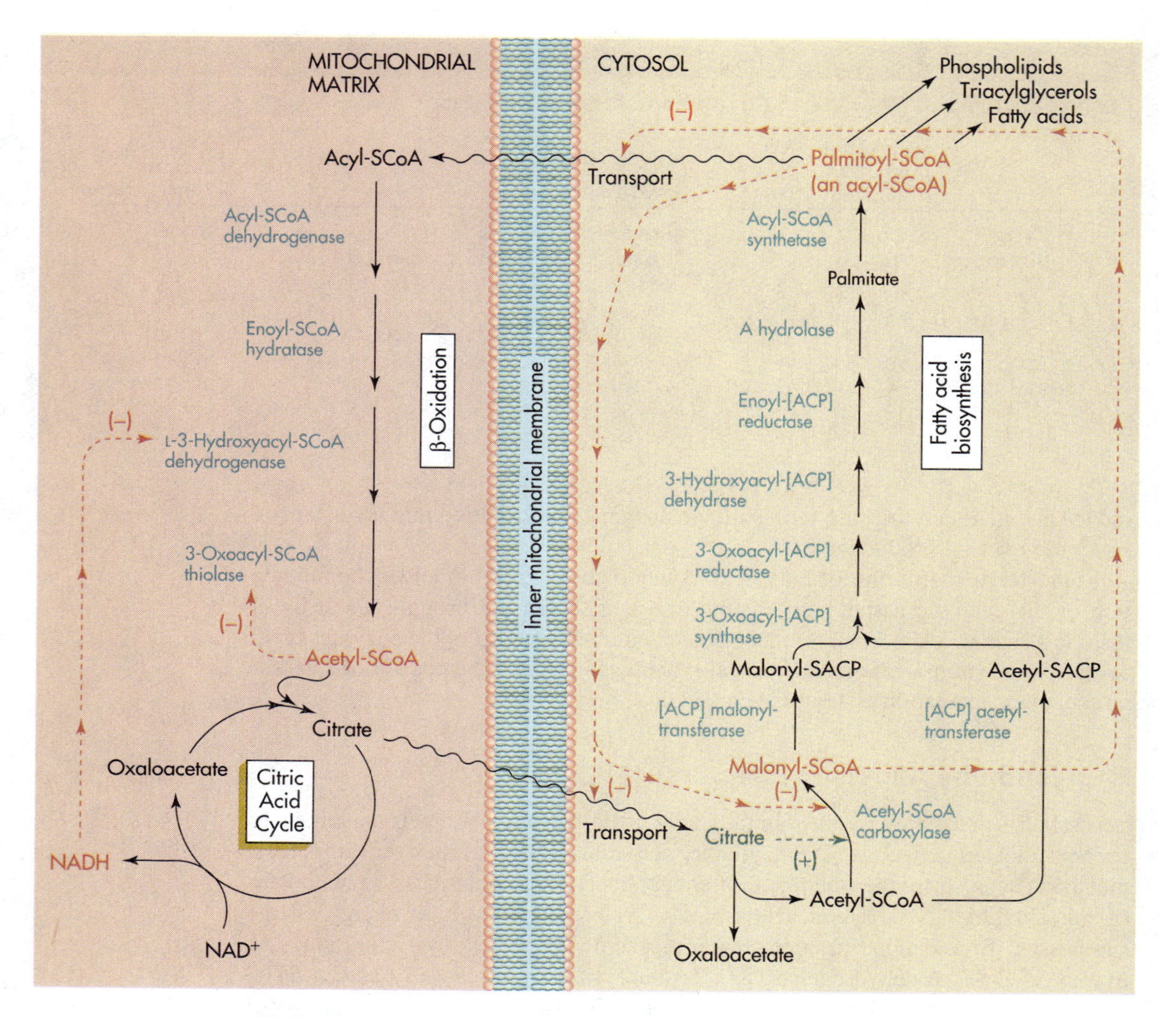

FIGURE 14.16
Allosteric regulation of fatty acid metabolism.

and inactivation of acetyl-SCoA carboxylase. This prevents ATP from being used for fatty acid production when the limited supply of ATP is needed for other purposes. The protein phosphatase that catalyzes the dephosphorylation and reactivation of acetyl-SCoA carboxylase is partially regulated by hormone-linked reaction cascades. The AMP-stimulated protein kinase is inhibited by ATP.

Hormone-Linked Regulation

Epinephrine, glucagon, and insulin, the three hormones encountered during the discussion of the regulation of carbohydrate metabolism, all exert a short-term impact on fatty acid metabolism by binding to target cell receptors and triggering intracellular reaction cascades that lead to the phosphorylation or dephosphorylation of spe-

TABLE 14.8

Hormonal Regulation of Fatty Acid Biosynthesis and β-Oxidation

	Glucagon	Epinephrine	Insulin
β-Oxidation (liver)	↑	↑	↓
Fatty acid biosynthesis (liver)	↓	↓	↑

cific enzymes. These covalent modifications either enhance or repress catalytic activity. Glucagon and epinephrine (released in response to low blood sugar levels) target liver cells where they initiate cAMP-dependent cascades to activate key enzymes in fatty acid oxidation and to inactivate key enzymes in fatty acid production (Table 14.8). Insulin, which normally signals high blood glucose levels and a well-fed and energy-abundant state, has the opposite effect; it activates key enzymes involved in fatty acid biosynthesis while repressing key enzymes involved in fatty acid oxidation. **[Try Problem 14.28]**

Regulation of Substrate Availability

The compartmentation of β-oxidation and fatty acid biosynthesis allows a cell to use membrane transport to help control substrate availability and, consequently, the relative rates of these pathways (Figure 14.16). Malonyl-SCoA accumulates in the cytosol of liver cells when there is an abundance of acetyl-ScoA, a high energy charge, and active fatty acid synthesis. It represses the simultaneous oxidation of fatty acids by inhibiting carnitine acyltransferase I, an enzyme required for the transport of fatty acids into mitochondria (Section 13.2). By blocking transport, malonyl-SCoA prevents fatty acids from reaching the compartmentalized enzymes required for their oxidation. High concentrations of cytosolic palmitoyl-SCoA, an intermediate in the incorporation of newly synthesized palmitate into fats, impede further fatty acid production by inhibiting both acetyl-SCoA carboxylase and the transport of acetyl groups (carried by citrate) from mitochondria to the cytosol.

Under certain circumstances, the rate of fatty acid biosynthesis is regulated by the availability of NADPH, a coenzyme that usually functions like a recycling substrate. After NADPH is converted to $NADP^+$ during one enzyme-catalyzed reaction, it is regenerated as $NADP^+$ is reduced in a distinct enzyme-catalyzed reaction. NADPH and $NADP^+$ do not remain associated with a single enzyme. A variety of enzymes compete for a limited quantity of these coenzymes.

Summary of Short-Term Regulation

The short-term regulation of fatty acid metabolism is summarized in Table 14.9. This table highlights the multitude of checks and balances that ultimately determine the relative rates of β-oxidation and palmitate biosynthesis. Since one pathway tends to be favored over the second at any given instance, fatty acid catabolism and anabolism seldom proceed at similar rates at the same instant. To a large extent, this avoids the utilization of energy to support a futile cycle.

TABLE 14.9

Short-Term Regulation of Fatty Acid Metabolism

	Compounds That Stimulate β-Oxidation or Inhibit Palmitate Biosynthesis	Compounds That Stimulate Palmitate Biosynthesis or Inhibit β-Oxidation
Allosteric effectors	Palmitoyl-SCoA	NADH Acetyl-SCoA Citrate
Hormones	Glucagon Epinephrine	Insulin
Membrane transport modulators	Palmitoyl-SCoA	Malonyl-SCoA
Cofactor availability	High $NADP^+$/NADPH ratio High NAD^+/NADH ratio	Low $NADP^+$/NADPH ratio Low NAD^+/NADH ratio

PROBLEM 14.27 What determines the maximum increase in reaction rate that can be achieved with short-term regulation of enzyme activity?

PROBLEM 14.28 List two anabolic processes in addition to fatty acid biosynthesis that are partially regulated by epinephrine or glucagon. For each listed process, indicate whether it is stimulated or inhibited by the presence of these hormones.

14.9 TRIACYLGLYCEROL BIOSYNTHESIS

Triacylglycerols, the major components of fat, function primarily as fuel molecules. They are storage forms of fatty acids and energy. A person of average weight possesses enough triacylglycerols to meet the basal energy needs of the human body for approximately 3 months.

Glycerol 3-phosphate and fatty acyl-SCoAs are the immediate building blocks for the assembly of triacylglycerols. Glycerol 3-phosphate is normally derived from dihydroxyacetone phosphate (an intermediate in glycolysis) or from glycerol through reactions catalyzed by **glycerol-3-phosphate dehydrogenase** and **glycerol kinase,** respectively.

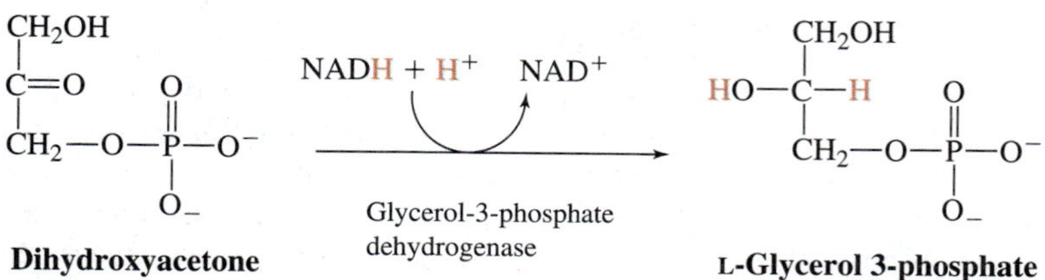

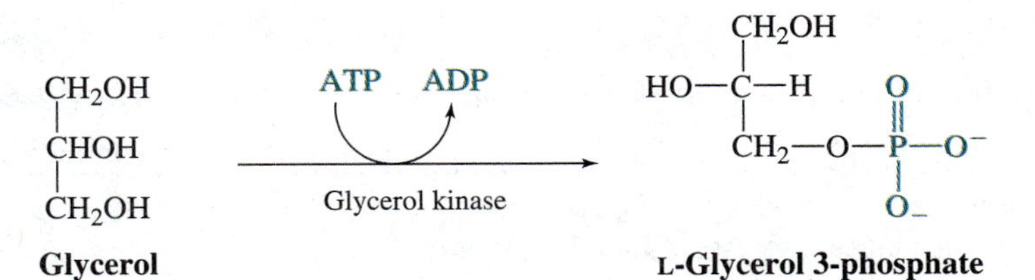

A fatty acyl group is transferred from an acyl-SCoA to the terminal hydroxyl group of glycerol 3-phosphate with the catalytic assistance of **glycerol-3-phosphate acyltransferase.**

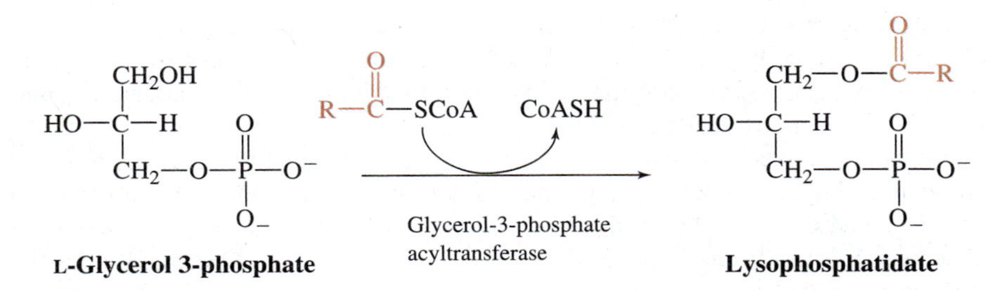

Lysophosphatidate is formed. The hydroxyl group in lysophosphatidate is acylated in a similar manner during a reaction catalyzed by **1-acylglycerol-3-phosphate acyltransferase.**

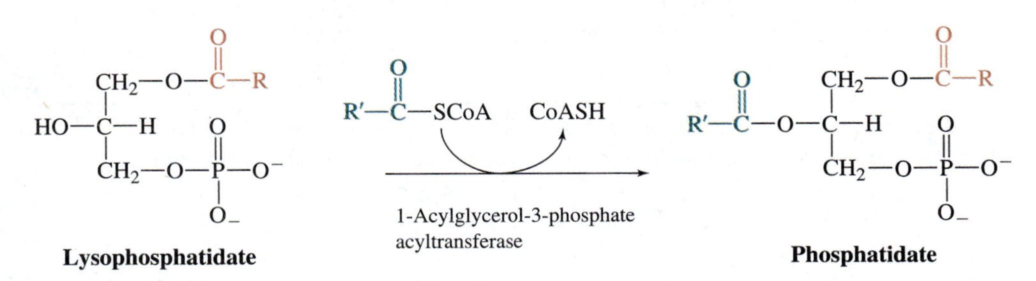

After the product, known as a phosphatidate, is hydrolyzed with the aid of **phosphatidate phosphatase,** a third acyl group is transferred from an acyl-SCoA to the diacylglycerol produced. **Diacylglycerol acyltransferase** catalyzes the latter reaction. **[Try Problems 14.29 and 14.30]**

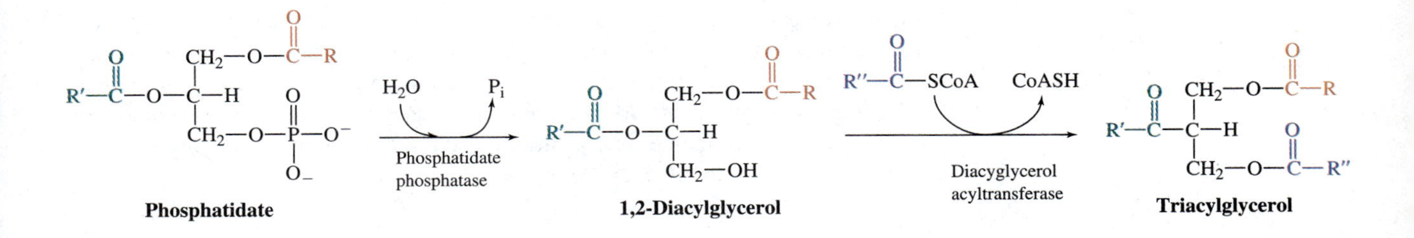

TABLE 14.10

Hormonal Regulation of the Biosynthesis and Hydrolysis of Triacylglycerols

	Glucagon	Epinephrine	Insulin
Triacylglycerol hydrolysis (adipose tissue)	↑	↑	↓
Triacylglycerol biosynthesis (adipose tissue)			↑

In humans, the catalytic activities responsible for the biosynthesis of triacylglycerols all reside within a multienzyme complex, known as the **triacylglyercol synthetase complex,** that is bound to the cytosolic surface of the membrane of the endoplasmic reticulum. The phosphatidates produced as intermediates during the assembly of triacylglycerols are also used as precursors for the synthesis of some phospholipid classes: the phosphatidylcholines; the phosphatidylethanolamines; and the phosphatidylinositols (Section 9.3).

Both triacylglycerol biosynthesis and triacylglycerol mobilization (hydrolysis) are regulated by hormones. Epinephrine and glucagon binding to adipocytes triggers cAMP-linked cascades that lead to the activation of a hormone-sensitive lipase, an enzyme that catalyzes triacylglycerol hydrolysis (Table 14.10). Although liver is the primary target of glucagon, this hormone does have a significant number of receptors on adipocytes. Insulin inhibits triacylglycerol hydrolysis and stimulates triacylglycerol production by mechanisms under investigation.

The next section turns to amino acid anabolism. This topic is closely related to the carbohydrate sections of this chapter, because amino acids are synthesized from compounds that are intermediates in carbohydrate metabolism.

PROBLEM 14.29 How many different triacylglycerols (for chiral triacylglycerols, include both D- and L-isomers in your count) can be constructed from a pool of metabolites containing glycerol, palmitoyl-SCoA, linoleyl-SCoA, and steryl-SCoA?

PROBLEM 14.30 List all the chiral compounds that are intermediates in the biosynthesis of triacylglycerols from glycerol and acyl-SCoA. The enzymes involved in metabolizing these intermediates are stereospecific.

14.10 AMINO ACID BIOSYNTHESIS

Humans are genetically programmed to synthesize adequate amounts of 11 of the 20 common protein amino acids. The other 9 are essential nutrients (Section 3.4, Table 14.11). Arginine is sometimes classified as essential because infants commonly exhibit a dietary requirement. Some authors also place tyrosine in the essential category because phenylalanine, one of the essential amino acids, is its only precursor within the body. **[Try Problem 14.31]**

All of the nonessential amino acids, except tyrosine, can be synthesized from one of four metabolites that are intermediates or products in the central metabolic pathways examined in previous chapters. These four metabolites are α-ketoglutarate, oxaloacetate, 3-phosphoglycerate, and pyruvate (Figure 14.17). Alanine, aspartate, and glutamate are each produced through a single-step transamination reaction (Figure 14.18). A small number of additional steps are required to synthesize each of the other nonessential amino acids. An examination of these additional steps can be found in some more encyclopedic texts.

The biosynthesis of the essential amino acids in plants and bacteria often proceeds through elaborate multistep pathways. Some organisms, for example, use nine distinct steps to produce leucine from pyruvate, and different organisms commonly use unique pathways to assemble the same amino acid. In contrast, the major pathways for carbohydrate and fat metabolism tend to occur universally. **[Try Problem 14.32]**

Now that the catabolism and anabolism of carbohydrates, fats, and amino acids have been examined, the next section summarizes these processes in an attempt to provide an integrated picture of the metabolism of these central classes of compounds.

TABLE 14.11

Essential and Nonessential Amino Acids

Essential Amino Acids[a]	Nonessential Amino Acids
Histidine	Alanine
Isoleucine	Arginine[b]
Leucine	Asparagine
Lysine	Aspartate
Methionine	Cysteine
Phenylalanine	Glutamate
Threonine	Glutamine
Tryptophan	Glycine
Valine	Proline
	Serine
	Tyrosine[c]

[a]Amino acids that are essential nutrients in humans
[b]Essential for infants but not adults
[c]Synthesized from phenylalanine, an essential amino acid

PROBLEM 14.31 Explain how the total absence of even a single essential amino acid within a cell completely blocks the assembly of most enzymes. What is the consequence of the total absence within a cell of a nonessential amino acid? Explain.

PROBLEM 14.32 Although the human body is not genetically programmed to synthesize the essential amino acids, it normally contains a source of at least small quantities of such amino acids. What is that source? Hint: Review Section 1.3.

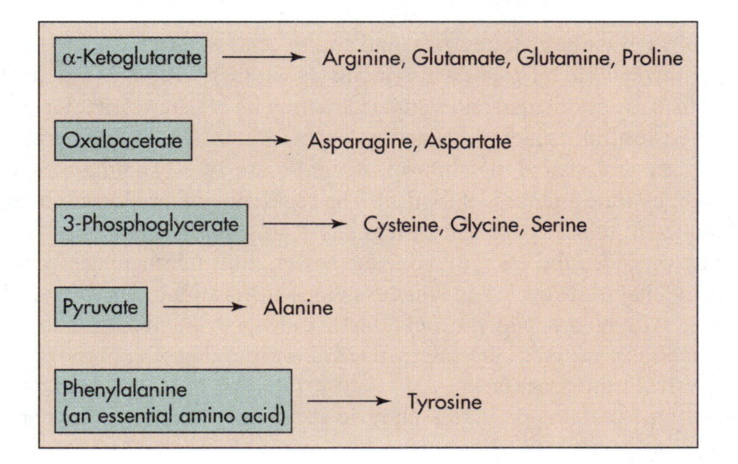

FIGURE 14.17
Biosynthetic precursors for the nonessential amino acids.

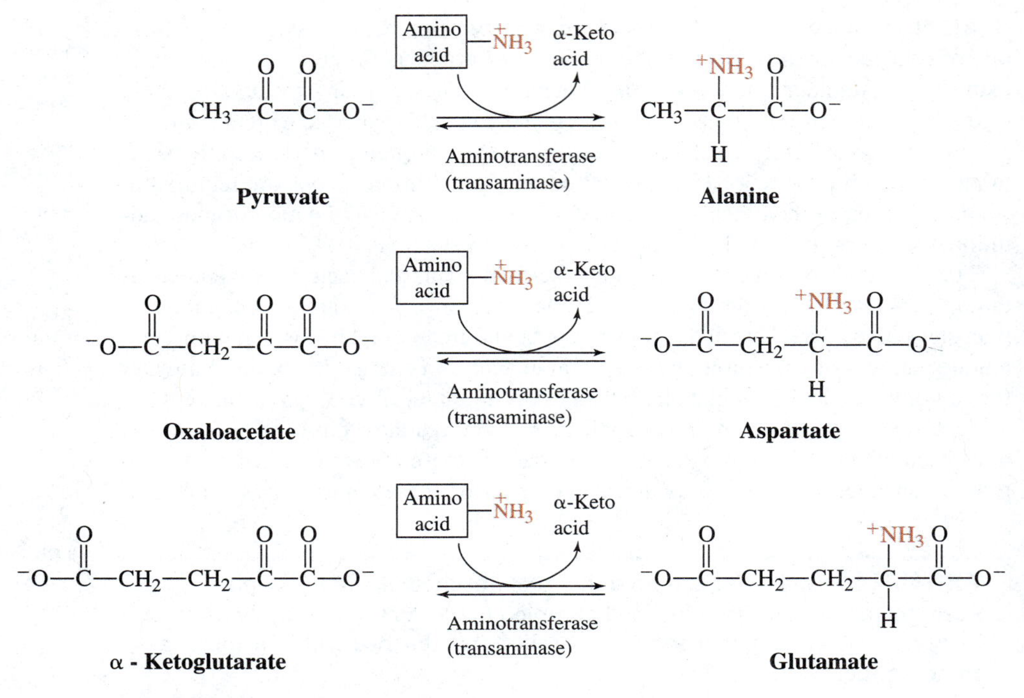

FIGURE 14.18
Biosynthesis of alanine, aspartate, and glutamate.

14.11 SUMMARY OF CARBOHYDRATE, FAT, AND AMINO ACID METABOLISM

The metabolism of carbohydrates, fats, and amino acids is interconnected by central metabolic pathways and by specific intermediates within these pathways. In humans, this metabolism is coordinated and regulated mainly by allosteric effectors and three key hormones: insulin; epinephrine, and glucagon. Figure 14.19 summarizes carbohydrate, fat, and amino acid metabolism, while Figure 14.20 summarizes the regulation of carbohydrate and fat metabolism. The regulation of amino acid metabolism is not examined in this text. Metabolic regulators, in general, ensure that catabolism continuously provides the energy, reducing power, and intermediates required for anabolism and that futile cycles, in which a compound is simultaneously synthesized and degraded with no associated benefit, are avoided. Metabolic regulators also ensure that catabolism provides the energy needed for muscle action, brain action, and other energy-requiring processes.

The next three sections provide a glimpse at the biosynthesis of the purines, the pyrimidines, and some related compounds. Because the precursors for purine and pyrimidine biosynthesis are intermediates in carbohydrates, fat, and amino acid metabolism, purine and pyrimidine anabolism is connected to the metabolism summarized in Figures 14.19 and 14.20.

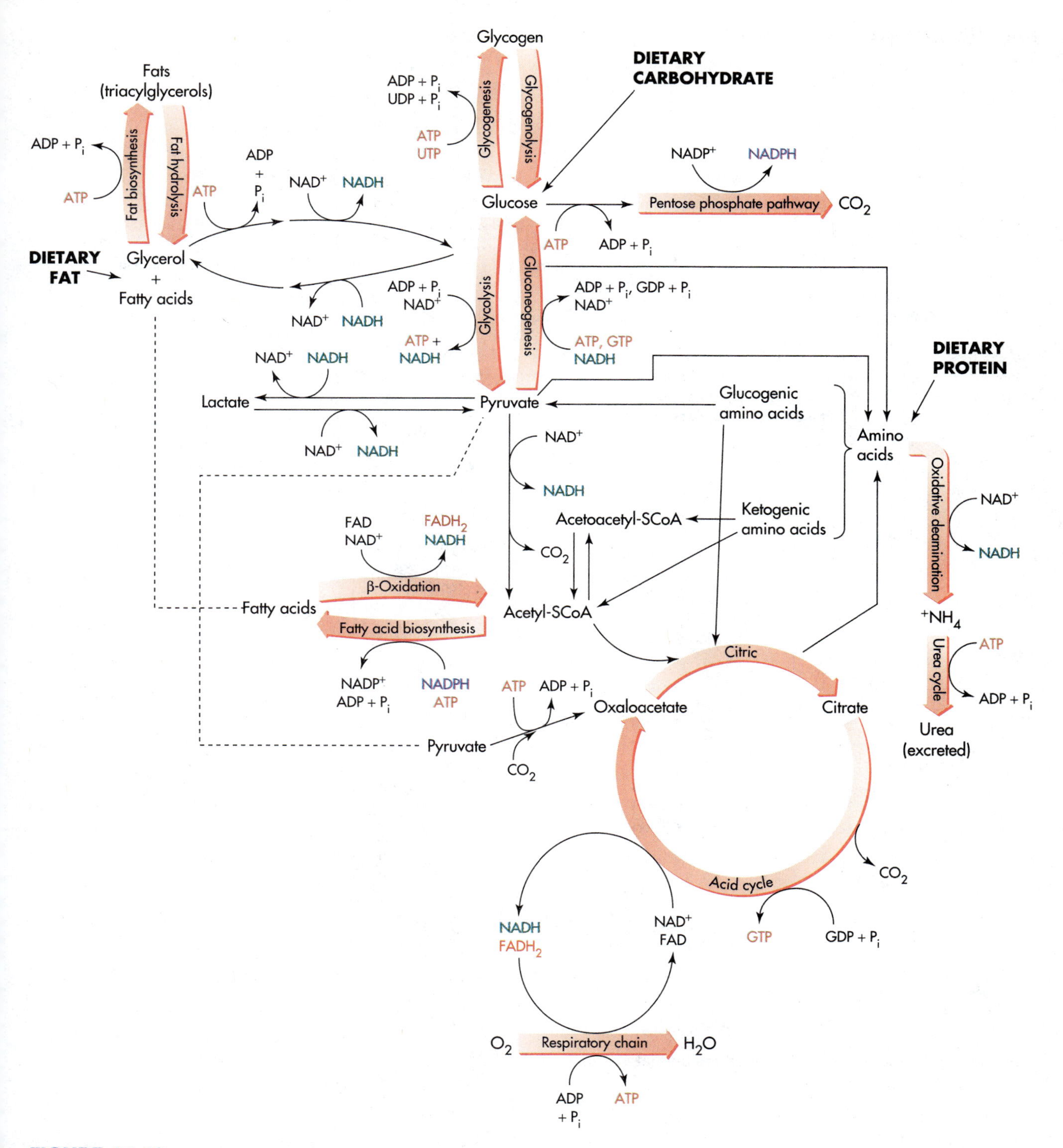

FIGURE 14.19
Summary of carbohydrate, fat, and amino acid metabolism. The human body can synthesize only the nonessential amino acids and the nonessential fatty acids.

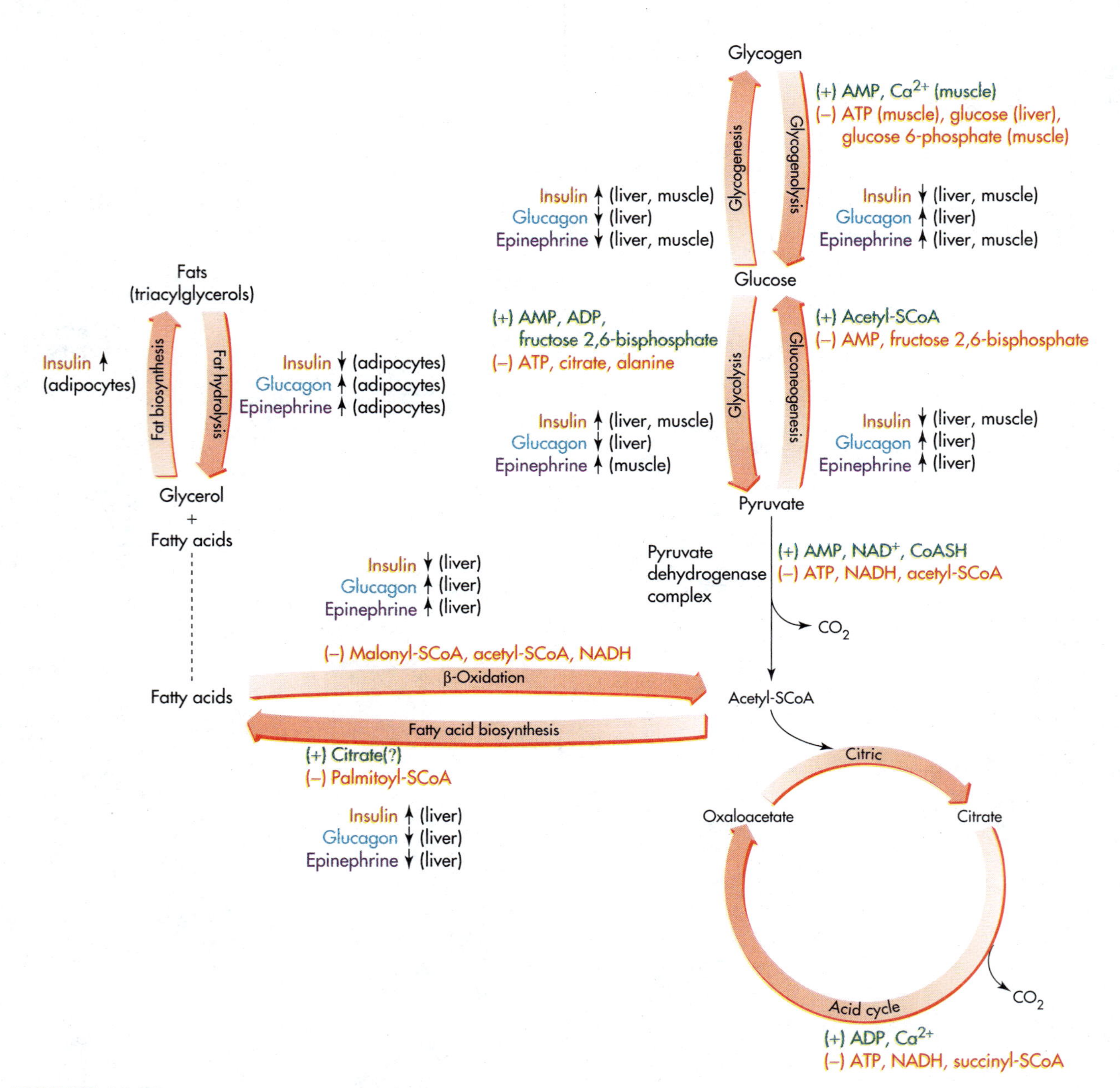

FIGURE 14.20
Summary of the regulation of carbohydrate and fat metabolism.

14.12 PURINE BIOSYNTHESIS

Purines are compounds with a purine ring (Section 7.1). **Pyrimidines** contain a pyrimidine ring.

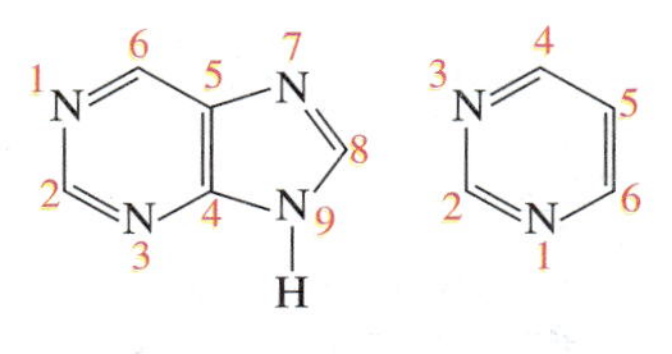

Purine **Pyrimidine**

Since specific purines and pyrimidines serve as building blocks for RNA and DNA, an adequate supply of these heterocyclic aromatic amines is essential for nucleic acid biosynthesis (Chapter 17).

5-Phosphoribosyl-1-pyrophosphate (PRPP) produced from ribose 5-phosphate generated within the pentose phosphate pathway (Section 12.2) is the initial reactant in the biosynthesis of purines. A sequence of 10 enzyme-catalyzed reactions converts PRPP to **inosine 5′-monophosphate (IMP,** inosinate), the first biosynthetic intermediate with an intact purine ring system (Figure 14.21). One to three atoms at a time are added to the growing purine ring, which is attached to a ribose phosphate residue during the ten step assembly process. Figure 14.22 summarizes the source of each atom within the purine ring system. Seven of the nine ring atoms can be traced to specific atoms contributed by one aspartate, two glutamine, one glycine, and one CO_2. The other two atoms are supplied by separate molecules of **N^{10}-formyltetrahydrofolate,** a coenzyme whose vitamin precursor is folic acid (Table 10.6, Figure 14.23). **[Try Problem 14.33]**

IMP can be converted to either adenosine 5′-monophosphate (AMP; adenylate) or guanosine 5′-monophosphate (GMP; guanylate) through separate two-step pathways. AMP and GMP can, in turn, be converted to ATP and GTP (Figure 14.24), which serve as the immediate source of energy for most energy requiring processes within the human body. These nucleoside triphosphates also represent activated monomers for RNA biosynthesis; they are substrates for RNA polymerases (Section 17.1). The total and relative amounts of ATP and GTP produced are regulated primarily through the feedback inhibition and the feed-forward activation of specific enzymes. *You should assume that all nucleoside monophosphates (nucleotides), nucleoside diphosphate, and nucleoside triphosphates encountered in this text are 5′-nucleoside derivatives unless specified otherwise.* **[Try Problems 14.34 and 14.35]**

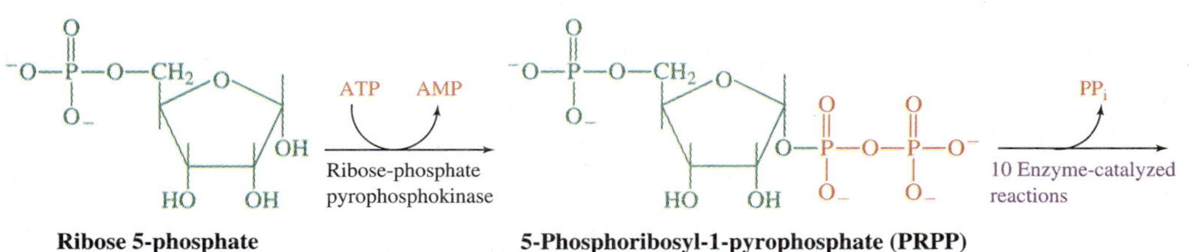

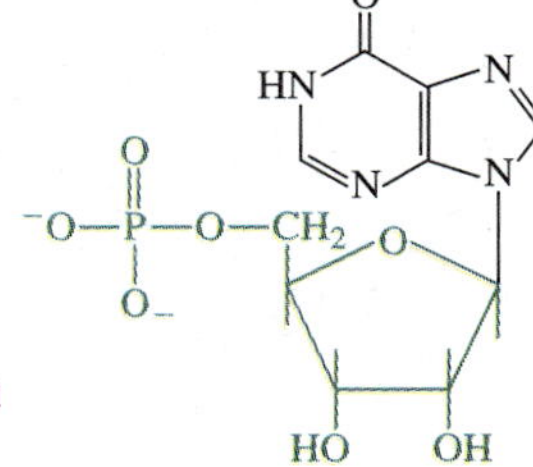

FIGURE 14.21
Biosynthesis of IMP.

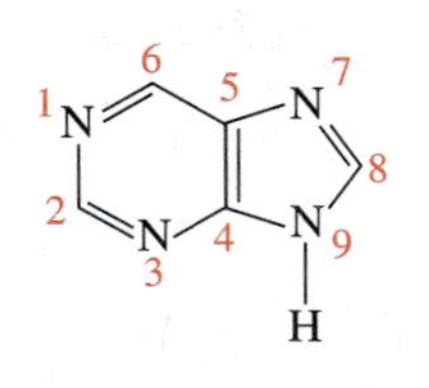

Atom	Source
1	Aspartate
2, 8	N^{10}-Formyltetrahydrofolate (two separate molecules)
3, 9	Glutamine (two separate molecules)
4, 5, 7	Glycine (a single molecule contributes all three atoms)
6	CO_2

FIGURE 14.22
Source of each atom within the purine ring.

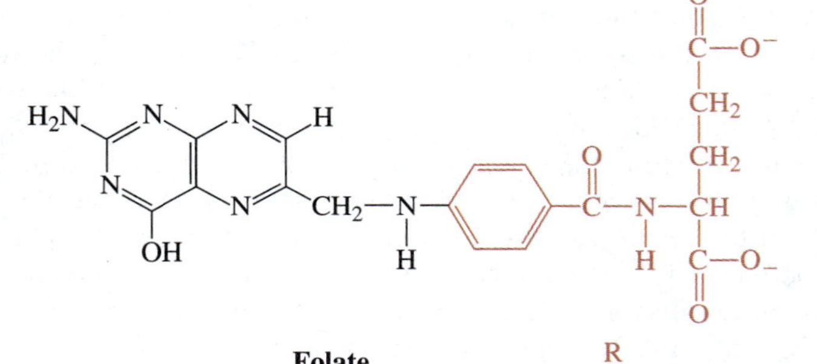

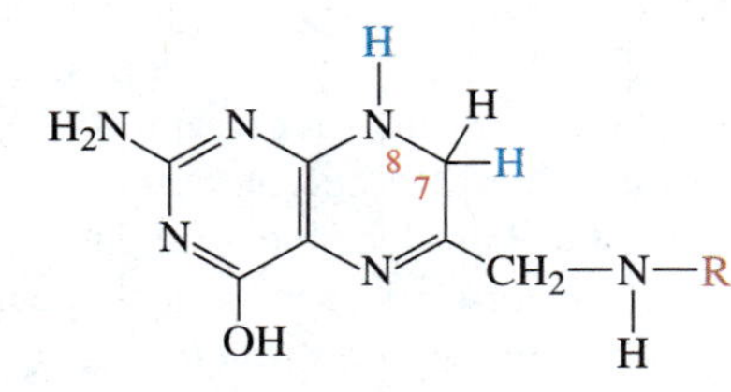

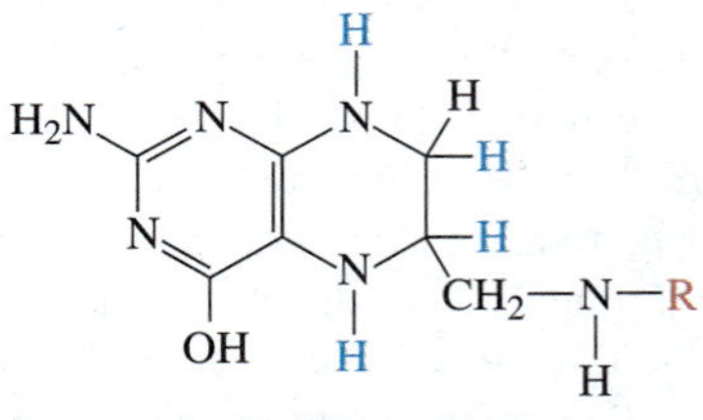

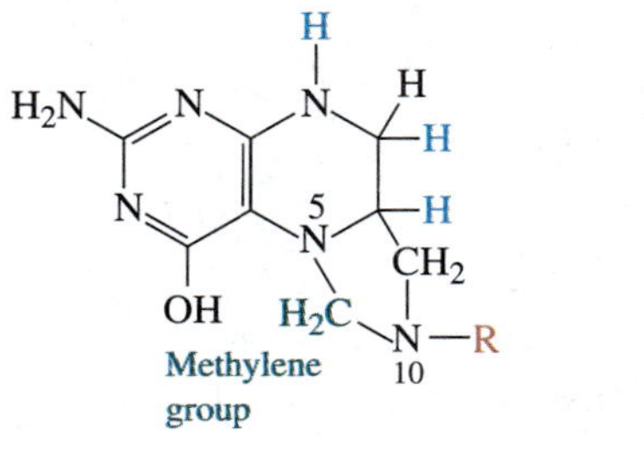

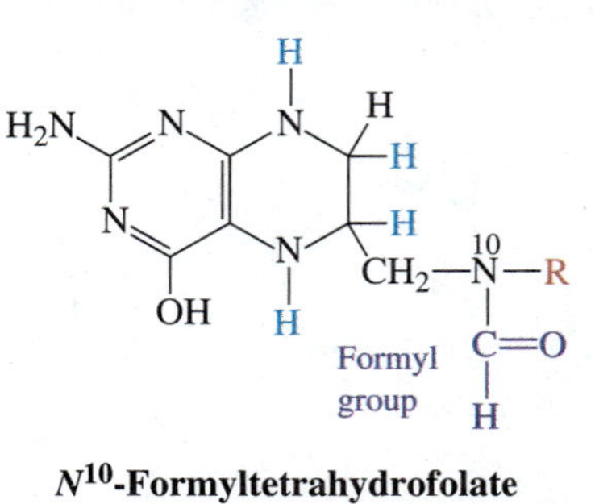

FIGURE 14.23
Structure of folate and some folate coenzymes.

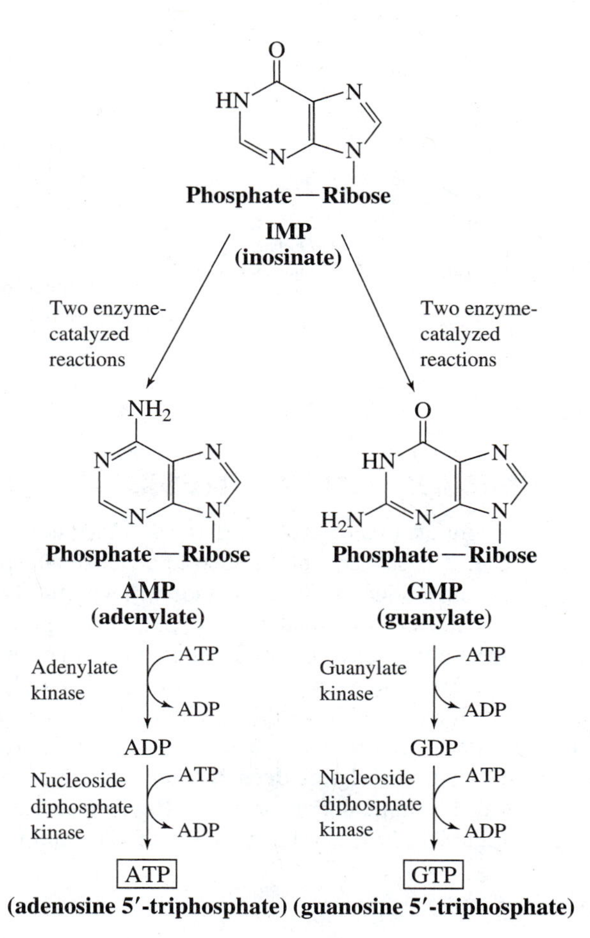

FIGURE 14.24
Biosynthesis of ATP and GTP. The nucleoside diphosphate kinase–catalyzed reaction is not the usual source of ATP. In heterotrophs, most ATP is synthesized in the respiratory chain from ADP and P_i.

PROBLEM 14.33 True or false? Aspartate is a substrate for one of the enzymes involved in the biosynthesis of the purine ring. Explain.

PROBLEM 14.34 Study Figure 14.24. Are the guanylate kinase- and nucleoside diphosphate kinase–catalyzed reactions reversible? Explain.

PROBLEM 14.35 What is meant by an "activated monomer?" Why do cells normally activate monomers before incorporating them into polymers? Give one other example, in addition to nucleoside triphosphates, of activated monomers.

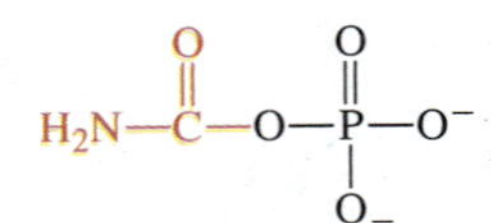

+

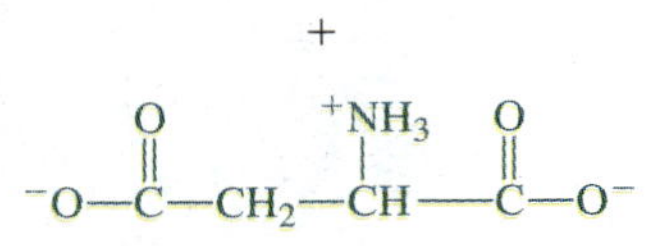

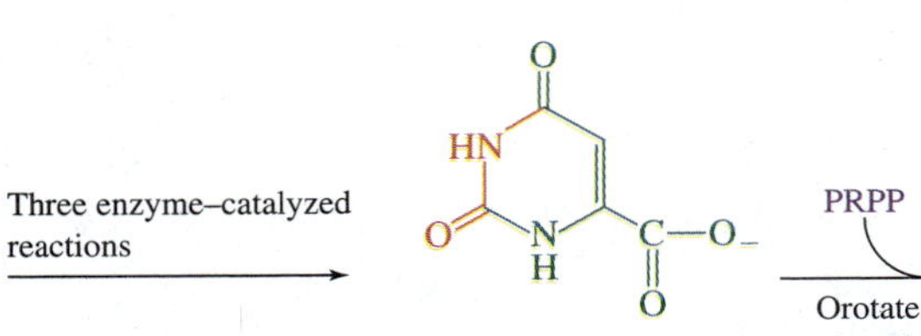

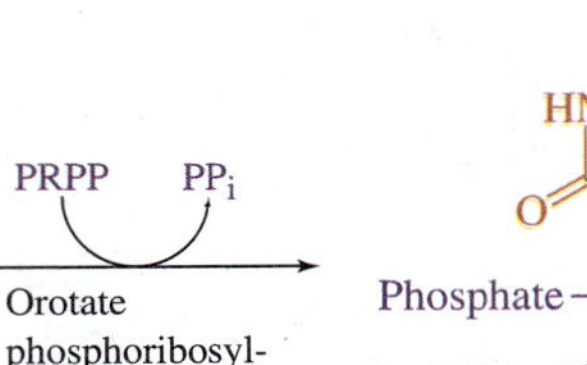

FIGURE 14.25
Biosynthesis of OMP.

14.13 PYRIMIDINE BIOSYNTHESIS

The pyrimidine ring, in contrast to the purine ring, is assembled in the absence of an attached sugar phosphate. It is subsequently linked to a ribose phosphate (donated by PRPP) to generate the nucleotide **orotidine 5′-monophosphate (OMP; orotidylate).** Carbamoyl phosphate and aspartate are the immediate precursors for the pyrimidine ring, and they supply all of the atoms ultimately found within this ring (Figure 14.25). Aspartate is a protein amino acid whereas carbamoyl phosphate is a reactant in the urea cycle; it reacts with ornithine to initiate the cycle (Figure 13.18). Uridine 5′-monophosphate (UMP), and ultimately UTP, are synthesized from OMP with the catalytic assistance of **orotidylate decarboxylase** and two kinases. In mammals, glutamine serves as the amino donor for the **CTP synthase**–catalyzed conversion of UTP to CTP (Figure 14.26). The biosynthesis of UTP and CTP is partially

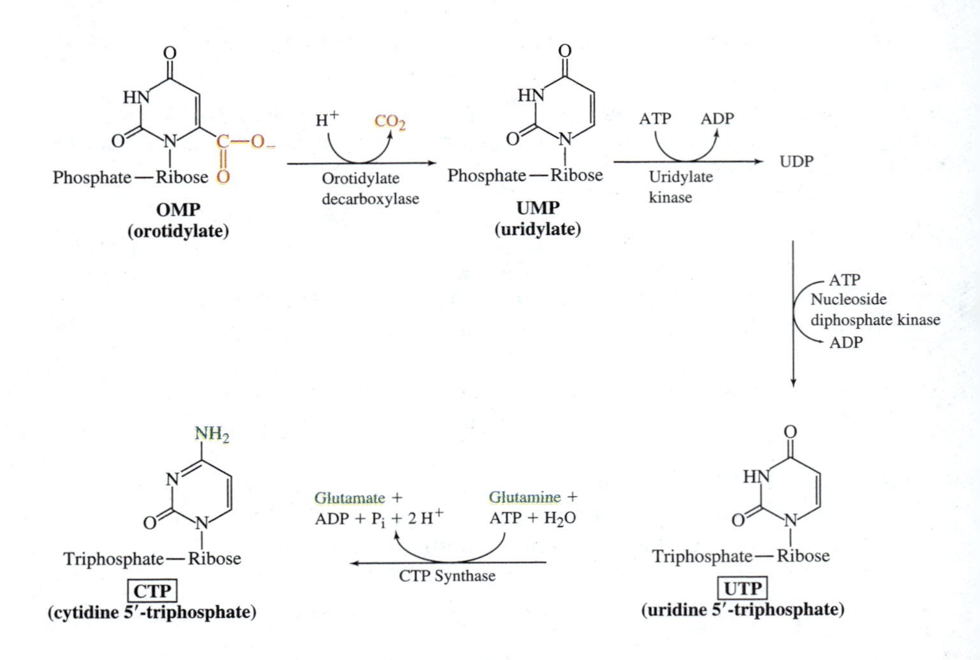

FIGURE 14.26
Biosynthesis of UTP and CTP.

controlled by the feedback inhibition of key regulatory enzymes. UTP and CTP, like ATP and GTP, represent activated forms of RNA monomers and serve as substrates for RNA polymerases (Section 17.1). **[Try Problem 14.36]**

PROBLEM 14.36 Write the structural formula for glutamine (one of the 20 protein amino acids). Study Figure 14.26, and then circle the group in glutamine that is transferred to UTP during the biosynthesis of CTP.

14.14 DEOXYRIBONUCLEOTIDE BIOSYNTHESIS

The two preceding sections have looked at the biosynthesis of AMP, GMP, UMP, and CMP, the ribonucleotides from which RNA is constructed, and the biosynthesis of ATP, GTP, UTP, and CTP. Ribonucleotides are the precursors for the production of deoxy-AMP (dAMP), deoxy-GMP (dGMP), deoxy-TMP (dTMP), and deoxy-CMP (dCMP), the deoxyribonucleotides that serve as building blocks for the assembly of DNA. *"Deoxy" refers to "2′-deoxy" unless specified otherwise.* In mammalian systems, ribonucleoside diphosphates produced from ribonucleoside monophosphates (ribonucleotides) serve as substrates for **ribonucleotide reductase** (also called **ribonucleoside-diphosphate reductase**), an enzyme that catalyzes the conversion of ribose residues to deoxyribose residues.

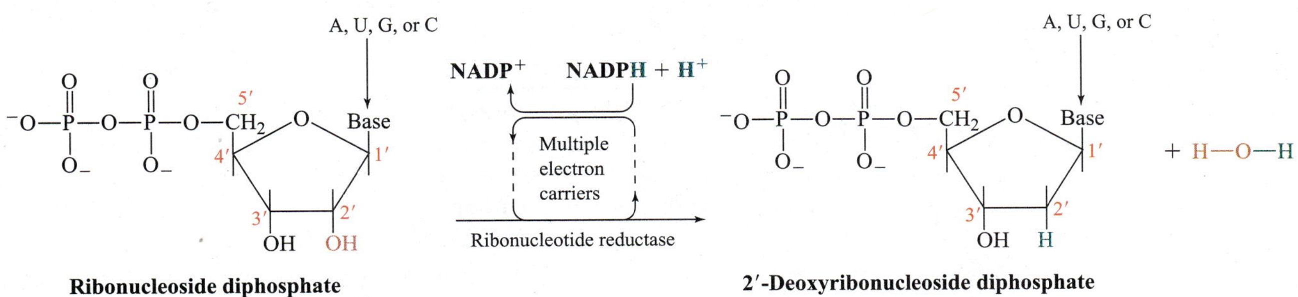

Although NADPH is the ultimate source of the reducing power, the electrons from NADPH flow through multiple carriers on the way to the ribose residue that is reduced. The immediate reducing agent is the ribonucleotide reductase itself; two thiol side chains are oxidized to a disulfide as the ribose residue is transformed into a deoxyribose residue.

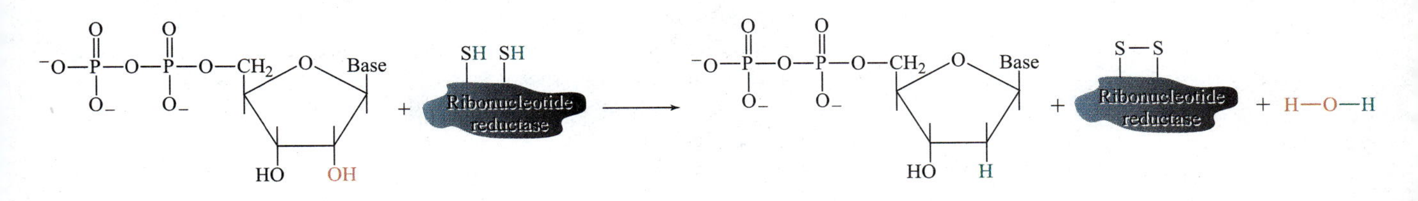

Deoxyribonucleoside diphosphates can be either hydrolyzed to yield deoxyribonucleotides or phosphorylated to generate deoxyribonucleoside triphosphates. [Try Problem 14.37]

For a cell to complete the assembly of the pyrimidine structures required for DNA synthesis, it must convert uracil residues to thymine residues. This is accomplished with the **thymidylate synthase**–catalyzed addition of a methyl group to the pyrimidine ring in dUMP.

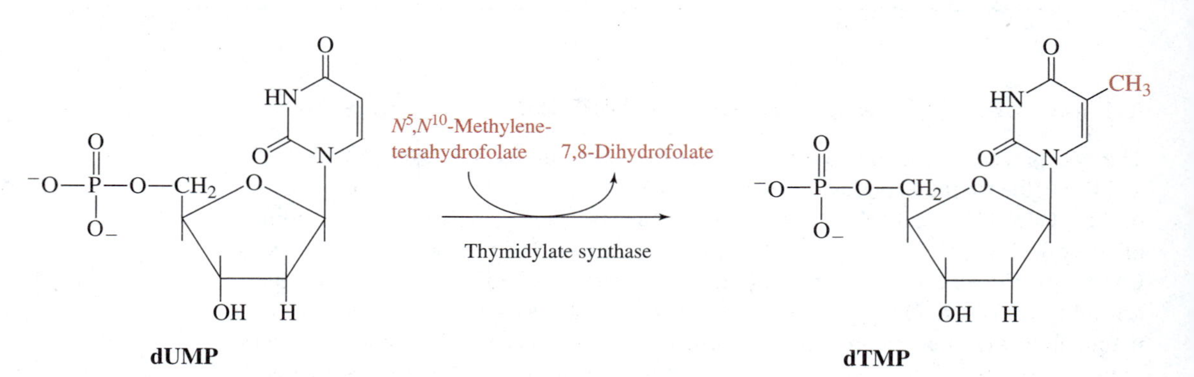

***N*^{5},*N*^{10}-methylenetetrahydrofolate** (Figure 14.23) serves as the methyl donor during this reaction. Figure 14.27 summarizes the biosynthesis of the four dexoyribonucleoside triphosphates that are the immediate reactants required for the DNA polymerase–catalyzed production of DNA (Section 17.4).

PROBLEM 14.37 Is the use of NADPH (rather than NADH) for the conversion of ribose residues to deoxyribose residues consistent with the previously encountered uses of NADH and NADPH in metabolism? Summarize the division of labor that usually exists between NADH and NADPH.

14.15 METABOLIC BASIS FOR CHEMOTHERAPY

Some of the chemicals commonly used in the treatment of cancer act by altering purine- or pyrimidine-linked metabolism, including DNA replication. This metabolism plays a central role in the rapid, uncontrolled **mitosis** (cell division) characteristic of malignant tumors (Section 19.8). Because each round of mitosis must be preceded by a round of DNA replication, a constant and abundant supply of deoxynucleoside triphosphates is required to support the rapid multiplication of cancer cells. Some compounds that inhibit DNA replication have proved to be effective chemotherapeutic agents. Because these drugs inhibit DNA replication in normal cells as well as in cancer cells, they commonly have serious side effects.

5-Fluorouracil, one of the first drugs specifically developed for the treatment of cancer, is converted to fluorodeoxyUMP (FdUMP) within the human body

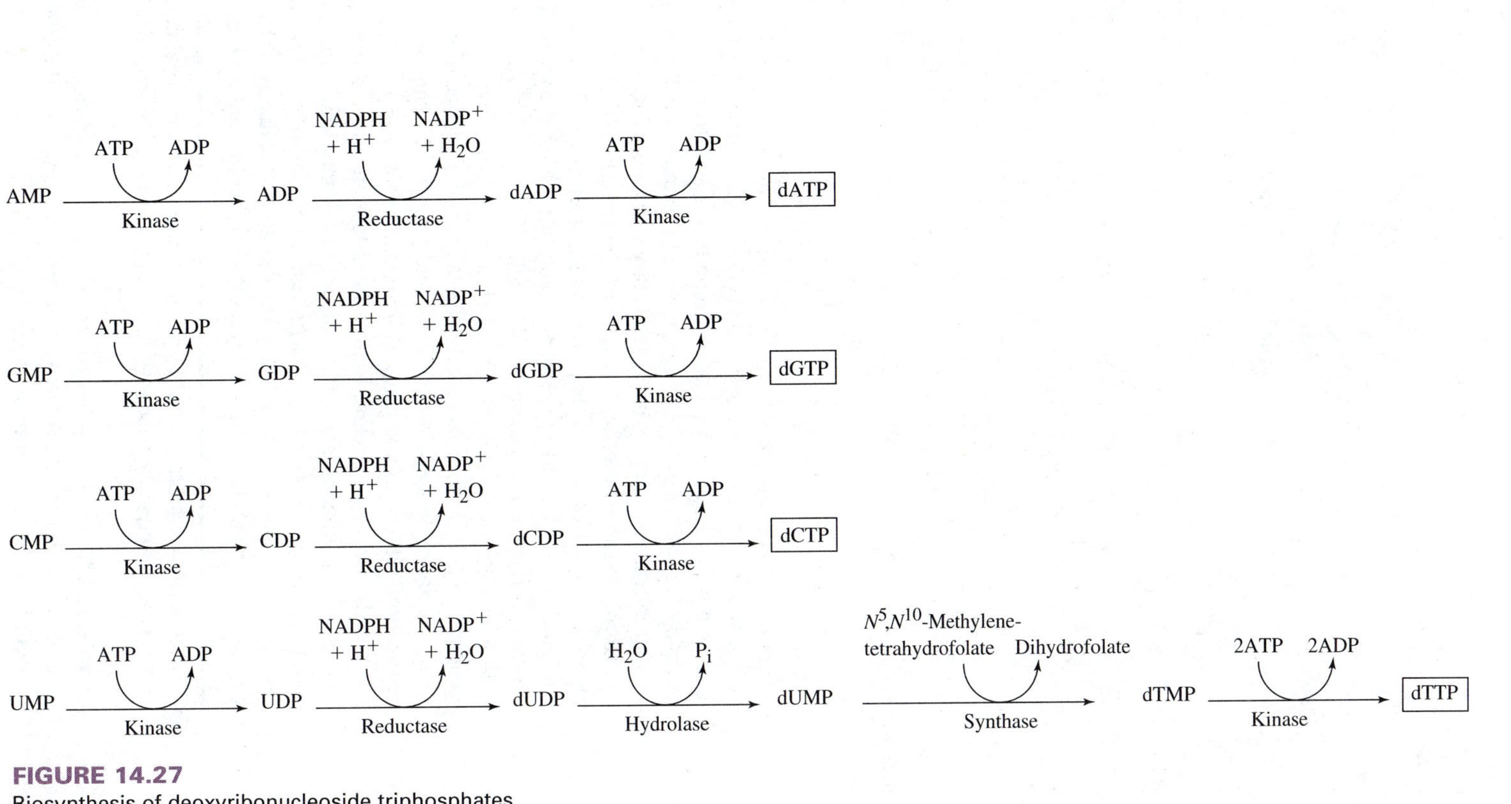

FIGURE 14.27
Biosynthesis of deoxyribonucleoside triphosphates.

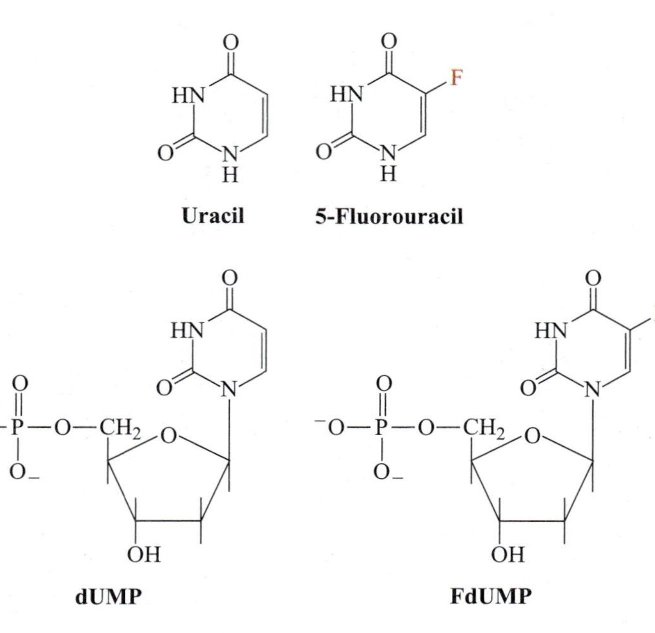

FIGURE 14.28
5-Fluorouracil and related compounds.

(Figure 14.28). This dUMP analog competes with dUMP for the active site on thymidylate synthase (the enzyme that converts dUMP to dTMP). Once bound, it becomes covalently and irreversibly joined to this active site. If enough copies of thymidylate synthase are inactivated in this manner, a cell runs low on dTMP and DNA replication becomes impossible. **[Try Problems 14.38 and 14.39]**

Methotrexate (amethopterin) and **aminopterin** (Figure 14.29) are two additional examples of widely employed anticancer drugs. These folate analogs block the production of dTMP by competitively inhibiting **dihydrofolate reductase,** an enzyme that helps regenerate N^5,N^{10}-methylenetetrahydrofolate from dihydrofolate following the production of dTMP (Figure 14.30). Thymidylate synthase cannot generate dTMP without its coenzyme (N^5,N^{10}-methylenetetrahydrofolate). Because DNA cannot replicate without dTMP, and cancer cells cannot divide without DNA replication, tumors cannot grow. **[Try Problems 14.40 and 14.41]**

A wide variety of bacterial and viral infections are also treated with chemicals that disrupt pathogen-linked metabolism. Many antibiotics, for example, selectively inhibit protein or nucleic acid biosynthesis in prokaryotes. Specific examples are examined in Chapters 17 and 18. AIDS therapies, including the use of AZT (a deoxynucleotide analog) to block viral RNA replication, are discussed in Section 17.5.

PROBLEM 14.38 Would you expect all of the normal cells in the human body to be equally susceptible to damage by 5-fluorouracil? Explain.

PROBLEM 14.39 5-Fluorouracil must be converted to FdUMP within a cell before it can block DNA replication. Suggest why 5-fluourouracil, rather than FdUMP, is used for cancer chemotherapy.

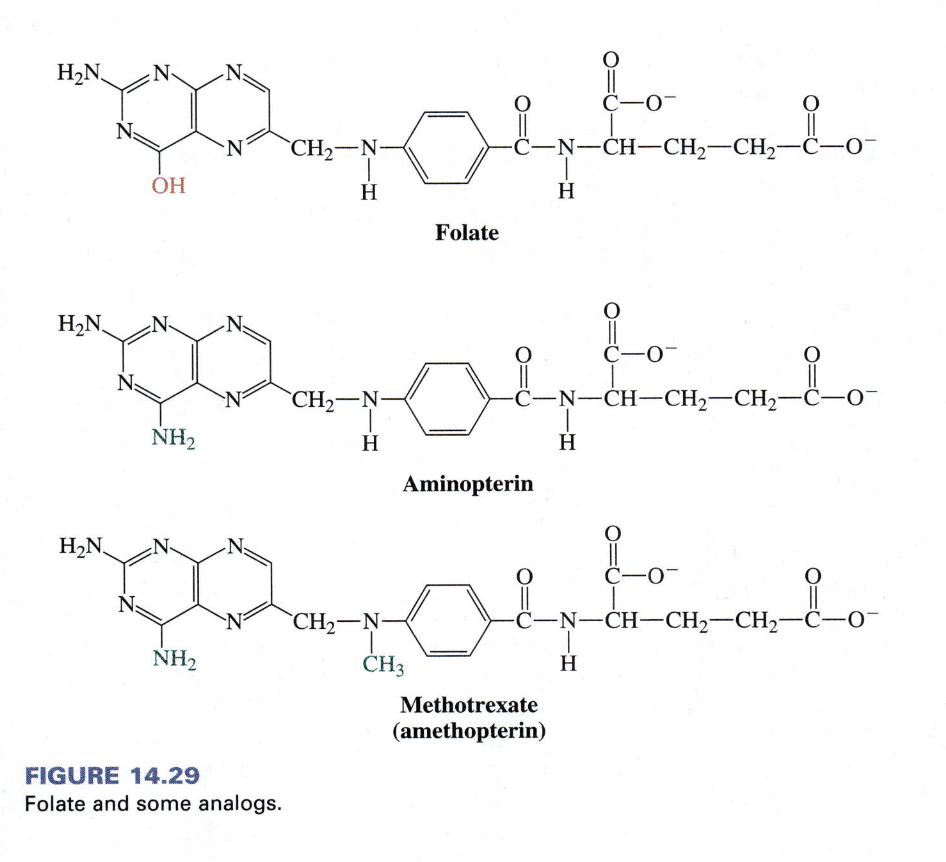

FIGURE 14.29
Folate and some analogs.

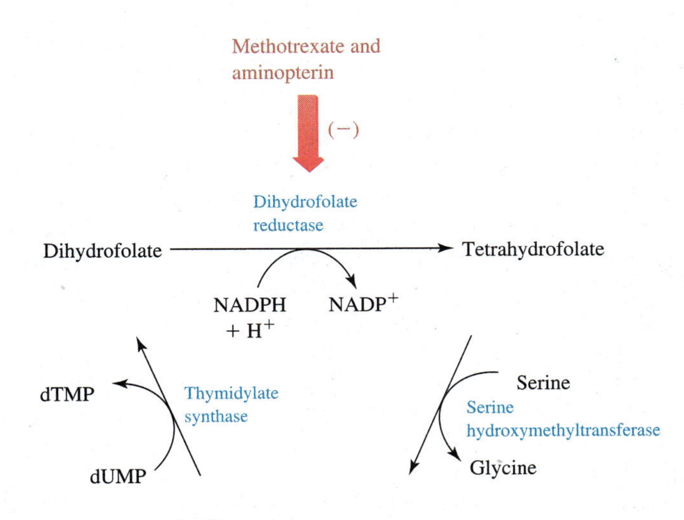

FIGURE 14.30
Methotrexate and aminopterin inhibition of dTMP biosynthesis.

PROBLEM 14.40 To what site on dihydrofolate reductase does methotrexate most likely bind? Explain.

PROBLEM 14.41 Study Figure 14.29 and then suggest a new compound that might be worth testing as an anticancer agent. Write the structural formula for this compound. Explain how you arrived at your answer.

SUMMARY

Humans, like all living organisms, contain thousands of compounds. Most of these are synthesized from dietary substances by the body itself. Collectively, the reactions involved constitute anabolism. The products of anabolism are, with few exceptions, continuously degraded and replaced (if needed) with newly synthesized copies of the same substance. This turnover eliminates some molecules that are no longer needed and tends to prevent the accumulation of damaged molecules that exhibit abnormal biological activity. There is growing evidence that the accumulation of alterations in DNA, one of the few substances in the body that does not completely turn over, contributes to biological aging.

Carbohydrate Metabolism

Gluconeogenesis is that anabolic pathway involved in the synthesis of glucose from noncarbohydrate precursors, including glycerol, lactate, and some protein amino acids. Humans are not genetically programmed to convert fatty acids to gluconeogenic precursors or carbohydrates. Most gluconeogenic precursors are converted to pyruvate in the process of incorporation into glucose. Pyruvate is transformed into glucose with the participation of the seven most readily reversible steps in glycolysis plus four unique reactions. Collectively, the unique reactions bypass the three irreversible (from a practical standpoint) steps in glycolysis. The hydrolysis of 2 ATP and 2 GTP helps generate a negative $\Delta G^{\circ\prime}$ for glucose production. When glycerol is the precursor for glucose synthesis, it is converted to dihydroxyacetone phosphate, an intermediate in the production of glucose from pyruvate. Liver is the primary site of gluconeogenesis.

Multiple allosteric effectors, including AMP, fructose 2,6-bisphosphate, ATP, citrate, and acetyl-SCoA, help avoid a futile cycle by ensuring that glycolysis and gluconeogenesis do not both proceed at similar rates at the same time. Phosphofructokinase-1 (glycolysis) and fructose-bisphosphatase (gluconeogenesis) are the primary targets for most of these effectors. Fructose 2,6-bisphosphate concentration is primarily controlled in liver by glucagon, a hormone that enhances the synthesis of glucose and its release into blood.

Lactate from anaerobic muscle is transported through the blood to the liver, where it is converted to pyruvate and then to glucose. The glucose is transported back to muscle, where it undergoes glycolysis to generate lactate and ATP. This cycle, known as the Cori cycle, helps shift part of the burden of ATP formation in anaerobic muscle from muscle to the liver. The same cycle helps return lactate levels to normal values after active muscle cells acquire an adequate supply of O_2.

Glycogenesis, the synthesis of glycogen from glucose, involves the UDPglucosyltransferase-catalyzed transfer of a glucosyl unit from UDP-glucose to a tyrosine side chain within a protein named glycogenin. After glycogenin catalyzes the stepwise $\alpha(1 \longrightarrow 4)$ attachment of roughly seven glucose residues to the initial glucose residue, glycogen synthase begins to catalyze the $\alpha(1 \longrightarrow 4)$ attachment of additional glucose residues to the growing polymer chain. Amylo-(1,4–1,6)-transglycosylase catalyzes the creation of $\alpha(1 \longrightarrow 6)$ branches.

Epinephrine, glucagon, and insulin play dominant roles in the reciprocal regulation of glycogenolysis and glycogenesis. Epinephrine (liver and muscle) and glucagon (liver) trigger reaction cascades that lead to the phosphorylation of both glycogen phosphorylase (the key enzyme in glycogenolysis) and glycogen synthase (the central enzyme in glycogenesis). Because the phosphorylase is activated while the synthase is inhibited, the anabolic/catabolic balance is shifted in favor of catabolism. The mechanism of action of insulin, which has an opposite effect, is under investigation.

Fat Metabolism

Most human fatty acids are assembled in the cytosol of hepatocytes and adipocytes from acetyl-SCoA generated in mitochondria. Most of the acetyl-SCoA is derived from the catabolism of dietary carbohydrates and proteins. The citrate-linked shuttle system, which moves acetyl-SCoA from mitochondria to the cytosol, provides part of the NADPH (reducing power) required for fatty acid biosynthesis. The remainder of the NADPH is produced within the pentose phosphate pathway.

In the cytosol, part of the acetyl-SCoA is converted to malonyl-SCoA, and then both the acetyl and malonyl groups are transferred from CoASH to separate molecules of acyl carrier protein (ACPSH). Fatty acid assembly begins with a condensation reaction between acetyl-SACP and malonyl-SACP that yields CO_2 and acetoacetyl-SACP. A round of reduction–dehydration–reduction generates butyryl-SACP. Additional cycles of condensation–reduction–dehydration–reduction add two carbon atoms at a time to the growing fatty acid chain. Even-numbered, saturated fatty acids with up to 16 carbons are produced through this process. In mammals, the synthesis is performed with the catalytic support of a multifunctional enzyme.

Part of the palmitate released by hydrolysis from palmitoyl-SACP (C_{16}-acyl-SACP) is used by the human body to generate longer chained fatty acids and certain unsaturated ones. Odd-numbered fatty acids are generated by microorganisms when propionyl-SACP, instead of acetyl-SACP, condenses with malonyl-SACP during the initial condensation step in fatty acid biosynthesis. The body is not genetically programmed to synthesize linoleate and linolenate, two of the polyunsaturated fatty acids that it requires. These essential fatty acids must be obtained in the diet.

The reciprocal regulation of β-oxidation and fatty acid biosynthesis involves both long-term and short-term modulation. Long-term modulation is linked to hormone-induced changes in enzyme concentrations. Short-term modulation is brought about by allosteric effectors, the covalent modification of existing enzymes, and the regulation of substrate and coenzyme availability. NADH and acetyl-SCoA allosterically inhibit β-oxidation, whereas citrate (+) and palmitoyl-SCoA (−) allosterically modulate fatty acid assembly. Epinephrine and glucagon initiate reaction cascades that stimulate β-oxidation and suppress fatty acid production in liver cells. Insulin has an opposing effect. Substrate availability is partially controlled by the modulation of the transport of fatty acids into mitochondria and the export of acetyl-SCoA from mitochondria. Collectively, the multiple regulatory strategies help cells avoid a futile cycle, where β-oxidation and fatty acid biosynthesis are proceeding at similar rates at the same time.

Fat biosynthesis entails the transfer of two acyl groups from acyl-SCoAs to glycerol 3-phosphate to produce a phosphatidate. Hydrolysis of the phosphatidate yields a diacylglycerol that can be acylated to generate a triacylglycerol. Phosphatidates are also intermediates in the production of some phospholipids.

Amino Acid Metabolism

Since the human body cannot synthesize adequate amounts of 9 of the 20 common protein amino acids, these essential amino acids must be acquired through the diet. All of the 11 nonessential amino acids, except tyrosine, can be synthesized from α-ketoglutarate, oxaloacetate, 3-phosphoglycerate, or pyruvate in reactions or reaction pathways that include an aminotransferase-catalyzed reaction.

Purine and Pyrimidine Metabolism

Purines and pyrimidines serve as building blocks for both RNA and DNA. The purine ring is constructed from atoms donated by aspartate, glutamine, glycine, CO_2, and N^{10}-formyltetrahydrofolate. IMP, the first purine-containing compound formed, is converted to AMP and GMP through separate two-step pathways. AMP and GMP are phosphorylated to form ATP and GTP. The pyrimidine ring acquires all of its atoms from carbamoyl phosphate and aspartate. OMP, generated by linking the purine ring to a ribose phosphate, is the precursor for UTP, which, in turn, is the precursor for CTP. ATP, GTP, CTP, and UTP are the normal substrates for RNA polymerases.

Deoxyribonucleoside diphosphates are created by reducing the 2′-hydroxyl group in ribonuceloside diphosphates. Deoxyribonucleoside diphosphates are hydrolyzed to yield deoxyribonucleotides or are phosphorylated to form deoxyribonucleoside triphosphates. Deoxy-UMP is converted to dTMP in a reaction where N^5,N^{10}-methylenetetrahydrofolate serves as a methyl donor. The substrates for DNA polymerases are dATP, dGTP, dCTP, and dTTP.

A variety of anticancer drugs block DNA replication in cancer cells by disrupting the assembly or functioning of deoxyribonucleoside triphosphates. Methotrexate (amethopterin), aminopterin, and 5-fluorouracil are specific examples. These chemotherapeutic agents are inhibitors of specific enzymes. Other metabolism-disrupting chemicals are used to treat both bacterial and viral infections.

EXERCISES

14.1 True or false? The intracellular concentration of a protein with a short half-life will always be less than the intracellular concentration of a protein with a much longer half-life. Explain.

14.2 Can you be confident that the activity of a pool of a cellular protein will remain constant as long as the rate of its production is exactly equal to its rate of degradation?

14.3 True or false? Any compound that can be converted to a four-carbon citric acid cycle intermediate can serve as a precursor for gluconeogenesis. Explain. List those compounds that are normally the major precursors for gluconeogenesis.

14.4 Which tissue or organ is the primary site for gluconeogenesis in humans? What is the primary fate of most of the glucose synthesized at this site?

14.5 Study Figure 14.3. To which of the six major international classes of enzymes does each of the enzymes in gluconeogenesis belong? Hint: Review Section 5.1.

14.6 Different species have relatively fixed and inherited maximum life spans. Construct a molecular aging hypothesis based on this observation.

14.7 From a thermodynamic standpoint, what determines whether a step in a metabolic pathway is reversible or irreversible?

14.8 True or false? Because the conversion of 1,3-bisphosphoglycerate and ADP to 3-phosphoglycerate and ATP is considered one of the reversible steps in glycolysis, 1,3-bisphosphoglycerate must be a high-energy compound. Explain.

14.9 Under anaerobic conditions, what prevents pyruvate from glycolysis from being used for gluconeogenesis?

14.10 Which, if any, of the enzymes involved in gluconeogenesis from pyruvate are located within the cytosol of a liver cell?

14.11 Once phosphorylated, glucose becomes trapped within a muscle cell. In contrast, phosphorylation is unable to prevent this monosaccharide from escaping from hepatocytes. Explain.

14.12 Glycerol from fat can be completely oxidized or converted to glucose. List the metabolic pathways that are involved in each process. Identify two metabolites that, when present in high concentrations, cause a cell to oxidize glycerol rather than convert it to glucose.

14.13 Phosphofructokinase-1 is a regulatory enzyme that is activated by AMP and fructose 2,6-bisphosphate and inhibited by citrate. Predict whether these three effectors will bind to different allosteric sites or compete for binding to the same allosteric site. Explain how you arrived at your answer.

14.14 List the five allosteric effectors described in this chapter that have direct impacts on the rate of glycolysis or gluconeogenesis. Circle those effectors which favor glycolysis when they are present at high concentrations. Which of the circled effectors, if present at high levels, signal a high energy charge within a cell?

14.15 Under what conditions is a substrate cycle a futile cycle?

14.16 What is the physiological importance of the Cori cycle?

14.17 What property of muscle cells makes the Cori cycle irreversible?

14.18 Explain why the Cori cycle tends to be nonfunctional during aerobic exercise.

14.19 List those factors that primarily determine the oxygen concentration within a cell at any given time.

14.20 Why is the body unable to use the acetyl-SCoA from β-oxidation to build glucose? Could acetyl-SCoA from glucose, itself, be used to rebuild glucose?

14.21 Summarize the reactions involved in converting dietary glucose to triacylglycerols. Use the summary to explain how the conversion of 10 g of glucose to fat leads to the production of less than 10 g of fat.

14.22 Biotin-containing enzymes are involved in both gluconeogensis and fatty acid biosynthesis. What is biotin? What types of reactions are usually catalyzed by biotin-containing enzymes?

14.23 Study Figure 14.13. List those classes of organic compounds that are intermediates during the biosynthesis of palmitate. For each class listed, record two reactions characteristic of that class. During this exercise, ignore the functional groups within ACPSH and CoASH.

14.24 Identify those intermediates of β-oxidation (if any) that are also intermediates in palmitate biosynthesis (Figures 14.13 and 14.16).

14.25 Anabolism, in general, is an energy-consuming process. Identify those steps in gluconeogenesis from pyruvate and in palmitate biosynthesis from acetyl-SCoA that represent the major energy-consuming steps within these pathways.

14.26 Mammalian fatty acid synthase is a multifunctional enzyme. What specific class of bonds covalently joins the multiple enzymes within mammalian fatty acid synthase?

14.27 Assign each of the enzymes involved in palmitate biosynthesis to one of the six major international classes of enzymes.

14.28 What specific participants in palmitate biosynthesis prevent NADH, rather than NADPH, from being used as the reducing agent within this metabolic pathway?

14.29 What major class of biochemicals is known to be produced from the essential fatty acids?

14.30 Individuals who exercise regularly tend to burn fat more readily than those who avoid regular exercise. Provide a molecular explanation for this fact.

14.31 Should the reciprocal regulation of gluconeogenesis and glycolysis described in this chapter be classified as short-term regulation, long-term regulation, or a combination of both? Explain.

14.32 What potential short-term or long-term risks, if any, would result if a cell allowed β-oxidation and fatty acid biosynthesis to proceed simultaneously at similar rates?

14.33 Draw a graph clearly illustrating the results obtained when initial velocity is plotted versus substrate concentration for a reaction catalyzed by a Michaelis-Menten enzyme. How does this graph relate to the claim that the rate of a metabolic pathway can be regulated by controlling substrate availability? Hint: Review Section 5.6.

14.34 How many net ATP are consumed (hydrolyzed) during the production of a triacylglycerol from glycerol and a pool of free fatty acids?

14.35 What specific bond is cleaved and what specific bond is formed each time an acyl group is transferred from a CoASH to a glycerol during triacylglycerol biosynthesis? Is the $\Delta G^{\circ\prime}$ for this reaction positive or negative? Explain.

14.36 What accounts for the ability of most bacteria to synthesize those amino acids that are essential nutrients for humans?

14.37 Clearly describe the structural relationship between each of the following: nucleotide, nucleoside, nucleoside diphosphate, nucleoside triphosphate, ATP, AMP, RNA, and nucleic acid. Hint: Review Chapter 7.

14.38 What specific compound is the source of the ribose phosphate residue in both purine and pyrimidine nucleotides? Explain why this compound represents an activated form of ribose phosphate. What metabolic pathway supplies ribose 5-phosphate?

14.39 List the two chemical modifications required to convert RNA monomers to DNA monomers.

14.40 What is the most likely explanation for the occasional incorporation of a dUMP residue into DNA at positions normally occupied by dTMP?

14.41 Suggest why intestinal disturbances and hair loss are common side effects of the anticancer drugs discussed in this chapter. Hint: Hair cells and the cells that line the intestine are normally undergoing continuous, relatively rapid division.

14.42 What enzyme is the ultimate target for the anticancer drug 5-fluorouracil? How does the inhibition of this enzyme block the replication of DNA?

14.43 What effect, if any, would a high intracellular dihydrofolate concentration have on the ability of methotrexate to inhibit the production of DNA?

14.44 Draw a graph clearly showing the results expected when dihydrofolate reductase activity (y-axis) is plotted versus intracellular methotrexate concentration (x-axis).

14.45 List three enzymes, in addition to dihydrofolate reductase and thymidylate synthase, that are appropriate targets for drugs designed to shut down DNA replication in cancer cells.

14.46 Describe the central role that enzymes play in the process called life.

SELECTED READINGS

Culotta E., and D. E. Koshland, Jr., DNA repair works its way to the top, *Science* 266, 1926–1929 (1994).

Discusses the "molecules of the year" and the "runners-up."

Gurr, M. I., and J. L. Harwood, *Lipid Biochemistry: An Introduction,* 4th ed. London: Chapman & Hall, 1991.

An introduction to lipid metabolism.

Lehninger, A. L., D. L. Nelson, and M. M. Cox, *Principles of Biochemistry,* 2nd ed. New York: Worth Publishers, 1993.

A thousand-page text with a thorough coverage of metabolism.

Moran, L. A., K. G. Scrimgeour, H. R. Horton R. S. Ochs, and J. D. Rawn, *Biochemistry,* 2nd ed. Englewood Cliffs, NJ, Neil Patterson/Prentice-Hall, 1994.

An encyclopedia-type biochemistry text containing a detailed discussion of metabolism. Includes a limited number of problems and some selected readings.

Smith, C. A., and E. J. Wood, *Energy in Biological Systems.* New York: Chapman & Hall, 1991.

A 171-page paperback that summarizes the basic metabolic pathways. Contains both questions and references.

Stryer, L., *Biochemistry,* 4th ed. New York: W. H. Freeman and Company, 1995.

Provides a thorough coverage of metabolism.

Werner, R., *Essential Biochemistry and Molecular Biology,* 2nd ed. New York: Elsevier Publishing Co., 1992.

A 463-page paperback that provides a concise and readable review of basic metabolism.

15 Photosynthesis and Nitrogen Fixation

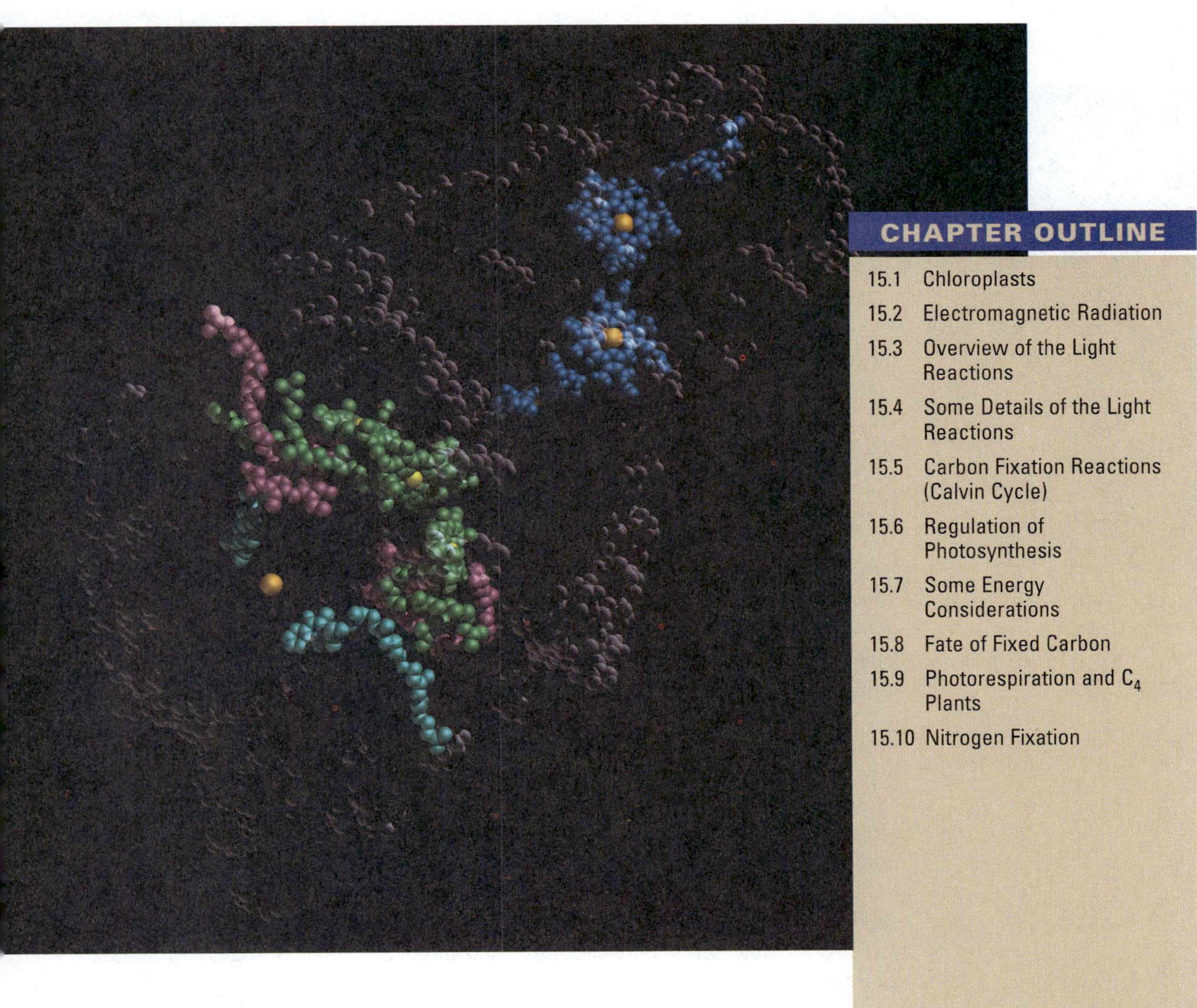

CHAPTER OUTLINE

Chapters 10 through 14 have provided an introduction to carbohydrate, fat, amino acid, and nucleotide metabolism. Although part of the metabolic pathways described are shared by all living organisms, other pathways are less widely distributed. The metabolic capabilities of an organism are determined by the DNA it has inherited, for this DNA encodes the enzymes responsible for catalyzing the individual reactions that, collectively, constitute metabolism. Heterotrophs and autotrophs differ because they contain distinct combinations of genes that encode distinct combinations of enzymes. Similarly, nitrogen-fixing organisms are programmed to synthesize certain enzymes that cannot be assembled from the genetic information within organisms that are unable to fix nitrogen.

Heterotrophs are living organisms that rely on organic compounds to meet their carbon and energy needs. In contrast, autotrophs acquire their carbon from atmospheric CO_2 and obtain energy from a source other than carbon compounds (Chapter 8). Autotrophs that employ light as a source of energy are called **phototrophs.** Phototrophs, including algae, plants, Cyanobacteria, and several groups of green and purple bacteria, are genetically programmed to synthesize compounds that enable them to perform photosynthesis, that process through which light energy is used to generate carbohydrates and O_2 from atmospheric CO_2 and water:

$$n\,CO_2 + n\,H_2O + \text{light} \rightleftharpoons (CH_2O)n + n\,O_2$$

Through the food chain, autotrophs provide humans, fish, and other heterotrophs with life-supporting organic compounds.

The carbohydrates synthesized through photosynthesis supply phototrophs with the carbon required to produce other classes of organic compounds. These carbohydrates can also be oxidized to release the energy they have acquired from sunlight.

Through the food chain, autotrophs provide humans and other heterotrophs with life-supporting organic compounds. Phototrophs also supply the O_2 that is used by most heterotrophs to oxidize these compounds:

$$\text{Organic compounds} + O_2 \rightleftharpoons CO_2 + H_2O + \text{energy}$$

The oxidation of organic compounds by heterotrophs constitutes the second half of the biological carbon cycle, the cycling of carbon between atmospheric CO_2 and organic compounds within living organisms (Figure 8.1). As autotrophs continuously "fix" carbon into organic compounds, heterotrophs constantly return the "fixed" carbon to the atmosphere. An oxygen cycle is coupled to the carbon cycle, since O_2 is consumed and then regenerated within this cycle. The production of O_2 by primitive phototrophs is thought to have created an O_2-rich and organic compound–rich environment that contributed to the evolution of oxidative heterotrophs and the beginning of the biological carbon and oxygen cycles.

A third biologically important elemental cycle, the **nitrogen cycle,** is superimposed upon the carbon and oxygen cycles (Figure 15.1). This cycle is initiated by a few species of bacteria that inherit the ability to convert atmospheric nitrogen (N_2) to ammonia (NH_3), a process called **nitrogen fixation.** Other participants in the biological nitrogen cycle are: bacteria that oxidize ammonia to nitrite and nitrate (a process called **nitrification**); bacteria and plants that reduce nitrate and nitrite to ammonia; bacteria that transform nitrate and nitrite into molecular nitrogen (a process called **denitrification**); and all the organisms that incorporate fixed nitrogen into proteins, nucleic acids, phospholipids, and numerous other nitrogen-containing organic compounds.

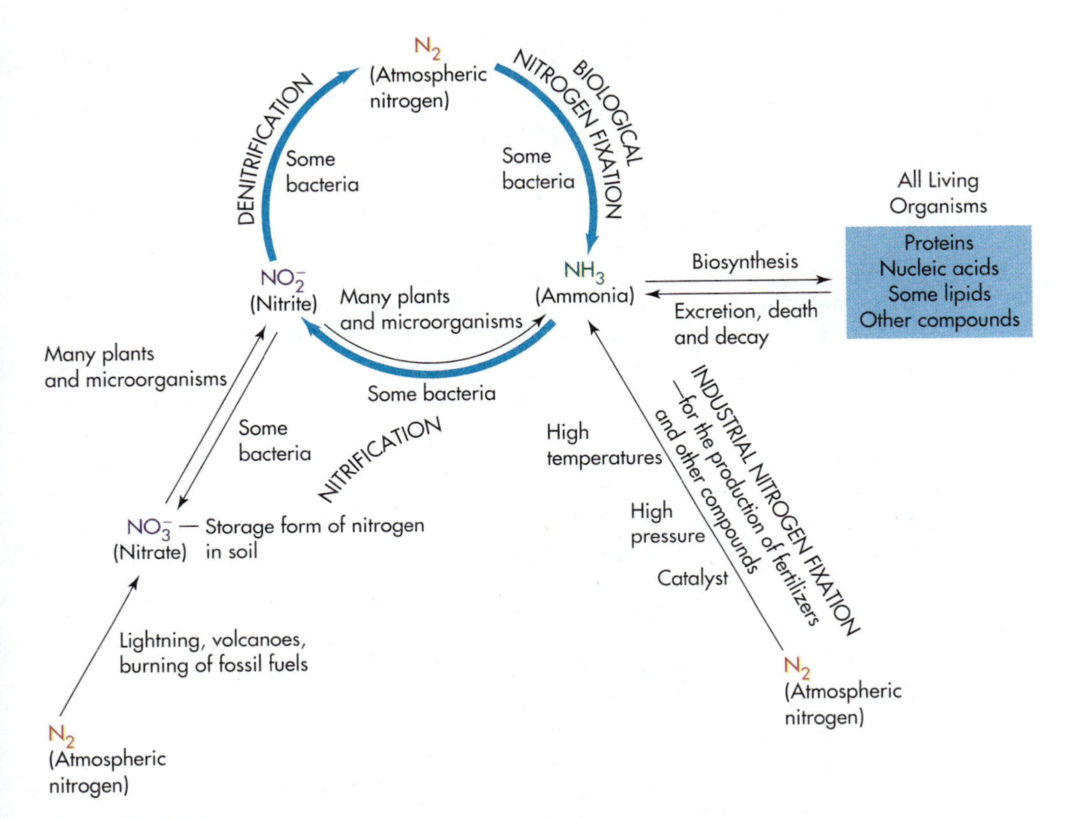

FIGURE 15.1
The nitrogen cycle.

Although metabolic reactions within living organisms are largely responsible for the removal of nitrogen from the atmosphere, a variety of other processes contribute to this removal. Lightning storms, volcanoes, and the burning of fossil fuels, for example, lead to the formation of nitrogen oxides that are converted to nitrite and nitrate ions by reactions that occur within our environment. Additional nitrogen is removed from the atmosphere as humans employ high temperatures, high pressures, and catalysts to fix nitrogen for use in fertilizers and other commercial products.

The nitrogen used by most heterotrophs comes from nitrogenous compounds in their diets. The nitrogen in these compounds can be traced to bacteria or plants further down the food chain. When organisms die, their fixed nitrogen is recycled by bacteria and other organisms or is transformed into N_2 through denitrification by bacteria. Since much of the fixed nitrogen is recycled, the continual biological fixation of nitrogen is said to "top off" the biological nitrogen supply. Chapters 13 and 14 have examined some of the nitrogen-linked metabolism in humans and other organisms.

This chapter continues the discussion of anabolism that was initiated in Chapter 14 by examining photosynthesis, nitrogen fixation, and some related topics. Featured biochemical themes include hypothesis construction, energy use and transformation, communication, compartmentation, specificity, and evolution and change. Chapter 16 moves into a discussion of two of the most important classes of nitrogen-containing compounds, RNA and DNA.

15.1 CHLOROPLASTS

Chloroplasts, a class of plastids (Section 1.3), are plant cell organelles that are responsible for photosynthesis. Chloroplasts share a number of properties with mitochondria: (a) They appear to have evolved from symbiotic prokaryotes; (b) they contain DNA whose genes are expressed by organelle-specific machinery; and (c) they are surrounded by a highly permeable outer membrane (allows the passage of most small molecules and ions) and a substantially less permeable inner membrane. In contrast to mitochondria (Section 11.4), chloroplasts lack cristae (their inner membrane is not invaginated) and they have a third membrane system, the **thylakoid membrane system,** within the inner membrane space. The thylakoid membrane folds to generate a single flattened, membrane-enclosed vesicle, called a **thylakoid.** The fluid-filled volume inside the inner membrane but outside the thylakoid membrane is called the **stroma,** and the fluid-filled volume within a thylakoid is known as the **lumen.** Different regions of a thylakoid stack together to form **grana.** Unstacked regions of the thylakoid make up the **stroma lamellae.** These structural features of chloroplasts, which are repeatedly encountered during the discussion of photosynthesis that follows, are summarized in Figure 15.2.

15.2 ELECTROMAGNETIC RADIATION

Since photosynthesis uses light as a source of energy for the synthesis of carbohydrates, a review of electromagnetic radiation is in order prior to the analysis of photosynthesis.

Properties of Light

Electromagnetic radiation has been introduced in Section 3.11 during the discussion of spectrophotometric analysis. This class of radiation consists of oscillating electric and magnetic fields that carry energy and travel at the speed of light. Electromagnetic radiation has both particulate and wavelike properties. A single particle or packet of radiation is known as a **photon.** The energy content of an electromagnetic wave is inversely proportional to its wavelength; the shorter the wavelength, the more energy its photon contains. This relationship is encompassed in the equation $E = hc/\lambda$ where $h = 6.63 \times 10^{-34}$ J · s, Planck's constant, c is the speed of light (3.0×10^8 m/s in a vacuum), and λ is the wavelength in meters.

Electromagnetic radiation is divided into eight major categories on the basis of differences in wavelength ranges (Table 3.10). Sunlight is a combination of three of these major classes: infrared (IR) radiation, visible light, and ultraviolet (UV) light. Most of the solar energy reaching the surface of the Earth falls within the visible spectrum. Visible light, with wavelengths from approximately 400 nm to 700 nm

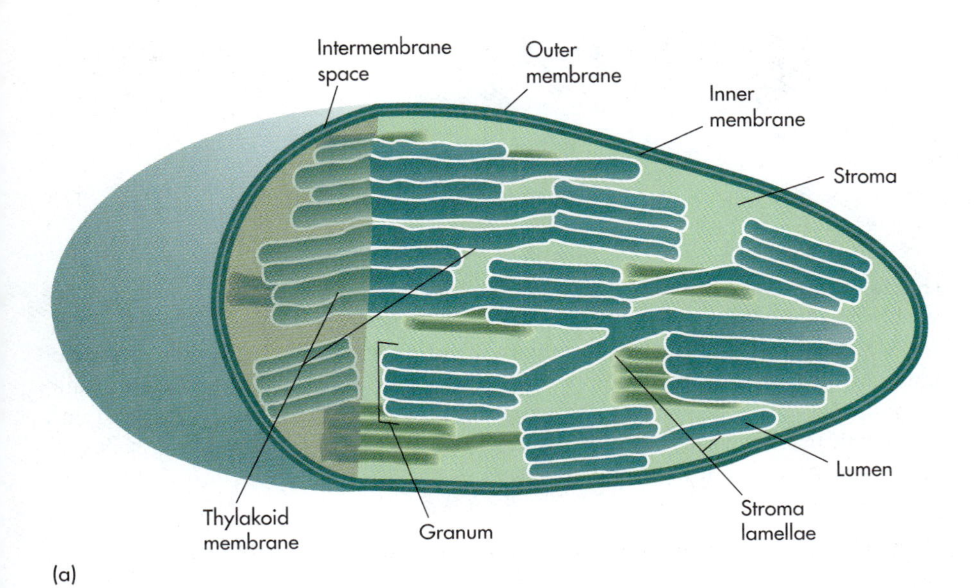

(a)

(b)

FIGURE 15.2
A chloroplast. (a) Schematic diagram. (Based on Moran et al., 1994) (b) Electron micrograph.

Wavelength (nm)	400	500	600	700
Energy (kJ/einstein)	300	240	200	170

FIGURE 15.3
The visible spectrum. One einstein = 1 mol of photons = 6.023×10^{23} photons.

(Figure 15.3), plays the central role in both photosynthesis and human vision. The visible portion of sunlight is perceived as white light because it possesses all of the wavelengths of visible electromagnetic radiation. Individual wavelengths of visible light correspond to different colors of light; the shortest wavelengths (that contain the most energy) appear violet, the longest wavelengths (that possess the least energy) appear red, and intermediate wavelengths are perceived as various shades of blue, cyan, green, yellow, or orange. **[Try Problem 15.1]**

Absorption and Emission of Electromagnetic Radiation

When visible light falls upon certain molecules, specific wavelengths of light are absorbed and other wavelengths are reflected or transmitted. This fact is explained by **quantum theory,** which proposes that very small particles, such as atoms and molecules, can exist in a limited number of discrete energy levels. To change from one "allowed" energy level to a second, a molecule must absorb or emit a packet of energy (**quantum**) exactly equal to the energy difference between the two energy levels. The transition is instantaneous and stepwise. Thus, for visible light to be absorbed by a molecule, its photon must possess exactly the right packet of energy to boost the molecule from one energy level to a second energy level of higher energy. A visible electromagnetic wave whose photon energy does not correspond to the energy difference between two energy levels cannot be absorbed.

The absorption of a photon of visible electromagnetic radiation by a molecule leads to an instantaneous, stepwise shift from a ground state energy level to an excited state energy level. The excited molecule tends to rapidly release the energy of excitation and to return to its ground state through one or a combination of mechanisms: part or all of the excitation energy can be converted to heat (molecular motion); part of this energy can be emitted as a photon of light, a process called **fluorescence;** an electron in a high energy orbital can be transferred to a neighboring molecule and replaced with an electron (that occupies a lower energy orbital) from an electron donor; or the excitation energy (without the electron) can be passed to a neighboring molecule, a process known as **resonance energy transfer** (Figure 15.4). The last two mechanisms are of prime importance during photosynthesis. The de-excitation routes utilized by an excited state molecule are determined by its structure and its environment. **[Try Problem 15.2]**

When a photon of ultraviolet light is absorbed by a molecule, the excited state molecule that is produced readily loses an electron. The ion that is formed tends to be unstable and very reactive. When such ions are created within living organisms, they often react with neighboring molecules and, in the process, alter their biological activities. The biological activity of the molecule absorbing the radiation is also altered. For this reason, UV radiation is harmful to most living organisms, including

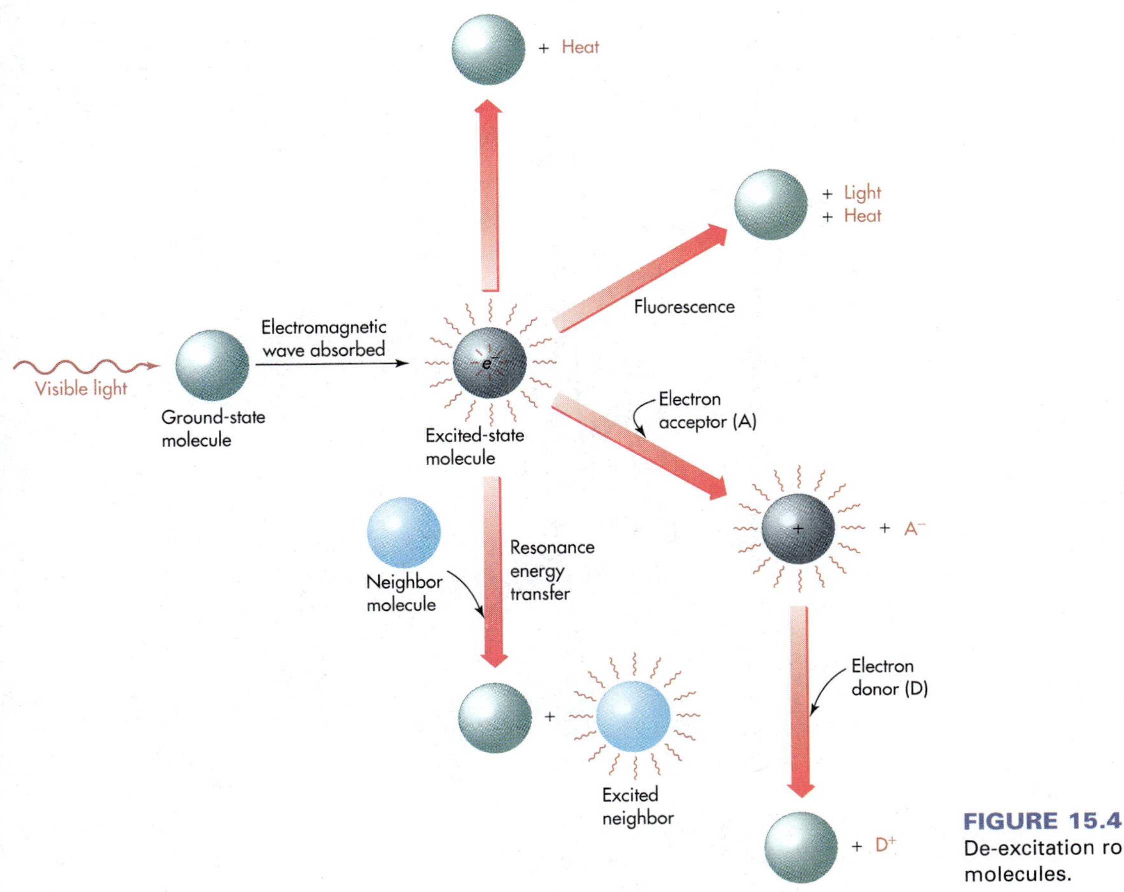

FIGURE 15.4
De-excitation routes for excited molecules.

plants and humans. In humans, UV radiation in sunlight is responsible for sunburns and most cases of skin cancer.

Absorbance and Absorption Spectra

Since different compounds tend to possess unique energy levels or distinct combinations of energy levels, different compounds tend to absorb at least some different wavelengths of electromagnetic radiation. The absorbance of a sample of a compound is a measure of the amount of light that it absorbs. Absorbance is determined by the wavelength of light absorbed, compound concentration, the compound's absorptivity (an experimental constant), and the distance the absorbed light travels through the sample of the compound (the pathlength) [Section 3.11]:

$$A = \epsilon bc; \text{ Beer's Law}$$

where A = absorbance (optical density)
ϵ = absorptivity (extinction coefficient)
b = pathlength
c = concentration of absorbing compound

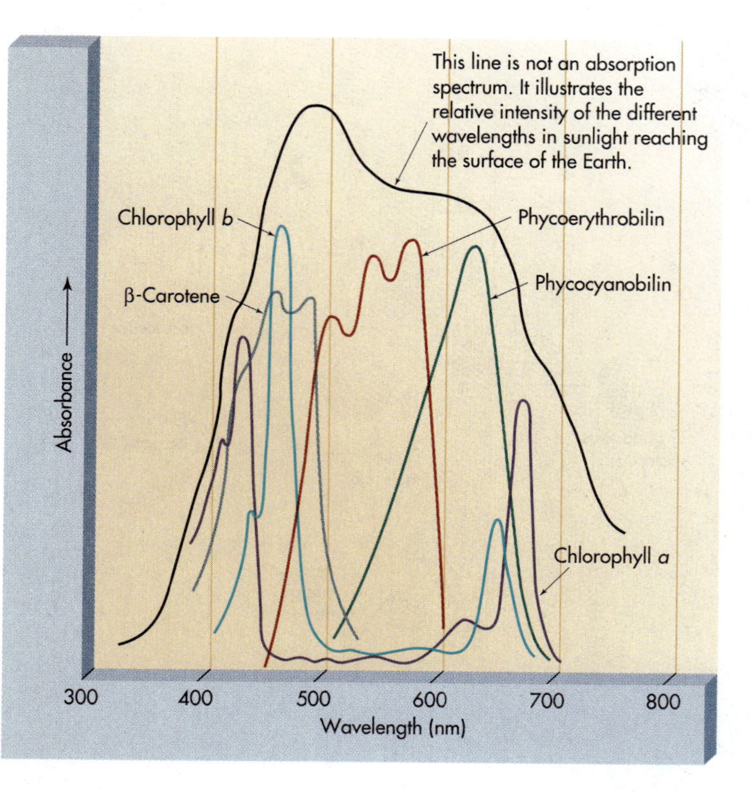

FIGURE 15.5
Absorption spectra for some photosynthetic pigments. (Based on Lehninger et al., 1993.)

Although the wavelength of the light absorbed determines the absorptivity, the value of the wavelength itself does not appear in the absorbance equation. The absorptivity is called molar absorptivity when the concentration of the absorbing compound is expressed as molarity. The absorbance of a mixture of compounds is equal to the sum of the absorbances of each individual compound. **[Try Problem 15.3]**

An **absorption spectrum** is the graph obtained when absorbance (y-axis) is plotted against the wavelength of light absorbed (x-axis). Figure 15.5 provides absorption spectra for some of the photosynthetic pigments (colored compounds) found in one or more classes of phototrophs. Different phototrophs usually possess unique combinations of pigments or unique ratios of pigments. Collectively, these compounds absorb efficiently most wavelengths of visible electromagnetic radiation. Distinct combinations of these pigments are largely responsible for the different colors associated with leaves and bacteria. Light absorbed by these pigments provides the energy for photosynthesis. **[Try Problems 15.4 and 15.5]**

PROBLEM 15.1 List the five major classes of electromagnetic radiation in addition to UV, IR, and visible light. Which class possesses the shortest wavelengths? Hint: Review Section 3.11.

PROBLEM 15.2 Explain why some molecules absorb no visible light. What is the color of such a compound in sunlight? Explain.

PROBLEM 15.3 Calculate the absorbance of a 0.1 *M* solution of a compound with a molar absorptivity of 0.8 $Lmol^{-1}cm^{-1}$. Assume that a standard cuvet (1-cm pathlength) is used for the absorbance measurement.

PROBLEM 15.4 Which one of the photosynthetic pigments in Figure 15.5 is best able to harvest (absorb) green light? orange light? Assume that similar pigment concentrations have been used to prepare all of the given curves. Why is this assumption important?

PROBLEM 15.5 On Figure 15.5, place the absorption spectrum obtained for a solution containing both β-carotene and phycoerythrin with each pigment present at the same concentration used to prepare its individual absorption spectrum.

15.3 OVERVIEW OF THE LIGHT REACTIONS

Photosynthesis is divided into two separate but coordinated sets of reactions known as the **light-induced electron transfer reactions (light reactions)** and the **carbon fixation reactions (dark reactions)**. The two sets of reactions are coupled to the extent that the ATP and NADPH produced during the light reactions are required for the carbon fixation reactions. The light reactions require light, whereas the carbon fixation reactions can proceed in either the presence or absence of light. *In vivo,* regulatory mechanisms usually shut down both sets of reactions when a phototroph moves from the light into the dark. This section and Section 15.4 explore the light reactions, and Section 15.5 discusses the carbon fixation reactions. **[Try Problem 15.6]**

The Z-Scheme

In green plants, the light-induced electron transfer reactions are carried out by two physically separated, thylakoid membrane–bound **photosystems** plus some accessory complexes and molecules. The photosystems, labeled **photosystem I** (the first system discovered) and **photosystem II,** are basic functional units for photosynthesis. Photosystem II precedes photosystem I in the electron transfer scheme. The existence of multiple photosystems was first revealed by experiments in which the **quantum yield** (number of product molecules formed per photon absorbed) during photosynthesis was measured in plants exposed to different combinations of visible wavelengths. When 680–700 nm light is combined with 520–600 nm light, the quantum yield is considerably greater than the sum of the quantum yields observed with each range of wavelengths alone.

The light reactions begin when a photon of visible light is absorbed by the **reaction center** of photosystem II (Figure 15.6). The absorbed energy boosts the reaction center pigment, **P680,** to an excited state (abbreviated **P680***) that has a much lower reduction potential than the ground state pigment. **$P680^+$** (not shown in Figure 15.6) and a negative counterion are created as P680* quickly transfers an electron to an appropriately positioned electron acceptor. At this point, light energy is converted to chemical energy. The electron extracted from P680* is passed from one

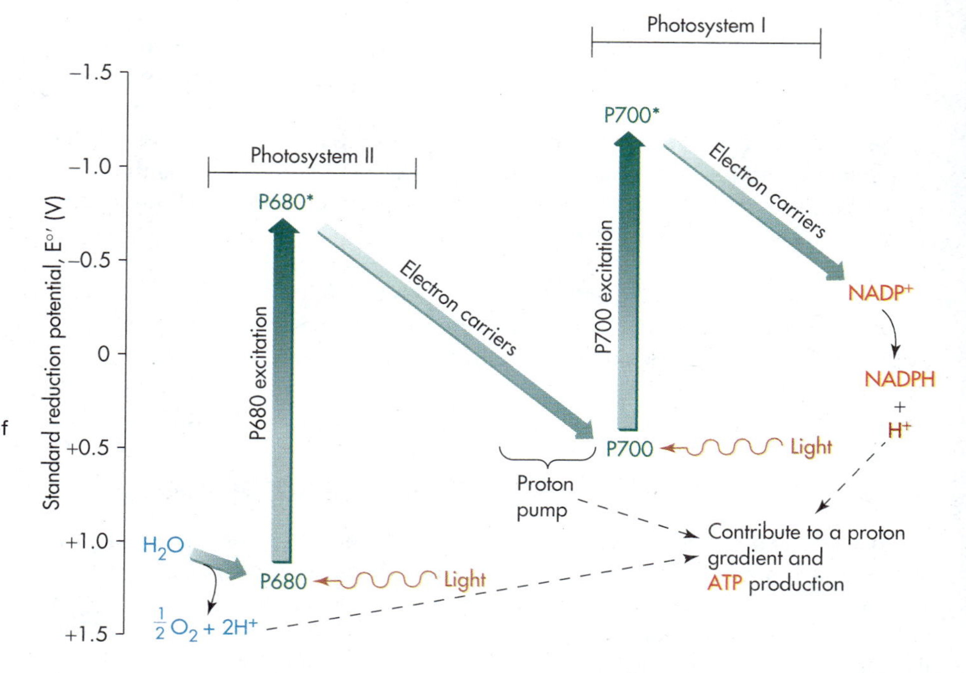

FIGURE 15.6
The Z-scheme. The green arrows depict the light-induced flow of electrons from water to $NADP^+$ through photosystem II, photosystem I, and a sequence of electron carriers. The flow of electrons is always from a carrier of lower reduction potential to one of higher reduction potential. Energy is released at each step. Part of the energy is used to create a proton gradient that possesses a proton-motive force. The proton-motive force is used by ATP synthase to produce ATP from ADP and P_i.

electron carrier to another within an electron transport system similar to the one involved in oxidative phosphorylation within mitochondria (Section 11.8). After it has traversed a number of carriers, the electron from photosystem II is delivered to photosystem I. **[Try Problems 15.7 and 15.8]**

Before the electron delivered to photosystem I can be accepted, the reaction center pigment in photosystem I, **P700,** must be transformed into a cation, **$P700^+$** (not shown in Figure 15.6). This is accomplished with the absorption of a photon of visible light. The light energy boosts P700 to an excited state, **P700*,** and substantially lowers its reduction potential. P700* promptly passes an electron to a properly positioned acceptor that functions as the next carrier in the electron bucket brigade that constitutes the light-induced electron transfer system. $P700^+$ returns to its ground state by accepting the electron delivered to it. The delivered electron is passed on down the transport system after P700 is converted to P700* through another round of photoexcitation. The flow of electrons is always from a carrier of lower reduction potential to one of higher reduction potential, and each transfer of an electron is an energy-releasing redox reaction in which the electron donor is oxidized and the acceptor is reduced. The $\Delta G^{\circ\prime}$ for each redox reaction is determined by the difference in standard reduction potentials for its two half-reactions (review Section 8.8):

$$\Delta G^{\circ\prime} = -n\mathfrak{F}\,\Delta E^{\circ\prime}$$

where n = Number of electrons transferred
$\mathfrak{F}$ = Faraday constant = 96.48 kJ/V·mol
$\Delta E^{\circ\prime}$ = Difference in reduction potential

The ultimate electron acceptor is $NADP^+$, which is converted to NADPH + H^+ following the acceptance of two electrons. The formation of NADPH + H^+ contributes to a proton gradient between the stroma and lumen. The flow of electrons is described as the **Z-scheme,** since a Z-shaped diagram is obtained when reduction potential is plotted against the flow of electrons within the transport system as in Figure 15.6. **[Try Problem 15.9]**

Photophosphorylation

Part of the energy released during the transfer of electrons within the Z-scheme is used to "pump" protons from the stroma into the lumen, an event that helps generate a proton-motive force between these compartments (Figure 15.7). NADPH production contributes to the same force (see previous paragraph). A Mg^{2+} gradient is created along with a proton gradient as Mg^{2+} ions move out of the lumen into the stroma to help balance the charges on the separate sides of the thylakoid membrane. Both the proton and the Mg^{2+} gradients are involved in the regulation of photosynthesis (Section 15.6), and the proton gradient provides energy for the production of ATP from ADP and P_i. During ATP synthesis, protons flow from the lumen to the stroma through ATP synthase, a membrane-spanning protein complex associated with the thylakoid membrane. The production of ATP in chloroplasts is described as **photophosphorylation** because, indirectly, photons are the source of the energy used for the phosphorylation of ADP. Photophosphorylation is considered part of the light reactions. **[Try Problem 15.10]**

Cyclic Electron Transport

When NADPH concentrations are high and $NADP^+$ concentrations are low, the rate of electron transport through the Z-scheme and the rate of the coupled production of ATP drop as a consequence of a low concentration of the terminal electron acceptor ($NADP^+$) for the Z-scheme. Under these conditions and certain other conditions,

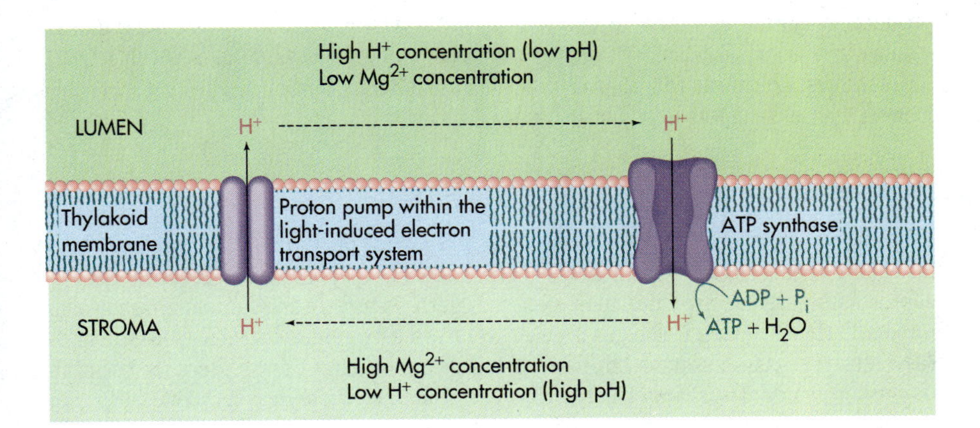

FIGURE 15.7
ATP production. The light-induced electron transport system leads to a net movement of protons (H^+ ions) from the stroma into the lumen. ATP is produced from ADP and P_i as the protons flow back into the stroma through ATP synthase. The creation of a H^+ gradient leads to the simultaneous production of a Mg^{2+} gradient as Mg^{2+} moves (movement not shown) from the lumen into the stroma to help balance charges on the two sides of the thylakoid membrane.

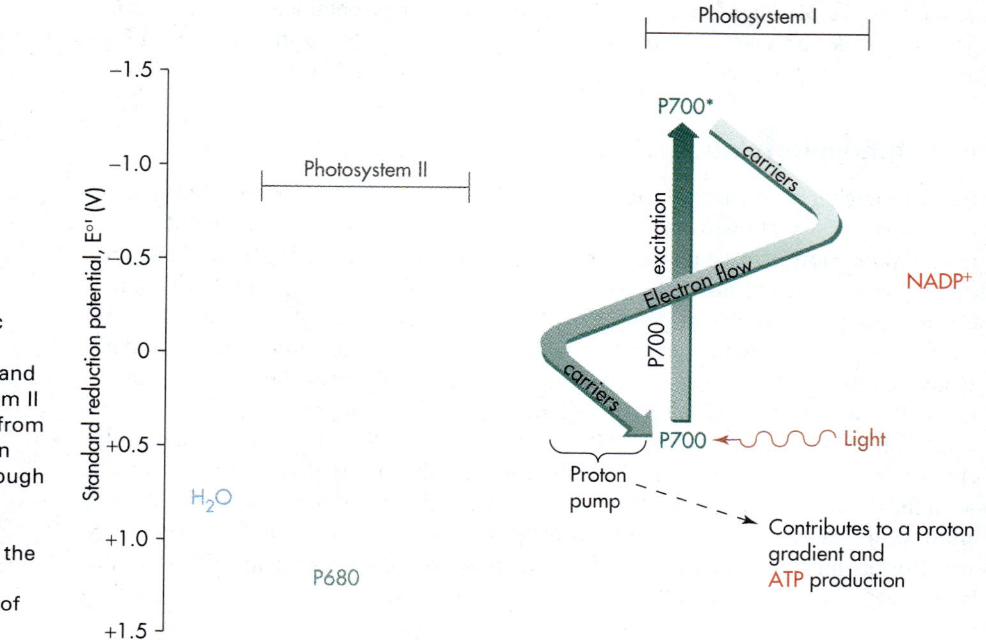

FIGURE 15.8
Cyclic electron transport. Cyclic electron transport leads to ATP synthesis in the absence of O_2 and NADPH production. Photosystem II is not involved. Electrons flow from P700* to $P700^+$ (produced when P700* donates an electron) through an electron transport system coupled to a proton pump. The proton-motive force created by the pump is used for the ATP synthase–catalyzed production of ATP from ADP and P_i.

photosystem I participates in a cyclic electron transport process that is independent of photosystem II and $NADP^+$ (Figure 15.8). During cyclic transport, electrons released by P700* are transported back to $P700^+$ (from which the electrons were originally acquired) through a sequence of electron carriers that bypass $NADP^+$. Since the electrons flow through a proton pump on the way to $P700^+$, light-powered ATP production continues in the absence of O_2 and NADPH formation. The ATP that is synthesized boosts the energy charge of the cell and helps maintain an ATP/NADPH ratio that is optimum for photosynthesis and other essential anabolic processes. Hence, cyclic transport is physiologically important.

O_2 Production

What becomes of the P680 that absorbed light energy to initiate the Z-scheme? As noted above, it is initially converted to $P680^+$ as P680* passes an electron to an acceptor within the electron transport system. $P680^+$ is transformed into a ground-state molecule (P680) when it accepts an electron from a protein associated with the photosystem II reaction center. Ultimately, this protein acquired the electron from an oxygen in water. Thus, water is the ultimate source of the electrons passed through the electron transfer system to $NADP^+$. Water molecules are converted to O_2 and H^+ as they lose electrons:

$$2\ H_2O \rightleftharpoons O_2 + 4\ H^+ + 4\ \text{electrons}$$

Since the H^+ are released into the lumen, they contribute to the proton gradient used to produce ATP from ADP and P_i (Figure 15.6).

Net Reaction

The following equations represent the net result of the light-induced electron transfer reactions:

$$2\ H_2O + 2\ NADP^+ + 8\ \text{photons} \rightleftharpoons O_2 + 2\ NADPH + 2\ H^+ + \text{proton gradient}$$

The proton gradient is used to synthesize ATP:

$$\text{Proton gradient} + ADP + P_i \rightleftharpoons ATP + H_2O$$

Although the stoichiometry surrounding the proton gradient and ATP production is uncertain, it appears that the noncyclic transfer of two electrons from water to $NADP^+$ creates a proton-motive force capable of supporting the production of 1 to 2 ATP. The ATP and NADPH generated contain part of the light energy harvested in photosystems I and II. Since two quanta (photons) are absorbed for each electron transferred, roughly four quanta are absorbed for every 1 to 2 ATP generated. Thus, the quantum yield for ATP production is 0.25 to 0.50. Four photons are absorbed for each NADPH produced. Photosynthesis, like so many other processes, entails energy transformations, one of the most pervasive central themes of biochemistry (Exhibit 1.4). Both the ATP and the NADPH produced during the light reactions are required for the carbon fixation reactions of photosynthesis and other anabolic processes. **[Try Problem 15.11]**

PROBLEM 15.6 Explain why the term "dark reaction" can be misleading. Some biochemists now recommend that this term be abandoned.

PROBLEM 15.7 True or false? In general, the transfer of an electron between two uncharged molecules leaves the donor with a 1+ charge and the acceptor with a 1− charge. Explain.

PROBLEM 15.8 Define standard reduction potential. Explain why P680* has a lower reduction potential than P680. Hint: Review Section 8.8.

PROBLEM 15.9 True or false? A compound with a low reduction potential is easily oxidized. Explain.

PROBLEM 15.10 Explain why ATP production is classified as a light reaction even though ATP synthase can operate in the dark.

PROBLEM 15.11 Explain why it takes eight photons to move four electrons from water to $NADP^+$ through the photosynthetic electron-transport system.

15.4 SOME DETAILS OF THE LIGHT REACTIONS

The details of the light-induced electron transfer reactions sometimes differ markedly from one phototroph to the next. This section examines certain aspects of the light reactions as they proceed in green plants.

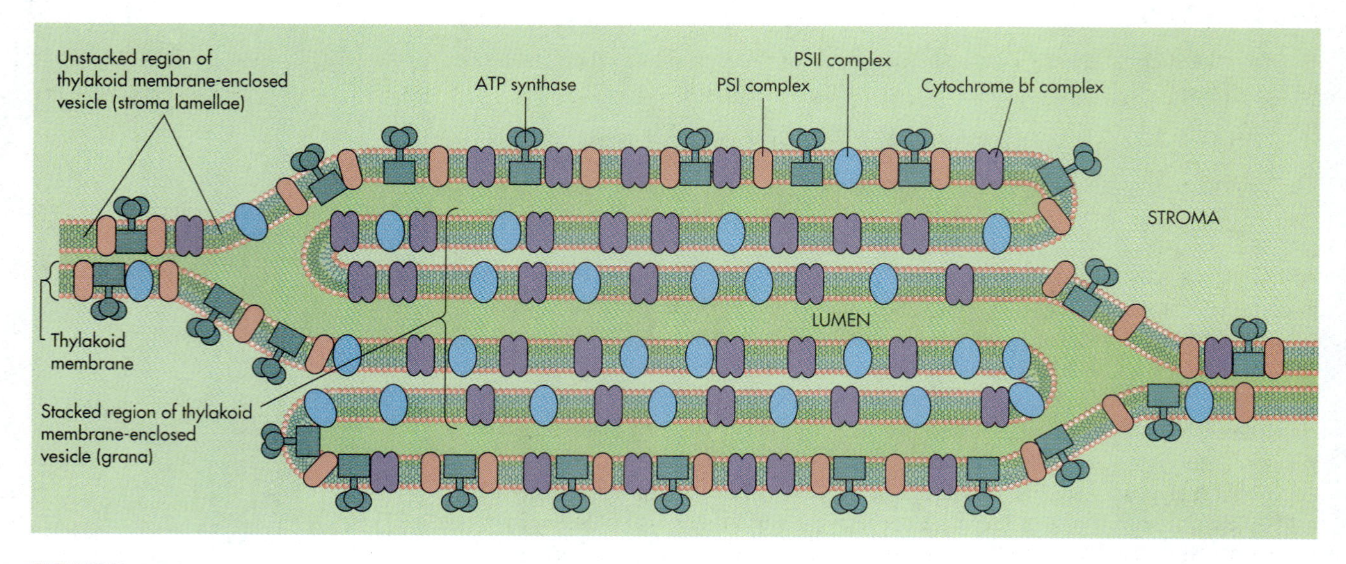

FIGURE 15.9
Membrane distribution of photosynthetic complexes. (Based on Voet & Voet, 1990.)

Light-induced ATP production *in green plants* involves four protein-containing, thylakoid membrane-bound complexes: photosystem I, photosystem II, the **cytochrome *bf* complex,** and ATP synthase. These separate, membrane-spanning structures are, with the exception of the cytochrome *bf* complex, unequally distributed between the stacked and unstacked regions of the thylakoid (Figure 15.9). Photosystem I, photosystem II, and the cytochrome *bf* complex are electron transport complexes whereas ATP synthase catalyzes the production of ATP. Mobile electron carriers transport electrons between the electron-transport complexes. Light-harvesting pigments are associated with both photosystems.

Photosynthetic Pigments

A typical chloroplast from a green plant contains hundreds of photosynthetic units, with each unit possessing 200 to 300 pigment molecules that participate in the harvesting of light. The most abundant pigments are two green compounds named **chlorophyll *a*** (Chl *a*) and **chlorophyll *b*** (Chl *b*) (Figure 15.10). In addition to chlorophyll, phototrophs contain a variety of accessory pigments, including β-carotene (in green plants), phycoerythrobilin (in certain algae and bacteria), and phycocyanobilin (also in algae and in some bacteria). β-Carotene, a precursor of vitamin A, is discussed in Sections 10.1 and 10.2. All of the photosynthetic pigment molecules possess extensive conjugated bonding (alternating single and double bonds) that primarily accounts for their ability to absorb visible light.

Chlorophylls contain a conjugated, planar ring system, the **chlorin ring,** that encompasses four pyrrole rings (numbered I through IV) plus one additional ring. The chlorin ring is similar to the protoporphyrin ring in the heme group of myoglobin and hemoglobin (Sections 4.8 and 4.9). However, in chlorins, one pyrrole ring (ring IV) is partially reduced. Furthermore Mg^{2+}, rather than Fe^{2+}, lies at the center of

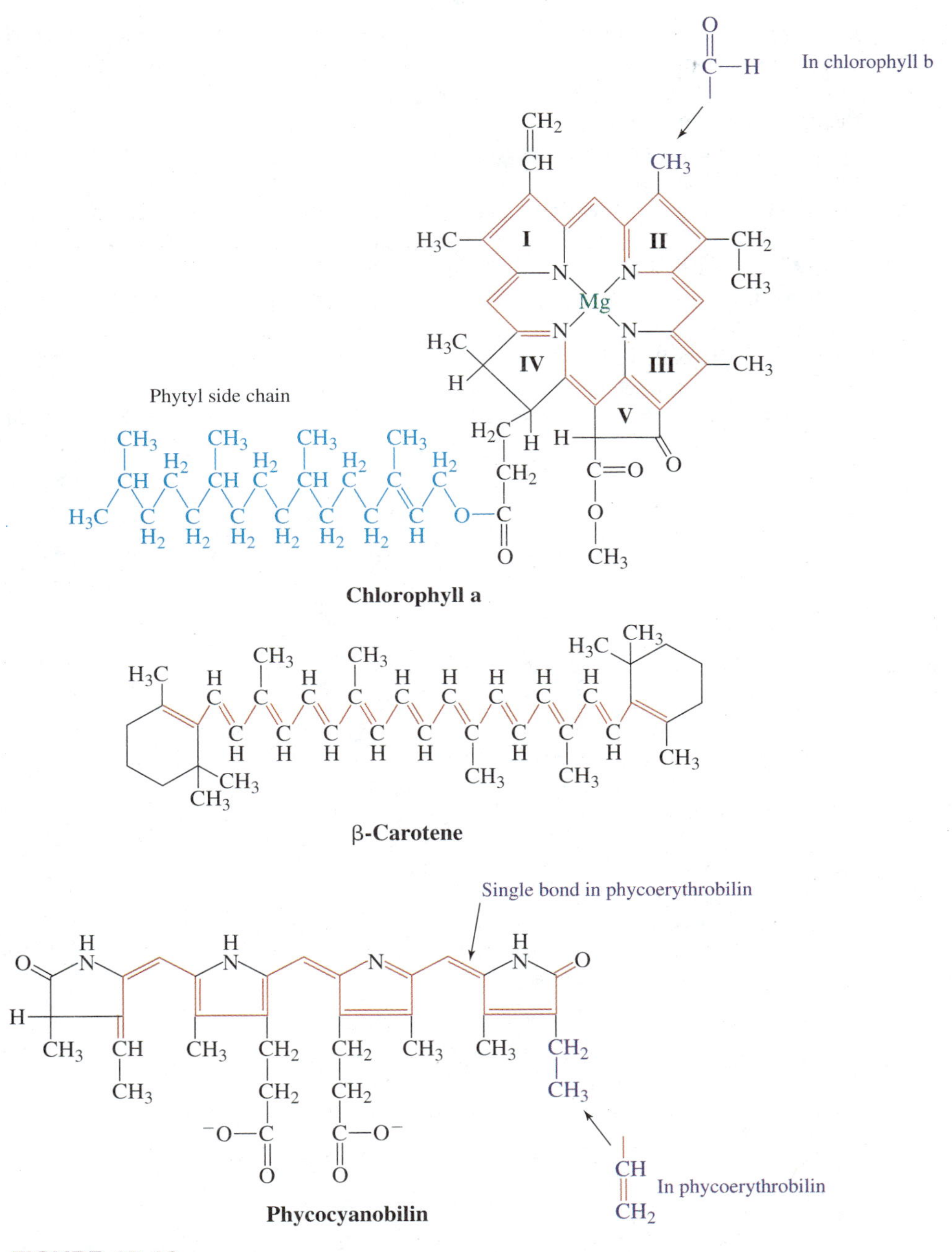

FIGURE 15.10
Some photosynthetic pigments. The red colored bonds identify regions of conjugation (alternating double and single bonds). These conjugated regions primarily account for the ability of these pigments to absorb visible light. (Based on Lehninger et al., 1993.)

the chlorin ring. The chlorophylls are also characterized by a hydrophobic **phytyl side chain.** Changes in the structure or the environment (or both) of the chlorin ring or its side chains account for the unique absorption spectrum associated with each of the distinctive chlorophylls in green plants and bacteria.

The reaction center pigments in photosystem I and photosystem II (P700 and P680, respectively) of green plants consist of a "special pair" of chlorophylls associated with specific proteins and additional pigments. The "special" chlorophylls are closely related to chlorophyll *a,* possibly a unique form of this chlorophyll. P680 has a long-wavelength absorbance maximum at 680 nm, whereas P700 absorbs maximally at 700 nm. The difference is attributed to the distinctive packaging of the "special pair" within each reaction center. The photosystem pigments outside the reaction centers are called **light-harvesting** or **antenna molecules.** Each photosystem has its own set of antenna molecules. These solar collectors feed light energy into the reaction center through resonance energy transfer (Figure 15.11). Not all light

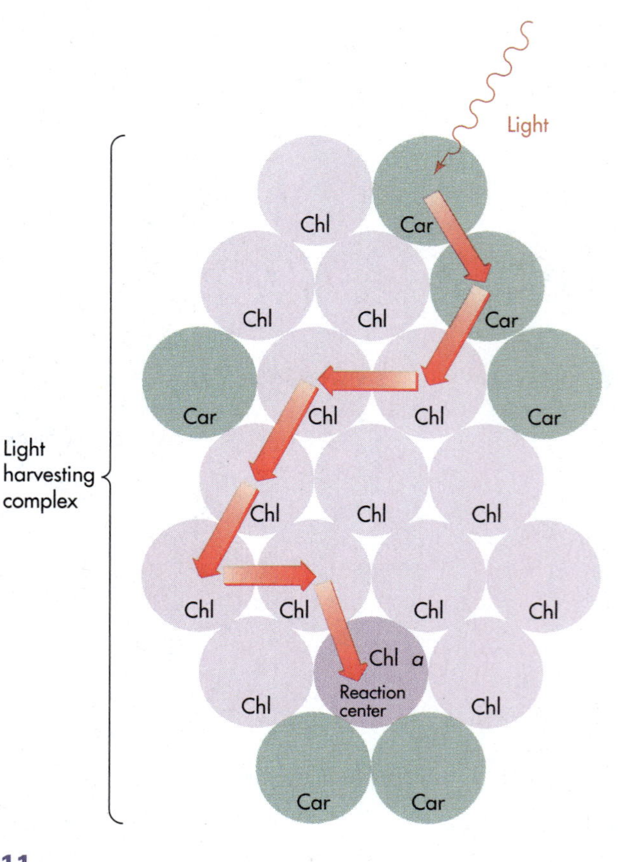

FIGURE 15.11
Absorption of light by a light-harvesting complex. Chl = chlorophyll, Car = β-carotene. The light energy absorbed takes a "random walk" to the reaction center. The entire "walk," which entails resonance energy transfer, typically occurs within a billionth of a second or less, with little energy "lost" as heat. (From Wolfe, 1993.)

energy is acquired from antenna molecules; P680 and P700 can also absorb photons directly. The use of multiple antenna molecules with distinctive absorption spectra allows a cell to harvest light efficiently throughout the visible spectrum. The harvested light initiates electron transfer reactions. **[Try Problem 15.12]**

Electron Transport in the Z-Scheme

Electron transfer begins with the absorption of light by P680 or the transfer of light energy to P680 to generate P680* (Section 15.3, Figure 15.6). P680*, within a few picoseconds (the prefix "pico" means 10^{-12}), transfers an electron to **pheophytin,** a membrane-associated pigment similar to chlorophyll but lacking Mg^{2+} (Figure 15.12). The electron flows from pheophytin to a membrane-bound **plastoquinone,** $\mathbf{Q_A}$ (Figure 15.13), which can accommodate two electrons. Q_A passes one electron at a time to $\mathbf{Q_B}$**,** a second plastoquinone, which is also a two-electron carrier. Q_B delivers its electrons to a mobile plastoquinone pool ($\mathbf{Q_{pool}}$) that passes the electrons to a cytochrome *bf* complex containing multiple electron carriers, including two unique cytochromes (different from those in mitochondria) and an iron–sulfur protein. The *b* cytochrome in the *bf* complex is named **cytochrome** $\mathbf{b_6}$**.** The electron-

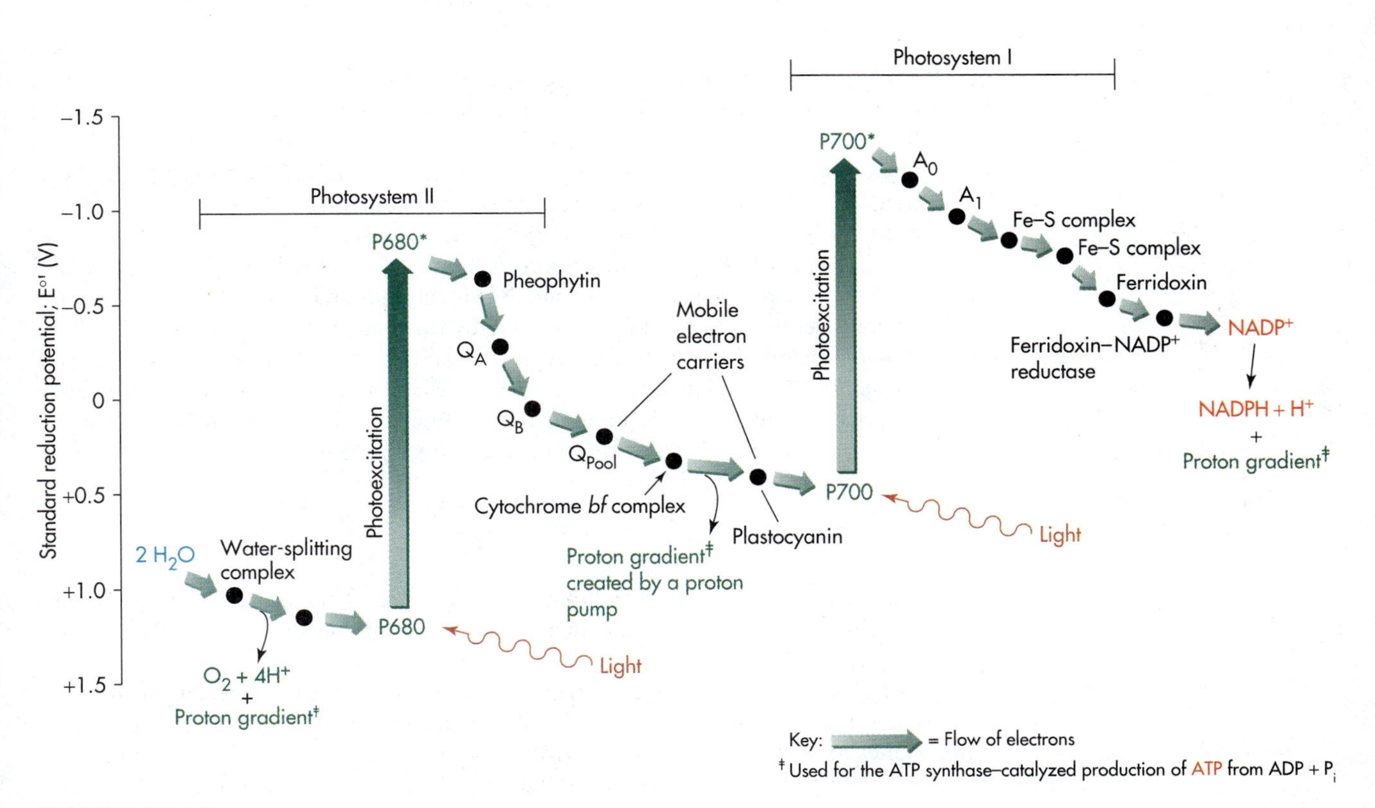

FIGURE 15.12
Electron carriers in the Z-scheme. (Based on Moran et al., 1994.)

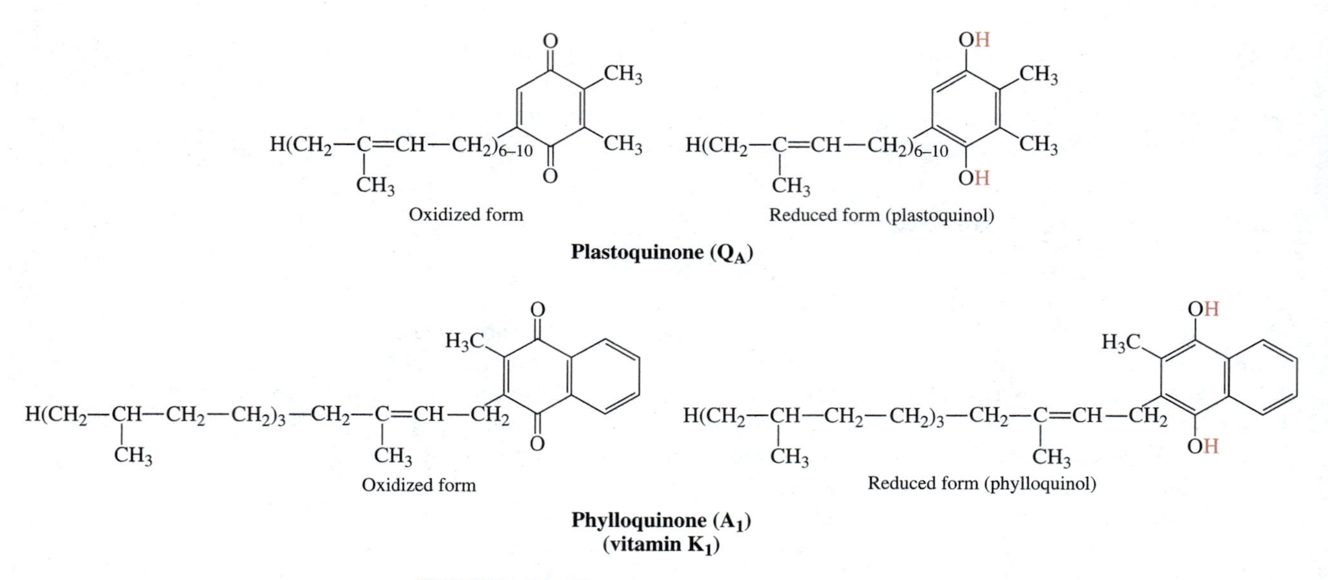

FIGURE 15.13
Structural formulas for plastoquinone (Q_A) and phylloquinone (A_1).

transport process continues as the cytochrome *bf* complex transfers an electron to a copper atom in a mobile protein called **plastocyanin** (Figure 15.14). Plastocyanin feeds electrons into photosystem I. The physical separation of photosystem II and photosystem I within the thylakoid membrane (Figure 15.9) prevents the direct transfer of energy (without an electron) from photosystem II to photosystem I, an event that would make photosystem II a light harvester for photosystem I.

The cytochrome *bf* complex is very similar to Complex III of the mitochondrial electron transport system: both contain cytochromes and iron–sulfur proteins, both contain a proton "pump" that is powered by the redox reactions associated with the flow of electrons, and both are coupled to the rest of the electron-transport system by mobile electron carriers. The nature of cytochromes and iron–sulfur proteins is discussed in Section 11.8. Particular iron atoms in these compounds oscillate between a +3 and a +2 oxidation state as electrons pass through:

$$e^- + Fe^{3+} \rightleftharpoons Fe^{2+}$$
$$Fe^{2+} \rightleftharpoons Fe^{3+} + e^-$$

The "pump" in both the cytochrome *bf* complex and Complex III of mitochrondria contributes to a proton gradient that powers ATP synthase–catalyzed ATP production (Figure 11.20 and 15.7).

For photosystem I to accept an electron from plastocyanin, its reaction center pigment must exist as a cation, $P700^+$. Acceptance of an electron returns the reaction center to its ground state, P700. After absorbing light energy, P700 is converted to P700*, a powerful reducing agent that quickly donates an electron to a cholorphyll *a*-like acceptor called A_0. A_0 passes the electron on to a molecule of **phylloquinone** (vitamin K_1) labeled $\mathbf{A_1}$ (Figure 15.13). During noncyclic light reactions, the electron gained by A_1 is transferred to $NADP^+$ through multiple iron–sulfur proteins

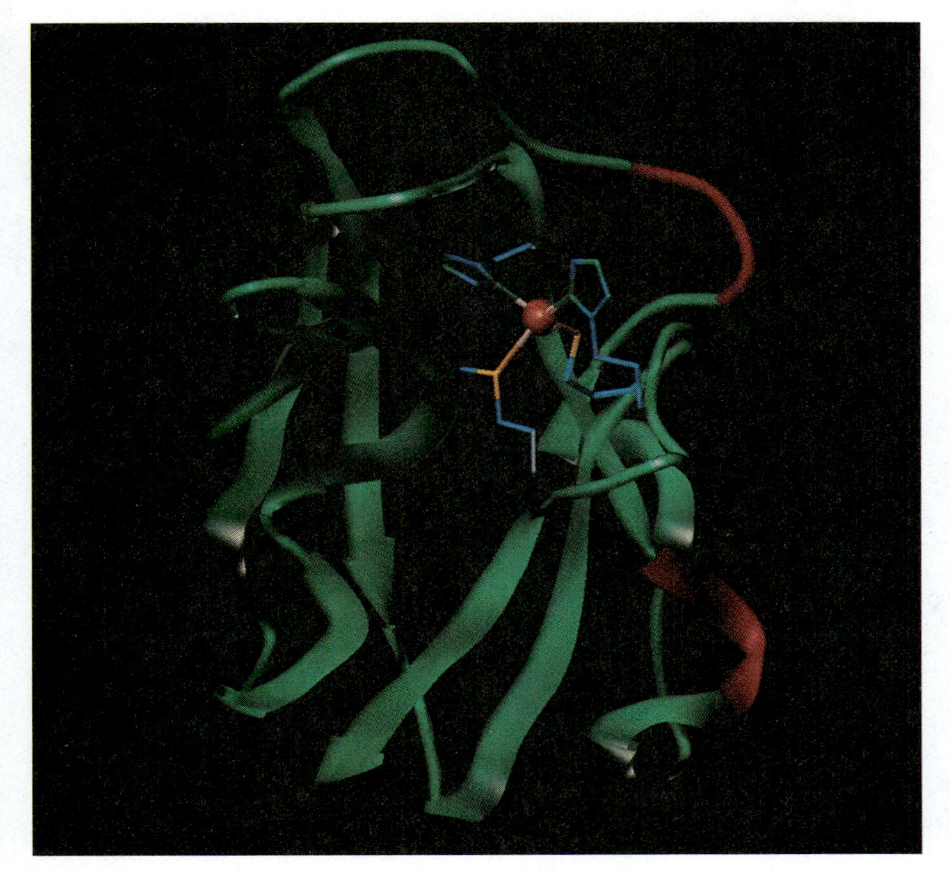

FIGURE 15.14
Plastocyanin from poplar leaves. Plastocyanin contains a single polypeptide chain with 99 amino acid residues. Its single copper (Cu) atom (red/orange sphere) alternates between +1 and +2 oxidation states as electrons pass through one at a time. The Cu is tetrahedrally liganded by the side chains of specific amino acid residues: His 37, Cys 84, His 87, and Met 92. The side chains are shown in stick form with their C, N, and S atoms blue, green, and yellow respectively. Negatively charged patches on the protein surface are red.

(the last called **ferredoxin**) and an FAD coenzyme within an enzyme named **ferredoxin-NADP$^+$ reductase** (also called **ferredoxin–NADP$^+$ oxidoreductase**). Since one proton (H^+) is removed from the stroma during the ferredoxin–NADP$^+$ oxidoreductase–catalyzed production of NADPH, the last step in noncyclic electron transfer contributes to the proton gradient used for ATP production.

During cyclic electron transport, the absorbance of light by P700 generates P700* which transfers an electron to ferredoxin through A_0, A_1, and the iron–sulfur proteins (Figure 15.15). P700$^+$ is generated in the process. At this point, the flow of the electron from P700* to NADP$^+$ is short circuited as ferredoxin passes the electron to the cytochrome *bf* complex. According to one hypothesis, ***cytochrome b*** (also called b_6) within the *bf* complex accepts the electron from ferredoxin during cyclic transport, whereas ***cytochrome f*** accepts an electron from Q_B during noncyclic transport. As the electron from ferredoxin flows through the cytochrome *bf* complex and plastocyanin to P700$^+$, the proton pump in the cytochrome *bf* complex contributes to the proton gradient used by ATP synthase for ATP production. The delivery of the electron to P700$^+$ regenerates the P700 that initiated the cycle.

The net reaction for cyclic electron transport and coupled ATP production is:

$$\text{ADP} + \text{P}_i + \text{light} \rightleftharpoons \text{ATP}$$

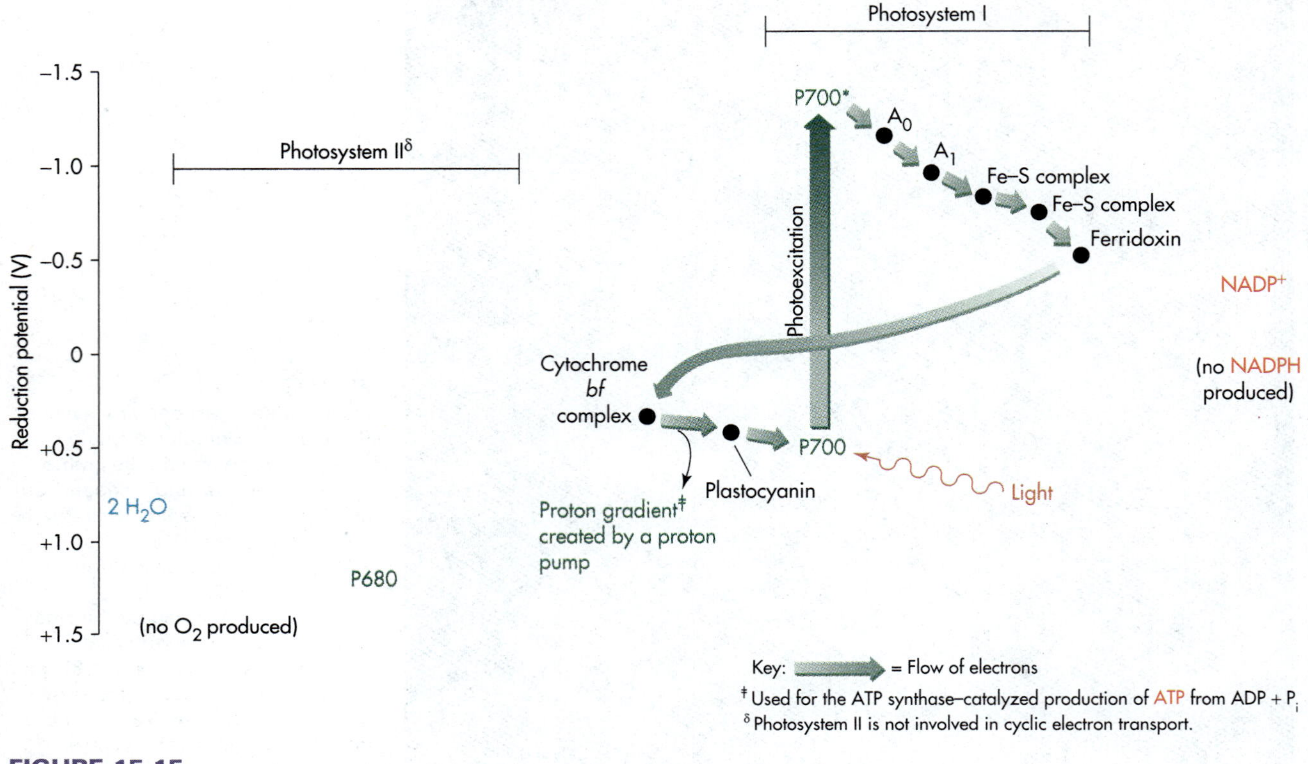

FIGURE 15.15
Participants in cyclic electron transport. (Based on Moran et al., 1994.)

This cyclic system, which leads to ATP synthesis in the absence of O_2 and NADPH production, is used by a cell to adjust its ATP/NADPH ratio to better meet its overall metabolic requirements. Photosynthesis itself uses more ATP than NADPH during the carbon fixation phase; three molecules of ATP are utilized for every two molecules of NADPH consumed (Section 15.5). The regulation of cyclic electron transport is under investigation. **[Try Problems 15.13 through 15.16]**

Chloroplast ATP Synthase

The ATP synthase associated with photophosphorylation in chloroplasts is structurally and functionally very similar to the ATP synthase (an isozyme) that catalyzes oxidative phosphorylation in mitochondria (Section 11.9). Both enzymes are membrane-spanning complexes that are powered by proton gradients. In chloroplasts, the energy for the creation of the proton gradient comes from light, whereas this energy is derived from the oxidation of organic compounds within mitochondria. The use of a proton-motive force for ATP synthesis is at the heart of chemiosmotic theory.

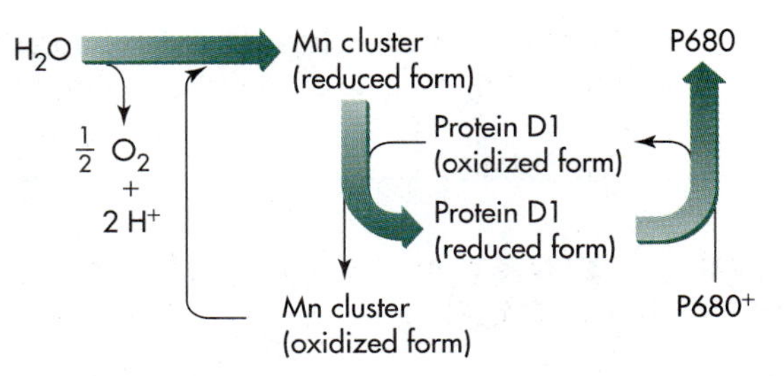

FIGURE 15.16
Reactions of the water-splitting complex. The green arrows depict the flow of electrons from water to P680.

Water-Splitting Complex

One final detail needs to be examined before we move on to the carbon fixation reactions. This is the extraction of electrons from water, the ultimate source of the electrons delivered to $NADP^+$ during noncyclic electron transport. The oxidation of water is accomplished by the **water-splitting complex** (also called the **oxygen-evolving complex**) which is considered part of photosystem II (Figure 15.16). A manganese cluster within the complex is the immediate recipient of electrons from water. Electrons are passed one at a time from this cluster to a protein called **D1** which then feeds the electrons to $P680^+$. Since the water-splitting reactions occur on the lumen side of the thylakoid membrane, the protons generated in the process ($2H_2O \rightleftharpoons O_2 + 4\ H^+ + 4$ electrons) contribute to the proton gradient that is central to ATP production. O_2 is released as a gas. **[Try Problems 15.17 and 15.18]**

PROBLEM 15.12 When a solution of purified cholorphyll is exposed to sunlight, the chlorophyll fluoresces; few, if any, chlorophyll cations are generated. In contrast, when P680 and P700 are exposed to sunlight their chlorophyll ionizes rather than fluoresces. What is the explanation for these differences?

PROBLEM 15.13 Assume that a plant evolves that contains all of the electron carriers in the standard Z-scheme plus one additional carrier. If the new carrier has a reduction potential of −1.0, where will this carrier reside within the newly evolved Z-scheme? Explain.

PROBLEM 15.14 True or false? Each transfer of electrons within the electron transport system creates or eliminates ions. Explain.

PROBLEM 15.15 Would cyclic electron transport be of any value if ferredoxin transferred electrons to plastocyanin rather than the cytochrome *bf* complex? Explain.

PROBLEM 15.16 Explain why P680 and P700 tend to fluoresce in sunlight when the flow of electrons through the electron transport system is blocked at a point after P700*.

PROBLEM 15.17 List the three separate steps in the light-induced electron transfer reactions that contribute to the proton gradient between the lumen and the stroma of chloroplasts.

PROBLEM 15.18 The lumen within chloroplasts occupies a relatively small volume. Explain how this facilitates the creation of a proton-motive force.

15.5 CARBON FIXATION REACTIONS (CALVIN CYCLE)

Photosynthetic carbon fixation is a cyclic process known as the **carbon fixation cycle, Calvin cycle** and **Calvin–Benson cycle.** The cycle, which begins with three molecules of ribulose 1,5-bisphosphate, can be conveniently divided into three stages: (1) the fixation of three carbons coupled to the formation of six 3-phosphoglycerates; (2) the reduction of each 3-phosphoglycerate to glyceraldehyde 3-phosphate; and (3) the transformation of five of the six glyceraldehyde 3-phosphates into three ribulose 1,5-bisphosphates (replace those that initiated the

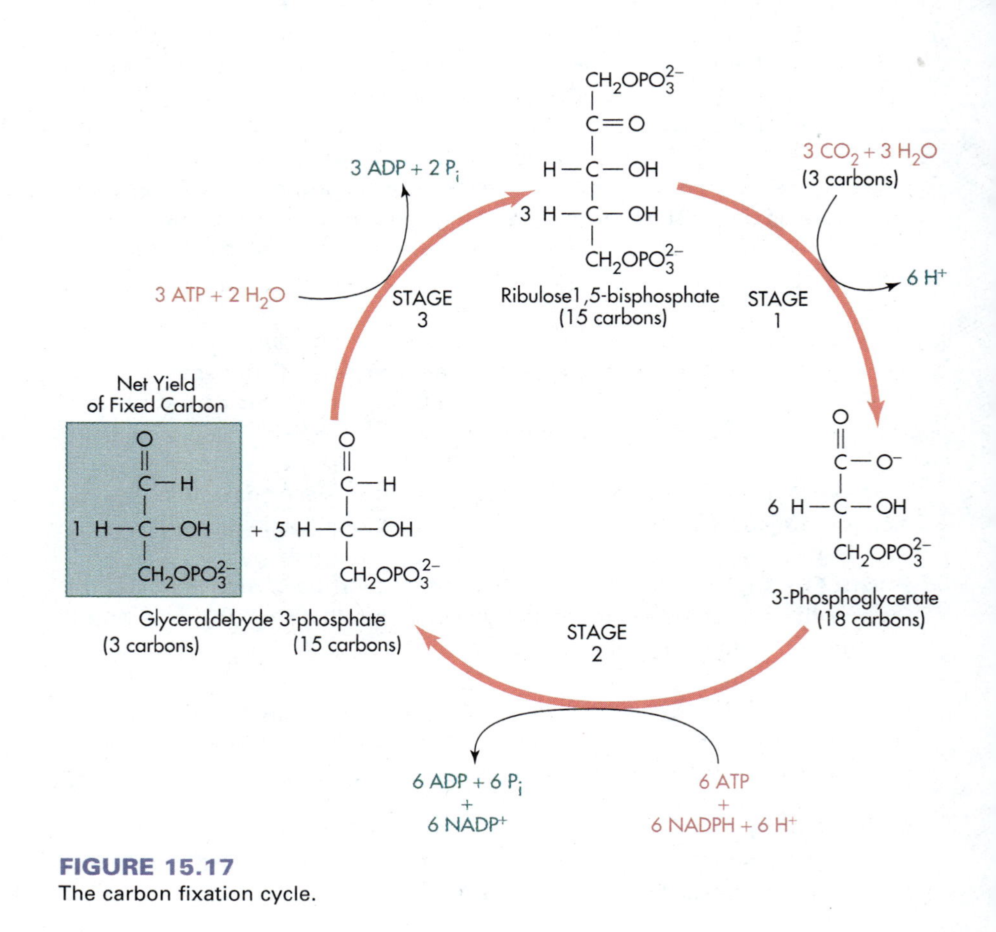

FIGURE 15.17
The carbon fixation cycle.

cycle) (Figure 15.17). The sixth glyceraldehyde 3-phosphate represents the net yield of the cycle in terms of carbon fixation. The net reaction for the cycle is:

$$3\ CO_2 + 9\ ATP + 6\ NADPH + 5\ H_2O \rightleftharpoons \text{triose phosphate} + 9\ ADP + 6\ NADP^+ + 8\ P_i$$

All of the energy used (in the form of ATP and NADPH) comes from photons harvested during the light reactions. The remainder of this section explores some of the details of the carbon fixation cycle.

Stage 1: Rubisco-Catalyzed Carbon Fixation

Rubisco (ribulose 1,5-bisphosphate carboxylase or **RuBP carboxylase/oxygenase)** (Figure 15.18) is the oligomeric enzyme directly responsible for catalyzing carbon fixation during photosynthesis. Part of the polypeptide subunits are encoded in nuclear DNA and part encoded in chloroplast DNA. Rubisco is the most abundant

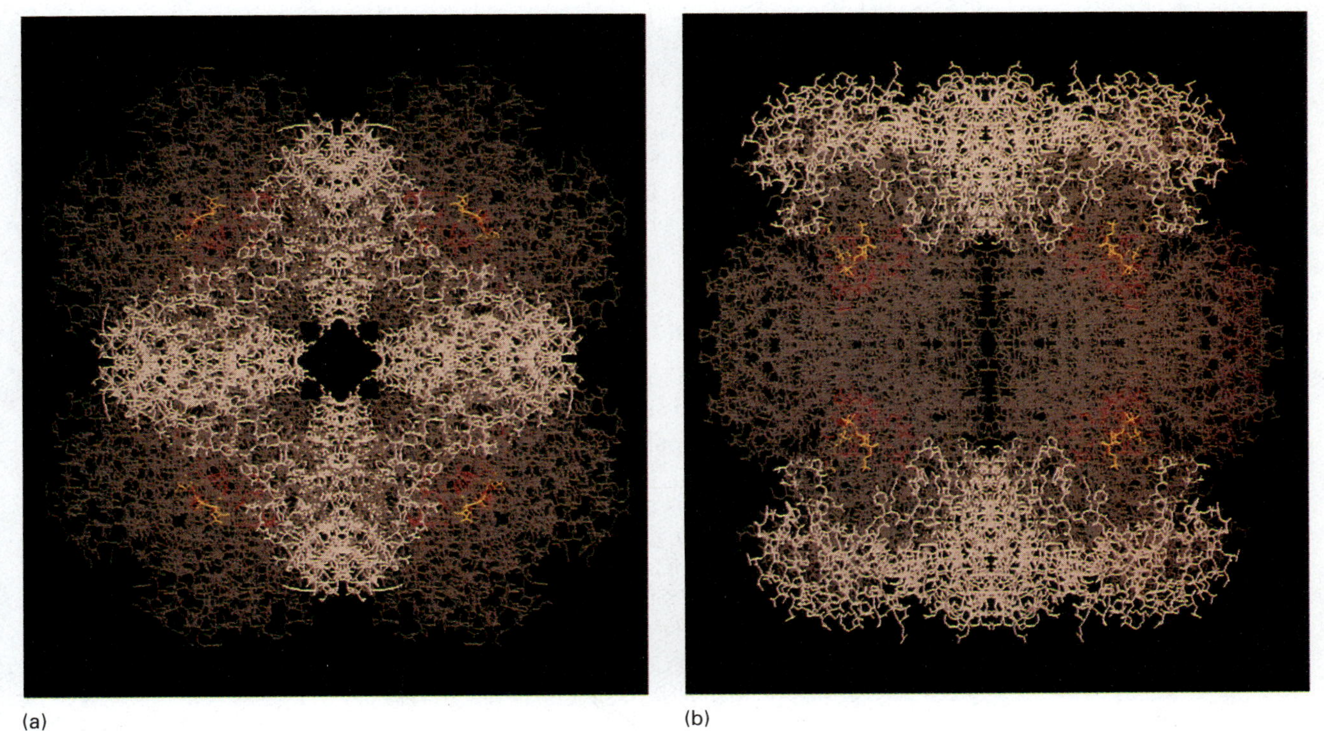

(a) (b)

FIGURE 15.18
Plant rubisco. (a) Top view of a stick model of plant rubisco. (b) Side view of the same stick model. Rubisco from higher plants contains eight identical large subunits (477 residues, encoded in chloroplast DNA) and eight identical small subunits (123 residues; encoded in nuclear DNA). The large subunits are gray and the small subunits are white. Active-site amino acid residues are red. Yellow sticks are sulfate molecules (not normal substrates) bound to active sites.

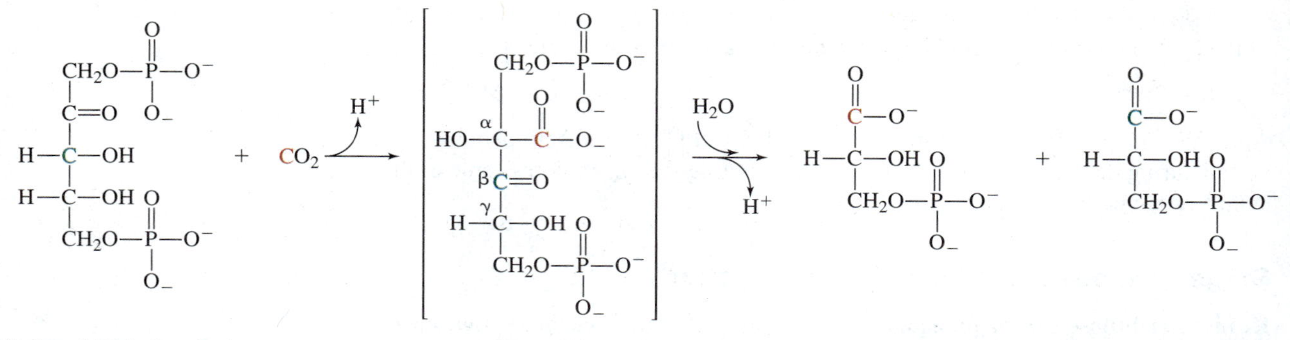

FIGURE 15.19
Rubisco-catalyzed reaction.

protein in chloroplasts and the most abundant enzyme in the biosphere. Located in the stroma of chloroplasts, it catalyzes the reaction of ribulose 1,5-bisphosphate with CO_2 to yield two molecules of 3-phosphoglycerate (Figure 15.19). This is a two-step process in which an initial carboxylation reaction generates an enzyme-bound β-keto acid intermediate. Hydrolysis of this intermediate releases final products. **[Try Problem 15.19]**

Stage 2: Glyceraldehyde 3-Phosphate Formation

Each of the two 3-phosphoglycerates produced from ribulose 1,5-bisphosphate and CO_2 is transformed into glyceraldehyde 3-phosphate during two steps catalyzed by 3-phosphoglycerate kinase and glyceraldehyde-3-phosphate dehydrogenase, respectively (Figure 15.20). These chloroplast-specific enzymes, which are isozymes of cytosolic enzymes that participate in glycolysis, catalyze what amounts to the reversal of two steps in glycolysis. In step one, stromal 3-phosphoglycerate kinase catalyzes the phosphorylation of 3-phosphoglycerate to yield 1,3-bisphosphoglycerate, an energy-rich mixed anhydride. This reaction is made thermodynamically favorable by the coupled cleavage of an anhydride link in ATP. During the second step, stromal glyceraldehyde-3-phosphate dehydrogenase catalyzes the NADPH-linked reduction of 1,3-bisphosphoglycerate to glyceraldehyde 3-phosphate. $NADP^+$ and inorganic phosphate are additional products. Note that, in contrast to its glycolytic counterpart, the stromal dehydrogenase employs NADPH (rather than NADH) as a reducing agent. This fact supports a previous generalization: NADPH is normally used for anabolic processes, whereas NADH usually passes electrons to O_2 through the respiratory chain, a process coupled to ATP production. The ATP and NADPH used for glyceraldehyde 3-phosphate synthesis are products of the light reactions. **[Try Problem 15.20]**

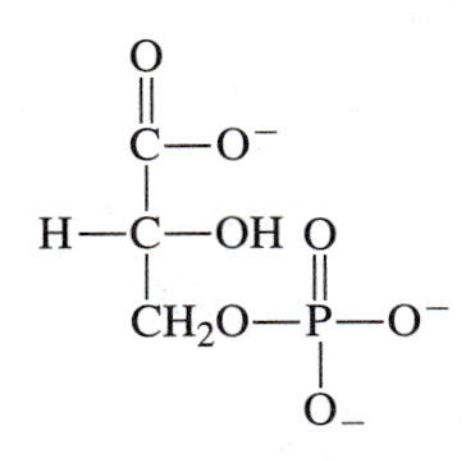

3-Phosphoglycerate

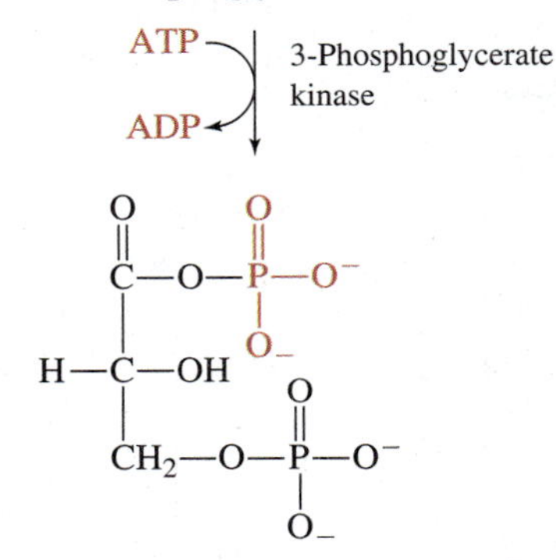

1,3-Bisphosphoglycerate

NADPH + H^+
$NADP^+$ + P_i
Glyceraldehyde-3-phosphate dehydrogenase

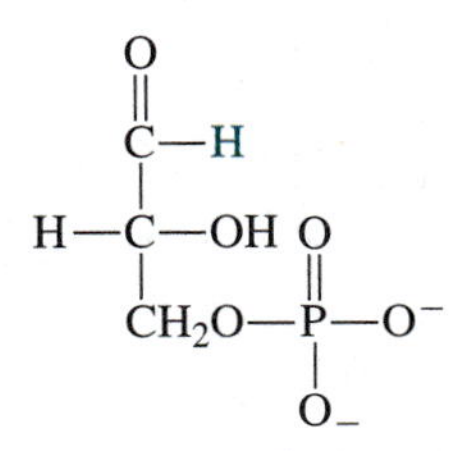

Glyceraldehyde 3-phosphate

FIGURE 15.20
Conversion of 3-phosphoglycerate to glyceraldehyde 3-phosphate.

Stage 3: Regeneration of Ribulose 1,5-Bisphosphate

Part of the glyceraldehyde 3-phosphate produced from ribulose 1,5-bisphosphate and CO_2 in the stroma must be utilized to regenerate three ribulose 1,5-bisphosphates if a cycle is to be created and maintained. On the average, five out of every six glyceraldehyde 3-phosphates formed are employed for this purpose. The reactions involved are, for the most part, a reversal of the reactions that convert pentose phosphates to hexose phosphates within the pentose phosphate pathway (review Section 12.2). During the regeneration of ribulose 1,5-bisphosphate, hexose phosphates produced from glyceraldehyde 3-phosphates are transformed into pentose phosphates. Figure 15.21 summarizes the entire regeneration process, and Figure 15.22 summarizes the flow of carbon that is involved.

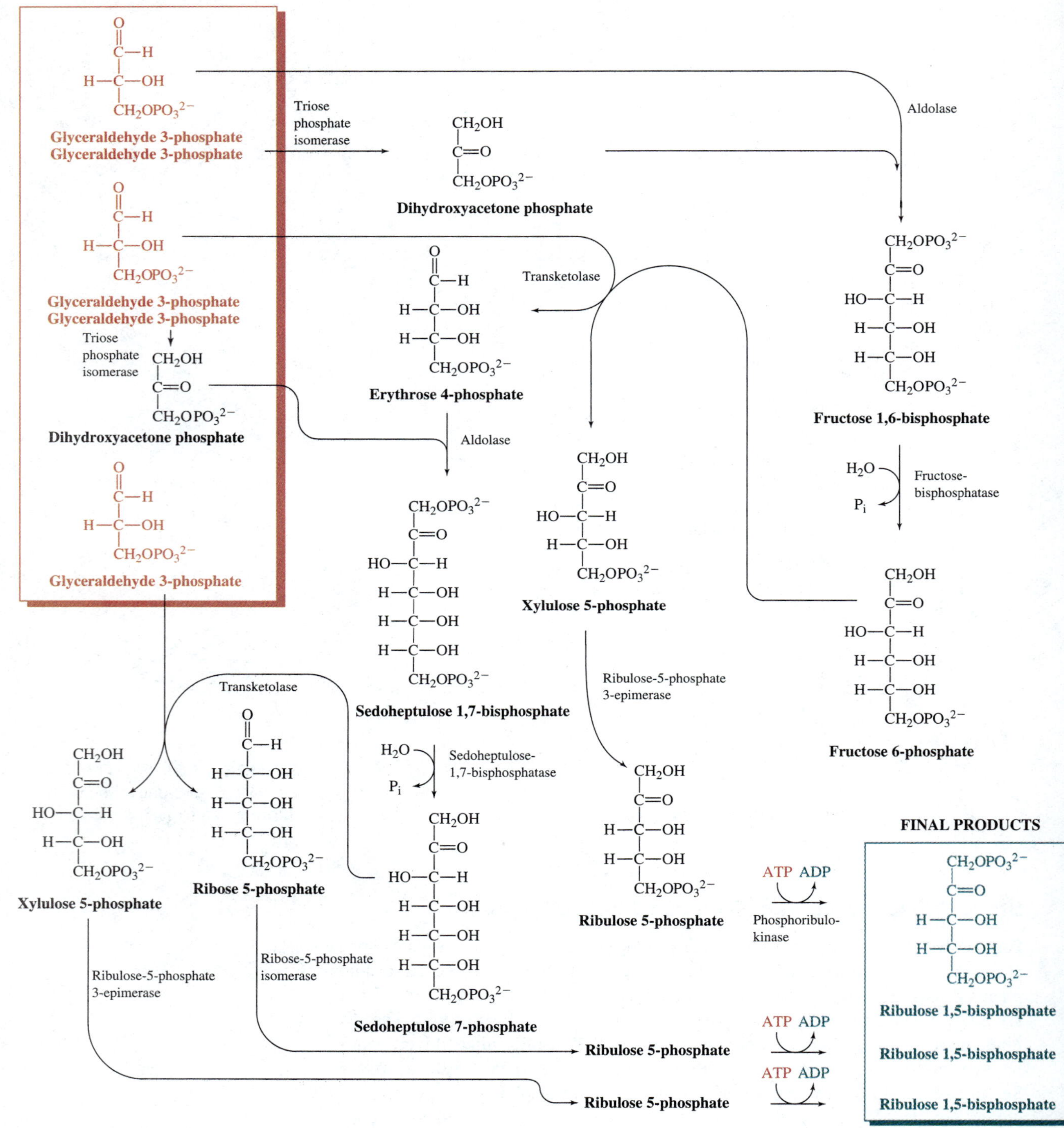

FIGURE 15.21
Regeneration of ribulose 1,5-bisphosphate from glyceraldehyde 3-phosphate. Five glyceraldehyde 3-phosphates are converted to three ribulose 1,5-bisphosphates.

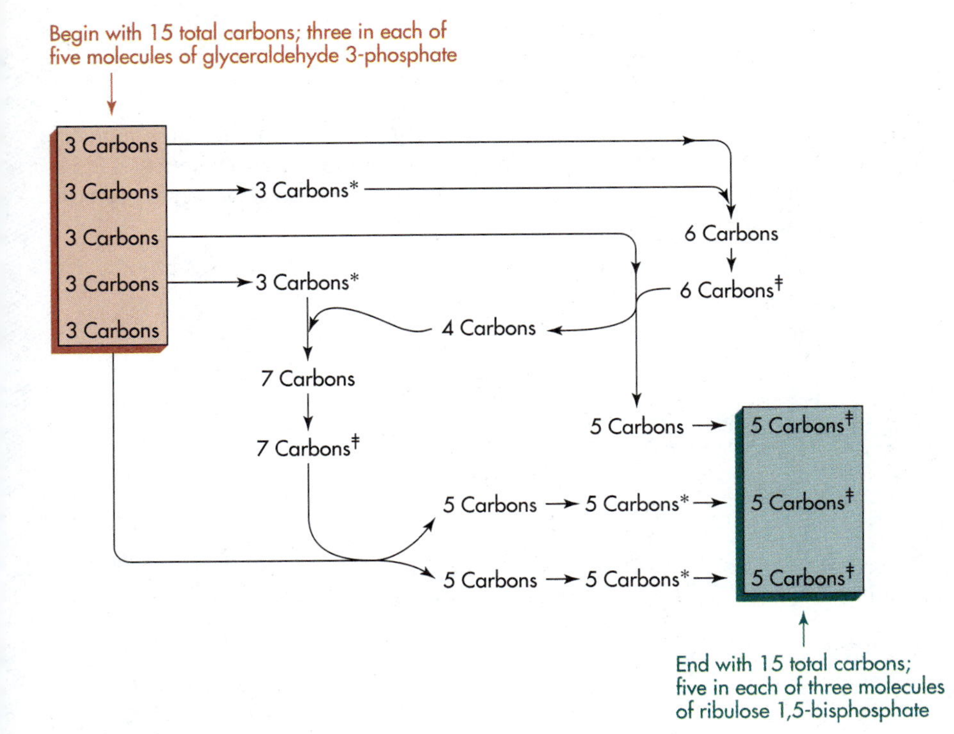

FIGURE 15.22
An accounting of carbon atoms during ribulose 1,5-bisphosphate regeneration. Intermediates, marked with asterisks, are products of isomerization reactions. Isomerization plays a central role in carbon flow. Compounds marked with ‡ are products of phosphorylation or dephosphorylation reactions.

The regeneration process begins with two stromal enzymes that are isozymes of cytosolic triose phosphate isomerase and aldolase, two glycolytic enzymes. The isomerase catalyzes the conversion of glyceraldehyde 3-phosphate to dihydroxyacetone phosphate, and aldolase catalyzes an aldol condensation between glyceraldehyde 3-phosphate and dihydroxyacetone phosphate to yield fructose 1,6-bisphosphate. These two reactions represent a direct reversal of two steps in glycolysis.

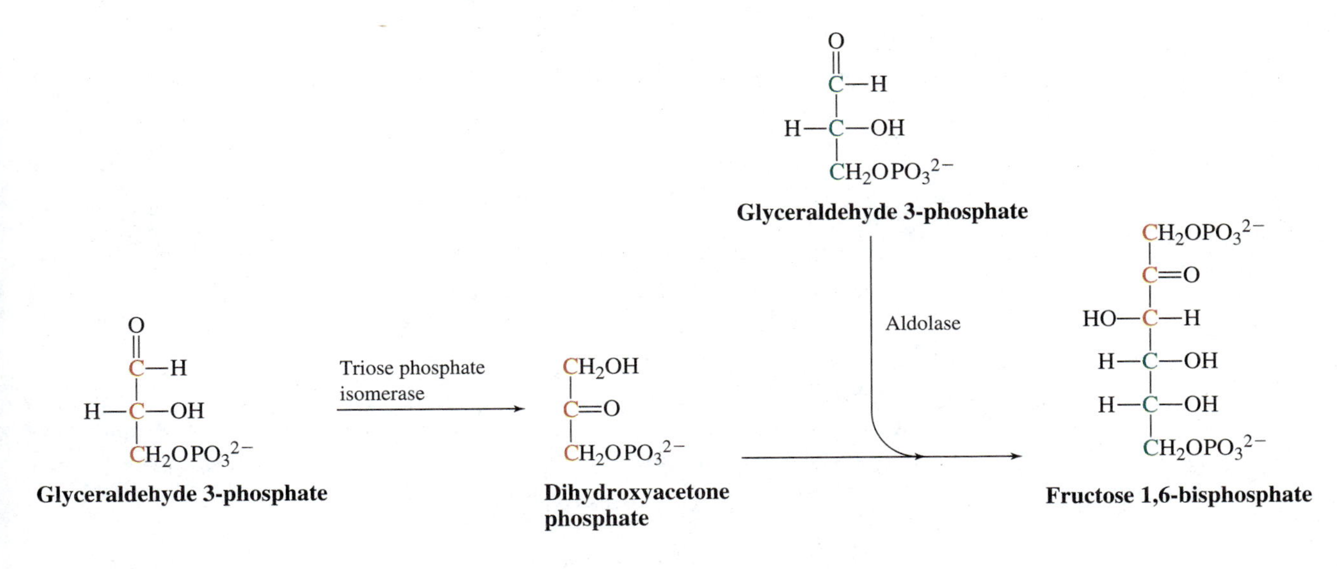

After the fructose 1,6-bisphosphate is hydrolyzed with the catalytic assistance of fructose-bisphosphatase (fructose-1,6-bisphosphatase), the fructose 6-phosphate that is generated transfers a ketol group ($CH_2OH—\overset{O}{\overset{||}{C}}—$) to another molecule of glyceraldehyde 3-phosphate (the third encountered in the regeneration scheme) in a reaction catalyzed by transketolase. Xylulose 5-phosphate and erythrose 4-phosphate are formed.

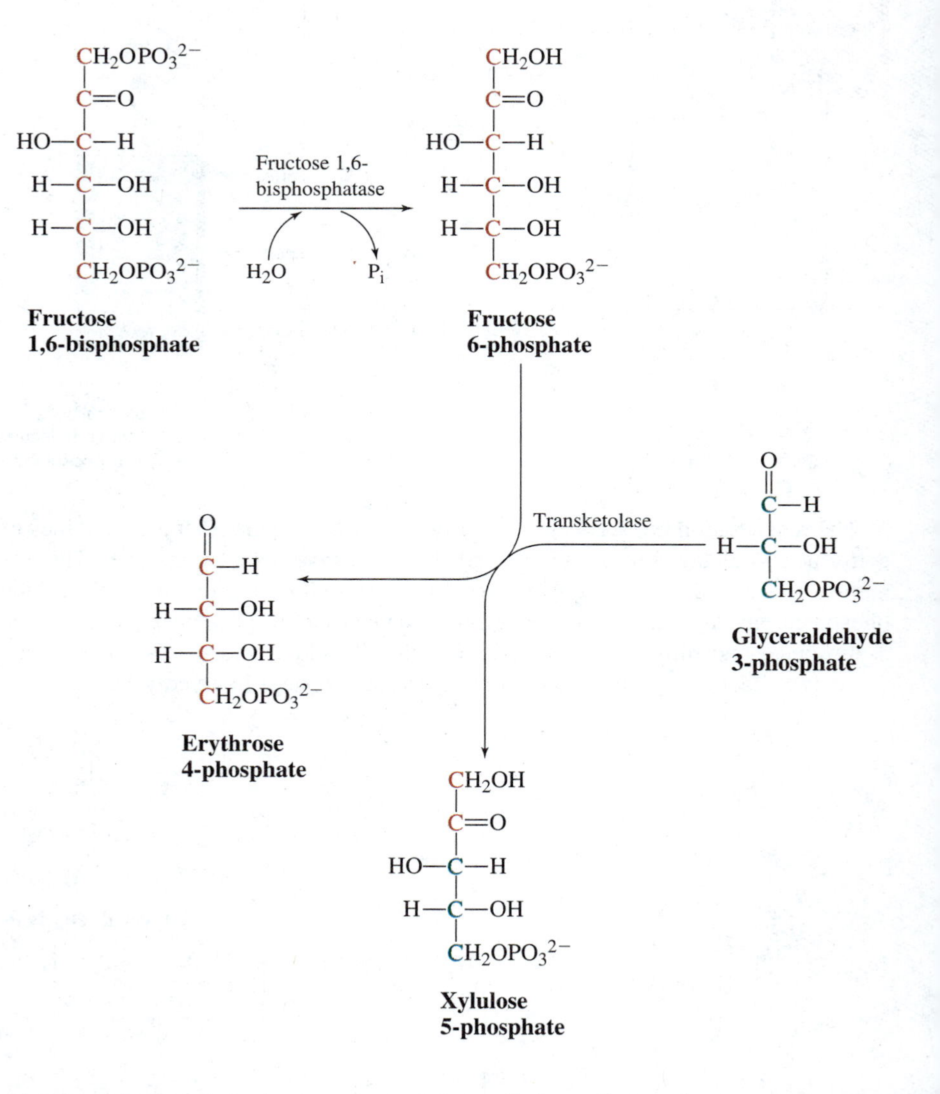

Xylulose 5-phosphate is transformed into ribulose 5-phosphate and then into ribulose 1,5-bisphosphate with the catalytic assistance of **ribulose-5-phosphate 3-epimerase** and **phosphoribulokinase,** respectively. ATP is the phosphoryl donor for the kinase reaction.

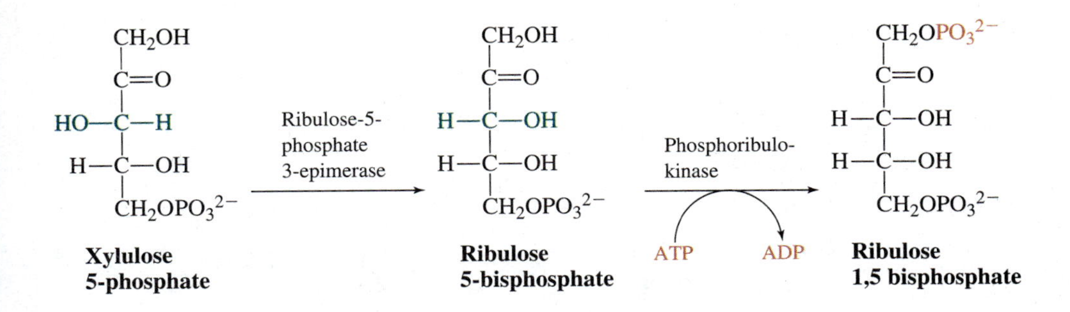

Erythrose 4-phosphate condenses with a molecule of dihydroxyacetone phosphate (produced from a fourth glyceraldehyde 3-phosphate) in a second aldolase-catalyzed reaction to yield sedoheptulose 1,7-bisphosphate. **Sedoheptulose-1,7-bisphosphatase (sedoheptulose-bisphosphatase)**–catalyzed hydrolysis leads to the formation of sedoheptulose 7-phosphate. **[Try Problem 15.21]**

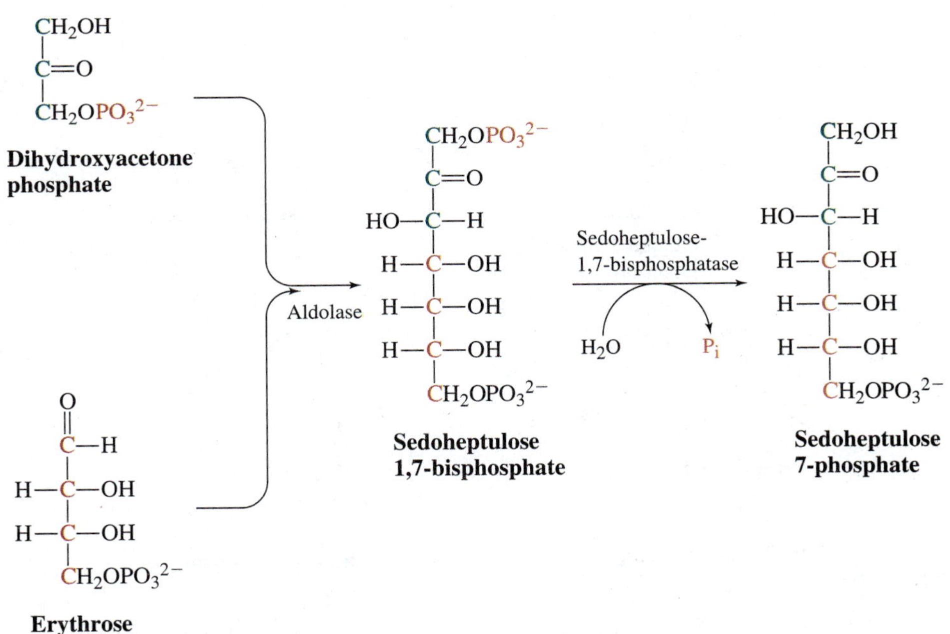

The sedoheptulose 7-phosphate transfers a ketol group to a fifth molecule of glyceraldehyde 3-phosphate in a second transketolase-catalyzed step. The xylulose 5-phosphate and ribose 5-phosphate produced are each converted to ribulose 5-phosphate in reactions catalyzed by ribulose-5-phosphate 3-epimerase and **ribose-5-phosphate isomerase,** respectively. The two ribulose 5-phosphate molecules are then transformed into ribulose 1,5-bisphosphate with the catalytic assistance of phosphoribulokinase. ATP is once again the phosphoryl donor (as shown on the next page).

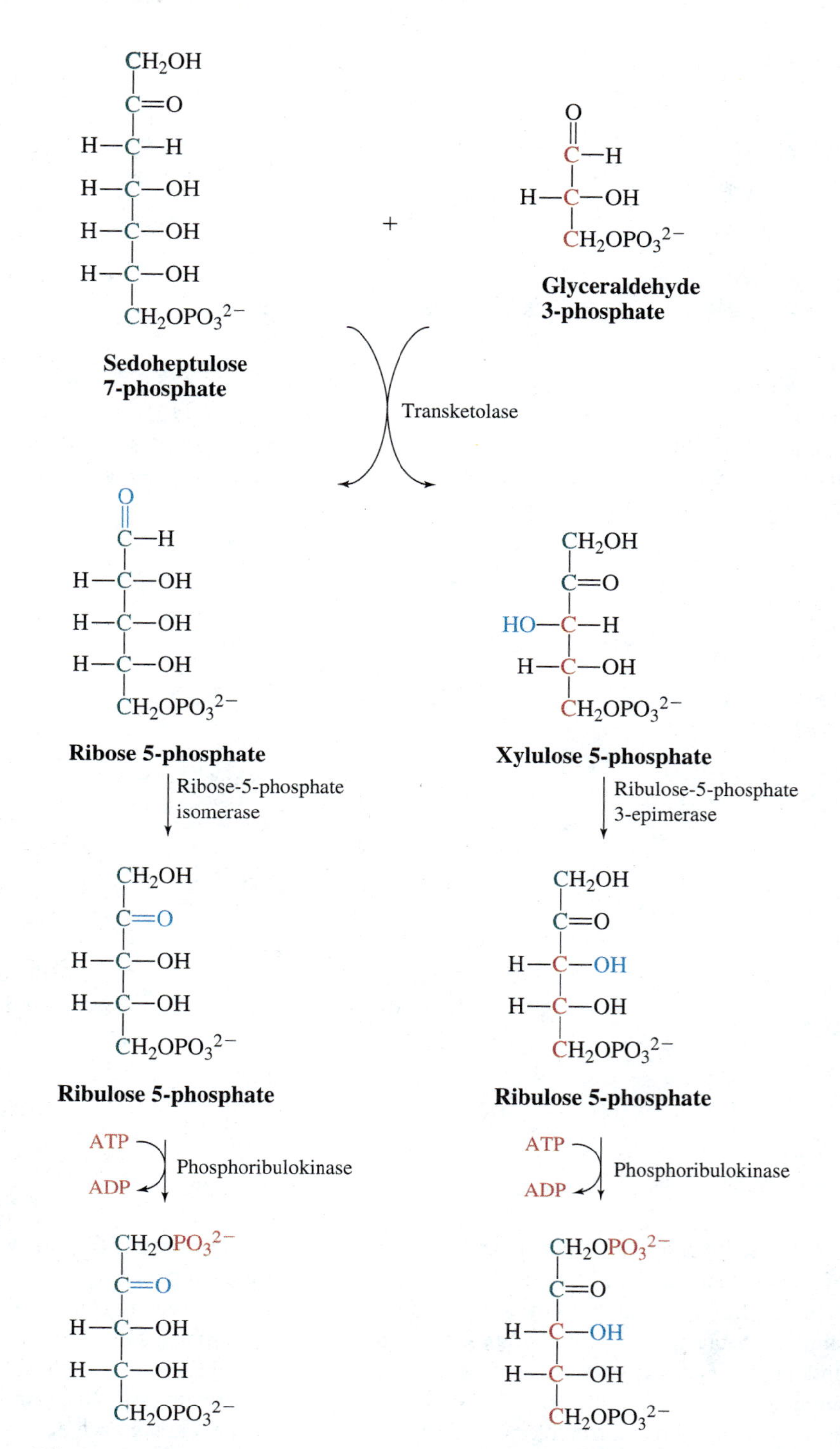

Sedoheptulose 7-phosphate
Glyceraldehyde 3-phosphate
Transketolase
Ribose 5-phosphate
Xylulose 5-phosphate
Ribose-5-phosphate isomerase
Ribulose-5-phosphate 3-epimerase
Ribulose 5-phosphate
Ribulose 5-phosphate
ATP
ADP
Phosphoribulokinase
ATP
ADP
Phosphoribulokinase
Ribulose 1,5-bisphosphate
Ribulose 1,5-bisphosphate

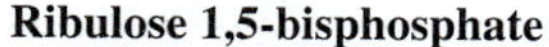

The net reaction for ribulose 1,5-bisphosphate regeneration from glyceraldehyde 3-phosphate is:

$$5\ \begin{matrix} O \\ \| \\ C{-}H \\ | \\ H{-}C{-}OH \\ | \\ CH_2OPO_3^{2-} \end{matrix} + 2\,H_2O + 3\,ATP \longrightarrow 3\ \begin{matrix} CH_2OPO_3^{2-} \\ | \\ C{=}O \\ | \\ H{-}C{-}OH \\ | \\ H{-}C{-}OH \\ | \\ CH_2OPO_3^{2-} \end{matrix} + 2\,P_i + 3\,ADP$$

Glyceraldehyde 3-phosphate **Ribulose 1,5-bisphosphate**

The next section examines how ribulose 1,5-bisphosphate regeneration and other photosynthetic processes are regulated. **[Try Problem 15.22 and 15.23]**

PROBLEM 15.19 Identify that bond in the β-keto acid intermediate in Figure 15.19 that is hydrolyzed during the second step of the rubisco-catalyzed reaction.

PROBLEM 15.20 Write the net reaction for the conversion of 3-phosphoglycerate to glyceraldehyde 3-phosphate. What is the minimum number of electrons that must be passed through the light-induced electron-transport system to support this net reaction? What is the minimum number of photons that must be absorbed?

PROBLEM 15.21 True or false? Ribulose-5-phosphate 3-epimerase is a transferase that catalyzes the interconversion of epimers. Explain.

PROBLEM 15.22 In the net reaction for ribulose 1,5-bisphosphate regeneration from glyceraldehyde 3-phosphate, three ATPs are among the reactants and three ADPs plus 2 P_i's are among the products. Are the two P_i's derived from the direct hydrolysis of ATP? If so, why aren't there three P_i's? If not, what is the source of the P_i's and what happened to the phosphate residues lost by the ATPs?

PROBLEM 15.23 For each enzyme involved in ribulose 1,5-bisphosphate regeneration from glyceraldehyde 3-phosphate, identify the major international class of enzymes (Table 5.1) to which it belongs.

15.6 REGULATION OF PHOTOSYNTHESIS

Because photosynthesis depends on light and CO_2, its rate is determined, in part, by the availability of these "substrates." The availability of ATP and NADPH from the light reactions helps set the pace of the carbon fixation cycle. Photosynthesis is further controlled at the level of specific enzymes that participate in the carbon fixation cycle. Several of the strategies used to regulate the catalytic activity of individual enzyme molecules are examined in this section.

The catalytic activities of rubisco depend on the binding of Mg^{2+} and the carbamylation (addition of CO_2 to form a carbamate) of one of its lysine side chains; the noncarbamylated enzyme, although able to bind ribulose 1,5-bisphosphate, is inactive. Although rubisco carbamylation, like hemoglobin carbamylation (Section 4.9), occurs nonenzymatically, the rubisco reaction is dependent on a light-activated enzyme. This enzyme, **rubisco activase,** catalyzes the displacement of noncovalently bound ribulose 1,5-bisphosphate from the noncarbamylated enzyme (Figure 15.23a).

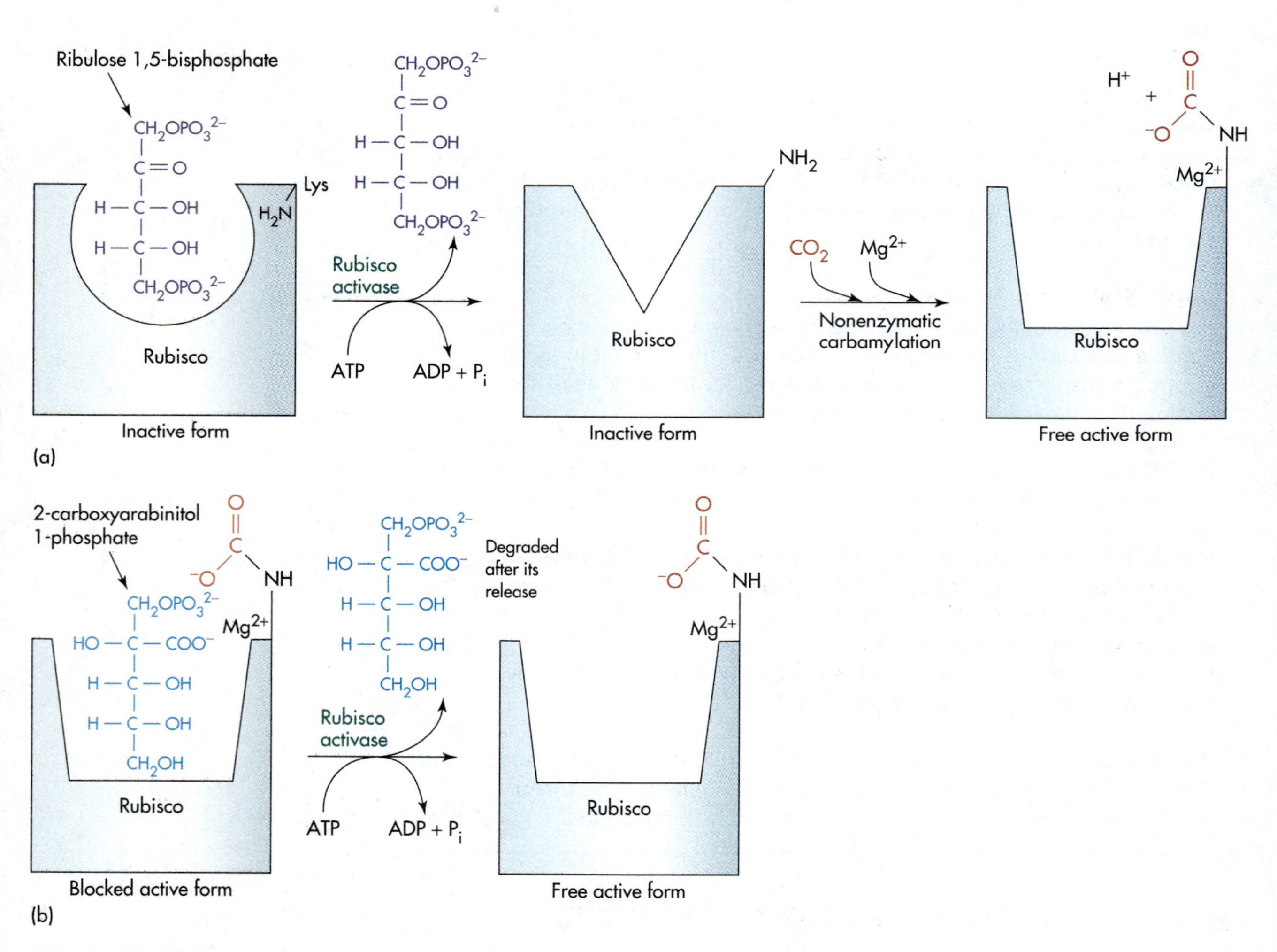

FIGURE 15.23
Regulation of rubisco by rubisco activase, CO_2, Mg^{2+}, and 2-carboxyarabinitol 1-phosphate. (a) Rubisco activase catalyzes the removal of ribulose 1,5-bisphosphate from the substrate binding site on the inactive form of rubisco. This allows rubisco to be activated through carbamylation and the binding of Mg^{2+}. CO_2 and Mg^{2+} concentrations help determine the concentration of active rubisco and the rate of carbon fixation. (b) Rubisco activase catalyzes the removal of 2-carboxyarabinitol 1-phosphate, a natural regulatory agent that operates as a competitive inhibitor, from the active site of the active form of rubisco. Rubisco is freed to bind ribulose 1,5-bisphosphate and catalyze carbon fixation.

Carbamylation cannot occur when ribulose 1,5-bisphosphate is bound. The activase-catalyzed release of the ribulose 1,5-bisphosphate is coupled to ATP hydrolysis and an alteration in the conformation of rubisco. Mg^{2+}, CO_2, and rubisco activase are positive regulators. **[Try Problem 15.24]**

Rubisco is also subject to negative regulation. In the dark, some plants produce a "nocturnal inhibitor," **2-carboxyarabinitol 1-phosphate** (Figure 15.24), which is structurally similar to the β-keto acid intermediate generated during the rubisco-catalyzed fixation of CO_2 (Figures 15.19 and 15.24). This inhibitor competes with ribulose 1,5-bisphosphate for the active site on carbamylated rubisco, a physiologically important example of competitive inhibition (Section 5.10). In the process of binding to the active site, 2-carboxyarabinitol 1-phosphate stabilizes the carbamylated active form of rubisco. When a plant is exposed to light, the 2-carboxyarabinitol 1-phosphate is released from rubisco with the catalytic assistance of the light-activated rubisco activase and then hydrolyzed with the catalytic assistance of a light-activated phosphatase (Figure 15.23b). Thus, regulation by 2-carboxyarabinitol 1-phosphate is coupled to regulation by both CO_2 (carbamylation) and rubisco activase. **[Try Problem 15.25]**

Rubisco, fructose-bisphosphatase (also called fructose-1,6-bisphosphatase), and sedoheptulose-1,7-bisphosphatase are all activated by the increases in stromal pH and Mg^{2+} concentration that are a result of the light-induced pumping of H^+ from the stroma to the lumen and the associated movement of Mg^{2+} from the lumen into the stroma (Figure 15.7). Certain other carbon fixation enzymes also have an alkaline pH optimum. Hence, an increase in the rate of the light reactions automatically increases the rate of key steps in the carbon fixation cycle (Figure 15.25). Since Mg^{2+} activates rubisco, in part, by complexing with the carbamate formed during the carbamylation reaction, CO_2 activation and Mg^{2+} activation are related processes (Figure 15.23a).

Fructose-bisphosphatase, sedoheptulose-1,7-bisphosphatase, phosphoribulokinase, and certain other carbon fixation cycle enzymes are each activated through the light-induced reduction of surface disulfide bonds (Figure 15.26). The reduction process begins when reduced ferredoxin coupled to photosystem I feeds electrons to

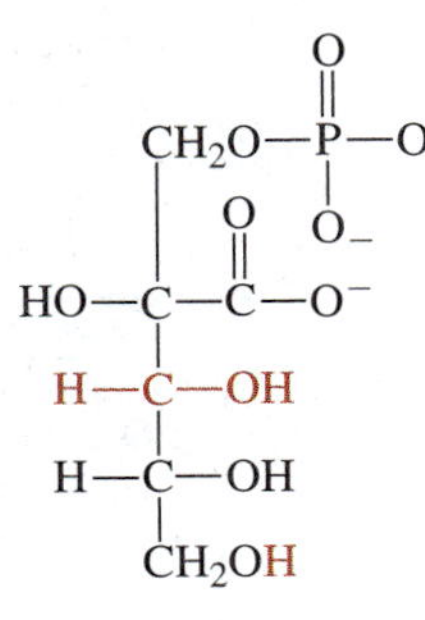

2-Carboxyarabinitol 1-phosphate; a competitive inhibition of rubisco.

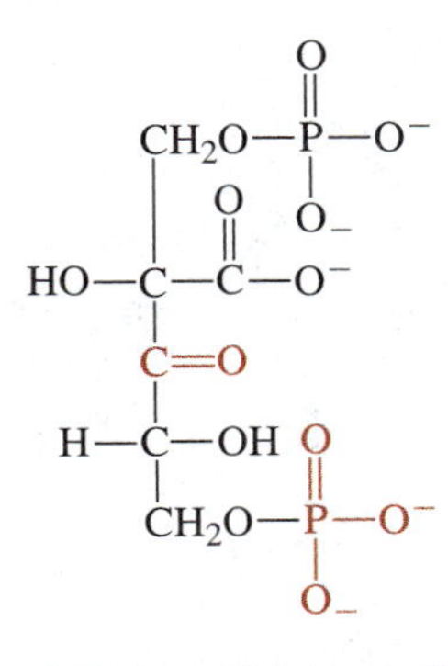

β-Keto acid intermediate in rubisco-catalyzed CO_2 fixation.

FIGURE 15.24
2-Carboxyarabinitol 1-phosphate.

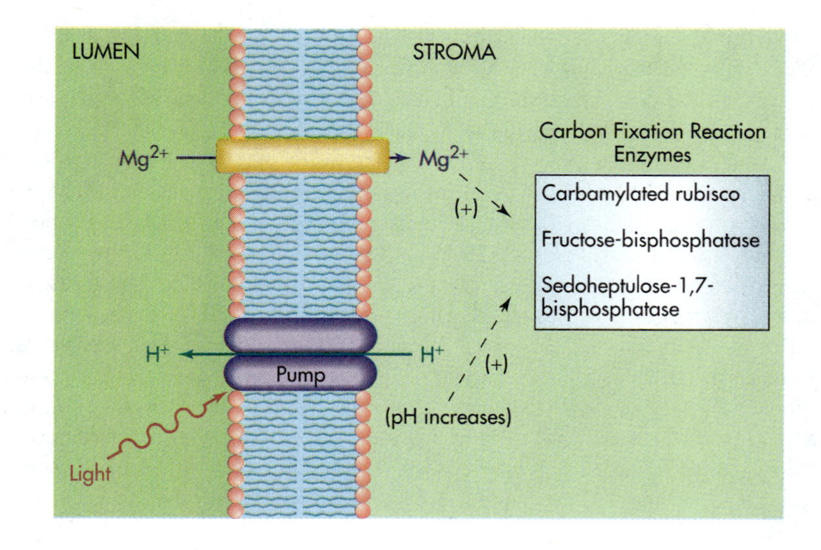

FIGURE 15.25
H^+ and Mg^{2+} regulation of photosynthesis. Light-induced shifts in pH and Mg^{2+} concentrations enhance the catalytic activity of several carbon fixation cycle enzymes.

a disulfide-containing protein named **thioredoxin.** A disulfide bond in thioredoxin is reduced to two thiol groups in the process. Thioredoxin is returned to its disulfide (oxidized) state as it reduces a disulfide bond in one of the regulated enzymes, a process described as **disulfide exchange.** This reaction, which is catalyzed by **thioredoxin reductase,** leads to alterations in the tertiary structure and to an enhancement of the catalytic capabilities of the regulated enzyme. **[Try Problem 15.26]**

The regulation of photosynthesis in chloroplasts is coupled to the regulation of glycolysis and gluconeogenesis in the cytosol. During active photosynthesis (in the light), dihydroxyacetone phosphate and glyceraldehyde 3-phosphate are exported from chloroplasts to the cytosol where they allosterically modulate cytosolic enzymes in-

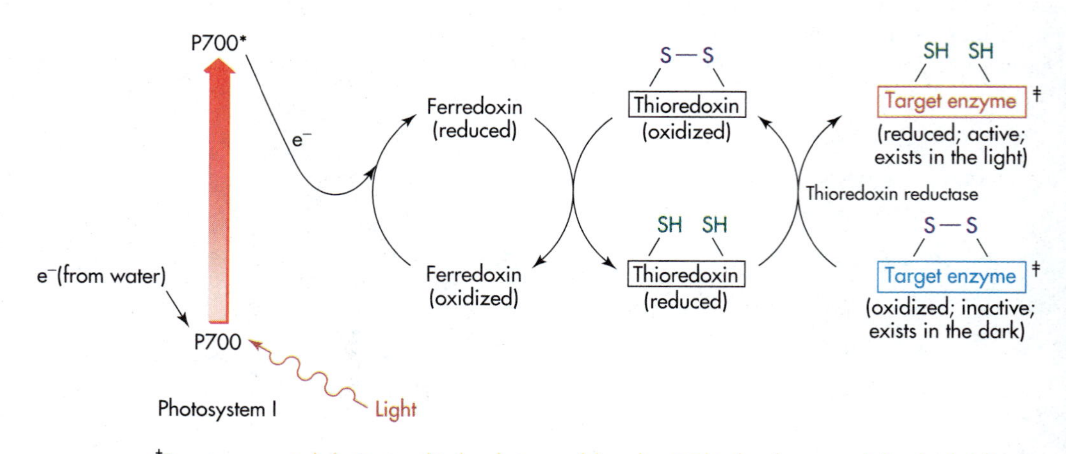

‡Target enzymes include: Fructose-bisphosphatase, sedoheptulose-1,7-bisphosphatase, and phosphoribulokinase.

FIGURE 15.26
Light-induced, reductive activation of carbon fixation cycle enzymes. Electrons from photosystem I are used to activate target enzymes by reducing disulfide bonds.

volved in the production and breakdown of fructose 2,6-bisphosphate. The decrease in fructose 2,6-bisphosphate that results leads to an increase in the rate of gluconeogenesis and decrease in the rate of glycolysis. Fructose 2,6-bisphosphate plays a similar role in the reciprocal regulation of glycolysis and gluconeogenesis in humans where fructose 2,6-bisphosphate concentrations are regulated primarily by hormones (Section 14.3). It makes sense to beef up glucose reserves when plants have an abundant supply of energy in the form of sunlight. In the dark, an increase in fructose 2,6-bisphosphate concentration in the cytosol of plants stimulates glycolysis while suppressing glucose production. Fructose 2,6-bisphosphate–stimulated glycolysis serves as an alternate source of essential energy when energy is not available from sunlight. Photosynthesis, glycolysis, and gluconeogenesis must all be coordinated to avoid futile cycles and to ensure that cells have a constant supply of useable energy in both the presence and absence of sunlight. **[Try Problems 15.27 through 15.30]**

PROBLEM 15.24 Write the equation for the carbamylation of a lysine side chain. Provide structural formulas for all reactants and products. Explain why carbamylation tends to change the conformation of an enzyme.

PROBLEM 15.25 Prepare a graph illustrating the type of results obtained when the rate of the rubisco-catalyzed reaction (y-axis) is plotted against 2-carboxyarabinitol 1-phosphate concentration (x-axis). Hint: Review Section 5.10.

PROBLEM 15.26 Write a balanced equation for the disulfide exchange reaction between $HSCH_2CH_2CH_2SH$ and $CH_3CH_2SSCH_2CH_2CH_3$.

PROBLEM 15.27 List those substances identified in this section that modulate rubisco activity by interacting directly with rubisco.

PROBLEM 15.28 Summarize all of the mechanisms described in this section through which the catalytic activity of chloroplast fructose-bisphosphatase is regulated.

PROBLEM 15.29 True or false? Simultaneous photosynthesis and gluconeogenesis is a futile cycle. Explain.

PROBLEM 15.30 Which major biochemical themes (Exhibit 1.4) are featured in this section?

15.7 SOME ENERGY CONSIDERATIONS

From a thermodynamic standpoint, how efficient is photosynthesis? What percentage of the light energy used by photosystems I and II is converted to chemical energy in carbohydrates? Consider the following reaction for the oxidation of glucose (discussed in Section 11.11):

$$C_6H_{12}O_6 + 6\,O_2 \rightleftharpoons 6\,CO_2 + 6\,H_2O \quad \Delta G^{\circ\prime} = -2823 \text{ kJ/mol}$$

Under standard conditions, the photosynthesis-linked production of glucose is basically the reverse of this process:

$$2823 \text{ kJ/mol (from light)} + 6\ CO_2 + 6\ H_2O \rightleftharpoons C_6H_{12}O_6 + 6\ O_2$$

Four electrons are transferred through the photosynthetic electron-transport system for each O_2 generated. The transfer of each electron requires two photons, one to activate P680 in photosystem II and a second to activate P700 in photosystem I. Thus, a total of eight photons (2 photons/electron × 4 electrons) are absorbed during the production of each O_2, and the generation of six O_2 (and one $C_6H_{12}O_6$) requires 48 photons.

The energy content of 1 einstein of visible light ranges from about 300 kJ for 400-nm blue light to roughly 170 kJ for 700-nm red light (Figure 15.3). One einstein equals 1 mol of photons (light waves) which equals 6.023×10^{23} photons. Since 48 einsteins are required to produce a mole of glucose, the minimum input energy (assuming that only long-wavelength red light is utilized) is close to 8160 kJ (48 einstein × 170 kJ/einstein). This means that, at best, around 35% [(2823 kJ/8160 kJ) × 100] of the light energy used for photosynthesis is transformed into chemical energy within glucose. A similar efficiency is associated with the trapping of energy in ATP during the oxidation of glucose (Section 11.11). **[Try Problem 15.31]**

PROBLEM 15.31 Calculate the approximate percentage of light energy trapped in glucose if photosynthesis is powered by 400 nm light rather than 700 nm light.

15.8 FATE OF FIXED CARBON

The previous section indicates that glucose can be produced from carbon fixed during photosynthesis. This raises the general question, "What is the fate of the carbon fixed during photosynthesis?" This section provides a partial answer to this question as it relates to green plants.

The fixation of three CO_2 within the carbon fixation cycle leads to the net production of one glyceraldehyde 3-phosphate (Figure 15.17). Normally, roughly half of the triose phosphate molecules are converted to starch within the stroma of chloroplasts. Most of the remainder are exported to the cytosol, where they are transformed into sucrose, the transport form of carbohydrates in plants.

The production of starch from glyceraldehyde 3-phosphate is summarized in Figure 15.27. After some glyceraldehyde 3-phosphate is converted to dihydroxyacetone phosphate, the dihydroxyacetone phosphate condenses with some remaining glyceraldehyde 3-phosphate to yield fructose 1,6-bisphosphate. Following the hydrolysis of fructose 1,6-bisphosphate, the fructose 6-phosphate generated is transformed into glucose 6-phosphate which, in turn, is converted to glucose 1-phosphate. Glucose 1-phosphate reacts with ATP to yield ADP-glucose, which transfers a glucosyl group to a nonreducing end of a starch molecule. Starch production is strictly analogous to glycogen biosynthesis (Section 14.5), except ADP-glucose, instead of UDP-glucose, serves as the glucosyl donor. The names of the enzymes involved in starch production are given in Figure 15.27. With the exception of **ADP-glucose pyrophosphorylase** and **starch synthase,** these enzymes, or isozymes of these enzymes, have been encountered previously. **[Try Problem 15.32]**

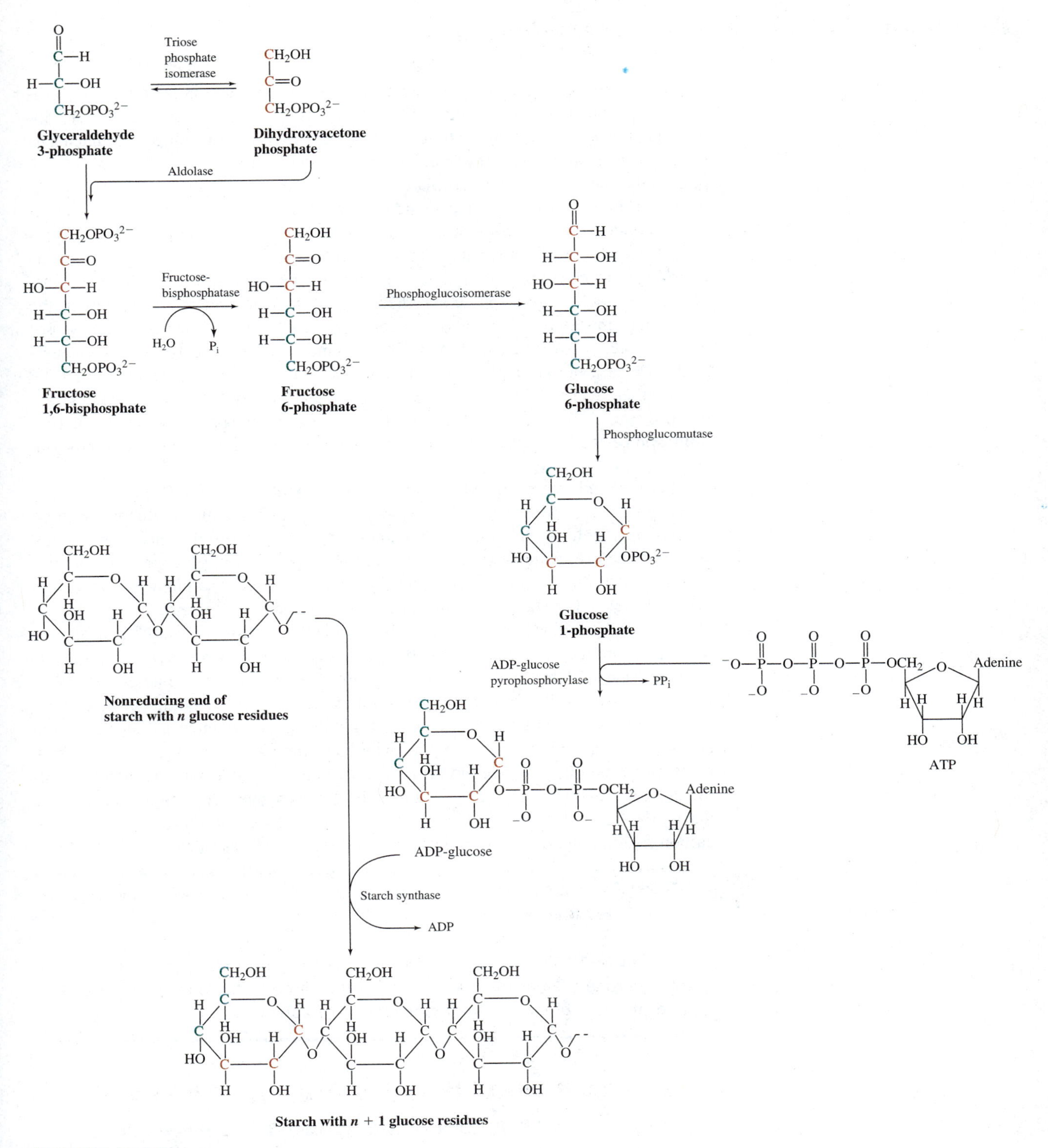

FIGURE 15.27
Conversion of glyceraldehyde 3-phosphate to starch.

Glyceraldehyde 3-phosphate destined for sucrose production is converted to dihydroxyacetone phosphate, which enters the cytosol through an antiport system; inorganic phosphate moves into cholorplasts as dihydroxyacetone phosphate makes its escape. This is an important exchange, because the entering phosphate replaces the phosphate that leaves the chloroplasts attached to the triose phosphate. Part of the cytosolic dihydroxyacetone phosphate is converted to glyceraldehyde 3-phosphate, which condenses with some of the remaining dihydroxyacetone phosphate to produce fructose 1,6-bisphosphate (Figure 15.28). After fructose 1,6-bisphosphate is hydrolyzed to fructose 6-phosphate, a portion of the fructose-6-phosphate is transformed to glucose 6-phosphate and then to glucose 1-phosphate. To this point, the initial steps in starch biosynthesis in chloroplasts have been duplicated in the cytosol with the catalytic assistance of cytosolic isozymes. The glucose 1-phosphate generated follows a unique path as sucrose biosynthesis continues. It reacts with UTP (rather than ATP) to form UDP-glucose (rather than ADP-glucose). The UDP-glucose transfers a glucosyl group to fructose 6-phosphate to form sucrose 6-phosphate. Hydrolysis of the sucrose 6-phosphate yields sucrose. **[Try Problem 15.33]**

Sucrose and starch produced from photosynthesis-derived glyceraldehyde 3-phosphate are storage forms of energy and carbon. Some of the glucose in these compounds, along with some photosynthesis-derived glucose not incorporated into starch or sucrose, is incorporated into cellulose, a polymer of glucose that is an important structural component in plants (Section 6.6). Cellulose biosynthesis, which is closely related to starch biosynthesis, is not examined in this text.

In the dark, photosynthetic plant cells mobilize the hexoses in starch and sucrose for energy production through oxidation, a process that involves glycolysis, the citric acid cycle, and the respiratory chain (Chapter 11). In plant cells incapable of photosynthesis (root cells, for example), carbohydrates from photosynthetically competent neighbors are oxidized both day and night. Some of the intermediates of starch and sucrose catabolism are used for anabolism, including the assembly of amino acids, proteins, fats, and nucleic acids. The carbon in sucrose and starch comes, ultimately, from atmospheric CO_2, whereas the energy comes from photons harvested during the light reactions of photosynthesis. The next section returns to the topic of carbon fixation and examines a puzzling catalytic activity associated with rubisco.

PROBLEM 15.32 What is the most likely source of the ATP used during starch biosynthesis?

PROBLEM 15.33 Is the energy required to incorporate a glucose residue into sucrose greater than, less than, or the same as the energy required to incorporate a glucose residue into starch? Explain.

15.9 PHOTORESPIRATION AND C_4 PLANTS

Rubisco is also called RuBP carboxylase/oxygenase. The "carboxylase" in the latter name is based on rubisco's ability to catalyze the fixation of CO_2. Why is "oxygenase" included in the name? Oxygen is an alternate substrate for rubisco; it

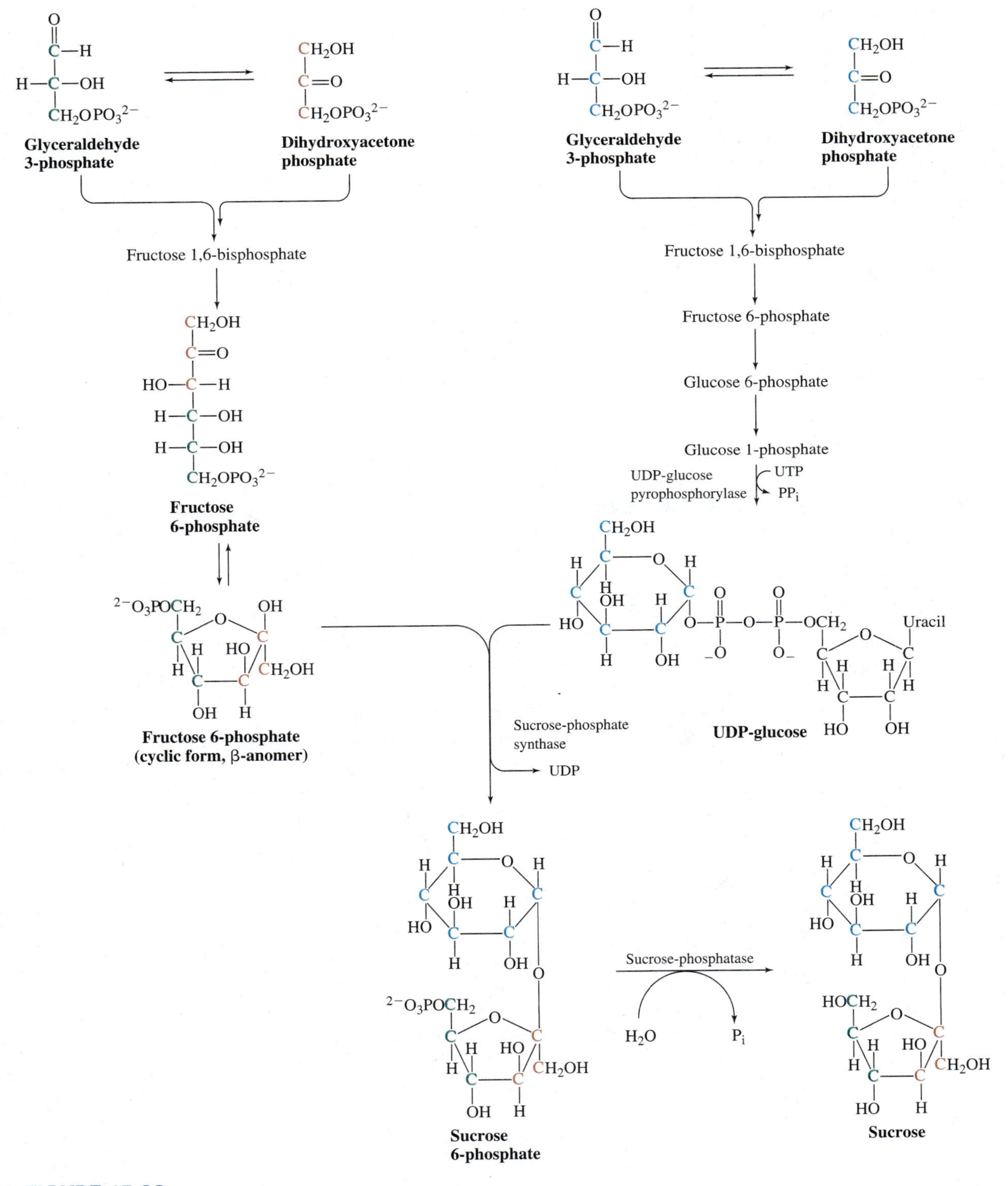

FIGURE 15.28
Conversion of glyceraldehyde 3-phosphate to sucrose.

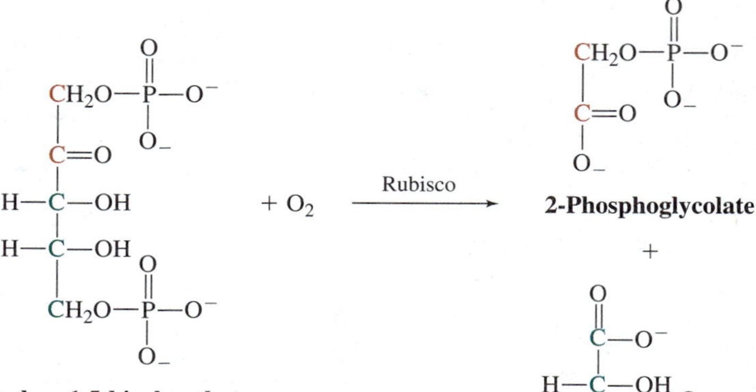

FIGURE 15.29
Oxygenase-catalyzed reaction.

competes with CO_2 for the same active sites, an example of substrate inhibition (Section 5.10). With O_2 at the active site, ribulose 1,5-bisphosphate is catalytically converted to 2-phosphoglycolate and 3-phosphoglycerate in the absence of carbon fixation (Figure 15.29). Since the 2-phosphoglycolate has no known function and energy is required to convert it to a useful metabolite, rubisco's oxygenase activity is indeed puzzling. Some speculate that it is a consequence of the initial evolution of rubisco in the absence of O_2, where there was no evolutionary pressure to select enzymes that could distinguish CO_2 from O_2. Others suggest that the oxygenase activity provides some benefit to plants, possibly regulating O_2 concentrations to help minimize oxidative damage. The construction and testing of hypotheses is one of the central themes in biochemistry and other natural sciences (Exhibit 1.4). **[Try Problem 15.34]**

The metabolic pathway that salvages 2-phosphoglycolate consumes additional O_2 and generates CO_2. For this reason, the production and salvage of 2-phosphoglycolate is known as **photorespiration.** The ratio of carbon fixation to photorespiration varies with growth conditions, including temperature. Higher temperatures and bright sunlight lead to a rapid rate of photosynthesis that tends to decrease the intracellular concentration of CO_2 (consumed during photosynthesis) and increase the concentration of O_2 (a product of photosynthesis). Any increase in the O_2/CO_2 ratio increases the rate of photorespiration relative to the rate of photosynthesis (Figure 15.30). Under some growth conditions, photorespiration may inhibit net biomass production by as much as 50%. **[Try Problem 15.35]**

Some plants of tropical origin have, through evolution, acquired mechanisms to increase the ratio of carbon fixation to photorespiration. These plants possess leaves whose design and operation greatly suppress photorespiration (Figure 15.31). A di-

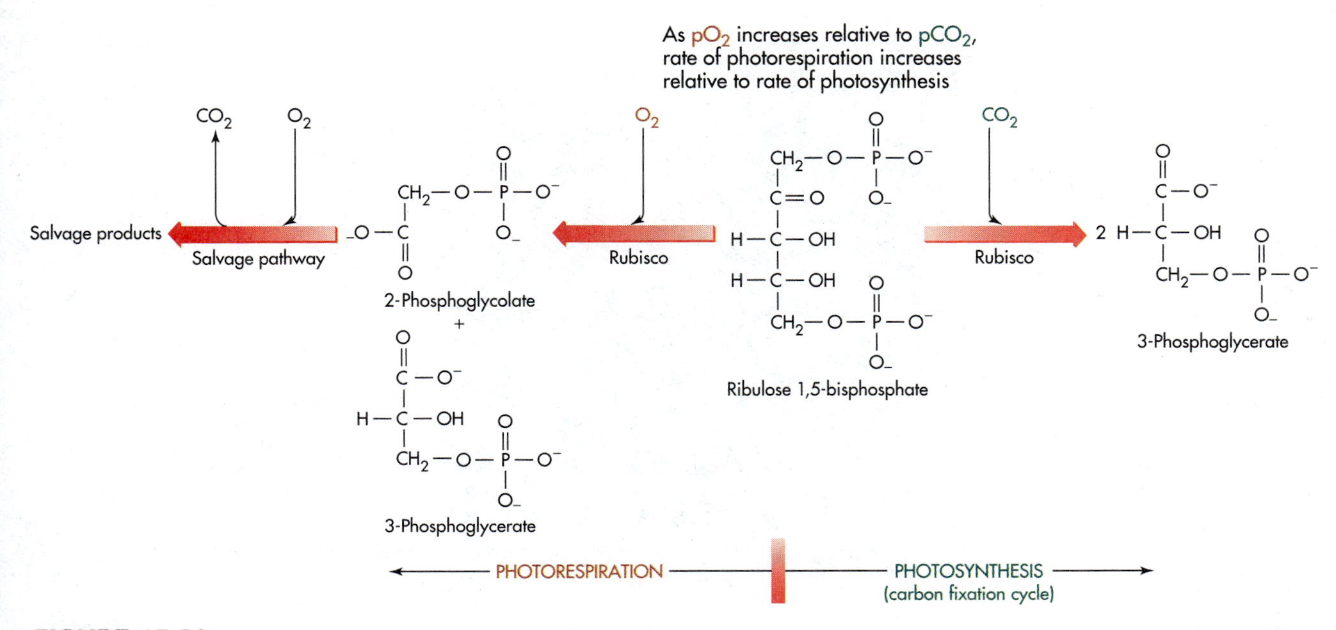

FIGURE 15.30
Impact of pO_2/pCO_2 ratio on photorespiration and photosynthesis.

vision of labor between outer **mesophyll cells** and inner **bundle-sheath cells** plays a crucial role in the suppression. Mesophyll cells lack rubisco but contain an enzyme, **phosphoenolpyruvate carboxylase (PEP carboxylase),** that catalyzes the carboxylation of phosphoenolpyruvate (PEP) to yield oxaloacetate (Figure 15.32):

$$\text{Phosphoenolpyruvate (PEP)} + HCO_3^- \rightleftharpoons \text{oxaloacetate} + P_i$$

The HCO_3^- used by PEP carboxylase comes from the reaction of atmospheric CO_2 with water:

$$CO_2 + H_2O \rightleftharpoons H_2CO_3 \rightleftharpoons HCO_3^- + H^+$$

PEP carboxylase–catalyzed carbon fixation is described as C_4 fixation because the fixation product (oxaloacetate) contains four carbons. Plants capable of C_4 fixation are known as **C_4 plants.** Since rubisco-catalyzed carbon fixation generates three-carbon products, it is called C_3 fixation. **C_3 plants** are only capable of C_3 fixation. **[Try Problem 15.36]**

The oxaloacetate produced during the PEP carboxylase–catalyzed reaction in mesophyll cells is converted to malate with the catalytic assistance of malate dehydrogenase. The malate is transported into underlying bundle-sheath cells where **malic enzyme** catalyzes its decarboxylation to yield CO_2 and pyruvate. The pyruvate is returned to mesophyll cells, where **pyruvate phosphate dikinase** catalyzes its transformation into PEP, which can be used for another round of C_4 carbon fixation. The CO_2 released during the decarboxylation of malate leads to a locally high concentration of this metabolite within bundle-sheath cells. The high CO_2 concen-

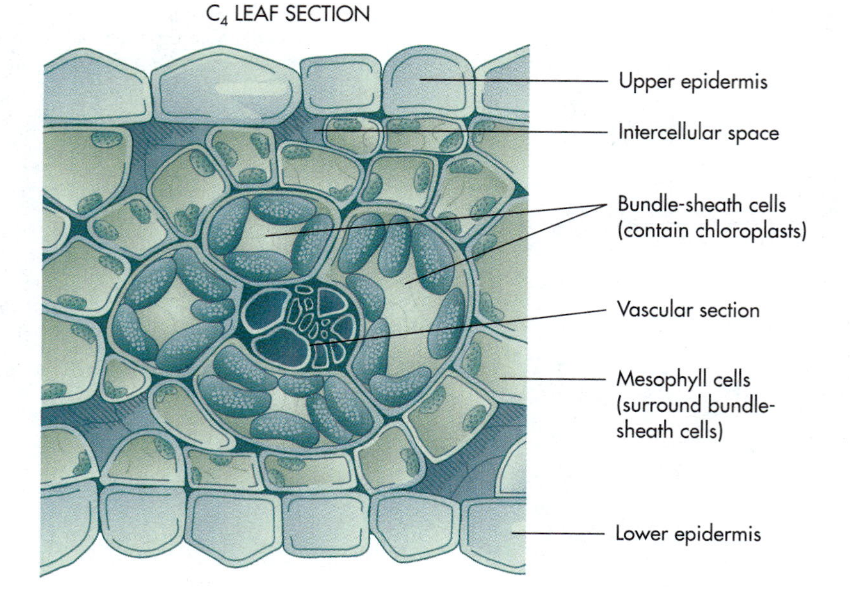

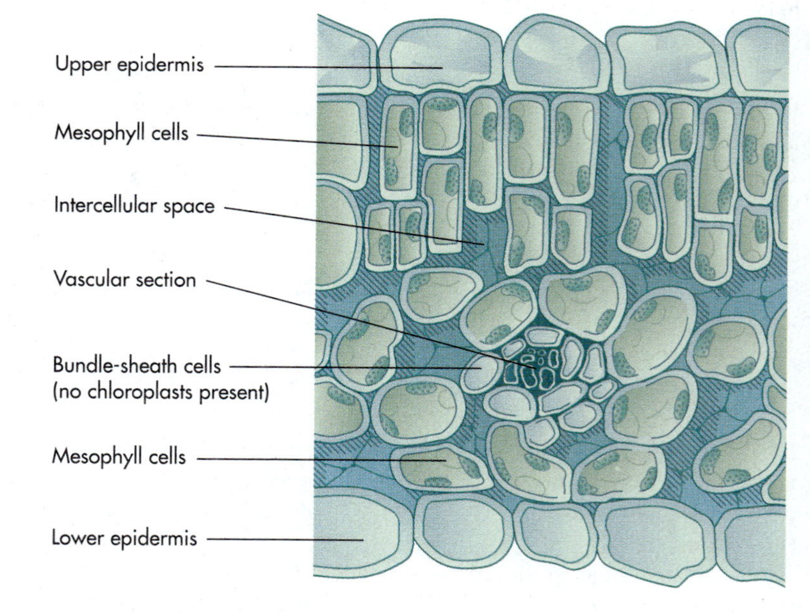

FIGURE 15.31
Leaf structure in C_4 and C_3 plants. In C_4 leaves, large bundle-sheath cells containing many chloroplasts are completely surrounded by mesophyll cells that lack rubisco and are unable to perform C_3 carbon fixation.

tration, coupled with a relatively low O_2 concentration, allows rubisco-catalyzed CO_2 fixation to proceed with little competition from photorespiration. Bundle-sheath cells possess rubisco plus all of the other enzymes that participate in the C_3 carbon fixation cycle (Figures 15.19 through 15.21). C_4 fixation is temporary fixation because its function is to deliver CO_2 to the bundle-sheath cells. In contrast, C_3 fixation is of a more permanent nature.

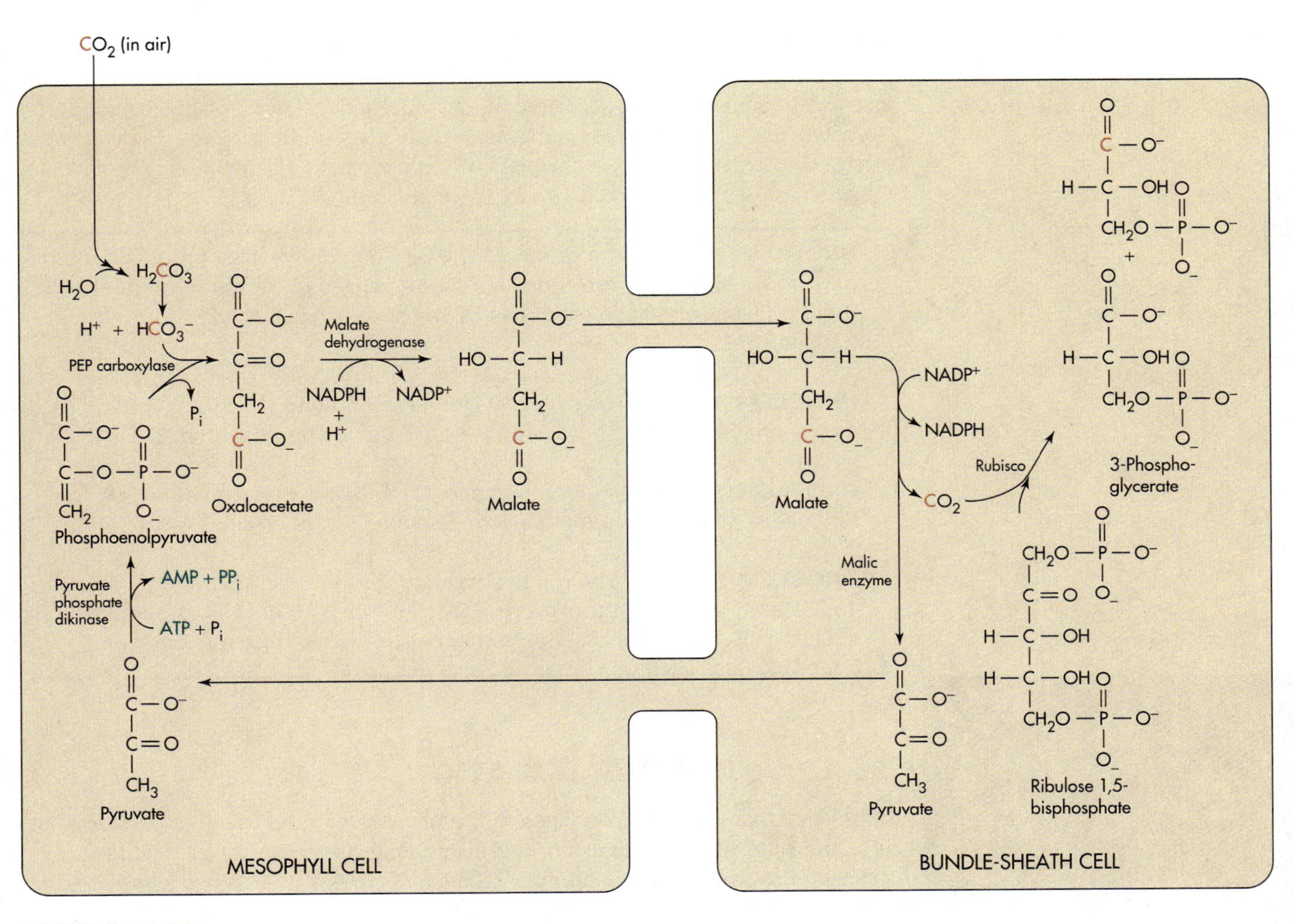

FIGURE 15.32
Carbon fixation in C_4 plants. CO_2 is fixed temporarily during a rubisco-independent reaction that is localized in outer mesophyll cells where the CO_2 to O_2 ratio is relatively low. The CO_2 is subsequently released in underlying bundle-sheath cells where a much higher CO_2 to O_2 ratio enhances rubisco-linked carbon fixation and suppresses photorespiration.

There are multiple varieties of C_4 plants, including some that operate through slightly different mechanisms than those just described. Permanent carbon fixation by all C_4 plants bears a higher cost than standard C_3 carbon fixation, because additional ATP is consumed during the metabolic cycle that delivers CO_2 fixed in the mesophyll cells to the C_3 system within bundle-sheath cells (Figure 15.32). However, under the conditions where C_4 plants usually grow, the higher energy cost is offset by a marked reduction in photorespiration and an increase in photosynthetic efficiency based on quantum yield. When light intensity and temperature are relatively high, as is common in the tropics much of the year and common in other regions during summer months, C_4 plants have a marked growth advantage over C_3 plants. C_4 plants may also have a growth advantage at very high altitudes where the ambient concentration of atmospheric CO_2 is relatively low. This possibility is presently of major interest to scientists attempting to predict the global consequences of an increasing concentration of atmospheric CO_2. C_4 plants include corn, sorghum, and sugarcane, plus crabgrass and a variety of other weeds. Genetic engineers may some day increase the yield of some crops by programming C_3 plants to function as C_4 plants (Section 19.12). [Try Problem 15.37]

PROBLEM 15.34 True or false? The 3-phosphoglycerate produced by the oxygenase activity of rubisco will be metabolized in the same manner as the 3-phosphoglycerate produced by the carboxylase activity of rubisco. Explain.

PROBLEM 15.35 True or false? Nocturnal inhibitor (2-carboxyarabinitol 1-phosphate) inhibits photorespiration as well as photosynthesis. Explain.

PROBLEM 15.36 Would you expect O_2 to be an alternate substrate for PEP carboxylase as it is for rubisco? Explain.

PROBLEM 15.37 Write the net reaction for the reaction cycle (Figure 15.32) in C_4 plants that transports CO_2 from mesophyll cells to bundle-sheath cells. Does this reaction represent a waste of energy? Explain.

FIGURE 15.33
Nitrogen-fixing nodules on soybean roots.

15.10 NITROGEN FIXATION

Nitrogen fixation, like carbon fixation, is a phenomenon of general importance that is restricted to a limited number of organisms. Only a few species of bacteria are genetically programmed to fix nitrogen (convert N_2 to NH_3). Although some of these bacteria are free living, others live as symbionts in the root nodules of leguminous plants, including beans, peas, alfalfa, and clover (Figure 15.33). The nitrogen-fixing bacteria provide the plants with nitrogen, and the plants supply the bacteria with organic compounds for energy production and other uses. In a symbiotic relationship, all participants must benefit.

Agricultural crop rotation commonly involves planting legumes every few years to build up fixed nitrogen in soils. Nonleguminous plants rely upon ammonia (NH_3), nitrite (NO_2^-), and nitrate (NO_3^-) in the soil to meet their nitrogen requirements. Nitrate is the predominant storage form of nitrogen in soil.

Nitrogenase Complex

Nitrogen fixation is performed with the catalytic assistance of two multisubunit enzymes that are packaged together in a complex known as **nitrogenase** or the **nitrogenase complex.** One enzyme, **dinitrogenase (nitrogenase) reductase,** is an iron–sulfur protein with a single Fe_4–S_4 redox cluster (center). The second enzyme, **dinitrogenase (nitrogenase),** is an iron–sulfur–molybdenum protein with multiple

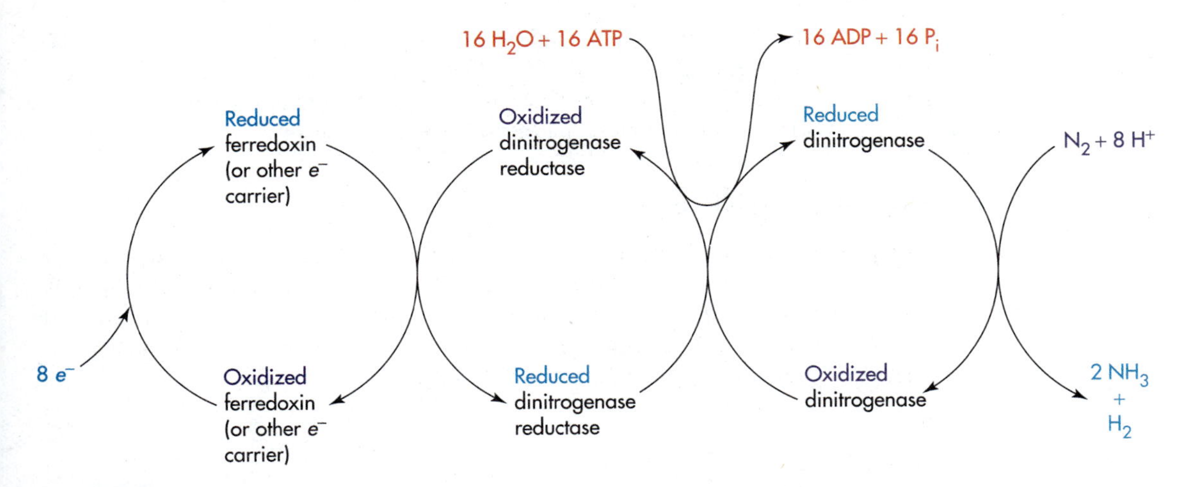

FIGURE 15.34
Nitrogen fixation by the nitrogenase complex. The eight electrons flow through the complex one at a time. Two ATPs are hydrolyzed for each electron delivered to N_2 or H^+. The ATPs bind to dinitrogenase reductase before they are hydrolyzed.

iron–sulfur clusters and two iron–sulfur–molybdenum clusters. In some dinitrogenases, vanadium replaces molybdenum. **[Try Problem 15.38]**

Electrons for the reduction of N_2 usually come from the oxidation of organic compounds or from photosynthesis. Ferredoxin is often the compound that delivers these electrons to the nitrogenase complex. When this is the case, most electrons flow from ferredoxin to dinitrogenase reductase to dinitrogenase to N_2 (Figure 15.34). At least part of the metal–sulfur clusters serve as electron carriers during the electron transport process. Two dinitrogenase reductase–bound ATP are hydrolyzed for each electron transferred. ATP binding and hydrolysis alter the conformation for the reductase, and possibly the conformation of the dinitrogenase as well. The changes in conformation facilitate the unidirectional flow of electrons through the nitrogenase complex. Thus, ATP contributes to catalysis; it does not provide a thermodynamic driving force for nitrogen fixation.

Although only six electrons are required to reduce N_2 to two NH_3, eight electrons must flow through the nitrogenase complex for one N_2 to be reduced. The two extra electrons are linked to the obligatory production of one H_2 during the course of NH_3 generation. The mechanism of this process is still under investigation. The net reaction catalyzed by the nitrogenase complex is:

$$N_2 + 8\,H^+ + 8e^- + 16\,ATP + 16\,H_2O \rightleftharpoons 2\,NH_3 + 16\,ADP + 16\,P_i + H_2$$

Nitrogen fixation is definitely a costly process from the standpoint of energy consumption.

The nitrogenase complex is extremely sensitive to O_2 inactivation. For this reason, N_2 fixation can proceed only when O_2 concentrations are extremely low. Aerobic organisms employ a variety of strategies to eliminate O_2 at nitrogen fixation sites. Some partially uncouple oxidation and phosphorylation to enhance the oxidation of organic fuels and "burn off" O_2. Others produce specialized cells with a low permeability for O_2. Legumes produce an O_2-binding protein, **leghemoglobin,** that removes O_2 from their symbiotic partners. **[Try Problem 15.39]**

Fate of Fixed Nitrogen

After ammonia from biological nitrogen fixation, biological nitrate reduction, or other sources (Figure 15.1) is converted to $^+NH_4$ through reaction with an acid, it is incorporated *into plants* with the catalytic assistance of glutamine synthetase and glutamate dehydrogenase. Glutamine synthetase catalyzes the ATP-dependent production of glutamine from glutamate and $^+NH_4$:

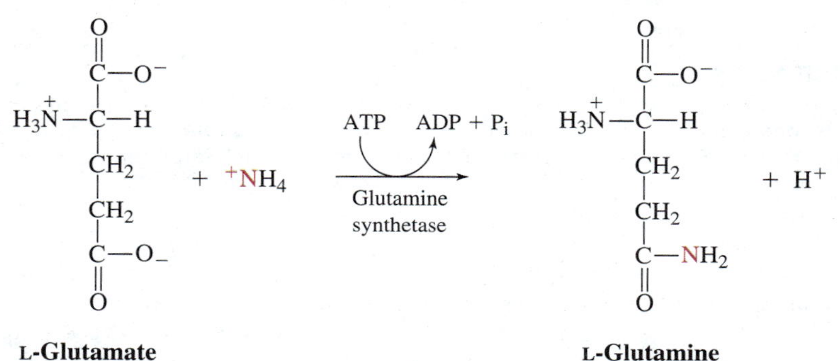

An isozyme of plant glutamine synthetase has been encountered in Chapter 13; it is used by mammals to convert toxic free $^+NH_4$ to glutamine, a transport form of $^+NH_4$ and amino groups in blood (Section 13.5). Glutamate dehydrogenase catalyzes the reductive amination of α-ketoglutarate:

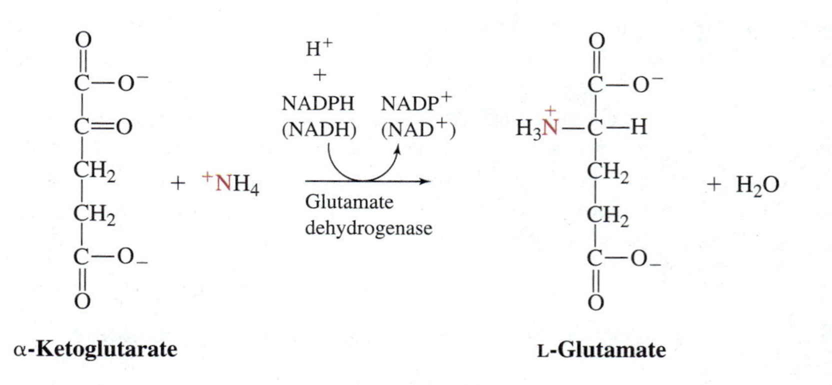

NADPH or NADH (depending on the organism involved) provides the reducing power and $^+NH_4$ provides the amino group for this process. In mammals, glutamate dehydrogenase catalyzes the reverse reaction during amino acid catabolism (Section

13.5). Glutamine and glutamate are used directly or indirectly as the source of nitrogen for the biosynthesis of other nitrogen-containing compounds, including other amino acids, purines, and pyrimidines (Chapter 14). **[Try Problem 15.40]**

Fertilizers

The nitrogen required for plant growth can be obtained from industrial nitrogen fixation as well as biological fixation. The Haber process employs 500°C, 300 atm of pressure, and an iron catalyst to transform a mixture of N_2 and H_2 into ammonia. Most of the fixed nitrogen is used in the manufacture of agricultural fertilizers. Since fixed nitrogen concentrations in soils frequently limit the growth rate of nonleguminous plants, nitrogen-rich synthetic fertilizers are widely utilized to enhance crop yields. Natural organic fertilizers containing animal excrements and dead plant material are commonly used for the same purpose. Biochemists may some day create self-fertilizing nonleguminous plants through genetic engineering (Chapter 19).

PROBLEM 15.38 Nitrogen fixation is a significantly exergonic reaction:

$$N_2 + 3\ H_2 \rightleftharpoons 2\ NH_3 \quad \Delta G^{\circ\prime} = -33.5\ \text{kJ/mol}$$

Why is a catalyst required for this reaction?

PROBLEM 15.39 Leghemoglobin is a monomeric protein with a single globin chain and an associated heme group. Does leghemoglobin bind O_2 more or less tightly than normal mammalian hemoglobin? Does leghemoglobin bind O_2 cooperatively? Explain. Hint: Review Section 4.9.

PROBLEM 15.40 During the biosynthesis of the purine ring, aspartate is the immediate source of the 1-nitrogen in this ring. Suggest how glutamate serves indirectly as the source of this nitrogen. Hint: Review Sections 14.10 and 14.12.

SUMMARY

Although photosynthesis and nitrogen fixation are of central importance to most living organisms, only certain organisms are genetically programmed to perform these metabolic feats. Most other organisms use O_2, organic compounds, and fixed nitrogen produced by the privileged few. Photosynthesis by autotrophs, coupled with the oxidation of organic compounds by heterotrophs, creates carbon and oxygen cycles. Nitrogen fixation and denitrification by bacteria lie at the heart of a biological nitrogen cycle.

Photosynthesis, which employs light energy to fix atmospheric carbon, can be summarized as follows:

$$n\ CO_2 + n\ H_2O + \text{light} \rightleftharpoons (CH_2O)_n + n\ O_2$$

It entails the light-induced electron-transport reactions (light reactions) plus the carbon fixation reactions, two separable processes that are normally tightly coordinated. During the light reactions in chloroplasts of green plants, light harvested by cholorphyll and other pigments within photosystems II and I is used to provide the reducing power that drives an electron-transport system containing a combination of membrane-bound and mobile carriers. Water is the ultimate electron donor and $NADP^+$ is normally the terminal electron acceptor. Chlorophylls, quinones, cytochromes, iron–sulfur proteins, and copper proteins all function as electron carriers during the electron transport process.

The flow of electrons through the electron carriers follows a Z-like pattern when standard reduction potential (*y*-axis) is plotted against electron carriers (*x*-axis). This flow of electrons is coupled to the production of a proton gradient across the thylakoid membrane. The resultant proton-motive force is utilized by a membrane-imbedded ATP synthase that catalyzes the production of ATP. This photophosphorylation proceeds through a mechanism very similar to that employed during oxidative phosphorylation in mitochondria.

The light reactions can be summarized as follows:

$$2\ H_2O + 2\ NADP^+ + 8\ \text{photons} \rightleftharpoons O_2 + 2\ NADPH + 2\ H^+ + \text{proton gradient}$$

$$\text{Proton gradient} + ADP + P_i \rightleftharpoons ATP + H_2O$$

Cyclic electron transport is used to adjust the ATP/NADPH ratio to best meet the ever changing metabolic needs of a cell. ATP and NADPH from the light reactions contain a significant fraction (possibly as much as one third) of the light energy captured by photosystems I and II.

Photosynthetic carbon fixation is a cyclic process that yields one net glyceraldehyde 3-phosphate for every three CO_2 fixed:

$$3\ CO_2 + 9\ ATP + 6\ NADPH + 5\ H_2O \rightleftharpoons \text{glyceraldehyde 3-phosphate} + 9\ ADP + 6\ NADP^+ + 8\ P_i$$

The cycle is conveniently divided into three stages. In stage one, rubisco catalyzes the reaction of three CO_2 with three ribulose 1,5-bisphosphates to yield six 3-phosphoglycerates. During the second stage, each of the six 3-phosphoglycerates is converted to glyceraldehyde 3-phosphate. In the final stage, five of the six glyceraldehyde 3-phosphates are transformed into three ribulose 1,5-bisphosphates, which replace those that initiated the cycle. The sixth glyceraldehyde 3-phosphate is the net yield in terms of carbon fixation. The required ATP and NADPH are provided by the light reactions.

The light reactions of photosynthesis are regulated primarily by the availability of light. The carbon fixation reactions are regulated by the availability of CO_2, ATP, and NADPH. They are further regulated at the level of specific carbon fixation cycle enzymes, including rubisco and fructose-bisphosphatase. Rubisco activity is a function of pH, Mg^{2+} concentration, extent of carbamylation (partially determined by CO_2 concentration), 2-carboxyarabinitiol 1-phosphate concentration, and rubisco activase activity. Disulfide exchange reactions, pH, and Mg^{2+} concentration all contribute to the modulation of the catalytic activity of fructose-bisphosphatase. Collectively, the regulatory mechanisms ensure that the carbon fixation cycle is activated in sunlight and that photosynthesis is coordinated with glycolysis and gluconeogenesis in order to meet the energy needs of a photosynthetic organism while avoiding futile cycles.

That portion of the glyceraldehyde 3-phosphate produced by the carbon fixation cycle that is not required to maintain the cycle is predominantly converted to starch in chloroplasts or to sucrose in the cytosol. During both starch and sucrose synthesis, glyceraldehyde 3-phosphate is transformed into glucose 1-phosphate. After glucose 1-phosphate reacts with a nucleoside triphosphate to form a nucleoside diphosphate–glucose, the glucosyl group is transferred to a nonreducing end of a growing starch molecule or to fructose 6-phosphate (during sucrose biosynthesis). Both starch and sucrose function as storage forms of carbon and energy in plants. The carbons in these compounds come from atmospheric CO_2, whereas their energy is derived from light harvested during the light reactions of photosynthesis.

Molecular oxygen (O_2) inhibits carbon fixation by serving as an alternate substrate for rubisco. The use of O_2 as a substrate makes rubisco an oxygenase and it leads to photorespiration, the production of phosphoglycolate, and the salvaging of this metabolite through a pathway that consumes O_2 and generates CO_2. C_4 plants suppress photorespiration by initially fixing CO_2 in a four-carbon compound in mesophyll cells that lack rubisco. The CO_2 is subsequently released in underlying bundle-sheath cells, where C_3 carbon fixation is performed under conditions that favor carbon fixation over photorespiration (locally high CO_2 concentrations and relatively low O_2 concentrations).

Some nitrogen-fixing bacteria live free, whereas others live as symbionts in the root nodules of legumes. Nitrogen fixation is catalyzed by the nitrogenase complex, which encompasses dinitrogenase reductase and dinitrogenase, two metal-containing enzymes. This complex catalyzes the transfer of electrons from the light reactions of photosynthesis, or an alternate source, to N_2. One H_2 is produced and 16 ATP are hydrolyzed for each N_2 reduced. The net reaction is:

$$N_2 + 8\,H^+ + 8\,e^- + 16\,ATP + 16\,H_2O \rightleftharpoons 2\,NH_3 + 16\,ADP + 16\,P_i + H_2$$

Since the nitrogenase complex is inactivated by O_2, nitrogen-fixing organisms employ various strategies to lower O_2 concentrations in these locations where nitrogen is fixed.

In plants, fixed nitrogen is normally incorporated into organic compounds by glutamine synthetase– and glutamate dehydrogenase–catalyzed reactions that generate glutamine and glutamate, respectively. Glutamine and glutamate serve directly or indirectly as the source of nitrogen for the biosynthesis of other nitrogen-containing compounds. Fertilizers containing industrially fixed nitrogen provide part of the nitrogen used by many nonleguminous crops.

EXERCISES

15.1 True or false? All (or virtually all) of the energy used by your body comes, indirectly, from sunlight. Explain.

15.2 What property of a wave of blue light accounts for its greater energy content relative to a wave of green light?

15.3 Suggest why evolution has led to organisms that use visible light, rather than UV or IR radiation, for photosynthesis. All three classes of radiation are present in sunlight.

15.4 True or false? If a compound absorbs no visible light, you can be confident that the compound will absorb no electromagnetic radiation of lower energy than visible light. Explain.

15.5 Place a curve on Figure 15.5 showing the absorption spectrum for a hypothetical compound that absorbs visible light more efficiently than any combination of the given photosynthetic pigments. What is the color of this compound?

15.6 What ions are created when a single electron is transferred from Fe^{2+} to Cu^{2+}?

15.7 What structural feature of photosynthetic pigments largely accounts for their ability to absorb visible light?

15.8 Why is it important that some antenna molecules have different absorption spectra than P680 and P700?

15.9 Explain why many compounds that absorb visible light efficiently would not make good antenna molecules.

15.10 If kept under low light conditions for prolonged periods, some phototrophs respond by increasing their production of chlorophyll and other photosynthetic pigments. How do these organisms benefit from this maneuver?

15.11 P680* of photosystem II has a low enough reduction potential to bypass the rest of the electron transfer system and deliver an electron directly to $NADP^+$. Suggest why this direct transfer of electrons never occurs. What impact would this direct transfer (if it occurred) have on the products of the light reactions?

15.12 The pH of the lumen is normally near 4.5 during photosynthesis, whereas the pH of the stroma is close to 8.0. What is the magnitude of the proton gradient under these conditions? This is the concentration gradient that drives ATP production.

15.13 Four different metals are components of one or more of the Z-scheme complexes. Identify where each of these metals appear within the Z-scheme and then describe its role or function.

15.14 Which favors cyclic electron transport, a high or a low ATP/NADPH ratio? Explain.

15.15 List the major similarities and differences between the electron-transport systems in mitochondria and those in chloroplasts.

15.16 Why is it important that green plants and other phototrophs be able to carry out oxidative phosphorylation as well as photophosphorylation?

15.17 Photosynthesis generates two substances, or classes of substances, that aerobic heterotrophs must acquire from their environment in order to survive. What are those two substances? Do heterotrophs produce any substances that phototrophs must acquire from their environment? If so, list these substances.

15.18 What would be the consequence of removing two molecules of glyceraldehyde 3-phosphate from a chloroplast following each round of the carbon fixation cycle? Explain.

15.19 Write the structural formula for the products formed during a transketolase-catalyzed reaction between ribose 5-phosphate and xylulose 5-phosphate. What is the significance of this reaction? Hint: Review Section 12.2.

15.20 Study Figure 15.21. Which of the reactions in this figure are, from a practical standpoint, irreversible? Explain. The irreversible steps in a metabolic pathway tend to be the points at which the pathway is regulated.

15.21 If an autotroph produced ATP and NADPH using energy from a nonorganic source other than light, could that autotroph use the same carbon fixation cycle employed by phototrophs? Explain.

15.22 Both rubisco activase and thioredoxin reductase are nonphotosynthetic enzymes involved in the regulation of photosynthesis. There are other nonphotosynthetic enzymes that, at least indirectly, are also involved in this regulation. Although not identified by their specific name in this chapter, their existence is noted or inferred. What reactions are catalyzed by these enzymes?

15.23 Suggest why the addition of 2-carboxyarabinitol 1-phosphate to a growth medium containing plant cells does not inhibit photosynthesis within these cells.

15.24 Does the 2-carboxyarabinitol 1-phosphate produced *in vivo* react stoichiometrically with rubisco? That is, does 1 mol of 2-carboxyarabinitol 1-phosphate inactivate 1 mol of rubisco? Explain.

15.25 List those mechanisms through which the light reactions of photosynthesis activate the carbon fixation reactions.

15.26 True or false? In general, an allosteric effector that inhibits glycolysis will probably inhibit photosynthesis as well (assuming that it is an effector for both pathways). Explain.

15.27 List two reasons why an increase in CO_2 concentration increases the rate of carbon fixation.

15.28 What is the net change in number of anhydride links during the transformation of two glyceraldehyde 3-phosphates into a glucose residue within starch?

15.29 If a plant cell is in need of extra energy, can the glyceraldehyde 3-phosphate generated during the carbon fixation cycle be oxidized directly without being converted to glucose, sucrose, or starch? Explain.

15.30 Can a single active site on rubisco simultaneously participate in both carbon fixation and photorespiration? Explain.

15.31 True or false? The O_2 produced during photosynthesis tends to inhibit further photosynthesis. Explain.

15.32 How does compartmentation (a major theme in biochemistry) contribute to the ability of C_4 cells to suppress photorespiration.

15.33 Prepare a graph illustrating the type of results obtained when rate of photosynthesis (y-axis) is plotted against O_2 concentration (x-axis) while holding CO_2 concentration constant.

15.34 What prevents C_3 carbon fixation from occurring in mesophyll cells of C_4 plants?

15.35 Would you expect both mesophyll cells and bundle-sheath cells of C_4 plants to produce nocturnal inhibitor (2-carboxyarabinitol 1-phosphate) when in the dark? Explain.

15.36 How many photons are required to provide the electrons needed to reduce one N_2 to two NH_3 when legume photosynthesis is the source of the electrons?

15.37 True or false? In phototrophs, nitrogen fixation occurs in chloroplasts. Explain.

15.38 Could a phototrophic bacterium use photorespiration to help reduce O_2 concentrations in preparation for nitrogen fixation? Explain.

15.39 NH_3 is a gas. What prevents it from escaping from a cell after nitrogen fixation?

SELECTED READINGS

Barber, J., Short-circuiting the Z-scheme, *Nature* 376, 388–389 (1995).

Describes a mutant green algae that can, under anaerobic conditions, reduce $NADP^+$, fix CO_2, and grow photoautotrophically in the absence of a functional photosystem I. Thus, under some circumstances, photosystem II alone can support photosynthesis.

Barber, J., and B. Andersson, Revealing the blueprint of photosynthesis, *Nature* 370, 31–34 (1994).

A brief, crisp review of photosynthesis.

Chapin, F. Stuart III, New cog in the nitrogen cycle, *Nature* 377, 199–200 (1995).

Reviews some of the evidence that organic nitrogen is a major and direct source of nitrogen for plants in infertile ecosystems.

Govindjee, and W. J. Coleman, How plants make oxygen, *Sci. Am.* 262(2), 50–58 (1990).

Explores the reactions of the water-splitting complex.

Hartman, F. C., and M. R. Harpel, Structure, function, regulation, and assembly of D-ribulose-1,5-bisphosphate carboxylase/oxygenase, *Annu. Rev. Biochem.* 63, 197–234 (1994).

A detailed analysis of rubisco.

Howard, J. B., and D. C. Rees, Nitrogenase: A nucleotide-dependent molecular switch, *Annu. Rev. Biochem.* 63, 235–264 (1994).

An examination of nitrogen fixation from the standpoint of structure–function relationships within the nitrogenase complex.

Kühlbrandt, W., Many wheels make light work, *Nature* 374, 497–498 (1995).

Reviews the atomic structure of a bacterial light-harvesting complex.

Lehninger, A. L., D. L. Nelson, and M. M. Cox, *Principles of Biochemistry,* 2nd ed. New York: Worth Publishers, 1993.

An encyclopedia-type text that covers both photosynthesis and nitrogen fixation. Contains a limited number of problems and some selected readings.

Moran, L. A., K. G. Scrimgeour, H. R. Horton, R. S. Ochs, and J. D. Rawn, *Biochemistry,* 2nd ed. Englewood Cliffs, NJ, Neil Patterson/Prentice-Hall, 1994.

An encyclopedia-type biochemistry text containing a chapter on photosynthesis and a separate discussion of nitrogen fixation. Includes selected readings.

Nisbet, E. G., J. R. Cann, and C. L. Van Dover, Origins of photosynthesis, *Nature* 373, 479–480 (1995).

Discusses observations supporting the hypothesis that photosynthesis was originally an accidental by-product of thermal detection in a chemotrophic organism.

Voet, D., and J. G. Voet, *Biochemistry,* 2nd ed. New York: John Wiley & Sons, 1995.

A comprehensive text with a chapter on photosynthesis and a separate discussion of nitrogen fixation. Encompasses both problems and selected readings.

Youvan, D. C., and B. L. Marrs, Molecular mechanisms of photosynthesis, *Sci. Am.* 256(6), 42–48 (1987).

Analyzes the reaction center of a photosynthetic bacterium.

16 Nucleic Acids

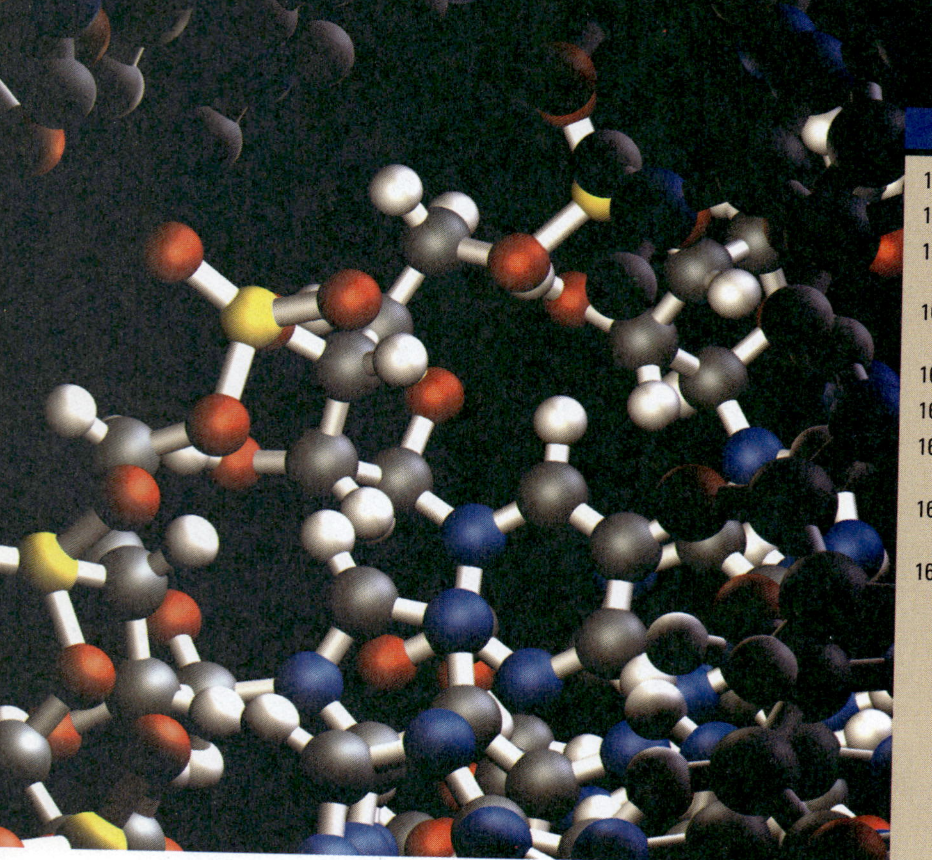

CHAPTER OUTLINE

Chapters 11 through 14 have provided an introduction to the metabolism of carbohydrates, fats, amino acids, and nucleotides. Chapter 15 examines photosynthesis and nitrogen fixation. With very rare exceptions, each of the numerous reactions encountered in these chapters is catalyzed by a separate, highly specific enzyme. To a very large extent, a cell controls its metabolism by regulating what enzymes it produces and then regulating the concentration and catalytic capabilities of these enzymes. The regulation of enzyme activity and reaction rates is one of the most pervasive central themes of biochemistry. Information for the synthesis of enzymes is carried in DNA, and the expression of that information entails other nucleic acids as well.

Nucleic acids are polymers of nucleotides (Section 7.4). They are divided into two major classes: deoxyribonucleic acid (DNA) and ribonucleic acid (RNA). DNA is the genetic material in all living organisms, including humans; it carries that information which is passed from generation to generation. The final expression of most genetic information consists of the production of an RNA or polypeptide. A majority of the RNAs participate directly or indirectly in polypeptide assembly. The polypeptides include the enzymes and enzyme subunits that participate in metabolism. It is crucial that these relationships be kept in focus as you attempt to develop a broad and integrated picture of those molecular reactions and interactions that create and maintain life.

This chapter examines the structure and chemical characteristics of nucleic acids and the general roles of DNA, mRNAs, tRNAs, and rRNAs in the storage and expression of genetic information. The major biochemical themes encountered include: experimentation and hypothesis construction; specificity; information transfer; structural hierarchy; evolution, diversity and change; and structure–function relationships. The overriding theme is the storage and transfer of information. This theme is developed further in Chapters 17 and 18, which explore some of the details of nucleic acid and protein biosynthesis.

16.1 DEOXYRIBONUCLEIC ACID

All of the DNA within a cell or organism is said to represent its genome (Table 16.1). In a eukaryotic cell, the vast majority of its DNA is located in nuclear DNA–protein complexes that resemble strings of beads. Such complexes are known as chromatin (Section 16.3). Individual strands of chromatin coil and fold into highly condensed structures known as chromosomes. Human cells contain 23 pairs of chromosomes with one member of each pair contributed by each parent.

Genes are segments of DNA that carry information for the synthesis of specific polypeptides or RNAs. Certain regulatory segments of DNA that do not encode polypeptides or RNA are also classified as genes by some authors. The DNA in a chromosome consists of a sequence of genes that is interspersed with polynucleotide segments that, for the most part, serve no known function. Surprisingly, the estimated 100,000 genes in the human genome account for less than 5% of the entire genome. At least a portion of the non-gene-containing part of the genome appears to participate in the regulation of gene expression and the maintenance of genome structure. **[Try Problems 16.1 and 16.2]**

TABLE 16.1

The Size of Genomes[a,b]

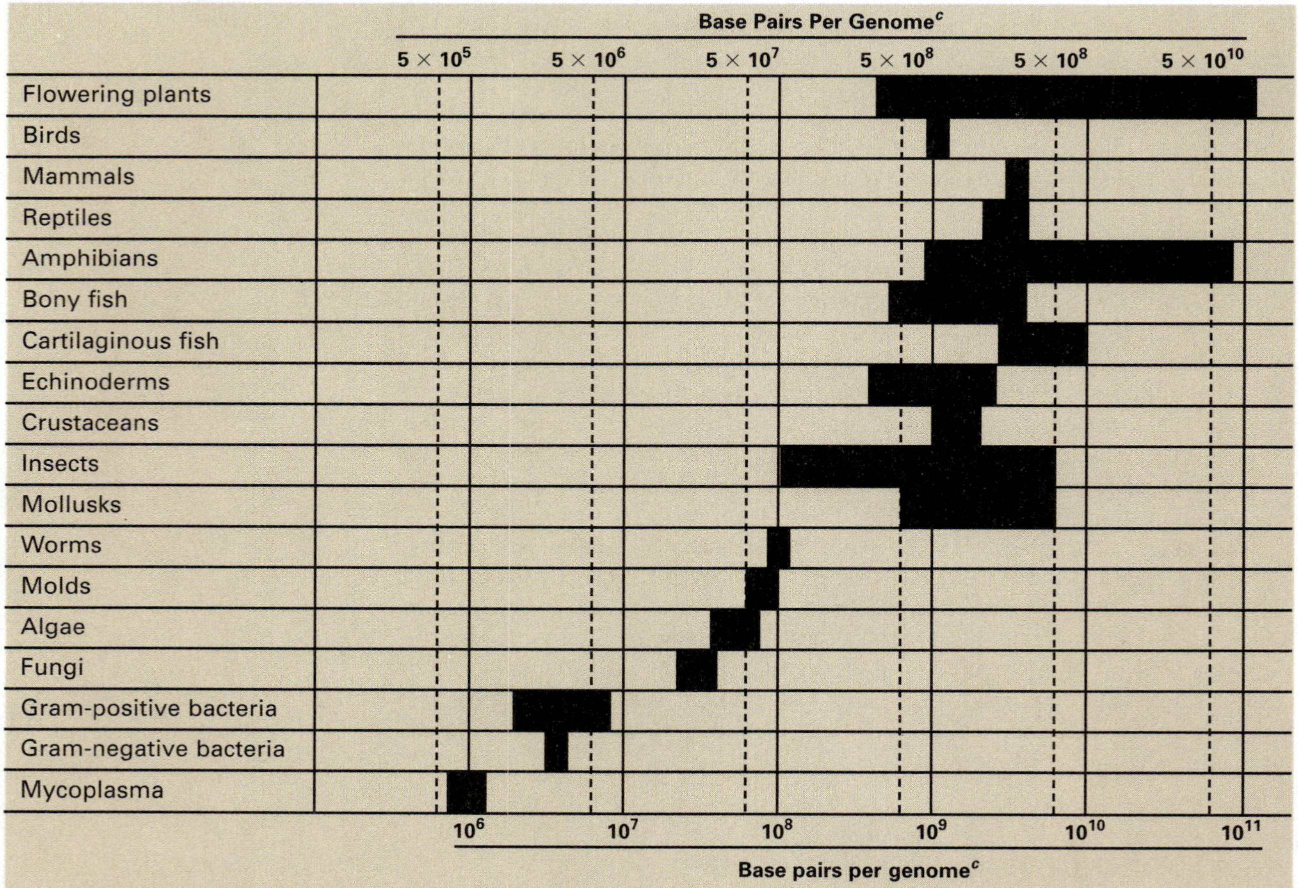

Base Pairs Per Genome[c]: 5×10^5 | 5×10^6 | 5×10^7 | 5×10^8 | 5×10^8 | 5×10^{10}

Organisms
Flowering plants
Birds
Mammals
Reptiles
Amphibians
Bony fish
Cartilaginous fish
Echinoderms
Crustaceans
Insects
Mollusks
Worms
Molds
Algae
Fungi
Gram-positive bacteria
Gram-negative bacteria
Mycoplasma

10^6 | 10^7 | 10^8 | 10^9 | 10^{10} | 10^{11}

Base pairs per genome[c]

[a]From Wolfe, 1993.
[b]The shaded areas represent the range of genome sizes within each group of organisms. The human genome contains approximately 3×10^9 base pairs.
[c]DNA contains two separate polynucleotide strands, with each nucleotide residue in one strand paired with a nucleotide residue in the second strand. The total number of nucleotide residues in a DNA equals two times the number of base pairs.

A Review of Nucleotides

Nucleotides are the basic building blocks for all nucleic acids, both RNA and DNA. Their structures and general properties are examined in Chapter 7. A common deoxyribonucleotide contains a phosphate residue, a 2-deoxy-β-D-ribose residue, and a residue of one of the common DNA bases, adenine (A), guanine (G), cytosine (C), and thymine (T). Each common ribonucleotide contains a phosphate residue, a β-D-ribose residue, and a residue of adenine, guanine, cytosine or uracil (U). Uracil, rather than thymine, is common to RNAs. Nucleotides are abbreviated with the same capital letters that are used to abbreviate their component bases. A, for example, is used as an abbreviation for both adenine and a nucleotide that contains adenine. Since the removal of the phosphate residue from a nucleotide generates a nucleoside, nucleotides are also called nucleoside monophosphates (NMPs). The attachment of a pyrophosphosphoryl group to the phosphate residue in a nucleotide generates a nucleoside triphosphate (NTP). The joining of many nucleotides through 3′,5′-phosphodiester links yields a polynucleotide with distinguishable 3′ and 5′ ends. Many of the facts reviewed in this paragraph are summarized in Figure 16.1 on page 646. This figure reviews some additional facts and terminology as well. Chapter 7 should be revisited if a more in-depth review is needed.

Base Pairing and Double Helix Formation

Two separate polynucleotide strands (or two segments of a single polynucleotide strand) that lie side by side lengthwise can be either **parallel** (both strands run 3′ to 5′ or 5′ to 3′) or **antiparallel** (one strand runs 5′ to 3′ while the second strand runs 3′ to 5′).

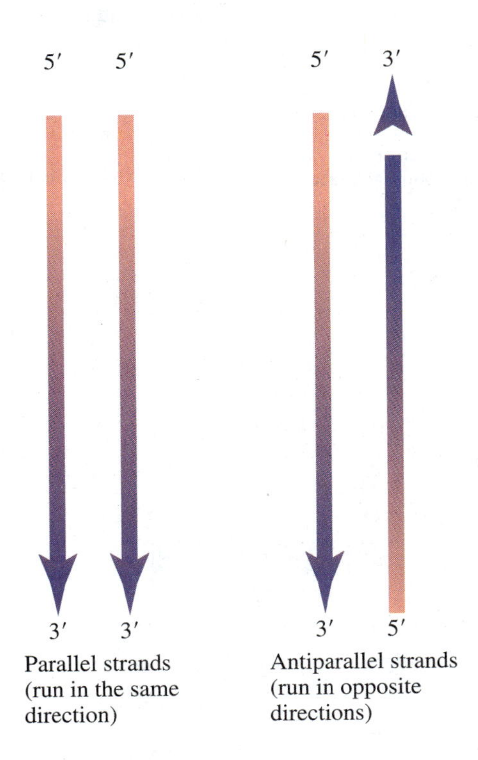

When two strands are antiparallel, certain purine and pyrimidine residues within the separate strands tend to H bond with one another to form planar **base pairs;** thymine and uracil pair with adenine, while guanine pairs with cytosine:

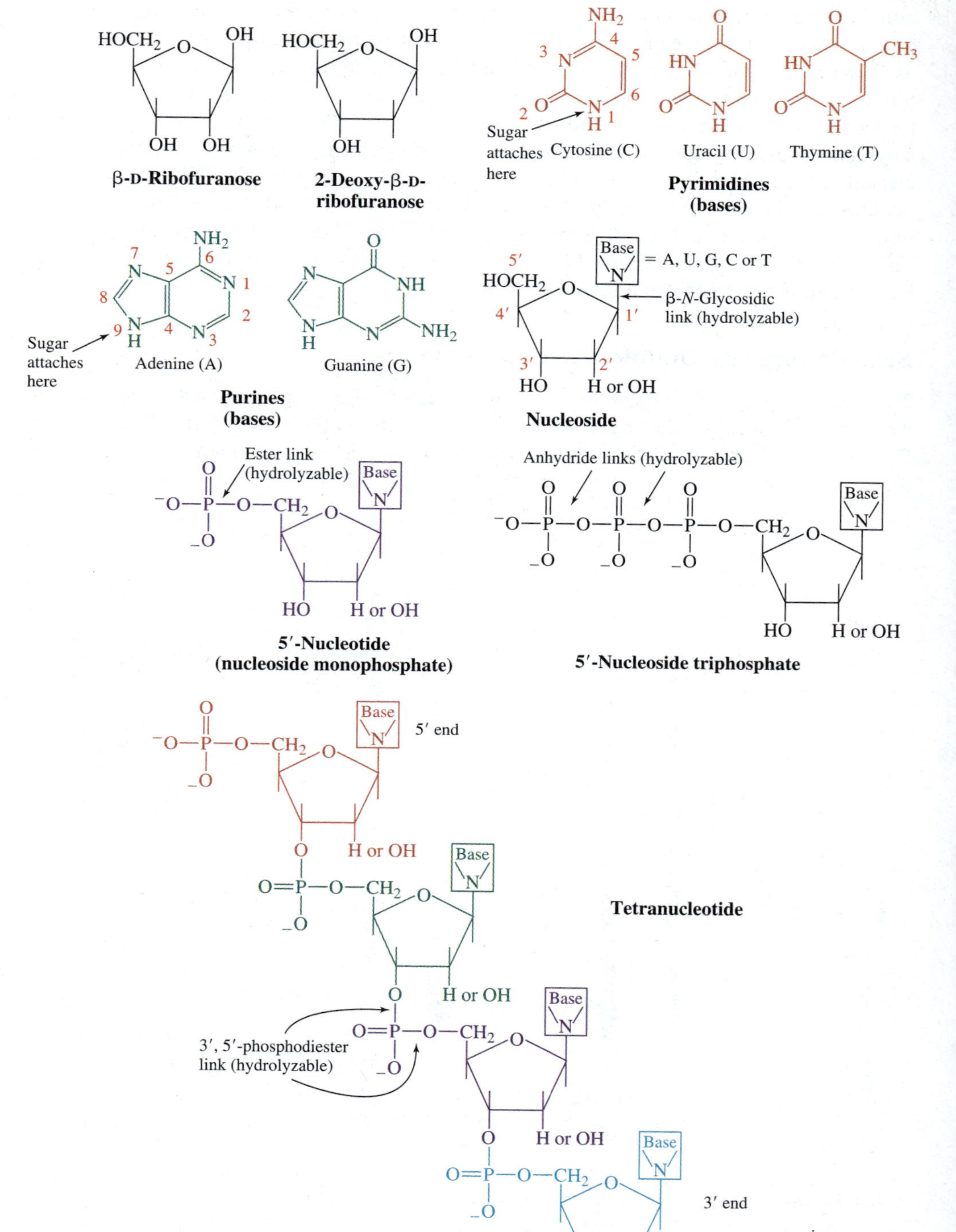

FIGURE 16.1
Nucleotides and some related compounds.

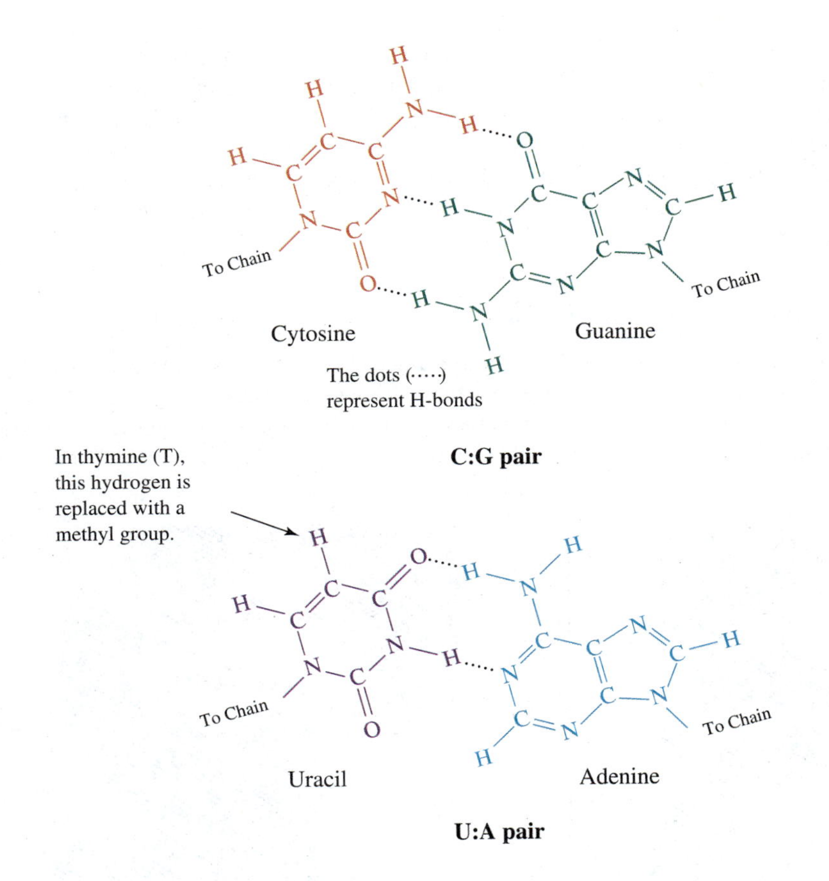

Bases that pair with one another are said to be complementary; both U and T are complementary to A, and G is complementary to C. Bases in **parallel** polynucleotide strands are unable to orient themselves in the manner necessary to form standard pairs. Although there are many documented examples of nonstandard base pairs (exceptions to the general pairing rules) within particular nucleic acids, individual nonstandard pairs tend to be less stable than their standard counterparts. **[Try Problems 16.3 and 16.4]**

A single base pair is relatively weak, and it is easily disrupted by the thermal energy that exists at body temperature. In contrast, multiple adjacent pairs can lead to positive cooperative bonding and to a relatively stable structural element within a nucleic acid molecule. When two antiparallel polynucleotide segments do interact to form multiple adjacent base pairs, they twist together into a **double helix** that resembles a spiraling staircase. Double helical regions are maintained through hydrophobic interactions between stacked base pairs as well as the H bonds within individual pairs. The twisting together of the two strands optimizes these interactions. The **secondary structure** of a nucleic acid refers to its base paired regions. Because complementary segments must exist before base paired regions can be formed, the **primary structure** (nucleotide sequence) of a nucleic acid determines its possible secondary structures.

A typical DNA molecule consists of two polydeoxyribonucleotide strands that are antiparallel, totally complementary, and base paired along their entire lengths. Because each A in one strand is normally paired with a T in the second strand and each G is normally paired with a C, the numbers of A's equals the number of T's and the G's equal the C's. DNA does not normally contain U. If present in DNA, U pairs with A and is considered a minor base (any base other than A, G, C, or T; Section 7.1).

The Helical Forms of DNA

The two strands in a duplex DNA molecule can twist together into a variety of helical arrangements. One of these structures, known as A-DNA (Figure 16.2), is comparable to the right-handed (coils clockwise) double helical arrangement common in RNAs (Section 16.5). A second, labeled B-DNA, is similar to A-DNA but has a dif-

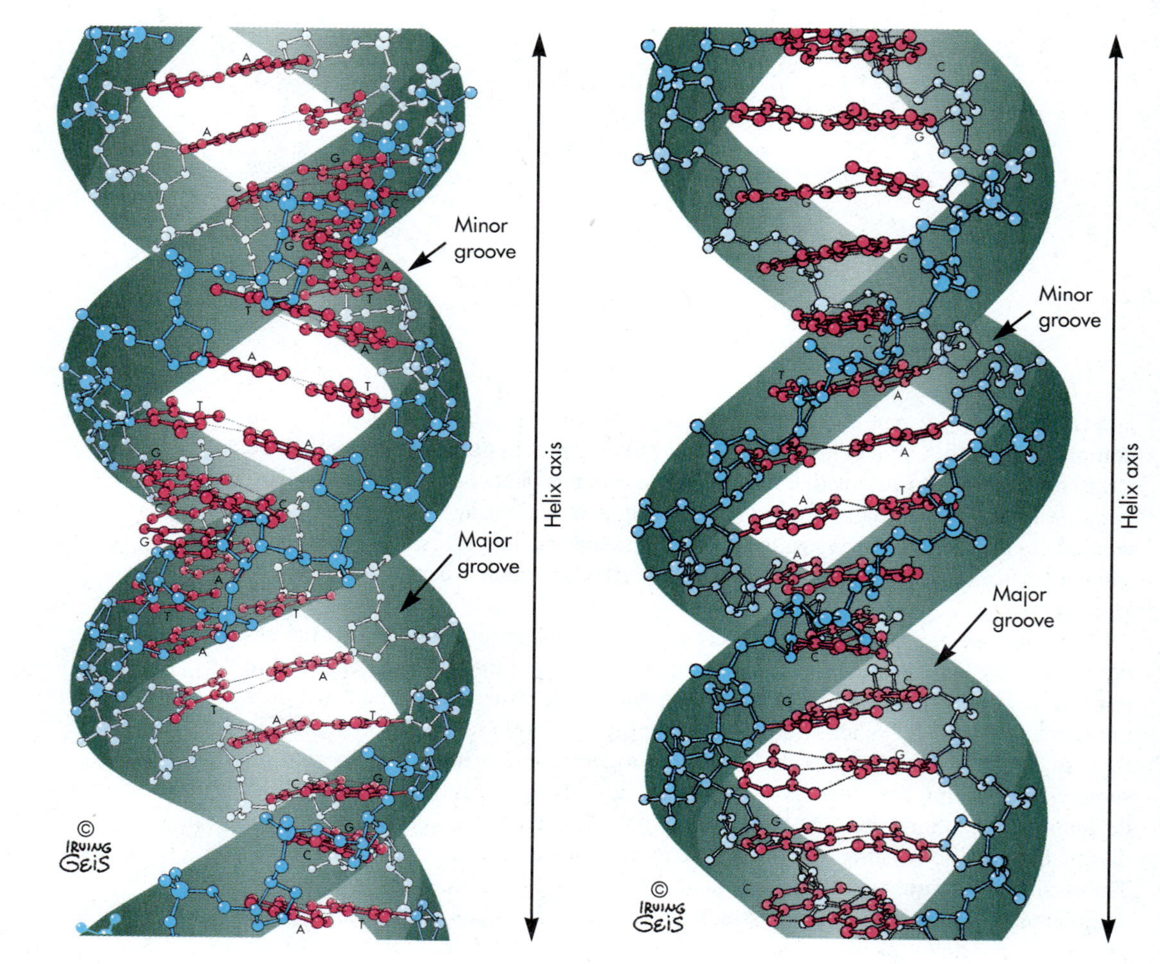

FIGURE 16.2
Ball-and-stick models of A-DNA (left) and B-DNA.

ferent number of base pairs per turn of the helix. The change in the number of base pairs per turn alters the diameter of the helix and the orientation of the stacked bases with respect to the helix axis. It also changes the width and depth of the **major** and **minor grooves** that are characteristic features of the helix surface. B-DNA corresponds to the famous, Noble Prize–winning model for DNA that was published by Watson and Crick in 1953. In B-DNA, there are approximately 10 base pairs per turn and the stacked base pairs are nearly perpendicular to the helix axis. **[Try Problems 16.5 and 16.6]**

In 1979 a new double helical form of DNA was identified that differed markedly from any previously characterized duplex. Since the sugar phosphate backbone in each strand followed a zig-zag path, it was labeled **Z-DNA.** It is left-handed and has a smaller diameter than A-DNA or B-DNA (Figure 16.3). Z-DNA, in contrast to most right-handed duplexes, is a rather potent antigen (stimulates antibody production). Antibodies against Z-DNA have been found in humans with certain **autoimmune diseases** (diseases in which the body is attacked by its own immune system) and they can be produced by injecting Z-DNA into experimental animals. Such an-

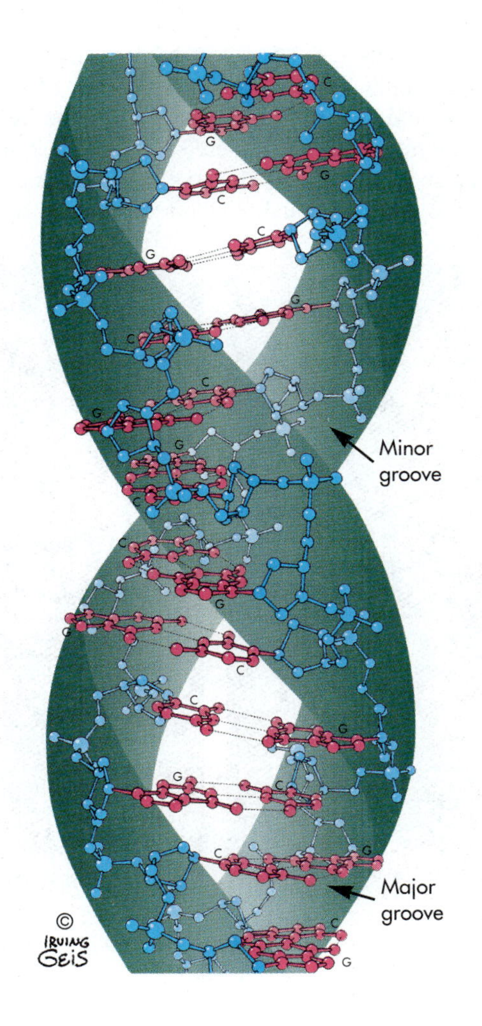

FIGURE 16.3
Ball-and-stick model of Z-DNA.

TABLE 16.2

Some Properties of A-, B-, and Z-DNA

	Handedness	Diameter (nm)	Base Pairs Per Turn of Helix	Rise (nm) Per Turn of Helix	Base Pair Tilt Relative to Helix Axis	Major[a] Groove	Minor[b] Groove
A-DNA	Right	~2.6	~11	2.5	19°	Narrow, deep	Wide, shallow
B-DNA	Right	~2.4	~10.4	3.5	1°	Wide, intermediate depth	Narrow, intermediate depth
Z-DNA	Left	~1.8	~12	4.6	9°	Flat	Narrow, deep

[a]Relative to other major grooves
[b]Relative to other minor grooves

tibodies are used as tools to identify and study Z-DNA. Table 16.2 summarizes some of the characteristics of A-DNA, B-DNA and Z-DNA. **[Try Problem 16.7]**

Figure 16.4 provides pictures of space-filling models of A-DNA, B-DNA, and Z-DNA. This figure should be carefully compared with the figures of the ball-and-stick models. The more frequently encountered ball-and-stick models can lead to serious misimpressions about the overall shape of DNA and the amount of space occupied by single atoms and individual nucleotide residues. In all of the described

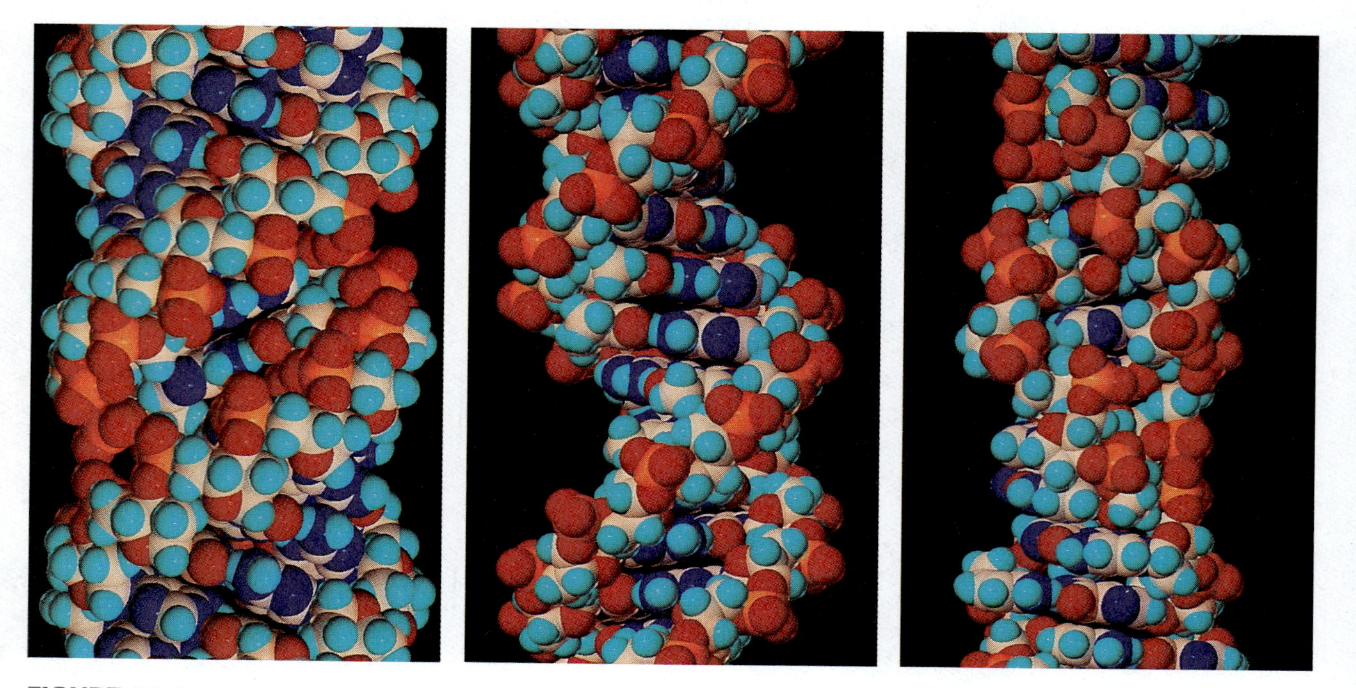

FIGURE 16.4
Space-filling models for A-DNA, B-DNA, and Z-DNA.

double helices, the two separate strands and the stacked base pairs are packed tightly together; there is very little free space within the core of a double helix. The actual construction of a model of DNA or a hands-on examination of such a model is highly recommended. Such an exercise can add significantly to your understanding and appreciation of the architectural features of these complex macromolecules. **[Try Problem 16.8]**

The multiple helical forms of DNA tend to exist in equilibrium, because localized base-paired regions are continuously separating and reforming. DNAs, like proteins, are said to breathe. Primary structure, supercoiling (next section), DNA methylation, salt concentration, protein binding, and other variables determine the relative abundance of the different helical forms. One segment of a DNA may exist in the B form and another segment along the same duplex exist in the Z form or A form. B-DNA appears to be the most common arrangement for DNA as it exists within chromatin and chromosomes (Section 16.3). Although there is some evidence that both the A and Z forms of DNA may be biologically important, their precise function is uncertain.

A variety of uncommon forms of DNA have been discovered that are architecturally distinct from the three antiparallel duplexes that have been described. These structural variants include parallel duplexes plus multiple triplexes and quadruplexes. All may exist *in vivo* and may be of biological importance. Suggested functions include the packaging of DNA into chromosomes, the stabilization of DNA, the regulation of gene expression, and the control of **recombination** (the exchange of DNA fragments between separate DNA molecules). These hypotheses must now be tested and retested with extensive research. Experimentation and hypothesis construction are at the heart of all the natural sciences, and they represent one of the central themes of biochemistry (Exhibit 1.4).

PROBLEM 16.1 Use common sense to describe or define the term "gene product."

PROBLEM 16.2 How many different genes must be expressed in order for a cell to produce a molecule of hemoglobin? Explain. Hint: Review Section 4.9.

PROBLEM 16.3 Will 3-methylcytosine (a minor base; Figure 7.1) pair with guanine in the same manner that cytosine pairs with guanine? Explain.

PROBLEM 16.4 What is the nucleotide sequence of the polyribonucleotide that is totally complementary to:

a. AUCUAAUCGCAUU
b. CGGGUUCGUUGGUGUGACG
c. AAACUCGCCCUUGUGGGCAAACCCCCGGUGUGC

Note: Section 7.4 describes multiple conventions used to abbreviate oligo- and polynucleotides. The 5′ end is to the left and 3′ end to the right unless specified otherwise.

PROBLEM 16.5 Study the diagram of B-DNA in Figure 16.2. Circle three G:C pairs in which the three H bonds are clearly visible. Put a rectangle around two A:T pairs in which the two H bonds are readily discernible.

PROBLEM 16.6 How many base pairs exist in the segments of A-DNA and B-DNA depicted in Figure 16.2?

PROBLEM 16.7 List one experimental observation which suggests that at least some Z-DNA actually exist within the human body.

PROBLEM 16.8 True or false? In B-DNA there is just enough space between stacked base pairs to accommodate a single layer of water molecules. Explain.

16.2 SUPERCOILED DNA

There is one additional structural variant of DNA that is worthy of special comment. Imagine that one end of a B-DNA molecule (contains two complementary polynucleotide strands wrapped into a double helix) is attached to a fixed support while the other end is twisted several times. A torsional stress will be generated that causes the entire linear duplex to form a coil. When this happens, the DNA is said to be **supercoiled** since it contains a coiled coil:

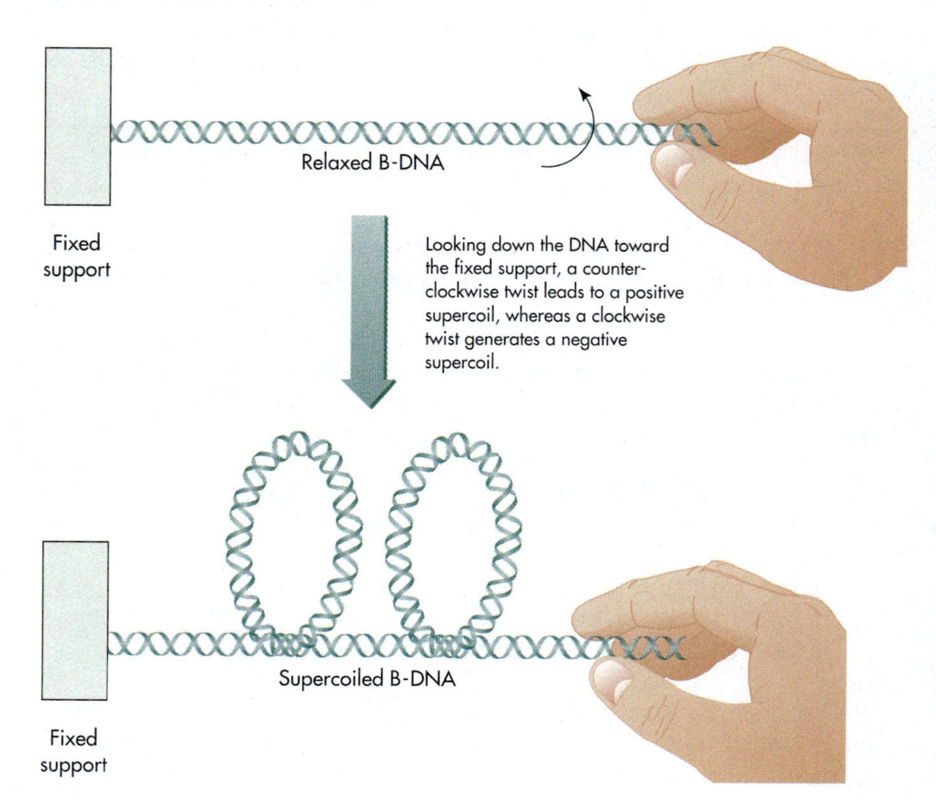

TABLE 16.3

Some Consequences of Supercoiling in B-DNA

	Linking Number	Energy Required to Unwind the Two Strands in B-DNA
Relaxed B-DNA	10.4	Depends on pH, ionic strength, and other variables
Negatively supercoiled B-DNA	<10.4	Decreased relative to relaxed DNA
Positively supercoiled B-DNA	>10.4	Increased relative to relaxed DNA

When the free end of the duplex is twisted counterclockwise (as you look down the double helix toward the immobilized end), the two polynucleotide strands within B-DNA tend to wind together more tightly. The supercoil that results is said to be a **positive supercoil.** If the free end of the duplex is twisted clockwise, the resultant torsional stress tends to unwind the two polymer strands and leads to a **negative supercoil.** A DNA molecule with no supercoils is said to be **relaxed.**

The number of times that one strand of a duplex DNA molecule winds around the other is the **linking number** for that DNA. The linking number for a relaxed B-DNA molecule is equal to the number of base pairs that it contains divided by 10.4, the approximate number of base pairs per turn of the B-helix under physiological conditions. Negative supercoiling decreases the linking number. Positive supercoiling increases it (Table 16.3). **[Try Problem 16.9]**

Supercoiling can be demonstrated with a piece of twine containing multiple strands that have been twisted together. Hold one end of the twine firmly, and then twist the other end. Twists in one direction will cause the strands within the twine to wrap together more tightly (the resultant supercoils will be positive). Twists in the opposite direction will cause the separate strands within the twine to partially unwind and will create negative supercoils. If the two ends of the twisted twine are tied together, the twine will be locked into a supercoiled arrangement. If one end of the twisted twine is released, the twine will tend to untwist spontaneously and return to its "relaxed" state. **[Try Problem 16.10]**

Once the twine experiment has been performed, it should become obvious that energy is required to create a supercoil and that the rotation of a DNA molecule must be restricted for a supercoiled structure to be generated and maintained. If both ends of a DNA molecule are free, twisting one end (takes energy) will cause the entire molecule to rotate. Rotation can be restricted by attaching one end of a linear DNA molecule to a protein, membrane, or other large particle. It can also be restricted by joining the two ends of a linear DNA molecule to form a **covalently closed, circular duplex DNA** (analogous to tying the two ends of a twisted piece of twine together).

Circular DNA molecules that have the same nucleotide sequence but different linking numbers are called **topoisomers** or **topological isomers.** Such isomers differ only in the extent or nature of their supercoiling. The fact that both prokaryotes and eukaryotes contain multiple enzymes, known as **topoisomerases,** that catalyze the interconversion of topological forms suggests that supercoiling is biologically

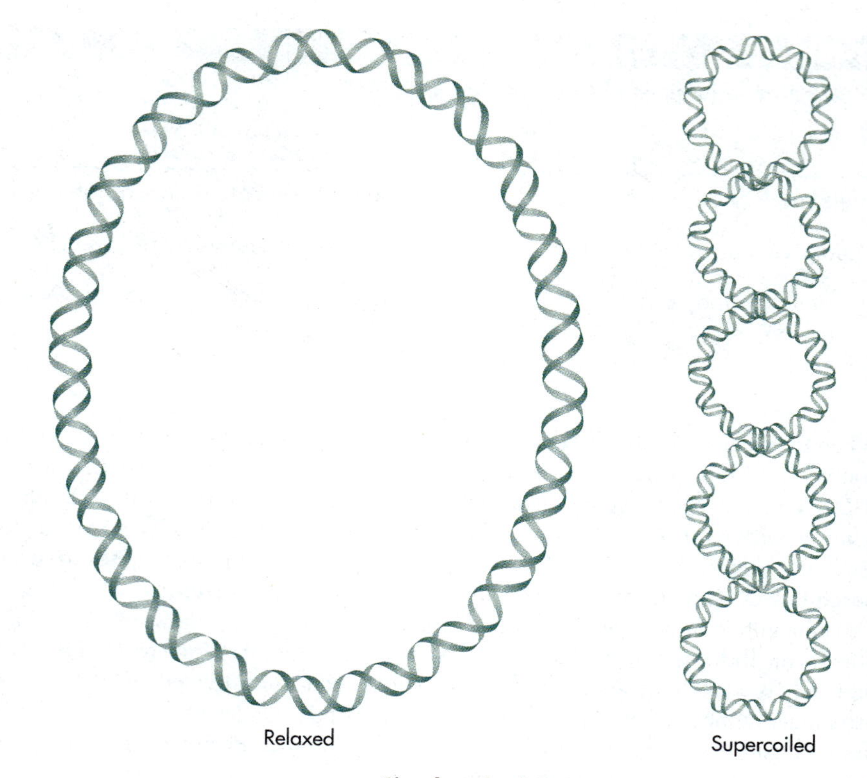

Circular, duplex DNA

important. Growing evidence indicates that supercoiled DNA is normally the metabolically active form of DNA. The DNA in nucleosomes is supercoiled, and supercoiling appears to play a role in DNA replication, DNA recombination, and transcription.

Some anticancer drugs, including **camptothecin** and its derivatives, bind and inhibit topoisomerases within cancer cells. The inhibition of these enzymes, it is hypothesized, kills the cells by blocking DNA replication and transcription. Camptothecin, like the highly publicized anticancer drug **taxol,** is a plant alkaloid. **Alkaloids** are a diverse group of cyclic, nitrogen-containing organic compounds found primarily in plants. Taxol, in contrast to camptothecin, is thought to kill cancer cells by blocking the disassembly of microtubules, protein networks involved in cell division.

PROBLEM 16.9 Does counterclockwise twisting increase or decrease the linking number in Z-DNA? Explain.

PROBLEM 16.10 Is the temperature required to unwind a segment of relaxed B-DNA greater than, less than, or the same as the temperature required to unwind a segment of negatively supercoiled B-DNA? Explain.

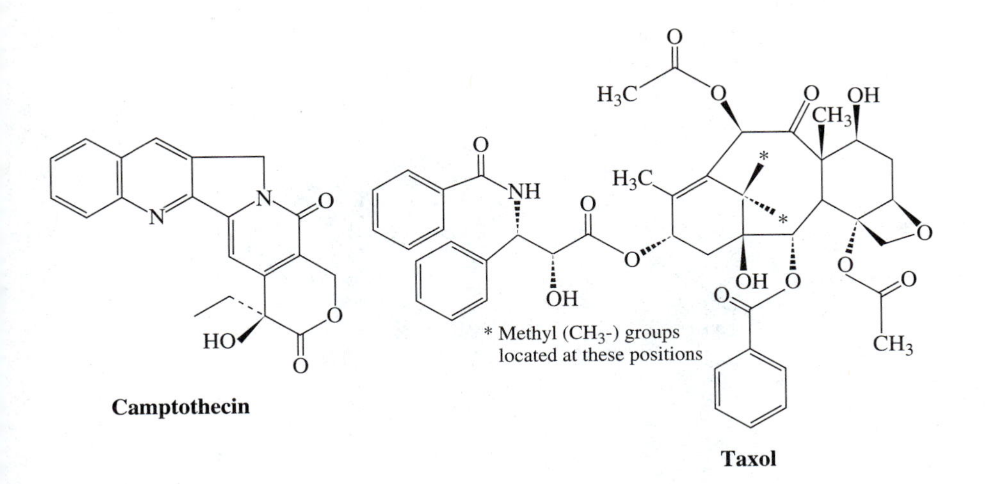

16.3 NUCLEOSOMES, CHROMATIN, AND CHROMOSOMES

The discussion of the topological forms of DNA leads logically into an examination of the topology and packaging of DNA as it normally exists *in vivo* (within living organisms). In eukaryotes, most DNA resides in basic structural units known as **nucleosomes** that consist of DNA packaged with basic proteins called **histones.** Because histones are rich in lysine or arginine or both (Table 16.4), they exist as polycations at physiological pH and bind tightly to DNA (a polyanion). The heart of a nucleosome, called a **core particle,** encompasses an octamer of histones (two copies each of **histones H2A, H2B, H3,** and **H4**) wrapped with a fragment of supercoiled duplex DNA containing approximately 150 base pairs. Additional DNA (around 50 base pairs) and a single copy of **histone H1** account for the remainder of a nucleosome (Figure 16.5).

Chromatin consists primarily of a coiled string of nucleosomes along a single duplex DNA molecule. Individual nucleosomes contain separate segments of this DNA. The DNA regions that join adjacent core particles are called **linker DNA.** Chromatin also encompasses nonhistone proteins and small amounts of RNA. Individual nucleosomes are produced in the laboratory by treating chromatin with **mi-**

TABLE 16.4

Some Characteristics of Calf Thymus Histones

Histone	% Arginine	% Lysine	Molecular Weight (daltons)	Copies per Nucleosome
H1	1	29	23,000	1
H2A	9	11	14,000	2
H2B	6	16	13,800	2
H3	13	10	15,300	2
H4	14	11	11,300	2

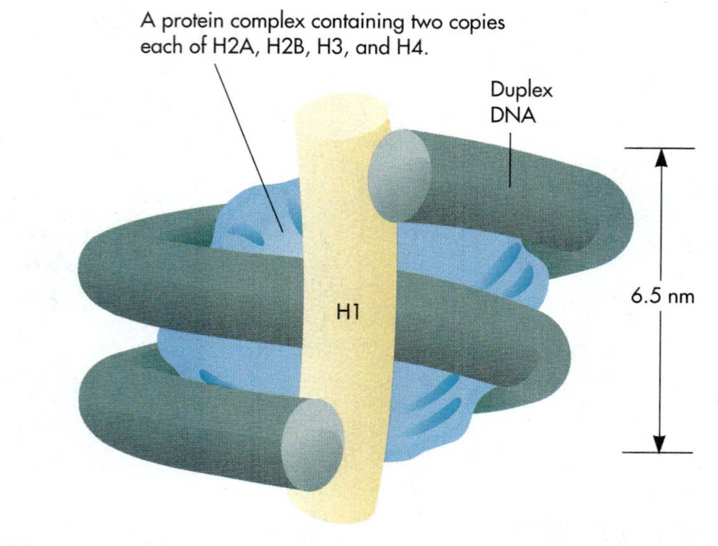

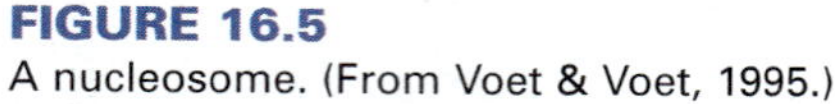

FIGURE 16.5
A nucleosome. (From Voet & Voet, 1995.)

crococcal nuclease, an enzyme that catalyzes the hydrolysis of a specific phosphate ester bond within linker DNA. **[Try Problems 16.11 and 16.12]**

Prior to the division of a eukaryotic cell, each chromatin molecule coils, folds, and packs around itself to form a dense nuclear body known as a **chromosome,** whose putative (reputed; supposed) structural hierarchy is depicted in Figure 16.6. In prokaryotic cells, the term *chromosome* is used to describe a cell's predominant DNA molecule regardless of how it is coiled or packaged. Prokaryotic cells lack histones, nucleosomes, and chromatin.

PROBLEM 16.11 Is micrococcal nuclease an endonuclease or an exonuclease? Is it an RNAse or a DNAse? Explain.

PROBLEM 16.12 Although linker DNA and free DNA are highly susceptible to nuclease-catalyzed hydrolysis, the DNA within intact core particles is relatively resistant to nuclease action. Suggest why this is the case.

16.4 RIBONUCLEIC ACIDS: AN OVERVIEW

The DNA within the chromosome(s) of an organism carries its blueprint for life. In DNA's nucleotide sequence resides all the information necessary for the construction and maintenance of a living organism. Much of this information encodes those polypeptides found within enzymes, the catalysts for metabolism. The assembly of polypeptides involves three major classes of RNA, messenger RNA (mRNA), transfer RNA (tRNA) and ribosomal RNA (rRNA), plus a number of minor classes. The

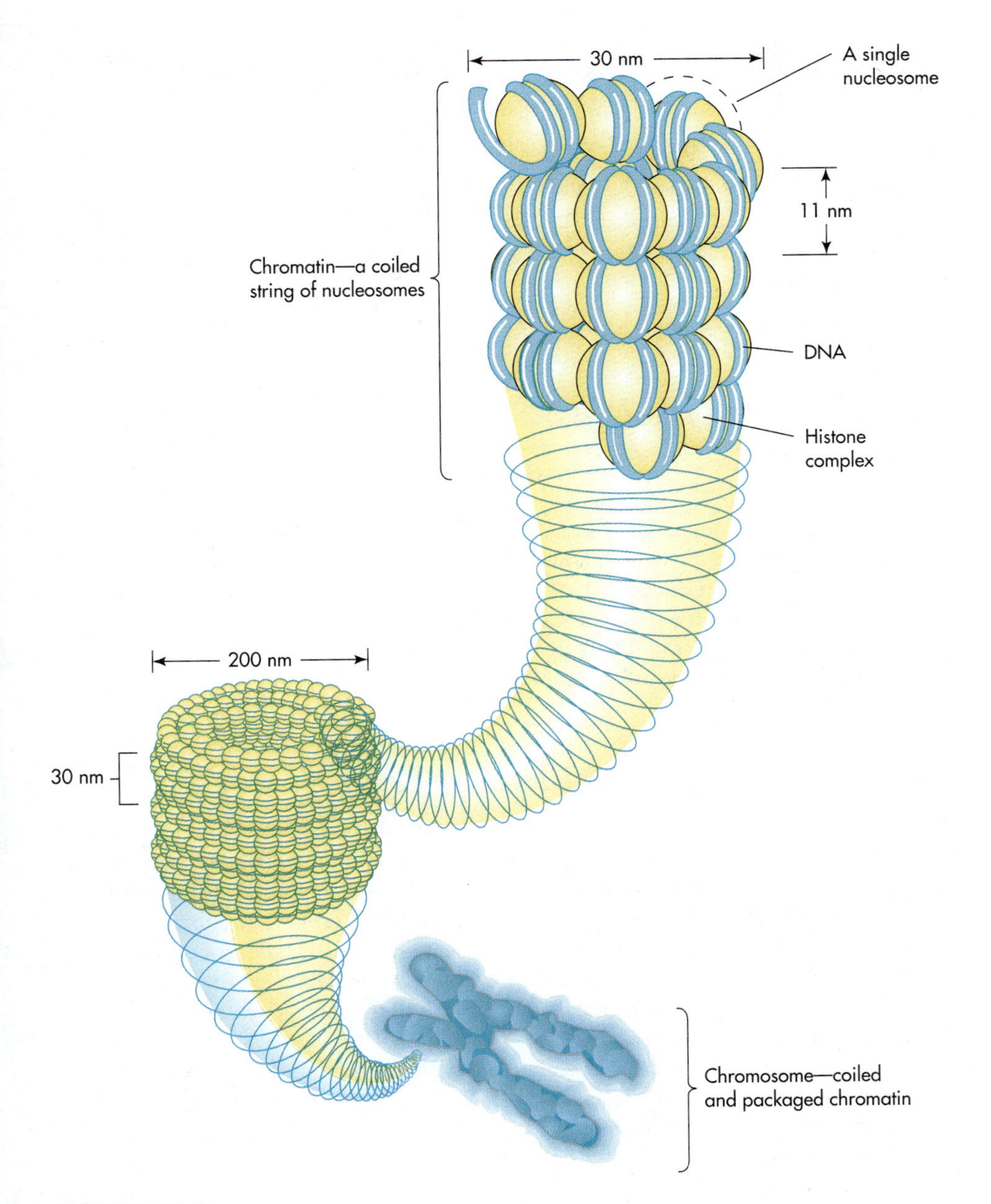

FIGURE 16.6
The putative relationship between nucleosomes, chromatin, and chromosomes. (Adapted from Zubay, 1993.)

next three sections provide an overview of the structure and function of the three major classes of RNA, while Chapter 18 examines the details of polypeptide assembly.

RNAs, like DNAs, are polynucleotides. RNA, in contrast to DNA, is single-stranded and contains β-D-ribose residues instead of 2-deoxy-β-D-ribose residues (Table 16.5). The common bases are A, G, C, and U. Thymine, if present, is consid-

TABLE 16.5

Some Differences Between RNA and DNA

Nucleic Acid	Sugar Residue	Common Bases	Usual Number of Nucleotide Residues Per Molecule	Usual Number of Polymer Strands Per Molecule	Major Function	Mode of Biosynthesis
Deoxyribonucleic acid (DNA)	Deoxyribose	ACGT	thousands to millions	2	Carries blueprints for life; genetic material	DNA replication
Ribonucleic acid (RNA)	Ribose	ACGU	60 to over 10,000	1	Involved in the expression of genetic information	Transcription

ered a minor base. RNAs are produced when cellular machinery uses a sequence of deoxyribonucleotide residues in DNA as a pattern or template to direct the joining of ribonucleotides, a process known as transcription. Thus, it is the sequence of deoxyribonucleotide residues in the DNA within a cell that determines the sequence of monomers in the cell's RNAs. Chapter 17 describes how this is accomplished. The flow of information from DNA to RNA is encompassed within one of the central themes of biochemistry, information transfer (Exhibit 1.4).

16.5 MESSENGER RNA

Messenger RNA (mRNA) molecules are single polynucleotide chains that carry information (a message from DNA) for the production of polypeptides. In eukaryotic cells, virtually all mRNAs bear instructions for the construction of a single polypeptide chain. In prokaryotes, many mRNAs are **polycistronic;** they contain messages for the assembly of multiple polypeptides. Section 1.3, which describes some of the differences between prokaryotes and eukaryotes, should be reviewed at this time, because the differences between these two classes of organisms will be of importance throughout the remainder of this text.

Messenger RNAs carry information in their nucleotide sequences in basically the same manner that information is carried in the sequence of letters in this text. The process through which a polypeptide is produced from information carried in mRNA is known as translation. A single mRNA molecule can be repeatedly translated to generate many copies of the polypeptide that it encodes.

Caps, Tails, and Untranslated Regions

All messenger RNA molecules contain a **coding (translated) sequence** that is preceded at its 5′ end by a **5′ untranslated region (5′-UTR; leader sequence)** and followed at its 3′ end by a **3′ untranslated region (3′-UTR; tailing sequence)** (Figure 16.7). The 5′ ends of most eukaryotic mRNAs are attached through a triphosphate bridge to the 5′ position of a 7-methyl guanosine residue:

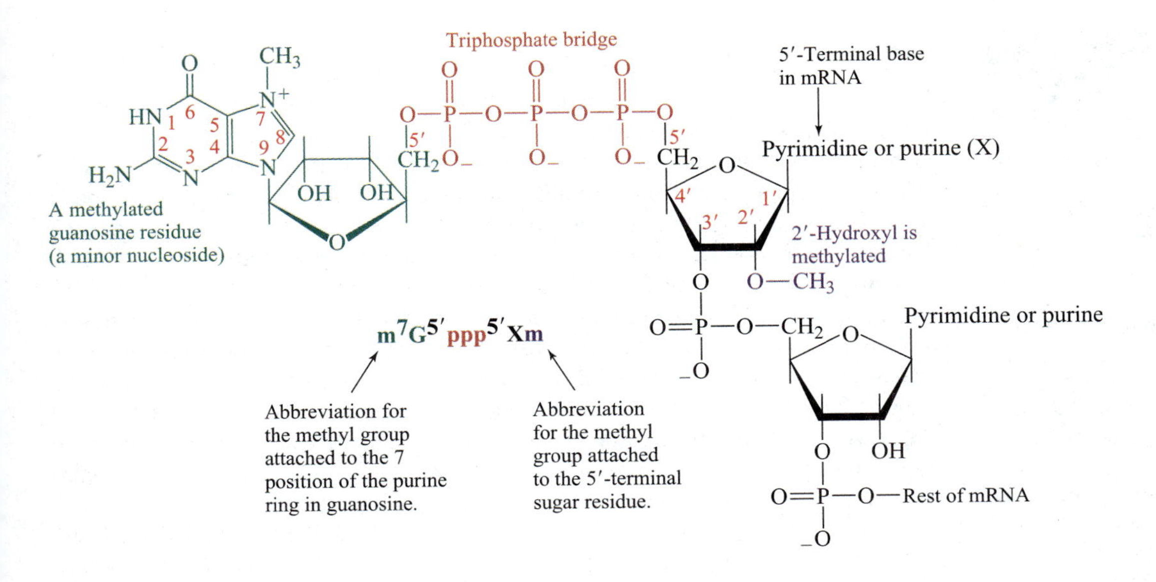

Such RNAs are said to be **capped.** The ribose residue within one or more of the 5′-terminal nucleotide residues in a capped RNA is commonly methylated. Any terminal nucleotide residues that contain a methylated ribose residue are considered part of the cap. The 3′ end of a eukaryotic mRNA usually contains an uninterrupted sequence of 50 to 200 AMP residues that were added to the 3′ untranslated region after the RNA was produced through the transcriptional process. Such mRNAs are said to contain a **poly-A tail.** Although a typical eukaroytic mRNA contains both a cap and a poly-A tail, most prokaryotic mRNAs lack these structural features. The caps, poly-A tails, and untranslated regions in mRNAs are hypothesized to be involved in mRNA transport, turnover, and translation. Some untranslated regions also appear to participate in the regulation of transcription (Section 17.1). **[Try Problem 16.13]**

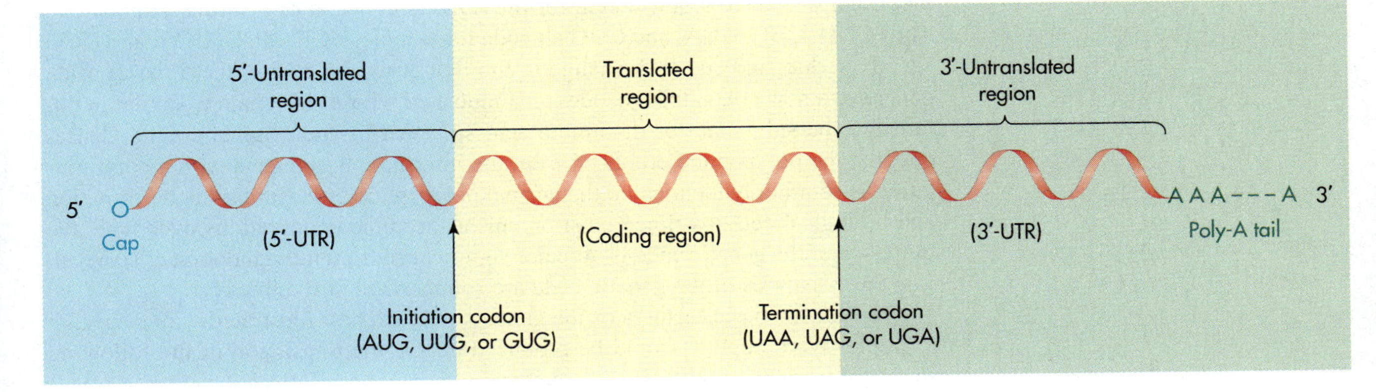

FIGURE 16.7
The features of a typical eukaryotic mRNA.

EXHIBIT 16.1
Basic Features of the Genetic Code

Four-letter alphabet: A, U, G, and C
Three-letter words (codons)
Sequential (with rare exceptions): Codons are read in sequence
Nonoverlapping: Each base is part of a single codon
Specific initiation codons
Specific termination codons
Nonambiguous (with rare exceptions): A given codon always specifies the same amino acid
Universal (with rare exceptions): All organisms use the same codon assignments
Buffered: Random base substitutions commonly lead to new codons for the same or similar amino acids

The Genetic Code

The polypeptide-encoding information in mRNAs is said to be carried in the form of the **genetic code.** Translated regions are read sequentially in a 5′ to 3′ direction with each monomer (nucleotide residue) representing a separate **code letter.** Because mRNAs are constructed from four different nucleotides (A, U, G, and C), the genetic code employs a four-letter alphabet (Exhibit 16.1). Within a translated region, each consecutive three-letter sequence usually represents a separate **codon** or **code word.**

With a four-letter alphabet and three-letter words, there are 64 possible codons (4^3; Table 16.6). Sixty-one of these codons normally specify one of the 20 common protein amino acids. The other three usually function as **termination codons;** they lead to the cessation of translation when encountered during the 5′ to 3′ reading of a translated region in an mRNA. Three of the 61 translatable codons, **AUG, UUG,** and **GUG,** serve as **initiation codons;** one of these (normally AUG; see footnotes to Table 16.6) is the first codon read. Thus, each translated region begins (at its 5′ end) with an initiation codon and closes with a termination codon.

There is a single codon for methionine and a single codon for tryptophan. There are six leucine codons and six serine codons. From two to four codons exist for each of the other common protein amino acids. The greater the number of codons for an amino acid, the more frequently that amino acid appears within proteins. This experimental correlation may make common sense, but it is not obvious why this would necessarily be the case. Since any given codon usually specifies only one amino acid, the genetic code is described as **nonambiguous.**

Although most organisms utilize exactly the same genetic code (often described as **universal**), a limited number of unique codon assignments have been documented in mitochondria and certain simple organisms that contain distinctive translational machinery. The near universality of the genetic code is consistent with the hypothesis that all life on Earth has evolved from a single ancestral cell. It also illustrates the general theme of evolution, diversity, and change (Exhibit 1.4).

One last feature of the genetic code is worthy of note. The code is buffered to minimize the impact of random point mutations (changes in single nucleotide residues in DNA) on an organism. A change in the third base in a codon, for example, usually leads to a new codon for the same amino acid or a similar amino acid. Thus CUU, CUC, CUA, and CUG all code for leucine, UCU, UCC, UCA, and UCG specify serine, and so on. A change in the first base in a codon (the 5′ base) often generates a new codon that encodes an amino acid whose side chain is similar in polarity to the side chain of the amino acid specified by the original codon. Consequently, the polypeptide produced from the information in a gene with a point mutation is commonly identical to the normal polypeptide or is similar in conformation and biological activity. Since most organisms accumulate mutations over time, the buffering of the genetic code is of major significance from the standpoint of survival. The basic features of the genetic code are summarized in Exhibit 16.1.

Some of the basic features of the genetic code are best illustrated with a specific example. What peptide would be produced during the translation of the following fragment from the middle of the translated region of an mRNA?

5′CUCGAUGAAAGAUGU3′

To solve this problem, you must first ascertain the **reading frame;** the 5′ nucleotide could be either the first, second, or last letter in a codon. The actual position of this 5′ nucleotide within a codon is determined by the **upstream** (toward the 5′ end, where translation begins) location of the initiation codon. Once the initiation codon is read, each consecutive three-nucleotide sequence usually represents another intact codon; it is very uncommon for the reading frame to shift once translation has begun. For the problem at hand, assume that the first nucleotide in the given fragment represents the 5′ end of a codon. If this is the case, the first three nucleotides represent one codon, the next three nucleotides a second codon, and so on. The fragment contains a total of five codons. When these codons are read consecutively from the

TABLE 16.6

The Genetic Code

First Base	Second Base: U		Second Base: C		Second Base: A		Second Base: G	
U	UUU	Phe	UCU	Ser	UAU	Tyr	UGU	Cys
U	UUC	Phe	UCC	Ser	UAC	Tyr	UGC	Cys
U	UUA	Leu	UCA	Ser	UAA	TERM[a]	UGA	TERM[a]
U	UUG[b]	Leu	UCG	Ser	UAG	TERM[a]	UGG	Trp
C	CUU	Leu	CCU	Pro	CAU	His	CGU	Arg
C	CUC	Leu	CCC	Pro	CAC	His	CGC	Arg
C	CUA	Leu	CCA	Pro	CAA	Gln	CGA	Arg
C	CUG	Leu	CCG	Pro	CAG	Gln	CGG	Arg
A	AUU	Ile	ACU	Thr	AAU	Asn	AGU	Ser
A	AUC	Ile	ACC	Thr	AAC	Asn	AGC	Ser
A	AUA	Ile	ACA	Thr	AAA	Lys	AGA	Arg
A	AUG[b]	Met	ACG	Thr	AAG	Lys	AGG	Arg
G	GUU	Val	GCU	Ala	GAU	Asp	GGU	Gly
G	GUC	Val	GCC	Ala	GAC	Asp	GGC	Gly
G	GUA	Val	GCA	Ala	GAA	Glu	GGA	Gly
G	GUG[b]	Val	GCG	Ala	GAG	Glu	GGG	Gly

[a]Termination codon.
[b]Initiation codon. With rare exceptions, AUG is the sole initiation codon in eukaryotes. In some prokaryotes, GUG and UUG are occasionally used for initiation.

5′ to the 3′ end during translation, the amino acids specified are joined in the sequence in which their codons are encountered:

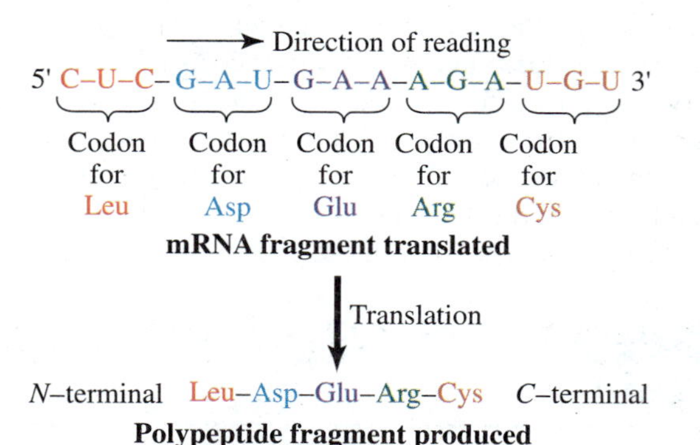

A polypeptide is assembled in the N-terminal to C-terminal direction. Chapter 18 explains why this is the case. **[Try Problems 16.14 through 16.17]**

mRNA Diversity

At any given time, a typical eukaryotic cell is probably in the process of translating the codons within the mRNAs for well over 1000 different proteins (Chapter 4). Since at least one unique mRNA molecule is normally involved in the production of each protein, a typical cell in the human body contains multiple copies of over a thousand mRNAs (often several thousand). Most of these mRNAs have half-lives (the time required for one-half the existing molecules to be inactivated) between one hour and several days (Table 16.7). Because new copies of an mRNA are commonly being produced at the same time that existing copies are being inactivated and degraded, mRNAs, like most molecules in living organisms, turn over (Section 14.1). At any instance, mRNAs usually account for 3% to 5% (by weight) of the total RNA within a cell. **[Try Problem 16.18]**

TABLE 16.7

Half-Lives for Some Mammalian mRNAs

Protein Encoded by mRNA	Half-Life of mRNA (h)
Dihydrofolate reductase	97
Fatty acid synthetase	48–96
Glyceraldehyde-3-phosphate dehydrogenase	75–130
Heat shock protein 70	2
Insulin receptor	9
Ornithine decarboxylase	0.5
Pyruvate kinase	30
Tyrosine aminotransferase	2

Structural Hierarchy

The total number of monomers in an mRNA molecule is determined by the number of amino acids in the polypeptide that it encodes and by the length of its untranslated regions. Most of the mRNAs in a eukaryotic cell possess 300 to 3000 nucleotide residues, with approximately 4% of these residues containing minor bases. Since efficient methods have been developed to sequence nucleic acids (Section 16.9), the primary structures (nucleotide sequences) of mRNAs are rapidly accumulating in computerized data banks around the world.

The single polynucleotide strand in an mRNA folds back upon itself in such a manner that certain complementary segments base pair in a classical antiparallel fashion. This tends to generate a complex pattern of base paired **(stem)** and **loop regions:**

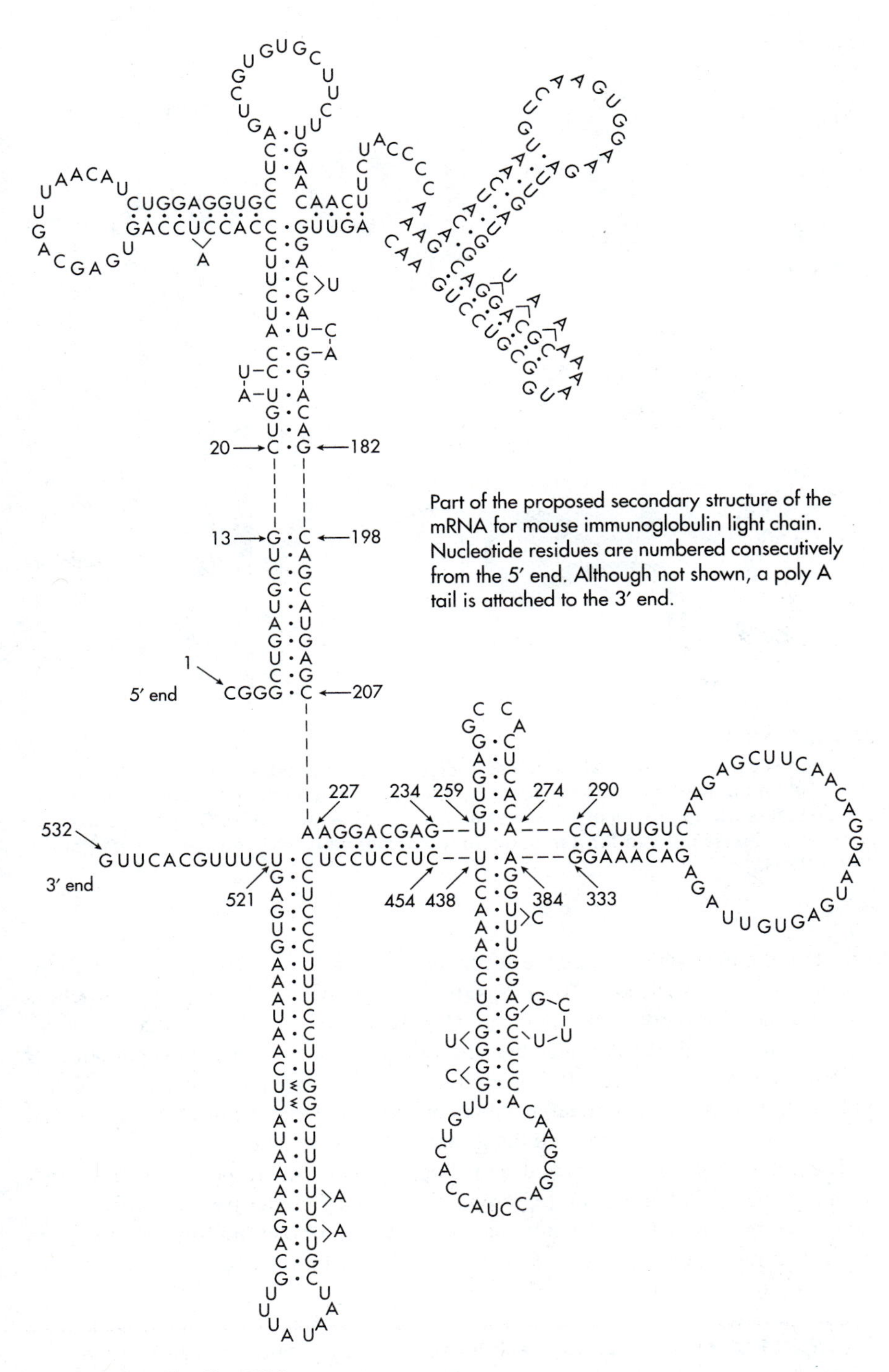

Part of the proposed secondary structure of the mRNA for mouse immunoglobulin light chain. Nucleotide residues are numbered consecutively from the 5′ end. Although not shown, a poly A tail is attached to the 3′ end.

Adapted from Devlin, 1992.

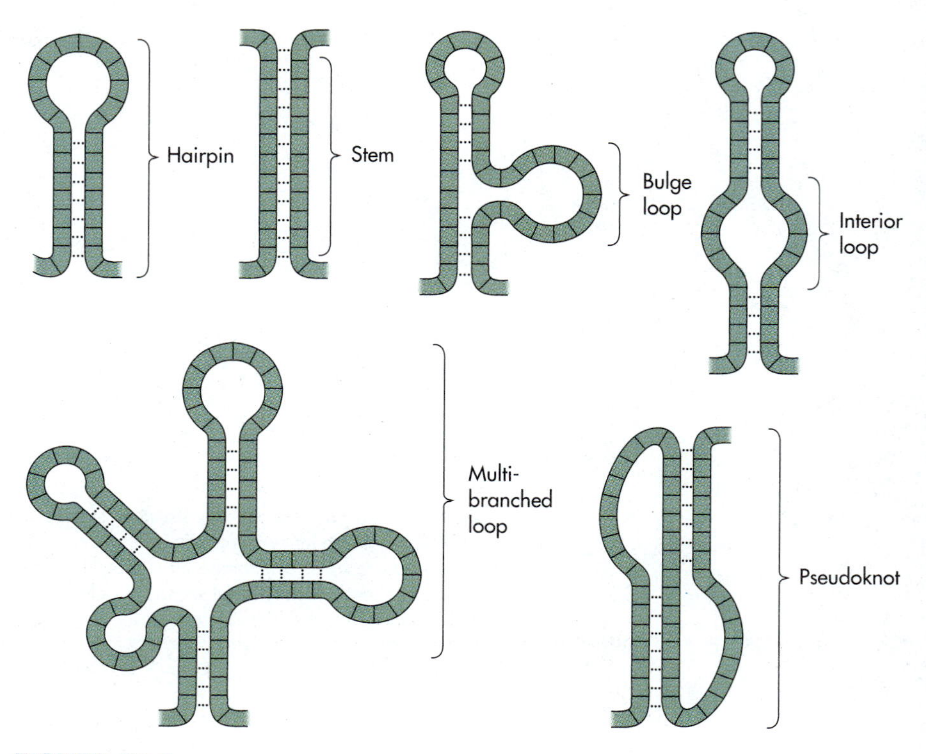

FIGURE 16.8
Common secondary structural elements in RNAs. In the paired regions, the RNA chain winds into a double helix in the A conformation. Flexibility in the RNA double helices and spatial rearrangements of the bases allow G–U, G–A, A–C, A–A, G–G, and other nonstandard base pairs to form in addition to the standard A–U and G–C pairs. (From Wolfe, 1993.)

The stem regions, which create the secondary structure of an RNA, are normally A-type double helices (Section 16.1). Some commonly occurring secondary structural elements have been given special names (Figure 16.8). Because the stems and loops often create a petal-like arrangement, the literature refers to the **flower model** for mRNA secondary structures. Because novel mRNAs have unique primary and secondary structures, the individual members of a group of distinct mRNAs can be said to represent different flowers, and the group to constitute a bouquet.

The folding of the base-paired and loop regions with respect to one another within a single RNA molecule represents the **tertiary structure** of that RNA. Although the precise tertiary structure has not been determined for any mRNA, there is considerable evidence that at least some mRNAs may possess biologically important tertiary structures.

PROBLEM 16.13 Identify each ester, anhydride, and ether link in the mRNA cap depicted in this section. Circle each nucleotide residue in this cap.

PROBLEM 16.14 How many nucleotide residues exist within the translated region of an mRNA that encodes a polypeptide with 527 amino acid residues? A eukaryotic mRNA normally possesses considerably more nucleotide residues than are found in its translated region. What structural features of mRNA account for this fact?

PROBLEM 16.15 What peptide is encoded in each of the following mRNA fragments? In each case, assume that the fragment is part of a translated region and that the first three nucleotide residues represent an intact codon.

a. CGUCCCGGACGCAAA
b. CCGUAUGCAGAAAACACAUAAUGU
c. GGGGGCGCGCGGCGCCCCGCG

PROBLEM 16.16 What peptides would be encoded in the mRNA fragments given in Problem 16.15, if, in each case, the first nucleotide residue represented the last letter in an intact codon?

PROBLEM 16.17 What oligopeptide is encoded in the following DNA sequence:

dTCTCCGACAGGTGAG

Assume that this segment of DNA serves as a template for the production of an mRNA fragment that resides within the coding region of an mRNA. Further, assume that the first nucleotide at the 5′ end of this mRNA fragment is the first nucleotide in an intact codon.

PROBLEM 16.18 The combination of mRNAs within a cell changes over time. Suggest why this is the case.

16.6 TRANSFER RNA

Transfer RNA (tRNA) is the second major class of RNA involved in polypeptide synthesis. tRNAs carry amino acids and transfer these (the basis for the "transfer" in transfer RNA) to a growing polypeptide chain during polypeptide biosynthesis. A typical eukaroytic cell contains 50 to 100 different tRNAs and, collectively, they account for 10 to 15% (by weight) of total cellular RNA. Normally, each tRNA accepts and carries one and only one of the 20 common protein amino acids. In addition, each tRNA molecule has a specifically positioned sequence of three nucleotide residues, known as an **anticodon,** that is complementary to one of the codons for the amino acid that it carries; a tRNA with an anticodon that is complementary to a codon for the amino acid alanine (abbreviated $tRNA^{Ala}$) will carry only alanine; a tRNA with an anticodon that is complementary to a codon for the amino acid serine

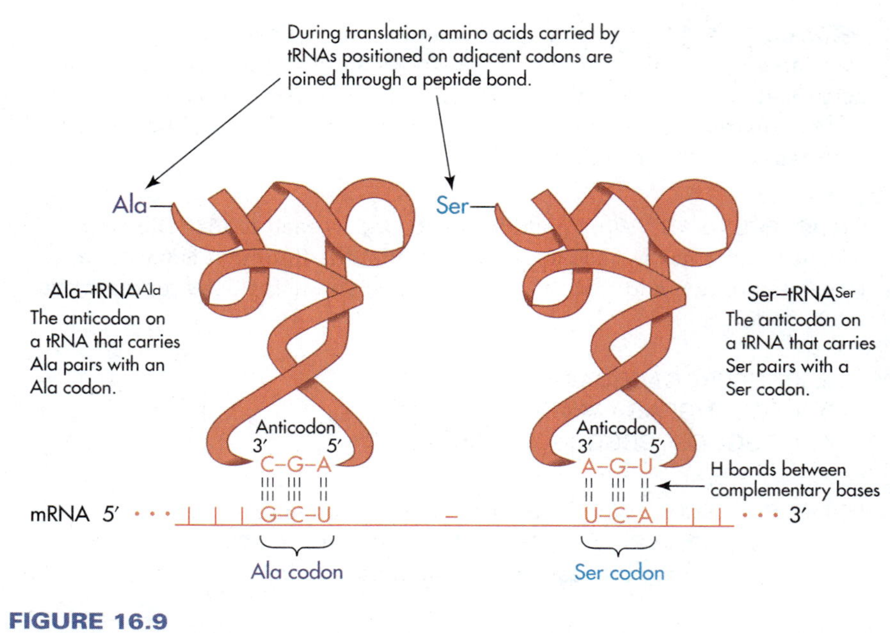

FIGURE 16.9
Codon–anticodon interactions.

($tRNA^{Ser}$) will always bind and carry serine; and so on (Figure 16.9). After an anticodon pairs with a complementary codon during the translation of an mRNA (see Chapter 18 for details), the amino acid attached to the tRNA is transferred to the growing polypeptide chain. Thus, it is through tRNAs that a sequence of codons in an mRNA determines the sequence of amino acids in a polypeptide; tRNAs, by pairing with select codons, directly **decode (translate)** the genetic code. **[Try Problems 16.19 and 16.20]**

Structural Hierarchy

The primary structure has been determined for several hundred different tRNAs isolated from a wide variety of organisms. Most tRNAs possess between 73 and 93 nucleotide residues, with up to 20% of these residues containing minor bases. The three-nucleotide sequence CCA is found at the 3′ end (also called the **CCA end** or **acceptor end**) of all tRNAs. Since tRNAs have base-paired and loop regions that give them a cloverlike appearance, they are said to have a **cloverleaf secondary structure.** The clover has four arms plus the acceptor system. The second arm from the 5′ end contains the **anticodon loop** which always possesses seven nucleotide residues. The anticodon consists of the three nucleotide residues in the center of this loop. Some of the structural features shared by most tRNAs are summarized in Figure 16.10. Although not shown in this figure, each base-paired region is twisted into a double helical arrangement similar to the A helix described for DNA (Section 16.1). The base-paired and loop regions in a tRNA fold together to generate an **L-shaped tertiary structure** (Figure 16.11). **[Try Problem 16.21]**

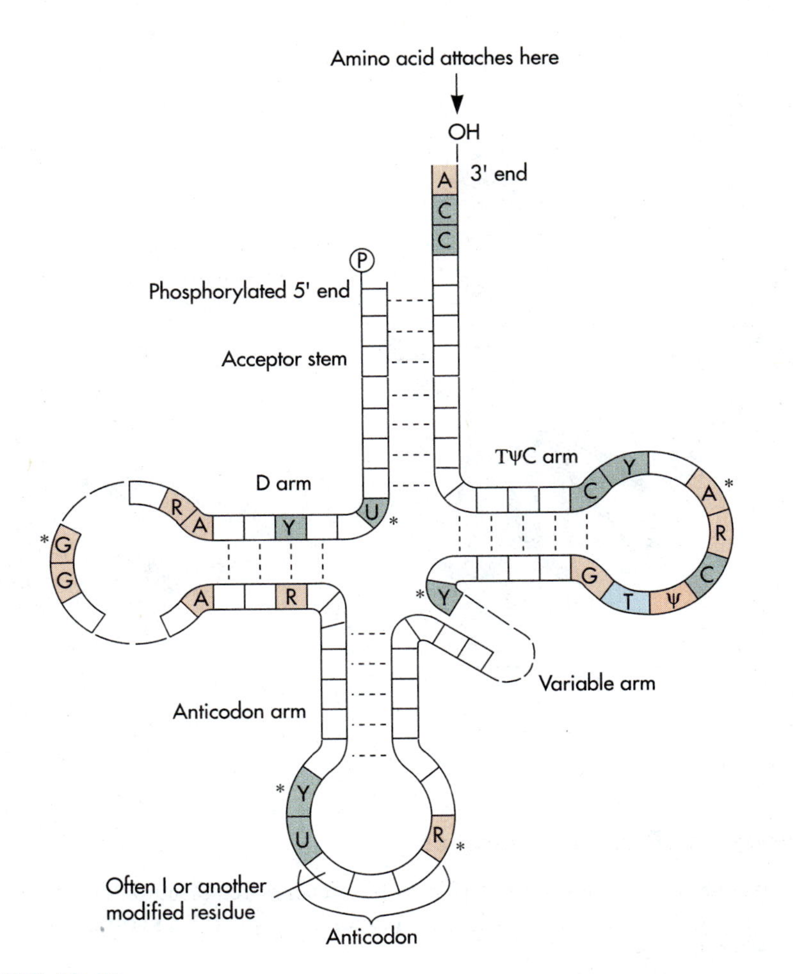

FIGURE 16.10
Some structural features shared by most tRNAs. Boxes depict nucleotide residues. Dashed lines represent hydrogen bonds between paired bases. The name for the D arm is based on the presence of dihydrouridine (D). The TΨC arm contains thymidine (T), pseudouridine (Ψ), and the sequence TΨC. Each arm, with the exception of the variable arm, always contains both a stem (based-paired region) and a loop. The number of nucleotide residues in each arm is more or less constant, except for the variable arm. Boxes containing letters identify conserved nucleotide residues, those that are the same in most tRNAs. R = purine-containing residue (A or G); Y = pyrimidine-containing residue (U or C); T = thymidine-containing residue; Ψ = pseudouridine-containing residue; and I = inosine-containing residue. The conserved residues marked with a star are commonly modified. (From Horton et al, 1993.)

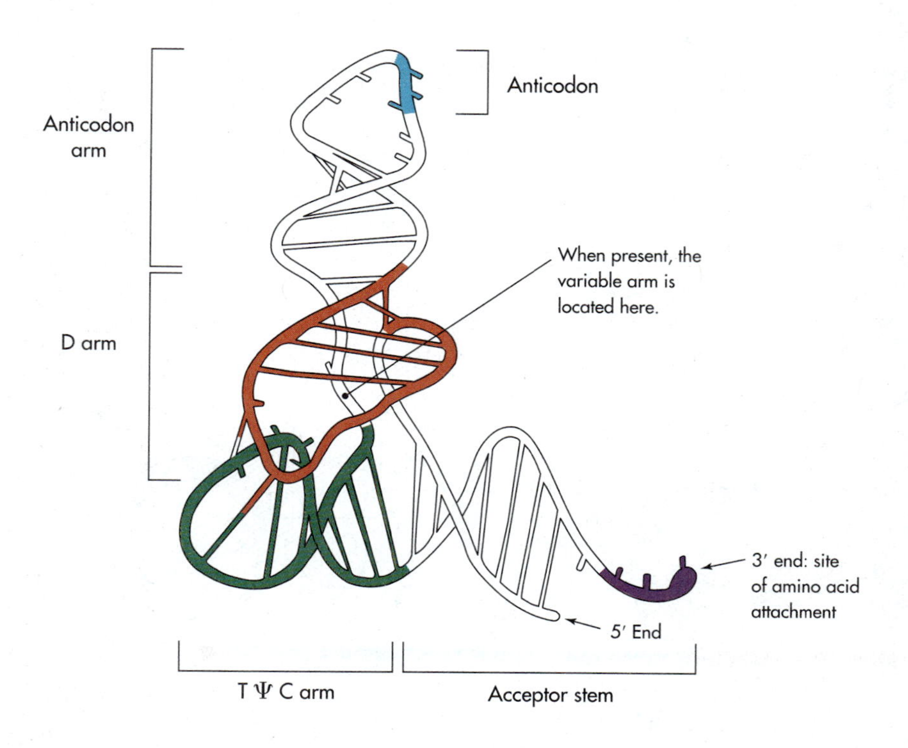

FIGURE 16.11
The tertiary structure of tRNA.

Aminoacyl–tRNA Synthetases

The attachment of amino acids to tRNAs is an energy-requiring process that is catalyzed by highly specific enzymes known as **aminoacyl–tRNA synthetases (amino acyl–tRNA ligases):**

$$\text{tRNA} + \text{amino acid} + \text{ATP} \xrightleftharpoons{\text{Aminoacyl–tRNA synthetase}} \text{aminoacyl–tRNA} + \text{PP}_i + \text{AMP}$$

The cleavage of ATP (generates PP_i plus AMP) provides a thermodynamic driving force for the reaction (Section 8.6). The PP_i produced is hydrolyzed to help "pull" the reaction (Le Châtelier's principle). Each cell contains at least one aminoacyl–tRNA synthetase for each of the 20 protein amino acids. An aminoacyl–tRNA synthetase for phenylalanine will only attach phenylalanine to those tRNAs whose anticodons are complementary to codons for phenylalanine; an aminoacyl–tRNA synthetase for methionine will only attach methionine to tRNAs whose anticodons pair with codons for methionine; and so on. Thus, any given aminoacyl–tRNA synthetase selectively recognizes both a specific amino acid and those specific tRNAs to which that amino acid should be attached (Figure 16.12). When an improper amino acid is joined to a tRNA, an abnormal polypeptide will be produced if the resultant aminoacyl–tRNA participates in translation. Fortunately, aminoacyl–tRNA synthetases rarely make a mistake. When they do, they often recognize the mistake and catalyze

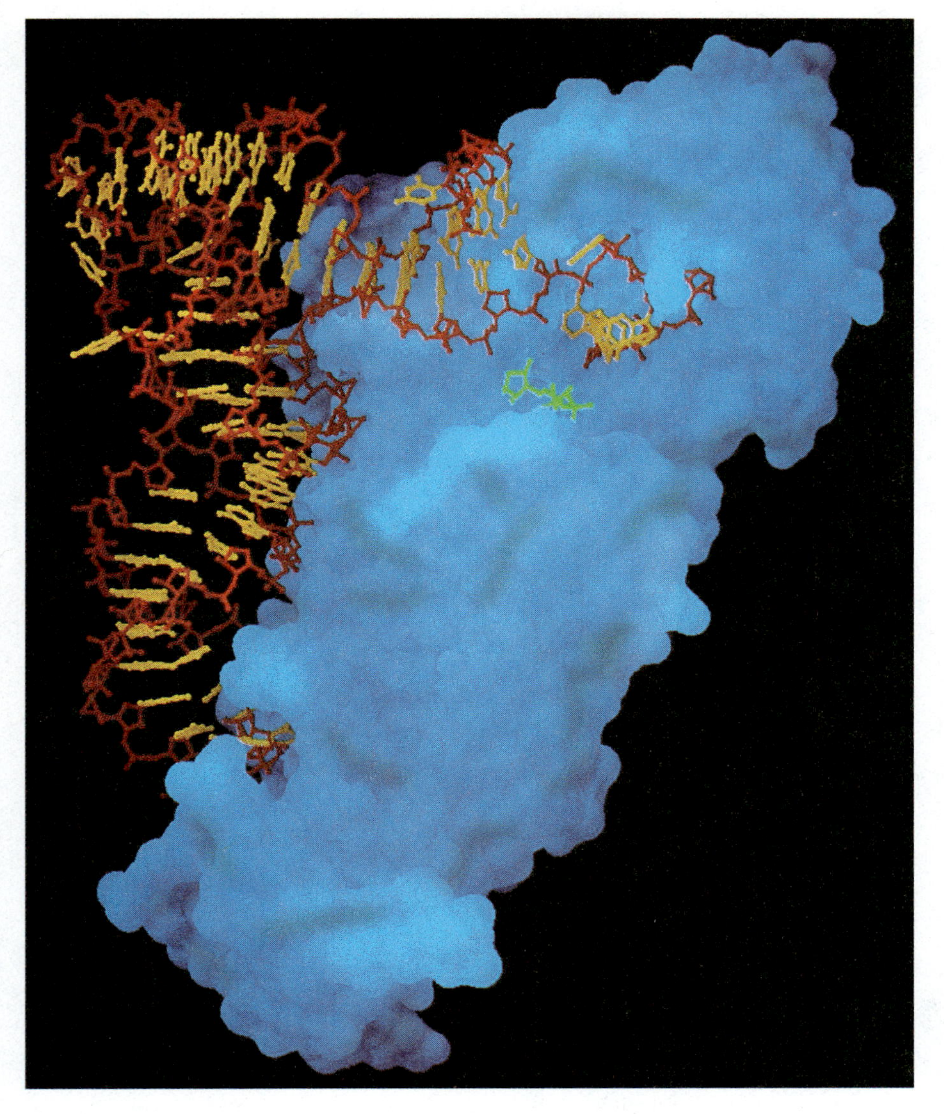

FIGURE 16.12
Binding of $tRNA^{Gln}$ by Gln-tRNA synthetase.

the hydrolytic removal of the improperly placed amino acid, a process described as **proofreading.** **[Try Problems 16.22 through 16.25]**

Amino acids are initially attached to tRNAs through an ester link to either the 2′ or 3′ position of the ribose residue within the 3′ terminal nucleotide residue (always an AMP residue) (Figure 16.13). Once attached, an amino acyl group can readily migrate between the 2′ and 3′ positions. A tRNA molecule to which an amino acid has been attached is said to be **charged** or **aminoacylated.** The amino acyl group carried by a tRNA is activated (relative to a free amino acid).

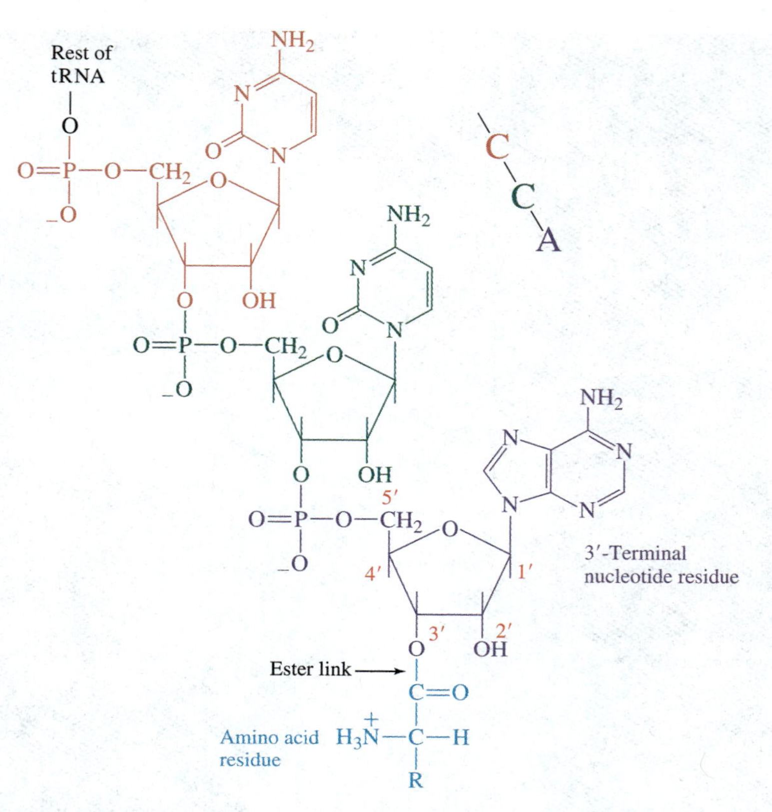

FIGURE 16.13
An amino acid esterified to the 3′ end (CCA end) of a tRNA. R is the amino acid side chain.

PROBLEM 16.19 What amino acid is normally carried by a tRNA molecule whose anticodon pairs with the codon CUU? Explain. What is the anticodon for this tRNA? Remember that complementary nucleic acid segments are antiparallel.

PROBLEM 16.20 Transfer-RNA_1 ($tRNA_1$) and $tRNA_2$ both accept and carry valine. Can you confidently conclude that these two tRNAs have the same anticodon? Explain.

PROBLEM 16.21 Can two tRNA molecules have the same secondary structure but different primary structures? Explain.

PROBLEM 16.22 Explain why aminoacyl–tRNA synthetases are classified as ligases.

PROBLEM 16.23 How many anhydride bonds are cleaved during the attachment of an amino acid to a tRNA? What is the significance of this cleavage?

PROBLEM 16.24 The equation for the attachment of an amino acid to a tRNA can be depicted as the sum of two coupled reactions. What are those two reactions?

PROBLEM 16.25 Aminoacyl–tRNA synthetases use L-amino acids, but not D-amino acids, as substrates. What does this tell you about the active sites of these enzymes? How does this explain the fact that nascent (newly synthesized) polypeptides contain only L-amino acid residues?

16.7 RIBOSOMAL RNA AND RIBOSOMES

The last major class of RNA is **ribosomal RNA (rRNA),** so named because rRNAs are integral components of ribosomes. **Ribosomes** are RNA–protein complexes that serve as factories for the assembly of polypeptides. Messenger RNAs are translated as they move through ribosomes, and an enzyme domain within ribosomes is directly responsible for the formation of peptide bonds (see Chapter 18 for details). A **prokaryotic ribosome,** called a **70S ribosome** on the basis of its sedimentation rate during centrifugation (see next paragraph), can dissociate into a **50S subunit** and a **30S subunit.** A cytosolic **eukaryotic ribosome (80S)** dissociates into a **60S subunit** and a **40S subunit:**

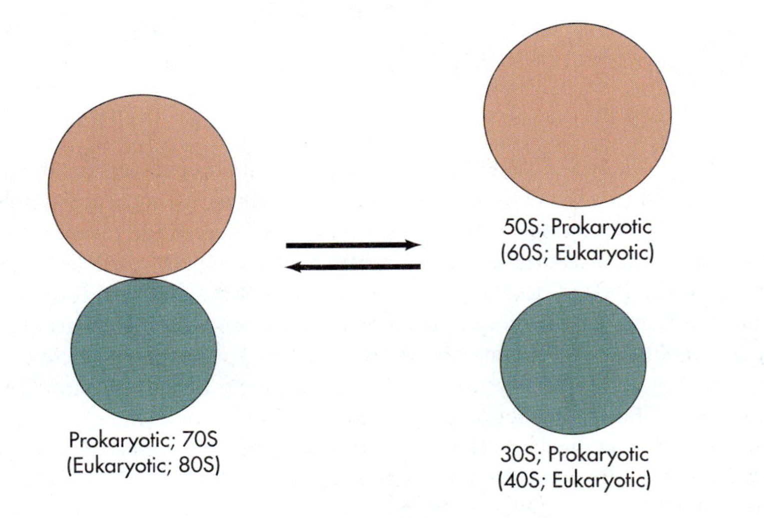

Each subunit contains from one to three rRNAs and between 20 and approximately 50 different polypeptides (Tables 16.8 and 16.9). **[Try Problems 16.26 and 16.27]**

TABLE 16.8

Some Characteristics of a Typical Prokaryotic Ribosome

Particle	S Value	Mass (daltons)	% RNA (by mass)	% Protein (by mass)	RNA Molecules Present	Number of Nucleotide Residues in rRNAs	Total Number of Polypeptides
Intact ribosome	70	2,520,000	66	34	16S rRNA		~55
					23S rRNA		
					5S rRNA		
Small subunit	30	930,000	60	40	16S rRNA	~1,540	~21
Large subunit	50	1,590,000	70	30	23S rRNA	~3,200	~34
					5S rRNA	~120	

TABLE 16.9

Some Characteristics of a Typical Eukaryotic Cytosolic Ribosome

Particle	S Value	Mass (daltons)	% RNA (by mass)	% Protein (by mass)	RNA Molecules Present	Number of Nucleotide Residues in rRNAs	Total Number of Polypeptides
Intact ribosome	80	4,220,000	60	40	18S rRNA		~82
					28S rRNA		
					5.8S rRNA		
					5S rRNA		
Small subunit	40	1,400,000	50	50	18S rRNA	~1,900	~33
Large subunit	60	2,820,000	65	35	28S rRNA	~4,700	~49
					5.8S rRNA	~160	
					5S rRNA	~120	

The **S** in 70S, 50S, 30S, 80S, 60S, and 40S is an abbreviation for the **Svedberg unit.** The Svedberg is a unit of **sedimentation coefficient,** a measure of how rapidly a particle sediments (moves toward the bottom of a centrifuge tube) during centrifugation. An **ultracentrifuge** is an instrument that subjects particles to high gravitational forces (50,000 to 500,000 times the normal gravitational force on the surface of the Earth) by spinning a tube containing a mixture of the particles at rapid speeds (20,000 to 70,000 rpm) (Figure 16.14). The S value of a particle is primarily determined by its mass, its shape, and its surface area. Because frictional forces retard sedimentation, compact particles tend to sediment more rapidly (have larger S values) than less compact particles of comparable mass. Similarly, a tiny metal ball and a feather of equal mass will settle to the ground at different rates when dropped from a tall building. Particles with approximately the same density tend to sediment more rapidly as their masses increase. However, since a 70S ribosome dissociates into a 50S subunit and a 30S subunit, one knows that S values are not proportional to mass. Transfer RNAs are 4S molecules while a typical mRNA sediments between 6S and 25S. **[Try Problem 16.28]**

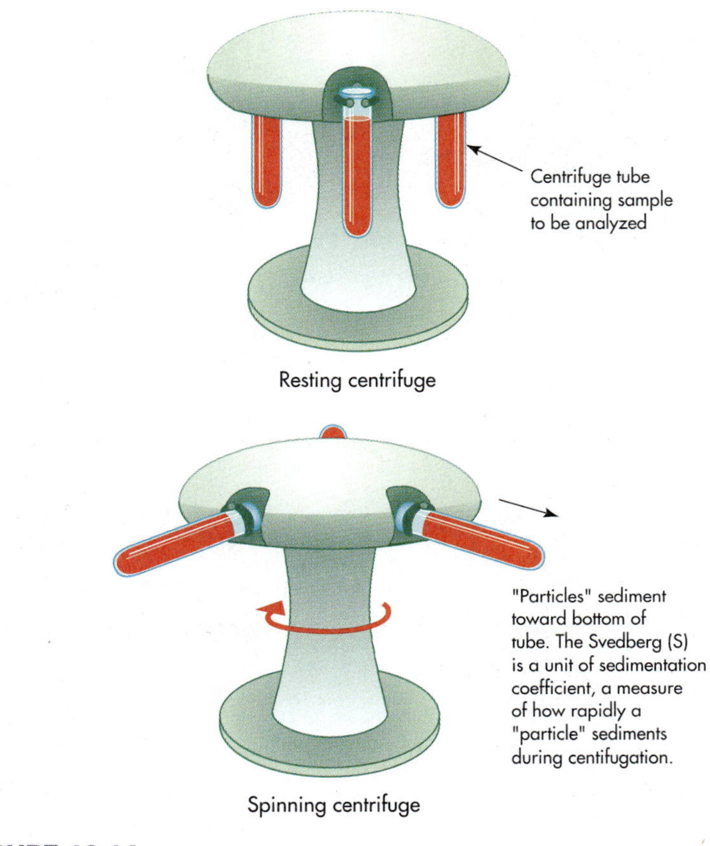

FIGURE 16.14
A centrifuge.

Although circles are commonly used to represent diagrammatically ribosomal subunits, it is known that such subunits are not really circular or spherical objects. Figure 16.15 contains a diagram of an *E. coli* ribosome and its subunits. Eukaryotic ribosomes and their subunits have similar shapes.

The eukaryotic ribosomes described in the above paragraphs are cytosolic ribosomes, those ribosomes found in the cytosol (Section 1.3) of cells. Mitochondria contain structurally unique ribosomes and a distinctive set of tRNAs that are used to assemble the proteins encoded in mitochondrial DNA. In plant cells, chloroplasts also contain distinctive ribosomes and a unique set of tRNAs that are involved in the expression of the genetic information carried in the DNA of these organelles.

Because most cells contain a large number of ribosomes, these protein-synthesizing factories account for about 80% (by weight) of the total RNA and 10% (by weight) of the total protein within a typical cell. All the ribosomes in a prokaryotic cell are probably identical. This may not be the case for cytosolic ribosomes in eukaryotic cells.

The primary structure has been determined for the rRNAs from many different types of cells, and all of these RNAs contain a small fraction of minor bases. Sec-

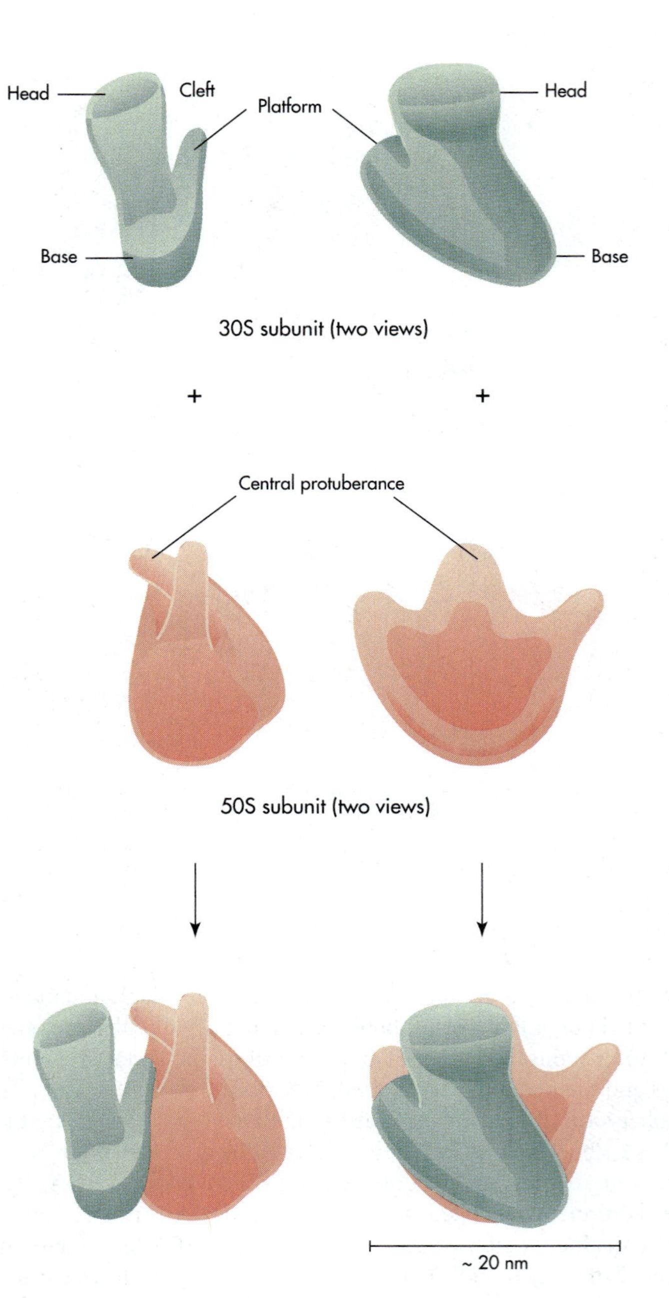

FIGURE 16.15
An *E. coli* ribosome. (From Voet & Voet, 1995.)

ondary structure models have been proposed for the sequenced rRNAs, and considerable progress has been made in determining the exact manner in which the rRNAs and ribosomal polypeptides interact within intact ribosomes and their subunits. Figure 16.16 depicts the secondary structure for the rRNA from the small subunit of each of five different ribosomes, including one eukaryotic cytosolic ribosome and one mitochondrial ribosome. Although the rRNAs differ in length and possess

(a)

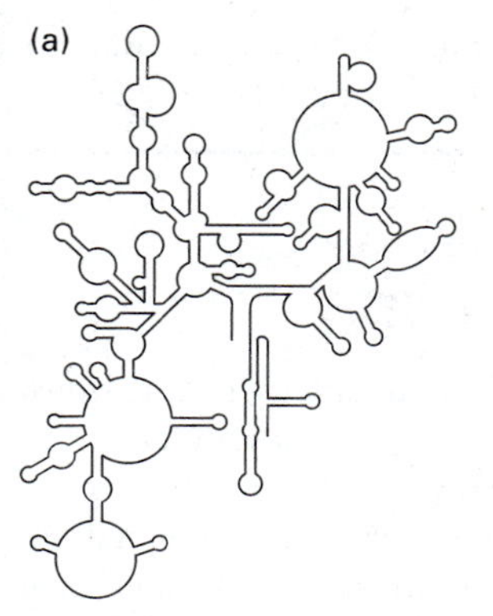

E. coli 16S rRNA: 1542 nucleotide residues

(b)

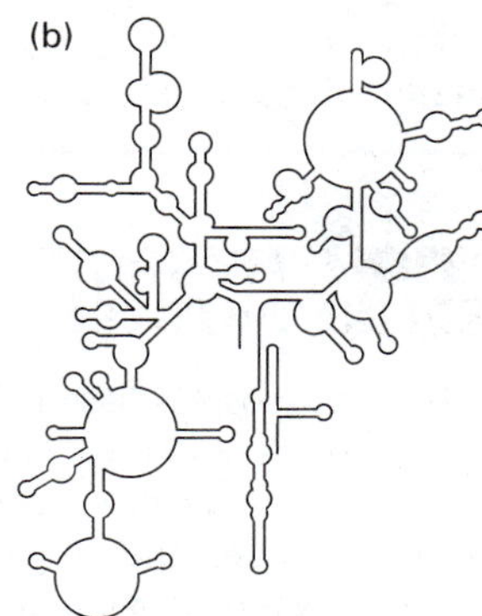

Xenopus laevis cytosolic 18S rRNA: 1825 nucleotide residues

(c)

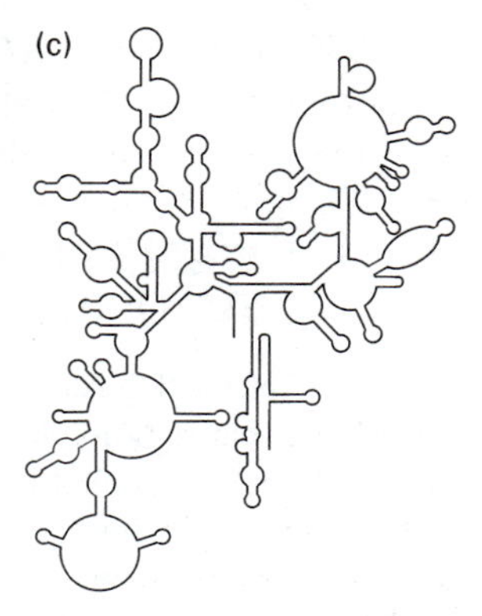

Saccharomyces cerevisiae mitochondrial 15S rRNA: 1640 nucleotide residues

(d)

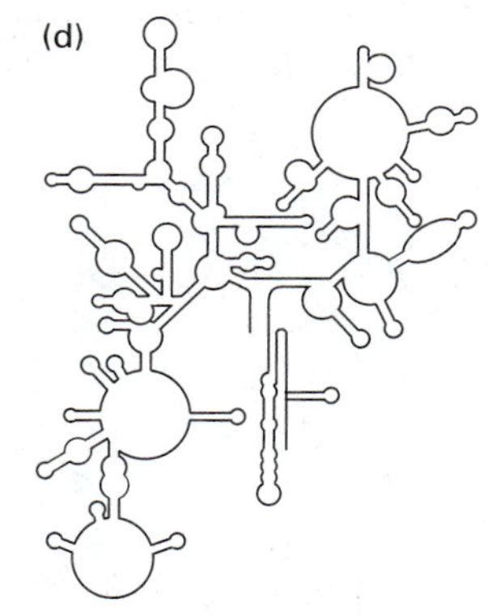

Maize chloroplast 16S rRNA: 1430 nucleotide residues

(e)

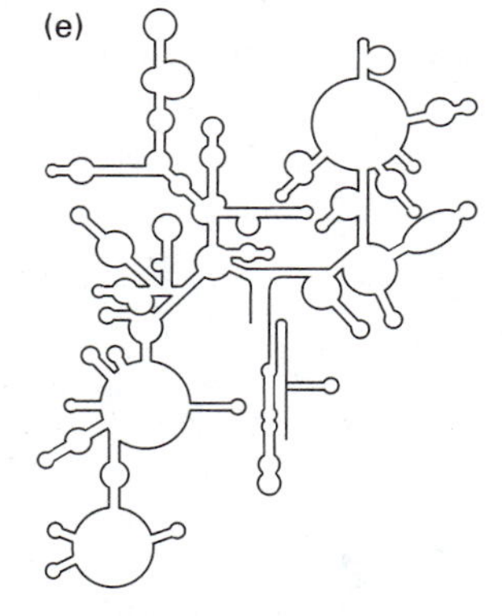

Halobacterium volcanii 16S rRNA: 1469 nucleotide residues

FIGURE 16.16
Proposed secondary structures for some small subunit rRNAs. The rRNAs from the small subunit of ribosomes from diverse organisms often differ significantly in size or primary structure or both, yet most have very similar secondary structures. (From Sadara, 1993.)

unique nucleotide sequences, they all have similar secondary structures. The similar secondary structures are claimed to be an example of convergent evolution. In this case, the convergent evolution is thought to be a consequence of structure–function relationships that are under investigation. Although not identified in many instances, the themes of structure–function relationships and evolution weave their way through virtually every chapter in this text.

PROBLEM 16.26 Approximately how many separate molecules does an intact prokaryotic ribosome contain? What is the approximate number of molecules in the 60S subunit of a typical cytosolic eukaroytic ribosome?

PROBLEM 16.27 What compounds are produced when a ribosome is completely hydrolyzed?

PROBLEM 16.28 Which has a larger S value, a 50,000 dalton fibrous protein or a 50,000 dalton globular protein? Explain.

16.8 DENATURATION, HYBRIDIZATION, AND ANTISENSE NUCLEIC ACIDS

DNA, mRNAs, tRNAs, and rRNAs all possess base-paired, double-helical regions. Because these regions are maintained by relatively weak noncovalent interactions (H bonds and hydrophobic forces), they can be disengaged without rupturing any covalent bonds. Nucleic acids in which base-paired regions have been disrupted are said to be **denatured.** Nucleic acids can be denatured by the same treatments that denature proteins (Section 4.6), since the native coilings and foldings of both classes of polymers are maintained primarily through similar noncovalent forces. When heat is used to denature a DNA molecule, the DNA is said to melt. The **melting point** of a DNA is that temperature at which half of its base pairs have been ruptured. **[Try Problem 16.29]**

Under appropriate conditions, denatured nucleic acids can reestablish base-paired regions, a process described as **renaturation** or **annealing.**

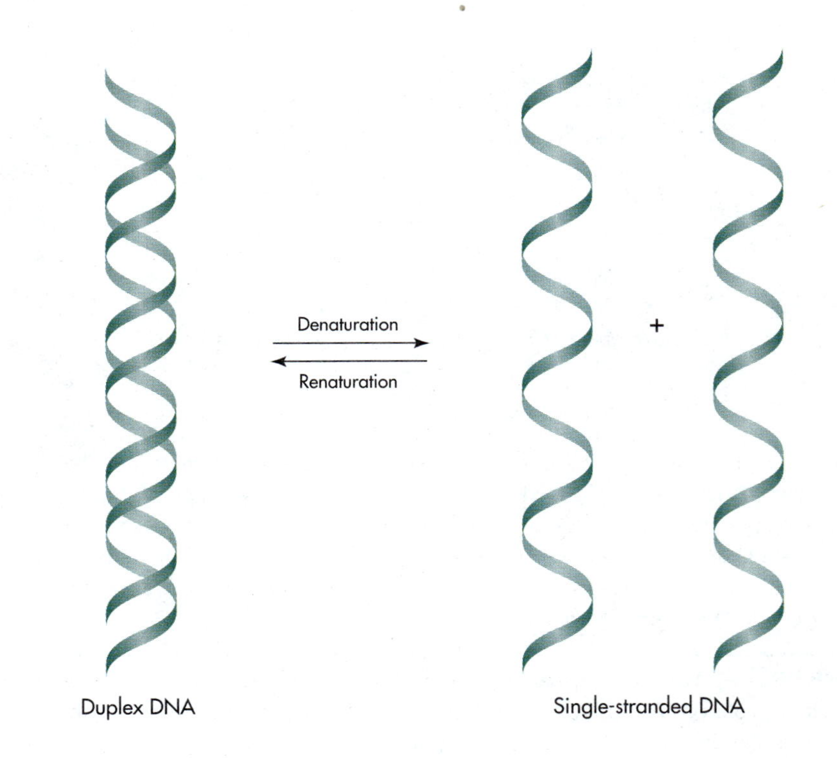

When a single-stranded DNA molecule anneals with a complementary RNA molecule, an **RNA:DNA hybrid** is formed and the process is known as **hybridization.**

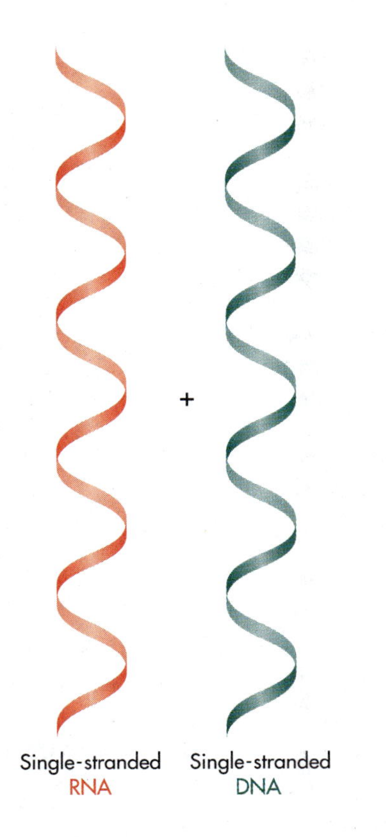

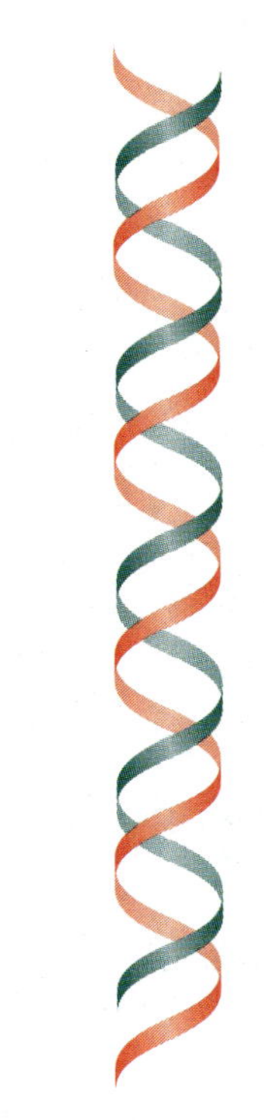

The term "hybridization" is also used by some authors to describe the pairing of two complementary single-stranded DNAs or two complementary single-stranded RNAs. **[Try Problem 16.30]**

The tendency for complementary polynucleotide strands to anneal can be put to practical use. **DNA probes,** for example, are single-stranded oligo- and polynucleotides that are used to search for (probe for) complementary nucleotide sequences in DNA. Such probes, which are discussed in more detail in Chapter 19, are employed to identify carriers of genetic diseases and to locate cells (including cancer cells and virus-infected cells) possessing abnormal or foreign DNA. Some DNA probes qualify as **antisense nucleic acids,** single-stranded nucleic acids that are complementary to biologically functional nucleic acid strands known as **sense strands.** Antisense nucleic acids are being used routinely to regulate sense strand utilization (Figure 16.17). How this is accomplished is best illustrated with a specific example.

A specific enzyme, **polygalacturonase,** plays a central role in the softening (senescence) of tomatoes. Information for the production of this enzyme is encoded in a tomato gene whose expression entails the production of a unique mRNA. Genetic engineers (Chapter 19) have programmed tomato plants to produce an antisense RNA that is complementary to the mRNA for polygalacturonase. Once produced, the antisense RNA anneals with the mRNA and blocks its translation. Polygalacturonase synthesis is shut down and tomato softening is retarded. Tomatoes from the genetically engineered plants taste better, last longer, and can be

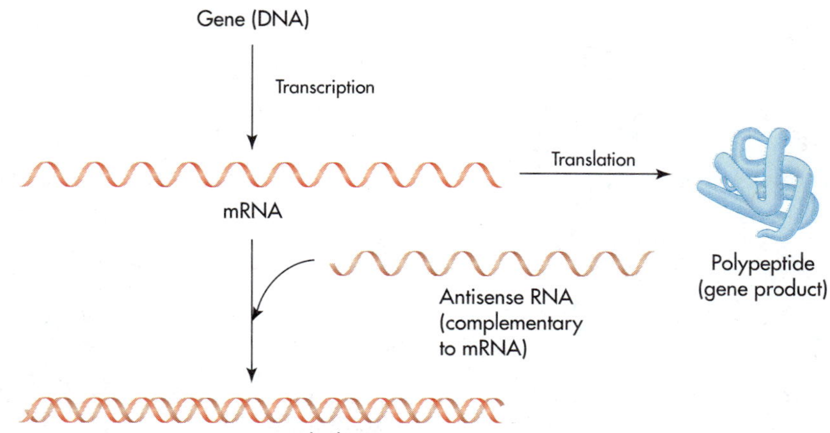

FIGURE 16.17
Control of gene expression with antisense RNA.

FIGURE 16.18
Genetically engineered tomatoes are shown with an electrophoresis gel in the background. Each band in the electrophoresis gel contains a DNA fragment that was separated from the other fragments during electrophoresis. Electrophoresis is used to analyze the genes in both normal and genetically engineered plants.

harvested when ripe (without turning to mush before they reach the supermarket) (Figure 16.18). Genetic engineers are using similar strategies to suppress the expression of specific genes in a wide variety of cells and organisms. Some organisms are naturally programmed to produce antisense nucleic acids that they use to help regulate the expression of their own genes.

Some antisense nucleic acids are presently being tested as drugs (sometimes described as genetic medicines; see selected readings). Much of their therapeutic potential lies in their ability to suppress overactive genes and their ability to block the expression of pathogen-specific genes whose products are essential for the survival of a pathogen. Clinical trials are in progress that are designed to evaluate the effectiveness of specific antisense nucleic acids in the treatment of myologenous leukemia (a form of blood cancer) and several different virus infections, including human immunodeficiency virus (HIV) infections. HIV is the virus associated with AIDS (Section 17.5).

PROBLEM 16.29 List three different conditions or treatments, in addition to heat, that denature nucleic acid molecules.

PROBLEM 16.30 True or false? All of the RNA within a cell should be able to hybridize with part of the denatured DNA from that cell. Explain.

16.9 NUCLEIC ACID SEQUENCE DETERMINATION

Any nucleic acid must be sequenced before its structure and function can be studied in detail. For this reason, an enormous amount of time and energy has been put into the development of sequencing technologies, and considerable progress has been made. With automated sequenators it is now possible to sequence even large DNA molecules relatively quickly and at a cost of less than a dollar a base pair.

The Sanger (Dideoxy) Method

Virtually all current DNA sequencing efforts are based on either the **Sanger (dideoxy) method** or the **Maxam–Gilbert procedure.** Both approaches utilize the same general strategy: four separate samples of the DNA strand to be analyzed are used to create four unique sets of complementary (to the analyzed strand) polynucleotide fragments. One set contains all possible fragments ending in A, one all possible fragments terminating in T, one all possible fragments ending in G, and one all possible fragments that conclude with C. After the fragments in each set are separated by size along adjacent strips of a flat electrophoresis gel, the sequence of the complement of the original polynucleotide is read from the relative positions of the fragments (each shows up as a separate band) on the gel. The sequence of the original polynucleotide can be determined from the sequence of its complement. The more frequently utilized Sanger methodology is examined in this section.

Assume that the following piece of a DNA molecule is to be sequenced:

3′dTGGCTATGGTTACCGC5′

A sample of this segment of DNA is incubated in a test tube with an appropriate primer (an oligonucleotide complementary to its 3′ end), dATP, dTTP, dCTP, dGTP, a small amount of **2′,3′-dideoxy ATP** (ddATP; Figure 16.19) and an enzyme named DNA polymerase. The primer and the deoxynucleoside triphosphates pair with nucleotide residues in the polynucleotide to be sequenced. The nucleotide residues within the nucleoside triphosphates are then added (by the catalytic activity of the polymerase) to the primer one at a time. The product is a nucleic acid strand complementary to the original polynucleotide (Figure 16.20). The original polynucleotide is said to serve as a template or pattern during this process. The same events are involved in DNA replication (Section 17.4). The primer is required because DNA polymerase cannot catalyze the joining of one free nucleotide to a second; it can only catalyze the addition of a nucleotide residue to a preexisting oligonucleotide or polynucleotide. To synthesize the appropriate primer, a separate procedure is used to determine the ordering of nucleotides at the 3′ end of the polynucleotide being sequenced.

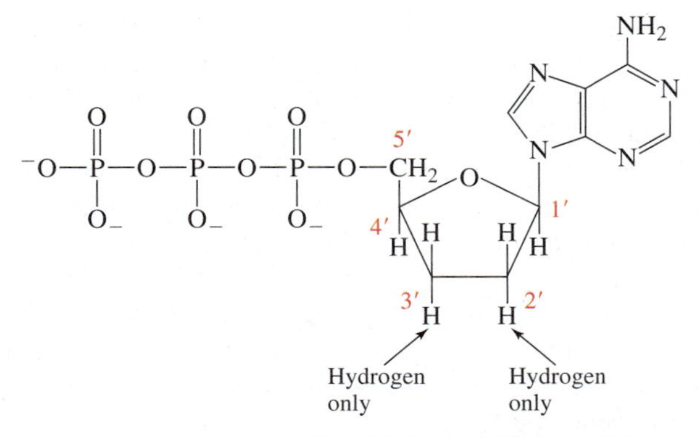

FIGURE 16.19
Dideoxy ATP. The oxygen has been removed from the hydroxyl groups that were at the 2′ and 3′ positions in a ribose residue; a dideoxy nucleoside triphosphate results.

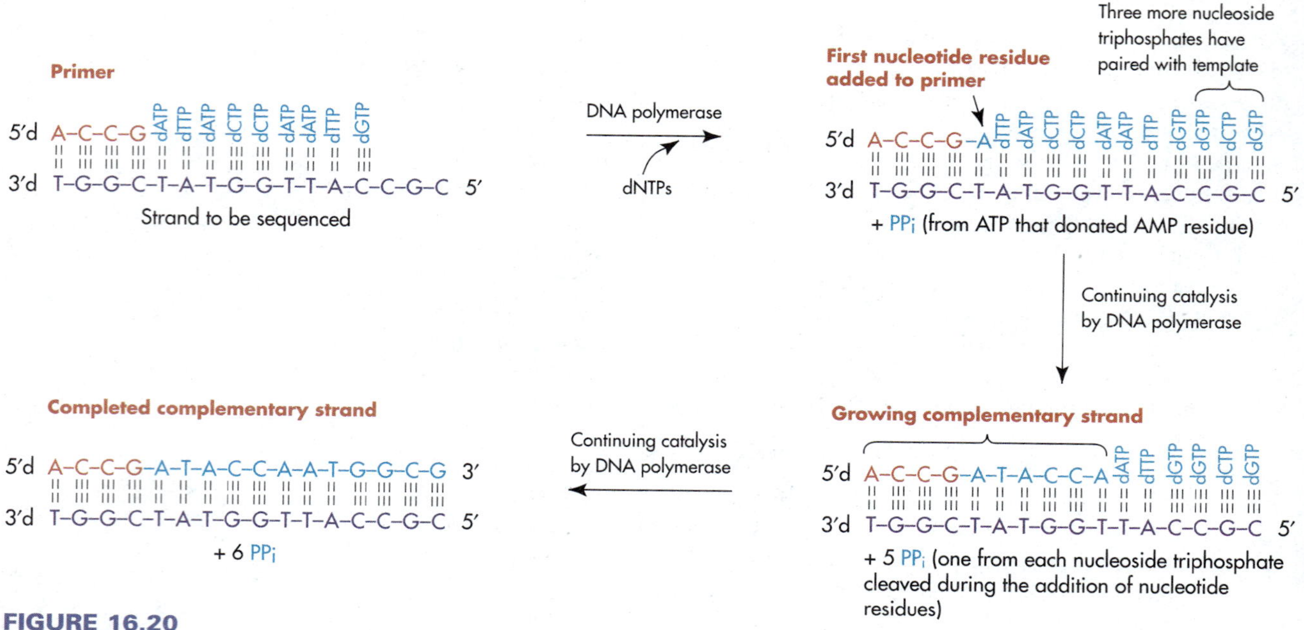

FIGURE 16.20
The production of a complementary nucleic acid strand.

In our reaction mixture, where an enormous number of complementary fragments are simultaneously being assembled, ddATP occasionally pairs with the template in place of dATP. The assembly of the complementary strand stops when ddAMP is added to the growing chain, since the ddAMP residue lacks the 3′-hydroxyl group required to add another nucleotide. The random nature of the ddATP pairing leads to the generation of a set of different length complementary fragments, with each fragment having a ddAMP residue at its 3′ end. The final set of fragments contains some of all of the possible fragments that would normally terminate in a dAMP residue (Table 16.10).

When we add ddTTP to the reaction mixture in place of ddATP, we generate a unique set of complementary (to the original polynucleotide) fragments, each with a ddTMP residue at its 3′ end. Similarly, incubation with ddCTP and ddGMP generates two additional sets of fragments (Table 16.10).

Four samples, one from each of the four reaction mixtures (one containing ddATP, one ddTTP, and so on), are placed side by side at the origin of an electrophoresis gel composed of polyacrylamide (most often) or agarose. Electrophoresis employing a polyacrylamide gel is known as **PAGE** (Section 3.10). Under the electrophoretic conditions employed, smaller oligo- and polynucleotides migrate through the gel more rapidly than larger ones, and molecules that differ in length by a single nucleotide can be separated from one another. Following electrophoresis, the separated fragments are located and visualized with fluorescent dyes or other techniques. When the deoxyribonucleoside triphosphates used to synthesize the complementary strand are labeled with a radioisotope, the electrophoretic bands can be detected with autoradiography (Section 3.12). Each fragment shows up as a separate band on the electrophoresis gel (Figure 16.21 on page 682).

TABLE 16.10

The Four Sets of Oligonucleotide Fragments Generated During the Sequencing of 3′dTGGCTATGGTTACCGC5′

Set Terminating In ddAMP	
A1	5′ d A–C–C–G–A[a] 3′
A2	5′ d A–C–C–G–A–T–A[a] 3′
A3	5′ d A–C–C–G–A–T–A–C–C–A[a] 3′
A4	5′ d A–C–C–G–A–T–A–C–C–A–A[a] 3′
Set Terminating In ddTMP	
T1	5′ d A–C–C–G–A–T[b] 3′
T2	5′ d A–C–C–G–A–T–A–C–C–A–A–T[b] 3′
Set Terminating In ddCMP	
C1	5′ d A–C–C–G–A–T–A–C[c] 3′
C2	5′ d A–C–C–G–A–T–A–C–C[c] 3′
C3	5′ d A–C–C–G–A–T–A–C–C–A–A–T–G–G–C[c] 3′
Set Terminating In ddGMP	
G1	5′ d A–C–C–G–A–T–A–C–C–A–A–T–G[d] 3′
G2	5′ d A–C–C–G–A–T–A–C–C–A–A–T–G–G[d] 3′
G3	5′ d A–C–C–G–A–T–A–C–C–A–A–T–G–G–C–G[d] 3′

[a]A = ddAMP.
[b]T = ddTMP.
[c]C = ddCMP.
[d]G = ddGMP.

The sequence of the newly synthesized complementary strand, minus the primer, can be read directly off of the four **DNA "ladders"** that exist side by side on the gel. Because the band that migrates the farthest (contains the shortest oligonucleotide) is in the ddATP column, dAMP must be the first nucleotide added to the primer. The next most rapidly moving band (contains the second shortest nucleotide) is in the ddTTP column, so dTMP is the second nucleotide added to the growing complement. Similar reasoning allows us to determine that dAMP is the third nucleotide added, dCMP is the fourth, and so on (study the DNA ladders in Figure 16.21). The sequence of the original strand can be determined from the sequence of its complement. Figure 16.22 presents some experimentally-prepared sequencing ladders, and Problem 16.31 calls for the interpretation of these ladders. **[Try Problem 16.31]**

The Human Genome Project

The United States government is presently supporting the **Human Genome Project (Initiative),** a massive research effort designed to provide the nucleotide sequence of the entire human genome, three to four billion base pairs. The project is scheduled to be completed soon after the turn of the century. When this goal is achieved, biochemists will have in their hands the sequences of the roughly 100,000 genes that are responsible for the production and maintenance of human life. Research will

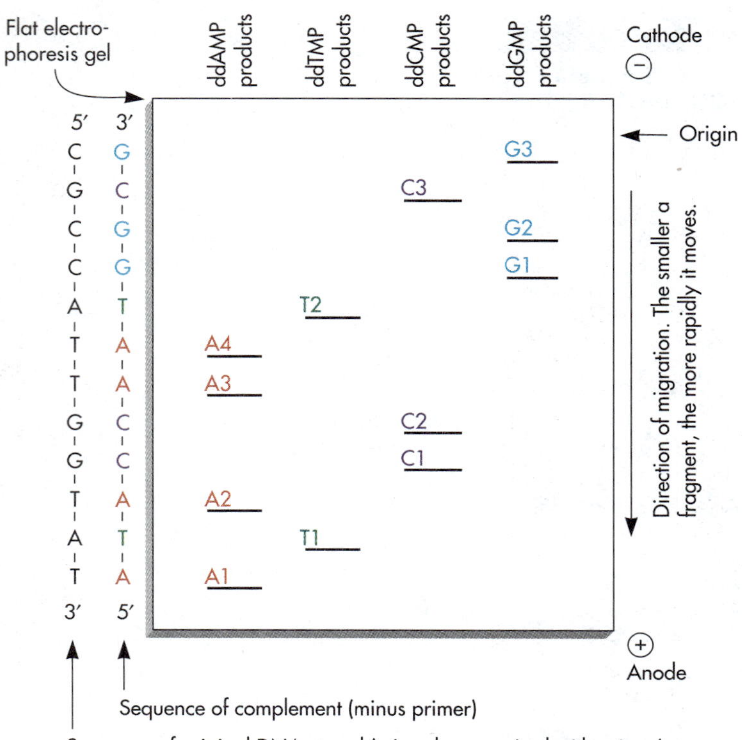

FIGURE 16.21
An idealized sequencing gel. After a sample of each of the four dideoxy reaction mixtures (one containing ddATP, one ddTTP, one ddCTP, and one ddGTP) is placed in a separate "trough" at the origin, an electric current is employed to separate the oligonucleotides in each reaction mixture according to size. Each band (visualized with the use of a fluorescent dye, a radioisotope, or some other procedure) represents a unique oligonucleotide. The numbering of the bands corresponds to the numbering used in Table 16.10. The farthest migrating band contains the primer with one added nucleotide. It is the smallest oligonucleotide generated during the production of the complementary strand. The first nucleotide added to the primer must have been AMP because the farthest migrating band resides in the ddAMP column. Since the next most rapidly migrating band is in the ddTTP column, the second nucleotide added to the growing complementary strand was TMP. AMP was the third nucleotide added, and CMP the fourth. The sequence of the rest of the complementary strand can be quickly read from the gel by continuing to employ the same reasoning. The reading and interpretation of sequencing gels has been automated. The sequence of the primer cannot be read from the gel. The primer normally has a predetermined sequence.

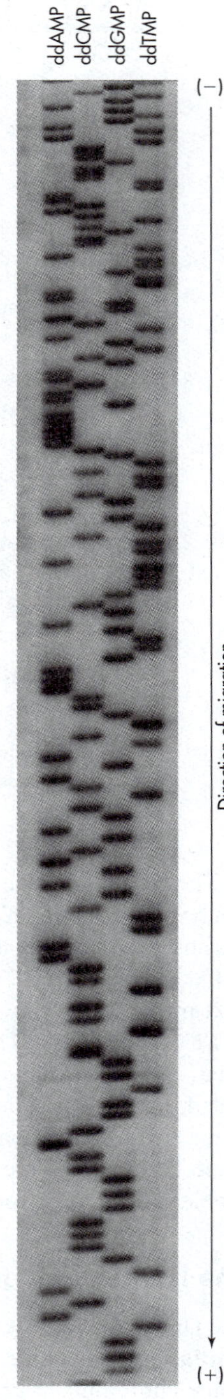

FIGURE 16.22
A Sanger sequencing gel.

then focus on the expression of individual genes and the functions of their products. The products of most genes are polypeptides, and the sequences of these polypeptides can, with the assistance of the genetic code, be read from the nucleotide sequences of their genes.

Spinoffs from the Human Genome Project are already having a major impact on the health sciences. The identification of genes linked to genetic diseases, cancer, and other health problems provide specific examples. As such genes are deciphered and analyzed, the door is opened for improved treatment or prevention. Chapter 19 expands on these observations.

The nucleotide sequence of the complete *Haemophilus influenzae* Rd genome (1,830,137 base pairs) was published in July 1995. It was the first genome of a free-living organism to be entirely sequenced. Other sequencing projects are divulging the ordering of base pairs within the genomes of a wide assortment of other organisms, both prokaryotic and eukaryotic (Exhibit 16.2). This information is being used to develop a better understanding of evolution and the basic chemistry of life. It is also of enormous practical value. The animal genome projects, for example, will lead to better animal models for the study of human diseases. The sequences of the genomes for human pathogens and parasites will facilitate drug and vaccine development. And all of the genome projects will enhance our ability to interpret and manipulate the gene content of organisms. The selective manipulation of gene content falls in the realm of genetic engineering, a topic discussed in some detail in Chapter 19.

EXHIBIT 16.2

Some Genomes That Have Been or Are Being Sequenced

Human
Mouse
Corn
Soybean
Wheat
Rice
D. melanogaster (a fruit fly)
S. cerevisiae (a yeast)
C. elegans (a nematode)
E. coli (a bacterium)
H. influenzae (a bacterium)

Jurassic Park

One sequencing effort, in particular, has captured the interest and the imagination of the general public. This is the sequencing of DNA from ancient organisms, including extinct plants and animals. *Jurassic Park,* a Steven Spielberg film released in 1993, is largely responsible for the public interest (Figure 16.23). In this film, dinosaur DNA was extracted from the gut of Mesozoic mosquitoes preserved in amber. This DNA, after being sequenced and patched, was employed to resurrect the age of the dinosaurs. In the real world, 2400-year-old fragments of DNA from an Egyptian mummy have been sequenced, DNA from the Otz Valley Iceman (~5000 years old) has been partially analyzed, and the ordering of the nucleotides in bits of DNA from amber-embedded leaves over 30 million years old has been published. Several sequenced fragments of DNA, isolated from an 80-million-year-old skeleton, are hypothesized to be from a dinosaur. One of the most ancient bits of DNA ever sequenced was extracted from a Jurassic wood beetle embedded in amber more than 120 million years ago.

FIGURE 16.23
A scene from the movie *Jurassic Park.*

The sequences of ancient DNAs must be viewed with caution, because multiple published sequences have already been traced to artifacts. If an entire ancient genome is ever isolated and sequenced, it will still be impossible (based on present knowledge and capabilities) to use that genome to create the organism encoded within it. In spite of the potential pitfalls, published ancient sequences are being utilized to compare present life forms with extinct organisms and to construct phylogenetic trees. If enough DNA can be sequenced from related mummified Egyptian kings, their family trees can be ascertained.

Although many biochemists have questioned the ability of ancient genomes to survive in a functional state for millions of years, 25 to 40 million-year-old bacterial spores were claimed to have been revived in May 1995. The spores were isolated from the abdominal contents of extinct bees preserved in buried amber from the Dominican Republic.

PROBLEM 16.31 What is the sequence of the first 25 residues (from the 3′ end) in the polynucleotide used to prepare the ladders illustrated in Figure 16.22? Assume that the primer is ATGCC and that this primer pairs with the terminal five bases at the 3′ end of the sequenced polynucleotide.

SUMMARY

DNA is the genetic material for all living organisms. Genes are segments of DNA that carry information for the production of specific RNAs or polypeptides. The final expression of most genetic information consists of the production of a functional polypeptide.

Most DNA molecules contain two distinct polynucleotide strands that are antiparallel, base paired along their entire length, and twisted into a double helix. Each duplex is capable of existing in multiple helical forms, including A, B, and Z forms, that differ in the number of base pairs per turn, the base-pair tilt, the direction of coiling, and surface topology. The multiple forms tend to exist in equilibrium, with the favored helix being determined by nucleotide sequence, extent of methylation, extent of supercoiling, protein–nucleic acid interactions, salt concentration and other variables. Under physiological conditions, most DNA probably exists as B-DNA. Several minor forms of DNA, including parallel duplexes, assorted triplexes, and multiple quadruplexes, are hypothesized to be involved in the maintenance of genome structure, DNA replication, transcription, and recombination.

The linking number for a duplex DNA is the number of times one strand wraps around the other. For relaxed B-DNA, the linking number is the number of base pairs in the DNA divided by 10.4, the number of base pairs per turn of the helix. As the relaxed linking number increases or decreases, a torsional stress is created that leads to a coiled coil arrangement known as a supercoil. A decrease in linking number generates a negative supercoil and weakens a duplex, whereas an increase in linking number leads to a positive supercoil and a stabilized duplex. DNA molecules that differ only in linking numbers are called topoisomers and enzymes that catalyze the interconversion of topoisomers are known as topoisomerases. Supercoiling is involved in packaging DNA into nucleosomes, and it is thought to be involved in DNA replication, transcription, and recombination.

In eukaryotes, segments of DNA are packaged with basic proteins, known as histones, to create nucleosomes. Chromatin consists of a string of nucleosomes along one continuous duplex DNA. Chromatin molecules coil and compress to generate chromosomes, packaged forms of DNA that usually contain a variety of nonhistone proteins and some RNA in addition to histones.

The three major classes of RNA all play central roles in the expression of the information carried in DNA. DNA directs the construction of these RNAs by serving as a pattern or template, a process known as transcription (examined in Section 17.1). The members of the major classes of RNA consist of single-stranded polymers that fold back upon themselves to form base-paired (stem) and loop regions. The base-paired regions contain two antiparallel segments of the polymer in which all of the purines and pyrimidines along one segment are H bonded to complementary bases along the second segment. Adjacent planar base pairs stack together and twist into a right-handed, spiraling staircase–type arrangement similar to the A-helix of DNA.

The primary structure of an RNA refers to its nucleotide sequence. Its secondary structure consists of its stem regions. The specific folding together of the multiple stem and loop regions generates a unique tertiary structure.

Messenger RNA, one of the three major classes of RNA, carries information (messages picked up from DNA) for the assembly of polypeptides. Messages are carried in the form of the genetic code, and the process through which a polypeptide is produced from the message in an mRNA is known as translation (examined in Chapter 18). As we move in a 5′ to 3′ direction along a typical eukaryotic mRNA, we encounter a cap, a 5′ untranslated region, a translated (coding) sequence, a 3′ untranslated region and a poly-A tail. Suggested functions of the untranslated regions include involvements in translation, mRNA transport, mRNA turnover, and the regulation of gene expression. The 300 to 3000 nucleotides in most mRNAs fold to generate flowerlike secondary structures. The tertiary structure has not yet been determined for any mRNA.

The genetic code is a four-letter code in which each nucleotide residue in the coding region of an mRNA serves as a code letter. Normally, each consecutive three-letter sequence within a coding region constitutes a separate code word (codon). Sixty-one of the 64 codons ordinarily specify one of the 20 common amino acids. The other three codons usually function as termination signals. The codons AUG, UUG, and GUG serve as initiation codons; one of these codons is the first codon read within the translated region of an mRNA. There are multiple codons for most amino acids and, with few exceptions, codon assignments are the same within all living organisms.

Transfer RNAs, which represent the second major class of RNA, are directly responsible for translating (decoding) the genetic code. Most of the dozens of tRNAs in a typical cell contain 73 to 93 nucleotide residues that fold into a cloverleaf secondary structure and an L-shaped tertiary structure. Normally, each tRNA accepts and carries one, and only one, amino acid, and each contains a sequence of three nucleotide residues known as an anticodon. During translation, the anticodon in a tRNA base pairs with one of the codons (on the mRNA being translated) for the amino acid that it carries, an event that decodes the paired codon; it is through tRNAs that the sequence of codons in an mRNA determines the order in which amino acids are linked during the assembly of a polypeptide (see Chapter 18 for details).

Aminoacyl–tRNA synthetases catalyze the attachment of specific amino acids to select tRNAs, a process in which an ester link is formed between the α-carboxyl group of an amino acid and the sugar residue at the 3′ end (CCA end) of a tRNA. ATP provides the energy for this key reaction, a reaction that activates amino acids in preparation for their incorporation into polypeptides.

Ribosomes are the factories where mRNAs are translated and polypeptides are assembled. Prokaryotic ribosomes are 70S complexes that dissociate into 30S and 50S subunits. Eukaryotic ribosomes (80S) are constructed from 40S and 60S subunits. Each subunit contains from around 20 to approximately 50 unique polypeptides and from one to three ribosomal RNAs (the third major class of RNA). Ribosomal RNAs range in size from about 120 nucleotides to close to 5000 nucleotides, and they fold into a variety of stem and loop regions that constitute their secondary structure. The packaging of rRNAs and polypeptides creates the functional domains of the ribosome.

The two strands in a duplex region of a nucleic acid can be separated, a process described as denaturation. Denaturation can be brought about by any treatment that ruptures the relatively weak H bonds and hydrophobic forces that maintain a duplex. Such treatments include heat, high salt concentrations, and organic solvents. When returned to nondenaturing conditions, separate complementary strands tend to repair with one another, a process called renaturation, annealing, or hybridization. Some authors reserve the term "hybridization" to refer to the pairing of complementary RNA and DNA strands to form an RNA:DNA hybrid.

Antisense nucleic acids are single-stranded polynucleotides that are complementary to biologically functional nucleic acid strands, called sense strands. Antisense nucleic acids are being used to locate and identify sense strands and to regulate sense strand utilization. When a single-stranded oligo- or polynucleotide is used to locate (probe for) its complement, that oligo- or polynucleotide is called a DNA probe.

The primary structure of a DNA strand can be determined with either the Sanger or the Maxam–Gilbert procedure. Each approach involves the creation of four distinct sets of complementary (to the DNA strand being sequenced) polynucleotide fragments. One set contains all possible fragments that terminate at the 3′ end in a dAMP residue. Another set contains all possible fragments that terminate at the 3′ end in a dTMP residue. All of the fragments in a third set terminate in dGMP, and those in the fourth set terminate in dCMP. The individual fragments in each of the four sets of fragments are separated from one another during electrophoresis under conditions where smaller fragments migrate more rapidly than larger ones. Each fragment shows up as a separate band or "rung" on a DNA ladder. When all four ladders (one from each of the four unique sets of fragments) are side by side on the same electrophoresis gel, the sequence of the polynucleotide strand complementary to original polynucleotide can be read from the gel. The sequence of the original polynucleotide can be determined from its complement. Biochemists are in the process of determining the sequence of the nucleotide residues within several entire genomes, including the human genome. Biochemists are also sequencing samples of ancient DNA.

EXERCISES

16.1 Summarize the major structural differences between DNA and RNA.

16.2 Identify the 3′ and the 5′ end of each polynucelotide strand in the ball-and-stick models of A-DNA and B-DNA depicted in Figure 16.2.

16.3 Describe the major structural differences between the three double-helical forms of DNA described in this chapter.

16.4 List the major classes of intermolecular bonds that are primarily responsible for maintaining the duplex forms of DNA. Do any covalent bonds exist between the two polynucleotide strands in a normal duplex DNA molecule?

16.5 Which, if any, of the following statements are correct? Explain.

- **a.** It is impossible to convert A-DNA to B-DNA or Z-DNA without breaking covalent bonds.
- **b.** A separate gene must exist for each member of a group of mRNAs that share no common sequences.
- **c.** A cell that is synthesizing 10,000 unique polypeptides must contain at least 10,000 genes.
- **d.** Messenger RNAs are constructed from two complementary polynucleotide strands that can be separated with denaturation.

16.6 Assume that an equilibrium exists between Z-DNA and B-DNA in a reaction mixture:

$$\text{Z-DNA} \rightleftharpoons \text{B-DNA}$$

Will the total amount of Z-DNA (bound plus unbound) in this reaction mixture increase, decrease, or remain the same if you add a protein that binds Z-DNA but does not bind B-DNA? Explain.

16.7 What peptide will be synthesized from the information carried in each of the following DNA fragments? Assume that each fragment codes for a translated region of an mRNA molecule and that the first three nucleotide residues at the 5′ end of each RNA fragment represent an intact codon.

- **a.** dAATGTTCTGTAAAGCCGCGCTGTCGTA
- **b.** dTTGCAGCGTAGCTAGTACGTAGCC

16.8 What is the linking number of a duplex DNA that contains 100 base pairs with 9 base pairs per turn of the double helix?

16.9 Which favors the conversion of B-DNA to Z-DNA: negative supercoiling or positive supercoiling? Explain.

16.10 Assume that the nuclear DNA in a human cell contains 3×10^9 base pairs. Approximately how many nucleosomes will the nucleus of this cell contain if all of its DNA is packaged into nucleosomes? Show your calculations.

16.11 What class of enzymes, besides nucleases, might reasonably be used in an attempt to disrupt or destroy the structure of chromatin? Explain.

16.12 A student suggests that each nucleosome contains a separate gene and that the number of nucleosomes in a cell is equal to the number of genes that it contains. Is this a reasonable suggestion? Explain.

16.13 Use a family tree-type diagram to illustrate the hierarchy among genes, genomes, nucleosomes, chromatin, core particles, and chromosomes.

16.14 How many separate polymer molecules are released when a nucleosome is completely denatured? Explain.

16.15 List the substrates for each of the following classes of enzymes, and then describe the catalytic activity of each class. To which of the six major international classes of enzymes does each of these classes of enzymes belong?

- **a.** Nucleases
- **b.** Proteinases
- **c.** Topoisomerases
- **d.** Aminoacyl–tRNA synthetases

16.16 What polypeptide fragments will be produced during the translation of the following mRNA fragments? Assume that each mRNA fragment is from a translated region and that the first three nucleotide residues represent an intact codon.

- **a.** CUGACCAAAAACUGGCGA
- **b.** GUUACGCACCCUCGGCGAUGACCCAUG
- **c.** UCGUUUCAACGCGCAGGUCCUAUU
- **d.** UUAUUGCUUCUCCUACUG

16.17 For each of the following terms and phrases list the major class or major classes of RNAs (if any) that it describes or characterizes.

- **a.** Cloverleaf secondary structure
- **b.** Is often capped
- **c.** Involved in translation
- **d.** Produced through transcription
- **e.** Contains codons
- **f.** Contains anticodons

g. Can be hydrolyzed
h. Can be "charged" with an amino acid
i. An integral part of ribosomes
j. A polyanion at physiological pH
k. Contains translated regions
l. A substrate for aminoacyl–tRNA synthetases
m. Often contains poly-A tails
n. Contains some minor bases
o. A polynucleotide
p. Directly translates the genetic code
q. A small protein
r. L-Shaped tertiary structure
s. Flower model secondary structure

16.18 Why would you expect brain cells and liver cells to contain at least some different mRNAs?

16.19 Explain why a rapidly growing cell usually contains more mRNAs than a dormant or relatively inactive cell.

16.20 What amino acid is carried by a tRNA that contains the anticodon AAG? ACA? GGC? UGU?

16.21 What is the minimum number of monomer residues in an mRNA that encodes a polypeptide with 370 amino acid residues?

16.22 Explain why an RNA molecule that contains only G and U (no A or C) will have no secondary structure.

16.23 What is the minimum number of mRNA molecules required for a cell to produce the polypeptide components within a single IgG molecule? Ten identical IgG molecules? Explain. Hint: Review Section 4.11.

16.24 Two mRNAs are totally complementary. Can you confidently conclude that both molecules encode the same polypeptide? Explain.

16.25 Let a line represent a polynucleotide chain, and then draw a diagram illustrating an RNA molecule that contains 8 separate base-paired regions.

16.26 Write abbreviations for two very different (in primary structure) mRNA fragments that would both code for Leu-Ala-Phe-Gln-Ser-Pro.

16.27 How does a cell benefit from the constant breakdown of its mRNA molecules?

16.28 How many different genes are needed to carry all of the information for the synthesis of a rat liver cytosolic ribosome? Explain.

16.29 Suggest why aminoacyl–tRNA synthetases are unable to catalyze the attachment of amino acids on denatured tRNAs.

16.30 A purified sample of 23S rRNA, 5S rRNA, and each of the 31 proteins normally found in a 50S *E. coli* ribosomal subunit is added to a reaction mixture. Thirty minutes later, intact 50S subunits are identified within this reaction mixture. What does this observation tell you about the assembly of the 50S ribosomal subunit?

16.31 Which one of the three major classes of RNA is present in largest amounts in a typical cell? Explain.

16.32 Write an abbreviation for the RNA fragment that will hybridize with:
a. dATTGCACATGATGCAATAGCA
b. dGGGCGGGCTGCGTCACGCGTA
c. dAACTTTAACTAAATATATTTT

16.33 Assume that each of the DNA fragments in Exercise 16.32 has paired with its complementary RNA fragment to yield a DNA:RNA hybrid. Which hybrid, (a), (b), or (c), will have the highest melting point? Explain. Hint: Study the A:T and G:C base pairs depicted in Section 16.1. The H bonds that maintain an A:T pair are the same as those that maintain an A:U pair.

16.34 What is the sequence of the DNA fragment used to prepare the following idealized Sanger sequencing gel?

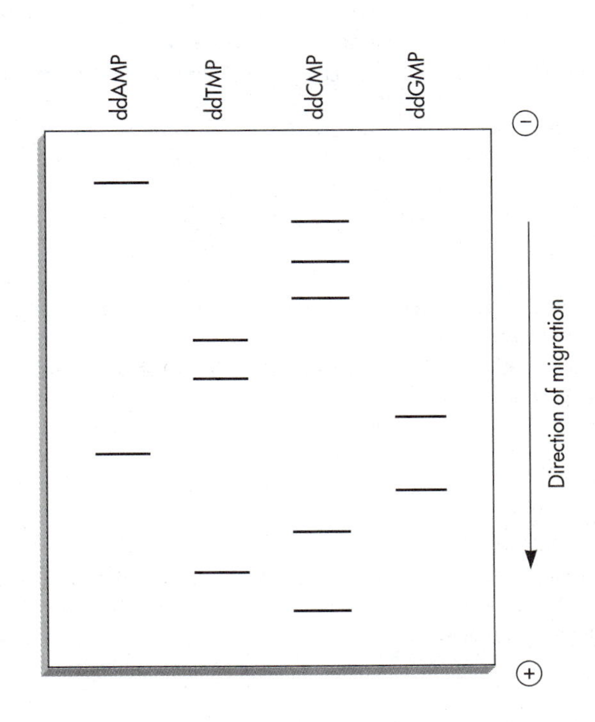

SELECTED READINGS

Aldhous, P., Managing the genome data deluge, *Science* 262, 502–503 (1993).

Describes the use of computers in the storage and analysis of the "growing flood of sequencing and mapping data" from numerous genome sequencing projects.

Blackburn, G. M., and M. J. Gait, Editors, *Nucleic Acids in Chemistry and Biology,* New York: Oxford University Press, 1990.

Approximately 400 pages in length, it is "designed to provide a compact, molecular perspective on the subject of nucleic acids." Beginning students should find it readable. Contains an adequate number of references.

Cano, R.J., and M.K. Borucki, Revival and identification of bacterial spores in 25- to 40-million-year-old Dominican amber, *Science* 268, 1060–1064 (1995).

Describes the culturing of bacterial spores from the abdominal contents of extinct bees preserved in 25- to 40-million-year-old amber. Examines some of the biochemical characteristics of a revived bacterium.

Cohen, J. S., and M. E. Hogan, The new genetic medicines, *Sci. Am.* 271(6), 74–82 (1994).

Explores the potential uses of antisense nucleic acids as drugs.

Gesteland, R. F., R. B. Weiss, and J. F. Atkins, Recoding: Reprogrammed genetic decoding, *Science* 257, 1640–1641 (1992).

Reviews some exceptions to the general rules that govern the translation of the information carried in mRNAs.

Gibbons, A., Possible dino DNA find is greeted with skepticism, *Science* 266, 1159 (1994).

Describes the isolation and analysis of DNA from an 80-million year old skeleton, and examines some of the controversy surrounding this research.

Lehninger, A. L., D. L. Nelson, and M. M. Cox, *Principles of Biochemistry,* 2nd ed. New York: Worth Publishers, 1993.

An encyclopedic text that provides adequate coverage of nucleic acid biochemistry. Contains some problems and further readings.

Marshall, E., A strategy for sequencing the genome 5 years early, *Science* 267, 783–784 (1995).

Explains how the complete sequence of the human genome can be determined as early as 2001.

Moffat, A. S., Triplex DNA finally comes of age, *Science* 252, 1374–1375 (1991).

Examines the structure and uses of triplex DNA.

Moran, L. A., K. G. Scrimgeour, H. R. Horton, R. S. Ochs, and J. D. Rawn, *Biochemistry,* 2nd ed. Englewood Cliffs, NJ: Neil Patterson/Prentice-Hall, 1994.

An encyclopedia-type biochemistry text that covers nucleic acid structure and function. Contains selected readings.

Morell, V., Dino DNA: The hunt and the hype, *Science* 261, 160–162 (1993).

Reviews efforts to isolate and sequence dinosaur DNA.

Nowak, R., Bacterial genome sequence bagged, *Science* 269, 468–470 (1995).

Describes the sequencing of the entire genome for H. influenzae, *and discusses the uses of genome sequence databases.*

Oeller, P. W., L. Min-Wong, L. P. Taylor, D. A. Pike, and A. Theologis, Reversible inhibition of tomato fruit senescence by antisense RNA, *Science* 254, 437–439 (1991).

Describes the use of antisense RNA to control the ripening and senescence of tomatoes.

Pääbo, S., Ancient DNA, *Sci. Am.* 269(5), 86–92 (1993).

Examines the sequencing of ancient DNA molecules and the use of sequence information to study evolution.

Rossl, P. E., Eloquent remains, *Sci. Am.* 266(5), 114–125 (1992).

Describes recent trends in molecular archaeology. Explains how DNA and protein analysis are being used in efforts "to trace the divisions, migrations, extinctions and expansions that have marked the biological history of humanity."

Singer, M., and P. Berg, *Genes & Genomes,* Mill Valley, CA: University Science Books, 1991.

Designed "to emphasize by depth and experimental approach" the science called molecular genetics. Intended for students with "limited prior knowledge of biochemistry, cell biology and genetics." Ample references.

Smith, L. M., The future of DNA sequencing, *Science* 262, 530–531 (1993).

Reviews recent developments in DNA sequencing technology.

Stein, C. A., and Y.-C. Cheng, Antisense oligonucleotides as therapeutic agents—Is the bullet really magical? *Science* 261, 1004–1012 (1993).

Examines attempts to use oligodeoxynucleotides to treat human diseases. Emphasis is placed on oligonucleotides that inhibit viral replication.

Voet, D., and J. G. Voet, *Biochemistry,* 2nd ed. New York: John Wiley & Sons, 1995.

A comprehensive text with thorough coverage of the structure and function of nucleic acids. Encompasses selected readings and some problems.

Williams, N., The trials and tribulations of cracking the prehistoric code, *Science* 269, 923–924 (1995).

Reviews some of the presentations and discussions at a conference titled Ancient DNA III in July 1995.

17 Nucleic Acid Biosynthesis

CHAPTER OUTLINE

Chapter 16 has examined the structure, properties and general functions of DNA and the three major classes of RNA. These nucleic acids play a central role in the storage, transmission, and expression of that information responsible for all hereditary characteristics. The final expression of most genetic information consists of the production of a polypeptide, a process that involves transcription (RNA synthesis directed by DNA) and translation (polypeptide assembly directed by mRNA). Although transcription and translation are commonly coupled in prokaryotes, these events are physically separated during the expression of nuclear genes in eukaryotes; transcription occurs in the nucleus while the translational machinery resides in the cytoplasm. The polypeptides synthesized during translation determine and control those reactions that occur within an organism. Enzymes play a central role in this regard (Chapter 5).

The biosynthesis of all nucleic acids, both RNAs and DNAs, involves the use of preexisting polynucleotide strands as patterns or templates (Figure 17.1). If no errors are made in the process, the nucleotide sequence of a newly assembled nucleic acid strand is totally complementary to its template. In most instances, the structure of a nucleic acid is subsequently modified through the alteration, addition, or deletion of specific nucleotide residues. In some cases, a newly synthesized nucleic acid is subjected to all three categories of modifications.

Transcription employs one strand of duplex DNA as a template. The immediate products include an RNA strand called a **primary transcript.** In eukaryotic cells, most primary transcripts are inactive precursors of mRNAs, tRNAs, or rRNAs. The conversion of primary transcripts to mature, biologically active RNAs is known as **processing.** It entails a variety of enzyme-catalyzed reactions. In prokaryotic cells, some primary transcripts are immediately active; no processing is required.

During DNA replication both strands of a **parent DNA** function as templates and two identical **daughter DNAs** are generated. Since each daughter contains one parent strand (present in the parent duplex) and one newly synthesized strand, the replication process is said to be **semiconservative.** In both prokaryotic and eukaryotic systems, DNA replication normally precedes each round of cell division, and each daughter cell ends up with the same genetic information that was present in the parent cell. The meiotic cell divisions that generate egg and sperm cells represent an important exception to this generalization.

The production of a DNA molecule from information carried in an RNA molecule is known as **reverse transcription.** In this instance, RNA serves as a template for the production of a complementary strand of DNA. This DNA subsequently functions as a template for the synthesis of its own complement, and the original RNA strand is degraded in the process.

This chapter explores some of the details of transcription, RNA processing, DNA replication, and reverse transcription. Several related topics, including retroviruses and the polymerase chain reaction, are examined as well. Most of the major biochemical themes are encountered along the way. The featured theme is information transfer, the flow of information from DNA to RNA, DNA to DNA, and RNA to DNA. Supporting themes include the regulation of gene expression, specificity, communication, structure–function relationships, catalysis and the regulation of reaction rates, experimentation and hypothesis construction, and the transformation and use of energy. Chapter 18 discusses translation.

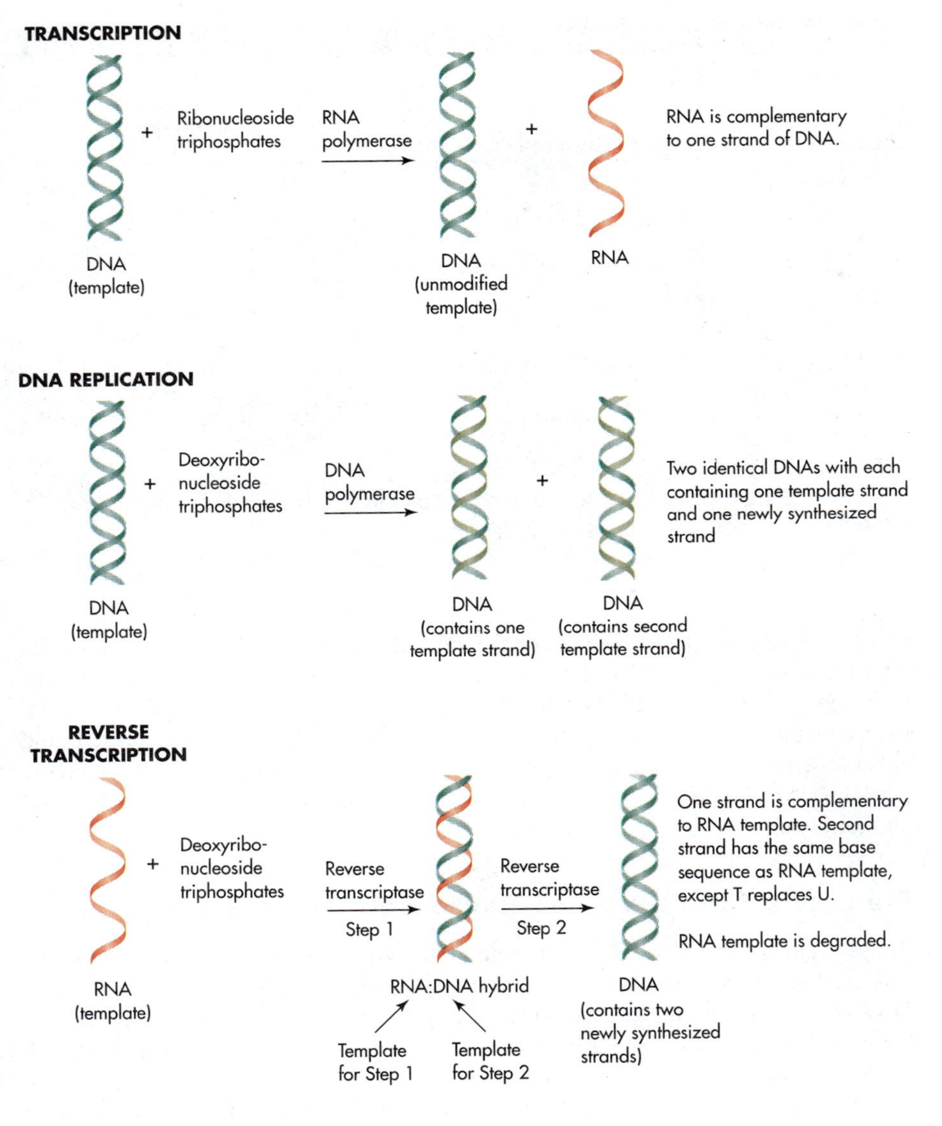

FIGURE 17.1
Nucleic acid biosynthesis.

17.1 TRANSCRIPTION

Transcription is catalyzed by enzymes, known as **DNA-dependent RNA polymerases** (often shortened to **RNA polymerases**) or **transcriptases.** These proteins contain multiple polypeptide subunits packaged into specific quaternary structures. In prokaryotes, a single RNA polymerase is usually responsible for the assembly of all classes of RNA. In eukaryotes, each of three distinct nuclear RNA polymerases, designated **I, II,** and **III,** catalyze the production of unique classes of RNAs (Table 17.1). Mitochondria and chloroplasts possess additional novel transcriptases that are involved in the expression of the genetic information within their genomes.

TABLE 17.1

Eukaryotic RNA Polymerases

Polymerase	Intracellular Location	Produces Primary Transcripts for:
RNA polymerase I	Nucleus	18S, 5.8S, and 28S rRNAs
RNA polymerase II	Nucleus	mRNAs Some snRNAs[a]
RNA polymerase III	Nucleus	tRNAs 5S rRNA Some snRNAs[a] Other small RNAs
Mitochondrial RNA polymerase	Mitochondria	All mitochondrion-produced RNAs
Chloroplast RNA polymerase	Chloroplasts	All chloroplast-produced RNAs

[a]snRNAs are small nuclear RNAs that participate in the processing of primary transcripts.

None of the RNA polymerases operate alone; each functions in association with one or more accessory proteins. The accessory proteins involved in eukaryotic transcription are known as **transcription factors.** In this section, emphasis will be placed on the biochemistry of prokaryotic transcriptases and RNA polymerase II, the eukaryotic enzyme responsible for nuclear mRNA biosynthesis. These polymerases participate in each of the three major stages of transcription: initiation, elongation, and termination.

Promoters, Start Sites, and Transcription Units

Those regions on duplex DNAs where RNA polymerases bind to initiate transcription are known as **promoters.** Prokaryotic RNA polymerases may initially dock to DNA at nonpromoter regions and then slide along the duplex until promoters are encountered. The mechanism through which eukaryotic polymerases reach promoters is under investigation.

A promoter is a patchwork of specific base pair sequences, known as **motifs,** that function as regulatory elements and protein binding sites. Most nucleic acid motifs are 6 to 10 base pairs in length, and many are palindromic sequences (described in Section 19.9). Nucleic acid motifs, like protein motifs, are frequently given distinctive names. The **TATA box** (A TA-rich region also called the **−10 sequence** and **Pribnow box**) and **−35 sequence,** for example, are classes of motifs involved in the recognition and binding of many prokaryotic promoters by prokaryotic RNA polymerases (Figure 17.2). A **CAAT box** (called a cat box) and a TATA box play similar roles in most of the promoters for RNA polymerase II from eukaroytes. Promoters for different RNA polymerases encompass unique motifs or distinct combinations or motifs.

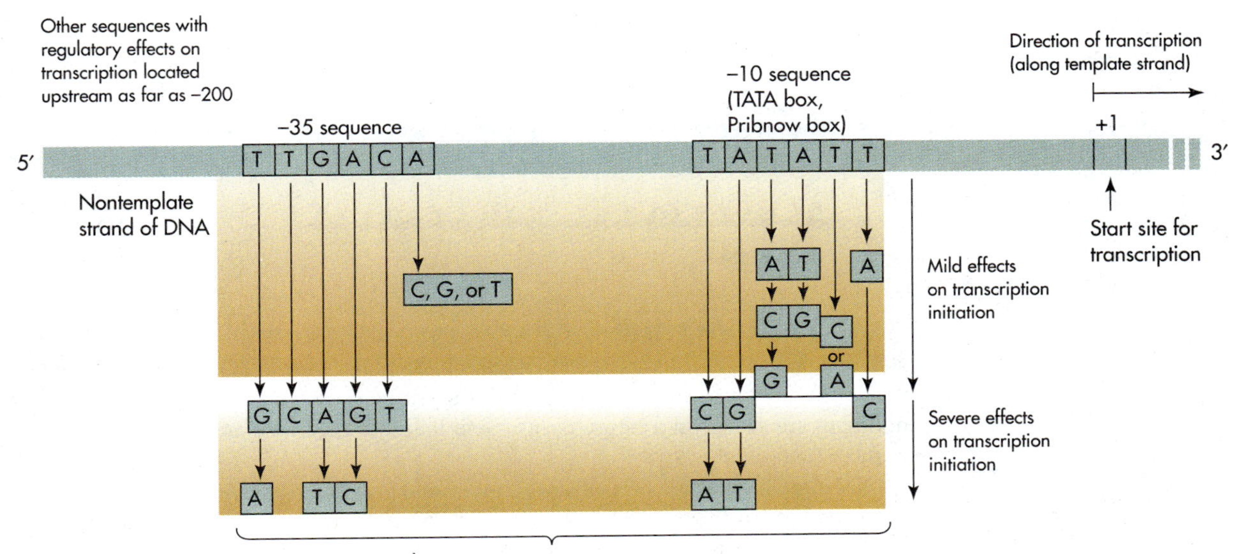

FIGURE 17.2
Effect of changes in −35 and −10 sequences on RNA polymerase binding to *E. coli* promoters. The top row of letters shows the consensus sequences for the −10 and −35 sequences of *E. coli* promoters that are recognized by RNA polymerase linked to a protein initiation factor called sigma factor 70 (σ^{70}). A consensus sequence is the most frequently encountered sequence when one compares corresponding segments of different nucleic acids. Promoters with "perfect" TTGACA and TATATT sequences at −35 and −10 positions bind the RNA polymerase:σ^{70} complex most strongly. Alterations in the consensus sequences weaken RNA polymerase binding to the extent indicated by the lengths of the arrows extending downward from the sequences. The rates at which genes are transcribed, which span a range of at least 1000, vary directly with the rate at which RNA polymerase forms a stable complex with the promoter. (From Wolfe, 1993.)

The "−10" in the term "−10 sequence" and the "−35" in the term "−35 sequence" refer to the approximate location of these sequences on DNA relative to the transcription **start site,** that site where transcription actually begins. That strand of DNA which functions as a template during transcription is called the **sense strand** or **template strand,** and its complement is labeled the **antisense strand** or **non-template strand.** The template strand is "read" in a 3′ to 5′ direction from the start site as a complementary RNA is assembled in a 5′ to 3′ direction (Figure 17.3). DNA base pairs on the 3′ side of the template strand start site are said to be **upstream** from the start site, and they are assigned consecutive negative numbers moving from the start site toward the 3′ end along the template strand. DNA base pairs on the 5′ side of the template strand start site are assigned positive numbers and are said to be **downstream** from the start site. The first nucleotide residue to be transcribed is the +1 residue, the second nucleotide residue to be transcribed is the +2 residue, and so on. Collectively, those base pairs containing nucleotide residues

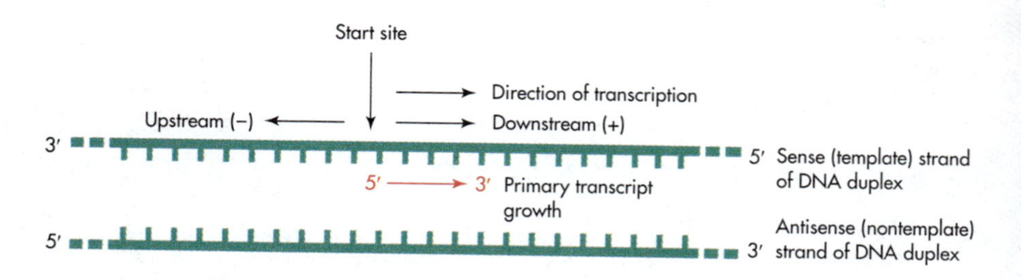

FIGURE 17.3
The transcription start site. The start site usually lies within the promoter, or a short distance outside the promoter on its downstream side.

that are transcribed constitute the gene(s) whose expression is initiated from the start site. Each promoter carries information that determines which strand of the duplex DNA serves as the template strand; the template strand in one gene may be the nontemplate strand in a neighboring gene that is transcribed from a separate promoter. **[Try Problem 17.1]**

Although nucleic acid motifs, promoters, and genes are segments of duplex DNA, the structures of these genetic elements are often depicted as the sequence of bases in their nontemplate strand (see Figure 17.2, for example). The nontemplate strand was selected for this purpose because, within a gene, its base sequence corresponds to the sequence of the primary transcript of the gene, except T replaces U. Both the nontemplate strand and the primary transcript are complementary to the template strand.

During the binding of RNA polymerases and other proteins to promoters, unique protein motifs selectively attach to particular nucleic acid motifs. The term "motif," as it applies to proteins, refers to certain recurring folding patterns that encompass α-helices or β-sheets or both (Section 4.4). Helix–turn–helix motifs and zinc finger motifs are common classes of DNA-binding motifs (Figures 17.4 and 17.5).

Each promoter is part of a separate **transcription unit,** a segment of DNA that is continuously transcribed to yield a single primary transcript. Prokaryotic RNA polymerases are capable of binding directly to promoters, and prokaryotic transcription units commonly contain multiple genes. In eukaryotes, transcription factors must attach to a promoter before polymerase binding, and transcription units normally possess a single gene. The proper positioning of a transcriptase (RNA polymerase) on a promoter determines which strand of the duplex DNA serves as the template strand.

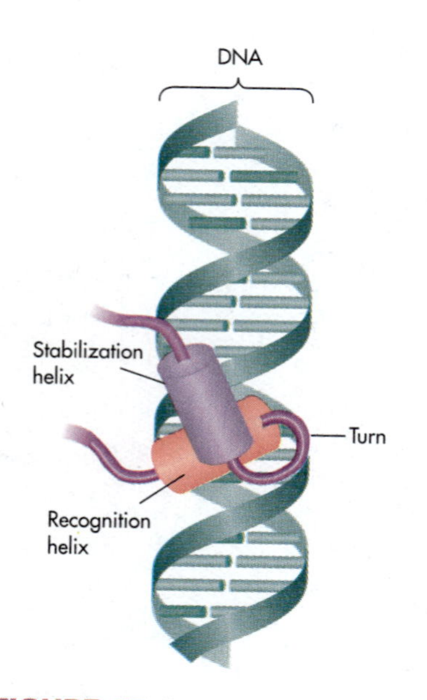

FIGURE 17.4
The helix–turn–helix motif is formed by two α-helical segments connected by a short turn. One of the two α-helices, termed the recognition helix, is positioned to fit precisely into the major groove of DNA. (From Wolfe, 1993.)

Initiation

Although a prokaryotic RNA polymerase is capable of binding directly to a compatible promoter, the initiation of transcription involves a protein **initiation factor** called a **sigma** (σ) **factor** or **sigma subunit.** The σ factor attaches to the core RNA polymerase and enhances its ability to distinguish promoters from nonpromoter regions on DNA. It does so by reducing the enzyme's affinity for nonpromoter regions while increasing its affinity for promoters. The σ factor–core polymerase complex is called a **holoenzyme** (*holo,* from Greek, means *whole* or *entire*). Most prokaryotic

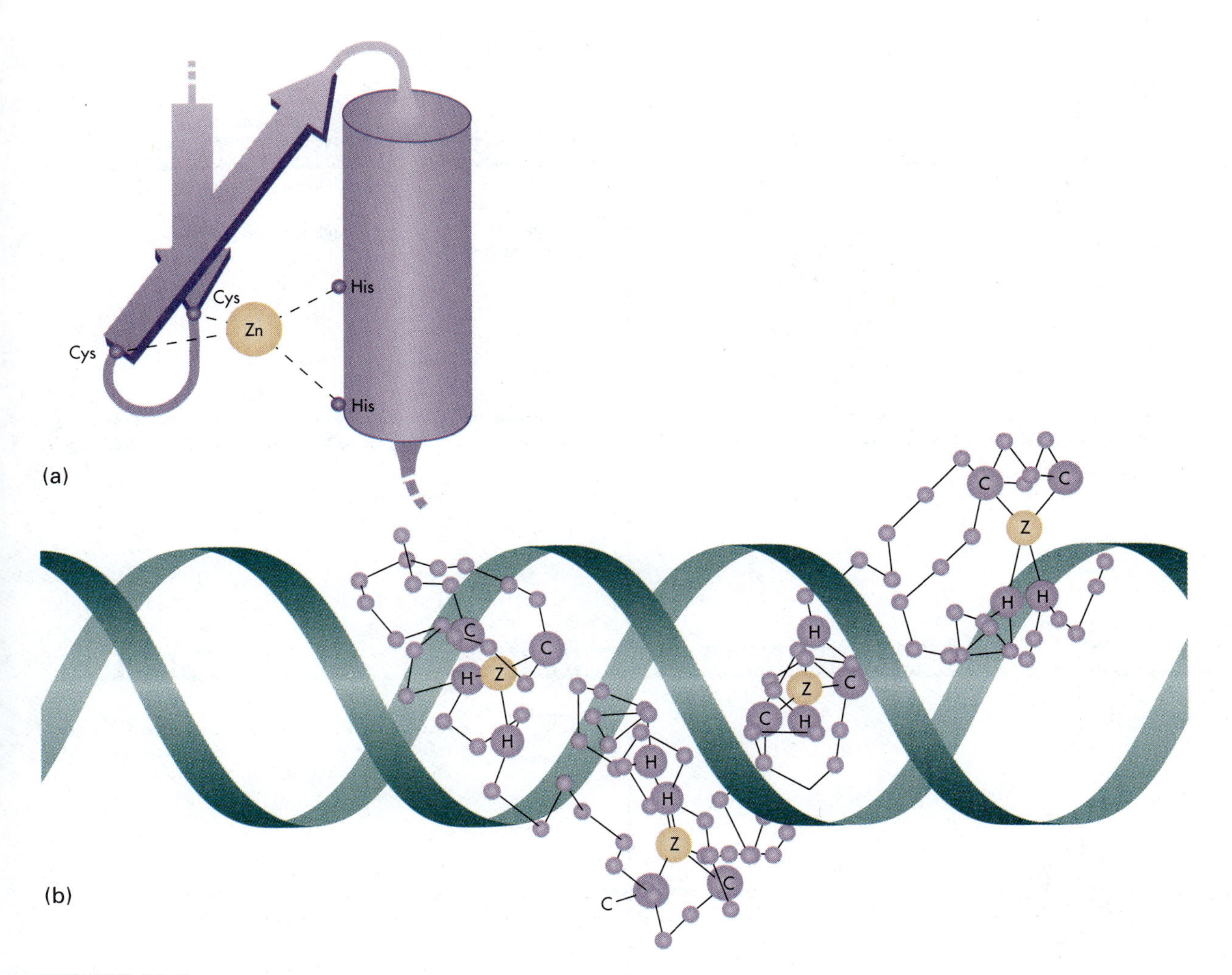

FIGURE 17.5
Zinc-finger motif. (a) Structure of the motif, which consists of two beta strands (arrows) and an α-helical segment (cylinder) stabilized by a zinc atom. (b) A series of four consecutive zinc fingers fitting into the major groove of a DNA molecule. H, histidine; C, Cysteine; Z, Zn. (a) From Wolfe, 1993. (b) From Berg, 1990.

cells contain multiple σ factors, with each factor involved in the recognition of distinct sets of promoters. After it has bound to a promoter, a σ factor–RNA polymerase complex facilitates the denaturation of a short segment (usually around 18 base pairs) of the duplex DNA at the start site and helps direct the pairing of nucleoside triphosphates with complementary bases on the sense (template) strand of the "opened" DNA. The holoenzyme then catalyzes the assembly of an oligoribonucleotide, called a **primer,** that is approximately 10 residues in length. At this point,

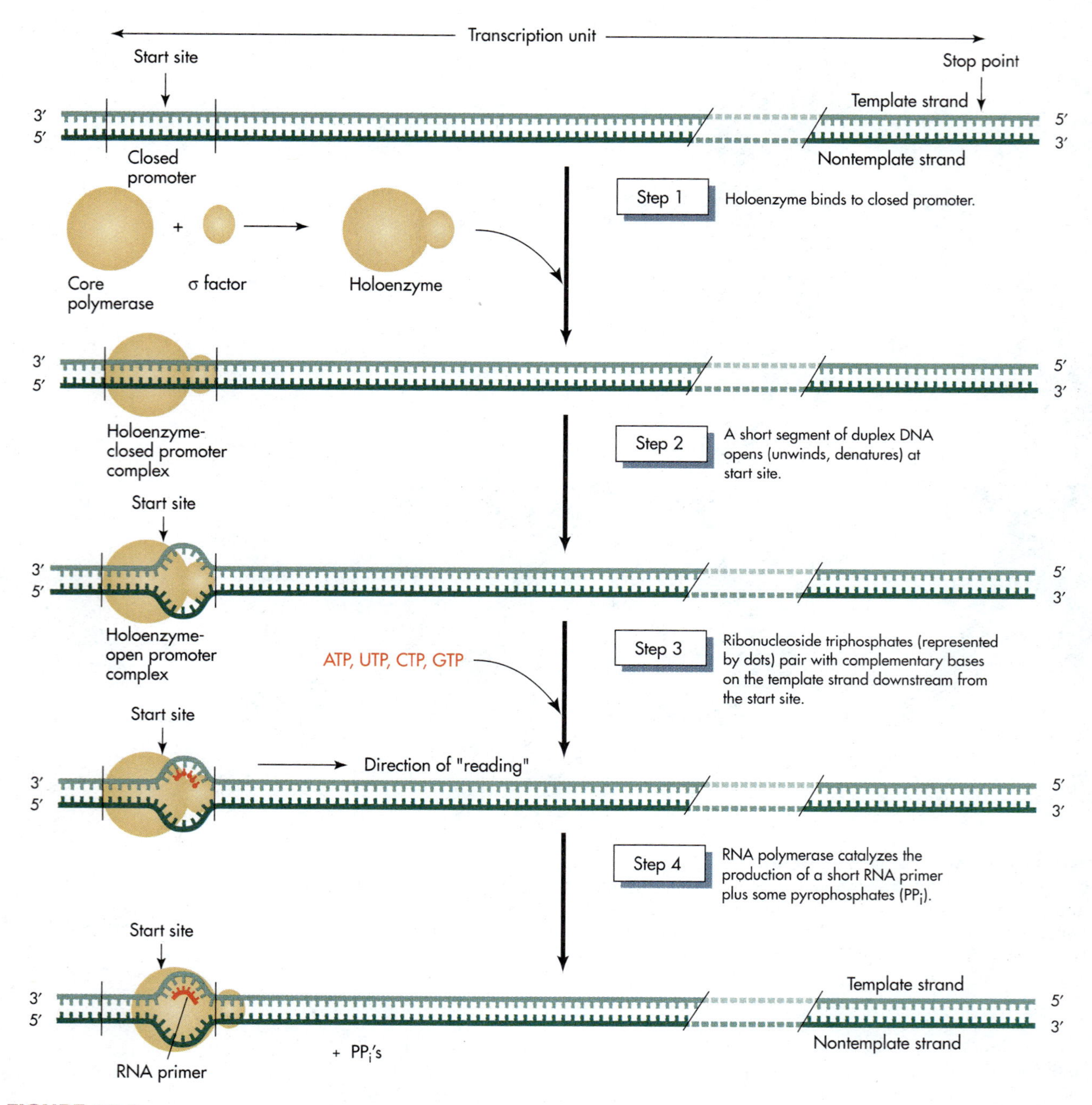

FIGURE 17.6
The initiation of transcription in *E. coli.*

the σ factor dissociates from the holoenzyme and the initiation phase of transcription is complete (Figure 17.6). **[Try Problems 17.2 and 17.3]**

The initiation of transcription in eukaryotes is considerably more complex; it involves multiple transcription factors and the stepwise assembly of an initiation complex. The transcription factors are divided into basal (general) factors (those required for transcription from most or all promoters) and promoter-specific factors (those that are more selective in their association with promoters). **TATA-binding protein (TBP)** is one of the most thoroughly studied basal transcription factors. It is involved in the binding of all three nuclear RNA polymerases to their promoters.

The initiation of transcription from most RNA polymerase II promoters begins with the attachment of TBP to an 8 base pair TATA box that is roughly 25 base pairs upstream from the start site (Figure 17.7). The attachment leads to dramatic changes in the structure of DNA: base pairing is disrupted within the box, compensating superhelical twists are created outside the box, and kinks are created at both ends of the box. Promoter-bound TBP possesses a wide variety of binding sites for other transcription factors, including multiple TBP-associated factors (TAFs) that adhere tightly to TBP. The TBP–TAF complex is labeled TFIID where TF stands for *transcription factor*.

The stepwise attachment of additional transcription factors and RNA polymerase II to the promoter–TFIID complex leads to the assembly of an active initiation complex. The additional transcription factors include the basal factors TFIIA, TFIIB, TFIIE, TFIIF, and TFIIH (Table 17.2). The initiation complex facilitates the denaturation of a segment of DNA at the start site and catalyzes the synthesis of an RNA primer during the final stage of the initiation process in eukaryotes.

The stepwise model for initiation complex assembly may not be applicable in all cases. In some eukaryotes, a holoenzyme containing RNA polymerase II and particular transcription factors appears to exist independently of promoters. In such cases, the binding of this holoenzyme to a promoter–transcription factor complex constitutes a key step in the initiation of transcription.

Regulation of Initiation

The initiation stage of transcription is frequently tightly regulated. In prokaryotes, this regulation is usually coupled to σ factor availability and, in some instances, to the concentration of specific protein **repressors** or protein **activators.** Prokaryotic repressors and activators usually modulate initiation by binding to specific DNA motifs within or near promoters. Genes whose activity is regulated by repressors tend to shut down in the presence of high repressor concentrations. They are said to be negatively regulated. Activators stimulate the initiation of transcription from positively regulated genes.

The regulation of transcription initiation in eukaryotes begins at the level of nucleosomes, the histone-DNA complexes described in Section 16.3. A nucleosome must be at least partially disrupted before the DNA within it can interact with transcription factors and RNA polymerase. The disruption of the nucleosome is regulated, at least in part, by specific nonenzyme proteins and by enzymes that catalyze the chemical modification of histones. Competition between histones and transcription factors for binding sites on DNA may also play a role in the disruption process.

The initiation of transcription is additionally controlled in eukaryotes by the availability of the basal transcription factors that make up the active basal initiation

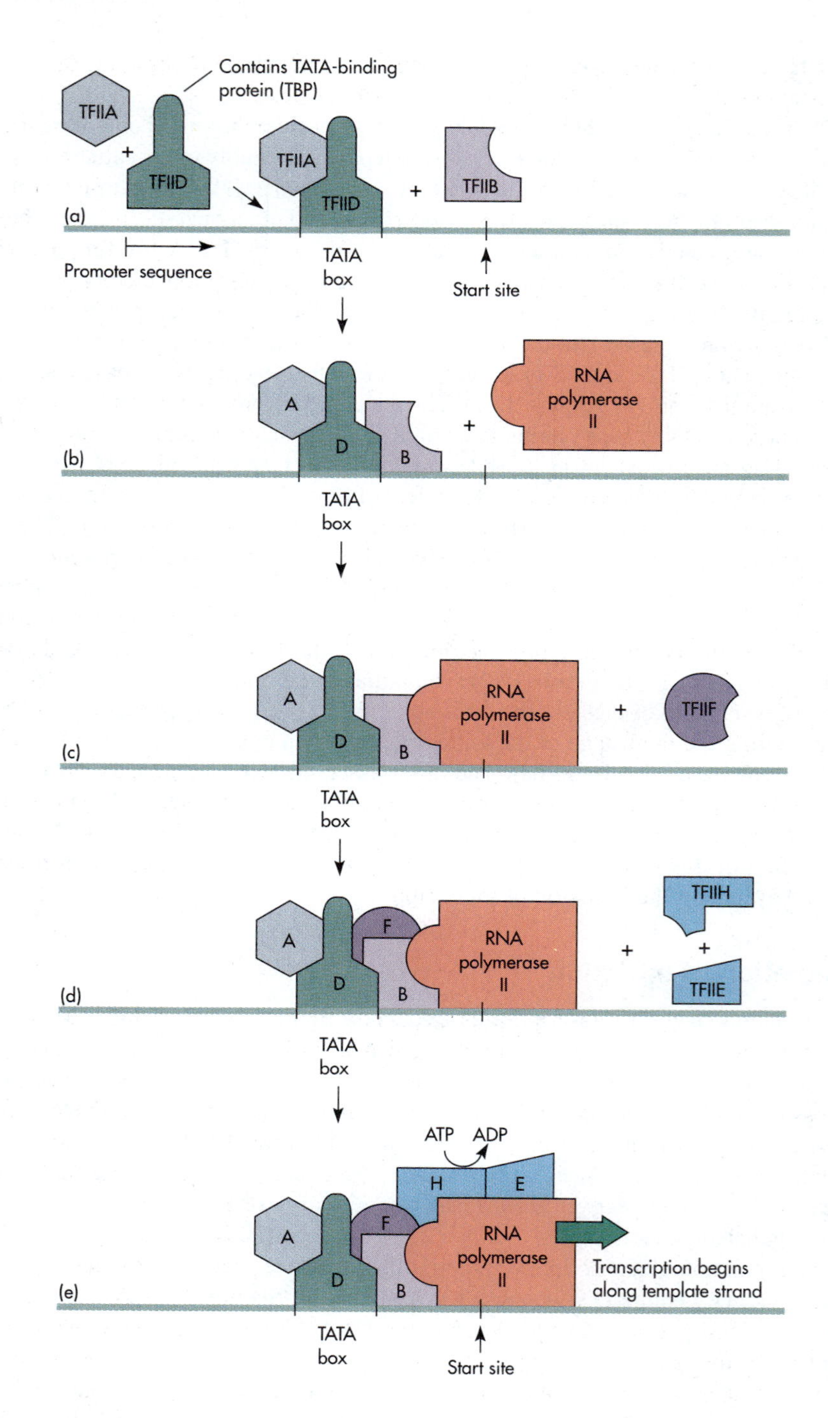

FIGURE 17.7
Initiation of transcription by eukaryotic RNA polymerase II. The probable interactions among the transcription factors, RNA polymerase II, and the promoter during the initiation of transcription. (a) The process begins as the transcription factor TFIID recognizes and binds the TATA box of an mRNA gene. The binding is accelerated by TFIIA, which may bind to TFIID before it attaches to the promoter. (b) After TFIID binds, TFIIB adds to the complex. (c) This creates a complex that binds RNA polymerase II and aligns it at the start site. (d) The assembly binds TFIIF. (e) After TFIIE and TFIIH join the complex, a kinase activity associated with TFIIH catalyzes the phosphorylation of RNA polymerase II. This activates the completed basal initiation complex; transcription can begin.

TABLE 17.2

Some Basal Transcription Factors

Factor	Some Characteristics and Proposed Roles During Transcription Initiation on RNA Polymerase II Promoters
TFIIA	Can bind TFIID even in the absence of DNA. May stabilize binding of TFIID to promoter.
TFIIB	Does not bind DNA. Required for selective binding of RNA polymerase II. May "measure" distance from TATA box to transcription start site.
TFIID	Contains TBP and tightly bound TAFs. Recognizes promoters and binds to the TATA box and neighboring regions. Possesses binding sites for other transcription activators.
TFIIE	Associates stably with RNA polymerase II even in solution. May not be essential for initiation from all promoters.
TFIIF	Does not bind to DNA. No catalytic activities. May promote selective binding of RNA polymerase II to initiation complex.
TFIIH	Possesses a protein kinase activity that can catalyze the phosphorylation of RNA polymerase II. Plays a crucial role in the assembly of the initiation complex. Also plays a role in DNA repair.

complex. Further control is exercised by distinct transcription factors that interact with regulatory segments of DNA known as **enhancers.** Enhancers, which contain unique patchworks of motifs, may be separated from the transcription units that they regulate by thousands of base pairs. After a transcription factor (a **eukaryotic activator**) binds an enhancer, the segment of DNA between the enhancer and the regulated transcription unit loops out to bring the enhancer–protein complex into close proximity of the promoter within the regulated unit (Figure 17.8). The enhancer–protein complex subsequently interacts with the initiation complex on the promoter and enhances the initiation process. Although enhancers were first discovered in eukaryotes, they are now known to exist in prokaryotes as well.

Initiation of the transcription of some eukaryotic genes is also regulated by DNA sequences, known as **silencers,** that, like enhancers, may be separated from targeted transcription units by thousands of base pairs. Selective interactions between a silencer, a specific transcription factor (a **eukaryotic repressor**), and a promoter-bound initiation complex can suppress the initiation of transcription at the promoter. However, the control of transcription in a eukaryotic cell is usually dominated by positive regulatory agents rather than negative ones. The initiation of transcription at most eukaryotic promoters is under the control of multiple regulatory sites, each of which interacts with one or more transcription factors. The average number of regulatory sites is probably at least five. The 5′-UTRs in some mRNAs (Section 16.5) also help modulate initiation from particular promoters. Such multifaceted regulation facilitates the coordination of the expression of distinct genes and it helps ensure that individual genes are expressed at the appropriate level. **[Try Problem 17.4]**

Because the gene products (proteins) within a cell determine its nature, characteristics, and capabilities (predominantly by determining what reactions occur within it), the control of gene expression is indeed central to life itself. A brain cell differs from a liver cell because a select combination of genes is active, not because it con-

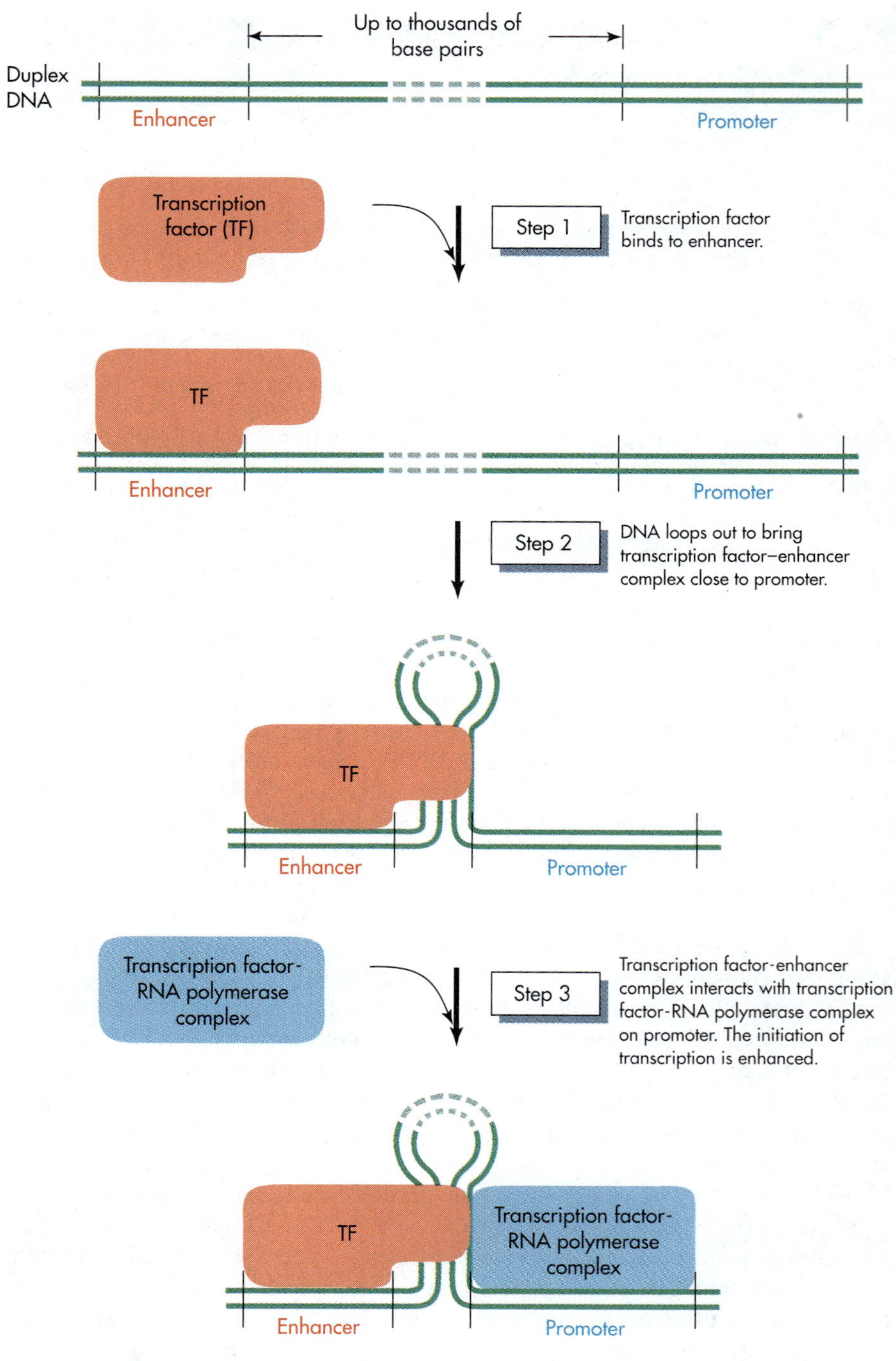

FIGURE 17.8
The regulation of transcription by enhancers.

tains a unique set of genes. Thus, the human body must carefully choreograph the activity of approximately 100,000 genes in each of more than 250 different types of cells. And even a given type of cell must constantly turn on and turn off the expression of specific genes in response to ever-changing needs or activities. In some instances, rather minor changes in gene expression can transform a normal cell into a cancerous one, further evidence of the importance of the regulation of gene activity. Since transcription is the first major step in gene expression, it is not surprising that several hundred human gene products have already been identified that are involved in the regulation of this event.

Primer Synthesis and Chain Elongation

Once an RNA polymerase and associated proteins have bound to a promoter and facilitated the denaturation of a short segment of DNA at the start site and the pairing of nucleoside triphosphates with the template strand, the polymerase begins to catalyze the assembly of an RNA molecule. The initial substrates for the polymerase are the two adjacent ribonucleoside triphosphates paired with the +1 and +2 bases on the template strand of DNA. The polymerase catalyzes the cleavage of a phosphate anhydride bond and the coupled formation of a 3′,5′-phosphodiester link that creates a dinucleotide (Figure 17.9). A pyrophosphate (PP_i) molecule is also formed. The enzyme then shifts on the template, repositions (on its active site) the ribonucleoside triphosphate that is base paired with the +3 nucleotide on the template strand of DNA, and generates a trinucleotide by catalyzing the splitting of another anhydride link and the formation of a second phosphodiester bond. The polymerase shifts again, repositions the next nucleoside triphosphate, and catalyzes the addition of a fourth monomer to the 3′ end of the growing RNA chain. And the process continues.

The polymer chain is extended in a 5′ to 3′ direction with the 5′ end containing a triphosphate residue. Since the splitting of an energy-rich compound provides a thermodynamic driving force for the addition of each monomer (the formation of each phosphodiester bond), transcription, like most anabolic pathways, is an energy-requiring event (Chapter 8). The enzyme-catalyzed hydrolysis of the anhydride link in the pyrophosphate molecules formed during the polymerization process "pulls" polymer formation toward completion (Le Châtelier's principle). At the same time, this hydrolysis reaction doubles the amount of energy "consumed" during the addition of each nucleotide residue to a growing RNA chain. **[Try Problems 17.5 through 17.7]**

The joining of the first 10 or so nucleotide residues in a growing RNA chain is viewed as part of the initiation stage of transcription, and the RNA fragment generated is considered a primer (Figure 17.6). The addition of a nucleotide residue to the primer marks the beginning of the elongation stage of transcription. One or more protein factors dissociate from the transcription complex and one or more accessory proteins usually join the complex as the shift is made from initiation to elongation. In prokaryotes, it is the σ factor that dissociates. The change in accessory factors alters the number of contacts between the RNA polymerase complex and the DNA that is being transcribed; during initiation in prokaryotes, about 70 base pairs on DNA are protected from enzyme-catalyzed hydrolysis, whereas only 30 or so base pairs are protected during elongation. The alteration in polymerase–DNA contacts is in-

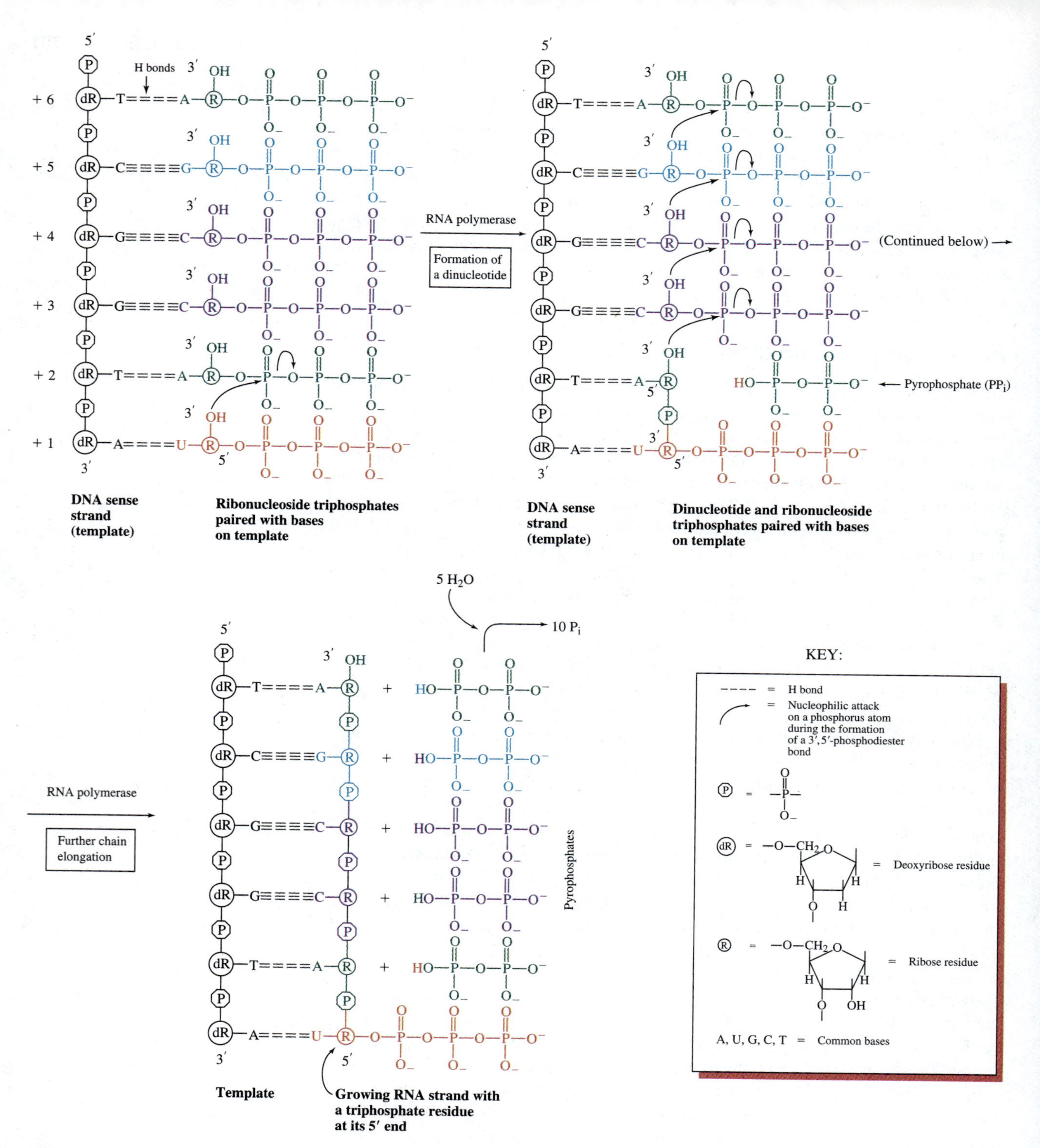

FIGURE 17.9
RNA assembly during transcription.

volved in **promoter clearance,** the movement of the polymerase off of the promoter as elongation begins. Once an RNA polymerase has moved off a promoter, a second RNA polymerase (with appropriate initiation factors) can bind to the promoter and initiate another round of transcription. The RNA polymerase that has moved off of the promoter continues to catalyze the synthesis of the primary transcript encoded in the transcription unit that encompasses the promoter. Hence, multiple segments of the same gene can be transcribed simultaneously by separate RNA polymerase complexes.

As an RNA is synthesized along one strand of the denatured segment of DNA, the shifting RNA polymerase complex leads to further unwinding of the template. Once the RNA polymerase complex has moved a certain distance along the DNA, the 5′ end of the growing RNA chain is continuously displaced from the template strand as the template and nontemplate strands reanneal (Figure 17.10). The net re-

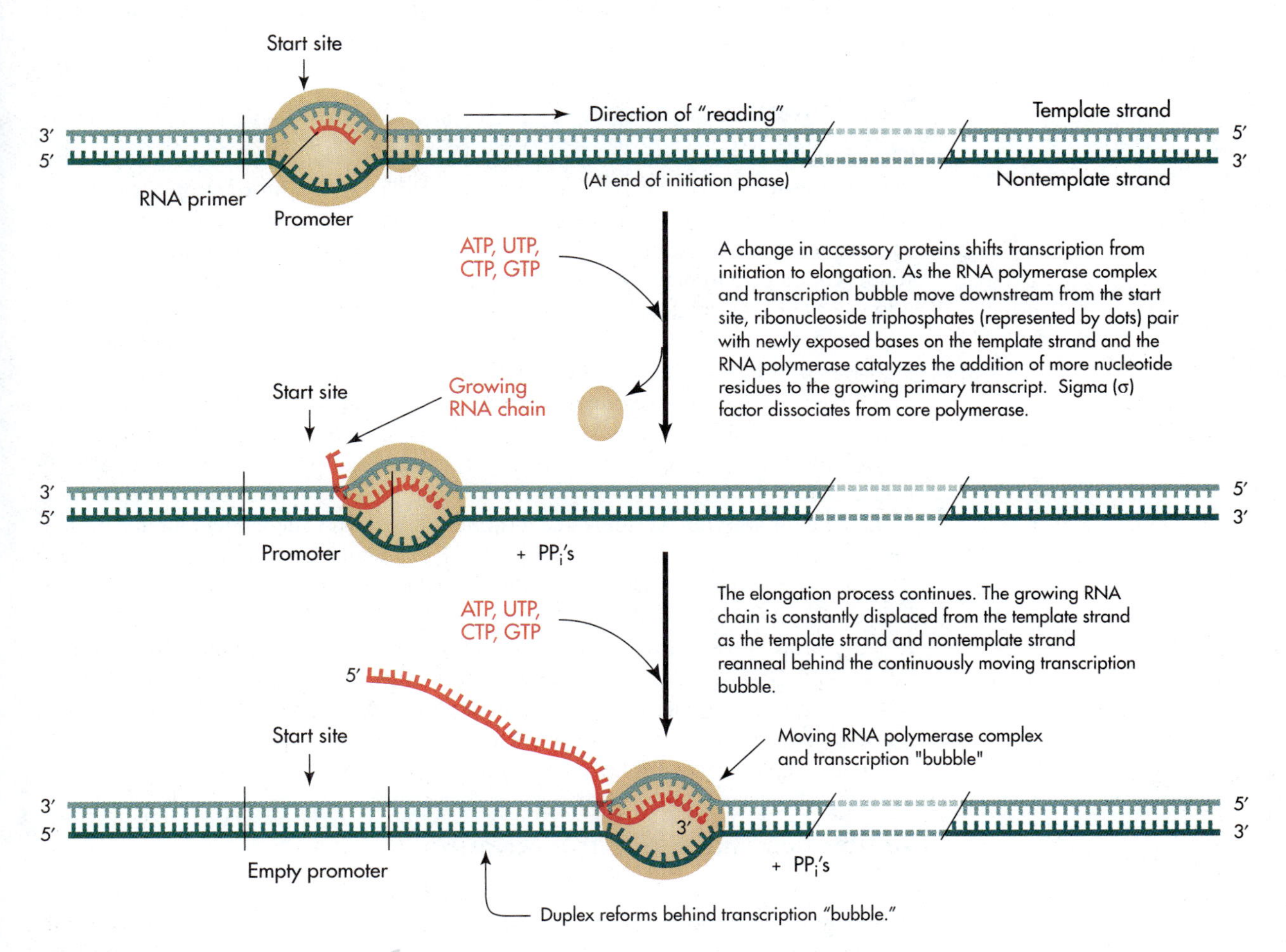

FIGURE 17.10
Movement of the RNA polymerase complex during the elongation phase of transcription in *E. coli.*

sult is a denatured "bulge" or "bubble" that moves along the DNA with the polymerase. The RNA:DNA hybrid within the bubble is approximately 12 base pairs in length. The supercoiling generated by the mobile bubble is relieved with topoisomerases (Section 16.2). Any nucleosomes encountered during elongation must be at least partially disrupted to allow the transcription complex to read through them. How this is accomplished is uncertain. **[Try Problems 17.8 and 17.9]**

The growing RNA is elongated at a rate of up to approximately 85 nucleotides per second until a terminator is encountered. Different template sequences tend to be transcribed at different rates with GC-rich sequences transcribed more slowly than AT-rich sequences. On the average, only one error is made for every 10^5 to 10^6 nucleotides incorporated into a primary transcript. An error is the addition of a nucleotide that is not complementary to the corresponding nucleotide on the template strand of DNA. Although the maintenance of this low error frequency is a commendable feat, it is paled by the fidelity exhibited during DNA replication (Section 17.4).

Termination

The termination stage of transcription, like the initiation stage, is a distinct process. A **terminator** is a base pair sequence in DNA that carries information for termination. Contrary to common-sense expectations, there are no protein factors that recognize and bind to terminators. In prokaryotes, the terminator often causes the RNA polymerase to pause on the template, an event that contributes to the dissociation of the newly synthesized RNA, the RNA polymerase, and any accessory proteins from the transcribed DNA (Figure 17.11). In some instances, the formation of a base-paired hairpin loop in the newly synthesized RNA also facilitates termination by destabilizing the segment of RNA:DNA hybrid within the transcription bubble. Distinct transcription units usually possess unique terminators. During termination, the transcription unit is returned to its original state; it can be repeatedly transcribed.

In some prokaryotic systems, the termination process is dependent upon a protein factor, called **rho** (ρ) [not shown in Figure 17.11], which is hypothesized to catalyze the energy-dependent unwinding of the RNA:DNA hybrid within the transcription bubble. Rho may also facilitate termination by wrapping around the single stranded RNA protruding from the transcription bubble. Rho action is coupled to the cleavage of ATP. Other prokaryotic protein factors, called **antiterminators,** can allow an RNA polymerase to **read** (transcribe) **through** select terminators and move on down the template to the next terminator. The details of the termination process in eukaryotic systems must still be unraveled.

Antibiotics That Block Transcription

A variety of substances, including some antibiotics, inhibit one or more steps in transcription. Antibiotics are biological organic compounds, predominantly of microbial origin, that are toxic to certain microorganisms. They have been encountered previously during the discussion of the chirality of amino acids in Section 3.3. **Actinomycin D** and the **rifamycins** are some of the antibiotics known to block

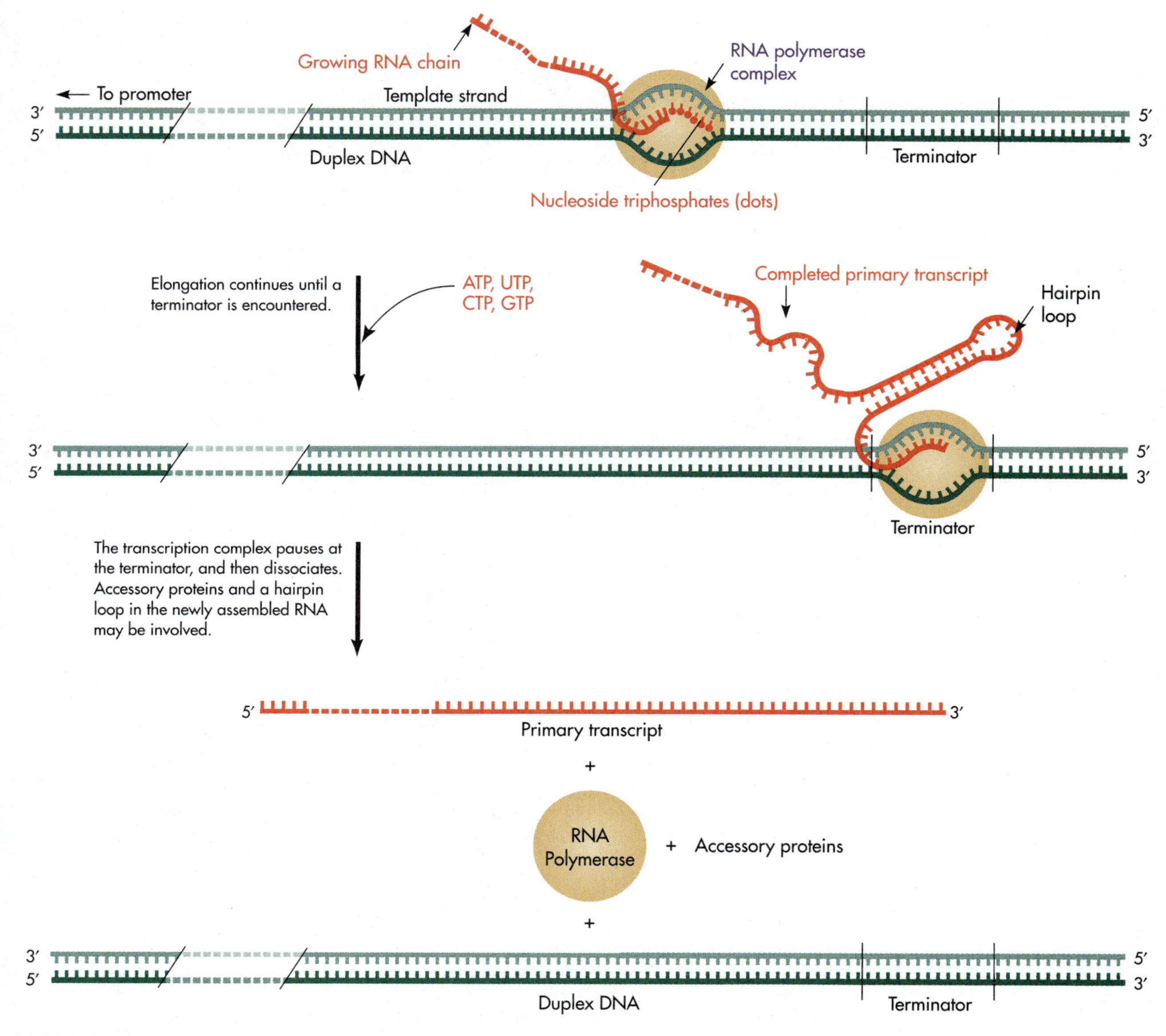

FIGURE 17.11
The termination of transcription in *E. coli.*

transcription (see structures at the top of page 706). Actinomycin D, an anticancer agent that contains some D-amino acid residues, inhibits both eukaryotic and prokaryotic transcription by binding to DNA. The rifamycins are prokaryotic specific; they selectively bind to a particular subunit within bacterial RNA poly-

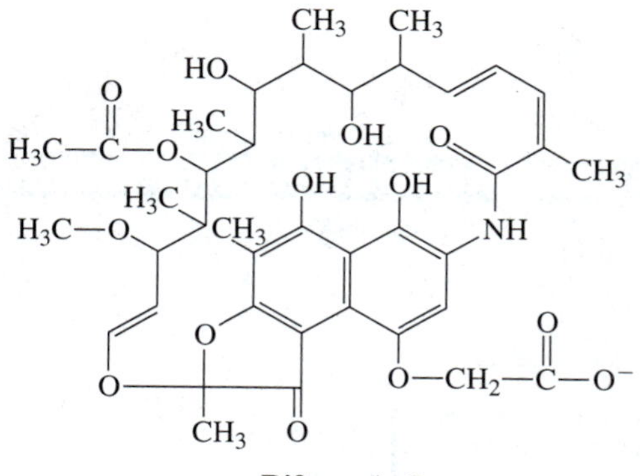

Rifamycin B

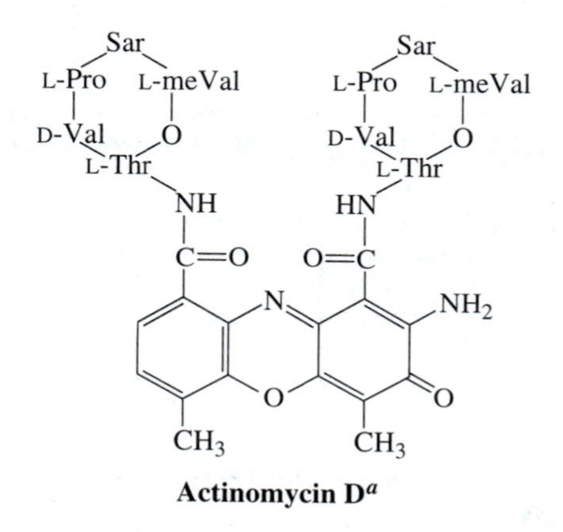

Actinomycin D[a]

[a]Sarcosine (Sar) is *N*-methylglycine; meVal is L-*N*-methylvaline; L-proline (Pro), L-threonine (Thr), and L-valine (Val) are common protein amino acids (Section 3.2); Actinomycin D contains two peptide ring systems. In each ring, the hydroxyl group in the side chain of a threonine residue is esterified to the carboxyl group in an L-*N*-methylvaline residue.

merases. The rifamycins are employed to treat certain gram-positive bacterial infections and tuberculosis.

Post-Transcriptional Regulation of Gene Expression

It has been emphasized that the final expression of most genetic information consists of the production of a biologically active polypeptide. In most instances, prokaryotes control gene expression by regulating transcription. Once a polypeptide-coding gene has been transcribed, the primary transcript is activated and then translated to yield a functional polypeptide. In eukaryotes, however, the major control point for gene expression is frequently post-transcriptional (after transcription) (Figure 17.12). In these cases, the rate-limiting event during active polypeptide production is primary transcript processing, translation of mRNA, or polypeptide processing (the conversion of a newly synthesized polypeptide into a biologically-active form; Section 18.11).

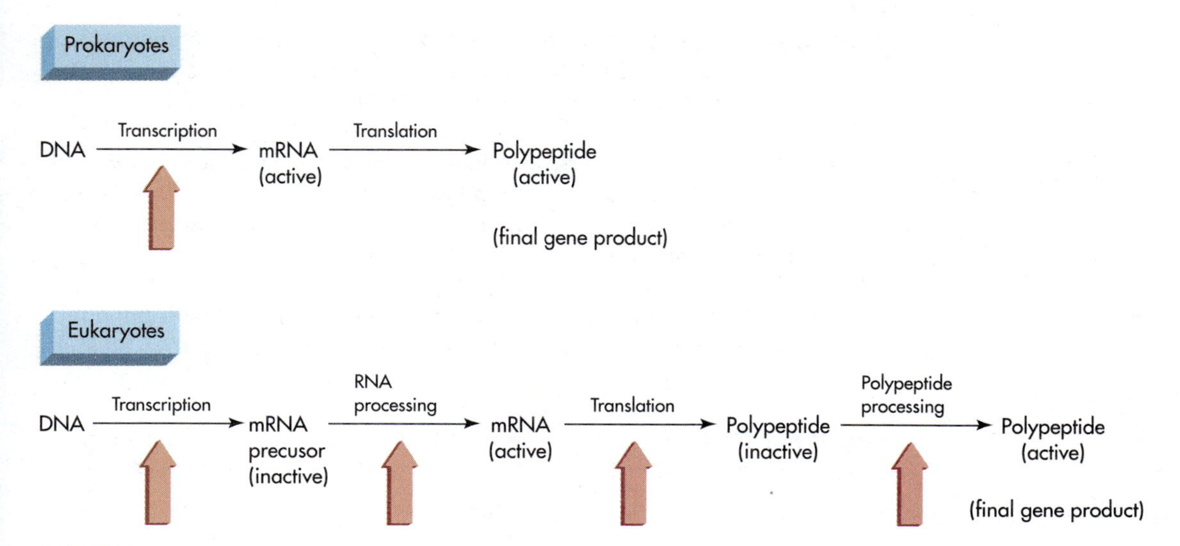

FIGURE 17.12
Control of gene expression. Colored arrows identify major sites of regulation. In prokaryotes, primary transcripts and nascent (newly synthesized) polypeptides are usually active, and gene expression is controlled primarily through the regulation of transcription. In eukaryotes, primary transcripts and nascent polypeptides are frequently inactive; they must be processed (chemically modified) to be activated. Each step in gene expression is a common site of regulation.

PROBLEM 17.1 The sequence of the TATA box is not exactly the same in all *E. coli* promoters. The nontemplate strand consensus sequence for this motif in a particular class of promoters is:

5′T(80)A(95)T(45)A(60)A(50)T(96)3′

T is the first base 80% of the time, A is the second base 95% of the time, and so on. The exact position of the TATA box from the start site can also vary by a few positions from promoter to promoter. Write an abbreviation for a segment of duplex DNA containing a consensus TATA box and a start site (label each) with a purine as the +1 base in the nontemplate strand. Assume that the first base pair in the box (contains the 5′ base in the nontemplate strand consensus sequence) is the −15 pair. In the abbreviation for DNA, use one-letter abbreviations for nucleotide residues. Although the base sequence between the TATA box and the start site may not, in reality, be random, use any base sequence desired outside the TATA box.

PROBLEM 17.2 Which is easier to unwind, an AT-rich region or a GC-rich region of DNA? Which region would you expect to find at a transcription start site? Explain.

PROBLEM 17.3 Why must a segment of DNA be unwound before that segment can serve as a template for RNA synthesis?

PROBLEM 17.4 List the three classes of motif-containing regulatory units within eukaryotic DNA that interact with transcription factors. What is the general role of each of these classes of regulatory elements?

PROBLEM 17.5 Use an arrow to identify each β-*N*-glycosidic link within the growing RNA strands depicted in Figure 17.9. Circle each nucleotide residue in each growing RNA strand. Put a box around the triphosphate residue at the 5′ end of each growing RNA strand.

PROBLEM 17.6 Use an arrow to identify each phosphate ester bond in all the diagrams to the right of (and below) the first reaction arrow in Figure 17.9. Put an X through each phosphate anhydride bond in these diagrams.

PROBLEM 17.7 What RNA fragment is produced when the following DNA fragment is transcribed?

dACTTTATCCCTGC

Recall that, unless specified otherwise, the 5′ end of an abbreviated polynucleotide is written to the left and the 3′ end to the right.

PROBLEM 17.8 List those specific interactions that hold the growing RNA chain and its DNA template together within a transcription bubble.

PROBLEM 17.9 Define topoisomerases and supercoiled DNA. Hint: Review Section 16.2

17.2 THE PROCESSING OF mRNA PRECURSORS

The final products of transcription are a primary transcript plus many pyrophosphates. Some primary transcripts in prokaryotes and most primary transcripts in eukaryotes are inactive precursors of biologically active RNAs. The term *processing* is used to describe the events involved in the production of a mature RNA from a primary transcript. This section summarizes some of what is known about the processing of mRNA precursors in eukaryotic systems. The primary transcripts for prokaryotic mRNAs rarely require processing; they tend to be biologically active without chemical modifications. Processing is catalyzed by specific enzymes, and it often begins before a primary transcript has been completely assembled.

hnRNAs and Transcript Domains

The primary transcripts for cytosolic, eukaryotic mRNAs are confined to the nucleus, where they are members of a class of nucleic acids known as **heterogeneous nuclear RNAs (hnRNAs).** Normally, each mRNA precursor is **monocistronic** (it carries the sequence of a single mature mRNA that encodes a single polypeptide) and several times longer than the mRNA whose sequence it encompasses. Most eukaryotic mRNA precursors are trimmed, spliced, capped, and polyadenylated within the nucleus during the maturation process. In addition, the common purines and pyrimidines in a small fraction of their nucleotide residues are converted to minor bases (Section 7.1) with the catalytic assistance of specific enzymes. The nucleotide sequence in some mRNA precursors is further modified through a process described as editing. These post-transcriptional or co-transcriptional events are not necessarily listed in the order in which they occur. Multiple processing enzymes may simultaneously catalyze the chemical modification of the same precursor. The nucleus within some eukaryotic cells contains discrete, localized RNA processing centers, termed **transcript domains,** where at least part of the molecules involved in the processing of primary transcripts are packaged into a highly organized complex.

Trimming

Trimming refers to the removal of nucleotide residues from the 3′ and 5′ ends of a primary transcript (Figure 17.13). In the case of mRNA precursors, only the 3′ end is normally trimmed. Trimming is performed by enzymes, called **nucleases,** that catalyze the hydrolysis of phosphodiester bonds within nucleic acids. Many nucleases are sequence specific; they catalyze the hydrolysis of particular bonds within specific nucleotide sequences. Most nucelases catalyze the hydrolysis of either single-stranded nucleic acids or double-stranded nucleic acids, but not both. Nucleases that

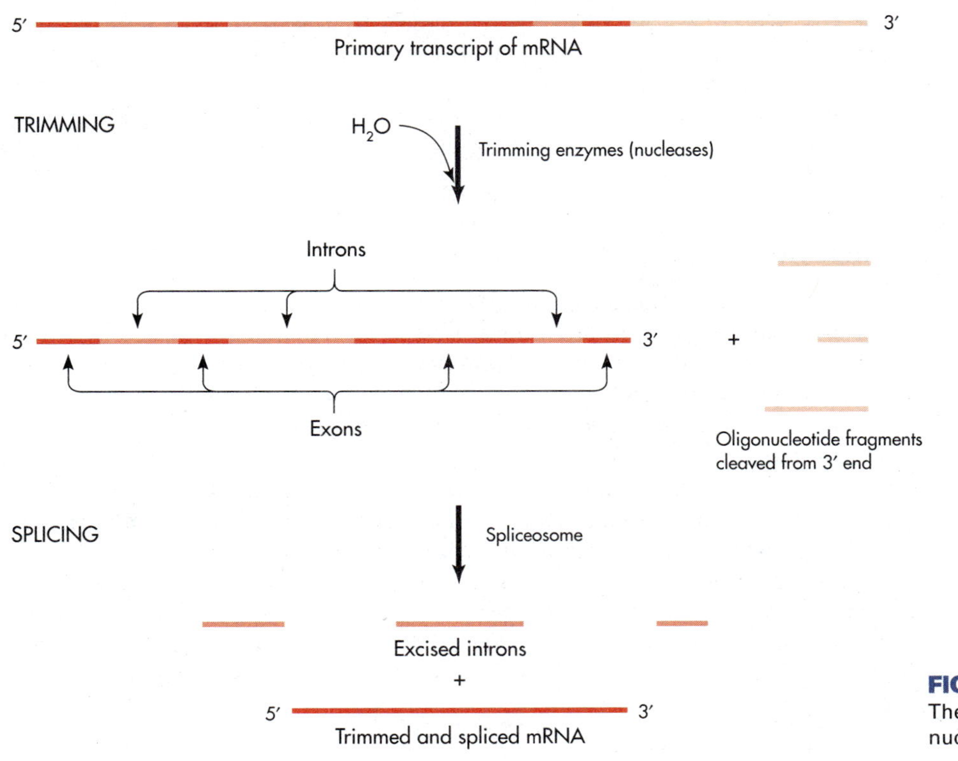

FIGURE 17.13
The trimming and splicing of a nuclear mRNA precursor.

catalyze RNA hydrolysis are called **RNases.** Those whose substrates include DNA are called **DNases.** An **exonuclease** will catalyze the hydrolysis of only terminal phosphodiester links (usually either 3′ or 5′, but not both), whereas **endonucleases** target internal bonds. **[Try Problem 17.10]**

Splicing

Splicing is a second common processing event in eukaryotes. The splicing story begins with the following observation: when the nucleotide sequence of a mature cytosolic mRNA is compared with the sequence of its nuclear primary transcript, one or more segments of the mature mRNA are usually separated within the primary transcript by sequences of bases that are not found within the mature mRNA. The extra sequences of bases are described as **intervening sequences** or **introns.** During processing, the introns are removed and the retained polynucleotide fragments, called **exons,** are **spliced** together with the catalytic assistance of **snRNPs** (pronounced "snurps"; an acronym for **small nuclear ribonucleoproteins**). Each snRNP encompasses one or more **small nuclear RNAs (snRNAs;** each contains from approximately 60 to 300 nucleotides) plus one or more polypeptide chains. The snRNAs help direct splicing reactions by pairing with bases at intron–exon junctions, and they may possess catalytic activities as well. The entire macromolecular complex that is responsible for splicing is known as a **spliceosome.** Detailed splicing mechanisms are presented in more encyclopedic texts.

Alternate splicing patterns are sometimes used to create two or more distinct mRNAs (usually encode unique polypeptides) from separate copies of the same mRNA precursor (Figure 17.14). Splicing factors known as **SR proteins** (not shown in Figure 17.14) help determine which splicing patterns predominate within a given cell at any given time. Thus, the SR proteins must be placed on the rapidly growing list of proteins known to help regulate the expression of genetic information. **[Try Problems 17.11 and 17.12]**

Ribozymes

Some of the mRNA precursors in mitochondria, chloroplasts, and certain bacteria, in contrast to most nuclear mRNA precursors, are **self-splicing.** These precursors remove their own introns without the assistance of snRNPs or other protein catalysts. Although self-splicing RNAs are frequently labeled catalytic RNAs, they are not genuine catalysts; they cannot be recovered unchanged at the end of the splicing reactions. Certain snRNAs may rank among the several authentic RNA catalysts that have been identified since the 1982 discovery of self-splicing RNAs. Since the discovery of RNA catalysts, the definition of enzymes has been changed from "protein catalysts" to "biological catalysts" (Chapter 5). **[Try Problem 16.13]**

RNA enzymes are called ribozymes, and they all appear to be metalloenzymes. Both their folding and their catalytic activity are dependent upon metal ions, principally divalent magnesium (Mg^{2+}). Many of the proteins that interact with nucleic acids, including some transcription factors and all polymerases, are also metal ion dependent.

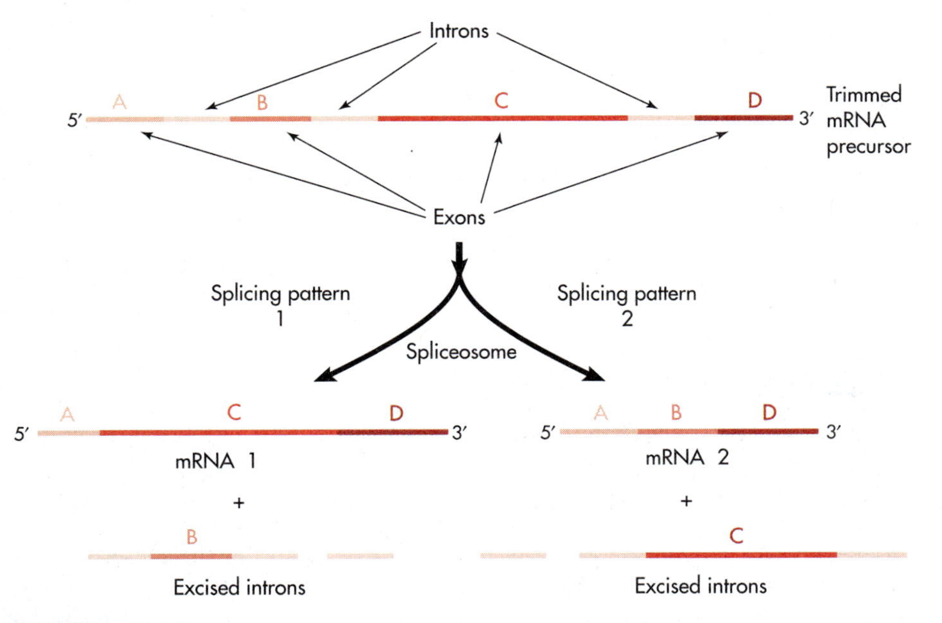

FIGURE 17.14
Alternate splicing patterns for an mRNA precursor. Exon B becomes part of an intron in splicing pattern 1. Exon C becomes part of an intron in splicing pattern 2.

The existence of ribozymes has generated considerable debate about the possible roles of RNAs in evolution and the origin of life. According to one popular hypothesis, RNAs evolved before proteins and DNAs. Numerous observations support the feasibility of this proposal. Initially, the living world may have been an RNA world.

Some current ribozyme research is rapidly leading into the clinical arena. Multiple laboratories have constructed RNAs that catalyze the hydrolysis of specific nucleotide sequences within other RNA or DNA molecules. A ribozyme that selectively catalyzes the hydrolysis of the RNA genome of HIV (the retrovirus involved in AIDS; see Section 17.5) is being developed as a potential therapeutic agent for the treatment of AIDS.

Split Genes

The introns in primary transcripts are encoded in the genes for these transcripts. Intron-containing genes are said to be **split genes.** Some genes and their primary transcripts possess 50 or more introns, with each intron containing from about 65 to over 20,000 nucleotides. This helps explain why the primary transcripts for many eukaryotic mRNAs are much longer than their mature mRNAs. Speculations about the evolutionary origin of introns and splicing make interesting reading. However, the biological function, if any, of most introns is still uncertain. Some exons encode protein motifs or domains, and, during evolution, new genes are sometimes created through the reshuffling of exons. **[Try Problem 17.14]**

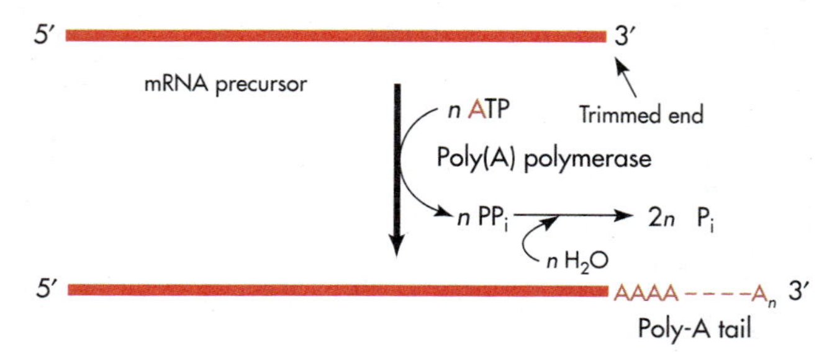

FIGURE 17.15
Poly-A tail formation.

Capping and Polyadenylation

Capping and **polyadenylation,** two other common processing events in eukaryotes, refer to the synthesis of a messenger's cap and poly-A tail, respectively (Section 16.5). Capping requires several enzymes, including the **methylases** that catalyze the addition of methyl groups to the guanine and ribose residues within caps. The enzyme **poly(A) polymerase** generates poly-A tails by catalyzing the consecutive addition of up to 200 AMP residues to the trimmed 3′ ends of mRNA precursors (Figure 17.15). ATPs donate the AMP residues, and the tail is assembled in the absence of any template. The hydrolysis of PP_i helps "pull" the reaction toward completion. Both the caps and tails enhance the stability of the mature mRNA by protecting its ends from exonucleases. The cap is also recognized by the eukaryotic translational machinery (Section 18.2).

Editing

Editing refers to the modification of the template-determined nucleotide sequence of an RNA through the interconversion of common bases or the addition or deletion of one or more nucleotide residues (Figure 17.16). Although common in plants and protozoa, editing is rare in mammalian systems. Genes whose primary transcripts are edited are called **cryptogenes.** Editing is catalyzed by specific enzymes within ribonucleoprotein complexes labeled **editosomes.** It is brought about with the assistance of **guide RNAs (gRNAs),** small RNAs that pair with those segments of the RNA precursor that are to be edited. Some mRNA precursors are edited to the point that they are no longer able to hybridize with their own genes, and editing is sometimes required to generate a translatable messenger. Although many mRNA precursors are never edited, the list of precursors known to be edited is growing constantly. Mitochondrial mRNA precursors are common on this list.

In humans, precursors for **apolipoprotein B (apoB)** mRNA and the mRNA of the B subunit of the **glutamate-activated receptor channel (GluR-B)** are edited. ApoB is an important component of some of the lipoproteins that transport lipids in blood (Section 9.7), whereas GluR-B is a component of an ion channel that plays a central role in the transmission of nerve impulses in the brain (Section 10.9). In both instances, the translation of edited mRNA leads to the production of a unique polypeptide, and the percentage of mature mRNA that is edited changes from less than 10% in the early embryo to about 90% in adults. The physiological significance of the

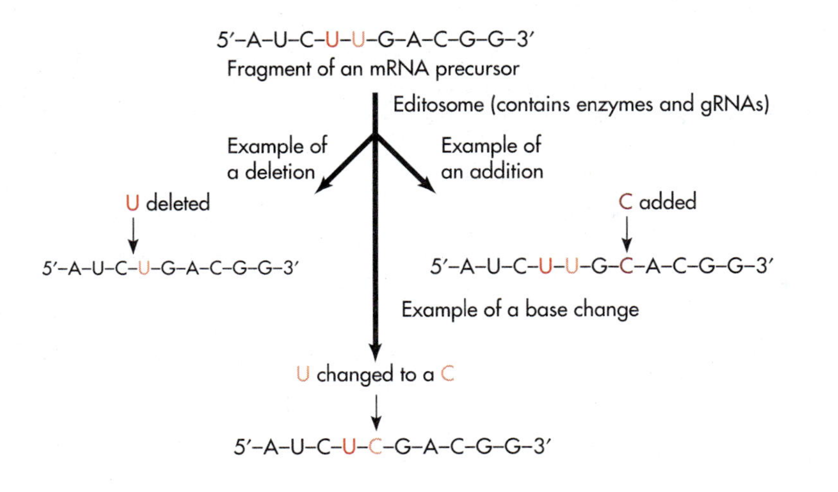

FIGURE 17.16
RNA editing.

editing is under investigation. An intron in the precursor for GluR-B mRNA plays an essential role in the editing process. This is one of a limited number of specific functions discovered for introns.

mRNA Processing Summary

The events involved in the processing of eukaryotic mRNA precursors are summarized in Table 17.3. Although the significance of some of the resultant modifications is still uncertain, other changes are known to have a marked impact on the activity or life span of the mature mRNAs ultimately created. **[Try Problems 17.15 through 17.18]**

TABLE 17.3

Events Involved in the Processing of mRNA Precursors

Processing Event[a]	Agent(s) Responsible	End Result
Trimming	Nucleases	Nucleotide residues are removed from the 3′ end
Splicing	Spliceosome (includes snRNAs) or mRNA precursor itself	Introns are removed and exons are joined
Capping	Capping enzymes	Cap is added
Tailing	Poly(A) polymerase	Poly-A tail is added
Minor base formation	Specific enzymes	Particular common bases are converted to minor bases
Editing	Editosome (includes gRNAs)	Nucleotide residues are added, deleted, or interconverted

[a]Most mRNA precursors are subjected to only a fraction of the listed events.

PROBLEM 17.10 Write an equation for the hydrolysis of a trinucleotide to form a mononucleotide and a dinucleotide. Provide structural formulas for all reactants and products. This is the type of reaction catalyzed by nucleases.

PROBLEM 17.11 How many separate splicing "events" are required to produce a mature mRNA from an mRNA precursor that contains 27 introns? How may separate exons does the mRNA precursor contain?

PROBLEM 17.12 The amino acid sequence of the polypeptide encoded in mRNA 1 (Figure 17.14) may be drastically different from the amino acid sequence in the polypeptide encoded in mRNA 2 (same figure). Explain. Hint: Review Section 16.5.

PROBLEM 17.13 Define "tertiary structure" as the term relates to RNAs. Would you expect ribozymes to possess unique tertiary structures and chiral active sites? Explain.

PROBLEM 17.14 Is it possible that an intron in the primary transcript for one mRNA could contain the nucleotide sequences for a second mRNA? Explain.

PROBLEM 17.15 Which of the events listed in Table 17.3 involve the cleavage or formation of phosphate ester bonds within mRNA precursors?

PROBLEM 17.16 True or false? The processing of the primary transcript for a eukaryotic mRNA normally leads to changes in its primary, secondary, and tertiary structure. Explain.

PROBLEM 17.17 A processing error leads to the production of an mRNA molecule that contains an extra 69 nucleotides at one unique site within that region normally translated. The polypeptide produced during the translation of this abnormal mRNA is found to differ from the polypeptide produced during the translation of the normal mRNA. How would you expect the normal and abnormal peptides to differ? Is the abnormal polypeptide necessarily longer than the normal one? Explain. Hint: Review Section 16.5.

PROBLEM 17.18 Most primary transcripts for eukaryotic mRNAs contain nucleotide residues not found in mature mRNAs, and most mature mRNAs contain nucleotide residues not found in their primary transcripts. What processing events account for this fact?

17.3 THE PROCESSING OF tRNA AND rRNA PRECURSORS

The primary transcripts for many eukaryotic tRNAs and rRNAs contain extra bases at both their 3′ and 5′ ends. Some also contain introns. Processing commonly involves trimming, splicing and minor base formation (Table 17.4). In addition, the nucleotide sequence CCA [found at the 3′ end of all mature tRNAs (Section 16.6)] must be added to those trimmed tRNA precursors that lack this sequence at their 3′ end. Although some eukaryotic tRNA precursors and rRNA precursors are also edited, the matured molecules lack caps and poly-A tails. **RNase P,** an enzyme involved in the trimming of the 5′ end of most tRNA precursors, is a ribonucleoprotein whose RNA component can accurately catalyze the trimming of primary transcripts in the absence of proteins. Thus, the RNA in RNase P is a genuine ribozyme. Some eukaryotic rRNA precursors, like some mRNA precursors, can remove their own introns.

The processing of the primary transcripts for prokaryotic tRNAs and prokaryotic rRNAs entails most of the same events involved in the processing of their eukaryotic counterparts. Trimming, minor base formation, and CCA addition (to tRNA precursors) are common occurrences. Introns and splicing, however, are rare in prokaryotes. In both prokaryotes and eukaryotes, rRNA maturation is coupled to ribosome assembly.

TABLE 17.4

Processing of Eukaryotic tRNA and rRNA Precursors

	Observed Events[a]
tRNA precursors	Trimming Addition of 3′ CCA Splicing Minor base formation Editing (rare)
rRNA precursors	Trimming Splicing Minor base formation Editing (rare)

[a]Most precursors are subjected to only a fraction of the listed events.

17.4 DNA REPLICATION

This section moves from RNA biosynthesis into DNA biosynthesis. The basic "logic" is the same in both processes. During DNA replication, a duplex DNA molecule unwinds, and then each polynucleotide strand serves as a pattern or template (through base pairing with deoxyribonucleoside triphosphates) for the synthesis of a complementary strand (Figure 17.1). The overall process can conveniently be divided into four stages: initiation, elongation, maturation, and termination. DNA replication is normally coupled to cell division; a cell does not replicate its DNA until it is ready to divide.

Origins

The sites on a duplex DNA molecule where replication begins are described as **origins.** Each origin is part of a discrete segment of DNA, called a **replicon,** that is capable of independent replication. The relatively small DNAs in most prokaryotes are usually closed circular duplex molecules containing a single origin. The larger DNAs in the chromosomes of eukaryotes are linear molecules that often possess dozens of origins. Since replication can proceed simultaneously in both directions from each origin, even very large DNAs can be rapidly replicated (Figure 17.17). Since all replication proceeds in a similar fashion, the examination of replication in one direction from a single origin provides a basic understanding of the overall process. This approach is used in the discussions that follow. **[Try Problem 17.19]**

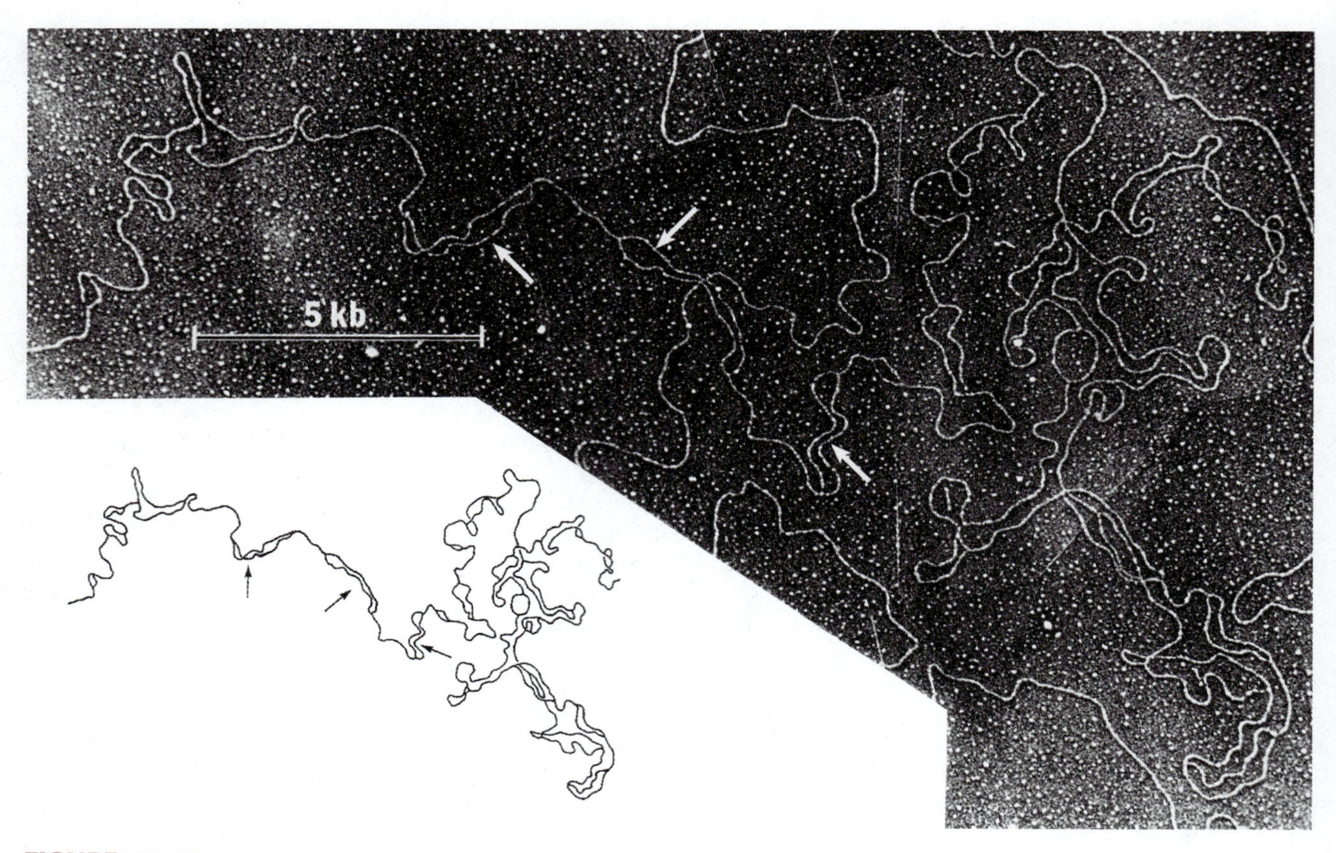

FIGURE 17.17
Bidirectional replication of *Drosophila* DNA from multiple origins. (a) Electron micrograph of replicating *Drosophila* (fruit fly) DNA. Each replication bubble (three are identified with white arrows) is a result of a bidirectional replication from a single origin. The bar, labeled 5 kb shows the length of a 5000-base pair segment of DNA. (b) Interpretation of the electron micrograph. (From Wolfe, 1993.)

DNA Unwinding

The first step in the initiation of DNA replication is the unwinding of a short segment of duplex DNA at an origin, an event that involves a variety of enzymes and other proteins (Figure 17.18). In *E. coli,* the process begins when a multiprotein complex binds to the origin and facilitates the separation of a small segment of the duplex. A topoisomerase may play a central role in this process by catalyzing the production of negative supercoils that weaken the right-handed duplex (Section 16.2). The transcription of adjacent genes may add to the negative supercoiling, since the creation of a transcription complex generates negative supercoiling behind the transcription bubble.

Once unwinding is initiated at the origin, further unwinding is driven by the **helicase**-catalyzed hydrolysis of ATP. Continuing strand separation is assisted in

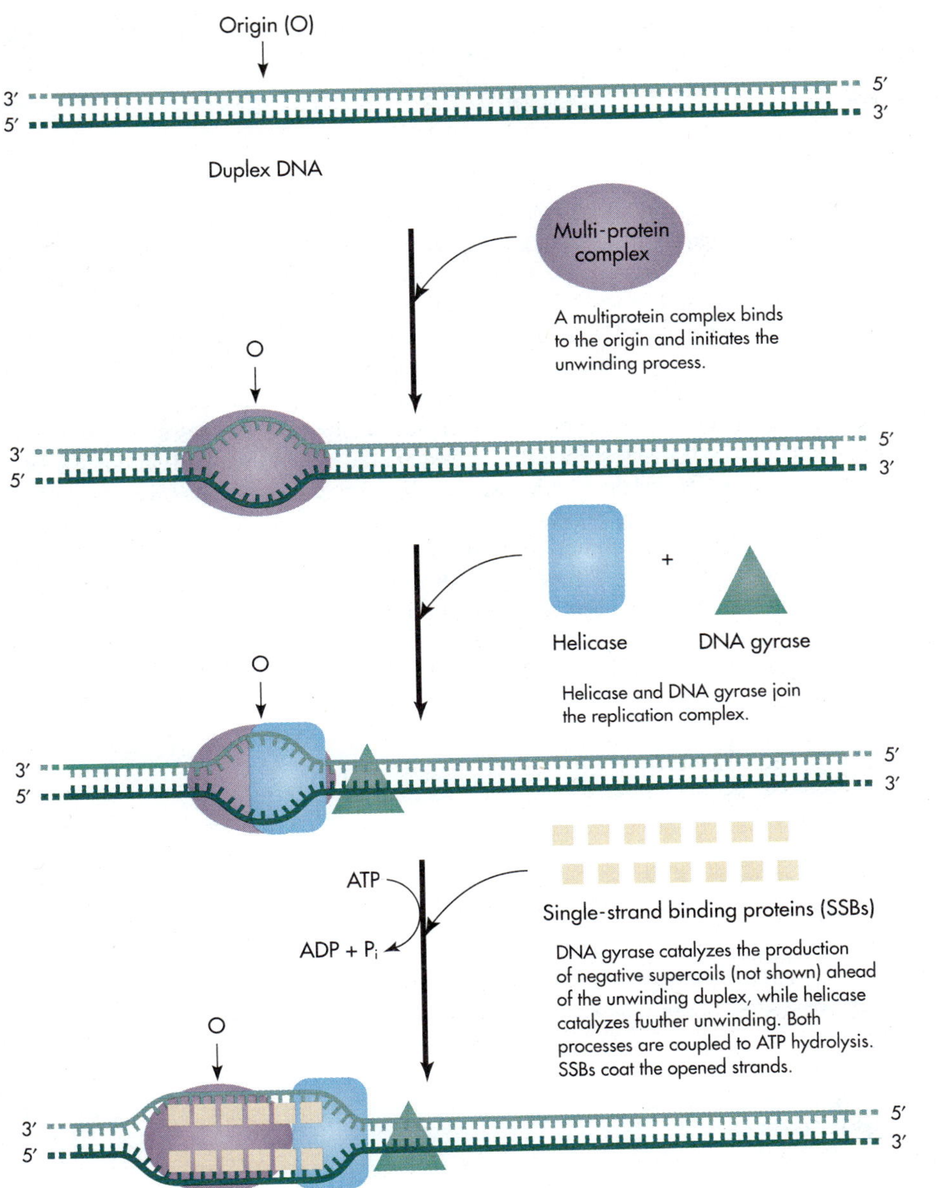

FIGURE 17.18
The unwinding of duplex DNA in *E. coli.*

prokaryotes by a topoisomerase, named **DNA gyrase,** which catalyzes the ATP-dependent generation of negative supercoils in the duplex DNA ahead of the site of strand separation. The gyrase-linked supercoiling helps relieve the torsional stress created by the unwinding of the parent duplex. **Single-strand binding proteins (SSBs)** reversibly coat (attach to) the separated parent strands to protect them from nucleases and to keep them from reannealing before they can be replicated. Although no enzyme strictly analogous to DNA gyrase has yet been detected in eukaryotes, unwinding appears to proceed in a similar manner in these more complex organisms. **[Try Problems 17.20 through 17.22]**

Primers and DNA Polymerases

To appreciate the second step in the initiation process, one needs some knowledge of **DNA polymerases,** those enzymes that catalyze the production of new DNA strands using preexisting strands of DNA or RNA as templates. Their substrates are deoxyribonucleoside triphosphates, and the overall process is strictly analogous to transcription (see Figure 17.9). DNA polymerases, like RNA polymerases, usually contain multiple polypeptide chains. However, in contrast to RNA polymerases, DNA polymerases cannot catalyze the production of a dinucleotide from two nucleoside triphosphates; they can only add deoxyribonucleotide residues to a preexisting oligonucleotide or polynucleotide that is base paired with a template (Figure 17.19). The preexisting oligo- or polynucleotide is called a primer, and it can be either RNA or DNA. Primers normally range in length from 6 to 60 nucleotides.

The second step in the initiation of DNA replication consists of the 5′ to 3′ production of an RNA primer (Figure 17.20) along one strand of the partially unwound parent duplex. Primer synthesis may actually be coupled to initial strand separation and not be a separate step. Although the primer initially synthesized at the origin may be assembled with the catalytic assistance of the same RNA polymerase that

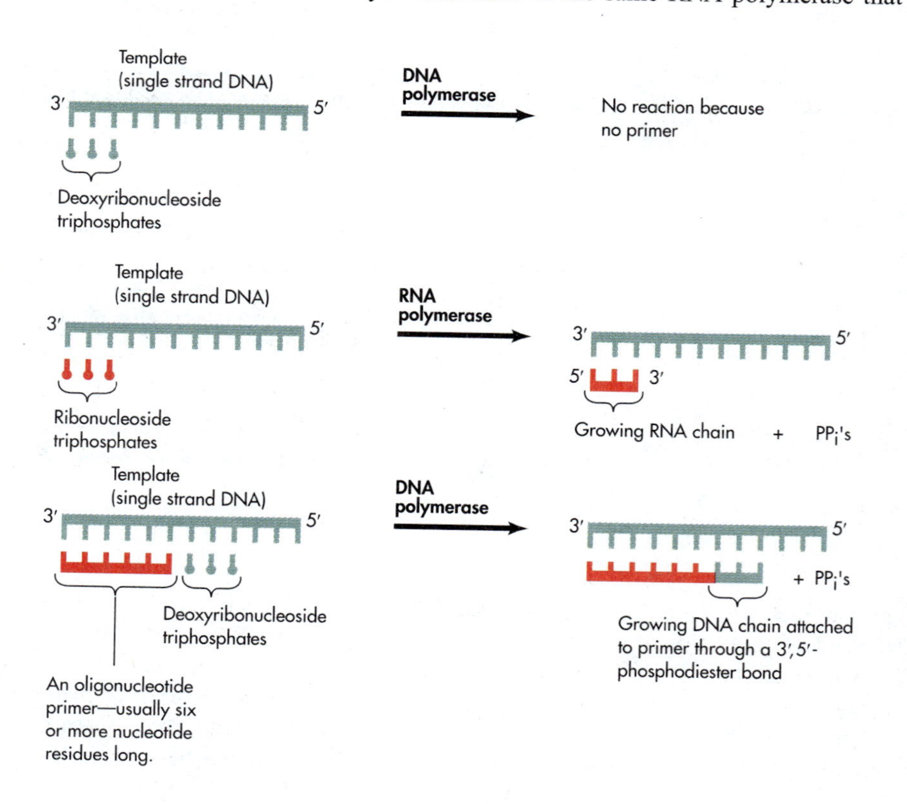

FIGURE 17.19
Primer requirements for DNA and RNA polymerases.

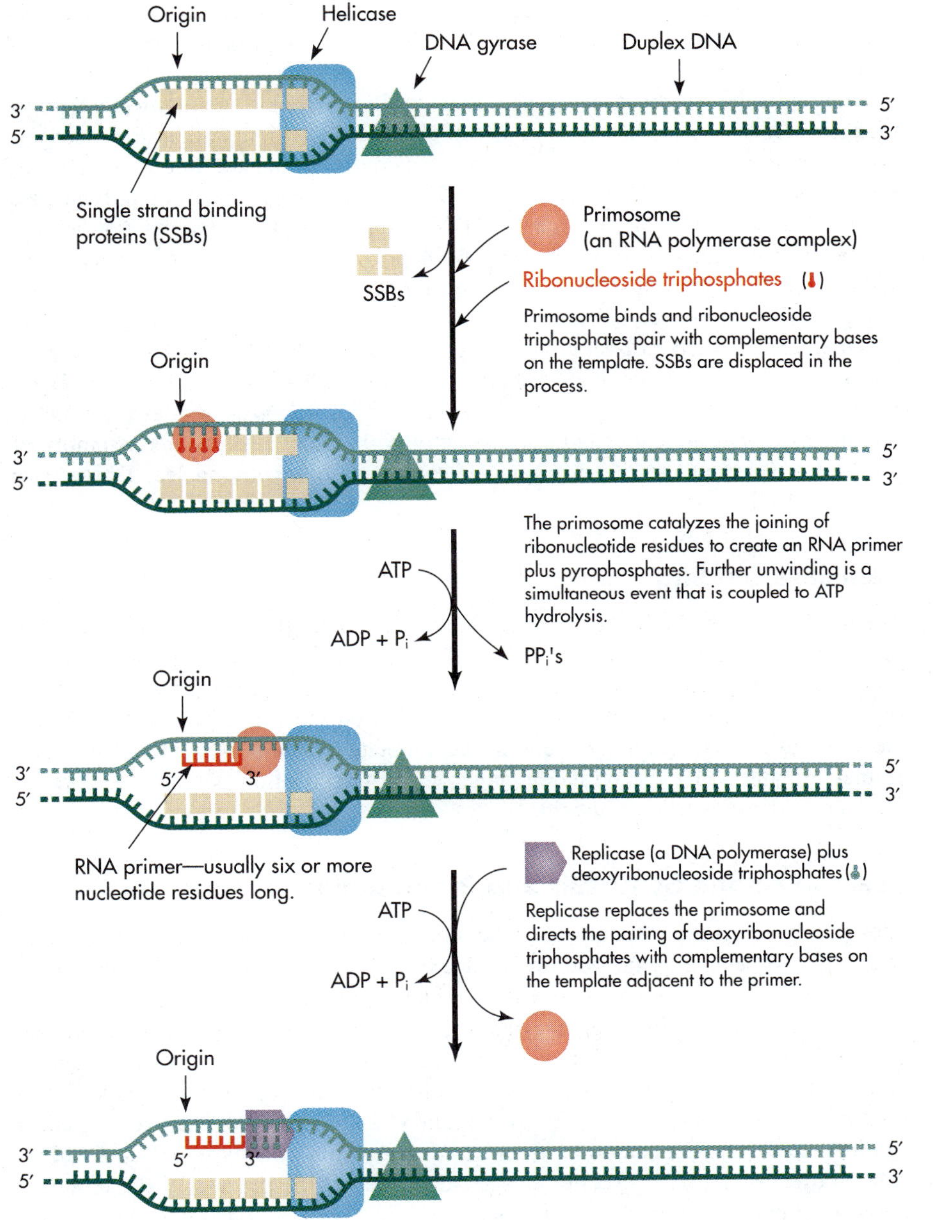

FIGURE 17.20
Primer production and the binding of a replicase in *E. coli.*

catalyzes the production of primary transcripts, the other primers involved in DNA replication (see discussion of lagging strand synthesis) are produced by a special RNA polymerase, called a **primase,** which is housed within a multiprotein complex called a **primosome.** Primases catalyze the assembly of primers through the normal transcriptional process (see Figure 17.9). Primosomes and the other participants in

DNA replication are all part of a giant complex called a **replisome.** In both eukaryotes and prokaryotes, replisome assembly is facilitated by protein chaperones (Section 4.4).

The replisomes in most cells encompass multiple **DNA-dependent DNA polymerases.** Only one of these, called a **replicase** catalyzes the addition of deoxyribonucleotide residues to the RNA primer generated during the initiation stage of replication. In *E. coli,* **DNA polymerase III** is the replicase and **DNA polymerase I** is involved in the maturation stage of DNA replication, DNA recombination, and DNA repair. A third *E. coli* polymerase, **DNA polymerase II,** appears to play a special role in DNA repair. Its function, if any, during DNA replication is uncertain. Although DNA polymerases II and III are constructed from multiple subunits, DNA polymerase I consists of a single polypeptide chain. The nuclei of eukaryotic cells contain at least two multisubunit DNA polymerases, **DNA polymerase** α and **DNA polymerase** δ**,** plus a large monomeric enzyme labeled **DNA polymerase** ϵ**.** DNA polymerase α has an associated primase activity that may catalyze the assembly of all the RNA primers involved in the replication of eukaryotic nuclear DNAs. **[Try Problem 17.23]**

Replicase Binding

The third step in the initiation process involves the binding of a replicase to the RNA primer:DNA complex (see Figure 17.20). This binding requires energy and involves multiple protein factors. The replicase, once bound, helps direct the pairing of deoxyribonucleoside triphosphates with those bases on the template that lie immediately downstream (toward the 5′ end of the template strand of DNA) from the RNA primer. At this point, the initiation stage of DNA replication is complete and the elongation phase is ready to begin.

Leading Strand Synthesis and Proofreading

Elongation begins immediately since the replicase is already bound to substrates and ready to start catalyzing the addition of deoxyribonucleotide residues to the primer. The process of chain extension is virtually identical to that illustrated for transcription in Figure 17.9. However, the replicase selectively uses deoxyribonucleoside triphosphates, rather than ribonucleoside triphosphates, as substrates. The daughter strand whose synthesis is initiated first is called the **leading strand.** This strand is continuously synthesized in a 5′ to 3′ direction as the replicase catalyzes the addition of one deoxyribonucleotide residue at a time while moving in a 3′ to 5′ direction along the template. SSB proteins are displaced in the process. If the replicase catalyzes the addition of a nucleotide that is paired with a noncomplementary nucleotide residue on the parent template, it pauses to catalyze the removal of the mismatched base before it continues to catalyze the synthesis of the leading strand. This process, described as **proofreading,** involves the hydrolysis of the phosphodiester bond that binds a mismatched nucleotide to the rest of the nascent polymer (Figure 17.21). Proofreading is brought about by a **3′ to 5′ exonuclease activity** associated with the replicase. Since a replicase adds nucleotide residues to a daughter strand at

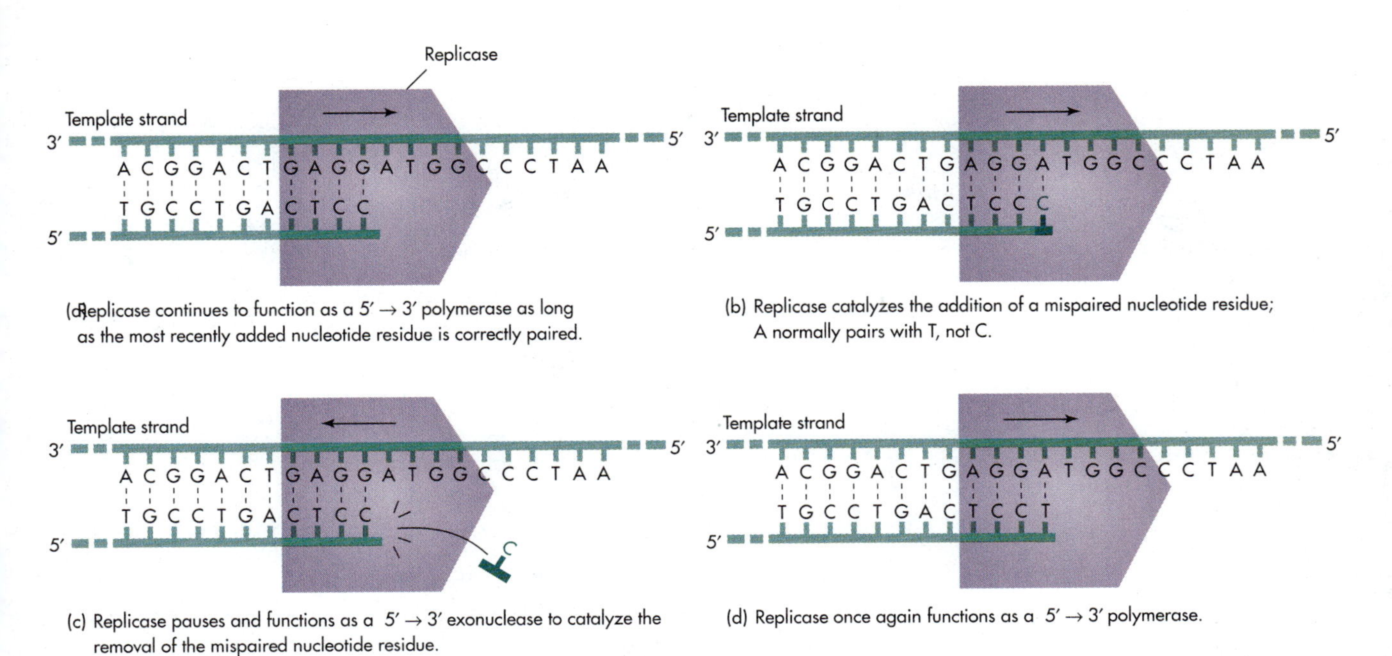

(a) Replicase continues to function as a $5' \rightarrow 3'$ polymerase as long as the most recently added nucleotide residue is correctly paired.

(b) Replicase catalyzes the addition of a mispaired nucleotide residue; A normally pairs with T, not C.

(c) Replicase pauses and functions as a $5' \rightarrow 3'$ exonuclease to catalyze the removal of the mispaired nucleotide residue.

(d) Replicase once again functions as a $5' \rightarrow 3'$ polymerase.

FIGURE 17.21
Proofreading by replicase (a DNA polymerase). (From Wolfe, 1993.)

a rate up to 50,000 per minute in prokaryotes and around 2000 per minute in eukaryotes, these enzymes literally zip along their templates during DNA replication. **[Try Problems 17.24 through 17.26]**

Lagging Strand Synthesis and Okazaki Fragments

While the leading strand of daughter DNA is being continuously synthesized along one strand of the parent DNA, the second daughter strand, called the **lagging strand** (because its production lags a bit behind the production of the leading strand), is synthesized in a discontinuous manner along the other parent strand. **Discontinuous synthesis** refers to the 5′ to 3′ production of many separate (not continuous), short fragments of DNA called **Okazaki fragments** (Figure 17.22). Each Okazaki fragment possesses a small RNA primer (typically less than 20 nucleotides long) plus approximately 1500 deoxyribonucleotide residues, in the case of prokaryotes, or around 150 deoxyribonucleotide residues, in the case of eukaryotes. Each primer is assembled by a primosome while the remainder of each Okazaki fragment is synthesized by a replicase. In *E. coli,* a single DNA polymerase III molecule, which has two active sites, may be involved in the simultaneous synthesis of both the leading strand and the lagging strand. According to this model, the lagging strand loops forward to interact with the replicase which is permanently associated with the continuously growing leading strand. **[Try Problem 17.27]**

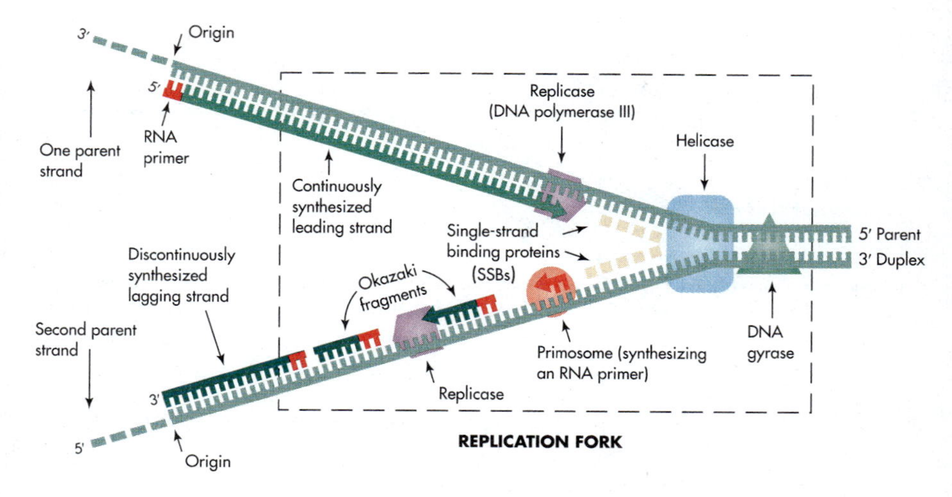

FIGURE 17.22
Elongation of the leading and lagging strands of daughter DNA in *E. coli.* Although replication is shown proceeding in only one direction from the origin, it normally proceeds in both directions simultaneously. With bidirectional replication, a separate replication fork moves off in both directions from the origin.

In eukaryotic cells, it seems that DNA polymerase δ is the replicase. It catalyzes the attachment of all the deoxyribonucleotides added to the leading strand and most of the deoxyribonucleotides added to the lagging strand. A primase subunit of DNA polymerase α is hypothesized to catalyze the assembly of the RNA primers in Okazaki fragments. A distinct subunit of α catalyzes the stepwise addition of about 35 deoxyribonucleotide residues to each primer. The segment of DNA assembled in this manner is labeled **initiator DNA (iDNA).** DNA polymerase δ replaces DNA polymerase α and catalyzes lagging strand extension after iDNA has been produced. In contrast to DNA polymerase δ, DNA polymerase α has no proofreading ability. **[Try Problems 17.28 and 17.29]**

The discovery that one daughter strand is synthesized discontinuously came as a total surprise because most biochemists had assumed that both daughter strands were simultaneously and continuously synthesized, each along a separate strand of the unwinding parent duplex. This model for DNA replication was once found in all biochemistry texts. Such simultaneous, continuous synthesis would require that one daughter grow in a 5′ to 3′ direction while the other daughter was growing in a 3′ to 5′ direction (Figure 17.23). However, no one has found a DNA polymerase that can catalyze 3′ to 5′ synthesis. The discontinous model for DNA replication explains why such a polymerase is not necessarily needed and may not exist. This bit of history provides a good example of the scientific method in action. Experimentation, observation, hypothesis construction, and hypothesis testing are major themes in biochemistry, and they are keys to the acquisition of scientific knowledge. When pondering the rapid changes in biochemistry, it is easier to keep the controversies and uncertainties in perspective if we remember this fact.

Replication Fork

The production of the two daughter DNA strands begins almost immediately after the two parent strands have started to unwind at an origin. As a consequence, replication is taking place along the unwound segments of the parent DNA at the same

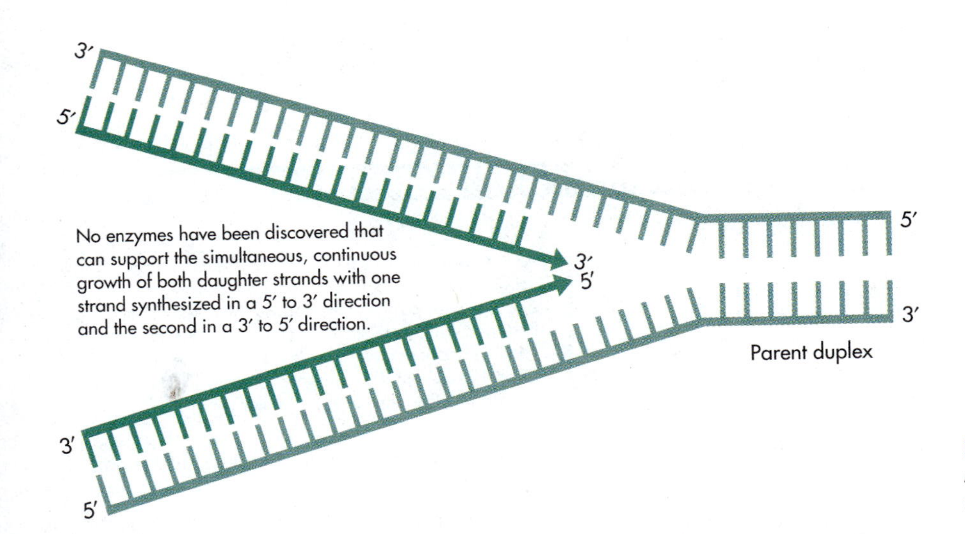

FIGURE 17.23
An abandoned model for DNA replication.

time that segments farther along this DNA are unwinding. In fact, replication normally keeps pace with unwinding and, at any instance, there are only short segments of unwound parent strands that are not paired with the growing daughter strands or in the process of being copied. Those sites on parent DNA where unwinding is occurring and where active replication is in progress are known as **replication forks** (see Figure 17.22). During the elongation phase of replication, the replication fork is in constant motion.

Maturation

The maturation stage of DNA replication is tightly coupled to chain elongation. **Maturation** refers to the removal of the RNA primers from the Okazaki fragments assembled during lagging strand production, the filling of the resultant gaps with deoxyribonucelotide residues, and the sealing together of the discontinuously assembled DNA fragments (Figure 17.24). Since this phase of replication begins immediately after Okazaki fragments start to appear, these fragments have short half-lives and do not accumulate to any significant extent. This helps explain why Okazaki fragments went undetected for many years.

In *E. coli,* DNA polymerase I has a 5′ to 3′ exonuclease activity that catalyzes the stepwise hydrolytic removal of the ribonucleotides from the RNA primer in each Okazaki fragment. The same enzyme possesses a 5′ to 3′ DNA polymerase activity that simultaneously catalyzes the addition of deoxyribonucleotides to the 3′ end of the adjacent Okazaki fragment to fill the gap left by the removal of the primer. This simultaneous removal of some ribonucleotides and addition of other nucleotides is known as **nick translation,** a term disliked by this author since nick translation is unrelated to translation as the term is most often used in biochemistry. *Translation* normally refers to the mRNA-directed production of a polypeptide. DNA polymerase I also possesses a 3′ to 5′ exonuclease activity (proofreading activity) that becomes involved in maturation if a mismatched deoxyribonucleotide residue is incorporated into an Okazaki fragment during nick translation. The maturation stage

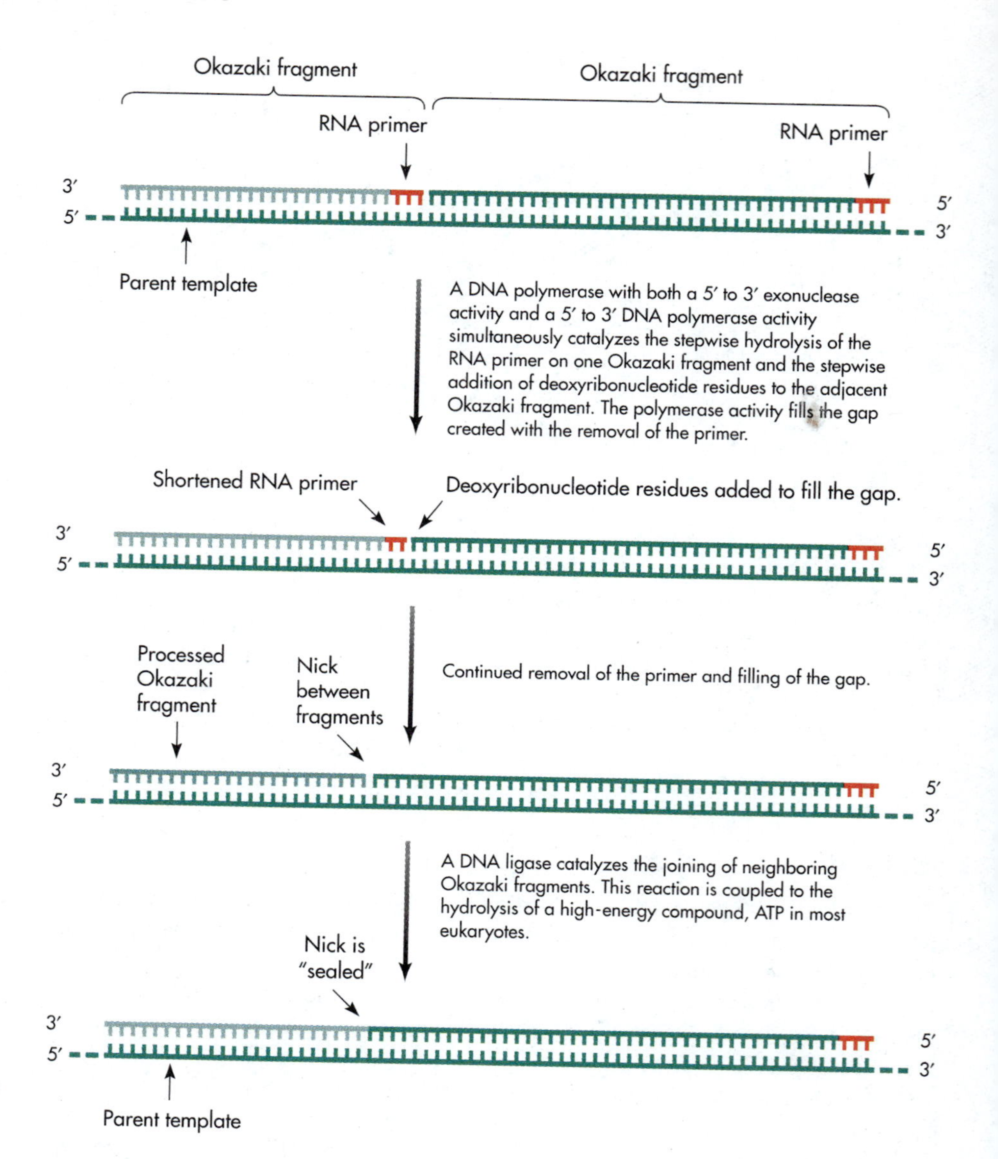

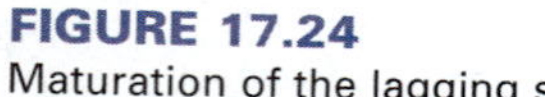

FIGURE 17.24
Maturation of the lagging strand of daughter DNA.

of DNA replication in eukaryotic cells may be catalyzed by DNA polymerase ϵ. Alternatively, DNA polymerase ϵ may be involved in DNA repair. **[Try Problem 17.30]**

The catalytic activities of DNA polymerase I or an analogous enzyme lead to processed Okazaki fragments (primers removed and resultant gaps filled) that lie side by side along their common template. Neighboring fragments are separated from one another only by nicks. In both prokaryotes and eukaroytes, an enzyme known as **DNA ligase** eliminates the nicks by catalyzing the formation of phosphodiester bonds between the processed Okazaki fragments. This reaction is coupled to the hydrolysis of an energy-rich compound, ATP in the case of most eukaryotes.

Termination

The processes of unwinding, elongation, and maturation simultaneously continue to the end of the replicon. The replisome then dissociates during a process described as **termination.** At least some of the various enzymes and protein factors within the replisome can be reused for the replication of other replicons. Most of the details of the termination process must still be unraveled. **[Try Problem 17.31]**

The termination of replication at the ends of a linear eukaryotic DNA molecule deserves special consideration. After the leading strand replicase has "copied" the 5′ end of its linear template (one strand of the parent DNA), replication stops before the 5′ end of the lagging strand has been completely assembled (Figure 17.25). The 5′ end of the lagging strand cannot be assembled through a mechanism so far described, because a terminal Okazaki fragment (if produced) could not be processed without an adjacent 5′ Okazaki fragment. Before the lagging strand is extended to its full length, the parent strand to which it is base paired is, itself, extended. This is accomplished with the catalytic assistance of an enzyme known as telomerase. Once the parent strand has been extended, lagging strand synthesis is completed. Since telomerase is a reverse transcriptase, a discussion of its proposed mechanism of action is reserved for Section 17.5 (Reverse Transcription and Retroviruses).

The Fidelity of Replication

The survival of individual organisms and species of organisms depends upon the accurate replication of the genetic information that is passed from cell to cell and generation to generation. Consequently, the fidelity of replication has evolved to a phenomenal degree; there is usually only one error for every 10^9 to 10^{10} nucleotides added to a growing daughter strand. On the average, this corresponds to less than three errors during the replication of the entire human genome (approximately 3×10^9 base pairs). Yet for some individuals, even a single error may be one too many. The possible consequences of abnormal changes in DNA sequence are examined in Chapter 19. The high fidelity of DNA replication is achieved through a combination of **error-avoidance** and **error-correction** mechanisms. **[Try Problem 17.32]**

Replicases play the central role in error avoidance by directing the pairing of deoxyribonucleoside triphosphates with complementary bases on parent templates and by using their proofreading ability to remove those mispaired nucleotide residues that are inadvertently inserted into a daughter strand (Table 17.5). The use of RNA primers, coupled with their automatic removal and replacement, is another error-avoidance mechanism. To understand why this is the case, we must first recognize that the error frequency is greater during the joining of the first few nucleotide residues in a new polynucleotide strand than it is during subsequent elongation. This fact is partially explained by the cooperative nature of the interactions between neighboring base pairs. A replicase also has difficulty proofreading a short oligonucleotide. Because the error-prone RNA primers are removed and the resultant gaps are filled under high fidelity conditions, most errors made during the initiation process are corrected during the maturation process.

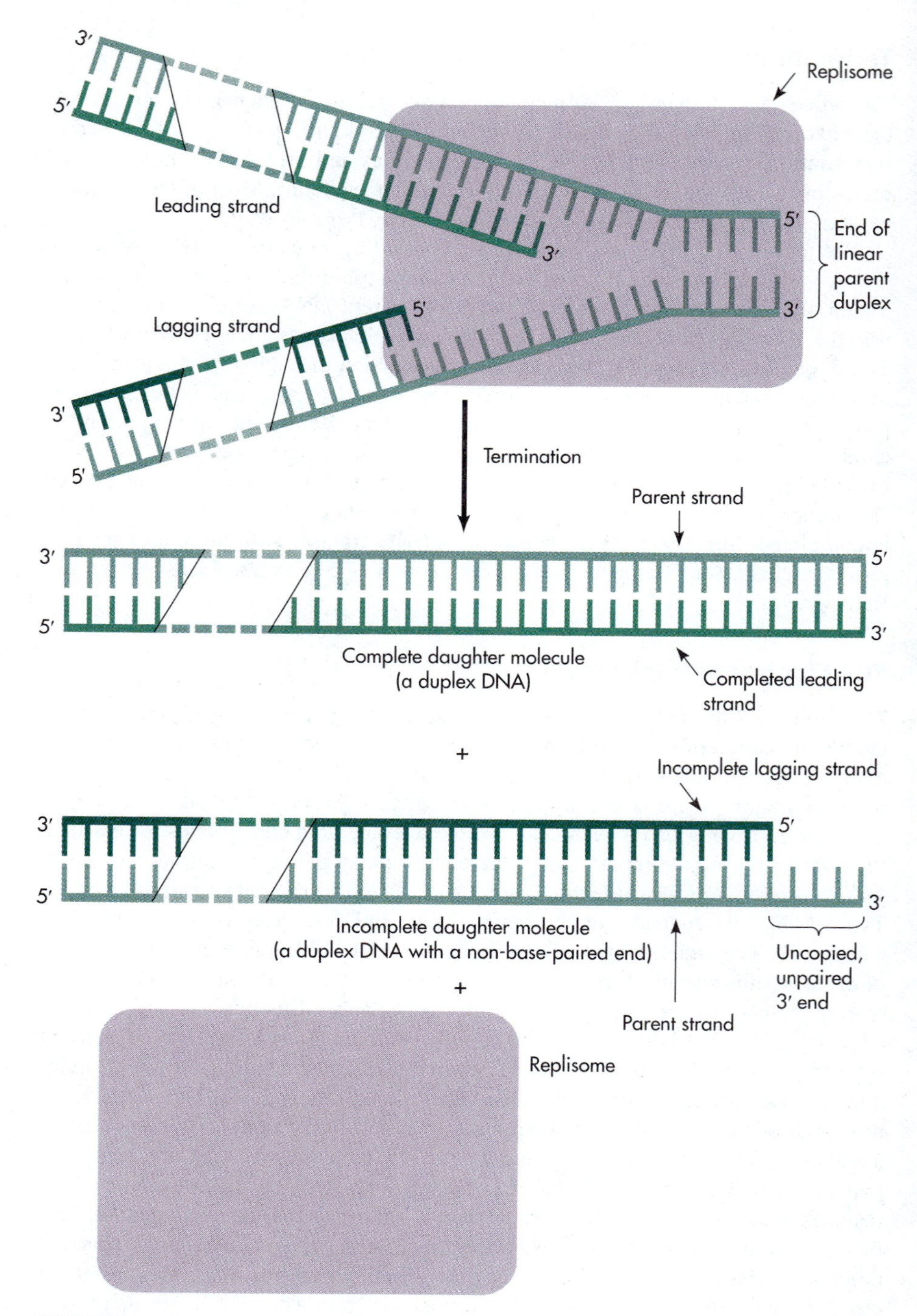

FIGURE 17.25
Incomplete synthesis of the lagging strand during termination of replication at the end of a linear parent duplex DNA.

TABLE 17.5

Some Mechanisms That Maintain the High Fidelity of DNA Replication

Mode of Action	Mechanism
Error avoidance	Replicase-assisted base pairing
	Proofreading
	Removal and replacement of primers under high-fidelity conditions
Error correction	Mismatch repair

The major error-correction mechanism is an energy-dependent process described as **mismatch repair.** It involves a multiprotein complex that recognizes mismatched bases and then employs multiple catalytic activities to replace the mispaired nucleotide residue in the daughter strand with the appropriate residue. In some cases, at least, this repair system distinguishes the daughter and parent strands by their extent of methylation (DNAs, like RNAs, contain a small fraction of minor nucleotides that are produced from the common nucleotides after the common nucleotides are incorporated into polymer strands. This means that newly replicated DNA must be processed). Parent–daughter discrimination is crucial because the replacement of the mismatched nucleotide residue in the parent strand would lead to error propagation rather than error correction.

Nucleosome Disruption

There is one additional aspect of DNA replication in eukaryotes that is worthy of mention. Replication, like transcription, must be accompanied by at least a partial disruption of nucleosome structure; the replisome cannot perform its tasks when DNA is tightly associated with histones. Both the disruption and the reassembly of nucleosomes is under investigation. Since the amount of DNA doubles during replication, twice as many nucleosomes are needed after replication. The synthesis of additional histones, described as **histone duplication,** is closely coordinated with DNA replication.

PROBLEM 17.19 How many replicons exist within a DNA molecule that contains 50 origins of replication? How many nucleotide residues does this DNA contain if the average replicon encompasses seven kilobase (kb) pairs?

PROBLEM 17.20 Why is energy required to unwind a duplex DNA? Is the energy required the same at all pH's? Explain. Hint: Review Sections 16.1 and 16.8.

PROBLEM 17.21 Although UTP contains the same amount of energy as ATP, UTP cannot be used in place of ATP during the helicase-catalyzed unwinding of DNA. What accounts for this fact?

PROBLEM 17.22 What classes of bonds most likely hold SSBs to single strands of DNA?

PROBLEM 17.23 Construct a family tree–type diagram illustrating the relationships among enzymes, replicases, primases, polymerases, transcriptases, DNA-dependent DNA polymerases, DNA-dependent RNA polymerases, DNA polymerase I, DNA polymerase III, DNA polymerase δ, and DNA ligase.

PROBLEM 17.24 What specific class of high-energy compound is cleaved each time a DNA polymerase catalyzes the addition of a monomer to a growing DNA chain? What is the significance of this cleavage reaction?

PROBLEM 17.25 Although proofreading (the hydrolytic removal of a mismatched nucleotide residue) is not, itself, an energy-requiring process, the temporary addition of mispaired nucleotide residues adds to the total energy cost of DNA replication. Explain.

PROBLEM 17.26 Approximately how long does it take a single eukaryotic DNA replicase molecule to catalyze the production of a leading strand of daughter DNA containing 10,000 nucleotides?

PROBLEM 17.27 What specific class of bond joins the RNA primer and the DNA within an Okazaki fragment?

PROBLEM 17.28 What is the minimum number of replicase molecules involved simultaneously in the replication of a eukaryotic DNA containing 50 active replicons? Explain.

PROBLEM 17.29 Assume that a replicon is replicated through bidirectional synthesis from a single origin. Will all of the Okazaki fragments produced during the replication process hybridize with the same strand of the parent DNA? Explain.

PROBLEM 17.30 List all of the substrates for DNA polymerase I of *E. coli.*

PROBLEM 17.31 List all of the enzymes identified in this section that will be present at an active *E. coli* replication fork and, therefore, be part of a replisome.

PROBLEM 17.32 Would you expect every uncorrected error in DNA replication to have an adverse impact on a cell or organism? Explain. Hint: Review Section 16.5.

17.5 REVERSE TRANSCRIPTION AND RETROVIRUSES

It has been known for many years that DNA can code for the synthesis of DNA (through replication) or RNA (through transcription) and that RNA can code for the production of proteins (through translation; Chapter 18). However, before 1970, it was generally accepted that information could not flow in the reverse direction; protein could not code for the production of RNA, and RNA could not serve as a template for DNA synthesis. This unidirectional hypothesis was known as the **central dogma.** In 1970, the discovery of an enzyme, called an **RNA-dependent DNA polymerase** or **reverse transcriptase,** which could use RNA as a template during the production of DNA, sent the biochemical community into a spin. The central dogma had been violated! The currently accepted view on the "allowed" flow of genetic information is summarized in Figure 17.26. This flow of information is one of the basic themes of biochemistry (Exhibit 1.4).

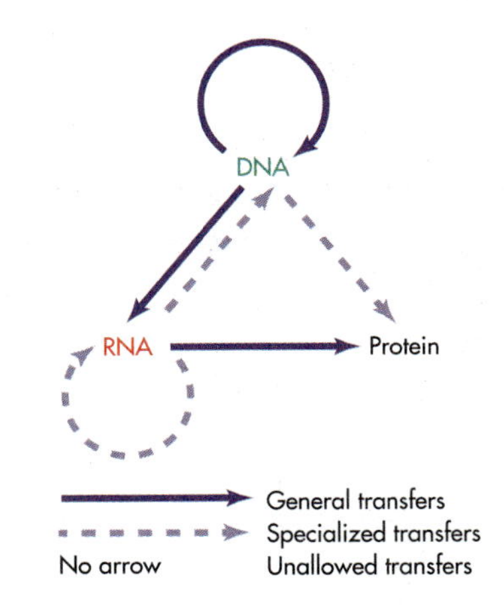

FIGURE 17.26
A summary of the "allowed" transfer of genetic information.

Reverse Transcriptases

Most reverse transcriptases possess three distinct catalytic activities that, collectively, allow them to catalyze the production of a duplex DNA molecule from the information in a single-stranded RNA. The three catalytic activities are: (1) an RNA-dependent DNA polymerase activity that uses RNA as a template and catalyzes RNA:DNA hybrid formation; (2) a nuclease activity, known as **RNase H,** that catalyzes the hydrolysis of the RNA in the RNA:DNA hybrid; and (3) a DNA-dependent DNA polymerase activity that, using the strand of DNA within the hybrid as a template, catalyzes the production of a complementary DNA strand (Figure 17.27). Since the original RNA template is degraded in the process, there is a one to one stoichiometry between original template and final product; a single duplex DNA is produced from the information in each RNA template. Once produced, the duplex DNA can be repeatedly replicated by an appropriate replisome if it contains an origin. **[Try Problem 17.33 through 17.35]**

Retroviruses

Reverse transcriptases were first discovered in RNA-containing viruses known as **retroviruses.** The prefix *retro-* means *backward.* It was selected to emphasize the fact that genetic information must flow backwards from RNA to DNA during the life cycle of these viruses. The most highly publicized retrovirus is the **human immunodeficiency virus (HIV),** a causative agent in the development of **acquired immune deficiency syndrome (AIDS).** A mature HIV, known as a **virion,** consists of two identical copies of an RNA genome (~10,000 nucleotide residues each), viral reverse transcriptase, additional viral enzymes, and a variety of other compounds, all enclosed in a viral envelope.

The life cycle of an HIV begins when the virus binds to a specific plasma membrane receptor, known as **CD4 antigen,** that marks those human cells susceptible to

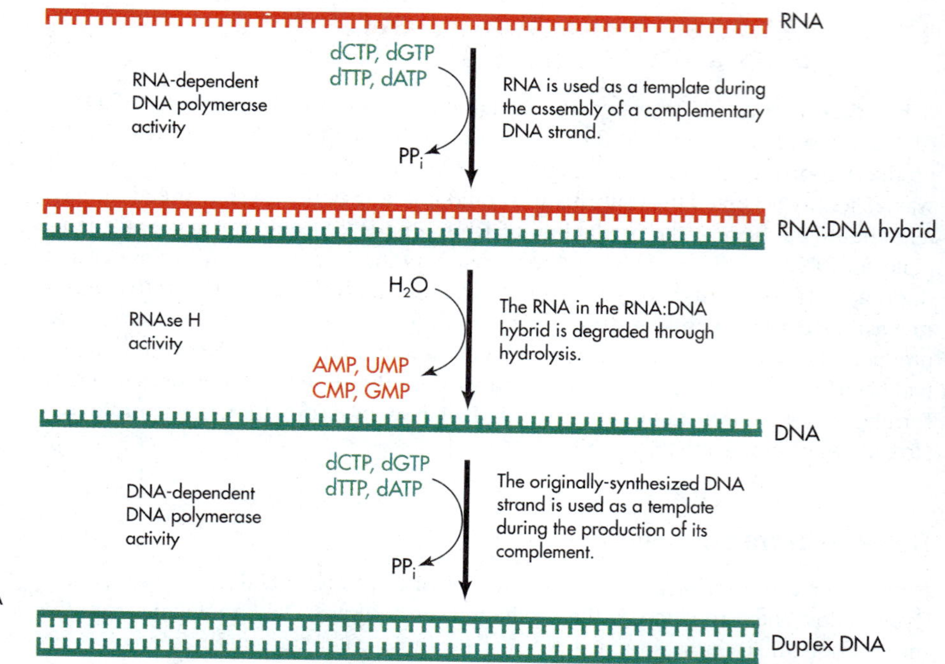

FIGURE 17.27
The reverse transcriptase–catalyzed production of duplex DNA from single-stranded RNA. Although not shown, primers are involved in the assembly of both strands of the DNA, and the second strand of DNA is synthesized at the same time the original RNA template is degraded.

infection (Figure 17.28). Host cells include the CD4-bearing T-lymphocytes (immune system cells abbreviated **CD4s**). Once a virus has attached, its RNAs and proteins invade the cytoplasm of the host cell, where the viral reverse transcriptase catalyzes the assembly of a duplex DNA using the viral RNA as a template. This DNA enters the nucleus, where it can be spliced into a host chromosome. The integrated DNA, known as a **provirus,** carries the blueprints for the construction of the proteins and RNAs required to complete the life cycle of the virus. A provirus may remain in an inactive (dormant) state for a prolonged period. When activated, it is transcribed, and the viral RNAs synthesized enter the cytoplasm to be translated. The viral proteins generated during translation are utilized, along with the viral RNAs, to assemble numerous virus particles. Once released from the infected cell, the new virions roam the body in search of other CD4-bearing cells whose depletion plays a central role in the ultimate collapse of the immune system in most AIDS patients. **[Try Problem 17.36]**

AIDS Therapies

Each step in the life cycle of HIV is a potential target for therapeutic agents. The controversial and widely employed AIDS drug, **3′-azido-2′,3′-dideoxythymidine (AZT),** is designed to block the reverse transcription of the viral RNA. The body converts AZT to AZT triphosphate, an analog of deoxythymidine triphosphate (dTTP; Figure 17.29) and a competitive inhibitor of reverse transcriptase (Section 5.10). Although several additional drugs target the same enzyme, novel therapeu-

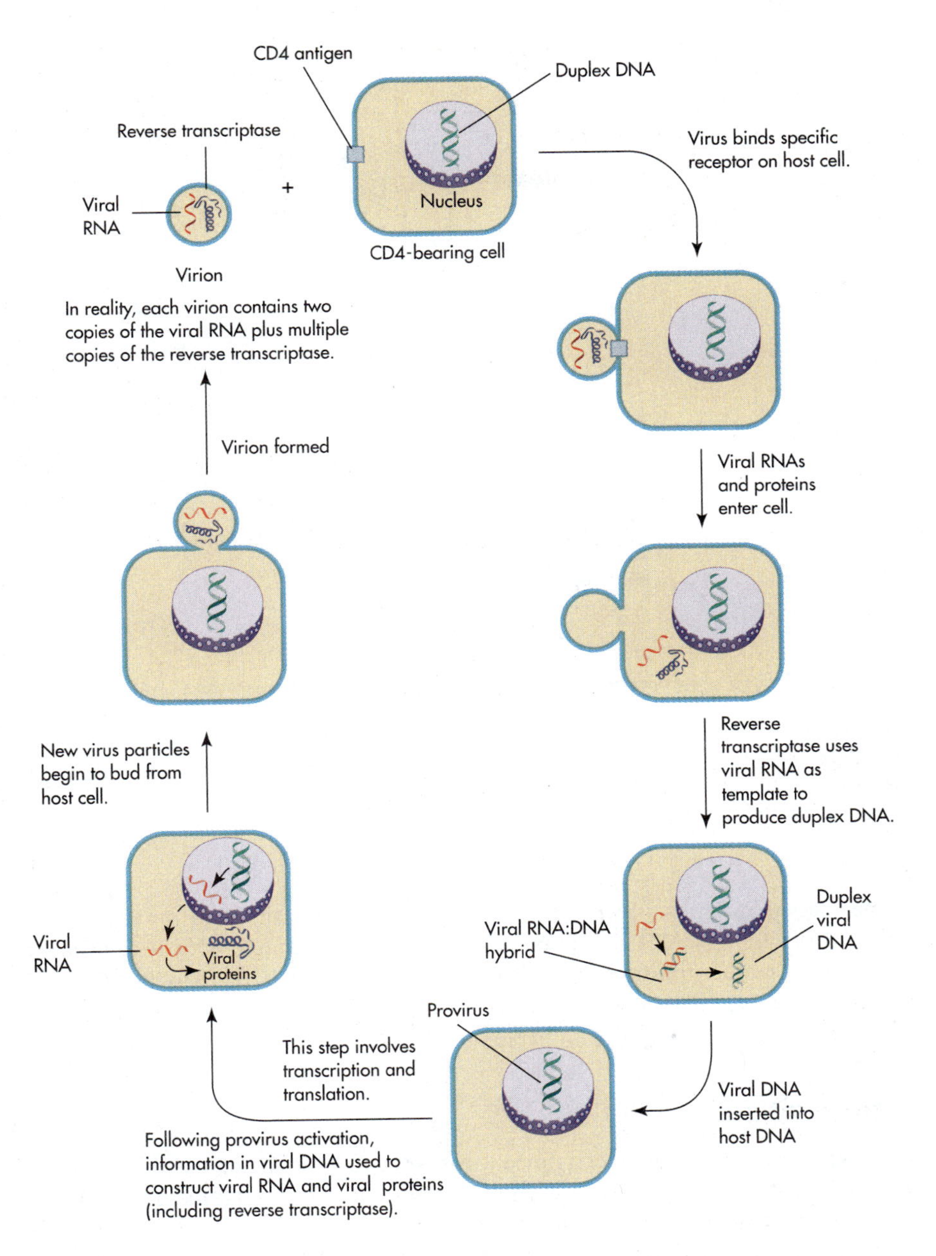

FIGURE 17.28
The life cycle of a retrovirus.

tic agents have been developed to inhibit virus attachment to membrane receptors, to interfere with the transport of viral RNA from the nucleus to the cytoplasm, or to disrupt other phases of the virus life cycle. Unfortunately, all of the existing, thoroughly tested drugs are relatively ineffective or have serious side effects. Because reverse transcriptases have no proofreading capability, retroviruses have relatively high mutation rates and tend to evolve drug resistance rapidly. The take-

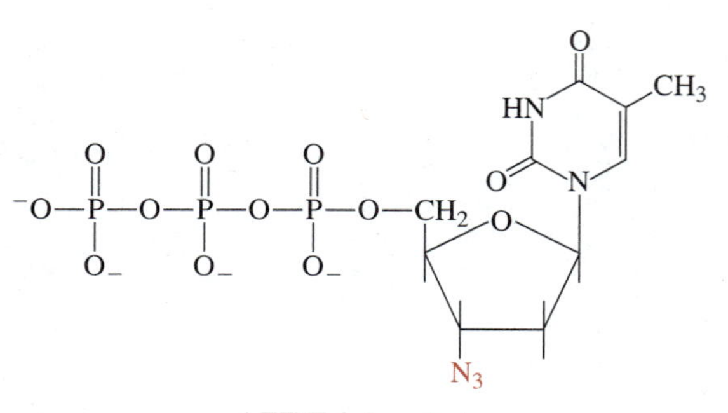

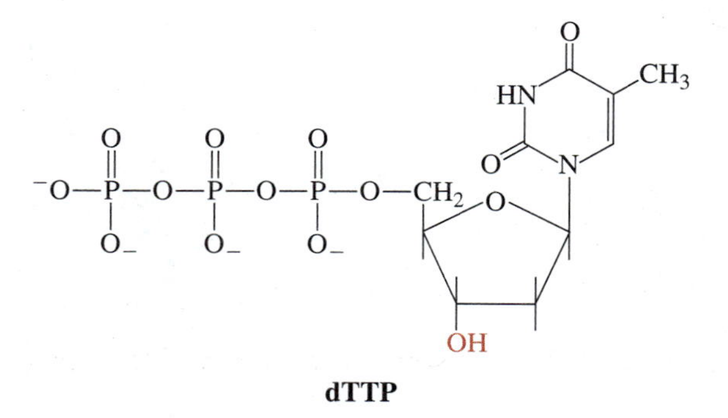

FIGURE 17.29
AZT triphosphate and dTTP.

home message? Think *prevention,* not treatment. When thinking *prevention,* be aware that although AZT is relatively ineffective in treating AIDS, it can reduce the risk of a mother transmitting HIV to her baby by almost 70%. Hopes for a cure lie in new drugs, new strategies for treatment, and combination therapies. **[Try Problem 17.37]**

Transposition

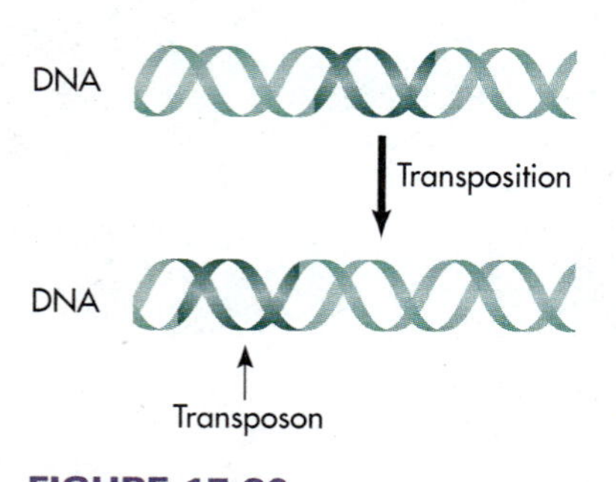

FIGURE 17.30
Transposition.

Reverse transcriptases are not confined to viruses and virus-infected cells. These enzymes have been discovered in a wide variety of normal cells, including human cells. Present evidence indicates that the blueprints for at least some of these reverse transcriptases are encoded in segments of DNA acquired during previous encounters with retroviruses. Eukaryotic reverse transcriptases appear to play a central role in **transposition,** the movement of segments of DNA, called **transposons** or "jumping genes," from one region in the genome to a second (Figure 17.30). Eukaryotic transposons are sometimes called **retrotransposons** since they are thought to move by means of RNA intermediates. Transpositions can lead to changes in gene expression, and, in some instances, they are involved in the transformation of normal cells into cancerous ones. **[Try Problem 17.38]**

Telomerase

Telomerase is a special reverse transcriptase that plays a vital role in the replication of linear DNA molecules (Section 17.4). To understand the full significance of this enzyme and how it operates, you need some knowledge of telomeres. **Telomeres,** which contain short repeated sequences of DNA plus associated protein, are found at the ends of eukaryotic chromosomes. Telomeres appear to have been highly conserved during evolution, because they are very similar in all organisms that have been examined. In most vertebrates, telomeres encompass many repeats of the sequence TTAGGG. A human cell possesses 92 telomeres, one at each end of each of its 46 chromosomes. Telomeres protect the ends of chromosomes and help prevent end-to-end fusion of neighboring chromosomes. They may also participate in other processes as well, including the regulation of gene expression and the regulation of cell division.

Before telomerase action is examined in more detail, a brief review is in order. Recall that the replication of the terminal replicons on linear duplex DNAs stops before lagging strand synthesis has been completed (see Figure 17.25). To complete the assembly of the lagging strand, the single strand end of its complementary parent strand is first extended. An RNA component within telomerase plays a central role in this process. After a complementary segment of this RNA base pairs with a telomere repeat at the end of the parent strand, an RNA-dependent DNA polymerase (reverse transcriptase) activity catalyzes the further addition of deoxyribonucleotides to this parent strand. The segment of telomerase RNA adjacent to the segment paired with the parent strand serves as a template (Figure 17.31). The telomerase shifts on the parent strand, a process called translocation, and then catalyzes the addition of still more nucleotide residues to this strand. Telomerase RNA is once again used as a template. Because the telomerase RNA also appears to contribute to the catalytic activity of this enzyme it may prove to be another example of a ribozyme (Section 17.2).

After the 3′ end of the parent strand has been extended to an appropriate length with the catalytic assistance of telomerase, synthesis of the lagging strand can be completed. How this is accomplished is uncertain. In the absence of telomerase, lagging strand synthesis is never completed and the length of the telomere on the daughter DNA is shortened relative to its length on the parent DNA.

Most normal cells lack telomerase activity. As a consequence, their telomeres become shorter and shorter with each round of cell division. This continual shortening may explain why most normal cells can divide a limited number of times; the shortening of telomeres beyond a certain point may lead to cell death. There is some evidence that the telomerases in cancer cells contribute to their uncontrolled growth. If true, selective inhibitors of these enzymes may be of therapeutic value in the treatment of cancer. A substantial amount of ongoing research is designed to explore this possibility. **[Try Problem 17.39]**

cDNA

Reverse transcriptase is not used only by normal cells and retroviruses, it is also employed as a tool by genetic engineers. This enzyme is used to construct genes for gene transfer experiments. When an mRNA serves as a template for reverse tran-

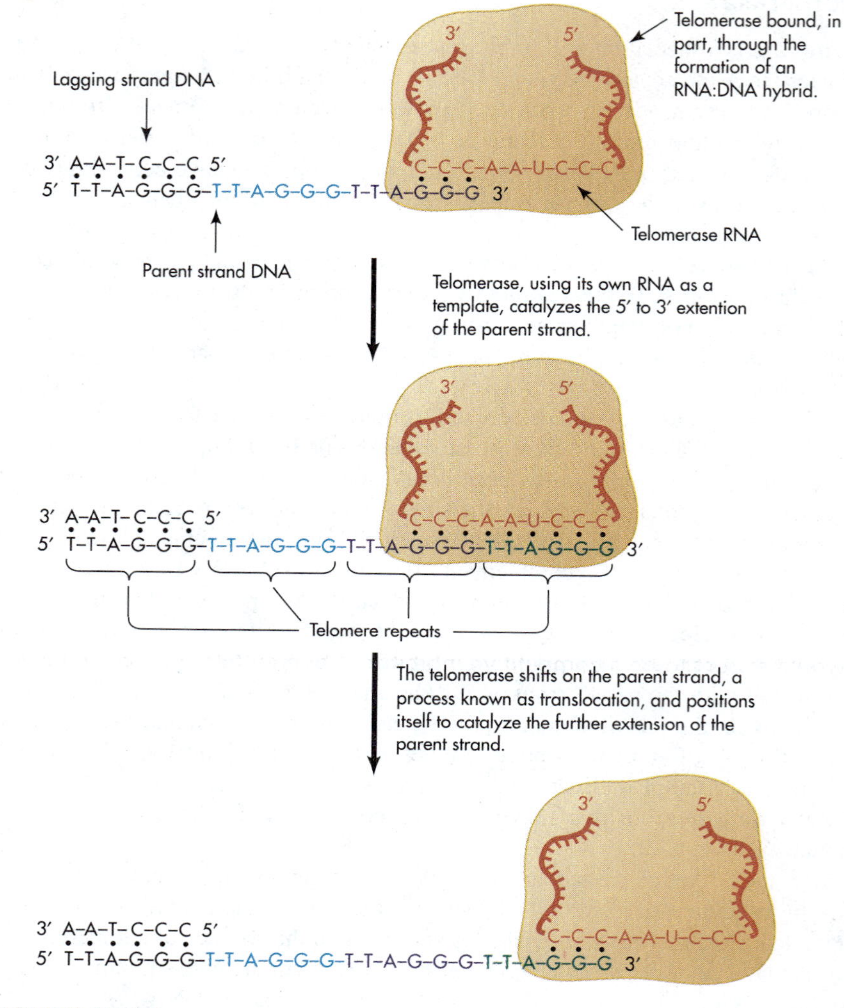

FIGURE 17.31
Telomerase action.

scription, the duplex DNA generated is called **cDNA** (complementary DNA), and it carries the information for the synthesis of the protein originally encoded in the mRNA. When coupled to appropriate regulatory elements, the cDNA construct can function as a gene. Genetic engineering and some of its accomplishments are discussed in Chapter 19.

PROBLEM 17.33 What duplex DNA molecule will be produced when the following RNA molecule serves as a template for reverse transcriptase?

AGAAAAAACGAGCUUU

Illustrate how each of the three catalytic activities of reverse transcriptase participates in the production of this duplex.

PROBLEM 17.34 List all of the molecules or classes of molecules that serve as substrates for reverse transcriptases. What is the number of distinct active sites that a single reverse transcriptase molecule most likely contains?

PROBLEM 17.35 To which of the six major classes of enzymes do reverse transcriptases belong?

PROBLEM 17.36 Suggest how a retrovirus benefits by inserting its genetic information into the chromosome of a host cell.

PROBLEM 17.37 True or false? You would expect AZT to inhibit DNA replication as well as reverse transcription. Explain.

PROBLEM 17.38 Suggest two mechanisms through which a transposition could alter the expression of a gene.

PROBLEM 17.39 Give the abbreviation of an oligonucleotide that you would expect to be a competitive inhibitor of human telomerase. Explain how you arrived at your answer.

17.6 THE POLYMERASE CHAIN REACTION

This chapter on nucleic acid biosynthesis closes with a glimpse at the **polymerase chain reaction (PCR),** a simple but novel process for the *in vitro* (outside of living cells) replication of DNA. PCR is patented and its value lies in its ability to produce an unlimited number of copies of a DNA molecule. The genetic material in a single cell can be amplified to generate enough DNA to fill a beaker. PCR has become an invaluable tool for basic researchers, clinicians, genetic engineers, forensic scientists, and others working with DNA.

PCR begins with the heat denaturation (98°C) of the DNA that is to be amplified (replicated). Small DNA primers that are complementary to the 3′ ends of the resultant single strands of DNA are added to the solution containing these DNAs. Some prior knowledge about the 3′-terminal sequences of the DNAs is normally utilized in the selection of appropriate primers. After the reaction mixture is cooled to about 60°C to allow the primers to anneal with complementary sequences on the single-stranded DNAs, DNA polymerase and deoxyribonucleoside triphosphates are added. DNA polymerase molecules, using the original DNA strands as templates, catalyze the attachment of one deoxyribonucleotide residue at a time to the annealed primers. The completion of this process leads to the duplication of the original DNA (Figure 17.32).

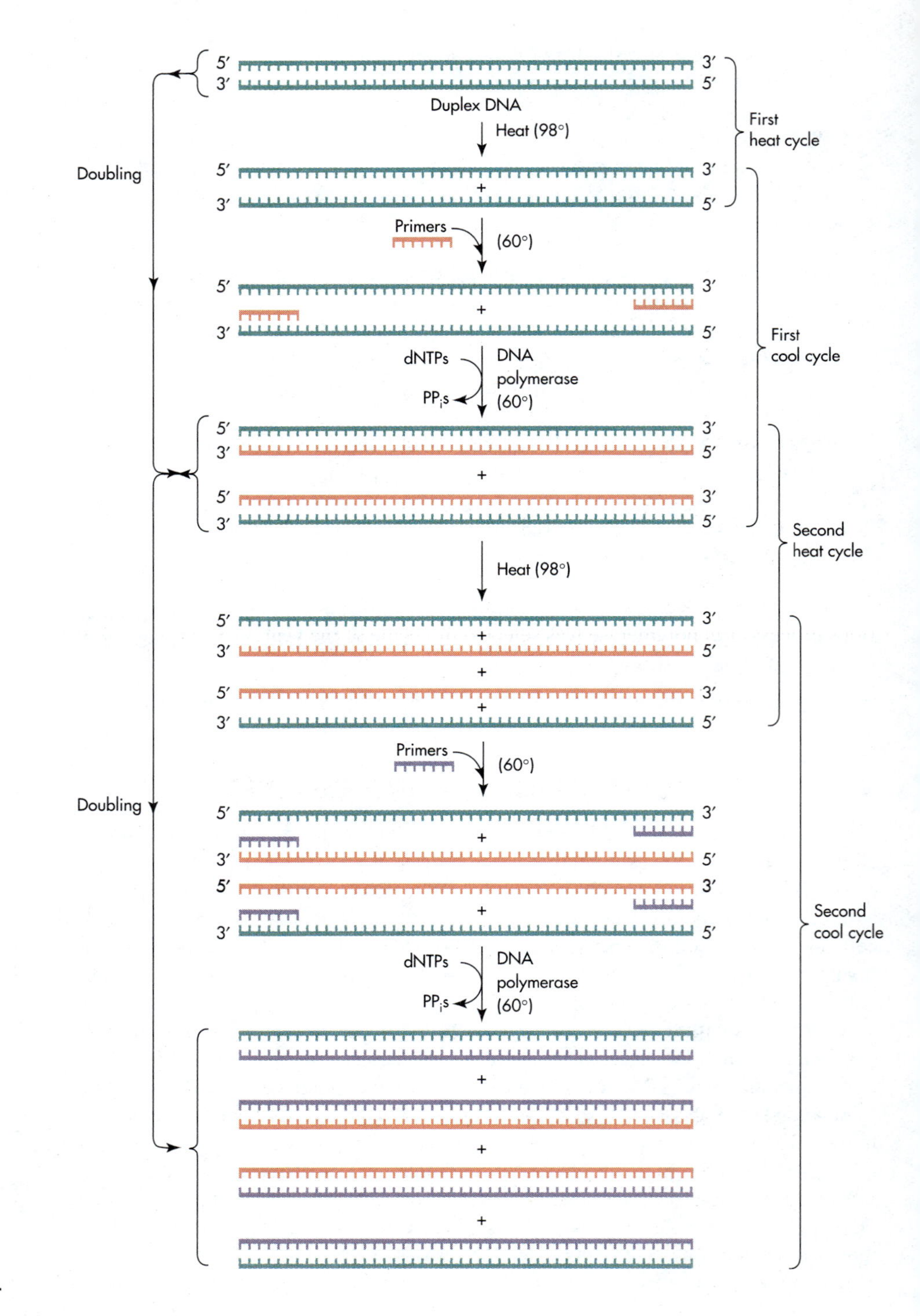

FIGURE 17.32
The polymerase chain reaction.

The solution containing the primers, deoxyribonucleoside triphosphates, DNA polymerase, and the duplicated DNA is once again heated to approximately 98°C to denature the newly synthesized duplex DNAs. A second cooling cycle allows DNA primers (a large excess is added originally) to pair with the single strands of DNAs and allows the catalytic activity of DNA polymerase to once again double the number of duplex DNA molecules present in the reaction mixture. At this point, the original DNA has been amplified four-fold. A third heat–cool cycle leads to an eight-fold amplification, a fourth to a 16-fold amplification, and so on. Each heat–cool cycle doubles the amount of DNA present at the beginning of that cycle. If needed, more primers and deoxyribonucleoside triphosphates can be added prior to any cycle. Since each heat–cool cycle can be as short as 1 or 2 minutes, a sample of DNA can readily be amplified 100 billion fold in an afternoon. The final magnitude of the amplification is 2^n where n equals the number of heat–cool cycles. **[Try Problems 17.40 through 17.42]**

Since enzymes and other globular proteins are usually denatured and inactivated at 98°C, how does DNA polymerase survive the repeated heat–cool cycles during PCR? The polymerase used (abbreviated ***Taq* polymerase**) is a heat-resistant enzyme from *Thermus aquaticus,* a thermophilic (heat-loving) bacterium. This microbe and its enzymes thrive in Yellowstone hot springs at temperatures close to the boiling point of water. The use of a heat-labile DNA polymerase would necessitate the separation of DNA and enzyme prior to each heat cycle, an expensive and laborious process. *Taq* polymerase was selected molecule of the year by *Science* in 1989, the first year that the magazine selected a molecule of the year.

PROBLEM 17.40 List the similarities and differences between PCR and normal DNA replication.

PROBLEM 17.41 What magnitude of DNA amplification is attained with PCR during 10 heat–cool cycles?

PROBLEM 17.42 Why are primers required for PCR? Can RNA primers be used in place of DNA primers? Explain.

SUMMARY

Transcription, DNA replication, and reverse transcription all involve the use of a preexisting nucleic acid strands as patterns or templates for the construction of new polynucleotide chains. In each instance, ribonucleoside triphosphates or deoxyribonucleoside triphosphates hydrogen bond with nucleotide residues in the template to form standard Watson–Crick base pairs, and, if no errors are made, the newly assembled polymer is antiparallel and totally complementary to its template. Each new polymer is extended in a 5′ to 3′ direction as a specific polymerase catalyzes the addition of one nucleotide residue at a time to the growing chain. The addition of each nucleotide residue entails the formation of a phosphate diester bond at the expense of an anhydride bond. The enzyme-catalyzed hydrolysis of the pyrophosphate that is formed in the process helps pull the reaction to completion (Le Châtelier's principle).

Transcription is catalyzed by DNA-dependent RNA polymerases (RNA polymerases; transcriptases) that require no preformed primers and utilize ribonucleoside triphosphates as substrates. A unique segment of one strand of a duplex DNA molecule serves as a template, and the immediate product is an RNA called a primary transcript. Each transcription unit in DNA begins with a sequence of base pairs, known as a promoter, that encompasses or is close to the transcription start site. Prokaryotic RNA polymerases bind to promoters with the assistance of a protein initiation factor labeled sigma (σ). In eukaryotic systems, a multitude of transcription factors collaborate to determine promoter utilization and to control the rate of initiation of transcription by RNA polymerase. Transcription factors sometimes interact with enhancers or silencers (regulatory elements on DNA) during this process. The binding of an RNA polymerase and appropriate accessory proteins to a promoter leads to the opening of a short segment of the duplex DNA at the start site, an event that frees the template strand so it can be transcribed. The RNA polymerase subsequently directs the pairing of ribonucleoside triphosphates with the template strand and catalyzes the production of a short RNA primer. This completes the initiation stage of transcription.

Once transcription has been initiated, a change in the accessory proteins associated with RNA polymerase shifts transcription to the elongation phase. A transcription bubble moves along the transcription unit while RNA polymerase catalyzes the addition of one residue at a time to the growing RNA chain. When a terminator is reached, transcription stops, the primary transcript is released, and the various components that remain within the transcription complex dissociate. The termination process is often regulated, in part at least, by one or more protein factors.

Many primary transcripts are biologically inactive precursors of mRNAs, tRNAs, or rRNAs. The processing of eukaryotic mRNA precursors involves one or more of the following events: trimming, splicing, capping, polyadenylation (creates poly-A tails), minor base formation, and editing. These chemical modifications are brought about by the catalytic activities of a variety of enzymes that may include some ribozymes. Ribonucleoprotein complexes, called spliceosomes and editosomes, are involved in splicing and editing functions, respectively. The processing of tRNA and rRNA precursors usually involves trimming, splicing, and minor base formation. Other events, including editing, may also be involved. Mature tRNAs and rRNAs lack caps and poly-A tails.

The main polymerases involved in DNA replication, known as DNA-dependent DNA polymerases or replicases, have four major characteristics: they use deoxyribonucleoside triphosphates as substrates; they can catalyze only the addition of deoxyribonucleotides to preexisting primers; they possess a proofreading ability; and the final products of catalysis are DNA and pyrophosphates. In eukaryotes, DNA replication usually proceeds simultaneously in both directions from multiple origins. Prokaryotic DNAs normally have a single origin.

The initiation of DNA replication encompasses three primary events: the separation of the parent duplex at an origin; the synthesis of an RNA primer; and the binding of a DNA polymerase. Strand separation involves several proteins, including a helicase, a topoisomerase, and single-strand binding proteins. Primer synthesis is accomplished with the catalytic assistance of an RNA polymerase. Elongation begins with the addition of a deoxyribonucleotide residue to the RNA primer and continues with the addition of one deoxyribonucleotide residue at a time to the growing DNA strand. The details of the termination process must still be ascertained.

The leading strand of daughter DNA is synthesized in a continuous fashion along one strand of the parent DNA. Only a single RNA primer is required. In contrast, the lagging strand is simultaneously assembled in a discontinuous manner; many short segments of DNA, called Okazaki fragments, are produced along adjacent segments of the second strand of the parent duplex. Each Okazaki fragment possesses its own RNA primer which is assembled by a primase housed within a primosome. A unique DNA polymerase catalyzes the removal of the primers from Okazaki fragments and the filling of the resultant gaps with deoxyribonucleotide residues. Once Okazaki fragments are processed, adjacent fragments are joined with the catalytic assistance of DNA ligases. In *E. coli,* DNA polymerase III is the replicase while DNA polymerase I is involved in the processing of the Okazaki fragments. DNA polymerase δ is the major eukaryotic replicase, whereas DNA polymerase α, which has an associated primase activity, catalyzes the production of RNA primers and iDNAs. DNA polymerase ϵ may be responsible for Okazaki fragment processing.

On the average, only one error is made for every 10^9 to 10^{10} nucleotides added to a growing daughter strand during DNA replication. This phenomenal fidelity, which is crucial to the survival of most organisms, is accomplished through a combination of error-avoidance and error-correction mechanisms. Proofreading by DNA polymerases plays a central role in error avoidance and mismatch repair helps correct errors that slip by the avoidance systems.

Most RNA-dependent DNA polymerases (reverse transcriptases) possess three distinct catalytic activities: an RNA-dependent DNA polymerase activity; an RNase activity; and a DNA-dependent DNA polymerase activity. Collectively, these three catalytic capabilities lead to the assembly of a duplex DNA molecule from information carried in an RNA template. In normal eukaryotic cells, reverse transcriptases appear to participate in transpositions, the movement of DNA fragments from one site in a genome to a second. A special reverse transcriptase, called telomerase, is involved in the maintenance of the telomeres found at the ends of eukaryotic chromosomes. Viral reverse transcriptases are key participants in the life cycle of HIV and other retroviruses. For this reason, multiple AIDS drugs, including AZT, have been designed to inhibit viral reverse transcriptases.

Polymerase chain reaction (PCR) is a patented process used to amplify the duplex DNA molecules in biological samples. A solution of the DNA to be amplified is heated, mixed with appropriate primers plus deoxynucleoside triphosphates and a heat resistant DNA polymerase, and then cooled. After the primers anneal with the 3′ ends of the separated strands of the heat-denatured DNA, DNA polymerase molecules catalyze the assembly of a complementary strand along each of the original polymer strands. The heat–cool cycle is repeated over and over again, with each cycle doubling the number of duplex DNA molecules present in the sample at the beginning of the cycle. PCR allows us to amplify the DNA content in minute biological samples, even single cells. This can provide enough DNA for chemical analysis and sequence determination.

EXERCISES

17.1 Describe the specific biological role, function, or activity of each of the following during transcription.

a. Enhancer
b. RNA polymerase
c. Antiterminator
d. Terminator
e. Sigma factor
f. TATA box
g. Promoter
h. Transcription factors

Clearly indicate whether each listed "item" is a segment of a nucleic acid, a protein, a carbohydrate, a nucleoprotein, or another class (please specify) of compound.

17.2 What impact, if any, would a marked drop in σ factor concentration have on the rate of protein production within an *E. coli* cell? Explain.

17.3 What is the minimum number of RNA polymerase molecules required to synthesize a primary transcript that contains 250 nucleotides? How many phosphate anhydride bonds are cleaved in the process? Approximately how much energy is released each time a phosphate anhydride bond is hydrolyzed?

17.4 If a eukaryotic cell suddenly stopped producing the mRNA for DNA ligase, approximately how long would it be before the concentration of this mRNA dropped to 10% of its initial value (value at the time mRNA production shut down)? Assume that the ligase mRNA has a half-life of 2 days.

17.5 Is there any theoretical limit to the number of times a single gene can be transcribed?

17.6 True or false? Since most eukaryotic mRNAs contain a poly-A tail with over 100 adjacent AMP residues, the template strands of the genes for most eukaryotic mRNAs must contain over one hundred adjacent dTMP residues. Explain.

17.7 Explain why a drug that preferentially blocks a unique promoter will tend to selectively inhibit the expression of a single gene or a single class of genes.

17.8 Compare the interactions of prokaryotic and eukaryotic transcriptases with their respective promoters.

17.9 In theory, the following DNA fragment contains information for the production of two different RNA fragments:

5′dGCAACGTTGCC3′
3′dCGTTGCAACGG5′

What is the monomer sequence within each of these two RNA fragments? In general, only one of the two possible RNA fragments is produced during the transcription of this DNA. What determines which RNA fragment is actually produced?

17.10 List the major steps involved in the processing of a typical eukaryotic mRNA precursor. Which steps are energy-requiring steps? Explain.

17.11 Describe the role, function, biological activity, and general significance of each of the following during the processing of an mRNA precursor:

a. snRNPs
b. hnRNA
c. Exons
d. RNases
e. Methylases

Clearly indicate whether each listed "item" is a protein, a nucleic acid, a nucleoprotein, or another class (please specify) of compound.

17.12 Although mRNAs, tRNAs, and rRNAs account for the majority of cellular RNA, several "minor" RNAs or "minor" classes of RNA also exist. List each minor RNA or minor class of RNA encountered in this chapter, and then briefly describe its biological role or function.

17.13 How would the primary transcript produced during the transcription of a mammalian cDNA differ from the primary transcript produced during the transcription of the normal gene for the mRNA used to construct the cDNA?

17.14 A biochemist isolates the template strand for the gene that codes for the primary transcript of a unique mRNA. She then hybridizes (Section 16.8) this sense strand DNA with the mature mRNA produced from the primary transcript and "photographs" the hybrid with an electron microscope. The following "picture" is obtained:

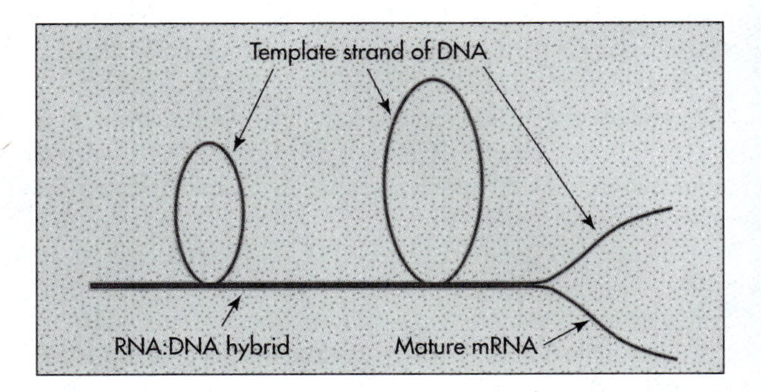

Explain the existence of the loops. What do the loops represent? Explain.

17.15 Draw a diagram illustrating a duplex DNA containing two different origins of replication with unidirectional replication in progress from one origin and bidirectional replication in progress from the second one. Use a separate line to represent each polynucleotide strand.

17.16 List all the specific proteins discussed in this chapter that participate in the initiation stage of DNA replication in *E. coli.* Briefly summarize the function of each listed protein.

17.17 List the properties of most DNA polymerases that distinguish them from RNA polymerases.

17.18 List the two specific events during DNA replication that involve an exonuclease activity.

17.19 List those specific steps or reactions during DNA replication where ATP or another high-energy compound is cleaved in order to make the process thermodynamically favorable.

17.20 List all of the events that are simultaneously occurring in the region of the replication fork during active DNA replication.

17.21 Explain why DNA replication involves a maturation stage whereas RNA synthesis (transcription) does not.

17.22 Describe the two distinct catalytic activities that are associated with DNA replicases.

17.23 Which of the following terms accurately describe or characterize a primase? A primer? A replicase? An Okazaki fragment? A replicon? A transcriptase? A reverse transcriptase? In each case, list all correct answers.

- **a.** Nuclease
- **b.** Oxidoreductase
- **c.** DNA polymerase
- **d.** RNA polymerase
- **e.** Isomerase
- **f.** Transferase
- **g.** Hydrolase
- **h.** Protein
- **i.** Enzyme
- **j.** Nucleic acid or nucleic acid fragment

17.24 Define *half-life.* What accounts for the short half-life of Okazaki fragments?

17.25 What processes, events, or reactions account for the extreme fidelity of DNA replication?

17.26 Explain why a single error in DNA replication can have lethal consequences for a cell or organism whereas a single error during transcription would virtually never have similar consequences.

17.27 Describe the three separate catalytic activities associated with most reverse transcriptase molecules.

17.28 This chapter has provided an introduction to DNA-dependent DNA polymerases, DNA-dependent RNA polymerases, and RNA-dependent DNA polymerases. A fourth class of nucleic acid polymerases exists. What is this fourth class? What are the substrates and products for this class?

17.29 Explain how a reverse transcriptase can lead to gene amplification (an increase in the number of copies of a gene) within a cell.

17.30 Explain why a human telomerase is unable to function in an organism whose telomeres contain repeats that differ from the repeats in human telomeres.

17.31 A tRNA molecule encoded by the RNA genome within RNA tumor viruses is thought to serve as a primer during the reverse transcriptase–catalyzed production of the RNA:DNA hybrid that contains the viral RNA. Draw a diagram illustrating an RNA:DNA hybrid containing a tRNA primer. Use separate lines to represent each polynucleotide strand. What feature of the tRNA allows it to serve as a primer?

17.32 Would you expect the error rates to be very similar or significantly different during PCR and normal DNA replication? Explain.

17.33 Can primers complementary to the 5′ends of single-stranded DNAs (rather than ones complementary to the 3′ ends) be used for PCR? Explain.

17.34 What specific chemical link is cleaved each time a nucleotide residue is added to a growing DNA strand during PCR?

SELECTED READINGS

Alberts, B., D. Bray, J. Lewis, M. Raff, K. Roberts, and J. D. Watson, *Molecular Biology of the Cell,* 3rd ed. New York: Garland Publishing, 1994.

An up-to-date and encyclopedic textbook for those wanting more details about nucleic acid biosynthesis. Numerous references are provided.

Barinaga, M., Ribozymes: killing the messenger, *Science* 262, 1512–1514 (1993).

Describes the possible clinical uses of ribozymes.

Beardsley, T., Smart genes, *Sci. Am.* 265(2), 86–95 (1991).

Examines the role of transcription factors in cellular differentiation.

Blackburn, G. M., and M. J. Gait, Editors, *Nucleic Acids in Chemistry and Biology,* New York: Oxford University Press, 1990.

Very readable. Contains some details not presented in most introductory-level biochemistry texts. Another good source of selected readings.

Carey, M., Simplifying the complex, *Nature* 402–403 (1994).

Reviews research indicating that a holoenzyme may be involved in the initiation of transcription in eukaryotes.

Chan, L., and P. H. Seeburg, RNA editing, *Sci. Am. Science & Medicine* 2(2), 68–77 (1995).

Reviews mRNA editing in mammalian systems.

Choy, B., and M. R. Green, Eukaryotic activators function during multiple steps of preinitiation complex assembly, *Nature* 366, 531–536 (1993).

Examines some of the details of transcription initiation in eukaryotes.

Conaway, R. C., and J. W. Conaway, General initiation factors for RNA polymerase II, *Ann. Rev. Biochem.* 62, 161–190 (1993).

A thorough review that contains 270 references.

Gray, M. W., Pan-editing in the beginning, *Nature* 368, 288 (1994).

Examines the distribution and possible evolutionary origin of editing.

Johnston, M. I., and D. F. Hoth, Present status and future prospects for HIV therapies, *Science* 260, 1286–1293 (1993).

A thorough review that contains 123 references.

Karran, P., Appropriate partners make good matches, *Science* 268, 1857–1858 (1995).

Reviews some of the details of the mismatch repair process in humans.

Krumm, A., and M. Groudine, Tumor suppression and transcription elongation: The dire consequences of changing partners, *Science* 269, 1400–1401 (1995).

Describes the central role of a protein named Elongin in the regulation of the elongation phase of transcription and discusses a dire consequence of a change in the rate of elongation.

Kunkel, T. A., DNA replication fidelity, *J. Biol. Chem.* 267, 18251–18254 (1992).

A review of two error discrimination processes that operate during DNA replication.

Lamond, A. I., Splicing in stereo, *Nature* 365, 294–295 (1993).

Describes the stereochemistry of the splicing mechanism for mRNA precursors.

Lingner, J., J.P. Cooper, and T.R. Cech, Telomerase and DNA end replication: No longer a lagging strand problem, *Science* 269, 1533–1534 (1995).

Suggests how telomerase may participate in the formation of the single stranded 3′ overhangs that are associated with many telomeres.

Marx, J., Chromosome ends catch fire, *Science* 265, 1656–1658 (1994).

Describes the probable functions of telomeres and telomerases.

Moran, L. A., K. G. Scrimgeour, H. R. Horton, R. S. Ochs, and J. D. Rawn, *Biochemistry,* 2nd ed. Englewood Cliffs, NJ, Neil Patterson/Prentice-Hall, 1994.

An encyclopedia-type biochemistry text that covers all aspects of nucleic acid biosynthesis.

Pabo, C. O., and R. T. Sauer, Transcription factors: structural families and principles of DNA recognition, *Ann. Rev. Biochem.* 61, 1053–1095 (1992).

Another comprehensive review.

Struhl, K., Duality of TBP, the universal transcription factor, *Science* 263, 1103–1104 (1994).

Reviews the roles of TATA-binding protein (TBP) in transcription initiation in eukaryotes.

Tjian R., Molecular machines that control genes, *Sci. Am.* 272(2), 54–61 (1995).

Examines the regulation of transcription in mammalian systems.

van Holde, K. E., D. E. Lohr, and C. Robert, What happens to nucleosomes during transcription? *J. Biol. Chem.* 267, 2837–2840 (1992).

A minireview that addresses its title.

Waga, S., and B. Stillman, Anatomy of a DNA replication fork revealed by reconstitution of SV40 DNA replication *in vitro, Nature* 369, 207–212 (1994).

A close look at the replication fork.

Wise, J. A., Guides to the heart of the spliceosome, *Science* 262, 1978–1979 (1993).

Examines the mechanism of spliceosome action.

18 Protein Biosynthesis

CHAPTER OUTLINE

Chapters 4 and 5 have emphasized the central roles of proteins in life processes. Chapters 16 and 17 have provided an overview of gene expression and examined some of the details of transcription and DNA replication. Several major points are worth reiterating.

A vast majority of all DNA of known function carries information for the production of mRNAs and, ultimately, polypeptides. Segments of DNA that encode a protein or an RNA are called genes. Translation is the process through which an mRNA is used to direct the joining of amino acids through peptide bonds. The order in which the amino acids are linked is determined by the sequence of codons within the translated region of the mRNA. Although some polypeptides are functional as originally assembled, many must be post-translationally modified (processed) to be biologically active. A biologically active polypeptide represents the final product of an mRNA-encoding gene.

The polypeptides present within an organism determine what reactions occur and how rapidly they occur. Collectively, these reactions determine the nature of the organism. This chapter examines translation and polypeptide processing, the two events directly involved in protein biosynthesis. The dominant theme is information transfer. Supporting themes include regulation of gene expression, structural hierarchy and compartmentation, evolution, specificity, and catalysis. Chapter 19 examines genetic diseases and genetic engineering, two topics closely associated with protein biosynthesis.

18.1 OVERVIEW OF TRANSLATION

To discuss the details of translation, we must have some knowledge about mRNAs, tRNAs, ribosomes, and the genetic code, topics examined in Chapter 16. A brief review is in order.

Role of mRNA

That segment of an mRNA which encodes a polypeptide is its translated (coding) region. This region begins at its 5′ end with an initiation codon and concludes at its 3′ end with a termination codon. Normally, codons are read sequentially in a 5′ to 3′ direction starting with the initiation codon. With rare exceptions, sixty-one of the 64 codons "specify" one of the 20 common protein amino acids, whereas the other three are termination codons (Table 18.1). The sequence of codons in the coding region of an mRNA determines the amino acid sequence of the polypeptide synthesized during translation.

Role of tRNA

Transfer RNAs are directly responsible for translating the genetic code. Each tRNA binds and carries a specific protein amino acid and contains an anticodon that pairs with (binds noncovalently to) one or more of the codons for that amino acid. The pairing of an anticodon in a tRNA with a codon in the translated region of an mRNA

TABLE 18.1

The Genetic Code

First Base	Second Base: U		C		A		G	
U	UUU	Phe	UCU	Ser	UAU	Tyr	UGU	Cys
	UUC	Phe	UCC	Ser	UAC	Tyr	UGC	Cys
	UUA	Leu	UCA	Ser	UAA	TERM[a]	UGA	TERM[a]
	UUG[b]	Leu	UCG	Ser	UAG	TERM[a]	UGG	Trp
C	CUU	Leu	CCU	Pro	CAU	His	CGU	Arg
	CUC	Leu	CCC	Pro	CAC	His	CGC	Arg
	CUA	Leu	CCA	Pro	CAA	Gln	CGA	Arg
	CUG	Leu	CCG	Pro	CAG	Gln	CGG	Arg
A	AUU	Ile	ACU	Thr	AAU	Asn	AGU	Ser
	AUC	Ile	ACC	Thr	AAC	Asn	AGC	Ser
	AUA	Ile	ACA	Thr	AAA	Lys	AGA	Arg
	AUG[b]	Met	ACG	Thr	AAG	Lys	AGG	Arg
G	GUU	Val	GCU	Ala	GAU	Asp	GGU	Gly
	GUC	Val	GCC	Ala	GAC	Asp	GGC	Gly
	GUA	Val	GCA	Ala	GAA	Glu	GGA	Gly
	GUG[b]	Val	GCG	Ala	GAG	Glu	GGG	Gly

[a]Termination codon.
[b]Initiation codon. With rare exceptions, AUG is the sole initiation codon in eukaryotes. In some prokaryotes, GUG and UUG are occasionally used for initiation.

constitutes the translation (decoding, reading) of that codon. Amino acids bound to tRNAs whose anticodons are paired with adjacent codons in mRNA are joined through a peptide bond during polypeptide assembly (Figure 18.1).

Role of Ribosomes

Ribosomes help direct and stabilize the interactions between tRNAs and mRNAs, and they provide the catalyst for peptide bond formation. Prokaryotic ribosomes encompass three rRNAs and around 55 proteins (Table 16.8), whereas a typical eu-

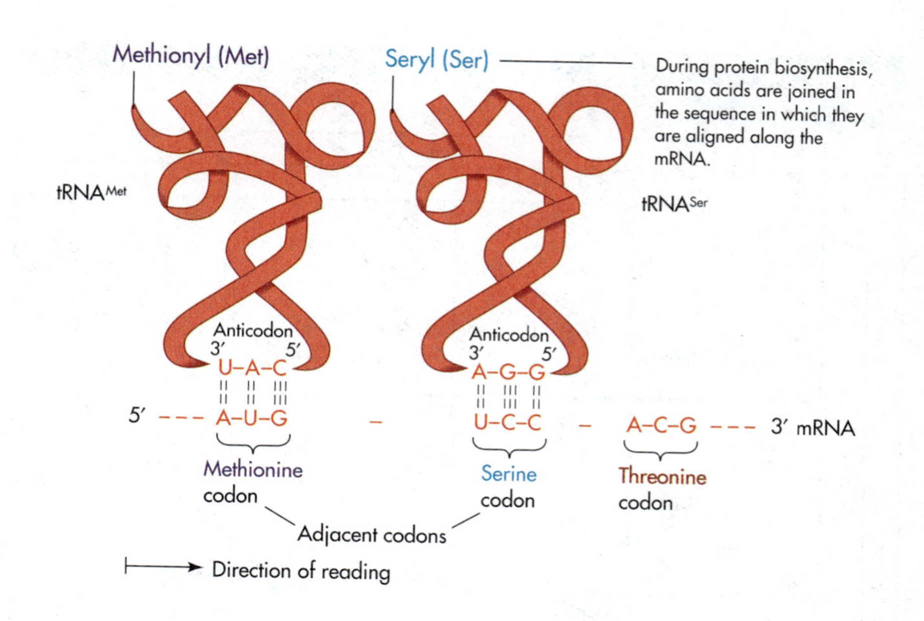

FIGURE 18.1
Codon decoding.

karyotic ribosome contains four rRNAs and roughly 80 proteins (Table 16.9). A prokaryotic ribosome dissociates reversibly into a 30S and a 50S subunit, a eukaryotic ribosome into a 40S and a 60S subunit (Figure 18.2).

Some General Considerations

Translation, like the biosynthesis of RNA, can be divided into three major stages: initiation, elongation, and termination. During the initiation stage, a ribosome binds to an mRNA in the region of the initiation codon and helps direct the binding of an aminoacylated initiator tRNA to the initiation codon. The elongation stage begins when an aminoacyl-tRNA pairs with the codon on the 3′ side of the initiation codon and the ribosome catalyzes the transfer of the amino acid on the initiator tRNA to the amino acid on the adjacent tRNA and the formation of a peptide bond. Elongation continues as the ribosome moves along the mRNA one codon at a time in a ratchet-like fashion, catalyzing the transfer of one amino acid residue at a time from a tRNA (paired with a codon) to the growing polypeptide. Termination entails the re-

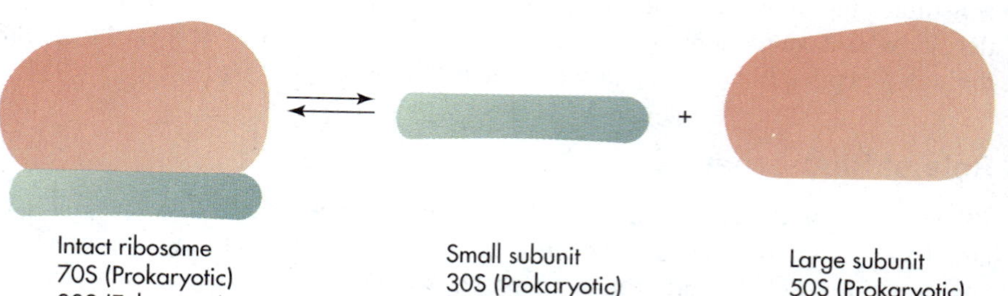

FIGURE 18.2
Dissociation of ribosomes.

lease of the completed polypeptide and the disassembly of the translational machinery. Accessory proteins participate at each stage.

Translation is very similar in both prokaryotes and eukaryotes; only the details differ. In this chapter, emphasis is placed on polypeptide synthesis as it occurs in *Escherichia coli,* the organism in which it has been most thoroughly studied. In the process, some of the distinguishing features of eukaryotic translation are identified.

18.2 INITIATION OF TRANSLATION

In *E. coli,* a unique **initiator tRNA ($tRNA_f^{Met}$),** a ribosome, an mRNA, GTP, three protein **initiation factors** (**IF1, IF2,** and **IF3**) and Mg^{2+} are all required to initiate translation. **[Try Problems 18.1 through 18.3]**

Initiator tRNAs

All initiator tRNAs, both prokaryotic and eukaryotic, are joined to the amino acid methionine before their participation in translation. The attachment entails the ATP-dependent, enzyme-catalyzed formation of an ester link between the carboxyl group in methionine and a hydroxyl group in the ribose residue at the 3′ end (—CCA end) of the initiator tRNA (Figure 18.3). The attached amino acid is activated, and the tRNA is said to be charged (not related to electrical charges) with the amino acid (Section 16.6). In prokaryotic systems, mitochondria, and chloroplasts, the preparation of the initiator tRNA for translation entails the additional enzyme-catalyzed placement of a formyl group (HCO—) on the methionine residue to generate **formylmethionyl-$tRNA_f^{Met}$ (fMet-$tRNA_f^{Met}$)**. Eukaryotic initiator tRNA is abbreviated $tRNA_i^{Met}$, rather than $tRNA_f^{Met}$, since the methionine residue that it carries is not formylated. **[Try Problem 18.4]**

Initiation Complex Formation in *E. coli*

Escherichia coli 70S ribosomes exist in equilibrium with a pool of 30S and 50S ribosomal subunits. *Step 1* in the initiation of polypeptide synthesis consists of the binding of IF3 to a 30S subunit in such a manner that this subunit no longer binds a 50S subunit (Figure 18.4). This reaction pulls the 30S subunit out of its equilibrium with 70S ribosomes and leads to the dissociation of additional ribosomes (Le Châtelier's principle). The 30S:IF3 complex subsequently attaches to a **ribosome binding site** on an mRNA *(step 2).* This binding site encompasses two key structural features: an **initiation codon** (usually **AUG,** but sometimes **GUG** or **UUG** in prokaryotes); and a **Shine–Dalgarno sequence.** The Shine–Dalgarno sequence base pairs with a complementary segment of the 16S rRNA that is part of the 30S subunit and, in the process, helps secure the mRNA to that subunit (not depicted in Figure 18.4). The pairing also determines which one of the multiple AUG, GUG, and UUG sequences normally found along an mRNA is to function as the initiation codon. The initiation codon is the first codon read during translation, and it sets the frame of reading for the remainder of the translated region of the mRNA (Section 16.5). **[Try Problem 18.5]**

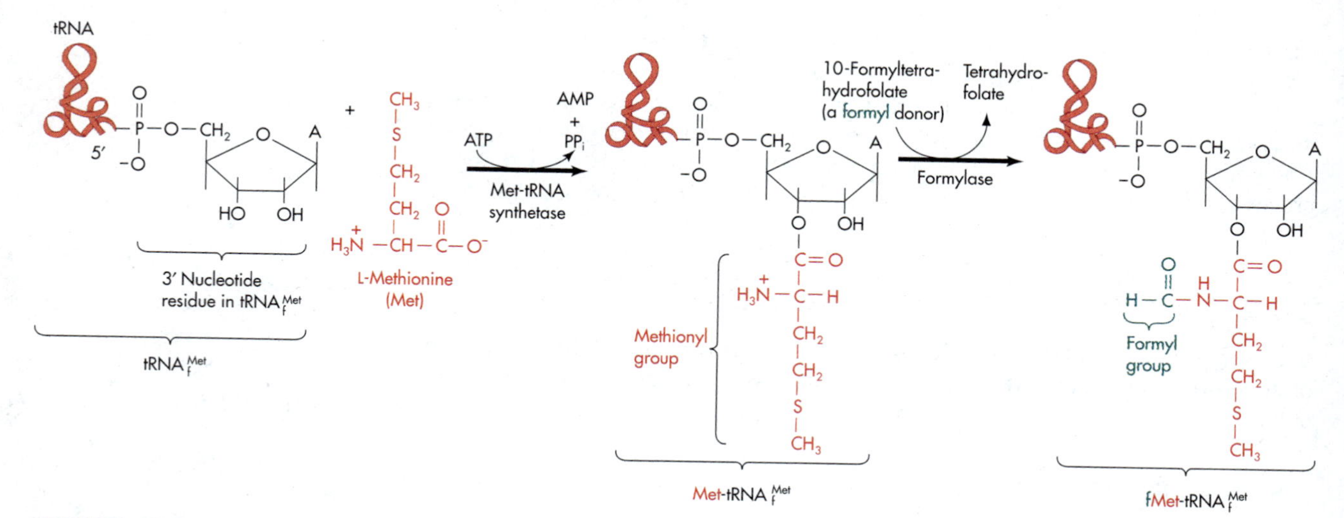

FIGURE 18.3
Preparation of a prokaryotic initiator tRNA for translation.

During *step 3* of initiation, an IF2:GTP complex binds fMet-tRNA$_f^{Met}$ and leads to the formation of a 30S:IF3:mRNA:IF2:GTP:fMet-tRNA$_f^{Met}$ complex. In the process, the anticodon of the charged and formylated initiator tRNA base pairs in an antiparallel manner with the initiation codon on the mRNA:

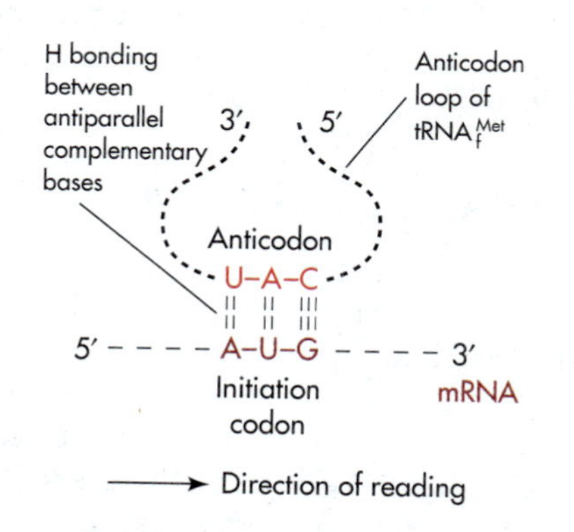

The final step in initiation *(step 4)* includes the following events: A 50S subunit joins the 30S:IF3:mRNA:IF2:GTP:fMet-tRNA$_f^{Met}$ complex to form a large preinitiation complex; GTP is hydrolyzed with the catalytic assistance of IF2; and IF3, P_i, and an IF2:GDP complex dissociate from the preinitiation complex. The initiation complex thus formed contains a 70S ribosome, an mRNA, and fMet-tRNA$_f^{Met}$. The third initiation factor, IF1, is known to enhance the activity of IF2 and IF3 *in vitro,* but its precise role and detailed mechanism of action *in vivo* are still uncertain. In the completed initiation complex, the initiation codon (still base paired with the anticodon in fMet-tRNA$_f^{Met}$) in mRNA is positioned at a tRNA binding site on the

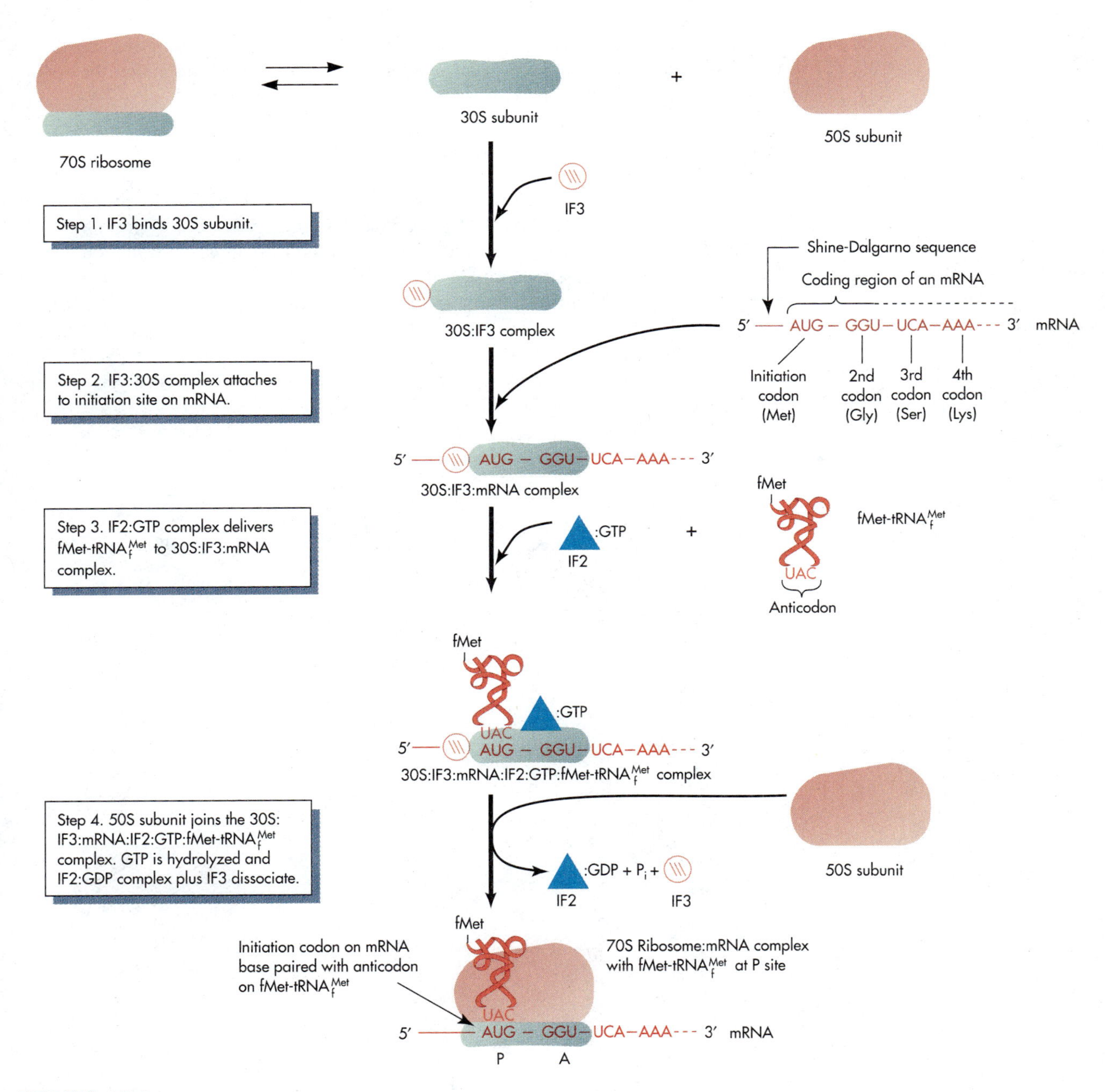

FIGURE 18.4
Initiation of polypeptide synthesis in *E. coli.*

ribosome known as the P (for "peptidyl") site. The second codon in the translated region of the mRNA is located at an adjacent tRNA binding site called the A (for "aminoacyl") site. **[Try Problems 18.6 and 18.7]**

Role of IF2

IF2 is a GTPase, an enzyme that catalyzes the hydrolysis of GTP. The properties of GTPases (also called GTP-binding proteins and, in some instances, G proteins) have been examined in Section 10.4 during the discussion of signal transduction across membranes. Such proteins normally possess other activities or functions in addition to their GTPase activities. IF2, for example, binds fMet-$tRNA_f^{Met}$ during the formation of the initiation complex. The non-GTPase activities depend on the presence of bound GTP. GTPases cycle between three states: GTP bound, "empty," and GDP bound. The cycling is always in the same direction because GTP hydrolysis is, from a thermodynamic standpoint, irreversible: **[Try Problem 18.8]**

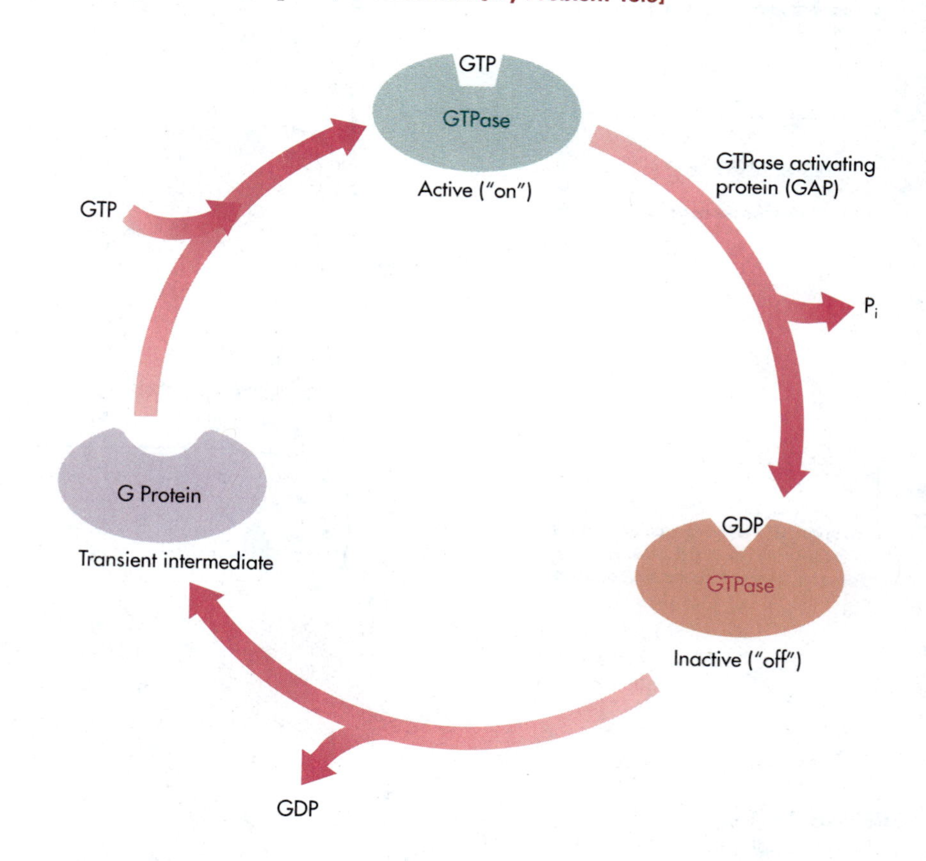

The GTPase cycle functions as a molecular switch with the GTP-bound form representing the "on" or active position (from the standpoint of the associated non-GTPase activity) and the GDP-bound state being the "off" or inactive position. The "empty" state is a transient intermediate. Proteins, called GTPase activating proteins

(GAPs), throw the switch from "on" to "off" by triggering the GTPase-catalyzed hydrolysis of GTP. A properly constructed and aligned preinitiation complex is probably the GAP that activates the GTPase activity of IF2 during the last step in the formation of the initiation complex.

IF2 and other GTPases monitor key steps in translation to ensure that these steps are completed correctly before polypeptide biosynthesis continues. Thus, IF2 does not catalyze the hydrolysis of GTP and complete the initiation stage of translation until the 70S ribosome:mRNA:fMet-$tRNA_f^{Met}$ complex has been properly assembled and is ready to participate in the elongation process.

Initiation in Eukaryotes

Although some of the details differ during the initiation of translation in eukaryotes, the initiation process terminates with the formation of an 80S ribosome:mRNA:Met-$tRNA_i^{Met}$ complex that is strictly analogous to the 70S ribosome:mRNA:fMet-$tRNA_f^{Met}$ complex of *E. coli.* Close to a dozen initiation factors (most abbreviated **eIF1, eIF2,** etc., where the "e" stands for "eukaryotic") participate in the formation of this complex, and a GTPase monitors the operation (Table 18.2). In contrast to prokaryotic systems, a 40S subunit:initiation factor complex first binds to the cap at the 5′ end of a eukaryotic mRNA (Section 16.5) and then scans along the messenger to the initiation codon (usually the first AUG encountered).

TABLE 18.2

Some Eukaryotic Initiation Factors

Factor	Possible Functions in Initiation
eIF1	May stabilize binding of initiator tRNA to small ribosomal subunit
eIF2	Forms complex with GTP and initiator tRNA; promotes binding of initiator tRNA to small subunit; monitors binding (a GTPase)
eIF2B	Promotes exchange of GTP for GDP associated with eIF2 to regenerate eIF2:GTP complex
eIF3	Promotes dissociation of ribosomal subunits; stabilizes ribosomal subunits in dissociated form
eIF4A	Binds to 5′ untranslated segment of mRNA and promotes its melting (denaturation)
eIF4B	May promote unwinding function of eIF4A
eIF4C	Promotes dissociation of ribosomal subunits; stabilizes small subunit in dissociated form
eIF4E	Recognizes and binds 5′ cap of mRNA; promotes binding of mRNA to small ribosomal subunit
eIF4F	Composite factor consisting of eIF4A, eIF4E, and p220
eIF5	Promotes binding of large ribosomal subunit to complete initiation
eIF6	May stabilize large ribosomal subunit in dissociated form
p220	May align eIF4E on 5′ cap of mRNA during binding

Modified from Wolfe, 1993.

PROBLEM 18.1 List the rRNAs within the 50S and 30S subunits of an *E. coli* ribosome. Are these rRNAs bonded covalently or noncovalently to the proteins within the ribosome? Hint: Review Section 16.7.

PROBLEM 18.2 What structural features do all tRNA molecules have in common? What term is used to describe the secondary structure of tRNAs? The tertiary structure of tRNAs? Hint: Review Section 16.6.

PROBLEM 18.3 List the similarities and differences between prokaryotic and eukaryotic mRNAs. Hint: Review Section 16.5.

PROBLEM 18.4 What specific class of chemical bond joins the formyl group to the methionine residue in fMet-$tRNA_f^{Met}$ (Figure 18.3)?

PROBLEM 18.5 In a reaction mixture where 70S ribosomes exist in equilibrium with 50S and 30S subunits, will the concentration of 70S ribosomes increase, decrease, or remain the same when (a) more 50S subunits are added to the reaction mixture and (b) some 50S subunits are removed from the reaction mixture? Explain.

PROBLEM 18.6 What specific bonds or interactions hold the initiation codon and its anticodon together within a 70S ribosome:mRNA:fMet-$tRNA_f^{Met}$ complex? Are these forces covalent or noncovalent?

PROBLEM 18.7 It is important to recognize that Figure 18.4 does not accurately represent the relative sizes or the actual shapes of the participants in polypeptide synthesis. In each of the following sets, circle the particle with the largest mass and put a rectangle around the particle with the smallest mass:

Set A: 70S ribosome, tRNA, GTP, 23S rRNA
Set B: IF2 (97,000 daltons), 50S subunit, 30S subunit, tRNA
Set C: Methionine, mRNA, IF3 (22,000 daltons), inorganic phosphate (P_i)

Assume that there are approximately 300 daltons per nucleotide residue. The sizes of tRNAs, mRNAs, and ribosomes are examined in Chapter 16. Mass is roughly proportional to size for most organic compounds.

PROBLEM 18.8 Use structural formulas to write the equation for the hydrolysis of GTP to yield GDP and inorganic phosphate. Why is this reaction virtually irreversible? Hint: Review Sections 8.2 and 8.4.

18.3 POLYPEPTIDE ELONGATION

The elongation stage of translation begins as soon as the initiation complex has been assembled. In *E. coli,* there are three protein **elongation factors** that participate in a three-step cycle that forms the first peptide bond and then adds one amino acid residue at a time to the growing polypeptide chain. Two of the elongation factors, EF-Tu ("Tu" denotes "temperature unstable") and EF-G (translocase), are GTPases and the third, EF-Ts ("Ts" for "temperature stable"), is involved in the cycling of EF-Tu between inactive and active forms.

Roles of EF-Tu and EF-Ts

During the first step in elongation, EF-Tu:GTP complexes deliver aminoacyl-tRNAs (aa-tRNAs) to the A site on the 70S:mRNA:fMet-$tRNA_f^{Met}$ initiation complex (Figure 18.5). An EF-Tu:GTP complex carrying an aminoacyl-tRNA whose anticodon is complementary to the resident codon at the A site attaches (attachment is temporary and not shown in Figure 18.5) to this site. Most of the exceptions to this generalization are explained by the **wobble hypothesis** (Section 18.5). According to this hypothesis, a "wobble" in the 5′ base of some anticodons allows these anticodons to pair with any one of multiple codons, including some that are not entirely complementary. The base pairing between the anticodon of the incoming aminoacyl-tRNA and the A site codon represents the translation or decoding of this codon, since the tRNA that binds always carries (if no mistakes have been made) that amino acid

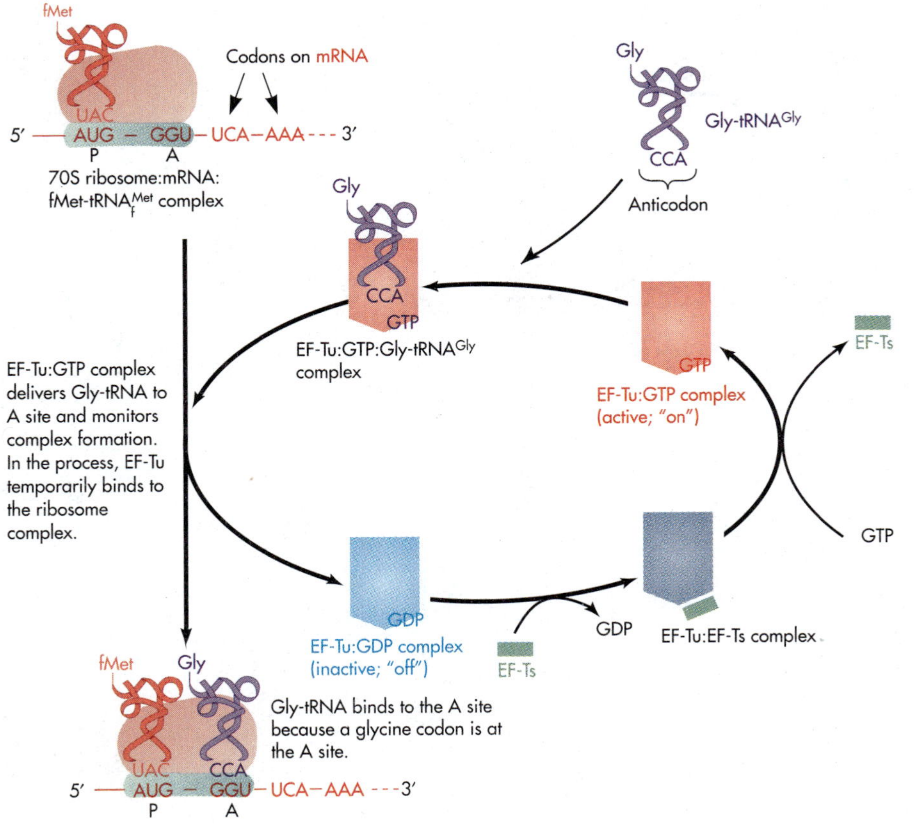

FIGURE 18.5
Roles of EF-Tu and EF-Ts in polypeptide elongation in *E. coli.* The EF-Tu:GTP complex is converted to an EF-Tu:GDP complex after the delivered glycyl-tRNA is properly positioned on the A site of the ribosome. EF-Ts helps replace the GDP bound to EF-Tu with GTP and, in the process, returns EF-Tu to its active ("on"; GTP-bound) state.

specified by the codon (Section 16.6). The 16S rRNA of the 30S subunit plays a central role in the decoding process; its decoding region binds to both mRNA and tRNA and helps stabilize the codon:anticodon duplexes at both the A and P sites. The GTP in the aminoacyl-tRNA:EF-Tu:GTP complex must be hydrolyzed with the catalytic assistance of EF-Tu before the elongation process can continue. **[Try Problems 18.9 and 18.10]**

The GTPase cycle associated with EF-Tu slows down elongation to allow mismatched anticodons to dissociate from codons before peptide bond formation. This GTPase-monitored delay, which is described as **kinetic proofreading,** increases the fidelity of polypeptide assembly up to 100-fold. The EF-Tu:GDP complex and inorganic phosphate (P_i) produced when GTP is hydrolyzed are released from the translational complex. EF-Ts resets the EF-Tu–linked molecular switch to the "on" position by triggering the replacement of the GDP in the EF-Tu:GDP complex with GTP. The resultant EF-Tu:GTP complex is ready to pick up another aminoacyl-tRNA and deliver it to a translating ribosome. **[Try Problem 18.11]**

Peptide Bond Formation

The second step in elongation consists of the transfer of the formylmethionine residue from the initiator tRNA at the P site on the ribosome to the amino acid on

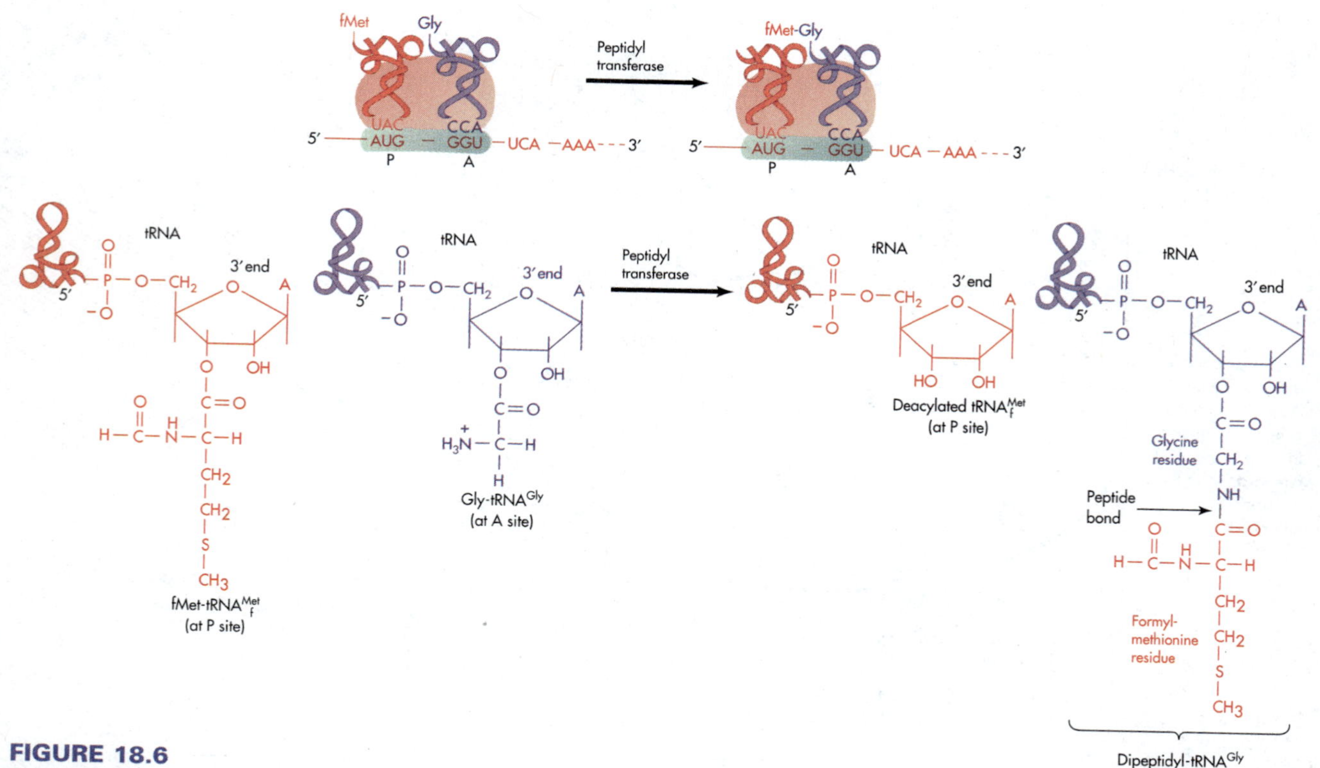

FIGURE 18.6
Peptidyl transferase–catalyzed reaction.

the tRNA at the A site and the formation of a peptide bond (Figure 18.6). The coupled cleavage of the ester link between the formylmethionyl and the initiator tRNA makes peptide bond formation thermodynamically favorable (leads to a negative $\Delta G^{\circ\prime}$). This explains why an amino acid attached to a tRNA is said to be activated (Section 16.6). The equilibrium constant for the joining of two free amino acids to form a dipeptide is relatively small, a fact emphasized in Section 3.5. **[Try Problem 18.12]**

Peptide bond formation is catalyzed by **peptidyl transferase,** an enzyme domain (independently folded region; Section 4.4) within the 50S subunit of the ribosome. There is now convincing evidence that this domain is constructed from 23S rRNA and is, therefore, a ribozyme (an RNA catalyst; see introduction to Chapter 5 and Section 17.2). After the peptidyl transferase–catalyzed reaction, the P site is occupied by a deacylated (empty) tRNA and the tRNA at the A site carries a dipeptidyl group (N-formylmethionylglycyl in the example in Figure 18.6); the actual assembly of a polypeptide has begun. In *E. coli,* and *N*-terminal end of a growing polypeptide chain initially contains a formylated methionine residue. This residue is frequently removed during polypeptide processing (Section 18.11). The second codon in the translated region of an mRNA determines which amino acid is attached to the *N*-terminal methionine residue. **[Try Problem 18.13]**

Polypeptide Synthesis

Elongation continues through an EF-G–mediated process known as **translocation.** During translocation, the ribosome shifts on the mRNA, causing the codon:dipeptidyl-tRNA complex at the A site of the ribosome to shift to the P site (Figure 18.7). The deacylated $tRNA_f^{Met}$ that was bound to the P site is simultaneously displaced from the ribosome and a new codon (the third codon within the coding region of the mRNA) is positioned at the A site. Translocation is triggered when EF-G (a GTPase) catalyzes the hydrolysis of the GTP within an EF-G:GTP complex. The energy released in the process induces the conformational changes required to activate the translocational domains of the ribosome. Displaced tRNAs are rapidly recharged (reacylated with appropriate amino acids) and can be reused. **[Try Problem 18.14]**

The second round of elongation begins when an aminoacyl-tRNA associated with an EF-Tu:GTP complex in the vicinity of the A site binds to the new codon at the A site. Correct codon–anticodon pairing leads to the binding of a tRNA that carries the amino acid specified by the new resident codon (a serine codon in Figure 18.7). Peptidyl transferase catalyzes the transfer of the dipeptidyl group in the peptidyl-tRNA at the P site to the aminoacyl group esterified to the tRNA at the A site; a second peptide bond is formed. Another round of translocation shifts the resultant tripeptidyl-tRNA (fMet-Gly-Ser-$tRNA^{Ser}$ in Figure 18.7) and its paired codon from the A site to the P site and brings the fourth codon in the translated region of the mRNA to the A site. An EF-Tu:GTP complex delivers the appropriate aminoacyl-tRNA to the new codon at the A site, and the cycles of elongation continue until a final round of translocation positions any one of the three **termination codons** (UAA, UAG, and UGA) on the A site. Elongation normally stops at this point, because most translational systems lack a tRNA capable of binding to a termination codon. **[Try Problems 18.15 and 18.16]**

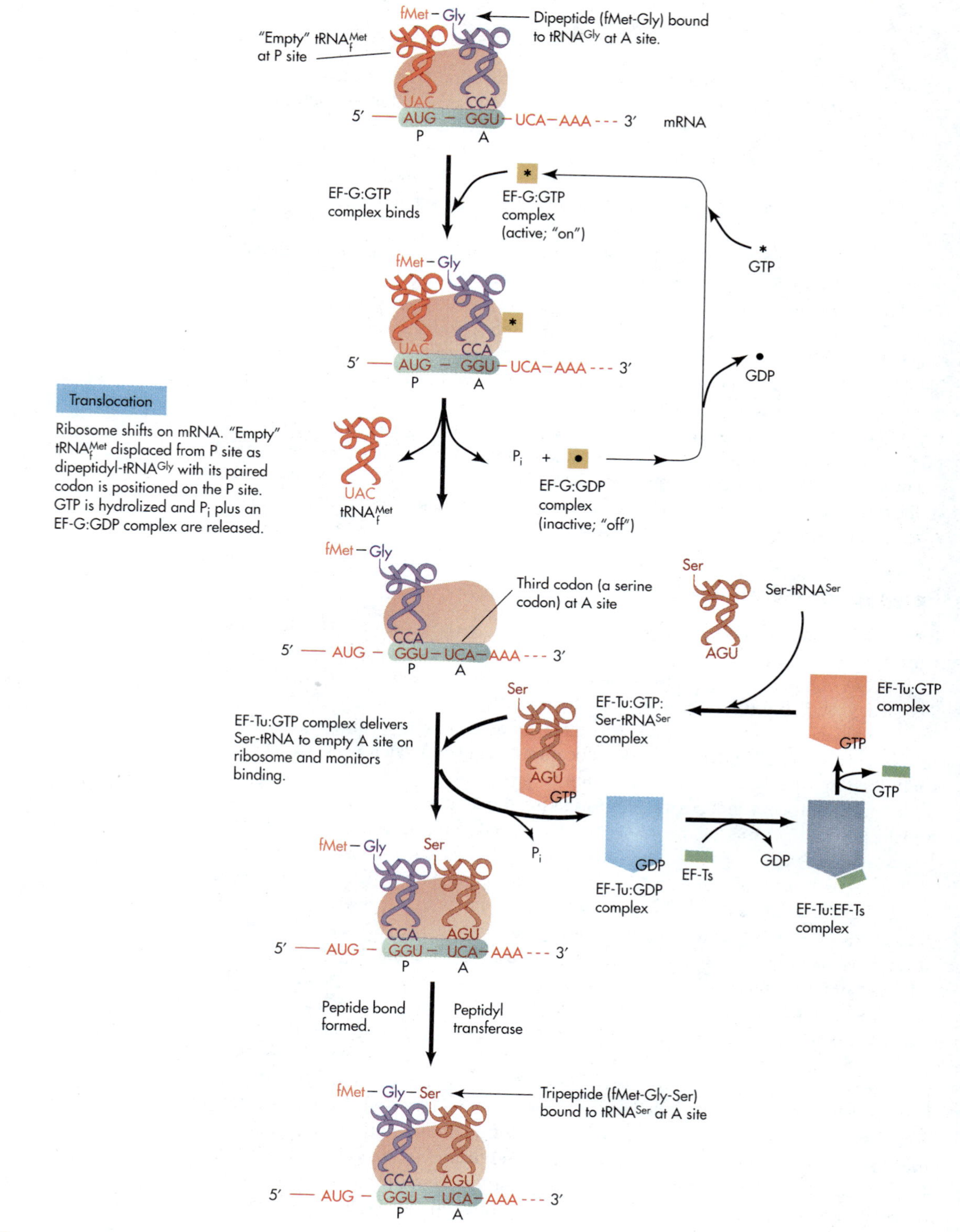

FIGURE 18.7
Translocation and the production of a tripeptide in *E. coli*.

Amino acids are joined at a rate up to about 18 per second during translation in *E. coli*. This phenomenal feat is made possible by a relatively high concentration of EF-Tu, the most abundant protein within an *E. coli* cell. The concentration of EF-Tu is approximately the same as the intracellular concentration of tRNA.

Two elongation factors, eEF1 and eEF2, have been identified in eukaryotic cells. The former is analogous to EF-Tu (or possibly EF-Tu plus EF-Ts) and the latter is analogous to EF-G. Thus, GTPase switches are coupled to aminoacyl-tRNA binding and to translocation in both eukaryotes and prokaryotes.

PROBLEM 18.9 If the second codon in the coding region of an mRNA is CCA, what amino acid is attached to the tRNA at the A site in Figure 18.5? Explain.

PROBLEM 18.10 Accurate translation of the genetic code requires that any tRNA that binds to a codon accept and carry only that amino acid specified by the codon. What specific molecules are responsible for the attachment of the proper amino acid to the proper tRNA? What properties or characteristics of these molecules make this process relatively error free? Hint: Review Section 16.6.

PROBLEM 18.11 What is the most likely GTPase activating protein (GAP) for EF-Tu? Explain.

PROBLEM 18.12 What compound provides energy for the aminoacyl-tRNA synthetase–catalyzed attachment of an amino acid to a tRNA? Indirectly, this compound provides the energy that drives peptide bond formation. Hint: Review Section 16.6.

PROBLEM 18.13 Identify each ester bond (including those in phosphate diesters), each amide bond, and each peptide bond in the structural formulas in Figure 18.6.

PROBLEM 18.14 How many different RNA molecules and how many different protein molecules are part of the giant ribonucleoprotein complex shown as the final product in Figure 18.7?

PROBLEM 18.15 In the absence of wobble, what is the minimum number of different tRNA molecules required to decode an mRNA whose coding region contains at least one copy of each of the translatable codons? Explain.

PROBLEM 18.16 A tRNA molecule with the anticodon AAA is mistakenly charged with the amino acid leucine. What polypeptide fragment will be produced if this mischarged tRNA participates in the translation of the following mRNA fragment?

5′-UGUCGUAGAUUUGAAACG-3′

Assume that the reading frame makes the first three nucleotide residues in this mRNA fragment an intact codon within the coding region of the mRNA. Clearly identify the N-terminal and the C-terminal end of the polypeptide fragment.

18.4 TERMINATION OF TRANSLATION

Three protein-**release factors** (**RF1, RF2,** and **RF3**) direct the termination process in *E. coli* (Figure 18.8). RF3 (another GTPase) binds GTP and stimulates the binding of RF1 or RF2 to the termination codon that is positioned at the A site during the last step in elongation. RF1 and RF2 are said to be codon specific because RF1 recognizes the codons UAA and UAG, whereas RF2 binds UGA and UAA. Once bound to a termination codon, a release factor leads to the hydrolysis of the ester link that joins the completed polypeptide to the tRNA on the P site and the release of the polypeptide and the deacylated tRNA from the ribosome. After the hydrolysis of GTP expels the release factors, the remaining mRNA:ribosome complex spontaneously dissociates to yield an mRNA plus a 70S ribosome. In this case, GTPase-linked monitoring ensures that the newly synthesized polypeptide is freed from tRNA before the translational complex falls apart. Ribosomes, mRNAs, tRNAs and release factors can all be recycled. [Try Problem 18.17]

The overall termination process is very similar in both prokaryotic and eukaryotic systems. Most eukaryotic systems, however, appear to contain only a single release factor (eRF), which catalyzes GTP hydrolysis and recognizes all of the termination codons. In the next section we examine the wobble hypothesis, a hypothesis introduced in Section 18.3 that is applicable to both prokaryotes and eukaryotes.

PROBLEM 18.17 What polypeptide fragment is produced during the translation of the following mRNA fragment? Assume that the first three nucleotide residues in the mRNA fragment represent an intact codon.

5′-AGACGUCGGAGUUAACCCUCC-3′

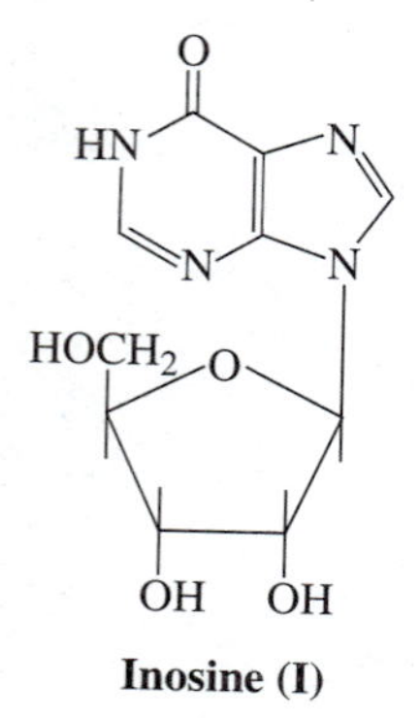

Inosine (I)

18.5 THE WOBBLE HYPOTHESIS

Although a given purine or pyrimidine will normally pair with a single unique partner, select bases at the 5′ position of an anticodon are able to "wobble" (change their orientations) in such a fashion that they can H bond with either standard partners or select nonstandard partners. This proposal, known as the **wobble hypothesis,** explains how it is possible for a single tRNA to decode (translate, bind to) more than one codon. G at the 5′ position of an anticodon can pair with either C or U at the 3′ position of a codon (Figure 18.9). Similarly, U can pair with A or G, and I (inosine; a minor nucleoside) can pair with U, C or A (Table 18.3).

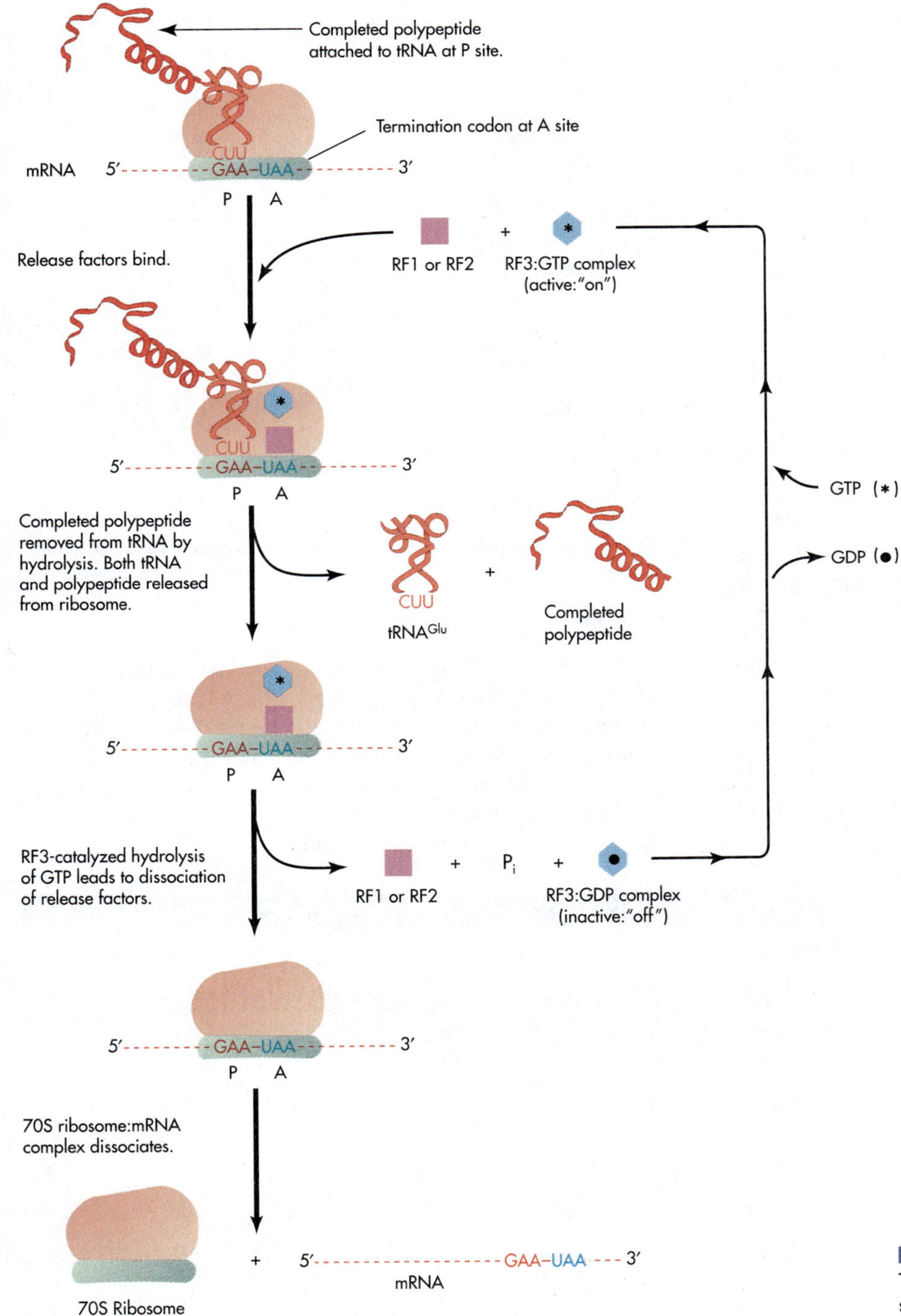

FIGURE 18.8
Termination of polypeptide synthesis in *E. coli.*

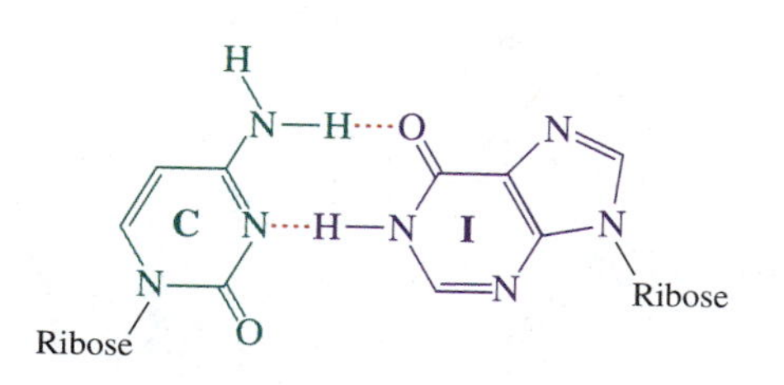

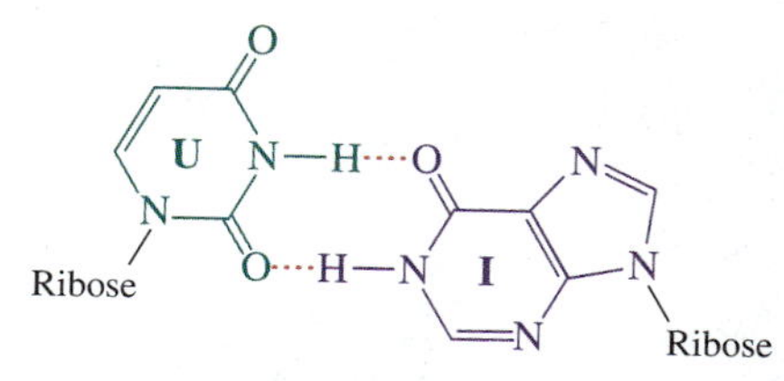

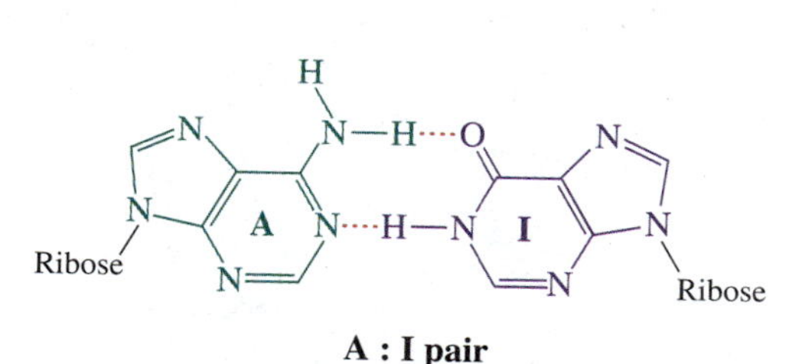

A : I pair

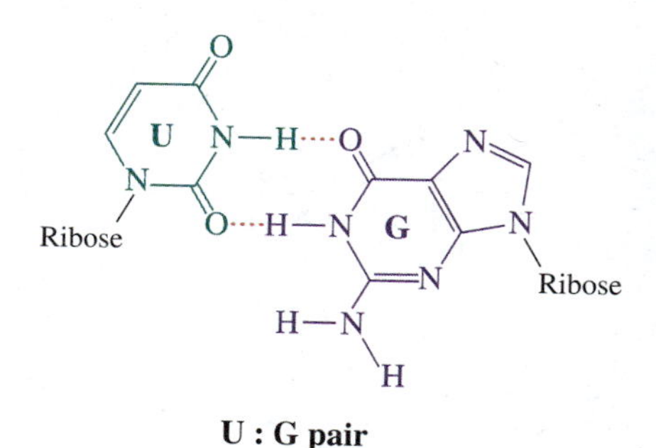

FIGURE 18.9
Wobble-induced base pairing.

An arginyl-tRNA with the anticodon ICG, for example, can attach to three distinct arginine codons: CGU, CGC, and CGA. When pondering this claim, recall that, in an oligonucleotide abbreviation, the 5′ end is to the left and the 3′ end to the right unless specified otherwise. The "wobbling" of A or C does not lead to the acquisition of alternative partners. In those instances where a single tRNA can pair with multiple codons, all the codons specify the same amino acid, that amino acid normally carried by the tRNA involved. **[Try Problem 10.18]**

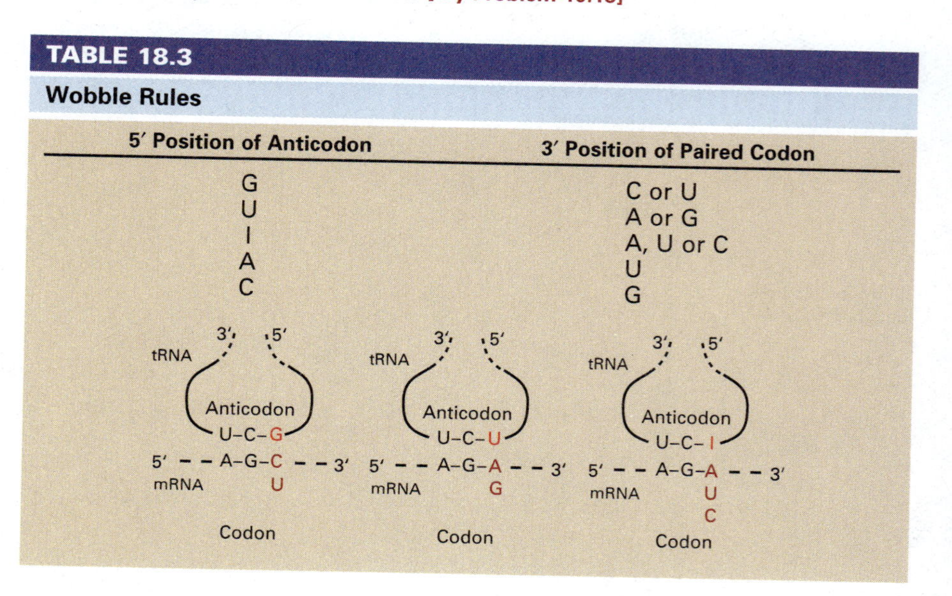

TABLE 18.3
Wobble Rules

5′ Position of Anticodon	3′ Position of Paired Codon
G	C or U
U	A or G
I	A, U or C
A	U
C	G

"Wobble" reduces the minimum number of tRNAs required to read the 61 translated codons (the three termination codons are not translated) from 61 to 32. By reducing the number of tRNAs required to express its genetic information, a cell can save energy (it takes energy to assemble tRNAs) and reduce its genome size (different genes are usually required to encode unique tRNAs). Lowering the number of tRNAs also lessens the time required to match a codon with an appropriate anticodon during translation. Unique ribosomes in mitochondria alter the traditional wobble rules and allow the genetic code in mitochondrial mRNAs to be decoded with 24 tRNAs.

PROBLEM 18.18 Which codons can be read (decoded) by each of the following anticodons:

a. GUG
b. UAA
c. IGU

What amino acid is carried by the tRNA associated with each of these anticodons?

18.6 RECODING THE GENETIC CODE

The translational machinery of a cell normally begins the decoding process with an initiation codon and then moves along the coding region of an mRNA reading each consecutive codon until a termination codon is reached. The initiation codon sets a frame of reading that is maintained throughout the entire process. Once again, we find exceptions to the general rule.

There are now documented cases of frame shifting, ribosome hopping, and the read-through (termination codon is read and translation continues past the normal termination site) of termination codons. In one extreme case involving the translation of a viral mRNA, a ribosome translates part of the messenger and then hops over 50 nucleotide residues before it resumes the decoding process. Such relatively rare anomalies are thought to result from the **recoding** of the genetic code by the mRNA that is being translated. That is, the mRNA appears to carry information that leads to the ribosomal gymnastics. **[Try Problem 18.19]**

PROBLEM 18.19 What peptide is encoded in the following mRNA fragment if, after the first two codons are read (assume the first three nucleotide residues represent an intact codon), the frame of reading shifts downstream (toward the 3′ end) one position:

5′ AUGCGUCCGACCCGUCUU 3′

As a consequence of the shift in reading frame, one nucleotide residue never serves as a code letter in a translated codon.

18.7 POLYSOMES

The discussion of recoding and ribosomal gymnastics leads conveniently into the topic of polysomes, another ribosome-linked phenomenon. After a ribosome has participated in the initiation of polypeptide synthesis at an initiation site on mRNA, it moves along the mRNA in a ratchet-like fashion toward the 3′ end of the translated region. Once the ribosome has moved off the initiation site, a second ribosome can bind to this site and simultaneously work its way along the same mRNA behind the first ribosome. Still other ribosomes can bind and move along the same mRNA in a similar manner (Figure 18.10). Complexes in which multiple ribosomes are moving along a single mRNA are called **polysomes.** Because each ribosome in a polysome is in the process of assembling a separate copy of the same polypeptide, many copies of a polypeptide can be produced very rapidly. Prokaryotic polysomes often contain over 50 ribosomes, whereas eukaryotic polysomes normally contain fewer than 12. **[Try Problem 18.20]**

PROBLEM 18.20 What ultimately determines the maximum number of ribosomes that can simultaneously bind to an mRNA molecule?

18.8 COUPLED TRANSCRIPTION AND TRANSLATION

In prokaryotic cells, polysomes commonly encompass segments on mRNA molecules that are still in the process of being synthesized. When the translation of an mRNA begins before the assembly of that mRNA has been completed, transcription and translation are said to be coupled. (Figure 18.11). Once translation is underway, it keeps pace with the continuing production of mRNA since the rate of decoding is approximately the same as the rate of codon synthesis. The coupling of transcription and translation is biologically important because bacterial mRNAs usually have short half-lives (typically a few minutes). Short-lived mRNAs must be quickly and efficiently translated before they are degraded.

The direct coupling of nuclear transcription with translation in eukaryotic cells is physically impossible, because a membrane separates the protein-synthesizing machinery in the cytoplasm from the mRNAs being produced in the nucleus. However, the compartmentation allows a cell to control the rate of gene expression by regulating the rate of transport of mRNAs from the nucleus into the cytoplasm. Mitochondria and chloroplasts produce unique mRNAs from information carried in their own DNAs. Because these mRNAs are translated by organelle-specific ribosomes and tRNAs, the coupling of transcription and translation is theoretically possible. The extent to which such coupling actually occurs is uncertain. **[Try Problem 18.21]**

PROBLEM 18.21 What aspect of mRNA biosynthesis in eukaryotes would hinder the coupling of transcription and translation even if a membrane barrier did not exist between the nucleus and the cytoplasm? Hint: Review Section 17.2.

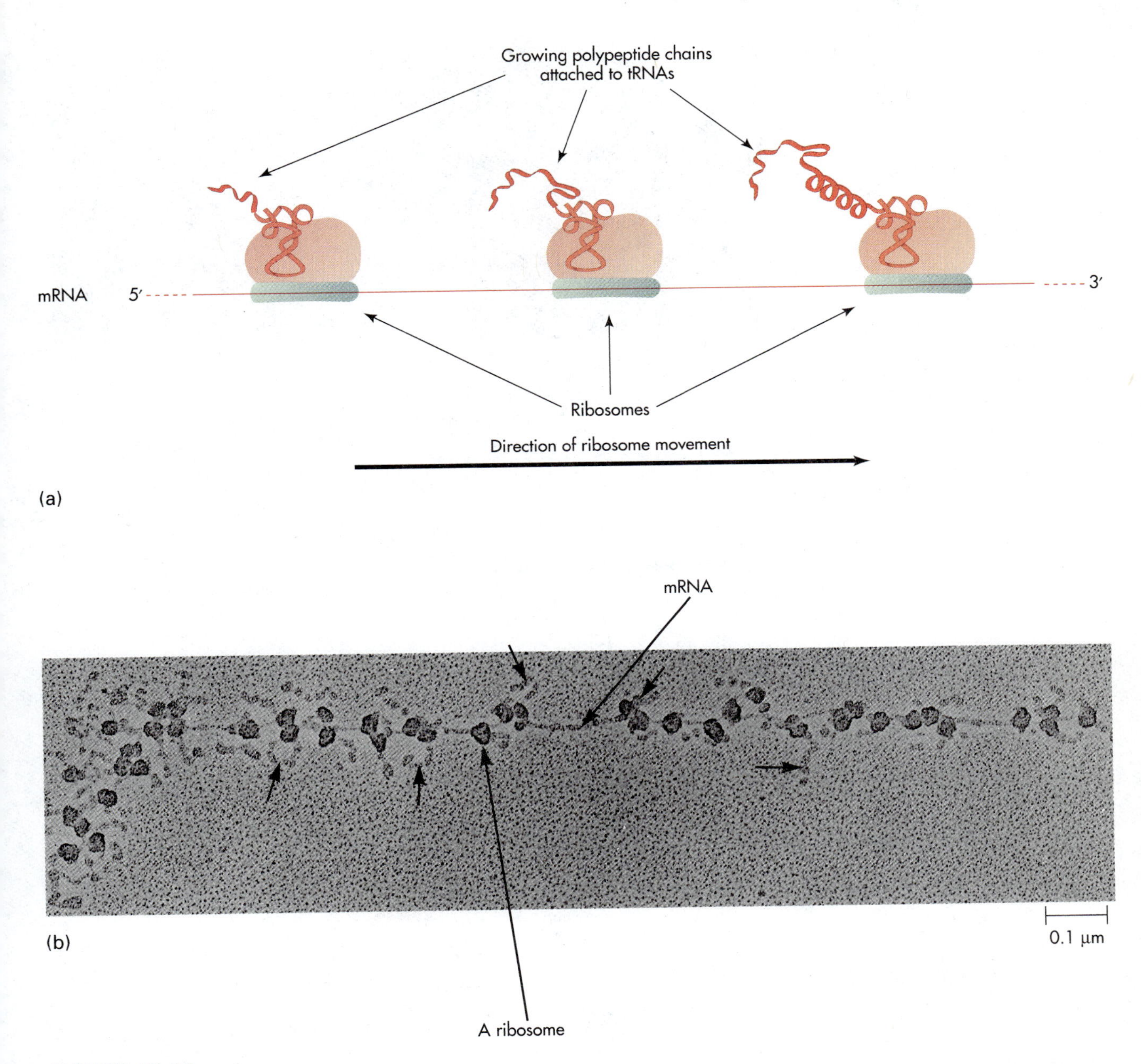

FIGURE 18.10
(a) A diagram of a polysome. (b) An electron micrograph of a polysome. The unlabeled arrows identify growing polypeptide chains.

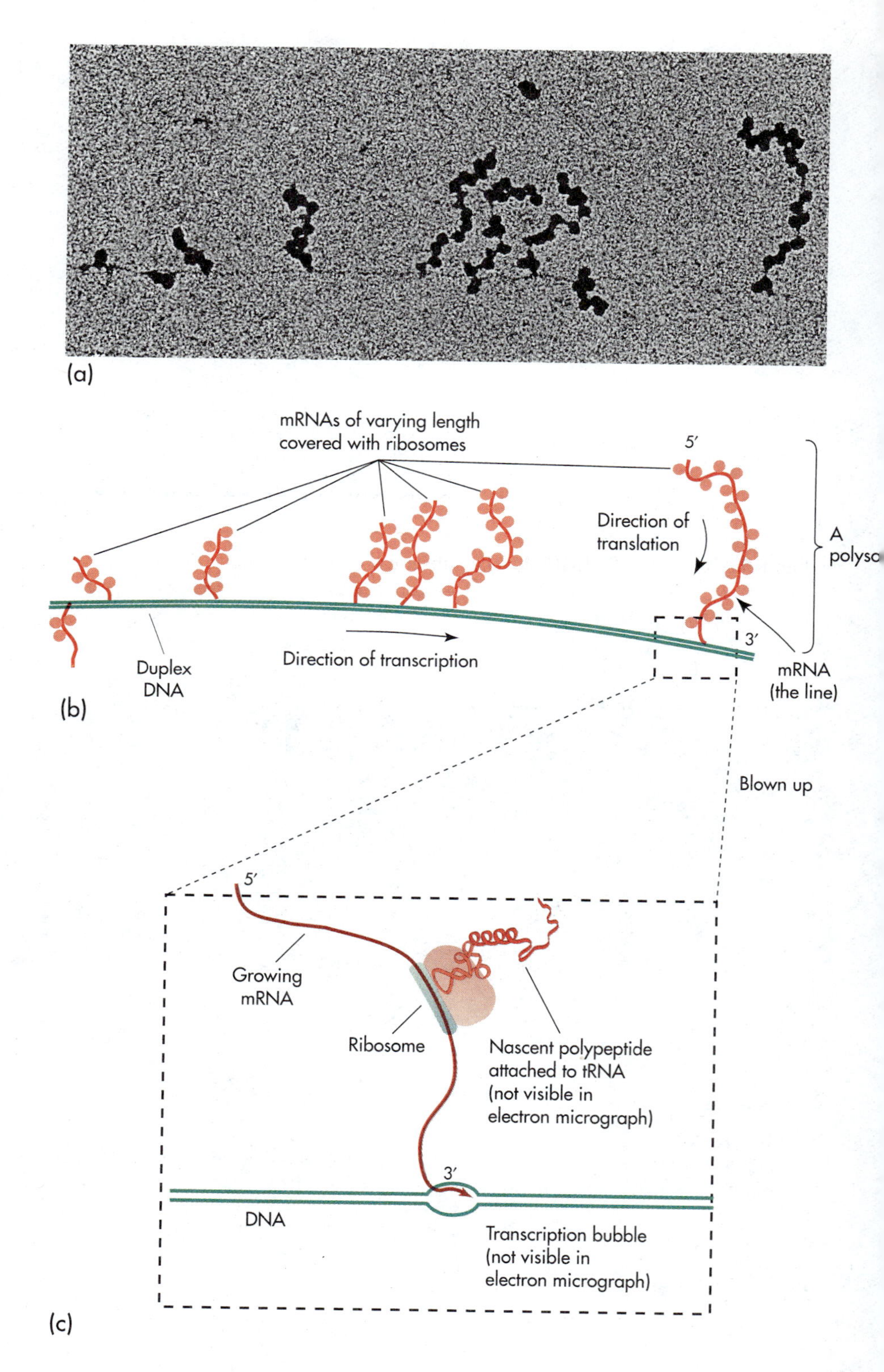

FIGURE 18.11
Coupled transcription and translation in a prokaryotic system. (a) An electron micrograph of coupled transcription and translation in *E. coli.* The nascent polypeptides are not visible. (b), (c) Interpretations of the electron micrograph.

18.9 SIGNAL PEPTIDES

The previous section emphasized the compartmentation of translation in the cytoplasm and transcription in the nucleus. In the cytoplasm itself, the translation of some nucleus-synthesized mRNAs is further compartmentalized; part of the cytoplasmic ribosomes are bound to the endoplasmic reticulum (ER), an extensive membrane network that is classified as an organelle (Section 1.3). The presence of ribosomes distinguishes the rough ER from the smooth ER. Although the ribosomes associated with the rough ER appear to be identical to the those that exist in the cytosol, the two categories of ribosomes are used by cells to synthesize different classes of polypeptides.

The polypeptides synthesized by the ER-bound ribosomes include certain membrane polypeptides, polypeptides destined for secretion, polypeptides used by the ER itself, and some of the polypeptides found within lysosomes, organelles rich in digestive enzymes (Section 1.3). The translation of the mRNAs for these polypeptides actually begins in the cytosol on free ribosomes. After a small segment of polypeptide known as a **signal peptide** has been assembled, the translational complex containing the newly synthesized signal peptide is picked up by a **signal recognition particle (SRP),** which shuts down further elongation of the signal peptide–containing polypeptide (Figure 18.12). SRP is constructed from six proteins and an RNA known as **7SL-RNA.** The subsequent attachment of SRP to one of the many **SRP receptors (docking proteins)** on the cytosolic surface of the ER delivers the polypeptidyl-tRNA : mRNA : ribosome complex to a ribosome receptor on the ER. After a GTPase in the SRP catalyzes the hydrolysis of an associated GTP, the SRP separates from its receptor and the elongation of the polypeptide resumes. The GTPase monitors the binding of the signal peptide and its associated ribosome to the ER and does not allow polypeptide growth to resume until all participants are properly positioned and correctly aligned.

The growing polypeptide is transported across the ER membrane with the assistance of a membrane-embedded **peptide translocation complex** (unrelated to the EF-G–assisted translocation of a peptidyl-tRNA from the A site to the P site on a ribosome during the elongation phase of translation). Although many of the details of the membrane translocation process remain to be unraveled, it appears that a protein conducting channel is involved. The signal peptide is normally removed by a membrane-bound peptidase during translocation. Once completed, the newly synthesized polypeptide is released into the ER. Such polypeptides are commonly subjected to glycosylation (Section 6.7) or other chemical alterations within the ER before their excretion or their ER-directed incorporation into membranes or lysosomes.

18.10 THE INHIBITION OF TRANSLATION BY ANTIBIOTICS

A variety of compounds have been discovered that inhibit polypeptide synthesis in prokaryotes without markedly inhibiting translation in eukaryotic cells. The ribosomes and protein factors that are unique to the simpler life forms usually serve as targets for these inhibitors. Such compounds have proved to be of enormous value

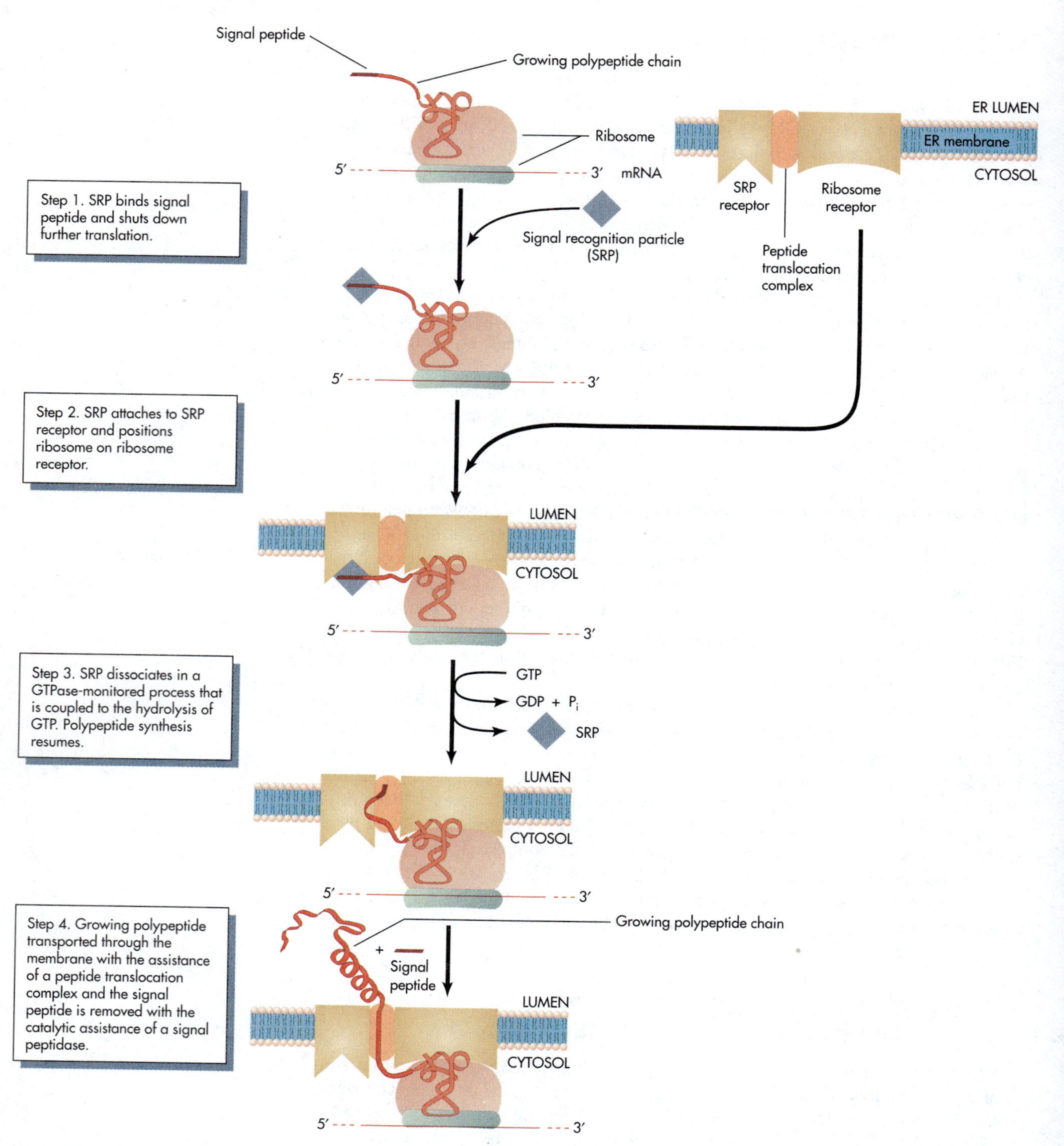

FIGURE 18.12
Synthesis of a polypeptide by the rough endoplasmic reticulum.

TABLE 18.4

Some Antibiotics That Inhibit Prokaryotic Translation

Antibiotic	Primary Mode of Action
Streptomycin	Binds to the 30S subunit of prokaryotic ribosomes. At low concentrations, induces abnormal codon–anticodon interactions and leads to the assembly of abnormal polypeptides. At high concentrations, inhibits the initiation of translation.
Tetracycline	Binds to the 30S subunit of prokaryotic ribosomes and blocks the attachment of aminoacyl-tRNAs.
Chloramphenicol	Binds to the 50S subunit of prokaryotic ribosomes and inhibits peptide bond formation. Also inhibits the peptidyl transferase of mitochondrial ribosomes.
Erythromycin	Binds to the 50S subunit of prokaryotic ribosomes and blocks translocation.

in the treatment of bacterial infections in humans. Many antibiotics, including **streptomycin** (an **aminoglycoside**), **tetracycline, chloramphenicol,** and **erythromycin,** provide specific examples (Table 18.4 and Figure 18.13). Unfortunately, the toxicity of these drugs is not as prokaryote specific as we would like; there are a range of po-

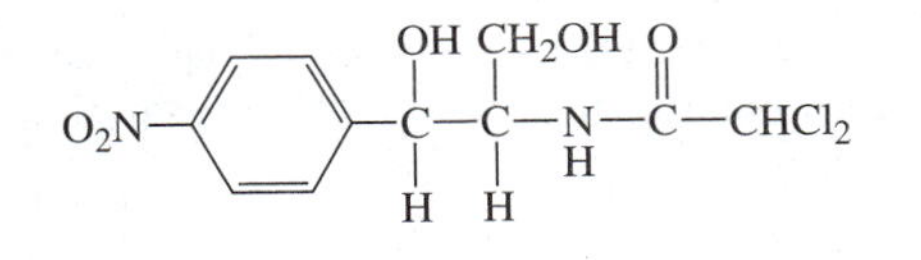

Chloramphenicol

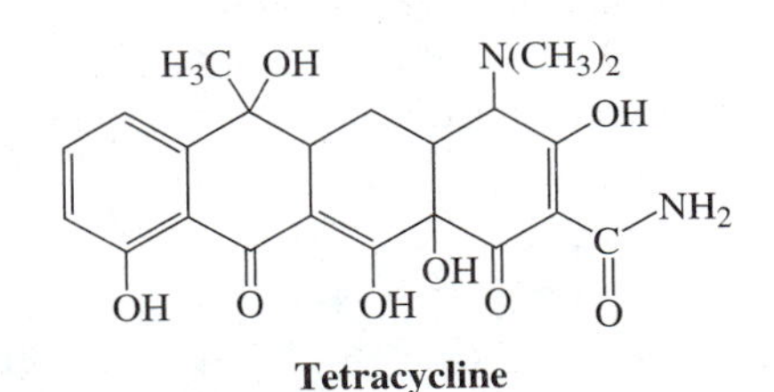

Tetracycline

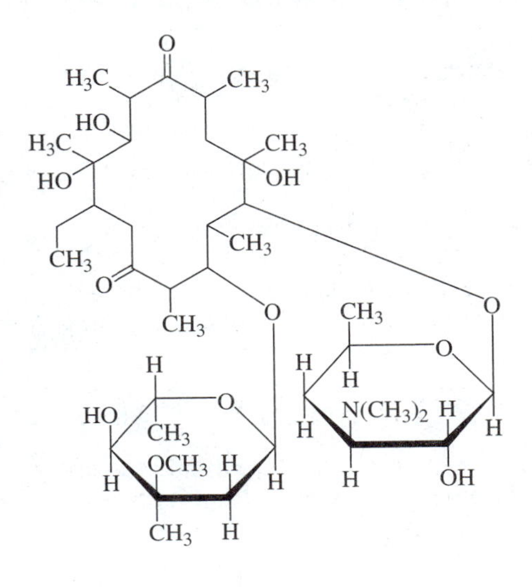

Erythromycin

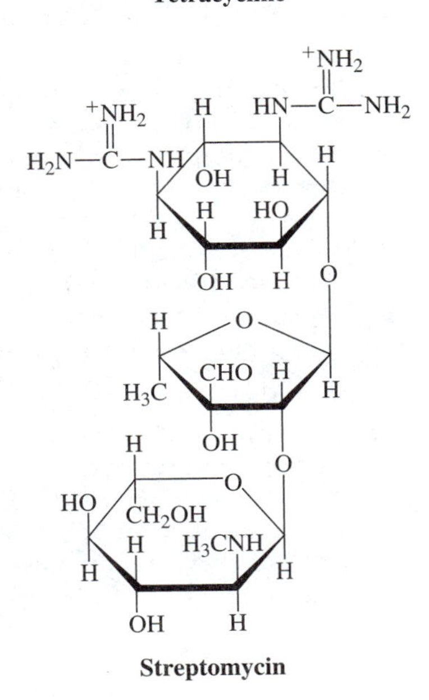

Streptomycin

FIGURE 18.13
Some antibiotics that inhibit prokaryotic translation.

tentially serious side effects. In some instances, the side effects result from the inhibition of mitochondrial polypeptide synthesis. **[Try Problem 18.22]**

PROBLEM 18.22 Some compounds inhibit polypeptide synthesis to the same extent in eukaryotic cells and prokaryotic cells. Are such compounds of value in the treatment of bacterial infections in humans? Explain.

18.11 POLYPEPTIDE PROCESSING

Many of the polypeptides encoded in mRNA lack normal biological activity immediately after they are released from tRNA during the termination stage of translation; they must be processed (chemically or conformationally altered, or both) to become functional. Processing may involve co-translational modifications, post-translational changes, or both. Exhibit 18.1 summarizes some common processing events. The hydrolysis of specific peptide bonds includes the removal of signal peptides (Section 18.9) and the activation of zymogens and other proproteins (Sections 3.6 and 5.14). The importance of prosthetic groups has been emphasized in Chapter 4 in the discussion of hemoglobin and myoglobin and in Chapter 5 during the discussion of coenzyme (organic prosthetic groups within enzymes). Coenzymes are essential participants in a large fraction of metabolism (Chapters 10 through 15).

The roles of chaperones and *cis-trans* isomerases in the generation of native tertiary and quaternary structures have been examined in Section 4.4. The discussion

EXHIBIT 18.1

Some Events Involved In Polypeptide Processing

COMMON

1. The hydrolysis of specific peptide bonds and the release, in some instances, of one or more peptide fragments
2. The addition of a prosthetic group
3. Chaperone-directed formation of native tertiary and quaternary structures
4. The conversion of specific *trans* peptide bonds to *cis* peptide bonds
5. The formation of disulfide bonds
6. The covalent modification of amino acid side chains, including phosphorylation, methylation, acylation, hydroxylation, and glycosylation

UNCOMMON

7. The enzyme-catalyzed conversion of specific L-amino acid residues to D-amino acid residues
8. The removal of an internal polypeptide fragment (an intein) and the covalent joining (through a peptide bond) of the retained fragments (exeins), an event called polypeptide splicing

of the structure of immunoglobulins in Section 4.11 has documented the importance of disulfide bonds in some proteins. Polypeptide acylation is introduced in Section 9.1 under the topic of fatty acids, and the importance of glycosylation is described in Section 6.7 of the carbohydrate chapter. Examples of protein hydroxylation were encountered during the discussion of collagen in Section 4.10.

The control of protein activity through phosphorylation has been introduced in Section 5.13 during the discussion of protein kinases and protein phosphatases. Numerous examples are encountered in Chapters 10 through 15. In mammalian systems, translation itself is partially regulated through the phosphorylation of several protein participants, including aminoacyl-tRNA synthetases, a ribosomal protein, initiation factors, and elongation factors.

Chapter 3 has emphasized the stereospecific nature of oligo- and polypeptides. With rare exceptions, these classes of compounds contain only L-amino acid residues. The basis for some of the exceptions was discovered in 1994 when an enzyme was isolated that catalyzes the conversion of a particular L-residue to a D-residue during the processing of the polypeptide precursor for a peptide in the venom of the funnel web spider *Agelenopsis aperta.* This interconversion has profound implications from the standpoint of protein structure and structure–function relationships. The mature peptide paralyzes spider prey by blocking certain voltage-gated calcium channels within nerve cell membranes. The D form of the peptide **(ω-agatoxin IVB)** is three to five times more active than the L form **(ω-agatoxin IVC).** At the moment, it is uncertain how many organisms possess isomerases capable of catalyzing such D to L interconversions.

Polypeptide splicing is analogous to RNA splicing; certain intervening monomer sequences within a polymer are removed, and the retained polymer sequences are spliced together. In RNAs, the intervening sequences are called introns and the retained sequences are labeled exons (Section 17.2). In polypeptides, the intervening sequences are named **inteins (protein introns)** and the retained sequences **exeins (protein exons).** Polypeptide splicing, like the much more common RNA splicing, has a marked impact on the structure and function of the polypeptide ultimately produced from the information carried in a gene. Inteins can catalyze their own removal and the splicing of the retained exeins. Similarly, some RNA precursors remove their own introns. Protein splicing has been observed in both prokaryotes and eukaryotes, and it may contribute to the pathogenicity of some bacteria.

The processing of a polypeptide normally leads to the alteration of individual amino acid side chains or to variations in the manner in which individual side chains fold together, or both. The net result is often the creation, elimination, or modification of active sites or binding sites. Some processing events are reversible, whereas others are irreversible. By helping to control protein activity and by helping to determine when and where proteins are activated, processing sometimes plays a central role in the regulation of individual metabolic pathways and crucial physiological events. Oncologists (those who study cancer) have even synthesized drugs designed specifically to block the processing of particular polypeptides that play key roles in the creation and maintenance of cancer cells. **[Try Problems 10.23 through 10.25]**

PROBLEM 18.23 What classes of compounds, besides peptides and proteins, are initially synthesized in inactive forms that must be processed? Hint: Review Chapter 17.

PROBLEM 18.24 Which, if any, of the events listed in Exhibit 18.1 can alter the tertiary structure of a polypeptide? Explain.

PROBLEM 18.25 Which of the events in Exhibit 18.1 involve the cleavage or formation of covalent bonds ? Explain.

SUMMARY

Translation refers to those events through which the information in consecutive codons along a translated region of an mRNA is used to direct the assembly of a polypeptide whose amino acid sequence is determined by the codon sequence. The overall process consists of three stages: initiation, elongation, and termination. The fidelity of polypeptide synthesis is enhanced by multiple GTPases (GTP-binding proteins) that function as molecular switches and monitor key steps along the way.

In *E. coli,* initiation involves three initiation factors, IF1, IF2 (a GTPase), and IF3. Collectively, these three proteins bring about the attachment of a 30S ribosome subunit to a ribosome binding site on an mRNA and the formation of a 70S ribosome:mRNA:fMet-$tRNA_f^{Met}$ initiation complex. GTP hydrolysis is coupled to the monitoring of complex formation by IF2. The assembled 70S ribosome has two tRNA binding sites, A and P. In the initiation complex, the anticodon of fMet-$tRNA_f^{Met}$ is bound to the initiation codon at the P site, and the codon to the 3′ side of the initiation codon sits at the A site. Close to a dozen initiation factors lead to the formation of an analogous 80S ribosome:mRNA:Met-$tRNA_i^{Met}$ complex in eukaryotic systems.

The elongation phase in *E. coli* involves two GTPases (EF-Tu and EF-G) and EF-Ts. These three proteins, along with peptidyl transferase, bring about a three-step cycle that adds one amino acid residue at a time to the growing polypeptide chain. Elongation begins when EF-Tu:GTP complexes deliver aminoacyl-tRNAs to the codon at the empty A site on the initiation complex. A tRNA with an appropriate anticodon attaches to the codon. This attachment represents the decoding of the genetic code, since the tRNA which pairs with the codon carries the amino acid specified by that codon. This step is monitored by EF-Tu; EF-Tu–catalyzed GTP hydrolysis is required before translation can proceed.

Peptidyl transferase, a ribozyme embedded in the 50S ribosomal subunit, catalyzes the transfer of a formylmethionyl group from the fMet-$tRNA_f^{Met}$ at the P site to the amino acid bound to the tRNA at the A site; a peptide bond is formed. EF-G–catalyzed hydrolysis of GTP triggers the translocation of the peptidyl-tRNA and its associated codon from the A site to the P site. After another aminoacyl-tRNA attaches to a new codon positioned at the A site, the dipeptidyl group on the tRNA at the P site is joined through a peptide bond to the amino acid attached to the tRNA at the A site. The tripeptidyl-tRNA at the A site is translocated to the P site and the cycle of aminoacyl-tRNA binding, peptide bond formation, and translocation is repeated until the entire coding region of the mRNA has been translated. Analogous reactions occur in eukaryotes.

Translation enters the termination phase when a final round of translocation positions a termination codon at the A site. In *E. coli,* three release factors, RF1, RF2, and RF3 (a GTPase), bring about the hydrolysis of the ester link between the completed polypeptide and the tRNA at the P site and the dissociation of the translational complex.

The wobble hypothesis explains how a tRNA with a single anticodon is able to pair with (decode) multiple codons. Select bases at the 5′ position of an anticodon "wobble" in such a fashion that they can H bond with either standard partners or certain nonstandard ones. Wobble reduces the minimum number of tRNAs required to translate the genetic code from 61 to 32 and it allows translation to proceed more rapidly.

Once the initiation codon sets the frame of reading, each consecutive codon is normally read in sequence until translation halts at the first termination codon encountered. However, information carried in the mRNA being decoded can lead to frame shifting, ribosome hopping, and the read-through of termination codons. In such instances, the mRNA is said to recode the genetic code.

In prokaryotes, transcription and translation are often coupled; the translation of an mRNA begins while it is still being assembled. This phenomenon, along with the ability of multiple ribosomes to translate simultaneously along the same mRNA (leading to polysome formation), allows a prokaryotic cell to produce many copies of the polypeptide encoded within the mRNA rapidly and efficiently. Although polysomes are also common in eukaryotes, the physical separation of transcription and translation makes direct coupling impossible.

In eukaryotic cells, certain classes of polypeptides, including those to be excreted or incorporated in lysosomes, are assembled by ribosomes attached to the ER. The synthesis of these polypeptides is initiated in the cytosol, where free ribosomes participate in the construction of a signal peptide. The signal peptide and the translational machinery that produced it bind to receptors on the cytosolic surface of an ER membrane. Continued translation leads to the translocation of the signal peptide–containing polypeptide chain into the ER. Completed polypeptides are processed within the ER and then distributed to appropriate destinations.

A variety of substances, including tetracycline, streptomycin, chloramphenicol and erythromycin, kill cells by inhibiting translation. Some of the prokaryotic-specific agents are important components in the chemical arsenal used to combat bacterial infections in humans.

Newly synthesized polypeptides frequently lack normal biological activity; they must be chemically or conformationally modified (processed) to become functional. Post-translational modifications commonly entail proteinase cleavage, the acquisition of prosthetic groups, chaperone and enzyme-assisted folding, acylation, hydroxylation, glycosylation and phosphorylation. Other chemical alterations, including splicing, are required in some instances.

EXERCISES

18.1 Describe the role played by rRNA during the attachment of a 30S ribosomal subunit to an mRNA.

18.2 List those protein "factors" that participate in the initiation stage of polypeptide synthesis in *E. coli*. What is the function of each listed factor?

18.3 AUG serves as an internal methionine codon as well as an initiation codon. When AUG is an internal codon, it is translated by a Met-$tRNA^{Met}$ that is distinct from fMet-$tRNA_f^{Met}$. The Met-tRNA that decodes internal AUG codons does not bind to AUG when it is the initiation codon, and fMet-$tRNA_f^{Met}$ does not bind to internal AUG codons. What accounts for this binding specificity?

18.4 Compound X is found to react with 30S subunits to form an X:30S complex. If the dissociation of 70S ribosomes into 30S and 50S subunits is initially at equilibrium, will the addition of compound X to a reaction mixture necessarily lead to the dissociation of more 70S ribosomes? Explain.

18.5 What is the minimum number of 70S ribosomes required to produce 10 identical copies of a polypeptide? Explain.

18.6 Predict what specific classes of bonds hold IF3 and a 30S subunit together within an IF3:30S complex? Explain the basis for your predictions.

18.7 Suggest how the release of an inorganic phosphate from the GTP in a GTPase:GTP complex turns off the non-GTPase function of the GTPase.

18.8 List the GTPases encountered in this chapter and then describe the non-GTPase function of each listed protein.

18.9 How many total GTPs are hydrolyzed for each amino acid added to a growing polypeptide chain during the elongation phase of translation? Explain. How does this expenditure of energy benefit the cell involved?

18.10 Does a polypeptide chain grow from its *N*-terminal toward its *C*-terminal or vice versa? Explain.

18.11 Is there any limit to the number of times a single mRNA molecule can be translated? Explain.

18.12 Is IF3 an enzyme? Explain.

18.13 A processing error leads to the production of a mature mRNA molecule with an intron in its 3′ untranslated region. Will the translation of this abnormal mRNA (assume that it can be translated) lead to the production of the same polypeptide assembled during the translation of the normal mRNA? Explain.

18.14 Draw a diagram similar to the ones in Figure 18.7 clearly showing how the translational complex will appear immediately after the sixth codon in the following mRNA fragment has been shifted to the P site during a round of translocation.

5′—AUGCCAGGGUUGAUAUGCCUUAAA—3′

Assume that the first three nucleotide residues in the given mRNA fragment constitute the initiation codon.

18.15 What specific class of bonds holds the peptide to the tRNA within a peptidyl-tRNA molecule?

18.16 What is accomplished during the translocation step in translation?

18.17 What determines the minimum number of unique tRNA molecules required to synthesize the following polypeptide fragment during translation?

Leu–Val–Glu–Leu–Phe–Leu–Met–Ser–Met

18.18 List all of the participants in translation that can be used repeatedly to produce many copies of a polypeptide.

18.19 Translation is often coupled to transcription in prokaryotes. What is the minimum number of nucleotides that must be joined per second in a growing mRNA chain in order for transcription to keep ahead of a translating ribosome that is moving along the growing mRNA? Explain.

18.20 What specific compounds recognize and bind termination codons during the termination phase of polypeptide synthesis in *E. coli?* What specific events follow the binding of these compounds to termination codons?

18.21 What impact, if any, does the length of the poly-A tail on an mRNA have on the maximum size of the polysome that can be produced during the translation of this mRNA? Explain.

18.22 *In vitro* translational systems are constructed by mixing purified components in test tubes. List all of the compounds that must be mixed to obtain an *E. coli* system that can (with the appropriate mRNAs) synthesize any desired polypeptides from free amino acids.

18.23 A point mutation (change in a single base pair in DNA) leads to the production of an mRNA that contains no termination codon. Can such an mRNA molecule be translated? Explain. If the mRNA molecule is translated, how will the polypeptide synthesized differ from the polypeptide synthesized during the translation of the normal mRNA?

18.24 The expression of most genetic information can be divided into four major processes: transcription, RNA processing, translation, and polypeptide processing. Which one of these processes represents the point at which gene expression is most often regulated in prokaryotes? Eukaryotes? Hint: Review Section 17.1.

18.25 How does a cell benefit from the production of inactive protein precursors?

18.26 Predict what major class of enzyme (Table 5.1) catalyzes the removal of the formyl group from the *N*-terminal methionine residue (a common event) within a newly synthesized prokaryotic polypeptide? In some instances, removal of the formyl group may be a co-translational process.

18.27 What is the role or function of each of the following in the assembly of polypeptides by the endoplasmic reticulum?

a. Signal peptide
b. SRP
c. Peptide translocation complex
d. GTP

SELECTED READINGS

Alberts, B., D. Bray, J. Lewis, M. Raff, K. Roberts, and J. D. Watson, *Molecular Biology of the Cell,* 3rd ed. New York: Garland Publishing, 1994.

An up-to-date and encyclopedic textbook for those wanting more details about protein biosynthesis. Numerous references are provided.

Blackburn, G. M., and M. J. Gait, Editors, *Nucleic Acids in Chemistry and Biology.* New York: Oxford University Press, 1990.

Very readable. Contains some details not presented in most introductory-level biochemistry texts. A good source of selected readings.

Bourne, H. R., D. A. Sanders, and F. McCormick, The GTPase superfamily: A conserved switch for diverse cell functions, *Nature* 348, 125–131 (1990).

Discusses molecular switches and the role of GTPase cycling. IF2 and EF-Tu are described as specific examples of GTPases.

Dobberstein, B., On the beaten pathway, *Nature* 367, 599–600 (1994).

Examines the role of GTPases in ER-linked polypeptide synthesis.

Draper, D.E., Protein-RNA recognition, *Annu. Rev. Biochem.* 64, 593–620 (1995).

Part of the specific protein-RNA interactions that are reviewed involve some of the participants in translation.

Genes that splice as proteins, *Cancer Watch* 4(1), 4–6 (1995).

A brief review of protein splicing.

Gesteland, R. F., R. B. Weiss, and J. F. Atkins, Recoding: Reprogrammed genetic decoding, *Science* 257, 1640–1641 (1992).

Reviews some exceptions to the general rules that govern the reading of the information carried in mRNAs.

Hershey, J. W. B., Protein phosphorylation controls translation rates, *J. Biol. Chem.* 264, 20823–20826 (1989).

A minireview that is appropriately titled.

Kreil, G., Conversion of L- to D-amino acids: a posttranslational reaction, *Science* 266, 996–997 (1994).

Examines some peptides that contain D-amino acid residues, and describes an enzyme that catalyzes the conversion of an L-amino acid residue to a D-amino acid residue within a polypeptide precursor for a spider venom peptide.

Lewin, B., *Gene V.* New York: John Wiley & Sons, 1994.

Chapters 7–10 are devoted to protein biosynthesis.

Moran, L. A., K. G. Scrimgeour, H. R. Horton, R. S. Ochs, and J. D. Rawn, *Biochemistry,* 2nd ed. Englewood Cliffs, NJ: Neil Patterson/Prentice-Hall, 1994.

An encyclopedic-type biochemistry text that covers all aspects of protein biosynthesis.

Samaha, R.R., R. Green, and H.F. Noller, A base pair between tRNA and 23S rRNA in the peptidyl transferase centre on the ribosome, *Nature* 377, 309–314 (1995).

Presents evidence that a specific base in E. coli *23S rRNA must pair with a C residue at the CCA end of a tRNA for the proper positioning of the tRNA on the ribosomal P site.*

Schroeder, R., Dissecting RNA function, *Nature* 370, 597–598 (1994).

A brief review of the role of RNA in translation.

Steiner, D. F., S. P. Smeekens, S. Ohagi, and S. J. Chan, The new enzymology of precursor processing endoproteases, *J. Biol. Chem.* 267, 23435–23438 (1992).

A minireview on the enzymes that catalyze peptide bond hydrolysis during the post-translational alteration of polypeptides.

Travis, J., Novel anticancer agents move closer to reality, *Science* 260, 1877–1878 (1993).

Describes the development of drugs that inhibit cancer growth by blocking polypeptide processing.

Voet, D., and J. G. Voet, *Biochemistry,* 2nd ed. New York: John Wiley & Sons, 1995.

A comprehensive text with a 64-page chapter devoted to translation. This chapter includes a good bibliography and 21 problems.

19 Genetic Diseases and Genetic Engineering

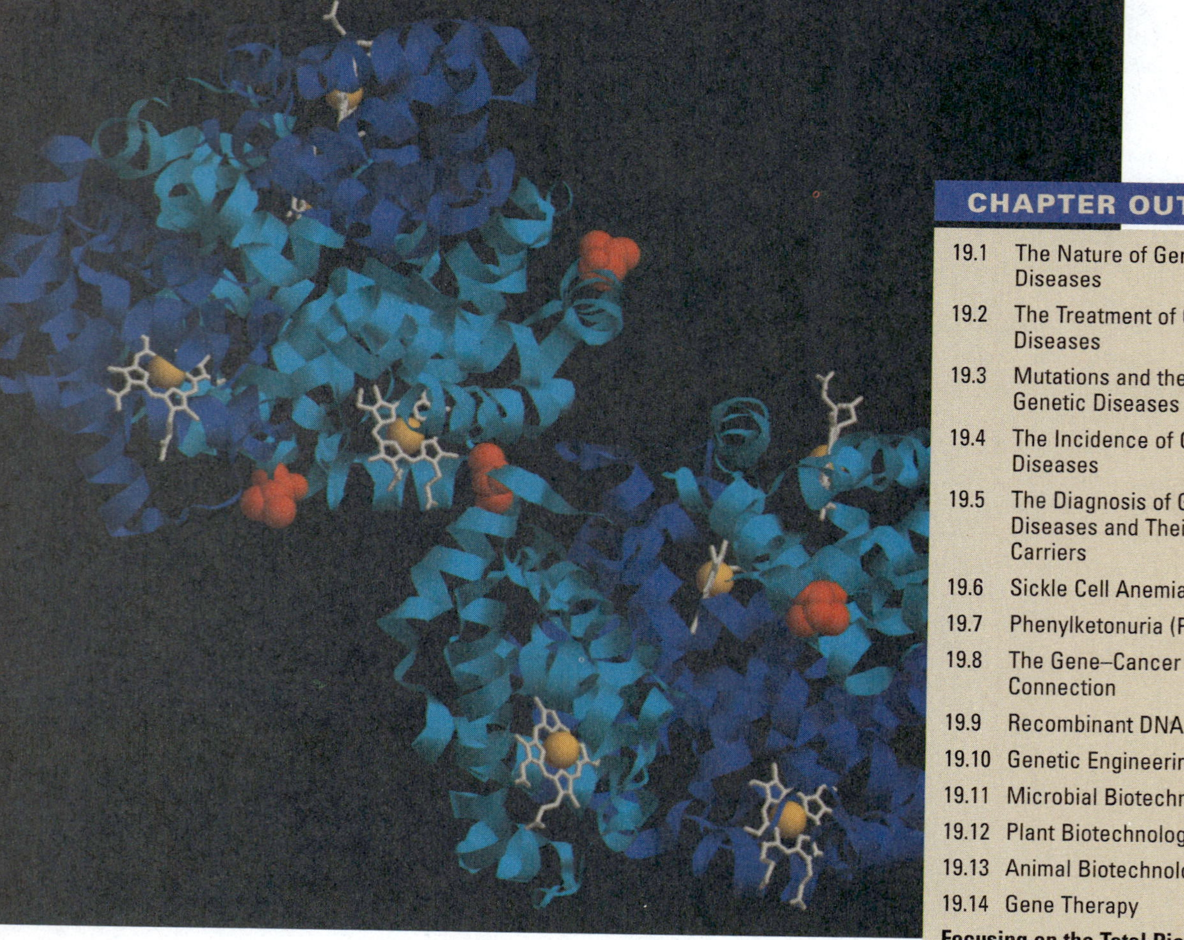

CHAPTER OUTLINE

Chapters 17 and 18 have examined the major processes involved in the expression of the genetic information encoded in the base pair sequence in DNA. Within the genetic information of an organism lies the inheritable blueprints for the production, maintenance, and reproduction of that organism. Errors in these blueprints can lead to **genetic diseases** (inherited diseases). Over 4000 human genetic diseases have been identified and, collectively, they have an enormous impact on individual families, our health care system, and our society. According to the March of Dimes Birth Defects Foundation, "one American in 100 is born with a serious genetic defect." Even more are born with less disabling genetic disorders. Over one fourth of all hospitalizations of children are due to inherited abnormalities.

Not all genetic disorders are apparent at birth or even during childhood. The symptoms of some defects do not appear for weeks, months, or even years. The late-appearing diseases include sickle cell anemia, cystic fibrosis, Tay–Sachs disease, and Huntington's disease (Table 19.1).

Although there is no satisfactory treatment for many genetic diseases, rapid advances in the ability to manipulate the gene content of cells have opened the door to new treatment strategies. The same technology is also having a profound impact on biochemistry, clinical chemistry, agriculture, the pharmaceutical industry, food processing, and many other disciplines. Much of this technology centers around the construction of recombinant DNA molecules, in which segments of DNA from different organisms have been covalently joined. The introduction of recombinant DNAs into host cells serves as the basis for genetic engineering.

TABLE 19.1

Some Genetic Diseases Whose Symptoms Are Not Usually Apparent at Birth

Genetic Disease	Symptoms	Age When Symptoms Usually Appear
Cystic fibrosis	Thick mucus clogs lungs and digestive tract	Birth to 4 years
Tay–Sachs disease	Brain damage leading to seizures, blindness, psychomotor regression, and death (usually within 3 years)	6 months to 1 year
Sickle cell anemia	Anemia, bouts of pain, damage to vital organs, susceptibility to infections	6 months on
Duchenne muscular dystrophy	Progressive wasting of muscles	2 to 5 years
Wilson's disease	Liver and neurological damage, skeletal changes	8 to 20 years
Huntington's disease	Uncontrollable movements, intellectual and emotional impairment, death	Late 30s on

This chapter begins by exploring genetic diseases, their nature, treatment, origin, incidence, and diagnosis. A discussion of two specific genetic diseases, sickle cell anemia and phenylketonuria (PKU), is then followed by an examination of the gene–cancer connection and a glimpse at genetic engineering and some of its accomplishments. The chapter closes with a section that provides an integrated summary of the molecular basis of life. Featured biochemical themes include information transfer, regulation of gene expression, evolution, structural hierarchy, association of like with like, catalysis, and specificity.

19.1 THE NATURE OF GENETIC DISEASES

All human genetic diseases are a consequence of abnormalities in the human genome. In some cases a single base pair change is involved, whereas in other instances a larger segment of DNA is modified, relocated, added, or deleted. **Down's syndrome (trisomy 21)** provides an extreme example, because this relatively common disease is linked to an extra copy of all of the DNA within an entire chromosome. Symptoms of Down's syndrome include mental retardation, growth retardation, and a distinct facial appearance.

An individual normally possesses two sets of DNA molecules, one set contained in the 23 chromosomes received from his or her mother and a second set contained in the 23 chromosomes acquired from his or her father. Since each set of DNA molecules contains a nearly complete set of human genes, a normal person possesses duplicate copies of most genes. Females have a pair of X chromosomes, whereas males possess one X chromosome and one Y chromosome. Hence, females lack Y chromosome–specific genes, and males contain a solitary copy of each X- and Y-specific gene. Members of gene pairs may both be normal, one may be normal and the other abnormal, or both may be abnormal. During the generation of egg or sperm cells, one member of each pair is randomly selected to be passed on to the next generation. A genetic disease that can result from an abnormality in only one of the two sets of genes is said to be **dominantly inherited.** A genetic disease that only exists when both sets of genes are abnormal is said to be **recessively inherited** (Figure 19.1 and Exhibit 19.1). **[Try Problem 19.1]**

The risk of acquiring or developing virtually any disease, even those that are not directly inherited, is impacted by genetics. Our susceptibility to heart disease, infections, and cancer (Section 19.8), for example, is partially determined by our gene content. In some instances, the genetic link is very strong, and changes in a single gene can markedly affect disease risk. **[Try Problem 19.2]**

Some authors now classify virus infections as **acquired genetic diseases.** This classification is based on the fact that a virus-infected cell acquires, and often incorporates into its own DNA (see retroviruses, Section 17.5), some form of the viral genome. If an egg or sperm cell is infected, it can pass genetic information from the virus on to the next generation.

No disease

DNA DNA
Two normal gene sets

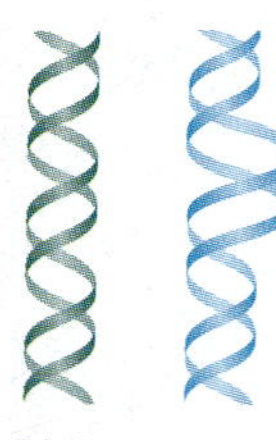

No disease with recessive inheritance

Disease with dominant inheritance

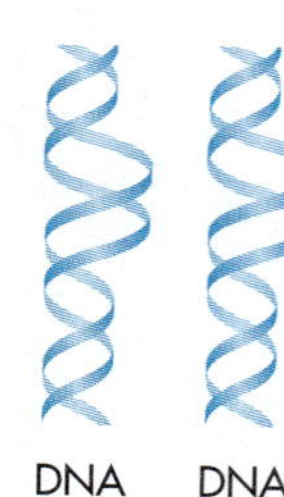

Disease with both dominant and recessive inheritance

FIGURE 19.1
Dominant and recessive inheritance of genetic diseases.

PROBLEM 19.1 Many, but not all, birth defects are a consequence of genetic diseases. Suggest another cause of birth defects.

EXHIBIT 19.1
Examples of Dominantly and Recessively Inherited Diseases

Dominantly Inherited Diseases
Achondroplasia—a form of dwarfism
Chronic simple glaucoma (some forms)—if untreated, leads to blindness
Huntington's disease—progressive nervous system degeneration
Hypercholesterolemia—high blood cholesterol levels and a propensity to heart disease
Polydactyly—extra fingers or toes

Recessively Inherited Diseases
Cystic fibrosis—thick mucus clogs lungs and digestive tract
Phenylketonuria (PKU)—if untreated, leads to mental retardation
Sickle cell anemia—anemia, bouts of pain, damage to vital organs, susceptibility to infections
Tay–Sachs disease—fatal brain damage
Gaucher's disease—enlarged spleen, skin pigmentation, bone lesions, neurological degeneration

PROBLEM 19.2 What system is primarily responsible for defending the body against infections? Explain how a change in a gene affecting a key component of this system could modify the risk of infections.

19.2 THE TREATMENT OF GENETIC DISEASES

A limited number of genetic defects respond well to surgery, diets, drugs, or other therapies. A cleft palate (a fissure in the roof of the mouth), for example, can be surgically sealed. PKU (Section 19.7) is controlled with diet, and hemochromatosis (Section 19.4) can be treated with bloodletting. The management of type I diabetes (Section 12.4) normally involves both insulin (a protein hormone) and diet. Relatively new treatment strategies for genetic diseases include gene therapy (Section 19.14) and the transplantation of normal cells, tissues, or organs. The use of pancreas transplants and pancreas cell implants to treat diabetes illustrates the transplantation strategy. Unfortunately, there is presently no satisfactory treatment for many of the most serious and most common genetic defects. **[Try Problem 19.3]**

There is no cure for any genetic disease if we define a "cure" as the correction of the abnormalities in DNA that are ultimately responsible for the disease. A true "cure" presents an enormous challenge because the defective genes exist within every cell of an affected individual. Future hopes for a "cure" lie in genetic engineering. **[Try Problem 19.4]**

PROBLEM 19.3 Everyone has had multiple contacts with individuals with genetic defects. List one defect for which there is no satisfactory treatment.

PROBLEM 19.4 Does an abnormal gene have an adverse effect on all the cells in which it resides? Does an abnormal hemoglobin gene, for example, have any direct impact on heart or liver cells? Explain. Hint: Review Section 4.9.

19.3 MUTATIONS AND THE ORIGIN OF GENETIC DISEASES

The origin of most genetic diseases can be traced to mutations in human egg or sperm cells or mutations in germ cells that divide to generate egg and sperm cells. A **mutation** is any change in the normal nucleotide sequence in DNA. Such alterations can result from errors during DNA replication, from **translocations** (the movement of a segment of DNA from one chromosome to another), from other categories of transpositions (Section 17.5), and from the action of chemical or physical **mutagens** (agents that lead to mutations; Exhibit 19.2). A change in a single base pair is called a **point mutation.** Some mutations are beneficial and play a role in evolution and the adaptation of species. Other mutations have no impact on an organism and are said to be **silent.** Still others are harmful to a cell, organism, or species. **[Try Problem 19.5]**

EXHIBIT 19.2

Some Frequently Encountered Mutagens[a]

Physical

Ultraviolet (UV) Light
X-Rays
Nuclear radiation

Chemical

Ozone (produced by lightning)
Some hydrazines (in edible mushrooms)
Some furocoumarins (in celery and figs)
Some pyrrolizidine alkaloids (in herbs and herbal teas)
Allyl isothiocyanate (in mustard and horseradish)
Aflatoxin (in mold that grows on peanuts and grains)
Methylglyoxal (in coffee)
Chlorogenic acid (in coffee)
Formaldehyde (a gas emitted by some building materials)
Nitrosamines (produced in the human body from nitrate and nitrite in vegetables and other foods)

[a]The risk associated with a mutagen is determined by many variables, including dose, frequency of exposure, other compounds present, route of exposure, diet, and DNA repair capabilities. Although risk is extremely difficult to quantitate, the risk associated with usual exposures to many of the listed mutagens is probably very low compared with many other health risks.

TABLE 19.2

Effect of Tautomerism on Base Pairing in DNA

Normal Pairing	Abnormal Pairing
Both Bases in Pair Exist in Most Stable Tautomeric Form	**Left Base in Pair Exists in Rare Tautomeric Form**
A–T	A–C
T–A	T–G
G–C	G–T
C–G	C–A

An organic chemistry phenomenon known as **tautomerism** accounts for part of the low spontaneous mutation rate observed in all organisms. This phenomenon is reviewed in Section 3.5 where it is used to help explain the planar nature of peptide bonds. It is revisited in Section 7.1 during the discussion of the chemical properties of purines and pyrimidines. Within normal duplex DNA molecules, adenine pairs with thymine, and guanine pairs and cytosine. The correct copying of parental DNA strands during DNA replication depends on the formation of these standard base pairs (Section 17.4). Tautomerism leads to a low frequency of mispairing, since a base in a rare tautomeric form will pair with a nonstandard partner (Table 19.2). The rare tautomer of adenine, for example, pairs with cytosine rather than thymine. The copying errors that result from tautomerism lead to abnormal nucleotide sequences (mutations) within newly synthesized strands of daughter DNA. Proofreading by DNA polymerases and corrective surgery by mismatch repair systems eliminate some of these errors (Section 17.4). Those errors that survive lead to mutations and, in some instances, genetic diseases. **[Try Problem 19.6]**

The importance of DNA repair systems extends well beyond the avoidance of genetic diseases. Many cancers and other noninherited ailments are linked to mutations, and even biological aging is affected by alterations in DNA (Section 14.1). The somatic mutation hypothesis of aging proposes that normal aging is due, in part at least, to the time-linked accumulation of mutations. Mutations do indeed accumulate in people over time, and, in mammals, an impressive correlation has been reported between maximum life span and DNA repair capabilities. Individuals with certain inherited disorders, including **xeroderma pigmentosum-A,** are subject to brain deterioration, immune system diseases, tumors, or other health problems as a consequence of deficiencies in DNA repair. Xeroderma pigmentosum-A leads to an increased risk of UV-induced skin cancer and corneal scarring.

PROBLEM 19.5 What feature of the genetic code (Section 16.5) explains why some mutations are silent?

PROBLEM 19.6 Write structural formulas for the two tautomeric forms of adenine. In each form, circle each hydrogen atom capable of hydrogen bonding. The difference in H bonding capabilities accounts for the differences in base pairing capabilities.

19.4 THE INCIDENCE OF GENETIC DISEASES

How do genetic diseases, including some that lead to death before puberty, persist within a population? Why do certain diseases exist at such high frequencies? Why doesn't natural selection eliminate most genetic diseases? In many instances, the frequency of a disease is so low that new mutations during one generation are able to replace the disease genes lost from the gene pool during the previous generation. **Duchenne-type muscular dystrophy** (Table 19.1) and **achondroplasia** (Exhibit 19.1) are among the diseases thought to be maintained solely through this mechanism. In other instances, the disease gene appears to offer some compensating advantages, not necessarily for the affected individuals but for the "population" as a whole.

Hemochromatosis, which may affect as many as one out of every 200 United States citizens, is one of the diseases in which the gene involved confers compensating advantages. Whereas normal individuals limit their absorption of iron from the contents of the small intestine, victims of hemochromatosis tend to absorb excess quantities of this essential nutrient and can succumb to involuntary, self-inflicted iron poisoning. At the same time, women who regularly lose large amounts of iron through menstrual bleeding, pregnancy, or nursing profit from increased iron absorption. Thus, the hemochromatosis gene benefits some women at the expense of men and certain other women. From the perspective of some women, hemochromatosis is not a disease at all; it is a survival mechanism. From an evolutionary standpoint, the death of some postreproductive men (the disease does not usually lead to death until relatively late in life) is a small price to pay in order to protect reproductive women from anemia. Section 19.6 examines sickle cell anemia and looks at a compensating advantage linked to this disease.

Modern medicine is contributing to an increase in the frequency of certain genetic diseases within technologically advanced societies. The reason? Medical advances have allowed us to escape, at least temporarily, from some evolutionary pressures. We are now keeping more and more diseased individuals alive long enough to allow them to pass their defective genes on to future generations. Without medical intervention, many of these genes would be eliminated from the gene pool. What fraction of the human population will have a genetic disease two generations from now? 10 generations from now? What will be the personal, social and economic costs of these diseases?

19.5 THE DIAGNOSIS OF GENETIC DISEASES AND THEIR CARRIERS

The diagnosis of some genetic diseases can be made quickly and confidently soon after birth (in some cases, even prenatally) based on the appearance of characteristic symptoms. For other diseases, symptoms are not disease specific (various other ailments lead to the same symptoms), or they do not surface for weeks, months, or years (Table 19.1). Individuals who carry a single copy of a gene for a recessively inherited disease are usually asymptomatic. How can diseased individuals and carriers of recessive genetic defects be identified in those cases where symptom-based diagnosis is impossible, and of what value is such identification? The latter question is addressed first.

Benefits of Diagnosis

The early detection of some genetic diseases opens the door for successful treatment. PKU (Section 19.7) provides an excellent example; early detection coupled with dietary restrictions can usually prevent the development of mental retardation and other symptoms. The value of early detection is debatable when there is no treatment for the genetic disease involved.

The identification of carriers of defective genes allows couples to make sometimes tough decisions regarding childbirth. Should a couple have a child if there is a significant probability (based on gene analysis) that it will be diseased? If both partners are carriers of the same recessively inherited disease, there is normally a 25% probability that any child of theirs will acquire two copies of the defective gene and be diseased (assuming that the disease is linked to a single gene). The expanding interest in carrier identification is documented by the growth in cystic fibrosis carrier testing between 1989 and 1992 (Figure 19.2). **[Try Problem 19.7]**

Diagnostic Strategies

The detection of carriers and asymptomatic diseased individuals is presently possible for only a small fraction of the known genetic diseases. Detection is commonly based on the analysis of tissue samples or body fluids for the presence of abnormal gene products (most are proteins) or the absence of a gene product. Sickle cell anemia (Section 19.6), for example, can be diagnosed by the identification of abnormal hemoglobin molecules within red blood cells. Alternatively, clinical laboratories sometimes check for abnormal concentrations of a substance that is linked to a genetic disease but is not encoded in a defective gene. This strategy is employed to identify PKU victims (Section 19.7). A relatively new approach to detection involves the direct analysis of a sample of DNA. Thanks to the polymerase chain reaction (PCR, Section 17.6), very small quantities of DNA can be amplified and analyzed.

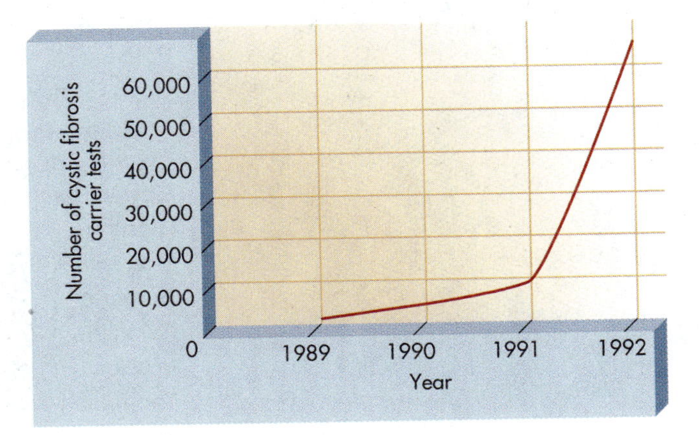

FIGURE 19.2
Growth in cystic fibrosis carrier testing in the U.S.

DNA Probes and RFLPs

Most DNA analysis employs enzymes known as **restriction endonucleases** that attach to specific base pair sequences within DNA, and then catalyzes the hydrolysis of both strands of the duplex. The DNA fragments produced are known as **restriction fragments.** When the restriction fragments from a specific DNA molecule are separated by size with agarose gel electrophoresis (Sections 3.10 and 16.9), the pattern of bands created on the gel by the separated fragments is called a **DNA fingerprint** (Figure 19.3). All restriction fragments of the same size (length) are found in the same band, whereas different sized fragments always appear in different bands. When corresponding segments of DNA from two different individuals are analyzed, many common restriction bands are usually found along with some bands that are unique to each individual. The unique bands, which arise from variations, among individuals, in the size of restriction fragments from equivalent regions of the genome, are called **restriction fragment length polymorphisms (RFLPs).** **[Try Problem 19.8]**

A variety of techniques are utilized to detect restriction fragment bands on agarose gels. One of the most common techniques employs **DNA probes,** small pieces of single-stranded DNA (usually 10 to 30 nucleotides long) that are complementary to sequences of nucleotides that we want to search (probe) for in a target DNA molecule. If a DNA molecule contains a strand that can base pair with a DNA

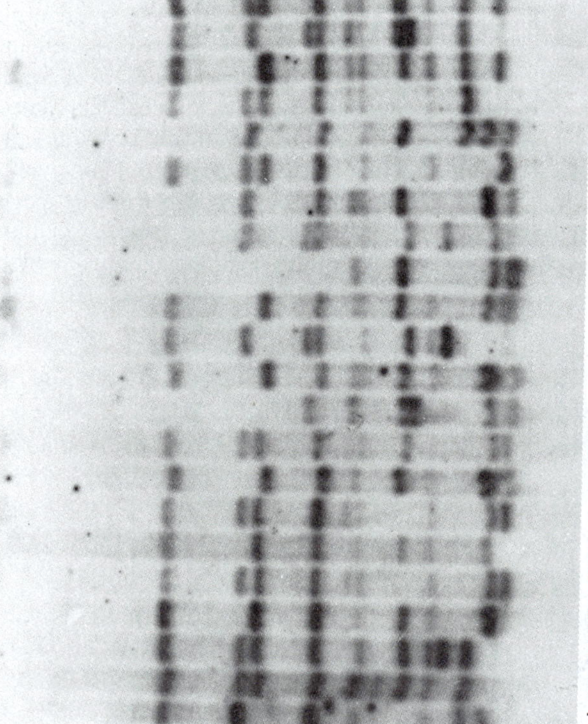

FIGURE 19.3 DNA fingerprints of 22 wine grape varieties. Petite sirah and durif, once thought to be the same, are distinct varieties because each has a unique fingerprint. Radioisotopes, fluorescent dyes, or other detection methods are required to locate the bands because DNA fragments are not directly visible on an agarose gel. All DNA fragments of the same size are found in the same band. Different sized fragments migrate in different bands.

probe along its entire length, that target DNA must contain the unique base sequence that is complementary to the probe. The base pairing of a probe with a nucleic acid strand is sometimes described as hybridization or annealing (Section 16.8). DNA probes can be used to detect RFLPs if the probes are complementary to sequences common to the different sized restriction fragments. The probes themselves are usually "tagged" with a radioisotope, a fluorescent dye, or an enzyme to allow them to be sensitively and easily detected and followed. **[Try Problem 19.9]**

When unique RFLPs are associated with diseased individuals, these RFLPs are said to be linked to the genetic disease, and they can be used to help identify both victims and carriers (Figure 19.4). Viewed from a slightly different perspective, spe-

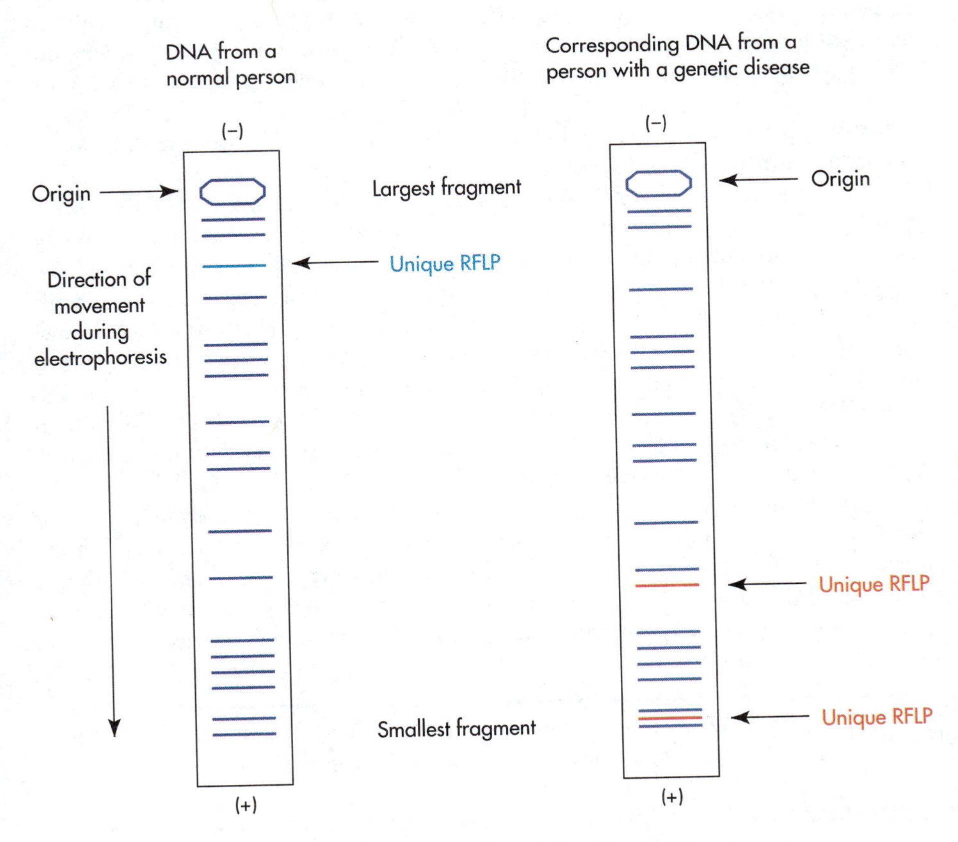

FIGURE 19.4
Disease-linked restriction fragment length polymorphisms. Each band represents a different sized restriction fragment. In this idealized example, the normal and the diseased individual share many restriction fragments that hybridize with the same probe. At the same time, each individual possesses one or more uniquely sized restriction fragments. The unique fragments represent RFLPs. If the unique RFLPs in the diseased individual are linked to the disease, their presence can be used to identify carriers of the disease-linked gene.

cific restriction fragments can serve as **markers** for a disease. Not all disease-linked RFLPs are due to the disease-linked gene itself; some are a consequence of genetic markers that lie close to the defective gene within the DNA of disease victims. One of the spinoffs of the Human Genome Project (Section 16.9) is the identification of genetic markers and genes for a growing number of genetic defects. Duchenne muscular dystrophy, Huntington's disease, cystic fibrosis, Lesch–Nyhan syndrome, PKU, hemophilia A, and sickle cell anemia are some of the many diseases that have already yielded to DNA analysis. **[Try Problem 19.10]**

The ideal DNA probe for a genetic disease would be complementary to part or all of a unique sequence within the segment of DNA responsible for the disease. The appropriate probe would hybridize with the DNA from diseased individuals and carriers but not with normal DNA. A disease-linked gene and its normal counterpart must both be sequenced (partially, at least) before such a probe can be constructed.

Additional Uses of DNA Probes

The diagnostic value of DNA probes extends well beyond classical genetic diseases. Because cancer cells and virus-infected cells usually contain nucleotide sequences and RFLPs that are not found in most normal cells, DNA probes are being developed and used for the diagnosis of cancer and virus infections. In the case of cancer, clinicians often probe for the presence of a specific oncogene (Section 19.8). Virus probes are usually complementary to unique sequences within the viral genome. DNA probes are also being used by dentists to detect and identify the bacteria responsible for periodontal disease.

DNA fingerprinting (sometimes called **DNA identity testing** and **DNA typing**) also has many applications outside the diagnostic arena. Since each individual (the exception is identical twins) tends to have a unique DNA fingerprint, DNA analysis is now routinely employed in forensic science, paternity testing, tissue typing, and the identification of missing persons. This fact made headlines during news coverage of the O. J. Simpson murder trial; DNA fingerprinting was used as evidence in an attempt to prove that O. J. Simpson's blood was present at the murder scene. DNA analysis of semen from a rape victim is another example of the forensic use of DNA technology. Many rapists have been convicted with the assistance of DNA analysis. **[Try Problem 19.11]**

Some in our society are increasingly concerned about the possible misuses of our rapidly expanding ability to detect genetic abnormalities. Will individuals found to contain certain genetic defects be denied employment, health insurance, or life insurance? Will they be given equal educational opportunities? Will they have difficulty finding mates? Who should be tested and who should have access to the final results?

PROBLEM 19.7 If both parents inherit no defective genes, can they be certain that their children will have no genetic disease? Explain.

PROBLEM 19.8 Explain how the DNA fingerprints for two individuals may be the same with one restriction endonuclease but different with a second restriction endonuclease.

PROBLEM 19.9 To use a DNA probe, one must first denature the DNA that is to be analyzed and then add the probe under conditions in which it can hybridize with complementary sequences. Define denaturation, and then explain the importance of the denaturation step during DNA probe analysis. Hint: Review Section 16.8.

PROBLEM 19.10 Several members in each of two unrelated families have the same genetic disease. DNA analysis of all members of one family identifies RFLPs that are present in all diseased members of the family but not present in any nondiseased members. When members of the second family are subjected to the same DNA analysis, none of them (not even the diseased individuals) contain the RFLPs linked to the diseased individuals in the first family. What is the most likely explanation for these observations? Assume that the disease is always due to changes in the same gene. Is the disease dominantly or recessively inherited? Explain.

PROBLEM 19.11 Cells growing in culture medium A produce some unique proteins and have a different size and shape than cells growing in culture medium B. Analysis of A-cells and B-cells indicates that their DNAs contain identical monomer sequences. Explain how it is possible for two cells with the same genetic information to look and behave differently. Can a DNA probe be used to distinguish A-cells and B-cells? Explain.

19.6 SICKLE CELL ANEMIA

Sickle cell anemia, one form of **sickle cell disease,** is a genetic disease associated with the production of abnormal hemoglobin molecules. Hemoglobin is a red blood cell protein whose major function is to transport O_2 and CO_2 (Section 4.9). The most abundant form of normal adult hemoglobin, hemoglobin A_1, contains two identical polypeptide chains known as α-chains and two identical polypeptide chains known as β-chains. Individuals with sickle cell anemia contain two abnormal copies of the gene for the β-chain, and they produce an abnormal hemoglobin known as **hemoglobin S.** Individuals with one normal β-chain gene and one abnormal β-chain gene have **sickle cell trait.** They produce some hemoglobin A_1 along with some hemoglobin S and have few, if any, symptoms of disease.

Molecular Basis

Glutamate is the sixth amino acid residue in the two β-chains of hemoglobin A_1, whereas valine occupies this position in the β-chains of hemoglobin S. This is the only difference in the primary structure of these two proteins. Hemoglobin A_1 and hemoglobin S have almost identical secondary, tertiary, and quaternary structures, and both are capable of binding and carrying O_2 and CO_2 when in solution. Hemoglobin S, however, has a hydrophobic surface patch that is not present in hemoglobin A_1. It is this surface patch, created by the replacement of a hydrophilic glutamate side chain with a hydrophobic valine side chain (Figure 19.5), that leads to health problems. [Try Problems 19.12 through 19.14]

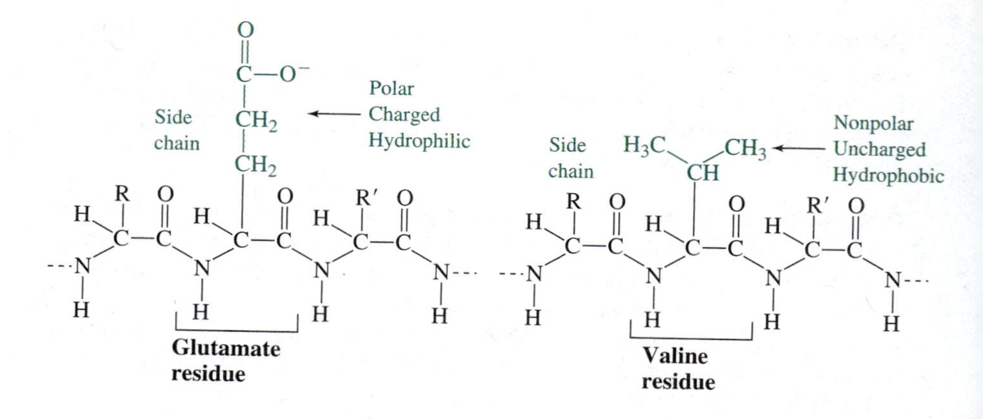

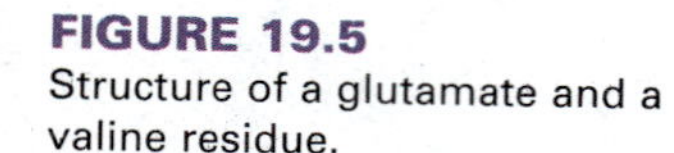

FIGURE 19.5
Structure of a glutamate and a valine residue.

The deoxy forms of both hemoglobin A_1 and hemoglobin S have a small hydrophobic pocket located at a specific site on their surfaces. The oxyhemoglobins do not contain this pocket because it is destroyed by the changes in conformation associated with the binding of O_2 (Section 4.9). In sickle cell patients, the hydrophobic valine side chains on the surface of hemoglobin S molecules tend to insert themselves into the hydrophobic pockets of neighboring deoxyhemoglobin molecules in order to avoid the polar aqueous environment in which hemoglobin molecules normally reside. When the concentration of deoxyhemoglobin S is high within red blood cells, many hemoglobin S molecules bind together through hydrophobic interactions in a process described as aggregation or polymerization; long hemoglobin-containing fibers are generated (Figure 19.6). These fibers cause red blood cells to **sickle** (become sickle shaped, Figure 19.7) and to lose deformability. The loss of deformability in the sickled cells prevents them from flexing and bending as they move through capillaries. This leads to pain and inflammation. Membrane damage in some sickled cells leads to cell rupture and to **anemia,** a deficiency in the ability to transport O_2. Worse yet, sickled cells can become trapped in capillaries and restrict the flow of blood and O_2 to tissues. Cell death can follow. Such capillary blockages contribute to the early and painful death of many individuals with sickle cell anemia. **[Try Problem 19.15]**

Diagnosis

The diagnosis of sickle cell anemia and sickle cell trait after birth normally entails the electrophoretic analysis of hemoglobin isolated from a sample of blood. Hemoglobin S has a characteristic and unique electrophoretic mobility under standard electrophoretic conditions. The electrophoretic separation of hemoglobin A_1 and hemoglobin S is made possible by the change in net charge associated with the valine (in sickle cell hemoglobin) for glutamate (in normal hemoglobin) substitution that is responsible for sickling. Prenatal (before birth) diagnosis of sickle cell diseases and their carriers is usually based on analysis of fetal DNA that has been amplified with PCR. The fetal DNA that is amplified is frequently acquired with **amniocentesis,** the surgical insertion of a hollow needle into the uterus of a pregnant woman to obtain amniotic fluid. Such fluid contains fetal cells.

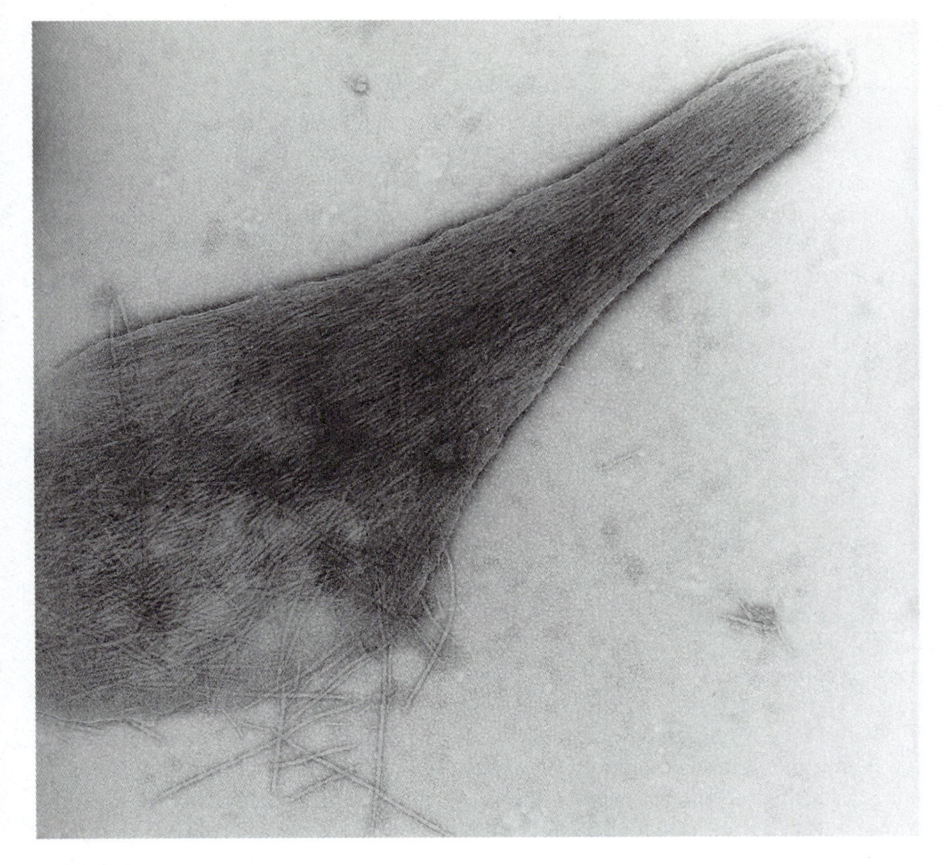

FIGURE 19.6
Electron micrograph of abnormal hemoglobin S fibers in a ruptured red blood cell from an individual with sickle cell anemia. Individual fibers are approximately 22 nm in diameter.

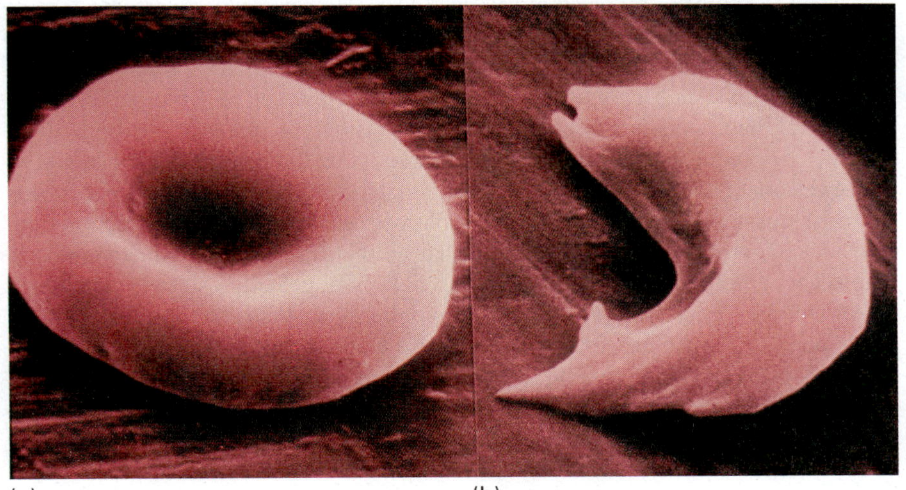

(a) (b)

FIGURE 19.7
(a) A normal and (b) a sickled red blood cell.

Treatment

A major treatment strategy employs drugs to reduce the frequency or extent of hemoglobin S aggregation within red blood cells. This can be accomplished by increasing deoxyhemoglobin solubility, by reducing hemoglobin S concentration, and by increasing the O_2 affinity of hemoglobin S. This strategy has met with limited success, since there are serious side effects associated with the drugs that have been tested. Other treatment strategies include exchange transfusions, bone marrow transplants (red blood cells are produced from stem cells in the bone marrow), vasodilation (to widen arteries), and gene therapy (Section 19.14). The most promising therapeutic option at the moment entails the daily administration of hydroxyurea.

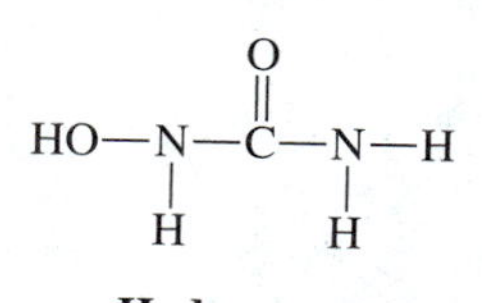

Hydroxyurea

The National Institutes of Health called for an early halt to a national hydroxyurea trial in January 1995 because interim results provided compelling evidence of the drug's value; it was decided that all patients should have immediate access to the drug. Hydroxyurea appears to function by elevating the concentration of fetal hemoglobin (hemoglobin F) within red blood cells, an event that reduces the probability that neighboring hemoglobin S molecules will collide and aggregate. How hydroxyurea increases hemoglobin F concentration is under investigation. Hydroxyurea is also used to treat some forms of leukemia, and its employment in the treatment of AIDS is under consideration.

Geographical Distribution

Sickle cell anemia is most common in malarial regions of the world, including Africa and South Asia, because the gene associated with this disease provides some resistance to **malaria,** a parasitic disease. The malaria parasite, a protozoan, must spend part of its life cycle within a red blood cell of its host. The red blood cells of individuals with sickle cell trait or sickle cell anemia contain an inhospitable environment. When a parasite enters such blood cells, these cells tend to sickle and to lose part of their potassium ions. The loss of potassium ions leads to the death of the parasite. Consequently, normal individuals are more likely than individuals with sickle cell anemia or sickle cell trait to die of malaria before they reach puberty. This means that in a malarial environment people with sickle cell trait have a reproductive advantage when compared with normal individuals. This reproductive advantage accounts for the high frequency of the sickle cell gene in resident populations. The sickle cell gene, like the hemochromatosis gene, benefits some (those with sickle cell trait) at the expense of others (those with sickle cell anemia). Sickle cell anemia is most common among African-Americans in the United States because many of their ancestors lived for generations in malaria-infested environments. Most states now test all newborn infants for this disease.

Sickle cell anemia clearly illustrates how changes in a single gene and a single protein can have serious health consequences. The symptoms of many genetic diseases have been traced to single gene/protein alterations. Phenylketonuria provides another specific example.

PROBLEM 19.12 Sickle cell anemia is due to a single nucleotide substitution (point mutation) in the genes for the β-chain of hemoglobin. Study Table 18.1 and then suggest what specific nucleotide substitution is involved.

PROBLEM 19.13 Explain why nonpolar amino acid side chains are usually folded into the core of globular proteins, whereas polar amino acid side chains are normally located on the exterior surface. Hint: Review Section 4.4.

PROBLEM 19.14 The single amino acid substitution involved in sickle cell anemia has no impact on hemoglobin's oxygen carrying ability. Explain. Would this also be the case for all other single amino acid substitutions within hemoglobin? Explain.

PROBLEM 19.15 Patients with sickle cell anemia are often advised to avoid vigorous exercise. Explain.

19.7 PHENYLKETONURIA (PKU)

Phenylketonuria (PKU) is a genetic disease caused, most often, by a defect in the gene that encodes phenylalanine hydroxylase, an enzyme that catalyzes the conversion of phenylalanine to tyrosine (Figure 13.22). The protein produced from the information carried in the abnormal gene is unable to catalyze this reaction. As a consequence, phenylalanine tends to accumulate in people with PKU. As phenylalanine concentrations increase, the phenylalanine is converted to phenylpyruvic acid, a **phenylketone,** and related compounds (Figures 13.23 and 19.8) that appear at elevated levels in blood and urine. Diagnosis in newborns normally involves an analysis of the concentrations of phenylalanine and phenylketones in blood. Prenatal diagnosis usually entails the analysis of fetal DNA amplified through PCR. Direct sequencing of segments of the abnormal gene is now common.

Toxic concentrations of phenylketones are thought to be responsible for the mental retardation and other symptoms associated with this disease. PKU is treated by limiting the dietary intake of phenylalanine and compounds, including aspartame (NutraSweet; Section 3.7), that contain releasable phenylalanine residues. Most aspartame-containing products bear a warning for PKU patients. Alternative treatment strategies and methods to improve compliance with dietary restrictions are under consideration because most adults treated for PKU since birth show subtle neuropsychological defects. Virtually all hospitals in the United States now screen all newborn infants for PKU. **[Try Problems 19.16 and 19.17]**

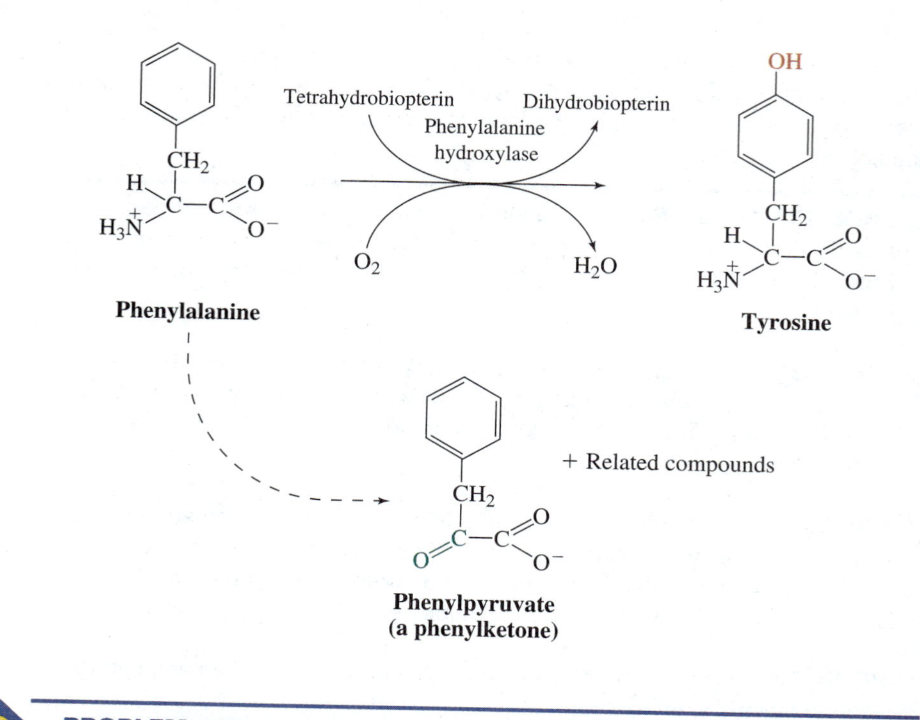

FIGURE 19.8
Phenylalanine metabolism. Most people with PKU are unable to produce a functional phenylalanine hydroxylase. As phenylalanine accumulates, part of it is converted to phenylpyruvate (a phenylketone) and related compounds (see Figure 13.23 for additional details).

PROBLEM 19.16 Explain why PKU is recessively inherited rather than dominantly inherited.

PROBLEM 19.17 True or false? Because phenylalanine is one of the protein amino acids, a PKU victim must limit protein intake. Explain.

19.8 THE GENE–CANCER CONNECTION

Although cancer, in contrast to sickle cell anemia and PKU, is not considered to be inherited in the classical sense, your risk of developing cancer is largely determined by genetics. Some knowledge of the differences between normal cells and cancer cells is needed in order to understand the molecular basis for the gene–cancer connection.

Nature of Cancer

The growth and division of normal cells is controlled by numerous compounds, including growth factors, growth factor receptors, and a variety of proteins involved in signal transduction following the binding of growth factors to receptors (Section 10.4). The action of these compounds must be carefully regulated throughout the body, since normal body function requires that distinctive types of cells divide at different, precisely controlled rates. A **cancer** is simply a group of cells whose growth and division are no longer under normal control. In contrast to normal cells, most cancer cells grow in random directions, divide rapidly, and serve no useful function (Figure 19.9). Their uncontrolled proliferation tends to disrupt normal body func-

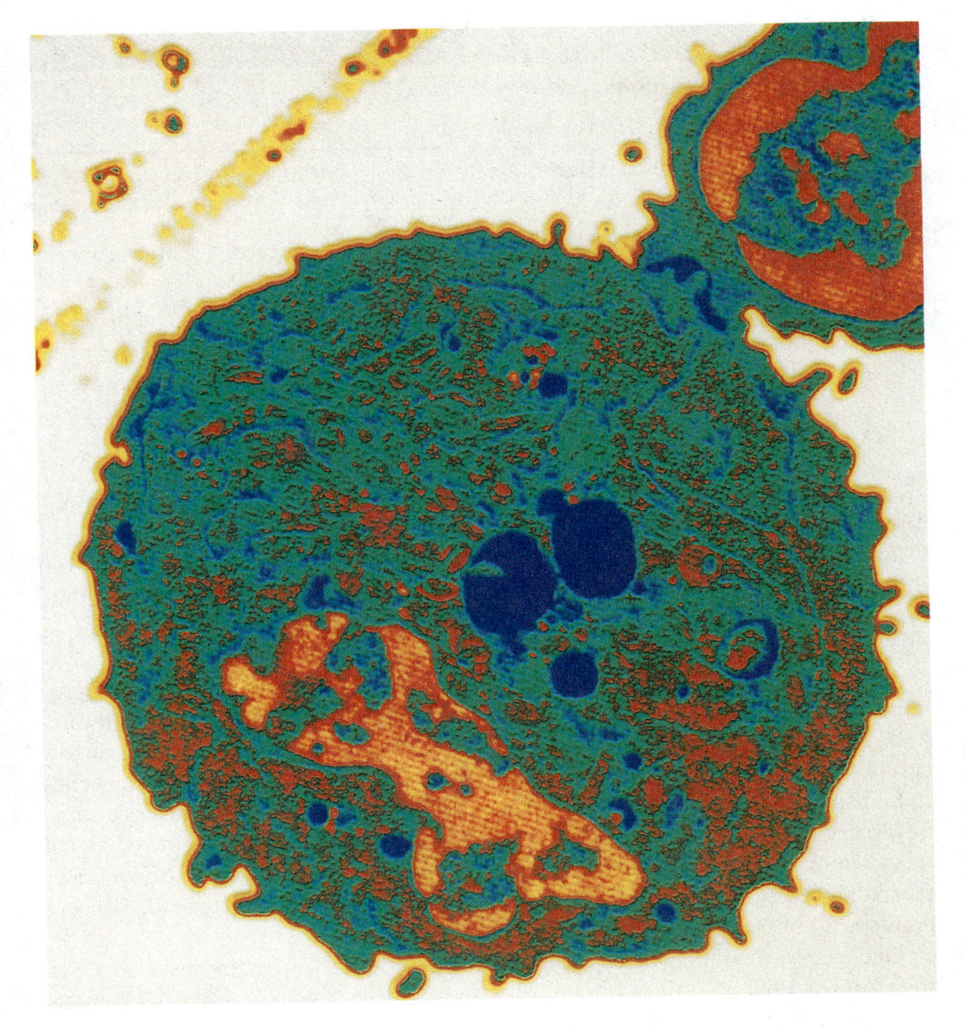

FIGURE 19.9
Colored transmission electron micrograph (TEM) of a cross-section through a human cancer cell. The typical uneven surface of the cell and its cytoplasmic projections can be seen. The cell has undergone a mitotic division and a daughter cell is partially visible in the upper right.

tions and to lead ultimately to death. **Oncology** is the study of cancer and an **oncologist** is one who studies cancer.

Protooncogenes and Oncogenes

Most cancer is a consequence of mutations in protooncogenes and tumor suppressor genes. Multiple genes must be altered to convert a normal cell into a cancerous one. Many colorectal cancers, for example, contain a total of seven to ten mutations in four or five distinct cancer-linked genes.

A **protooncogene** is a normal gene that can be mutated to create an **oncogene,** a gene that contributes to the **transformation** of a normal cell into a cancerous one. Because the conversion of a protooncogene to an oncogene leads to alterations in the activities of gene products, not to a loss or inactivation of gene products, oncogenes act in a dominant fashion; a mutation in only one member of a pair of protooncogenes leads to progression toward cancer. Most protooncogenes encode proteins that

TABLE 19.3

Some Classes of Protooncogene Proteins

Class of Proteins	Function
Growth factors	Stimulate cell growth and division
Growth factor receptors	Bind growth factors and initiate signal transduction
GTP-binding proteins	Involved in signal transduction following growth factor binding to receptor
Protein kinases	Involved in signal transduction following growth factor binding to receptor; catalyze the phosphorylation of proteins
DNA-binding proteins	Involved in controlling gene expression; some are transcription factors

are involved in the positive regulation of normal cell growth and division (Table 19.3). The mutations that generate oncogenes normally lead to the overproduction of growth stimulators or to the production of overactive stimulators. Since close to 100 different oncogenes have been identified, the message seems clear: Changes in the proteins that exert positive control over cell growth and division can lead to a loss of control and, ultimately, to cancer. **[Try Problem 19.18]**

Some of the most thoroughly studied protooncogenes are the genes for Ras proteins, the G-proteins that play a central role in growth factor-linked signal transduction (Section 10.4). The abnormal Ras proteins encoded in *Ras* oncogenes cause a cell to overreact to growth factors. *Ras* oncogenes are the most common oncogenes in humans, contributing to the development of roughly 30% of all human cancers. Several other protooncogenes encode proteins that participate in Ras-linked signal transduction. These protooncogenes include genes for growth factors, growth factor receptors, GAPs, Raf protein, and several protein kinases (study Figure 10.10). **[Try Problem 19.19]**

Oncogenes were first discovered in **oncogenic viruses,** retroviruses that can transform normal cells into tumor cells. Retroviruses, including the AIDS virus (Section 17.5), can reversibly insert a duplex DNA form of their genome into cellular chromosomes following the infection of host cells. In the past, some retroviruses acquired oncogenes from host cells when the viral genome was imprecisely "released" from the host chromosome. Thus, viral oncogenes and cellular oncogenes are really one and the same.

Tumor Suppressor Genes

Tumor suppressor genes are a second major class of cancer-linked genes. By definition, a tumor suppressor gene is one that, when inactivated, increases cancer risk. Viewed from a different perspective, the introduction of a normal tumor suppressor gene into a cancer cell can often suppress its wild and unruly growth. A mutation can inactivate a tumor suppressor gene by either blocking its expression or converting it to a gene that encodes a nonfunctional gene product. In contrast to oncogenes, tumor suppressor genes act recessively; both copies of a gene pair must be mutated before

TABLE 19.4

Some Known or Suspected Tumor Suppressor Genes

Gene	Proposed Function(s) of Gene Product
Rb	Regulates transcription factors and the cell cycle
p53	Transcription factor; regulates cell cycle, DNA repair, and apoptosis (cell suicide)
BRCA1	Transcription factor
APC	Communicates between cell surface proteins and microtubules
WT1	Transcription factor
NF1	GTPase-activating protein (GAP) for Ras protein
NF2	A cytoskeletal protein
DCC	Cell-surface protein that helps cells stick together
MTS1	Regulates the cell cycle

the genes contribute to the progression towards malignancy (cancer). However, the inheritance of one defective tumor suppressor gene greatly increases the probability that some cells will eventually end up with both members of the gene pair defective. Such inheritance accounts for a significant fraction of the inherited predisposition to cancer. Approximately one dozen tumor suppressor genes have been identified, including the breast cancer susceptibility gene labeled ***BRCA1*** (Table 19.4). It is estimated that up to 85% of the women who inherit a single mutated *BRCA1* gene will develop breast cancer by age 65. Several tumor suppressor genes, including *BRCA1*, appear to encode transcription factors (Section 17.1) or proteins that regulate transcription factors.

The most highly publicized tumor suppressor gene, named ***p53*** ("p" for "protein" and "53" for "53,000" daltons), was selected molecule of the year by *Science* in 1993. More than 50 types of human cancers carry *p53* mutations, and defects in *p53* may be associated with over 50% of all human cancers. The expression of the *p53* gene is enhanced by DNA damage. Once produced, p53 protein leads to the inhibition of DNA replication and cell division and the stimulation of DNA repair, allowing the damaged cell to repair its DNA before that DNA is replicated. This repair is important, because the replication of damaged DNA increases the mutation rate and the chance that an oncogene will be produced or a tumor suppressor gene will be inactivated.

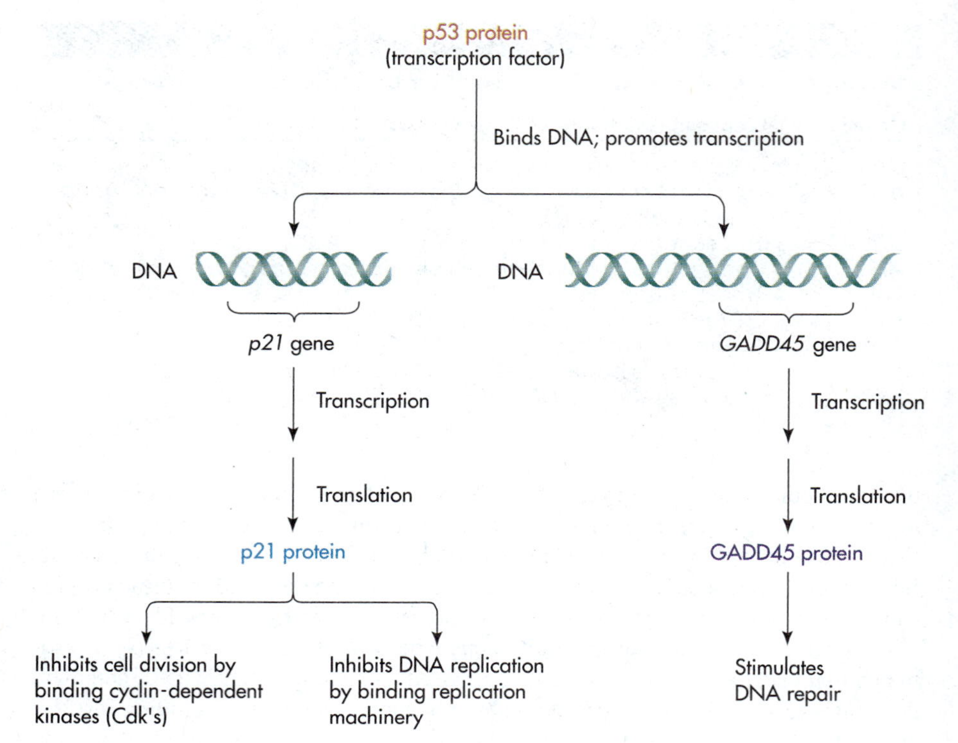

FIGURE 19.10
Mode of action for p53 protein. p53 protein is a transcription factor that inhibits cell division and DNA replication and stimulates DNA repair by promoting the expression of particular genes, including *p21* and *GADD45.*

The normal p53 protein is a transcription factor that promotes the expression of multiple genes, including ***p21*** that encodes a protein called **p21** ("p" for "protein" and "21" for "21,000 daltons"), **WAF1,** or **Cip1** and ***GADD45*** that encodes a protein labeled **GADD45 protein** (Figure 19.10). "GADD" is an acronym for "growth-arrested and DNA-damage-inducible." GADD45 protein stimulates DNA repair, while p21 protein inhibits both DNA replication and cell division. p21 protein's regulation of DNA replication entails its direct interaction with replication machinery. p21 protein blocks the cell cycle by binding to and inhibiting **cyclin-dependent kinases (Cdk's)** that play a central role in driving cell division. The **cyclins** are proteins that regulate cyclin-dependent kinases. The p53 protein encoded in many mutated *p53* genes lacks the ability to stimulate p21 protein and GADD45 protein production. p53 protein's modulation of transcription appears to involve direct interactions with TATA-binding protein (TBP), one of the basal transcription factors (Section 17.1). **[Try Problem 19.20]**

In some types of cells and under some circumstances, normal p53 protein not only halts cell division, stops DNA replication, and stimulates DNA repair, but it also triggers **apoptosis** (programmed cell death; cell suicide). If the DNA in apoptosis-prone cells acquires severe damage, the cells set out on a pathway of self-destruction. This ensures that the damaged DNA is not passed to future generations of cells. By functioning as a damage control specialist, normal p53 protein helps the body avoid the accumulation of cells in which protooncogenes have been converted to oncogenes or tumor suppressor genes have been inactivated. The accumulation of other genetic abnormalities is suppressed as well.

Mice with their *p53* genes experimentally inactivated appear normal at birth, but tumors begin to develop within weeks. By the end of 6 months, all of the mice have tumors or have died.

Mutator Genes

The central role of mutations in the development of cancer is now clear; a cancer normally arises and evolves as a consequence of the accumulation of mutations in multiple genes within the same cell over time. Protooncogenes and tumor suppressor genes are altered by these cancer-linked mutations. Given this fact, it is not surprising that a third class of genes, called **mutator genes,** also impacts one's cancer risk. Mutator genes are genes whose products help determine the mutation rate within a cell. The products of these genes help control the fidelity of DNA replication or participate in the correction of errors made during replication. A mutation in a mutator gene can lead to an increased rate of mutation with a coupled increase in the rate at which oncogenes are generated and tumor suppressor genes are inactivated.

An understanding of the gene–cancer connection has opened the door to new treatment strategies. One of these strategies, known as gene therapy, employs recombinant DNA technology. The same strategy is applicable to the treatment of genetic diseases. Recombinant DNA and genetic engineering (the basis for gene therapy) are the next topics of discussion.

PROBLEM 19.18 Explain how a mutation in the promoter of a protooncogene could modify the frequency of expression of the gene and transforms it into an oncogene. Hint: Review Section 17.1.

PROBLEM 19.19 If a mutation in a *Ras* gene leads to the production of a Ras protein with reduced GTPase activity, will the abnormal protein inhibit, enhance, or have no impact on signal transduction following the binding of a growth factor to a Ras-linked receptor? Explain. Hint: Review Section 10.4.

PROBLEM 19.20 True or false? Based on Figure 19.10, the gene that encodes p21 protein is likely a tumor suppressor gene. Explain.

19.9 RECOMBINANT DNA

Webster's Dictionary defines **genetic engineering** as "the techniques by which genetic material can be altered by recombinant DNA so as to change or improve the hereditary properties of microorganisms, animals, plants, etc." This section examines the construction of recombinant DNA and lays the groundwork for a discussion of genetic engineering, including gene therapy.

A **recombinant DNA (rDNA)** is a double-stranded (duplex) DNA molecule that contains segments of DNA from two or more sources, often distinct species. The construction of such molecules is sometimes described as **gene splicing,** because specific genes are commonly joined (spliced together) in the process. Although multiple approaches have been developed to reshuffle segments of DNA, only the restriction endonuclease–based method will be examined in this text.

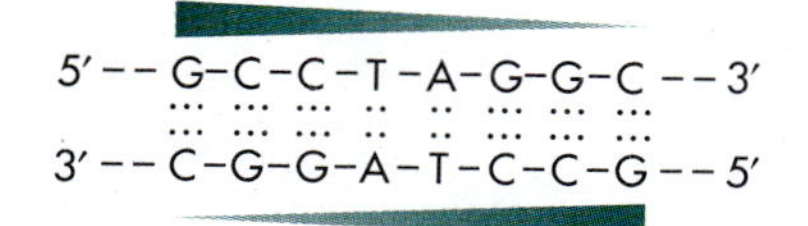

FIGURE 19.11
A palindromic sequence within duplex DNA.

Restriction endonucleases, as noted in Section 19.5, are sequence-specific DNAses. They are produced by microorganisms attempting to protect themselves from foreign DNA; they restrict the expression of the foreign DNA by catalyzing its degradation. Restriction endonucleases catalyze the hydrolysis of duplex DNAs at **palindromic sequences (palindromes),** segments of DNA (typically four to eight base pairs in length) in which the base sequence read in a 5′ to 3′ direction along one strand is the same as the base sequence read in a 5′ to 3′ direction along the complementary strand (Figure 19.11). Segments of DNA containing a palindrome may exist in equilibrium with a **cruciform** structure (Figure 19.12). Cruciforms are created when the two halves of each strand of a palindrome bulge out from the rest of the duplex and pair with one another. Although it is uncertain whether cruciforms exist *in vivo,* one hypothesis suggests that they serve as recognition and attachment sites for a variety of DNA-binding proteins. It is known that palindromic sequences serve as recognition and attachment sites for some proteins. Over 100 different restriction endonucleases, each recognizing a different palindromic sequence, are now commercially available. This pool of enzymes allows genetic engineers to selectively cut (hydrolyze) a DNA molecule at any of a wide variety of different sites. **[Try Problems 19.21 and 19.22]**

Restriction endonucleases not only serve as tools for excising genes, they also facilitate the joining of excised genes to create recombinant DNAs. After some restriction endonucleases bind to a unique palindromic sequence, they catalyze the hydrolysis of one strand of the DNA toward one end of this sequence while catalyzing

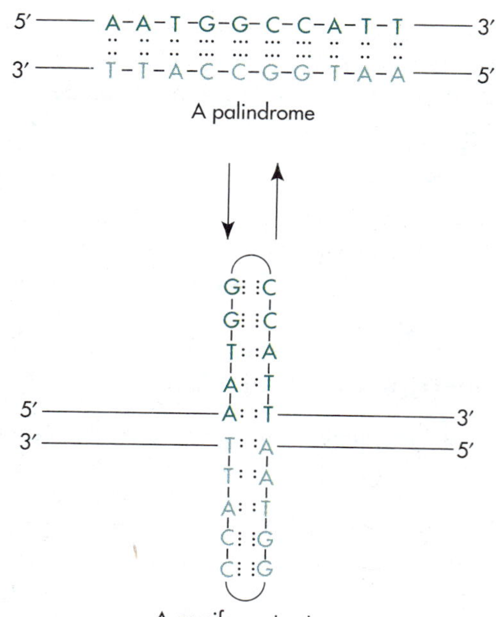

FIGURE 19.12
A cruciform structure in DNA.

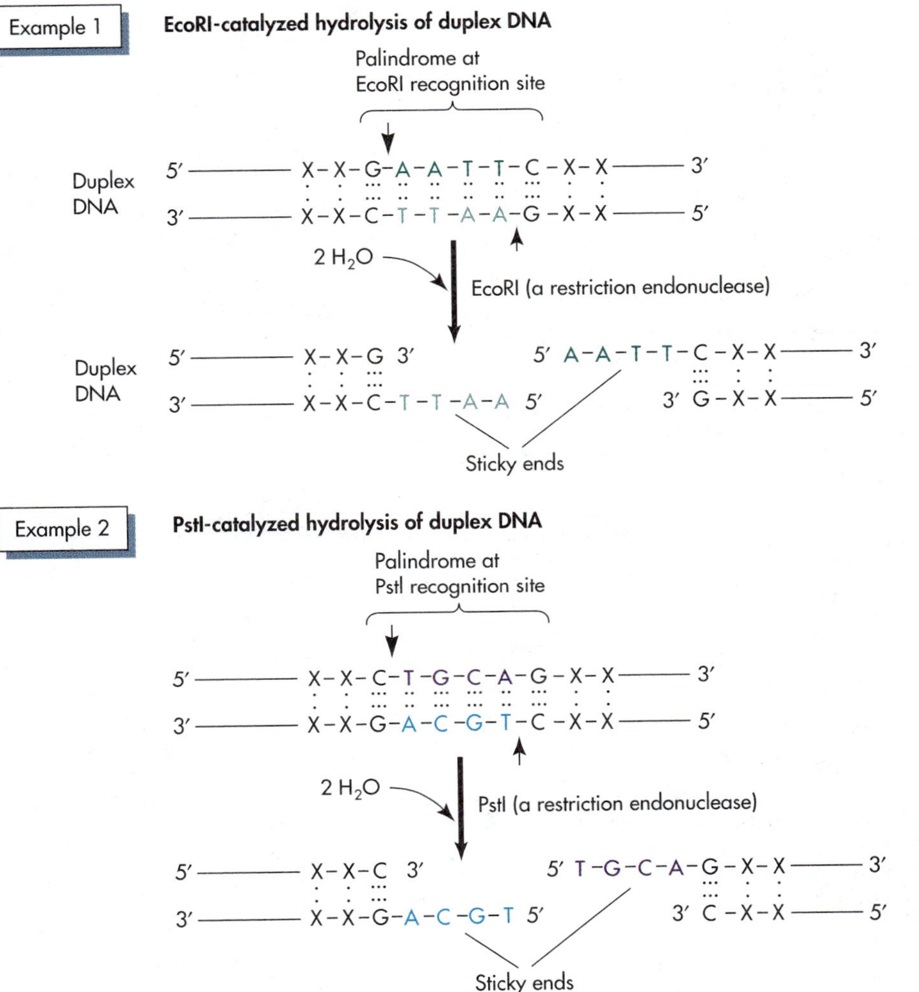

FIGURE 19.13 The generation of sticky ends through the restriction endonuclease–catalyzed hydrolysis of duplex DNA. Arrows (↑,↓) identify specific sites of hydrolysis within palindromic sequences.

the hydrolysis of the second strand of the DNA toward the opposite end of the palindrome (Figure 19.13); they do not catalyze the hydrolysis of both strands of the duplex in the middle of the palindrome. The two DNA fragments produced tend to cling together through noncovalent base pairing, since each fragment contains a short single strand end (tail) that is complementary to the single strand end of the other fragment. Single strand ends that can base pair along their entire lengths are said to be sticky or cohesive. The same **sticky ends** are produced each time a given restriction endonuclease catalyzes the hydrolysis of any duplex DNA molecule. Sticky ends play a key role in the joining of DNA fragments to produce recombinant DNAs. **[Try Problems 19.23 and 19.24]**

Assume that a bacterial DNA molecule contains one copy of the palindromic sequence recognized by the restriction endonuclease called EcoRI and a human DNA molecule contains two copies of the same palindrome. If these DNAs are separately

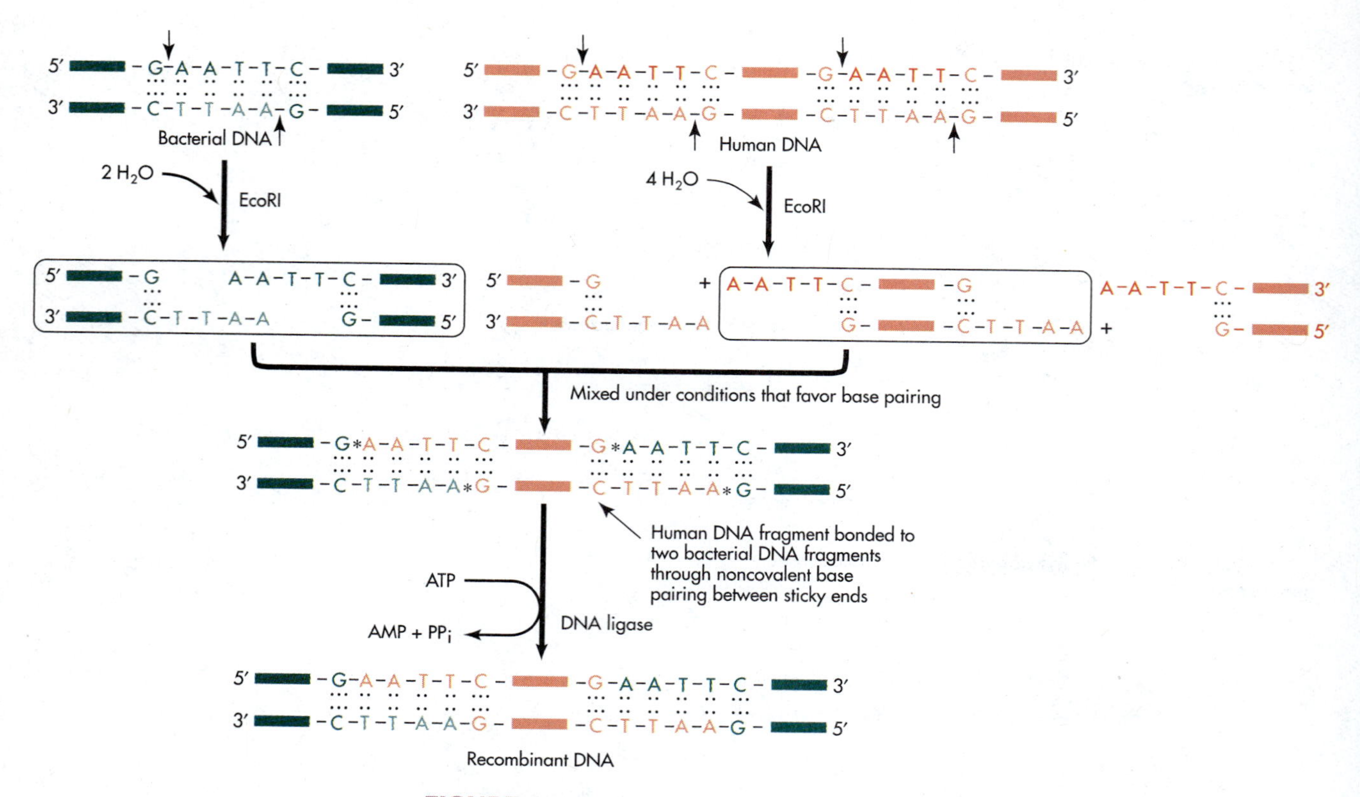

FIGURE 19.14
Construction of a recombinant DNA molecule. The final product is a duplex DNA molecule containing a human DNA fragment spliced into a bacterial DNA. Arrows (↑,↓) locate sites of EcoRI-catalyzed hydrolysis. Asterisks show sites where DNA ligase catalyzes the formation of phosphodiester bonds to covalently join (ligate) the three DNA fragments.

digested with EcoRI, the bacterial DNA will be hydrolyzed at its single EcoRI recognition site while the human DNA will be cut at both of its EcoRI recognition sites (Figure 19.14). The single-strand ends of the two bacterial DNA fragments will be complementary to some of the single-strand ends of the three human DNA fragments and vice versa. If appropriate DNA fragments are mixed under conditions where sticky ends will base pair, some constructs will be formed in which bacterial DNA fragments are noncovalently attached to a human DNA fragment through sticky end interactions. **DNA ligase,** an enzyme encountered during the discussion of DNA replication in Section 17.4, can then be used to catalyze the joining (sealing) of these fragments through covalent phosphate diester links. The product is a recombinant DNA. Most recombinant DNAs are constructed in this fashion. **[Try Problems 19.25 and 19.26]**

PROBLEM 19.21 Write the structural formula for a dinucleotide and then write the equation for the hydrolysis of this dinucleotide to produce two nucleotides. What specific bond is cleaved (cut) during this hydrolysis reaction? The same bond is cleaved during the restriction endonuclease–catalyzed hydrolysis of duplex DNA.

PROBLEM 19.22 Identify the largest palindrome in the following duplex DNA molecule, and then rewrite the abbreviation for this DNA in a cruciform structure.

5′ dGACCTCACCGTACGGTCGCA 3′
3′ dCTGGAGTGGCATGCCAGCGT 5′

PROBLEM 19.23 Identify each PstI recognition site (see Figure 19.13) within the following DNA molecule. How many duplex DNA fragments are produced when this molecule is hydrolyzed with the catalytic assistance of PstI?

5′ dATGCTGCAGTTATACTGCAGGCACACTGCAGTTATAA 3′
3′ dTACGACGTCAATATGACGTCCGTGTGACGTCAATATT 5′

PROBLEM 19.24 Write an abbreviated structure for a duplex DNA molecule that has a single strand end that will stick to (pair with) the single strand end in the following DNA molecule:

5′ dATAGCTGCATAGCTCGACCCGTACGATGACAGTTGCGG 3′
3′ dTATCGACGTATCGAGCTGGGCATGCTAC 5′

PROBLEM 19.25 In the experiment illustrated in Figure 19.14, a mixture of "spliced" products is actually obtained. The product shown must be isolated from this mixture. Describe or illustrate one other product that is formed along with the product shown.

PROBLEM 19.26 Can recombinant DNA technology be used to rearrange the genes in a human chromosome? Explain.

19.10 GENETIC ENGINEERING

Most of the genetic engineering that is making headlines today involves the transfer of specific genes into selected host cells. There are typically three steps involved: (1) Restriction endonucleases are used as molecular scissors to cut (catalyze the excision of) the desired gene out of the donor DNA; (2) the excised gene is covalently joined to a **carrier DNA (vector);** and (3) the resultant recombinant DNA is introduced into a host cell (Figure 19.15). The gene to be transferred is not always cut out of a DNA. It is often synthesized from the mRNA that it encodes with the catalytic assistance of a reverse transcriptase (Section 17.5). Alternatively, a gene can be chemically synthesized. **[Try Problem 19.27]**

EXHIBIT 19.3
Functions of Gene Vectors

- Transport genes into host cells.
- Help genes survive and replicate.
- Help genes integrate into host chromosome.
- Regulate expression of genes.

Gene vectors are designed to do one or more of the following: transport genes into host cells; help genes survive and replicate once inside host cells; help genes integrate into host chromosomes; and regulate gene expression (Exhibit 19.3). The most common vectors are plasmids and viruses. **Plasmids** are extrachromosomal, circular, duplex DNA molecules that are capable of replicating independently of a

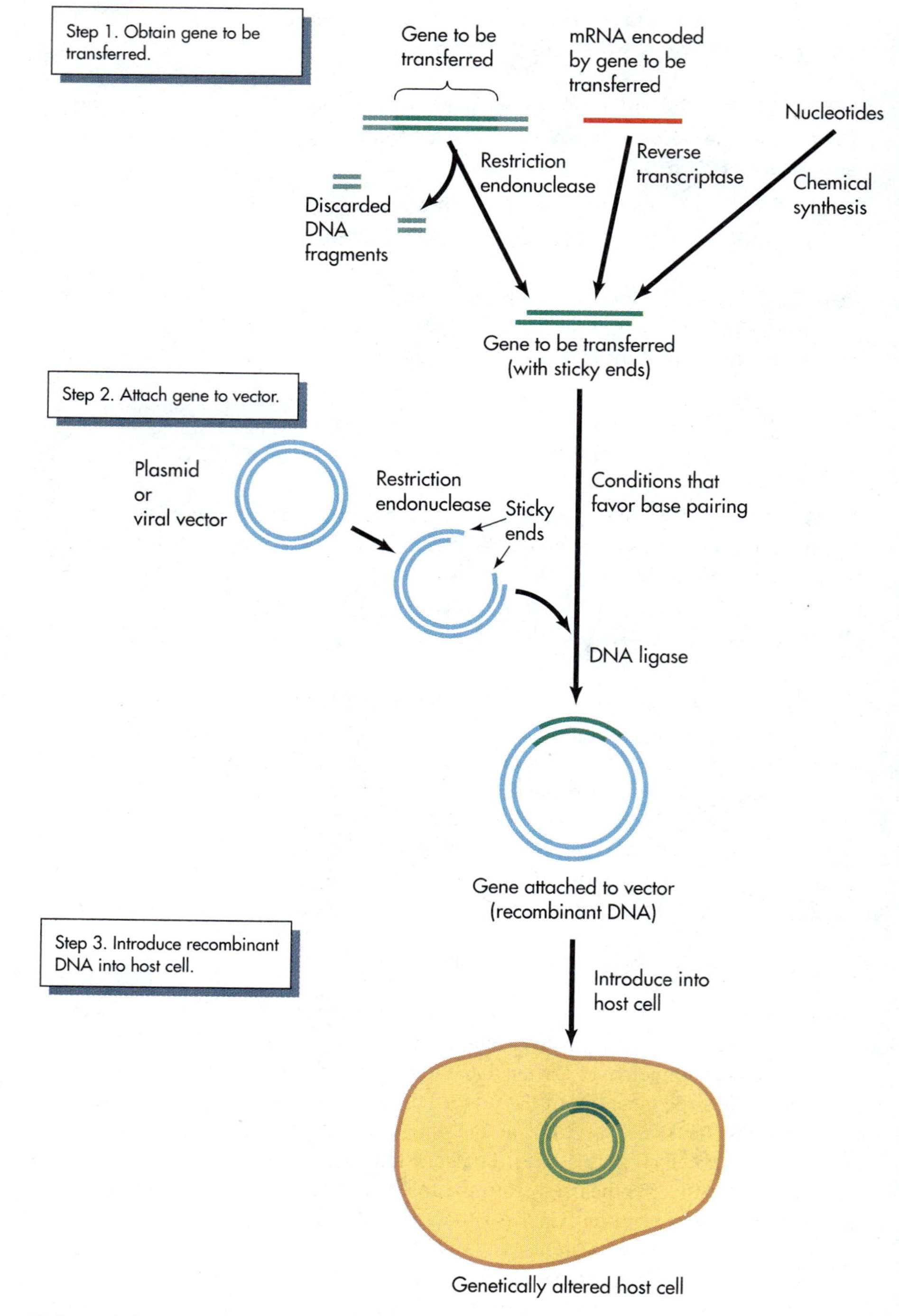

FIGURE 19.15
Steps in a typical gene transfer experiment.

cell's chromosomes. They exist naturally in a wide variety of bacteria, plants and animals, and some can be reversibly inserted into cellular chromosomes. Cells may contain multiple plasmids and multiple copies of the same plasmid.

Viruses contain a nucleic acid core (the viral chromosome or viral genome) surrounded by a protein coat (Section 17.5). A virus normally infects a host cell by binding to specific receptors on the plasma membrane, and then introducing its chromosome into that cell. The viral nucleic acid (usually a duplex DNA) takes control of certain cellular events in order to support its own replication and the reproduction of the virus. The host cell is frequently damaged or killed in the process. Some viral chromosomes, like some plasmids, can be reversibly inserted into cellular chromosomes. Clever genetic engineers have created crippled viruses that can be used to introduce foreign genes into host cells without damaging the cells.

A variety of other vectors have also been developed. Some, such as **yeast artificial chromosomes (YACs),** allow much larger chunks of DNA to be transferred than is possible with the usual viral and plasmid vectors. A YAC contains, in addition to the DNA to be transferred, control elements from yeast which make it look, and replicate, like a yeast chromosome. YACs are facilitating the study of genetic diseases and normal traits that involve multiple genes. Because of the importance of vectors in gene therapy (Section 19.14), the National Institutes of Health (NIH) established three *national gene vector laboratories* (University of Indiana, University of Michigan, and University of Pennsylvania) in 1995 to create and produce high-quality gene transfer agents for use by physicians around the country.

Most purified DNA molecules, including recombinant DNAs, have no built-in mechanism for entering cells. Under normal conditions, their entry is determined by the ability of host cells to passively take up any extracellular DNA at a very low efficiency. For this reason, genetic engineers have developed methods to enhance the entry of DNAs into host cells. Some of these methods are summarized in Table 19.5.

TABLE 19.5

Methods for Enhancing the Entry of Recombinant DNAs into Host Cells

Method	Description
1. Calcium phosphate	Host cells are incubated with DNA that has been precipitated as an insoluble calcium phosphate complex.
2. Cationic liposomes	DNA is incorporated into artificial, lipid bilayer–enclosed particles known as liposomes or lipid vesicles that are able to fuse with plasma membranes and release their contents into host cells.
3. Electroporation	An electric field is used to increase host cell permeability to DNA by reversibly disrupting local regions of the plasma membrane.
4. Bioblaster	Microscopic, DNA-coated spheres are fired at host cells at speeds that allow them to pass through plasma membranes without seriously damaging the cells.
5. Microinjection	A microsyringe is used to inject a DNA solution into host cells.

When a viral vector is used, virus particles can be employed to introduce attached genes into host cells.

Why would we want to introduce a foreign gene into a cell or organism? The reasons are many and varied. Some of these reasons are explored in the next four sections.

PROBLEM 19.27 List the three separate steps catalyzed by a reverse transcriptase during the production of a duplex DNA from an mRNA. Hint: Review Section 17.5.

19.11 MICROBIAL BIOTECHNOLOGY

Biotechnology is the application of biological systems and organisms to technical and industrial processes. This discipline has been booming since the development of recombinant DNA technology and a 1980 U.S. Supreme Court ruling that allows new life forms created through genetic engineering to be patented. There are presently hundreds of biotechnology companies that, collectively, have sought thousands of patents on genetically modified organisms. The world's first patent on a genetically altered mammal (a mouse) was issued in the United States in 1988, and by mid-1989 approximately 50 animal patent applications had accumulated in the U.S. Patent Office. Initially, the patenting of mammals generated considerable debate within both the public press and the U.S. Congress.

One of the first practical applications of recombinant DNA technology involved the introduction of specific foreign genes into bacteria, yeast, and other microorganisms. Some of these organisms incorporated the genes into their chromosomes and started to produce foreign proteins. In the process, they were functioning as small factories, often assembling large quantities of a protein that was in high demand and short supply. Proteins synthesized by such genetically engineered organisms are now being used in biochemistry, clinical laboratories, medicine, agriculture, food processing, and other disciplines. **[Try Problem 19.28]**

Human insulin was the first microbially produced human protein to reach the market in the United States. Sold under the trade name **Humulin,** this product is now being used by over half of the millions of diabetics in the United States. In the past, some of these patients were forced to use animal insulins because of a shortage of the human hormone. These animal proteins lead to adverse side reactions in some users.

Microbially produced **human growth hormone (hGH)** quickly followed human insulin to the pharmaceutical marketplace. It was approved for clinical use by the Food and Drug Administration in the fall of 1985, and it is now being used to treat dwarfism in those children unable to produce enough of this essential protein. Before 1985, hGH was difficult to obtain and always in short supply. Since 1985, drug companies and scientists have been looking into novel uses for this now abundant protein. In 1987 the National Institutes of Health began a very controversial clinical trial designed to determine whether the drug can be used to treat prospective height deficiency by making healthy short children grow taller. The effects of the protein on weight control and aging are also being examined. Sales of the drug are tightly monitored since it is abused by some athletes who hope that it will increase muscle mass and improve performance. The possible health consequences of excess hGH are uncertain.

The list of foreign proteins now being produced by genetically engineered microorganisms or cultured cells is already quite long and it continues to grow. Additional examples are provided in Table 19.6. One of these examples, **recombinant bovine growth hormone (rbGH),** is used to increase milk production in dairy cattle. It has been highly publicized in both Europe and the United States because of its potential economic, social, and health impacts. Some claim that the use of rbGH will increase milk supply, cause a drop in milk prices, and force small dairy farmers out of business. Others have suggested that the use of rbGH may lead to an increased use of antibiotics by farmers (treated cattle have a higher than normal incidence of udder infection) and an increase in antibiotic residues in milk. Animal rights activists are concerned about the impact the hormone may have on the cattle themselves. The Food and Drug Administration approved the use of rbGH (marketed by Monsanto under the trade name **Posilac**) in dairy cattle in 1994. **[Try Problems 19.29 and 19.30]**

Microorganisms have been programmed to produce a variety of valuable products in addition to proteins. *Escherichia coli,* for example, has been engineered to build an unusually large amount of tryptophan (one of the 20 common protein amino acids; Chapter 3) and to convert part of that tryptophan to indigo, a dye used to treat denims. Such bioengineering feats normally involve the transfer of specific enzyme-

TABLE 19.6

Some Foreign Proteins Being Produced by Genetically Engineered Microorganisms or Eukaryotic Cells

Protein(s)	Activity, Use, or Proposed Use
Alpha-1-antitrypsin	Treat emphysema
Bone morphogenetic proteins	Stimulate bone and cartilage formation
Bovine growth hormone (bGH)	Increase milk production in cows
Calf prochymosin	Manufacture cheese
Colony stimulating factors	Stimulate growth of immune cells
Erythropoietin	Stimulate red blood cell production
Factor VIII	Treat hemophilia
Foot and mouth vaccine	Immunize cattle against foot and mouth disease
Human growth hormone (hGH)	Treat dwarfism
Insulin	Treat diabetes
Insulin-like growth factor	Treat osteoporosis and a rare form of dwarfism
Interferons	Treat cancer and viral infections
Tissue plasminogen activator (TPA)	Treat heart attacks; dissolves blood clots

encoding genes to the target organism. After the genes are expressed, the resultant enzymes may catalyze reactions that are foreign to the host organism or enhance existing reactions. Indigo-producing *E. coli* are of commercial interest since most indigo used today is synthesized in factories from toxic chemicals, including aniline, formaldehyde, and sodium cyanide.

The genetic engineering of microorganisms does pose some potential risks, and there is the possibility that this technology will some day be misused. What, for example, would happen if a new pathological life form were accidentally or intentionally (possibly for biological warfare) created in a lab and then escaped (or was intentionally released) into our environment? Could genetically altered organisms impact an ecosystem even if they were not pathological to humans? For obvious reasons, certain types of gene transfer experiments are carried out under rigorous containment conditions.

PROBLEM 19.28 Once a bacterium is genetically programmed to produce a human protein, can that ability be passed on to future generations of bacteria? Explain.

PROBLEM 19.29 True or false? We presently have the technology to program bacteria to produce virtually any human protein. Explain.

PROBLEM 19.30 Some of the polypeptides synthesized in the human body must be processed to become biologically active (Section 18.11). What problems does mandatory processing create for the genetic engineer who wants to program a bacterium to produce the biologically active form of a protein?

19.12 PLANT BIOTECHNOLOGY

The genetic engineering of plants may well be the most rapidly advancing area of biotechnology. A wide variety of plants have been genetically altered, and hundreds of field trials, involving dozens of crops, have been carried out. The bioengineering of plants is relatively simple because most plant cells are **totipotent;** each cell can be used to generate a mature, reproductively competent plant. In mammals, only zygotes and early embryo cells can as yet be used to grow new organisms, and this requires a host uterus.

A **transgenic organism** (also called **intergeneric organism**) carries one or more foreign genes that can be transmitted to offspring. Field trials with transgenic plants require a permit, and the test plots must normally be surrounded by moats, fences, and vegetation-free areas to ensure that genetically altered plants do not escape and lead to possible ecological or environmental problems. In spite of all the precautions, a half acre of experimental transgenic corn was washed down Beaver Creek in Iowa by floods during the summer of 1993.

Many transgenic plants have been created that possess improved agronomic traits such as pest and disease resistance (Table 19.7). Virus resistance, for example, has been programmed into a wide variety of crops by introducing genes for the coat

TABLE 19.7

Some Transgenic Plants with Improved Agronomic Traits

Plant	Traits[a]
Apple	Insect resistance
Oilseed rape (canola)	Herbicide tolerance, insect resistance, modified seed oils
Corn	Herbicide tolerance, insect resistance, virus resistance
Potato	Herbicide tolerance, insect resistance, virus resistance, increased starch production
Rice	Insect resistance, modified seed protein storage
Strawberry	Insect resistance
Tomato	Herbicide tolerance, insect resistance, virus resistance, modified ripening
Walnut	Insect resistance

[a]Herbicide tolerance, insect resistance, and virus resistance does not indicate a resistance to all herbicides, all insects, or all viruses. In most cases, resistance is limited to a small number of herbicides, insects, or viruses.

protein of the target virus. The virus proteins produced from information in the viral genes serve as "vaccines"; they "immunize" the plants against the virus. Various plants have been made insect resistant by introducing a gene (from the bacterium *Bacillus thuringiensis*) that encodes a protein toxic to some insects but apparently harmless to mammals and most other organisms. Researchers are also attempting to create crops that are drought resistant, frost resistant, heat resistant, salt resistant, and self-fertilizing.

Another area of plant biotechnology entails efforts to design plants that produce fruits with delayed spoilage, enhanced flavor, and improved nutritional value. Progress is being made. The use of genetic engineering to create a slow-ripening, stay-hard tomato was described in Section 16.8 during the discussion of antisense nucleic acids. That tomato, marketed by Calgene under the name Flavr Savr, began appearing on supermarket shelves in May 1994. Potatoes have been engineered to produce more starch, in part because high starch potatoes are cheaper and easier to process into french fries and potato chips. **[Try Problem 19.31]**

Plants, like microorganisms, are also being converted into **bioreactors** that produce specialty chemicals and novel biopolymers. Turnips have been programmed to make human interferon (Table 19.6). Some tobacco plants are building mouse antibodies, called **plantibodies,** that may some day be used for diagnostic purposes. Other tobacco plants are programmed to produce human serum albumin (employed in fluid replacement solutions and for other purposes) and the starch digesting enzyme α-amylase (Chapter 6; uses include the production of breads and low calorie beers). The oilseed rape plant is now a source of Leu-enkephalin, one of the human opiate peptides discussed in Section 3.6. The list goes on.

PROBLEM 19.31 Describe two genes or classes of genes that could be used to improve the nutritional value of plants.

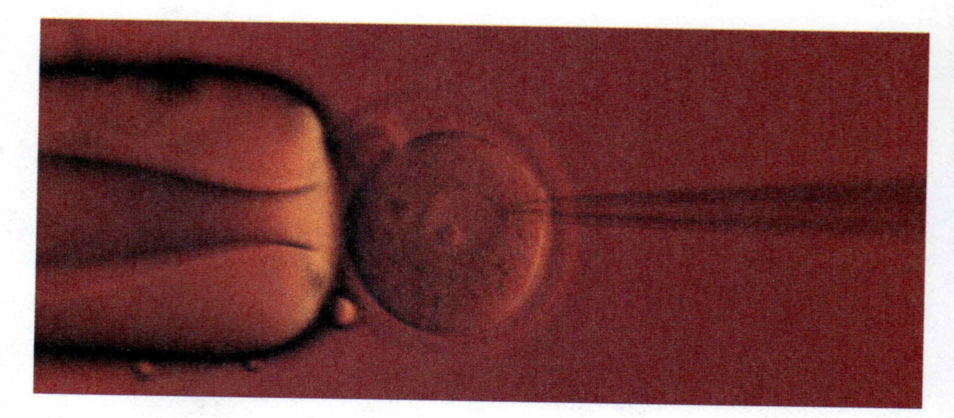

FIGURE 19.16
The microinjection of a mouse egg cell.

19.13 ANIMAL BIOTECHNOLOGY

FIGURE 19.17
A super mouse with a normal mouse.

A large number of both human and nonhuman genes have been introduced into a variety of animals, including mammals. Some mice, pigs, and rabbits, for example, now carry human growth hormone (hGH) genes and make detectable amounts of this human protein. These animals were produced by injecting recombinant DNA molecules into the nuclei of fertilized egg cells (Figure 19.16), and then implanting these egg cells into the uteruses of the animals that had donated them. Some of the mature organisms that developed from the genetically engineered egg cells incorporated the hGH gene into their chromosomes and were able to pass this gene on to their offspring. Some of the transgenic mice have grown to nearly twice the size of normal mice as a consequence of the hGH they have produced. These giant mice, sometimes described as **super mice** or **mighty mice,** have made headlines around the world (Figure 19.17). Although pigs engineered to produce hGH have not grown to be much larger than their normal siblings, further research may eventually lead to giant hams and giant slices of bacon. Some fish engineered to produce recombinant fish growth hormone do grow much more rapidly than their normal siblings (Figure 19.18). **[Try Problem 19.32]**

Transgenic farm animals, like microorganisms and plants, are being used as bioreactors. Human alpha-1-antitrypsin (Table 19.6) and human blood clotting fac-

FIGURE 19.18
"Super" fish. Nontransgenic (left) and growth hormone transgenic (right) coho salmon siblings at 14 months of age. Length of fish at far right is 41.8 cm.

tor IX are both being isolated in the laboratory from the milk of sheep containing the corresponding human genes. Some of these critters would make good targets for rustlers; a single animal can produce over $100,000 worth of antitrypsin each year. Human tissue plasminogen activator (Table 19.6) has been found in the milk of transgenic goats, and specific human genes have been successfully introduced into dairy cattle. Even pigs have been programmed to secrete human proteins in their milk. The use of transgenic farm animals to produce pharmaceutical agents is called **gene pharming.**

The creation of animal models for human diseases is another hot area of animal biotechnology. In fact, the first mammalian patent was issued on a mouse (called an oncomouse) designed to be particularly susceptible to cancer. Such animals are extremely valuable in cancer research, and shortly after patenting in 1988, an oncomouse was selling for around $50. Research on any human disease is facilitated if animals can be found or created that possess analogous ailments. These animals can then be employed to unravel the molecular interactions associated with the disease and to test treatment strategies. Table 19.8 lists some of the animal models presently being used to study human diseases.

TABLE 19.8

Some Animal Models Used To Study Human Diseases

Animal	Gene Defect	Human Disease Equivalent
Watanabe rabbit	Low-density lipoprotein (LDL) receptor	Familial hypercholesterolemia (propensity to heart disease)
"Cone-head" mouse	β-Glucuronidase	Sly syndrome (a lysosomal storage disease; brain and skeletal damage)
mdx Mouse	Dystrophin	Duchenne muscular dystrophy (progressive wasting of muscle)
Mouse	Glucocerebrosidase	Gaucher disease (enlarged spleen, bone lesions, cerebral degeneration)
Mouse	Hypoxanthine–guanine phosphoribosyl transferase (HGPRT)	Lesch–Nyhan syndrome (mental retardation, spasticity, and compulsive biting of lips and fingers)
Golden retriever	Dystrophin	Duchenne muscular dystrophy (progressive wasting of muscle)
Irish setter	Factor VIII	Hemophilia A (delayed blood clotting)
Mouse	Cu_1Zn superoxide dismutase (SOD)	Amyotrophic lateral sclerosis (ALS) (a neurodegenerative disease also known as Lou Gehrig's disease; always fatal)
Mouse	β-Amyloid precursor protein (APP)	Alzheimer's disease (neurodegenerative disorder leading to senile dementia and ultimately death)

One of the common approaches to the creation of an animal model involves the inactivation of specific genes within totipotent embryo stem cells of experimental animals. This is accomplished by isolating copies of the targeted genes and then inactivating them through the *in vitro* insertion of an additional segment of DNA that disrupts their normal sequence. After the modified genes are introduced into the stem cells whose genes are to be inactivated, they selectively interact with their normal counterparts by pairing with complementary sequences. A small fraction of the treated cells replace the normal genes with the disrupted ones (a process called **homologous recombination**) or insert the abnormal genes into the normal ones. Either way, the function of the normal genes is knocked out. The modified stem cells are implanted into the uteruses of host animals where they may grow into a mature organism unable to express the knocked-out genes. If they fail to do so, this may mean that the knocked-out genes are essential for development or survival. Designer **knockout mutants** are used to help discover the roles and functions of the genes that are knocked out. Some also serve as animal models for human genetic diseases.

Genetically engineered animals are being developed for an enormous number of purposes in addition to those identified in this glimpse at animal biotechnology. Pigs, for example, are being programmed to produce leaner meat, and researchers are attempting to bioengineer mosquitoes that are unable to serve as hosts for the protozoan parasite that leads to malaria. [Try Problem 19.33]

PROBLEM 19.32 Some might consider the production of a giant mouse a misuse of genetic engineering. However, these mice, along with many related transgenic animals, are of enormous value to researchers studying basic biochemistry. Explain this fact.

PROBLEM 19.33 Can (in theory, at least) the eye color, intelligence, and size of an individual be influenced by manipulating the gene content of the fertilized egg cell from which that individual develops? Explain.

19.14 GENE THERAPY

The first federally authorized introduction of a foreign gene into a human occurred in 1989. At that time, white blood cells carrying a bacterial gene were injected into a terminally ill cancer patient. The foreign gene allowed researchers to distinguish the injected blood cells, known as **tumor-infiltrating lymphocytes,** from preexisting ones. This ability made it possible to determine the life span of the injected cells, to study their movement throughout the body and to monitor their interactions with cancer cells.

Gene therapy is the transfer of new genetic material to the cells of an organism with resulting therapeutic benefit. The first federally authorized attempts at human gene therapy were initiated in 1990. White blood cells were removed from two girls, aged 4 and 9 years, and grown in culture (Figure 19.19). Both girls lacked a functional **adenosine deaminase (ADA)** gene, a defect that leads to **severe combined immunodeficiency (SCID)** and frequent life-threatening bouts of infection. After

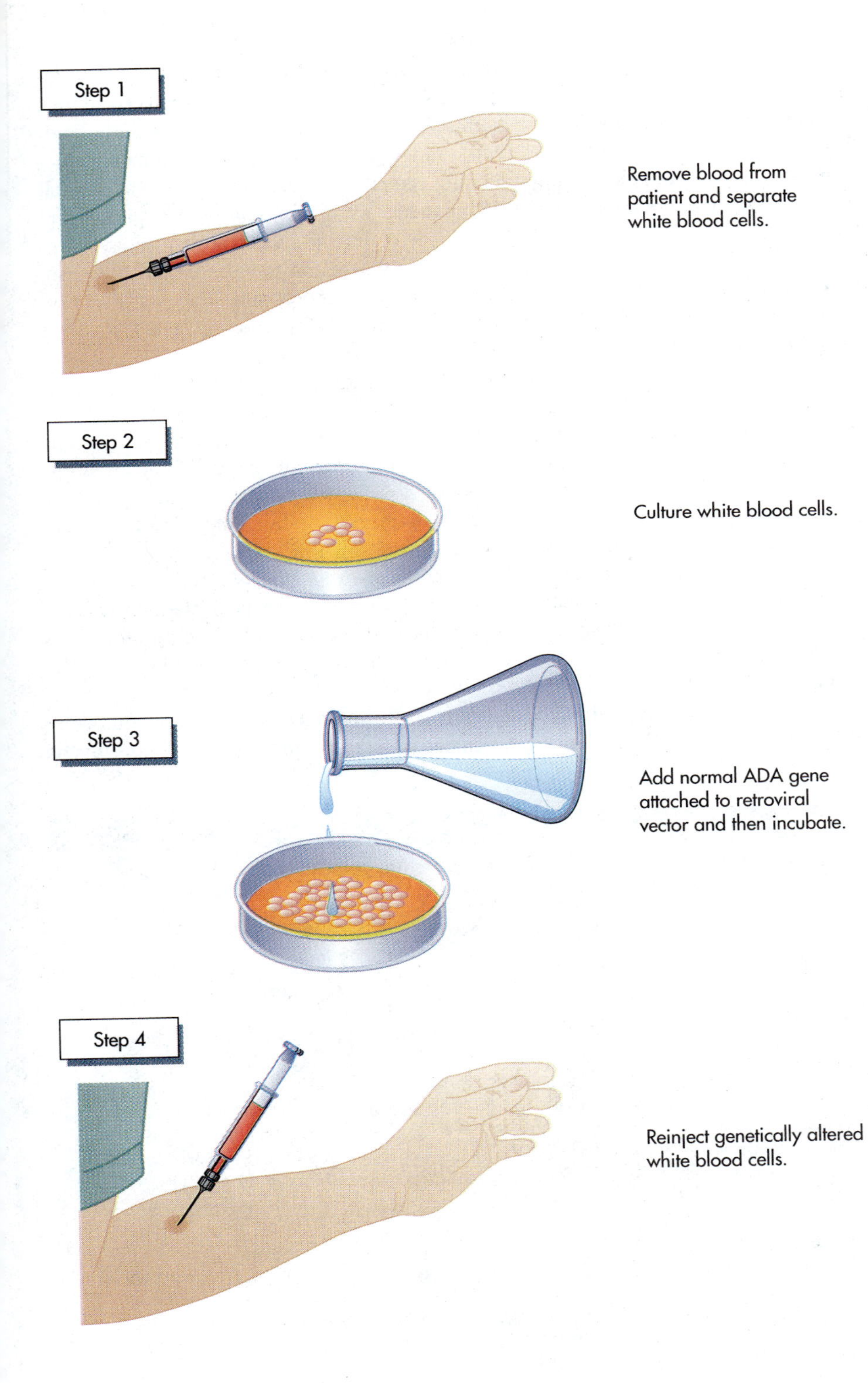

FIGURE 19.19
Gene therapy for ADA deficiency.

normal copies of the ADA gene were attached to a retroviral vector, they were introduced into the cultured white blood cells. The genetically altered cells were then injected back into the blood of the donors. This treatment protocol increased the total number of blood lymphocytes (white blood cells) and improved immune function. In one girl, ADA levels in circulating lymphocytes rose to 25% of the normal value. Soon after treatment began, both patients were attending public schools and subject to no more than the average number of infections. Treatment must be repeated periodically since the engineered lymphocytes do have a limited life span. Once researchers are able to isolate and genetically reprogram bone marrow stem cells (the cells from which lymphocytes and other blood cells develop), a single treatment may suffice. [Try Problems 19.34 and 19.35]

The introduction of ADA genes into SCID patients is an example of **gene augmentation therapy,** therapy designed to introduce normal copies of defective genes into patients. The ideal candidates are those genetic diseases that are caused by a single abnormal gene whose product is inactive or only partially active. Table 19.9 provides a partial list of those diseases considered prime targets for such therapy. Clinical trials are underway for several of these diseases including hemophilia B, cystic fibrosis, and familial hypercholesterolemia.

1991 saw the first attempt to use gene therapy to treat cancer. That year the first of several terminally ill malignant melanoma (a form of skin cancer) patients was injected with tumor-infiltrating lymphocytes that had been genetically programmed to synthesize **tumor necrosis factor (TNF),** a protein shown to cause the shrinkage of tumors in mice. Since tumor-infiltrating lymphocytes are known to infiltrate and destroy tumors, it was hypothesized that any TNF produced by these immune cells would help them eliminate the targeted melanoma cells. Unfortunately, the results of

TABLE 19.9

Some Genetic Diseases Considered Prime Targets for Gene Therapy

Disease	Product of Defective Gene	Symptoms
Severe combined immunodeficiency (SCID)	Adenosine deaminase (ADA)	Immunodeficiency; susceptibility to infection
Emphysema	Alpha-1-antitrypsin	Progressive degeneration of lungs
Cystic fibrosis	Cystic fibrosis transmembrane regulator	Thick mucus clogs lungs and digestive tract
Hemophilia A	Factor VIII	Delayed blood clotting
Hemophilia B	Factor IX	Delayed blood clotting
Gaucher disease	Glucocerebrosidase	Enlarged spleen, bone lesions, cerebral degeneration
β-Thalassemia	β-Chain of hemoglobin	Anemia
Sickle cell anemia	β-Chain of hemoglobin	Anemia, damage to vital organs
Familial hypercholesterolemia	Low-density lipoprotein (LDL) receptor	Heart disease

TABLE 19.10

Some Gene Therapies in Clinical Trials

Disease	Therapeutic Gene
ADA-deficient SCID	Adenosine deaminase (ADA)
Cystic fibrosis	Cystic fibrosis transmembrane regulator
Familial hypercholesterolemia	Low-density lipoprotein (LDL) receptor
Ovarian cancer	Herpes simplex thymidine kinase
Brain tumor	Herpes simplex thymidine kinase
Malignant melanoma	Interleukin-2
Hemophilia B	Factor IX
AIDS	Human immunodeficiency virus envelope

this trial were not encouraging, and the trial became embroiled in controversy. In spite of this fact, gene therapy is now an accepted therapeutic option for the treatment of cancer and numerous other ailments in addition to genetic diseases. Dozens of clinical trials are in progress around the world and hundreds of individuals have received exogenous functional genes with therapeutic intent. Table 19.10 identifies some of the diseases being targeted by these trials.

PROBLEM 19.34 The introduction of normal copies of a defective gene into the cells of an individual with a genetic disease does not guarantee that there will be any therapeutic benefit. Explain this fact.

PROBLEM 19.35 Suggest which genes should be knocked out to create a mouse model for SCID? If the knockout mouse did not develop SCID, what conclusions could be drawn?

SUMMARY

Genetics plays a central role in determining our susceptibility to virtually all diseases. Genetic diseases are those that are inherited as a consequence of an abnormal human genome. Most can be traced to mutations in egg or sperm cells or their germ cell precursors. To acquire a recessively inherited disease, we must receive an abnormal set of genes from both parents. A disease that results from the presence of a single abnormal set of genes is said to be dominantly inherited. Although the symptoms of a limited number of human genetic diseases respond well to treatment, there is presently no cure for any genetic disease if *cure* means the correction of the molecular defect ultimately responsible.

The frequency of a genetic disease is at least partially maintained through the acquisition of new mutations during each generation and the compensating advantages associated with some disease-linked genes. DNA repair systems play a key role in determining mutation rates and controlling the frequency of inherited defects.

Although some genetic defects can be identified on the basis of unique symptoms soon after birth, symptom-based diagnosis is not immediately possible for other diseases. Carriers of most recessively inherited defects generally do not have symptoms of disease. For this reason, the identification of diseased individuals and carriers is often based on chemical analysis. This analysis commonly entails the use of DNA probes to search for disease-linked RFLPs. Cancer cells, viruses, and bacteria can also be located and identified on the basis of the unique genetic markers they usually possess.

Sickle cell anemia is a recessively inherited genetic disease linked to a single amino acid change in the β-chain of hemoglobin. The abnormal hemoglobin molecules aggregate within red blood cells to form fibers that cause these cells to become sickle shaped. The misshapen red blood cells are easily ruptured and tend to scar capillaries. They can also clog capillaries, cutting off the flow of blood to tissues. The disease is particularly common in malaria-afflicted regions of the world, where the gene involved tends to protect carriers from malaria.

PKU, another recessively inherited disease, is caused by defective genes for phenylalanine hydroxylase, an enzyme involved in the conversion of phenylalanine to tyrosine. In diseased individuals, blood concentrations of phenylketones (produced following the buildup of unused phenylalanine) rise to toxic levels (the basis for diagnosis) that lead to mental retardation and other symptoms. Early detection and dietary restrictions can prevent the development of the most severe symptoms.

Cancer is usually a consequence of mutations in protooncogenes and tumor suppressor genes. Protooncogenes encode proteins involved in the positive regulation of normal cell growth and division. Mutations in protooncogenes create oncogenes whose products lead to the overstimulation of cell growth and division. The products of normal tumor suppressor genes suppress the growth of both normal and cancerous cells. Mutations that block the expression of these genes or inactivate the gene products lead to a significant increase in cancer risk.

Mutations in a tumor suppressor gene named *p53* are linked to a wide variety of tumors, and they may be involved in the development of over 50% of all human cancers. The normal p53 protein is a transcription factor that promotes the production of two proteins, p21 and GADD45, that (directly or indirectly) block cell division, inhibit DNA replication, and stimulate DNA repair. In some cell types, normal p53 protein also triggers apoptosis. Collectively, the activities of the normal p53 protein tend to resist the accumulation of abnormal DNA, including oncogenes and inactivated tumor suppressor genes.

Genetic engineering is the use of recombinant DNA to change or improve the hereditary properties of an organism. It can be used to make genetic alterations not possible with traditional breeding and selection techniques. Most genetic engineering involves the splicing of specific genes into either virus or plasmid vectors and the introduction of the resultant recombinant DNAs into host cells. The splicing is accomplished with the catalytic assistance of restriction endonucleases and DNA ligases, key tools of the trade. The entry of recombinant DNAs into host cells can be enhanced with a variety of techniques, including Ca^{2+} incubation, electroporation, cationic liposomes, and microinjection.

Genetic engineering is at the heart of the biotechnology industry, an industry that utilizes biological systems and living organisms to tackle technical and industrial problems. A wide variety of microorganisms, plants, and animals have been genetically programmed to serve as bioreactors for the production of pharmacologically active human proteins and other specialty chemicals. Recombinant DNAs have been used to create plants with improved agronomic traits, including virus and insect resistance. Progress is being made in the development of crops with delayed spoilage, enhanced flavors, and improved nutrition. The traits of farm animals are also being altered. The list of transgenic organisms is already quite long and is growing rapidly.

Genetic engineering contributes to improvements in the treatment of genetic diseases in two distinct ways: it produces animal models that are used to develop and test new treatment strategies; and it allows us to introduce therapeutically beneficial genes into diseased individuals, a process described as gene therapy. Gene therapy is also being utilized to treat a variety of noninherited diseases, including cancer and AIDS.

FOCUSING ON THE TOTAL PICTURE

This text has provided an introduction to the major building blocks of life and some basic metabolism. From this overview, it is possible to construct a general picture of biochemistry and the molecular basis for life. As the text draws to an end, it seems appropriate to bring that picture into focus and to take a brief look into the future.

The instructions for the production and maintenance of human life are carried in the DNA that is initially present in every cell of the body. Half of that DNA can be traced to a sperm cell and the other half to an egg cell. Most DNA of known function encodes information for the construction of the body's proteins, which provide structure and support; move the body; defend the body; transport other molecules; provide vision; coordinate reactions occurring throughout the body (serve as chemical messengers); and, most importantly, serve as catalysts (enzymes) (Figure 19.20). The coordination and control of the production and the activity of enzymes is central to life, because it is these polymers that kinetically control all metabolism, including the replication of the DNA that encodes them. The replication of DNA allows life's blueprints to be passed from cell to cell and generation to generation.

During its lifetime, the human body hosts probably 50,000 to 100,000 distinct, genetically programmed chemical transformations. Each, with rare exceptions, is catalyzed by a separate, highly selective enzyme. Although some reactions are involved only in embryonic development, many are required to sustain all stages of life. Most transformations can clearly be placed into one of two major categories: catabolic or anabolic. Catabolism is involved in degrading compounds, whereas anabolism is responsible for the construction of molecules.

A significant portion of human catabolism is directed toward the release of chemical energy through the O_2-linked oxidation of carbohydrates, fats, proteins, and other components in the diet. Figures 10.24, 10.25, 10.26, 14.19 and 14.20 and Tables 10.11 and 10.12 provide an overview of much of this metabolism, and they should be revisited at this time. The released energy, primarily in the form of ATP and NADH, is used for a variety of processes including muscle action, brain action, signal transduction, and anabolism. In addition to providing energy, catabolism generates intermediates and reducing power for anabolism, and it plays a central role in the turnover of the body's own molecules, those it has produced through anabolism.

Dietary substances and their catabolic intermediates are the principal starting materials for anabolism. The products of anabolism account for the shape, form, and function of the human body. Although many of these products have not yet been purified and studied, they are central to the various levels of reactivity, structure, organization, communication, and coordination that characterize life itself.

Metabolism can be divided into numerous discrete reactions and reaction sequences, known as metabolic pathways, that are interconnected through a variety of mechanisms. Each sequence of coupled reactions has a single rate-limiting step that determines the rate of the entire pathway. Strategies for controlling reaction rates include the regulation of numerous variables, including enzyme concentration, pH, effector concentration, substrate availability, and cofactor concentration. Enzyme concentrations are themselves determined by the rates of gene expression and enzyme degradation. An additional strategy for controlling reaction rates entails the covalent modification of existing enzymes.

Hormones and allosteric effectors play key roles in the communication that takes place among discrete pathways, cells and tissues. This communication helps keep individual cells responsive to both their own needs and the needs of the body as a whole. Human metabolism is also regulated through compartmentation and metabolic channeling. Compartmentation includes the division of labor between organelles, cells, and tissues.

The body is unable to construct adequate amounts of a small number of required organic compounds: vitamins, essential amino acids, and essential fatty acids (Section 10.1). The diet must provide these substances along with water, an abundant source of energy and carbon, and approximately two dozen chemical elements. Eleven essential elements are known as the bulk elements because they are required in relatively large amounts and, collectively, account for over 99.9% of the atoms within the human body. The other essential elements, called trace elements, are required in much smaller quantities.

At the frontiers of human biochemistry, researchers are studying nerve and brain action, including memory; the regulation of gene expression; cell division and differenti-

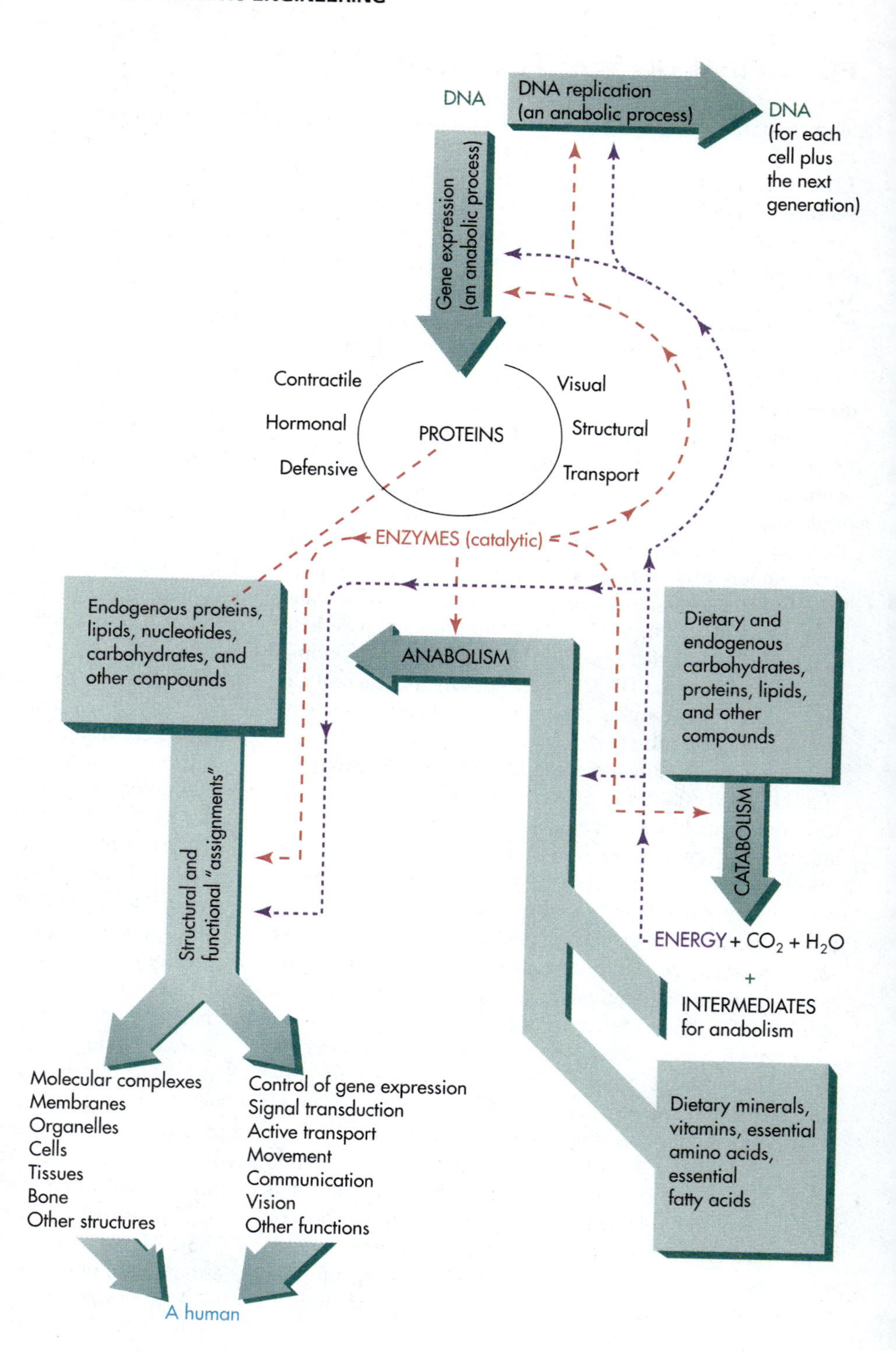

FIGURE 19.20
Molecular basis for human life.

ation; enzyme catalysis; protein folding; genetic and nongenetic diseases; signal transduction; cellular structure and organization; virus action; metabolic checks and balances; metabolic turnover; aging; and numerous other topics. In a variety of these endeavors, the Human Genome Project and spinoffs from this project have opened many doors. Basic biochemical research holds the key to understanding and treating cancer, heart disease, arthritis, genetic diseases, mental illness, and other medical problems, possibly even aging. More selective and more potent drugs are already on the horizon. Novel treatment strategies, including gene therapy, may improve the prognosis for a number of human ailments in the near future.

In 1995, *Science* magazine asked 50 renowned scientists at the frontier what they see in the future for science. There answers were published in an article titled "Through the Glass Lightly" (see selected readings). Those answers make interesting reading.

At present, much of the excitement in biochemistry lies in the growing ability to selectively manipulate genetic information. This fact is highlighted by the thousands of patents that have been sought for genetically altered organisms since this became possible in 1980. Plants, animals, and microorganisms have been programmed to serve as factories for the production of a wide variety of useful biochemicals. Plants and microorganisms are being engineered to help clean up the environment. Knockout mutants have been bioengineered for medical and biochemical research. Crops are being created that are pest resistant, drought resistant, frost resistant, salt resistant, spoilage resistant, and more energy efficient. Genetic engineers are attempting to improve the traits of farm animals, and dozens of gene therapy trials are in progress around the world. The list goes on.

Over the next several decades, there is no scientific discipline that will have a greater impact on human health and well being than biochemistry. It may, however, all be for naught if the global population explosion destroys our life support system. Biochemistry may even assist in combating this problem; further research could lead to improved methods for regulating reproduction. These are indeed interesting times for biochemistry and biochemists.

EXERCISES

19.1 A student claims that she developed a genetic disease as a consequence of an exposure to a chemical mutagen in an undergraduate chemistry lab. Is there any possibility that this is a valid claim? Explain.

19.2 True or false? Some genetic diseases are contagious. Explain.

19.3 Is there any chance that future generations will have genetic diseases that are nonexistent today? Explain.

19.4 Both a husband and his wife have the same dominantly inherited genetic disease. Will all of their children have the same genetic disease? Explain.

19.5 Explain why individuals with certain genetic diseases are unable to pass those diseases on to future generations.

19.6 Assume that the base A in the following DNA fragment is in its rare tautomeric form when this fragment replicates. Show the nucleotide sequence in each of the two daughter duplexes that result from this round of replication. Show the nucleotide sequence in each of the four granddaughter duplexes that result from the replication of the daughter DNAs in the absence of any rare tautomeric forms.

5′ CCGCGACGGCGCGG 3′
3′ GGCGCTGCCGCGCC 5′

19.7 Gene *X* normally encodes an enzyme that catalyzes the methylation of specific nucleotide residues in an RNA. A mutation in gene *X* leads to the production of an enzyme that catalyzes the methylation of inappropriate nucleotide residues. These abnormal methylations lead to illness and death in the affected individuals. Is the genetic disease associated with the mutation in gene *X* dominantly or recessively inherited? Explain.

19.8 Hemochromatosis can be treated with bloodletting. Suggest another simple treatment for this disease that does not entail gene therapy.

19.9 True or false? The incidence of genetic diseases tends to be higher in families with ineffective DNA repair systems than in other families. Explain.

19.10 Explain why the gonads of patients are commonly covered with a lead apron during diagnostic X-rays, whereas no attempt is usually made to similarly shield the head and many other parts of the body.

19.11 Can an appropriate DNA probe be used to diagnose individuals infected with the AIDS virus? Explain. Hint: Review Section 17.5.

19.12 Explain why oxyhemoglobin S molecules, unlike deoxyhemoglobin S molecules, will not polymerize (aggregate, clump together) within red blood cells to form fibers.

19.13

a. Would you expect deoxyhemoglobin molecules with an aspartic acid residue at the 6 position of their β-chains to aggregate (clump together) in the same manner that deoxyhemoglobin S molecules aggregate?

b. Would aggregation be expected if this number 6 residue were isoleucine? Phenylalanine? Explain. Hint: Review amino acid structures in Section 3.2.

19.14 Are the red blood cells in a sickle cell patient more or less likely to sickle when the patient is breathing pure oxygen rather than air? Explain.

19.15 When the DNA from a patient with two copies of an abnormal gene is examined with a unique DNA probe, the probe binds to a single restriction fragment. However, DNA from a patient with a single abnormal gene yields two different restriction fragments that bind to the probe, indicating that restriction fragment length polymorphisms exist. How many different sized restriction fragments will bind to the same DNA probe when DNA from an individual with two normal (no abnormal) genes is analyzed? Explain.

19.16 Could knockout mutants be used as animal models for sickle cell anemia? Explain.

19.17 Although neither parent has sickle cell anemia, their first child inherits this genetic disease. What is the probability that their second child will have the same disease? Explain.

19.18 Based on the information given in this chapter, could PKU, like sickle cell anemia, be the consequence of a change in a single codon in a single gene? Explain.

19.19 Is there any chance that the red blood cells in a fetus with sickle cell anemia will sickle before birth? Explain. Hint: Review Section 4.9.

19.20 True or false? Since a point mutation can convert a protooncogene to an oncogene, a point mutation should be able to convert the oncogene back into a protooncogene. Explain.

19.21 Explain why a correlation exists between a person's DNA repair capabilities and that person's cancer risk?

19.22 True or false? A tumor suppressor gene mutation that leads to an overactive gene product increases the probability that the mutant cell will become cancerous. Explain.

19.23 Although tumor suppressor genes are commonly inherited, you will never encounter a person who has inherited an oncogene. Suggest why this is the case.

19.24 Illustrate two different palindromic sequences that each contain 10 base pairs. Use single-letter abbreviations for nucleotide residues.

19.25 Look up the term *palindrome* in a dictionary and then give two examples of word palindromes.

19.26 True or false? A palindromic sequence will always contain an even number of base pairs. Explain. If you answer *false,* give an example to justify your answer.

19.27 Plasmid X is hydrolyzed with the catalytic assistance of the restriction endonuclease EcoRI, whereas plasmid Y is not a substrate for this enzyme. What does this observation reveal about the nature of each of these two plasmids?

19.28 A linear duplex DNA molecule is "cut" into 10 fragments by EcoRI-catalyzed hydrolysis and "cut" into 5 fragments when digested in the presence of PstI.

a. How many fragments will be produced when this DNA is digested in the presence of both EcoRI and PstI?

b. If a fingerprint is prepared for the original DNA after digestion with PstI alone, how many electrophoretic bands will the fingerprint most likely contain?

c. Explain why the actual number of bands in the fingerprint (see part b) may differ from this most likely number.

19.29 If the genetic code used by a bacterium were markedly different from the genetic code used by humans, could that bacterium produce human insulin after it was provided a human insulin gene? Explain.

19.30 Suggest two useful traits that could be genetically programmed into plants in addition to those traits identified in this chapter.

19.31 Describe those techniques used to enhance the entry of recombinant DNA molecules into host cells.

19.32 Human promoters (for transcription) differ from those used by bacteria. Why does this create a problem for genetic engineers attempting to program bacteria to produce human proteins? Describe one possible solution to this problem.

19.33 Explain why the random splicing of a harmless gene into a chromosome could have extremely serious consequences for an organism.

19.34 Human growth hormone (hGH) causes mice to grow to a larger than normal size. Is it likely that normal mice produce a mouse growth hormone (mGH)? If so, what is the probable structural relationship between hGH and mGH?

19.35 If a person is a carrier of three lethal genetic diseases, which specific cells in this person must be genetically altered to ensure that the defective genes will not have an adverse impact on future generations?

19.36 Human cells use three distinct reactions, each catalyzed by a separate enzyme, to convert compound M to compound N. Mice are unable to metabolize compound M. What is the minimum number of human genes that a mouse must acquire to synthesize N from M? Explain why the actual number of required genes (cannot be determined from information given) might be larger than the minimum number.

SELECTED READINGS

Alberts, B., D. Bray, J. Lewis, M. Raff, K. Roberts, and J. D. Watson, *Molecular Biology of the Cell,* 3rd ed. New York: Garland Publishing, 1994.
Includes a chapter devoted to DNA technology.

Biotech Special Report, *Science* 256, 766–813 (1992).
Contains 12 articles on human diseases, gene therapy and biotechnology.

Capecchi, M. R., Targeted gene replacement, *Sci. Am.* 270(3), 52–59 (1994).
Describes how researchers can now create mice bearing any chosen mutations in any known gene. This technology is revolutionizing the study of mammalian biology.

Cavenee, W. K., and R. L. White, The genetic basis of cancer, *Sci. Am.* 272(3), 72–79 (1995).
Tracks some of the research that has led to our present understanding of the gene–cancer connection.

Culotta, E., and D. E. Koshland, Jr., p53 sweeps through cancer research, *Science* 262, 1958–1961 (1993).
Reviews the biochemistry of p53 and explains why p53 was selected "molecule of the year" by Science.

Cummings, M. R., *Human Heredity: Principles and Issues,* 3rd ed. New York: West Publishing Co., 1994.
Good coverage of genetic diseases.

Dickson, D., Pig heart transplant 'breakthrough' stirs debate over timing of trials, *Nature* 377, 185–186 (1995).
Describes transgenic pigs that have been developed to provide hearts for implantation into humans. Hearts from the transgenic pigs block the action of a key protein responsible for transplant rejection.

Duff, K., and J. Hardy, Mouse model made, *Nature* 373, 476–477 (1995).
Describes a transgenic mouse that shows promise as an animal model for the study of Alzheimer's disease.

The emerging world of plant science: Frontiers in biotechnology, *Science* 268, 653–691 (1995).
A special section that includes nine articles on plant biotechnology.

Fackelmann, K., Drug wards off sickle-cell attacks, *Science News* 147, 68 (1995).
Describes the reasons for the early termination of clinical trails involving hydroxyurea.

Gasser, C. S., and R. T. Fraley, Transgenic crops, *Sci. Am.* 266(6), 62–69 (1992).
Describes the construction of transgenic plants and discusses the properties of some of the plants that have been produced.

Kareiva, P., Transgenic plants on trial, *Nature* 363, 580–581 (1993).
Addresses the ecological concerns associated with transgenic plants.

Lehrman, S., Challenge to growth hormone trial, *Nature* 364, 179 (1993).
Describes part of the controversy over the use of recombinant human growth hormone in normal children.

Levine, A. J., Tumor suppressor genes, *Sci. Am. Science Medicine* 2(1), 28–37 (1995).

A very readable review of tumor suppressor genes.

Marshall, E., Gene therapy's growing pains, *Science* 269, 1050–1055 (1995).

Reviews the state of gene therapy during the summer of 1995. "With more than 100 clinical trials started and hundreds of millions of dollars at stake, the field is struggling to meet expectations."

Marx, J., New link found between p53 and DNA repair, *Science* 266, 1321–1322 (1994).

Describes evidence that p53 activates a gene involved in DNA repair.

Marx, J., Oncogenes reach a milestone, *Science* 266, 1942–1944 (1994).

Reviews the milestones in oncogene research since they were discovered in normal cells in 1975.

Marx, J., Mouse model found for ALS, *Science* 264, 1663–1664 (1994).

Describes the production of a mouse model for ALS, also known as Lou Gehrig's disease.

Moran, L. A., K. G. Scrimgeour, H. R. Horton, R. S. Ochs, and J. D. Rawn, *Biochemistry,* 2nd ed. Englewood Cliffs, NJ: Neil Patterson/Prentice-Hall, 1994.

Contains a 36-page chapter entitled "Recombinant DNA Technology."

Morgan, R. A., and W. F. Anderson, Human gene therapy, *Annu. Rev. Biochem.* 62, 191–217 (1993).

A thorough review of human gene therapy, with over 200 references.

Nowak, R., Breast cancer gene offers surprises, *Science* 265, 1796–1799 (1994).

Describes the isolation of the breast cancer susceptibility gene BRCA1 and explores some BRCA1-associated questions that remain to be answered.

Rennie, J., Grading the gene tests, *Sci. Am.* 270(6), 89–97 (1994).

Examines genetic testing and some of the ethical issues associated with these tests.

Rodgers, G. P., C. T. Noguchi, and A. N. Schechter, Sickle cell anemia, *Sci. Am. Science Medicine* 1(4), 48–57 (1994).

A review of the nature, diagnosis, and treatment of sickle cell anemia.

Singer, M. and P. Berg, *Genes & Genomes,* Mill Valley, CA: University Science Books, 1991.

Outstanding coverage of gene vectors and recombinant DNAs. Ample references.

Steller, H., Mechanisms and genes of cellular suicide, *Science* 267, 1445–1449 (1995).

A brief review of apoptosis and its molecular basis.

Thompson, C. B., Apoptosis in the pathogenesis and treatment of disease, *Science* 267, 1456–1462 (1995).

Explores the role of apoptosis in disease and the possibility that some diseases can be treated through the regulation of apoptosis.

Through the glass lightly, *Science* 267, 1609–1618 (1995).

Provides the responses when a collection of scientists at the frontier were asked what they see in the future for science.

Travis, J., Scoring a technical knockout in mice, *Science* 256, 1392–1394 (1992).

Reviews the production and uses of knockout mutants.

Ward, M., Corn crosses last hurdle for genetically modified crops, *Nature* 376, 544 (1995).

Describes transgenic corn seeds that have been approved for marketing by the U.S. Environmental Protection Agency. Plants produced from the seeds will be resistant to the European corn borer, an insect that causes losses of around one billion dollars annually in the United States.

CREDITS

This page constitutes an extension of the copyright page. We have made every effort to trace the ownership of all copyrighted material and to secure permission from copyright holders. In the event of any question arising as to the use of any material, we will be pleased to make the necessary corrections in future printings. Thanks are due to the following authors, publishers, and agents for permission to use the material indicated.

Chapter-Opening Images

All the computer-generated models of the chapters were created by Ken Eward/Biografx, NYC. The structural coordinates for some of the images were determined by the researchers listed along with the description of the image below.

Chapter 1, page 1 Model of the hormone insulin, displaying the molecule's backbone. In the background is a false-color micrograph of the human pancreas. Insulin-producing cells have been selectively stained and appear magenta in this micrograph. (Based on structural coordinates determined by A. Wlodawer & H. Savage.)

Chapter 2, page 21 The transfer of a proton from lactic acid to water to generate H_3O^+. Water molecules are seen faintly in the background to remind the viewer that this reaction occurs in an aqueous environment.

Chapter 3, page 61 Glutamate (glutamic acid), lysine, and methionine at pH 6.0.

Chapter 4, page 112 Ribbon model of a human class I histocompatibility protein, a polypeptide involved in recognition of foreign proteins by the immune system. Two types of protein secondary structure, the alpha helix (pink) and the beta sheet (yellow) are readily apparent in this model, as is the tertiary structure of the protein. (Based on structural coordinates determined by D.R. Madden, D.N. Garbocz, & D.C. Wiley.)

Chapter 5, page 169 A tyrosine kinase (Lck): polypeptide backbone is white, molecular surface is shown as a blue grid. (Based on structural coordinates determined by M.J. Eck, S.K. Atwell, S.E. Shoelson, & S.C. Harrison.)

Chapter 6, page 236 Computer model of the polysaccharide amylose (starch), showing its helical shape.

Chapter 7, page 272 DNA: sugar-phosphate backbone is white, stacked base pairs are blue.

Chapter 8, page 288 NADPH (yellow stick model) bound to glutathione reductase (molecular surface). (Based on structural coordinates determined by P.A. Karplus & G.E. Schultz.)

Chapter 9, page 324 Computer model of a phospholipid bilayer composed of glycerophospholipids and sphingomyelins. (Based on structural coordinates determined by R. Pastor & R. Venable.)

Chapter 10, page 375 Ras protein (ribbons) and bound GDP (red molecular surface). (Based on structural coordinates determined by P.J. Kraulis, P.J. Domaille, S.L. Campbell-Burk, T. Vanaken, & E.D. Laue.)

Chapter 11, page 427 Hexokinase monomer (blue molecular surface) and bound glucose (yellow molecular surface). (Based on structural coordinates determined by T.A. Steitz, C.M. Anderson, & R.E. Stenkamp.)

Chapter 12, page 479 Transketolase: ribbons model with bound TPP prosthetic groups (white spheres). Color code is based on the polarity of amino acid side chains. (Based on structural coordinates determined by Y. Lindqvist, G. Schneider, & M. Kikkola.)

Chapter 13, page 502 Lipase: ribbons model with bound calcium ions (white spheres). Color code is based on atomic mobility. (Based on structural coordinates determined by Y. Bourne & C. Cambillau.)

Chapter 14, page 537 Thymidylate synthase (homodimer): polypeptide backbone is white, molecular surfaces are shown as grids. (Based on structural coordinates determined by A. Kamb, J. Finer-Moore, & R.M. Straud.)

Chapter 15, page 591 Photosynthetic reaction center of the purple bacterium *Rhodopseudomonas viridis* with attached prosthetic groups. Apoprotein—translucent shell; heme groups (4)—blue; bacteriochlorophylls (4)—green; bacteriopheophytins (2)—purple; quinones (2)—cyan; iron—orange; magnesium—yellow. (Based on structural coordinates determined by J. Deisenhofer, O. Epp, K. Miki, R. Huber, & H. Michel.)

Chapter 16, page 642 A close up view of a small portion of $tRNA^{Phe}$ (yeast). (Based on structural coordinates determined by A. Jack, J.E. Ladner, & A. Klug.)

Chapter 17, page 689 The backbone of tramtrack protein (yellow), a DNA-binding protein with two zinc finger motifs, bound to duplex DNA (space filling model with one strand red and the second strand blue). Zinc ions (white) are shown at 1.25 times true size for clarity. (Based on structural coordinates determined by L. Fairall, J.W.R. Schwabe, L. Chapman, J.T. Finch, & D. Rhodes.)

Chapter 18, page 743 Ser-tRNA (magenta) bound to seryl-tRNA synthetase (pale purple and medium blue subunits). Molecular surfaces are shown as grids. (Based on structural coordinates determined by V. Biou, A. Yaremchuk, M. Tukalo, & S. Cusak.)

Chapter 19, page 774 Sickle cell hemoglobin dimer, showing alpha (blue) and beta (cyan) chains, the organic component of the heme groups (white), heme irons (orange), and the valine side chains (red) associated witht the number 6 residues of the beta chains. A valine side chain on one hemoglobin molecule is inserted into a hydrophobic pocket in a neighboring hemoglobin molecule. (Based on structural coordinates determined by E.A. Padlan & W.E. Love.)

Photographs and Computer Models

Chapter 1: 3: Figure 1.1 (inset) © Geoff Tompkinson/SPL/Photo Researchers. **8:** Figure 1.2a © CRNI/SPL/Photo Researchers. **10:** Figure 1.3a © Chin Ho Lin/Courtesy of Wadsworth Publishing. **11:** Figure 1.4 © Don W. Fawcett/Visuals Unlimited. **12:** Figure 1.5 (left) © K.R. Porter/Photo Researchers. **13:** Figure 1.6 © R. Bolender & D. Fawcett/ Visuals Unlimited; Figure 1.7a © Martha J. Powell/Visuals Unlimited. **14:** Figure 1.8a © W. R. Hargreaves/Courtesy of Wadsworth Publishing. **15:** Figure 1.9 © M. Schliwz/Visuals Unlimited. **Chapter 2: 23:** © Stern/Black Star. **27:** © Vandystadt/Photo Researchers. **40:** © Richard Megna/Fundamental Photos. **Chapter 3: 73:** Figure 3.2 © Renee Lynn/Photo Researchers. **94:** Figure 3.13 Courtesy of Beckman Instruments. **104:** Figure 3.18 (inset) © SIU/Visuals Unlimited. **Chapter 4: 122:** Figure 4.4b © Tripos, Inc. **141:** Figure 4.20 © Tripos, Inc. **152:** Figure 4.30b © Tripos, Inc. **157:** Courtesy of Peck Ritter. **158:** Figure 4.35 Courtesy of Sir John Kendrew/FRS/ Medical Research Council/ Cambridge, England. **159:** Figure 4.36 Courtesy of Sir John Kendrew/FRS/Medical Research Council/Cambridge, England. **162:** © Will & Deni McIntyre/Photo Researchers. **Chapter 5: 213:** Figure 5.15a & bCourtesy of Dr. Lester Reed/Biochemical Institute/TX. **Chapter 6: 240:** © Martin/Custom Medical Stock. **251:** Courtesy of the Agricultural Research Service/USDA. **255:** © Lisa Quinones/Black Star. **262: top,** © Leonard Lessin/ Peter Arnold; **bottom,** Courtesy of BEANO. **Chapter 7: 277:** © Biophoto Associates/Photo Researchers. **Chapter 8: 291:** © Dallas & John Heaton/Westlight. **292:** © Wesley Boxce/Photo Researchers. **301:** Courtesy of Peck Ritter. **Chapter 9: 332:** Figure 9.6 © P. Motta/Dept. of Anatomy/Univ. La Sapienza Rome/SPL/Photo Researchers. **348:** Figure 9.19 © Cabisco/Visuals Unlimited. **352:** Figure 9.21a Courtesy Dr. Margaret Perry/Roslin Institute, Scotland. **367:** Figure

9.33a & b Courtesy of Ron Milligan & Jerry Hinshaw/Scripps Research Institutes. **Chapter 10: 386:** Figure 10.2 © Biophoto Associates/Photo Researchers. **389:** © Black Star. **407:** Figure 10.17 © Tripos, Inc. **Chapter 11: 429:** © Visuals Unlimited. **439:** © Andrew Wood/Photo Researchers. **471:** © Chad Ehlers/Tony Stone Images. **Chapter 12: 484:** Figure 12.4a © Tripos, Inc. **486:** © Chase Swift/Westlight. **487:** Figure 12.6 © Tripos. **498:** © Mark Clarke/SPL/Photo Researchers. **Chapter 13: 520:** © Adamsmith Productions/Westlight. **533:** Figure 13.24 © Jane Thomas/Visuals Unlimited. **Chapter 14: 540:** © Carl Purceli/Photo Researchers. **Chapter 15: 592:** Courtesy of Peck Ritter. **595:** Figure 15.2b © George B. Chapman/Priscilla Deradoss/Visuals Unlimited. **609:** Figure 15.14 © Tripos, Inc. **613:** Figure 15.18a & b © Tripos, Inc. **634:** Figure 15.33 Courtesy of the Agricultural Research Service/USDA. **Chapter 16: 650:** Figure 16.4 © Tripos, Inc. **669:** Figure 16.12 Courtesy of T.A. Steitz, Yale University/From *Science Magazine, Vol 246*(1989). **678:** Figure 16.18 Courtesy of the Agricultural Research Service/USDA. **682:** Figure 16.22 Courtesy of G.L. Costa/Stratagene. **683:** Figure 16.23 © 1993 ILM/Universal/Courtesy of Industrial Light & Magic. **Chapter 17: 716:** Figure 17.17 © D.S. Hogness/Courtesy of Wadsworth Publishing. **Chapter 18: 763:** Figure 18.10b Courtesy of S.L. McKnight & Oscar Miller, Jr./University of Virginia. **764:** Figure 18.11a Courtesy of Oscar Miller, Jr., University of Virginia/From *Science Magazine, Vol 169*(1970). **Chapter 19: 782:** Figure 19.3 Courtesy of Carole Meredith/Journal of Enology and Viticulture. **787:** Figure 19.6 Courtesy of R. Josephs/From *Journal of Molecular Biology Vol 135*(1979), pp. 651–674, T.E. Willems & R. Josephs; Figure 19.7a & b © Stanley Flegler/Visuals Unlimited. **791:** Figure 19.9 © Alfred Pasieka/SPL/Photo Researchers. **806:** Figure 19.16 Courtesy of Douglas Hanahan/U.C. San Francisco; Figure 19.17 © Jackson Lab/Visuals Unlimited; Figure 19.18 Courtesy of Robert H. Devlin/Oceans & Fisheries, Canada.

Illustrations

Chapter 4: 121: Figure 4.3a & b © Irving Geis. **122:** Figure 4.4a © Irving Geis. **140:** Figure 4.19 © Irving Geis. **143:** Figure 4.21 © Irving Geis. **152:** Figure 4.30a © Irving Geis. **154:** Figure 4.32 from *Biochemistry,* Second Edition, by Moran, Scrimgeour, Horton, Ochs, and Rawn, p. 5–15. Copyright © 1994 Prentice-Hall, Inc. Reprinted by permission of Prentice Hall, Upper Saddle River, New Jersey. **Chapter 5: 215:** Figure 5.16 © Irving Geis. **Chapter 6: 257:** Figure 6.4(bottom, left & right) © Irving Geis. **Chapter 9: 333:** Figure 9.7 (b) copyright © 1994 Oldways Preservation & Exchange Trust. Reprinted by permission. **347:** Figure 9.18 from *Textbook of Biochemistry: With Clinical Correlations,* Third Edition, by T. M. Devlin, p. 69. Copyright © 1992 Wiley-Liss. Reprinted by permission of John Wiley & Sons, Inc. **350:** Figure 9.20 appeared in "Are You Headed for a Heart Attack?", *Consumer Reports on Health,* 6(10), October 1994, p. 113. Reprinted by permission of Ted Pass, Ph.D. and StrateCision, Inc. **361:** Figure 9.28 from "How Cells Absorb Glucose," by G. E. Lienhard, J. W. Slot, D. E. James, and M. M. Mueckler, *Scientific American,* January 1992. Copyright © 1992 by Scientific American. Reprinted by permission. **363:** Figure 9.29 adapted from "How Cells Absorb Glucose," by G. E. Lienhard, J. W. Slot, D. E. James, and M. M. Mueckler, *Scientific American,* January 1992. Copyright © 1992 by Scientific American. Reprinted by permission. **Chapter 10: 383:** Table 10.5 from *Recommended Dietary Allowances,* 10th Edition, by the Food and Nutrition Board, National Research Council-National Academy of Sciences, 1989. **Chapter 11: 469:** Figure 11.20(b) reprinted with permission from "Our Primary Source of ATP," by R. L. Cross, *Nature, 370,* 25 August 1994, p. 594. Copyright © 1994 Macmillan Magazines Ltd. **Chapter 15: 606:** Figure 15.11 from *Molecular and Cellular Biology,* by S. L. Wolfe, p. 374. Copyright © 1993 Wadsworth Publishing Company. Reprinted by permission. **Chapter 16: 644:** Table 16.1 from *Molecular and Cellular Biology,* by S. L. Wolfe, p. 760. Copyright © 1993 Wadsworth Publishing Company. Reprinted by permission. **648:** Figure 16.2 © Irving Geis. **649:** Figure 16.3 © Irving Geis. **656:** Figure 16.5 from *Biochemistry* by Donald Voet and Judith G. Voet. Copyright © 1995 John Wiley & Sons. Reprinted by permission of John Wiley & Sons, Inc. **657:** Figure 16.6 adapted from Geoffrey Zubay, *Biochemistry,* 3rd Edition. Copyright © 1993 Wm. C. Brown Communications, Inc., Dubuque, Iowa. Reprinted by permission of Times Mirror Higher Education Group, Inc., Dubuque, Iowa. All Rights Reserved. **664:** Figure 16.8 from *Molecular and Cellular Biology,* by S. L. Wolfe, p. 575. Copyright © 1993 Wadsworth Publishing Company. Reprinted by permission. **667:** Figure 16.10 from *Principles of Biochemistry,* by H. R. Horton, L. A. Moran, R. S. Ochs, J. D. Rawn, and K. G. Scrimgeour, p. 22–25. Copyright © 1993 Neil Patterson Publishers/Prentice-Hall, Inc. Reprinted by permission of Prentice Hall, Englewood Cliffs, New Jersey. **674:** Figure 16.15 from *Biochemistry* by Donald Voet and Judith G. Voet. Copyright © 1995 John Wiley & Sons. Reprinted by permission of John Wiley & Sons, Inc. **675:** Figure 16.16 from *Cell Biology: Organelle Structure,* by David E. Sadava. Copyright © 1993 Jones & Bartlett Publishers, Inc. Reprinted by permission. **Chapter 17: 693:** Figure 17.2 from *Molecular and Cellular Biology,* by S. L. Wolfe, p. 641. Copyright © 1993 Wadsworth Publishing Company. Reprinted by permission. Data from P. H. von Hippel et al., *Ann. Rev. Biochem. 53:*389 (1984). **694:** Figure 17.4 from *Molecular and Cellular Biology,* by S. L. Wolfe, p. 544. Copyright © 1993 Wadsworth Publishing Company. Reprinted by permission. **695:** Figure 17.5 (a) from *Molecular and Cellular Biology,* by S. L. Wolfe, p. 544. Copyright © 1993 Wadsworth Publishing Company. Reprinted by permission; (b) redrawn with permission, from the *Annual Review of Biophysics and Biophysical Chemistry, Volume 19:*405, © 1990, by Annual Reviews, Inc. **716:** Figure 17.17 (b) from *Molecular and Cellular Biology,* by S. L. Wolfe, p. 977. Copyright © 1993 Wadsworth Publishing Company. Reprinted by permission. **721:** Figure 17.21 from *Molecular and Cellular Biology,* by S. L. Wolfe, p. 964. Copyright © 1993 Wadsworth Publishing Company. Reprinted by permission. **Chapter 18: 751:** Table 18.2 from *Molecular and Cellular Biology,* by S. L. Wolfe, p. 652. Copyright © 1993 Wadsworth Publishing Company. Reprinted by permission.

INTERNET RESOURCES*

Locations	Types of Information
Telnet Site	
uwin.u.washington.edu	A good place to begin that is easy to use. Offers many of the Internet resources.
Anonymous FTP Sites	
ftp.dartmouth.edu	Includes many classical texts and some software.
ftp.ds.internic.net	path /pub/the-scientist To read *The Scientist.*
ftp.ucs.ubc.ca	path /pub/mac/info-mac/ A good source of Macintosh software.
ftp.uu.net	path /index Contains one of the most extensive archives of programs and information files. Use "index" (path name) to get lists of available files.
ftp.wustl.edu *also* http://wuarchive.wustl.edu	Holds extensive collection of Macintosh and PC software.
oak.oakland.edu	path /pub A good site to retrieve software.
sunsite.unc.edu	path /pub Includes academic information and software for computers manufactured by SUN Systems.
ftp.bio.indiana.edu	Site for academically-useful software in biological sciences.
ftp.cso.uiuc.edu	A large, general-purpose site that includes a variety of software for different PCs.
World Wide Web Sites	
http://www.nih.gov:80/science/	National Institutes of Health scientific resources.
http://www.gene.com:80/ae/	A National Education Program sponsored by a biotech corporation. Contains resources for science teachers.
http://www.asap.unimelb.edu.au/hstm/hstm_ove.htm	Relates to history of science and technology and also has e-mail facility for various Listservs and electronic journals.
http://golgi.harvard.edu	Offers access to databases for molecular biology and biochemistry, instructional resources in biology, and EC enzymes. Also includes information on Harvard University.
http://golgi.harvard.edu/journals.html	Provides a list of electronic journals. Journals have either full text, abstracts only or index page only. Check out your favorite journal here.
http://www.imb-jena.de/IMAGE.html	An image library of biological molecules.
http://www.chem.ucla.edu/chempointers.html	Links to various chemistry sites around the world, including FTP sites.
http://www.ebi.ac.uk/biocat/biocat.html	A directory that lists various software packages for molecular biology and genetics.
http://www.missouri.edu/~c621913/links.html	A collection of 3-D images of DNA, RNA, and selected enzymes.
http://www.cchem.berkeley.edu/Table/index.html	Includes the periodic table and associated information.
http://www.biochem.abdn.ac.uk/	Gives 3-D graphics.

*This list was compiled by Dr. Prakash H. Bhuta, Eastern Washington University.

Locations	Types of Information
http://www.acs.org/	The home page of the American Chemical Society.
http://highwire.stanford.edu/jbc	The *Journal of Biological Chemistry* online (still in development stage and free).
http://www.csc.fi/molbio/progs/	Software and databases of interest to molecular biologists.
http://www.yahoo.com/	An excellent site to search using key words or subject categories.
http://lcweb.loc.gov:80/homepage/	Another useful site for searching with key words.
http://www.brookscole.com/brookscole.html	Provides information about Brooks/Cole Company's texts and technology products.
Gopher Sites	
gopher.micro.umn.edu	A site at the University of Minnesota—one of the first sites. It can link to other gopher sites.
gopher.nsf.gov	A clearinghouse for many scientific reports and documents. It also can lead you to other US government gopher services.
gopher.hs.jhu.edu	Administered by the History of Science Department at Johns Hopkins University. Contains information on the history of science.

Gene Bank and Protein Data Banks

The National Center for Biotechnology Information (NCBI) provides access to GenBank and several other sequence databases through two e-mail servers:

RETRIEVE (for record retrieval from GenBank and other databases)
BLAST (for sequence similarity searching)

These services are free of charge and only require the ability to send and receive Internet electronic mail. The addresses are:

retrieve@ncbi.nlm.nih.gov (RETRIEVE server)
blast@ncbi.nlm.nih.gov (BLAST server)

For information on how to use either server, send a message to the server address with HELP as the only word in the body of the message. You will receive instructions for formatting a query in the return message. The NCBI also has World Wide Web page:

http://www.ncbi.nlm.nih.gov/

Note: Failing to connect to one of these sites may be caused by one of these three problems: (1) a typographical error in the address (try retyping), (2) the host is too busy (try connecting during non-rush hours), or (3) the address has changed (try using the search function on the browsing tool).

Answers to Problems

CHAPTER 1

1.1 The number of possible questions is enormous. For any individual, that number is determined, in part, by his or her curiosity, imagination, and scientific background. Some sample questions follow.

What specific reactions or molecular interactions are responsible for curiosity? Imagination? Intelligence? Memory? Learning? Sexual preference? Vision? Depression? Joy? Hatred? Hearing? Smell? Pain? Hunger? Muscle action? High blood pressure? Sleep? And so on.
How does the immune system destroy viruses and bacteria?
How does dietary fiber lower the risk of colon cancer?
How do dietary saturated fats increase the risk of heart disease?
How does sunlight lead to skin cancer?
How does radiation therapy destroy cancer cells?
How does carbon monoxide kill a person? Why is oxygen used to treat carbon monoxide poisoning?
Why is mercury so toxic?
Why does iron deficiency lead to anemia?
How does acupuncture work?
How do anesthetics work?
How do antibiotics kill bacteria and what is the molecular explanation for their side effects?
What is the molecular explanation for the fact that muscles get sore following exercise?
How do steroid supplements modify the chemistry of athletes?
What is the molecular composition of a tooth and what leads to tooth decay?
What chemicals are found in the mouth and how do these impact oral health?
What causes normal cells to become cancerous?
What is the nature of viruses and how do they impact their target cells?
How do plants fix nitrogen?
How do plants capture and use sunlight?
How do plants communicate with one another?
How do organisms survive in extreme environments (high temperature, high salt concentrations, low pH, etc.)?
What determines the ultimate size and life span of an organism?
How do salmon recognize their spawning streams?
What accounts for the toxicity of snake and spider venoms?

1.2 There are limits. From a theoretical standpoint, these limits are primarily set by the fixed rules or laws under which nature apparently operates and by the limited resources available in our world. Our imperfect understanding of natural laws and natural processes also limits what we are able to accomplish. From a practical standpoint, accomplishment limits are at least partially determined by costs, politics, social values, and other factors not directly related to science research.

CHAPTER 2

2.1 Nonpolar. The bond angle in a water molecule is approximately 105°, and water is highly polar. If, however, the bond angle were 180°, the two identical and very polar O—H bonds would cancel one another; one end of the molecule would be no more positive or negative than the opposite end. This imaginary molecule would be nonpolar:

H—O—H (bond dipole arrows point toward O from each H)

2.2 **a.** δ− C—H δ+   C—H (arrow toward C)

b. Nonpolar

c. δ− N—H δ+   N—H (arrow toward N)

d. δ+ C—O δ−   C—O (arrow toward O)

e. δ+ C—F δ−   C—F (arrow toward F)

2.3 The C—F bond is most polar. It is the electronegativity difference of the bonded atoms that primarily determines the polarity of a covalent bond. The greater the difference, the more polar the bond. Table 2.2 indicates that the electronegativity difference for the C—F bond is 1.4 (4.0 minus 2.6), considerably greater than that for any of the other bonds in Problem 2.2.

2.4

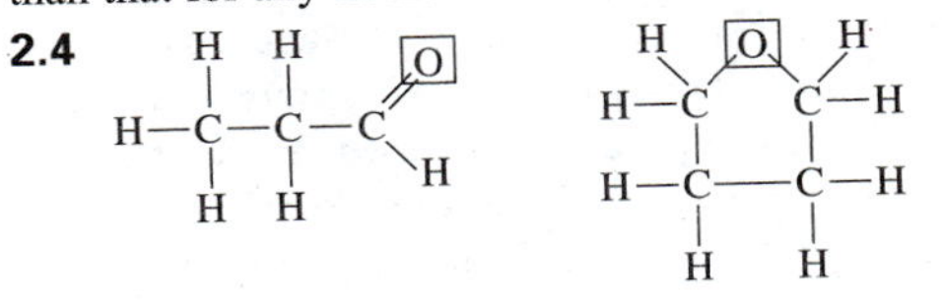

(Continued)

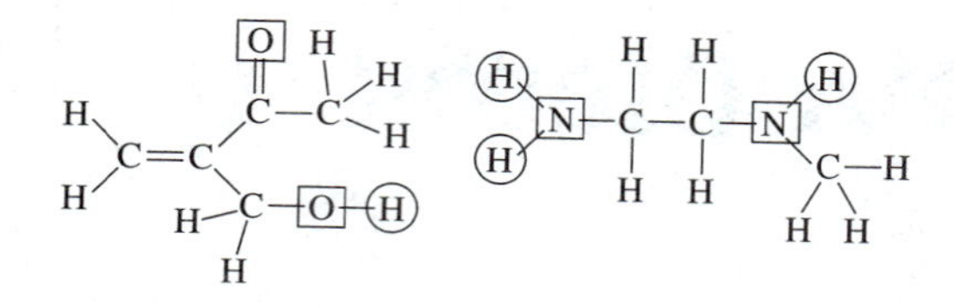

Each circled hydrogen is covalently bonded to a nitrogen or an oxygen and has a significant partial positive charge. Each oxygen and nitrogen with a square around it has a significant partial negative charge and is able to H bond with any hydrogen (other than one to which it is covalently bonded) capable of forming an H bond.

2.5 a.

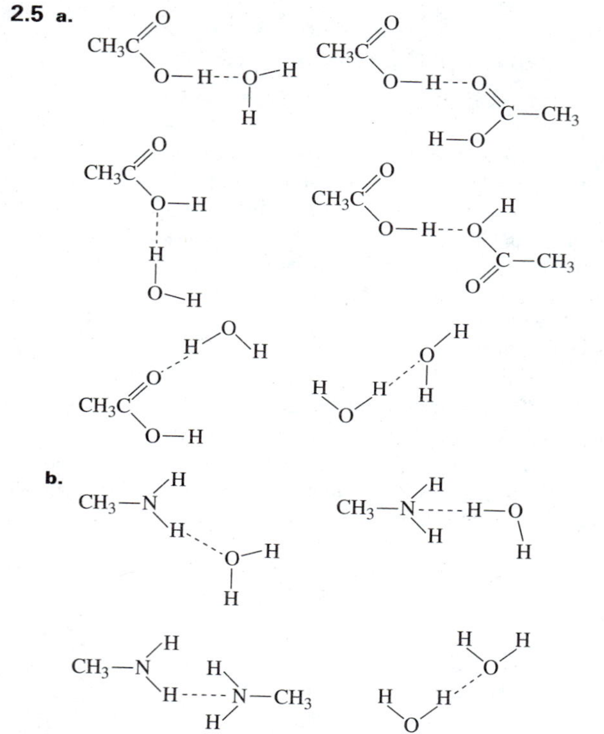

Solute-solute, solute–solvent, and solvent–solvent H bonds exist in both solutions. Although there are no examples given here, it is common for a single molecule to be simultaneously H bonded to multiple other molecules. An H bond is simply an attraction between oppositely charged centers. Any positive charge or partial positive charge will be attracted to any and all negative charges or partial negative charges in the same vicinity.

2.6 Any sample of a substance is in its most random state when the molecules of that substance are uniformly distributed throughout the volume available to it. There is more order and less entropy when more solute molecules reside in one half of a solution than reside in the other half. Similarly, a more random situation exists when the boys and girls at a dance are uniformly mixed than when the most of the boys are at one end of the dance floor and most of the girls at the other end.

2.7 a. van der Waals forces. Since cyclohexane is a nonpolar hydrocarbon, the only attractions between neighbors would result from temporary dipoles.

b. van der Waals plus hydrophobic. In an aqueous mixture, nonpolar molecules tend to aggregate and to become surrounded by polar water molecules. An increase in entropy provides the driving force for this phenomenon, and the interactions are said to be hydrophobic. In reality, it is difficult to get water and cyclohexane to mix. Cyclohexane tends to separate from water to form a two-phase system. The hydrocarbon phase floats on the water.

2.8 In pure dimethyl ether, the predominant intermolecular bonds will be polar forces (dipole–dipole interactions) other than H bonds. Dimethyl ether has a permanent dipole but possesses no hydrogens capable of H bonding.

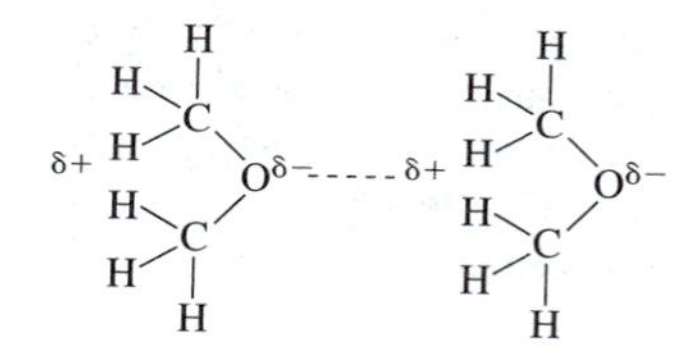

Water will H bond with dimethyl ether in an aqueous solution:

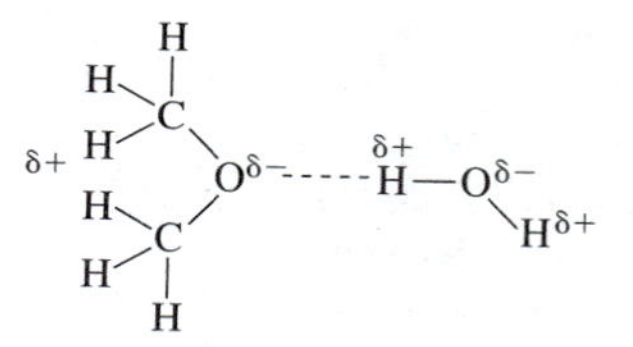

The van der Waals forces that also hold neighboring molecules together in both samples are significantly weaker than the electrostatic attractions.

2.9 All answers are rounded off to two significant figures:

a. $[H^+] = K_w/[OH^-]$

$[H^+] = 1.0 \times 10^{-14}/2.0 \times 10^{-11} = 5.0 \times 10^{-4}\ M$

b. $[H^+] = 1.0 \times 10^{-14}/6.3 \times 10^{-5} = 1.6 \times 10^{-10}\ M$

c. $[H^+] = 1.0 \times 10^{-14}/8.4 \times 10^{-8} = 1.2 \times 10^{-7}\ M$

2.10 All answers are rounded off to two significant figures:

a. $[OH^-] = K_w/[OH^-]$

$[OH^-] = 1.0 \times 10^{-14}/1.4 \times 10^{-10} = 7.1 \times 10^{-5}\ M$

b. $[OH^-] = 1.0 \times 10^{-14}/7.9 \times 10^{-7} = 1.3 \times 10^{-8}\ M$

c. $[OH^-] = 1.0 \times 10^{-14}/6.6 \times 10^{-5} = 1.5 \times 10^{-10}\ M$

2.11

Solution	$[H^+]^a$	$[OH^-]^a$	pH^b	pOH^b	Acidic, Basic or Neutral?
1	2.5 *M*	4.0×10^{-15} *M*	−0.4	14.4	Acidic
2	1.0×10^{-7} *M*	1.0×10^{-7} *M*	7.0	7.0	Neutral
3	5.0×10^{-13} *M*	2.0×10^{-2} *M*	12.3	1.7	Basic
4	6.3×10^{-8} *M*	1.6×10^{-7} *M*	7.2	6.8	Basic

[a]Answers are rounded off to two significant figures.
[b]Answers are expressed to the nearest tenth of a unit.

The following formulas and definitions were used to complete this table:

$K_w = [H^+][OH^-] = 1 \times 10^{-14}$;
$[H^+] = K_w/[OH^-]$; $[OH^-] = K_w/[H^+]$

$pH = -\log [H^+]$; $[H^+] = \text{antilog}(-pH) = 10^{-pH}$

$pOH = -\log [OH^-]$; $[OH^-] = \text{antilog}(-pOH) = 10^{-pOH}$

$pH + pOH = 14$

Acidic solution: $[H^+] > [OH^-]$

Basic solution: $[OH^-] > [H^+]$

Neutral solution: $[H^+] = [OH^-]$

2.12 Phosphoric acid is the strongest. Hydrogen phosphate is the weakest. The larger the equilibrium constant for the dissociation of an acid, the stronger the acid. Since $pK_a = -\log K_a$, pK_a decreases as K_a increases.

2.13

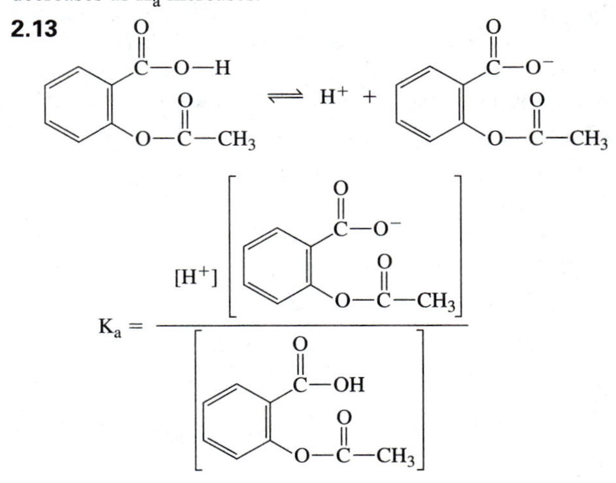

2.14 Strong. Since 3.7×10^{17} is an extremely large number (370,000,000,000,000,000), virtually all of the molecules will exist in a dissociated form at equilibrium. For most purposes, the dissociation can be considered complete.

2.15 There are many possible answers. Some examples follow:

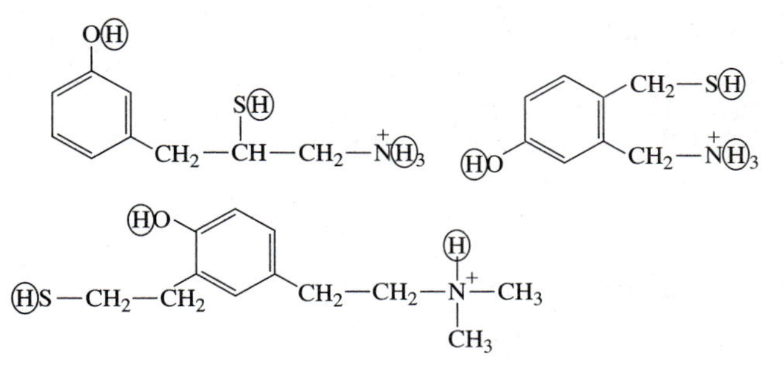

The acidic hydrogens are circled in each molecule. In the first two examples, only one of the three hydrogens attached to nitrogen is able to dissociate. Phenols, thiols, and protonated amines are all weak acids.

2.16

Weak Acid	Conjugate Base
H_3PO_4	$H_2PO_4^-$
$H_2PO_4^-$	HPO_4^{2-}
HPO_4^{2-}	PO_4^{3-}
H_2CO_3	HCO_3^-
HCO_3^-	CO_3^{2-}
CH_3COOH	CH_3COO^-
$CH_3CHOHCOOH$	$CH_3CHOHCOO^-$
NH_4^+	NH_3
C_6H_5—OH	C_6H_5—O^-
$CH_3NH_3^+$	CH_3NH_2

2.17 Pairs a, c, and d each represent an acid:conjugate base pair. An acid always contains one more H^+ than its conjugate base. Consequently, the net charge on an acid is always +1 greater than the net charge on its conjugate base. Although the net charge on an acid is one unit more positive, it may still be negative (consider $H_2PO_4^-$ and HPO_4^{2-}, for example).

One needs the dissociation constant for an acid to determine its strength.

2.18 With spectator ions:

$$HClO_4 + NaOH \rightleftharpoons H_2O + NaClO_4$$

or

$$H^+ + ClO_4^- + Na^+ + OH^- \rightleftharpoons H_2O + Na^+ + ClO_4^-$$

Since $HClO_4$ is a strong acid and NaOH a strong base, both will (for all practical purposes) be completely dissociated in an aqueous solution. Dissolved salts, including $NaClO_4$, also dissociate into their component ions in aqueous solvents. These facts are illustrated in the second equation given above. The spectators should be considered spectators only from the standpoint of acid–base reactions. These ions may be biologically active and physiologically important.

Without spectators:

$$H^+ + OH^- \rightleftharpoons H_2O$$

2.19 Without spectators:

$$OH^- + H^+ \rightleftharpoons H_2O$$

With spectators:

$$H_2SO_4 + 2\ KOH \rightleftharpoons 2\ H_2O + K_2SO_4$$

or

$$2\ H^+ + SO_4^{2-} + 2\ K^+ + 2\ OH^- \rightleftharpoons 2\ H_2O + 2\ K^+ + SO_4^{2-}$$

2.20 Without spectators:

$$C_6H_5OH + OH^- \rightleftharpoons H_2O + C_6H_5O^-$$

$$CH_3CHOHCOOH + OH^- \rightleftharpoons H_2O + CH_3CHOHCOO^-$$

With spectators:

$$C_6H_5OH + KOH \rightleftharpoons H_2O + C_6H_5OK$$

or

$$C_6H_5OH + K^+ + OH^- \rightleftharpoons H_2O + C_6H_5O^- + K^+$$

$$CH_3CHOHCOOH + KOH \rightleftharpoons H_2O + CH_3CHOHCOOK$$

or

$$CH_3CHOHCOOH + K^+ + OH^- \rightleftharpoons H_2O + CH_3CHOHCOO^- + K^+$$

Once again, spectators should only be considered spectators from the standpoint of the acid–base reactions.

2.21 **a.** $OH^- + H_3PO_4 \rightleftharpoons H_2O + H_2PO_4^-$

b. $2\ OH^- + H_3PO_4 \rightleftharpoons 2\ H_2O + HPO_4^{2-}$

c. $3\ OH^- + H_3PO_4 \rightleftharpoons 3\ H_2O + PO_4^{3-}$

2.22 With spectators:

$$HCl + NaHCO_3 \rightleftharpoons H_2CO_3 + NaCl$$

or

$$H^+ + Cl^- + Na^+ + HCO_3^- \rightleftharpoons H_2CO_3 + Na^+ + Cl^-$$

Without spectators:

$$H^+ + HCO_3^- \rightleftharpoons H_2CO_3$$

2.23

Carboxylic acid ester

Phophoric acid ester

Amides

Carboxylic acid anhydrides

Carboxylic acid–phosphoric acid anhydrides

Acetals

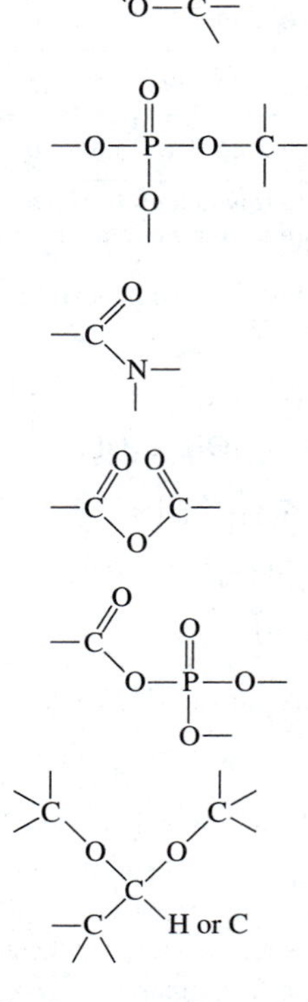

2.24 Catalyst: any substance that increases the rate of a reaction yet can be recovered unchanged at the end of the reaction.

Dehydrating agent: any agent that can remove water. Dehydration, in general, refers to the removal or loss of water.

2.25 **a.** Alka-Seltzer with spectators:

$$HNO_3 + NaHCO_3 \rightleftharpoons H_2CO_3 + NaNO_3$$

or

$$H^+ + NO_3^- + Na^+ + HCO_3^- \rightleftharpoons H_2CO_3 + Na^+ + NO_3^-$$

Alka-Seltzer without spectators:

$$HCO_3^- + H^+ \rightleftharpoons H_2CO_3$$

b. TUMS with spectators:

$$2\ HNO_3 + CaCO_3 \rightleftharpoons H_2CO_3 + Ca(NO_3)_2$$

or

$$2\ H^+ + 2\ NO_3^- + Ca^{2+} + CO_3^{2-} \rightleftharpoons H_2CO_3 + Ca^{2+} + 2\ NO_3^-$$

TUMS without spectators:

$$2\,H^+ + CO_3^{2-} \rightleftharpoons H_2CO_3$$

2.26 **i.** $OH^- + HPO_4^{2-} \rightleftharpoons H_2O + PO_4^{3-}$;

$H^+ + PO_4^{3-} \rightleftharpoons HPO_4^{2-}$

ii. $OH^- + CH_3NH_3^+ \rightleftharpoons H_2O + CH_3NH_2$;

$H^+ + CH_3NH_2 \rightleftharpoons CH_3NH_3^+$

iii. $OH^- + HCN \rightleftharpoons H_2O + CN^-$;

$H^+ + CN^- \rightleftharpoons HCN$

iv. $OH^- + H_2CO_3 \rightleftharpoons H_2O + HCO_3^-$;

$H^+ + HCO_3^- \rightleftharpoons H_2CO_3$

The spectator ions (the Na^+ from the NaOH and the Cl^- from the HCl) have been eliminated in each case.

2.27 Phosphoric acid, 1.15–3.15; dihydrogen phosphate, 6.20–8.20; hydrogen phosphate, 11.4–13.4; carbonic acid, 5.35–7.35; hydrogen carbonate, 9.3–11.3; acetic acid, 3.76–5.76; lactic acid, 2.86–4.86; ammonium, 8.25–10.25; phenol, 9.0-11.0; protonated form of methyl amine, 9.5–11.5.

$$\text{Buffer range} = pK_a \pm 1$$

2.28 At pH = 7.4: [CB]/[WA] = antilog (7.4 − 7.2) = antilog (0.2) = 1.6/1 or (multiplying both numerator and denominator by 10) 16/10. Thus, in a phosphate buffer at pH 7.4, there will be approximately 16 HPO_4^{2-} for every 10 $H_2PO_4^-$.

At pH = 6.8: [CB]/[WA] = antilog (6.8 − 7.2) = antilog (−0.4) = 0.4/1 or (multiplying both numerator and denominator by 10) 4/10.

At pH = 7.8: [CB]/[WA] = antilog (7.8 − 7.2) = antilog (0.6) = 4/1.

At pH = 9.2: [CB]/[WA] = antilog (9.2 − 7.2) = antilog (2.0) = 100/1.

pH = pK_a + log ([conjugate base]/[weak acid]); rearranged, log ([CB]/[WA]) = pH − pK_a; [CB]/[WA] = antilog (pH − pK_a). The pK_a for $H_2PO_4^-$ is 7.2 (Table 2.8).

2.29 The amount of conjugate base in a buffer pair determines the maximum amount of acid it is able to consume.

2.30 0.49 *M*

By definition, [buffer] = [conjugate base] + [weak acid] = 0.37 *M* + 0.12 *M* = 0.49 *M*.

$$pH = pK_a + \log([CB]/[WA]) = 4.76 + \log(0.37/0.12) = 5.25$$

2.31 17 to 100 = 17/100. pH = pK_a + log ([conjugate base]/[weak acid]); rearranged, log ([CB]/[WA]) = pH − pK_a and [CB]/[WA] = antilog (pH − pK_a). For acetic acid (pK_a = 4.76, see Table 2.8) at pH = 4.0, [CB]/[WA] = antilog (4.0 − 4.76) = antilog (−0.76) = 0.17/1 or (multiplying both numerator and denominator by 100) 17/100.

2.32 Le Châtelier's principle: when the equilibrium for a chemical reaction is destroyed, that reaction undergoes either a net forward or a net reverse reaction until its equilibrium is once again restored. The net movement tends to counteract the change initially responsible for the loss of equilibrium.

At equilibrium, the rate of the forward reaction equals the rate of the reverse reaction, so there is no net reaction and no change in reactant or product concentration over time. A reaction that is not at equilibrium spontaneously and continuously moves (through a net forward or net reverse reaction) toward equilibrium.

2.33 Alkalosis exists. pH = pK_a + log ([conjugate base]/weak acid]); since the pK_a for H_2CO_3 is 6.35 (Table 2.8), pH = 6.35 + log (50/1) = 6.35 + 1.70 = 8.05, considerably above the normal pH range for blood. In fact, a person with a blood pH of 8 would most likely be dead.

2.34 Respiratory acidosis. Respiratory diseases decrease the efficiency with which the body eliminates CO_2 from the lungs. This leads to an accumulation of CO_2, which reacts with water to form H_2CO_3 ($H_2O + CO_2 \rightleftharpoons H_2CO_3$). The partial dissociation of the H_2CO_3 leads to an increase in H^+ concentration and a decrease in pH.

2.35 NH_4^+:NH_3. There are many possible answers. The protonated form of most amines is a weak acid with a +1 charge. Its conjugate base is uncharged. The $CH_3NH_3^+$:CH_3NH_2 pair provides another example.

2.36 $H_2PO_4^-$ and HCO_3^- (Table 2.8) are two examples. In all instances, the net charge on the conjugate base will be −2. The loss of an H^+ from any −1 species will lead to a −2 ion, since the departing H^+ leaves an electron (negatively charged) behind. An acid is always one unit more positive (it may still be negative or have no charge) than its conjugate base. There are many other examples of weak acids with a −1 net charge, including $CH_3OPO_3H^-$ and other phosphate esters.

2.37 H_3PO_4, $H_2PO_4^-$, H_2CO_3, CH_3COOH, $CH_3CHOHCOOH$

Since [conjugate base]/[weak acid] = antilog (pH − pK_a) and the antilog of a positive number is always greater than 1, the conjugate base (dissociated) form of a weak acid is favored whenever the pH is above the pK_a for that weak acid.

2.38 (a) 0, (b) +1, (c) −2. At pH 7.0: pH > 0.5 above the pK_a of the carboxyl group, so the dissociated form ($-COO^-$) is highly favored and contributes a −1 charge (to the nearest half charge unit); pH > 0.5 below the pK_a of the —SH group, so its protonated form is favored and it contributes a 0 charge (to the nearest half charge unit); pH > 0.5 below the pK_a of the $-NH_3^+$, so it exists predominantly in its weak acid form and contributes a +1 charge (to the nearest half charge unit). Summing the charges contributed by the three functional groups in cysteine, one obtains: (−1) + (0) + (+1) = 0.

At pH 1.0: —COOH contributes a 0 charge (protonated form highly favored with pH > 0.5 below pK_a); —SH contributes a 0 charge (protonated form strongly favored with pH > 0.5 below pK_a); $—NH_3^+$ contributes a +1 charge (pH once again well below pK_a, so weak acid form predominant). Net charge to the nearest half charge unit = (0) + (0) + (+1) = +1.

At pH 12: —COOH contributes a −1 charge (pH way above pK_a), —SH contributes a −1 charge (exists mainly as $—S^-$ since pH well above pK_a); $—NH_3^+$ contributes a 0 charge (dissociated form [$—NH_2$] highly favored with pH > 0.5 above pK_a). Net charge to the nearest half charge unit = (−1) + (−1) + (0) = −2.

2.39 (a) +2 (b) +1 (c) −1. At pH 1.0: since the pH is more than 0.5 below the pK_a for every functional group, we assume (when calculating net charge to the nearest half unit) that each group exists entirely in its protonated form. Net charge = +1 (from α-NH_3^+) + 1 (from ε-NH_3^+) + 0 (from —COOH) = +2.

At pH 7.0: since the pH is more than 0.5 below the pK_a for both amino groups and more than 0.5 above the pK_a of the carboxyl group, the amino groups are highly protonated and the carboxyl group almost entirely dissociated. Net charge to the nearest half charge unit = (+1) + (+1) + (−1) = +1.

At pH 12.0: all functional groups are considered to be entirely dissociated since the pH is more than 0.5 above the pK_a for each. Net charge to the nearest half charge unit = (0) + (0) + (−1) = −1.

2.40

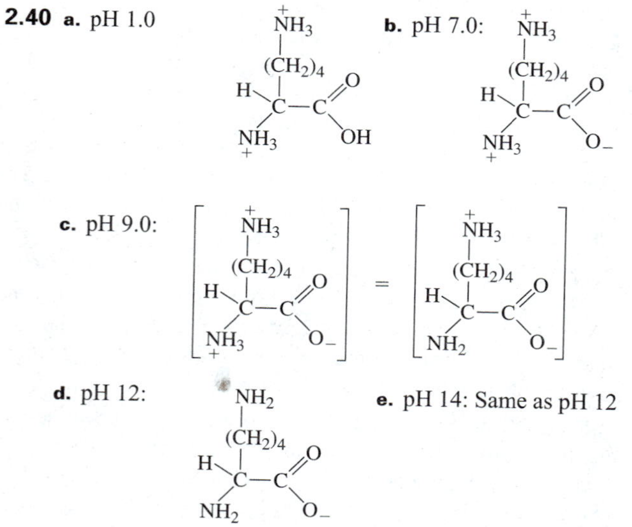

If the pH is above the pK_a of a functional group, the functional group primarily exists in its nonprotonated form. At pH's below the pK_a, the protonated form is favored. When pH = pK_a, protonated and nonprotonated forms are present in equal concentrations.

2.41 a. pH 5.5

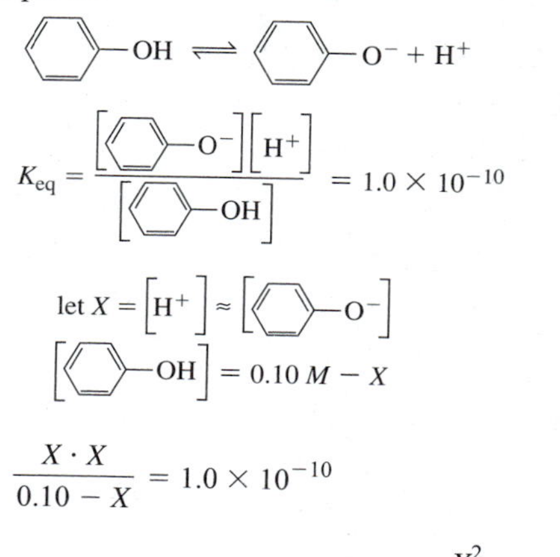

$$\frac{X \cdot X}{0.10 - X} = 1.0 \times 10^{-10}$$

If one assumes $0.10 - X \approx 0.10$; $\frac{X^2}{0.10} = 1.0 \times 10^{-10}$

$$X^2 = 1.0 \times 10^{-11}$$

$$X = [H^+] = \sqrt{1.0 \times 10^{-11}} = 3.2 \times 10^{-6}\,M$$

$$pH = -\log [H^+] = -\log (3.2 \times 10^{-6}) = 5.5$$

The simplifying assumption is valid in this instance since, for most purposes, 3.2×10^{-6} is very small compared to 0.10. That is, $0.10 - (3.2 \times 10^{-6})$ is very close to 0.10.

b. pH 4.6. NH_4Cl is an ionic compound that dissociates when dissolved in water to yield NH_4^+ and Cl^- ions. The NH_4^+ is a weak acid with a pK_a = 9.25 (Table 2.8).

$$NH_4^+ \rightleftharpoons NH_3 + H^+$$

$$K_{eq} = \frac{[NH_3][H^+]}{[NH_4^+]} = 5.62 \times 10^{-10}$$

Let $X = [H^+] \approx [NH_3]$

$$[NH_4^+] = 1.0 - X$$

Assuming $1.0 - X = 1.0$; $\frac{X \cdot X}{1.0} = 5.62 \times 10^{-10}$

$$X^2 = 5.62 \times 10^{-10}$$

$$X = [H^+] = \sqrt{5.62 \times 10^{-10}} = 2.4 \times 10^{-5}\,M$$

$$pH = -\log [H^+] = -\log (2.4 \times 10^{-5}) = 4.6$$

Since $1.0 - (2.4 \times 10^{-5})$ is very close to 1.0, the simplifying assumption is, for most purposes, valid.

2.42 a. $CH_3CHOHCOOH$, conjugate acid of lactate

NH_4^+, conjugate acid of NH_3

b. $H_2O + CH_3CHOHCOO^- \rightleftharpoons OH^- + CH_3CHOHCOOH$

$H_2O + NH_3 \rightleftharpoons OH^- + NH_4^+$

c. $$K_b \text{ (for lactate)} = \frac{[OH^-][CH_3CHOHCOOH]}{[CH_3CHOHCOO^-]}$$

$$K_b \text{ (for } NH_3) = \frac{[OH^-][NH_4^+]}{[NH_3]}$$

The $[H_2O]$ that appears in the equilibrium constant expressions for the reactions in part b is incorporated into the value of K_b, so it does not appear in the K_b expressions.

d. $K_a \times K_b = K_w$

$K_b = K_w/K_a = 1.0 \times 10^{-14}/1.38 \times 10^{-4}$ (for lactate)

K_b (for lactate) $= 7.2 \times 10^{-11}$

pK_b (for lactate) $= -\log K_b = -\log (7.2 \times 10^{-11}) = 10.1$

K_b (for NH_3) $= 1.0 \times 10^{-14}/5.62 \times 10^{-10} = 1.8 \times 10^{-5}$

pK_b (for NH_3) $= -\log (1.8 \times 10^{-5}) = 4.7$

CHAPTER 3

3.1 Smallest amino acid is glycine with 10 atoms (at pH 7.0). There is a tie for largest amino acid; both tryptophan and arginine have 27 atoms (at pH 7.0).

You need count only the side chain atoms to identify the largest and smallest amino acids. The largest is close to three times the size of the smallest.

3.2 Required elements: H, C, N, O, and S.
All belong to an A family.
All are nonmetals.
Valence: hydrogen, 1; O and S, 2; N, 3 or 4; C, 4.
H (least electronegative), C, S, N, O (most electronegative)
H (smallest), O, N, C, S (largest)

Each row in the periodic table is known as a period. Each vertical column in the periodic table represents a group or family. Families are identified with specific names and numbers. The A families consist of the two families to the far left and the six families to the far right. These families are numbered from I to VIII as one moves from left to right. Elements in the same family tend to have similar chemical properties.

Elements that are poor conductors, are nonmalleable, are nonductile, and lack luster are classified as nonmetals. They are clustered in the upper right-hand portion of the periodic table.

Most isolated atoms (noble gases represent a notable exception) are unstable and reactive; they tend to react with other atoms to form compounds. The valence of an atom is the number of bonds that it forms when it does react. For an A family element, its most common valence is determined by the number of electrons that it must gain, lose, or share to obtain a stable, noble gas electron configuration. Nonmetals tend to gain or share electrons when they react.

Electronegativity tends to increase from left to right across a period and from bottom to top in a family. For the A-family elements, atomic radius tends to decrease from left to right in a period and from bottom to top in a family.

3.3 Carboxyl group (strongest acid), guanidino group (weakest acid).

K_a = dissociation constant for a weak acid and $pK_a = -\log K_a$. The larger the K_a, the stronger the acid and the smaller the pK_a; the larger the pK_a, the weaker the acid.

3.4 The presence of the α-NH_3^+ group stabilizes the negatively charged dissociated form of the neighboring α-carboxyl group (a positive charge tends to "neutralize" a negative one; isolated charges are unstable) and, in the process, increases the K_a of the carboxyl group. Carboxyl groups further removed from the α-amino group are stabilized to a lesser extent.

In general, the equilibrium constant for a reaction is determined by the relative stability (as measured by difference in free energy) of reactants and products. The more stable the products, relative to reactants, the larger the equilibrium constant. This prerequisite concept will be reviewed in more detail in Chapter 8.

3.5 pH 1: Lys, Arg, and His.
pH 12: Arg.

A weak acid functional group exists predominantly in its protonated form at pH's below its pK_a and exists predominantly in its dissociated form at pH's above its pK_a The exact ratio of the two forms can be calculated with the Henderson–Hasselbalch equation (Section 2.7):

$$pH = pK_a + \log \frac{[\text{dissociated form}]}{[\text{protonated form}]}.$$

Since pH 1 is below the pK_a for all amino acid functional groups, each functional group exists primarily in its protonated form at this pH. The protonated forms of the Lys, Arg, and His side chains are positively charged. At pH 12, all amino acid functional groups, except the guanidino group, exist mainly in a dissociated form that is either uncharged or negatively charged. Only the guanidino group exists primarily in a positively charged protonated form.

3.6 pH 1: none
pH 12: Asp, Glu, Tyr, and Cys

Since pH 1 is below the pK_a for all the functional groups in amino acids, each functional group will exist primarily in its protonated form. None of the protonated forms has a negative charge. At pH 12, all functional groups, except the guanidino group, will exist mainly in their dissociated forms. The dissociated forms of the Asp, Glu, Tyr, and Cys side chains are negatively charged (Table 3.4, Figure 3.1).

3.7 pH 1: His +2; Tyr +1
pH 12: His −1; Tyr −2

A functional group is given the charge of its protonated form if the pH is 0.5 or more below its pK_a. It is given the charge of its dissociated form if the pH is 0.5 or more above its pK_a. At pH's within 0.5 of the pK_a, a weak acid functional group is given a charge equal to the average of the charges on its protonated and nonprotonated forms. The net charge on a molecule is determined by summing the charges assigned to its individual functional groups. This process was examined in detail in Section 2.7.

At pH 1, the imidazole group and the α-amino group of His are each assigned a +1 charge, whereas the carboxyl group is given a zero charge. Net charge = (+1) + (+1) + (0) = +2.

At pH 12, the imidazole and α-amino group of His are both predominantly uncharged, whereas the carboxyl group is assigned a −1 charge. Net charge = (0) + (0) + (−1) = −1.

At pH 1, the α-amino group of Tyr is positively charged, whereas the carboxyl group and the hydroxyl group are both uncharged. Net charge = (+1) + (0) + (0) = +1.

At pH 12, the α-amino group of Tyr is uncharged, whereas each of the other two functional groups possess a −1 charge. Net charge = (0) + (−1) + (−1) = −2.

All amino acids have net negative charges at very high pH's and net positive charges at very low pH's.

3.8

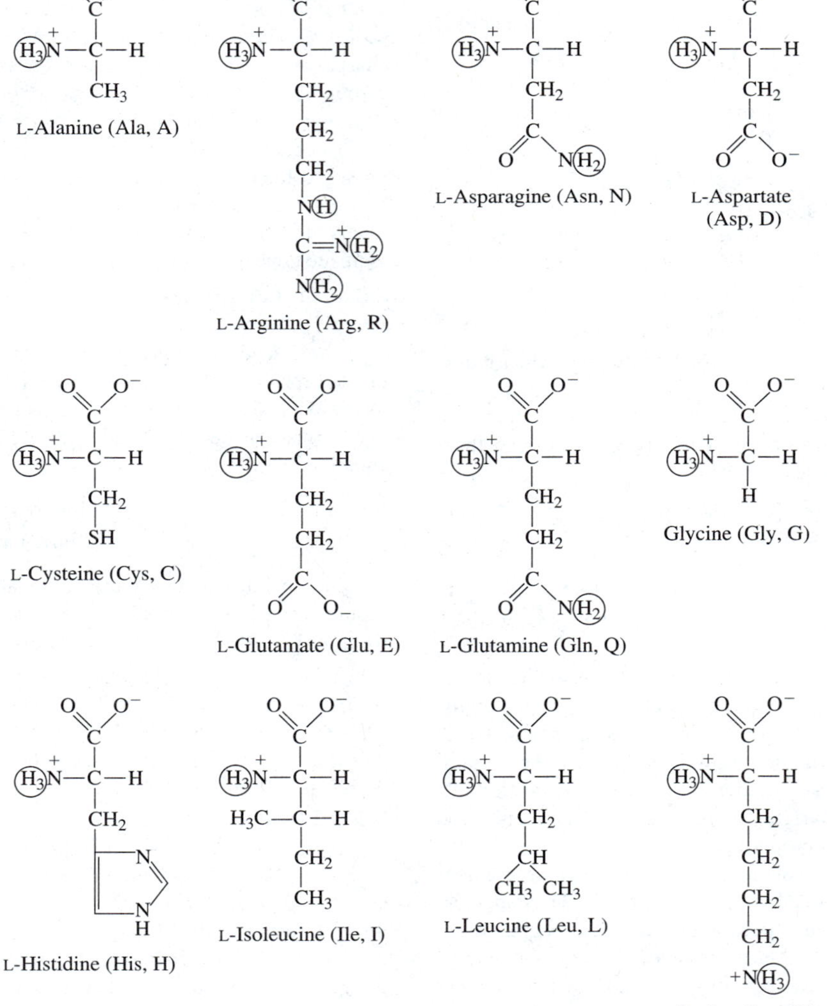

(Continued)

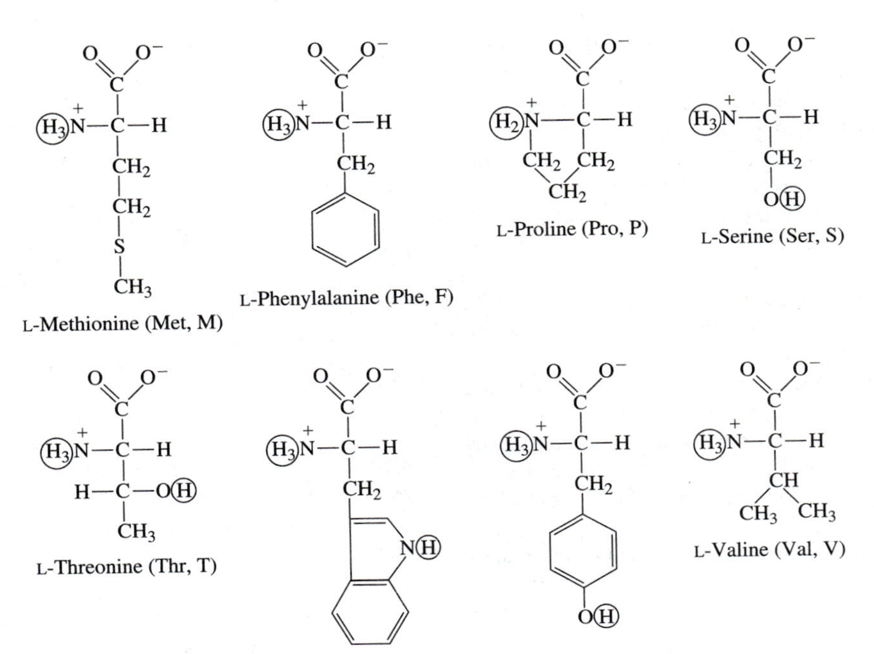

3.9 Polar: b, c, d, f, h, j, k, and l.
Nonpolar: a, e, g, and i.

A polar bond is one in which there exists a separation of charge; one end of the bond is partially negative, and the other end is partially positive. The polarity of a bond is primarily determined by the electronegativity difference between the bonded atoms. Electronegativity is a measure of an atom's electron attracting ability. Technically, any difference in electronegativity leads to a charge separation, but if the difference is small (less than about 0.4) the polarity is considered insignificant for many purposes. Large electronegativity differences (greater than about 1.7) lead to bonds that are predominantly ionic in nature. A periodic table with electronegativity values is provided on back inside cover.

For a molecule to be polar there are two conditions that must be satisfied: (a) it must contain one or more polar bonds; and (b) if it does contain more than one polar bond, the polarities must not cancel. To determine whether polar bonds will cancel, we must consider the magnitude of the polarity of each bond and their relative orientations. The relative orientation of the bonds within a molecule is determined by the bond angles about the molecule's central atoms. A hypothetical example of the canceling of polar bonds is examined in Problem 2.1.

A comparison of compounds (a) through (e) will further illustrate the approach you should use to determine whether or not a molecule is polar. Each of these five compounds has a tetrahedral molecular geometry; all bond angles are approximately 109°.

Compound (a) is nonpolar because it has no significantly polar bonds. Compound (b) is polar because it has only one significantly polar bond (the C—F bond) and this bond is not canceled by opposing polar bonds. In virtually all cases, a compound with only one significantly polar bond will be polar; opposing nonpolar or weakly polar bonds rarely cancel the polar bond's polarity. However, a large molecule with a single polar bond and many nonpolar bonds will have a more nonpolar than polar character, and that molecule may behave as though it were nonpolar. Compounds (c) and (d) are also polar because the 109° bond angles make it impossible for their multiple polar bonds to cancel; in both cases there is one "side" of the molecule that has a partial negative charge and an opposite "side" that has a partial positive charge. This is best seen by building and studying molecular models. Keep in mind that each fluorine atom tends to pull electrons (negatively charged) away from the carbon. Compound (e) is nonpolar because the four highly polar bonds do cancel; one "side" of this molecule is no more negative than any other side because of its overall molecular symmetry. If four equal forces 109° apart pulled on a central carbon atom, that central atom would not move. Similarly, the electron cloud around this carbon atom remains spherically symmetrical as the four fluorine atoms pull (at 109° angles) with equal force on the shared electrons.

3.10

a. *Plane of symmetry*: A plane (flat surface) that divides an object into two components that are mirror images.

b. *Plane-polarized light*: A beam of light in which all of the electric fields oscillate in the same plane. Each wave of light contains an oscillating electric field and an oscillating magnetic field. In a beam of nonpolarized light, the electric fields associated with separate waves usually oscillate in different planes. When such light is funneled into a polarizer, only those waves whose electric fields oscillate in the same plane will pass through; all other waves of light are removed.

c. *Chiral center* (also known as a stereogenic center): An atom that is not contained in a plane of symmetry. The most common example is a carbon atom with four different groups attached to it.

d. *Chiral compound*: A compound that does not contain a plane of symmetry. The presence of one of more chiral centers accounts for the chirality of most chiral compounds. In large molecules, however, chirality can also result from the stable, nonsymmetrical folding of achiral centers.

e. *Achiral compound*: A compound that is not chiral; it has a plane of symmetry.

f. *Isomers*: different compounds with the same molecular formula.

g. *Stereoisomers*: Isomers in which atoms have the same connectedness but differ in their relative spatial orientation. Normally, stereoisomerism is a consequence of the restricted rotation about one or more bonds or the existence of one or more chiral centers.

h. *Geometric isomers*: Stereoisomers that exist because of restricted rotation about one or more bonds. Geometric isomers come in pairs: a cis isomer and a trans isomer.

i. *Structural isomers*: Isomers that are a consequence of differences in the connectedness of atoms.

j. *Enantiomers*: Stereoisomers that are nonsuperimposable mirror images of one another. Enantiomers always come in pairs: a D-isomer and an L-isomer.

k. *Diastereomer*: Stereoisomers that are not mirror image isomers. Geometric isomers are one class of diastereomer.

3.11

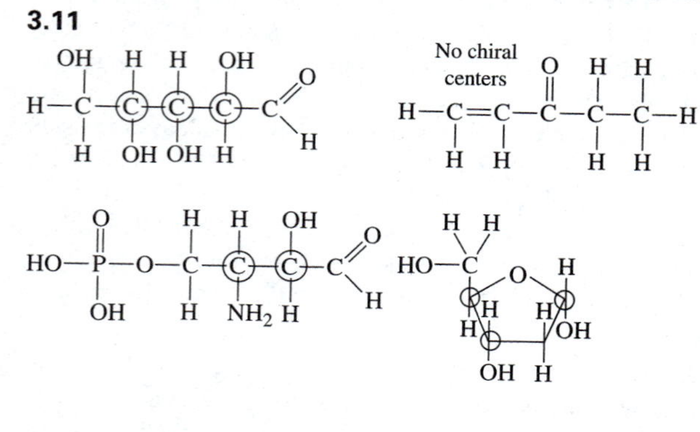

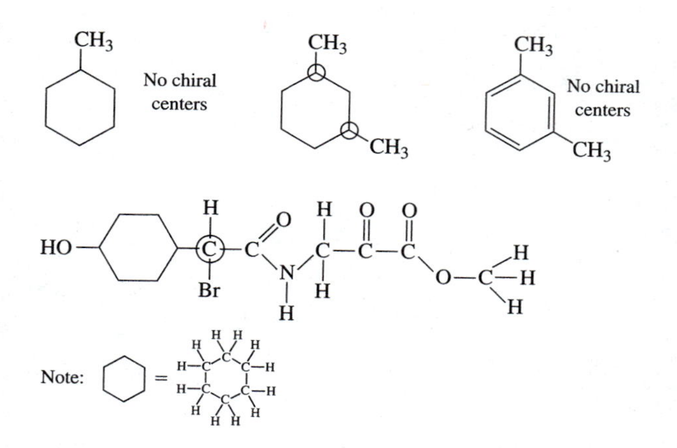

3.12 Threonine and isoleucine. The α-carbon of every protein amino acid, except glycine, is a chiral center. In both threonine and isoleucine, the first carbon into the side chain is also a chiral center.

3.13

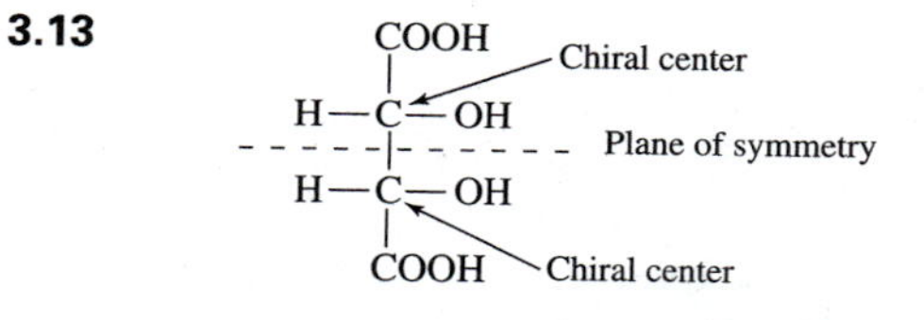

Such compounds are called **meso compounds**. There are many of them.

3.14

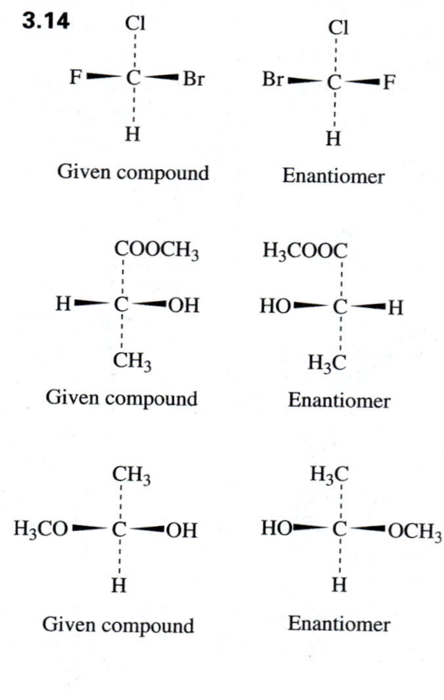

3.15 Best buy = 75 g of D,L-amino acids
$15/50 g L-amino acids = 30 cents/g of L-amino acids
$10/75 g D,L-amino acids = $10/37.5 g L-amino acids = 27 cents/g of L-amino acids

Although the 75 g of D,L-amino acids would contain only 37.5 g of the L-amino acids (those that can be utilized by the human body), these L-amino acids cost less per gram. Since D-amino acids cannot be metabolized by the body, they have no nutritional value. The answer assumes that both supplements contain the same ratios of the 20 protein amino acids. The answer might be different if the 30 cent/g L-amino acid supplement contained a higher percentage or better balance of the essential amino acids.

3.16 Acid chlorides and acid anhydrides are two well-described activated carboxylic acid derivatives.

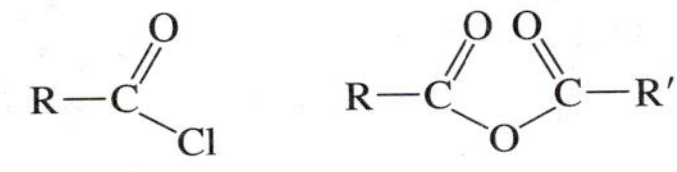

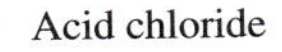

Carboxylic acid anhydride

The hydrolysis of either an acid chloride or an acid anhydride leads to the release of a substantial amount of energy. By coupling this hydrolysis to a second reaction, the energy can be used to drive the second reaction (Section 8.6).

3.17 $$\Delta G^\circ = -2.3RT \log K_{eq}$$

ΔG° = free energy of products minus free energy of reactants
R = gas constant
T = absolute temperature (K)
K_{eq} = equilibrium constant

A reaction is spontaneous ($K_{eq} > 1$) when the products are more stable (have less free energy) than the reactants. The more stable the products, relative to reactants, the larger the equilibrium constant (Section 8.2).

3.18

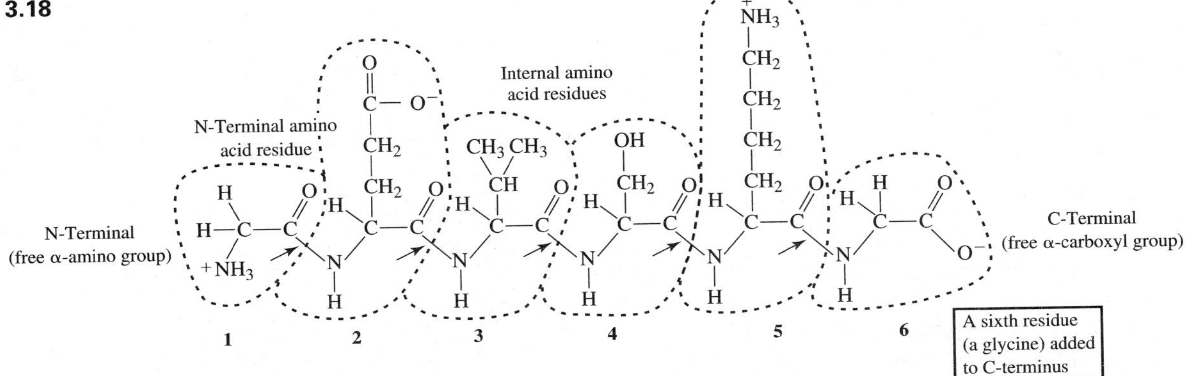

There are 40 possible hexapeptides. Each of the 20 protein amino acids could be added to either end of the pentapeptide. The arrows (→) identify peptide bonds. A pentapeptide contains four peptide bonds, a hexapeptide five, a heptapeptide six, and so on.

3.19 Residues 2, 4, and 5. Residue 2 (asparate) contains a side chain carboxyl group that could be joined through an amide link to the α-amino group of any amino acid. The α-carboxyl group of the amino acid attached to the side chain could then be bonded to additional amino acids to create a peptide branch. In a similar manner, the side chain hydroxyl group in residue 4 (serine) could be joined through an ester link to the carboxyl group in any amino acid. The free α-amino group in the attached amino acid could then be linked through peptide bonds to additional amino acids. The amino group in the side chain of residue 5 (lysine) could also be joined to the carboxyl group in any amino acid. In this case, an amide link would be formed.

3.20

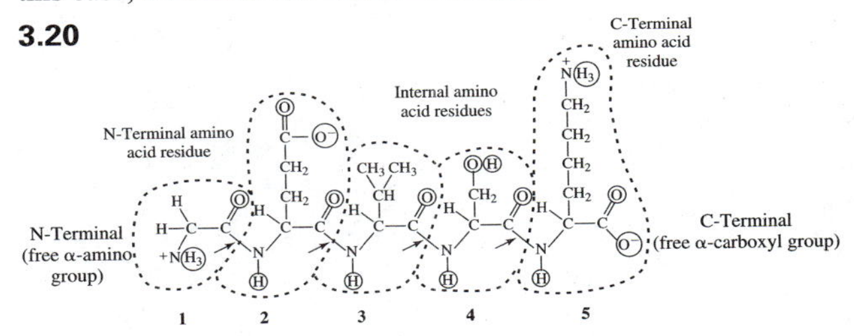

Note: The nitrogen in a peptide bond does not participate in H bonding since it acquires a partial positive charge through resonance:

3.21

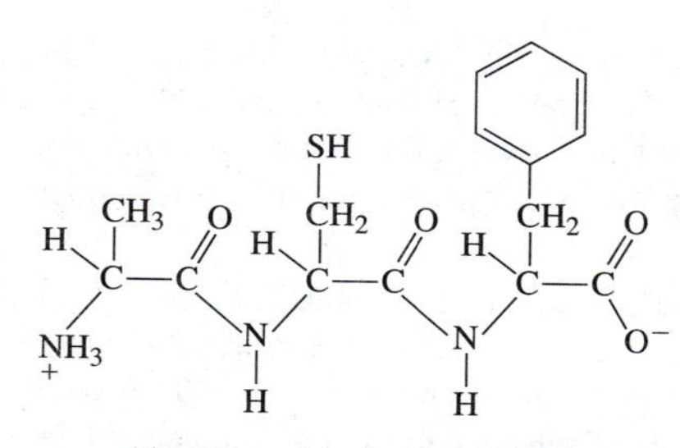

Alanyl - cysteinyl - phenylalanine

There are six possible answers: alanyl-cysteinyl-phenylalanine, alanyl-phenylalanyl-cysteine, cysteinyl-alanyl-phenylalanine, cysteinyl-phenylalanyl-alanine, phenylalanyl-alanyl-cysteine, and phenylalanyl-cysteinyl-alanine.

3.22

a. pH 1:

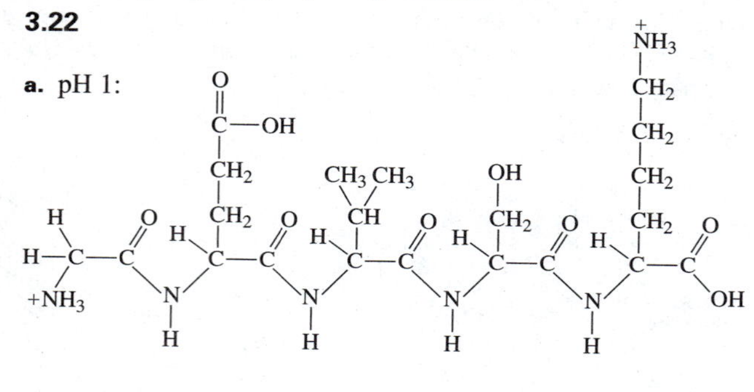

b. pH 6:

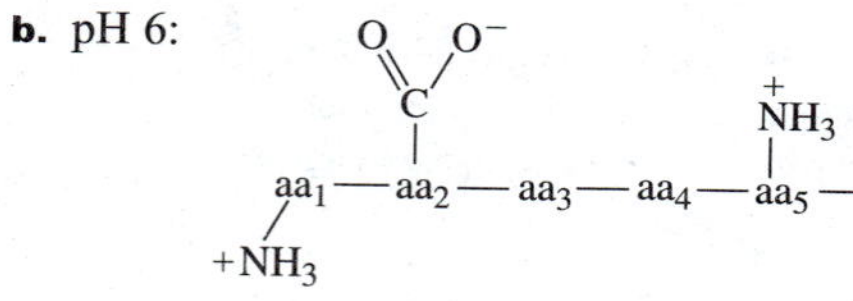

c. pH 10.5:

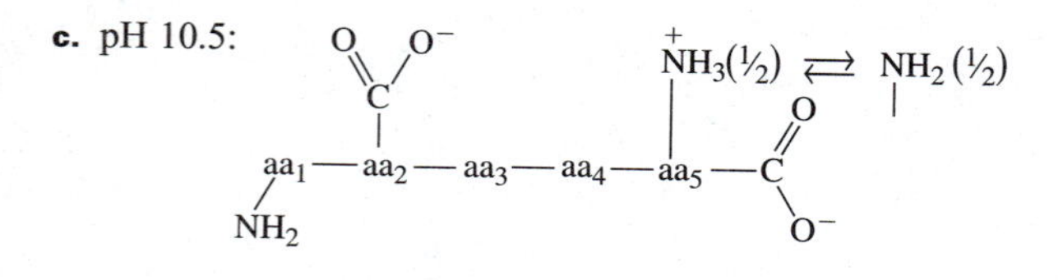

d. pH 14:

The Henderson–Hasselbalch equation indicates that a weak acid functional group is primarily protonated when the pH is below its pK_a and primarily nonprotonated at pH's above its pK_a. The same equation predicts that [conjugate base] = [weak acid] whenever pH = pK_a. Since the pK_a for the lysine side chain is 10.5, it will be half protonated and half nonprotonated at pH 10.5.

3.23 pH 1, +2; pH 4.3, +0.5; pH 8, 0; pH 13, −2

If the pH is 0.5 or more below the pK_a of a functional group, that group is assigned the charge of its weak acid form. If the pH is 0.5 or more above the pK_a of a functional group, that group is assigned the charge of its conjugate base form. When the pH is within ±0.5 of the pK_a for a functional group, the group is assigned a charge equal to the average of the charge on its weak acid (protonated) form and the charge on its conjugate base form. The net charge on a compound is equal to the sum of the charges on all of its charged functional groups. These rules are derived and illustrated in Section 2.7.

At pH 1, net charge = (+1) + (0) + (+1) + (0) = +2

At pH 4.3, net charge = (+1) + (−0.5) + (+1) + (−1) = +0.5

At pH 8, net charge = (+1) + (−1) + (+1) + (−1) = 0

At pH 13, net charge = (0) + (−1) + (0) + (−1) = −2

3.24

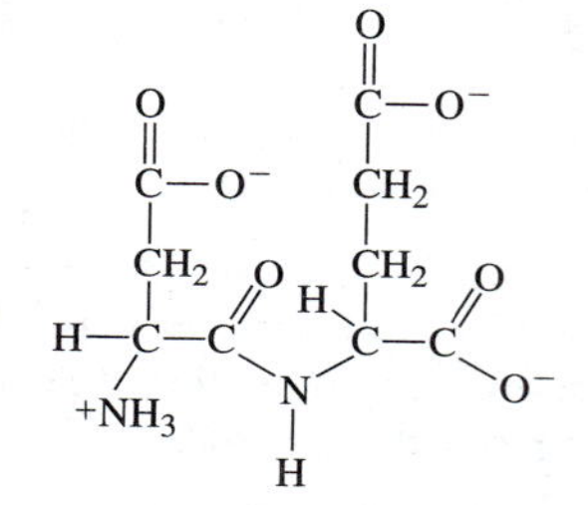

There are other possible answers. Each of the two residues can be either glutamic acid or aspartic acid, the only two amino acids with negatively charged side chains at pH 7. Since the charge on the N-terminal α-amino group cancels the charge on the C-terminal α-carboxyl group, each side chain in the peptide must have a negative charge in order to end up with a net charge of −2.

3.25 A net negative charge. An increase in pH removes protons from acidic functional groups. The removal of a proton either eliminates a positive charge or creates a negative charge (study Table 3.4). In either case, the net charge on a peptide will become less positive, zero, or more negative (depending on its initial net charge and the number of protons removed). If the net charge is initially zero, as it is when pH = pI, the net charge will become negative.

3.26

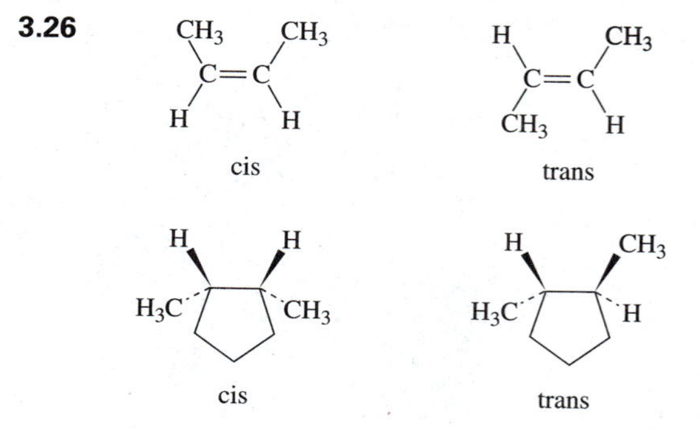

Cis means on the same side (of a double bond or ring).

Trans means on opposite sides (of a double bond or ring).

Geometric isomers represent one class of stereoisomers. They always come in pairs with one member of the pair designated cis and the second member designated trans. Geometric isomers are examples of diastereomers (see the answer to Problem 3.10).

3.27

Delocalized π-Bonding

A sigma (σ) bond results from the end to end overlap of a variety of orbitals, including *s* orbitals, *sp* hybrid orbitals, sp^2 hybrid orbitals, and sp^3 hybrid orbitals:

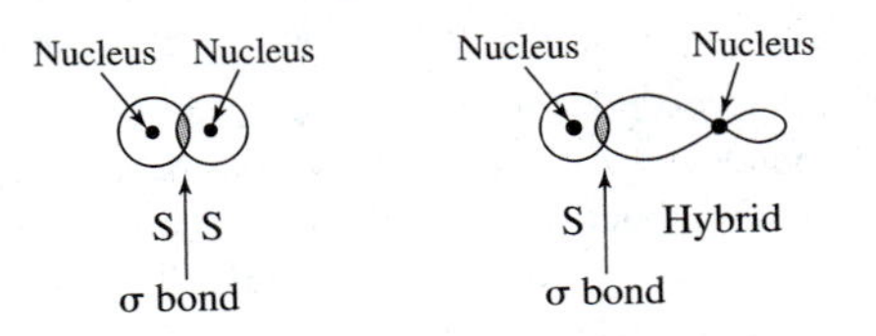

Since there is a symmetrical distribution of the electron cloud about a σ bond, there is free rotation. A pi (π) bond is a product of the side-by-side overlap of two nonhybridized *p* orbitals:

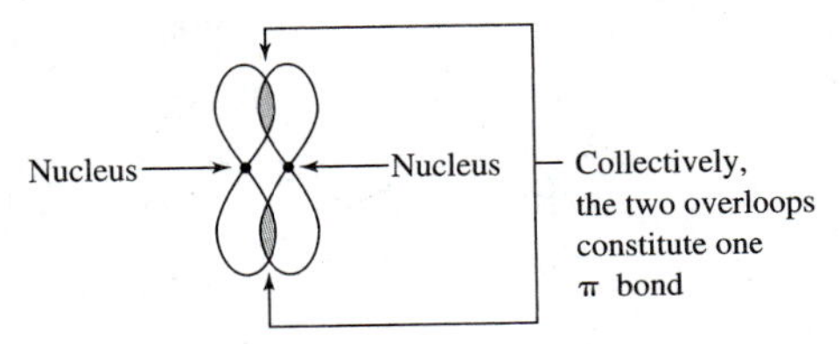

Since rotation would destroy the overlap that is responsible for the bond, rotation is restricted.

Delocalized π bonding involves the simultaneous side-by-side overlap of more than two nonhybridized *p* orbitals (each associated with a separate atom):

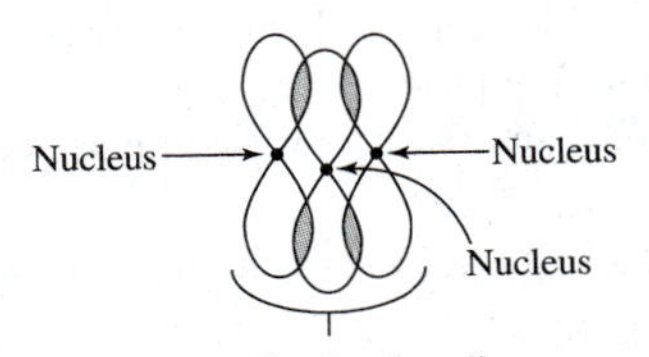

Delocalized π bonding

There is normally restricted rotation about all the π bonds to the participating atoms.

Resonance

Any time multiple formulas can be written for a compound that differ only in the position of electrons (the connectedness and spatial arrangement of atoms unchanged), each formula is said to represent a resonance form of the compound and the compound is, in reality, a resonance hybrid of all its resonance forms. This concept is normally introduced in organic chemistry during the discussion of the structure of benzene:

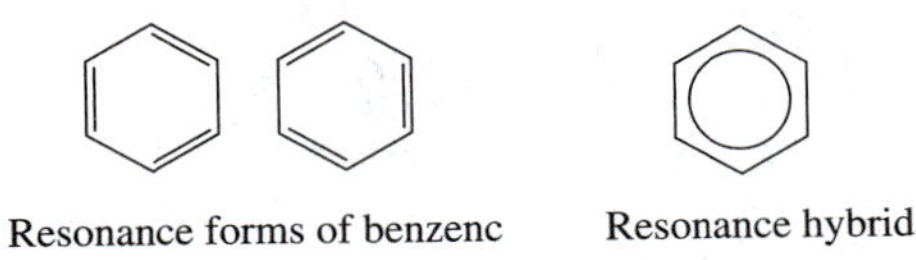

Resonance forms of benzenc Resonance hybrid

Benzene does not contain alternating double and single bonds as its resonance structures imply; each bond in the ring is identical and each has partial double bond and partial single bond character.

Tautomerism

Tautomers are compounds that exist in equilibrium and differ in the location of a single hydrogen and a single double bond. A variety of functional groups, including those of aldehydes, ketones, and amides, exist in multiple tautomeric forms:

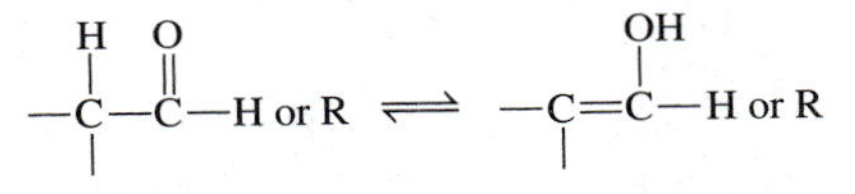

Keto form Enol form

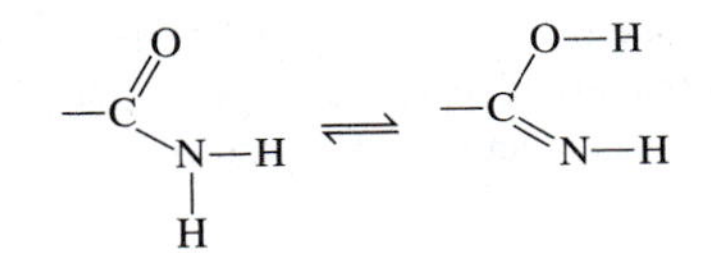

Additional Examples

Specific additional examples of delocalized π-bonding, resonance, and tautomerism follow:

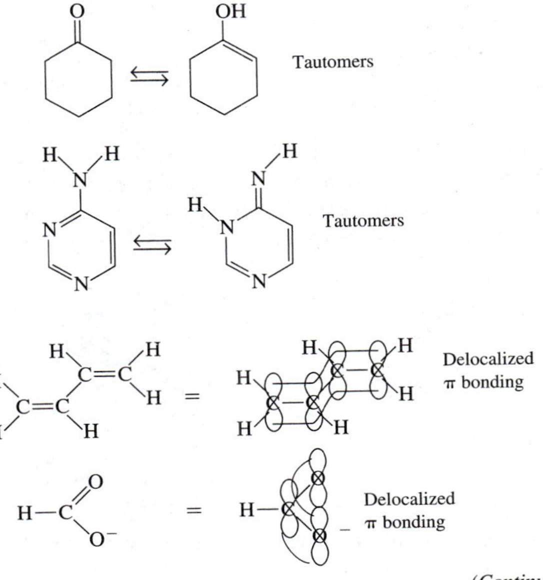

(Continued)

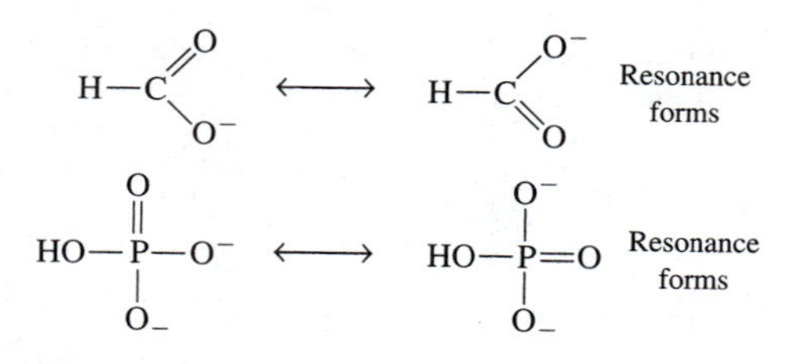

3.28 120°. Most A-family elements have four outer bonding orbitals: one *s* and three *p*. According to the hybridization model for bonding, these orbitals can be hybridized in one of three ways at the time they overlap with orbitals in other atoms to form bonds. All four outer orbitals can merge and rearrange themselves to generate four identical sp^3 hybrid orbitals that are oriented 109° apart and point toward the corner of a tetrahedron:

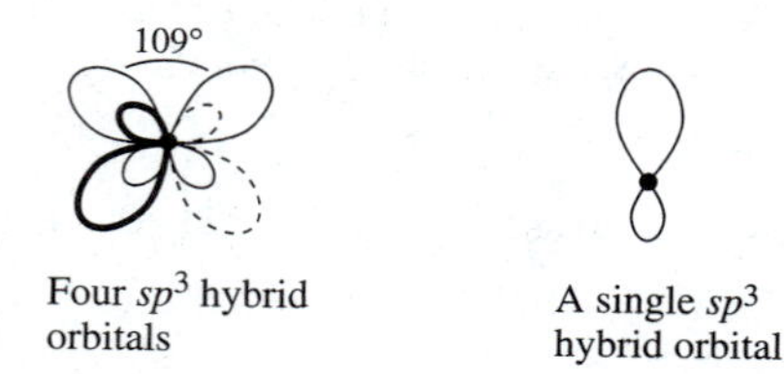

Four sp^3 hybrid orbitals

A single sp^3 hybrid orbital

When two *p* and one *s* orbital hybridize, the result is three sp^2 hybrid orbitals that are planar and 120° apart. The single nonhybrid *p* orbital is perpendicular to the plane in which the hybrid orbitals lie:

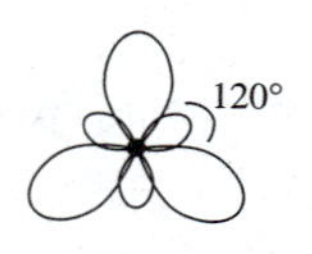

Three sp^2 hybrid orbitals

120°

p orbital

Three hybrid orbitals and one nonhybridized *p* orbital

The third hybridization option involves the creation of two *sp* hybrid orbitals by merging one *s* orbital with one *p* orbital. These two orbitals are 180° degrees apart and are perpendicular to each of the two nonhybridized *p* orbitals:

Two *sp* hybrid orbitals

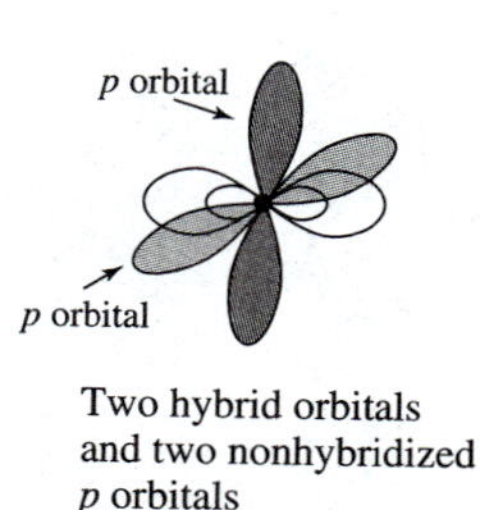

Two hybrid orbitals and two nonhybridized *p* orbitals

The sp^3, sp^2, and *sp* hybridization of a central atom leads to 109°, 120°, and 180° bond angles, respectively, because the angle between two orbitals at the time they overlap with orbitals of other atoms to form bonds determines the bond angle that results. In this context, the following theoretical generalizations are worth remembering:

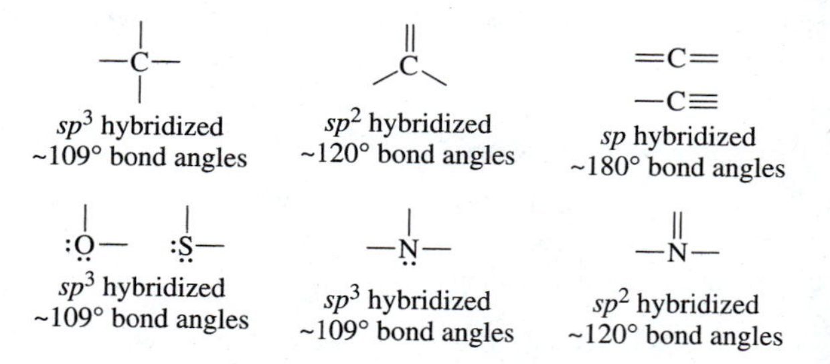

3.29 109°. Any carbon (or other A-family element) that forms four single bonds is sp^3 hybridized and will have approximately 109° bond angles about it. A double bond and two single bonds to a carbon leads to sp^2 hybridization and 120° bond angles. A triple and a single bond or two double bonds to a carbon leads to *sp* hybridization and a 180° bond angle (see answer to Problem 3.28).

3.30

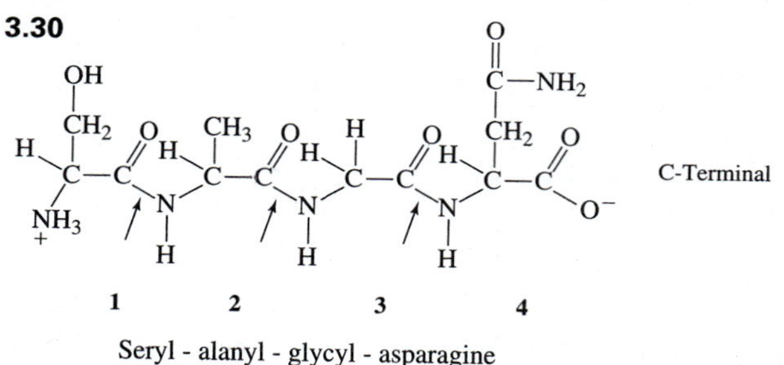

Seryl - alanyl - glycyl - asparagine

3.31 SAGN

3.32 See answer at top of page A-15.

3.33 See answer at top of page A-15.

3.34 Oxytocin. When compared with vasopressin, oxytocin contains Leu (nonpolar) in place of Arg (polar and charged at physiological pH values) and Ile (nonpolar) in place of Phe (also nonpolar).

3.35 Vasopressin, +2; Oxytocin +1. Vasopressin has three weak acid functional groups: a phenol, a protonated guanidino group (guanidinium), and a protonated amino group. Amide groups have little tendency to either accept or donate an H^+. Oxytocin contains a phenol and a protonated amino group but lacks a guanidinium group. Because all three of these functional groups have pK_a's above 9, they will all be almost entirely protonated at pH 7. The protonated form of the phenol group is uncharged, whereas the protonated forms of the guanidino and amino groups have a +1 charge.

3.36 If multiple peptides are present in a sample, it is impossible to tell how much of each amino acid came from which peptide. It

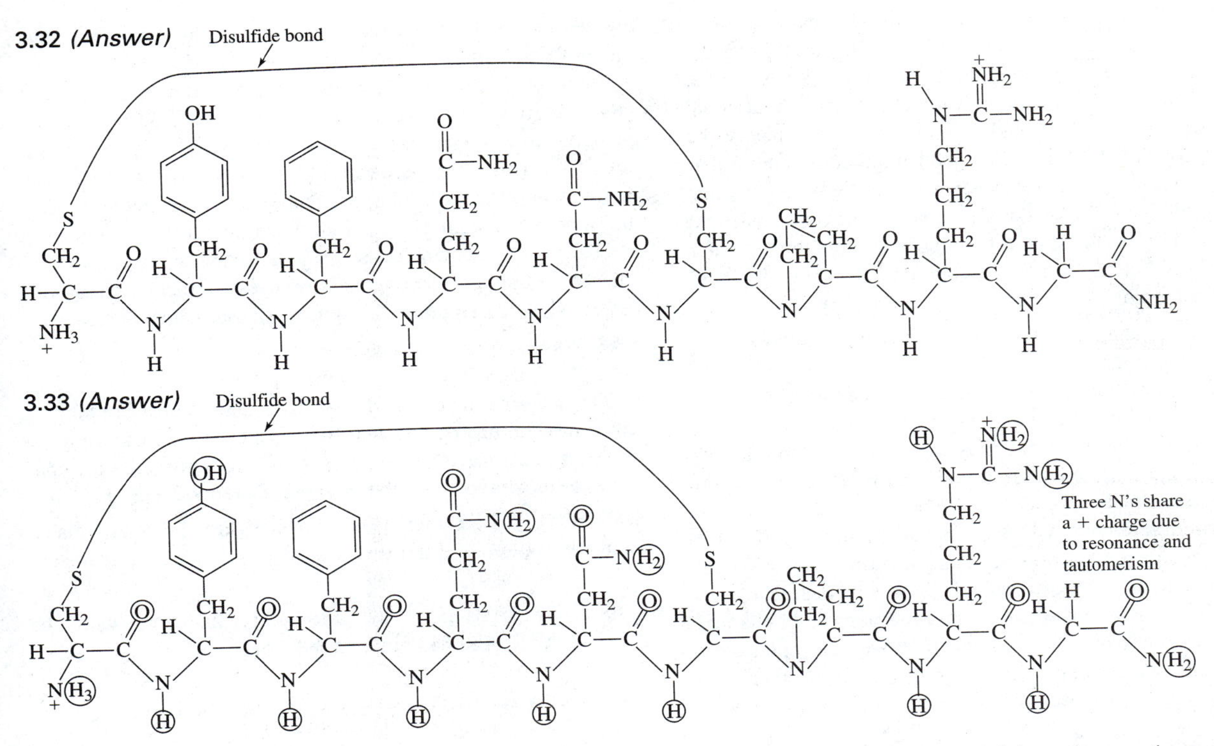

is possible that nonpeptide components in a mixture could coelute with an amino acid and impact peak height. This would be of no concern if the detection method were amino acid specific. Impurities might react chemically with the peptide or with amino acids.

3.37 No. A peptide is completely hydrolyzed before the separation of its amino acids. The hydrolysate carries no information about the order in which the amino acids had originally been linked.

3.38 (a) Glu-Glu. (b) Ala-Lys and Lys-Lys would tend to emerge together before the Glu-Glu. (c) pH 7, yes; pH 1, yes; pH 14, no.

a. A cation exchange resin (negatively charged) binds all cations, including Ala-Lys and Lys-Lys (both have a net positive charge at pH 7). Glu-Glu, which is negatively charged at pH 7, is repelled by the negative functional groups on the resin and passes directly through the column.

b. An anion exchange column (positively charged) binds negatively charged molecules, including Glu-Glu (approximately −2 net charge at pH 7). Because both Ala-Lys and Lys-Lys have a net positive charge at pH 7, they are repelled by the resin and tend to move rapidly through the column together.

c. At pH 7, Glu-Glu does not bind at all [see (a), above], whereas Ala-Lys and Lys-Lys bind with different affinities. Because the Ala-Lys has a smaller net positive charge than Lys-Lys, it binds less tightly to the negative resin than the Lys-Lys. Glu-Glu elutes first, followed by Ala-Lys, and, finally, Lys-Lys.

At pH 1, each of the three peptides has a different net positive charge. Lys-Lys (net charge of +3) is more difficult to displace than Ala-Lys (net charge +2), which is more difficult to displace than Glu-Glu (net charge +1).

At pH 14, all of the peptides have a net negative charge and are repelled by the negative resin. They all tend to run directly through the column and elute together.

3.39 It can. The shape of a molecule can have an impact on its ability to enter the pores of the stationary phase beads. Those molecules whose shapes make it difficult for them to enter the pores move through the column ahead of similarly-sized but differently-shaped molecules that readily enter the pores.

3.40 True. For many organic compounds, size is roughly proportional to molecular weight, because many organic compounds contain roughly equal proportions of the same elements. Since larger molecules find it more difficult to enter gel filtration pores, they move more rapidly through a gel filtration column.

3.41 Insulin, the largest, will tend to emerge first. The smallest compound, an amino acid, will tend to emerge last.

The relative rate of movement of compounds through a gel filtration column is primarily determined by the sizes of the pores within the gel filtration beads selected for the stationary phase. If the pores are extremely small, none of the compounds enter the beads, and all tend to elute together. If the pores are very large, all of the compounds enter and exit the beads with about equal ease and, once again, they tend to move at approximately the same rate through the column. The five compounds could probably be separated from one another if we were to use a pore size that just excluded insulin. All of the remaining compounds could access the beads, but the smallest compound would tend to enter more readily and penetrate more deeply than the others; it would be the last compound to emerge from the column.

3.42 pH 7: Lys and methyl amine migrate to the cathode, others to the anode.
pH 2: All but lactic acid migrate to the cathode.
pH 10.5: All but methyl amine migrate to the anode.

To answer this question one must calculate the approximate net charge on each compound at each of the three pH's. A compound with a net positive charge will move toward the cathode (−), whereas compounds with a net negative charge will move toward the anode (+) (opposite charges attract). Precise net charge calculations are not required since any degree of net charge leads to some movement toward the oppositely charged pole. If we are concerned about the rate of movement, precise net charge calculations are important. The net charge (whether + or −) of each compound at each pH is summarized in the following table:

Compound	Net Charge pH 7	pH 2	pH 10.5
Glu	−	+	−
Lys	+	+	−
Ala-Val-Cys-Asp	−	+	−
Lactic acid	−	−[a]	−
Methyl amine	+	+	+

[a]Net charge is very small since pH 2 is almost 2 units below the pK_a for lactic acid. At pH 2, lactic acid will predominantly exist in its uncharged protonated form; its time averaged charge to the nearest half of a charge unit is zero.

A weak acid functional group predominantly exists in its protonated form at pH's below its pK_a and mainly exists in its nonprotonated form at pH's above its pK_a. These facts can be used to quickly estimate the charge contribution of any given functional group. The net charge on a compound is equal to the sum of the charges contributed by its individual functional groups (Section 2.7).

3.43 Sets a, b, and c. Two compounds can normally be separated electrophoretically if a pH can be found at which there is a significant difference in their charge to size ratios. If the support material interacts differently with each of two compounds, it may be possible to separate them even when they have the same charge to size ratio. In this problem, however, the support material is assumed to be inert.

Because all of the compounds in this problem are roughly the same size, the ability to create significant charge differences (by changing the pH) primarily determines whether the two compounds in a pair can be electrophoretically separated. Because both Gly and Ala have the same functional groups with virtually the same pK_a's, there is no pH at which they differ significantly in charge; they cannot be separated. Likewise, methanol and ethanol are both uncharged unless we use extremely high pH values, in which case both will acquire a similar negative charge.

3.44 Wave A: 2.38 cm; 0.0238 m
Wave B: 3.79 cm; 0.0379 m

The answer is determined by using a ruler to measure the distance, in centimeters, between two adjacent peaks or two adjacent dips in the wave. Since 1 cm = 10^{-2} m, centimeters are converted to meters by multiplying the measured value by 10^{-2}.

3.45 Wave A is most energetic. Wavelength is inversely related to energy content; the shorter the wavelength, the greater the energy.

3.46 $\epsilon_{552} = 370\ \text{L mol}^{-1}\ \text{cm}^{-1}$. More likely than not, ϵ would be different at 552 nm, 643 nm, and 390 nm.

$$\epsilon = \frac{A}{bc} = \frac{0.37}{(1\ \text{cm})(1 \times 10^{-3}\ \text{mol L}^{-1})} = 370\ \text{L mol}^{-1}\ \text{cm}^{-1}$$

Because the absorbance of a solution usually changes as the wavelength being absorbed is modified, ϵ is normally a function of wavelength. If a solution of drug MX-2 happened to have the same absorbance at two different wavelengths, ϵ would be the same at these wavelengths.

3.47 $9.74 \times 10^{-4}\ M$.

$$c = \frac{A}{\epsilon b} = \frac{0.19}{(195\ \text{L mol}^{-1}\ \text{cm}^{-1})(1\ \text{cm})} = 9.74 \times 10^{-4}\ M$$

Pathlength is assumed to be 1 cm (considered standard) unless specified otherwise.

3.48 $552\ \text{nm} \times \frac{10^{-9}\ \text{m}}{1\ \text{nm}} = 5.52 \times 10^{-7}$ m; **visible**

$$258\ \text{nm} \times \frac{10^{-9}\ \text{m}}{1\ \text{nm}} = 2.58 \times 10^{-7}\ \text{m}$$; **UV**

$$10\ \text{nm} \times \frac{10^{-9}\ \text{m}}{1\ \text{nm}} = 10^{-8}\ \text{m}$$; **UV/X-ray border**

$$10^{-5}\ \text{nm} \times \frac{10^{-9}\ \text{m}}{1\ \text{nm}} = 10^{-4}\ \text{m}$$; **microwave/IR border**

The answers are read from Table 3.10 after wavelengths are converted to meters.

3.49 $A_{218} = \epsilon_{218}bc = (33{,}500 \text{ L mol}^{-1} \text{ cm}^{-1})(1 \text{ cm})(2.2 \times 10^{-5} \text{ M}) = 0.737$

Because it is not specified, the pathlength is once again assumed to be 1 cm.

3.50 ${}^{35}_{17}Cl$, ${}^{36}_{17}Cl$, ${}^{34}_{17}Cl$, ${}^{37}_{17}Cl$, ${}^{38}_{17}Cl$ and others. There are many possible answers. Although the subscript must be 17, the superscript can, in theory, be any whole number greater than 16.

3.51 10 μCi = 6.2×10^3 dpm with both isotopes. The prefix μ means 10^{-6}. One curie (Ci) equals 3.7×10^{10} dps.

$$10\ \mu\text{Ci} \times \frac{10^{-6}\ \text{Ci}}{1\ \mu\text{Ci}} \times \frac{3.7 \times 10^{10}\ \text{dps}}{1\ \text{Ci}} \times \frac{1\ \text{min}}{60\ \text{s}} = 6.2 \times 10^3\ \text{dpm}$$

By definition, 10 μCi of any radioisotope undergoes 6.2×10^3 dpm; the unit μCi is not a function of the radioisotope involved.

3.52 (a) No; (b) Yes. Alpha-rays emitted inside a patient lack the energy necessary to escape from the body. In contrast, most of the γ-rays can escape and expose anyone in the vicinity of the patient.

3.53 1/1024 of the unstable atoms remain after 10 half-lives.

$$1 \rightarrow \frac{1}{2} \rightarrow \frac{1}{4} \rightarrow \frac{1}{8} \rightarrow \frac{1}{16} \rightarrow \frac{1}{32} \rightarrow \frac{1}{64} \rightarrow \frac{1}{128} \rightarrow \frac{1}{256} \rightarrow \frac{1}{512} \rightarrow \frac{1}{1024}$$

3.54
1. Limit time spent around the radiation source.
2. Stay as far away from the source as possible. Exposure is inversely proportional to the square of the distance (X) from the source; exposure is proportional to $1/X^2$. By doubling your distance from a source, you reduce your exposure fourfold.
3. Use shielding.

3.55 $\frac{29000}{98000} \times 100 = 29.6\ \%$ counting efficiency

3.56 No. Some nonapeptides, including those containing no glutamate residues and those synthesized before the injection of ^{14}C-glutamate, will contain no ^{14}C (assuming that none of the glutamate is metabolized to another amino acid). Any nonapeptides that contain multiple glutamate residues and are synthesized in large amounts after the injection of the ^{14}C glutamate will contain relatively large amounts of ^{14}C.

3.57
1. The imaging isotope should have a short half-life so as to minimize the radiation exposure associated with diagnosis.
2. The types of radiation emitted by the imaging isotope should lead to as little risk as possible.
3. The imaging isotope should be chemically nontoxic.

3.58 No. Exposure to an external beam of nuclear radiation leads to chemical and physical changes within a patient, but it does not create any radioisotopes. Thus, exposure to an external beam of nuclear radiation does not cause a patient to emit nuclear radiation.

CHAPTER 4

4.1 The hydrolysis of a simple protein yields a mixture of free amino acids and some ammonia (if the protein contains glutamine or asparagine). The free amino acids are products of the hydrolysis of the peptide bonds that hold amino acids together within proteins. Ammonia is generated during the hydrolysis of the amide links within Gln and Asn side chains.

The hydrolysis of a metalloprotein yields the same products plus one or more metal ions. The *metallo*-prefix means that the protein contains at least one metal ion as a prosthetic group.

4.2 A peptide must have over 50 amino acid residues and a specific three-dimensional shape under physiological conditions to be considered a protein. The 50 residue cutoff is not written in stone, but it seems to be accepted by most authors. Whereas all proteins are peptides, not all peptides are proteins.

4.3 A protein with 500 amino acid residues will contain roughly 8000 atoms (500 × 16). The actual number of atoms will depend upon the ratio of large to small amino acids within one molecule of the protein. The precise count will also be a function of pH and prosthetic groups (if present). The pH impacts the count by determining the extent of dissociation of weak acid functional groups. Regardless of the exact number of atoms, most proteins are truly giant molecules.

4.4 No. The number of atoms per amino acid ranges from 10 to approximately 27. A protein rich in small amino acids will have a much lower molecular weight than one that contains predominantly large amino acid residues, even if the two proteins contain the same total number of residues. Many of the possible amino acid ratios will lead to different, but similar, molecular weights.

4.5 Arg–Asp–Val–Leu–Val–Gly–Pro–Ser–Met–His–Tyr–Thr–Ile–Cys–Lys–Glu–Ser–Phe–Ala–Gly–Phe–Gly–Gly–Asp–Glu.

For discussion purposes, number the peptides in Group 1 from 1 to 4 (top to bottom) and letter the peptides in Group 2 from A to E (top to bottom). Since Peptide Y contains only one Pro and one Met, the Pro–Ser–Met sequence in Peptide A indicates that the C-terminal Pro–Ser sequence in peptide 2 must have been joined (in peptide Y) to the N-terminal Met in peptide 4. Similarly, the Thr-Ile sequence in peptide C indicates that the C-terminal Thr of peptide 4 must have been linked to the N-terminal Ile of peptide 1. Finally, the Glu–Ser–Phe sequence in peptide E leads to the conclusion that the C-terminal Glu–Ser in peptide 1 must have been joined to the N-terminal Phe in peptide 3.

4.6 The α-helix in Figure 4.3 contains just over 6 residues, including 5 complete residues and two partial ones. With 3.6 residues per turn, there is one complete turn and close to two thirds of a second turn.

A residue contains all of those atoms contributed to a molecule by an individual amino acid. Each internal residue, except proline residues, contains a carbonyl, an α-carbon, an α-hydrogen, a side chain, and an α- nitrogen with one hydrogen attached. No hydrogen is attached to the α-nitrogen in an internal proline residue. An N-terminal residue will contain a free α-amino group, whereas a C-terminal residue will possess a free α-carboxyl group. One simple way to count the residues within a peptide fragment is first to circle an easily identifiable internal residue, and then to circle one residue at a time while working in both directions from the one initially circled. The peptide fragments illustrated in Figure 4.3 contain some partial residues.

4.7 In each helix, the C-terminus is at the top and the N-terminus is at the bottom. Take any peptide bond within a peptide fragment and hydrolyze it. The free carboxyl group that results will be pointing toward the C-terminal end and the free amino group will be pointing toward the N-terminus.

4.8

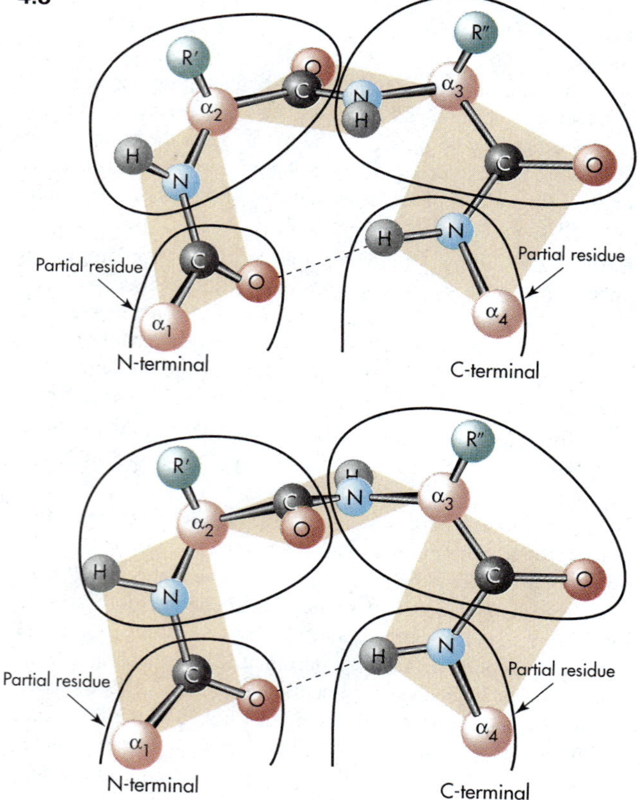

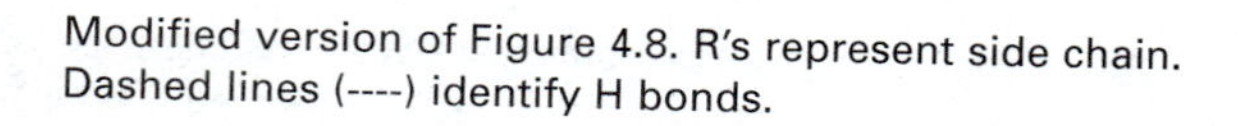
Modified version of Figure 4.8. R's represent side chain. Dashed lines (----) identify H bonds.

4.9

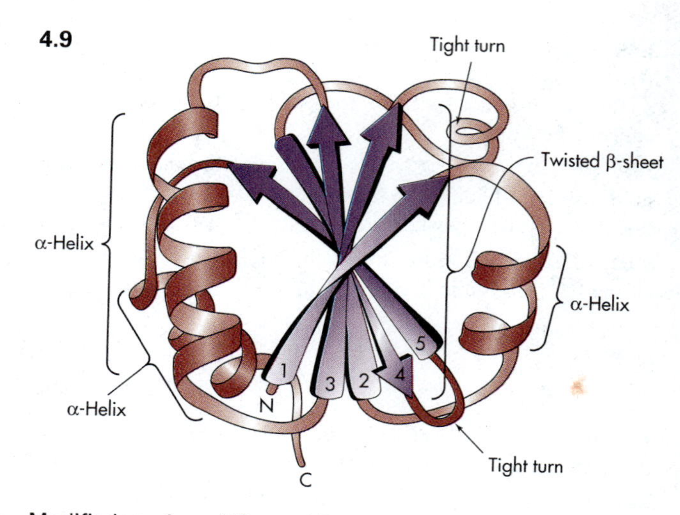

Modified version of Figure 4.7.

For a turn to be considered a tight turn, a chain must *abruptly* reverse its direction. An abrupt reversal is usually classified as one that involves no more than five amino acid residues.

4.10 Gly, Asn, and Ser.
Study Table 4.4.

4.11 Since the pK_a of the glutamic acid side chain is approximately 4.2, each side chain along a polyglutamic acid molecule will be uncharged at pH 1 but negatively charged at pH 7. The repulsion between like-charged side chains prevents α-helix formation at pH 7.

4.12 Helical region.

This prediction, which is shaky at best, is based on the information given in Table 4.4. Eight of the 11 residues are commonly found in α-helices, whereas 5 of the 11 are seldom found in β-sheets.

4.13 Steric hindrance is the repulsion (between like-charged electron clouds) that results when two groups of atoms on the same molecule are brought too closely together; two atoms cannot occupy the same point in space at the same time. Certain amino acid sequences are incompatible with an α-helix because, when they are coiled into such a helix, bulky side chains are packed too closely together (steric hindrance results). In a similar manner, steric hindrance can disrupt a β-sheet.

4.14 Polar in, nonpolar out.

In a nonpolar environment (solvent), polar amino acid residues tend to fold into the middle of a protein in order to escape from that environment. This relates to the major biochemical theme, "like associates with like"; polar associates with polar, nonpolar associates with nonpolar, but polar and nonpolar do not associate.

4.15 Since the peptide has only a single polar side chain and multiple nonpolar side chains, it will most likely be buried inside a folded protein in an aqueous environment. Nonpolar in, polar out.

4.16 **a.** **b.**

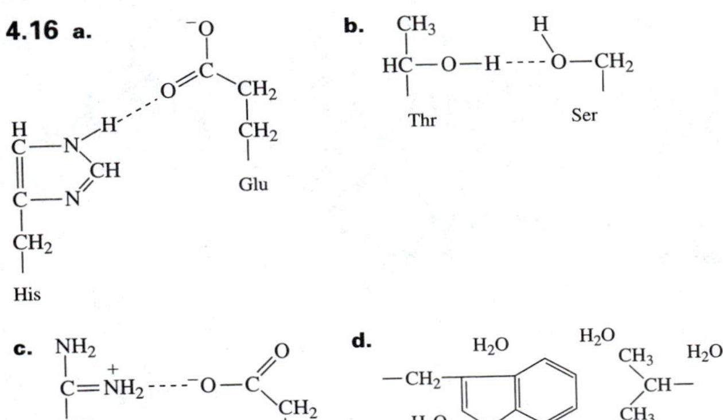

c. **d.**

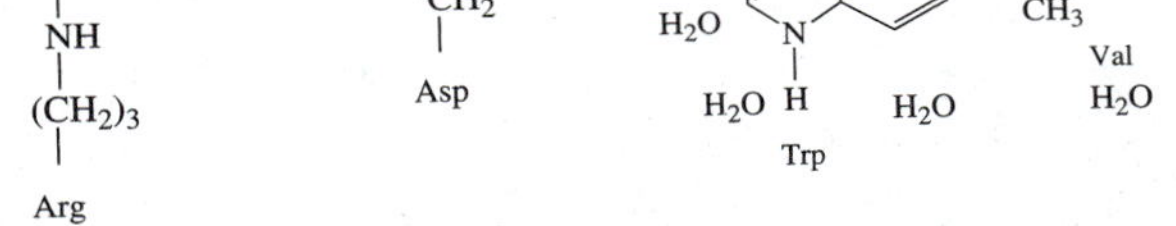

Water is a key component of diagram d. Hydrophobic forces are, by definition, those forces that hold nonpolar groups together in an aqueous environment. Polar water molecules, in clinging together, tend to exclude nonpolar groups. An overall increase in entropy results (Section 2.2).

4.17 Each motif contains roughly 23 amino acid residues.

This estimate is based on the fact that there are 3.6 amino acid residues per turn of an α-helix. In the DNA-binding motif, each of two α-helices contains approximately 3 turns and there are 1 to 3 residues in the bend between the helices (based on the length of the connecting ribbon relative to the length of the ribbon within 1 turn of an α-helix). Total residues = (3.6×6) + about 2 = roughly 23. A similar approach leads to a similar answer for the calcium-binding motif.

4.18 Approximately 60° in DNA-binding motif; roughly 75° in calcium-binding motif.

The angle between the two α-helices and the length of the connecting peptide fragment both play roles in determining the fit between each motif and its partner in binding (that substance with which it interacts and binds).

4.19 Domains often contain multiple motifs plus a variety of unique loops and coils. They maintain their distinctive conformations (foldings) when cut out of a polypeptide chain. A domain can, for some purposes, be viewed as a protein within a protein; different domains within the same protein often serve different functions. Domains are sometimes very large and frequently possess an intricate and complex tertiary structure.

Motifs, being parts of domains, are a step down in the structural hierarchy of a protein. Since the precise folding of a motif is maintained with the assistance of amino acid residues that are external to the motif itself, a motif cannot exist in the absence of other peptide components. Motifs, in contrast to domains, are relatively simple structural elements that usually contain a limited number of amino acid residues.

4.20 Immunoglobulin VL domain: 2 sheets, no helixes.
Rhodanese domain 2: 1 sheet, 6 helixes.
Subtilisin: 2 sheet, 8 helices.
Cytochrome c_3: 1 helix.
Elastase: 2 sheets, 2 helixes.
Pyruvate kinase domain 1: 1 sheet, 8 helixes.
Pyruvate kinase domain 2: 2 sheets, 1 helix.
Pyruvate kinase domain 3: 1 sheet, 5 helixes.

See answer structure on page A-20.

4.21 An oligomer with two polypeptide chains is called a dimer. A tetramer possesses four chains.

This terminology is consistent with the previously encountered uses of the prefixes oligo-, di-, and tetra-. See dipeptide, tetrapeptide, and oligopeptide, for example (Table 3.1).

4.22 Four. Each subunit contains at least one domain (independently folded region). Single-domain subunits fold in such a manner that the removal of any peptide fragment has an impact on the secondary or tertiary structure of the polypeptide that remains.

4.23 Denaturation exposes nonpolar (hydrophobic, water hating) side chains that are normally buried in the core of a native protein. In an aqueous environment, the general rule for protein folding is "nonpolar in: polar out." The general rule for solubility is "like dissolves like"; polar dissolves polar, nonpolar dissolves nonpolar, but polar and nonpolar do not mix. Once the nonpolar core of a globular protein is exposed to an aqueous environment, the protein tends to precipitate as a consequence of the association of nonpolar side chains on neighboring molecules.

4.24 **a.** Changes in pH primarily alter salt bridges (by adding or removing charges from participating side chains) and H bonds (by adding or removing participating hydrogens).

b. High ethanol concentrations primarily disrupt hydrophobic bonds (by mixing with the hydrophobic core of a protein) and H bonds (by H bonding with compatible protein functional groups and, in the process, weakening the H bonds that already exist to these functional groups). Ethanol is a good general solvent; it is able to mix, to a significant extent, with both polar and nonpolar substances.

c. High temperatures tend to disrupt all classes of bonds except for the covalent disulfide bonds. At high enough temperatures, even covalent bonds can be broken; proteins decompose before they melt.

4.25 To say that something is planar means that it will fit in or on a flat surface or plane; a planar object is a flat object. Techni-

4.20 *(Answer)*

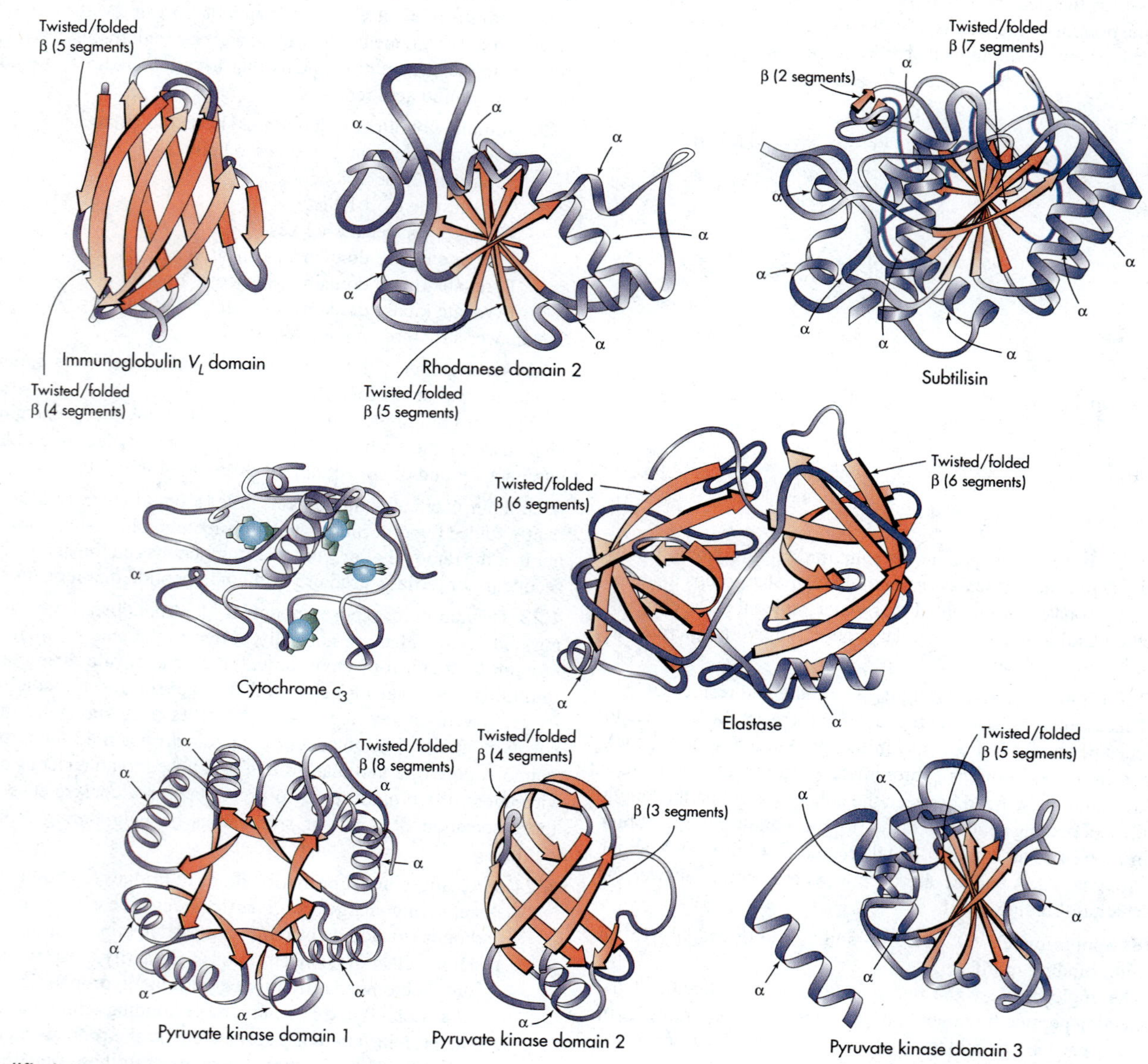

Modified version of Figure 4.13.

cally, the heme group as a whole is not planar, because part of the side chains attached to the porphyrin ring are not planar. For some purposes, however, the relatively slight deviation from a strict planar geometry is considered insignificant; the heme group as a whole is often described as planar.

4.26 The answers lie in Figures 4.18 and 4.19. There are seven bends or loops in myoglobin (not counting the folding of the short N-terminal and C-terminal fragments that lie outside of α-helical regions).

There is not a single amino acid residue between either the B and C helices or the D and E helices. In both of these instances, the bend is between the last residue in one helix and the first residue in the second (as one moves along the polypeptide from N-terminal to C-terminal).

There are eight residues in the loop between the C and D helices and an equal number between the E and F helices.

Proline plays a role in the creation of four of the bends or loops: the BC bend (Pro is second residue in C helix), the EF loop (Pro is third residue in F helix), the FG loop (Pro is first residue in G helix) and the GH loop (Pro is a loop residue). Since a Pro residue has no hydrogen on its α-nitrogen (required for H bonding along the helix) and there is restricted rotation about each of the bonds within its ring system, Pro is incompatible with a stable α-helix, except when one of the first three residues from the N-terminus. With restricted rotation about its α-carbon, a Pro residue does not have the rotational flexibility of other residues; the possible values for ϕ and ψ are limited (Section 3.5).

4.27 No. Since myoglobin contains no cysteine residues, it lacks the sulfhydryl groups (-SH) required to form a disulfide bond. Disulfides are produced by oxidatively coupling two thiols.

4.28

$$K_{eq} = \frac{[MbO_2]}{[Mb][O_2]}$$

When myoglobin's O_2 affinity increases, the K_{eq} for O_2 binding also increases. At any given concentration of O_2, a larger fraction of the myoglobin will be bound to O_2 at equilibrium.

If more O_2 is added to the reaction mixture, the equilibrium will be temporarily destroyed, but the equilibrium constant itself will be unchanged. The equilibrium constant is determined by the difference in free energy between reactants and products and by temperature ($\Delta G° = -RT \ln K_{eq}$), not by reactant or product concentration (Section 8.2).

4.29 B, E, F, and G form the walls. A and H provide a floor.

The heme pocket is created and maintained by a very specific folding (tertiary structure). The precise fit of the heme group into this pocket can only be fully appreciated by working with computer graphics or space-filling models.

4.30 The noncovalent binding of a molecule to a protein can create stearic or electronic strains that lead to changes in conformation. Such binding can also disrupt or modify some of the noncovalent interactions involved in maintaining secondary, tertiary, or quaternary structure. The binding of O_2 to a hemoglobin subunit, for example, modifies the precise positioning of the Fe^{2+} within the protoporphyrin ring of the heme group and alters the interactions between the heme group and neighboring histidine side chains. The "ripple effect" that results leads to a change in subunit conformation which is propagated to neighboring subunits through alterations in quaternary structure interactions.

The strong covalent bonds (peptide bonds) that maintain primary structure are not cleaved or rearranged by the noncovalent binding of a ligand.

4.31 896 million. If fully oxygenated, 280 million hemoglobin molecules will carry 1120 million O_2 (280 million × 4), since each hemoglobin can bind 4 O_2. When 80% oxygenated, the hemoglobin will carry 80% of this number (1120 million × 0.8) which equals 896 million.

4.32 Cooperative binding, by definition, refers to the effect that the binding of one ligand has on the binding of additional ligands. A compound must possess multiple binding sites to exhibit cooperative binding. Myoglobin has only a single O_2 binding site.

4.33 Removal of Hb will trigger a net reverse reaction that will lead to an increase in O_2 concentration and a drop in $Hb(O_2)_4$ concentration. LeChatelier's principle states that when a reaction at equilibrium is disturbed, that reaction will move (through either a net forward reaction or a net reverse reaction) to counter the disturbance and restore equilibrium.

$$Hb + H^+ \rightleftharpoons Hb{-}H^+$$

$$Hb + 2,3\text{-BPG} \rightleftharpoons Hb{-}2,3\text{-BPG}$$

4.34 a, c, and d.

Any condition that leads to the prolonged reduction of O_2 concentration within the lungs leads to an increase in red blood cell 2,3-BPG levels. The delivery of O_2 to tissues will be improved in the process. Lung cancer, pneumonia, and a sinus infection all have an adverse impact on the functioning of the lungs and the acquisition of O_2 by the lungs. The O_2 pressure within the lungs is reduced as a consequence.

4.35 To the left. With its binding curve lying to the left, myoglobin will be more fully oxygenated than fetal hemoglobin at any given O_2 concentration. In the fetus, as well as in the mother, O_2 must flow from hemoglobin in the blood to myoglobin within muscle cells. When the myoglobin O_2 binding curve lies to the left of that for fetal hemoglobin, this desired direction of O_2 movement is facilitated.

4.36 Collagen is rich in Gly, Pro, and Hyp, and it contains many copies of the sequence Gly–Pro–Hyp. Pro and Hyp are both incompatible with an α-helix (except when one of the first three residues from the N-terminus). Glycine also tends to destabilize an α-helix (Table 4.4).

4.37 Cross-linked collagen contributes to: changes in skin texture, elasticity, and appearance; more rigid and less efficient arterial walls (impacts the entire circulatory system); less flexible lungs, with a lower vital capacity; more brittle bones; and stiff joints and tendons.

In general, the structure and function of any body component that contains collagen can be adversely affected by chemical modifications in this protein. Since many tissues and organs contain some collagen, the cross-linking of collagen has numerous consequences.

4.38 Since the protein molecules in hair exist in an α-helical arrangement, they can be stretched without breaking the relatively strong, covalent peptide bonds that maintain the polymer backbone. Only the H bonds maintaining the helix are disrupted during the stretching process. A good analogy is the stretching of a slinky.

Since the polypeptide chains in a collagen molecule exist in an almost fully extended arrangement initially, there is very little room for stretch. Although a collagen molecule tends to break rather than stretch, considerably more force is required to break a collagen molecule than is required to stretch a hair fiber. To break a polypeptide strand, one must rupture covalent bonds.

4.39

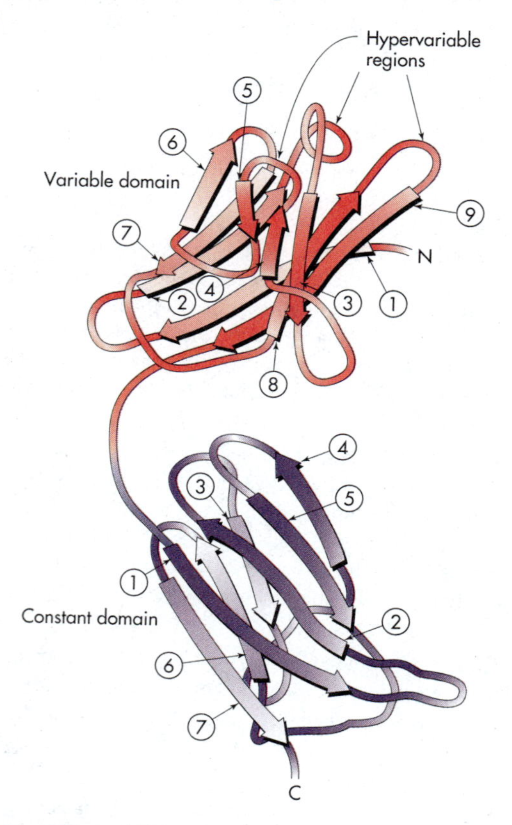

Modified version of Figure 4.34.

The number 3 β-strand in the constant domain is in the lower sheet (contains three strands), whereas the number 4 β-strand in the same domain is a component of the upper four-stranded sheet. β-Strands 2 and 3 in the constant domain are also in separate sheets, as are 5 and 6. Other examples can be found within the variable domain.

4.40 The immunoglobulin fold would probably be more resistant to denaturation because it is partially maintained through a covalent disulfide bond. The globin fold is totally maintained through noncovalent interactions. This prediction cannot be made with full confidence if we consider that denaturation is not an all or nothing process. Part of the immunoglobulin fold could be unfolded more readily than the globin fold. Complete unfolding, however, would tend to occur more readily in the case of the globin fold.

4.41

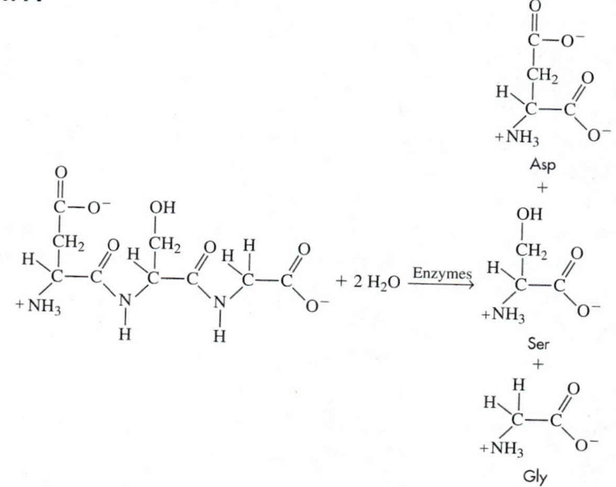

4.42 Yes. Since muscle is mainly protein (Table 4.1), the human body needs amino acids to build muscle. The typical diet in the United States, however, contains more than enough protein to meet the amino acid requirements for even a serious weight lifter. Excess dietary protein does not help an athlete build muscle mass, and it does not improve strength, power, or endurance. For most people, dietary amino acid or protein supplements are a waste of money and can lead to diarrhea and other health problems.

4.43 Quantum theory proposes that very small particles (electrons, protons, atoms, molecules, etc.) can exist in a limited number of discrete energy levels. To change from one "allowed" level to a second, a very small particle must do so in an instantaneous, stepwise fashion and must absorb or emit a packet of energy (quantum) exactly equal to the energy difference between the two levels.

4.44 A low-energy nucleus is "flipped" to a high-energy spin state when it absorbs a radiowave with an energy content exactly equal to the energy difference between the low and high energy spin states. A nucleus will absorb only those radiowaves that bring about an "allowed" energy level transition.

4.45 A separate peak will be observed for each chemically unique set of ^{1}H atoms.

- **a.** One peak. All of the hydrogens in acetone are chemically equivalent.
- **b.** Three peaks. Each of the three carbons is in a unique environment. Consequently, all of the hydrogens attached to the same carbon are in a common unique environment.

This leads to three hydrogens in each of two classes and two hydrogens in a third class.

c. Four peaks. Each of the four carbons is in a unique environment. Once again, this leads to four different classes of hydrogens. Two of the classes have three hydrogens each, whereas the other two classes have two hydrogens each.

CHAPTER 5

5.1 LDH is an oxidoreductase since it catalyzes an oxidation–reduction reaction.

Enzymes that catalyze cis–trans interconversions are isomerases. Geometric isomers come in pairs with one member of the pair having a cis configuration and the other member a trans configuration.

5.2 **a.** Rate = 10 mmol D/min = 1.67×10^2 μmol D/s:

Since 5 mmol represents a relatively small amount of product, the calculated velocity is approximately equal to the initial velocity (the velocity before a significant amount of product has accumulated).

$$5 \text{ mmol}/0.5 \text{ min} = 10 \text{ mmol/min}$$

$$\frac{10 \text{ mmol}}{\text{min}} \times \frac{1 \ \mu\text{mol}}{10^{-3} \text{ mmol}} \times \frac{1 \text{ min}}{60 \text{ s}} = \frac{1.67 \times 10^2 \ \mu\text{mol}}{\text{s}}$$

b. Since 2 mol of E are produced for each mol of D generated (see balanced equation), 10 mmol of E will be produced in 0.5 min:

$$5 \text{ mmol D} \times \frac{2 \text{ mmol E}}{1 \text{ mmol D}} = 10 \text{ mmol E}$$

c. Rate = 3.33×10^2 μmol E/s:

$$\frac{10 \text{ mmol E}}{0.5 \text{ min}} \times \frac{1 \ \mu\text{mol}}{10^{-3} \text{ mmol}} \times \frac{1 \text{ min}}{60 \text{ s}} = 3.33 \times 10^2 \ \mu\text{mol E/s}$$

d. Since 1 mol of B is consumed for each mol of D formed (see balanced equation), B disappears at a rate of 5 mmol/0.5 min = 10 mmol/min.

5.3

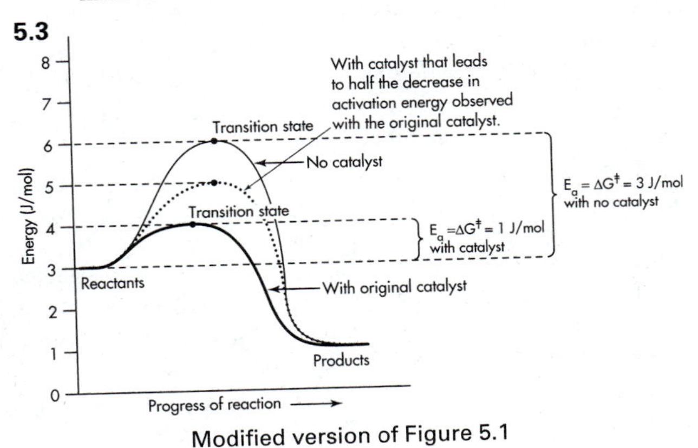

Modified version of Figure 5.1

5.4

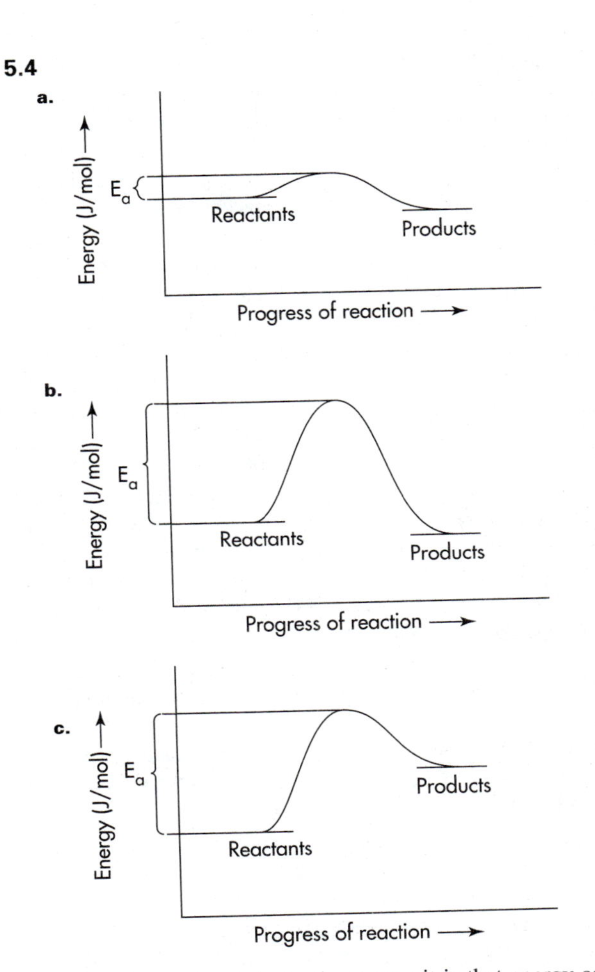

5.5 Heat is one form of kinetic energy; it is that energy associated with atomic and molecular motion. Temperature is a measure of the intensity of heat energy. The higher the temperature, the more rapid, on the average, the motion of the atoms and molecules in a sample of matter. At body temperature (37°C), molecules are bouncing around and colliding at tremendous speeds. This is true in solids, liquids and gases. The freedom of movement is greater in a gas than in a liquid and is greater in a liquid than in a solid.

5.6

$$K_{eq} = \frac{[HCO_3^-][H^+]}{[H_2CO_3]}$$

$$K_{eq} = \frac{[CH_3COOCH_3][H_2O]}{[CH_3OH][CH_3COOH]}$$

$$K_{eq} = \frac{[C]^2[D]^2}{[A]^3[B]^2}$$

The equilibrium constant for a reaction is usually abbreviated K_{eq}. It is, by definition, equal to the product of the equilibrium concentrations of reaction products divided by the product of the equilibrium concentrations of reactants, with each concentration raised to a power corresponding to the compound's coefficient in the balanced equation for the reaction (see Section 8.2 for a more in-depth review). Equilibrium constants were first encountered, utilized, and reviewed (in problems) in Section 2.3

5.7 True. At equilibrium, the rate of the forward reaction equals the rate of the reverse reaction. The equilibrium constant is a measure of the extent to which a reaction will occur before the net reaction stops (equilibrium is attained). A large K_{eq} indicates that product concentration is much greater than reactant concentration at equilibrium. The K_{eq} is small when reactant concentrations are considerably greater than product concentrations at equilibrium. To understand why this is the case, study the answer to Problem 5.6. The K_{eq} expressions reveal that product concentrations (at equilibrium) must be large relative to reactant concentrations (at equilibrium) in order for the K_{eq} to be large. The value of a quotient increases as the value of the numerator increases relative to the value of the denominator.

5.8 False. Rates of reactions and equilibrium constants are independent variables. Rates are determined by activation energies, collision frequencies, and collision forces, whereas, at a fixed temperature, K_{eq}'s are determined by the free energy difference between reactants and products. An in-depth review of free energy and equilibrium is provided in Chapter 8.

5.9 The initial concentration of reactants has no impact on K_{eq}, since it has no impact on the relative free energy content of reactants and products. At a constant temperature, it is the difference in free energy content between reactants and products that determines K_{eq} (Section 8.2). The free energy content of a molecule is the same regardless of the number of other molecules present in the same vicinity.

5.10 Enzyme X concentration in the serum sample is 30,000 U/mL. This corresponds to 5.01×10^{-4} kat/mL.

$$\frac{50\ \mu\text{mol}}{\text{s}} \times \frac{60\ \text{s}}{\text{min}} = 3000\ \mu\text{mol/min}$$

Since 1 U = 1 μmol/min, 3000 μmol/min = 3000 U.

Since the 0.1 mL of serum contains 3000 U, 1 mL of serum contains 30,000 U:

$$\frac{3000\ \text{U}}{0.1\ \text{mL}} \times 1\ \text{mL} = 30{,}000\ \text{U}$$

Since 1 U = 1.67×10^{-8} kat, 30,000 U/mL = 5.01×10^{-4} kat/mL:

$$\frac{30{,}000\ \text{U}}{\text{mL}} \times \frac{1.67 \times 10^{-8}\ \text{kat}}{1\ \text{U}} = 5.01 \times 10^{-4}\ \text{kat/mL}$$

5.11 A 1060-fold purification was achieved.

$$\frac{33{,}300\ \text{(specific activity of final preparation)}}{12.5\ \text{(specific activity of crude extract)}} = 1060$$

5.12 2×10^8 U.

$$\frac{10^5\ \text{U}}{\text{mg}} \times \frac{1000\ \text{mg}}{1\ \text{g}} \times 2\ \text{g} = 2 \times 10^8\ \text{U}$$

5.13 1.0 U/mL.

Since 1 U = 1 μmol/min, 0.01 μmol/min = 0.01 U (in 10 μL of enzyme preparation added to reaction mixture)

$$\frac{0.01\ \text{U}}{10\ \mu\text{L}} \times \frac{10^3\ \mu\text{L}}{1\ \text{mL}} = 1.0\ \text{U/mL}$$

5.14 One. A catalyst, by definition, can be used over and over again. Although enzymes are rather fragile catalysts with limited lifespans, a single enzyme molecule can, in theory, convert 10^6 molecules of reactant to product.

5.15 **a.** Peptide (amide) bonds.
b. H bonds.
c. H bonds, disulfide bonds, hydrophobic forces, salt bridges, and van der Waals forces.
d. H bonds, hydrophobic forces, salt bridges, and van der Waals forces.

This question is designed to drive home the fact that protein enzymes have all of the characteristics that were described for proteins in Chapter 4. You may want to review the properties of proteins at this time, and you should maintain a mental picture of proteins when studying enzymes.

5.16 2500. There tends to be a separate enzyme for each reaction that occurs within a living organism.

5.17 Since brain cells and liver cells serve different functions, each cell type must encompass some unique reactions. In general, a distinctive set of reactions requires a novel set of enzymes (see answer to Problem 5.16).

5.18 False. Since cis–trans isomers do not differ in chirality (many are achiral), a chiral binding site is not needed to distinguish these isomers. Chiral binding sites are required to distinguish stereoisomers that differ in chirality; it takes a chiral binding site to recognize a chiral compound.

5.19 Hydrogen bonds, other interactions involving permanent dipoles, salt bridges, hydrophobic forces, and van der Waals forces. In an aqueous environment, one or more of these forces is primarily responsible for holding any two molecules together (Sections 2.1 and 2.2). Enzymes, like other proteins, possess a wide variety of functional groups and are capable of participating in the formation of all classes of noncovalent bonds.

5.20 True. The heme group is held in the highly selective heme pocket of myoglobin by noncovalent interactions, predominantly hydrophobic bonds, salt bridges, and H bonds. Under appropriate conditions, the heme group can be reversibly removed. The heme group must have a binding site on those enzymes responsible for its production and degradation, and these binding sites are most likely similar to myoglobin's heme pocket.

5.21 High temperatures, extremes in pH, high salt concentrations, mechanical agitation, organic solvents, and surface-active agents such as soaps and detergents. Since an enzyme must have a highly specific conformation to be biologically active, any treatment that unfolds a protein will inactivate an enzyme.

5.22 Denatured. Denaturation, by definition, is the unfolding of a protein without rupturing covalent bond (Section 4.6).

5.23 The biological activity of an enzyme is determined by its highly specific folding (its conformation), since it is this folding that brings together those amino acid side chains that constitute the active site. Since the primary structure of a protein determines its secondary, tertiary, and quaternary structures, a change in primary structure can impact both conformation and biological activity. That impact can, in theory, be either positive or negative as far as catalytic activity is concerned. Since evolution tends to lead to the selection of primary structures that confer a high degree of catalytic activity, a random change in primary structure seldom improves upon what has been selected during roughly 4 billion years of evolution.

5.24 A change in pH leads to either the net addition or net removal of H^+ from all weak acid functional groups. Such changes tend to disrupt or lead to the formation of H bonds and salt bridges. Hydrophobic interactions may also be impacted as functional groups are made more or less polar through the transfer of protons (H^+). Thus, pH shifts lead to alterations in the very bonds that determine and maintain the conformations of enzymes and other proteins.

5.25 No. Pepsin is in its native, biologically active conformation at its pH optimum of 1.5. A shift to pH 7.0 would have a drastic impact on its side chain interactions and, consequently, its overall folding and catalytic activity.

5.26 Autoclaves kill bacteria by denaturing (either partially or totally) enzymes and other proteins. The denaturation of an enzyme shuts down a reaction. In some cases, the loss of a single reaction can have lethal consequences. A cell will definitely die whenever very many of its reactions are shut down.

5.27 See answer at top of next column.

5.28 Since conformation is a function of pH (see answer to Problem 5.24), both the catalytic site and binding site on an enzyme tend to change as the pH is altered. Changes in the catalytic site usually lead to alterations in V_{max}, whereas changes in the binding site are reflected by changes in K_m.

5.29 Substrate C binds most tightly since it has the smallest K_m. Since we are assuming that the K_m is equal to the K_d (dissociation

5.27 *(Answer)*

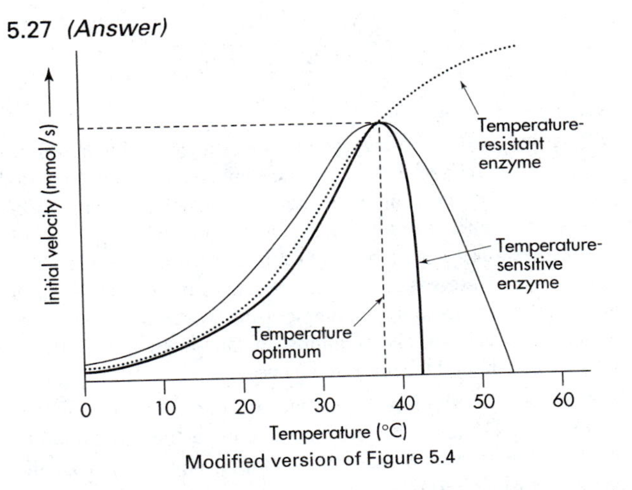

Modified version of Figure 5.4

constant) for the enzyme–substrate complex (a good assumption in many instances), the smaller the K_m, the smaller the K_d and the greater the amount of substrate bound to enzyme when the dissociation of the enzyme–substrate complex is at equilibrium.

K_m is primarily determined by the nature and geometric arrangement of the amino acid side chains at the binding site on the enzyme. Since it is the interactions of substrates with the catalytic site that determine the rate of an enzyme-catalyzed reaction, the K_m is no measure of V_{max} (even when there is substantial overlap between binding and catalytic sites). An enzyme may have a large K_m and a small V_{max} and vice versa.

5.30 $K_m = 1.2 \times 10^{-3}\ M$:

Given $v_o = V_{max}[S]/(K_m + [S])$, multiply both sides of this equation by $K_m + [S]$:

$$v_oK_m + v_o[S] = V_{max}[S]$$

Subtract $v_o[S]$ from both sides of the new equation:

$$v_oK_m = V_{max}[S] - v_o[S]$$

Divide both sides of this equation by v_o:

$$K_m = \frac{V_{max}[S] - v_o[S]}{v_o}$$

$$K_m = \frac{(35 \times 0.0002) - (5 \times 0.0002)}{5}$$

$$K_m = 1.2 \times 10^{-3}\ M$$

5.31 $v_o = 0.4\ \mu$mol/min:

$$v_o = \frac{V_{max}[S]}{K_m + [S]}$$

$$v_o = \frac{2 \times (2 \times 10^{-5})}{(8 \times 10^{-5}) + (2 \times 10^{-5})}$$

$$v_o = 0.4\ \mu\text{mol/min}$$

5.32 At V_{max}, all of the enzyme molecules are bound to substrate and in the process of catalyzing the conversion of reactants to products at every moment; all of the enzyme molecules are working all of the time. It follows that to attain one-half V_{max}, half of the enzymes must be bound to substrate and working. If the number of active workers is cut in half, the rate of production is cut in half. Similarly, when only one-tenth of the workforce is working, one-tenth as much work is accomplished. One-tenth of the enzyme molecules are bound to substrate whenever $v_o = (1/10)\ (V_{max})$.

5.33 Since, by definition, K_m is that substrate concentration that leads to (1/2) (V_{max}), half of the enzyme molecules are bound to substrate whenever substrate concentration equals K_m (see the answer to Problem 5.32).

5.34 Double the substrate concentration and then remeasure initial velocity. If the remeasured velocity is greater than 3 mmol/min then V_{max} must be greater as well. If the remeasured velocity is still 3 mmol/min, the enzyme must have been saturated at the original substrate concentration and must have been operating at V_{max}.

5.35

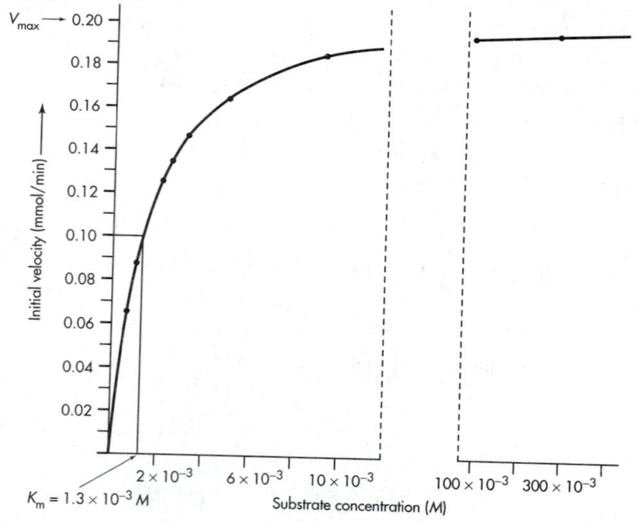

a. V_{max} is approximately 0.2 mmol/min (rate when hyperbola levels off).

b. K_m is approximately 1.3×10^{-3} M ([S] that yields one half V_{max}).

c. Since a substrate concentration of 2×10^{-1} M is well off the graph, this concentration must be saturating. The initial velocity equals V_{max} or approximately 0.2 mmol/min. Once at saturating substrate concentration, a further increase in substrate concentration has no effect on initial velocity. Restated, since at saturating substrate concentrations all of the enzyme molecules are already working all the time, a further increase in substrate concentration cannot enhance the amount of work accomplished.

d. There is a significant degree of self-regulation at substrate concentrations up to about 4×10^{-3} M. At higher concentrations, the curve starts to level off and the degree of self-regulation drops markedly.

5.36 The reaction is second order, since the substrate concentration is raised to the second power. A doubling of substrate concentration leads to a four-fold increase in v_o; 2 squared is 4.

5.37 The K_{eq} for $E + S \rightleftharpoons ES$ is k_1/k_{-1} (the reciprocal of the K_d for the ES complex).

The K_{eq} for $ES \rightleftharpoons E + P$ is k_2/k_{-2}.

When considering the reaction $ES \rightleftharpoons E + P$:

$$\text{Rate of forward reaction} = k_2[\text{ES}]$$

$$\text{Rate of reverse reaction} = k_{-2}[\text{E}][\text{P}]$$

At equilibrium, $k_2[\text{ES}] = k_{-2}[\text{E}][\text{P}]$. Dividing both sides of this equation by [ES] and k_{-2} yields the equilibrium constant expression which is found to be equal to k_2/k_{-2}:

$$\frac{k_2}{k_{-2}} = \frac{[\text{E}][\text{P}]}{[\text{ES}]} = K_{eq}$$

5.38 d. By definition, the turnover number is the amount of work (the number of substrate molecules catalytically converted to product) that can be accomplished by one worker (one enzyme molecule) during a specified time period. The most efficient worker accomplishes the most work during any fixed period of time.

5.39

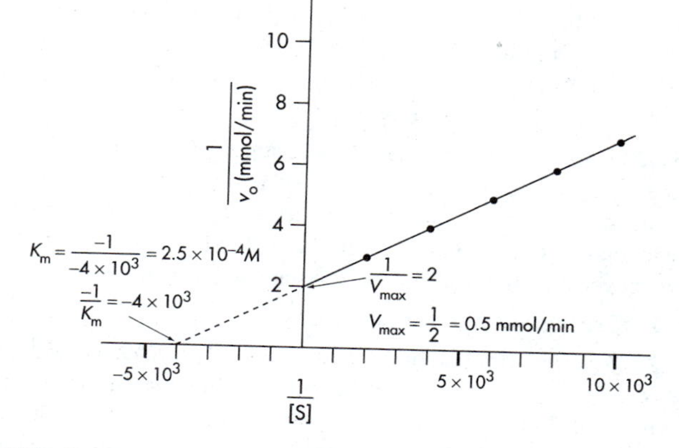

5.40 Greater. Since the substrate must pull (Le Châtelier's principle) the enzyme away from the inhibitor, a higher substrate concentration is required to force all of the enzyme into an enzyme–substrate complex.

5.41 Most competitive inhibitors bind to the same site as a normal substrate. Since such sites typically have enormous discriminatory powers, any compound significantly different from a substrate will tend to be excluded from the binding site or will normally bind with such low affinity that it will be of no significance from the standpoint of competitive inhibition.

Since noncompetitive inhibitors bind to sites that are structurally distinct from substrate binding sites, there are, in theory, no constraints on the nature of these sites or the type of molecules that they can accommodate. Any given site, however, will normally bind a single compound or a limited number of closely related compounds.

5.42 **a.**

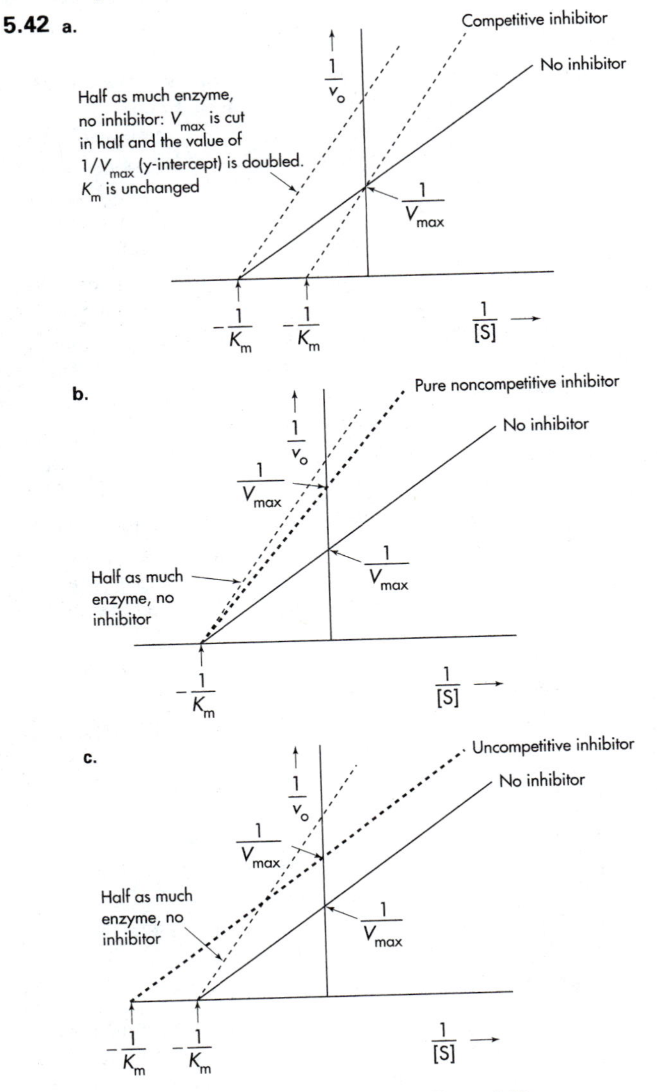

Modified version of Figure 5.10

Pure noncompetitive inhibitors are mimicked by a reduction in enzyme concentration because they continuously inactivate a specific fraction of the total enzyme. Enzyme not bound to inhibitor is normal. V_{max} decreases ($1/V_{max}$ increases) while K_m remains the same.

5.43 **a.**

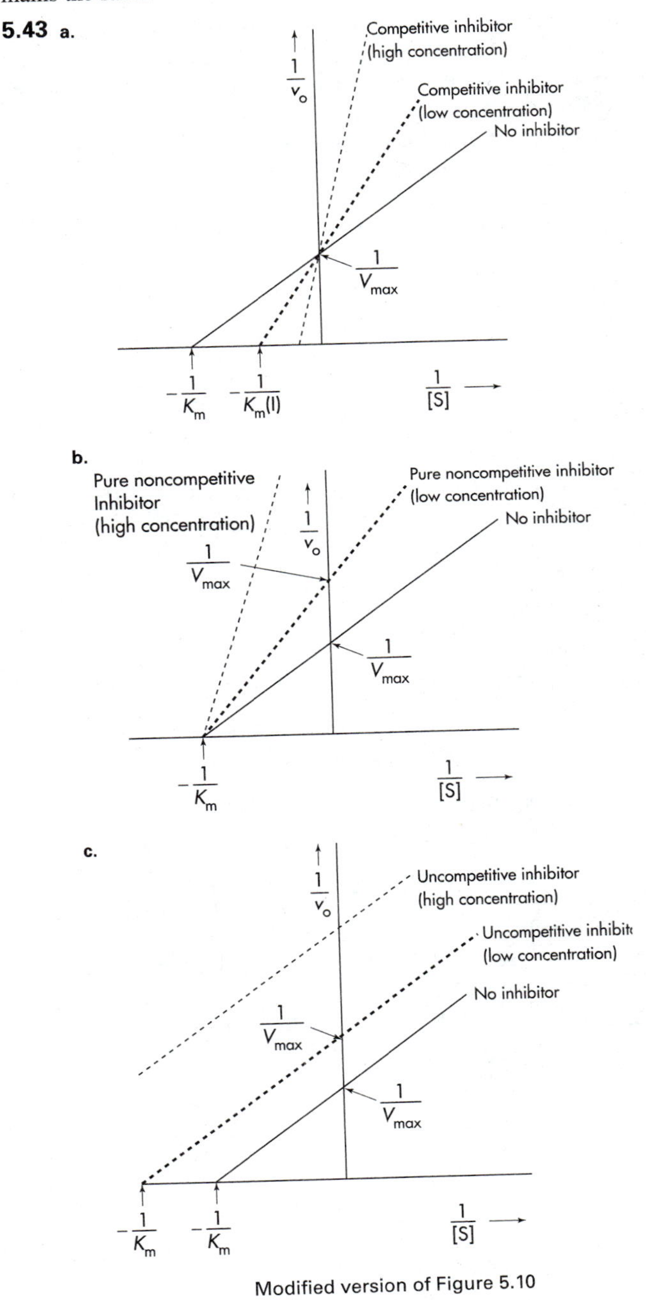

Modified version of Figure 5.10

5.44 With three subunits per isozyme, there are four possible isozymes: AAA, AAB, ABB, and BBB.

With five subunits per isozyme, there are six possible isozymes: AAAAA, AAAAB, AAABB, AABBB, ABBBB, and BBBBB.

5.45 Yes, it is possible. The positive effectors cancel the action of the negative effectors. The effector binding sites are still important because they allow the enzyme to respond almost instantly to changes in the concentration of any of its effectors.

5.46 Hemoglobin is not an enzyme because it is not a catalyst; it is a transport agent that does not increase the rate of any reaction. The four O_2 binding sites on each hemoglobin molecule are both structurally and functionally analogous to substrate binding sites on enzymes. The active sites of enzymes are unique because they contain a catalytic site in addition to a binding site.

5.47

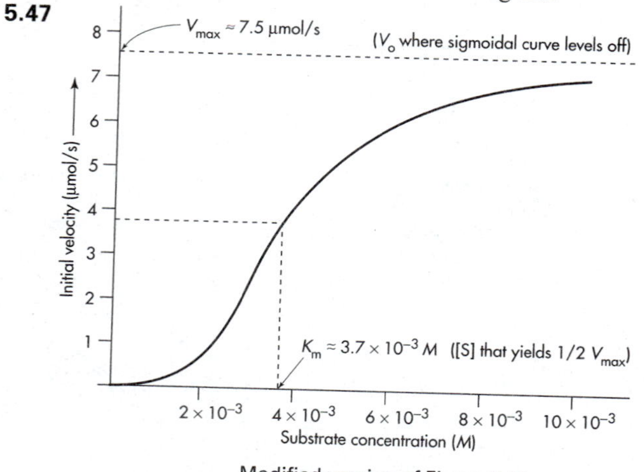

Modified version of Figure 5.12

V_{max} is approximately 8.0 µmol/s.

K_m is close to 3.3×10^{-3} *M*.

The theoretical interpretations of K_m and V_{max} differ for allosteric enzymes and classical Michaelis–Menten enzymes, because the allosteric enzymes fail to adhere to several of the assumptions involved in the derivation of the Michaelis–Menten equation.

5.48

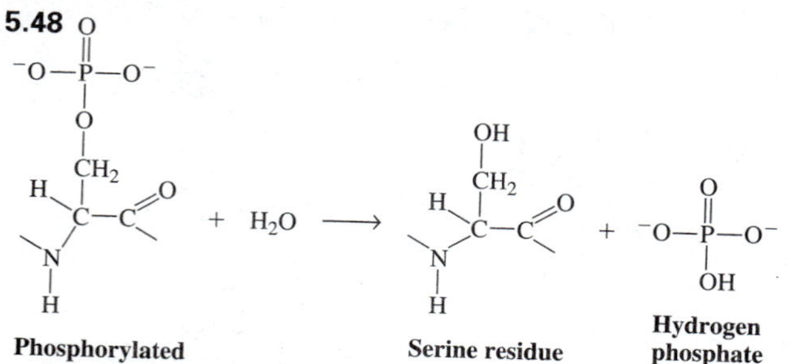

5.49 The phosphorylation or dephosphorylation of a single serine residue can:

a. Directly alter a binding site, a catalytic site or both. A phosphorylated serine side chain interacts differently with a substrate molecule than does a free serine side chain.

b. Disrupt or modify some of the interactions that maintain a specific conformation and, in the process, indirectly alter an active site. A specific conformation is required to create both the catalytic site and the binding site.

c. Block or expose an active site.

5.50 Trypsin is most specific from the standpoint of which amino acid residues are recognized by the enzyme. It preferentially catalyzes the hydrolysis of peptide bonds at the C-terminal side of lysine and arginine residues.

Carboxypeptidase is most specific from the standpoint of which specific peptide bond is cleaved. It will only catalyze the hydrolysis of that peptide bond closest to the C-terminal end.

Pepsin, trypsin, and chymotrypsin are endopeptidases (they are capable of catalyzing the hydrolysis of internal peptide bonds), whereas carboxypeptidase is an exopeptidase (it can only catalyze the hydrolysis of a C-terminal peptide bond).

5.51 Proteinases are susceptible to digestion (enzyme-catalyzed hydrolysis) by neighboring proteinases; proteinases can serve as substrates for proteinases. In addition, proteinases move through the digestive system (and out of the body) along with certain other materials.

5.52 The abnormal amino acid sequence in trypsinogen is associated with a peptide fragment that is "cut out" of this zymogen during its conversion to trypsin.

5.53 False. A multifunctional enzyme can lead to metabolic channeling, because its separate domains can fold together to trap intermediate reactants and products. A soluble multienzyme system is unable to participate in metabolic channeling, because its intermediate reactants and products equilibrate with metabolite pools in the surrounding medium as they diffuse between separate enzymes in the system.

5.54 Ligand binding, including the binding of allosteric effectors.

Covalent modification, including phosphorylation and peptide bond cleavage.

Isozymes with distinctive K_m's and V_{max}'s.

Compartmentation with distinct enzymes and unique environments (including pH) within separate compartments.

Construction of enzymes that operate in self-regulating regions of velocity versus substrate concentration plots.

5.55 See answer at top of page A-29.

5.56 Any amino acid with a nonpolar side chain can be used to help line a hydrophobic pocket. The amino acids with nonpolar side chains are Ala, Ile, Leu, Met, Phe, Pro, and Val (Table 3.6). Borderline (polar/nonpolar) amino acids are Cys, Gly, and Trp.

5.55 *(Answer)*

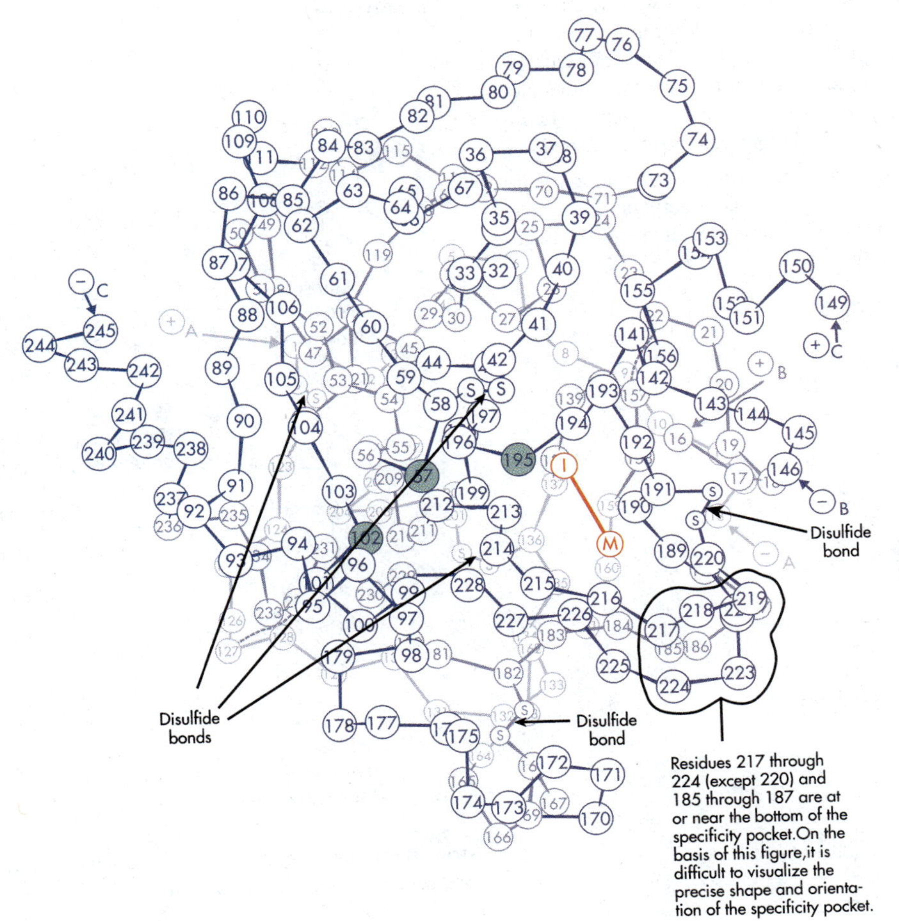

5.57 Since opposite charges attract, we would expect the Asp-containing binding pocket to preferentially accommodate amino acids with positively charged side chains at physiological pH's. Thus, the peptide bonds to the C-terminal side of lysine and arginine residues would be preferentially hydrolyzed. Trypsin, which has an aspartate at the bottom of its specificity pocket, exhibits a strong preference for lysine and arginine (Table 5.7).

5.58 The transition state can be stabilized by positive or partially positive groups that neutralize the negative charge on the oxygen. Isolated charges are inherently unstable. Experimentally, the positive charge on His 57 and H bonding to the negative oxygen by particular enzyme functional groups appear to account for the stabilization of the transition state.

5.59 There are 7 bonds between the enzyme and its substrates in the enzyme·tyrosine·ATP complex. There are 14 H bonds between the enzyme and the transition state for the first step of the catalyzed reaction. The increase in H bonding is hypothesized to stabilize the transition state and, in the process, lower the activation energy for the first step of the reaction. One of the 14 H bonds in the transition state is atypical, it involves a hydrogen attached to sulfur in a cysteine side chain. It is questionable whether or not this interaction should be classified as an H bond.

5.60 The covalent bond between tyrosine and AMP is a mixed anhydride bond. A carboxylic acid residue and phosphate ester residue are joined in the anhydride. Anhydrides are normally high-energy compounds (Section 8.5) and highly reactive. The cleavage of an anhydride bond commonly provides a thermodynamic driving force for coupled reactions (Section 8.6). The cleavage of the anhydride bond in tyrosyl–AMP drives the second step of the reaction catalyzed by tyrosyl–tRNA synthetase.

5.61 During an EIA, an antigen becomes attached to an antibody–enzyme complex. Since the enzyme acts catalytically during the conversion of reactant to colored product, an enzyme molecule in a single antigen–antibody–enzyme complex can be used time and time again to generate product. A single molecule of a catalyst can, in theory, be used to produce an infinite number of molecules of the same product. In practice, catalyst molecules tend to be inactivated or to run out of reactant before this is accomplished.

5.62 The stereospecificity must arise from the immunoglobulin since it is this protein, not the enzyme, that directly binds the antigen. The chirality of the antigen binding sites on an antibody accounts for its stereospecificity.

5.63

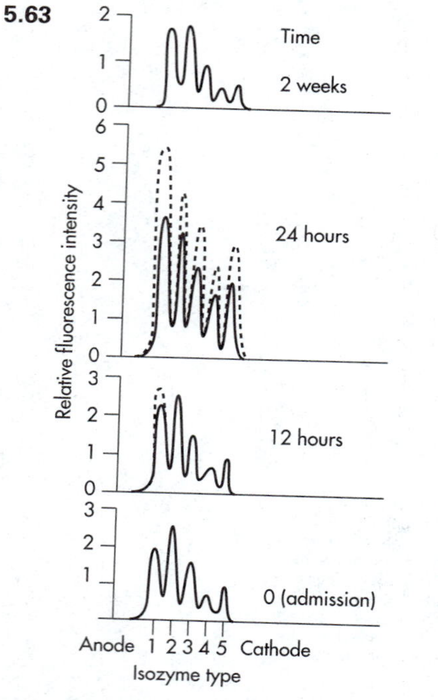

The dashed lines (---) depict the results expected if the heart attack and associated tissue damage had been twice as severe. The *increase* in isozyme levels would be roughly twice as great.

5.64 Orally administered asparaginase would not be as effective since most of it would be digested in the stomach and intestine. Little, if any, active enzyme would enter the blood. The ingested asparaginase could not be distinguished from other dietary protein by the digestive enzymes in the body.

CHAPTER 6

6.1 See answer at top of next column.

6.2 Configuration: the unique spatial arrangement of atoms that identifies a specific stereoisomer. Stereoisomers are isomers whose atoms exhibit the same connectedness but are locked into different spatial orientations. Stereoisomerism can result from restricted rotation about one or more bonds or from chirality. See rest of answer at top of page A-31.

6.1 *(Answer)*

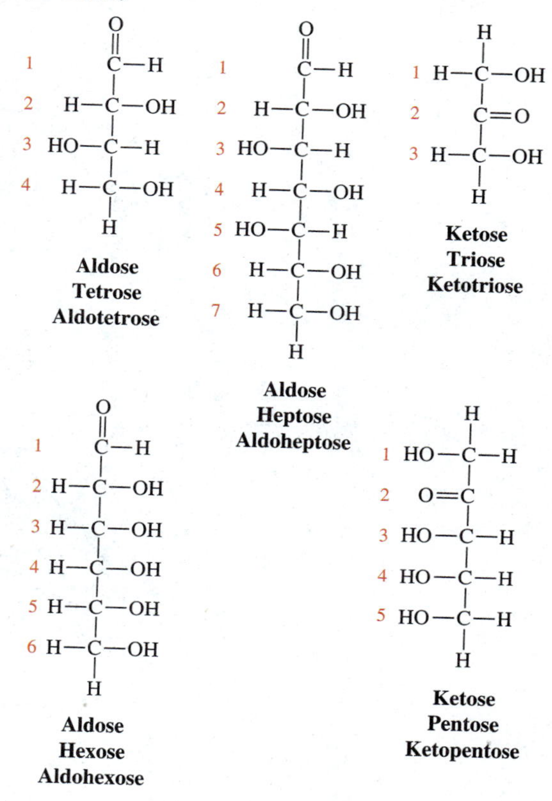

6.3

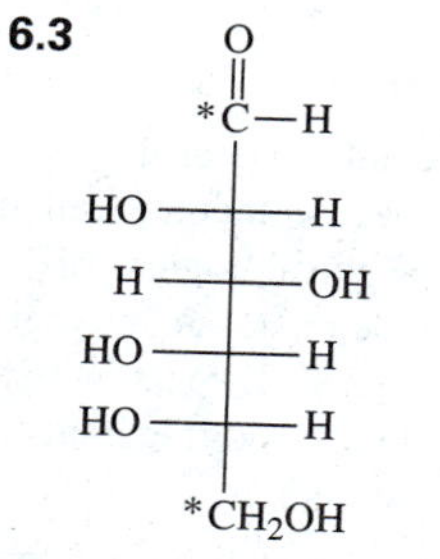

An * indicates an achiral carbon; attached atoms can be written in any desired orientation. L-Glucose is the complete mirror image of D-glucose.

6.4 Any chiral compound, including D-ribose, has only one enantiomer, its mirror image. Since D-ribose contains 3 chiral centers, it is one of a group of 8 (2^3) stereoisomers. Thus, there are 7 stereoisomers of D-ribose ($2^3 - 1$). One is its enantiomer and the other 6 are diastereomers.

6.2 *(Answer)*

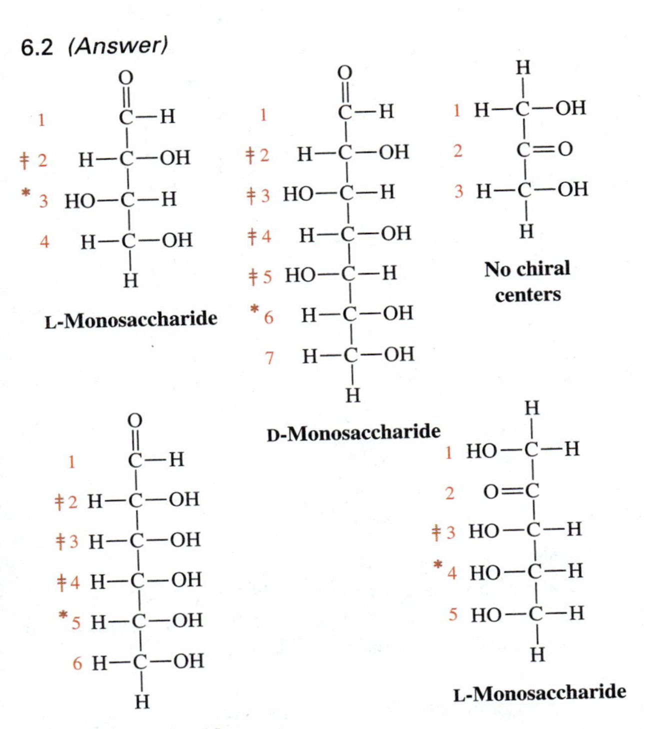

An * identifies the highest numbered chiral centers. An —OH to the left of a highest numbered chiral center depicts an L configuration and an L-monosaccharide. An —OH to the right denotes a D configuration and a D-monosaccharide.

A ‡ identifies chiral centers in addition to the highest numbered chiral centers.

6.5 Dihydroxyacetone has no chiral centers and no bonds with restricted rotation. Consequently, it is not a stereoisomer. Because the attachment of a phosphate to either hydroxyl fails to generate a chiral carbon, dihydroxyacetone phosphate is not a stereoisomer.

6.6 Ribose is chemically more reactive than deoxyribose. In organic compounds, the functional groups are the reactive centers. Ribose contains one more functional group than deoxyribose. This extra hydroxyl group, being adjacent to other functional groups, tends to be more reactive than it would be otherwise. It may also make its neighboring functional groups more reactive. The greater stability of deoxyribose may account for its use (rather than the use of ribose) as a building block for DNA; it is essential that the genetic material in an organism be relatively stable.

6.7 See answer at top of next column.

6.8 Three- and four-membered rings tend to be unstable as a consequence of ring (angle) strain. The atoms in the ring, being sp^3 hybridized and having a tetrahedral orbital geometry, are

6.7 *(Answer)*

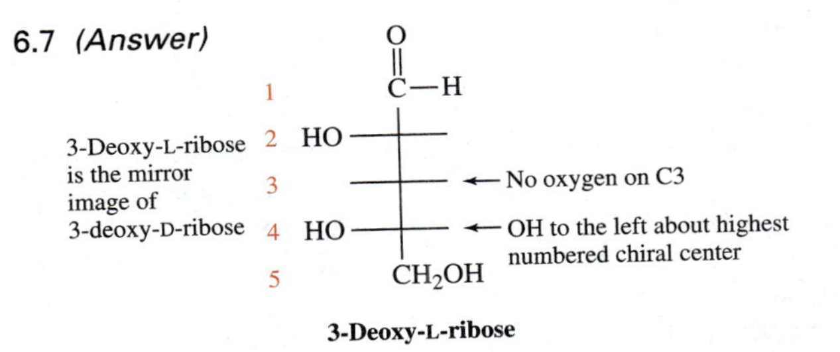

most stable with bond angles near 109°. The 60° bond angles in three-membered rings and the 90° bond angles in four-membered rings create considerable strain (instability).

6.9

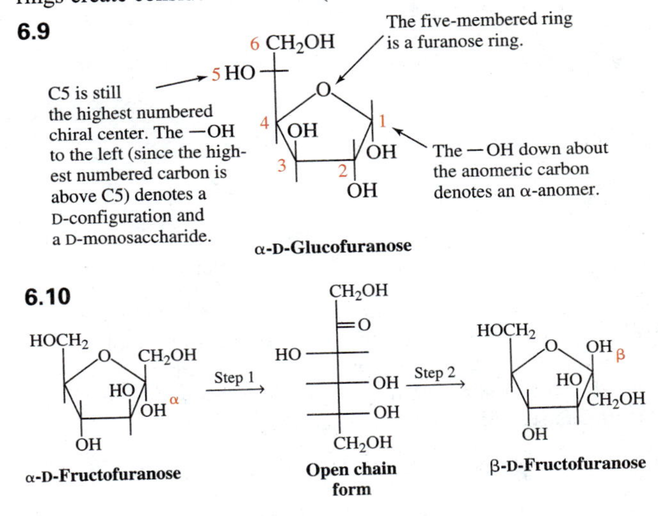

Anomers must pass through the open chain form to be interconverted.

6.11

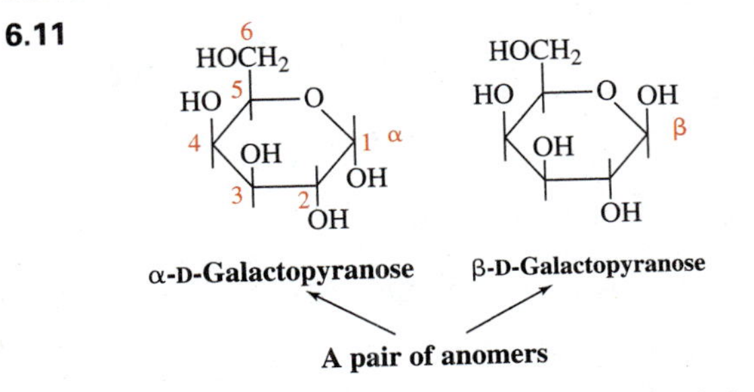

Note: The configuration about the C4 is all that distinguishes glucose and galactose.

6.12 Oxidation refers to a loss of electrons. In biochemistry, oxidation is often associated with the addition of oxygen or the removal of hydrogen.

Reduction entails a gain of electrons. In biochemistry, reduction is frequently linked to the removal of oxygen or the addition of hydrogen. Both oxidation and reduction are reviewed in detail in Section 8.7.

Dehydration, in general, refers to the removal of water (H_2O). When related to chemical reactions, two hydrogens and one oxygen are removed from a compound to produce water.

Alkylation is the addition of an alkyl (alkane) group to a compound. Methyl (CH_3—), ethyl (CH_3CH_2—), and propyl ($CH_3CH_2CH_2$—) groups are examples of alkyl groups. In organic chemistry, an R is commonly used to represent a generic alkyl group. This text uses the same convention.

6.13 Primary (1°) alcohols can be oxidized to aldehydes, which can be further oxidized to yield carboxylic acids:

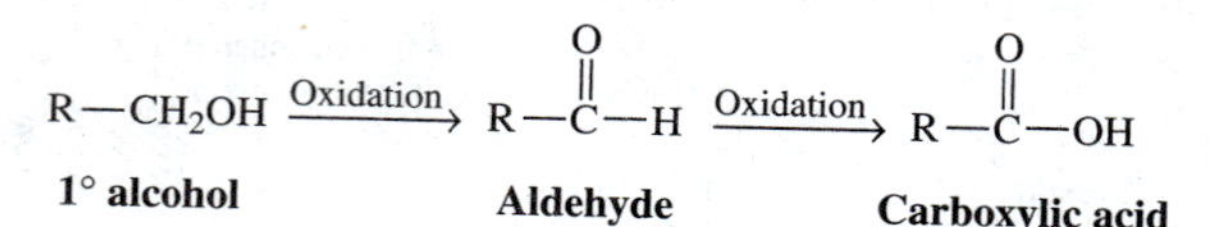

Secondary (2°) alcohols are oxidized to ketones:

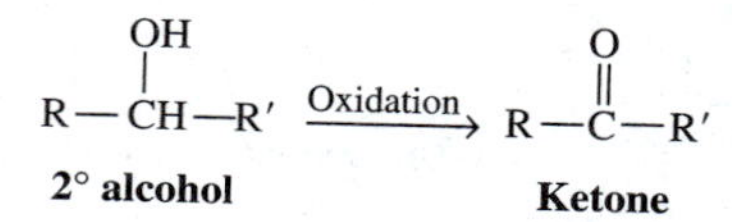

Tertiary alcohols are not oxidized under the relatively mild conditions used to oxidize primary and secondary alcohols. Under more severe conditions, all classes of alcohols can be oxidized to a variety of products. Any alcohol, for example, can be burned (an oxidative process) to generate CO_2 and H_2O:

$$\text{Alcohol} + O_2 \rightleftharpoons CO_2 + H_2O + \text{energy}$$

Primary, secondary, and tertiary alcohols can be dehydrated to yield alkenes:

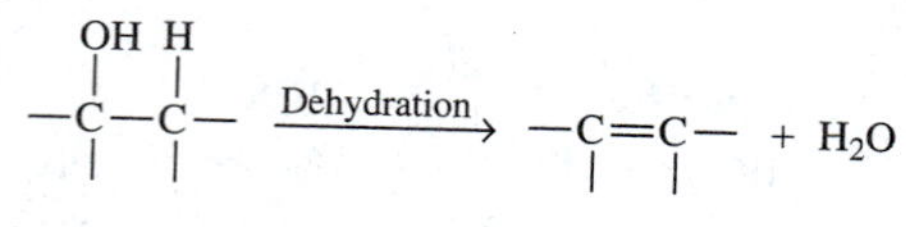

The alkylation of an alcohol leads to the production of an ether:

$$R{-}OH \xrightarrow{\text{Alkylation}} R{-}O{-}R'$$

Any alcohol **Ether**

6.14 Aldehydes are readily oxidized to carboxylic acids and can be reduced to primary alcohols:

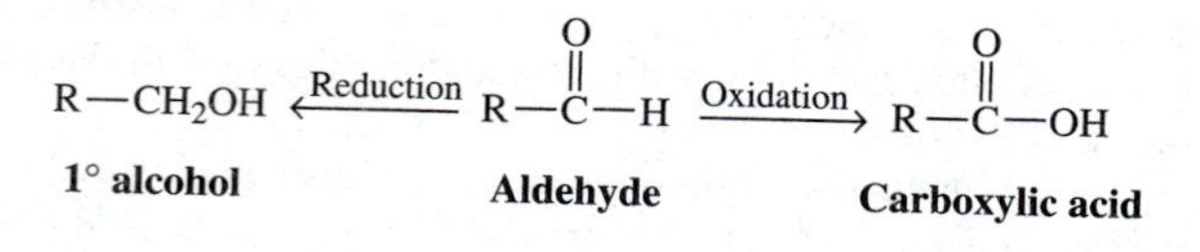

6.15

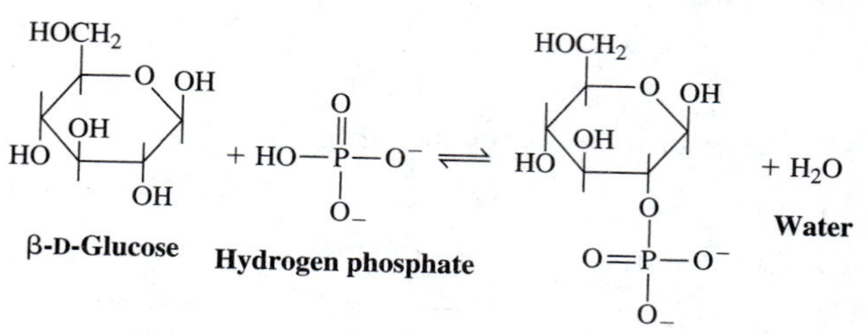

6.16 Three.

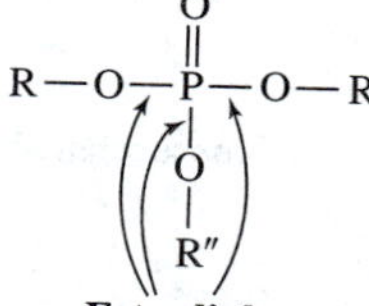

6.17 None.

All of the carbohydrates encountered through Section 6.3 can exist in an open-chain form that possesses a carbonyl group and is readily oxidizable. A compound that is readily oxidized is a good reducing agent.

6.18 Glycosylation modifies free amino groups in hemoglobin, amino groups that would otherwise be protonated and positively charged at physiological pH's. Thus, the net charge on glycosylated hemoglobin is more negative than the net charge on normal hemoglobin at appropriate pH values. These charge differences cause free hemoglobin and glycosylated hemoglobin to have unique electrophoretic mobilities (see electrophoresis, Section 3.10). Electrophoretic mobility is determined mainly by charge to size ratio.

6.19 Virtually all proteins are susceptible to nonenzymatic reaction with blood glucose and other monosaccharides, because most proteins contain one or more free and accessible amino groups. Some free amino groups are buried within the tertiary structure of proteins and resistant to chemical modification under nondenaturing conditions.

6.20 See answer at top of page A-33.

6.21 Five different disaccharides can be produced.
See rest of answer on page A-33.

6.22 One glucose and one galactose.

Lactose is a disaccharide containing one residue of each of these two monosaccharides.

6.23 True.

When two monosaccharides are joined through their anomeric carbons, neither residue is able to exist in a readily oxidizable open-chain form.

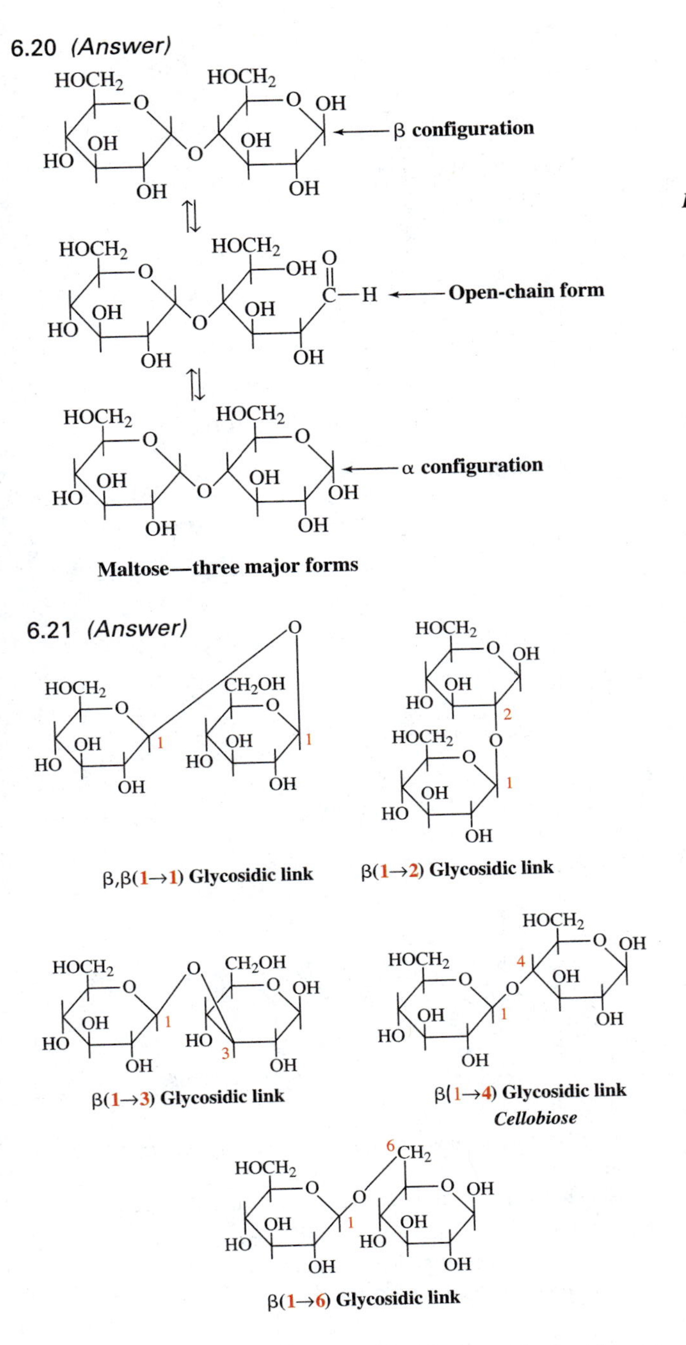

6.24 Acylation of an amino group yields an amide. Amides can be hydrolyzed.

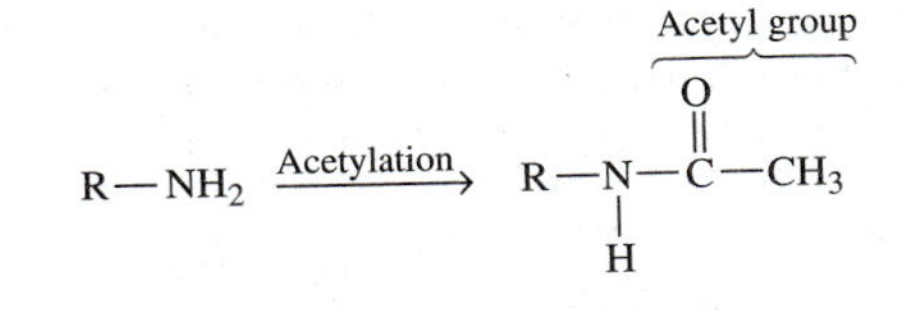

Hydrolysis of an amide:

Amide bond cleaved

R—C(=O)—N(R')—H

Oxygen attaches here

Hydrogen attaches here

H—O—H

Bond cleaved

Amide

$$\rightleftharpoons R{-}C({=}O){-}OH + H_2N{-}R'$$

Carboxylic acid **1° Amine**

Note: The arrows from the water molecule do not indicate a flow of electrons or a reaction mechanism. They depict where the components of water attach during the course of the reaction. The carboxylic acid and amine generated will react together to form a salt:

$$R{-}C({=}O){-}OH + H_2N{-}R' \longrightarrow R{-}C({=}O){-}O^- + H_3\overset{+}{N}{-}R'$$

Acetylation refers to the addition of an acetyl group (CH_3CO—) to a compound. An acetyl group is a specific example of an acyl group (RCO—).

6.25

HOCH2 HOCH2 HOCH2 HOCH2 HOCH2

·· O 4 OH 1 O 4 OH 1 O 4 OH 1 O 4 OH 1 O 4 OH 1 O ··

OH OH OH OH OH

HOCH2 HOCH2

·· O OH 1 O OH 1 O — α(1→6) Glycosidic link

OH OH

HOCH2 HOCH2 CH2 HOCH2 HOCH2

·· O OH 1 O OH 1 O OH 1 O OH 1 O OH 1 O ··

OH OH OH OH OH

In each glucose residue, C1 is the anomeric carbon. Each residue contains one anomeric carbon.

6.26 (1 ⟶ 6), (1 ⟶ 2), and (1 ⟶ 3) branches are theoretically possible, because each glucose residue along a α(1 ⟶ 4)-linked chain contains a free hydroxyl group at positions 2, 3, and 6. The configuration about the anomeric carbons within the branch can be either α or β, so β branches as well as α branches are possible.

6.27

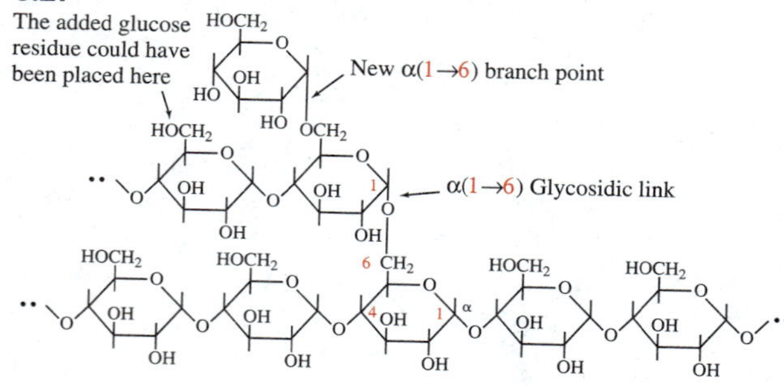

6.28 Chitin is converted to chitosan by hydrolyzing its amide bonds and, in the process, generating free amino groups. Whereas chitin is uncharged, chitosan is a polycation at physiological pH. Each amino group ($-NH_2$) exists predominantly in its protonated form ($-\overset{+}{N}H_3$) at pH's near 7.0

6.29

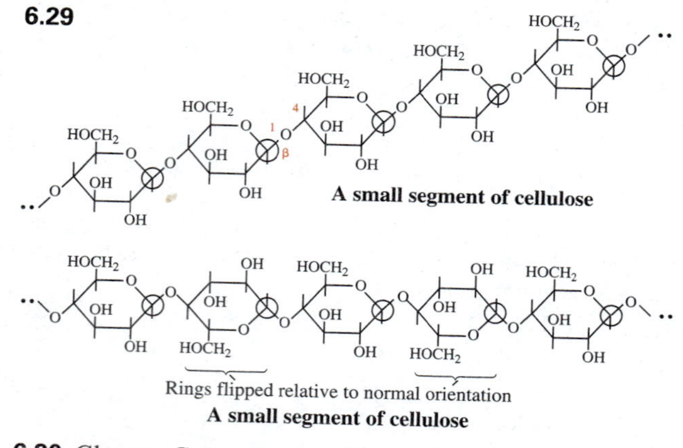

6.30 Glucose. Cotton is primarily constructed from cellulose, a polymer of glucose. Since the only hydrolyzable bonds in cellulose are the glycosidic bonds that hold neighboring monomers together, the hydrolysis of cellulose yields glucose. Concentrated solutions of strong acids "burn" holes in cotton by catalyzing the hydrolysis of the glycosidic bonds.

6.31

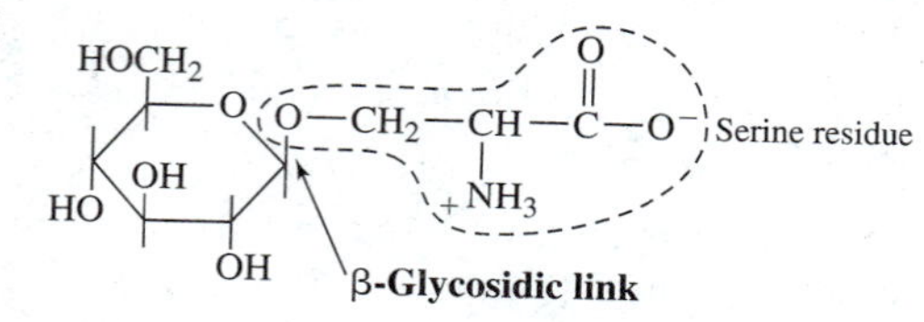

6.32 See answer at top of next column.

6.33 β(1 ⟶ 3) and β(1 ⟶ 4).
See rest of answer in next column.

6.34 Minus 12,500.

6.32 *(Answer)*

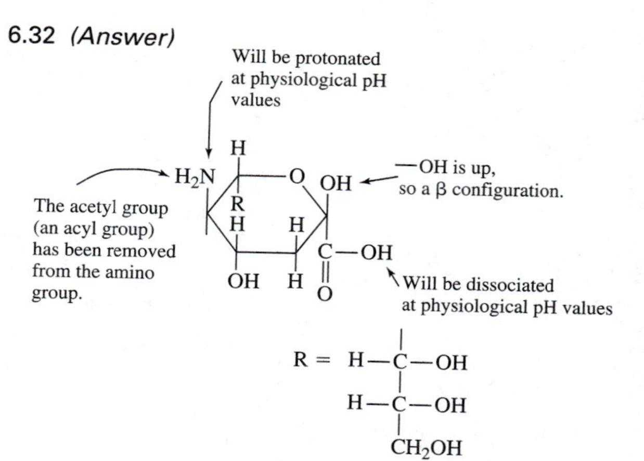

6.33 *(Answer)*

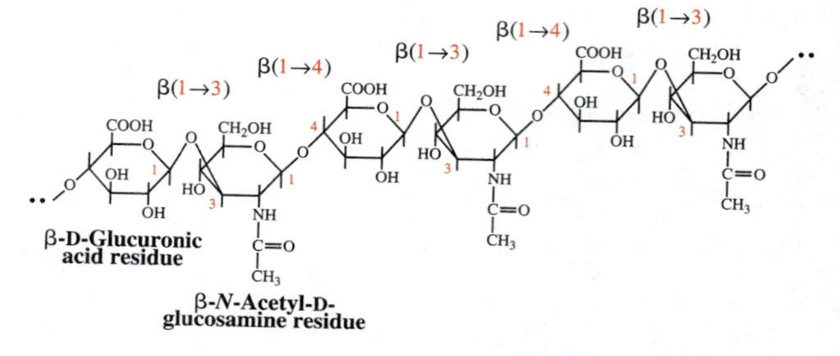

In hyaluronic acid every other residue is a D-glucuronic acid residue, and the carboxyl group in each glucoronic acid residue has a −1 charge at pH 7.0.

6.35 β(1 ⟶ 3) and β(1 ⟶ 4).

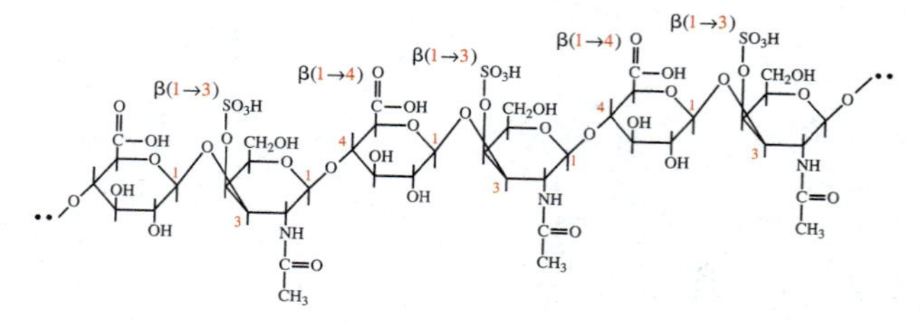

6.36 D-Glucuronic acid, sulfuric acid, acetic acid, and D-galactosamine.

Acetal, ester, and amide bonds can all be hydrolyzed.

6.37 Because starch and sucrose are the major carbohydrates in a typical diet, glucose and fructose are the predominant monosaccharides generated during the digestion of dietary carbohydrates.

Starch is a polymer of glucose, whereas sucrose is a disaccharide containing one glucose and one fructose residue.

6.38 Add lactase.

6.39 Active transport.

6.40 Not all carbohydrates are fattening. Since some cannot be digested, the calories that they contain are inaccessible. The major carbohydrate in green leafy vegetables is cellulose, a nondigestible, nonabsorbable polymer that tends to pass intact through the human digestive system.

6.41 Glucose (dextrose), maltose, and the disaccharide containing two glucose joined through an $\alpha(1\rightarrow6)$ glycosidic link.

6.42 Yes. The purification of any compound will separate it from all other compounds, including vitamins, amino acids, minerals, and other nutrients.

CHAPTER 7

7.1 Net charge on phosphate: at pH 5 = −1; at pH 7.2 = −1.5. Net charge on methyl phosphate: at pH 5 = −1; at pH 7.2 = −2. Net charge on dimethyl phosphate: at pH 5 = −1; at pH 7.2 = −1.

To calculate net charge to the nearest half charge unit: A functional group is assumed to exist entirely in its protonated form if the pH is more than 0.5 below its pK_a; it is assumed to exist entirely in its dissociated form if the pH is more than 0.5 above its pK_a; and it is visualized as existing in its protonated form half the time and in its dissociated form half the time at pH's within 0.5 of its pK_a. These rules, and the basis for them (the Henderson–Hasselbalch equation) are examined in detail in Section 2.7.

The calculation of the net charge on phosphate will be used to illustrate these rules. Phosphoric acid (pK_a's of 2.2, 7.2, and 12.4) is assumed to have one of its acidic hydrogens completely dissociated and two hydrogens entirely associated at pH 5, since this pH is more than 0.5 units above one of its pK_a's and more than 0.5 units below two pK_a's:

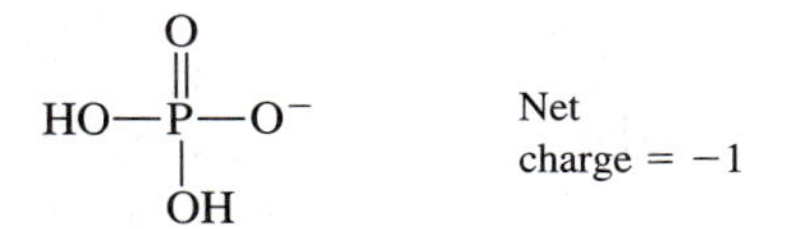

Phosphoric acid at pH 5.0

At pH 7.2, one hydrogen in phosphoric acid is assumed to be permanently lost, a second hydrogen is assumed to be dissociated half the time, and the third is nondissociated:

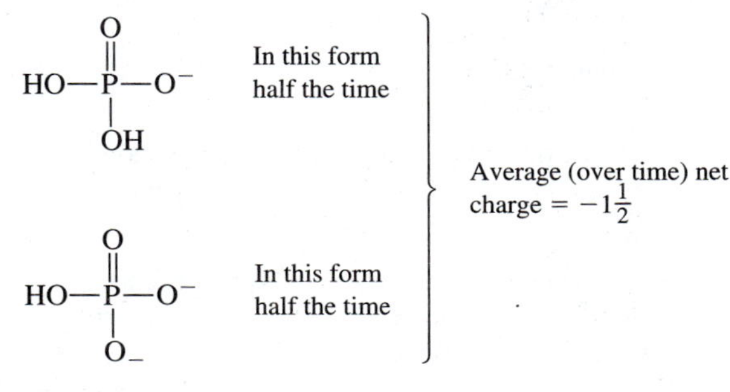

Phosphoric acid at pH 7.2

Rework the problems in Section 2.7 for a further review of net charge calculations.

7.2

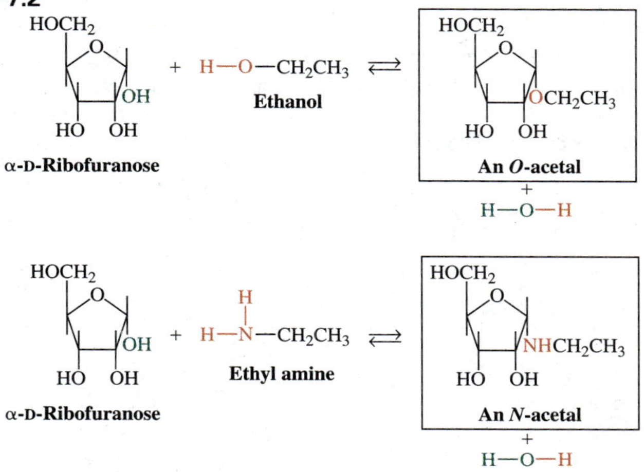

When a cyclic form of a monosaccharide reacts to form an acetal, the acetal is called a glycoside and the acetal link is said to be a glycosidic link. The reaction with ethanol yields an O-glycoside whereas the reaction with ethyl amine generates an N-glycoside (Section 6.5).

7.3

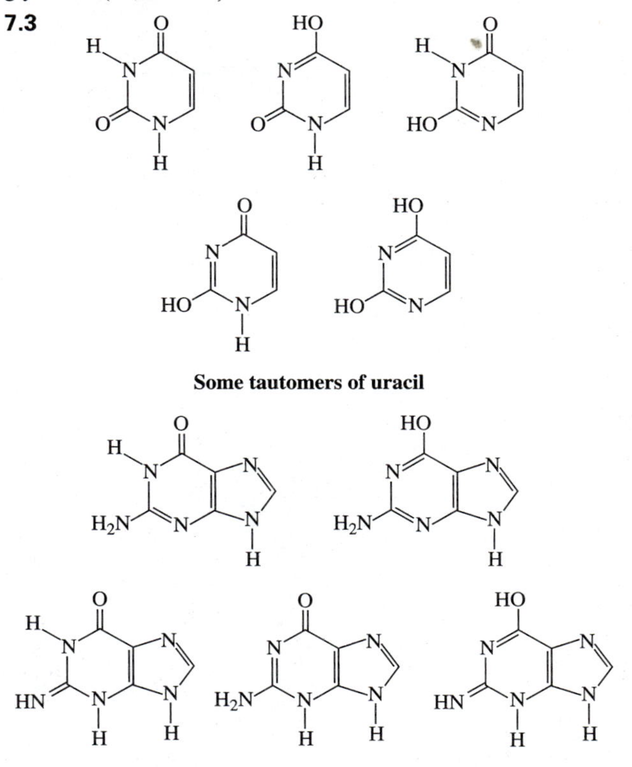

7.4 The enzymes that an organism assembles determine what other compounds it produces since the enzymes in an organism determine what reactions occur (at a biologically significant rate) in that organism. Each step of every synthetic pathway tends to be catalyzed by a separate, highly specific enzyme. Information for the synthesis of enzymes is carried in DNA, the genetic material in all cells. Each organism is genetically programmed to synthesize a limited number of enzymes.

7.5

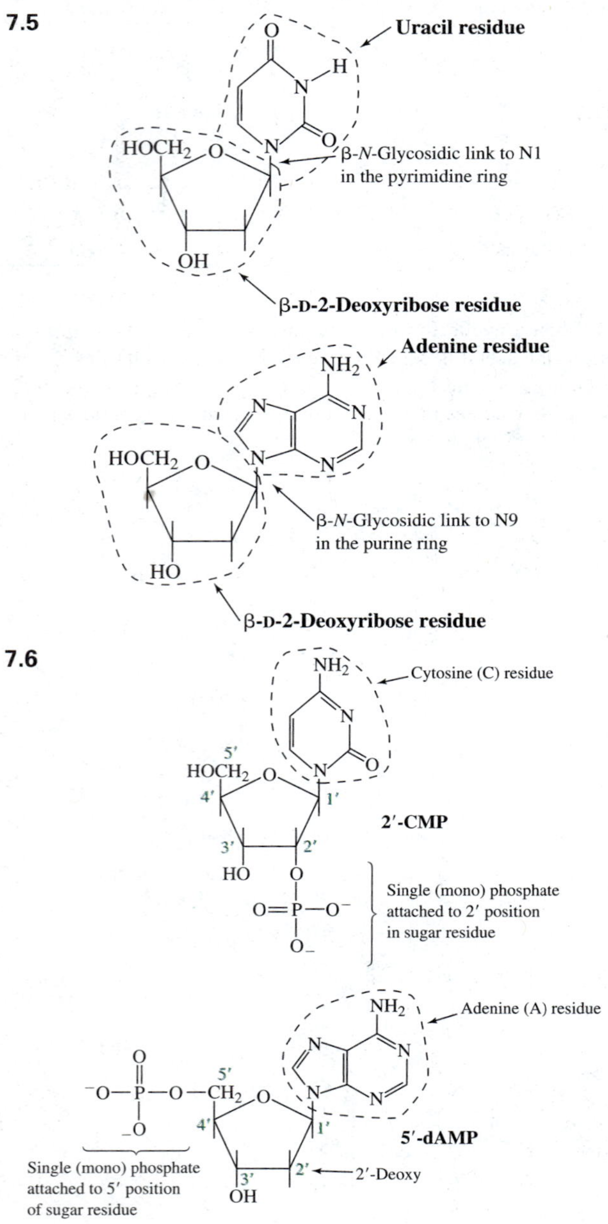

7.6

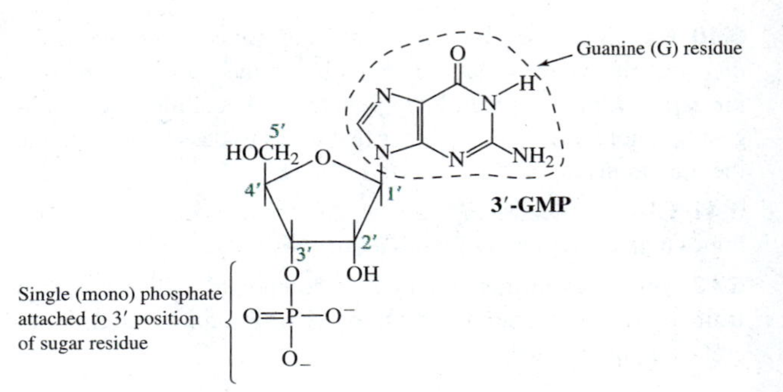

(Continued)

7.7 The net charge on a nucleotide: at pH 6.0 = −1.5; at pH 7.0 = −2; and at pH 8.0 = −2.

Nucleotides are phosphate monoesters that have phosphate pK_a's close to 1 and 6. If the pH is more than 0.5 below a pK_a, the functional group involved is assumed to exist in its protonated from. If the pH is more than 0.5 above a pK_a, the functional group is assumed to reside in its dissociated form. When the pH is within 0.5 of a pK_a, there are roughly equal amounts of the associated and dissociated forms (see Problem 7.1 and Section 2.7). In a nucleotide at pH 6.0, for example, one acidic hydrogen on the phosphate residue is entirely dissociated, whereas the second acidic hydrogen is dissociated approximately half the time:

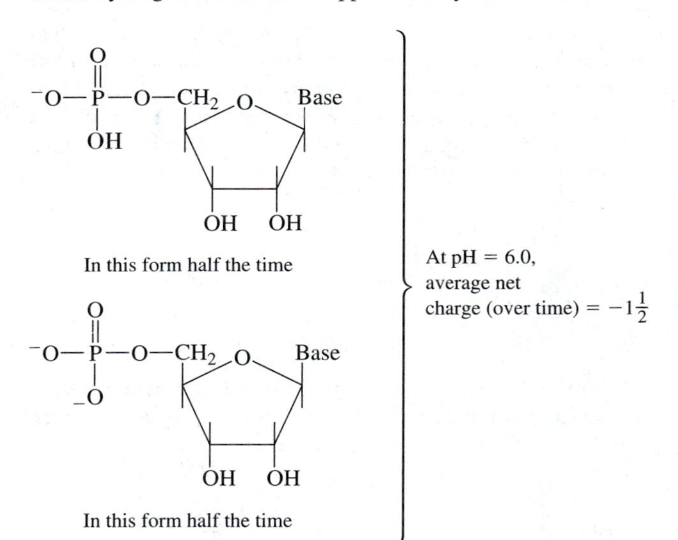

At pH values below 4.0 (approximately), purine and pyrimidine residues begin to accept protons and to acquire positive charges.

7.8 See answer on top of page A-37.

7.9 The binding of an effector to an allosteric enzyme modulates its catalytic activity. Some effectors are positive (stimulatory) and others are negative (inhibitory) (Section 5.12).

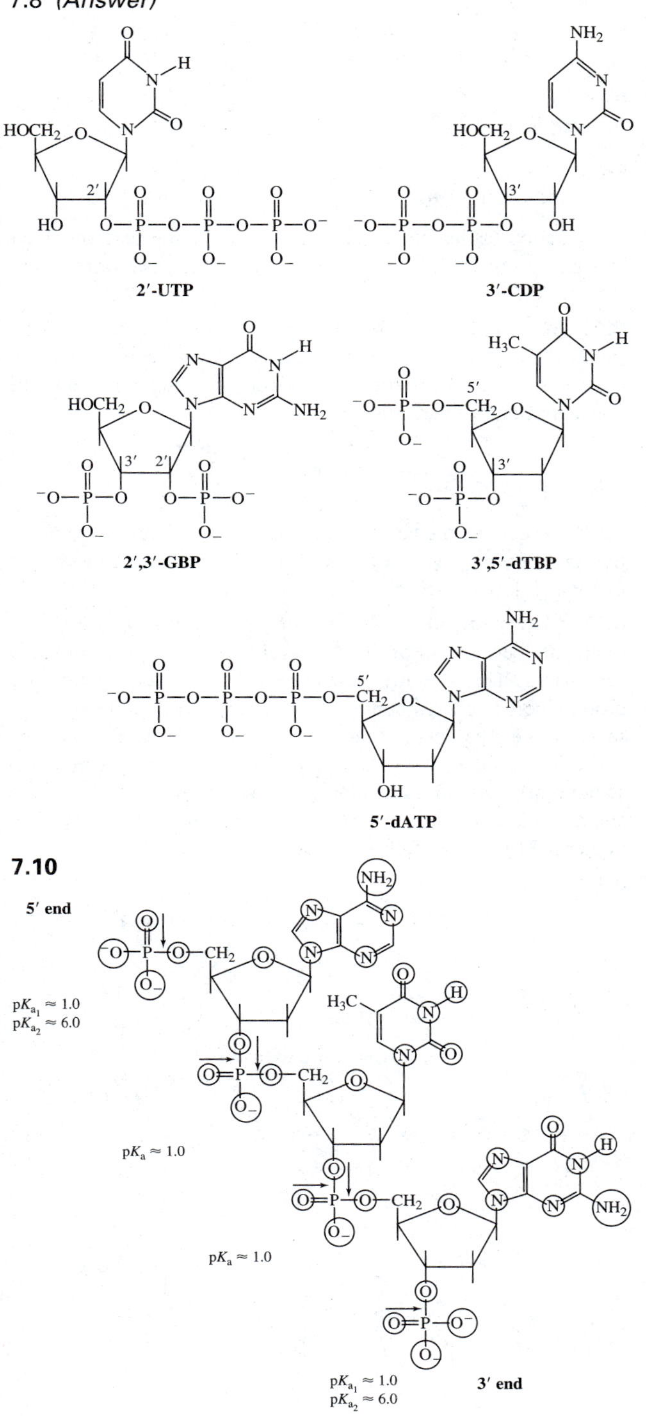

At pH 7.0, each phosphate hydroxyl group will be predominantly dissociated, since the pH is 1 (approximately) or more units above each pK_a (see answers to Problems 7.1 and 7.7). This leads to a net charge of roughly −6. At pH values below 4.0 (approximately), purine and pyrimidine residues become significantly protonated and begin to acquire a significant time-averaged positive charge.

CHAPTER 8

8.1 Without O_2, organic fuel molecules (the dietary source of the energy used by heterotrophs) cannot be oxidized to release energy. Since many life-sustaining processes, including muscle contraction and the transmission of nerve impulses, require a constant supply of energy, O_2 deprivation leads to death.

8.2 $CH_3CH_2OH + 3\ O_2 \rightleftharpoons 2\ CO_2 + 3\ H_2O$ + energy

$$C_6H_{12}O_6 + 6\ O_2 \rightleftharpoons 6\ CO_2 + 6\ H_2O + \text{energy}$$

To balance an equation, first write formulas for all reactants to the left of $\rightleftharpoons$ and formulas for all products to the right. Place the smallest possible whole number coefficients before the formulas so that the number of hydrogen atoms is the same on both sides of the equation, the number of oxygen atoms is the same on both sides, and so on.

8.3 Heat is the energy associated with atomic and molecular motion. Temperature is a measure of the intensity of heat energy. As the temperature rises, the average rate of movement of atoms and molecules increases.

8.4 Gravitational potential energy (water behind the dam) is transformed into kinetic energy as water flows over the dam and through turbines. The kinetic energy of the rotating turbines is transformed into the kinetic energy of an electric current (the flow of electrons). At each step, some energy is also transformed into heat.

8.5 Since the molecular weight of water is 18 amu, 1 mol of water weighs 18 g. Molarity is moles per liter and 1 L = 1000 mL. Since 1 L of pure water weighs 1000 g (1000 mL × 1 g/mL) and contains 55.5 mol (1000 g × 1 mol/18 g), its molarity is 55.5. Under standard biological conditions we are dealing with solutions whose $[H_2O]$ is approximately the same as that of pure water.

8.6 Standard conditions and standard biological conditions differ only in $[H^+]$ and $[H_2O]$. Under both sets of conditions, all other reactants and products are present at 1 *M* concentration or 101.3 kPa (if gases). Thus, in the absence of H^+ and H_2O (as reactants or products), reaction conditions are the same.

8.7 $$K_{eq} = \frac{[\text{ethyl acetate}][H_2O]}{[\text{ethanol}][\text{acetic acid}]}$$

$$K_{eq} = \frac{0.4 \times 5}{2 \times 2} = 0.5$$

In this instance, water is not a solvent and $[H_2O]$ is not constant, so $[H_2O]$ cannot be incorporated into the value of K_{eq}; the $[H_2O]$ must be left in the K_{eq} expression.

8.8 The addition of a reactant or product to a reaction at equilibrium temporarily destroys the equilibrium, but it has no impact on K_{eq} or $\Delta G°$. $\Delta G°$ is the free energy difference between reactants and products *under standard conditions.* If conditions are not standard, $\Delta G \neq \Delta G°$, but $\Delta G°$ is unchanged. At a constant temperature, K_{eq} is determined by $\Delta G°$.

8.9 No effect. $\Delta G^{°}$ is the free energy difference between reactants and products under standard conditions. A catalyst has no impact on the free energy of reactants or products. Catalysts increase reaction rates, in part, by lowering activation energies.

8.10 Within many regions of a cell, buffers hold $[H^+]$ relatively constant and the amount of water produced or consumed during a reaction is, in the short run at least, negligible compared with the amount of water present originally. Most cells also employ multiple mechanisms to maintain a relatively constant water content. Water can readily pass though cell membranes.

8.11 False. When $RT \ln ([C]^c[D]^d/[A]^a[B]^b)$ is more positive than $\Delta G^{°'}$ is negative, the sum is positive. This occurs whenever a net conversion of products to reactants is required to reach equilibrium.

8.12 $\Delta G_p = \Delta G^{°'} + RT \ln \dfrac{[C]^c[D]^d}{[A]^a[B]^b}$

$$\Delta G_p = -31{,}000 \text{ J/mol} + (8.315 \text{ J/mol} \cdot \text{K})(298 \text{ K}) \ln \frac{(0.2 \times 10^{-3})(1.6 \times 10^{-3})}{2.35 \times 10^{-3}}$$

$$\Delta G_p = -31{,}000 + 2{,}480 \ln (1.36 \times 10^{-4}) = -31{,}000 + (-22{,}100) = -53{,}100 \text{ J/mol}$$

Since ΔG_p is negative, a net forward reaction will be required to reach equilibrium.

8.13 $\Delta G° = 0$ when, under standard conditions, $\Delta H°$ is equal to $T\Delta S°$. When $\Delta G° = 0$, $RT \ln K_{eq} = 0$ and $K_{eq} = \text{antiln } 0 = 1$. The reaction is at equilibrium under standard conditions, where all reactants and products are present at 1 *M* concentration.

8.14 True. If ΔH is negative, ΔG will be negative (reaction will be spontaneous) at all temperatures (given that ΔS is positive and $\Delta G = \Delta H - T\Delta S$). On the other hand, if ΔH is positive, ΔG is positive at low temperatures. However, as the temperature rises, a point is reached where $T\Delta S$ becomes larger than ΔH. At this point, ΔG becomes negative (reaction becomes spontaneous) since $\Delta G = \Delta H - T\Delta S$.

8.15 Positive. $\Delta S = S_{products} - S_{reactants}$. When a compound dissociates into two or more products, the randomness of the system increases. Since $S_{products}$ is greater than $S_{reactants}$, ΔS is positive.

8.16

a.

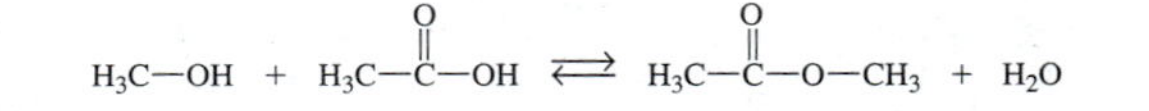

b. $H_3C{-}OH + H_3C{-}C(=O){-}O{-}C(=O){-}CH_3 \rightleftharpoons H_3C{-}C(=O){-}O{-}CH_3 + H_3C{-}C(=O){-}OH$

c. $H_3C{-}OH + H_3C{-}C(=O){-}Cl \rightleftharpoons H_3C{-}C(=O){-}O{-}CH_3 + HCl$

Reaction (a) has the smallest K_{eq} because the reactants have less free energy relative to products than is the case for reactions (b) and (c).

8.17 ΔG_p will increase (become less negative) because the reaction will be moving closer to equilibrium. In general, ΔG_p increases as the concentration of products increases relative to the concentration of reactants.

$$\Delta G_p = \Delta G^{°'} + RT \ln \frac{[ADP][P_i]}{[ATP]}$$

The logarithm in the above equation becomes larger (less negative or more positive) as the ADP to ATP ratio rises (assuming that the $[P_i]$ increases or remains the same).

8.18 At very low pH, ATP experiences little, if any, like-charge repulsion, because all of the hydroxyl groups are predominantly protonated. The loss in like-charge repulsion contributes to a drop in the free energy content of ATP so that ATP no longer possesses more free energy than its hydrolysis products. Since the hydrolysis products are also fully protonated (from a practical standpoint), they do not exhibit the same resonance stabilization and dissociation stabilization (relative to reactants) that they do at higher pH (when not fully protonated).

8.19

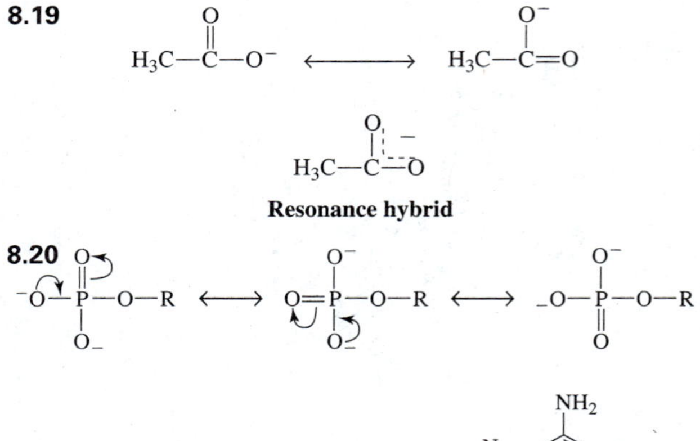

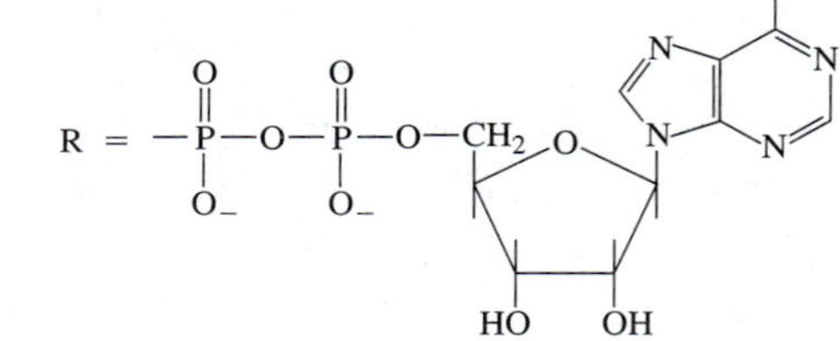

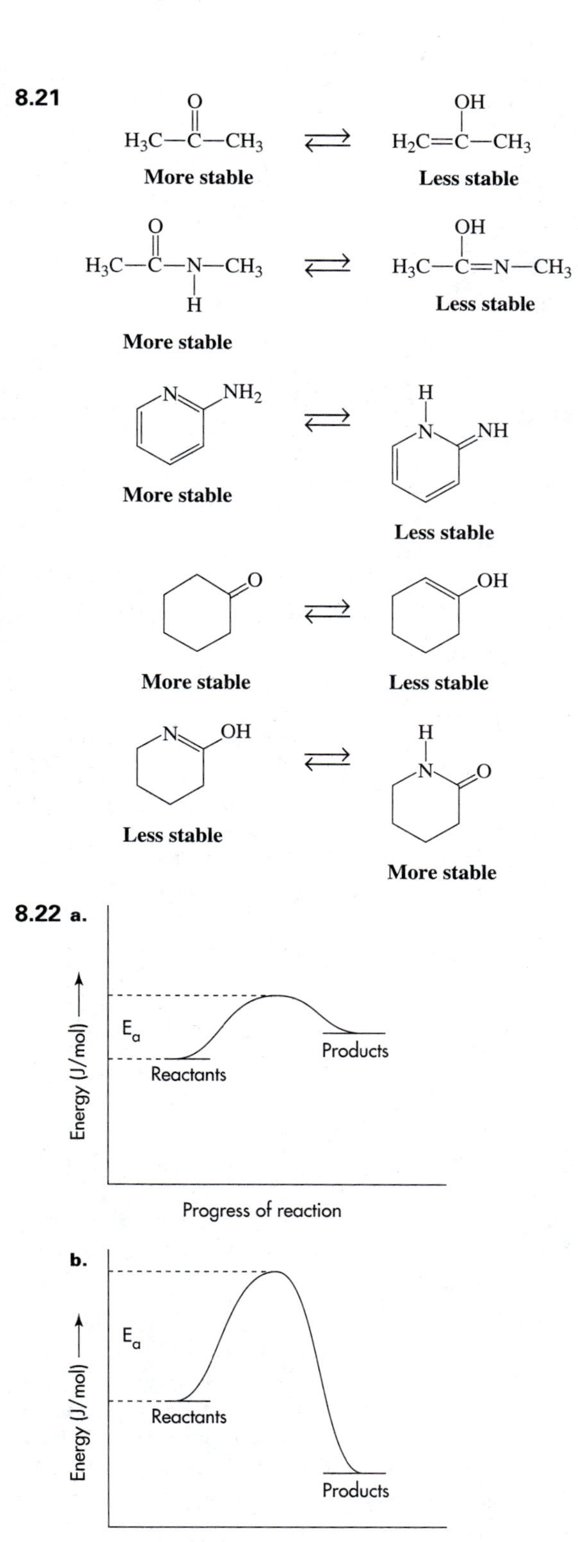

Reaction b has the greatest thermodynamic driving force, since products possess less free energy than reactants.

Reaction a will tend to proceed most rapidly, since it has the lowest activation energy. A drop in activation energy tends to increase the fraction of the total collisions (between reactants) that are productive (lead to product formation).

8.23 Because the second reaction has a $\Delta G^{\circ\prime}$ of + 31 kJ/mol, the $\Delta G^{\circ\prime}$ for the first reaction must have a value more negative than −31 kJ/mol for the $\Delta G^{\circ\prime}$ of the net reaction to be negative. A positive driving force (under standard conditions) is always associated with a negative $\Delta G^{\circ\prime}$. The $\Delta G^{\circ\prime}$ for the net reaction is equal to the sum of the $\Delta G^{\circ\prime}$ values for the coupled reactions.

8.24 False.

$\Delta G^{\circ\prime}$ may differ markedly from ΔG_p. It is ΔG_p that determines the spontaneity of a reaction under physiological conditions.

8.25

$$2\ CH_3SH + CH_3CH_2S—SCH_2CH_3 \rightleftharpoons CH_3S—SCH_3 + 2\ CH_3CH_2SH$$

$2\ CH_3SH$	$CH_3CH_2S—SCH_2CH_3$
Reducing agent	Oxidizing agent
Electron donor	Electron acceptor
Oxidized	Reduced

$$CH_3CHO + 2\ CuO \rightleftharpoons CH_3COOH + Cu_2O$$

CH_3CHO	$2\ CuO$
Reducing agent	Oxidizing agent
Electron donor	Electron acceptor
Oxidized	Reduced

$$H_2O_2 + CH_3OH \rightleftharpoons 2\ H_2O + CH_2O$$

H_2O_2	CH_3OH
Oxidizing agent	Reducing agent
Electron acceptor	Electron donor
Reduced	Oxidized

8.26 Alkanes, alcohols, (aldehydes, ketones), carboxylic acids.

As we move from left to right along this list, oxygen is added or hydrogens are removed. Normally, both of these events lead to oxidation.

8.27 Within Set A, cytochrome *c* (Fe^{3+}) has the greatest electron affinity (because it has the largest standard reduction potential) and is most readily reduced (because it has the greatest affinity for electrons). Reduction is the gain of one or more electrons. The greater the electron affinity of an agent, the more readily it gains electrons.

Within Set B, ubiquinone has the greatest electron affinity and is most readily reduced.

8.28 $$E' = E^{\circ\prime} + \frac{0.026\ \text{V}}{n} \ln \frac{[\text{electron acceptor}]}{[\text{electron donor}]}$$

$$E' = -0.320\ \text{V} + \frac{0.026\ \text{V}}{2} \ln \frac{0.10 \times 10^{-3}\ M}{5.0 \times 10^{-3}\ M}$$

$$E' = -0.371\ \text{V}$$

8.29 O_2, Fe^{3+}, and the oxidized forms (Fe^{3+} forms) of all the cytochromes will be spontaneously reduced by ubiquinol under standard conditions, because these oxidizing agents have a greater electron affinity than ubiquinone. A compound is reduced when it accepts an electron. The electron acceptor is an oxidizing agent, because it removes an electron from a donor (which is oxidized in the process).

8.30 For reaction a:

$$\Delta G^{\circ\prime} = -n\mathfrak{F}\Delta E^{\circ\prime}, \text{ where } \Delta E^{\circ\prime} = (-0.166) - (-0.185) = 0.019 \text{ V}$$

$$\Delta G^{\circ\prime} = -2(96.48 \text{ kJ/V} \cdot \text{mol})(0.019 \text{ V}) = -3.7 \text{ kJ/mol}$$

$$\Delta G^{\circ\prime} = -RT \ln K_{eq}'; \ \ln K_{eq}' = \frac{\Delta G^{\circ\prime}}{-RT}$$

$$K_{eq}' = \text{antiln} \frac{\Delta G^{\circ\prime}}{-RT}$$

$$K_{eq}' = \text{antiln} \frac{-3700 \text{ J/mol}}{-(8.315 \text{ J/mol} \cdot \text{K})(298 \text{ K})} = 4.5$$

For reaction b:

$$\Delta G^{\circ\prime} = -2(96.48 \text{ kJ/V} \cdot \text{mol})(0.197 \text{ V}) = -38.0 \text{ kJ/mol}$$

$$K_{eq}' = \text{antiln} \frac{-38000}{-(8.315)(298)} = 4.57 \times 10^6$$

CHAPTER 9

9.1 Polar molecules dissolve in polar solvents, nonpolar molecules dissolve in nonpolar solvents, but polar and nonpolar substances do not mix to form solutions. "Like associates with like" is one of the major themes of biochemistry (Exhibit 1.4). Sections 2.1 and 2.2 provide an in depth review of solubility.

9.2 Technically, any nonpolar compound is an acceptable answer, since solvents and solutions can be solids, liquids, or gases. In practice, however, the term "solvent" normally implies a liquid unless specified otherwise. There are many nonpolar or relatively nonpolar compounds that are liquids at room temperature. Some examples follow:

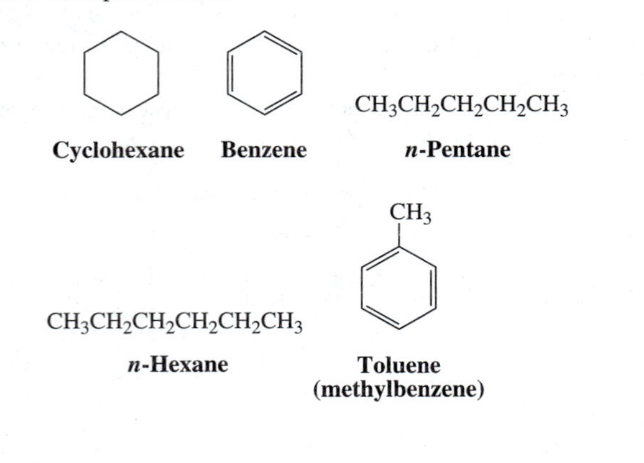

9.3

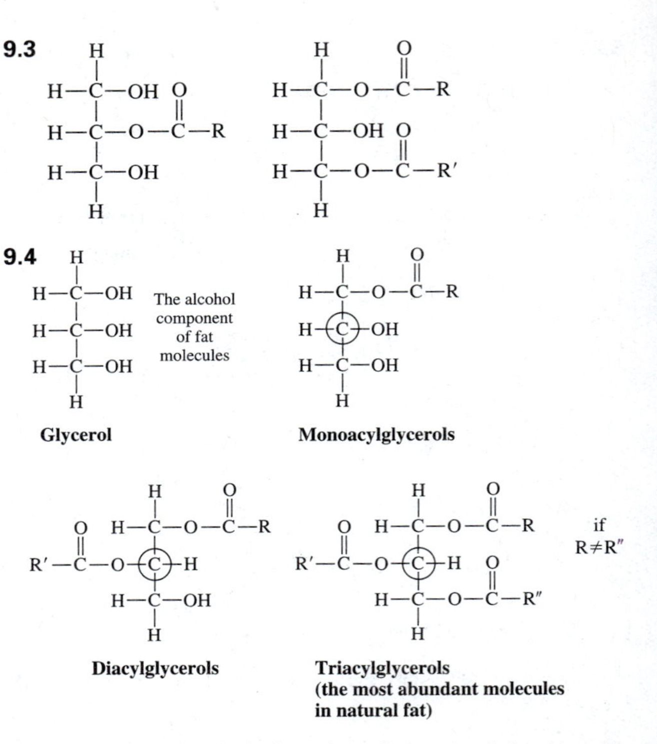

R, R′ and R″ are usually hydrocarbon chains containing 11 to 19 carbons. Table 9.1 provides specific examples.

Triacylglycerols, like the other chiral compounds in living organisms, are synthesized with the catalytic assistance of stereospecific enzymes.

9.5 The many possible answers include:

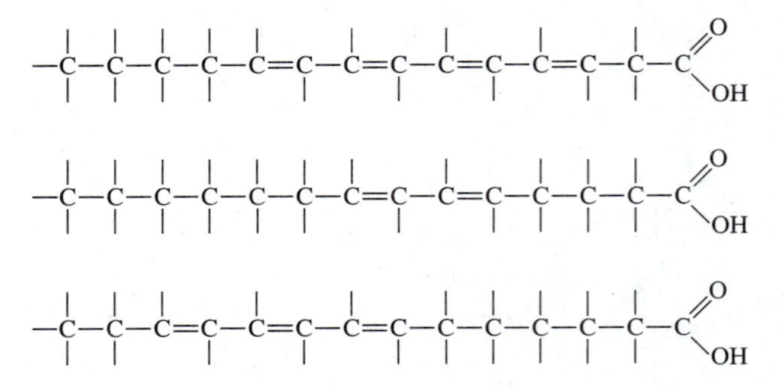

Most fatty acids have a cis configuration about their carbon–carbon double bonds and a methylene group (CH_2) between adjacent carbon-carbon double bonds, so it is very unlikely that any of the given answers represent an acid found in human fat. Not all *cis*-unsaturated carboxylic acids are fatty acids, only those found within fat molecules. Human fatty acids are identified by hydrolyzing human fat, and then separating and characterizing the carboxylic acids produced.

9.6 Linoleic acid, γ-linolenic acid, and arachidonic acid are ω-6 fatty acids. Oleic acid is an ω-9 fatty acid.

9.7 A fatty acid can be attached to the hydroxyl group in a serine, a threonine, or a tyrosine side chain through an **ester link,** and it can be joined to a lysine side chain or an *N*-terminal α-amino group to form an **amide.** A fatty acid is also capable of forming a **thioester link** with a cysteine side chain and an **anhydride link** with any protein carboxyl group. Living organisms are genetically programmed to form a limited number of these possible bonds.

9.8 True. Vegetable oils are classified as unsaturated because of the carbon–carbon double bonds they contain. Complete hydrogenation would eliminate all of the carbon–carbon double bonds.

9.9 Since fats contain long hydrocarbon chains, they are in a more reduced state than proteins and carbohydrates. Viewed from a different perspective, proteins and carbohydrates contain a larger percentage of oxygen than fats and are in a more oxidized state. The less oxidized a compound initially, the more energy is released upon complete oxidation. Since oxidation is required to convert a fat to a protein or carbohydrate, some energy is released in the process.

9.10 50 g carbohydrate × 4 kcal/g = 200 kcal
25 g fat × 9 kcal/g = 225 kcal
5 g protein × 4 kcal/g = 20 kcal
445 kcal total

% calories from fat = (225/445) × 100 = 50.6%.

Note: A "calorie," as the term is normally used in nutrition, is really a kilocalorie. A kilocalorie is sometimes abbreviated Cal with a capital C.

9.11 True. Although both fats and waxes are carboxylic acid esters, most fat molecules are triacylglycerols that contain three adjacent ester links while a wax molecule contains a single ester link. Di- and monoacylglycerols contain one and two hydroxyl groups, respectively. In addition, fats tend to contain more carbon–carbon double bonds than waxes.

9.12 See answer on page A-42.

9.13 See answer in next column.

9.14 See answer on page A-43.

The glycerol residues and the ester links can be included in the polar heads. However, these components are not as polar as the charged "heads" that are circled.

9.15 Lipid bilayers are maintained predominantly by noncovalent hydrophobic interactions. Section 2.2 explains why this is the case.

9.16 See answer in next column.

9.17 See answer in next column.

9.18 Glycosphingolipids are lipids that contain both a sphingosine residue and a carbohydrate residue. *Glyco-* refers to carbon hydrate and *sphingo-* indicates the presence of a sphingosine residue. Sphingomyelins are not glycosphingolipids because they contain no carbohydrate residue(s).

9.19 A β-glycosidic link (Section 6.5) joins the two residues.

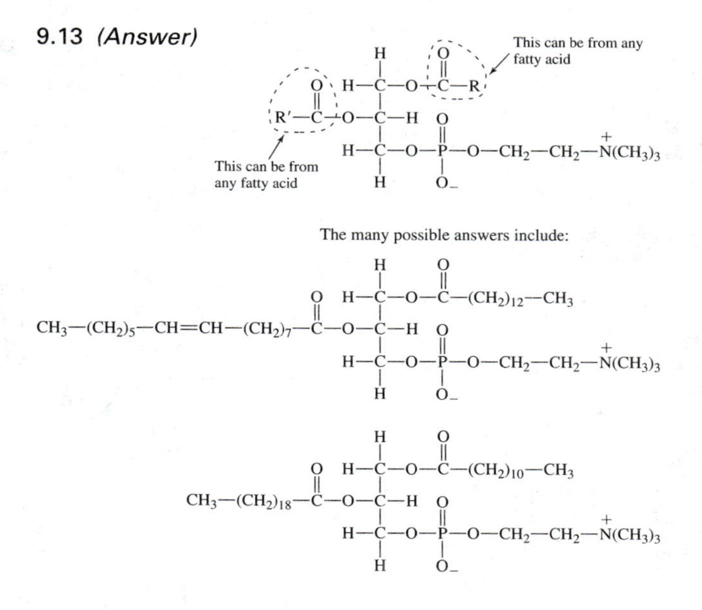

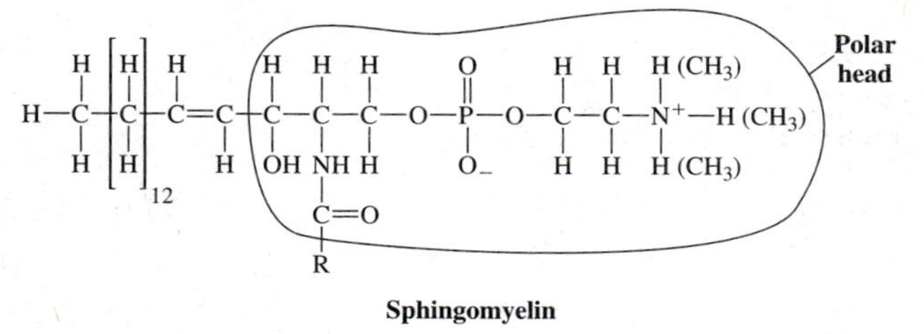

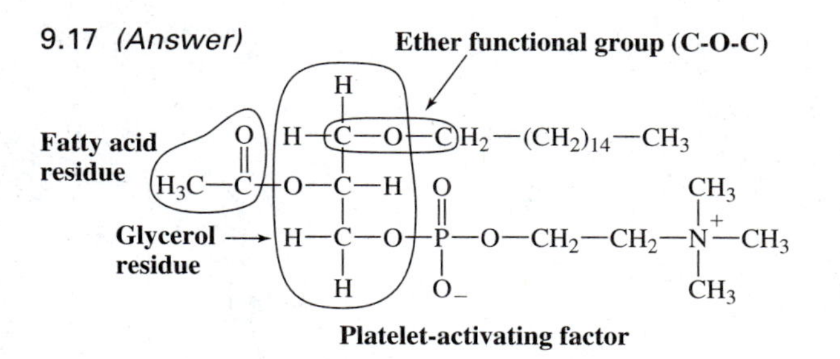

9.12 *(Answer)*

Some Glycerophospholipids (Modified version of Table 9.4)

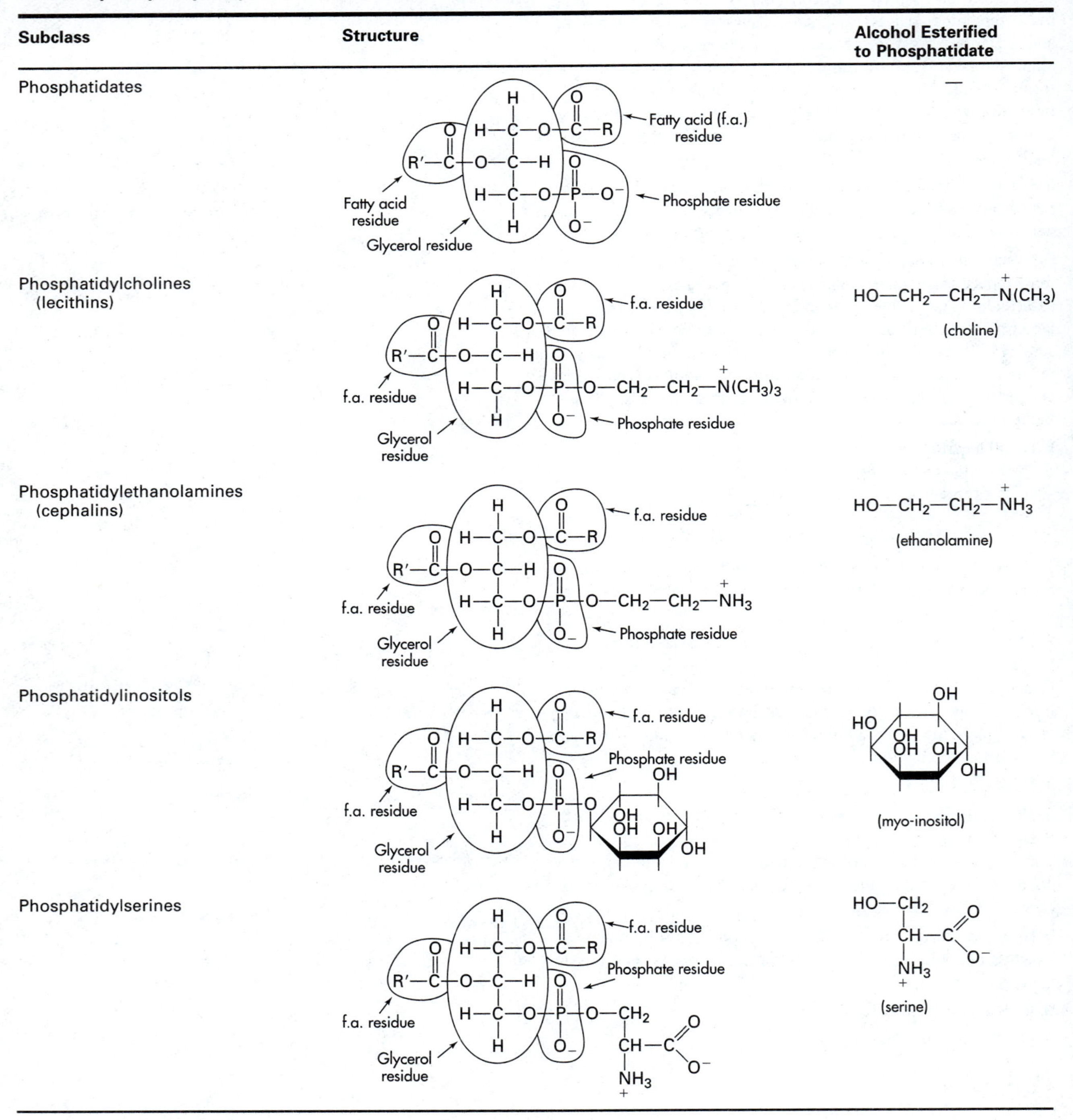

Subclass	Structure	Alcohol Esterified to Phosphatidate
Phosphatidates		—
Phosphatidylcholines (lecithins)	$-O-CH_2-CH_2-\overset{+}{N}(CH_3)_3$	$HO-CH_2-CH_2-\overset{+}{N}(CH_3)_3$ (choline)
Phosphatidylethanolamines (cephalins)	$-O-CH_2-CH_2-\overset{+}{N}H_3$	$HO-CH_2-CH_2-\overset{+}{N}H_3$ (ethanolamine)
Phosphatidylinositols		(myo-inositol)
Phosphatidylserines		$HO-CH_2-CH(\overset{+}{N}H_3)-COO^-$ (serine)

9.14 *(Answer)*

Glycerophospholipids (Modified version of Table 9.4)

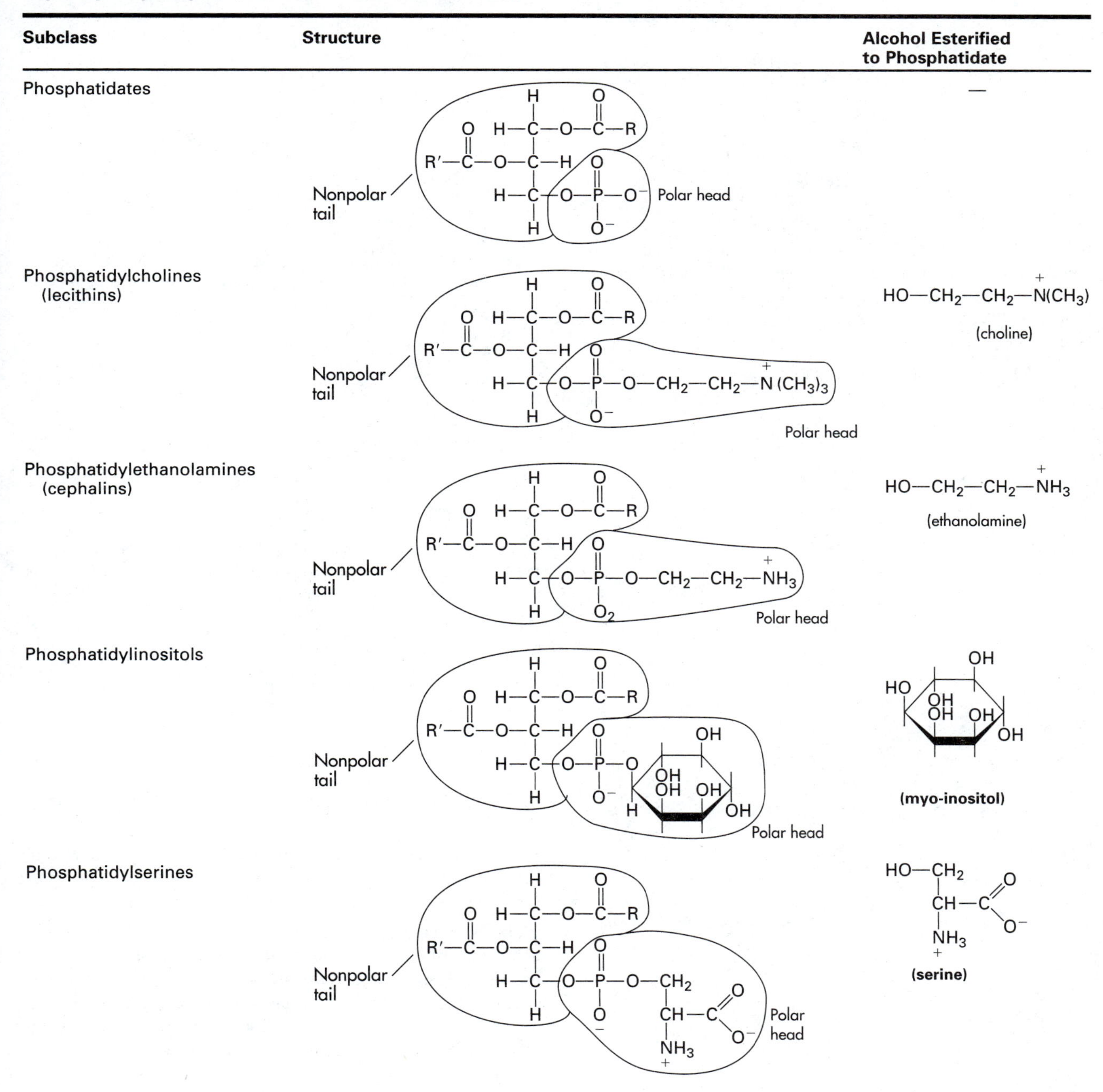

Subclass	Structure	Alcohol Esterified to Phosphatidate
Phosphatidates	Nonpolar tail; Polar head	—
Phosphatidylcholines (lecithins)	Nonpolar tail; Polar head	$HO{-}CH_2{-}CH_2{-}\overset{+}{N}(CH_3)$ (choline)
Phosphatidylethanolamines (cephalins)	Nonpolar tail; Polar head	$HO{-}CH_2{-}CH_2{-}\overset{+}{N}H_3$ (ethanolamine)
Phosphatidylinositols	Nonpolar tail; Polar head	**(myo-inositol)**
Phosphatidylserines	Nonpolar tail; Polar head	**(serine)**

9.20 This compound is an acetal, alcohol, amide and alkene.

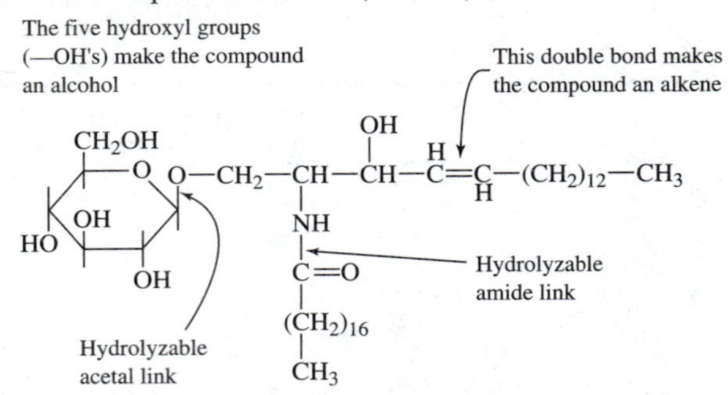

9.21 TXB_3 is a thromboxane that contains three carbon–carbon double bonds outside of the oxane ring. The B indicates that the substituents on the thromboxane backbone are the same as those shown for TXB_2 in Figure 9.14. A change in these substituents leads to a change in the letter used to designate the thromboxane.

9.22 PGE_1 is a carboxylic acid, ketone, alcohol, and alkene.

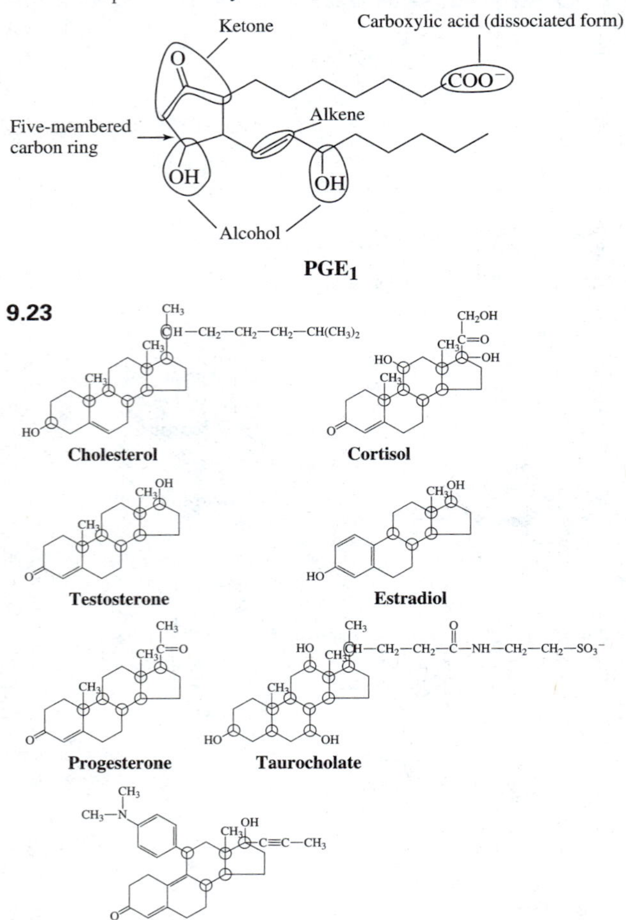

Steroids, like the other chiral compounds produced by living organisms, are constructed through stereospecific, enzyme-catalyzed reactions.

9.24

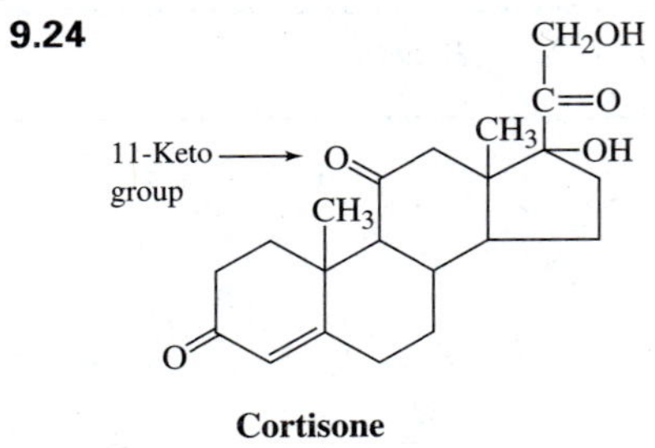

9.25 Secondary sexual characteristics include: hair distribution, fat distribution (breasts, etc.), deepness of voice, and muscle mass.

9.26

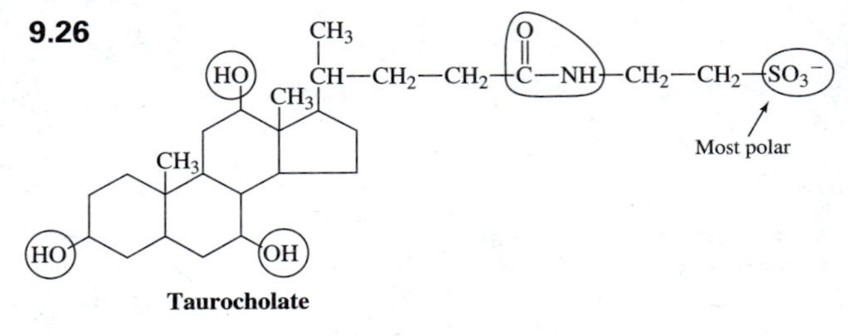

The $—SO_3^-$ group is most polar. An ionic bond can be considered an extreme case of a polar covalent bond. The hydroxyl groups are the second most polar functional groups.

9.27

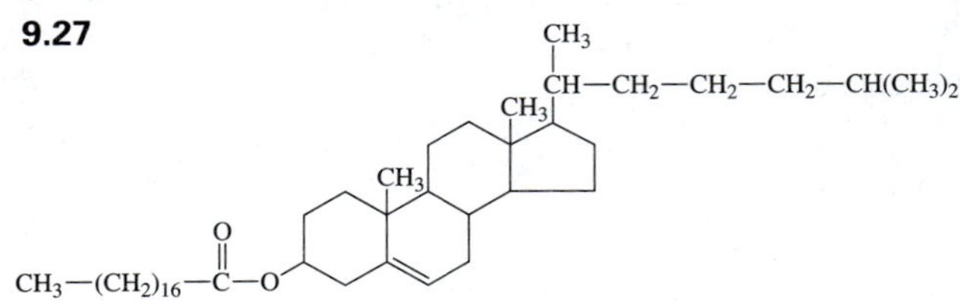

9.28 Percentage protein (dry weight) is the best predictor of density.

9.29 Useable energy. The energy used by most cells comes from the oxidation of fuels, primarily carbohydrates, fats, and proteins. Oxygen (O_2) is required for the release of this energy.

9.30 Blood cholesterol level is only one of several risk factors, and the distribution of cholesterol between multiple lipoproteins has a significant impact on the risk associated with cholesterol itself; two people with the same blood cholesterol level may have markedly different heart disease risks because of differences in other risk factors or differences in the lipoprotein distribution of their cholesterol.

9.31 In most cases, heart disease is not considered to be inherited, since an appropriate life-style can greatly reduce one's risk. It is the predisposition to heart disease, not the heart disease itself, that is inherited. The risk of developing any disease, even infectious diseases, is partly determined by genetics (Chapter 19). Although some people are genetically susceptible to colds, they do not inherit colds. Some genetic diseases, including familial hypercholesterolemia, lead to heart disease regardless of lifestyle. In these instances, heart disease is usually considered to be inherited.

9.32 Peripheral proteins, which contain many polar functional groups, interact primarily with the polar heads of the lipids in the bilayer. Consequently, the predominant interactions are salt bridges, H bonds, and other polar forces.

9.33 The mobility of individual membrane components is possible because the interactions that predominantly maintain membrane structure are noncovalent. If the separate membrane components were linked by covalent bonds, membranes would be rigid and inflexible, and individual membrane components would be locked into a fixed position relative to their neighbors.

9.34 The half-life of a compound is the time required for one half the molecules or formula units in a sample of the compound to be degraded or eliminated. The components in membranes are constantly removed or broken down and replaced with new components. This *tends to prevent* the accumulation of damaged molecules and allows a membrane to change its composition over time. A membrane is a dynamic structural and functional element that must sometimes respond quickly to changes in its environment.

9.35 Like dissolves like (see Problem 9.1). The nonpolar core of the lipid bilayer plays a central role in determining membrane permeability. Small nonpolar molecules can mix with this core and pass through it, whereas polar and ionic substances are unable to do so.

9.36 K_m increases, V_{max} remains the same. A competitive inhibitor competes with the normal substrate for its binding site on the transporter. With inhibitor present, it takes a higher substrate concentration to half saturate (determines K_m) the transporter. Hence, *K_m increases.* Although it also takes a higher substrate concentration to reach V_{max} (saturate the transporter), *V_{max} is unchanged;* at high enough substrate concentration, a transporter is constantly bound to substrate and in the process of transporting that substrate. Think Le Châtelier's principle. Section 5.10 provides a thorough discussion of competitive inhibition.

9.37 In general, the rate of free diffusion is determined by the magnitude of the concentration gradient, the temperature, the viscosity of the solvent, the size and shape of the diffusing solute, and the nature of any permeability barrier.

9.38 4.3 ATP or 5 ATP. The hydrolysis of ATP to yield ADP and phosphate releases roughly 31 kJ/mol (7.4 kcal/mol) under standard conditions (Section 8.4). If we assume that the same amount of energy is released during ATP-driven active transport, a minimum of 4.3 ATP (134 kJ/mol divided by 31 kJ/mol) must be hydrolyzed for each molecule of substance transported. Since you cannot hydrolyze a fraction of an ATP molecule, you could argue that a minimum of 5 ATP must be hydrolyzed to transport one molecule of substrate. Alternatively, the 4.3 can be viewed as the average number of ATP that must be hydrolyzed to transport each of a large number of substrates.

9.39 The initial magnitude of the concentration gradient. The greater the magnitude of a concentration gradient, the greater the entropy difference of the two sides of a membrane. As the entropy difference increases, so does the energy required to move a substance against a concentration gradient.

9.40 The major biochemical themes are specificity, structural hierarchy and compartmentation, transformation and use of energy, catalysis and regulation of reaction rates, and structure–function relationships.

9.41 All of the α-helices are designed to span the lipid bilayer, which is of fairly uniform thickness, around 5 nm. Although distinct α-helical regions could extend different distances out of the bilayer, this is not the case for GluT1 (Figure 9.28).

9.42 Since D-mannose can be transported by GluT1, it competes with glucose for the monosaccharide binding site within this protein; it is a competitive inhibitor. L-Mannose does not function in a similar capacity, since binding sites on proteins are stereospecific; a binding site that accommodates a chiral compound normally will not accommodate its enantiomer.

9.43 The Na^+K^+ transporter plays a central role in the maintenance of multiple concentration gradients that impact the relative solute concentration inside and outside a cell. Osmotic pressure is determined by these relative solute concentrations.

9.44 The Na^+K^+ transporter does not function in reverse under normal physiological conditions; the phosphorylation step in the transport cycle tends to be irreversible (from a practical standpoint) because of the cleavage of a phosphate anhydride bond in ATP. When ADP and P_i concentrations are high (relative to [ATP]) and the normal K^+ and Na^+ gradients are reversed, the Na^+K^+ transporter can function as an ATP synthase. The major ATP synthases in cells can be viewed as H^+ transporters functioning in reverse (Section 11.9).

9.45 A Na^+ channel would most likely be lined with hydrophilic, uncharged amino acid side chains, answer b. A Na^+ ion would be repelled by both a nonpolar channel (polar and nonpolar do not mix well) and a positively charged channel (like charges repel).

CHAPTER 10

10.1 Carbon, hydrogen, nitrogen, and oxygen.

10.2 Nonpolar fat deposits tend to accumulate nonpolar compounds, including the fat-soluble vitamins. Like associates with like.

10.3 Polar. Like dissolves like. A compound must be polar to dissolve in polar solvents such as water.

10.4 Alcohols: thiamine, riboflavin, pyridoxine, pantothenic acid, ascorbic acid, cobalamin, vitamin A, vitamin D
Carboxylic acids: niacin, pantothenic acid, biotin, folic acid
Amines: thiamine, riboflavin, niacin, pyridoxine, folic acid, cobalamin (some are heterocyclic aromatic amines)
Aromatic compounds: thiamine, riboflavin, niacin, pyridoxine, folic acid, cobalamin, vitamin E, vitamin K
Sulfides: thiamine, biotin (both cyclic sulfides)
Ethers: vitamin E (cyclic ether)
Ester: ascorbic acid (cyclic ester), cobalamin (phosphate ester)
Thiols: none
Phenols: pyridoxine, vitamin E

Class of Compound	Defining Functional Group
Alcohol	Aliphatic —OH
Carboxylic acid	R—C(=O)—OH
Amines	R—NH₂ (1°); R—NHR′ (2°); R—NR′R″ (3°)
Aromatic compounds	Benzene ring (⌬) or similar ring system
Sulfides	R—S—R′
Ethers	R—O—R′
Esters	R—C(=O)—O—R′ (carboxylic acid ester); R—C(=O)—S—R′ (thioester); R—O—P(=O)(O⁻)—O⁻ (phosphate ester)
Thiols	R—SH
Phenols	Aromatic —OH

10.5 Niacin, with two polar functional groups.

10.6 Pyridoxine, biotin, folic acid, ascorbic acid, and vitamin A (see Tables 10.3 and 10.4).

10.7 True. The gene content of an organism determines what enzymes it is capable of producing. The enzymes in an organism determine what reactions occur within it and how rapidly these reactions occur. For these reasons, genetics impacts the ability of the human body to absorb, store, and utilize vitamins. It also impacts the life span of a vitamin and the compounds produced from the vitamin. In addition, genetics helps determine the intestinal flora, the amount of 7-dehydrocholesterol in the skin, and numerous other variables that affect dietary vitamin requirements.

10.8 Half-life (within the body): the time required for one half of the molecules in a sample of a compound to be modified or eliminated from the body. The half-life of a compound depends upon how rapidly it is enzymatically altered; how rapidly it is eliminated in feces and/or urine; and how rapidly it is nonenzymatically hydrolyzed, oxidized, or modified in other ways. Distinctive forms of a vitamin often react with different enzymes and usually differ in susceptibility to nonenzymatic chemical alterations. In general, the shorter the half-life of a vitamin, the more frequently it needs to be consumed.

10.9 Since the requirements of a cell or organism can change abruptly, a quick-reversal option is beneficial in hormone action. Duel regulation provides this option; production of an inhibitory hormone can promptly suppress the action of a stimulatory hormone and vice versa.

10.10 Diacylglycerols contain no acidic or basic functional groups and are uncharged at all pH's. Since they are predominantly nonpolar in nature, they are lipids and are virtually insoluble in water. The rest of the second messengers depicted are polar, charged, and water soluble (like dissolves like) at physiological pH. With the exception of Ca^{2+}, the charges are due to the phosphate residues that have one or more acidic hydrogens with a pK_a below 7. At pH values above its pK_a, a functional group exists primarily in its dissociated form (Section 2.7).

10.11 The noncovalent interactions are H bonds, other polar forces, salt bridges, hydrophobic forces, and van der Waals forces. One or more of these interactions are responsible for holding neighboring molecules together in all samples of matter including hormone-receptor complexes. (Sections 2.1 and 2.2). Hormone receptors are proteins, and hormone binding sites are analogous to substrate binding sites on enzymes (Sections 5.4 and 5.5). Hormone binding sites, in contrast to substrate binding sites on enzymes, are not associated with a catalytic site.

10.12 Yes. Hormone receptors are proteins, and binding sites on proteins are normally chiral and stereospecific (they can distinguish stereoisomers; Section 5.4).

10.13 Second messengers normally function as allosteric effectors (Section 5.12). Their binding to an enzyme may alter its active site by changing its conformation. Alternatively, effector binding may shift an equilibrium that exists between active and inactive forms of the enzyme.

A second messenger could also bind to and block or modify an active site.

10.14 Transcription. DNA participates in transcription and is confined to the nucleus of cells. It is physically separated from the translational machinery, which is localized in the cytoplasm. When transcription is altered, translation is indirectly altered because of changes in mRNA concentration.

10.15 Most of the reactions that occur within cells have high activation energies and do not occur at a significant rate in the absence of enzymes (Section 5.2).

10.16

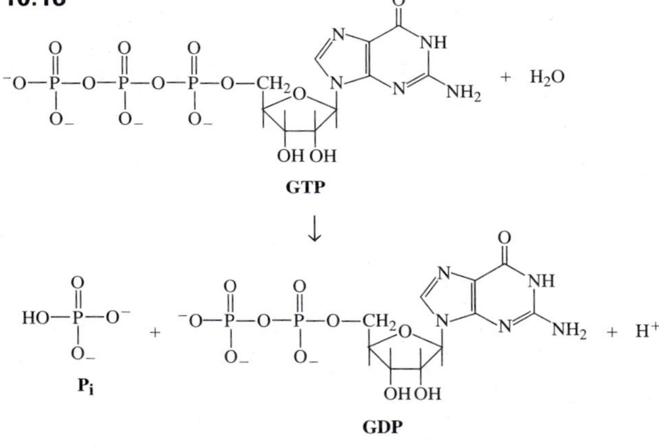

This reaction has a large negative $\Delta G^{\circ\prime}$ and is virtually irreversible because an energy-rich phosphate anhydride is hydrolyzed (see Section 8.4).

10.17 Decrease. GDP and GTP compete for a common binding site on Ras proteins. The more tightly GDP binds, the more difficult it is to activate a protein by replacing GDP with GTP. Ras protein, like other G proteins, is activated by GTP binding.

10.18 Phosphorylation normally converts an uncharged hydroxyl group to a phosphate ester that is negatively charged (at physiological pH) and bulkier than the hydroxyl group (Section 5.13). Such alterations impact side chain interactions and lead to conformational changes in a protein. Conformational changes, in turn, tend to modify active sites. A phosphate group can also directly block an active site or alter its chemistry.

10.19 Receptor protein, GDP, ATP, adenylate cyclase, G protein, any regulators of G protein, and any regulators of adenylate cyclase, including Ca^{2+}. There are other possible answers as well.

10.20 See answer at top of next column.

10.21 See answer in next column.

10.22 Epinephrine receptor, G_s protein, adenylate cyclase, cAMP-dependent kinase, phosphorylase b kinase, and glycogen phosphorylase b.

10.20 *(Answer)*

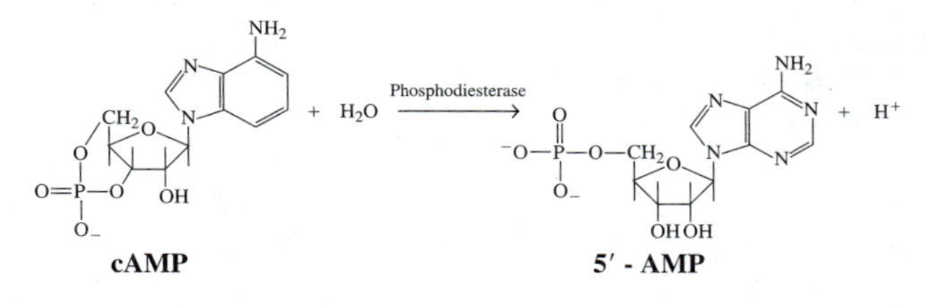

10.21 *(Answer)*

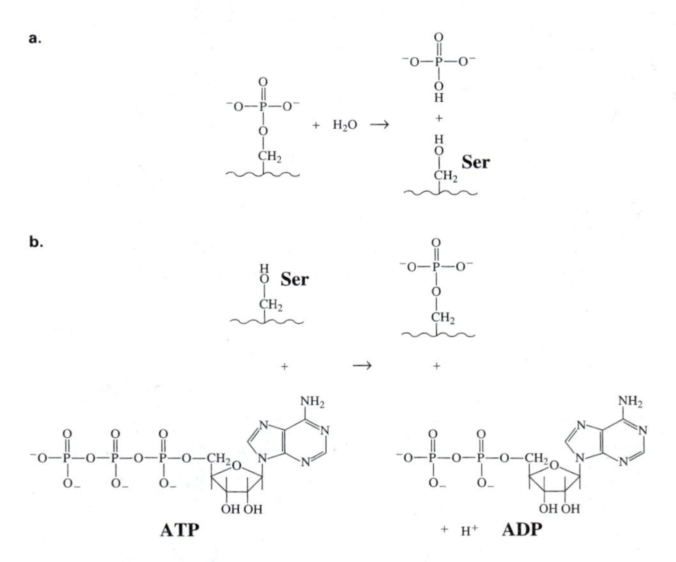

10.23 The burning of fossil fuels, including the combustion of gasoline in automobiles. No carbon monoxide is produced when burning is complete, but complete combustion is rare within human-engineered devices.

10.24 The diacylglycerol, being a nonpolar lipid, mixes well with the lipid bilayer in the membrane but has little tendency to pass into the aqueous environment in the cytosol. The reverse is true for the polar, charged inositol trisphosphate.

10.25

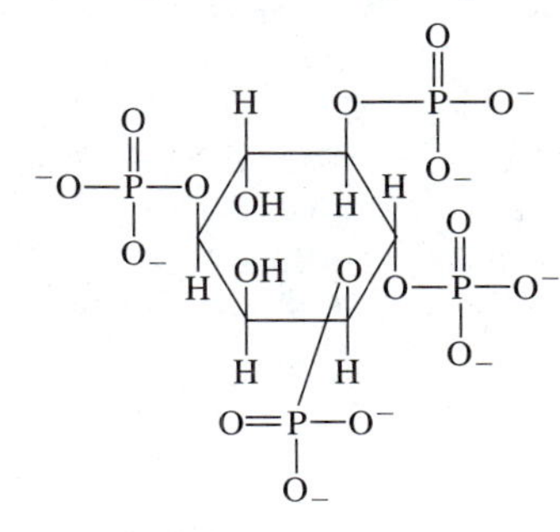

10.26 Although Na^+ plays a central role in the transmission of a nerve impulse, it is not a second messenger. A second messenger binds to one or more specific targets, usually enzymes or other proteins. Na^+ has no such targets. Ca^{2+} or a Ca^{2+}–calmodulin complex binds to specific target molecules during Ca^{2+}-linked signaling.

10.27 Considerable energy is required to operate the pumps that create and maintain the ion gradients involved in the transmission of a nerve impulse; it takes energy to create electrical and chemical gradients. A smaller amount of energy is also required to synthesize and package neurotransmitters and to produce and maintain other participants in nerve transmission.

10.28 Lipids: none

Carboxylic acids: aspartate, glutamate, neuropeptides, 4-aminobutyrate, glycine.

Esters: acetylcholine, ATP (phosphate ester)

Amines: aspartate, ATP, dopamine, epinephrine, glutamate, histamine, 5-hydroxytryptamine, neuropeptides, norepinephrine, 4-aminobutyrate, glycine. All except acetyl choline, which is a quaternary ammonium salt.

Amides: neuropeptides

Alcohols: ATP, epinephrine, some neuropeptides, norepinephrine.

Note: The answer to Problem 10.4 reviews some functional groups and classes of organic compounds.

10.29 There are a number of acceptable answers, including: an altered receptor that does not bind hormone normally or does not respond normally after it binds a hormone; a defective G protein that does not bind GTP and/or GDP with normal affinity or does not function normally after it is "turned on" by GTP docking; a modification in the membrane-associated enzyme coupled to a G protein; a defect in one of the kinases in a reaction cascade; and so on. In general, a change in any of the participants in signal transduction could alter the response triggered by hormone binding.

10.30 Hormones, agonists, and antagonists compete for the same binding sites on protein receptors. Since binding sites on proteins are normally highly selective, most ligands that bind have kindred structures. Similarly, most competitive inhibitors of enzymes are structurally similar to normal substrates (Section 5.10).

10.31 One redox reaction is $2\ H_2 + O_2 \rightleftharpoons 2\ H_2O$. The O_2 is the oxidizing agent, and H_2 is the reducing agent. The H_2 is oxidized (from the standpoint of oxidation–reduction, hydrogen atoms lose electrons to oxygen atoms that have a greater electron affinity) while the O_2 is reduced (it gains electrons from hydrogen). Another redox reaction is: $CH_4 + 2\ O_2 \rightleftharpoons CO_2 + 2\ H_2O$. Methane is oxidized while O_2 is reduced. O_2 is the oxidizing agent and methane the reducing agent. From the standpoint of oxidation–reduction, the carbon loses electrons to oxygen which has a greater electron affinity.

A nonredox reaction is: $HCl + NaOH \rightleftharpoons H_2O + NaCl$. Each atom "claims" the same number of electrons before and after the reaction; none of the atoms gain or lose electrons.

Note: There are an enormous number of possible answers, since every reaction falls into one of these two categories. Section 8.7 and its associated problems provide additional examples.

10.32 NADPH and NADH are equally powerful reducing agents because their half-reactions have the same low $E^{\circ\prime}$. FAD (in flavoproteins) is the strongest oxidizing agent, because its half-reaction has the largest $E^{\circ\prime}$.

10.33 A multienzyme complex allows a metabolic pathway to be carried out quickly and efficiently, since the product of one enzyme-catalyzed reaction can be passed directly to the next enzyme in the pathway. The metabolic channeling that results may also be important (Section 5.15).

10.34
- **a.** Since enzyme activity is a function of pH, temperature, ionic strength, and other variables, reaction rates can be controlled, to some extent at least, by regulating these variables. Lysosomal enzymes, for example, tend to be less active at cytosolic pH than at the more acidic pH that exists in lysosomes. This helps protect a cell from autodigestion if a lysosome should rupture (Section 5.5).
- **b.** The K_m of an enzyme helps determine how rapidly a reaction occurs at any given substrate concentration.
- **c.** The V_{max} for an enzyme and enzyme concentration set the upper limit for the rate of an enzyme-catalyzed reaction.
- **d.** Substrate inhibition helps regulate the relative rate of utilization of alternate substrates.
- **e.** Allosteric effectors and other activators or inhibitors modulate the activities of their target enzymes, a ligand-based regulatory mechanism.
- **f.** Some enzymes are initially synthesized in an inactive form (zymogen) that is activated only when and where the enzyme is needed.
- **g.** Reaction rates can be regulated by controlling how rapidly enzymes are degraded, an event that determines the half-life of the enzyme and helps determine enzyme concentration.
- **h.** The compartmentation of enzymes helps determine where certain reactions occur.
- **i.** Isozymes allow the same reaction to proceed at different rates in different cells, even when enzyme and substrate concentrations are the same.
- **j.** Covalent modification, including phosphorylation, commonly modulates enzyme activity.
- **k.** Multienzyme complexes facilitate reactions and help regulate the flow of metabolites through alternate metabolic pathways.

10.35 Nucleus, mitochondrion, endoplasmic reticulum, Golgi apparatus, and lysosome (Table 1.2). This is not a comprehensive listing of organelles, but it includes all of those organelles that you will encounter as you move on through this text.

10.36

$$\text{(glucose)} + 6\,O_2 \longrightarrow 6\,CO_2 + 6\,H_2O + \text{energy}$$

α-D-Glucose

$$CH_3(CH_2)_{16}COOH + 26\,O_2 \longrightarrow 18\,CO_2 + 18\,H_2O + \text{energy}$$

The oxidation of stearic acid leads to the release of the most calories per gram because this fatty acid is in a more reduced state than glucose initially. See Section 9.1 for a more in-depth explanation.

10.37 The brain is the command center for the entire body. You can lose an arm, a leg, or even a kidney and still survive, but brain loss or severe brain damage is fatal in every case. Through evolution special barriers and protection have been acquired by the brain because of its unique, central, and essential role. The blood–brain barrier shields the brain from many chemicals (some toxic or potentially toxic) that have access to other tissues and organs.

10.38 True. The typical American meal is high in carbohydrate, primarily starch and sucrose. The digestion of these substances releases glucose that enters the blood and temporarily elevates blood glucose levels.

10.39 The glucose cross-linking reactions are bimolecular reactions in which glucose must collide with the substance to which it becomes cross-linked before it can react. In any bimolecular reaction, reaction rate increases with reactant concentration because of an increase in the number of collisions per unit time (Section 5.2).

10.40

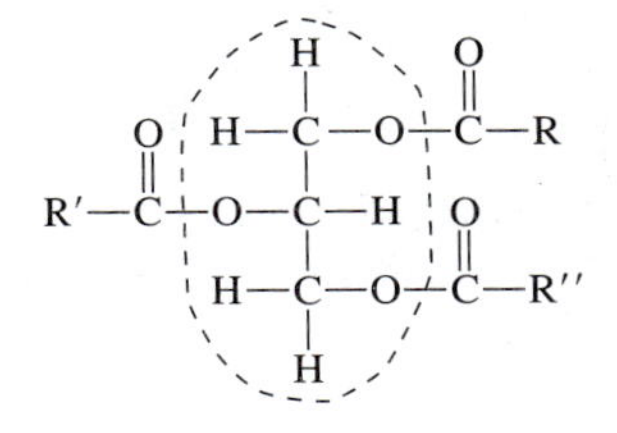

Glycerol residue

A triacylglycerol is a fat molecule. Fat is composed predominantly of triacylglycerols plus smaller amounts of mono- and diacylglycerols.

10.41

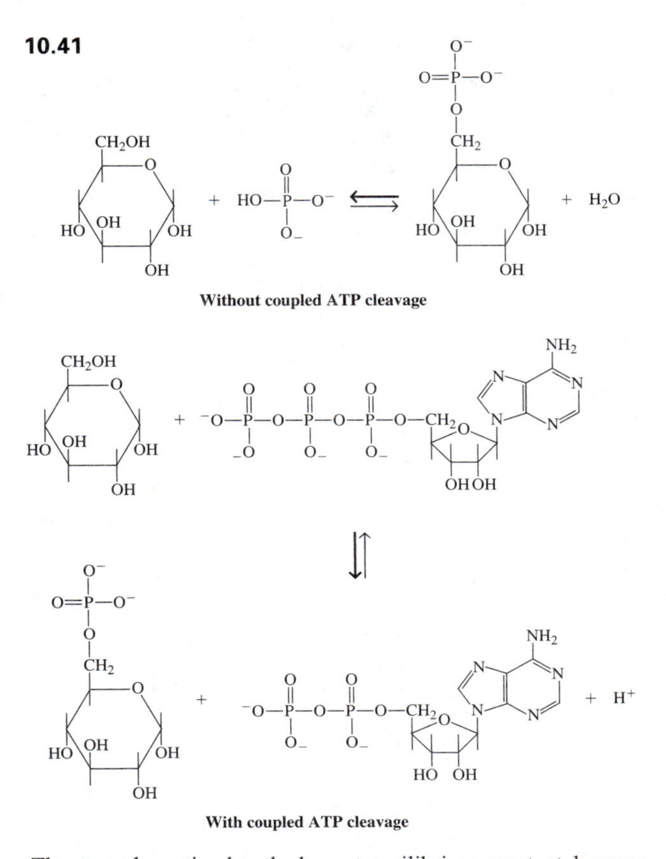

The second reaction has the largest equilibrium constant, because the cleavage of the anhydride bond in ATP provides a thermodynamic driving force. Since the $\Delta G^{\circ\prime}$ for the hydrolysis of ATP is approximately −31 kJ/mol, the $\Delta G^{\circ\prime}$ for the overall reaction is also negative. From a thermodynamic standpoint, the phosphorylation of glucose by ATP can be viewed as occurring in two steps: the hydrolysis of ATP; and the reaction of the released phosphate with glucose to yield glucose 6-phosphate. The coupling of reactions, a concept central to metabolism, is examined in detail in Section 8.6.

10.42 Hormone concentration can be adjusted by altering the rate at which a hormone is eliminated or degraded.

CHAPTER 11

11.1 Starch, a polymer of glucose, yields many molecules of glucose on hydrolysis. Sucrose, a disaccharide, yields one glucose and one fructose. Lactose, another disaccharide, yields glucose plus galactose. The structural formulas for these carbohydrates are presented in Sections 6.5 and 6.6.

11.2

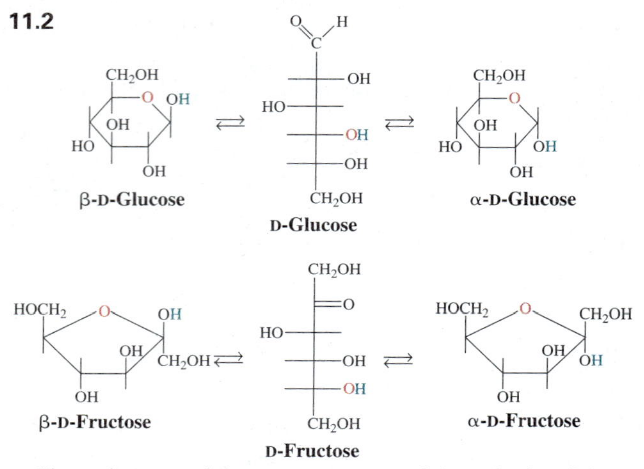

Since glucose and fructose are structural (constitutional) isomers, their direct interconversion entails an isomerization reaction. Enzymes that catalyze isomerization reactions are called isomerases (Section 5.1).

11.3 *Pyruvate* refers to the dissociated or salt form of pyruvic acid; it is the conjugate base of pyruvic acid. Similarly, citrate is the conjugate base of citric acid, lactate is the conjugate base of lactic acid, and so on. At physiological pH, a typical carboxylic acid exits predominantly in its dissociated form, because the pK_a values for carboxylic acids are normally at least 2 units below physiological pH. The [conjugate base] = [weak acid] when pH = pK_a. The conjugate base to weak acid ratio increases by 10 for each 1 unit rise in pH (Section 2.7).

11.4 In glucose 6-phosphate, the phosphoryl group is joined to glucose through a phosphate ester link. Glucose 6-phosphate is negatively charged at pH 7, since phosphate monoesters have two acidic hydrogens with pK_a values near 1 and 6, respectively. A weak acid exists predominantly in its dissociated form at pH values above its pK_a (Section 2.7).

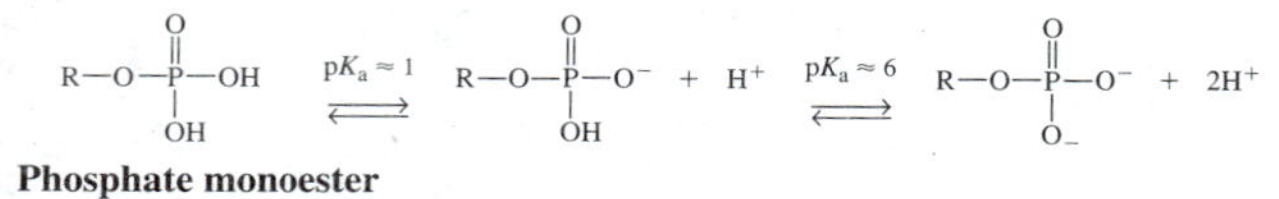

11.5 The six major classes of enzymes are oxidoreductases, transferases, hydrolases, lyases, isomerases, and ligases (Table 5.1). Hexokinase and other kinases are transferases; they catalyze the transfer of a phosphoryl group from ATP to a second substrate (glucose in the case of hexokinase).

11.6 The concentration of glucose 6-phosphate will increase—think "Le Châtelier's principle." Although a net conversion of fructose 6-phosphate to glucose 6-phosphate is required to restore equilibrium, the equilibrium constant itself is not changed. Equilibrium constants are determined by temperature and the free energy difference between reactants and products, not by the initial concentration of reactants and products (Section 8.2). Catalysts do not alter equilibrium constants; they simply reduce the time required to attain equilibrium.

11.7 The equilibrium constant is > 100 since $\Delta G^{\circ\prime} = -14.2$ kJ/mol (Table 8.3). The more negative the $\Delta G^{\circ\prime}$ for a reaction, the larger its equilibrium constant and the less reversible the reaction under standard biological conditions.

11.8

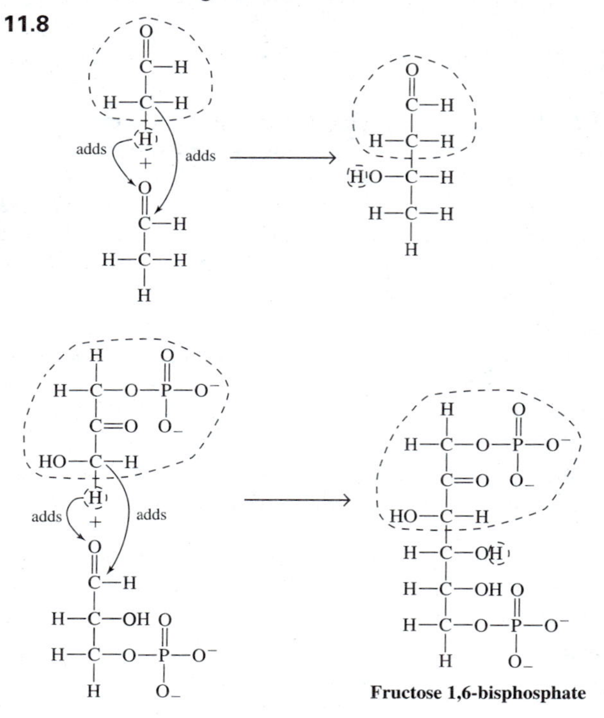

Both reactions involve the making and breaking of similar covalent bonds; both reactions are aldol condensations. Note: the arrows in the equations do not represent reaction mechanisms.

11.9 See answer on page A-51.

In both reactions, an aldehyde is converted to a ketone (an aldose to a ketose) as a carbonyl group ($-\overset{O}{\overset{\|}{C}}-$) shifts one position.

11.10 See answer on page A-51.

11.9 *(Answer)*

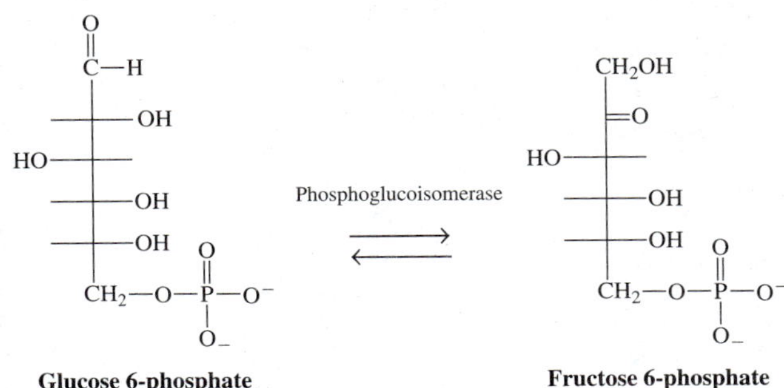

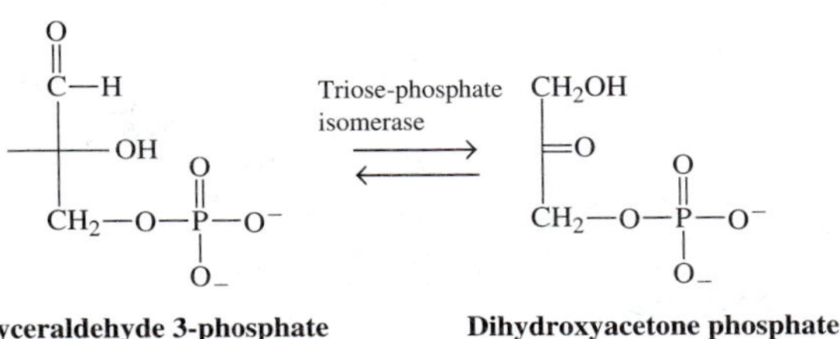

11.10 *(Answer)*

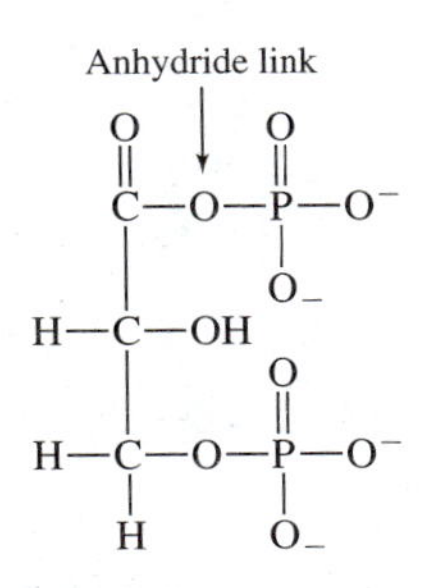

11.11 Ten steps are involved in glycolysis. Since each step is catalyzed by a separate enzyme, there are 10 unique enzymes. At least 10 genes must be expressed, because each enzyme contains one or more distinctive polypeptides and distinctive polypeptides are normally encoded in separate genes. The actual number of genes required to construct the enzymes for glycolysis is greater than 10, since some of the enzymes contain more than one polypeptide chain. Section 10.3 provides an overview of genes and gene expression, and Chapters 16 through 18 examine these topics in more detail.

11.12 Lactate production is triggered by the accumulation of NADH, a substrate for lactate dehydrogenase. In the presence of O_2, the NADH from glycolysis is rapidly converted to NAD^+ as electrons (from the NADH) are shuttled into mitochondria. In the absence of O_2, the flow of electrons from NADH into mitochondria stops, since O_2, the terminal electron acceptor, is required to maintain this flow. Hence, NADH levels rise in the cytosol.

11.13 Glycolysis slows down under conditions where lactate is reconverted to pyruvate, since the NAD^+ needed for glycolysis is being used to oxidize lactate.

11.14 Positive. High concentrations of NAD^+ signal a low energy charge and the need for increased energy production. By stimulating phosphofructokinase-1 and glycolysis, NAD^+ would help a cell meet its energy needs.

11.15 False. A single binding site could not accommodate all of the structurally diverse effectors. Binding sites on proteins are normally highly selective, even stereospecific.

11.16 K_m equals the substrate concentration that leads to an initial velocity equal to one half V_{max}. Since glucokinase has a large K_m for glucose, a high glucose concentration is required to reach one half V_{max}. At substrate concentrations far below the K_m, the rate of the glucokinase-catalyzed reaction is very slow (Section 5.6).

11.17 Decrease. An increase in pyruvate concentration leads to an increase in the rate of formation of citrate, alanine, and ATP, three compounds that allosterically inhibit glycolysis.

11.18 Since no energy is required for an erythrocyte to perform its biological task (O_2 and CO_2 transport), it doesn't need powerhouses. By eliminating mitochondria (the powerhouses of cells), a mature red blood cell can accommodate more hemoglobin. This increases its O_2 and CO_2 carrying capacity.

11.19 During both reactions, NADH reduces a ketone to a secondary alcohol. The reactions are catalyzed by oxidoreductases.

11.20 The decarboxylation of an α-keto acid yields an aldehyde and CO_2. The oxidation of an aldehyde leads to a carboxylic acid. The oxidative decarboxylation of pyruvate generates acetic acid plus CO_2, with acetaldehyde being an intermediate. When oxidative decarboxylation is catalyzed by the pyruvate dehydrogenase complex, the acetic acid generated ends up bonded to CoASH through a thioester link.

11.21 A catalyst is a substance that increases the rate of a reaction and is unchanged at the end of the reaction. TPP, lipoic acid, and FAD are said to function catalytically within the pyruvate dehydrogenase complex, because they contribute to an increase in the rate of oxidative decarboxylation and are unchanged at the end of the reaction.

11.22 High acetyl-SCoA concentrations are normally associated with the rapid production of ATP within the citric acid cycle and respiratory chain. Under these conditions, it is wasteful to oxidize pyruvate. The feedback inhibition of the pyruvate dehydrogenase complex by acetyl-SCoA inhibits this oxidation and conserves the pyruvate so it can be put to better use.

High NAD^+ concentrations signal a low energy charge (low [ATP]) and a need to increase the rate of ATP production. By stimulating the pyruvate dehydrogenase complex, NAD^+

contributes to an increased rate of fuel oxidation and the rate of oxidative phosphorylation (ATP production).

11.23

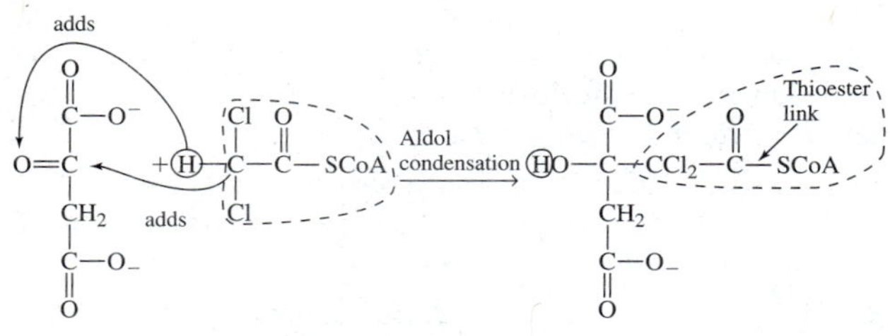

Note: The curved arrows do not represent a reaction mechanism.

11.24 True. Although citrate and isocitrate are structural (constitutional) isomers, they are not interconverted directly. The interconversion entails two steps: a dehydration reaction followed by a hydration reaction. Hence, aconitase is a lyase (Table 5.1).

11.25 10 ATPs. Although only one GTP (an ATP equivalent) is directly generated within the citric acid cycle, 9 ATPs can be produced when the three NADHs and one $FADH_2$ formed within the cycle pass electrons through the respiratory chain to O_2. Therefore, 9 of the 10 ATPs are synthesized within the respiratory chain.

11.26 The citric acid cycle stops under anaerobic conditions because mitochondria run out of NAD^+ and FAD, two reactants required for cycling. In the absence of O_2, the NAD^+ and FAD in mitochondria are quickly and permanently (until O_2 is restored) converted to NADH and $FADH_2$, since there is no terminal acceptor for the electrons carried by the reduced forms of these coenzymes. In the presence of O_2, NAD^+ and FAD are regenerated as NADH and $FADH_2$ pass electrons to O_2. Under anaerobic conditions, the cytosol regenerates the NAD^+ needed for glycolysis by converting pyruvate to lactate.

11.27 Positive. A high concentration of FAD would signal a low energy charge, since FAD only builds up when the citric acid cycle is relatively inactive and there is a low potential for generating ATP within the respiratory chain. When the energy charge is low, a cell needs to turn up the rate of fuel oxidation to produce more energy. The stimulation of the citric acid cycle by FAD would help accomplish this.

11.28 The joining of each inorganic phosphate to an ADP to generate ATP leads to the release of a water molecule. See Problem 11.32.

11.29 O_2 is a substrate for both NO synthase and heme oxygenase, the enzymes that catalyze the production of NO and CO, respectively. NO and CO are gaseous biological messengers.

11.30 False. When $\Delta E^{\circ\prime}$ is negative, $\Delta G^{\circ\prime}$ is positive. $\Delta G^{\circ\prime}$ must be negative for a reaction to be spontaneous and to have a $K_{eq} > 1$ under standard biological conditions (Section 8.2).

11.31 Myoglobin and hemoglobin are heme proteins featured in Chapter 4. In both proteins, the heme iron is involved in binding O_2 (Sections 4.8 and 4.9).

11.32

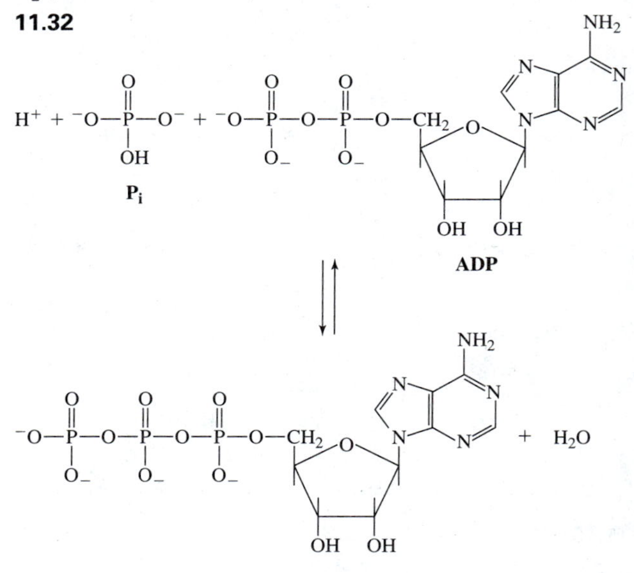

Under standard conditions, the K_{eq} for this reaction is <1000, since $\Delta G^{\circ\prime} = +31$ kJ/mol (Table 8.3).

11.33 True. A concentration gradient contains more entropy than a solution of uniform concentration. All other factors being equal, the greater the entropy of a system, the greater its free energy. Consequently, energy is required to create a concentration gradient (Section 8.3).

11.34 Since H^+ are constantly being pumped out of the matrix into the intermembrane space, the $[H^+]$ is lower inside the matrix. Since pH = –log $[H^+]$, the pH increases as [H+] decreases (Section 2.3). Hence, the pH of the matrix is greater than that of the intermembrane space. Since $[OH^-][H^+] = 10^{-14}$, $[OH^-]$ rises as [H+] drops. Therefore, the $[OH^-]$ is greater in the matrix.

11.35

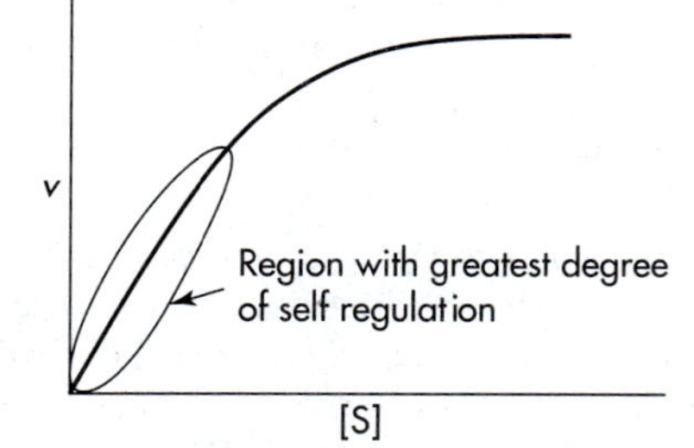

11.36 28 ATP. Two (net) are produced in glycolysis and two in the citric acid cycle (Table 11.4).

11.37 1600 kJ (50 kJ/mol × 32 mol)

11.38 $\Delta G^{\circ\prime} \approx -62$ kJ/mol. Since the hydrolysis of each phosphate anhydride link releases roughly 31 kJ/mol, the hydrolysis of two such links yields around 62 kJ/mol (Section 8.4).

CHAPTER 12

12.1 Glycogen is an alcohol, a hemiacetal, and an acetal (as depicted in Figure 12.1). When its reducing end exists in an open chain form, it is an aldehyde rather than a hemiacetal.

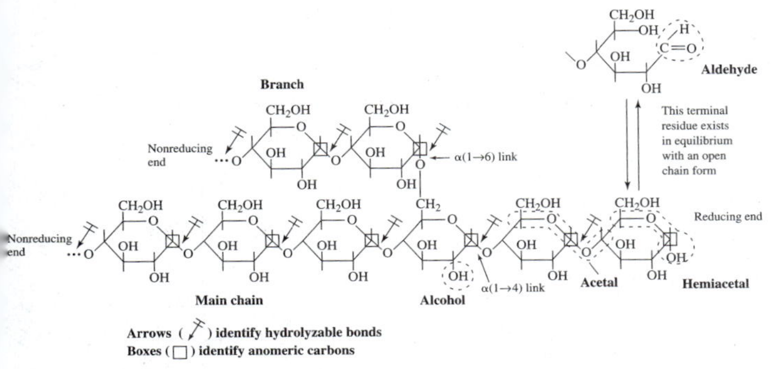

12.2 A "buffer" is a weak acid:conjugate base pair that, when present in an aqueous solution, resists changes in pH when an acid or base is added. Buffers help keep the $[H^+]$ of most biological fluids within specific ranges. Since the liver helps hold blood glucose levels within a normal range, it is said to "buffer" glucose concentrations. The term "buffer" has a similar meaning in both instances.

12.3 True. With 10 reducing ends, 10 glycogen phosphorylase molecules can simultaneously catalyze the phosphorolysis of the same molecule (one attached to each reducing end). In contrast, only one glycogen phosphorylase at a time can catalyze the phosphorolysis of a linear glycogen molecule. In any given time period, 10 workers can do 10 times as much work as a single worker.

12.4 Protein kinases catalyze the formation of phosphate esters. To form an ester, phosphoric acid (phosphate) must react with an alcohol or phenol (Section 5.13). Serine and threonine are alcohols, whereas tyrosine is a phenol. The side chains of these amino acids are the targets for protein kinases.

The phosphorylation of an amino acid side chain can alter a binding site or catalytic site at the active site of an enzyme either directly or indirectly. Phosphorylation of a side chain at the active site itself alters this site by directly changing the size, shape, charge, and intermolecular bonding capabilities of one of its components. Phosphorylation of a side chain at a site distinct from the active site can indirectly alter the active site by leading to a change in the conformation (folding) of the polypeptide chain(s) that make up the enzyme. The active site on an enzyme is created and maintained with a highly specific folding of the polypeptide chain(s) within it. Phosphorylation tends to modify the conformation of a protein by altering the side chain interactions that maintain secondary, tertiary, and quaternary structure.

12.5 **a.** Increase. An antagonist will reduce the amount of epinephrine bound to muscle cell receptors. This, in turn, leads to a decrease in the magnitude of the cAMP-linked cascades that catalyze the conversion of glycogen phosphorylase b to glycogen phosphorylase a and the inactivation of phosphorylase a phosphatase. Consequently, glycogen phosphorylase b tends to accumulate. Study Figures 12.4 and 12.5.

b. Decrease. As receptor concentration rises, the number of epinephrine bound per cell tends to increase. This increases the concentration of the participants in the cAMP-linked cascade that catalyzes the transformation of glycogen phosphorylase b into glycogen phosphorylase a. A coupled cascade also catalyzes the inactivation of phosphorylase a phosphatase, which provides catalytic assistance for the conversion of phosphorylase a back to phosphorylase b.

c. Decrease. Ca^{2+}, by allosterically activating phosphorylase b kinase, stimulates the conversion of glycogen phosphorylase b to glycogen phosphorylase a.

d. Decrease. Since glycogen phosphorylase a phosphatase catalyzes the conversion of glycogen phosphorylase a to glycogen phosphorylase b, its inhibition will block the regeneration of phosphorylase b and lead to the accumulation of phosphorylase a.

e. Decrease. Phosphorylase b kinase catalyzes the conversion of glycogen phosphorylase b to glycogen phosphorylase a.

12.6 Although a cell or organism may need to mobilize fuel reserves (glycogen and fat) at one instance, it may benefit by conserving these reserves a short time later. The hormones that regulate fuel metabolism must turn over rapidly so that an organism can respond quickly to changes in fuel and energy requirements. If epinephrine or glucagon remained in the blood for a long period, muscle or liver cells, or both, would continue to mobilize fuel when that fuel was no longer needed.

12.7 See answer on page A-54.
The oxidized and reduced forms of nicotinamide adenine dinucleotide phosphate are the same as those for nicotinamide adenine dinucleotide (Figure 11.1), except for the additional phosphate.

12.8 In Step 1, a hemiacetal is oxidized to a lactone (cyclic ester). In Step 3, the oxidation of an alcohol to a ketone is coupled to decarboxylation.

12.9 See answer on page A-54.
Since a lactone is a cyclic carboxylic acid ester, hydrolysis of a lactone yields an alcohol and a carboxylic acid.

12.7 *(Answer)*

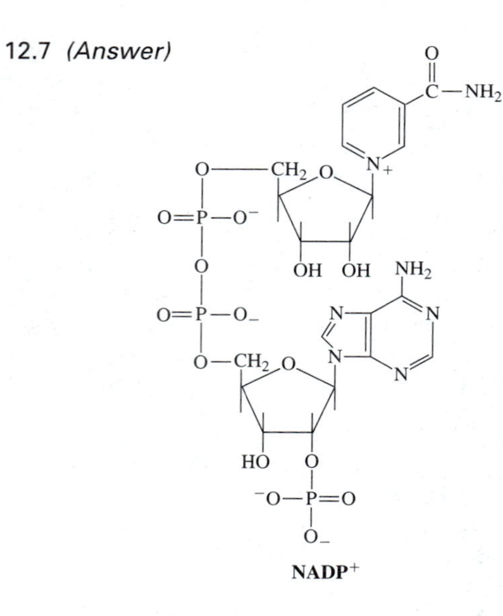

NADP⁺

12.9 *(Answer)*

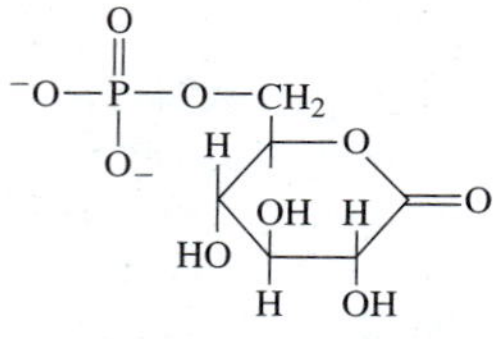

6-Phosphogluconolactone

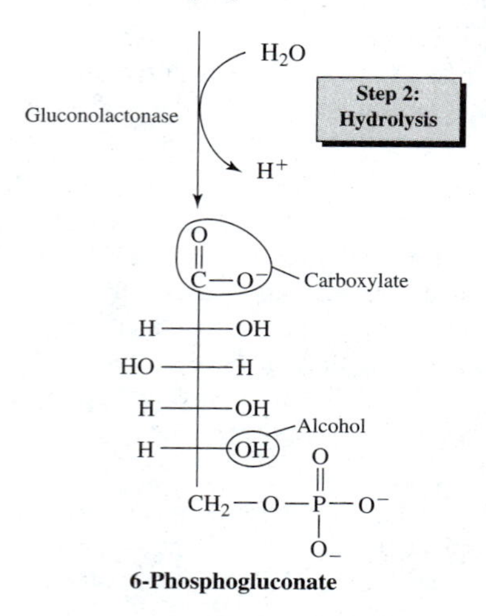

6-Phosphogluconate

12.10 Ribose 5-phosphate, ribulose 5-phosphate, and xylulose 5-phosphate. Ribulose 5-phosphate and xylulose 5-phosphate are diastereomers (non–mirror image stereoisomers). Both are structural isomers of ribose 5-phosphate.

12.11

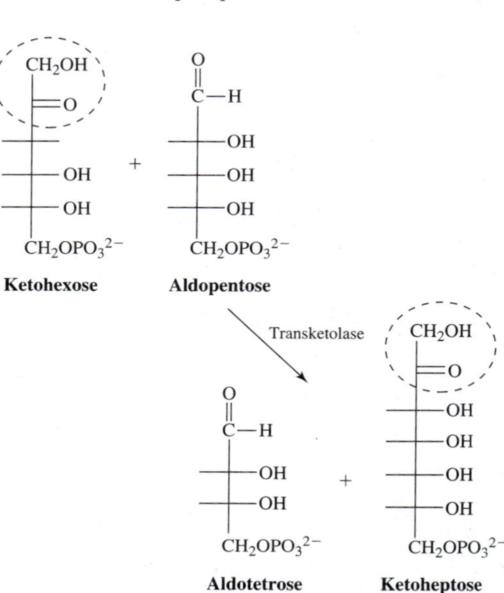

There are a number of possible answers. The transfer of a terminal carbonyl-containing two-carbon fragment from any 2-ketose to any aldose is an acceptable answer.

12.12 The 24 electrons removed from glucose during the pentose phosphate pathway–linked oxidation of glucose are initially acquired by $NADP^+$, which is converted to NADPH + H^+ in the process. These electrons ultimately end up in the products of the anabolic pathways that utilize the reducing power provided by NADPH.

12.13 See answer at the top of page A-55.

12.14 Under anaerobic conditions, all of the mitochondrial NAD^+ is converted to NADH; mitochondrial NAD^+ cannot be regenerated from NADH in the absence of O_2, because O_2 is the terminal acceptor of the electrons carried by NADH. The production of acetyl-SCoA from pyruvate requires NAD^+ (Study Figure 12.11).

12.15

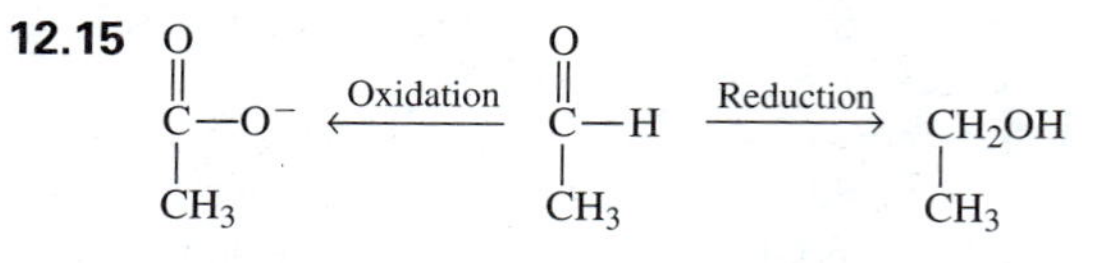

Oxidation of acetaldehyde yields acetic acid (acetate), whereas its reduction generates ethanol.

12.13 *(Answer)*

TABLE 12.1
Role of the Pentose Phosphate Pathway in the Oxidation of Glucose 6-Phosphate (Modified to show carbon balance)

Oxidative phase	12 $NADP^+$ + 6 H_2O + 6 glucose 6-phosphate ⟶ 12 NADPH + 12 H^+ + 6 CO_2 + 6 ribulose 5-phosphate glucose 6-phosphate: (6 × 6 = 36 carbons); CO_2: (6 × 1 = 6 carbons); ribulose 5-phosphate: (6 × 5 = 30 carbons)
Isomerization reactions	6 Ribulose 5-phosphate ⟶ 2 ribose 5-phosphate + 4 xylulose 5-phosphate (6 × 5 = 30 carbons) (2 × 5 = 10 carbons) (4 × 5 = 20 carbons)
Transaldolase and transketolase reactions	2 Ribose 5-phosphate + 4 xylulose 5-phosphate ⟶ 4 fructose 6-phosphate + 2 glyceraldehyde 3-phosphate (2 × 5 = 10 carbons) (4 × 5 = 20 carbons) (4 × 6 = 24 carbons) (2 × 3 = 6 carbons)
Part of gluconeogenesis	4 Fructose 6-phosphate + 2 glyceraldehyde 3-phosphate + H_2O ⟶ 5 glucose 6-phosphate + P_i (4 × 6 = 24 carbons) (2 × 3 = 6 carbons) (5 × 6 = 30 carbons)
Summary of the pentose phosphate cycle	12 $NADP^+$ + 7 H_2O + 6 glucose 6-phosphate ⟶ 12 NADPH + 12 H^+ + P_i + 5 glucose 6-phosphate + 6 CO_2 (6 × 6 = 36 carbons) (5 × 6 = 30 carbons) (6 × 1 = 6 carbons)
Net reaction	12 $NADP^+$ + 7 H_2O + glucose 6-phosphate ⟶ 12 NADPH + 12 H^+ + P_i + 6 CO_2 (1 × 6 = 6 carbons) (6 × 1 = 6 carbons)

12.16 True. A defective insulin gene, like the loss of pancreatic islet β-cells, can lead to a loss of biologically active insulin. A defective gene (a segment of DNA) normally leads to a defective gene product (usually a protein).

12.17 Too much insulin can lead to a dangerous drop in blood sugar levels by stimulating the uptake of too much glucose by adipocytes and muscle cells (Table 10.11). Severe hypoglycemia can lead to a coma and even death as brain cells run low on fuel and energy (Section 10.12).

CHAPTER 13

13.1

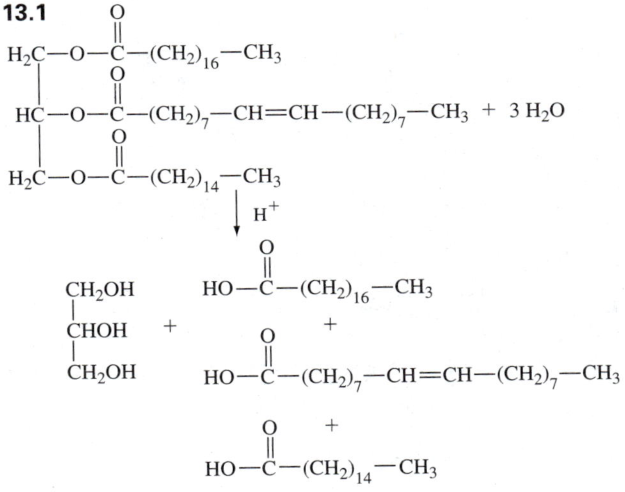

$$\begin{array}{l} H_2C-O-\overset{O}{\overset{\|}{C}}-(CH_2)_{16}-CH_3 \\ HC-O-\overset{O}{\overset{\|}{C}}-(CH_2)_7-CH{=}CH-(CH_2)_7-CH_3 \quad + \; 3\,H_2O \\ H_2C-O-\overset{O}{\overset{\|}{C}}-(CH_2)_{14}-CH_3 \end{array}$$

$$\downarrow H^+$$

$$\begin{array}{l} CH_2OH \\ CHOH \\ CH_2OH \end{array} + \begin{array}{l} HO-\overset{O}{\overset{\|}{C}}-(CH_2)_{16}-CH_3 \\ + \\ HO-\overset{O}{\overset{\|}{C}}-(CH_2)_7-CH{=}CH-(CH_2)_7-CH_3 \\ + \\ HO-\overset{O}{\overset{\|}{C}}-(CH_2)_{14}-CH_3 \end{array}$$

There are many possible answers because there are many unique triacylglycerol molecules. In the organic chemistry laboratory, the reaction can be base catalyzed instead of acid catalyzed. Salts of fatty acids, rather than free fatty acid, are obtained as products. The reaction is catalyzed by lipases within the human body.

13.2 $CH_3(CH_2)_{16}COOH + 26\ O_2 \longrightarrow 18\ CO_2 + 18\ H_2O + \text{energy}$

13.3 18.5. The oxidation process generates 6 NADH, 1 $FADH_2$, 2 ATP, and 1 GTP (an ATP equivalent). It consumes one ATP (Figure 13.2). When coupled to the respiratory chain, each NADH yields around 2.5 ATP and each $FADH_2$ about 1.5 ATP.

$$(6\ \text{NADH} \times 2.5) + (1\ \text{FADH}_2 \times 1.5) + 2 + 1 - 1 = 18.5\ \text{ATP (net)}$$

13.4 A carboxylic acid ester link joins the acyl group to carnitine in an acyl–carnitine molecule. The equilibrium constant for the carnitine acyltransferase I–catalyzed reaction is most likely of intermediate value since a thioester is cleaved as an oxyester is formed. $\Delta G°' \approx 30$ kJ/mol for the hydrolysis of a thioester (Section 8.5), whereas the corresponding figure for a typical oxyester is closer to 20 kJ/mol. Being more stable, oxyester formation is favored, but not to a large extent. The reaction is reversible. Neighboring functional groups, such as the ones in carnitine, can have a significant impact on ester stability. The information presented in this text does not allow you to assess that impact.

13.5 The additional alcohols are the D-isomer of L-3-hydroxyacyl-SCoA plus both the D and L isomers of 2-hydroxyacyl-SCoA:

$$\mathrm{R{-}CH_2{-}CH_2{-}\underset{\displaystyle }{\overset{\displaystyle OH}{\overset{|}{C}}}H{-}\overset{\displaystyle O}{\overset{\|}{C}}{-}SCoA}$$

2-Hydroxyacyl-SCoA
(can exist as D- and L-isomers)

The specificity of the active site on enoyl-SCoA hydratase accounts for the single product formed during β-oxidation.

13.6

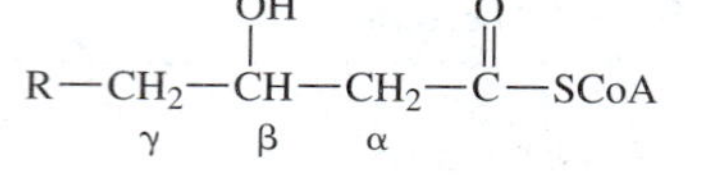

L-3-Hydroxyacyl-SCoA

13.7

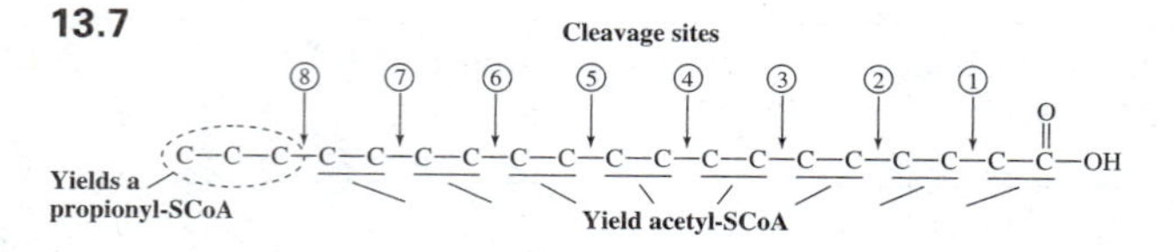

Eight rounds of β-oxidation release eight acetyl-SCoAs plus one propionyl-SCoA.

13.8 120 net ATP. When compared to palmitate oxidation (Table 13.1), stearate oxidation requires one additional round of β-oxidation and yields one additional acetyl-SCoA. The additional $FADH_2$ and NADH + H^+ yield 4 ATP, while the additional acetyl-SCoA yields 10 ATP when oxidized within the citric acid cycle–respiratory chain. Thus, the net yield of ATP is 14 greater for the oxidation of stearate than for the oxidation of palmitate; 106 + 14 = 120.

13.9 β-oxidation is involved mainly in the oxidation of fatty acids for energy production. When the energy charge of a cell is already high, as signaled by a high NADH concentration, β-oxidation should be shut down to conserve fuel molecules.

13.10 Both acetoacetic acid (pK_a = 3.62) and 3-hydroxybutyric acid (pK_a = 4.39) are weak acids, a fact predictable from the single carboxyl group present in each compound. Carboxylic acids are normally weak acids with pK_a values between 1 and 5. At physiological pH, acetoacetic acid and 3-hydroxybutyric acid exist predominantly as acetoacetate and β-hydroxybutryate (their dissociated forms), respectively. Hence, these ketone bodies are depicted in their dissociated forms in Figure 13.11.

13.11 Carbohydrate reserves are quickly depleted during starvation. The body then turns to fats (predominantly) and protein to meet its energy needs. When fat reserves are depleted, proteins become the predominant fuel. If starvation continues, death is close at hand.

13.12 pH = 2.31

$$\mathrm{CH_3{-}\overset{O}{\overset{\|}{C}}{-}CH_2{-}\overset{O}{\overset{\|}{C}}{-}OH \rightleftharpoons CH_3{-}\overset{O}{\overset{\|}{C}}{-}CH_2{-}\overset{O}{\overset{\|}{C}}{-}O^- + H^+}$$

$$K_a = \frac{\left[\mathrm{CH_3{-}\overset{O}{\overset{\|}{C}}{-}CH_2{-}\overset{O}{\overset{\|}{C}}{-}O^-}\right][\mathrm{H^+}]}{\left[\mathrm{CH_3{-}\overset{O}{\overset{\|}{C}}{-}CH_2{-}\overset{O}{\overset{\|}{C}}{-}OH}\right]} = 2.40 \times 10^{-4}$$

$pK_a = 3.62 = -\log K_a$ $\qquad K_a = \text{antilog} \ -3.62$

Let $[H^+] = x$. Since one $\mathrm{CH_3{-}\overset{O}{\overset{\|}{C}}{-}CH_2{-}\overset{O}{\overset{\|}{C}}{-}O^-}$ is formed for each H^+, $\left[\mathrm{CH_3{-}\overset{O}{\overset{\|}{C}}{-}CH_2{-}\overset{O}{\overset{\|}{C}}{-}O^-}\right] = x$.

$\left[\mathrm{CH_3{-}\overset{O}{\overset{\|}{C}}{-}CH_2{-}\overset{O}{\overset{\|}{C}}{-}OH}\right] = 0.100 - x$, since one $\mathrm{CH_3{-}\overset{O}{\overset{\|}{C}}{-}CH_2{-}\overset{O}{\overset{\|}{C}}{-}OH}$ is lost for each H^+ and $\mathrm{CH_3{-}\overset{O}{\overset{\|}{C}}{-}CH_2{-}\overset{O}{\overset{\|}{C}}{-}O^-}$ generated during dissociation. Therefore,

$$\frac{x \cdot x}{0.100 - x} = 2.40 \times 10^{-4}$$

Assume $0.100 - x \approx 0.100$:

$$\frac{x^2}{0.1} = 2.40 \times 10^{-4}$$

$$x^2 = 2.40 \times 10^{-5}$$

$$x = \sqrt{2.40 \times 10^{-5}} = [H^+] = 4.90 \times 10^{-3}\ M$$

$$\mathrm{pH} = -\log [H^+] = -\log (4.90 \times 10^{-3}) = 2.31$$

13.13

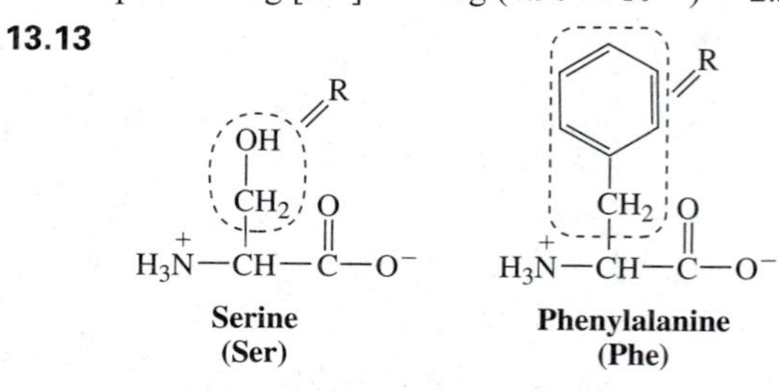

Serine (Ser) **Phenylalanine (Phe)**

13.14 True. Since glutamate is in a more reduced state than α-ketoglutarate, the conversion of glutamate to α-ketoglutarate entails oxidation and the reverse process involves reduction. The α-carbon in glutamate has an oxidation number of zero, whereas the oxidation number of the α-carbon in α-ketoglutarate is +2.

13.15

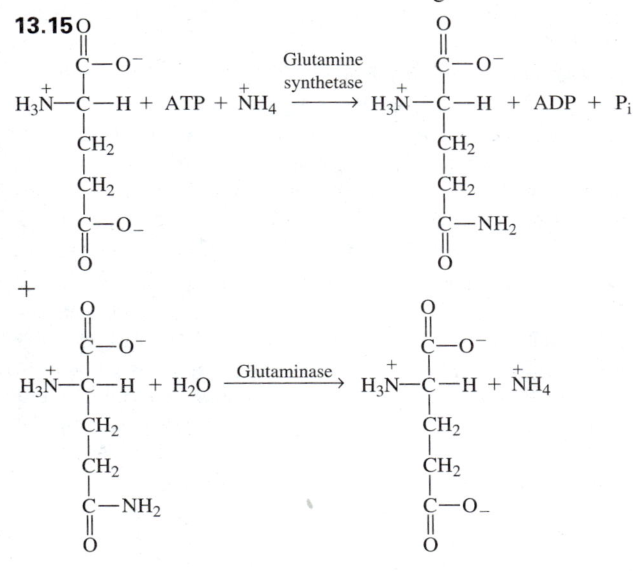

Net reaction: $ATP + H_2O \longrightarrow ADP + P_i$

This two-step reaction sequence would represent a waste of ATP if the two reactions occurred at the same location one right after the other. However, this is not the case. The first reaction occurs in extrahepatic tissue and the second occurs in the liver. The ATP is not wasted; it is involved in the conversion of $^+NH_4$ to a non-toxic form for transport.

13.16 Ammonium ($^+NH_4$) is an ion and a weak acid. Ammonia (NH_3) is its conjugate base:

$$^+NH_4 \rightleftharpoons NH_3 + H^+$$

The ratio of a conjugate base to its weak acid is determined by pH and the pK_a of the weak acid (Section 2.7):

$$pH = pK_a + \log \frac{[\text{conjugate base}]}{[\text{weak acid}]}$$

(The Henderson–Hasselbalch equation)

13.17 One of the nitrogens in urea is contributed by aspartate (see the argininosuccinate synthase–catalyzed reaction within the urea cycle). Aspartate can acquire its amino group from other amino acids through aminotransferase–catalyzed reactions. The amino group is never released as $^+NH_4$ along the way.

13.18 Restrict the dietary intake of proteins and amino acids. In practice, the feeding of α-keto acids produced from essential amino acids is also therapeutic. These α-keto acids scavenge amino groups that would otherwise be converted to $^+NH_4$. The essential amino acids generated in the process tend to be used by the body to build its own proteins.

13.19 12.5 mol ATP. Alanine is converted to pyruvate during the aminotransferase-catalyzed reaction. Pyruvate is oxidized with the catalytic assistance of the pyruvate dehydrogenase complex (generates 1 NADH) and the citric acid cycle (yields 3 NADH, 1 $FADH_2$, and 1 GTP). Each NADH yields approximately 2.5 ATP and each $FADH_2$ around 1.5 ATP when the electrons they acquired from pyruvate are fed to O_2 through the respiratory chain. GTP is an ATP equivalent.

$$\text{Total ATP} = (4\ \text{NADH} \times 2.5) + (1\ \text{FADH}_2 \times 1.5) + 1 = 12.5$$

13.20 Dihydropteridine reductase. When a cell is unable to catalyze the regeneration of tetrahydrobiopterin, it does not matter whether phenylalanine hydroxylase is active or not. Phenylalanine accumulates since phenylalanine hydroxylase lacks the tetrahydrobiopterin required for it to catalyze the oxidation of phenylalanine to tyrosine.

13.21 Six or seven. The minimum number of genes corresponds to the number of distinct enzymes involved (6 if dihydropteridine reductase is not included and 7 if it is included). Each different enzyme contains at least one unique polypeptide chain, and each unique polypeptide chain is normally encoded in a separate gene. The number of genes involved will be greater than the minimum if any enzyme contains multiple distinct polypeptide chains with each encoded in a separate gene. Restated, more than one gene is required to encode an enzyme when the enzyme contains more than one type of polypeptide chain.

CHAPTER 14

14.1 Short. Since the metabolic needs of the human body can change quickly, the alterations triggered by a hormone may be beneficial at one time but detrimental a short time later. Short half-lives allow the body to respond quickly to fluctuating needs and requirements.

14.2 Mutations can alter genes, and alterations in genes can lead to altered gene products. Gene products include enzymes, protein hormones, membrane receptors, immunoglobulins, and the rest of the body's proteins. These proteins play central roles in muscle action, brain action, immune function, and other processes that enable an organism to cope with environmental stresses and challenges. When alterations in gene products lead to changes in biological activities, brain action, muscle action, immune function, and other processes are compromised and coping ability is reduced.

14.3 Collagen is an extracellular support element in skin, tendon, cartilage, bone, blood vessels, teeth, and certain other components of the body. As collagen becomes cross-linked, skin, tendons, cartilage, bones, joints, and blood vessels become less flexible. It also becomes more difficult for nutrients and other materials to move in and out of cells as the cells become surrounded by cross-linked polymers, and it is more difficult for materials to enter and leave the blood through blood vessels. As blood vessels lose flexibility, there is an increased chance that a blood clot will clog an artery and lead to a stroke or a heart attack. Collagen cross-linking leads to other changes as well.

14.4 A high NAD^+/NADH ratio favors the oxidative transformation of lactate to pyruvate, because NAD^+ is the oxidizing agent for this reaction. The reverse reaction, which employs NADH as a reducing agent, is favored by a high NADH/NAD^+ ratio.

Under aerobic conditions, little lactate is formed because the NAD^+/NADH ratio is high. NADH that is formed tends to be converted rapidly to NAD^+ as it passes electrons to O_2 through the respiratory chain. The pyruvate to lactate reaction is the only source of lactate (Section 11.2)

14.5 Pyruvate carboxylase is a ligase, whereas PEPCK is a lyase. See Table 5.1.

14.6 Fructose-bisphosphatase is a hydrolase. See Table 5.1.

14.7 False. The more negative the $\Delta G°'$, the greater the thermodynamic driving force and the larger the equilibrium constant. $\Delta G°' = -RT \ln K_{eq}$ (Section 8.2).

14.8 The conversion of 2 glycerols to glucose yields +3 ATP net. The glycerol kinase–catalyzed reaction (occurs twice, once for each glycerol) consumes 2 ATP, whereas the glycerol phosphate dehydrogenase–catalyzed reaction yields 5 ATP when the 2 NADH produced pass electrons to O_2 through the respiratory chain. No ATP is generated or consumed during the production of glucose from dihydroxyacetone phosphate.

The conversion of 2 glycerol to 2 pyruvate yields +12 ATP net: +3 from the formation of 2 dihydroxyacetone phosphate and +9 ATP as the 2 dihydroxyacetone phosphates are converted to pyruvate (study Figure 14.3).

14.9 Active sites bind substrates and catalyze the transformation of the substrates to products. Allosteric sites (usually distinct from active sites) bind effectors. Once bound, the effectors modulate the catalytic activity of the enzyme by altering its conformation or shifting an equilibrium that exists between multiple forms of the enzyme.

14.10 Both ATP and NADH would most likely be positive effectors. When the energy charge of a cell is high, as signaled by either high ATP or high NADH concentrations, it makes sense to suppress glycolysis (generates energy) and use some of the high energy charge to synthesize fuel (glucose) for future use. The activation of fructose-bisphosphatase (a regulatory enzyme in gluconeogenesis) enhances fuel production.

14.11 It is released as heat. Recall that when oxidation and phosphorylation are uncoupled within the respiratory chain, virtually all of the energy of oxidation is converted to heat (see discussion on uncouplers and brown fat in Section 11.9). The substrate cycling–bumblebee story also supports this conclusion.

14.12 Under aerobic conditions, high NADH levels lead to high ATP levels as electrons are passed from the NADH to O_2 through the respiratory chain. When the respiratory chain is shut down in the absence of O_2, NADH from glycolysis is used to convert pyruvate to lactate. Although the NAD^+ produced in the process keeps glycolysis functional, relatively few ATP are generated.

14.13 Under the anaerobic conditions that exist in active muscle, the NAD^+ produced during the lactate dehydrogenase–catalyzed reaction is rapidly converted to NADH during glycolysis. The rate of glycolysis increases under anaerobic conditions, partially as a consequence of an increase in the concentration of positive allosteric effectors (including AMP) and a drop in the concentration of some negative modulators (including ATP).

14.14

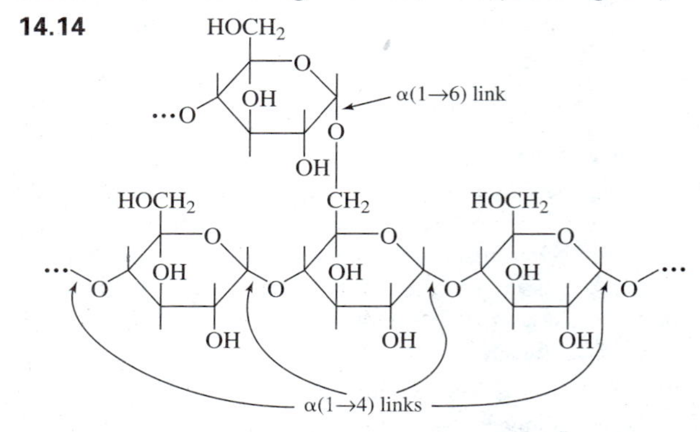

14.15 None. The would-be reducing end is attached to glycogenin through a glycosidic link. Consequently, none of the glucose residues exists in equilibrium with a readily oxidizable, open-chain aldehyde form. A substance that is readily oxidized is a reducing agent, because the compound that oxidizes it is reduced.

14.16 True. Each branch is attached to its "main chain" (which may itself be a branch) by a single $\alpha(1 \rightarrow 6)$ link.

14.17 184 net. Each of the 92 glucoses added to the octomer primer (92 + 8 = 100) must be converted to a UDP-glucose. For each UDP-glucose produced, three anhydride links are cleaved (assuming that the PP_i from the UDP-glucose pyrophosphorylase–catalyzed reaction is hydrolyzed) and one anhydride link is formed (when glucose 1-phosphate is joined to UMP). Thus, the net cleavage of anhydride links is 2 for each UDP-glucose generated: $2 \times 92 = 184$. The transfer of a glucose residue from UDP-glucose to a growing glycogen chain involves the cleavage of a phosphate ester link, not an anhydride link.

There are four enzymes involved directly in the synthesis: glucokinase (or hexokinase); phosphoglucomutase; UDP-glucose pyrophosphorylase, and glycogen synthase. The catalytic activity of glycogenin is not required for the synthesis, although it was involved in the production of the performed octomer. The enzyme count increases to five if one includes inorganic pyrophosphatase, the enzyme that catalyzes the hydrolysis of PP_i to "pull" the UDP-glucose pyrophosphorylase–catalyzed reaction (Figure 14.8).

14.18 No. The modulation of the activity of an enzyme has a similar impact on the rate of both the forward and the reverse reaction; it cannot enhance the forward reaction while suppressing the reverse reaction or vice versa.

14.19 Most of the ATP used for anabolism is generated as electrons from NADH are passed to O_2 through the respiratory chain.

14.20 Acetyl-SCoA carboxylase is a ligase. See Table 5.1.

14.21 See answer at bottom of page.

14.22 $CH_3CH_2CH_2CH_2CH_2CH_2CH_2CH_2CH_2CO$—. A butyryl group ($C_4$-acyl-ACP) is produced during the first cycle (Figure 14.13). Each additional cycle adds two more carbons to the alkane chain.

14.23 Seven malonyl-SCoAs. A single malonyl-SCoA is required to produce C_4-acyl-SACP. An additional malonyl-SCoA is required for each additional two-carbon fragment added. Palmitate has 16 carbons. See Figure 14.13.

14.24 Seven different enzymes. Each enzyme is identified with a separate name in Figure 14.13. Each enzyme, with the exception of [ACP] acetyltransferase, must function repeatedly during palmitoyl-SACP biosynthesis.

14.25 Metabolic channeling by fatty acid synthase is partly responsible. In addition, the enzymes for β-oxidation are confined to mitochondria whereas fatty acid biosynthesis occurs in the cytosol. Furthermore, the enzyme that initiates β-oxidation uses acyl-SCoA as a substrate, not acyl-SACP.

14.26 See answer at top of next column.

14.26 *(Answer)*

$$CH_3CH_2\overset{O}{\overset{\|}{C}}CH_2\overset{O}{\overset{\|}{C}}—$$

From propionyl-SACP (CH_3CH_2C); From malonyl-SACP (CH_2C—)

14.27 The maximum increase in rate is determined by the rate initially and the maximum rate for the enzyme catalyzed reaction (attained when substrate is saturating and the enzyme is in its most active form). The maximum increase in rate is the difference between these two values.

14.28 Glucagon and epinephrine both inhibit glycogenesis in the liver, and epinephrine inhibits glycogenesis in muscle as well. Both glucagon and epinephrine stimulate gluconeogenesis in the liver. Figure 14.20 summarizes the effect of these hormones on several central metabolic pathways.

14.29 27

P=palmitoyl
L=linoleyl
S=steryl

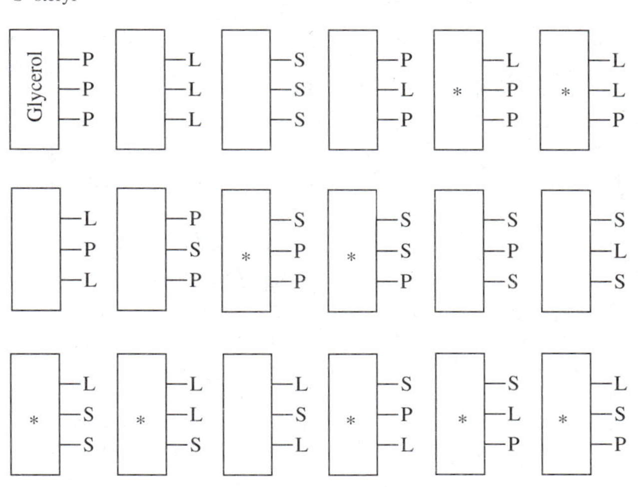

The middle carbon in glycerol is a chiral center in those triacylglycerols marked with asterisks. When this is the case, both a

14.21 *(Answer)*

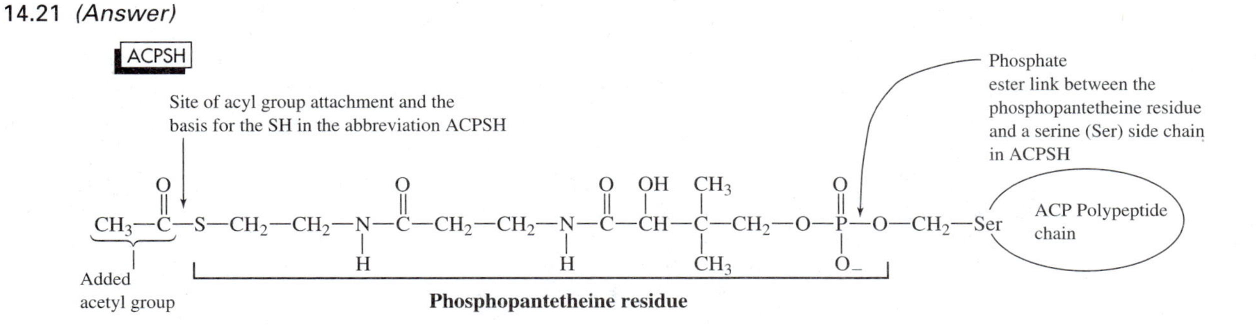

D- and an L-isomer (not shown in the schematic diagrams) can exist. Thus, the triacylglycerols marked with an asterisk must be counted twice.

14.30 The chiral intermediates are glycerol 3-phosphate, lysophosphatidates, phosphatidates, and 1,2-diacylglycerols. A triacylglycerol is, itself, chiral in some instances (see answer to Problem 14.29).

14.31 Most enzymes contain at least one copy of each of the 20 common protein amino acids. If any one of these amino acids (either essential or nonessential) is missing within a cell, most enzymes cannot be completely assembled and the cell will die. The "essential" and "nonessential" designations relate to nutritional requirements that are a consequence of the body's inability to synthesize some amino acids. From the standpoint of polypeptide synthesis, the "nonessential" amino acids are just as essential as the "essential" ones.

14.32 Bacteria living in the intestine (Section 1.3) are able to synthesize the essential amino acids. They supply the body with small amounts of these nutrients.

14.33 True. Every step in purine biosynthesis and most other anabolic pathways is catalyzed by a specific enzyme. Hence, every compound that contributes atoms during purine assembly must serve as a substrate for one or more of the enzymes involved.

14.34 Yes. The reactions are reversible since one anhydride link is formed as a second anhydride link is cleaved; reactants and products have a similar free energy content (see Section 8.2).

14.35 An "activated monomer" is one whose transfer (from the carrier to which it is attached) to a growing polymer is thermodynamically favored over the attachment of a free monomer unit to the same polymer. An activated monomer possesses more free energy than a free monomer. Monomers are activated to provide a thermodynamic driving force for polymer formation. A reaction is favored (has a large equilibrium constant) whenever the reactants have more free energy than the products; $\Delta G°' = -RT \ln K_{eq}$ (Section 8.2). UDP-glucose contains the activated monomer (glucose) involved in the production of glycogen (Section 14.5).

14.36

$O=C(-NH_2)-CH_2-CH_2-CH(-\overset{+}{N}H_3)-C(=O)-O^-$

NH_2 ← This must be the NH_2 transferred because glutamine is converted to glutamate during CTP production.

Glutamine

14.37 Yes. NADPH is normally used to provide reducing power for anabolism, while NADH carries electrons to the respiratory chain. Within the respiratory chain, the flow of electrons from NADH to O_2 is coupled to the production of ATP (see Table 14.5).

14.38 No. Cells that are dividing rapidly and continuously replicating DNA are most susceptible. A cell that is not replicating DNA is not susceptible to damage by drugs that selectively block DNA replication.

14.39 FdUMP, which is negatively charged at physiological pH, is unable to pass through biological membranes.

14.40 Methotrexate binds to the dihydrofolate binding site on dihydrofolate reductase. Since methotrexate is structurally very similar to dihydrofolate (see Figures 14.29 and 14.23), it is mistaken for the normal substrate of the enzyme; a classic example of competitive inhibition.

14.41 A possible new drug:

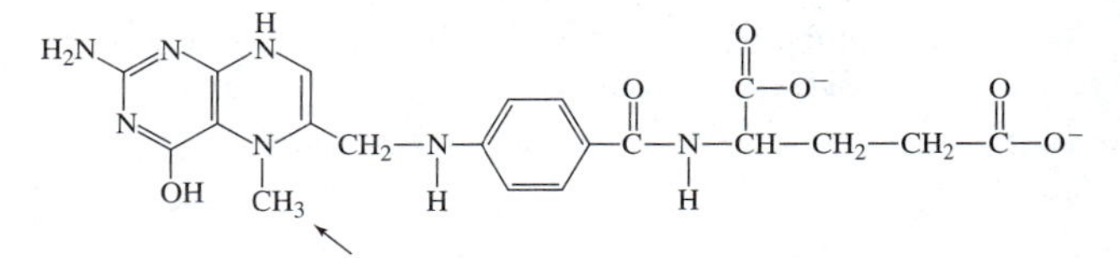

There are many possible answers. Any compound structurally very similar to dihydrofolate but possessing distinct chemical properties could be considered a potential drug.

CHAPTER 15

15.1 Extremely low-frequency electromagnetic fields, radio waves, microwaves, X-rays, and γ-rays (Table 3.11). Gamma (γ) rays possess the shortest wavelengths and the largest amount of energy.

15.2 For a compound to absorb visible light, it must possess "allowed" energy levels whose difference in energy content corresponds to the energy in a photon of visible light. The energy levels in many compounds do not satisfy this requirement. A compound that does not absorb visible light will reflect or transmit this light and will appear white if viewed in sunlight.

15.3 $A = \epsilon bc = 0.8\ \text{Lmol}^{-1}\text{cm}^{-1} \times 0.1\ \text{molL}^{-1} \times 1\ \text{cm} = 0.08$

15.4 A comparison of Figures 15.3 and 15.5 reveals that phycoerythrobilin is best at absorbing green wavelengths (around 550 nm), whereas phycocyanobilin is best at harvesting orange wavelengths (around 600 to 640 nm). The concentration assumption is important because, at high enough concentration, a solution of any of the pigments would absorb more green light than a very dilute solution of phycoerythrobilin; absorbance is proportional to concentration (Beer's law).

15.5 See answer at top of page A-61.
At each wavelength, one simply adds the absorbance of each of the two dyes. The absorbance of the mixture will only differ from the absorbance of the individual dyes at those wavelengths where both dyes absorb.

15.5 *(Answer)*

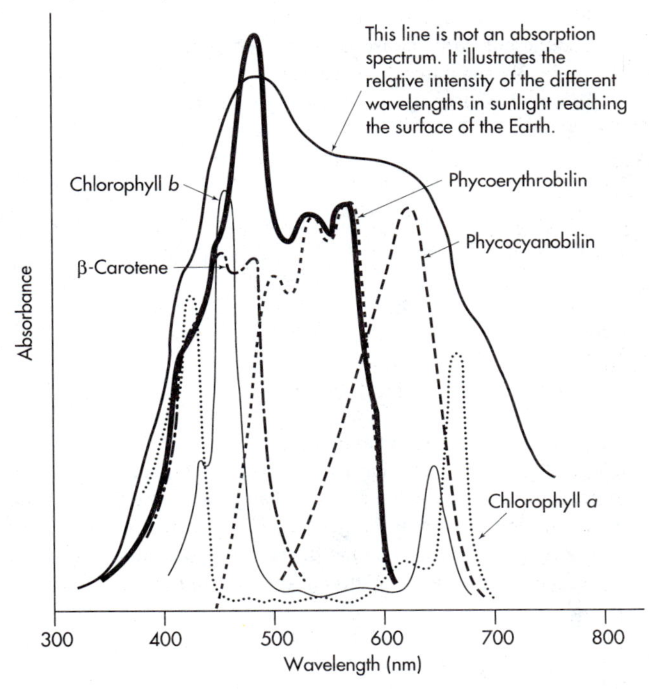

15.6 The term "dark reaction" implies that a reaction occurs only in the dark. However, the dark reactions of photosynthesis proceed in the presence of light and are normally shut down when a plant moves from the light into the dark. These reactions do not depend on light; given enough ATP and NADPH, they can proceed in the dark. In contrast, the light reactions are photoreactions that cannot proceed in the absence of light; they require the absorption of a photon.

15.7 True. In an uncharged molecule, the number of protons (1+ charge each) equals the number of electrons (1− charge each). If an electron is lost, a net positive charge results. If an electron is acquired, a negative ion is created.

15.8 The standard reduction potential is a quantitative measure of the electron affinity of an electron acceptor in a half-reaction (Section 8.8). A positive reduction potential means that, under standard conditions, an acceptor has a greater electron affinity than H^+, an internationally agreed upon "reference" acceptor. A negative potential signals that, under standard conditions, an electron acceptor will hold onto an accepted electron less tightly than H^+ will. Thus, the lower the reduction potential, the more readily an acceptor will part with an electron it has accepted.

An excited molecule will part with an electron more readily than a ground-state molecule, because it contains an electron that has been pulled partially away from the positive nucleus during the process of excitation. Because P680* is the excited state of P680, it will relinquish an electron more readily than will the P680 ground state. Thus, P680* has a lower reduction potential. Viewed from a different perspective, because P680 binds electrons more tightly than does P680*, P680 has a higher reduction potential.

15.9 True. Oxidation corresponds to a loss of electrons, and a low reduction potential indicates that a compound readily gives up an electron. A low reduction potential (a low potential for being reduced [accepting an electron]) is always associated with a high oxidation potential.

15.10 Although ATP synthase is capable of operating in the dark, the proton gradient used by this enzyme is created and maintained by the light reactions. The proton-motive force that drives ATP production is quickly lost when a chloroplast is deprived of light.

15.11 A separate photon is required to move an electron through each photosystem. Both P680 and P700 must absorb a photon before they can pass an electron they have accepted to the next electron carrier in the photosynthetic electron transport system.

15.12 In a solution of purified chlorophyll, chlorophyll molecules lack properly positioned neighbors with appropriate reduction potentials that can serve as electron acceptors. Therefore, the energy of excitation is lost through alternate mechanisms, primarily fluorescence and conversion to heat.

15.13 The new carrier will fall between A_0 and A_1 in the Z-scheme, since the spontaneous flow of electrons is always from compounds with lower reduction potentials to compounds with higher reduction potentials. If the new carrier were positioned between any other carriers, this requirement would not be met.

15.14 True. The transfer of an electron corresponds to the transfer of a negative charge (see Problem 15.7). The transfer of a negative charge to an anion will create a new anion with a charge one unit more negative than the original anion. The transfer of a negative charge to an uncharged molecule will generate a 1− ion. The transfer of a negative charge to a cation will reduce the positive charge on the cation by one unit; if the cation is a 1+ ion, it will become an uncharged molecule. The molecules or ions that donate electrons will undergo related changes in charge.

15.15 No. The cyclic electron transport scheme is of value because it contributes to the proton gradient used to produce ATP. The cytochrome *bf* complex contains the only proton pump within the cyclic electron transport system. If this component of the cycle is bypassed, the cycle becomes futile.

15.16 When electron transport is blocked, electron acceptors behind (in the direction of the electron source) the block become saturated with electrons and can no longer serve as acceptors. When there is no acceptor for an electron from an excited state chlorophyll (P680* or P700*), chlorophyll must release its excitation energy through some other mechanism. This energy is commonly lost as fluorescence. See Problem 15.12.

15.17 (a) The water splitting reaction releases H^+ from water into the lumen. (b) Protons are pumped into the lumen as electrons flow through the cytochrome *bf* complex. (c) The ferredoxin–$NADP^+$ oxidoreductase–catalyzed production of NADPH removes H^+ from the stroma.

15.18 It takes fewer protons to change the proton concentration in a small-volume solution than in a large-volume solution. One hundredth mole of H^+ in 10 mL yields a 1 *M* solution. One hundredth mole of H^+ in one liter yields a 0.01 *M* solution.

15.19

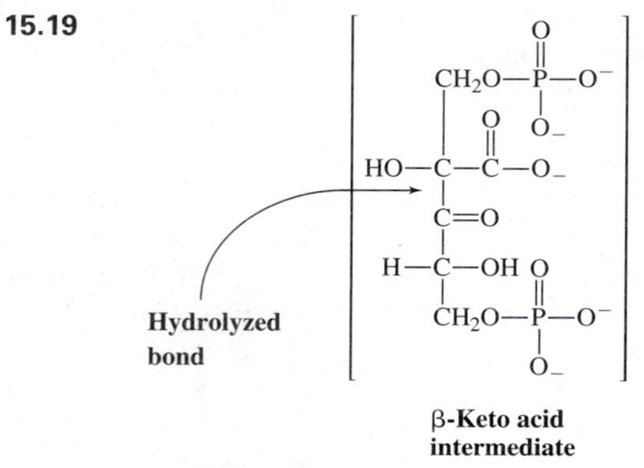

15.20 3-Phosphoglycerate + ATP + NADPH + H^+

$$\rightleftharpoons$$

glyceraldehyde 3-phosphate + ADP + P_i + $NADP^+$

A minimum of two electrons must flow from water to $NADP^+$ through the electron transport system. This is the number of electrons required to convert one $NADP^+$ to NADPH. It is also the number of electrons thought to be required to generate a proton-motive force capable of producing 1 to 2 ATP. Four photons are required to move two electrons through the transport system. For each electron that traverses the entire electron transport system, each of the two photosystems must absorb one photon.

15.21 False. Epimerases do catalyze the interconversion of epimers, but such reactions are classified as isomerization reactions, not group transfer reactions. Epimers are monosaccharides that differ only in the configuration about a single chiral center. Xylulose 5-phosphate and ribulose 5-phosphate differ only in the configuration about C3.

15.22 The two P_i's are from hydrolysis reactions catalyzed by fructose-bisphosphatase and sedoheptulose-1,7-bisphosphatase (Figure 15.21). The phosphoryl groups from the ATPs are transferred to ribulose 5-phosphates to produce ribulose 1,5-bisphosphates.

15.23

Triose phosphate isomerase; an isomerase
Aldolase; a lyase
Transketolase; a transferase
Fructose-bisphosphatase (fructose-1,6-bisphosphatase); a hydrolase
Ribulose-5-phosphate 3-epimerase; an isomerase
Phosphoribulokinase; a transferase
Sedoheptulose-bisphosphatase (sedoheptulose-1,7-bisphosphatase); a hydrolase
Ribose-5-phosphate isomerase; an isomerase

15.24

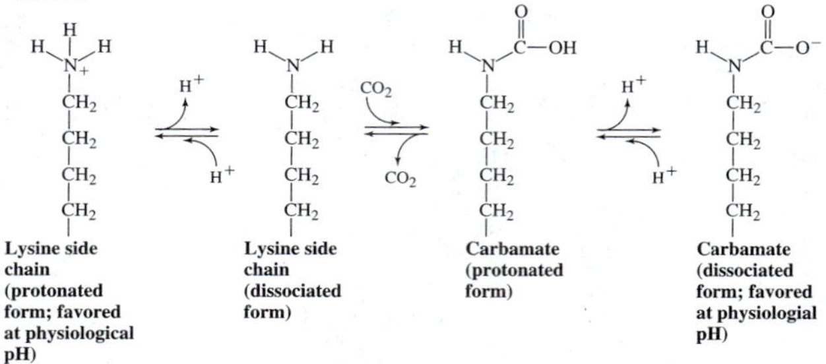

Carbamylation changes an amino group with a 1+ charge at physiological pH values to a carbamate with a 1− charge (at physiological pH values). Such a marked change normally has a drastic impact on how the altered functional group interacts with neighboring functional groups within the tertiary structure of a protein. Consequently, the conformation of the protein is modified.

15.25

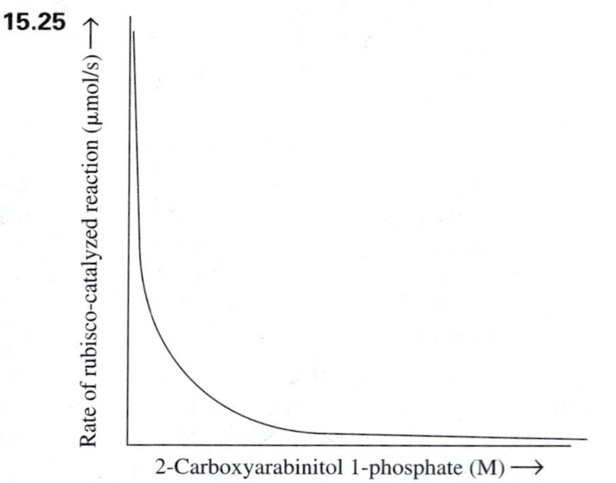

The curve obtained is an upside down hyperbola. The 2-carboxyarabinitol 1-phosphate gradually wins the competition with ribulose 1,5-bisphosphate for the enzyme. At extremely high inhibitor concentrations, rubisco will be saturated with inhibitor and will be inactive.

15.26

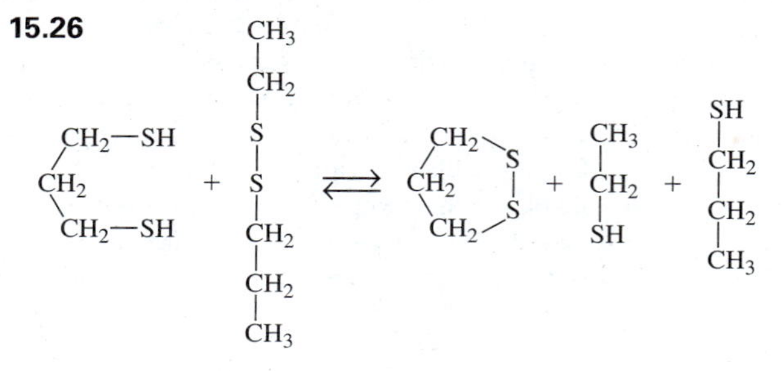

15.27 H^+, Mg^{2+}, 2-carboxyarabinitol 1-phosphate, CO_2, rubisco activase. The fact that an increase in pH activates rubisco indicates that H^+ interacts with this enzyme. Each enzyme tends to have a specific H^+ concentration at which it is most active (Section 5.5). CO_2 is both a modulator and a substrate. As a modulator, it forms a carbamate with a lysine side chain. As a substrate, it binds to the active site and is incorporated into 3-phosphoglycerate. The CO_2 residue attached to the lysine side chain remains attached to this side chain during rubisco-catalyzed reactions.

15.28 The catalytic activity of chloroplast fructose-bisphosphatase is regulated by pH, [Mg^+], and disulfide exchange.

15.29 False. Both of these metabolic pathways are involved in building carbohydrates. A futile cycle encompasses one pathway that degrades a compound at the same time a second pathway produces it. The energy released during the degradation is required for the synthesis. Nothing is gained.

15.30 The featured themes are communication, specificity, and catalysis and the regulation of reaction rates. The loss of the light-induced reactions is communicated to the carbon fixation reactions through a variety of mechanisms. This communication leads to changes in the rates of specific reactions.

15.31 20% [(2823 kJ/14,400 kJ) × 100]. The approximate input of light energy with 400 nm light is 14,400 kJ (48 einstein × 300 kJ/einstein).

15.32 Photosystem I-linked cyclic electron transport is the most likely source of the ATP. Starch is only assembled in the presence of light when precursors are being made through photosynthesis. Under these conditions, both cyclic and noncyclic electron transport generates ATP within the same intracellular compartments where starch is being synthesized. The use of ATP from oxidative phosphorylation would create a futile cycle.

15.33 The energy required is the same. The mechanism of incorporation is very similar in both instances. Study Figures 15.27 and 15.28.

15.34 True. Because the two molecules are identical and both are released at exactly the same site within a cell, neighboring molecules, including enzymes, could not distinguish them.

15.35 True. Nocturnal inhibitor inhibits the binding of ribulose 1,5-bisphosphate to rubisco, the enzyme whose catalytic activities are responsible for initiating both photorespiration and the carbon fixation reactions of photosynthesis. Because ribulose 1,5-bisphosphate is a substrate for both the oxygenase-catalyzed reaction and the carboxylase-catalyzed reaction, both reactions come to a standstill when this substrate is unable to bind rubisco.

15.36 No. The substrate of PEP carboxylase is HCO_3^- not CO_2. It is highly unlikely that the HCO_3^- binding site on PEP carboxylase would also have a significant affinity for O_2. O_2 is a nonpolar molecule while HCO_3^- is a polar charged molecule.

15.37 $H_2O + ATP \rightleftharpoons AMP + PP_i$. The PP_i is hydrolyzed in a separate reaction to "pull" the cycle. The hydrolysis of the ATP and PP_i does not represent wasted energy, because it is used to help deliver CO_2 to the bundle-sheath cells and to create (in the process) a locally high CO_2 concentration that favors carbon fixation over photorespiration.

15.38 N_2 is relatively unreactive as a consequence of a large activation energy attributed to the very strong triple bond between the two nitrogens. Catalysts, including enzymes, lower activation energies for reactions. The rate of a reaction is not determined by its $\Delta G°'$. Rate is determined primarily by activation energy, collision frequency, collision force and reactant orientation at collision. These facts lie at the heart of catalysis and the regulation of reaction rates, a central biochemical theme.

15.39 Leghemoglobin binds O_2 more tightly than does mammalian hemoglobin, and it binds O_2 noncooperatively. Extremely tight binding between leghemoglobin and O_2 is required to reduce O_2 concentrations to levels that do not significantly inhibit N_2 fixation. If human hemoglobin bound O_2 as tightly, this protein would pull O_2 out of tissues already low in O_2. With a single protein chain and a single heme group, leghemoglobin possesses a single O_2 binding site. Cooperative binding requires multiple binding sites.

15.40 The aspartate used for purine biosynthesis acquires its nitrogen from glutamate through a transaminase (aminotransferase)-catalyzed reaction:

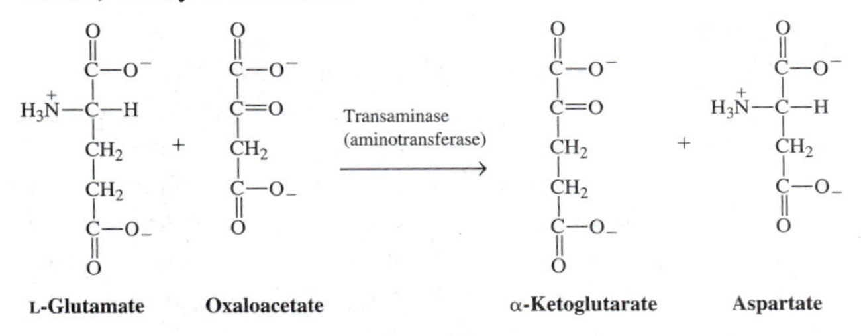

CHAPTER 16

16.1 "Gene product" refers to that substance (either an RNA or polypeptide) which is ultimately produced from the information carried within a gene. Most gene products are polypeptides.

16.2 Since hemoglobin contains two different polypeptide chains (Section 4.9), two genes must be expressed, one for each distinct polypeptide. Although a hemoglobin molecule possesses four polypeptide chains, the two α-chains are encoded in the same gene and the two β-chains in a single separate gene. A single gene can be repeatedly expressed to produce many copies of its product. In general, each unique polypeptide is encoded in a separate gene.

16.3 A 3-methyl group on cytosine prevents the formation of a hydrogen bond, creates like-charge repulsion, and sterically interferes with G:C pairing. Hence, 3-methylcytosine will not pair with guanine.

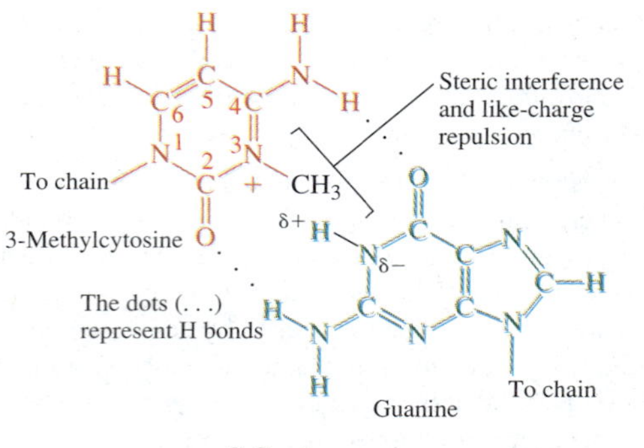

C:G pair

16.4 **a.** 3′UAGAUUAGCGUAA5′
b. 3′GCCCAAGCAACCACACUGC5′
c. 3′UUUGAGCGGGAACACCCGUUUGGGGGCCA CACG5′

Each A in one strand pairs with a U or a T in its complementary strand. The base G is always paired with a C. The complements given above contain U rather than T since the problem specified that the complements should be poly**ribo**nucleotides not poly**deoxyribo**nucleotides. It is important that the directionality of the complements be indicated; complementary base pairing requires that the paired strands be antiparallel.

16.5 See answer in the next column.

16.6 There are 17 base pairs depicted in the A-DNA fragment and 14 base pairs shown in the B-DNA fragment. This is best determined by counting the total number of bases (identified with an A, T, C, or G in Figure 16.2) and dividing by 2. Although most base pairs are readily discernible, some are difficult to pick out. The pairs at the crossover sites in A-DNA are particularly difficult to identify. The answer to Problem 16.5 pinpoints some of the specific base pairs in the B-DNA fragment.

16.7 Persons with certain autoimmune diseases possess antibodies against Z-DNA. The presence of a specific antibody indicates a prior exposure to a specific antigen (Z-DNA in this case).

16.8 False. The diagram of the space-filling model of B-DNA (Figure 16.4) clearly reveals that there is not even enough room for a single hydrogen atom between normally stacked base pairs. Molecules can slip between the stacked bases during the breathing process, but such intercalation disrupts the normal geometry of the double helix.

16.9 Counterclockwise twisting would decrease the linking number of Z-DNA, since it is a left-handed helix (B-DNA is right handed). The counterclockwise twisting of a left-handed helix reduces the number of times the two strands wrap around one another and tends to unwind the helix.

16.5 *(Answer)*

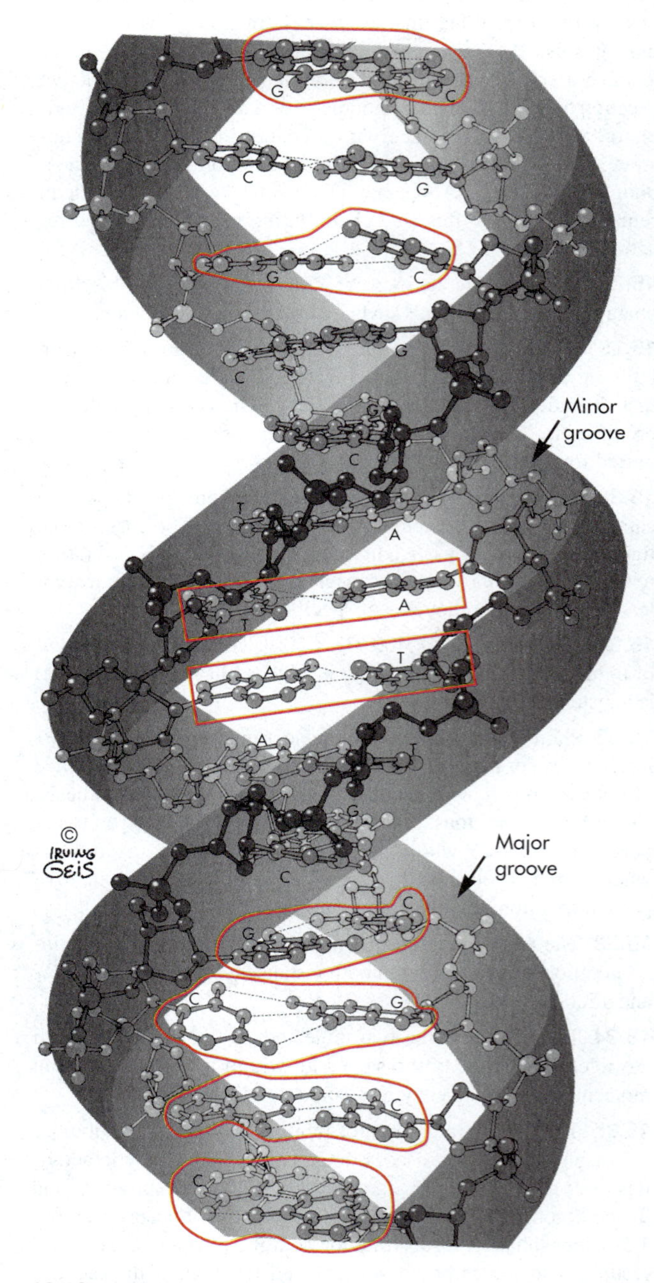

16.10 A higher temperature is required to unwind relaxed B-DNA. Negative supercoiling tends to unwind DNA and, in the process, weaken the interactions between complementary strands. With weakened interactions, less heat energy is required to separate the two strands. Chapter 17 explains how negative supercoiling is employed to help achieve strand separation during DNA replication.

16.11 Micrococcal nuclease is an endonuclease (it catalyzes the hydrolysis of internal phosphodiester bonds) and a DNAse (it catalyzes the hydrolysis of DNA).

16.12 The proteins in the core particles block nuclease access to the DNA; DNA bound to histones does not fit very well into the active sites of nucleases.

16.13

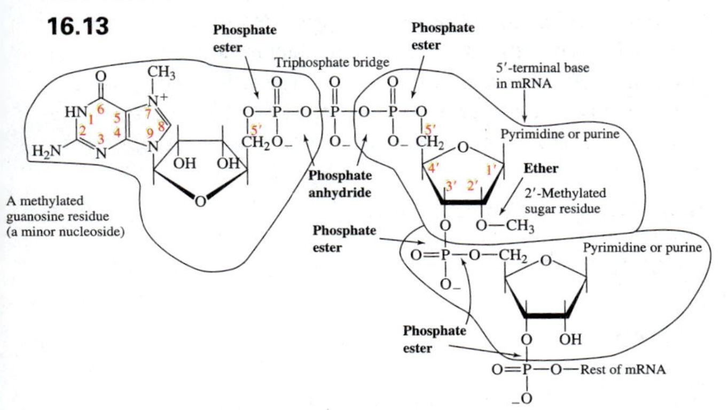

16.14 Since three separate nucleotides are required to code for each of the 527 amino acids, 1581 (3 × 527) nucleotides exist within the translated region of the mRNA. The answer is 1584 if we consider the termination codon part of the translated region. An mRNA contains a 5′-UTR and a 3′-UTR in addition to the translated region. These untranslated regions, which are thought to play a role in translation and other processes, may add a large number of nucleotides to the mRNA. Most eukaryotic mRNAs also contain a 3′-poly-A tail with 50 to 200 AMP residues.

16.15

(a) 5′ C–G–U–C–C–C–G–G–A–C–G–C–A–A–A 3′
N-Terminal Arg – Pro – Gly – Arg – Lys *C*-Terminal

(b) 5′ C–C–G–U–A–U–G–C–A–G–A–A–A–A–C–A–C–A–U–A–A–U–G–U 3′
N-Terminal Pro – Tyr – Ala – Glu – Asn – Thr (*C*-Terminal) — Termination codon not translated

(c) 5′ G–G–G–G–G–C–G–C–G–C–G–G–C–G–C–C–C–C–G–C–G 3′
N-Terminal Gly – Gly – Ala – Arg – Arg – Pro – Ala *C*-Terminal

The codons in each mRNA fragment are read consecutively in a 5′ to 3′ direction. Each codon, with the exception of the 3 termination codons, specifies one of the 20 common protein amino acids (Table 16.6). The ordering of codons along the translated region of an mRNA determines the ordering of amino acids within the encoded peptide. A peptide is assembled in an N-terminal to C-terminal direction.

16.16

(a) 5′ C–G–U–C–C–C–G–G–A–C–G–C–A–A–A 3′
N-Terminal Val – Pro – Asp – Ala *C*-Terminal

(b) 5′ C–C–G–U–A–U–G–C–A–G–A–A–A–A–C–A–C–A–U–A–A–U–G–U 3′
N-Terminal Arg – Met – Gln – Lys – Thr – His – Asn *C*-Terminal

(c) 5′ G–G–G–G–G–C–G–C–G–C–G–G–C–G–C–C–C–C–G–C–G 3′
N-Terminal Gly – Ala – Arg – Gly – Ala – Pro *C*-Terminal

A comparison of these answers to the answers to Problem 16.15 reveals the importance of the reading frame. A shift in reading frame normally has a marked impact on the amino acid sequence within the polypeptide encoded in an mRNA.

16.17 5′ dT–C–T–C–C–G–A–C–A–G–G–T–G–A–G 3′

↓ Transcription

3′ A–G–A–G–G–C–U–G–U–C–C–A–C–U–C 5′

mRNA fragment

The mRNA fragment is redrawn with its 5′ end to the left and then read:

5′ C–U–C–A–C–C–U–G–U–C–G–G–A–G–A 3′

***N*-Terminal** Leu – Thr – Cys – Arg – Arg ***C*-Terminal**

The expression of the information in the DNA strand entails transcription followed by translation. During transcription, an mRNA fragment is produced that is complementary and antiparallel to the DNA template. The amino acid sequence within the peptide encoded in the mRNA fragment is read from the genetic code table (Table 16.6) after the mRNA fragment is redrawn with its 5′ end to the left and 3′ end to the right; mRNAs are always read in a 5′ to 3′ direction.

16.18 The needs and activities of a cell are constantly changing as O_2 concentrations fluctuate, as nutrients are acquired, as environmental conditions are modified, and so on. Because the enzymes in a cell primarily determine what reactions occur within it, certain new enzymes must be assembled and some existing enzymes must be eliminated or inactivated in order to alter ongoing reactions in response to these changing needs and activities. The construction of novel enzymes involves the production of novel mRNAs; to synthesize a new protein, a cell must first transcribe the gene that encodes it. To eliminate an enzyme, a cell normally degrades existing copies of the enzyme and its mRNA(s) (multiple mRNAs will be involved in the production of an enzyme if an enzyme contains two or more different polypeptides) and shuts down the transcription of its gene(s). Thus, specific mRNAs come and go as genes are turned on and off in response to the immediate requirements of a cell.

16.19 Leucine. A tRNA always carries that amino acid specified by the codon to which it binds through anticodon:codon pairing. The anticodon would be 3′GAA5′.

16.20 No. There are multiple codons for valine and, in general, each codon pairs with a distinct anticodon. The wobble hypothesis (Section 18.5) explains why there are some common exceptions to this generalization.

16.21 Yes. Secondary structure is determined by the size and the location of the complementary segments along an RNA chain, not by the specific base sequences within these complementary segments. RNAs with totally different nucleotide sequences could possess complementary segments of exactly the same length that are located in exactly the same positions along a polymer chain. Viewed from a slightly different perspective, the replacement of a base pair in a stem region of an RNA with another base pair alters the primary structure, but not the secondary structure, of this RNA.

16.22 Aminoacyl–tRNA synthetases are ligases, since they catalyze the joining of two molecules with the concomitant cleavage of ATP (see Table 5.1).

16.23 One anhydride bond is cleaved during the conversion of ATP, an amino acid, and a tRNA to an aminoacyl–tRNA, AMP, and PP_i. The pyrophosphate produced is subsequently hydrolyzed to yield two phosphates, an event that entails the cleavage of a second anhydride link. The coupled cleavage of ATP provides a thermodynamic driving force for the initial reaction. The hydrolytic removal of the PP_i helps pull the reaction to completion (Le Châtelier's principle).

16.24 tRNA + amino acid $\rightleftharpoons$ aminoacyl–tRNA + H_2O

$$+ \qquad H_2O + \text{ATP} \rightleftharpoons \text{AMP} + PP_i$$

$$\text{tRNA} + \text{amino acid} + \text{ATP} \rightleftharpoons \text{aminoacyl–tRNA} + \text{AMP} + PP_i$$

16.25 Because it takes a chiral catalyst to distinguish chiral reactants, aminoacyl–tRNA synthetases must have chiral active sites. Because an amino acid must be attached to a tRNA before it is added to a growing polypeptide chain, only L-amino acids are used to build proteins.

16.26 An intact prokaryotic ribosome contains around 58 molecules, close to 55 polypeptides, plus 3 rRNAs (Table 16.8). The 60S subunit of a eukaryotic ribosome contains about 52 molecules, close to 49 polypeptides, plus 3 rRNAs (Table 16.9).

16.27 Because ribosomes are composed of RNAs and polypeptides, complete hydrolysis yields ribose, phosphate, purines (adenine, guanine, and some minor bases), pyrimidines (uracil, cytosine, and some minor bases), NH_3 (from asparagine and glutamine), and a mixture of free amino acids. Although not mentioned to this point in the text, some of the ribosomal polypeptides are subjected to covalent modifications that may lead to additional hydrolysis products as well.

16.28 The globular protein will sediment more rapidly and have a large S value, because it is more compact and there is less friction associated with its sedimentation.

16.29 Denaturing conditions for nucleic acids include extremes in pH, high temperatures, organic solvents, high salt concentrations, and detergents. These conditions disrupt the noncovalent H bonds and hydrophobic forces that maintain the base-paired regions.

16.30 True. Because all RNAs are produced through transcription, each RNA within a cell is complementary to that segment of DNA that served as a template for its production. The posttranscriptional addition of poly-A tails (Section 16.5), mRNA editing (Section 17.2), and minor base formation lead to exceptions to this generalization.

16.31 See answer on page A-67.

CHAPTER 17

17.1

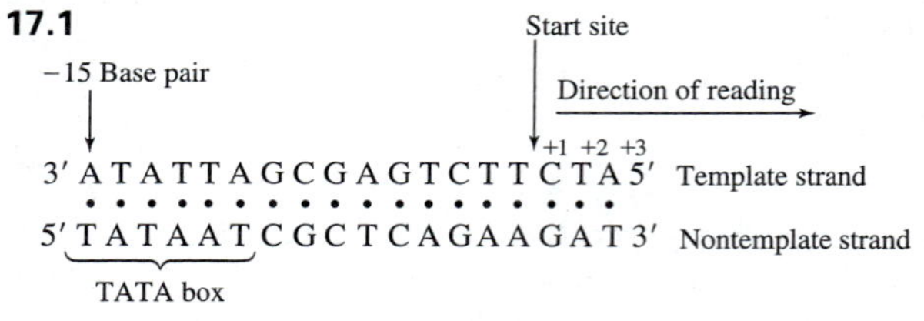

The +1 base in the nontemplate strand is usually A or G.

17.2 An AT-rich region is easier to unwind since there are three H bonds within each GC pair but only two H bonds within each AT pair (Section 16.1). The region of a promoter that initially opens during transcription (a region near the start site) is usually AT rich, presumably because such regions facilitate unwinding.

17.3 RNA synthesis involves the formation of standard base pairs between the template strand of DNA and ribonucleoside triphosphates. The formation of such pairs is impossible when the sense strand (template strand) is paired with the nontemplate strand.

17.4 Promoters, enhancers, and silencers are the regulatory segments of DNA that interact with transcription factors. Promoters are involved in transcriptase binding and transcription initiation. Both enhancers and silencers, in association with appropriate transcription factors, modulate the utilization of promoters.

17.5 See answer on page A-67.

17.6 See answer on page A-67.

17.7 5′ ACTTTATCCCTGC3′ DNA template
3′ UGAAAUAGGGACG5′ RNA fragment

The RNA produced during transcription is complementary and antiparallel to its template.

16.31 *(Answer)*

ddAMP ddCMP ddGMP ddTMP

Direction of migration

3′ Complement CACCGGGCCCCGTACATGGGCCCCGTA

5′ Sequence of original polynucleotide GTGGCCCCGGGCATGTACCCCGGGCAT

primer

17.5 *(Answer)*

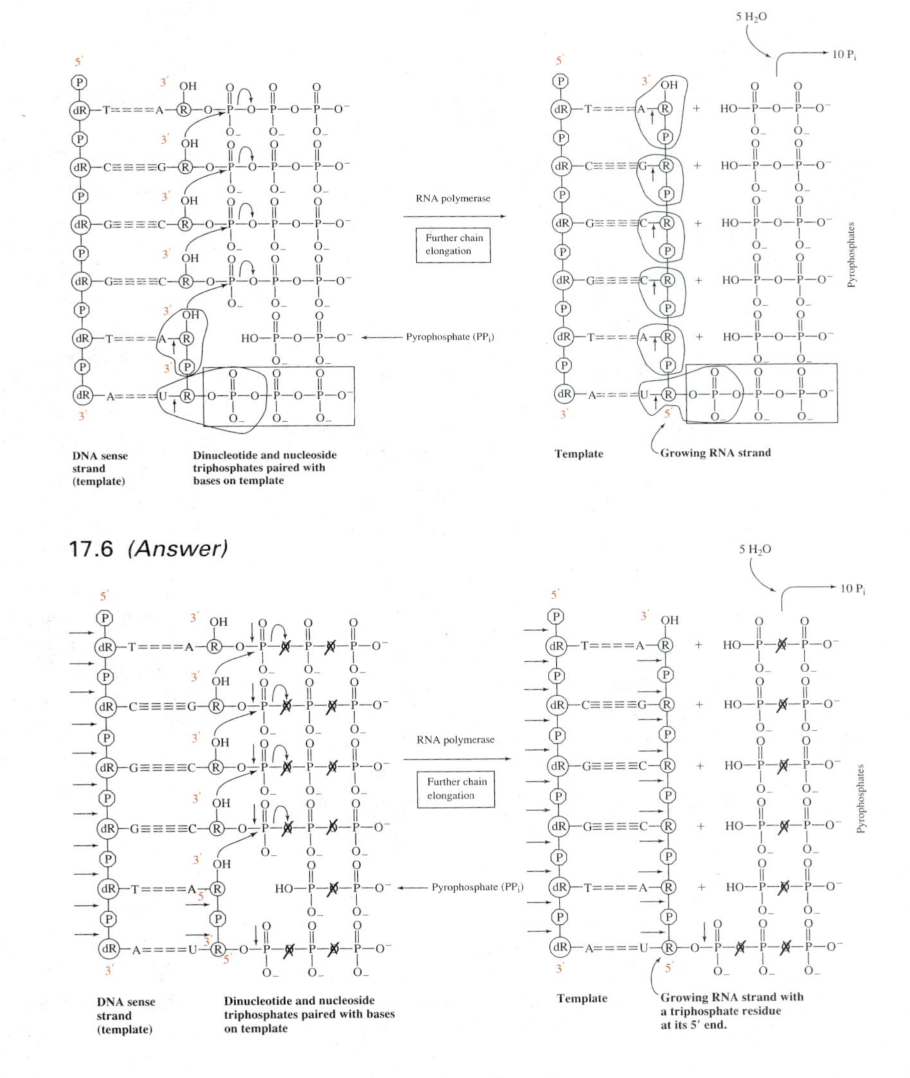

17.8 The RNA:DNA hybrid within the transcription bubble is maintained by H bonds within individual base pairs and by hydrophobic interactions between adjacent stacked pairs. These are the same forces that hold separate polymer strands together within duplex DNAs and maintain the base paired (stem) regions within RNAs (Chapter 16). RNA:DNA hybrids are described in Section 16.8.

17.9 Topoisomerases are enzymes that catalyze the interconversion of topological forms of DNA, forms of DNA that differ only in linking number. Supercoiled DNA has a linking number different from that of relaxed DNA, the favored form of DNA when the ends of a duplex are free to rotate.

17.10

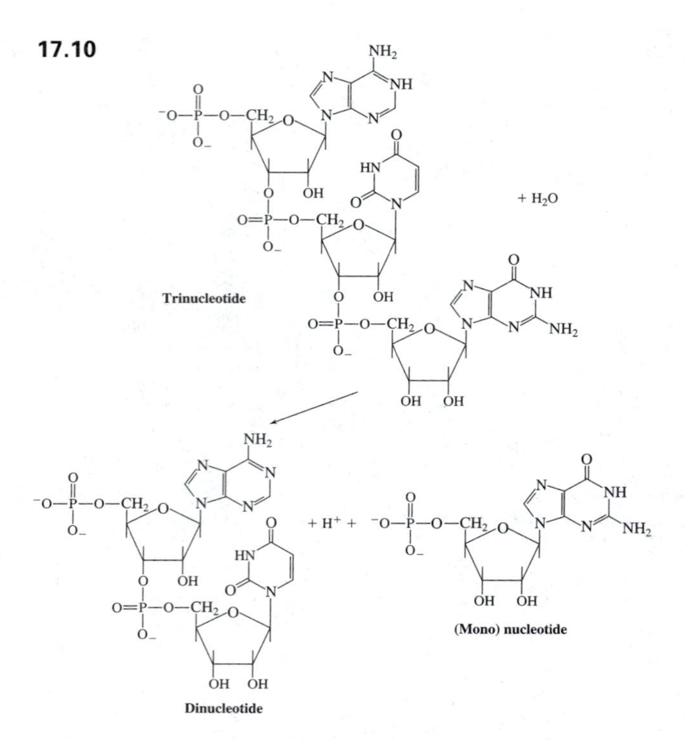

17.11 27 distinct splicing events are required, one to remove each intron. The mRNA precursor contains 28 exons, since the number of exons is always one greater than the number of introns.

17.12 Differences in reading frames and differences in exons can both contribute to differences in the encoded polypeptide. To review the concept of reading frame, assume that the sequence 5′ACGGC3′ lies within the translated region of an mRNA. ACG represents an intact codon within one imaginable reading frame. A second reading frame leads to the recognition of CGG as an intact codon. GGC is a codon within the last possible reading frame. A shift in reading frame markedly changes the codon sequence within a translated region even when the sequence of nucleotide residues is unchanged.

17.13 Tertiary structure; the folding together of base-paired regions (stems) and loop regions to create a molecule with a unique three dimensional shape and geometry.

Since ribozymes are highly selective, stereospecific catalysts, they must possess unique, chiral active sites. Such sites require a distinct tertiary structure.

17.14 Yes. This possibility exists for introns large enough to encode an mRNA. There are documented cases of this phenomenon.

17.15 All events except minor base formation. Although the creation of some minor bases may involve phosphate ester formation, this is not usually the case. One or more phosphate ester bonds are broken or formed whenever a polynucleotide is hydrolyzed, a nucleotide residue is added, a nucleotide residue is removed, and a polynucleotide is spliced. Splicing reactions involve transesterification reactions in which one ester bond is broken while a second is being formed.

17.16 True. Trimming, splicing, poly-A tail additions, and editing all lead to changes in the sequence of nucleotide residues (primary structure). Minor base formation can be viewed as doing the same, although some minor bases have the same partners (within base pairs) as the common bases from which they are constructed. Marked changes in primary structure normally lead to changes in secondary and tertiary structure as well, since primary structure determines the base pairing and folding options for a polynucleotide. The thermodynamically most stable secondary and tertiary structures are usually the biologically active ones. The same is true for polypeptides (Chapter 4).

17.17 If inserted between separate codons, the polynucleotide will lead to an extra 23 amino acid residues (69 divided by 3; it takes three nucleotides to code for one amino acid) at a specific site within the encoded polypeptide or lead to premature termination (if it contains an in-phase termination codon). If inserted between nucleotides within a codon, the polynucleotide (if it does not contain an in-phase termination codon) will change the reading frame for all preexisting nucleotide residues to the 3′ side of the disrupted codon. Twenty-three additional codons will be added as well. A change in reading frame usually impacts the length of an encoded polypeptide by altering the location of termination codons.

17.18 Splicing and trimming remove residues contained in mRNA precursors (primary transcripts), while capping and tailing add residues not found in primary transcripts. Editing can add or remove particular residues. Minor base formation modifies residues but does not change the residue count.

17.19 50 replicons, since each origin is part of a separate replicon. The DNA contains a total of 350 kb pairs (50×7) and 700,000 total nucleotide residues (350,000 pairs $\times$ 2 residues/pair).

17.20 The separate strands in a duplex DNA are actively held together by numerous H bonds and hydrophobic forces. Although individual bonds are quite weak, the total bond energy is substantial because of the large number of bonds and the cooperative nature of the bonding. Any conditions that weaken H bonds or hydrophobic forces tend to reduce the energy required to denature (separate the strands in; unwind) duplex DNA. These conditions include extremes in pH.

17.21 Enzyme specificity. Most enzymes that use ATP as a substrate can distinguish ATP from GTP, CTP, UTP, and their deoxy counterparts. The substrate binding sites on these enzymes must interact selectively with the adenine residue in ATP. The binding sites may interact with other sites on ATP as well.

17.22 The attachment of SSB to DNA is undoubtedly noncovalent because SSB binding does not require an enzyme and it is readily reversible. Since both nucleic acids and proteins are capable of H bonding, hydrophobic interactions, and salt bridge formation, we would expect these noncovalent forces to be predominantly responsible for the attachment of any protein to a nucleic acid. Experimentally, this is indeed the case.

17.23

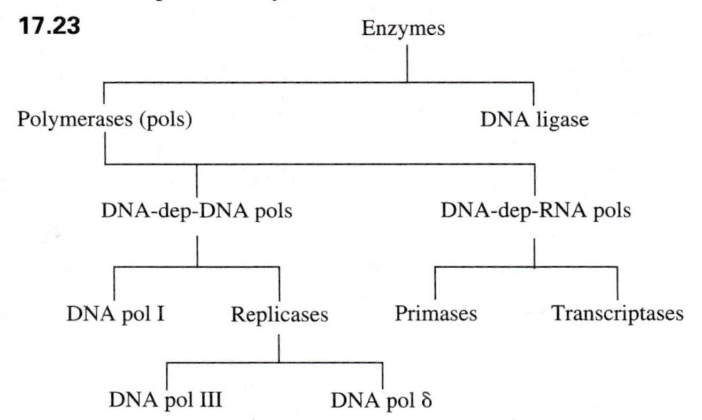

17.24 The addition of each monomer is coupled to the cleavage of a phosphate anhydride. The cleavage of an anhydride bond leads to products that are thermodynamically more stable than the reactants (the products have less free energy). This provides a thermodynamic driving force for the reaction; it increases the equilibrium constant for the reaction (Section 8.2).

17.25 The hydrolytic removal of a mispaired nucleotide residue is an exergonic process that requires no energy. However, energy is consumed during the addition of a mispaired nucleotide residue initially. This energy, which adds to the total energy burden for replication, is wasted from the standpoint of polymer extension.

17.26 Five minutes, since eukaryotic replicases catalyze the addition of nucleotide residues to growing daughter strands at a rate of approximately 2000/min (10,000 nucleotides ÷ 2000 nucleotides/min = 5 min).

17.27 A phosphate ester bond. After the primer has been assembled, the replicase extends the chain by catalyzing the addition of one deoxyribonucleotide residue at a time in the same manner that RNA polymerase adds ribonucleotide residues (see Figure 17.9).

17.28 200. Since replication is usually bidirectional within each replicon, you should assume that there are two replication forks per replicon. A minimum of two replicase (DNA polymerase δ) molecules are active at each fork, one continuously synthesizing the leading strand and the second involved in the assembly of Okazaki fragments to be incorporated into the lagging strand. This leads to a minimum of four replicase molecules per replicon and 200 per 50 replicons If a single replicase molecule with two active sites is involved in the simultaneous synthesis of leading and lagging strands (see discussion of *E. coli* replicase), the minimum number of replicase molecules required is reduced to 100.

17.29 No. The strand of DNA that is discontinuously copied when a replication fork moves off in one direction from the origin differs from the strand that is discontinuously copied as a second replication fork moves in the opposite direction. During bidirectional replication, the Okazaki fragments produced at one replication fork will hybridize with one of the parent strands while those generated at the other replication fork will hybridize with the second parent strand.

17.30 dATP, dCTP, dGTP, dTTP, Okazaki fragments, the growing lagging strand, and H_2O. DNA polymerase I has three catalytic activities: a 5′ to 3′ DNA polymerase activity, a 5′ to 3′ exonuclease activity involved in the removal of RNA primers during the maturation of the lagging strand, and a 3′ to 5′ exonuclease activity responsible for proofreading. The DNA polymerase activities employ dATP, dCTP, dGTP, and dTTP (analogous to the use of ribonucleoside triphosphates by RNA polymerase, Figure 17.9) and Okazaki fragments as substrates, whereas the exonuclease activity catalyzes the hydrolytic removal (involves H_2O) of nucleotide residues from Okazaki fragments and the growing lagging strand.

17.31 Helicase, DNA gyrase, primase (within primosome), DNA polymerase I, DNA polymerase III, and DNA ligase. Although single-strand binding protein is part of the replisome, it is not an enzyme; it acts stoichiometrically, not catalytically.

17.32 No. Some mutations (abnormal changes in DNA) are silent (have no effect on an organism) and others are beneficial. The buffered genetic code (Section 16.5) and the presence of noncoding DNA provide a partial explanation for why some mutations are silent. Evolution is driven by mutations; a mutation can give an organism a survival advantage by changing a protein (DNA carries information for the synthesis of proteins) so that it better fits a structural or functional niche.

17.33 See answer on page 70.

17.34 Deoxyribonucleoside triphosphates, the RNA in the RNA:DNA hybrid, primers, the growing DNA strands, and H_2O are all substrates for reverse transcriptases. Since a reverse transcriptase contains three distinct catalytic activities (an RNA-dependent DNA polymerase activity; an RNase H activity; and a DNA-dependent DNA polymerase activity), one would expect to find three distinct active sites. Deoxyribonucleoside triphosphates, primers, and the growing DNA strands would be substrate for both polymerase active sites, whereas the RNA in the RNA:DNA hybrid and H_2O would be the reactants for the RNase H–catalyzed hydrolysis of the RNA in the hybrid (see Figure 17.27; primers not shown).

13.33 *(Answer)*

5′ A G A A A A A A C G A G C U U U 3′ (original RNA)

↓ RNA-dependent DNA polymerase activity of reverse transcriptase; original RNA serves as template during the production of a complementary strand of DNA. Although not shown, primers are involved.

5′ A G A A A A A A C G A G C U U U 3′ (original RNA)
3′d T C T T T T T T G C T C G A A A 5′ (one strand of duplex DNA)

↓ RNAse H activity of reverse transcriptase; original RNA degraded.

3′d T C T T T T T T G C T C G A A A 5′ (one strand of duplex)

↓ DNA-dependent DNA polymerase activity of reverse transcriptase; the newly synthesized DNA serves as a template for the synthesis of its own complement. The second DNA strand is assembled while the original RNA is degraded. Although not shown, primers are involved.

3′d T C T T T T T T G C T C G A A A 5′
5′d A G A A A A A A C G A G C T T T 3′

Final duplex

17.35 Reverse transcriptases are hydrolases (they catalyze the hydrolysis of the RNA in RNA:DNA hybrids) and transferases (they catalyze the transfer of nucleotide residues from deoxyribonucleoside triphosphates to the growing polymer chains). Table 5.1 summarizes the six major classes of enzymes.

17.36 Once integrated into a host chromosome, the viral genetic information is replicated along with the host DNA. Each time a cell divides, the viral information is automatically passed on to each daughter cell. Viral integration probably reduces the risk of degradation (by nuclease-catalyzed hydrolysis) and other insults, since the viral DNA, like normal host DNA, is presumably packaged in nucleosomes. One would also expect the integrated viral DNA to be maintained by the repair systems that care for host DNA.

17.37 True. If AZT triphosphate is mistaken for dTTP by reverse transcriptase, the same error is likely to be made by other DNA polymerases. However, the inhibition constant (a measure of how tightly an inhibitor binds an enzyme) may be different from one polymerase to the next.

17.38 A transposition could remove a promoter from a transcription unit; modify a promoter; create a new promoter; create, eliminate or alter an enhancer or a silencer; remove a portion of a gene from a transcription unit; create an intact gene from a partial gene; and so on.

17.39 One would expect the oligonucleotide corresponding to the human telomere repeat (TTAGGG) to compete with the ends of telomeres for binding to telomerase RNA.

17.40 Similarities: Primers and parent templates are involved. 5′ to 3′ polymerization is catalyzed by DNA polymerases. Differences: Heat, rather than enzymes and other proteins, is used to separate (unwind) parent strands for PCR. The primers for PCR are DNA, rather than RNA. Since the primers are presynthesized, no primase is required for PCR. Since there are no origins for PCR, no origin-recognition factors are required. Since both daughter strands are synthesized continuously, PCR entails no Okazaki fragments and no maturation phase (requiring DNA ligase and other enzymes). There are no replication forks and no SSBs during PCR. The products of PCR are not subjected to mismatch repair. Telomerase is not required for PCR.

17.41 A 1024-fold amplification is attained.

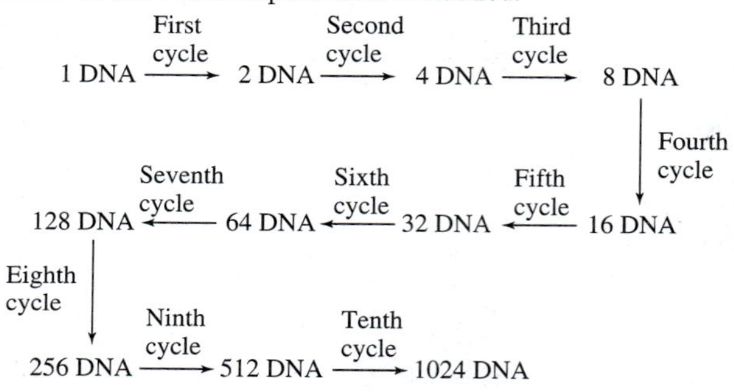

Each heat–cool cycle doubles the DNA that is present at the start of the cycle. In general, the magnitude of amplification is 2^n where n = number of heat–cool cycles:

$$2^{10} = 2 \cdot 2 \cdot 2 \cdot 2 \cdot 2 \cdot 2 \cdot 2 \cdot 2 \cdot 2 \cdot 2 = 1024$$

17.42 Primers are required for PCR because no DNA polymerase has been discovered that can replicate DNA without a primer. Although RNA primers can be used, they are usually unacceptable, because the primers are permanently incorporated into the daughter strands. PCR lacks the enzymes required to remove RNA primers and replace them with DNA fragments.

CHAPTER 18

18.1 The 50S subunit contains a 23S and a 5S rRNA, whereas the 30S subunit encompasses a 16S rRNA (Table 16.8). The three rRNAs and 55 different proteins in a typical prokaryotic ribosome are all associated through noncovalent interactions.

18.2 All tRNAs have a cloverleaf secondary structure, an L-shaped tertiary structure, and the nucleotide sequence CCA at their 3′ end (Figures 16.10 and 16.11). Certain other nucleotides are also located at the same site within the primary, secondary or tertiary structures of most tRNAs. All tRNAs contain a 7-nucleotide anticodon loop with the anticodon located in the middle of this loop.

18.3 All mRNAs, both prokaryotic and eukaryotic, contain a coding (translated) sequence along with a 5′ untranslated region and a 3′ untranslated region. They all possess a complex combination of base-paired (stem) and loop regions that create a flower-like secondary structure. Most eukaryotic mRNAs, in contrast to most prokaryotic mRNAs, have a 5′ cap plus a 3′ poly-A tail (Figure 16.7). Prokaryotic mRNAs tend to be polycistronic (encode multiple polypeptides), whereas eukaryotic mRNAs are usually monocistronic (encode a single polypeptide). Review Section 16.5.

18.4 Amide bond.

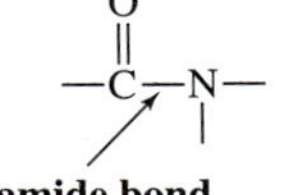

18.5 (a) 70S concentration increases; (b) 70S concentration drops. The explanation lies in Le Châtelier's principle: If a chemical reaction at equilibrium is disturbed, there will be a spontaneous net forward or net reverse reaction that counters the disturbance and tends to restore equilibrium.

18.6 Codons and anticodons are joined through noncovalent base pairing, which entails H bonding and hydrophobic base stacking (Section 16.1). The noncovalent attachment of mRNAs and tRNAs to ribosomes helps stabilize and maintain codon–anticodon duplexes.

18.7 Set A: 70S ribosome, largest; GTP, smallest
Set B: 50S subunit, largest; tRNA, smallest
Set C: mRNA, largest; inorganic phosphate, smallest

Chapter 16 provides some specific information on the relative and absolute sizes of part of these translation participants. To answer this question, you must convert the number of nucleotides in the listed RNAs to approximate total mass in daltons. Since most tRNAs contain from 73 to 93 nucleotide residues and there are around 300 daltons/nucleotide residue, weights of tRNAs range from roughly 22,000 to 28,000 daltons. Messenger RNAs and the larger rRNAs contain a substantially larger number of nucleotide residues and have a considerably greater mass. However, an rRNA cannot have a greater mass than a ribosome because it is part of a ribosome.

18.8 See answer at top of next column.

The hydrolysis of GTP is virtually irreversible because GTP contains substantially more free energy than GDP plus P_i (Section 8.2). The structural basis for this free energy difference is explained in Section 8.4.

18.9 Proline. A tRNA carries that amino acid specified by the codon to which it normally attaches. Table 18.1 reveals that the codon CCA specifies proline (Pro).

18.10 Aminoacyl-tRNA synthetases catalyze the attachment of amino acids to tRNAs. The fidelity of the process is enhanced by the proofreading ability of these ligases; they catalyze the hydrolytic removal of amino acids attached to inappropriate tRNAs.

18.8 *(Answer)*

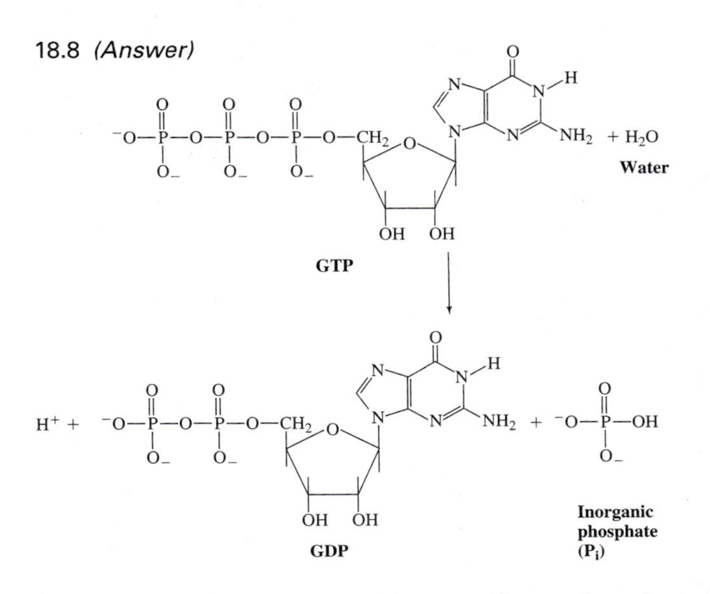

18.11 The most likely GAP is a ribosome with an aminoacyl-tRNA properly positioned (requires proper codon-anticoden pairing) at the A site. GTP hydrolysis, which triggers the move to the next step in translation, should not be allowed until the aminoacyl-tRNA delivered to the A site is properly positioned on this site.

18.12 The cleavage of ATP to yield AMP plus PP_i provides the thermodynamic driving force for the formation of aminoacyl-tRNAs. The PP_i is hydrolyzed to help "pull" the reaction to completion.

18.13 See answer at top of page A-72.

18.14 Six RNAs; two tRNAs, one mRNA, and three rRNAs. Fifty-five proteins, all part of the 70S ribosome (Table 16.8). The tripeptide residue attached to the tRNA at the A site does not qualify as a protein.

18.15 Sixty-one, a separate tRNA for each of the 61 translatable codons. In the absence of wobble, each codon will pair with a unique anticodon.

18.16 *N*-terminus Cys–Arg–Arg–Leu–Glu–Thr *C*-terminus

Codon read as Leu instead of
Phe by mischarged tRNA
↓
5′ –UGU–CGU–AGA–UUU–GAA–ACG– 3′
N-Terminus – Cys – Arg – Arg – Leu – Glu – Thr – *C*-terminus

A tRNA with the anticodon AAA will pair with the codon UUU and will normally carry the amino acid phenylalanine (Phe). If the tRNA is charged with leucine (Leu), the charging error will cause the UUU codon to be read (decoded) as leucine rather than phenylalanine.

18.13 *(Answer)*

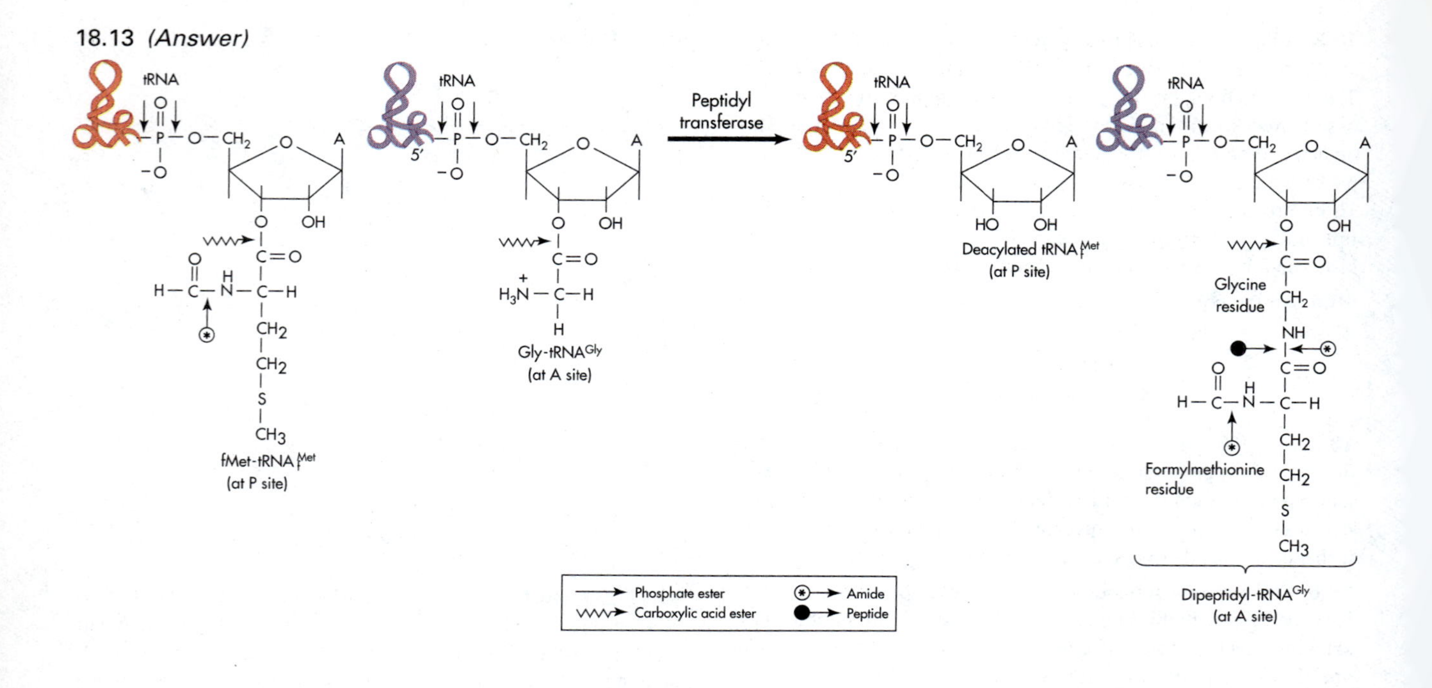

Polypeptides always grow in an *N*-terminal to *C*-terminal direction. Figure 18.7 reveals why this is the case. Messenger RNAs are always read 5′ to 3′, another detail of importance.

18.17 *N*-terminus Arg–Arg–Arg–Ser *C*-terminus

Termination codon
↓
5′–AGA–CGU–CGG–AGU–UAA–CCCUCC–3′
N-terminus Arg – Arg – Arg – Ser *C*-terminus

A genetic code table is used to decode the individual codons. The nucleotide residues to the 3′ side of the termination codon are part of the 3′ untranslated region of the mRNA.

18.18 a. the anticodon GUG pairs with CAC and CAU, both histidine codons.

b. the anticodon UAA pairs with UUA and UUG, both leucine codons.

c. the anticodon IGU pairs with ACU, ACC and ACA, three threonine codons.

In the absence of mistakes, a tRNA always carries that amino acid specified by the codon(s) to which its anticodon binds. Thus, a tRNA with the anticodon GUG will carry histidine, one with the anticodon UAA will carry leucine, and one with the anticodon IGU will carry threonine. Codons are always read in a 5′ to 3′ direction. Anticodons bind in an antiparallel manner to those codons with which they pair.

18.19 *N*-terminus Met–Arg–Arg–Pro–Val *C*-terminus

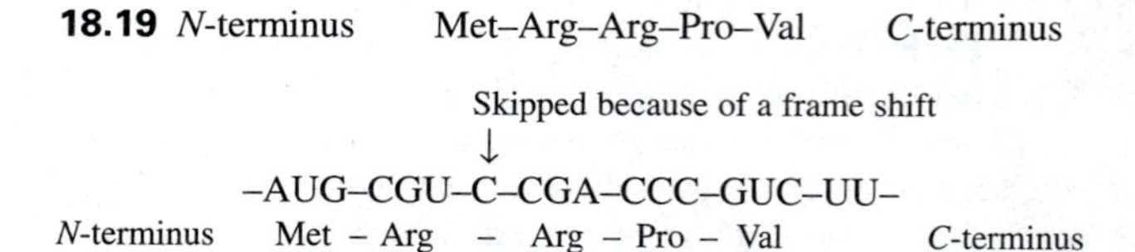

18.20 The maximum number of ribosomes that can simultaneously bind to an mRNA is determined by stearic hindrance and the length of the coding region of the mRNA. A new ribosome cannot attach to the ribosome binding site on an mRNA until a previously bound ribosome has moved out of the way. For obvious reasons, the larger the coding region of an mRNA, the more ribosomes it can accommodate. The maximum density observed for ribosomes on mRNA is around one ribosome per 80 nucleotide residues. At this density, the ribosomes are not packed together as tightly as theoretically possible.

18.21 Eukaryotic mRNAs are initially synthesized as inactive precursors that must be processed to be accurately translated. Although some processing events are probably co-transcriptional, others are post-transcriptional. The proper decoding of most eukaryotic mRNA precursors requires capping, splicing, and possibly trimming. In some cases, other processing events are required as well.

18.22 No. These compounds are of no value in treating bacterial infections, because humans are as much at risk as bacteria.

18.23 In eukaryotes, tRNAs, rRNAs, and mRNAs are all synthesized as inactive precursors that must be processed. Certain prokaryotic RNAs are also produced as inactive precursors that undergo processing. From one perspective, certain lipids and carbohydrates must also be processed; they must bind to other compounds (often proteins) to be functional.

18.24 Although all of the events listed in Exhibit 18.1 are capable of altering tertiary structure, some do not necessarily do so. The attachment of a prosthetic group and disulfide bond formation, for example, may stabilize an existing structure rather than alter it. In most instances, however, all of the events lead to at least minor changes in tertiary structure by altering side chain interactions.

18.25 Each event, except 3, potentially entails the making or breaking of covalent bonds within the polypeptide. Because peptide bonds are covalent, their hydrolysis leads to covalent bond cleavage in every case. Some, but not all, prosthetic groups become covalently linked to their polypeptide partners. Disulfide bonds are covalent bonds. The interconversions of isomers (both *cis* and *trans* and D and L) entails the making and breaking of bonds. Covalent modifications of amino acid side chains, by definition, involve the making or breaking of covalent bonds. Covalent peptide bonds are cleaved and formed during splicing.

CHAPTER 19

19.1 Damage to the embryo or fetus. Possible causes include malnutrition, chemical imbalances, toxic chemicals, physical insults (including ionizing radiation), and infections.

19.2 The immune system. Antibodies (immunoglobulins), central components of the immune system, were examined in Section 4.11. A mutation affecting antibody production or any other aspect of immune function could reduce the efficiency with which the body recognizes and destroys infectious agents.

19.3 With over 4000 genetic diseases, there are many possible answers including Down syndrome, sickle cell anemia, cystic fibrosis, Tay–Sachs disease, Huntington's disease, and Duchenne muscular dystrophy. How we define "satisfactory treatment" determines, in part, what specific diseases should be on this list. We could claim, for example, that there is no satisfactory treatment for diabetes, because so many victims develop serious complications, including the loss of kidney function and vision. However, most diabetics who adhere strictly to treatment protocols avoid or delay complications. Is the treatment for diabetes satisfactory?

19.4 No. Because red blood cells are normally the only cells that express hemoglobin genes (Section 4.9), heart and liver cells are not affected by abnormalities in the hemoglobin genes they contain. Restated, abnormal genes "turned off" in a cell during normal differentiation do not usually harm the cell.

19.5 The genetic code is buffered. A nucleotide residue substitution (point mutation) at the 3′ end of a codon often creates to a new codon for the same amino acid or a codon for a chemically-similar amino acid.

19.6

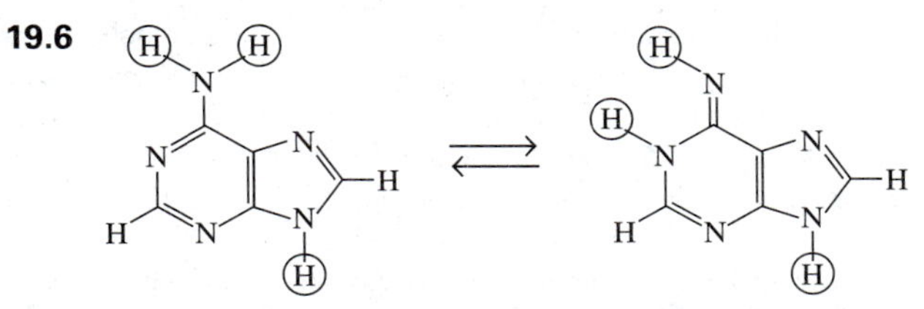

19.7 No. Mutations accumulate continuously during the life span of all organisms. A noninherited mutation in an egg or sperm cell can be passed to the next generation and lead to a genetic disease.

19.8 Distinct restriction endonucleases catalyze the hydrolysis of DNA at different palindromes. If copies of the palindrome recognized by one restriction endonuclease are distributed in an identical fashion in two corresponding DNAs, both DNAs will yield the same sized restriction fragments. If copies of the palindrome recognized by a second restriction endonuclease are positioned at distinctive sites in the two DNAs, enzyme-catalyzed hydrolysis will normally lead to RFLPs.

19.9 Denaturation is the separation of the base paired regions in a nucleic acid. In the case of double-stranded DNA, this corresponds to the separation of the two strands in the duplex. Since all of the bases in duplex DNA are already paired with bases in a complementary strand, they are inaccessible to a DNA probe. Once DNA is denatured, each strand is free to pair with a probe.

19.10 The disease-linked RFLPs in the one family are not due to the defective gene itself; they are a consequence of a neighboring genetic element that runs in the one family but not the second. Alternatively, although the same gene is defective in both families, the specific defect in the gene is different in the two families. The defect in the second family does not alter restriction endonuclease cleavage sites in the same manner as the defect in the first family. The disease is probably dominantly inherited. If it were recessively inherited, some of the nondiseased family members would most likely be carriers and exhibit disease-linked RFLPs.

19.11 Liver and brain cells are thought to have the same genetic information. They differ because some unique genes are expressed in each of the two cell types (Chapter 1). Similarly, gene activity in cells in culture is partially determined by culture conditions. A cell, for example, may only express the genes needed to metabolize galactose when this sugar is present. A DNA probe cannot not distinguish the two cell types, because a probe cannot distinguish genes that are being expressed from those that are silent (inactive). Probe binding simply denotes the presence of a specific base sequence in the probed DNA.

19.12 The codons for glutamic acid are GAA and GAG, whereas the valine codons are GUU, GUC, GUA, and GUG. The replacement of the middle A in either glutamic acid codon with U creates a valine codon.

19.13 Like associates with like, a central theme in biochemistry (Exhibit 1.4): polar associates with polar, nonpolar associates with nonpolar, but polar and nonpolar do not readily mix. This association rule explains the experimentally determined rule for the folding of proteins in an aqueous environment: "nonpolar in, polar out" (Section 4.4). The nonpolar side chains tend to fold into the core of a protein in order to avoid the polar aqueous environment. Similarly, HbS molecules clump together so valine side chains on their surfaces can escape from surrounding water molecules.

19.14 The amino acid substitution in HbS has no significant impact on O_2 binding, because it has no significant impact on the secondary, tertiary, or quaternary structures that create and maintain the O_2 binding sites. The substitution occurs on the surface at a site removed from the heme pocket. Many substitutions lead to changes in O_2 binding sites and, consequently, to alterations in O_2 binding.

19.15 Only deoxyhemoglobin S molecules (not oxyhemoglobin S molecules) aggregate to form the fibers that cause red blood cells to sickle. Vigorous exercise tends to reduce O_2 concentration within red blood cells and to increase deoxyhemoglobin S concentration (Le Châtelier's principle; Section 4.9). As deoxyhemoglobin S concentration increases, the probability that neighboring molecules will aggregate increases as well.

19.16 A single normal copy of the phenylalanine hydroxylase gene leads to the production of enough of this enzyme to prevent phenylalanine and phenylketones from building up to dangerous levels.

19.17 True. Most dietary amino acids come from dietary proteins (Section 4.12), and most proteins contain phenylalanine. Technically, a PKU victim could consume all the protein desired if that protein lacked phenylalanine.

19.18 Promoters are transcriptase binding sites that play a central role in the initiation of transcription. Since promoters help determined how often a gene is transcribed, a change in a promoter could lead to either an increase or decrease in gene utilization. An increase in frequency of gene expression could lead to the overproduction of a growth stimulator and the progression toward malignancy.

19.19 Reduced Ras-linked GTPase activity enhances signal transduction, since it increases the length of time the Ras protein (a G protein that functions as a molecular switch) remains in the "on" position following GTP binding. It has been discovered that some oncogenes do encode Ras proteins with reduced GTPase activities.

19.20 True. The loss or inactivation of p21 protein has many of the same consequences as the loss or inactivation of p53 protein, since the normal action of p53 depends upon normal p21. When p21 protein is inactive, the rate of cell division and DNA replication increases (relative to the rate when p21 is active), even when p53 protein is normal. The *p21* gene is indeed a tumor suppressor gene.

19.21

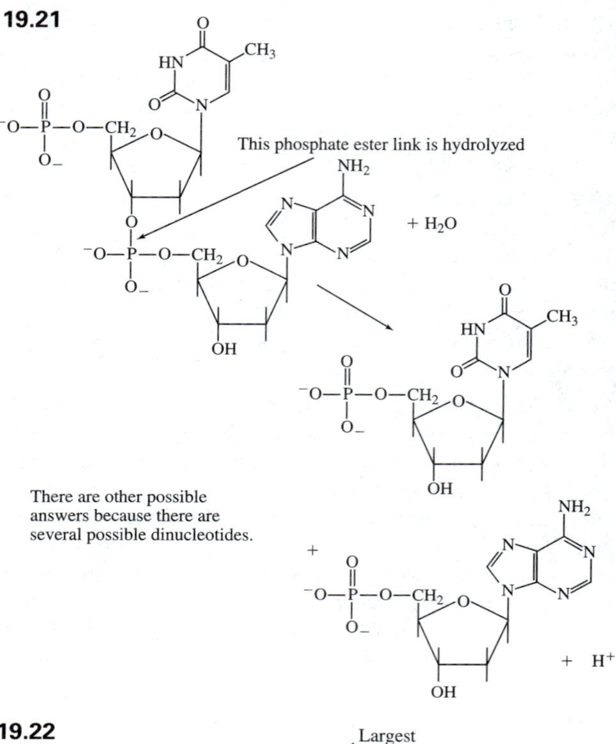

19.22

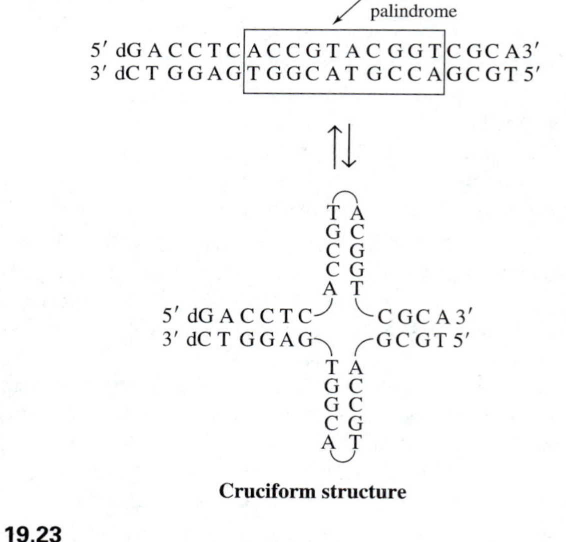

19.23

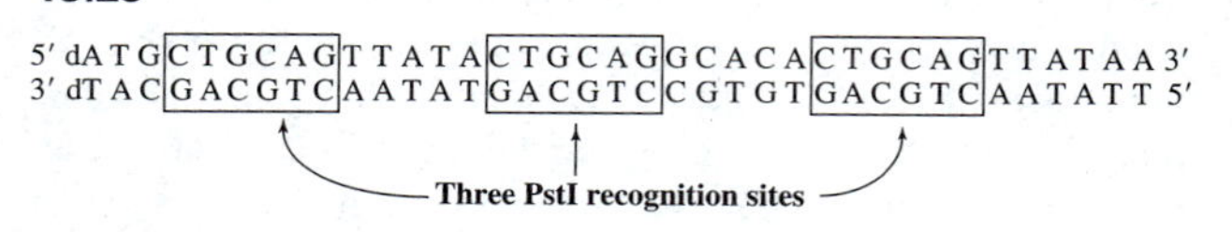

PstI-catalyzed hydrolysis yields four fragments; if you cut a string at three sites, you get four pieces.

19.24

5′ T G A A A A C T 3′
3′ T G T C A A C G C C A C T T T T G A 5′

Any base pair sequence is acceptable

19.25 There are an enormous number of theoretically possible constructs. The original bacterial DNA could be reformed. Any number of copies of the human DNA fragment could all be spliced together and a bacterial fragment could be attached to one or both ends. Two copies of the human DNA fragment could be spliced together to form a closed circular duplex DNA molecule. And so on. Some examples follow:

5′ A-A-T-T-C▭G-A-A-T-T-C▭G 3′
3′ G▭C-T-T-A-A-G▭C-T-T-A-A 5′

Two copies of the human fragment spliced together to give a linear "dimer" with sticky ends.

Two copies of the human fragment spliced into a closed circular duplex molecule.

▬G-A-A-T-T-C▭G-A-A-T-T-C▭G-A-A-T-T-C▬
▬C-T-T-A-A-G▭C-T-T-A-A-G▭C-T-T-A-A-G▬

Two copies of the human fragment spliced into the bacterial DNA.

19.26 Yes. Recombinant DNA technology can be used to "cut" genes out of a chromosome and to rejoin (splice) them in any desired order or sequence.

19.27 See Figure 17.27.

19.28 Yes, if the gene is incorporated into the host chromosome or to an appropriate vector. Copies of some plasmids are passed to each daughter cell during cell division.

19.29 True. The techniques described in this chapter can be used to attach a functional form of any gene to a vector. At the moment, however, the genes for many human proteins have not been located (on chromosomes) or studied.

19.30 Bacteria are not naturally programmed to perform some of the processing required to activate a nascent human polypeptide. Although genes for the processing enzymes can be isolated or produced and introduced into bacteria, there is no guarantee that the enzymes produced from the genes will be functional in the host environment and that they will be synthesized in the appropriate concentrations and at the correct time. Many processing enzymes may be involved, some multimeric. In some instances, part of the processing enzymes and their genes have not been identified. In addition, bacteria may lack the proper concentrations of the substrates required for an appropriate reaction rate. There are other considerations as well.

19.31 The nutritional value of a plant can be improved by introducing genes that encode:

a. proteins with a proper balance of essential amino acids
b. proteins that enhance essential amino acid production
c. proteins that enhance vitamin production
d. proteins that enhance essential fatty acid production
e. proteins that enhance the production of beneficial fiber
f. proteins that regulate fat production
g. proteins that reduce the concentration of toxic substances, including the natural mutagens found in most plants
h. proteins that make plant materials more readily digestible
i. etc.

19.32 The giant mice are an animal model for the study of growth and hGH. They are used to study the chemical reactions and interactions that lead to the physiological changes brought about by hGH. They are also used to study the control of gene expression and the numerous variables that determine how rapidly hGH is assembled and degraded.

19.33 Yes. One can, in theory, modify any inherited trait by altering the gene content of a fertilized egg cell, because an inherited trait is, by definition, determined by gene content. To modify a trait selectively, we must first identify the gene(s) involved. Some traits are determined by interactions between many different gene products, each encoded in a distinct gene. The genes responsible for many human traits have not been identified.

19.34 For the introduced gene to provide a therapeutic benefit it must survive, and it should ideally be replicated and passed on to daughter cells if the host cell divides. It may also be necessary that the gene be expressed in a specific group of cells, at a precise level, and at precisely the right point in time. The introduced gene and its carrier should not disrupt the normal functioning of another gene or any other molecule. The vector and its regulatory elements play a central role in determining the therapeutic value of a gene.

19.35 Based on the information in this chapter, the ADA genes should be knocked out. In reality, there are multiple forms of SCID, some linked to other genes. If the knockout mice did not develop SCID, this could indicate that: (1) there are differences in mouse and human immune systems; (2) mice possess enzymes or metabolic pathways that serve as alternate sources of the products of the ADA-catalyzed reaction; (3) there are differences in the ways in which mice and humans utilize or interact with the reactants or products of the ADA-catalyzed reaction; and so on.

The Genetic Code*

First Base	Second Base: U		C		A		G	
U	UUU	Phe	UCU	Ser	UAU	Tyr	UGU	Cys
	UUC		UCC		UAC		UGC	
	UUA	Leu	UCA		UAA	TERM[a]	UGA	TERM[a]
	UUG[b]		UCG		UAG		UGG	Trp
C	CUU	Leu	CCU	Pro	CAU	His	CGU	Arg
	CUC		CCC		CAC		CGC	
	CUA		CCA		CAA	Gln	CGA	
	CUG		CCG		CAG		CGG	
A	AUU	Ile	ACU	Thr	AAU	Asn	AGU	Ser
	AUC		ACC		AAC		AGC	
	AUA		ACA		AAA	Lys	AGA	Arg
	AUG[b]	Met	ACG		AAG		AGG	
G	GUU	Val	GCU	Ala	GAU	Asp	GGU	Gly
	GUC		GCC		GAC		GGC	
	GUA		GCA		GAA	Glu	GGA	
	GUG[b]		GCG		GAG		GGG	

*This is shown with color coding as Table 16.6 on page 661.

[a]Termination codon.

[b]Initiation codon. With rare exceptions, AUG is the sole initiation codon in eukaryotes. In some prokaryotes, GUG and UUG are occasionally used for initiation.

D0150903